Graphing Calculator Supplement
For Texas Instruments Calculator

Carolyn L. Meitler

AF692216

Finite Mathematics

for Business, Economics, Life Sciences, and Social Sciences

Eighth Edition

Raymond A. Barnett

Michael R. Ziegler

Karl E. Byleen

PRENTICE HALL, Upper Saddle River, NJ 07458

Executive Editor: Sally Simpson
Editorial Assistant: Sara Beth Newell
Special Projects Manager: Barbara A. Murray
Production Editor: Michele Wells
Supplement Cover Manager: Paul Gourhan
Supplement Cover Designer: Liz Nemeth
Manufacturing Buyer: Alan Fischer

© 1999 by Prentice-Hall, Inc.
Simon & Schuster / A Viacom Company
Upper Saddle River, NJ 07458

All rights reserved. No part of this book may be reproduced, in any form or by any means, without permission in writing from the publisher

Printed in the United States of America

10 9 8 7 6 5 4 3 2

ISBN 0-13-932195-0

Prentice-Hall International (UK) Limited, *London*
Prentice-Hall of Australia Pty. Limited, *Sydney*
Prentice-Hall Canada, Inc., *London*
Prentice-Hall Hispanoamericana, S.A., *Mexico*
Prentice-Hall of India Private Limited, *New Delhi*
Prentice-Hall of Japan, Inc., *Tokyo*
Simon & Schuster Asia Pte. Ltd., *Singapore*
Editora Prentice-Hall do Brazil, Ltda., *Rio de Janeiro*

Table of Contents

PREFACE

Graphing calculators are powerful tools that can help in learning mathematics. This manual includes examples that illustrate how to use the calculator with topics covered in a finite mathematcs course. A detailed solution to each example shows how a graphing calculator can be used to solve the problem.

Appendices

Four appendices are included. Each appendix contains keystrokes necessary to solve problems for the following ™Texas Instruments calculators:

Appendix A	TI-82 Operations
Appendix B	TI-83 Operations
Appendix C	TI-85 Operations
Appendix D	TI-86 Operations

More than one method of solution using the calculator is shown for many examples. Neither programming nor parametric equations are used.

The appendices correspond to one another. The same example is used in each appendix. This allows instruction to be given to a class of students in which more than one calculator model is being used.

Calculations

Default settings on the calculator have been used for all calculations in the problem solutions, unless otherwise noted. Numbers are rounded to the desired number of significant digits or number of decimal places as the problem requires after all calculations have been completed. Directions on how to set the calculator to a specified number of significant digits, or a given number of decimal places, before performing the calculations are given in Appendices A-17, B-17, C-17 and D-17. Setting a specified number of decimal places was not done in the examples of this manual but might be convenient in many problems.

Built-in Calculator Functions

The built-in functions (features) of the calculator are used in some of the examples. Discussion of some of the built-in features for a particular calculator can also be found in the appendix pertaining to that calculator.

Graphs

The curves shown in the figures of this manual are not smooth but rather are intended to represent the graphing calculator display.

ACKNOWLEDGMENTS

The completion of this project would not have occurred without the support and assistance of several individuals. First, and foremost, is Michael R. Ziegler. His suggestions were most helpful in forming the philosophy, layout, and content of this manual. Also I would like to thank James Briselden who helped in checking for accuracy and consistency. Finally, I thank my family for their encouragement and support.

Carolyn L. Meitler

Chapter 1

A Beginning Library of Elementary Functions

This chapter contains examples using the calculator to illustrate:

- Plotting points and drawing lines
- Graphing functions
- Graphing piecewise-defined functions
- Evaluating algebraic expressions
- Determining the domain of a function from a graph
- Evaluating functions for a given value of x
- Investigating the change in the value of an expression as the value of a variable changes
- Investigating the change in the graph of a function when constant terms are changed
- Finding intercepts
- Solving an equation in one variable
- Solving an inequality in one variable

EXAMPLE 1 ***TEXTBOOK SECTION 1-1***

(A) Plot the points (-5, 2) and (1, 3). (B) Draw a line segment with these points as the endpoints.

Solution (A): First viewing window variables need to be set. For this example, set Xmin=-10, Xmax=10, Xscl=1, Ymin=-10, Ymax=10, and Yscl=1.

Get the [Pt-On(] function on the calculator. Enter the values -5 and 2 and plot the first point. Get the [Pt-On(] function again and plot the second point. (See Appendix Section A-3, B-3, C-3 or D-3 of this manual.)

Solution (B): Get the Line(function of the calculator and draw the line. (See Appendix Section A-3, B-3, C-3 or D-3 of this manual.)

EXAMPLE 2 — *TEXTBOOK SECTION 1-1*

(A) Find the y coordinate of the points on the graph of $y = 3 - x^2$ for $x = -3, -1, 0, 1, 3$. (B) Plot the resulting ordered pairs. (C) Graph the function.

Solution (A): Store $3 - x^2$ in the function list in your calculator as: 3 - X ^ 2. (See Appendix Section A-4, B-4, C-4 or D-4.) One method to obtain the y coordinates is to store the first value for x in the calculator and evaluate the function. Repeat for the other four values of x. The resulting ordered pairs are (-3, -6), (-1, 2), (0, 3), (1, 2), and (3, -6).

A second method is to store the values for x as a list and to evaluate the function for the entire list. First store $3 - x^2$ in the function list. Then store the list and evaluate it.

<u>TI-82</u>

2nd { (-) 3 , (-) 1 , 0 , 1 , 3 2nd } STO▸ 2nd L1 ENTER

2nd Y-VARS 1 <Function...> 1 <Y1> (2nd L1) ENTER

<u>TI-83</u>

2nd { (-) 3 , (-) 1 , 0 , 1 , 3 2nd } STO▸ 2nd L1 ENTER

VARS ▶ 1 <Function...> 1 <Y1> (2nd L1) ENTER

<u>TI-85 & TI-86</u>

2nd LIST F1 <{> (-) 3 , (-)

1 , 0 , 1 , 3 F2 <}> STO▸ x-VAR ENTER

2nd ALPHA y 1 ENTER

Solution (B): Refer to Example 1 above on how to plot points on the calculator. Set the viewing window variables to [-10, 10]1 by [-10, 10]1 (see Appendix Section A-3, B-3, C-3 or D-3 of this manual).

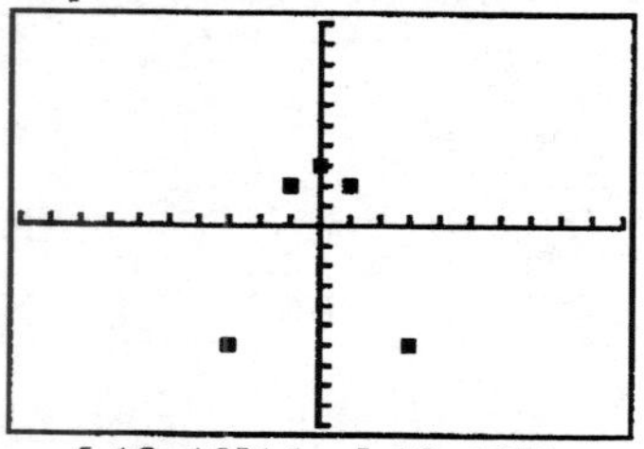

[-10, 10]1 by [-10, 10]1

Method 1 Plot each point separately

Get the Pt-On(function on the calculator. Enter the values -3 and -6 and plot the first point. Get the \x(Pt-On() function again, enter the values of the coordinates to plot the second point. Repeat for the other points. The result is shown to the right. (See Example 1 above and Appendix Section A-3, B-3, C-3, or D-3 of this manual.)

Method 2 Plot all the points at one time

Store the points in the lists for the statistics function of the calculator. Draw a scatter diagram. (See Appendix Section A-16, B-16, C-16 or D-16 of this manual.) Be sure to clear or deselect all functions in the function list. (See Appendix Section A-3, B-3, C-3 or C-3 of this manual.)

TI-82

STAT 4 <ClrList> 2nd L1 , 2nd L2 ENTER STAT
1 <Edit...> (-) 3 ENTER (-) 1 ENTER 0 ENTER 1 ENTER 3 ENTER
► (-) 6 ENTER 2 ENTER 3 ENTER 2 ENTER (-) 6 ENTER
2nd STAT PLOT 1 ENTER ▼ ENTER ▼ ENTER ▼ ► ENTER GRAPH

TI-83

STAT 4 <ClrList> ENTER 2nd L1 , 2nd L2 ENTER
STAT 1 <Edit...> (-) 3 ENTER (-) 1 ENTER 0 ENTER 1 ENTER
3 ENTER ► (-) 6 ENTER 2 ENTER 3 ENTER 2 ENTER (-) 6
ENTER 2nd STAT PLOT 1 ENTER ▼ ENTER ▼ 2nd L1 ▼
2nd L2 GRAPH

TI-85

STAT F2 <EDIT> F1 <xStat> ENTER F2 <yStat> ENTER F5 <CLRxy>
(-) 3 ENTER (-) 6 ENTER (-) 1 ENTER 2 ENTER
0 ENTER 3 ENTER 1 ENTER 2 ENTER 3 ENTER
(-) 6 EXIT F3 <DRAW> F2 <SCAT> CLEAR

TI-86

2nd STAT F2 <EDIT> ▲ CLEAR ▼ (-) 3 ENTER (-) 1 ENTER
0 ENTER 1 ENTER 3 ENTER ► ▲ CLEAR ▼ (-) 6 ENTER
2 ENTER 3 ENTER 2 ENTER (-) 6 ► ▲ CLEAR ▼ 1 ENTER
1 ENTER 1 ENTER 1 ENTER 1 ENTER 2nd QUIT 2nd STAT
F3 <PLOT>

F1 <PLOT1> ENTER ▼ F1 <SCAT> ▼ F1 <xStat> ▼ F2 <yStat>
2nd QUIT GRAPH F5 <GRAPH>

Solution (C): Set the viewing window variables to [-10,10]1 by [-10, 10]1 (see Appendix Section A-6, B-6, C-6, or D-6 of this manual). Store $3 - x^2$ in the function list in your calculator if it has not already been stored. Select the function if it has been deselected.

Clear or deselect all other functions. (See Appendix Section A-6, B-6, C-6, or D-6 of this manual.)

Clear all drawings. (See Appendix Section A-3, B-3, C-3, or D-3.)

Turn all statistical plots off if you are using the TI-82, TI-83 or TI-86 calculators. (See below and Appendix Section A-16, B-16, or D-16 of this manual.) If you do not turn the statistical plots off the points will plot as well as the function.

TI-82

2nd STAT PLOT 4 ENTER

TI-83

2nd STAT PLOT 1 ENTER

TI-85

STAT F3 <DRAW> F5 <CLRDW>

TI-86

2nd STAT F3 <PLOT> F5 <PlOff> ENTER

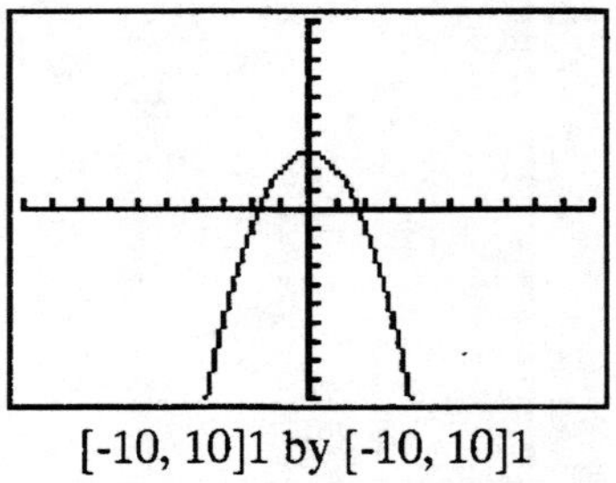

[-10, 10]1 by [-10, 10]1

Graph the function (see Appendix Section A-6, B-6, C-6 or D-6 of this manual).

EXAMPLE 3 ***TEXTBOOK SECTION 1-1***

Graph $y = \sqrt{4 - x}$ and determine the domain by examining the graph.

Solution: Store the function in the function list in the calculator as: √ (4 - X). Set the viewing window variables to [-5, 5]1 by [-5, 5]1 (see Appendix Section A-7, B-7, C-7 or D-7 Example 2 of this manual). Graph the function.

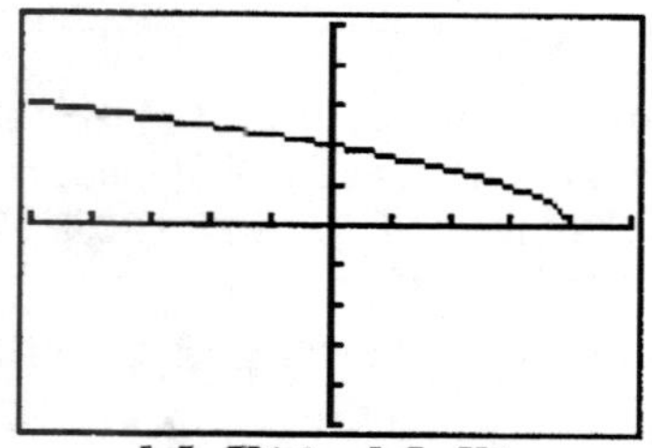

[-5, 5]1 by [-5, 5]1

Use the built-in function root or zero to find the x intercept (or root) of this function. (See Appendix Section A-7, B-7, C-7, or D-7 of this manual.)

The x intercept is 4. Hence the domain is $x \le 4$.

EXAMPLE 4 ***TEXTBOOK SECTION 1-1***

(A) Fill in the chart below for $f(x) = 2x^2 + 3$.

(B) How does the quantity $\frac{f(2+h) - f(2)}{h}$ change as h assumes values .2, .1, .01, and .001.

(C) Make a conjecture as to what happens to this quantity when h gets closer and closer to zero.

h	.2	.1	.01	.001	→ 0
$\frac{f(2+h) - f(2)}{h}$					

Solution (A): $f(2+h) = 2(2+h)^2 + 3$ and $f(2) = 11$. Store $2(2+h)^2 + 3$ as 2 (2 + [ALPHA] H) ^ 2 + 3 as Y1 and 11 as Y2 in the function list on your calculator. Store (Y1-Y2)÷H as Y3 by entering:

TI-82

[(] [2nd] [Y-VARS] [1] [1] <Y1> [-] [2nd] [Y-VARS] [2] [2] <Y2> [)] [÷] [ALPHA] [X]

TI-83

[(] [VARS] [►] [1] [1] <Y1> [-] [VARS] [►] [1] [2] <Y2> [)] [÷] [ALPHA] [X]

TI-85 & TI-86

[(] [2nd] [ALPHA] [Y] [1] [-] [2nd] [ALPHA] [Y] [2] [)] [÷] [ALPHA] [H]

Store .2 as H and evaluate Y3 by recalling it (see Appendix Section A-4, B-4, C-4 or D-4 of this manual). Repeat by storing each of the other values of h and evaluating Y3. The results are shown below.

h	.2	.1	.01	.001	→	0
$\frac{f(2+h) - f(2)}{h}$	8.4	8.2	8.02	8.002	→	8

Note that the list feature of the calculator could be used by storing {.2, .1, .01, .001} as a list and then evaluating the function for this list. To do this you must use X in place of H in the TI-82 and TI-83 calculators and store the list as L1. To evaluate the function use y3(L1). The list can be stored as H on the TI-85 and TI-86 calculators. To evaluate y3 enter [2nd] [ALPHA] [Y] [3].

Solution (B) & (C): The quantity gets closer and closer to 8 as h gets closer and closer to 0.

EXAMPLE 5 ***TEXTBOOK SECTION 1-1***

A price-demand function that models a set of data is $p(x) = 94.8 - 5x$.

(A) Evaluate the price-demand function at $x = 2, 5, 8,$ and 12 million.

(B) Find the error between the result of using the function to calculate the price and the actual data shown below.

x	p
2	87
5	68
8	53
12	37

(C) Plot the points and graph the price-demand function that models the data.

Solution (A): Store the function $p(x) = 94.8 - 5x$ as Y1 as 94.8 - 5 X. Store 2 as X. Recall the function Y1 and evaluate. Repeat for the other values of x. The list feature of the calculator could also have been used. (See Appendix Section A-4, B-4, C-4, or D-4 of this manual and Example 2 above.) The result are shown in the second column of the table to the right.

x	$p(x)$	p	*Error* $p(x)-p$
2	84.8	87	-2.2
5	69.8	68	1.8
8	54.8	53	1.8
12	34.8	37	-2.2

Solution (B): Calculate $p(x)$-p to find the error. The results are shown in the table above.

Solution (C): Set the viewing window variables at [0, 15]1 by [0, 100]10. (See Appendix Section A-8, B-8, C-8, or D-8 of this manual on how to choose and set the viewing window variables.) Plot the points (see Example 2 above). Store the price-demand function as Y1 as 94.8 - 5 X and graph.

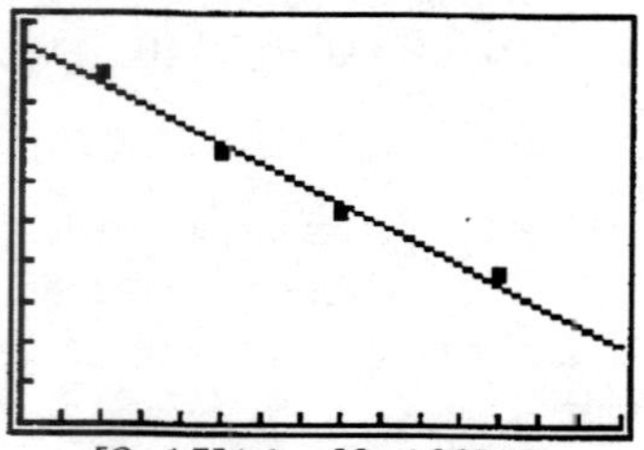

[0, 15]1 by [0, 100]10

EXAMPLE 6 ***TEXTBOOK SECTION 1-2***

Graph $f(x) = x^2$, $f(x) = x^2+4$, and $f(x) = x^2 - 7$ on the same coordinate axes. Compare the graphs and describe the effect of adding or subtracting a constant from $f(x)=x^2$.

Solution: Store the functions as three separate functions in the calculator as: X^2, X^2+4, and X^2-7. Graph using [-15, 15]1 by [-20, 20]1. We see that the second graph has the same shape but is shifted vertically upward 4 units from $f(x) = x^2$. The third graph has the same shape but is shifted vertically downward 7 units from $f(x) = x^2$.

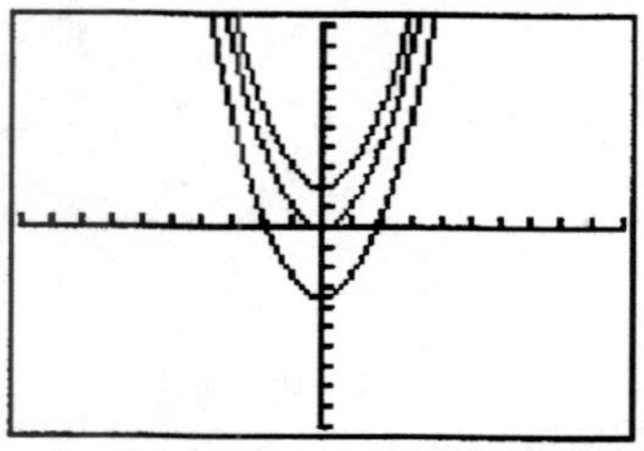

[-15, 15]1.5 by [-20, 20]2

EXAMPLE 7 ***TEXTBOOK SECTION 1-2***

(A) Find the value(s) of x for which $f(x) = |x|-5$ is positive.

(B) Discuss how the graph of this function differs from $y = |x|$.

Solution (A): Store $f(x)$ as Y1 as ABS (X) - 5 where ABS is the built-in absolute value function in your calculator (see Appendix Section A-2, B-2, C-2 or D-2 Example 1 of this manual). Set the viewing window variables to [-10, 10]1 by [-10, 10]1. Graph the function.

Zoom and trace, or use the root finder capability of your calculator, to find the x intercepts. (See Appendix Section A-7, B-7, C-7 or D-7 of this manual.) These roots are -5 and 5. The x intercepts are the cut-off values for the set of values for which $f(x)$ is positive.

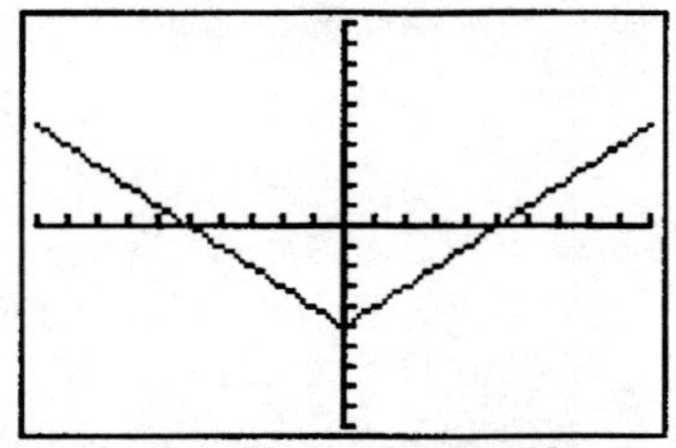

[-10, 10]1 by [-10, 10]1

Examine the graph to see where the graph is above the y axis. This is where the function is positive. The function is positive for x values greater than 5 and for x values less than -5. Hence, the solution is the set of all real numbers such that $x < -5$ or $x > 5$.

Solution (B): The graph of $f(x)$ differs from $y=|x|$ in that the graph of $f(x)$ is shifted 5 units downward from that of $y = |x|$.

EXAMPLE 8 ***TEXTBOOK SECTION 1-2***

$$\text{Graph } f(x) = \begin{cases} 2x & x \le 1 \\ x^2 & 1 < x \le 2 \\ 1 - x & 2 < x \end{cases}.$$

Solution: Store the function in the calculator as three separate pieces using an inequality statement to restrict the domain for each piece. The function will not be graphed for values of x where the function is undefined. Hence, we will divide by the inequality statement which causes the function to be only graphed for x values where it is defined. (See Appendix Section A-6, B-6, C-6, and D-6, of this manual.)

Y1: (2X)/(X≤1)	[(] [2] [X] [)] [÷] [(] [X] [2nd] [TEST] [≤] [1] [)]
Y2: (X^2)/(1<X)/(X≤2)	[(] [X] [^] [2] [)] [÷] [(] [1] [2nd] [TEST] [<] [X] [)] [÷] [(] [X] [2nd] [TEST] [≤] [X] [)]
Y1: (1-X^2)/(2<X)	[(] [1] [-] [X] [)] [÷] [(] [2] [2nd] [TEST] [<] [X] [)]

Graph the function using the standard viewing window.

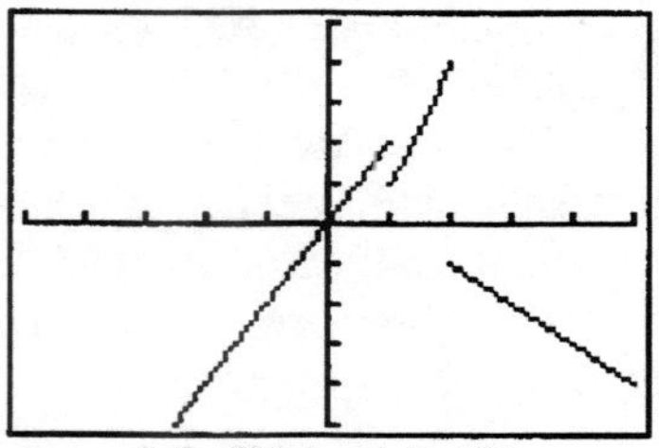

[-5, 5]1 by [-5, 5]1

EXAMPLE 9 **TEXTBOOK SECTION 1-2 & 1-3**

Use graphing techniques to explore whether or not the two lines $3.15x + 2.79y = 5.38$ and $3.05x + 2.70y = 5.30$ are parallel.

Solution: Graph both lines on the same set of coordinate axes using [-10, 10]1 by [-10, 10]1. At first glance there appears to be only one line. However, use either the zoom box or zoom in features of the calculator, or set the graph screen at [.5, 1.5]0 by [.5, 3]0 to see that there are, in fact, two lines. The lines now appear to be parallel. However, their slopes vary slightly and hence are actually two intersecting lines. (The point of intersection is $(58, -63.\overline{5})$).

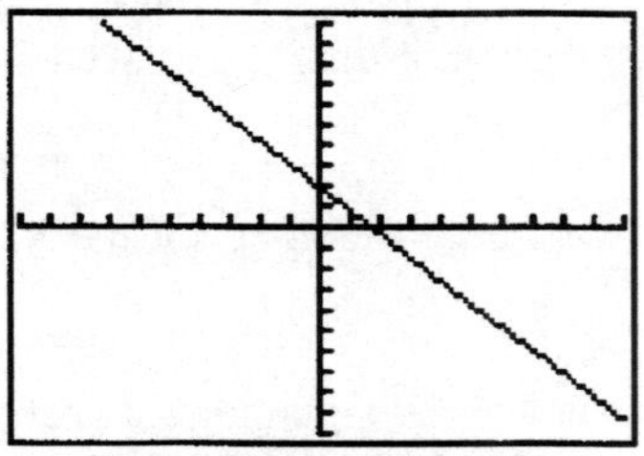

[-10, 10]1 by [-10, 10]1

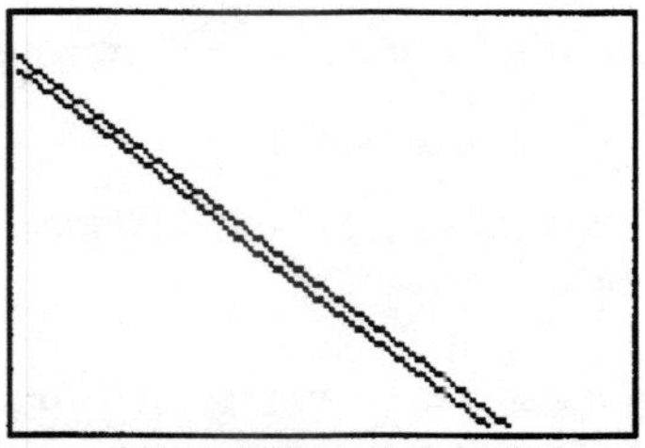

[.5, 1.5]0 by [.5, 1.5]0

EXAMPLE 10 **TEXTBOOK SECTION 1-2 & 1-3**

(A) Using a graphing utility, simultaneously graph $y = x + b$ for $b = -5, -3, 0, 3$, and 5 in a standard viewing window. Discuss the significance of the constant term.

(B) Using a graphing utility, graph $y = mx - 1$ for $m = -2, -1, 0, 1$, and 2 simultaneously in a standard viewing window. Discuss the significance of the slope.

Solution (A): Store the functions in the function list. Use [-10, 10]1 by [-10, 10]1 as the viewing window variables (see Appendix Section A-6, B-6, C-6 or D-6 of this manual). Graph. By examining the graph we see that the lines appear parallel. The slope of the line is 1 (the coefficient on x). The lines have y intercept equal to the constant term.

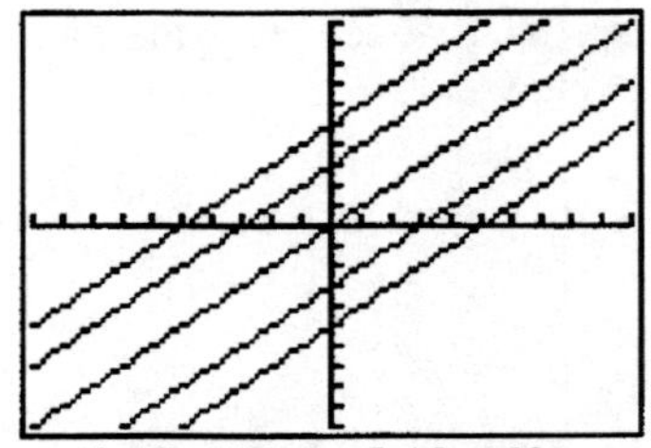

[-10, 10]1 by [-10, 10]1

Solution (B): Store the functions in the function list. Use [-10, 10]1 by [-10, 10]1 as the viewing window variables (see Appendix Section A-6, B-6, C-6 or D-6 of this manual). Graph. By examining the graph we see that the lines appear to have the same y intercept of -1 but have different slopes. The coefficient of x is the slope of the line.

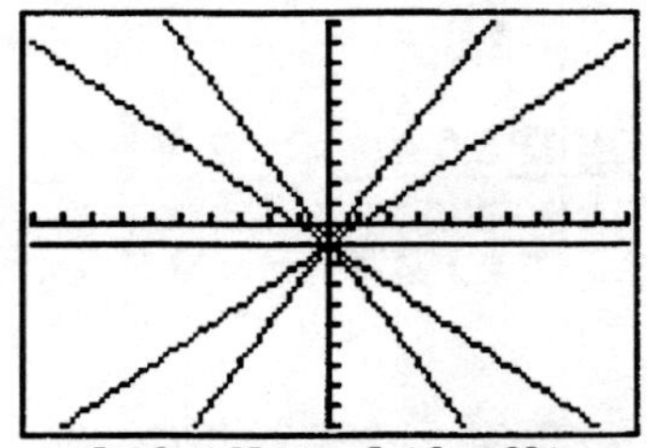

[-10, 10]1 by [-10, 10]1

EXAMPLE 11 ***TEXTBOOK SECTION 1-2 & 1-4***

(A) Graph $f(x) = -x^2 + 5x + 3$.

(B) Conjecture what you think will happen to the graph if the 3 is replaced by 6 in the function.

(C) Conjecture what you think will happen to the x and y intercepts if the 3 in $f(x)$ is replaced by a 6.

(D) Check your conjecture by graphing $g(x) = -x^2 + 5x + 6$ on the same set of coordinate axes.

Solution (A): Store $f(x)$ as Y1 in the calculator as - X ^ 2 + 5 X + 3. Set the viewing window variables at [-10, 10]1 by [-10, 10]1. Graph.

Solution (B): Let the new function be named $g(x) = -x^2 + 5x + 6$. A good conjecture is that the graph of $g(x)$ will be shifted upward 3 units from $f(x)$.

Solution (C): A good conjecture is that the least x intercept for $g(x)$ will be less than the least x intercept for $f(x)$ and that the greatest x intercept for $g(x)$ will be greater than the greatest x intercept for $f(x)$.

Furthermore, the y intercept of $g(x)$ will be greater than the y intercept for $f(x)$.

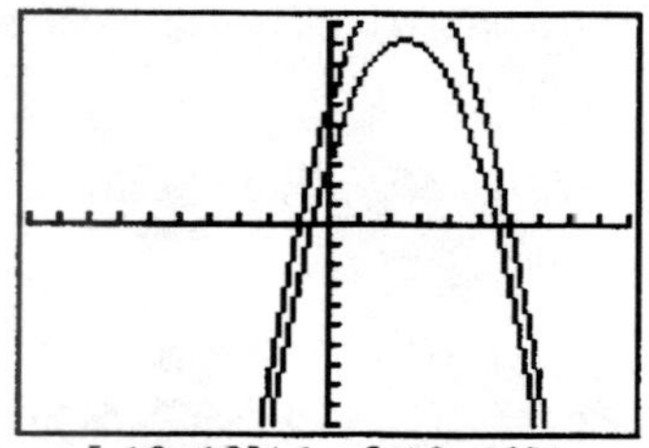

[-10, 10]1 by [-10, 10]1

Solution (D): The graph of $f(x)$ and $g(x)$ are shown to the right. See Appendix Section A-7 & A-8, B-7 & B-8, C-7 & C-8, or D-7 & D-8.

EXAMPLE 12 ***TEXTBOOK SECTION 1-2 & 1-4***

Consider the functions $f(x) = 2(x - 2)^2 + 3$ and $g(x) = 2(x - 4)^2 - 2$. Discuss the relationship of the graph of $f(x)$ to that of $g(x)$.

Solution: The graph of $g(x)$ will be two units to the left and 5 units downward from $f(x)$. The graphs are shown to the right.

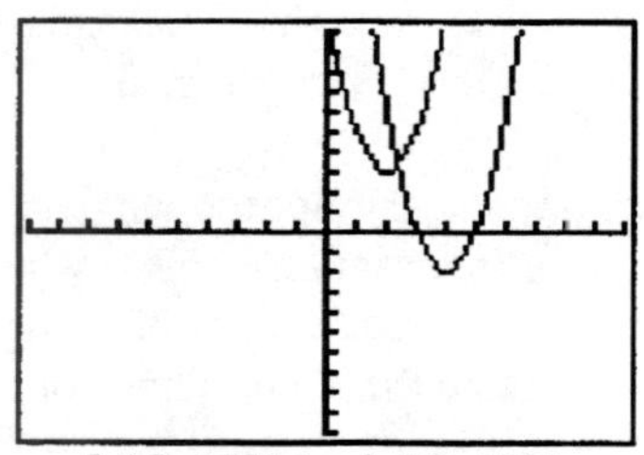

[-10, 10]1 by [-10, 10]1

EXAMPLE 13 ***TEXTBOOK SECTION 1-4***

The market research department of a company recommended to management that the company manufacture and market a promising new product. After extensive surveys, the research department backed up the recommendation with the price-demand equation $x = 4600 - 29p$ where x is the number of units that retailers are likely to buy per month at $\$p$ per unit for $0 \le p \le 150$, and where \$71,000 is the fixed cost and \$60 is the variable cost per unit.

(A) Find the cost function in terms of the price p.

(B) Find the revenue function in terms of the price p.

(C) Find the price p at which the company will break even. This price is where the cost is equal to the revenue.

Solution (A): The cost function is $C = 71000 + 60x = 71000 + 60(4600 - 29p)$.

Solution (B): The revenue equation is $R = xp = (4600 - 29p)p$.

Solution (C): We wish to find the price p at which the cost equals the revenue. Since the calculator uses X as the variable, change p to X and store the cost equation and the revenue equation in the calculator as separate functions as: 71000 + 60 (4600- 29 X) and (4600- 29 X) X .

The following chart will help in determining the viewing window variables.

x	Cost	Revenue
0	341000	0
50	254000	152500
100	167000	160000
150	80000	22500

A first choice for the viewing window variables could be [0, 150]10 by [100000, 200000]10000. The graph is shown to the right. (See Appendix Section A-8, B-8, C-8 or D-8 of this manual.)

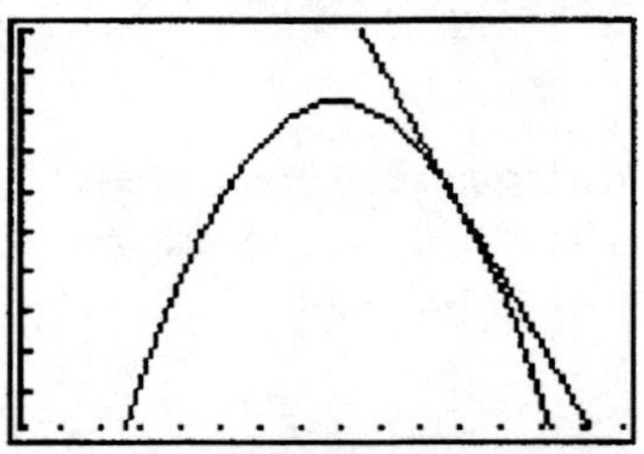

[0, 150]10 by [100000, 200000]100000

Examining the graph we see there appears to be one point of intersection. This is where the cost function and the revenue function have the same value. We want to find the X coordinate of this point.

Use the zoom feature of the calculator or change the viewing window variables to [104, 114]0 by [150000, 160000]0 to see that the graphs really do not intersect. This means that the revenue will never equal the cost. Hence, there is no break-even point. Since the revenue will always be less than the cost, this would not be a good product to produce based on this analysis.

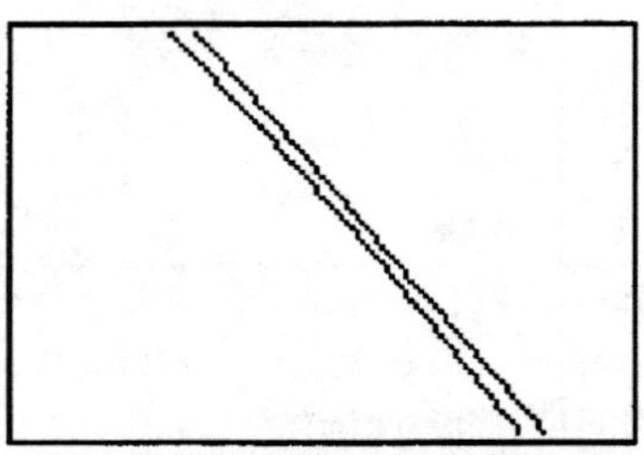

[104, 114]0 by [150000, 160000]0

This can be verified by attempting to solve the equation $71000 + 60(4600 - 29p) = (4600 - 29p)p$. This can be done algebraically by hand or using the built-in features (see Appendix Section A-9, B-9, C-9 or D-9 of this manual).

Exercise Set 1

1. Graph $f(x) = \sqrt{x^2 - 1}$. Use the graph to approximate the domain.

2. Graph $f(x) = \dfrac{\sqrt{x + 4}}{x - 1}$. Use the graph to approximate the domain.

3. Evaluate the $Q(x) = x^2 - 5x + 1$ for $x = 2.5, 2.1, 2.01,$ and 2.001. Examine the results to determine what value $Q(x)$ gets close to as x gets close to 2.

4. Repeat Problem 3 above for $x = 1, 1.5, 1.9, 1.99,$ and 1.999. Examine the results to determine what value $Q(x)$ gets close to when x gets close to 2.

5. Let $f(x) = x^2 - 2.3x + 3$. How does the quantity $\dfrac{f(1.5+h) - f(1.5)}{h}$ change as h assumes values .2, .1, .01, and .001? Fill in the chart and make a conjecture as to what happens to this quantity when h gets closer and closer to zero.

h	.2	.1	.01	.001	→	0
$\dfrac{f(1.5+h) - f(1.5)}{h}$						

6. Let $f(x) = 3x^2 - 1.3x + 3$. How does the quantity $\dfrac{f(.8+h) - f(.8)}{h}$ change as h assumes values .2, .1, .01, and .001. Fill in the chart and make a conjecture as to what happens to this quantity when h gets closer and closer to zero.

h	.2	.1	.01	.001	→	0
$\dfrac{f(.8+h) - f(.8)}{h}$						

7. Repeat Example 5 above for $p(x) = 1190 - 36x$ and the values given to the right. Find the error between the price-demand function $p(x)$ that models the data and the price p for each value of x.

x	p
2	87
5	68
8	53
12	37

8. Graph the following functions and explain how each is related to the graph of one of the six basic functions.

	Transformed Function	Basic Function
(A)	$g(x) = -\|x + 3\|$	$f(x) = \|x\|$
(B)	$g(x) = (x - 4)^2 - 3$	$f(x) = x^2$
(C)	$g(x) = 7 - \sqrt{x}$	$f(x) = \sqrt{x}$
(D)	$g(x) = -0.4x^2$	$f(x) = x^2$

9. (A) Graph $f(x) = 1.2x - 4.2$ in a standard viewing window.

   (B) Find the x and y intercepts to one decimal place using trace and zoom or an appropriate built-in function of your calculator.

   (C) Find the x and y intercepts algebraically to one decimal place.

   (D) Using the results of parts (A)-(C), find the solution set for the linear inequality $1.2x - 4.2 > 0$.

   (E) Conjecture, then check, what the solution to Part (D) would be if the function was $g(x) = 2.2x - 4.2$.

10. Repeat Problem 9 using $f(x) = 1.2x - 3.2$.

11. Graph the following equations simultaneously in a standard viewing window:

    $3x + 2y = 6 \qquad 3x + 2y = 3 \qquad 3x + 2y = -6 \qquad 3x + 2y = -3.$

    Conjecture, then check, if the graph of $3x + 2y = 0$ is in this family of lines.

12. Graph the following two equations simultaneously in a standard viewing window:

    $3x + 4y = 12 \qquad 4x - 3y = 12.$

    Conjecture, then check, how the graphs of $3x + 4y = 0$ and $4x - 3y = 0$ are related.

13. Graph the piecewise-defined function and find any points of discontinuity:

$$\text{Graph } f(x) = \begin{cases} 2x - 1 & x \le 0 \\ x^2 + 2 & 0 < x \le 2 \\ 4 + x & 2 < x \end{cases} .$$

14. Graph the piecewise-defined function and find any points of discontinuity:

$$\text{Graph } f(x) = \begin{cases} -3x & x \le -1 \\ \sqrt{x^2+8} & -1 < x \le 1 \\ 4 + x & 1 < x \end{cases}.$$

15. Use graphing techniques to explore whether or not the two lines $5.23x - 8.92y = 6.34$ and $5.21x - 8.89y = 6.30$ are parallel.

16. Use graphing techniques to explore whether or not the two lines $5.23x - 8.92y = 6.34$ and $y = .60x + 6.38$ are parallel.

17. Graph $f(x) = x^2$, $f(x) = 3x^2$, and $f(x) = -1.5x^2$ on the same coordinate axes. Compare the graphs and describe the effect of multiplying $f(x)=x^2$ by a constant.

18. Graph $f(x) = x^2$, $f(x) = (x-1)^2$, and $f(x) = (x+2)^2$ on the same coordinate axes. Compare the graphs and describe the effect of adding/subtracting a constant from the x term of $f(x)=x^2$.

NOTES

NOTES

Chapter 2

Additional Elementary Functions

This chapter contains examples using the calculator to illustrate:

- Exploring asymptotic behavior of a function
- Approximating zeros or roots of a polynomial
- Finding regression models
- Determining the domain of a function
- Exploring discontinuities of a function
- Exploring vertical asymptotes of a function
- Exploring horizontal asymptotes of a function
- Evaluating exponential and logarithmic functions
- Graphing exponential and logarithmic functions
- Solving problems involving growth or decay
- Graphing inverse functions
- Using the change of base formula to evaluate logarithms

EXAMPLE 1 ***TEXTBOOK SECTION 2-1***

How does the graph of $f(x) = x^4 - 3x^2 + x + 2$ compare to the graph of $y = x^4$ as we "zoom out?"

Solution:

Graph both functions using a viewing window of [-10, 10]1 by [-10, 10]1. [-10, 10]1 by [-40, 40]4, and [--10, 10]1 by [-100, 100]10. We see that $f(x)$ gets close to y for larger values of x.

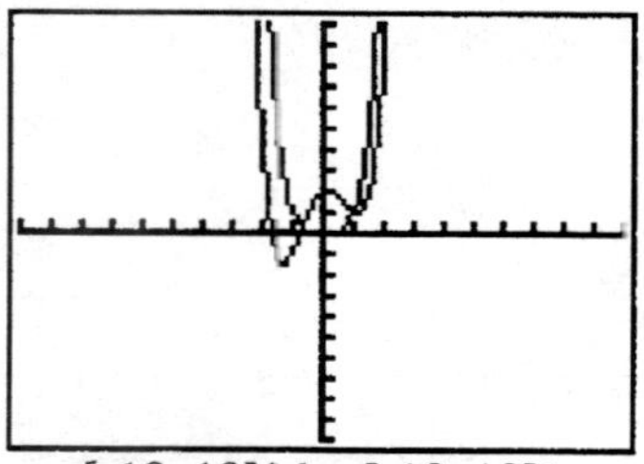

[-10, 10]1 by [-10, 10]1

Repeat using [-10, 10]1 by [-40, 40]4. and [-10, 10]1 by [-100, 100]10. We see that $f(x)$ gets close to y for larger values of x.

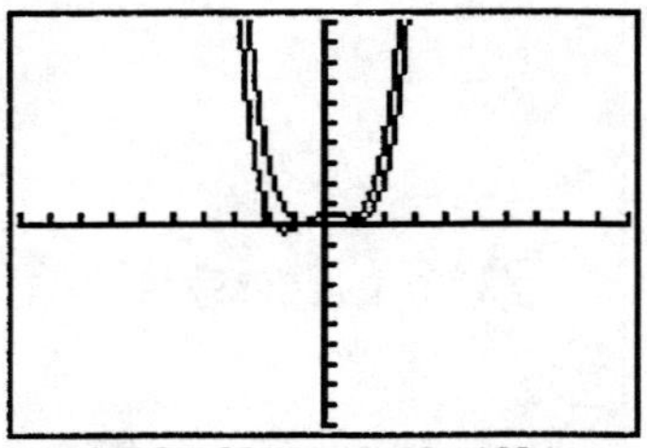

[-10,10]1 by [-40, 40]4

Repeat using [-10, 10]1 by [-100, 100]10.

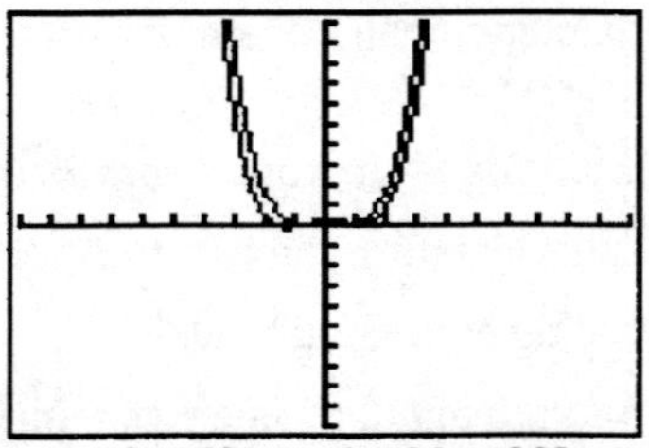

[-10, 10]1 by [-100, 100]1

We see that $f(x)$ gets close to y as x gets larger. In the graphs shown at the right, we needed to increase the range (values of y) to see how the graphs behaved as x increased since the values of the function were so large for smaller values of x.

EXAMPLE 2 ***TEXTBOOK SECTION 2-1***

Approximate, to two decimal places, the zeros of $f(x) = 2x^4 - 5x^3 - 4x^2 + 3x + 6$.

Solution: Graph the function by storing it as Y1 in the calculator. Set the viewing window variables at [-10, 10]1 by [-10, 10]1. Use the root finder feature of the calculator, or store the function as Y1, store 0 as Y2, and use the intersect feature of the calculator. (See Appendix Section A-7, B-7, C-7 or D-7 of this manual.)

The x intercepts are 1.15 and 2.89.

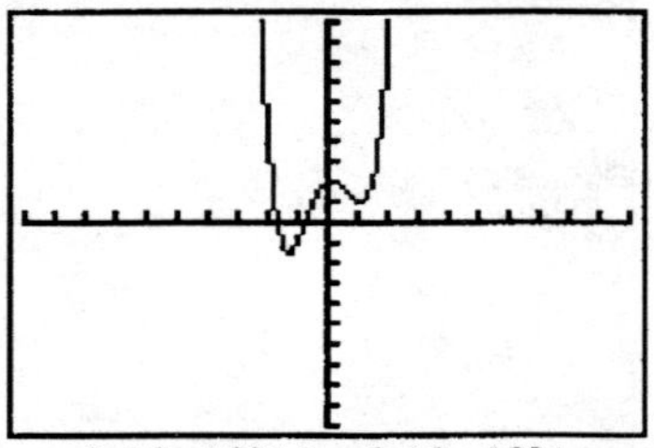

[-10, 10]1 by [-10, 10]1

EXAMPLE 3 ***TEXTBOOK SECTION 2-1***

Given the following set of data:

x	10	14	18	22	26	30	34	38	44
y	5	12	26	56	96	152	226	326	536

(A) Find a cubic polynomial regression model.

(B) Graph the data points and the regression model on the same set of coordinate axes.

(C) Predict the value of y for $x = 50$ using this regression model.

Solution (A): Enter the data into the statistical lists in the calculator. (See Appendix Section A-16, B-16, C-16 or D-16 Example 2 of this manual.) Calculate the cubic regression model. The result is $y = .01x^3 - .21x^2 + 3.23x - 17.54$ where the coefficients have been rounded to two decimal places.

Solution (B): Store the regression equation in the Y= list. Set the viewing window variables at [0, 50]5 by [0, 600]100. Graph the function and the points at the same time. (See Appendix Section A-16, B-16, C-16, or D-16 Example 2(C) of this manual.)

Solution (C): Store 50 as x in the calculator. Recall the function from the Y= list or RegEq from the VARS list and evaluate. The result is 816.38 rounded to two decimal places.

EXAMPLE 4 ***TEXTBOOK SECTION 2-1***

Consider the function $f(x) = \dfrac{x^2 - 4}{x - 2}$.

(A) Graph using viewing window variables of [-4.7, 4.7]1 by [-1, 5.2]1 on the TI-82 and TI-83, or [-6.3, 6.3]1 by [-1, 5.2]1 on the TI-85 and TI-86. Note that by using these viewing window variables the coordinates for x and y at the bottom of the viewing window are in steps of .1 when you use the arrow keys to find screen locations on the calculator. This is because on the TI-82 and TI-83 the screen is 95 pixels across by 63 pixels down; and on the TI-85 and TI-86 the screen is 127 pixels across by 63 pixels down.

(B) Find the domain.

Solution (A): Enter the function in the function list as (X^2-4)÷(X-2). Set the viewing window variables and graph. Note that there is a hole in the graph at $x = 2$. Use the trace feature of the calculator to see this.

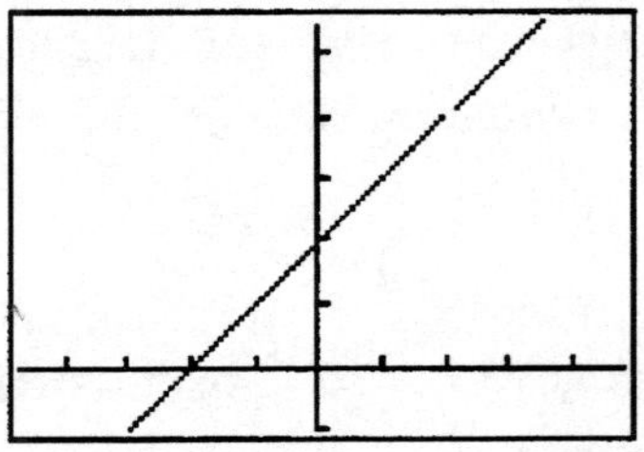

Solution (B): When you use the trace feature of the calculator you see that when the x coordinate is 2 there is no coordinate for y at the bottom of the screen. This indicates that 2 is not in the domain of the function.

Another way to see this is to return to the home screen, store 2 as X, and evaluate the function. An error message occurs indicating that the function cannot be evaluated for this value of x. This can also be seen be examining the denominator of $f(x)$. The denominator is equal to zero when $x = 2$. Hence the domain of $f(x)$ is $D = \{x \mid x \neq 2\}$.

EXAMPLE 5 ***TEXTBOOK SECTION 2-1***

Consider the function $f(x) = \dfrac{3x^2 + 5x - 8}{x^2 - 1}$.

(A) Graph using [-4.7, 4.7]1 by [-12.4, 12.4]1 on the TI-82 and TI-83, or [-6.3, 6.3]1 by [-12.4, 12.4]1 on the TI-85 and TI-86. Note that by using this range the coordinates for x at the bottom of the viewing window are in steps of .1 and y is in steps of .4 when you use the arrow keys to find screen locations (not using the trace feature) on the calculator This is because the screen on the TI-82 and TI-83 it is 95 pixels across by 63 pixels down; and on the TI-85 and TI-86 the screen is 127 pixels across by 63 pixels down.

(B) Find the domain.

Solution (A): Enter the function in the function list as (3X^2+5X-8)÷(X^2−1). Set the viewing window variables and graph the function. Note that there is unusual behavior when $x = -1$ (a vertical asymptote) and $x = 1$ (a hole in the graph).

Solution (B): Use trace to see that there is no y coordinate when the x coordinate is 1 or -1. This indicates that -1 and 1 are not in the domain of the function.

Another way to see this is to return to the home screen, store 1 as X, and evaluate the function. An error message occurs indicating that the function cannot be evaluated for this value of x. This can also be seen be examining the denominator of $f(x)$. The denominator is equal to zero when $x = 1$. Similarly for $x = -1$.

Hence the domain of $f(x)$ is $D = \{x \mid x \neq 1, x \neq -1\}$.

EXAMPLE 6 ***TEXTBOOK SECTION 2-1***

(A) Investigate the behavior of $f(x) = \dfrac{3x^2 + 5x - 8}{x^2 - 1}$ when x is close to 1 and -1.

(B) Write summary statements to Part (A) symbolically by filling in the blanks in the notation $f(x) \to$ ___ as $x \to 1$ and $f(x) \to$ ___ as $x \to -1$.

(C) Write the equations of any vertical asymptotes.

(D) Investigate the graph of f as x increases or decreases without bound.

Solution: We shall conduct our investigation by constructing a table for $f(x)$ for x values near but not equal to the given value and by examining the graph of the function using the trace and zoom features of the calculator.

Solution (A): Store the function in the calculator as (3 X ^ 2 + 5 X - 8) ÷ (X ^ 2 - 1). (See Appendix Section A-4, B-4, C-4 or D-4 of this manual.) Store .9 as X and evaluate the expression. The result is 5.632, rounded to the nearest thousandth. Repeat for the other values of x. [NOTE: The list feature could be used in evaluating this function.]

x	.900	.990	.999	→ 1 ←	1.001	1.010	1.100
$f(x)$	5.632	5.513	5.501→ ?		?←5.499	5.488	5.381

The table shows that the function is approaching 5.5 as x approaches 1 from both the right and the left. Hence, $\lim_{x \to 1} f(x) = 5.5$. However, as we saw in Example 3 above, when 1 is stored for x and the function is evaluated we get an error message. The denominator has value 0 when $x = 1$ and the calculator is attempting to divide by zero. This indicates that the function is not defined for $x = 1$. $x = 1$ is not in the domain of the function.

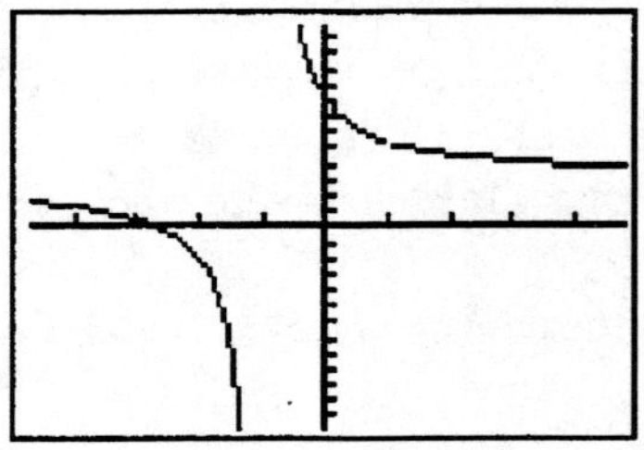

Set the viewing window variables as in Example 3 above graph the function that is stored. Use the trace feature of the calculator and move the cursor along the curve for values of x near 1. Note the y values are near 5.5. Now zoom in on the point of the graph where x is close to 1 or change the viewing window variables to [.90, 1.10].01 by [5.3, 5.7].1 and trace again.

We do not see a hole in the graph displayed on the calculator screen since the x values on the screen do not equal 1 exactly. The graph will show a hole at x = 1 if the viewing window variables are set as in Example 3 above.

Repeat the investigation for values of x near -1. Make sure the function is still stored in your calculator.

x	-1.100	-1.010	-1.001	→ -1 ←	-.999	-.990	-.900
$f(x)$	-47	-497	-4997→ ?		?←-5003	503	53

The table shows that the function values are getting greater and greater negatively for values of x close to but a little less than -1. The function values are getting greater and greater positively for values of x close to but a little greater than -1.

Set the viewing window variables to [-10, 10]1 by [-10, 10]1 and graph the function that is stored. Use the trace feature of the calculator and move the cursor along the curve for values of x near -1. We see that the y values get very large in the negative direction as x gets close to but is less than -1. Then as the cursor is moved from a little less than -1 to x values a little greater than -1 the y values jump to very large positive numbers. However, as we saw in Example 2 above, when -1 is stored for x and the function is evaluated we get an error message. The denominator has value 0 when $x = -1$ and the calculator is attempting to divide by zero. This indicates that the function is not defined for $x = -1$. $x = -1$ is not in the domain of the function.

Solution (B): Symbolically we write $f(x) \to 4$ as $x \to 1$ and $f(x) \to -\infty$ as $x \to -1^-$ $f(x) \to \infty$ as $x \to -1^+$.

Solution (C): The line $x = -1$ is a vertical asymptote.

Solution (D): To investigate the function as x increases without bound let us make a table. Store the function as Y1 in the calculator. Store the value of 10 as X and evaluate the function. Repeat for each value of X or use the list feature of your calculator. (See Appendix Section A-4, B-4, C-4 or D-4 of this manual.)

x	10	100	1,000	10,000	→	∞
$f(x)$	3.4545	3.0495	3.0050	3.0005	→	3

The graph of $y = f(x)$ approaches the line $y=3$ from above as x increases without bound. This can also be seen by using the trace feature of the calculator to exam the graph. Use the right arrow key to move the cursor far to the right. Watch the values of the y coordinate as you do this.

We repeat the process to investigate the function as x decreases without bound. We complete the table below.

x	-10	-100	-1,000	-10,000	→	-∞
$f(x)$	2.4444	2.9495	2.9950	2.9995	→	3

The graph of $y = f(x)$ approaches the line $y = 3$ from below as x decreases without bound. This can also be seen by using trace on the calculator to exam the graph. Use the left arrow key to move the cursor far to the left. Watch the values of the y coordinate as you do this.

EXAMPLE 7 ***TEXTBOOK SECTION 2-1***

(A) Investigate the behavior of $f(x) = \frac{|x - 2|}{x - 2}$ as x is near 2.

(B) Does $f(x)$ have a discontinuity at $x = 2$?

Solution (A): We shall conduct our investigation by constructing a table for $f(x)$ for x values near but not equal to 2 and by examining the graph of the function for x near 2 using the trace and zoom features of the calculator.

Store the function in the calculator as ABS (X - 2) ÷ (X - 2) where ABS is the built-in absolute value function in your calculator. (See Appendix Section A-2 and A-4, B-2 and B-4, C-2 and C-4, or D-2 and D-4 of this manual.) Store each value of x in the calculator and evaluate the stored function or use the list feature of the calculator.

x	1.900	1.990	1.999	→ 2 ←	2.001	2.010	2.100
$f(x)$	-1	-1	-1 → ?	?←	1	1	1

We see that the function values are -1 for values of x less than 2. The function values are 1 for values of x greater than 2.

Set the calculator in dotted mode. Set the viewing window variables to [-10, 10]1 by [-10, 10]1 and graph. Use the trace feature of the calculator and move the cursor near $x = 2$. Watch the y values as x moves from a little less than 2 to a little greater than 2. The y value jumps from -1 to 1.

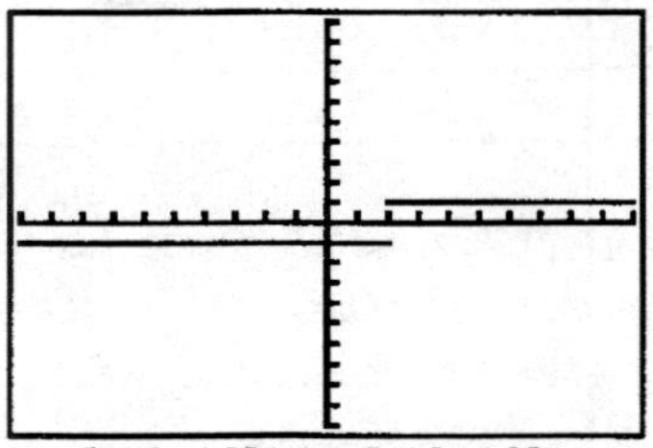

[-10, 10]1 by [-10, 10]1

Solution (B): Substitute 2 for x in the calculator and evaluate the function. We get an error message. This is because the calculator is trying to evaluate $\frac{0}{0}$. Hence we see that $f(x)$ is discontinuous at $x = 2$.

EXAMPLE 8 ***TEXTBOOK SECTION 2-2***

Graph $f(x) = 1.2^x$, $g(x) = 2.5^x$, $h(x) = -4^x$, and $k(x) = 4^x$ on the same set of coordinate axes. Describe the differences between the graphs.

Solution: Store the four functions in the calculator as 1.2^X, 2.5^X, -4^X, and 4^X. Note that -4^x is not the same function as $(-4)^x$. In fact, -4 cannot be the base of an exponential function.

Graph first using [-10, 10]1 by [-10, 10]1. It is hard to see the difference between the graphs. Change the viewing window variables to [-3, 3]1 by [-5, 5]1 to get the graph shown at the right.

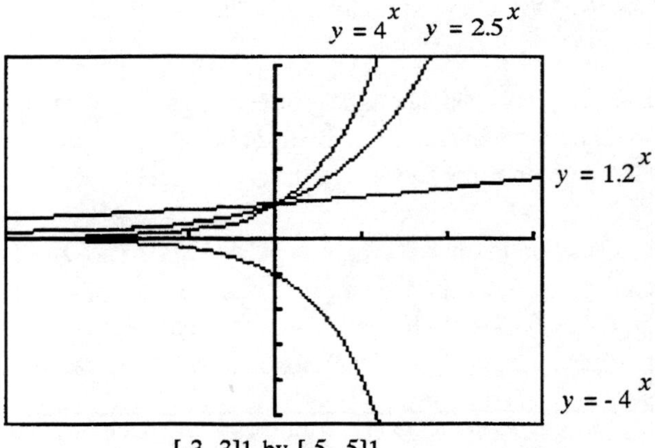

[-3, 3]1 by [-5, 5]1

The graph $f(x) = 1.2^x$ is lower than $g(x) = 2.5^x$ when $x > 0$, equal to $g(x) = 2.5^x$ when $x = 0$

and higher than $g(x) = 2.5^x$ when $x < 0$. A similar comparison can be made between $g(x) = 2.5^x$ and $k(x) = 4^x$. The graph of $h(x)$ is reflected about the x axis from $k(x)$.

EXAMPLE 9 ***TEXTBOOK SECTION 2-2***

A group of 400 parents, relatives, and friends are waiting anxiously at an airport for a student charter flight to return after a four-week trip to Europe. It is stormy and the plane is late. A particular parent thought he had heard that the plane's radio had gone out and related this news to some friends, who in turn passed it on to others, and so on. Sociologists have studied rumor propagation and have found that a rumor tends to spread according to: $N(t) = \dfrac{400}{1+399e^{-0.4t}}$ where t is measured in minutes.

(A) Graph this function.

(B) How many people will have heard the rumor after 5 minutes?

(C) Approximate the amount of time for 50% of the people to have heard the rumor.

(D) Approximate the amount of time for 75% of the people to have heard the rumor.

(E) Approximate the amount of time it will take for everyone to hear the rumor.

Solution: (A) Change t to X and store the function in the calculator as: 400 ÷ (1 + 399 **e**^ (⁻ .4 X)). (Recall e^ is the inverse ln function.)

We need to determine the viewing window variables. Evaluate the function for some values of X. Round numbers up since $N(x)$ represents the number of people having heard the rumor. For example, $N(10) = 48.146$ means that 49 people heard the rumor.

x	$N(x)$
0	1
10	49
20	353
30	400

From this table we see a good starting viewing window is [0, 50]5 by [0, 500]100.

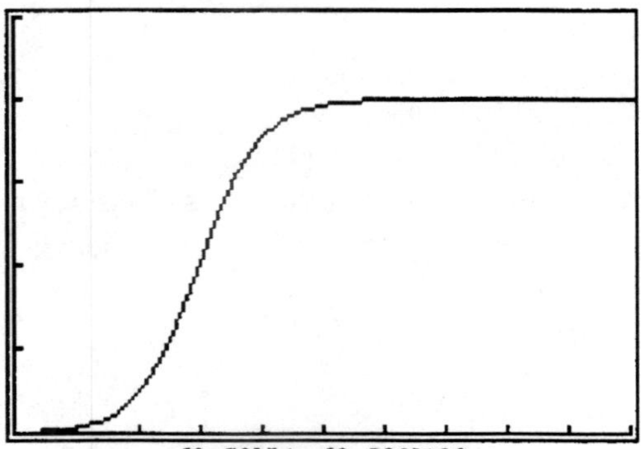

[0, 50]5 by [0, 500]100

(B) Evaluate the function for $x = 5$. The result is 8 people.

(C) Use the trace feature of the calculator and move the cursor to a position on the curve where the y value is close to 200 since 200 is 50% of 400. Zoom in on this value until accuracy is to one decimal place and round up. The value of x is approximately 15. It takes approximately 15 minutes for one-half of the people to have heard the rumor.

(D) Use the trace feature of the calculator and move the cursor to a position on the curve where the y value is close to 300 since 300 is 75% of 400. Zoom in on this value until accuracy is to one decimal place and round up. The value of x is approximately 18. It takes approximately 18 minutes for three-fourths of the people to have heard the rumor.

(E) Change the viewing window variables back to [0, 50]5 by [0, 500]100. Now use trace and watch the y variable as x increases. We see that as x gets larger and larger the y value gets closer and closer to 400. This makes sense since all people will hear a rumor eventually. We also see that 400 is reached when $x \approx 32$ which means that everyone will have heard the rumor in about 32 minutes.

EXAMPLE 10 — *TEXTBOOK SECTION 2-3*

(A) Graph $y = 3^x$, its inverse, and $y = x$ on the same set of coordinate axes. (B) Calculate the y coordinate for $y = 3^x$ when $x = 2$. Calculate the y coordinate for $y = \log_3 x$ when $x = 9$. Mark the points on the respective graphs. Discuss the relationship between these two pairs of coordinates.

Solution (A): First we need to write the inverse function of $y = 3^x$. We want to graph $y = \log_3 x$. The calculator can only calculate or graph functions having base 10 or base *e*. However, there is a formula that allows us to change from one base, say *b*, to another base, say *c*. This formula is: $\log_b x = \dfrac{\log_c x}{\log_c b}$. In our situation we have:

$\log_3 x = \dfrac{\log_{10} x}{\log_{10} 3}$. Hence we will store the three functions separately in the calculator as

3^X, X, and ([LOG] X) ÷ ([LOG] 3). Graph using [-6, 6]1 by [-4, 4]1. Note the y axis limits are 2/3 those of the x axis. This will make the scale marks about the same distance apart on both axes. The graph is shown at the right below.

Solution (B): For $y = 3^x$, $y = 9$ when $x = 2$. So the first coordinate pair is (2, 9).

For $\log_3 x$, $y = 2$ when $x = 9$. The second coordinate pair is (9, 2).

The coordinates are interchanged between the function and its inverse. The points are shown on the graph above. They are equidistant from the line $y = x$.

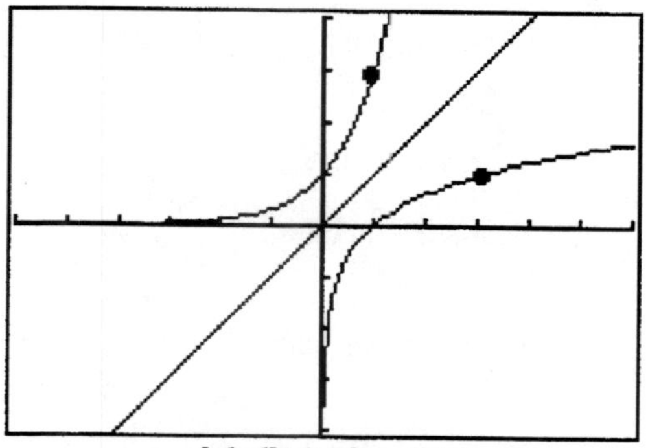

[-6, 6]1 by [-4, 4]1

Exercise Set 2

1. Approximate the zeros of $f(x) = x^5 - 12x^4 + 8x^3 + 11$ to two decimal places.

2. Approximate the zeros of $f(x) = x^5 + 14x^4 - 12x^3 - 13$ to two decimal places.

3. The following table shows the total national expenditures (in billion \$) and the percapita expenditures (in dollars) for selected years since 1960 (U. S. Census Bureau).

Date	1960	1965	1970	1975	1980	1985	1990
Total Expenditures (billion \$)	27.1	41.6	74.4	132.9	247.2	428.2	697.5
Per Capita Expenditures	143	204	346	592	1002	1666	2588

(A) Let x represent the number ofyears since 1960 and find a cubic regression polynomial for the total national expenditures.

(B) Use the regression polynomial found in Part (A) to estimate the total national expenditures (to the nearest hundred thousand) for 1995.

4. Use the data from Problem 3 to find a cubic regression polynomial for the per capita expenditures and estimate the percapita expenditures (to the nearest dollar) for 1995.

5. Given the rational function $f(x) = \dfrac{2x^2 - 7x + 5}{x^2 - 2x + 1}$.

   (A) Find any vertical and horizontal asymptotes for the graph.

   (B) Graph the function.

6. Given the rational function $f(x) = \dfrac{2x^2 - 3x - 5}{x^2 - 4x + 3}$.

   (A) Find any vertical and horizontal asymptotes for the graph.

   (B) Graph the function.

7. Graph $f(x) = 2.1^{x-1}$, $g(x) = 2.1^{x-2}$, $h(x) = 2.1^{x+1}$, and $j(x) = 2.1^x$ on the same graphing screen. Describe the differences between the graphs.

8. Graph $f(x) = 2.1^x$, $g(x) = 2.1^{3x}$, and $h(x) = 2.1^{-2x}$ on the same graphing screen. Describe the differences between the graphs.

9. Evaluate to five decimal places $\log_{1.3} 368.93$.

10. Evaluate to five decimal places $\log_{12.3} 368.93$.

11. Graph $y = 4^x$, its inverse, and $y = x$ on the same set of coordinate axes.

12. Graph $y = 1.5^x$, its inverse, and $y = x$ on the same set of coordinate axes.

NOTES

Chapter 3

Mathematics of Finance

This chapter contains examples using the calculator to illustrate:

- Calculation of simple interest, compound interest, future value of an ordinary annuity, and present value of an ordinary annuity
- Graphing of the future value with simple interest and compound interest
- Approximation of the interest rate in future value or amortization
- Evaluation of an algebraic expression to find future value and amortization schedules

EXAMPLE 1 ***TEXTBOOK SECTION 3-1***

$38,597 is invested in an account earning simple interest of 3.57%.

(A) How much interest will be earned if the money is left in the account for one year? Two years? Five years?

(B) Graph the future value as a function of time.

The replay key allows rapid calculation of expressions that are similar. This key will replay the last executed expression so you can change a quantity and recalculate. (Press [2nd] [ENTRY].)

Solution (A): Method 1 Each calculation can be done separately on the home screen without storing a formula. Enter 38597×.0357×1 in the calculator and evaluate. The result is $1,377.91. Use the replay key and change the 1 to a 2. Evaluate again. The result is $2,755.83. Use the replay key again and change the 2 to a 5. Evaluate again. The result is $6,889.56.

Method 2 Return to the home screen by pressing [2nd] [QUIT]. The future value of this loan after *t* years is A = 38597(1+.0357t). Change the *t* to an X and store the function as Y1 in the calculato as 38597 (1 + .0357 * X) as Y1 in the function list. Store the numbers 1, 2, and 5 in a list and evaluate the expression (see Appendix Section A-4, B-4, C-4 or D-4 of this manual).

Solution (B): The future value of this loan after t years is A = 38597(1+.0357t). Change the t to an X and store the function as Y1 in the calculator. An appropriate setting for the viewing window variables would be [0, 10]1 by [0, 100000]10000. The graph is shown at the right.

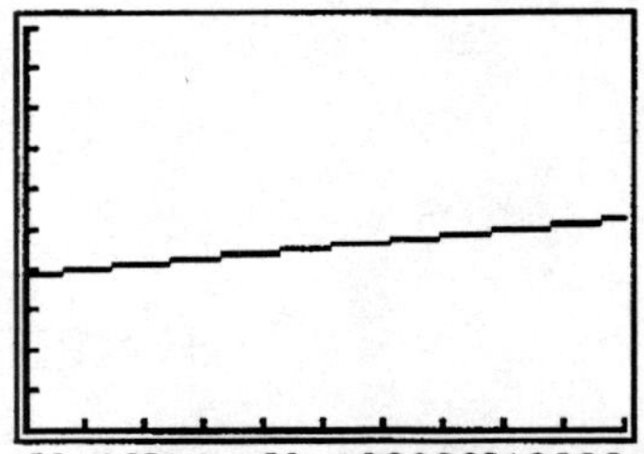

[0, 10]1 by [0, 100000]10000

Note that this graph implies interest is accumulating continuously when in fact it is calculated at the end of each time period.

EXAMPLE 2 ***TEXTBOOK SECTION 3-1 & 3-2***

$10000 is deposited in an account.

(A) Calculate the amount in the account at the end of 30 years if the account earns 8% simple interest.

(B) Calculate the amount in the account at the end of 30 years if the account earns 8% interest compounded yearly.

(C) Calculate the amount in the account at the end of 30 years if the account earns 8% interest compounded monthly.

(D) Calculate the amount in the account at the end of 30 years if the account earns 8% interest compounded daily.

(E) Graph the growth of the accounts for Part (B), (C) and (D) using viewing window variables of [0, 30]5 by [0, 111000]0. Compare the graphs. To see the difference between the graphs for monthly and daily compounding, set the viewing window variables at [29.9, 30].1 by [109000, 110400]0.

Solution: Set the number of decimal places to two. (See Appendix Section A-17, B-17, C-17 or D-17 of this manual.)

Solution (A): The formula for calculating future value for simple interest is $A = P(1+rt)$ where A =future value, P=principal, r=yearly interest rate, and t=number of years. For this situation we have 10000(1+.08×30) = 34000. Enter this on the home screen and calculate by pressing ENTER. $34,000 is in the account after 30 years.

Solution (B): The formula for calculating the future value for compound interest is $A = P(1 + \frac{r}{n})^{nt}$ where A =future value, P=principal, r=yearly interest rate, n=number of compounding periods in one year, and t=number of years. For this situation we have $10000\left(1 + \frac{.08}{1}\right)^{1\times30} = 100626.57$. Enter this on the home screen as 10000 (1 + .08 ÷ 1) ^ (1 * 30) and calculate by pressing [ENTER]. There is $100,626.57 in the account after 30 years.

Solution (B) on the TI-83 calculator:
The TI-83 calculator has financial functions that will calculate compound interest when certain variables are stored in the calculator. These are explained below for Solution (B).

[2nd] [FINANCE] [1] <TVM Solver...>	To get the FINANCE CALC menu and TVM Solver
[30] [ENTER]	To set N=30, the number of years.
[8] [ENTER]	To set I%=8, the interest rate.
[(-)] [10000] [ENTER]	To set PV=-10000, as a cash outflow which is the present value deposited.
[0] [ENTER]	To set PMT=0 since there is no payment each period.
[ENTER]	Do not set FV since this is what we are looking for.
[1] [ENTER]	To set P/Y=1, the number of payment periods per year.
[1] [ENTER]	To set C/Y=1, the number of compounding periods per year.
[ENTER]	So the calculation of interest will be done at the end of the compounding period.
[▲] [▲] [▲] [ALPHA] [SOLVE]	To place the cursor on FV and solve for the future value.

Solution (C): Using the formula in Part (B) we have $10000\left(1 + \frac{.08}{12}\right)^{12\times30} = 109357.30$. Use the above calculator instructions but set P/Y=1 and C/Y=12. Solving for PV we find that there is $109,357.30 in the account after 30 years.

Solution (D): Using the formula in Part (B) we have $10000\left(1 + \frac{.08}{365}\right)^{365\times30} = 110202.78$. Use the above calculator instructions but set P/Y=1 and C/Y=365. Solving for PV we find that there is $110,202.78 in the account after 30 years.

Solutions (C) and (D) on the TI-83 calculator: These can be found by changing the number of compounding periods per year, P/Y, to the appropriate value: P/Y=1 and C/Y=12 for Solution (C) and P/Y=1 and C/Y=365 for Solution (D).

Solution (E): Store the following in the function list in the calculator as:

Y1	10000 (1 + .08 ÷ 1) ^ (1 X)
Y2	10000 (1 + .08 ÷ 12) ^ (12 X)
Y3	10000 (1 + .08 ÷ 365) ^ (365 X)

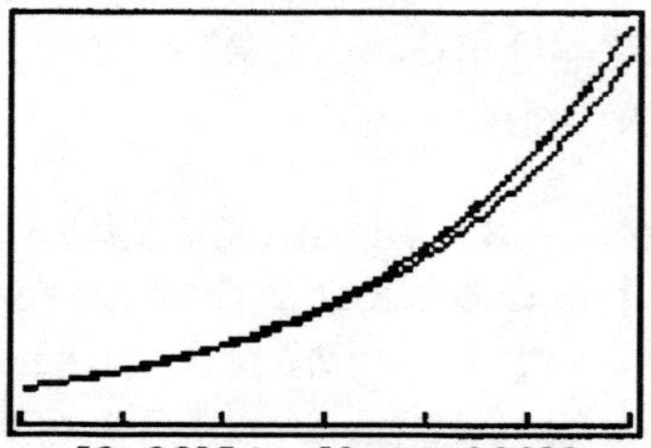
[0, 30]5 by [0, 111000]0

Set the viewing window variables to [0, 30]5 by [0, 111000]0 and graph. Watch very closely as Y2 and Y3 are graphed. Since the range is so wide for the y variable we see very little difference between these curves for the smaller values of x.

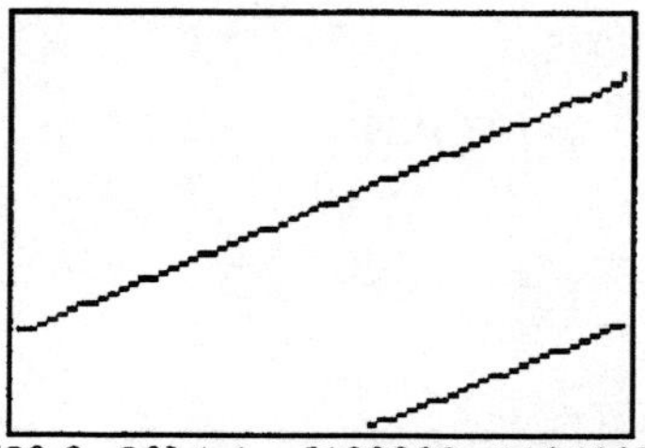
[29.9, 30].1 by [109000, 110400]0

Change the viewing window variables to [29.9, 30].1 by [109000, 110400]0 to see the difference between the graphs for the last tenth of the thirtieth year. We see there is little difference between result of monthly and daily compounded interest.

Try changing the viewing window to other settings to see these graphs. One suggestion is [25, 30]1 by [90000, 111000]0.

EXAMPLE 3 ***TEXTBOOK SECTION 3-2***

$4,570 is invested at 7.8% compounded monthly. (A) Make a chart showing the amount in the account at the end of each period for eight months. (B) How much interest is earned each month? (C) How much interest is earned overall? (D) How does the interest in the first month relate to that in each of the succeeding months? (E) Repeat Parts (A) through (D) under the conditions that the interest is withdrawn from the account at the end of each month. (F) Graph the future value as a function of time for each of the situations in this example. Why is the graph for the situation in Part (E) a horizontal line?

Solution (A) & (B): TI-81, TI-82, and TI-85 Store the expression $P(1+i)^n$ in the calculator as a function in the calculator as P(1+I)^N. Store P×I as a separate function in the calculator. Store 4570 as P, .078÷12 as I, and 1 as N.

Calculate the second expression P×I. The result is $29.71.
Calculate the first expression $P(1+i)^n$. The result is $4599.71.
Store this result as P.

Calculate the second expression again. The result is $29.90
Calculate the first expression again. The result is $4629.60.
Store this result as P.

Repeat this procedure until the chart has been filled in.

Period Number	Principal at the start of the month	Amount at the end of the month	Amount of interest earned each period
1	4570	4599.71	29.71
2	4599.71	4629.60	29.90
3	4629.60	4659.70	30.09
4	4659.70	4689.98	30.29
5	4689.98	4720.47	30.48
6	4720.47	4751.15	30.68
7	4751.15	4782.03	30.88
8	4782.03	4813.12	31.08

Sum = $243.12

TI-83 Use the built-in financial functions of the calculator to calculate the amount at the end of each month. Find the interest by subtraction.

Keystrokes	Explanation
[4570] [STO▸] [ALPHA] [A]	To store the principal at the beginning of the time period as A.
[2nd] [FINANCE] [1] <TVM Solver...>	To get the FINANCE CALC menu and TVM Solver.
[1] [ENTER]	To set N=1, the number of months.
[7.8] [÷] [12] [ENTER]	To set I%=8/12, the interest rate per month.
[(-)] [4570] [ENTER]	To set PV=-4570, as a cash outflow which is the present value deposited.
[0] [ENTER]	To set PMT=0 since there is no payment each period.
[ENTER]	It is not necessary to enter a value for FV since this is what we are looking for.
[1] [ENTER]	To set P/Y=1, the number of payment periods per month.
[1] [ENTER]	To set C/Y=1, the number of compounding periods per month.
[ENTER]	So the calculation of interest will be done at the end of the compounding period.

Keystrokes	Explanation
▲ ▲ ▲ ALPHA SOLVE	Place the cursor on FV and solve for the future value. The future value for the first month is $4599.71.
2nd QUIT	To exit the finance menu.
2nd FINANCE 6 <tvm_FV> - ALPHA A ENTER	To find the amount of interest earned this period.
2nd FINANCE 6 <tvm_FV> STO▸ ALPHA A	To store the principal for the next time period.

Repeat the above from the second line beginning 2nd FINANCE ... and store 2 as N. Repeat for each value of N until the chart has been completed.

Solutions (C) and (D) can be found by changing the number of compounding periods per year, P/Y, to the appropriate value: P/Y=12 in Solution (C). P/Y=365 in Solution (D).

Solution (C): The sum of the interest amounts should be equal to the value of the expression $P(1+i)^n - P$ for P=4750, I=.078/12, and N=8. Make certain you still have the expression $P(1+i)^n$ stored in the calculator as P(1+I)^N. Store 4570 as P, .078÷12 as I, and 8 as N. Evaluate the expression. The result is $4813.12 Now subtract P. The result is $243.12. This is the same amount as the sum of the "Amount of interest earned each period" column in the table on the preceding page. [NOTE: The financial function of the TI-83 could be used as shown in Solution (B) and using N = 8.

Solution (D): The interest on each succeeding month is the interest from the previous month plus the interest on this amount. For example: The interest for month 2 is $29.71 + 29.71 \times \frac{.078}{12}$. This demonstrates that the interest also earns interest each succeeding month.

Solution (E): The chart is now:

Period Number	Principal at the start of the month	Amount at the end of the month	Amount of interest earned each period
1	4570	4599.71	29.71
2	4570	4599.71	29.71
3	4570	4599.71	29.71
4	4570	4599.71	29.71
5	4570	4599.71	29.71
6	4570	4599.71	29.71
7	4570	4599.71	29.71
8	4570	4599.71	29.71

Sum = $29.71x8 = $237.68

When the interest earned during a month is withdrawn at the end of the month, the principal remains the same each month and hence, the interest earned in each month is the same.

Solution (F): The expression for the first situation is $4570\left(1 + \frac{.078}{12}\right)^x$ for integer values of x so that the interest is calculated and added to the principal at the beginning of the next month. The graph is shown to the right. The graph can be displayed on the calculator by entering (4570 (1 + .078 ÷ 12) ^ iPart X) as Y1,

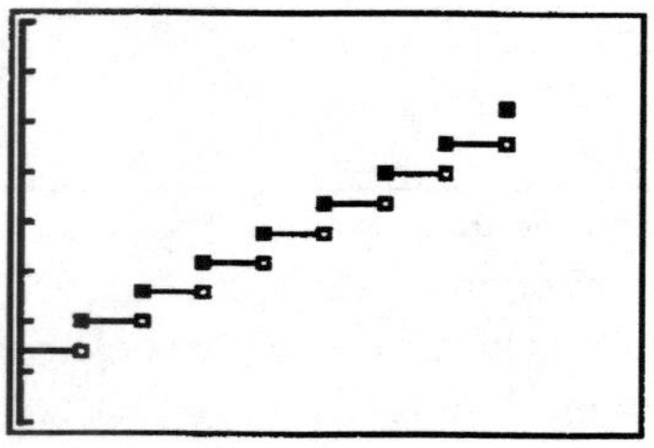

setting the graph mode to dot and using [0, 9.5]1 by [4500, 4900]0 on the TI-82 and TI-83, and [0, 12.6]1 by [4500, 4900]0 on the TI 85 and TI-86. iPart is found on the MATH menu. The open and closed dots are not displayed on the calculator. (See Appendix Section A-11, B-11, C-11, or C-11 to see how to get iPart on your calculator.)

In the second situation, the amount in the account is $4,570 for the entire month until the end of the month when the interest is calculated. The interest is added to the account at the beginning of the next month but then immediately withdrawn. This graph can be displayed using the same viewing window variables. Store 4570 (1 + .078 ÷ 12) (X = iPart X) as Y1 and 4570 ÷ (X ≠ iPart X) as Y2 where the = and ≠ signs are from the TEST menu and iPart is from the MATH menu. The open and closed dots are not displayed on the calculator.

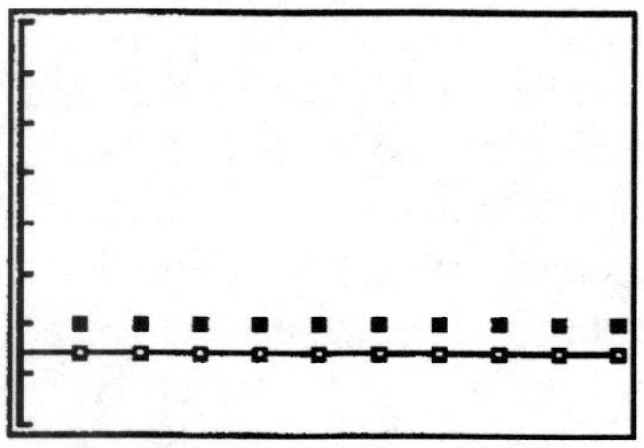

[0, 10]1 by [4500, 4900]0

We usually graph the above two functions as continuously increasing functions. To do this enter 4570 (1 + .078 ÷ 12) ^ X as Y1 in the calculator for the first situation. Enter 4570 ÷ (X < 1) as Y2 and (4570 (1 + .078 ÷ 12)) ÷ (X > 1) as Y3 and for the second situation. The inequality signs are found on the TEST menu. Set the viewing window dimensions at [0, 10]1 by [4500, 4900]0. Graph. The open and closed dots are not displayed on the calculators. (See graph to the right.)

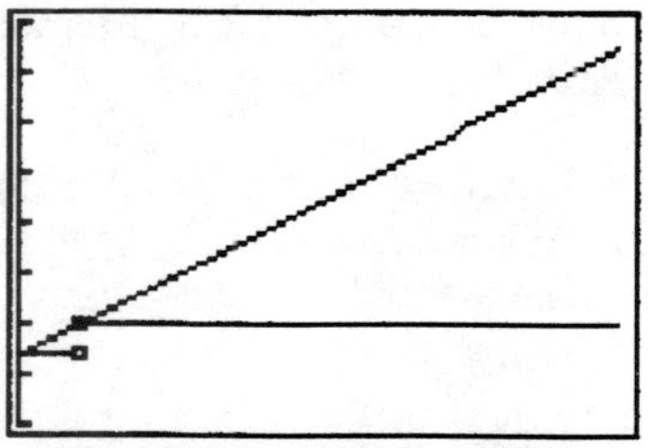

[0, 10]1 by [4500, 4900]0

Note that the future value starts at $4570 since both situations start with the same principal. The graph in the second situation is a horizontal line since the future value for each month is $4599.71 as the principal does not change each month since the interest is withdrawn from the account at the end of each month.

EXAMPLE 4 — *TEXTBOOK SECTION 3-2*

Graph the growth of an investment of $5,000 at 10% compounded semiannually and the growth of an investment of $4,500 at 12% compounded daily. (A) Which investment would be better if you plan to leave the money invested for a period of 10 years? (B) How much time will have elapsed before the investments yield the same amount?

Solution (A): We must first store the appropriate expressions in the functions list in the calculator. Enter 5000 (1 + .10 ÷ 2) ^ (2 x X) as Y1 for the first investment. Enter 4500 (1 + .12 ÷ 365) ^ (365 x X) as Y2 for the second investment. Set the viewing window variables at [0, 10]1 by [5000, 25000]500. The graphs are shown to the right. The investment of $4500 at 12% is the better investment if money is left in the account for 10 years.

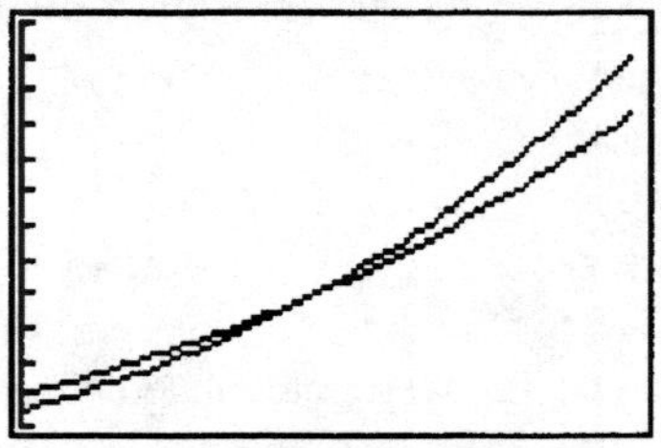

[0, 10]1 by [4000, 16000]0

Solution (B): Trace and zoom, or use other calculator methods, to find that the point of intersection occurs at x=4.7 years. Since the first investment calculates interest each 6-months, the investments would be said to be equal after 5 years.

EXAMPLE 5 — *TEXTBOOK SECTION 3-3*

$150 is deposited at the end of each month in an account earning 5.6% interest compounded monthly. Deposits were started five years ago. How much interest was earned during the third year?

Solution: We need to find the amount that was in the account at the end of the second year. Then we need to find the amount that was in the account at the end of the third year. The amount of interest earned during the third year is then given by

$$\left(\begin{array}{c}\text{the amount in the account}\\ \text{at the end of the 3rd year}\end{array}\right) - \left(\begin{array}{c}\text{the amount in the account}\\ \text{at the end of the 2nd year}\end{array}\right) - \left(\begin{array}{c}\text{the amount deposited}\\ \text{the third year}\end{array}\right)$$

To find the amount in the account at the end of the second year calculate

$$S = 150\,\frac{\left(1 + \frac{.056}{12}\right)^{24} - 1}{\frac{.056}{12}}$$

. Enter this into the calculator as 150 ((1 + .056 ÷ 12) ^ 24 - 1) ÷ (.056 ÷ 12). The result is $3,799.98.

To find the amount in the account at the end of the third year calculate

$$S = 150\,\frac{\left(1 + \frac{.056}{12}\right)^{36} - 1}{\frac{.056}{12}}$$

. The result is $5,865.25. The amount deposited during the third year is $150 x 12 = $1,800. Hence the total interest earned in the third year is $5,865.25 – 3,799.98 – 1,800 = $265.27.

TI-83

2nd FINANCE 1 <TVM Solver...>	To get the FINANCE CALC menu and TVM Solver.
24 ENTER	To set N=24, the number of payment periods.
5.6 ENTER	To set I%=5.6, the interest rate per year.
0 ENTER	To set PV=0, as there is no present value at the start.
(-) 150 ENTER	To set PMT=-150 since the payment is $150/period.
ENTER	It is not necessary to enter a value for FV since this is what we are looking for.
12 ENTER	To set P/Y=12, the number of payment periods per year.
12 ENTER	To set C/Y=12, the number of compounding periods per year.
ENTER	So the calculation of interest will be done at the end of the compounding period.
▲ ▲ ▲ ALPHA SOLVE	Place the cursor on FV and solve for the future value. The future value is $3799.98.
2nd QUIT	To exit the finance menu.
2nd FINANCE 6 <tvm_FV> STO▸ ALPHA A	To store the amount in the account at the end of 2 years.

Repeat using 36 as N to find the amount in the account at the end of 3 years.

2nd QUIT	To exit the finance menu.
2nd FINANCE 6 <tvm_FV> - ALPHA A - 150 ENTER	To find the amount of interest earned during the third year.

EXAMPLE 6 **TEXTBOOK SECTION 3-3**

A sinking fund has been accumulating money for five years. $250 is deposited at the end of each month. Currently there is $17353.15 in the account. What is the interest rate?

Solution: The formula is $S = R\,\frac{(1+i)^n - 1}{i}$. We are looking for $r = 12i$. We wish to find i from this equation. However, the formula cannot be solved for i algebraically.

i can be approximated graphically. Substituting the numbers into the equation for S, R, and n, we have $17353.15 = 250\,\frac{(1+i)^{60} - 1}{i}$. Store the left side of this equation as a function as 17353.15 in the calculator. Store the right side of this equation as a second function as: 250 ((1 + X) ^ 60 - 1) ÷ X where X represents i.

We now need to determine values for the viewing window variables. We know i will be between 0 and 1 since it is the monthly interest rate. So the viewing window variables for the horizontal axis will be [0, 1]1.

Since the first function stored is 17353.15 we must scale the y axis to accommodate this large number. Hence graph using [0, 1].1 by [0, 20000]1000. However, when this viewing window is used, nothing of the second function is seen on the graph. Hence we must adjust the variables. We suspect that perhaps the first screen dimensions are too big. Changing the viewing window to [0, .1].01 by [0, 20000]1000 gives the graph on the left below (see next page).

Using trace we see that the point of intersection is near (.005, 17500). Zoom or change the range to [0, .05].01 by [17000, 18000]1000, to get the graph on the right. (Note: It may take the calculator a little time to graph.)

Continue tracing and zooming, or changing the range, until you find $x \approx 0.00483$. This is the value of i. The rate we are seeking is $12i$ = .058 or 5.8%.

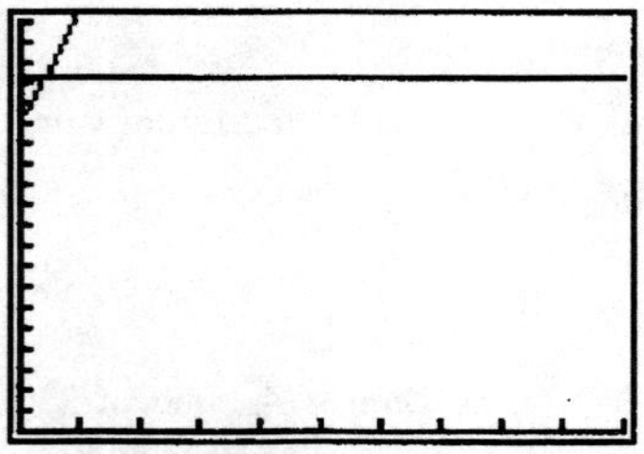

[0, .1].01 by [0, 20000]1000

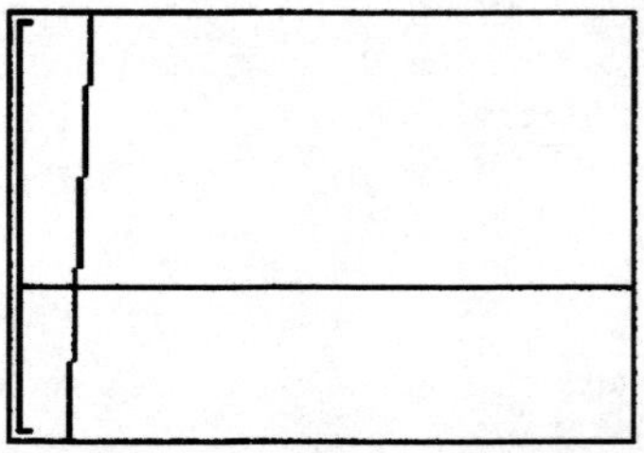

[0, .05].01 by [17000, 18000]1000

TI-83

Keystrokes	Explanation
2nd FINANCE 1 <TVM Solver...>	To get the FINANCE CALC menu and TVM Solver.
60 ENTER	To set N=60, the number of payment periods.
ENTER	You do not need to enter any particular value for I% since this is what we are looking for.
0 ENTER	To set PV=0, as there is no present value at the start.
(-) 250 ENTER	To set PMT=-250 since the payment is $250/period.
17353.15 ENTER	To set FV=17353.15, the future value.
12 ENTER	To set P/Y=12, the number of payment periods per year.
12 ENTER	To set C/Y=12, the number of compounding periods per year.
ENTER	So the calculation of interest will be done at the end of the compounding period.
▲ ▲ ▲ ▲ ▲ ▲	Place the cursor on I% and solve for the interest rate.
ALPHA SOLVE	
	The interest rate is 5.8%.

EXAMPLE 7 ***TEXTBOOK SECTION 3-4***

Suppose you have purchased a car for $11,500. The rate of financing is 9.8% compounded monthly.

(A) What are your monthly payments if you finance for one year?

(B) Make an amortization schedule.

(C) How much interest do you pay if you do not pay off the loan early?

(D) Revise the table to reflect the situation if you made $1,500 payments instead those calculated in Part (A).

(E) How much interest is paid in the situation of Part (D)?

(F) Find the payment number after which no more than one-half of the original unpaid balance is due.

Solution (A): Since we are dealing with money in this problem, it may be helpful to set the number of decimal places in the calculator to two. (See Appendix Section A-17, B-17, C-17 and D-17 of this manual.)

The formula for finding the payment is $P = 11500 \dfrac{\frac{.098}{12}}{1 - \left(1 + \frac{.098}{12}\right)^{-12}}$. Enter this into the calculator as: 11500 ((.098 ÷ 12) ÷ (1 - (1 + .098 ÷ 12) ^- 12)).

The result of the calculation is 1009.96. The payments are $1,009.96.

<u>TI-83</u>

Keystrokes	Explanation
[2nd] [FINANCE] [1] <TVM Solver...>	To get the FINANCE CALC menu and TVM Solver.
[12] [ENTER]	To set N=12, the number of payment periods.
[9.8] [ENTER]	To set the interest at 9.8%.
[11500] [ENTER]	To set PV=11500, the present value (cost of the car).
[ENTER]	Do not need to set this to a particular value since this is what we are looking for.
[0] [ENTER]	To set FV=0, the future value.
[12] [ENTER]	To set P/Y=12, the number of payment periods per year.
[12] [ENTER]	To set C/Y=12, the number of compounding periods per year.
[ENTER]	So the calculation of interest will be done at the end of the compounding period.
[▲] [▲] [▲] [▲]	Place the cursor on PMT and solve for the payment.
[ALPHA] [SOLVE]	The result is negative since this is a payment. The payment is $1009.96.

Solution (B): The amortization schedule can be found as follows:

Store the functions $P \times I$, $A - P \times I$, and $P - (A - P \times I)$ as three separate functions in the calculator. Store 11500 as P, .098÷12 as I, and 1009.96 as A.

Evaluate the three stored functions. The results are the interest, unpaid balance reduction, and unpaid balance, respectively. The results of the first calculation are 93.92, 916.04, and 10583.96. Store Ans as P. Repeat the instructions in this paragraph, recording the results as you go until the following table is complete.

Payment Number	Payment	Interest on Unpaid Balance	Unpaid Balance Reduction	New Unpaid Balance
1	1009.96	93.92	916.04	10583.96
2	1009.96	86.44	923.52	9660.43
3	1009.96	78.89	931.07	8729.37
4	1009.96	71.29	938.67	7790.70
5	1009.96	63.62	946.34	6844.36
6	1009.96	55.90	954.06	5890.30
7	1009.96	48.10	961.86	4928.44
8	1009.96	40.25	969.71	3958.73
9	1009.96	32.33	977.63	2981.10
10	1009.96	24.35	985.61	1995.48
11	1009.96	16.30	993.66	1001.82
12	1009.96	8.18	1001.78	.04

Note that the unpaid balance is not 0. This is because of rounding that takes place during the calculations. This amount probably would be added to your last payment.

<u>TI-83</u>

The unpaid balance column can be filled in by using the finance feature of the calculator as shown below:

2nd FINANCE 1 <TVM Solver...>	To get the FINANCE CALC menu and TVM Solver.
1 ENTER	To set N=1, the number of payment periods.
9.8 ENTER	To set I%=9.8, the interest rate per year.
11500 ENTER	To set PV=15000, as this is the amount owed at the beginning of the first period.
(-) 1009.96 ENTER	To set PMT=-1009.96, the payment each period.
ENTER	This does not need to be set as this is what we are looking for, the future value at the end of the first payment period.
12 ENTER	To set P/Y=12, the number of payment periods per year.
12 ENTER	To set C/Y=12, the number of compounding periods per year.
ENTER	So the calculation of interest will be done at the end of the compounding period.
▲ ▲ ▲ ALPHA SOLVE	Place the cursor on FV and solve for the future value.

The result is negative since this is what you still owe.
The result is -10583.96.
The new unpaid balance is $10,583.96.

▲ ▲ 2nd FINANCE (-) 6 <TVM_FV>	To store the unpaid balance as the new present value.
ENTER	Leave the payment at $1009.96
ENTER ALPHA SOLVER	To get the next unpaid balance.

The interest on the unpaid balance for the next period is found by multiplying the new unpaid balance by the interest rate for one period, .098÷12 = .00081667.

The unpaid balance reduction is found by subtracting the interest from the payment.

Solution (C): The amount of interest is $1009.96×12 - $11500 = $619.52.

Solution (D): The payment schedule is:

Payment Number	Payment	Interest on Unpaid Balance	Unpaid Balance Reduction	New Unpaid Balance
1	1500.00	11500.00x.098/12= 93.92	1500-93.92= 1406.08	11500.00-1406.08 = 10093.92
2	1500.00	82.43	1417.57	8676.35
3	1500.00	70.86	1429.14	7247.21
4	1500.00	59.19	1440.81	5806.40
5	1500.00	47.42	1452.58	4353.82
6	1500.00	35.36	1464.44	2889.38
7	1500.00	23.60	1476.40	1412.98
8	1500.00	11.54	1488.46	-75.48
9				
10				
11				
12				

We notice that the last value in the "New Unpaid Balance" column is -75.48. This means that the last payment need only be 1500.00-75.48 or $1,424.52.

Solution (E): The amount of interest is (1500.00×7+1424.52) - 11500 = 424.52. This illustrates the value in making payments in excess of those required if your contract will allow you to do so.

Solution (F): Half of the original unpaid balance of \$11,500 is \$5,750. We wish to find x so that $5750 = 1009.96 \dfrac{1 - \left(1 + \frac{.098}{12}\right)^{-(12-x)}}{\left(\frac{.098}{12}\right)}$.

We see from the table in Part (A) that this occurs between the sixth and seventh payment. Enter 5750 as Y1 in the calculator. Enter 1009.96 ((1 - (1 + .098 ÷ 12) ^ - (12 - X)) ÷ (.098 ÷ 12)) as Y2 in the calculator. Set the viewing window variables to [0, 8]1 by [4000, 7000]1000. Graph. The result is shown to the right.

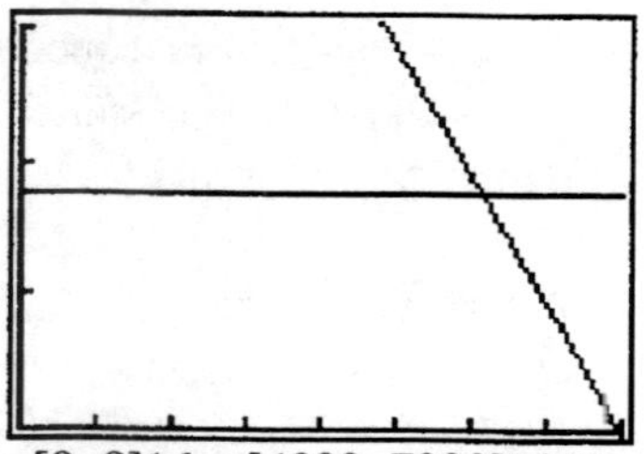

[0, 8]1 by [4000, 7000]1000

Trace and zoom, or use other calculator methods, to find the x coordinate of the point of intersection. Since this is not an integer, we round up. Hence the number of payments needed is seven. The unpaid balance after the seventh payment will be a little less than one half of the original unpaid balance.

Exercise Set 3

1. \$43,865 is invested in an account earning simple interest of 4.96%. How much interest will be earned if the money is left in the account for one year? Two years? Five years?

2. \$8,392 is invested in an account earning simple interest of 7.83%. How much interest will be earned if the money is left in the account for one year? Two years? Five years?

3. \$6,925 is invested at 6.5% compounded quarterly. (A) Make a chart showing the amount in the account at the end of each quarter for 18 months. Include a column on the chart showing how much interest is earned each quarter. (B) Make a chart showing the amount of the investment at the end of each quarter for 18 months if the interest is withdrawn after it is calculated. (C) Make a graph comparing these two situations.

4. \$200 is invested at 5.8% compounded monthly. (A) Make a chart showing the amount in the account at the end of each month for 9 months. Include a column on the chart showing how much interest is earned each quarter. (B) Make a chart showing the amount of the investment at the end of each quarter for 18 months

if the interest is withdrawn after it is calculated. (C) Make a graph comparing these two situations.

5. $50 is deposited each month in an account earning 4.8% interest compounded monthly. Deposits were started five years ago. How much interest was earned during the third year?

6. $375 is deposited each month in an account earning 7.2% interest compounded daily. Deposits were started five years ago. How much interest was earned during the third year?

7. A sinking fund has been accumulating money for five years. $200 is deposited each month. Currently there is $13532 in the account. What is the interest rate?

8. You bought a new car for $12,500. Your payments are $743.25 for 18 months. Find the interest rate?

9. A stereo system has been purchased for $3500. The rate of financing is 9.8% compounded monthly. What are your payments if you finance for one year? Make an amortization schedule. How much interest do you pay?

10. Repeat Example 5 above if you finance at 8.5% for 18 months. Do you pay more interest under this plan?

NOTES

Chapter 4

Systems of Linear Equations; Matrices

This chapter contains examples using the calculator to illustrate:

- Solving of systems of equations in two variables using graphing
- Solving of systems of equations using Gauss-Jordan reduction
- Solving systems of equations using the inverse coefficient matrix
- Solving problems using matrix operations

EXAMPLE 1 ***TEXTBOOK SECTION 4-1***

Given the system of equations $\begin{aligned} 11x + 2y &= 1 \\ 2x + 5y &= -24 \end{aligned}$.

(A) Solve to two decimal places using graphing techniques.

(B) Conjecture, then check, the effect on the solution of changing the 11 in the first equation to 20.

(C) Conjecture, then check, the effect on the solution of changing the 1 in the first equation of the original system to 6.

Solution (A): Each of these equations must be solved algebraically for y in order to enter it into the calculator.

The system becomes:	Store in the calculator as two separate functions:
$y = \dfrac{1-11x}{2}$	(1 - 11 X) ÷ 2
$y = \dfrac{-24-2x}{5}$	(-24 - 2 X) ÷ 5

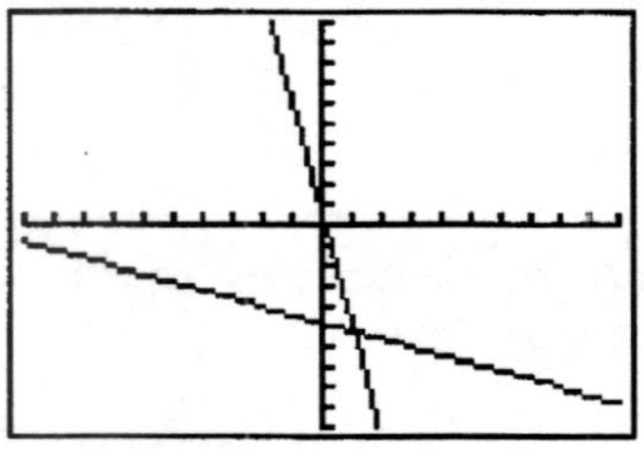

[-10, 10]1 by [-10, 10]1

Graph the lines using [-10,10]1 by [-10,10]1. Trace and zoom, or use other calculator techniques, to find the coordinates of the point of intersection. The solution is $x = 1.04$, $y = -5.22$.

Solution (B): Change the 11 in the first equation to 20. Store this in the calculator as Y3 as (1 - 20 X) ÷ 2. Changing the 11 in the first equation to 20 changes the slope of the line from $-\frac{11}{2} = -5.5$ to $-\frac{20}{2} = -10$. This means the line has steeper slope but the same y intercept.

Hence the x coordinate of the point of intersection will move to the left and the y coordinate will move upward since the slope of the second line is negative. The coordinates of the point of intersection are (.55, -5.02).

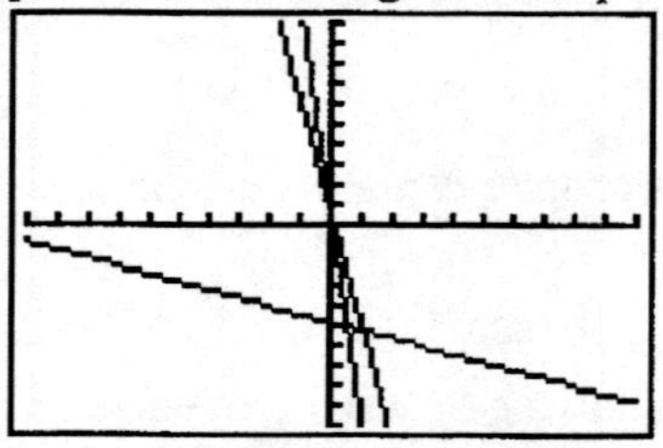

[-10, 10]1 by [-10, 10]1

(All three lines are graphed to the right.)

Solution (C): Change the 1 to 6 in the original first equation. Store this in the calculator as Y3 as (6 - 11 X) ÷ 2. Changing the 1 to 6 in the original first equation changes the y intercept from $\frac{1}{2}$ to $\frac{6}{2} = 3$. This shifts the first line upward (or to the right). Hence the point of intersection will move to the right and downward. The x coordinate will be greater and the y coordinate will be less than the coordinates of the original intersection point. The solution is $x = 1.53$, $y = -5.41$.

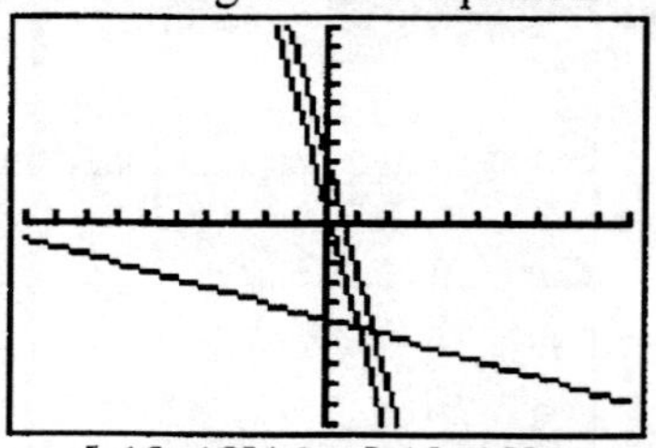

[-10, 10]1 by [-10, 10]1

EXAMPLE 2 ***TEXTBOOK SECTION 4-1***

Solve
$$\begin{aligned} 9x - 3y &= 24 \\ 11x + 2y &= 1 \\ 2x + 5y &= -24 \end{aligned}$$
using graphing techniques.

Solution: Each of these equations must be solved algebraically for y in order to enter it into the calculator.

The system becomes:

$$y = \frac{24-9x}{-3}$$

$$y = \frac{1-11x}{2}$$

$$y = \frac{-24-2x}{5}$$

Store in the calculator as three separate functions:

(24 - 9 X) ÷ -3

(1 - 11 X) ÷ 2

(-24 - 2 X) ÷ 5

Graph these three lines using [-10, 10]1 by [-10, 10]1. The lines appear to intersect in a single point. Zoom or change the viewing window variables to [.55, 1.55]1 by [-5.8, -4.6]1 to see that the three lines do not intersect in a single point (see graphs below). Consequently there is no one ordered pair that satisfies all three equations at the same time. Hence there is no solution to this system of equations.

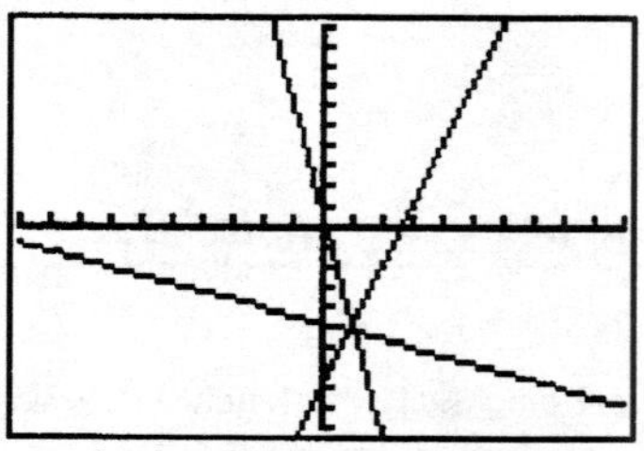

[-10, 10]1 by [-10, 10]1

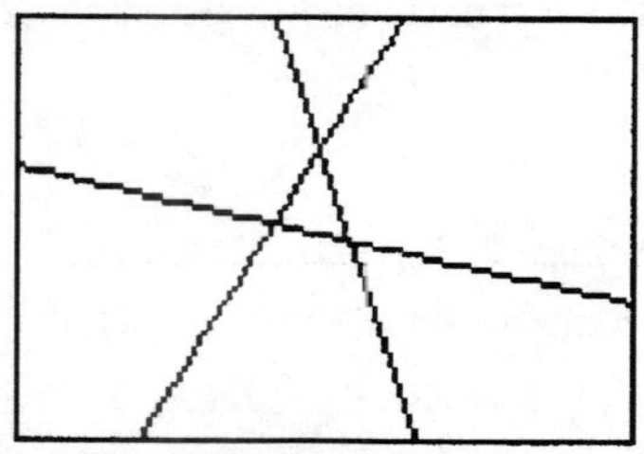

[.55, 1.55]1 by [-5.8, -4.6]1

EXAMPLE 3 ***TEXTBOOK SECTION 4-2 & 4-3***

Given the system of equations:

$$\begin{aligned} 7x - 2y + z &= 3 \\ 3x + y - z &= 6 \\ x - 3y + 2z &= 0 \end{aligned}$$

(A) Solve to two decimal places using Gauss-Jordan Elimination.

(B) Conjecture, then check, how the solution for x would be changed if the 3 in the first equation was changed to a 0.

(C) Conjecture, then check, what the solution to the system would be if both the 3 in the first equation and the 6 in the second equation are changed to 0.

Solution (A): Enter the augmented coefficient matrix into the calculator. The particular keystrokes for your calculator are described in Appendix Section A-12, B-12, C-12 or D-12 Example 4 of this manual. Use the built-in features of your calculator to perform row operations to get the reduced form.

$$\left[\begin{array}{ccc|c} 7 & -2 & 1 & 3 \\ 3 & 1 & -1 & 6 \\ 1 & -3 & 2 & 0 \end{array}\right] \sim \left[\begin{array}{ccc|c} 1 & 0 & 0 & -1 \\ 0 & 1 & 0 & -19 \\ 0 & 0 & 1 & -28 \end{array}\right].$$

Hence the solution to this system is $x = -1$, $y = -19$, and $z = -28$.

Solution (B): If the 3 is changed to a 0 the value for x will be less than -1. In fact, the solution for x is - 2.

Solution (C): The solution would be $x=0$, $y=0$, and $z=0$ since there would be no values added or subtracted in the constant column.

EXAMPLE 4 ***TEXTBOOK SECTION 4-4 & 4-5***

Given the matrices below find: (A) AB (B) BA (C) $A + A^{-1}$ (D) $2.18A$ (E) A^2 and A^3. Round all resulting numbers to two decimal places after calculations have been completed.

$$A = \begin{bmatrix} .28 & .31 & .45 \\ .06 & -.02 & -.40 \\ .09 & -.42 & -.93 \end{bmatrix} \qquad B = \begin{bmatrix} .47 \\ -6.38 \\ .81 \end{bmatrix}$$

Use two decimal place accuracy in all answers.

Solution: Store matrices A and B in the calculator. (See Appendix Section A-12, B-12, C-12 or D-12 of this manual.)

(A) Find the product AB. The result is $\begin{bmatrix} -1.48 \\ -.17 \\ 1.97 \end{bmatrix}$.

(B) Find the product BA. An error message appears on the calculator screen. This is because the dimensions are not correct for multiplication. You cannot multiply a 3×1 matrix by a 3×3 matrix.

(C) Find the sum $A + A^{-1}$. The result is $\begin{bmatrix} 3.51 & -1.84 & 2.94 \\ -.37 & 6.49 & -3.41 \\ .60 & -3.57 & -.41 \end{bmatrix}$.

(D) Find 2.18A . The result is $\begin{bmatrix} .61 & .68 & .98 \\ .13 & -.04 & -.87 \\ .20 & -.92 & -2.03 \end{bmatrix}$.

(E) Find A^2 by recalling matrix A and multiplying it by itself or using the $\boxed{x^2}$ key.

The result is $\begin{bmatrix} .14 & -.11 & -.42 \\ -.02 & .19 & .41 \\ -.08 & .43 & 1.07 \end{bmatrix}$.

Now recall the answer using [2nd] [ANS] and multiply it by A.

The result is $\begin{bmatrix} -.01 & .22 & .49 \\ .04 & -.18 & -.46 \\ .10 & -.49 & -1.21 \end{bmatrix}$.

EXAMPLE 5 ***TEXTBOOK SECTION 4-4***

The time requirements for the manufacture of slalom and trick water skis is given by :

	Fabricating department	Finishing department	
Trick ski	6.0 hr	1.5 hr	= A
Slalom ski	4.0 hr	1.0 hr	

Now suppose the company has two manufacturing plants, X and Y, in different parts of the country and that their hourly rates for each department are given in the following matrix:

	Plant X	Plant Y	
Fabricating department	\$8/hr	\$7/hr	= B
Finishing department	\$6/hr	\$4/hr	

(A) Find the total labor costs for each ski at each factory.

(B) Conjecture, then check, how the total labor costs change if the hourly rates at Plant X in the fabricating department and the finishing department increased to \$8.50/hr and \$6.35/hr respectively?

Solution (A): Enter the two matrices into the calculator. (See Appendix Section A-12, B-12, C-12 or D-12 of this manual.) Find AB. The result is $\begin{bmatrix} \$57 & \$48 \\ \$38 & \$32 \end{bmatrix}$.

Solution (B): A good conjecture would be that the total labor costs would increase since both the hourly rates increased. Change the matrix elements in matrix B as indicated above and recalculate. The result is now is $\begin{bmatrix} \$60.53 & \$48 \\ \$40.35 & \$32 \end{bmatrix}$. The labor costs at Plant X increase \$3.53 per ski for the trick ski and \$2.35 per ski for the slalom ski.

EXAMPLE 6 ***TEXTBOOK SECTION 4-6***

Solve to two decimal places using inverse matrix methods:

$$\begin{aligned} 7x - 2y + z &= 3 \\ 3x + y - z &= 6 \\ x - 3y + 2z &= 0 \end{aligned}$$

Solution: Enter the coefficient matrix into the calculator. The particular keystrokes for your calculator are described in Appendix Section A-12, B-12, C-12 or D-12 Example 2 of this manual. Use the built-in features of your calculator to get the inverse matrix shown below.

$$A^{-1} = \begin{bmatrix} .3333 & -.3333 & -.3333 \\ 2.3333 & -4.3333 & -3.3333 \\ 3.3333 & -6.3333 & -4.3333 \end{bmatrix}$$

Find the matrix product of A^{-1} and B:

$$\begin{bmatrix} .3333 & -.3333 & -.3333 \\ 2.3333 & -4.3333 & -3.3333 \\ 3.3333 & -6.3333 & -4.3333 \end{bmatrix} \begin{bmatrix} 3 \\ 6 \\ 0 \end{bmatrix} = \begin{bmatrix} -1 \\ -19 \\ -28 \end{bmatrix}$$

. Hence $x = -1$, $y = -19$, and $z = -28$.

Exercise Set 4

1. Solve $\begin{aligned} 7x - 5y &= 14 \\ 11x + 2y &= 1 \\ 2x + 5y &= -10 \end{aligned}$ using graphing techniques.

2. Solve $\begin{aligned} 7x - 5y &= 14 \\ 11x + 2y &= 1 \\ 2x + 3.5y &= -7.5 \end{aligned}$ using graphing techniques.

3. Conjecture, then check, how the solution to Problem 1 above will change if the constant in the first equation is 16 instead of 14.

4. Conjecture, then check, how the solution to Problem 2 above will change if the coefficient on x in the first equation is 6 instead of 7.

5. Solve to two decimal places using Gauss-Jordan reduction:

$$\begin{aligned} 6x - 3y + 2z &= 3 \\ 3x + 2y - 3z &= 6.1 \\ x - 3y + z &= 5.2 \end{aligned}$$

6. Solve to two decimal places using Gauss-Jordan reduction:

$$\begin{aligned} 5x + 3y + 2z &= 4.5 \\ 3.1x - 2y + z &= 6.2 \\ 3x - 3.7y + 2.2z &= 9 \end{aligned}$$

7. Solve to two decimal places using inverse matrix methods:

$$\begin{aligned} 5x + 3y + 2z &= 4.5 \\ 3.1x - 2y + z &= 6.2 \\ 3x - 3.7y + 2.2z &= 9 \end{aligned}$$

8. Solve to two decimal places using inverse matrix methods:

$$\begin{aligned} 6x - 3y + 2z &= 3 \\ 3x + 2y - 3z &= 6.1 \\ x - 3y + z &= 5.2 \end{aligned}$$

9. Repeat all parts of Example 5 in this chapter of the manual if the hourly rate for the fabricating department at Plant X begins at $7.50/hr instead of $8/hr. Again, describe what happens when Plant X costs rise to $8.50/hr and $6.35/hr for fabricating and finishing, respectively.

10. Suppose a third type of ski was to be manufactured needing 8 hours in the fabricating department and 2 hours in the finishing department. Repeat Example 5 in this manual above under these conditions.

NOTES

Chapter 5

Linear Inequalities and Linear Programming

This chapter contains examples using the calculator to illustrate:

- Solving of systems of inequalities in two variables using graphing
- Finding corner points graphically
- Solving a linear programming problem in two variables using graphing
- Solving linear programming problems using matrix methods

EXAMPLE 1 ***TEXTBOOK SECTION 5-1***

Consider the system of linear inequalities listed below:

$$\begin{aligned} 2.3x + y &\geq 10 \\ 3x + y &\leq 14 \\ 1.8x + 4.8y &\leq 48.3 \\ x &\geq 0 \\ y &\geq 0 \end{aligned}$$

(A) Solve the system graphically.

(B) Find the corner points.

(C) Conjecture, then check, the effect on the solution if the 10 in the first inequality was changed to 5.

(D) Conjecture, then check, the effect on the solution if the 2.3 in the original system of inequalities was changed to a 1.

Solution (A): <u>*Step 1*</u> Change the inequality signs to equal signs and graph the lines. This results in graphing the lines:

$$y = 10 - 2.3x$$
$$y = 14 - 3x$$
$$y = \frac{48.3 - 1.8x}{4.8}$$
$$x = 0$$
$$y = 0$$

Store the first three of these expressions in the calculator as separate functions: 10 - 2.3 X, 14 - 3 X, and (48.3 - 1.8 X) ÷ 4.8 . Graph using [0,15]1 by [0,15]1. We know the graph is in the first quadrant because of the last two inequalities.

<u>*Step 2*</u> Change the expressions stored as Y1, Y2, and Y3 into statements with the appropriate inequality sign. (See Appendix Section A-5, B-5, C-5 or D-5 of this manual.) To do this use the INS function on the calculator and get the inequality signs from the [TEST] menu. The expressions are now stored in Y1, Y2, and Y3 as: Y ≥ 10 - 2.3 X, Y ≤ 14 - 3 X, and Y ≤ (48.3 - 1.8 X) ÷ 4.8.

<u>*Step 3*</u> Choose a point not on any of the lines. We will use (1, 1). Enter 1 for X and 1 for Y in the calculator. Evaluate the inequalities by recalling each of the expressions stored in the calculator. If the result is 1 the statement is true and the point is in the solution region of that inequality. If the result in the calculator is 0, then the statement is not true and the point is not in the solution region. In this case we want the side of the line on the opposite side from (1, 1).

We see that the first inequality is false for x=1 and y=1. The second and third inequalities are true. So we want the region above the first line and below the second and third lines. This is shown in the figure to the right.

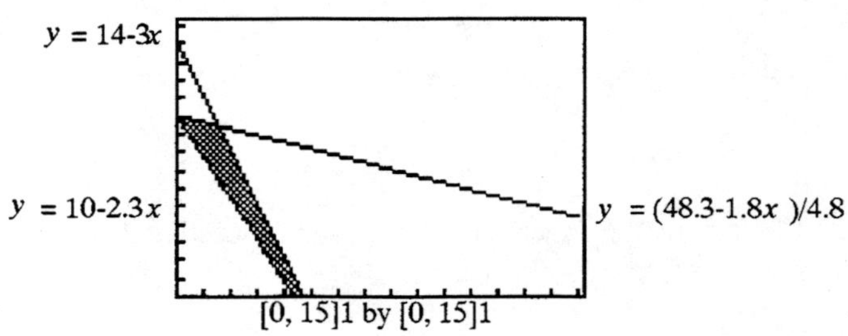

The solution of this system of linear inequalities is the set of ordered pairs corresponding to points in the shaded region and including the line segments that bound it. The shading cannot be done on all graphing calculators. It is shown here to indicate the solution region.

Solution (B): Change the expressions stored as Y1, Y2, and Y3 back to those in Step 1 above by deleting the Y and the inequality signs. Graph the lines. The corner points are found by tracing and zooming, or using other calculator methods (see Appendix Section A-7, B-7, C-7 or D-7 of this manual). The two corner points on the y axis are (0, 10) and (0, 10.06). The two corner points on the x axis are (4.35, 0) and (4.67, 0). The other corner point is (1.50, 9.50).

Solution (C): If the first inequality was $2.3x + y \geq 5$, the y intercept of the boundary line would be 5 instead of 10. This would shift the line downward. Since the slope of the line would be the same, the x intercept would shift to the left.

This situation is shown in the graph to the right. The corner points are now (0.00, 5.00), (0.00, 10.06), (2.17, 0.00), (4.67, 0.00) and (1.50, 9.50).

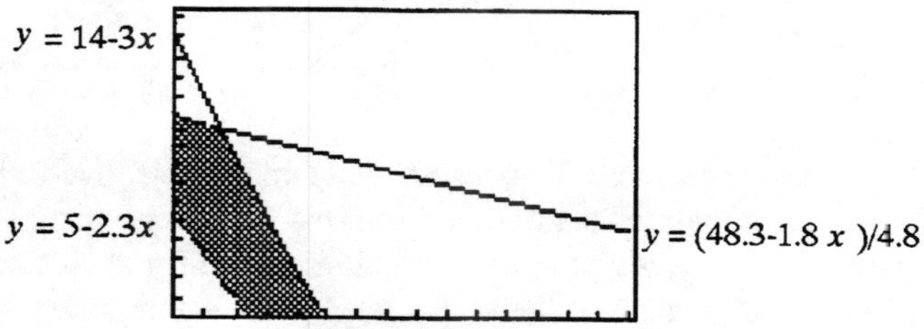

[0, 15]1 by [0, 15]1

Solution (D): If the first inequality was $x + y \geq 10$, the slope of this boundary line is not as steep as that of the original boundary line. The y intercept remains the same. Hence the x intercept moves to the right. In fact, the x intercept is now 10. Since the x intercept of y=14-3x is 4.67, the corner point is now the intersection of the two lines y=14-3x and y=10 - x.

The corner points of the solution region are now (0.00, 5.00), (0.00, 10.06), (2.00, 8.00), and (1.50, 9.50).

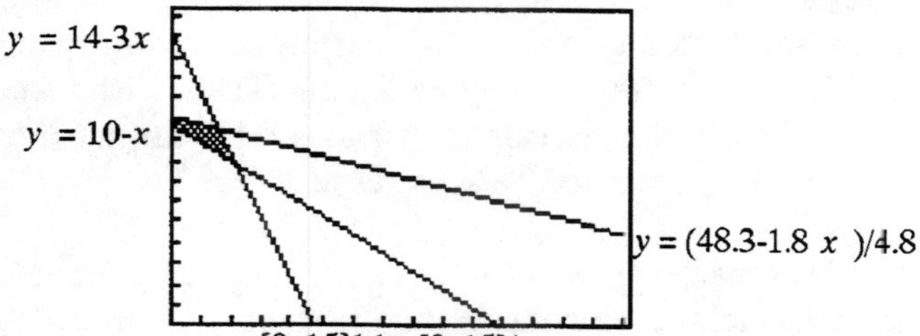

[0, 15]1 by [0, 15]1

EXAMPLE 2 — *TEXTBOOK SECTION 5-2*

Maximize and minimize the function $z = 1.2x + y$ over the region graphed in Example 1 above.

Solution: We wish to see what the objective function line looks like. Graph the line $1.2x + y = 15$. (The value of 15 was chosen because the largest y intercept was 10.06.) Enter this in the calculator as 15-1.2X. Also graph 14-1.2X, 13-1.2X, 7-1.2X and 3-1.2X on the graph.

You may not be able to get all of the lines graphed on the screen at the same time. (See Appendix Sections A-6, B-6, C-6 or D-6 of this manual.) Sketch a graph on paper and add the parallel lines.

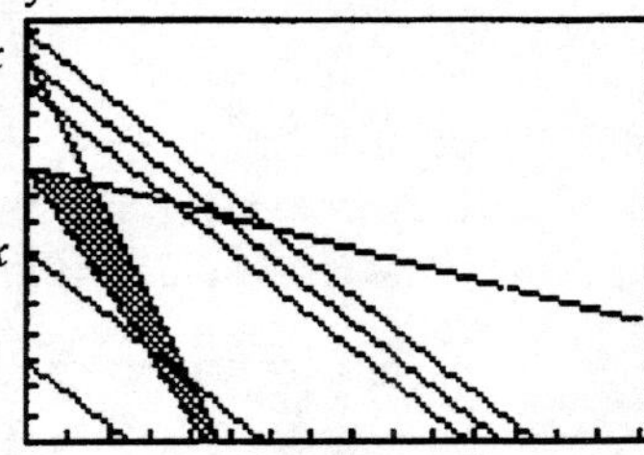

[0, 15]1 by [0, 15]1

We see that as we decrease the constant value in the objective function line, the first corner point encountered is (1.50, 9.50). This will be the point that gives the maximum value to z. Similarly, we see that the smallest value for z will occur at the corner point (4.35, 0). Evaluating z at these points results in: A maximum value of z occurs at (1.5, 9.5). The maximum value of z is 11.3. A minimum value of z occurs at (4.35, 0). The minimum value of z is 5.22.

EXAMPLE 3 — *TEXTBOOK SECTION 5-4*

A manufacturer of lightweight mountain tents makes a standard model and an expedition model for national distribution. Each standard tent requires 1 labor hour from the cutting department and 3 labor hours from the assembly department. Each expedition tent requires 2 labor hours from the cutting department and 4 labor hours from the assembly department. The maximum labor hours available per day in the cutting department and the assembly department are 32 and 84, respectively.

If the company makes a profit of \$50 on each standard tent and \$80 on each expedition tent, how many tents of each type should be manufactured each day to maximize the total daily profit (assuming all tents can be sold)?

Solution: The mathematical model for this problem is:

$$\begin{aligned} \text{Maximize} \quad & P = 50x + 80y \\ \text{Subject to} \quad & x + 2y \le 32 \\ & 3x + 4y \le 84 \\ & x, y \ge 0 \end{aligned}$$

This problem could be done graphically since there are only two variables. However, we will use the Simplex Method.

The initial system is:

$$\begin{aligned} x + 2y + s_1 \qquad\qquad &= 32 \\ 3x + 4y \qquad + s_2 \qquad &= 84 \\ -50x - 80y \qquad\qquad + P &= 0 \end{aligned}$$

So the initial tableau is:

$$\begin{array}{c} \\ S1 \\ S2 \\ P \end{array} \begin{array}{c} \begin{array}{ccccc} x_1 & x_2 & S_1 & S_2 & P \end{array} \\ \left[\begin{array}{ccccc|c} 1 & 2 & 1 & 0 & 0 & 32 \\ 3 & 4 & 0 & 1 & 0 & 84 \\ \hline -50 & -80 & 0 & 0 & 1 & 0 \end{array}\right] \end{array}.$$

Use matrix row operations to perform the pivot operations. (See Appendix Section A-12, B-12, C-12 or D-12 of this manual.) The final tableau is:

$$\begin{array}{c} \\ S1 \\ S2 \\ P \end{array} \begin{array}{c} \begin{array}{ccccc} x_1 & x_2 & S_1 & S_2 & P \end{array} \\ \left[\begin{array}{ccccc|c} 0 & 1 & 1.5 & -.5 & 0 & 6 \\ 1 & 0 & -2 & 1 & 0 & 20 \\ \hline 0 & 0 & 20 & 10 & 1 & 1480 \end{array}\right] \end{array}.$$

So the maximum profit of \$1480 is achieved if 20 standard tents and 6 expedition tents are manufactured and sold. Both slack variables have a value of zero. To see this, substitute the values for x and y into the slack variable equations. This means that all the hours available in the two departments are used.

Exercise Set 5

1. Solve the following system of linear inequalities graphically, and find the corner points:

$$\begin{aligned} 4.5x + 2.1y &\geq 15 \\ 3x + 7.2y &\geq 18.4 \\ 1.8x + 4.8y &\leq 48.3 \\ x &\geq 0 \\ y &\geq 0 \end{aligned}$$

2. Solve the following system of linear inequalities graphically, and find the corner points:

$$\begin{aligned} 5.5x + 3.1y &\geq 13.5 \\ 2x + 6.2y &\leq 19.2 \\ 1.8x + 4.8y &\leq 48.3 \\ x &\geq 0 \\ y &\geq 0 \end{aligned}$$

3. Maximize and minimize the function $z = 8.2x + 2.1y$ over the region graphed in Problem 1 above.

4. Maximize and minimize the function $z = 3.1x - y$ over the region graphed in Problem 2 above.

5. Repeat the first part of Example 3 above if the profit on standard tents is $90 and on expedition tents $70.

6. Repeat Example 3 if a new type of tent is introduced requiring 3 labor hours from the cutting department and 4.5 labor hours from the assembly department. The profit on each new tent is $60. What is the effect of having 38 hours and 90 hours available per day in the cutting department and the assembly department, respectively?

NOTES

Chapter 6

Probability

This chapter contains examples using the calculator to illustrate:

- Finding permutations and combinations
- Calculating probabilities and expected values
- Generating random numbers
- Calculating conditional probabilities
- Applying Bayes' Formula
- Finding expected values

EXAMPLE 1 ***TEXTBOOK SECTION 6-1 & 6-2***

A group of 12 people are meeting.

(A) In how many ways can 3 be chosen for one committee and 5 be chosen for a second committee? No person can serve on both committees.

(B) How would this result be affected if there were 5 people chosen for the first committee and 3 for the second committee?

(C) How would the result be affected if there were only 11 people from which to choose?

(D) Suppose the 12 people are numbered from 1 to 12. Use the random number generator in the calculator with a seed number of 6 to pick 7 of these 12 randomly.

Solution (A): This is a problem involving the fundamental theorem of counting and combinations. Use the built-in features of your calculator to find $C_{12,3}C_{9,5}$. The result is 27,720 ways.

Solution (B): The calculation would be $C_{12,5}C_{7,3}$ if 5 people were chosen for the first committee and 3 people were chosen for the second committee. The result is the same, 27,720 ways.

Solution (C): Press [2nd] [ENTRY] to replay the last entry. Change the last entry 12 to 11 and 7 to 6. Recalculate. The result is 9,240. By reducing the number of people you start with by one, the number of ways is about one-third as many.

Solution (D): Store the seed number in the storage location RAND (see Appendix Section A-11, B-11, C-11 or D-11 of this manual). [IPart] (12 x [Rand]) will generate integers x such that $0 \le x \le 11$. Since we want integers $1 \le x \le 12$, we add 1 to this expression. So we enter [IPart] (12 x [Rand]) +1 in the calculator to get the numbers:

6	2	5	10	2	7	6	9	10
8	10	9	9	1	1	12	12	5
11	7	5	8	2	11	10	1	5

Since we cannot have repeats, the people picked are numbered

6 2 5 10 7 9 8.

EXAMPLE 2 ***TEXTBOOK SECTION 6-3***

(A) Use the random number generator to get a sample of 40 tosses of a nickel and a dime.

(B) How do the results compare to what you expected from an equally likely sample space?

Solution (A): Choose a seed number (say 3). Store this in the storage location RAND (see Appendix Section A-11, B-11, C-11 or D-11 of this manual). Suppose 0 represents no heads, 1 represents a head on the nickel and a tail on the dime, 2 represents a tail on the nickel and a head on the dime, and 3 represents two heads. We wish to generate random numbers x, where x is an integer and $0 \le x \le 3$. A method that works on all calculators is to enter [IPart] (4 x [Rand]) [ENTER]. The resulting numbers are:

0	2	2	1	0	1	0	3	1	3
1	3	1	0	0	1	3	0	1	1
2	1	2	1	3	0	0	1	1	2
0	3	2	3	2	2	3	2	3	0

Counting we see the results are: Ten 0's, twelve 1's, nine 2's, and nine 3's.

Solution (B): The expected result was 10 in each category.

EXAMPLE 3 — *TEXTBOOK SECTION 6-4*

Simulate 36 repetitions of rolling a pair of dice and recording their sum. Display a histogram of the results. The generated data and the histograms will be different for different calculators and different seed values.

Solution: Choose a seed number (say 5). Store this in the storage location RAND (see Appendix Section A-11, B-11, C-11 or D-11 of this manual).

TI-82

[5] [STO▸] [MATH] [▶] [▶] [▶] [1] <rand> [ENTER]
[MATH] [▶] [2] <iPart> [(] [10] [×] [MATH] [▶] [▶] [▶] [1] <rand> [)] [+]
[MATH] [▶] [2] <iPart> [(] [10] [×] [MATH] [▶] [▶] [▶] [1] <rand> [)]
[ENTER] [ENTER] [ENTER] ... [ENTER] [ENTER]

Repeat pressing the [ENTER] key until you have 36 numbers that are between 2 and 12 inclusive.

The resulting numbers are:

9	4	8	5	9	~~16~~	12	4	5	4
~~13~~	12	11	~~15~~	~~1~~	11	10	~~16~~	11	9
8	6	10	10	6	~~15~~	2	~~1~~	3	~~13~~
3	3	2	~~17~~	~~1~~	8	~~14~~	3	11	6
10	8	~~13~~	10	9	6	7	8		

The crossed out numbers above are discarded because they are not between 2 and 12. Store these numbers in the list L1.

Deselect or clear all functions stored in the Y= list. Set the viewing window variables to [0, 14]1 by [0, 10]1 and graph the histogram.

[2nd] [STAT PLOT] [ENTER] [ENTER] [▼] [▶] [▶] [▶] [ENTER]
[▼] [ENTER] [▼] [ENTER] [GRAPH]

TI-83

[5] [STO▸] [MATH] [▶] [▶] [▶] [1] <rand> [ENTER]
[MATH] [▶] [▶] [▶] [5] <randInt> [1] [,] [6] [,] [36] [)] [+]
[MATH] [▶] [▶] [▶] [5] <randInt> [1] [,] [6] [,] [36] [)] [STO▸] [2nd] [L1] [ENTER]

This generates 36 numbers between 1 and 6 inclusive and stores the numbers in the L1 list.

The resulting numbers are:

9	6	2	9	5	7	6	3	8	5
11	8	6	8	5	9	3	4	2	4
5	9	9	9	6	6	6	8	2	3
8	12	3	6	10	7				

Deselect or clear all functions stored in the Y= list. Set the viewing window variables to [0, 14]1 by [0, 10]1 and graph the histogram.

[2nd] [STAT PLOT] [ENTER] [ENTER] [▼] [►] [►] [ENTER]

[▼] [ENTER] [▼] [ENTER] [GRAPH]

<u>TI-85</u>

[5] [STO►] [2nd] [MATH] [F2] <PROB> [F4] <rand> [ENTER]

[EXIT] [F1] <NUM> [F2] <iPart> [(] [10] [×] [EXIT] [F2] <PROB> [F4] <rand> [)] [+]

[EXIT] [F1] <NUM> [F2] <iPart> [(] [10] [×] [EXIT] [F2] <PROB> [F4] <rand> [)]

[ENTER] [ENTER] [ENTER] ... [ENTER] [ENTER]

Repeat pressing the [ENTER] key until you have 36 numbers that are between 2 and 12 inclusive.

The resulting numbers are:

9	4	8	5	9	~~16~~	12	4	5	4
~~13~~	12	11	~~15~~	~~1~~	11	10	~~16~~	11	9
8	6	10	10	6	~~15~~	2	~~1~~	3	~~13~~
3	3	2	~~17~~	~~1~~	8	~~14~~	3	11	6
10	8	~~13~~	10	9	6	7	8		

The crossed out numbers above are discarded because they are not between 2 and 12. Store these numbers in the list xStat with 1 as the yStat values.

Deselect or clear all functions stored in the y= list.
Set the viewing window variables to [0, 14]1 by [0, 10]1 and graph the histogram.

[STAT] [F3] <DRAW> [F1] <HIST> [CLEAR]

<u>TI-86</u>

[5] [STO►] [MATH] [►] [►] [►] [1] <rand> [ENTER]

[MATH] [►] [►] [►] [5] <randInt> [1] [,] [6] [,] [36] [)] [+]

[MATH] [►] [►] [►] [5] <randInt> [1] [,] [6] [,] [36] [)] [STO►] [2nd] [L1] [ENTER]

This generates 36 numbers between 1 and 6 inclusive and stores the numbers in the L1 list.

The resulting numbers are:

9	6	2	9	5	7	6	3	8	5
11	8	6	8	5	9	3	4	2	4
5	9	9	9	6	6	6	8	2	3
8	12	3	6	10	7				

Deselect or clear all functions stored in the y= list.
Store 1 as the frequency for each of the 36 pieces of data entered in the xStat list.
Set the viewing window variables to [0, 14]1 by [0, 10]1 and graph the histogram.
[2nd] [STAT] [F4] <DRAW> [F1] <HIST> [CLEAR]

EXAMPLE 3 ***TEXTBOOK SECTION 6-5***

In a study to determine employee voting patterns in a recent strike election, employees were selected at random and the following tabulation was made:

		SALARY CLASSIFICATION		
		Hourly(H)	Salary(S)	Salary+Bonus(B)
TO	Yes(Y)	394	175	20
STRIKE	No(N)	142	111	123

(A) Convert this table to a probability table.

(B) What is the probability of a randomly selected employee from this group voting to strike (Y) given that the person is paid hourly (H)?

(C) Are the events H and Y independent?

Round all answers to three decimal places.

Solution (A): Store this data as a 2x3 matrix A. Then calculate the sum of the elements. This can be done by recalling the elements using the features of the calculator, or simply adding the six numbers in the calculation mode. The result is 965. Store this answer as T using Ans→T. Then calculate $\frac{1}{T}[A]$ and store as matrix B. This is the probability matrix:

$$B = \frac{1}{T}\begin{bmatrix} 394 & 175 & 20 \\ 142 & 111 & 123 \end{bmatrix} = \begin{bmatrix} .408 & .181 & .021 \\ .147 & .115 & .127 \end{bmatrix}$$

Solution (B): We wish to calculate $P(Y|H) = \frac{B(1,1)}{B(1,1)+B(2,1)}$. Do this by recalling the elements of matrix B. Reentering the values of matrix B will cause round-off errors. In the calculation mode of the calculator enter **[B](1,1)** ÷ (**[B](1,1)** + **[B](2,1)**) where [B](1,1) represents the way to recall the element in the first row and first column of matrix B. It is shown in bold here to indicate that a single number is stored in the calculator in each location. The result is .735 .

Solution (C): We wish to see if P(Y|H) = P(H). P(Y|H) was calculated in Part (B). Calculate P(H) by calculating (**[A](1,1)** + **[A](2,1)**) ÷ T. The result is .555 . Hence the events H and Y are not independent since .555 ≠ .735 .

EXAMPLE 4 ***TEXTBOOK SECTION 6-6***

A new inexpensive skin test is devised for detecting tuberculosis. To evaluate the test before it is put into use, a medical researcher randomly selects 1,000 people. Using precise but more expensive methods already available, it is found that 8% of the 1,000 people tested have tuberculosis. Now each of the 1,000 subjects is given the new skin test and the following results are recorded: The test indicates tuberculosis in 96% of those who have it and in 2% of those who do not. Based on these results, what is the probability of a randomly chosen person having tuberculosis, given that the skin test indicates the disease?

Solution: A tree diagram is shown below that represents this situation.

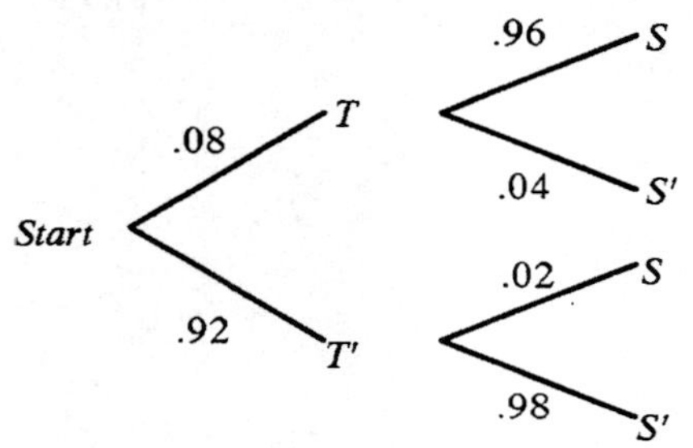

T = Tuberculosis
T' = No tuberculosis
S = Skin test indicates tuberculosis
S' = Skin test indicates no tuberculosis

We are interested in finding P(T|S). This is given by:

$$P(T|S) = \frac{P(T \cap S)}{P(T \cap S) + P(T' \cap S)} = \frac{.08 \times .96}{.08 \times .96 + .92 \times .02} = .81$$

To calculate this, enter the following into the calculator:

(.08 × .96) ÷ (.08 × .96 + .92 × .02).

EXAMPLE 5 ***TEXTBOOK SECTION 6-7***

A spinner device is numbered from 1 to 6 and each of the six numbers has a probability of occurring according to the chart below. Any winning bet of $1 will pay $5. A player always bets $1 on the 5. If the spinner lands on the 5 the player wins $4 plus gets the $1 bet back; otherwise, the $1 bet is lost. What is the expected value of the game to this player (long-run average gain or loss per game)? Use two decimal place accuracy. Is this the best number for this player to choose?

Numbers	*1*	*2*	*3*	*4*	*5*	*6*
Probability	$\frac{1}{21}$	$\frac{2}{21}$	$\frac{3}{21}$	$\frac{4}{21}$	$\frac{5}{21}$	$\frac{6}{21}$

Solution: The payoff table for this player is:

x_i	\$4	-\$1
p_i	$\frac{5}{21}$	$\frac{16}{21}$

To calculate the expected value enter 4×(5÷21) -1×(16÷21). The result is .19. In the long run this player will win $.19 each game on the average.

Use the replay key to calculate the other possibilities:

Number bet on	*1*	*2*	*3*	*4*	*5*	*6*
Expected value	-.76	-.52	-.29	-.05	.19	.43

The best choice for this player would be to bet on the 6 each game. In the long run this player would then win $.43 each game on the average.

Exercise Set 6

1. Consider a group of 18 people:

   (A) In how many ways can 4 be chosen for one committee and 3 be chosen for a second committee? A person can serve on both committees.

   (B) How would the result from Part (A) be affected if there were 5 people chosen for the first committee and 3 for the second committee?

   (C) How would the result from Part (A) be affected if there were only 15 people?

   (D) How would the result from Part (B) be affected if there were only 15 people?

2. From a group of 20 books, in how many ways can 5 be chosen for one bookrack and 10 be chosen for a second bookrack and 3 be chosen for a third bookrack. How would this result be affected if there were 25 books to start with?

3. Twenty-five paintings are available for purchase at an art show. The paintings are numbered from 0 to 24. The artist wishes to choose five randomly to sell for charity. What are the painting numbers that he should choose?

4. A group of 30 people are meeting. We want to randomly choose 10 of these people to work on a specific task. If the people are numbered from 1 to 30, who should be chosen?

5. Repeat Example 3 above if a third category, Undecided (U), was added to the vote a person would cast, where the Undecided (U) category had 102 Hourly, 87 Salary, and 187 Salary+Bonus votes.

6. Repeat Example 1 if the company did not have a Salary+Bonus category but did have the third category of Undecided (U) from Problem 1.

7. A company's brand (X) has 22% of the market. A market research firm finds that if a person currently uses brand X, the probability is .61 that he or she will buy it the next time (represented by X_c). On the other hand, if a person currently does not use X (represented by $X^{¿}_{c}$), the probability is .53 that he or she will switch to X the next time (represented by X_n).

   (A) Draw a tree diagram representing this situation.

   (B) What is the probability that a person is presently using brand X if he/she purchases brand X on the next purchase?

8. Repeat Problem 3 if the company only had 15% of the market.

9. A gaming device is numbered from 1 to 10 and each of the ten numbers has a probability of occurring according to the chart below. If the device lands on the chosen number the player wins $8 and gets the bet back; otherwise, the $2 bet is lost. Suppose a player always bets $2 on the 5.

Numbers	*1*	*2*	*3*	*4*	*5*	*6*	*7*	*8*	*9*	*10*
Probability	.05	.05	.06	.06	.21	.21	.08	.08	.10	.10

   (A) What is the expected value of the game to this player (long-run average gain or loss per game)? Use two decimal place accuracy.

   (B) Is this the best number for this player to choose?

10. Repeat Problem 9 if the bet is $4 and the player wins $5 and gets the bet back.

Chapter 7

Data Description and Probability Distributions

This chapter contains examples using the calculator to illustrate:

- Graphing quantitative data
- Calculating measures of central tendency
- Calculating measures of dispersion
- Calculating the probability of x successes in n Bernoulli Trials

EXAMPLE 1 ***TEXTBOOK SECTION 7-1***

(A) Create a bar graph for the data shown in the table at the right.

(B) Create a broken-line graph for the data.

US. Public Debt

Year	Billions of Dollars
1950	$ 256.1
1960	284.1
1970	370.1
1980	907.7
1990	3233.3

Solution (A): Assign a number to each year and express the debt as a whole number as shown below:

Year	1950	1960	1970	1980	1990
x	1	2	3	4	5
y	2561	2841	3701	9077	32333

Clear the functions from the function list and any drawings (see Appendix Section A-6, B-6, C-6 or D-6 of this manual). Clear the statistical registers in the calculator (see Appendix Section A-17, B-17, C-17 or D-17 of this manual).

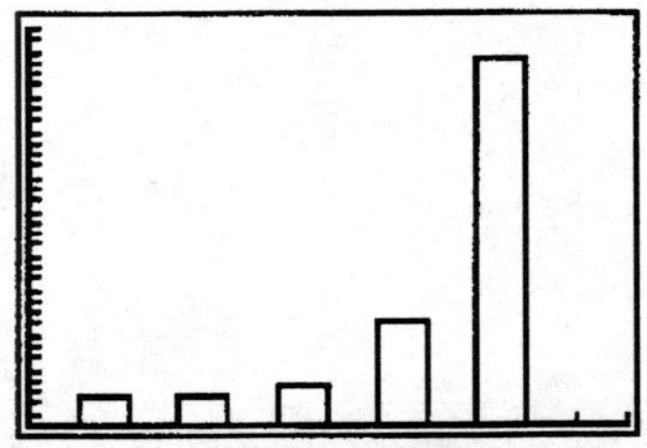

[0, 1].5 by [0, 35000]1000

Store the data using the number as the x variable and the amount as the y variable. Set the viewing window variables at [0, 6].5 by [0, 35000]1000. The bars will be .5 in width and there will be a gap of .5 between bars because an Xscl of .5 will graph data as $0 \le x < .5$, $.5 \le x < 1$, $1 \le x < 1.5$, etc. and there is no data for $.5 \le x < 1$, $1.5 \le x < 2$, etc. The center of the bars is not at the x value, but is at the center of the bar, at $x+.25$

Solution (B): Sort the data, if it is not already sorted. Clear the drawings (see Appendix Section A-6, B-6 or C-6 of this manual). Now use the xy-line option on the statistical draw menu to get the graph at the right. See Appendix Section A-16, B-16, or C-16 Example 2(E) of this manual.

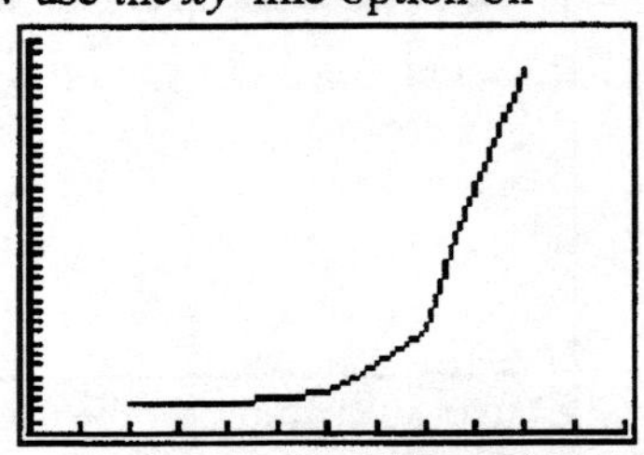

[0, 6].5 by [0, 35000]1000

The line segments are drawn from (1, 2561) to (2, 2841) to (3, 3701) to (4, 9077) to (5, 32333) which are the top left corners of the bars shown in Solution (A) above.

EXAMPLE 2 ***TEXTBOOK SECTION 7-1***

The following table shows price-earnings ratios of common stocks chosen at random from the New York Stock Exchange:

Price-Earnings (PE) Ratios

7	11	6	6	10	6	31	28	13	19
6	18	9	7	5	5	9	8	10	6
10	3	4	6	7	9	9	19	7	9
17	33	17	12	7	5	7	10	7	9
18	17	4	6	11	13	7	6	10	7

Construct a histogram using 0 as the lower limit, 40 as the upper limit, and class intervals of 5. Use the arrow keys on your calculator to move the cursor and explore the graph. Where do you think the "center" of the data is?

Solution: (See Appendix Section A-16, B-16, C-16 or D-16 of this manual on how to enter data and get a histogram for your calculator.) Set the viewing window variables to [0,40]5 by [-5,30]5. The result is shown to the right. Remember that the Xscl is the class width.

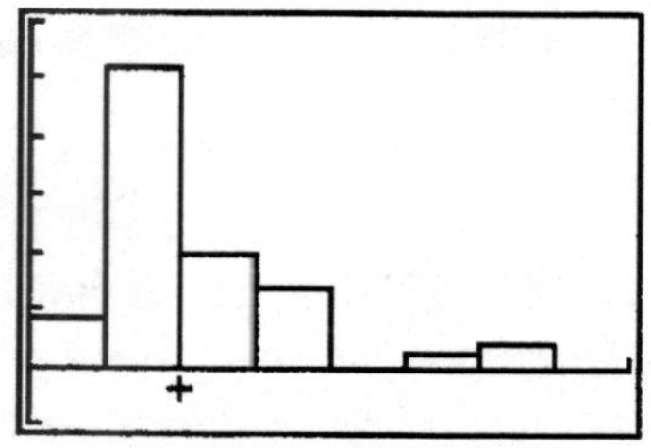

[0,40]5 by [-10,30]5

The "center" of the data is usually thought of as being the balance point or the mean. The cursor shown on the graph to the right is probably approximately where you placed your cursor. (Now you see the reason for setting the limits for the y axis to start below zero. This gives room on the graph to see the cursor and the coordinates of its location. The coordinates are not shown on the graph above because each calculator model will have different coordinates.)

EXAMPLE 3 ***TEXTBOOK SECTION 7-2***

Given the data below:

(A) Create a frequency polygon for this data.

(B) Create a relative cumulative frequency ogive for this data.

Class Interval	Midpoint x	Frequency y	Relative Cumulative Frequency
299.5-349.5	324.5	1	.01
349.5-399.5	374.5	2	.03
399.5-449.5	424.5	5	.08
449.5-499.5	474.5	10	.18
499.5-549.5	524.5	21	.39
549.5-599.5	574.5	20	.59
599.5-649.5	624.5	19	.78
649.5-699.5	674.5	11	.89
699.5-749.5	724.5	7	.96
749.5-799.5	774.5	4	1.00

Solution (A): Add a line to the table both at the beginning and at the end.

249.5-299.5	274.5	0	.00
799.5-849.5	824.5	0	1.00

Store the data in the calculator so that x=midpoint and y=frequency.

Sort the data on x. Set the viewing window variables at [250, 850]50 by [0, 25]1. Draw the xy-line. The TI-82 will show the dots if you draw the scatter graph first before the xy-line. (See Appendix Section A-16, B-16, C-16 or D-16 of this manual.)

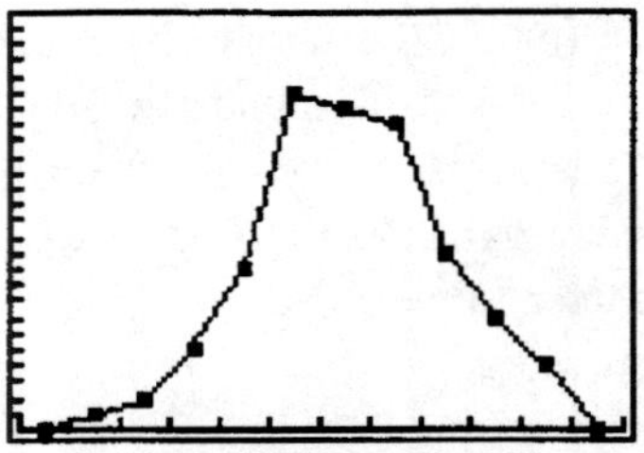

[250, 850]50 by [0, 25]1

Solution (B): Add a line at the beginning of the original table:

249.5-299.5	274.5	0	.00

The calculator needs to have a whole number for the y variable in order to draw the xy-line. Change the relative cumulative frequency to whole numbers by multiplying each by a number, say 100. Store the data in the calculator so that x=upper class boundary and y=whole number representing the relative cumulative frequency. Set the viewing window variables at [250, 850]50 by [0, 150]10. Draw the xy-line. The TI-82, TI-83, or TI-86 will show the dots if you draw the scatter graph first before the xy-line.

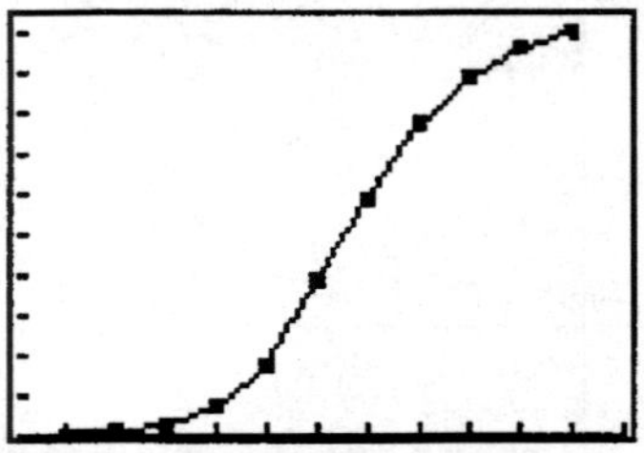

[250, 850]50 by [0, 150]10

EXAMPLE 4 ***TEXTBOOK SECTION 7-3***

(A) Find the median and mode of the data in Example 2 above.

(B) Find the median and mode of the data in Example 3 above.

Solution (A): Clear the statistical registers in the calculator and enter the data as was done in Example 2 above.

<u>TI-82, TI-83, and TI-86</u>

These calculator models have a built-in function that will find the median for you. See Appendix Section A-16, B-16 or D-16 of this manual. The median function is on the 1-Var Stats list. (Use the down arrow to see the last part of the list.) The median is 9 and the mode is 7.

TI-85

Use the calculator to sort the data on the *x* variable. (See Appendix Section C-16 of this manual.) There are 50 pieces of data so the median value is between the 25th and 26th value. Use the arrow keys while in the edit mode to examine the data to find the values in the 25th and 26th position. Take the average of these two values. The median is 9.

The mode is 7.

Solution (B): Clear the statistical registers in the calculator and enter the data as was done in Example 3(A) above.

TI-82, TI-83 and TI-86

These calculator models have a built-in function that will find the median for you. See Appendix Section A-16, B-16 or D-16 of this manual. The median function is on the 1-Var Stats list. (Use the down arrow to see the last part of the list.) The median is listed is 574.5. This is the midpoint of the class of the 50.5th value. However, if we want the number that is 11.5/20ths into the class, we must use the method shown for the TI-85 calculators below.

TI-85

Use the calculator to sort the data on the *x* variable. (See Appendix Section C-16 of this manual.) There are 100 pieces of data so the median value is between the 50th and 51st value. Use the arrow keys in the edit mode and count down the frequencies of the *y* variable) until you find the class having the 50th piece of data. This is the sixth class. Since there are 39 numbers before this class, we need to find the numbers that is 11.5th position into the sixth class. We find the median by taking the fraction of the class width and adding it on to the lower boundary of the class: $\frac{11.5}{\text{number in class}}$ x (class width) + (lower class boundary) = $\frac{11.5}{20}$ x 50 + 549.5 = 578.25. Hence the median is 578.25.

The modal class has midpoint 524.5. This can be observed in the tabled data or by inspecting the histogram with range [249.5, 824.5]50 by [-5, 25]1.

EXAMPLE 5 ***TEXTBOOK SECTION 7-3 & 7-4***

Find the mean and standard deviation to two decimal places for the data in Example 2 above.

Solution: Store the data in the calculator with x=the data value and y=the frequency (this is 1 if entering each piece of data separately). (See Appendix Section A-16, B-16, C-16 or D-16 of this manual on how to get the one-variable statistics on your calculator.)

The results are: $\bar{x} = 10.52$ and $s = 6.63$. Since this is a sample of data, we use the sample standard deviation and not population standard deviation.

EXAMPLE 6 *TEXTBOOK SECTION 7-3 & 7-4*

Find the mean and standard deviation to two decimal places for the data in Example 3 above.

Solution: Store the data using the midpoint of the class as the x variable and the frequency as the y variable as was done in Example 3(A) above. (See Appendix Section A-16, B-16, C-16 or D-16 of this manual on how to get the one-variable statistics on your calculator.) The results are: $\bar{x} = 579$ and $s = 94.31$. Since this is a sample of data, we use the sample standard deviation and not population standard deviation.

EXAMPLE 7 *TEXTBOOK SECTION 7-5*

The probability of recovering after a particular type of operation is .95. Find the probability that at least 5 of the next 50 patients having this operation will not recover.

Solution: At least 5 not recovering is the same as at most 45 will recover. So we want to find the probability of at least five patients not recovering is equivalent to less than 45 patients recovering.

$$\begin{aligned} P(X \le 45) &= P(1) + P(2) + \ldots + P(45) \\ &= 1 - P(X > 45) \\ &= 1 - P(X \ge 46) \\ &= 1 - [P(46) + P(47) + P(48) + P(49) + P(50)] \end{aligned}$$

We need to find P(46) + P(47) + P(48) + P(49) + P(50) and subtract the result from 1.

This is a binomial distribution situation with $n = 50$, $p = .95$, and $q = .05$ when a recovery is considered a success. The probability of x people recovering is given by: $P(x) = C_{50,x}\,(.95)^x(.05)^{50-x}$. Store this expression in the calculator as: (50 **nCr** X) (.95 ^ X) .05 ^ (50 - X) where nCr is found on the probability menu (see Appendix

Section A-11, B-11, C-11 or D-11 of this manual). Store 46 for X and evaluate the expression. Store the result in memory M. Store 47 for X and reevaluate the expression. Add the result to the contents of memory M and store again as memory M. Continue in this manner until the sum of the probabilities is in memory M.

(50 nCr X) (.95 ^ X) .05 ^ (50 - X)	
46 → X	
Recall the expression and evaluate.	
Ans → M	Result is .136
47 → X	
Recall the expression and evaluate.	Result is .220
Ans + M → M	Result is .356
48 → X	
Recall the expression and evaluate.	Result is .261
Ans + M → M	Result is .617
49 → X	
Recall the expression and evaluate.	Result is .202
Ans + M → M	Result is .819
50 → X	
Recall the expression and evaluate.	Result is .077
Ans + M → M	Result is .896

Find 1 - M. The result is .104. Thus the probability that at least 5 people out of 50 do not survive the operation is .104 or a little more than 1 chance in 10.

EXAMPLE 8 — *TEXTBOOK SECTION 7-6*

Given the normal probability distribution having mean 15.4 and standard deviation 2.3:

(A) Graph using [6, 24]1 by [0, .3].1.

(B) Find the value of c so that $P(9.2 \le x \le c) = .8534$.

(C) Conjecture, then check by graphing, the location and shape of a normal probability distribution having mean 17 and standard deviation 1.5.

Solution (A): The formula for this situation is: $f(x) = \frac{1}{2.3\sqrt{2\pi}} e^{-\frac{(x - 15.4)^2}{2(2.3^2)}}$.

Store this in the calculator as:

1 ÷ (2.3 x √ (2 x π)) x [e^] (- ((X - 15.4) ^ 2) ÷ (2 x 2.3 ^ 2)).

(Recall that [e^] is found using [2nd] [LN].)

Set the viewing window variables to [6, 24]1 by [0, .3].1. Graph.

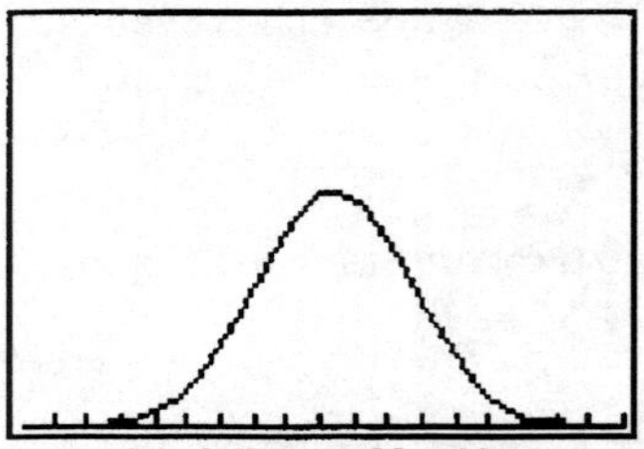

[6, 24]1 by [0, .3].1

Solution (B): The z value for x=9.2 is $z = \frac{9.2-15.4}{2.3} = -2.70$. From the table we find the area under the curve for z between -2.70 and 0 is .4965. Hence we know that the x value we want is on the right side of the mean which is 15.4. .8534-.4965=.3569 is the area on the right side of x=15.4. This corresponds to a z of .98.

The c value is found by solving $.98 = \frac{c - 15.4}{2.3}$. $c = 17.65$.

Solution (C): The graph will be shifted to the right 1.6 units and will be taller and skinnier since the standard deviation is smaller.

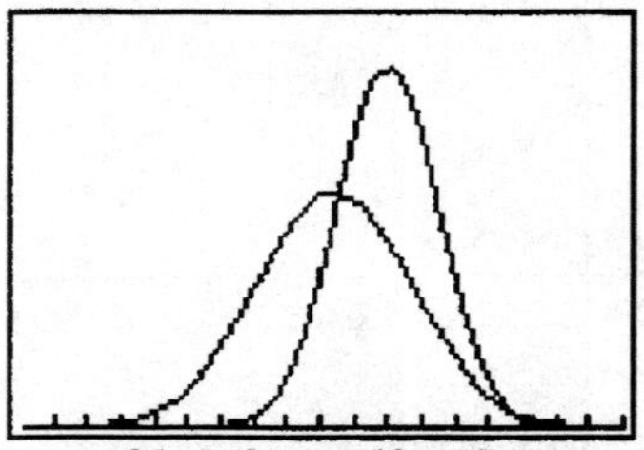

[6, 24]1 by [0, .3].1

Exercise Set 7

1. The following table shows scores of students on a 35 question true/false test.

Scores on True/False test

27	31	16	26	10	16	31	28	13	19	16	28	29	17
25	25	29	18	10	26	20	13	24	26	27	19	29	19
17	19	27	33	17	12	27	25	17	10	7	19	18	27
34	26	11	13	17	26	10	27						

Construct a histogram using 0 as the lower limit, 35 as the upper limit, and class intervals of 5.

Use the arrow keys on your calculator to move the cursor and explore the graph. Where do you think the "center" of the data is?

2. Suppose the first column of scores in Problem 7 were all perfect papers with scores of 35. Repeat Problem 1 and discuss how this change of data effects the histogram.

3. Find the mean and standard deviation of the sample data in Problem 1.

4. Find the mean and standard deviation of the sample data in Problem 2.

5. Repeat Example 7 but find the probability that 3, 4 or 5 will not recover.

6. Repeat Example 7 but change the probability of recovering to .97.

NOTES

NOTES

Chapter 8

Games and Decisions

This chapter contains examples using the calculator to illustrate:

- Finding the expected value of a game
- Finding the solution to a 2x2 nonstrictly determined matrix game

EXAMPLE 1 ***TEXTBOOK SECTION 8-2***

Find the expected value of the two-finger Morra game for R if strategies

$P = [.248 \quad .752]$ and $Q = \begin{bmatrix} .381 \\ .619 \end{bmatrix}$ are used by R and C, respectively, and the game

matrix is $M = \begin{bmatrix} a & b \\ c & d \end{bmatrix} = \begin{bmatrix} 2 & -3 \\ -3 & 4 \end{bmatrix}$.

Solution: The expected value of the game for R is given by PMQ. Store the matrix P as a 1x2 matrix A, and M as a 2x2 matrix B in the calculator. Store matrix Q as a 2x1 matrix C in the calculator. Multiply the matrices $A*B*C$ to find the product PMQ. (See Appendix Section A-12, B-12, or C-12 of this manual.) The result is .731. This means that in the long run, with R and C using the indicated strategies, R will win \$.73 per game and C will lose \$.73 per game, on the average.

EXAMPLE 2 ***TEXTBOOK SECTION 8-2***

(A) Solve the two-finger Morra game of Example 1 above.

(B) Determine if the game is fair.

Solution (A): We need to find the optimal strategy for R and C. The optimal strategy for R is given by $P^* = \begin{bmatrix} \frac{d-c}{D} & \frac{a-b}{D} \end{bmatrix}$ and the optimal for C is given by $Q^* = \begin{bmatrix} \frac{d-b}{D} \\ \frac{a-c}{D} \end{bmatrix}$ where $D = (a + d) - (b + c)$. Calculate D and store in memory location D. The result is 12. Now calculate the other quantities as (4-(-3))÷D, (2-(-3))÷D, etc. The result is $P^* = [.583 \quad .417]$ and $Q^* = \begin{bmatrix} .583 \\ .417 \end{bmatrix}$.

Solution (B): We need to check to see if the expected value $\frac{\det M}{D}$ is 0. Calculate determinant of M using the calculator. (See Appendix Section A-12, B-12, or C-12 of this manual.) The result is -1. The expected value for R is $-\frac{1}{12}$. Since the expected value of the game is not zero, this is not a fair game.

Exercise Set 8

1. Find the expected value of the two-finger Morra game for R if strategies

   $P = [.187 \quad .813]$ and $Q = \begin{bmatrix} .381 \\ .619 \end{bmatrix}$ are used by R and C, respectively, and the game matrix is $M = \begin{bmatrix} a & b \\ c & d \end{bmatrix} = \begin{bmatrix} 2.5 & -3.2 \\ -3.4 & 4.8 \end{bmatrix}$.

2. Find the expected value of the two-finger Morra game for R if strategies

   $P = [.683 \quad .317]$ and $Q = \begin{bmatrix} .381 \\ .619 \end{bmatrix}$ are used by R and C, respectively, and the game matrix is $M = \begin{bmatrix} a & b \\ c & d \end{bmatrix} = \begin{bmatrix} 3.2 & -2.7 \\ -2.8 & 6.2 \end{bmatrix}$.

3. Find the optimal strategies for the two-finger Morra game of Problem 1 and determine if the game is fair.

4. Find the optimal strategies for the two-finger Morra game of Problem 2 and determine if the game is fair.

Chapter 9

Markov Chains

This chapter contains examples using the calculator to illustrate:

- Finding next state and stationary state matrices in Markov chains
- Finding powers of the transition matrix
- Determining if a transition matrix is regular
- Finding a limiting matrix of a transition matrix

EXAMPLE 1 ***TEXTBOOK SECTION 9-1, 9-2, and 9-3***

A given plant species has red, pink, or white flowers according to the genotypes RR, RW, and WW, respectively. If each of these genotypes is crossed with a pink-flowering plant (genotype RW), then the transition matrix P is:

$$P = \text{This generation} \begin{matrix} \text{Red} \\ \text{Pink} \\ \text{White} \end{matrix} \overset{\begin{matrix} \textbf{Next generation} \\ \text{Red} \quad \text{Pink} \quad \text{White} \end{matrix}}{\begin{bmatrix} .50 & .50 & .00 \\ .25 & .50 & .25 \\ .00 & .50 & .50 \end{bmatrix}}$$

Assuming the plants of each generation are crossed only with pink plants to produce the next generation:

(A) Find, to two decimal places, the percentage of each color plant after four generations and after 10 generations if the initial generation is all red plants.

(B) Show that, regardless of the makeup of the first generation, the genotype composition will eventually stabilize at 25% red, 50% pink, and 25% white.

(C) Repeat Parts (A) and (B) if the initial generation is all white plants.

(D) Conjecture, then check, what the fourth generation, tenth generation, and stationary matrix will be if the initial matrix is $S_0 = [0 \quad 1 \quad 0]$.

(E) Is P a regular matrix?

(F) Find the limiting matrix $\overline{P}$.

Solution (A): We need to find the fifth state matrix with initial matrix of $S_0 =$ $[1 \quad 0 \quad 0]$. Store the transition matrix P in the calculator. The fourth state matrix is found by calculating $S_0P^4 = [1 \quad 0 \quad 0]\begin{bmatrix} .50 & .50 & .00 \\ .25 & .50 & .25 \\ .00 & .50 & .50 \end{bmatrix}^4$. To do this, store the 3x3 matrix $P = \begin{bmatrix} .50 & .50 & .00 \\ .25 & .50 & .25 \\ .00 & .50 & .50 \end{bmatrix}$ as matrix A in the calculator. Also, store the 1x3 matrix $S_0 = [1 \quad 0 \quad 0]$ as matrix B in the calculator. (See Appendix Section A-12, B-12, C-12, or D-12 of this manual.) Find the fourth power of A and store it as matrix C. The result is $\begin{bmatrix} .28125 & .50000 & .21875 \\ .25000 & .50000 & .25000 \\ .21875 & .50000 & .28125 \end{bmatrix}$.

Now multiply the initial state matrix by matrix C. The result is [.28125 .50000 .21875].

An alternative method would be to find BA^4 directly using matrix multiplication without finding A^4 first.

We also need to find $S_0P^{10} = BA^{10}$. The result is [. 25049 .50000 .24951]. We will see that this is very close to the stationary matrix.

Solution (B): We need to find the stationary matrix. This matrix is found by solving the matrix equation:

$$[m \quad n \quad p]\begin{bmatrix} .50 & .50 & .00 \\ .25 & .50 & .25 \\ .00 & .50 & .50 \end{bmatrix} = [m \quad n \quad p]$$. This is the same as

$$[m \ n \ p]\begin{bmatrix} .50 & .50 & .00 \\ .25 & .50 & .25 \\ .00 & .50 & .50 \end{bmatrix} = [m \ n \ p]\begin{bmatrix} 1 & 0 & 0 \\ 0 & 1 & 0 \\ 0 & 0 & 1 \end{bmatrix}$$

$$[m \ n \ p]\left\{\begin{bmatrix} .50 & .50 & .00 \\ .25 & .50 & .25 \\ .00 & .50 & .50 \end{bmatrix} - \begin{bmatrix} 1 & 0 & 0 \\ 0 & 1 & 0 \\ 0 & 0 & 1 \end{bmatrix}\right\} = [0 \ 0 \ 0]$$

$[m \ n \ p]\begin{bmatrix} -.50 & .50 & .00 \\ .25 & -.50 & .25 \\ .00 & .50 & -.50 \end{bmatrix} = [0 \ 0 \ 0]$. This result is equivalent to

$$\begin{aligned} -.50m + .25n \quad &= 0 \\ .50m - .50n + .50p &= 0. \\ .25n - .50p &= 0 \end{aligned}$$

This is a dependent system.

Only two of these equations can be used. The third equation to use in finding the stationary matrix is $m+n+p=1$. Hence we need

to solve the matrix equation $\begin{bmatrix} -.50 & .25 & .00 \\ .50 & -.50 & .50 \\ 1.00 & 1.00 & 1.00 \end{bmatrix}\begin{bmatrix} m \\ n \\ p \end{bmatrix} = \begin{bmatrix} 0 \\ 0 \\ 1 \end{bmatrix}$. Enter the 3×3 coefficient matrix into the calculator as matrix A and find its inverse. This result is

$\begin{bmatrix} -2.00 & -.50 & .25 \\ .00 & -1.00 & .50 \\ 2.00 & 1.50 & .25 \end{bmatrix}$. Store this as matrix *C*. Store the constant 3×1 matrix as matrix B in the calculator. Multiply the inverse matrix by the constant matrix, *CB*. The

result is $\begin{bmatrix} .25 \\ .50 \\ .25 \end{bmatrix}$. The stationary matrix is then [.25 .50 .25] to two decimal place accuracy. The expected genotype composition will eventually stabilize at 25% red, 50% pink, and 25% white.

Solution (C): We need to find the product $S_0P^4 = BA^4 =$

$[0 \ 0 \ 1]\begin{bmatrix} .50 & .50 & .00 \\ .25 & .50 & .25 \\ .00 & .50 & .50 \end{bmatrix}^4$. Store the 1×3 matrix S_0 as matrix *B* in the calculator. Store the 3×3 matrix *P* as matrix *A* in the calculator. Find the matrix product BA^4. The result is [.21875 .50000 .28125].

The tenth state matrix is [.24951 .50000 .25049].

The steady-state matrix is the same as found in Part (B) above.

Solution (D): A good conjecture would be that the fourth state matrix would be close to [.24 .52 .24], that the tenth state matrix would be close to the stationary matrix, and that the stationary matrix is the same as in Parts (B) and (C).

As it turns out, $S_0P^4 = BA^4 = [0\ \ 1\ \ 0]\begin{bmatrix} .50 & .50 & .00 \\ .25 & .50 & .25 \\ .00 & .50 & .50 \end{bmatrix}^4 = [\ .25\ .50\ .25\ ]$ = stationary matrix.

Solution (E): P is a regular matrix since A^2 has all positive elements.

Solution (F): Find successive powers of P to find $\bar{P}$. Do this on the calculator using matrix A since this is where we stored the elements of matrix P.

$$\bar{P} = \begin{bmatrix} .25 & .50 & .25 \\ .25 & .50 & .25 \\ .25 & .50 & .25 \end{bmatrix}$$

. We need to find A^{34} (A^{40} on the TI-85) to see the limiting matrix. An easy way to do this is to recall matrix A. Then enter Ans × [A]. Pressing the ENTER key repeatedly will continue to perform this multiplication. Count the number of times you push the ENTER key.

Exercise Set 9

1.. A company's brand X has 22% of the market. A market research firm finds that if a person uses brand X now, represented by X_0, the probability is .61 that he or she will buy it the next time, represented by X_1. On the other hand, if a person does not use brand X now (represented by X_0'), the probability is .53 that he or she will switch to brand X the next time, represented by X_1'.

   (A) Draw a probability tree diagram representing this situation.

   (B) What is the probability that a person is currently using brand X, represented by X_0, if he/she purchases brand X on the next purchase, represented by X_1?

2. Repeat Problem 1 if the company only had 15% of the market.

3. Refer to Problem 1 above. (A) What share of the market will the company have after 5 purchases? (B) What share of the market will the company have in the long run?

4. Repeat Problem 1 if in Problem 1 the probability is .75 that a person currently using brand X, represented by X_0, will purchase it the next time.

5. Repeat Problem 1 if the company introduced a new product Y so that the initial matrix is [.35 .21 .44] and the transition matrix is:

$$\begin{array}{r} \\ \\ X_0 \\ Y_0 \\ \textit{neither } X_0 \textit{ nor } Y_0 \end{array} \begin{array}{ccc} X_1 & Y_1 & \textit{neither} \\ & & X_1 \textit{ nor } Y_1 \\ \left[\begin{array}{c} .58 \\ .35 \\ .13 \end{array} \right. & \begin{array}{c} .29 \\ .45 \\ .17 \end{array} & \left. \begin{array}{c} .13 \\ .20 \\ .70 \end{array} \right] \end{array}.$$

6. Repeat Problem 5 where $P(X_1|X_0) = .63$, $P(Y_1|X_0) = .21$, $P(\textit{neither}|X_0) = .16$ with the remaining probabilities the same. In other words, the first row of the matrix in Problem 9 is .63, .21, .16.

NOTES

NOTES

Appendix A

TI-82 Operations

A-1 Operations

Press [ON] to turn on the calculator.

Press [2nd] [+] to get the **MEMORY** screen (shown at the right).

```
MEMORY
1: Check RAM...
2: Delete...
3: Reset...
```

Choose 3:Reset... by pressing [3] or use the down-arrow key [▼] to highlight 3:Reset... and press [ENTER].

The display shows the **RESET MEMORY**.

```
RESET MEMORY
1: No
2: Reset
```

Choose 2: Reset or use the down-arrow key to highlight 2:Reset... and press [ENTER]. You may need to adjust the contrast (directions follow).

`Mem cleared` should appear on the screen.

If it doesn't, press [2nd] and then hold the [▲] (to increase the contrast) or [2nd] [▼] (to decrease the contrast) until you see it on the screen.

Press [CLEAR] to clear the screen.

Press [2nd] [OFF] to turn the calculator off.

The [2nd] Key

Press this key to access the operation listed on the face of the calculator above and to the left of a key. The operations above the key are in a light blue color on the face of the calculator. Note that a flashing up arrow [↑] appears on the screen after [2nd] key is pressed.

> In this manual, the functions on the face of the calculator above a key will be referred to in square boxes just as if the function was printed on the key cap. For example, [ANS] is the function above the [(−)] key.

The [ALPHA] and [A-LOCK] Keys

The operation above and to the right of a key is accessed by first pressing [ALPHA] or [A-LOCK]. A flashing [A] is displayed as the cursor on the screen after the [ALPHA] key is pressed. [A-LOCK] locks the calculator into alpha mode until the [ALPHA] is pressed.

Home Screen

Calculations are done and commands are entered on the Home Screen. You can always get to this screen (aborting any calculations in progress) by pressing

QUIT
[2nd] [MODE]. From here on, this will be referred to as [2nd] [QUIT] in this appendix.

The [MODE] Key

When you press [MODE] you see the screen shown below. The highlighted items are active.

To change an item, select it using the arrow keys and press [ENTER] to activate the selection.

Type of notation for display of numbers.	Normal Sci Eng
Number of decimal places displayed.	Float 0123456789
Type of angle measure.	Radian Degree
Function, parametric, polar or sequence graphing.	Func Par Pol Seq
Connected/dot plotted points on graphs.	Connected Dot
Graphs functions separately or all at once.	Sequential Simul
Allows a full screen or split screen to be used.	FullScreen Split

Menus

The TI-82 Graphics calculator uses menus for selection of specific functions. The items on the menus are identified by numbers followed by a colon. There are two ways to choose menu items:

1. Press the number corresponding to the menu item.
2. Use the arrow keys to highlight the selection and then pressing [ENTER].

In this manual the menu items will be referred to using the key to be pressed followed by the meaning of the menu. For example, on the ZOOM menu, [1] <ZBox> refers to the first menu item ZBox.

Editing

It is easy to correct errors when entering data into the calculator by using the arrow keys, [DEL], and [INS] keys.

[◄] or [►]	Moves the cursor to the left or right one position.
[▲]	Moves the cursor up one line.
[▼]	Moves the cursor down one line.
[DEL]	Deletes one or more characters at the cursor position.
[2nd] [INS]	Inserts one or more characters at the cursor position.
[2nd] [ENTRY]	Replays the last executed line of input.

A-2 Calculating

EXAMPLE 1: Calculate $|-4^2 - \sqrt{34}|$

Enter numbers and functions in the calculator in the same order as you would read an expression. In the examples that follow, do not press [ENTER] unless specifically instructed to do so. Enter all the keystrokes without pressing the ENTER key until [ENTER] is listed in the example.

The subtraction sign and the negative sign are different.

The **subtraction sign** is a dark blue key on the lower right side of the calculator. This is used when subtracting one quantity from another.
The **negative sign** (-) is a light gray key in the bottom row of keys on the calculator. This is used when entering a negative quantity.

Solution: [2nd] [QUIT] [2nd] [ABS] [(] [(-)] [4] [x^2] [-] [2nd] [√] [34] [)] [ENTER]. The result displayed is 21.83095189.

A-3 Plotting Points and Drawing Line Segments

EXAMPLE 1: (A) Plot the points (-3, 6) and (1,5). (B) Draw the line segment having these points as the endpoints.

Solution (A):

Keystrokes	Explanation
[Y=] [CLEAR] [▼] [CLEAR] etc.	Clear all functions from the function list.
[2nd] [QUIT]	Set the viewing window dimensions.
[ZOOM] [6] <ZStandard> [CLEAR]	[ZOOM] [6] <ZStandard> automatically sets the dimensions to Xmin=-10, Xmax=10, Xscl=1, Ymin=-10, Ymax=10, and Yscl=1. This will be denoted as [-10, 10]1 by [-10, 10]1 in this manual. Press [CLEAR] to get back to the home screen.
[2nd] [DRAW] [1] <ClrDraw> [ENTER]	Get the [DRAW] menu and clear the drawings from the calculator screen.
[2nd] [DRAW] [▶] [1] <PT-On(> [(-)] [3] [,] [6] [)] [ENTER]	Get the [DRAW] menu again. Choose the point on option and enter the coordinates of the point with a comma separating them. The point will be plotted.
[2nd] [QUIT]	
[2nd] [DRAW] [▶] [1] <PT-On(> [1] [,] [5] [)] [ENTER]	Enter the coordinates of the second point and plot.

Solution (B):

Keystrokes	Explanation
[2nd] [QUIT]	Return to the Home Screen. Get the [DRAW] menu and choose the line option.
[2nd] [DRAW] [2] <Line(>	Enter the coordinates of the points separated by commas.
[(-)] [3] [,] [6] [,] [1] [,] [5] [)] [ENTER]	The line segment will be drawn between the two points.

A-4 Storing and Evaluating an Expression, Lists and Tables

EXAMPLE 1: Store the expression $\frac{-b+\sqrt{b^2-4ac}}{2a}$ and evaluate it to solve $3x^2+4x-5=0$.

Solution:

Keystrokes	Explanation
[Y=] [CLEAR]	You will see a list of function names Y1, Y2, Y3, etc. We can store ten different functions or algebraic expressions in this list.
[(] [(-)] [ALPHA] [B] [+] [2nd] [√]	
[(] [ALPHA] [B] [x^2] [-] [4]	Then we can recall them and evaluate or graph them. We will use Y1. Press these keys to enter the expression.
[ALPHA] [A] [×] [ALPHA] [C] [)] [)]	
[÷] [(] [2] [ALPHA] [A] [)]	(-B+√(B²-4AC))/(2A)
[ENTER] [2nd] [QUIT]	Get back to the home screen.
[3] [STO▸] [ALPHA] [A] [ENTER]	Store 3 as A, 4 as B and -5 as C.
[4] [STO▸] [ALPHA] [B] [ENTER]	
[(-)] [5] [STO▸] [ALPHA] [C]	
[ENTER]	
[2nd] [Y-VARS] [1] <Function>	Recall the expression and evaluate.
[1] <Y1> [ENTER]	The result is .79 to two decimal places. So one solution to this quadratic equation is x=.79.
[Y=] [▶] ... [▶] [-] [ENTER] etc.	Get the function list. Now change the addition sign to a subtraction sign and evaluate again. The other solution is x=-2.12.

> Throughout this manual, when using menus, the number of the desired selection will be shown followed by what the selection is. Thus, [Y-VARS] [1] <Function> means to pick the first selection (that is, Function) off the Y-VARS menu.

EXAMPLE 2: Given $f(x) = \sqrt[3]{x^4+2x}$ and $g(x) = 4^{2x-90}$, find $f(x) + 3\sqrt{g(x)}$ at $x = 45.2$.

Solution:

Keystrokes	Explanation
Y= CLEAR	Clear the old expression for Y1 by pressing CLEAR
(X,T,θ ^ 4 + 2 X,T,θ)	Enter $f(x)$ as Y1. `(X^4+2X)^(1/3)`
^ (1 ÷ 3) ENTER	Note that the X,T,θ key displays the letter X. (The letter T is displayed if parametric mode is being used, and θ is displayed if in polar graphing mode.)
Y= CLEAR	Clear the function stored as Y2.
4 ^ (2 X,T,θ	Enter $g(x)$ as Y2. `4^(2X-90)`
- 90) 2nd QUIT	
45.2 STO▸ X,T,θ ENTER	Store 45.2 as X.
2nd Y-VARS 1 <Function>	Form $f(x) + 3\sqrt{g(x)}$ and evaluate.
1 <Y1> + 3 2nd √ 2nd	
Y-VARS 1 <Function>	The result is 164.97 to two decimal places.
2 <Y2> ENTER	

EXAMPLE 3: Evaluate the expression $\frac{3x^2 + 5x - 8}{x^2 - 1}$ to three decimal places for $x = .900$, .990, .999, 1.001, 1.010, and 1.100 by: (A) evaluating the expression for each value of x individually; and (B) using a list.

Solution (A): First store the expression in the Y= list. Then store the value for x. Recall the expression and evaluate it.

Keystrokes	Explanation
Y= CLEAR	Clear the old expression for Y1 by pressing CLEAR.
(3 X,T,θ ^ 2 + 5 X,T,θ	Store the expression as Y1 and return to the home screen.
- 8) ÷ (X,T,θ ^ 2 -	
1) 2nd QUIT	
.9 STO▸ X,T,θ ENTER	Store the value of x.
2nd Y-VARS 1 <Function>	Recall the expression and evaluate.
1 <Y1> ENTER	

Store the next value for x and reevaluate the expression. Repeat for the other values of x. The corresponding results are: 5.632, 5.513, 5.501, 5.499, 5.488, and 5.381.

Solution (B): Store the values of x in a list and evaluate the expression for each value of x simultaneously.

Y= CLEAR — Clear the old expression for Y1 by pressing CLEAR.

(3 X,T,θ ^ 2 + 5 X,T,θ − 8) ÷ (X,T,θ ^ 2 − 1) 2nd QUIT — Store the expression as Y1 and return to the home screen.

2nd { .9 , .99 , .999 , 1.001 , 1.01 , 1.1 2nd } STO▸ 2nd L1 ENTER — Store the values of x in the list L1.

2nd Y-VARS 1 <Function> 1 <Y1> (2nd L1) ENTER — Simultaneously calculate the value of the expression stored as Y1 for the values of x in list L1. The results are: 5.632, 5.513, 5.501, 5.499, 5.488, and 5.381. To view the results, use the ◄ and ► keys.

EXAMPLE 4: Evaluate the expression $\dfrac{3x^2 + 5x - 8}{x^2 - 1}$ to three decimal places for values of x at each tenth from 0 to 1 using a table.

Solution: First store the expression in the Y= list. Set the table parameters to begin at $x = 0$ and to have an increment of .1. Get the table.

Y= CLEAR — Clear the old expression for Y1 by pressing CLEAR.

(3 X,T,θ ^ 2 + 5 X,T,θ − 8) ÷ (X,T,θ ^ 2 − 1) 2nd QUIT — Store the expression as Y1 and return to the home screen.

2nd TblSet 0 ENTER .1 ENTER — Set the table to begin evaluating the expression at x = 0 with a step size of .1

ENTER ▼ ENTER — Set the calculator to automatically display values of x and Y1.

2nd TABLE ▼ ... ▼ — Get the table. Arrow down to see the rest of the table. The highlighted value will appear at the bottom of the table.

X	Y1	
0	8	
.1	7.5455	
.2	7.1667	
.3	6.8462	
.4	6.5714	
.5	6.3333	
.6	6.125	
X=0		

A-5 Evaluating Inequalities

EXAMPLE 1: Given the inequality $x^2 - 3t < 5x + 9t^2$. Test this inequality for $x = 1.89$ and $t = -4.52$.

Solution:

Keystrokes	Explanation
[1.89] [STO▶] [X,T,θ] [ENTER]	Store the value of the variables.
[(-)] [4.52] [STO▶] [ALPHA]	
[T] [ENTER]	Enter the expression.
[X,T,θ] [^] [2] [-] [3] [ALPHA]	X^2-3T<5X+9T^2
[T] [2nd] [TEST] [5] << > [5]	
[X,T,θ] [+] [9] [ALPHA]	The result is 1. This means the inequality is true for these values of the variables. If the result was a 0 then the inequality would not have been true.
[T] [^] [2] [ENTER]	

A-6 Graphing

EXAMPLE 1: Graph $y = 3x^2-6$ and $y=3x^2+1$ on the same graphing screen.

Solution:

Keystrokes	Explanation
[Y=] [CLEAR]	Get the function list and clear the existing function for Y1. Enter the first function as Y1. Now clear the second function location and enter the second function as Y2.
[3] [X,T,θ] [^] [2] [-] [6] [ENTER]	
[CLEAR]	
[3] [X,T,θ] [^] [2] [+] [1] [ENTER]	Clear or deselect all others by placing the cursor over the = sign using the left arrow and pressing enter. Now this function will not graph. Note that the = sign is no longer highlighted.
[2nd] [DRAW] [1] <ClrDraw>	Clear drawings.
[ENTER]	
[2nd] [STAT PLOT] [1] [▶] [ENTER] etc.	Turn off all statistical plots.
[ZOOM] [6] <ZStandard>	The sixth option on the ZOOM menu will automatically set the graphing screen to have $-10<x<10$ and $-10<y<10$. Also, the functions will be automatically graphed.

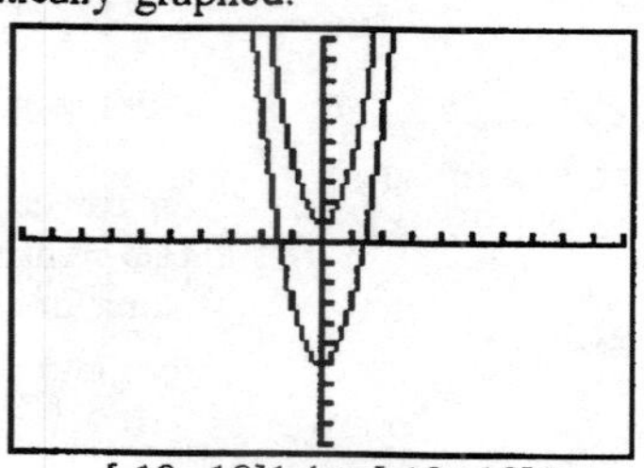

[-10, 10]1 by [-10, 10]1

ZOOM **6** automatically sets the graphing screen limits to $-10 < x < 10$ and $-10 < y < 10$ with scale marks every unit for both the x and y axes. The graph screen limits will be denoted in this manual by the notation [-10, 10]1 by [-10, 10]1. The graph screen limits can be set to other values, if desired, by pressing WINDOW and entering the desired values, then pressing GRAPH.

6 **<ZStandard>** means to choose the sixth option from the ZOOM menu.

The function will <u>not</u> graph if you move the cursor over the = sign and press ENTER. You notice the highlighting disappears from the = sign. You can still evaluate this expression for stored values of the variables. To select the function, place the cursor on the = sign and press ENTER again.

EXAMPLE 2: Graph $f(x) = \begin{cases} .5x + 1 & x \le 2 \\ x^2 + 2 & x > 2 \end{cases}$.

Solution: The calculator will only graph functions for values of x for which the function is defined. So we write this piecewise-defined function in a form where the pieces are defined only for the values of x that we want. Enter the function as two separate functions with (.5 X + 1) ÷ (X ≤ 2) as Y1 and (X ^ 2 + 2) ÷ (X > 2) as Y2. The first function is defined when $x \le 2$ and the second function is defined when $x > 2$. (See Appendix Section A-5 on how to get the inequality symbols from the TEST menu, and see Appendix Section A-6 Example 1 on how to store functions.)

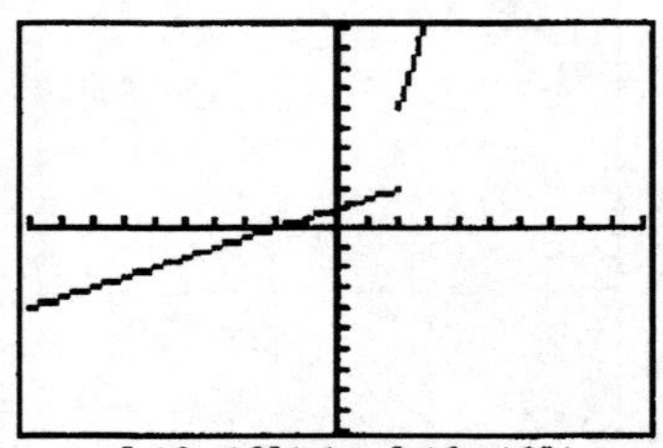

[-10, 10]1 by [-10, 10]1

A-7 TRACE, ZOOM, WINDOW, ROOT, and INTERSECT

EXAMPLE 1: Find the x intercept of $y = 3x^2 - 6$.

Solution:

First graph $y = 3x^2 - 6$ on the standard graphing screen (see Appendix Section A-6 Example 1 above).

2nd CALC 2 <root>

After you get the root finder, you will see

Lower Bound?

on the calculator screen.

Use right and left arrow keys and move the cursor along the graph until the cursor is to the left of the root and press ENTER.

◄ or ► ENTER

◄ or ► ENTER

◄ or ► ENTER

▲ or ▼

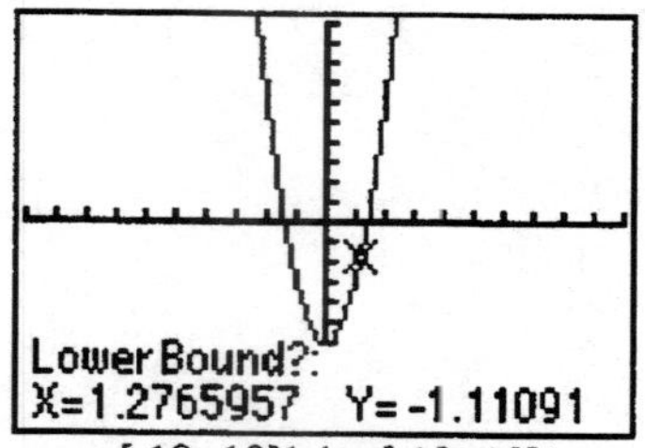

[-10, 10]1 by [-10, 10]1

Now move the cursor to the right of the room and press ENTER. Now move the cursor near the root and press ENTER. The calculator will display the *x* and *y* value of the root. Notice that the *y* value is very close to zero.

If we had two or more functions graphed, we could use the up and down arrow keys to move between the functions.

There are five ways to get a closer look at the intercept and one way to use the calculator to find the root:

1. Change the WINDOW dimensions.
2. Set the Zoom Factors and zoom in.
3. Use the Zoom Box feature of the calculator.
4. Use the Root Finder feature of the calculator.
5. Use the Intersect feature of the calculator.

EXAMPLE 2: Find the x intercept for $x>0$ of $y = 3x^2-6$. Use three decimal place accuracy.

Solution: Store the function in the calculator. Graph using [-10, 10]1 by [-10, 10]1.

<u>METHOD 1 Set the WINDOW dimensions.</u>

Tracing the function graphed above we see the x intercept is near 1.5. So we shall set the graph window close to that value.

WINDOW ENTER 1 ENTER
2 ENTER .1 ENTER
(-) .5 ENTER .5
ENTER .1 GRAPH

Set the WINDOW to [1, 2].1 by [-.5, .5].1. Notice we enter the minimum, maximum and scale values for the *x* axis, and then do the same for the *y* axis.

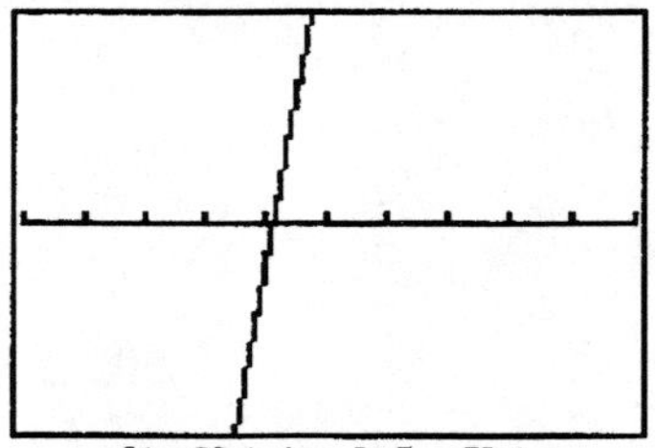

[1, 2].1 by [-.5, .5].1

Use trace again to get a better approximation. We see that the *x* intercept is between 1.41 and 1.42. Set the WINDOW to [1.41, 1.42].01 by [-.01, .01].001. We now find the *x* intercept to be 1.414 accurate to three decimal places.

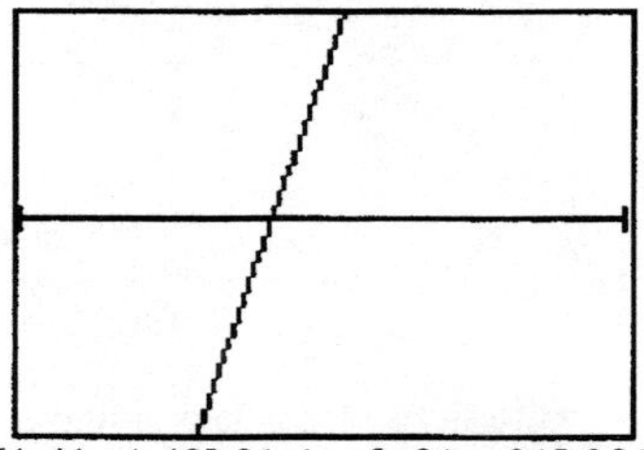

[1.41, 1.42].01 by [-.01, .01].001

<u>METHOD 2 Zoom In</u>

ZOOM 6 <ZStandard>
ZOOM ► 4 <Set Factors>
5 ENTER 5 2nd QUIT

Get the [-10, 10]1 by [-10, 10]1 display as shown in Section A-6 Example 1. Choose the MEMORY from the ZOOM menu. Choose 4 <Set Factors>.
Set the Zoom Factors to 5 and 5. After zooming in the new Xmin will be $\frac{1}{5}$ the previous Xmin.

ZOOM 2 <Zoom In>.
◄ or ► or ▲ or ▼
ENTER

Use the arrow keys repeatedly to get the cursor close to the *x* intercept, say at x=1.4893617 y=.64516129. The new display will be centered where you placed the cursor.

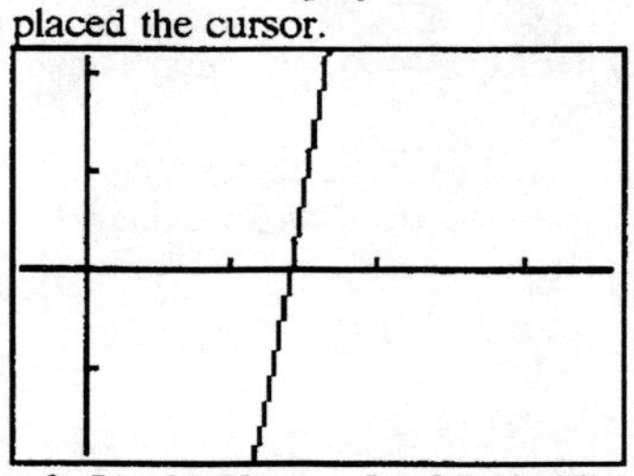

[-.51, 3.49]1 by [-1.35, 2.65]1

Now use TRACE again to get a new approximation to the *x* intercept.
Repeat zooming in as many times as necessary to get three decimal place accuracy.
The *x* intercept is 1.414.

METHOD 3 Zoom Box

[ZOOM] [6] <ZStandard> — Get the [-10, 10]1 by [-10, 10]1 display as shown in Section A-6 Example 1.

[ZOOM] [1] <ZBox> — Get the Zoom Box option.

[◄] or [►] or [▲] or [▼]
[ENTER] — Use the arrow keys repeatedly to get the cursor close to but a little above and to the left of the x intercept, say at X=1.0638298 Y=.64516129. Press [ENTER] to place the upper corner of the Zoom Box at this point.

[◄] or [►] or [▲] or [▼]
[ENTER] — Now move the cursor down and to the right of the x intercept, say at X = 1.7021277 Y =-.3225806. Press [ENTER] again. The graph will be drawn again with the box as the outside of the new display.

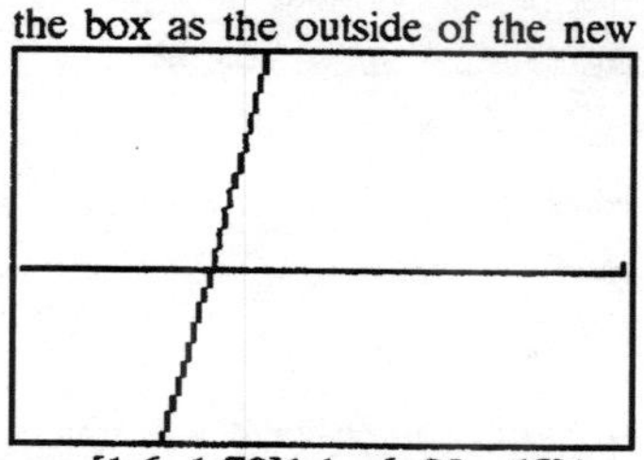

[1.6, 1.70]1 by [-.32, .65]1

Trace to get an estimate of the x intercept. Use Zoom Box a second time to get an even better approximation of the x intercept. The x intercept is 1.414.

METHOD 4 Root Finder

[2nd] [CALC] [2] <root>

[◄] or [►] [ENTER] — To place the cursor at a point on the graph to the left of the x intercept.

[◄] or [►] [ENTER] — To place the cursor at a point on the graph to the right of the x intercept.

[ENTER] — For the guess.

The value of the root will appear at the bottom of the screen. The root is (1.41, 0) rounded to two decimal places. Repeat for the other root.

METHOD 5 Intersect

[Y=] [▼] [0] [2nd] [QUIT] — To enter $y = 0$ as the second function.

[2nd] [CALC] [5] <intersect>

[◄] or [►] [ENTER] — To place the cursor at a point on the graph of the first curve near one of the x intercepts.

[◄] or [►] [ENTER] — To place the cursor at a point on the graph of the second curve near the same intercept.

[◄] or [►] [ENTER] — For the guess.

The coordinates of the point of intersection will appear at the bottom of the screen. The coordinates of the rightmost x intercept is (1.41, 0) rounded to two decimal places. Repeat for the other x intercept.

A-8 Choosing the WINDOW Variables Values

There is no one right way to determine the window variables that should be used for the graph of a function. Three methods to try are:

1. Graph using the standard screen and then zoom in or zoom out, if necessary.
2. Analyze the features of the function and then choose the viewing window dimensions.
3. Evaluate the function at several values and then choose the viewing window dimensions.

EXAMPLE 1 Graph $y = .03x^3 + .02x^2 - 5x + 10$.

Solution:

METHOD 1 Graph using the standard screen and then zoom out

Enter the function as .03 X ^ 3 + .02 X ^ 2 - 5 X + 10. Press [ZOOM] [6] to get the [-10, 10]1 by [-10, 10]1 display. (See Section A-6 of this manual.) We have too small a display to see the behavior of the graph clearly. Press [ZOOM] [▶] [4] and set the Zoom Factors to XFact=4 and YFact=4. Press [ZOOM] [3], move the cursor so it is where you want the center of the new screen to be, (we will leave it at the origin for this example) and press [ENTER]. The result is shown below. Note the double lines as the axes. This is because the Xscl and Yscl values do not change when zooming in. Since the scale values for x and y are small compared to the distance between Xmin and Xmax, and Ymin and Ymax, the scale marks appear to run together. You can fix this by choosing a more appropriate scale (say at 5 for Xscl and Yscl for this example). [Note if Xscl or Yscl are set to 0, no scale marks will be drawn.]

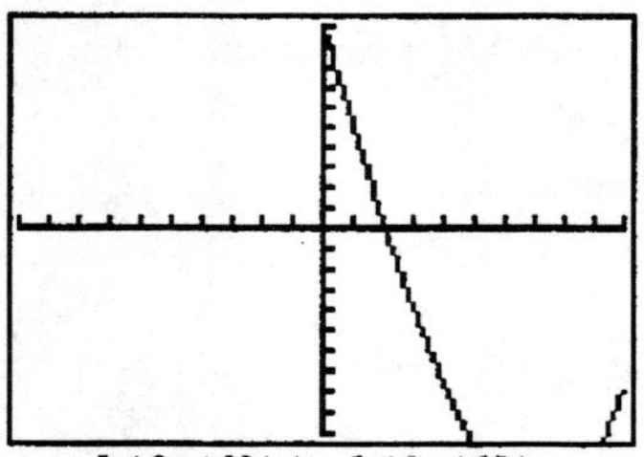

[-10, 10]1 by [-10, 10]1

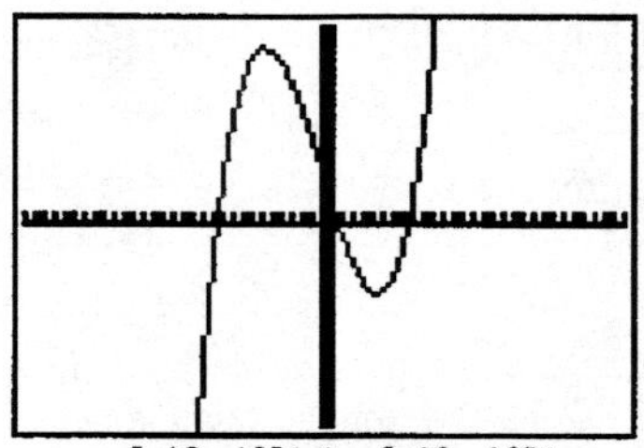

[-40, 40]1 by [-40, 40]1

METHOD 2 Analyze the function and then choose the window variable values

We will analyze the term with highest power on x since this is the dominant term of the function.

We see that the leading coefficient is .03. This means that the vertical change in the graph will be approximately .03 unit for every time x^3 increases 1 unit. An equation that represents this is $.03x^3 = y$ or $x^3 = \frac{1}{.03}y$ or $x^3 \approx 30y$. So if y increases by 1 unit, x will increase by about $\sqrt[3]{30}$ or about 3 units. This indicates that we want the x axis three times as wide as the y axis. Hence, a good first choice for the horizontal axis is -30 to 30, and a good first choice for the vertical axis is -10 to 10. Set Xmin = -30, Xmax = 30 and Xscl = 3. We choose 3 as the Xscl because this will put 20 tick marks on the axis.

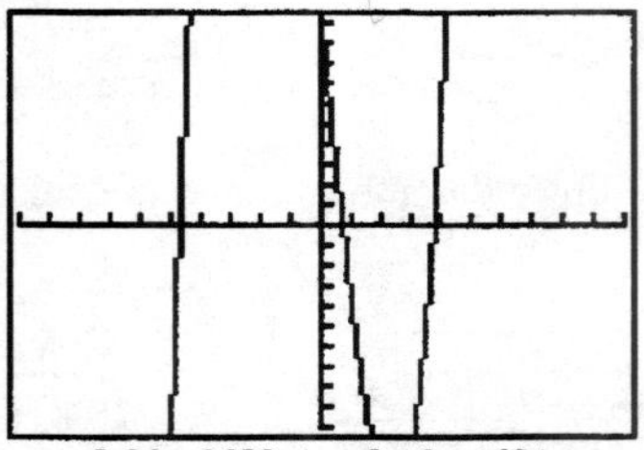

[-30, 30]3 by [-10, 10]1

The values for the vertical variables needs some adjusting since the turning points of the graph are not shown on the screen. Window variable values of [-30, 30]1 by [-40, 40]5 would show the turning points of the graph.

<u>METHOD 3 Evaluate the function at several values and then choose the viewing window dimensions</u>

Evaluate the function by storing a value for x and evaluating the function (see Section A-4). Repeat for each value of x. The choice of x values is arbitrary.

x	$f(x)$
-25	-321.25
-10	32
0	10
10	-8
25	366.25

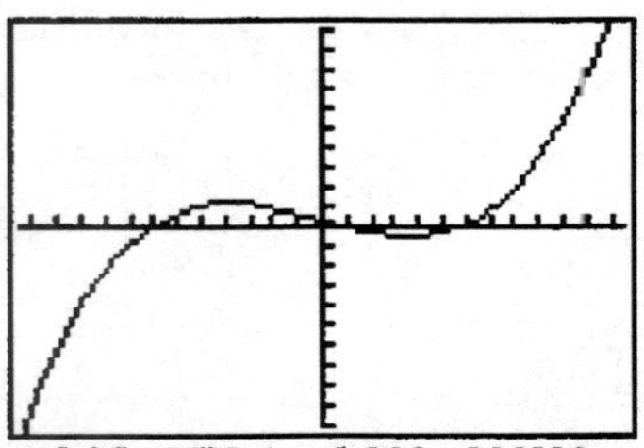

[-25, 25]2 by [-300, 300]30

Choose the Xscl and Yscl so there are about 20 marks on each axis (see METHOD 2 above). So a good first choice for the WINDOW could be [-25, 25]2 by [-300, 300]30.

A-9 Equations in One Variable

An equation in one variable can be solved graphically in two ways:

1. Algebraically move all terms to one side of the equation. Create a function $f(x)$=(the expression). Graph the function. Zoom and trace, or use other calculator features, to find the x intercepts. (See Appendix Section A-7 of this manual.) The set of x intercepts is the solution to the equation.

2. Create two functions: $f(x)$ = (the left side of the equation) and $g(x)$ = (the right side of the equation). Graph both functions. Zoom and trace, or use other calculator features, to find the x value of the points of intersection of the two graphs. (See Appendix Section A-7 of this manual.) The set of x values is the solution to the equation.

EXAMPLE: Solve $\sqrt[3]{x+2} = 4x^2 - 3\sqrt{x+4}$. Use two decimal places.

Solution:

<u>METHOD 1</u> Write the equation as $\sqrt[3]{x+2} - (4x^2 - 3\sqrt{x+4}) = 0$. Create the function $f(x) = \sqrt[3]{x+2} - (4x^2 - 3\sqrt{x+4})$. Enter this function into the calculator and graph.

[Y=] [(] [X,T,θ] [+] [2] [)] [^] [(] [1] [÷] [3] [)] [-]
[(] [4] [X,T,θ] [^] [2] [-] [3] [2nd] [√] [(] [X,T,θ]
[+] [4] [)] [)] [ENTER]
[ZOOM] [6] <Standard>

Enter the function and graph.

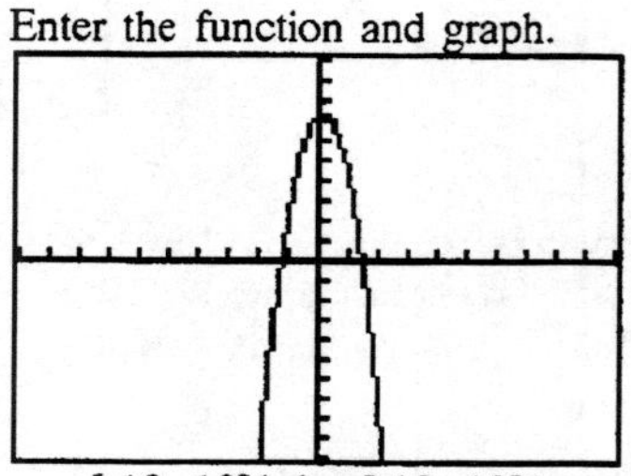

[-10, 10]1 by [-10, 10]1

We see the function crosses the x axis in two places. Trace and zoom, or use other calculator features, to find these x values. The solution to this equation is $x = -1.22$ and $x = 1.46$.

<u>METHOD 2</u> Create the two functions $f(x) = \sqrt[3]{x+2}$ and $g(x) = 4x^2 - 3\sqrt{x+4}$. Graph these and find the x values of the intersection points.

[Y=] [(] [X,T,θ] [+] [2] [)] [^] [(] [1] [÷] [3] [)]
[ENTER]
[4] [X,T,θ] [^] [2] [-] [3] [√] [(] [X,T,θ] [+] [4] [)]
[ENTER]
[ZOOM] [6] <Standard>

Enter the functions and graph.

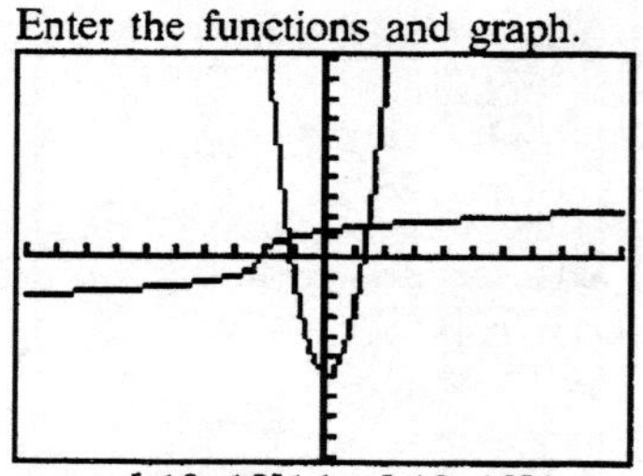

[-10, 10]1 by [-10, 10]1

We see the functions intersect in two places. Trace and zoom to find the coordinates of these points. The intersection points are approximately at (-1.22, .92) and (1.46, 1.51). So the solution to this equation is $x = -1.22$ and $x = 1.46$.

<u>A-10 Inequalities in One Variable</u>

An inequality in one variable can be solved graphically in two ways:

1. Algebraically move all terms to one side of the inequality and change the sign to an equal sign. Create a function $f(x)$ = (the expression). Graph the function. Trace and zoom to find the x intercepts. The set of x intercepts are the cutoff values for the solution. Now determine the intervals over which the inequality is true and write the solution. Pick a value either greater or less than the x intercept. Test the inequality using the method shown in Appendix Section A-5.

2. Create two functions: $f(x)$ = (the left side of the inequality) and $g(x)$ = (the right side of the inequality). Graph both functions. Trace and zoom to find the x value of the points of intersection of the two graphs. The set of x coordinates are the cutoff values for the solution. Now determine the intervals over which the inequality is true and write the solution. Pick a value either greater or less than the x intercept. Test the inequality using the method shown in Appendix Section A-5.

EXAMPLE: Solve $\sqrt[3]{x+2} < 4x^2 - 3\sqrt{x+4}$. Use two decimal places

Solution:

<u>METHOD 1</u> Write the inequality as: $\sqrt[3]{x+2} - (4x^2 - 3\sqrt{x+4}) < 0$. Change the sign to an equal sign: $\sqrt[3]{x+2} - (4x^2 - 3\sqrt{x+4}) = 0$. Create the function $f(x) = \sqrt[3]{x+2} - (4x^2 - 3\sqrt{x+4})$, graph, and find the x intercepts. This was done above in Appendix Section A-9 METHOD 1 of this manual. The cutoff values are $x = -1.22$ and $x = 1.46$. Observing the graph we see it is below the x axis for $x < -1.22$ and $x > 1.46$. Thus for these values of x we have $f(x)<0$ and hence $\sqrt[3]{x+2} < 4x^2 - 3\sqrt{x+4}$. Hence the solution to this inequality can be written as $(-\infty, -1.22)\cup(1.46, \infty)$.

<u>METHOD 2</u> Create the two functions $f(x) = \sqrt[3]{x+2}$ and $g(x) = 4x^2 - 3\sqrt{x+4}$. Graph these and find the x values of the intersection points. This was done above in Appendix Section A-9 METHOD 2 of this manual. The intersection points are at (-1.22, .92) and (1.46, 1.52). Inspecting the graph we see that $f(x) < g(x)$ for $x < -1.22$ and $x > 1.46$. Hence the solution to this inequality can be written as $(-\infty, -1.22)\cup(1.46, \infty)$.

A-11 Permutations, Combinations, and Random Numbers

EXAMPLE 1: Find $P_{19,13}$ the number of permutations of 19 distinct objects taken 13 at a time without repetition.

Solution:

[19] [MATH] [►] [►] [►] [2] <nPr> [13] [ENTER]	Enter the numbers and get the built-in permutation function from the MATH menu. The result is displayed in scientific notation as 1.689515283E14.

EXAMPLE 2: Find $C_{19,13}$ the number of combinations of 19 distinct objects taken 13 at a time without repetition.

Solution:

[19] [MATH] [►] [►] [►] [3] <nCr> [13] [ENTER]	Enter the numbers and get the built-in function from the MATH menu. The result is 27132.

EXAMPLE 3: Generate ten unique random numbers at least 0 and less than 50.

Solution:

5 STO▸ MATH ▸ ▸ ▸
1 <rand> ENTER

We need to store a "seed" number in the storage location RAND. This tells the random number generator where to start.

MATH ▸ 2 <iPart> (
50 x MATH ▸ ▸ ▸
1 <rand>) ENTER ENTER
ENTER ENTER ENTER ENTER
ENTER ENTER ENTER ENTER

RAND generates a number greater than 0 and less than 1. If we multiply by 50 and take the integer part, the result is a number greater than or equal to 0 and less than 50.

The 10 random numbers are:

36 13 6 17 21 ~~21~~ 20 5 31 16 48

Note that repetitions can occur but 21 cannot be used twice.

A-12 Matrices

Given the matrices $A = \begin{bmatrix} 1 & -4 & 6 \\ 2 & -1 & 0 \\ 5 & -3 & 6 \end{bmatrix}$ and $B = \begin{bmatrix} 4 & 15 \\ -2 & 6 \\ 2 & 5 \end{bmatrix}$.

EXAMPLE 1: Find -2.5AB.

Solution:

MATRX ▶ ▶ 1 <[A]> 3
ENTER 3 ENTER

Choose matrix A and enter the dimensions.

1 ENTER (-) 4 ENTER
6 ENTER 2 ENTER
(-) 1 ENTER 0 ENTER
5 ENTER (-) 3
ENTER 6 ENTER

Enter the matrix elements.

MATRX ▶ ▶ 2 <[B]>
3 ENTER 2 ENTER

Choose matrix B and enter the dimensions.

4 ENTER 15 ENTER
(-) 2 ENTER 6 ENTER
2 ENTER 5

Enter the matrix elements.

2nd QUIT

Return to the Home Screen. Note that [A] is the matrix A. This is different than A which represents the variable A.

Keystrokes	Explanation
(-) 2.5 MATRX 1 <[A]> x MATRX 2 <[B]> ENTER	Enter the matrix operations. The result is $\begin{bmatrix} -60 & -52.5 \\ -25 & -60 \\ -95 & -217.5 \end{bmatrix}$.

EXAMPLE 2: Find A^{-1} using matrix A from Example 1.

Solution:

Keystrokes	Explanation
MATRX 1 <[A]> x^{-1} ENTER	The operations are done from the Home Screen. The result rounded to three decimal places is: $A = \begin{bmatrix} -.167 & .167 & .167 \\ -.333 & -.667 & .333 \\ -.028 & -.472 & .194 \end{bmatrix}$
STO▸ MATRX 2 <[B]> ENTER	To inspect the result, store it as another matrix, say [B], recall it and use the arrow keys to see the elements.

EXAMPLE 3: Solve the system of equations using the inverse of the coefficient matrix. Round answers to two decimal places.

$$\begin{aligned} x - 4y + 6z &= 2.5 \\ 2x - y &= 3.2 \\ 5x - 3y + 6z &= -3.4 \end{aligned}$$

Solution: Note that the coefficient matrix is matrix A from Section A-12 Example 1 of this manual. The method of finding the inverse of a matrix was shown above in Section A-12 Example 2. Enter the constants as a 3×1 matrix B and find $A^{-1}B$.

Keystrokes	Explanation
MATRX ▶ ▶ 2 <[B]> 3 ENTER 1 ENTER	Enter the matrix dimensions.
2.5 ENTER 3.2 ENTER (-) 3.4	Enter the matrix elements.
2nd QUIT	Return to the Home Screen.
MATRX 1 <[A]> x^{-1} MATRX 2 <[B]> ENTER	Perform the matrix operations. The result is $\begin{bmatrix} -.45 \\ -4.10 \\ -2.24 \end{bmatrix}$. The solution to the system of equations to two decimal places is $x = -.45$, $y = -4.10$, $z = -2.24$.

EXAMPLE 4: Find the solution to the system of Example 3 above using Gauss-Jordan reduction. In other words, find the reduced form of the matrix $\begin{bmatrix} 1 & -4 & 6 & 2.5 \\ 2 & -1 & 0 & 3.2 \\ 5 & -3 & 6 & -3.4 \end{bmatrix}$.

Solution:

Keystrokes	Explanation
MATRX ▶ ▶ 1 <[A]>	Name the matrix A. Enter the matrix dimensions.
3 ENTER 4 ENTER	
1 ENTER (-) 4 ENTER	Enter the matrix elements row by row.
6 ENTER 2.5 ENTER	
2 ENTER (-) 1 ENTER	
0 ENTER 3.2 ENTER	
5 ENTER (-) 3 ENTER	
6 ENTER (-) 3.4	
2nd QUIT	Return to the home screen.

After a calculation, some numbers may look like decimals. However upon storing, recalling and examining the matrix using the arrow keys you will find they are decimals that are very close to zero. This situation can be eliminated by setting the number of decimal places to a number, say 3 (see Section A-3 of this manual).

Keystrokes	Explanation
MATRX 1 <[A]> ENTER	Display the matrix.
MATRX ▶ ALPHA	We wish to multiply by -2, matrix A, row 1 and add it to row 2. The result will be put in row 2.
A <*row+(>	`*row+(-2,[A],1,2)`
(-) 2 , MATRIX 1 <[A]>	
, 1 , 2) ENTER	The result is $\begin{bmatrix} 1 & -4 & 6 & 2.5 \\ 0 & 7 & -12 & -1.8 \\ 5 & -3 & 6 & -3.4 \end{bmatrix}$.
STO▸ MATRIX 1 <[A]> ENTER	Store the result as matrix A so you can perform the next operation.
MATRX ▶ ALPHA	Now multiply by -5, matrix A, row 1 and add it to row 3.
A <*row+(>	`*row+(-5,[A],1,3)`
(-) 5 , MATRIX 1 <[A]> ,	
1 , 3) ENTER	The result is $\begin{bmatrix} 1 & -4 & 6 & 2.5 \\ 0 & 7 & -12 & -1.8 \\ 0 & 17 & -24 & -15.9 \end{bmatrix}$. Store the result as matrix A.
STO▸ MATRIX 1 <[A]> ENTER	

Keystrokes	Explanation
[MATRX] [▶] [0] <*row(> [1] [÷] [7] [,] [MATRIX] [1] <[A]> [,] [2] [)] [ENTER]	Multiply by 1/7, matrix A, row 2. `*row(1÷7,[A],2)` The result is $\begin{bmatrix} 1 & -4 & 6 & 2.5 \\ 0 & 1 & -1.71 & -.26 \\ 0 & 17 & -24 & -15.9 \end{bmatrix}$ rounded to two decimal places.
[STO▶] [MATRIX] [1] <[A]> [ENTER]	Store the result as matrix A.

Continue in this manner until you get $\begin{bmatrix} 1 & 0 & 0 & -.45 \\ 0 & 1 & 0 & -4.10 \\ 0 & 0 & 1 & -2.24 \end{bmatrix}$. The solution to the system of equations is $x = -.45$, $y = -4.10$, $z = -2.24$.

> Note: To swap rows of a matrix use [8] <rowSwap(> from the [MATRX] MATH menu. The command rowSwap([A],2,3) will swap rows 2 and 3 in matrix A.
>
> Note: To add one row to another use [9] <row+(> from the [MATRX] MATH menu. The command row+([A],2,3) will add row 2 to row 3 in matrix A.

A-13 Logarithmic, Exponential, and Hyperbolic Functions

EXAMPLE 1: Graph $f(x) = 10^{3.2x}$ for $-.5 < x < .5$.

Solution: Set the WINDOW to [-.5, .5].1 by [-10, 10]1. (See Appendix B Section A-7 Example 2 Method 1.)

[Y=] [10] [^] [(] [3.2] [X,T,θ] [)] [GRAPH]   Enter the functions and graph.

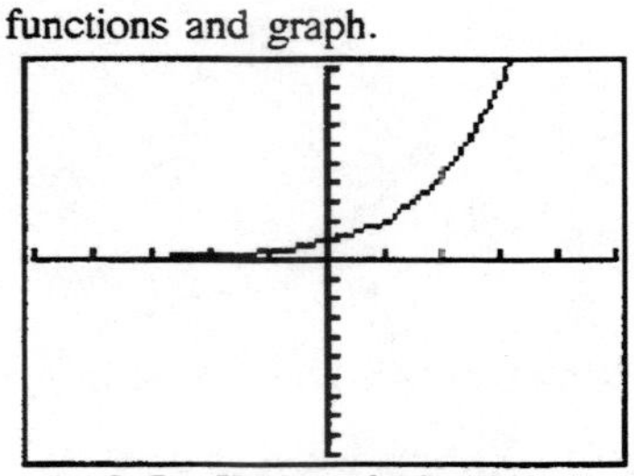

[-.5, .5].1 by [-10, 10]1

EXAMPLE 2: Graph $y = \dfrac{e^{x} + e^{-x}}{2}$.

Solution: This could be entered using [(] [2nd] [eˣ] [X,T,θ] [+] [2nd] [eˣ] [(-)] [X,T,θ] [)] [÷] [2]. However this is the hyperbolic cosine function. The following uses this built-in function. Recall that [eˣ] represents the built-in exponential function above the [LN] key.

[Y=] [CLEAR] [MATH] [►] [►]
[2] <cosh> [X,T,θ]
[ZOOM] [6] <ZStandard>

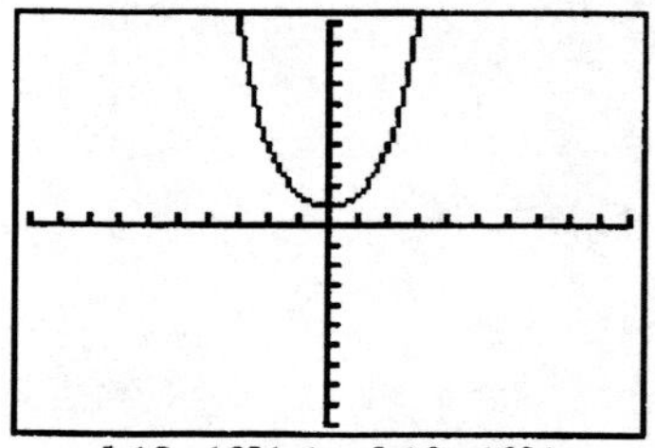

[-10, 10]1 by [-10, 10]1

EXAMPLE 3: Solve $\ln(x^2 + x) = 3x + 2$.

Solution:

[Y=] [CLEAR] [LN] [(] [X,T,θ] [^] [2] [+] [X,T,θ] [)] [ENTER] — Enter the function $f(x) = \ln(x^2 + x)$ as Y1 in the function list.

[CLEAR] [3] [X,T,θ] [+] [2] [ENTER] — Enter the function $g(x) = 3x + 2$ as Y2 in the function list.

[ZOOM] [6] <ZStandard>

Graph them on the same graphing screen. Trace and zoom to find the x value of the point of intersection. This will be the solution to the equation.

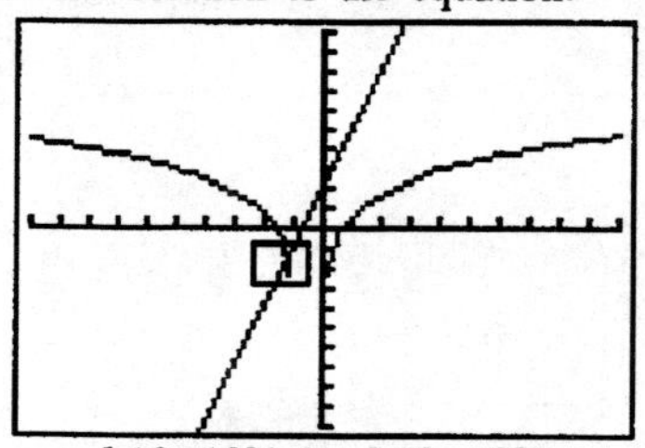

[-10, 10]1 by [-10, 10]1

The solution is -1.18, rounded to two decimal places.

A-14 Trigonometric Functions

EXAMPLE 1: Evaluate $f(x) = \cos x + \csc x$ at: (A) 53°45'18" and (B) 78.52 radians

Solution: (A)

[Y=] [CLEAR] [COS] [X,T,θ] [+] [1] [÷] [SIN] [X,T,θ] [ENTER] — Store the function.

[MODE] [▼] [▼] [►] [ENTER] — Change the calculator to degree angle measure.

[2nd] [QUIT]

[53] [+] [45] [÷] [60] [+] [18] [÷] [3600] [STO►] [X,T,θ] — Change the angle to decimal degrees while storing it as the variable X.

[2nd] [Y-VARS] [1] [1] <Y1> [ENTER] — Evaluate the function. The result to two decimal places is 1.83.

Solution:* *(B)

[MODE] [▼] [▼] [ENTER] — Change the calculator to radian angle measure.

[2nd] [QUIT] — Since the function was stored in Part (A) above, we need only to store the angle measure and evaluate the stored function.

[78.52] [STO▸] [X,T,θ] [ENTER] — Store 78.52 as X.

[2nd] [Y-VARS] [1] [1] <Y1> [ENTER] — Evaluate the function. The result to two decimal places is 49.47.

EXAMPLE 2: Graph $f(x) = \cos x + \csc x$.

Solution: The function was entered in the calculator in Part (A) above.

[MODE] [▼] [▼] [▼] [▼] [▶] — Change the graph mode to Dot. This will eliminate the "vertical lines."

[ENTER] [2nd] [QUIT]

[ZOOM] [7] <ZTrig> — This option automatically sets the viewing window variables to [-6.28, 6.28]1.57 by [-3, 3].25.

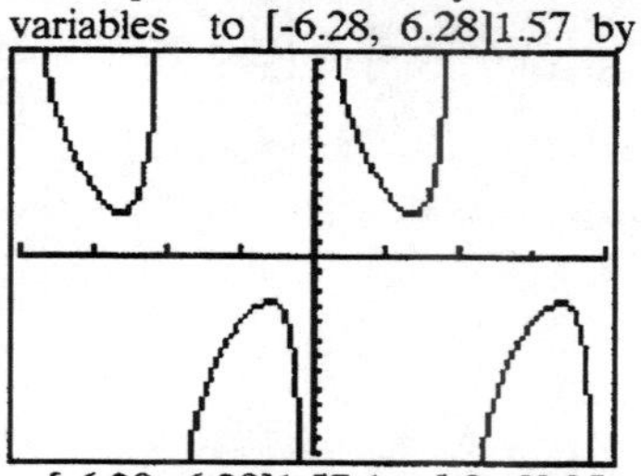

[-6.28, 6.28]1.57 by [-3, 3].25

A-15 Numeric Differentiation, Integration, and Tangent Lines

EXAMPLE 1: (A) Find the equation of the tangent line to $f(x) = \sqrt{x^5 - \sqrt[3]{x}}$ at $x = 3.54$. Use two decimal place accuracy. (B) Sketch the function and its tangent line using [2, 5]1 by [0, 60]5.

Solution (A): The function and its derivative can be evaluated at this value of x.

[Y=] [CLEAR] — Enter the function and store it as Y1.

[2nd] [√] [(] [X,T,θ] [^]

[5] [-] [X,T,θ] [^] [(] [1] [÷] [3] [)] [)]

Keystrokes	Explanation
2nd QUIT	Get the Home Screen and store the value of x.
3.54 STO▸ X,T,θ ENTER	
2nd Y-VARS 1 1 <Y1>	Evaluate the function. $f(3.54) = 23.55$ to two decimal places.
ENTER	
MATH 8 <nDeriv(>	Get the numeric derivative built-in function. `nDeriv(Y1,x,3.54,.001)`
2nd Y-VARS 1 1 <Y1>	This will evaluate the derivative with respect to x of
, X,T,θ , 3.54 , .0001)	the function stored in Y1 at $x = 3.54$ with .0001 tolerance. If the tolerance is not entered, the calculator will use .001.
ENTER	The result is 16.67.

The equation of the tangent line is $y - 23.55 = 16.67(x - 3.54)$.

Solution (B): Method 1 Enter the function as Y1 in the calculator and the equation of the tangent line as Y2 in the calculator and graph.

Method 2 Enter the function as Y1 in the calculator. Set the viewing rectangle to [2, 5]1 by [0, 60]5. Use the tangent drawing feature of the calculator by entering

2nd QUIT 2nd DRAW 5 <Tangent(> 2nd 1 <Function>

1 <Y1> , 3.54) ENTER . The function and its tangent will be graphed as shown to the right.

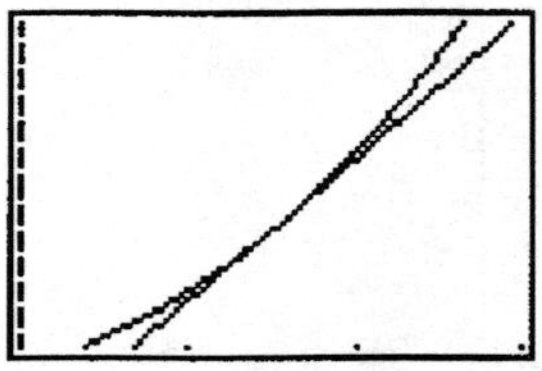

[2, 5]1 by [10, 60]1

EXAMPLE 2: Find $\int_{1.32}^{2.58} \sqrt{x^5 - \sqrt[3]{x}}\, dx$. Use two decimal place accuracy.

Solution:

Keystrokes	Explanation
Y= CLEAR 2nd √ (X,T,θ ^	Enter the function and store it.
5 - X,T,θ ^ (1 ÷ 3))	
2nd QUIT	Get the numeric integration function built into the calculator function:
MATH 9 <fnInt(>	`fnInt(Y1,X,1.32,2.58)`.
2nd Y-VARS 1 1 <Y1>	This will evaluate the definite integral. The result is 6.95. The tolerance in calculating
, X,T,θ , 1.32 , 2.58)	the definite integral is .00001.

A-16 Statistics

EXAMPLE 1: Given the following data find: (A) the mean; (B) the sample standard deviation ; (C) an histogram with first class beginning at 0 and width 4; (D) repeat Parts A and B after deleting the data value 89 from the list. How did eliminating this outlier effect the mean and standard deviation?

23	56	73	12	89	52	45	19	40	57	39
38	18	21	15	37	42	51	38	54	60	

Use one decimal place in answers.

Solution (A) and (B):

Keystrokes	Explanation
[STAT] [4] <ClrList> [2nd] [L1] [ENTER]	Clear the list named L1.
[STAT] [1] <Edit>	Get the data entry menu.
[23] [ENTER] [38] [ENTER] Etc.	Enter the data in the L1 list. Enter all data. There are 21 values. If you make a mistake, use the arrow keys to highlight the number and type it in again. To remove a data value, use the arrow keys to highlight number and press [DEL].
[2nd] [QUIT]	Exit the data entry screen.
[STAT] [▶] [3] [ENTER] [2nd] [QUIT] [STAT] [▶] [1] [ENTER]	Get the statistics menu again. Press [3] to see the setup. Under 1-Var Stats the Xlist should be L1 and the Freq should be 1. The mean is 41.86 and the sample standard deviation is 19.82.

Solution (C):

Keystrokes	Explanation
[WINDOW] [ENTER] [0] [ENTER] [100] [ENTER] [4] [ENTER] [(-)] [5] [ENTER] [5] [ENTER] [1]	The smallest value is 12 so let Xmin=0. The greatest value is 89. Let us set Xmax=100. Let Xscl=4 since this is the class width. Let Ymin=-5 to get some room below the histogram. We do not know the frequencies yet so let Ymax=5 for a first try. Let Yscl=1.
[Y=] [CLEAR] ...	Clear all functions on the function list.
[2nd] [DRAW] [1] <ClrDraw> [ENTER]	Clear all drawings from the display.

Keystrokes	Explanation
[2nd] [STAT PLOT] [1] [ENTER]	Set PLOT 1 to ON.
[▼] [▶] [▶] [▶] [ENTER]	Select Type as Histogram.
[▼] [ENTER] [▼] [ENTER]	Select Xlist as L1 and Freq as 1.
[2nd] [QUIT]	
[GRAPH] [ENTER]	Graph the histogram.

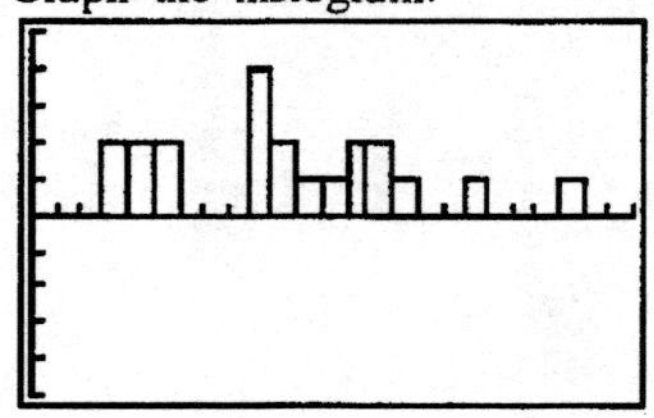

Solution (D):

Keystrokes	Explanation
[STAT] [1] [▼] [▼]...[▼] [▼] [DEL]	Get the data list and arrow down to the 89. Press [DEL].
[STAT] [▶] [1] [ENTER]	Get the statistics menu and the one variable sample statistics. The mean is now 39.5 and the sample standard deviation is 17.05. Notice how much influence an outlier has!

EXAMPLE 2: Given the following bivariate data: (A) Find the correlation coefficient. (B) Find the least squares line. (C) Graph the least squares line and the scatter plot on the same graph. (D) Plot the scatter diagram without the least squares line. (E) Sort the data on x and join the points using straight line segments.

x	5	8	10	29	9	24
y	3	4	8	20	3	19

Solution (A):

Keystrokes	Explanation
[STAT] [4] <ClrList> [2nd] [L1] [,] [2nd] [ENTER]	Clear list L1 and list L2.
[STAT] [1] <Edit>	Get the lists.
[5] [ENTER] [8] [ENTER] [10] [ENTER] [29] [ENTER] [9] [ENTER] [24] [ENTER]	Enter the data in list L1.
[▶] [3] [ENTER] [4] [ENTER] [8] [ENTER] [20] [ENTER] [3] [ENTER] [19] [ENTER]	Move to list L2. Enter the data in list L2. The data for y in L2 must correspond pairwise to the data for x in L1.
[STAT] [▶] <Calc> [3] <Set Up> [▼] [▼] [ENTER] [▼] [▶] [ENTER]	Set up the calculator to use L1 as the x variable and L2 as the y variable in the calculations.
[STAT] [▶] <Calc> [9] <LinReg(a+bx)> [ENTER]	This give you the linear correlation coefficient, the constant term and the coefficient on x term for the least squares line. r = .98.

Solution (B):

Keystrokes	Explanation
VARS 5 <Statistics> ► ► 7 <RegEQ>	The least squares line equation is displayed. It is $\hat{y}$ = -1.72 + .79*x* with the coefficients rounded to two decimal places.

Solution (C):

Keystrokes	Explanation
WINDOW ENTER 0 ENTER 30 ENTER 1 ENTER (-) 5 ENTER 30 ENTER 1	Set the graph WINDOW to [0, 30]1 by [-5, 30]1. Observe the data to determine the WINDOW values. Set the *y* value lower than needed so if you trace, the coordinates at the bottom of the screen do not cover up the graph.
Y= CLEAR ...	Clear all functions on the function list.
2nd DRAW 1 <ClrDraw> ENTER	Clear all drawings from the display.
	Turn the other STAT PLOTS off.
Y= VARS 5 ► ► 7 <RegEQ>	The least squares line equation will be put into the function list so you can graph it.
2nd STAT PLOT 1 <Plot 1> ENTER ▼ ENTER ▼ ENTER ▼ ► ENTER ▼ ENTER GRAPH	Set up the calculator to get the scatter plot with the line.

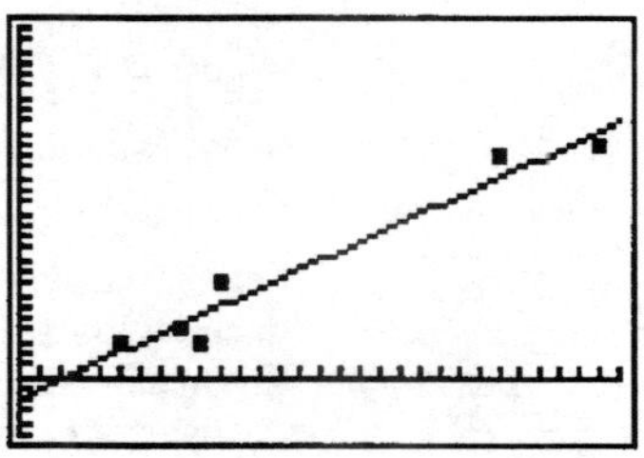

Solution (D):

Keystrokes	Explanation
Y= CLEAR or Y= ◄ ENTER	Clear the functions from the list or deselect them so they won't graph.
GRAPH	Plot the points on the graph.

Solution (E):

Keystrokes	Explanation
[Y=] [CLEAR] or [Y=] [◄] [ENTER]	Clear the functions from the list or deselect them so they won't graph.
[STAT] [2] [2nd] [L1] [,] [2nd] [L2] [)] [ENTER]	Sort the data on the x variable. The list L1 will be sorted in ascending order. The list L2 will sort along with L1 as pairs.
[2nd] [STAT PLOT] [1] <Plot 1> [ENTER]	Set the calculator to draw line segments joining the points.
[▼] [►] [ENTER] [▼] [ENTER] [▼] [►] [ENTER] [▼] [ENTER]	This will join the points together with straight line segments in the order they appear in the data list. This is why we sorted the data on x first.
[GRAPH]	

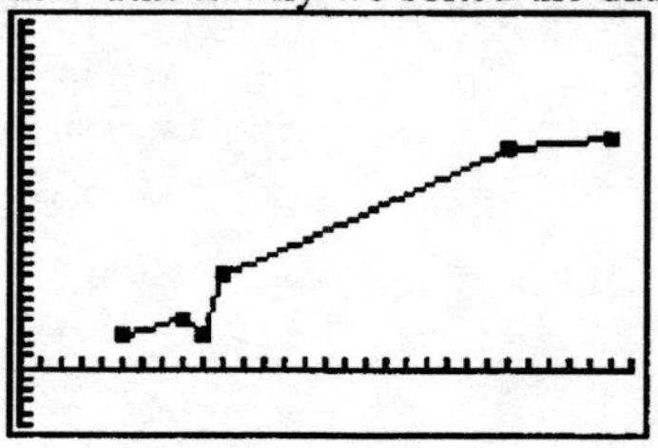

A-17 Setting the Display Mode for Numbers

Numbers can be entered into the calculator in scientific notation, a set number of significant digits, or a fixed number of decimal places.

EXAMPLE 1: Calculate $(9.35417\times 10^4)(-3.64389\times10^{-6})$ using:
(A) scientific notation, (B) two significant digits, (C) four decimal place accuracy.

Solution (A):

Keystrokes	Explanation
[MODE] [►] [ENTER] [2nd] [QUIT]	Set the calculator to scientific mode with floating decimal point.
[9.35417] [2nd] [EE] [4] [×] [(-)] [3.64389] [2nd] [EE] [(-)] [6] [ENTER]	The result displayed on the calculator is -3.408556652E-1. This means $-3.408556652\times10^{-1}$.

Solution (B):

Keystrokes	Explanation
[MODE] [ENTER] [▼] [►] [►] [ENTER]	Set the calculator to scientific notation with 1 fixed decimal point instead of a floating decimal. This will give two significant digits.
[2nd] [QUIT]	Return to the Home Screen.
[2nd] [ENTRY] [ENTER]	Use the replay key to replay the last calculation done in Solution (A). The result to two significant digits is -3.4E-1.

Solution (C):

Keystrokes	
[MODE] [▶] [ENTER]	Set the calculator to normal mode with 4 digits.
[▼] [▶] [▶] [▶] [▶] [▶] [ENTER]	
[2nd] [QUIT]	Return to the Home Screen. Use the replay key to replay the last calculation.
[2nd] [ENTRY] [ENTER]	The result to four decimal places is -.3409.

You may now wish to return your calculator to its default settings (all items on the mode screen having the left item highlighted). The examples throughout this manual are worked using the default settings unless otherwise noted.

NOTES

NOTES

Appendix B

TI-83 Operations

B-1 Operations

Press [ON] to turn on the calculator.

Press [2nd] [+] to get the **MEMORY** screen (shown at the right).

Choose 5:Reset... by pressing [5] or use the down-arrow key [▼] to highlight 5:Reset... and press [ENTER] to get the **RESET MEMORY** screen.

Choose 2: Defaults... by pressing [2] or use the down-arrow key [▼] to highlight 2: Defaults... and press [ENTER] to get the **RESET DEFAULTS** screen.

Choose 2: Reset or use the down arrow key to highlight 2:Reset... and press [ENTER]. The screen will display Defaults set. If the screen looks blank it is because the contrast setting was also reset.

Press [2nd] and then hold the [▲] (to increase the contrast) or [▼] (to decrease the contrast) until you see the message on the screen.

Press [CLEAR] to clear the screen.

Press [2nd] [OFF] to turn the calculator off.

```
MEMORY
1: Check RAM..
2: Delete..
3: Clear Entries
4: ClrAllLists
5: Reset..
```

```
RESET MEMORY
1: All Memory..
2: Defaults..
```

```
RESET DEFAULTS
1: No
2: Reset
```

```
Defaults set
```

[2nd]

Press this key to access the operation above and to the left of a key. The operations above the key are in a gold color on the face of the calculator. Note that a flashing up arrow [↑] appears on the screen after [2nd] key is pressed.

> In this manual, the functions on the face of the calculator above a key will be referred to in square boxes just as if the function was printed on the key cap. For example, [ANS] is the function above and to the left of the [(-)] key.

[ALPHA] and [A-LOCK]

The operation above and to the right of a key is accessed by first pressing [ALPHA]. A flashing [A] is displayed as the cursor on the screen after the [ALPHA] key is pressed.

[2nd] [A-LOCK] locks the calculator into alpha mode until the [ALPHA] is pressed again.

Home Screen

Calculations are done and commands are entered on the Home Screen. You can always get to this screen (aborting any calculations in progress) by pressing

QUIT
[2nd] [MODE]. From here on, this will be referred to as [2nd] [QUIT] in this appendix.

[MODE]

When you press [MODE] you see the screen shown below. The highlighted items are active.

To change an item select it using the arrow keys and press [ENTER] to activate the selection.

Type of notation for display of numbers.	Normal Sci Eng
Number of decimal places displayed.	Float 0123456789
Type of angle measure.	Radian Degree
Function, parametric, polar or sequence graphing.	Func Par Pol Seθ
Connected/not connected plotted points on graphs.	Connected Dot
Graphs functions separately or all at once.	Sequential Simul
Allows a full screen or split screen to be used.	Rea a+bi re^θi
	Full Horiz G-T

Menus

The TI-83 Graphics calculator uses menus for selection of specific functions. The items on the menus are identified by numbers followed by a colon. There are two ways to choose menu items:

1. Press the number corresponding to the menu item.
2. Use the arrow keys to highlight the selection and then pressing [ENTER].

> In this manual the menu items will be referred to using the key to be pressed followed by the meaning of the menu. For example, on the ZOOM menu, [1] <ZBox> refers to the first menu item ZBox.

Editing

It is easy to correct errors when entering data into the calculator by using the arrow keys, [DEL], and [INS] keys.

[◄] or [►]	Moves the cursor to the left or right one position.
[▲] or [▼]	Moves the cursor up one line or down one line.

DEL	Deletes one or more characters at the cursor position.
2nd INS	Inserts one or more characters at the cursor position.
2nd ENTRY	Replays the last executed line of input.

B-2 Calculating

EXAMPLE 1: Calculate $|-4^2 - \sqrt{34}|$

Enter numbers and functions in the calculator in the same order as you would read an expression. In the examples that follow, do not press ENTER unless specifically instructed to do so. Enter all the keystrokes without pressing the ENTER key until ENTER is displayed in the example.

The subtraction sign and the negative sign are different.

The **subtraction sign** is a blue key on the lower right side of the calculator. This is used when subtracting one quantity from another.

The **negative sign** (-) is a gray key in the bottom row of keys on the calculator. This is used when entering a negative quantity.

Solution: 2nd QUIT MATH ► 1 <abs(> (-) 4 x^2 - 2nd √ 34) ENTER.
The result displayed is 21.83095189.

> Throughout this manual, when using menus, the number of the desired selection will be shown followed by what the selection is. Thus, VARS 1 <Window> means to pick the first selection (that is, Window) off the VARS menu.

B-3 Plotting Points and Drawing Line Segments

EXAMPLE 1: (A) Plot the points (-3, 6) and (1,5). (B) Draw the line segment having these points as the endpoints.

Solution (A):

Y= CLEAR ▼ CLEAR etc.	Clear all functions from the function list.
2nd QUIT	
ZOOM 6 <ZStandard>	Set the viewing window dimensions. ZOOM 6 <ZStandard> automatically sets the dimensions to Xmin=-10, Xmax=10, Xscl=1, Ymin=-10, Ymax=10, and Yscl=1. This will be denoted as [-10, 10]1 by [-10, 10]1 in this manual.
CLEAR	Press CLEAR to get back to the home screen.

Keystrokes	Explanation
[2nd] [DRAW] [1] <ClrDraw> [ENTER]	Get the [DRAW] menu and clear the drawings from the calculator screen.
[2nd] [DRAW] [►] <POINTS> [1] <PT-On(> [(-)] [3] [,] [6] [)] [ENTER]	Get the [DRAW] menu again. Choose the point on option and enter the coordinates of the point with a comma separating them. The point will be plotted.
[2nd] [QUIT]	
[2nd] [DRAW] [►] <POINTS>	
[1] <PT-On(> [1] [,] [5] [)] [ENTER]	Enter the coordinates of the second point and plot.

Solution (B):

Keystrokes	Explanation
[2nd] [QUIT]	Return to the Home Screen.
[2nd] [DRAW] [2] <Line(>	Get the [DRAW] menu and choose the line option.
[(-)] [3] [,] [6] [,] [1] [,] [5] [)] [ENTER]	Enter the coordinates of the points separated by commas. The line segment will be drawn between the two points.

B-4 Storing and Evaluating an Expression, Lists and Tables

EXAMPLE 1: Store the expression $\frac{-b+\sqrt{b^2-4ac}}{2a}$ and evaluate it to solve $3x^2 + 4x - 5 = 0$.

Solution:

Keystrokes	Explanation
[Y=] [CLEAR]	You will see a list of function names Y1, Y2, Y3, etc. We can store ten different functions or algebraic expressions in this list.
[(] [(-)] [ALPHA] [B] [+] [2nd] [√] [ALPHA] [B] [x^2] [-] [4] [ALPHA] [A] [×] [ALPHA] [C] [)] [)] [÷] [(] [2] [ALPHA] [A] [)] [ENTER]	Then we can recall them and evaluate or graph them. We will use Y1. Press these keys to enter the expression. `(-B+√(B²-4AC))/(2A)`
[2nd] [QUIT]	Get back to the home screen.
[3] [STO►] [ALPHA] [A] [ENTER] [4] [STO►] [ALPHA] [B] [ENTER] [(-)] [5] [STO►] [ALPHA] [C] [ENTER]	Store 3 as A, 4 as B and -5 as C.
[VARS] [►] <Y-VARS> [1] <Function> [1] <Y1> [ENTER]	Recall the expression and evaluate. The result is .79 to two decimal places. So one solution to this quadratic equation is x = .79.

[Y=] [▶] ... [▶] [-] [ENTER] etc.	Get the function list. Now change the addition sign to a subtraction sign and evaluate again. The other solution is x = -2.12.

EXAMPLE 2: Given $f(x) = \sqrt[3]{x^4+2x}$ and $g(x) = 4^{2x-90}$. Find $f(x) + 3\sqrt{g(x)}$ at x = 45.2.

Solution:

[Y=] [CLEAR]	Clear the old expression for Y1 by pressing [CLEAR]
[(] [X,T,θ,n] [^] [4] [+] [2] [X,T,θ,n] [)] [^] [(] [1] [÷] [3] [)] [ENTER]	Enter $f(x)$ as Y1. `(X^4+2X)^(1/3)` Note that the [X,T,θ,n] key displays the letter X. (The letter T is displayed if parametric mode is being used, and θ is displayed if in polar graphing mode, and n is displayed if in the sequence mode.)
[CLEAR]	Clear the function stored as Y2.
[4] [^] [(] [2] [X,T,θ,n] [-] [90] [)] [2nd] [QUIT]	Enter $g(x)$ as Y2. `4^(2X-90)`
[45.2] [STO▶] [X,T,θ,n] [ENTER]	Store 45.2 as X.
[VARS] [▶] <Y-VARS> [1] <Function> [1] <Y1> [+] [3] [2nd] [√]	Form $f(x) + 3\sqrt{g(x)}$ and evaluate.
[VARS] [▶] <Y-VARS> [1] <Function> [2] <Y2> [)] [ENTER]	The result is 164.97 to two decimal places.

EXAMPLE 3: Evaluate the expression $\dfrac{3x^2 + 5x - 8}{x^2 - 1}$ to three decimal places for x = .900, .990, .999, 1.001, 1.010, and 1.100 by: (A) Evaluating the expression for each value of x individually; and (B) using a list.

Solution (A): First store the expression in the Y= list. Then store the value for x. Recall the expression and evaluate it.

[Y=] [CLEAR]	Clear the old expression for Y1 by pressing [CLEAR].
[(] [3] [X,T,θ,n] [^] [2] [+] [5] [X,T,θ,n] [-] [8] [)] [÷]	Store the expression as Y1 and return to the home screen.

[(] [X,T,θ,*n*] [^] [2] [−]
[1] [)] [2nd] [QUIT]
[.9] [STO▸] [X,T,θ,*n*] [ENTER] Store the value of x.
[VARS] [▶] <Y-VARS> Recall the expression and evaluate.
[1] <Function> [1] <Y1> [ENTER]

Store the next value for x and reevaluate the expression. Repeat for the other values of x. The corresponding results are: 5.632, 5.513, 5.501, 5.499, 5.488, and 5.381.

Solution (B): Store the values of x in a list and simultaneously evaluate the expression for each value of x.

[Y=] [CLEAR] Clear the old expression for Y1 by pressing [CLEAR].
[(] [3] [X,T,θ,*n*] [^] [2] [+] Store the expression as Y1 and return to the home screen.
[5] [X,T,θ,*n*] [−] [8] [)] [÷]
[(] [X,T,θ,*n*] [^] [2] [−]
[1] [)] [2nd] [QUIT]
[2nd] [{] [.9] [,] [.99] [,] [.999] [,] Store the values of x in list L1.
[1.001] [,] [1.01] [,] [1.1] [2nd] [}]
[STO▸] [2nd] [L1] [ENTER]
[VARS] [▶] <Y-VARS> Simultaneously calculate the value of the expression stored as Y1 for the values of x in list L1.
[1] <Function> [1] <Y1> [(] [2nd] The results are: 5.632, 5.513, 5.501, 5.499, 5.488,
[L1] [)] [ENTER] and 5.381. Use the [◀] and [▶] keys to view the entire list.

EXAMPLE 4: Evaluate the expression $\frac{3x^2 + 5x - 8}{x^2 - 1}$ to three decimal places for values of x at each tenth from 0 to 1 using a table.

Solution: First store the expression in the Y= list. Set the table parameters to begin at $x = 0$ and to have an increment of .1. Get the table.

[Y=] [CLEAR] Clear the old expression for Y1 by pressing [CLEAR].
[(] [3] [X,T,θ] [^] [2] [+] [5] [X,T,θ] Store the expression as Y1 and return to the home screen.
[−] [8] [)] [÷] [(] [X,T,θ] [^] [2] [−]
[1] [)] [2nd] [QUIT]
[2nd] [TblSet] [0] [ENTER] Set the table to begin evaluating the expression at x = 0 with a step size of .1
[.1] [ENTER]

Keystrokes	Explanation
[ENTER] [▼] [ENTER]	Set the calculator to automatically display values of x and Y1.
[2nd] [TABLE] [▼]...[▼]	Get the table. Arrow down to see the rest of the table. The highlighted value will appear at the bottom of the table.

X	Y1	
0	8	
.1	7.5455	
.2	7.1667	
.3	6.8462	
.4	6.5714	
.5	6.3333	
.6	6.125	
X=0		

B-5 Evaluating Inequalities

EXAMPLE 1: Given the inequality $x^2 - 3t < 5x + 9t^2$. Test this inequality for $x = 1.89$ and $t = -4.52$.

Solution:

Keystrokes	Explanation
[1.89] [STO▸] [X,T,θ,*n*] [ENTER]	Store the value of the variables.
[(-)] [4.52] [STO▸] [ALPHA]	
[T] [ENTER]	
[X,T,θ,*n*] [^] [2] [-] [3] [ALPHA]	Enter the expression.
[T] [2nd] [TEST] [5] << > [5]	X^2-3T<5X+9T^2
[X,T,θ,*n*] [+] [9] [ALPHA]	
[T] [^] [2] [ENTER]	The result is 1. This means the inequality is true for these values of the variables. If the result was a 0 then the inequality would not have been true.

B-6 Graphing and Style of Graph

EXAMPLE 1: Graph $y = 3x^2 - 6$ and $y = 3x^2 + 1$ on the same graphing screen using a solid thin line for the first function and a solid thick line for the second function.

Solution:

Keystrokes	Explanation
[Y=] [CLEAR] [◄] [◄] [ENTER]... [►] [►]	Get the function list and clear the existing function for Y1. Move the cursor to the left of the Y1. Press enter until you see the solid thin line symbol \.
[3] [X,T,θ,*n*] [^] [2] [-] [6] [ENTER]	Now move the cursor to the right of the equal sign and enter the first function as Y1.
[Y=] [CLEAR] [◄] [◄]	

Keystrokes	Explanation
[ENTER] ... [►] [►] [3] [X,T,θ,n] [^] [2] [+] [1] [ENTER]	Clear the second function location. Set the thick line for the graph and enter the second function as Y2. Clear or deselect all others by placing the cursor over the = sign using the left arrow and pressing enter. Now this function will not graph. Note that the = sign is no longer highlighted.
[2nd] [DRAW] [1] <ClrDraw>	Clear drawings.
[ENTER]	
[2nd] [STAT PLOT] [4] <PlotsOff>	
[ENTER]	Turn off all statistical plots. The sixth option on the ZOOM menu will automatically set the graphing screen to have $-10 < x < 10$ and $-10 < y < 10$. Also, the functions will be automatically graphed.
[ZOOM] [6] <ZStandard>	

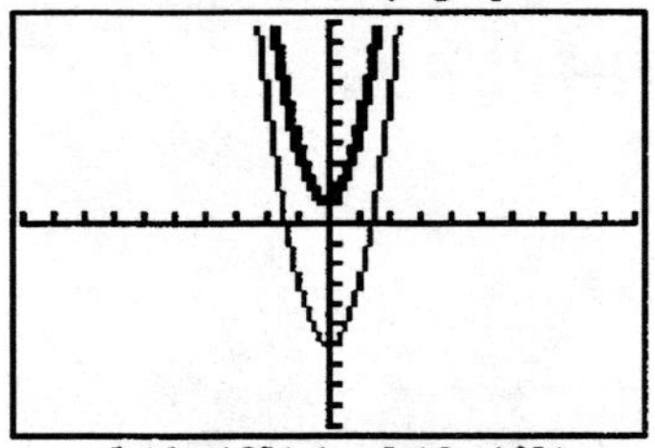

[-10, 10]1 by [-10, 10]1

> **[ZOOM] [6]** automatically sets the graphing screen limits to $-10 < x < 10$ and $-10 < y < 10$ with scale marks every unit for both the x and y axes. The graph screen limits will be denoted in this manual by the notation [-10,10]1 by [-10,10]1. The graph screen limits can be set to other values, if desired, by pressing [WINDOW] and entering the desired values, then pressing [GRAPH].

> **[6] <ZStandard>** means to choose the sixth option from the ZOOM menu.

> The function will <u>not</u> graph if you move the cursor over the = sign and press [ENTER]. You notice the highlighting disappears from the = sign. You can still evaluate this expression for stored values of the variables. To select the function, place the cursor on the = sign and press [ENTER] again.

EXAMPLE 2: Graph $f(x) = \begin{cases} .5x + 1 & x \le 2 \\ x^2 + 2 & x > 2 \end{cases}$.

Solution: The calculator will only graph functions for values of x for which the function is defined. So we write this piecewise-defined function in a form where the pieces are defined only for the values of x that we want. Enter the function as two separate functions with (.5 X + 1) ÷ (X ≤ 2) as Y1 and (X ^ 2 + 2) ÷ (X > 2) as Y2. The first function is defined when $x \leq 2$ and the second function is defined when $x > 2$. (See Appendix Section B-5 on how to get the inequality symbols from the [TEST] menu, and see Appendix Section B-4 on how to store functions.)

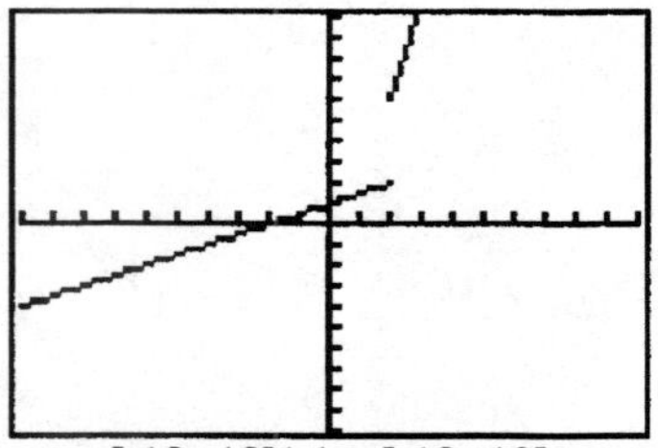

[-10, 10]1 by [-10, 10]1

B-7 TRACE, ZOOM, WINDOW, Zeroes and Intersect

EXAMPLE 3: Find the x intercept of $y = 3x^2 - 6$. Use three decimal place accuracy.

Solution: First graph $y = 3x^2 - 6$ on the standard graphing screen (see Appendix Section B-6 Example 1 above).

[TRACE]

[◄] or [►]

[▲] or [▼]

When you push [TRACE] you see a new cursor near the center of the graph screen and the coordinates of that point on the graph displayed at the bottom of the screen. Use right and left arrow keys and move the cursor along the graph.

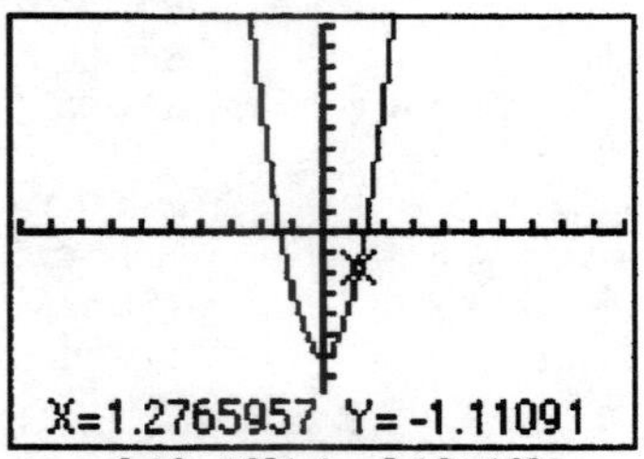

[-10, 10]1 by [-10, 10]1

There are several ways to get a closer look at the intercept:

1. Change the WINDOW dimensions.
2. Set the Zoom Factors and zoom in.
3. Use the Zoom Box feature of the calculator.
4. Use the Zero feature of the calculator.
5. Use the Intersect feature of the calculator.

<u>METHOD 1 Set the WINDOW dimensions.</u>

Using trace on the function graphed above we see the x intercept is near 1.5. So we shall set the graph window close to that value.

WINDOW 1 ENTER
2 ENTER .1 ENTER
(-) .5 ENTER .5
ENTER .1 GRAPH

Set the WINDOW to [1, 2].1 by [-.5, .5].1. Notice we enter the minimum, maximum and scale values for the x axis, and then do the same for the y axis.

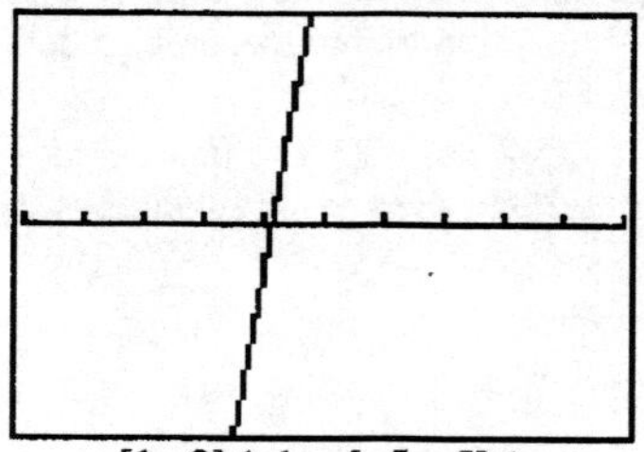

[1, 2].1 by [-.5, .5].1

Use trace again to get a better approximation. We see that the x intercept is between 1.41 and 1.42. Set the WINDOW to [1.41, 1.42].01 by [-.01, .01].001. We now find the x intercept to be 1.414 accurate to three decimal places.

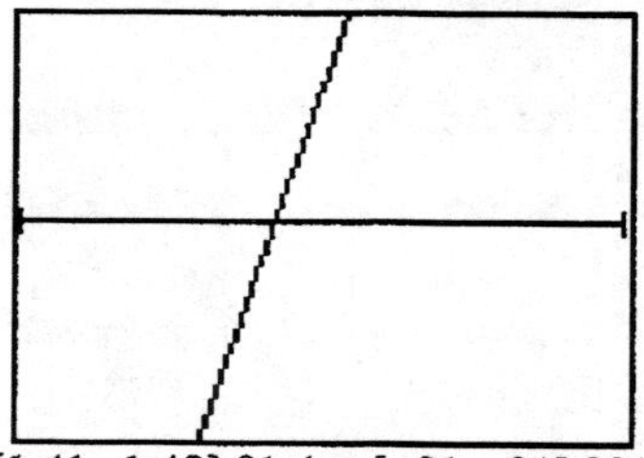

[1.41, 1.42].01 by [-.01, .01].001

<u>METHOD 2 Zoom In</u>

ZOOM 6 <ZStandard>
ZOOM ► <MEMORY>
4 <SetFactors>
5 ENTER 5 2nd QUIT
ZOOM 2 <Zoom In>.
◄ or ► or ▲ or ▼
ENTER

Get the [-10, 10]1 by [-10, 10]1 display as shown in Section B-6 Example 1. Choose the MEMORY from the ZOOM menu. Choose 4 <Set Factors>.

Set the Zoom Factors to 5 and 5. After zooming in the new Xmin will be $\frac{1}{5}$ the previous Xmin.

Use the arrow keys repeatedly to get the cursor close to the x intercept, say at x = 1.4893617 y = 0. The new display will be centered where you placed the cursor.

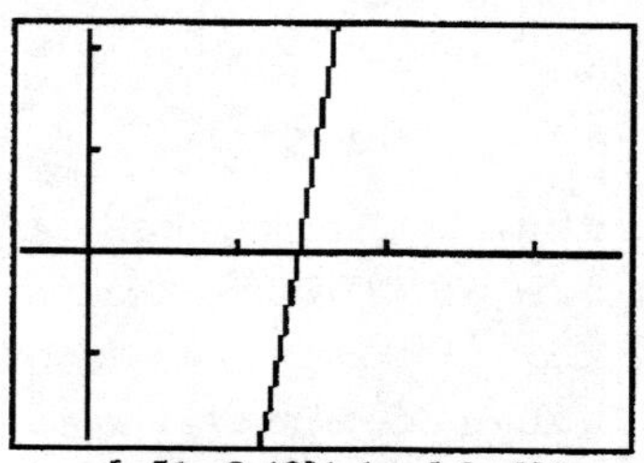

[-.51, 3.49]1 by [-2, 2]1

Now use TRACE again to get a new approximation to the x intercept.
Repeat the zooming in process as many times as necessary to get three decimal place accuracy.
The x intercept is 1.414.

<u>METHOD 3 Zoom Box</u>

[ZOOM] [6] <ZStandard>	Get the [-10, 10]1 by [-10, 10]1 display.
[ZOOM] [1] <ZBox>	Get the Zoom Box option.
[◄] or [►] or [▲] or [▼]	Use the arrow keys repeatedly to get the cursor close to but a little above and to the left of the x intercept, say at X = 1.2765957 Y = .64516129.
[ENTER]	Press [ENTER] to place the upper corner of the Zoom Box at this point.
[◄] or [►] or [▲] or [▼] [ENTER]	Now move the cursor down and to the right of the x intercept, say at X = 1.7021277 Y = -.3225806. Press [ENTER] again. The graph will be drawn again with the box as the outside of the new display.

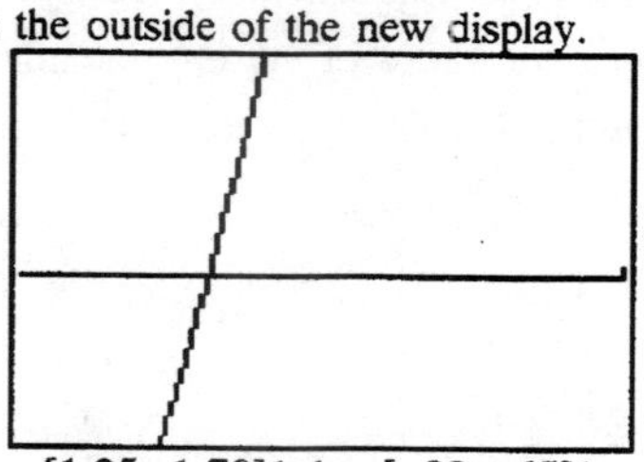

[1.25, 1.70]1 by [-.32, .65]1

Use trace to get an estimate of the x intercept. Use the Zoom Box feature of the calculator a second time to get an even better approximation of the x intercept. The x intercept is 1.414.

<u>METHOD 4 zero</u>

[ZOOM] [6] <ZStandard>	Get the [-10, 10]1 by [-10, 10]1 display.
[2nd] [CALC] [2] <zero>	Get the zero feature.
[◄] or [►] [ENTER]	To place the cursor at a point on the graph to the left of the x intercept.
[◄] or [►] [ENTER]	To place the cursor at a point on the graph to the left of the x intercept.
[ENTER]	For the guess.

The value of the root will appear at the bottom of the screen. The zero is (1.41, 0) rounded to two decimal places. Repeat for the other root.

<u>METHOD 5 Intersect</u>

[ZOOM] [6] <ZStandard>	Get the [-10, 10]1 by [-10, 10]1 display.
[Y=] [▼] [0] [2nd] [QUIT]	To enter $y = 0$ as the second function.
[2nd] [CALC] [5] <intersect>	
[◄] or [►] [ENTER]	To place the cursor at a point on the first graph near one of the x intercepts.

◄ or ► ENTER	To place the cursor at a point on the second graph near the same intercept.
ENTER	For the guess.

The coordinates of the point of intersection will appear at the bottom of the screen. The coordinates of the rightmost x intercept is (1.41, 0) rounded to two decimal places. Repeat for the other x intercept.

B-8 Choosing the WINDOW Variables Values

There is no one right way to determine the window variables that should be used for the graph of a function. Three methods to try are:

1. Graph using the standard screen and then zoom in or zoom out, if necessary.
2. Analyze the features of the function and then choose the viewing window dimensions.
3. Evaluate the function at several values and then choose the viewing window dimensions.

EXAMPLE 1 Graph $y = .03x^3 + .02x^2 - 5x + 10$.

Solution:

METHOD 1 Graph using the standard screen and then zoom out

Enter the function as .03 X ^ 3 + .02 X ^ 2 - 5 X + 10. Press ZOOM 6 to get the [-10,10]1 by [-10,10]1 display. (See Section B-6 of this manual.) We have too small a display to see the behavior of the graph clearly. Press ZOOM ► 4 and set the Zoom Factors to XFact = 4 and YFact = 4. Press ZOOM 3, move the cursor so it is where you want the center of the new screen to be, (we will leave it at the origin for this example) and press ENTER. The result is shown below. Note the double lines as the axes. This is because the Xscl and Yscl values do not change when zooming in. Since the scale values for x and y are small compared to the distance between Xmin and Xmax, and Ymin and Ymax, the scale marks appear to run together. You can fix this by choosing a more appropriate scale (say at 5 for Xscl and Yscl for this example).

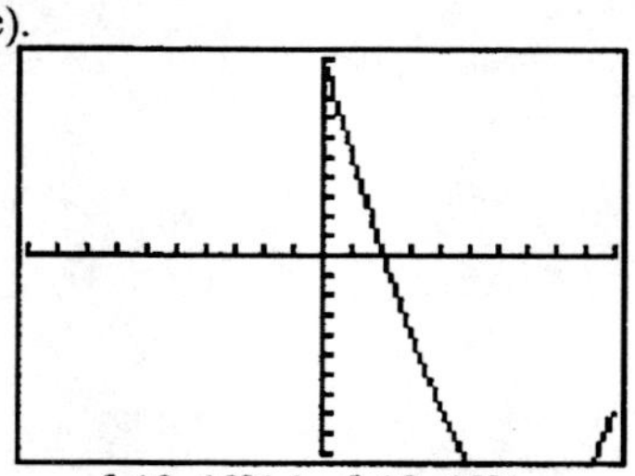

[-10, 10]1 by [-10, 10]1

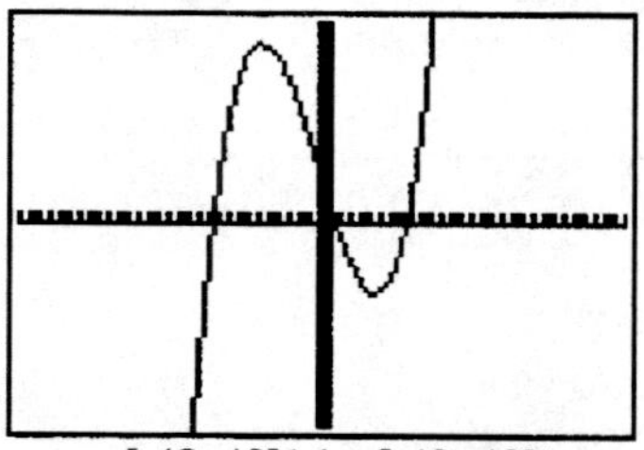

[-40, 40]1 by [-40, 40]1

METHOD 2 Analyze the features of the function and then choose a WINDOW

We see that the leading coefficient is .03. This means that the vertical change in the graph will be approximately .03 unit for every time x^3 increases 1 unit. An equation that represents this is $.03x^3 = y$ or $x^3 = \frac{1}{.03}y$ or $x^3 \approx 30y$. So if y increases by 1 unit, x will increase by about $\sqrt[3]{30}$ or about 3 units. This indicates that we want the x axis three times as wide as the y axis. Hence, a good first choice for the horizontal axis is -30 to 30, and a good first choice for the vertical axis is -10 to 10. Set Xmin = -30, Xmax = 30 and Xscl = 3. We choose 3 as the Xscl because this will put 20 tick marks on the axis.

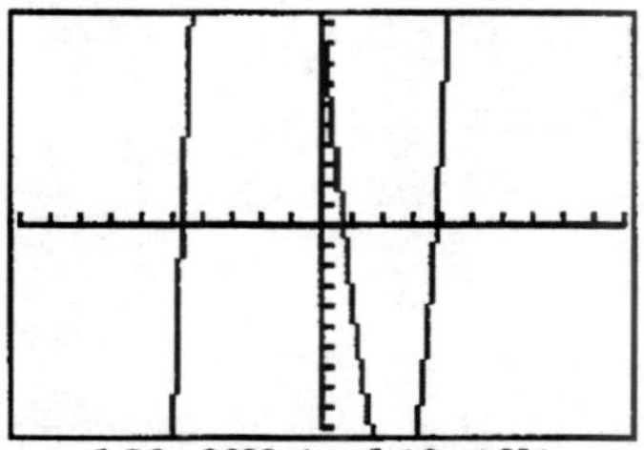

[-30, 30]3 by [-10, 10]1

The values for the vertical variables needs some adjusting since the turning points of the graph are not shown on the screen. Window variable values of [-30, 30]1 by [-40, 40]5 would show the turning points of the graph.

<u>METHOD 3 Evaluate the function at several values and then choose the viewing window dimensions</u>

Evaluate the function by storing a value for x and evaluating the function (see Section B-4). Repeat for each value of x. The choice of x values is arbitrary.

x	$f(x)$
-25	-321.25
-10	32
0	10
10	-8
25	366.25

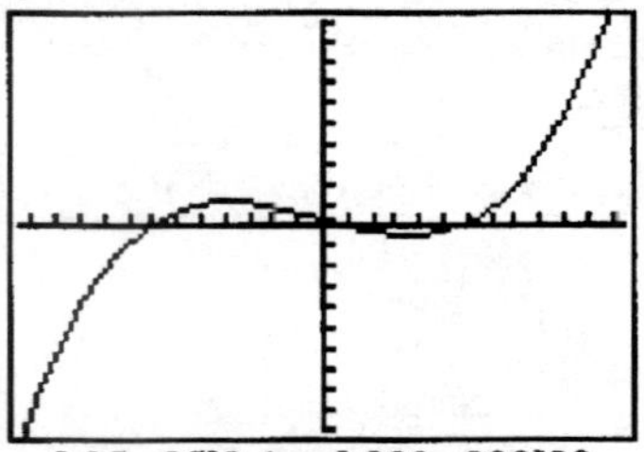

[-25, 25]2 by [-300, 300]30

Choose the Xscl and Yscl so there are about 20 marks on each axis (see METHOD 2 above). So a good first choice for the WINDOW could be [-25, 25]2 by [-300, 300]30.

B-9 Equations in One Variable

An equation in one variable can be solved graphically in two ways:

1. Algebraically move all terms to one side of the equation. Create a function $f(x)$ = (the expression). Graph the function. Zoom and trace to find the x intercepts. The set of x intercepts is the solution to the equation. (See Appendix Section B-7 of this manual.)
2. Create two functions: $f(x)$ = (the left side of the equation) and $g(x)$ = (the right side of the equation). Graph both functions. Zoom and trace to find the x value of the points of intersection of the two graphs. The set of x values is the solution to the equation.

EXAMPLE: Solve $\sqrt[3]{x + 2} = 4x^2 - 3\sqrt{x + 4}$. Use two decimal places.

Solution:

METHOD 1 Write the equation as $\sqrt[3]{x + 2} - (4x^2 - 3\sqrt{x + 4}) = 0$. Create the function $f(x) = \sqrt[3]{x + 2} - (4x^2 - 3\sqrt{x + 4})$. Enter this function into the calculator and graph.

Y= (X,T,θ,n + 2) ^ (1 ÷ 3) -
(4 X,T,θ,n ^ 2 - 3 2nd √ X,T,θ,n
+ 4)) ENTER
ZOOM 6 <Standard>

Enter the function and graph.

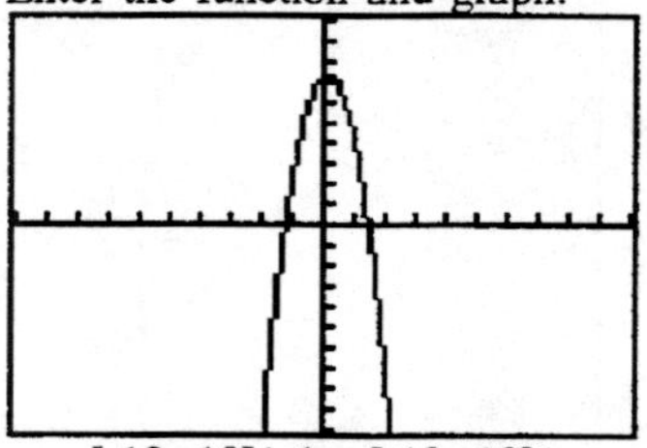

[-10, 10]1 by [-10, 10]1

We see the function crosses the x axis in two places. Trace and zoom, or use other features of the calculator, to find these x values. (See Appendix Section B-7 of this manual.) The solution to this equation is $x = -1.22$ and $x = 1.46$.

METHOD 2 Create the two functions $f(x) = \sqrt[3]{x + 2}$ and $g(x) = 4x^2 - 3\sqrt{x+4}$. Graph these and find the x values of the intersection points. (See Appendix Section B-7 of this manual.)

Y= (X,T,θ,n + 2) ^ (1 ÷ 3)
ENTER
4 X,T,θ,n ^ 2 - 3 √ X,T,θ,n
+ 4) ENTER

ZOOM 6 <Standard>

Enter the functions and graph.

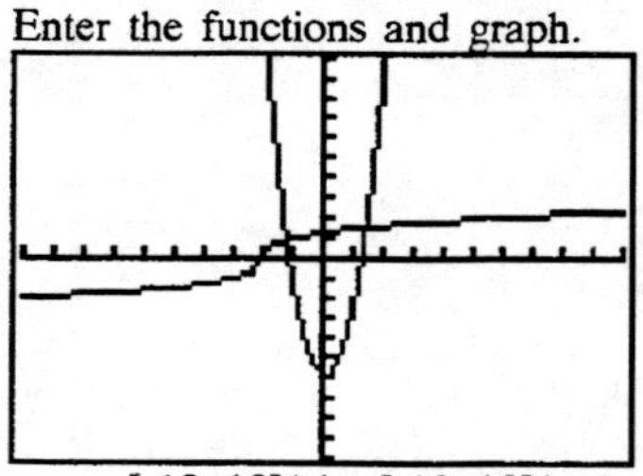

[-10, 10]1 by [-10, 10]1

We see the functions intersect in two places. Zoom and trace, or use other calculator methods, to find the coordinates of these points of intersection. (See Appendix Section B-7 of this manual.) The intersection points are approximately at (-1.22, .92) and (1.46, 1.51). So the solution to this equation is $x = -1.22$ and $x = 1.46$.

B-10 Inequalities in One Variable

An inequality in one variable can be solved graphically in two ways:

1. Algebraically move all terms to one side of the inequality and change the sign to an equal sign. Create a function $f(x)$ = (the expression). Graph the function. Zoom and trace to find the x intercepts. The set of x intercepts are the cutoff values for the solution. Now determine the intervals over which the inequality is true and write the solution. Pick a value either greater or less than the x intercept. Test the inequality using the method shown in Appendix Section B-5.

2. Create two functions: $f(x)$ = (the left side of the inequality) and $g(x)$ = (the right side of the inequality). Graph both functions. Zoom and trace to find the x value of the points of intersection of the two graphs. The set of x coordinates are the cutoff values for the solution. Now determine the intervals over which the inequality is true and write the solution. Pick a value either greater or less than the x intercept. Test the inequality using the method shown in Appendix Section B-5.

EXAMPLE: Solve $\sqrt[3]{x+2} < 4x^2 - 3\sqrt{x+4}$. Use two decimal places

Solution:

<u>METHOD 1</u> Write the inequality as: $\sqrt[3]{x+2} - (4x^2 - 3\sqrt{x+4}) < 0$. Change the sign to an equal sign: $\sqrt[3]{x+2} - (4x^2 - 3\sqrt{x+4}) = 0$. Create the function $f(x) = \sqrt[3]{x+2} - (4x^2 - 3\sqrt{x+4})$, graph, and find the x intercepts. This was done above in Appendix Section B-9 METHOD 1 of this manual. The cutoff values are $x = -1.22$ and $x = 1.46$. Observing the graph we see it is below the x axis for $x < -1.22$ and $x > 1.46$. Thus for these values of x we have $f(x)<0$ and hence $\sqrt[3]{x+2} < 4x^2 - 3\sqrt{x+4}$. Hence the solution to this inequality can be written as $(-\infty, -1.22) \cup (1.46, \infty)$.

<u>METHOD 2</u> Create the two functions $f(x) = \sqrt[3]{x+2}$ and $g(x) = 4x^2 - 3\sqrt{x+4}$. Graph these and find the x values of the intersection points. This was done above in Appendix Section B-9 METHOD 2 of this manual. The intersection points are at (-1.22, .92) and (1.46, 1.52). Inspecting the graph we see that $f(x) < g(x)$ for $x < -1.22$ and $x > 1.46$. Hence the solution to this inequality can be written as $(-\infty, -1.22) \cup (1.46, \infty)$.

B-11 Permutations, Combinations, and Random Numbers

EXAMPLE 1: Find $P_{19,13}$ the number of permutations of 19 distinct objects taken 13 at a time without repetition.

Solution:

Keystrokes	Explanation
[19] [MATH] [►] [►] [►] <PRB> [2] <nPr> [13] [ENTER]	Enter the numbers and get the built-in permutation function from the MATH menu. The result is displayed in scientific notation as 1.689515283E14.

EXAMPLE 2: Find $C_{19,13}$ the number of combinations of 19 distinct objects taken 13 at a time without repetition.

Solution:

Keystrokes	Explanation
[19] [MATH] [►] [►] [►] <PRB> [3] <nCr> [13] [ENTER]	Enter the numbers and get the built-in function from the MATH menu. The result is 27132.

EXAMPLE 3: Generate ten unique random numbers at least 0 and less than 50.

Solution:

5 STO▸ MATH ▸ ▸
▸ <PRB> 1 <rand> ENTER

We need to store a "seed" number in the storage location RAND. This tells the random number generator where to start.

MATH ▸ <NUM> 3 <iPart>
50 x MATH ▸ ▸ ▸ <PRB>
1 <rand>) ENTER ENTER

RAND generates a number greater than 0 and less than 1. If we multiply by 50 and take the integer part, the result is a number greater than or equal to 0 and less than 50.

ENTER ENTER ENTER ENTER
ENTER ENTER ENTER ENTER

The 10 random numbers are:

36 13 6 17 21 ~~21~~ 20 5 31 16 48

Note that repetitions can occur but 21 cannot be used twice.

B-12 Matrices

Given the matrices $A = \begin{bmatrix} 1 & -4 & 6 \\ 2 & -1 & 0 \\ 5 & -3 & 6 \end{bmatrix}$ and $B = \begin{bmatrix} 4 & 15 \\ -2 & 6 \\ 2 & 5 \end{bmatrix}$.

EXAMPLE 1: Find -2.5AB.

Solution:

MATRX ▸ ▸ <EDIT> 1 <[A]>

Choose matrix A and enter the dimensions.

3 ENTER 3 ENTER

1 ENTER (-) 4 ENTER
6 ENTER 2 ENTER
(-) 1 ENTER 0 ENTER
5 ENTER (-) 3
ENTER 6 ENTER

Enter the matrix elements.

MATRX ▸ ▸ <EDIT> 2 <[B]>

Choose matrix B and enter the dimensions.

3 ENTER 2 ENTER

4 ENTER 15 ENTER
(-) 2 ENTER 6 ENTER
2 ENTER 5

Enter the matrix elements.

2nd QUIT

Return to the Home Screen. Note that [A] is the matrix A. This is different than A which represents the variable A.

(-) 2.5 MATRX 1 <[A]>
x MATRX 2 <[B]>
ENTER

Enter the matrix operations.

The result is $\begin{bmatrix} -60 & -52.5 \\ -25 & -60 \\ -95 & -217.5 \end{bmatrix}$.

EXAMPLE 2: Find A^{-1} using matrix A from Example 1.

Solution:

MATRX 1 <[A]> x^{-1} ENTER

The operations are done from the Home Screen. The result rounded to three decimal places is:

$$A = \begin{bmatrix} -.167 & .167 & .167 \\ -.333 & -.667 & .333 \\ -.028 & -.472 & .194 \end{bmatrix}.$$

STO▸ MATRX 2 <[B]>
ENTER

To inspect the result use the arrow keys to see the elements. The mode can be set to three decimal places. See Appendix Section B-17 of this manual.

EXAMPLE 3: Solve the system of equations using the inverse of the coefficient matrix. Round answers to two decimal places.

$$\begin{aligned} x - 4y + 6z &= 2.5 \\ 2x - y &= 3.2 \\ 5x - 3y + 6z &= -3.4 \end{aligned}$$

Solution: Note that the coefficient matrix is matrix A from Section B-12 Example 1 of this manual. The method of finding the inverse of a matrix was shown above in Section B-12 Example 2. Enter the constants as a 3×1 matrix B and find $A^{-1}B$.

MATRX ▶ ▶ <EDIT> 2 <[B]> — Enter the matrix dimensions.

3 ENTER 1 ENTER

2.5 ENTER — Enter the matrix elements.

3.2 ENTER (-) 3.4

2nd QUIT — Return to the Home Screen.

MATRX 1 <[A]> x^{-1}
MATRX 2 <[B]> ENTER

Perform the matrix operations.

The result is $\begin{bmatrix} -.45 \\ -4.10 \\ -2.24 \end{bmatrix}$. The solution to the system of equations to two decimal places is $x = -.45$, $y = -4.10$, $z = -2.24$.

EXAMPLE 4: Find the solution to the system of Example 3 above using Gauss-Jordan reduction. In other words, find the reduced form of the matrix $\begin{bmatrix} 1 & -4 & 6 & 2.5 \\ 2 & -1 & 0 & 3.2 \\ 5 & -3 & 6 & -3.4 \end{bmatrix}$.

Two different methods can be used:

1. Use the matrix math row operation features.
2. Use the reduced row-echelon form matrix math feature.

Each method is described below. First we need to enter the matrix elements into the calculator.

[MATRX] [►] [►] <EDIT> [1] <[A]> — Name the matrix A. Enter the matrix dimensions.

[3] [ENTER] [4] [ENTER]

[1] [ENTER] [(-)] [4] [ENTER] — Enter the matrix elements row by row.

[6] [ENTER] [2.5] [ENTER]

[2] [ENTER] [(-)] [1] [ENTER]

[0] [ENTER] [3.2] [ENTER]

[5] [ENTER] [(-)] [3] [ENTER]

[6] [ENTER] [(-)] [3.4]

[2nd] [QUIT] — Return to the home screen.

After a calculation, some numbers may look like decimals. However, upon storing, recalling, and examining the matrix using the arrow keys you will find they are decimals that are very close to zero. This situation can be eliminated by setting the number of decimal places to a number, say 3 (see Section B-3 of this manual).

METHOD 1

The reduced form of the matrix can be found step by step using the matrix math commands in the calculator.

[MATRX] [1] <[A]> [ENTER] — Display the matrix.

[MATRX] [►] <MATH> [ALPHA] — We wish to multiply by -2, matrix A, row 1 and add it to row 2. The result will be put in row 2.

[F] <*row+(>

`*row+(-2,[A],1,2)`

[(-)] [2] [,] [MATRIX] [1] <[A]>

[,] [1] [,] [2] [)] [ENTER] — The result is $\begin{bmatrix} 1 & -4 & 6 & 2.5 \\ 0 & 7 & -12 & -1.8 \\ 5 & -3 & 6 & -3.4 \end{bmatrix}$.

[STO►] [MATRIX] [1] <[A]> — Store the result as matrix A so you can perform the next operation.

[ENTER]

MATRX ▶ <MATH> ALPHA F <*row+(> (-) 5 , MATRIX 1 <[A]> , 1 , 3) ENTER

Now multiply by -5, matrix A, row 1 and add it to row 3.

```
*row+(-5,[A],1,3)
```

The result is $\begin{bmatrix} 1 & -4 & 6 & 2.5 \\ 0 & 7 & -12 & -1.8 \\ 0 & 17 & -24 & -15.9 \end{bmatrix}$.

STO▶ MATRIX 1 <[A]> ENTER

Store the result as matrix A.

MATRX ▶ <MATH> E <*row(> 1 ÷ 7 , MATRIX 1 <[A]> , 2) ENTER

Multiply by 1/7, matrix A, row 2.

```
*row(1÷7,[A],2)
```

The result is $\begin{bmatrix} 1 & -4 & 6 & 2.5 \\ 0 & 1 & -1.71 & -.26 \\ 0 & 17 & -24 & -15.9 \end{bmatrix}$
rounded to two decimal places.

STO▶ MATRIX 1 <[A]> ENTER

Store the result as matrix A.

Continue in this manner until you get $\begin{bmatrix} 1 & 0 & 0 & -.45 \\ 0 & 1 & 0 & -4.10 \\ 0 & 0 & 1 & -2.24 \end{bmatrix}$. The solution to the system of equations is $x = -.45$, $y = -4.10$, $z = -2.24$.

> NOTE: To swap rows of a matrix use ALPHA C <rowSwap(> from the MATRX MATH menu. The command rowSwap([A],2,3) will swap rows 2 and 3 in matrix A.
>
> Note: To add one row to another use ALPHA D <row+(> from the MATRX MATH menu. The command row+([A],2,3) will add row 2 to row 3 in matrix A.

METHOD 2

The reduced row-echelon form can be found using the rref command in the calculator.

MATRX ▶ <MATH> ALPHA B <rref> MATRX 1 <[A])

B-13 Logarithmic, Exponential, and Hyperbolic Functions

EXAMPLE 1: Graph $f(x) = 10^{3.2x}$ for $-.5 < x < .5$.

Solution: Set the WINDOW to [-.5,.5].1 by [-10,10]1. (See Appendix C Section B-7 Example 1 Method 1.)

[Y=] [10] [^] [(] [3.2] [X,T,θ,n] [)]
[GRAPH]

Enter the function and graph.

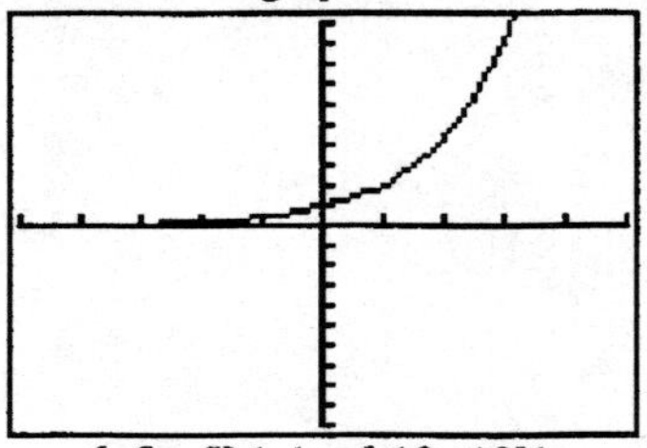

[-.5, .5].1 by [-10, 10]1

EXAMPLE 2: Graph $y = \frac{e^{x} + e^{-x}}{2}$.

Solution: This could be entered as [(] [2nd] [e^x] [X,T,θ,n] [+] [2nd] [e^x] [(-)] [X,T,θ,n] [)] [)] [÷] [2]. Recall that [e^x] represents the built-in exponential function above the [LN] key. However this is the hyperbolic cosine function. The following uses this built-in function.

[Y=] [CLEAR] — Clear the existing functions from the Y= list.

[2nd] [CATALOG] [▼] ... [▼] [ENTER] — Get the catalog of functions, arrow down until the cursor is at the cosh function, and press ENTER.

[X,T,θ,n] [)]

[ZOOM] [6] <ZStandard>

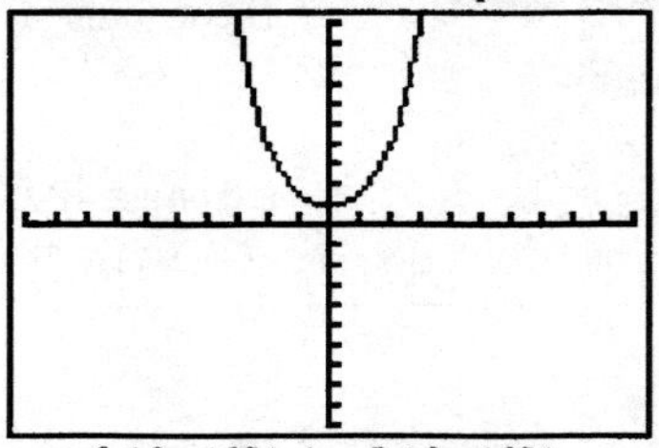

[-10, 10]1 by [-10, 10]1

EXAMPLE 3: Solve $\ln(x^2 + x) = 3x + 2$

Solution:

[Y=] [CLEAR] [LN] [X,T,θ,n] [^] [2] [+] [X,T,θ,n] [)] [ENTER] — Enter the function $f(x) = \ln(x^2 + x)$ as Y1 in the function list.

[CLEAR] [3] [X,T,θ,n] [+] [2] [ENTER] — Enter the function $g(x) = 3x + 2$ as Y2 in the function list.

`ZOOM` `6` <ZStandard>

Graph them on the same graphing screen. Trace and Zoom, or use other calculator methods, to find the x value of the point of intersection. (See Appendix Section B-7 of this manual.) This will be the solution to the equation.

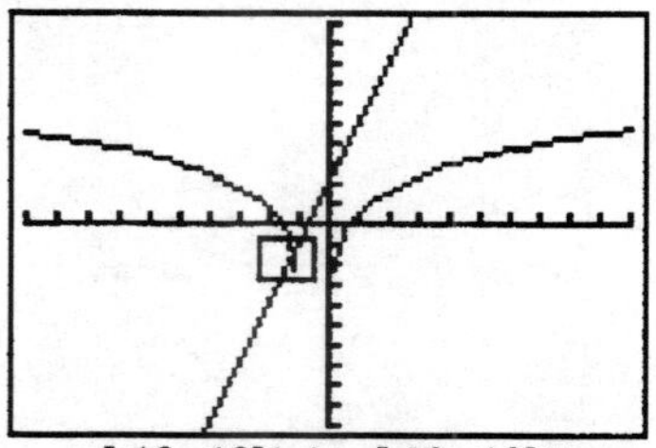

[-10, 10]1 by [-10, 10]1

The solution is -1.18, rounded to two decimal places.

B-14 Trigonometric Functions

EXAMPLE 1: Evaluate $f(x) = \cos x + \csc x$ at (A) 53°45'18" and (B) 78.52 radians.

Solution: (A)

`Y=` `CLEAR` `COS` `X,T,θ,n` `)`
`+` `1` `÷` `SIN` `X,T,θ,n` `)`

Store the function.

`VARS` `▶` <Y-VARS>
`2nd` `QUIT`

Change the calculator to degree angle measure.

`MODE` `▼` `▼` `▶` `ENTER`
`2nd` `QUIT`

Change the angle to decimal degrees while storing it as the variable X.

`53` `+` `45` `÷` `60` `+` `18` `÷`
`3600` `STO▶` `X,T,θ,n` `ENTER`

`1` <FUNCTION> `1` <Y1>
`ENTER`

Evaluate the function. The result to two decimal places is 1.83.

Solution: (B)

`Y=` `CLEAR` `COS` `X,T,θ,n` `)`

Store the function.

`+` `1` `÷` `SIN` `X,T,θ,n` `)`
`2nd` `QUIT`
`MODE` `▼` `▼` `ENTER`
`2nd` `QUIT`

Change the calculator to radian angle measure. Since the function was stored in Part (A) above, we need only to store the angle measure and evaluate the stored function.

[78.52] [STO▶] [X,T,θ,*n*] [ENTER] — Store 78.52 as X.

[VARS] [▶] <Y-VARS> [1] <FUNCTION> [1] <Y1> [ENTER] — Evaluate the function. The result to two decimal places is 49.47.

EXAMPLE 2: Graph $f(x) = \cos x + \csc x$.

Solution: The function was entered in the calculator in Part (A) above.

[MODE] [▼] [▼] [▼] [▼] [▶] [ENTER] [2nd] [QUIT] — Change the graph mode to Dot. This will eliminate the "vertical lines."

[ZOOM] [7] <ZTrig> — This option automatically sets the viewing window variables to [-6.28, 6.28]1.57 by [-3, 3].25.

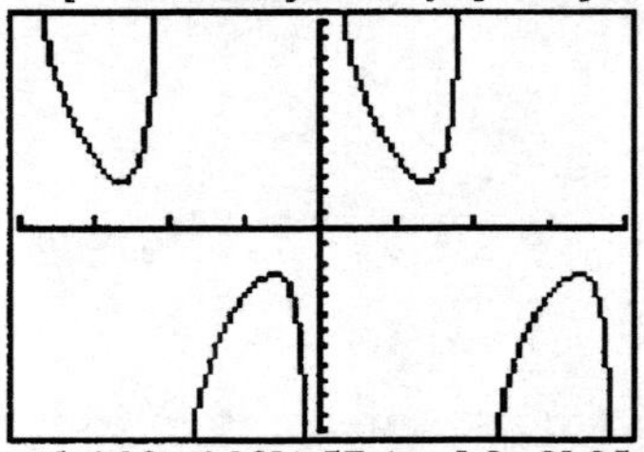

[-6.28, 6.28]1.57 by [-3, 3].25

B-15 Numeric Differentiation and Integration, and Tangent Lines

EXAMPLE 1: (A) Find the equation of the tangent line to $f(x) = \sqrt{x^5 - \sqrt[3]{x}}$ at $x = 3.54$. Use two decimal place accuracy. (B) Graph the function and the tangent line at $x = 3.54$ using [0, 6]1 by 0, 50]2.

Solution (A): The function and its derivative can be evaluated at this value of x.

[Y=] [CLEAR] [2nd] [√] [X,T,θ,*n*] [^] [5] [-] [X,T,θ,*n*] [^] [(] [1] [÷] [3] [)] [)] — Enter the function and store it as Y1.

[2nd] [QUIT] [3.54] [STO▶] [X,T,θ,*n*] [ENTER] — Get the Home Screen and store the value of x.

VARS ▶ <Y-VARS>
1 <FUNCTION> 1 <Y1>
ENTER

Evaluate the function. $f(3.54) = 23.55$ to two decimal places.

MATH 8 <nDeriv(>
VARS ▶ <Y-VARS>
1 <FUNCTION> 1 <Y1>
, X,T,θ,*n* , 3.54 , .0001)
ENTER

Get the numeric derivative built-in function.

```
nDeriv(Y1,X,3.54,.0001)
```

This will evaluate the derivative with respect to x of the function stored in Y1 at $x = 3.54$ with .0001 tolerance. If the tolerance is not entered, the calculator will use .001.

The result is 16.67.

The equation of the tangent line is $y - 23.55 = 16.67(x - 3.54)$.

Solution (B): The tangent line can be drawn from the Home Screen. Set the WINDOW dimensions to [0, 5]1 by [0, 50]2.

2nd DRAW 5 <Tangent>
VARS ▶ <Y-VARS>
1 <FUNCTION>
1 <Y1> , 3.54) ENTER

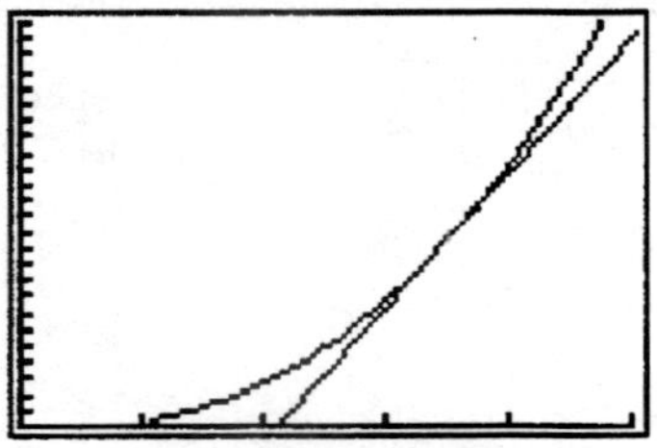

[0, 5]1 by [0, 50]2

EXAMPLE 2: Find $\int_{1.32}^{2.58} \sqrt{x^5 - \sqrt[3]{x}}\,dx$. Use two decimal place accuracy.

Solution:

Y= CLEAR 2nd √ X,T,θ,*n* ^
5 - X,T,θ,*n* ^ (1 ÷ 3))

Enter the function and store it as Y1.

2nd QUIT
MATH 9 <fnInt(>
VARS ▶ <Y-VARS> 1 <FUNCTION>
1 <Y1> , X,T,θ,*n* , 1.32 , 2.58
, .00001) ENTER

Get the numeric integration function built into the calculator function:

```
fnInt(Y1,X,1.32,2.58)
```

This will evaluate the definite integral. The result is 6.95. The tolerance in calculating the definite integral is .00001.

B-16 Statistics

EXAMPLE 1: Given the following data find: (A) the mean, (B) the sample standard deviation, (C) an histogram with first class beginning at 0 and width 4, and (D) repeat Parts A and B after deleting the data value 89 from the list. How did eliminating this outlier effect the mean and standard deviation?

23	56	73	12	89	52	45	19	40	57	39
38	18	21	15	37	42	51	38	54	60	

Use one decimal place in answers.

Solution (A) and (B):

Keystrokes	Explanation
[STAT] [4] <ClrList> [2nd] [L1] [ENTER]	Clear the list named L1.
[STAT] [1] <Edit>	Get the data entry menu.
[23] [ENTER] [56] [ENTER] Etc.	Enter the data in the L1 list. Enter all data. There are 21 values. If you make a mistake, use the arrow keys to highlight the number and type it in again. To remove a data value, use the arrow keys to highlight number and press [DEL].
[2nd] [QUIT]	Exit the data entry screen.
[STAT] [▶] <CALC> [1] <1-Var Stats> [2nd] [L1] [ENTER]	Get the statistics menu again. Press [1] to get the one variable statistics. Since each piece of data was entered individually, we need only to list L1 after 1-Var Stats. The calculator will use 1 as the frequency for each number in the L1 list. The mean is 41.86 and the sample standard deviation is 19.82.

Solution (C):

Keystrokes	Explanation
[WINDOW] [0] [ENTER] [100] [ENTER] [4] [ENTER] [(-)] [5] [ENTER] [5] [ENTER] [1]	The smallest value is 12 so let xMin=0. The greatest value is 89. Let us set Xmax=100. Let Xscl=4 since this is the class width. Let Ymin=-5 to get some room below the histogram. We do not know the frequencies yet so let Ymax=5 for a first try. Let yScl=1.
[Y=] [CLEAR] ...	Clear all functions on the function list.
[2nd] [DRAW] [1] <ClrDraw> [ENTER]	Clear all drawings from the display.

2nd STAT PLOT 1 <Plot1...OFF>	Select Plot1
ENTER	Set Plot1 to ON.
▼ <Type:> ► ► < ⌊⌋ > ENTER	Select Type as Histogram.
▼ <Xlist:> ENTER ▼ ENTER	Select Xlist as L1 and Freq as 1.
GRAPH	Graph the histogram.

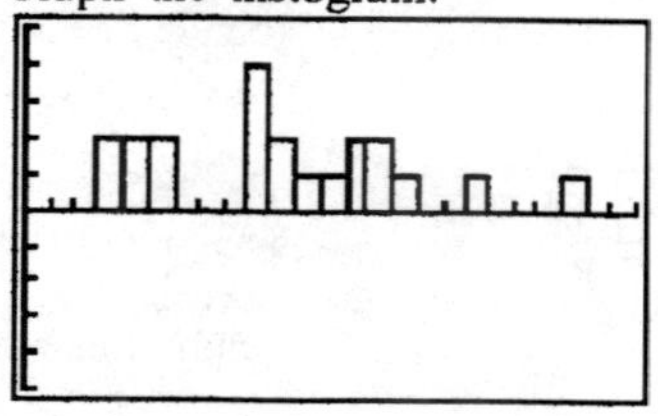

Solution (D):

STAT 1 ▼ ▼ ... ▼ ▼ DEL	Get the data list and arrow down to the 89. Press DEL.
STAT ► <CALC> 1 <1-Var Stats> 2nd L1 ENTER	Get the statistics menu and the one variable sample statistics. The mean is now 39.5 and the sample standard deviation is 17.05. Notice how much influence an outlier has!

EXAMPLE 2: Given the ungrouped frequency data, shown at the right: (A) Find the mean. (B) Find the sample standard deviation. (C) Draw a boxplot.

x	*f*
2.4	4
3.5	6
4.9	8
5.3	5
6.8	2

Solution (A) & (B):

STAT 4 <ClrList> 2nd L1 , 2nd L2 ENTER	Clear the lists named L1 and L2.
STAT 1 <Edit>	Get the data entry menu.
2.4 ENTER 3.5 ENTER Etc.	Enter the data in the L1 list. Enter all data. If you make a mistake, use the arrow keys to highlight the number and type it in again. To remove a data value, use the arrow keys to highlight number and press DEL.

Keystrokes	Explanation
[▶] [4] [ENTER] [6] [ENTER] Etc.	Enter the frequencies in the L2 list.
[2nd] [QUIT]	Exit the data entry screen.
[STAT] [▶] <CALC> [1] <1-Var Stats> [2nd] [L1] [,] [2nd] [L2] [ENTER]	Get the statistics menu again. Press [1] to get the one variable statistics. The calculator will use the L1 list for the values of x and the L2 as the frequency for each number in the L1 list. The mean is 4.40 and the sample standard deviation is 1.27, rounded to two decimal places.

Solution (C):

Keystrokes	Explanation
[WINDOW] [ENTER] [0] [ENTER] [10] [ENTER] [1] [ENTER] [(-)] [1] [ENTER] [4] [ENTER] [1] [ENTER] [1]	The smallest value is 2.4 so let Xmin=0. The greatest value is 6.8 so let Xmax=10. Let Xscl=1. Let Ymin=-1 to get some room below the graph. Let Ymax=4 and Yscl=1. Leave Xres at 1.
[Y=] [CLEAR] ...	Clear all functions on the function list.
[2nd] [DRAW] [1] <ClrDraw> [ENTER]	Clear all drawings from the display.
[2nd] [STAT PLOT] [1] <Plot1...OFF>	Select Plot1
[ENTER]	Set Plot1 to ON.
[▼] <Type:>	Select Type as Boxplot.
[▶] [▶] [▶] [▶] < > [ENTER]	
[▼] <Xlist:> [2nd] [L1] [▼] [2nd] [L2]	Select Xlist as L1 and Freq as L2.
[GRAPH]	Graph the boxplot.

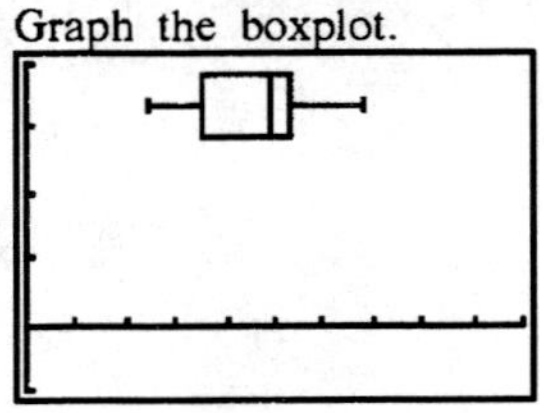

EXAMPLE 2: Given the following bivariate data: (A) Find the correlation coefficient. (B) Find the equation of the least squares line. (C) Graph the least squares line and the scatter plot on the same graph. (D) Plot the scatter diagram without the least squares line. (E) Sort the data on x and join the points using straight line segments.

x	5	8	10	29	9	24
y	3	4	8	20	3	19

Solution (A):

Keystrokes	Explanation
STAT 4 <ClrList> 2nd L1 , 2nd L2 ENTER	Clear list L1 and list L2.
STAT 1 <Edit>	Get the lists.
5 ENTER 8 ENTER 10 ENTER 29 ENTER 9 ENTER 24 ENTER	Enter the data in list L1.
► 3 ENTER 4 ENTER 8 ENTER 20 ENTER 3 ENTER 19 ENTER	Move to list L2. Enter the data in list L2. The data for y in L2 must correspond pairwise to the data for x in L1.
STAT ► <CALC> 8 <LinReg(a+bx)> 2nd L1 , 2nd L2 ENTER	Get the linear regression coefficients using L1 as the x variable and L2 as the y variable in the calculations.
STAT ► ► <Tests> ALPHA E <LinRegTTest>	This gives you many pieces of information. The information we want is: a = -1.72, b = .79, r = .98.
2nd L1 ENTER 2nd L2 ENTER ALPHA 1 ENTER ▼ ENTER ENTER ▼ ▼ ▼ ▼	The linear correlation coefficient is r = .98. The equation of the line is $\hat{y} = -1.72 + .79x$.

Solution (B):

Keystrokes	Explanation
VARS 5 <Statistics> ► ► 7 <RegEQ>	The least squares line equation is displayed. It is $\hat{y} = -1.72 + .79x$ with the coefficients rounded to two decimal places.

Solution (C):

Keystrokes	Explanation
WINDOW 0 ENTER 30 ENTER 1 ENTER (-) 5 ENTER 30 ENTER 1	Set the graph WINDOW. Observe the data to determine the WINDOW values. Set the y value lower than needed so if you trace, the coordinates at the bottom of the screen do not cover up the graph.
Y= CLEAR ...	Clear all functions on the function list.
2nd DRAW 1 <ClrDraw> ENTER	Clear all drawings from the display.

Keystrokes	Explanation
[Y=] [CLEAR] [VARS] [5] [▶] [▶] <EQ>	Clear or deselect any existing functions in the list. Clear Y1.
[1] <RegEQ>	
[2nd] [STAT PLOT]	The least squares line equation will be put into the function list so you can graph it.
[4] <PlotsOff> [ENTER]	
[2nd] [STAT PLOT] [ENTER]	Set up the calculator to get the scatter plot with the line.
[ENTER] [▼] [ENTER] [▼]	
[2nd] [L1] [ENTER]	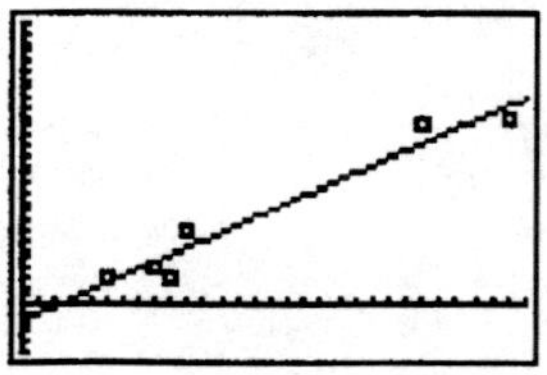
[2nd] [L2] [ENTER]	
[GRAPH]	

Solution (D):

Keystrokes	Explanation
[Y=] [CLEAR] or [Y=] [◀] [ENTER]	Clear the functions from the list or deselect them so they won't graph.
[GRAPH]	Plot the points on the graph.

Solution (E):

Keystrokes	Explanation
[Y=] [CLEAR] or [Y=] [◀] [ENTER]	Clear the functions from the list or deselect them so they won't graph.
[STAT] [2] [2nd] [L1] [,] [2nd] [L2] [)]	Sort the data on the *x* variable. The list L1 will be sorted in ascending order. The list L2 will sort along with L1 as pairs.
[ENTER]	
[2nd] [STAT PLOT] [1] <Plot 1> [ENTER]	Set the calculator to draw line segments joining the points.
[ENTER]	This will join the points together with straight line segments in the order they appear in the data list. This is why we sorted the data on *x* first.
[▼] [▶] [ENTER] [▼] [2nd] [L1] [ENTER]	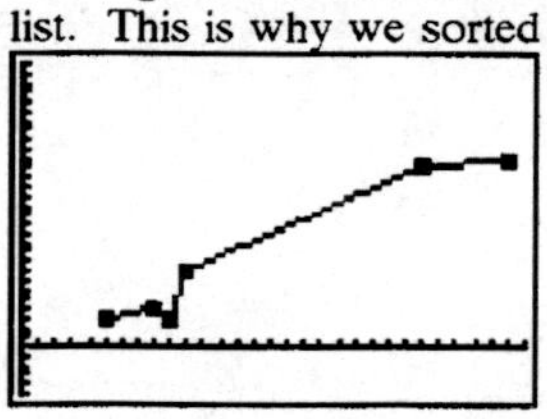
[2nd] [L2]	
[GRAPH]	

B-17 Setting the Display Mode for Numbers

Numbers can be entered into the calculator in scientific notation, a set number of significant digits, or a fixed number of decimal places.

EXAMPLE 1: Calculate $(9.35417\times 10^4)(-3.64389\times10^{-6})$ using:
(A) scientific notation (B) two significant digits (C) four decimal place accuracy.

Solution (A):

Keystrokes	Explanation
[MODE] [►] [ENTER] [2nd] [QUIT]	Set the calculator to scientific mode with a floating decimal point.
[9.35417] [2nd] [EE] [4] [×] [(-)] [3.64389] [2nd] [EE] [(-)] [6] [ENTER]	The result displayed on the calculator is -3.408556652E-1. This means $-3.408556652 \times 10^{-1}$.

Solution (B):

Keystrokes	Explanation
[MODE] [►] [ENTER] [▼] [►] [►] [ENTER]	Set the calculator to scientific notation with 1 fixed decimal point instead of a floating decimal. This will give two significant digits.
[2nd] [QUIT]	Return to the Home Screen.
[2nd] [ENTRY] [ENTER]	Use the replay key to replay the last calculation done in Solution (A). The result to two significant digits is -3.4E-1.

Solution (C):

Keystrokes	Explanation
[MODE] [ENTER] [▼] [►] [►] [►] [►] [►] [ENTER]	Set the calculator to normal mode with 4 digits.
[2nd] [QUIT] [2nd] [ENTRY] [ENTER]	Return to the Home Screen. Use the replay key to replay the last calculation. The result to four decimal places is -.3409.

You may now wish to return your calculator to its default settings (all items on the mode screen having the left item highlighted). The examples throughout this manual are worked using the default settings unless otherwise noted.

NOTES

NOTES

Appendix C

TI-85 Operations

C-1 Operations

Press ON to turn on the calculator.

Press CLEAR to clear the screen.

Press EXIT to clear any menus displayed at the bottom of the screen.

Press 2nd ▲ to make the display darker.

Press 2nd ▼ to make the display lighter.

Press 2nd + to get the RESET screen.

Then press F3 <RESET> F1 <ALL> to clear all the memories, and press F4 <YES> to clear. You must now adjust the display so you can see it. Enter a few numbers and adjust the display contrast as described above.

RAM	DELET	RESET		
ALL	MEM	DFLTS		

Mem Cleared
Defaults set

Press 2nd OFF to turn off the calculator.

Home Screen

The screen where calculations are done is called the Home Screen. You can always get to the Home Screen by pressing 2nd QUIT. Note: QUIT is above the EXIT key.

The item above the key will be referenced as if it is on the key. For example, QUIT, which is above the EXIT key, will be referenced from now on as 2nd QUIT.

2nd

This key must be used to access the calculator functions above and to the left of a key cap. A new cursor is displayed, ↑. These functions are in yellow on the calculator face.

ALPHA and 2nd ALPHA

This key is used to access the calculator functions and memory locations above and to the right of a key cap. These functions are in light blue on the calculator face.

The cursor changes to an A when ALPHA is pressed. Pressing ALPHA and then a letter will display an upper case letter.

The cursor changes to a a when 2nd ALPHA is pressed. Pressing 2nd ALPHA will display a lower case letter.

There is a separate memory location for the upper case letter and for the lower case letter.

MODE

This is found above the MORE key and is used to set the operating mode of the calculator. When the highlighted items are on the left, the calculator is in default mode. To choose a different mode: Use the arrow keys to choose the item you want to change to, then press the ENTER key. The new selection will now be highlighted. Press 2nd QUIT to return to the Home Screen.

Menus

Menus are listed across the bottom of the screen. To choose a menu item press the function key immediately below the menu item. In this manual, these functions will be referred to as the key to press in a box followed by the function in angle brackets. For example, the trace function of the calculator is found on the GRAPH menu above the F4 key is denoted by F4 <TRACE>.

Error Correction

Use the 2nd INS key to insert characters. (INS is above the DEL key.)

Use the DEL key to delete characters.

Use the arrow keys ◄ or ► to move to the left or right.

Use the up arrow key ▲ to move up one line.

Use the down arrow key ▼ to move down one line.

2nd ENTRY is used to replay the last executed operation. (ENTRY is above the ENTER key.)

C-2 Calculating

Many calculator functions are found above key caps or in menus. For example the square root function is above the x^2 key but the absolute value is found in the MATH menu.

EXAMPLE 1: Calculate $|4^2 - \sqrt{34}|$.

Solution:

2nd MATH F1 <NUM> F5 <abs> (4 x^2 − 2nd √ 34) ENTER. ENTER tells the calculator to calculate. The result displayed is 10.1690481052.

C-3 Plotting Points and Drawing Line Segments

EXAMPLE 1: (A) Plot the points (-3, 6) and (1, 5). (B) Draw the line segment having these points as the endpoints.

Solution (A):

Keystrokes	Explanation
[GRAPH] [F3] <ZOOM> [F4] <ZSTD>	We first need to set the viewing window dimensions. [ZOOM] [F4] <Standard> automatically sets the dimensions to xMin=-10, xMax=10, xScl=1, yMin=-10, yMax=10, yScl=1 and xRes=1. This will be denoted as [-10, 10]1 by [-10, 10]1 in this manual.
[EXIT] [EXIT]	
[GRAPH] [MORE] [F2] <DRAW> [MORE] [F5] <CLDRW>	Get the [DRAW] menu and clear the drawings from the calculator screen.
[2nd] [QUIT]	
[2nd] [CATALOG] [▼]...[▼] [ENTER]	Get the [CATALOG] listing all of the functions. Use the down arrow, or [F1] <Page↓>, until the pointer points to [PtOn(]. Press [ENTER].
[(-)] [3] [,] [6] [)] [ENTER]	Enter the coordinates of the first point.
[2nd] [ENTRY] [◄]...[◄] [1] [,] [5] [)] [ENTER]	Use the replay feature of the calculator to get the last entry. Use the left arrow and change the coordinates to those of the second point. Delete any unwanted characters using [DEL].

Solution (B):

Keystrokes	Explanation
[2nd] [QUIT]	Return to the Home Screen.
[2nd] [CATALOG] [▼]...[▼] [ENTER]	Get the [CATALOG] listing all of the functions. Use the down arrow or [F1] <PAGE↓> until the pointer points to [Line(] 1. Press [ENTER].
[(-)] [3] [,] [6] [,] [1] [,] [5] [)] [ENTER]	Enter the coordinates of the first point and the second point separated by commas. Press [ENTER] to draw the line segment.

C-4 Storing and Evaluating an Expression, and Using Lists

EXAMPLE 1: Store the expression $\frac{-b+\sqrt{b^2-4ac}}{2a}$ and evaluate it to find one solution of $3x^2+4x-5=0$.

Solution:

GRAPH F1 <y(x)=> CLEAR
((-) ALPHA B + 2nd √ (
ALPHA B x^2 – 4 ALPHA
A × ALPHA C))
÷ (2 ALPHA A)

We can store up to 99 functions or algebraic expressions in this list. Then we can recall them and evaluate or graph them.

Press these keys to enter the expression.

2nd QUIT

Get back to the home screen.

3 STO▸ A ENTER 4 STO▸ B
ENTER (-) 5 STO▸ C ENTER

Store 3 as A, 4 as B and -5 as C.

METHOD 1

2nd VARS F1 <ALL>
F1 <PAGE↓>... F1 <PAGE↓> ▼ ENTER
ENTER

Get the variables list. Press F1 <PAGE↓> and/or the arrow keys until the cursor is pointing at **y1**. Press ENTER ENTER
The result is .786299647847.

METHOD 2

2nd ALPHA Y 1 ENTER

Enter y1 on the screen and press ENTER.
The result is .786299647847.

EXAMPLE 2: Given $f(x) = \sqrt[3]{x^4+2x}$ and $g(x) = 4^{2x-90}$. Find $f(x) + 3\sqrt{g(x)}$ at $x = 45.2$.

Solution:

GRAPH F1 <y(x)=> CLEAR (x-VAR
^ 4 + 2 x-VAR) ^ (1 ÷ 3)
ENTER

Enter the functions as y1 and y2. Press CLEAR to delete the old y1 function.

CLEAR 4 ^ (2 x-VAR - 90)

Note that lower case x is used in functions. Upper case and lower case letters are different. Values can be stored for each of these.

2nd QUIT

45.2 STO▸ x-VAR ENTER

Store 45.2 as x.

2nd VARS F1 <ALL> ▼ ... ▼ ENTER
+ 3 2nd √ 2nd VARS F1 <ALL>
▼ ... ▼ ENTER ENTER

Form $f(x) + 3\sqrt{g(x)}$ as y1 + 3√ y2 getting y1 and y2 from the variables list. The expression will be evaluated when you press ENTER.
The result is 164.97 to two decimal places.

EXAMPLE 3: Evaluate the expression $\dfrac{3x^2 + 5x - 8}{x^2 - 1}$ to three decimal places for $x = .900$, .990, .999, 1.001, 1.010, and 1.100 by: (A) evaluating the expression for each value of x individually; and (B) using a list.

Solution (A): First store the expression in the Y= list. Then store the value for x. Recall the expression and evaluate it.

Keystrokes	
GRAPH F1 <y(x)=> CLEAR	Clear the old expression for y1 by pressing CLEAR.
(3 x-VAR ^ 2 +	Store the expression as y1 and return to the home screen.
5 x-VAR – 8) ÷	
(x-VAR ^ 2 –	
1) 2nd QUIT	
.9 STO▸ x-VAR ENTER	Store the value of x.
2nd alpha Y 1 ENTER	Recall the expression and evaluate.

Store the next value for x and reevaluate the expression. Repeat for the other values of x. The corresponding results are: 5.632, 5.513, 5.501, 5.499, 5.488, and 5.381, rounded to three decimal places.

Solution (B): Store the values of x in a list and evaluate the expression for each value of x simultaneously.

Keystrokes	
GRAPH F1 <y(x)=> CLEAR	Clear the old expression for y1 by pressing CLEAR.
(3 x-VAR ^ 2 +	Store the expression as Y1 and return to the home screen.
5 x-VAR – 8) ÷	
(x-VAR ^ 2 –	
1) 2nd QUIT	
2nd LIST F1 <{>	Store the values of x in the list x.
.9 , .99 , .999 , 1.001	
, 1.01 , 1.100 F2 <}>	
STO▸ x-VAR	
ENTER	Simultaneously calculate the value of the expression stored as Y1 for the values stored in list x. The results are: 5.632, 5.513, 5.501, 5.499, 5.488, and 5.381.
2nd alpha Y 1 ENTER	

C-5 Evaluating Inequalities

EXAMPLE 1: Given the inequality $x^2 - 3t < 5x + 9t^2$. Test this inequality for $x = 1.89$ and $t = -4.52$.

Solution:

Keystrokes	Explanation
[1.89] [STO▸] [x-VAR] [ENTER] [(-)] [4.52] [STO▸] [2nd] [alpha] [T] [ENTER]	Store the value of the variables. Lower case letters are used.
[x-VAR] [^] [2] [-] [3] [2nd] [alpha] [T] [2nd] [TEST] [F2] :<< > [5] [x-VAR] [+] [9] [2nd] [alpha] [T] [^] [2] [ENTER]	Enter the expression and evaluate. The result is 1. This means the inequality is true for these values of the variables. If the inequality was false, the result would have been 0.

C-6 Graphing

EXAMPLE 1: Graph $y = 3x^2 - 6$ and $y = 3x^2 + 1$ on the same graphing screen.

Solution:

Keystrokes	Explanation
[STAT] [F3] <DRAW> [F5] <CLDRW>	Clear drawings.
[GRAPH] [F1] <y(x)=> [CLEAR] [3] [x-VAR] [^] [2] [-] [6] [ENTER] [CLEAR] [3] [x-VAR] [^] [2] [+] [1] [ENTER]	Store the functions. Remember to clear out the old function whenever entering a new one. Clear or deselect all others by placing the cursor at the expression for the function and pressing [F5]. Now this function will not graph. Note that the = sign is no longer highlighted.
[EXIT]	Exit the $y(x)$= menu.
[F3] <ZOOM> [F4] <ZSTD>	Press [F3] and then [F4] to get the standard viewing window. This will automatically set the graphing screen to have $-10 < x < 10$ and $-10 < y < 10$ and graph the functions.

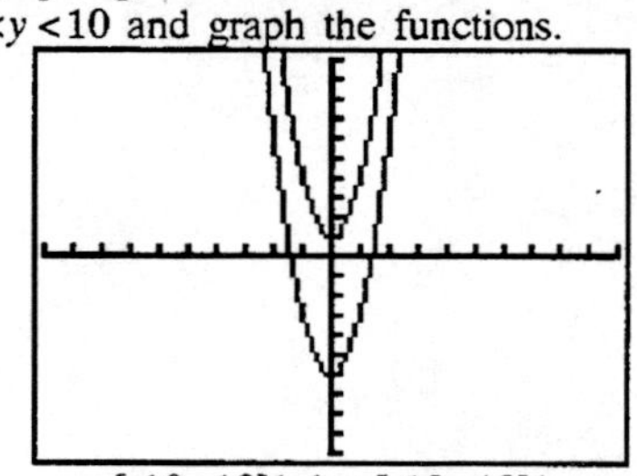

[-10, 10]1 by [-10, 10]1

Keystrokes	Explanation
[CLEAR]	Note that pressing [CLEAR] will remove the menus from the bottom of the screen.
[2nd] [QUIT]	

| ZSTD | automatically sets the graphing screen limits to $-10 < x < 10$ and $-10 < y < 10$. These graph screen limits will be denoted in this manual by the notation [-10, 10]1 by [-10, 10]1. The graph screen limits can be set by pressing F2 <RANGE> on the GRAPH menu and entering the desired values.

F2 <RANGE> means to choose the RANGE option from the menu on the screen above the F2 key.

The function will <u>not</u> graph if you choose a function on the F1 <y(x)=> menu and then press F5 <SELCT>. You will notice that the highlighting disappears from the = sign. You can still evaluate this expression for stored values of the variables. To select the function, press F5 <SELCT> again.

EXAMPLE 2: Graph $f(x) = \begin{cases} .5x + 1 & x \le 2 \\ x^2 + 2 & x > 2 \end{cases}$.

Solution:
The calculator will only graph functions for values of x for which the function is defined, so we write this piecewise-defined function in a form where the pieces are defined only for the values of x that we want. Enter two functions as $y1=(.5x+1)\div(x\le2)$ and $y2=(x^2+2)\div(x>2)$. The first function is defined with $x \le 2$ and the second function is defined when $x > 2$.

Keystrokes	
GRAPH F1 <y(x)=> CLEAR	Clear the existing function.
(.5 x-VAR + 1) ÷ (x-VAR 2nd TEST F4 <≤> 2) ENTER	Enter the first piece of the piecewise-defined function.
CLEAR (x-VAR ^ 2 + 2) ÷ (x-VAR 2nd TEST F3 <>> 2) ENTER	Clear the existing function. Enter the second piece of the piecewise-defined function.
EXIT EXIT F3 <ZOOM> F4 <ZSTD>	Graph.

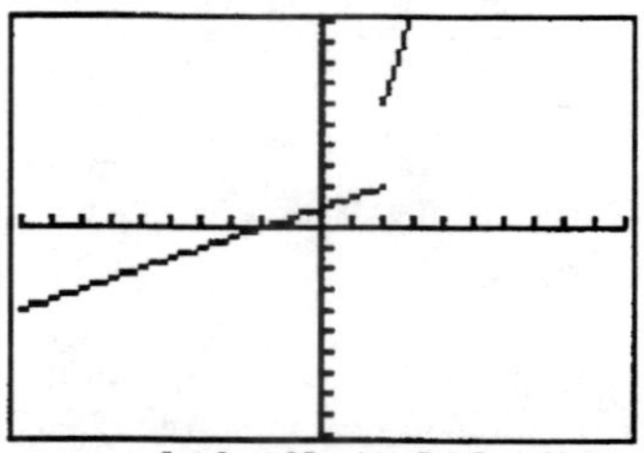

[-10, 10]1 by [-10, 10]1

C-7 TRACE, ZOOM, RANGE and SOLVER

EXAMPLE 1: Find the x intercept(s) for x>0 for $y = 3x^2 - 6$. Use three decimal place accuracy.

Solution: Store the function as 3 X ^ 2 - 6 as y1. Graph using the [-10, 10]1 by [-10, 10]1. (see Example 1 from Section D-6).

F4 <TRACE> — You see a new cursor near the center of the graph screen and the coordinates of that point on the graph displayed at the bottom of the screen.

◄ or ►

Use right and left arrow keys and move the cursor along the graph. If we had the graph of two or more functions displayed, we could use the up and down arrow keys to move between the functions. The 1 in the upper right-hand corner indicates that the cursor is on the graph of y1.

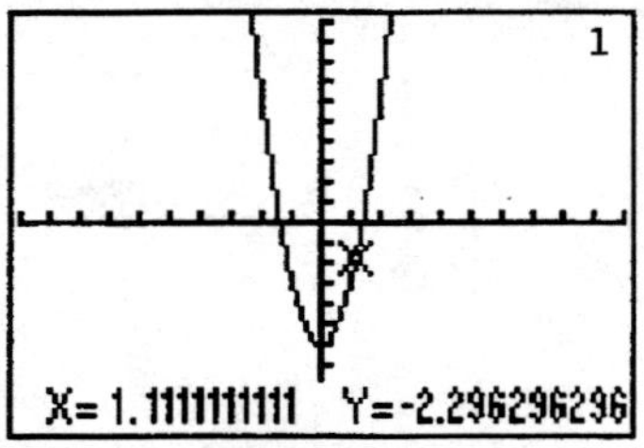

[-10, 10]1 by [-10, 10]1

There are four methods we can use to get an approximation to the x intercepts.

1. We can zoom in on the point by changing the graph screen parameters values.
2. We can zoom in on the point by settin the Zoom Factors and zoom in.
3. We can zoom in on the point by using the Zoom Box feature.
4. We can use the Solver feature .

METHOD 1 Set the RANGE

Using the trace feature of the calculator, trace the functions graphed above. We see the x intercept is near 1.5. So we shall set the RANGE closer to that value.

GRAPH F2 <RANGE> 1
ENTER 2 ENTER .1 ENTER
(-) .5 ENTER .5 ENTER .1
GRAPH F4 <TRACE>

Set the RANGE to [1, 2].1 by [-.5, .5].1.

Press EXIT to restore the menu at the bottom of the screen.

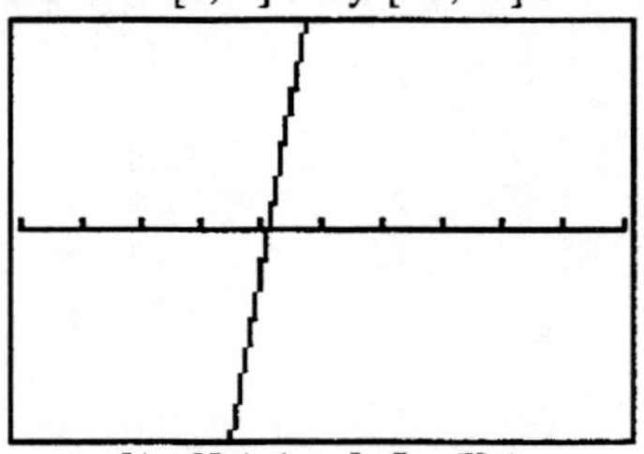

[1, 2].1 by [-.5, .5].1

Use TRACE again to get a better approximation. The trace cursor always starts at the middle x value of the RANGE. It may not be on the display. Watch the values for x at the bottom of the screen.

We see that the x intercept is between 1.41 and 1.42. Set the RANGE to [1.41, 1.42].01 by [-.01, .01].001 and graph again. We now find the x intercept to 1.414 accurate to three decimal places.

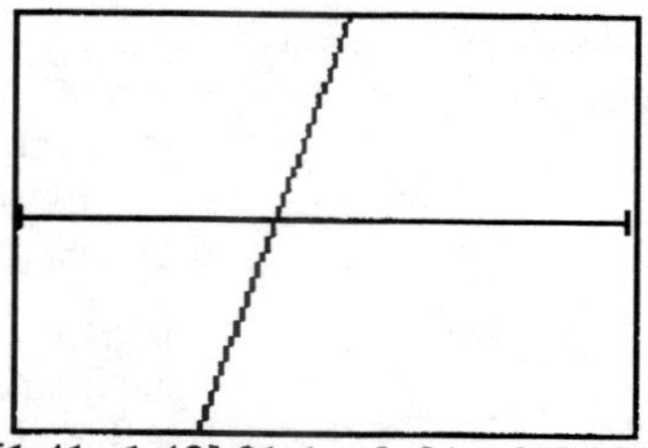

[1.41, 1.42].01 by [-.01, .01].001

<u>METHOD 2 Zoom In</u>

GRAPH F3 <ZOOM>
F4 <ZSTD>

Get the [-10, 10]1 by [-10, 10]1 display as shown at the beginning of this section.

MORE MORE F1 <ZFACT>
5 ENTER 5

Set the Zoom Factors to 5 and 5. The graph will be 5 times as large in both the x and y directions.

ENTER F3 <ZOOM> F2 <ZIN>
◄ or ► or ▲ or ▼

Use the ◄ or ► arrow keys to get the cursor to where you want the center of the new display, say at (1.42, 0).

EXIT

Press EXIT to restore the menu at the bottom of the screen.

ENTER

The new display will be centered where you placed the cursor.

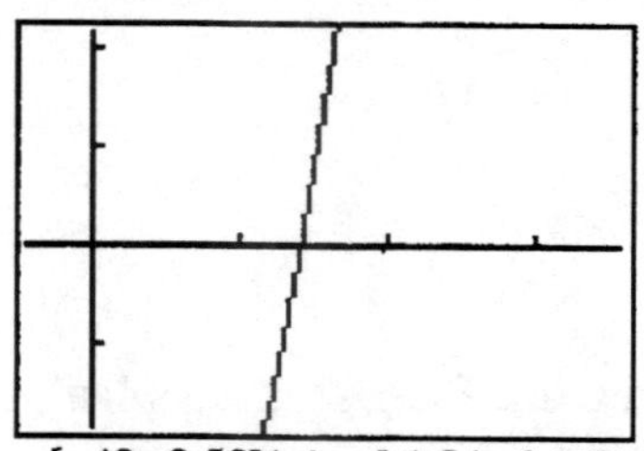

[-.42, 3.58]1 by [-1.84, 2.16]1

Now trace again to get a new approximation to the x intercept.
Repeat the zooming in process as many times as necessary to get accuracy to three decimal places. The x intercept is 1.414 .

<u>METHOD 3 Zoom Box</u>

Keystrokes	Description
[GRAPH] [F3] <ZOOM> [F4] <ZSTD>	Get the [-10, 10]1 by [-10, 10]1 display as shown at the beginning of this section.
[F1] <BOX>	Get the Zoom Box option.
[◄] or [►] or [▲] or [▼] [ENTER] [◄] or [►] or [▲] or [▼] [ENTER]	Use the arrow keys repeatedly to get the cursor close to but a little above and to the left of the x intercept, say at X = 1.1111111111, Y = .9677419355. Press [ENTER] to place the upper corner of the Zoom Box at this point. Now move the cursor down and to the right of the x intercept, say at X = 2.0634920635 Y = -.6451612903. Press [ENTER] again. The graph will be drawn again with the box as the outside of the new display.

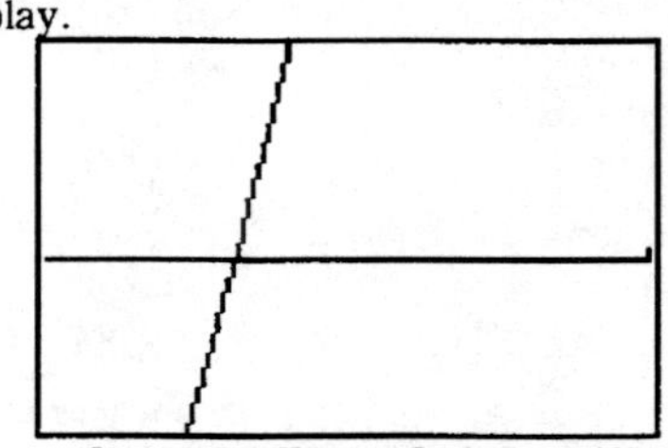

[1.11, 2.06]1 by [-.645, .968]1

Use Trace again to get another estimate of the x intercept. Use Zoom Box a second time to get a better approximation of the x intercept. The x intercept is 1.414 .

<u>METHOD 4 Solver</u>

Keystrokes	Description
[2nd] [SOLVER] [F1] <y1> [alpha] [=] [0] [ENTER]	Get the Solver feature.
[1.5] [F5] <SOLVE>	Enter an estimate of the x intercept and solve. The x intercept is 1.414.

C-8 Choosing the RANGE Variables Values

There is no one right way to determine the RANGE variables values that should be used for the graph of a function. Three methods to try are:

1. Graph using the standard screen and then zoom out.
2. Analyze the features of the function and then choose a RANGE.
3. Evaluate the function at several values.

EXAMPLE 1 Graph $y = .03x^3 + .02x^2 -5x + 10$.

Solution:

<u>METHOD 1 Graph using the standard screen and then zoom out</u>

Enter the function as .03 X ^ 3 + .02 X ^ 2 - 5 X + 10.

Press [EXIT] [F3] <ZOOM> [F4] <ZSTD> to get the [-10, 10]1 by [-10, 10]1 display.

Press [F3] <ZOOM> [MORE] [MORE] [F1] <ZFACT> and set the Zoom Factors to xFact = 4 and yFact=4. Press [F3] <ZOOM> [F3] <ZOUT>.

Move the cursor so it is where you want the center of the new screen to be, and press [ENTER]. The result is shown on the next page. The screen has been zoomed out to give us a better view of the graph. Note the double lines as the axes. This is because the Xscl and Yscl values are too close together. They do not change when zooming in.

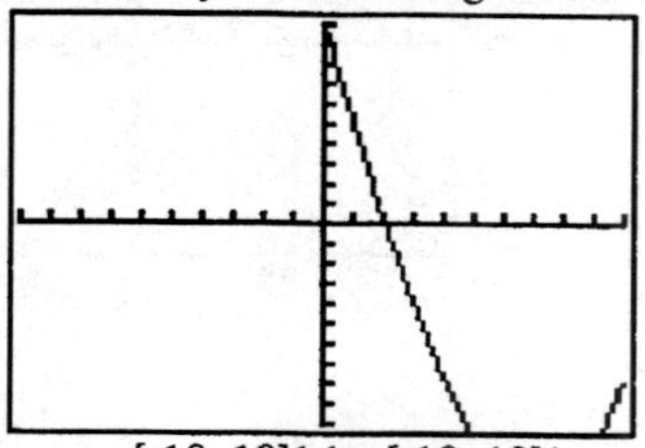

[-10, 10]1 by [-10, 10]1

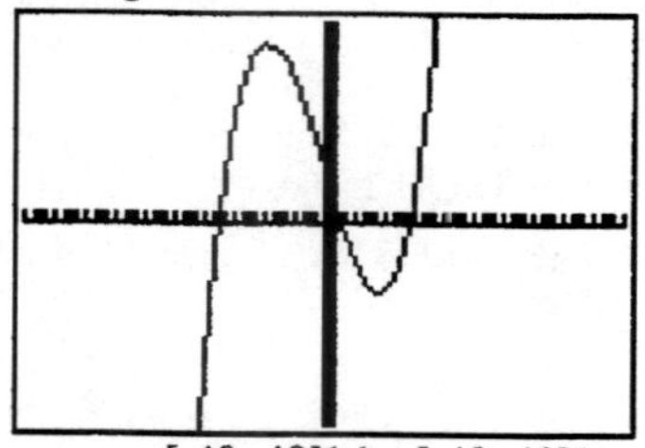

[-40, 40]1 by [-40, 40]1

<u>METHOD 2 Analyze the function and then choose the window variable values</u>
We will analyze the term with highest power on x since this is the dominant term of the function.

We see that the leading coefficient is .03. This means that the vertical change in the graph will be approximately .03 unit for every time x^3 increases 1 unit. An equation that represents this is $.03x^3 = y$ or $x^3 = \frac{1}{.03}y$ or $x^3 \approx 30y$. So if y increases by 1 unit, x will increase by about $\sqrt[3]{30}$ or about 3 units. This indicates that we want the x axis three times as wide as the y axis. Hence, a good first choice for the horizontal axis is -30 to 30, and a good first choice for the vertical axis is -10 to 10. Set Xmin = -30, Xmax = 30 and Xscl = 3. We choose 3 as the Xscl because this will put 20 tick marks on the axis.

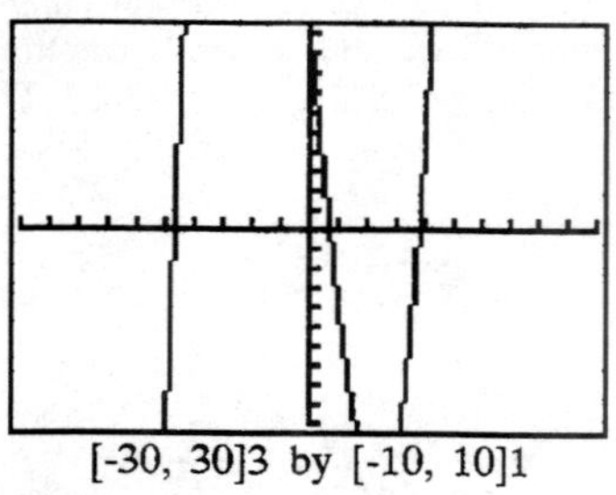

[-30, 30]3 by [-10, 10]1

The values for the vertical variables needs some adjusting since the turning points of the graph are not shown on the screen. Window variable values of [-30, 30]1 by [-40, 40]5 would show the turning points of the graph.

METHOD 3 Evaluate the function at several values of x and then choose a RANGE. The choice of x values is arbitrary. Store the value of x in memory location x, then evaluate the function. Repeat for all values of x.

x	$f(x)$
-25	-321.25
-10	32.00
0	10.00
10	-8.00
25	366.25

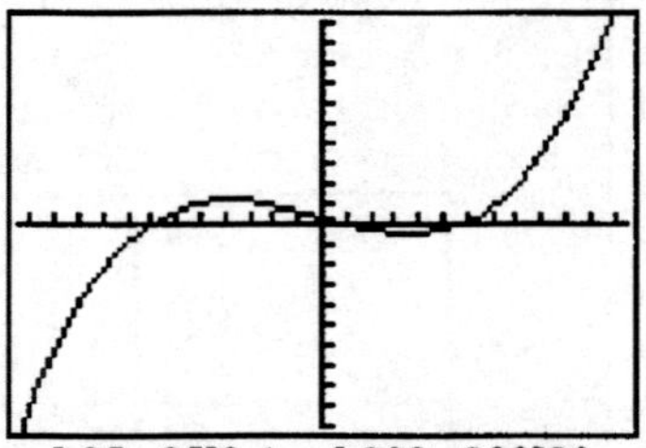

[-25, 25]2 by [-300, 300]30

Choose the Xscl and Yscl so there are about 20 marks on each axis (see METHOD 1 above).

So a good first choice for the RANGE could be [-25, 25]2 by [-300, 300]30.

C-9 Equations in One Variable

An equation in one variable can be solved graphically in two ways:

1. Algebraically move all terms to one side of the equation so that you have an expression equal to zero. Create a function $f(x)$ = (the expression). Graph the function.

   There are three ways to solve an equation in the form $f(x) = 0$ or to find the x intercepts of the graph of $f(x)$:

   a. Trace and zoom

   b. Use the solver

   c. Use the root finding feature

   The set of x intercepts is the solution to the equation.

2. Create two functions: $f(x)$ = (the left side of the equation) and $g(x)$ = (the right side of the equation). Graph both functions. Trace and zoom, or use the solver, to find the x coordinate of the points of intersection of the two graphs. The set of x coordinates is the solution to the equation.

EXAMPLE: Solve $\sqrt[3]{x + 2} = 4x^2 - 3\sqrt{x+4}$. Use two decimal places.

Solution:

METHOD 1 Write the equation as $\sqrt[3]{x + 2} - (4x^2 - 3\sqrt{x+4}) = 0$. Create the function

$f(x) = \sqrt[3]{x + 2} - (4x^2 - 3\sqrt{x+4})$. Enter this function into the calculator and graph.

GRAPH F1 <y(x)=> CLEAR
(x-VAR + 2) ^ (1 ÷ 3)
- (4 x-VAR ^ 2 - 3
2nd √ (x-VARS + 4))
EXIT F3 <ZOOM> F4 <ZSTD>

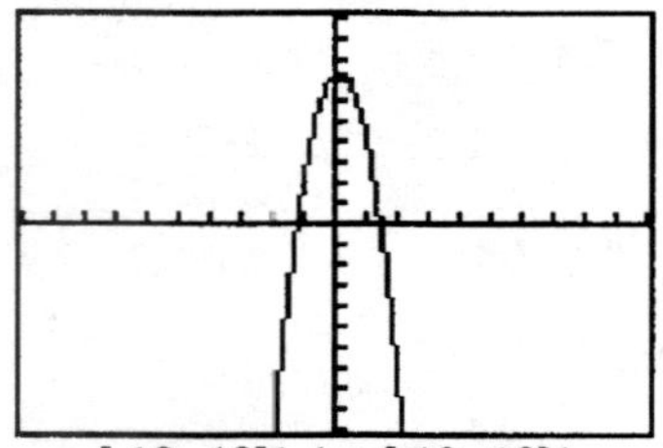

[-10, 10]1 by [-10, 10]1

We see the function crosses the x axis in two places.

a. Trace and zoom. This method is described in section C-8.

b. Solver.

GRAPH F1 <$y(x)$> ▼ 0 EXIT
2nd SOLVER F1 <y1> ALPHA =
F2 <y2)> ENTER 2
F5 <SOLVE>
-2 F5 <SOLVE>

Get the y(x)= menu and enter 0 as one of the functions. This is the equation of the x axis. Now set Now set y1 equal to y2. Enter a number near the x intercept you are finding. We used $x = 2$. Solve.
Enter another value of x near the other x intercept. We used -2. Solve.
The two intercepts are $x = -1.22$ and $x = 1.46$.

c. Root.

GRAPH MORE F1 <MATH> F3 <ROOT>
◀ or ▶ ENTER

Get the MATH menu from GRAPH menu at the bottom of the screen after you press GRAPH. Choose ROOT and move the cursor close to the x intercept. Press ENTER to get the root.

METHOD 2 Create the two functions $f(x) = \sqrt[3]{x + 2}$ and $g(x) = 4x^2 - 3\sqrt{x+4}$. Graph these and find the x coordinates of the intersection points. This can be done using trace and zoom or using the solver. The solver was used in the description below.

GRAPH F1 <y(x)=> CLEAR (x-VAR
+ 2) ^ (1 ÷ 3)
ENTER 4 x-VAR ^ 2
- 3 2nd √ (x-VAR + 4)
EXIT F3 <ZOOM> F4 <ZSTD>

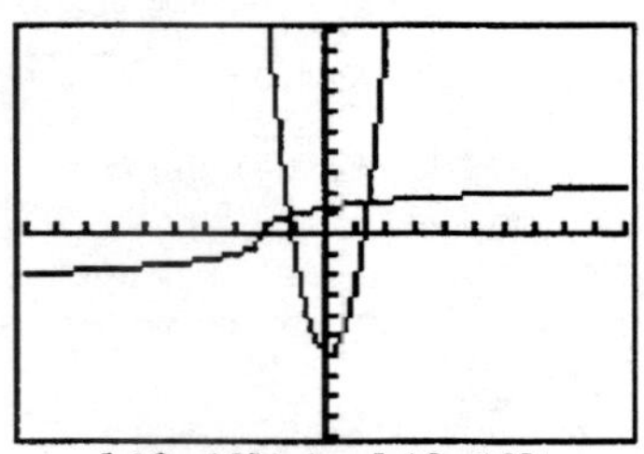

[-10, 10]1 by [-10, 10]1

[2nd] [SOLVER] [F1] <y1>
[ALPHA] [=] [F2] <y2)> [ENTER] [-2]
[F5] <SOLVE>

You'll need to enter a value of x near the x coordinate of one intersection point so the calculator knows which intersection point you are solving for. We chose 2. The result is $x = 1.46$, $y = 1.51$.

[2] [F5] <SOLVE>

Repeat for the other intersection point. We used -2 as the approximate value of this x coordinate. The result is $x = -1.22$, $y = .92$.

So the solution to this equation is $x = -1.22$ and $x = 1.46$.

C-10 Inequalities in One Variable

An inequality in one variable can be solved graphically in two ways:

1. Algebraically move all terms to one side of the equation so that you have an expression equal to zero. Create a function $f(x)$ = (the expression). Graph the function.

   There are three ways (as described in Section C-9) to solve an equation in the form $f(x) = 0$ or to find the x intercepts of the graph of $f(x)$:

   a. Trace and zoom

   b. Use the solver

   c. Use the root finding feature

   The set of x intercepts are the cutoff values for the solution. Now determine the intervals over which the inequality is true and write the solution.

2. Create two functions: $f(x)$ = (the left side of the equation) and $g(x)$ = (the right side of the equation). Graph both functions. Find the x coordinate of the points of intersection of the two graphs. The set of x coordinates are the cutoff values for the solution. Now, determine the intervals over which the inequality is true and write the solution.

EXAMPLE: Solve $\sqrt[3]{x+2} < 4x^2 - 3\sqrt{x+4}$. Use two decimal places.

Solution: METHOD 1 Write the inequality as: $\sqrt[3]{x+2} - (4x^2 - 3\sqrt{x+4}) < 0$.
Change the sign to an equal sign: $\sqrt[3]{x+2} - (4x^2 - 3\sqrt{x+4}) = 0$. Create the function $f(x) = \sqrt[3]{x+2} - (4x^2 - 3\sqrt{x+4})$, graph, and find the x intercepts. This was done above in Appendix Section C-9 Method 1 of this manual. The cutoff values are $x = -1.22$ and $x = 1.46$. Observing the graph, we see it is below the x axis for $x < -1.22$ and $x > 1.46$. Hence the solution to this inequality can be written as
$(-\infty, -1.22) \cup (1.46, \infty)$.

METHOD 2 Create the two functions $f(x) = \sqrt[3]{x+2}$ and $g(x) = 4x^2 - 3\sqrt{x+4}$. Graph these and find the x values of the intersection points. This was done above in Appendix Section C-9 Method 2 of this manual. The intersection points are at (-1.22, .92) and (1.46, 1.51). Inspecting the graph we see that $f(x) < g(x)$ for $x < -1.22$ and $x > 1.46$. Hence the solution to this inequality can be written as $(-\infty, -1.22) \cup (1.46, \infty)$.

C-11 Permutations, Combinations, and Random Numbers

EXAMPLE 1: Find: (A) $P_{19,13}$ and (B) $C_{19,13}$.

Solution (A): This is a direct calculation.

Keystrokes	Explanation
[19] [2nd] [MATH] [F2] <PROB> [F2] <nPr> [13] [ENTER]	Enter the numbers and get the built-in function from the MATH menu. The result is displayed in scientific notation as 1.68951528346E14.

Solution (B): This is a direct calculation.

Keystrokes	Explanation
[19] [2nd] [MATH] [F2] <PROB> [F3] <nCr> [13] [ENTER]	Enter the numbers and get the built-in function from the MATH menu. The result is 27132.
[EXIT] [EXIT]	To return to the home screen.

EXAMPLE 2: Generate ten unique random numbers at least 0 and less than 50.

Solution:

Keystrokes	Explanation
[5] [STO▸] [2nd] [MATH] [F2] <PROB> [F4] <rand> [ENTER]	We need to store a "seed" number in the storage location RAND. This tells the random number generator where to start.
[EXIT] [F1] <NUM> [F2] <IPart> [(] [50] [x] [EXIT] [F2] <PROB> [F4] <rand> [)] [ENTER] [ENTER]	RAND generates a number greater than 0 and less than 1. If we multiply by 50 and take the integer part, the result is a number greater than or equal to 0 and less than 50.
[ENTER] [ENTER] [ENTER] [ENTER] [ENTER] [ENTER] [ENTER] [ENTER]	The 10 random numbers are: 36 13 6 17 21 ~~21~~ 20 5 31 16 48 Not that repetitions can occur but 21 cannot be used twice if want unique random numbers.

C-12 Matrices

Given the matrices $A = \begin{bmatrix} 1 & -4 & 6 \\ 2 & -1 & 0 \\ 5 & -3 & 6 \end{bmatrix}$ and $B = \begin{bmatrix} 4 & 15 \\ -2 & 6 \\ 2 & 5 \end{bmatrix}$.

EXAMPLE 1: Find -2.5AB.

Solution:

Keystrokes	Explanation
2nd MATRX F2 <EDIT> A ENTER	Choose matrix A and enter the dimensions.
3 ENTER 3 ENTER	
1 ENTER (-) 4 ENTER 6 ENTER	Enter the matrix elements row by row.
2 ENTER (-) 1 ENTER 0 ENTER	
5 ENTER (-) 3 ENTER 6 EXIT	
2nd MATRX	Choose matrix B and enter the dimensions.
F2 <EDIT> B ENTER	
3 ENTER 2 ENTER	
4 ENTER 15 ENTER	
(-) 2 ENTER 6 ENTER	Enter the matrix elements.
2 ENTER 5	
EXIT	Return to the Home Screen.
(-) 2.5 2nd MATRX	Enter the matrix operations.
F1 <NAMES> F1 <A> × F2 <B>	The result is $\begin{bmatrix} -60 & -52.5 \\ -25 & -60 \\ -95 & -217.5 \end{bmatrix}$.
ENTER	

EXAMPLE 2: Find A^{-1}.

Keystrokes	Explanation
2nd MATRX F1 <NAMES> F1 <A>	The operations are done from the Home Screen. The result rounded to three decimal places is:
2nd x^{-1} ENTER	

$$A = \begin{bmatrix} -.167 & .167 & .167 \\ -.333 & -.667 & .333 \\ -.028 & -.472 & .194 \end{bmatrix}.$$

Some numbers may look like decimals. However upon examining the matrix using the arrow keys you will find they are decimals that are very close to zero. This situation can be eliminated by setting the number of decimal places to a number, say 3 (see Section C-3 MODES and/or C-17 of this manual).

EXAMPLE 3: Solve the system of equations using the inverse of the coefficient matrix. Round answers to two decimal places.

$$\begin{aligned} x - 4y + 6z &= 2.5 \\ 2x - y &= 3.2 \\ 5x - 3y + 6z &= -3.4 \end{aligned}$$

Solution: The inverse of the coefficient matrix was found above in Section C-12 Example 2. Enter the constants as matrix B and find $A^{-1}B$.

Keystrokes	Explanation
2nd MATRX F2 <EDIT> 2 :<B> ENTER 3 ENTER 1 ENTER 2.5 ENTER 3.2 ENTER (-) 3.4 EXIT	Enter the matrix dimensions and enter the matrix elements.
EXIT 2nd MATRX F1 <NAMES> F1 <A> 2nd x^{-1} × F2 <B> ENTER	Return to the Home Screen. Perform the matrix operations. The result is $\begin{bmatrix} -.45 \\ -4.1 \\ -2.24 \end{bmatrix}$. The solution to the system of equations is $x = -.45$, $y = -4.1$, $z = -2.24$.
EXIT EXIT	Return to the home screen.

EXAMPLE 4: Find the solution to the system of Example 3 above using Gauss-Jordan reduction. In other words, find the reduced form of the matrix $\begin{bmatrix} 1 & -4 & 6 & 2.5 \\ 2 & -1 & 0 & 3.2 \\ 5 & -3 & 6 & -3.4 \end{bmatrix}$.

Solution:

<u>METHOD 1</u> Use row operations on the matrix to get the reduced form.

Keystrokes	Explanation
2nd MATRX F2 <EDIT> A ENTER 3 ENTER 4 ENTER	Enter the matrix dimensions.
1 ENTER (-) 4 ENTER 6 ENTER 2.5 ENTER 2 ENTER (-) 1 ENTER 0 ENTER 3.2 ENTER 5 ENTER (-) 3 ENTER 6 ENTER (-) 3.4 EXIT EXIT	Enter the matrix elements. Notice that if you are continuing from Example 1 - 3 of this section, you need only enter the elements in column 4. Use the down arrow to move from one row to the next.
2nd MATRX F4 <OPS> MORE F5 <mRAdd> (-) 2 , EXIT F1 <NAMES> F1 <A> , 1 , 2) ENTER	We wish to multiply by -2, matrix A, row 1 and add it to row 2. The result will be put in row 2. `mRadd(-2,A,1,2)` The result is $\begin{bmatrix} 1 & -4 & 6 & 2.5 \\ 0 & 7 & -12 & -1.8 \\ 5 & -3 & 6 & -3.4 \end{bmatrix}$.
STO▸ F1 <A>	Store the result in matrix A.
ENTER 2nd MATRX F4 <OPS> MORE F5 <mRAdd> (-) 5 , EXIT F1 <NAMES> F1 <A> , 1 , 3) ENTER	Now multiply row 1 by -5 and add it to row 3 putting the result in row 3. `mRadd(-5,A,1,3)` The result is $\begin{bmatrix} 1 & -4 & 6 & 2.5 \\ 0 & 7 & -12 & -1.8 \\ 0 & 17 & -24 & -15.9 \end{bmatrix}$.

Keystrokes	Explanation
[STO▸] [A] [ENTER]	Store the result in matrix A.
[2nd] [MATRX] [F4] <OPS> [MORE] [F4] <multR> [1] [÷] [7] [,] [ALPHA] [A] [,] [2] [)] [ENTER] [STO▸] [A]	Multiply row 2 by 1/7 . The result is $\begin{bmatrix} 1 & -4 & 6 & 2.5 \\ 0 & 1 & -1.71 & -.26 \\ 0 & 17 & -24 & -15.9 \end{bmatrix}$ rounded to two decimal places. Store the result in matrix A.
	Continue in this manner until you get $\begin{bmatrix} 1 & 0 & 0 & -.45 \\ 0 & 1 & 0 & -4.1 \\ 0 & 0 & 1 & -2.24 \end{bmatrix}$. The solution to the system of equations is $x = -.45, y = -4.1, z = -2.24$.

> Note: To swap rows of a matrix use [F4] <OPS> [MORE] [F2] <rSwap> from the [MATRX] menu. The command rSwap(A,2,3) will swap rows 2 and 3 in matrix A.

<u>METHOD 2</u> Use the built-in capabilities of the calculator to get the reduced form.

Keystrokes	Explanation
[2nd] [MODE] [▼] [▶] [▶] [▶] [ENTER]	Set the mode to two decimal places.
[2nd] [MATRX] [F2] <EDIT>	Enter the matrix dimensions.
[A] [ENTER] [3] [ENTER] [4] [ENTER] [1] [ENTER] [(-)] [4] [ENTER] [6] [ENTER] [2.5] [ENTER] [2] [ENTER] [(-)] [1] [ENTER] [0] [ENTER] [3.2] [ENTER] [5] [ENTER] [(-)] [3] [ENTER] [6] [ENTER] [(-)] [3.4] [EXIT]	Enter the matrix elements. Notice that if you are continuing from Example 1 - 3 of this section, you need only enter the elements in column 4. Use the down arrow to move from one row to the next.
[2nd] [MATRX] [F4] <OPS> [F5] <rref> [EXIT] [F1] <NAMES> [F1] <A> [ENTER]	We wish to get the reduced form of this matrix. Use the matrix operation rref to get this on the calculator.
	The result is $\begin{bmatrix} 1.00 & 0.00 & 0.00 & -.45 \\ 0.00 & 1.00 & 0.00 & -4.10 \\ 0.00 & 0.00 & 1.00 & -2.24 \end{bmatrix}$. Use the right and left arrow keys to view the entire matrix.
	Hence, the solution to the system of equations, rounded to two decimal places, is $x = -.45, y = -4.10, z = -2.24$.

C-13 Logarithmic, Exponential, and Hyperbolic Functions

EXAMPLE 1: Graph $f(x) = 10^{3.2x}$ for $-.5 < x < .5$.

Solution: Set the RANGE to [-.5,.5].1 by [-10,10]1.

[GRAPH] [F1] <y(x)=>
[CLEAR] [10] [^] [(] [3.2] [x-VAR] [)] [EXIT]
[F2] <RANGE> [(-)] [.5] [ENTER]
[.5] [ENTER] [.1] [ENTER] [(-)] [10] [ENTER]
[10] [ENTER]
[1] [F5] <GRAPH>

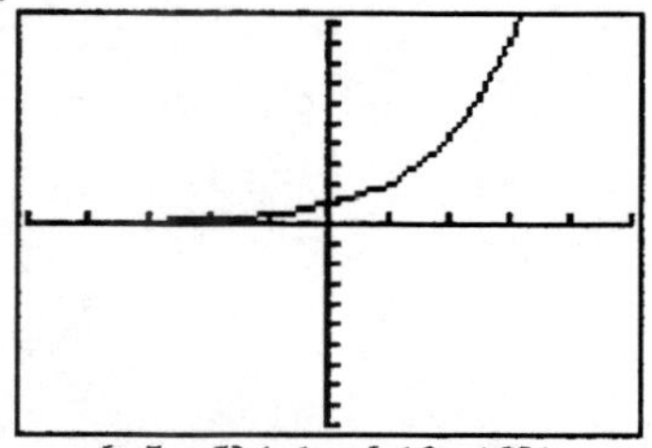

[-.5, .5].1 by [-10, 10]1

EXAMPLE 2: Graph $y = \dfrac{e^{x} + e^{-x}}{2}$.

Solution: This could be entered using [(] [2nd] [e^x] [x-VAR] [+] [2nd] [e^x] [(-)] [x-VAR] [)] [÷] [2]. However the built-in hyperbolic cosine function can also be used.

[GRAPH] [F1] <y(x)=> [CLEAR]
[2nd] [MATH] [F4] <HYP> [F2] <cosh>
[x-VAR] [EXIT] [F3] <ZOOM>
[F4] <ZSTD>

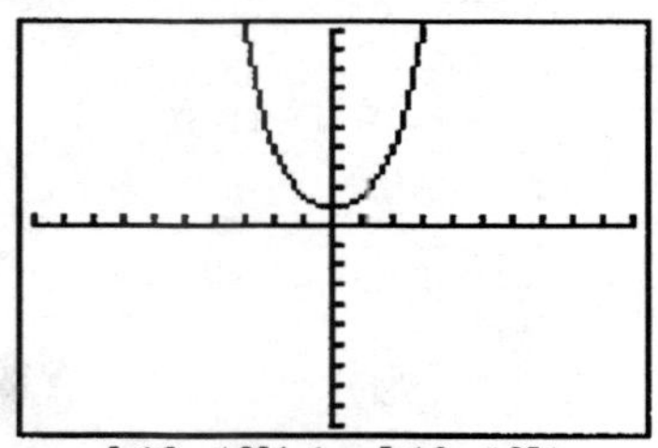

[-10, 10]1 by [-10, 10]1

EXAMPLE 3: Solve $\ln(x^2 + x) = 3x + 2$

Solution:

[GRAPH] [F1] <y(x)=> [CLEAR] — Enter the function $f(x) = \ln(x + x^2)$ as Y1 in the function list.
[LN] [(] [x-VAR] [^] [2]
[+] [x-VAR] [)] [ENTER] — Enter the function $g(x) = 3x + 2$ as Y2 in the function list.
[CLEAR] [3] [x-VAR] [+] [2]
[ENTER]

Keystrokes	Explanation
EXIT F3 <ZOOM> F4 <ZSTD>	Graph them on the same graphing screen. Zoom and trace to find the x value of the point of intersection. This will be the solution to the equation.

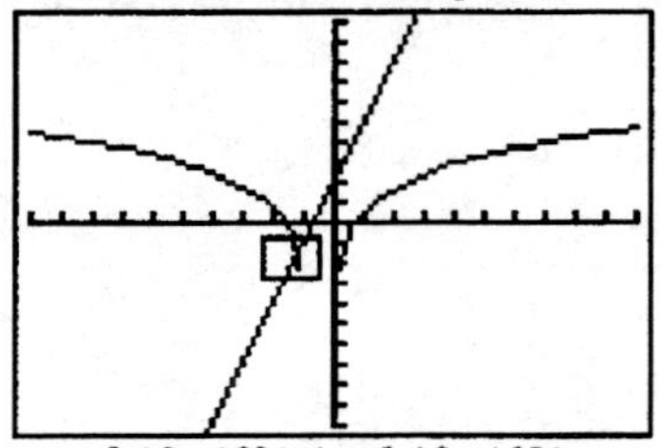

[-10, 10]1 by [-10, 10]1

The solution is -1.18, rounded to two decimal places.

C-14 Trigonometric Functions

EXAMPLE 1: Evaluate $f(x) = \cos x + \csc x$ at (A) 53°45'18" and (B) 78.52 radians.

Solution: (A)

Keystrokes	Explanation
GRAPH F1 <y(x)=> CLEAR COS x-VAR + 1 ÷ SIN x-VAR EXIT EXIT	Store the function.
2nd MODE ▼ ▼ ► ENTER	Change the calculator to degree angle measure.
CLEAR 53 + 45 ÷ 60 + 18 ÷ 3600 STO► x-VAR ENTER	Change the angle to decimal degrees while storing it as the variable x. Store 54.755° as x.
2nd ALPHA y 1 ENTER	Evaluate the function. The result to two decimal places is 1.83.

Solution: (B)

Keystrokes	Explanation
2nd MODE ▼ ▼ ENTER	Change the calculator to radian angle measure.
2nd QUIT	Since the function was stored in part (A) above, we need only to store the angle measure and evaluate the stored function.
78.52 STO► x-VAR ENTER	Store 78.52 as x.
2nd ALPHA y 1 ENTER	Recall the function. The result to two decimal places is 49.47.

EXAMPLE 2: Graph $f(x) = \cos x + \csc x$.

Solution: The function was entered in the calculator in Example 1 above.

Keystrokes	Explanation
GRAPH MORE F3 <FORMT>	Change the graph mode to Dot. This will eliminate the "vertical lines."
▼ ▼ ▶ ENTER	
F5 <GRAPH> F3 <ZOOM>	This option automatically sets the graph RANGE to [-8.24,8.24]1.57 by [-4,4]1.
MORE F3 <ZTRIG>	

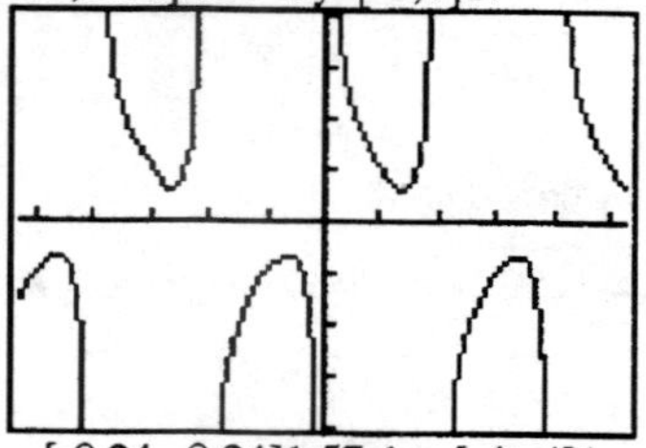

[-8.24, 8.24]1.57 by [-4, 4]1

C-15 Numeric Differentiation and Integration, and Tangent Lines

EXAMPLE 1: (A) Find the equation of the tangent line to $f(x) = \sqrt{x^5 - \sqrt[3]{x}}$ at $x = 3.54$. Use two decimal place accuracy. (B) Sketch the function and its tangent line using [2, 5]1 by [10, 60]5.

Solution (A): The function and it's derivative can be evaluated at this value of x .

Keystrokes	Explanation
GRAPH F1 <y(x)=> CLEAR 2nd	Enter the function and store it as y1.
√ (x-VAR ^ 5 - x-VAR ^	
(1 ÷ 3))	
2nd QUIT 3.54 STO▶	Get the Home Screen and store the value of x.
x-VAR ENTER	
2nd VARS F1 <ALL>	Use the arrow keys to highlight the function.
▼ ... ▼ ENTER ENTER	Press ENTER to evaluate the function. $f(3.54)$ = 23.55.
2nd CALC F3 <der1> 2nd alpha	Get the numeric derivative built-in function: `der1(y1,x,3.54)`
y 1 , x-VAR , 3.54)	This will evaluate the derivative with respect to x of the function stored in y1 at x = 3.54. The result is 16.67. (The default step size used in x increments is .001.)
ENTER	

The equation of the tangent line is $y - 23.55 = 16.67(x - 3.54)$.

Solution (B): Method 1 Enter the function as Y1 in the calculator and the equation of the tangent line as Y2 in the calculator and graph.

Method 2 Enter the function as y1 in the calculator. Set the viewing rectangle to [2, 5]1 by [10, 60]5 and graph the function.

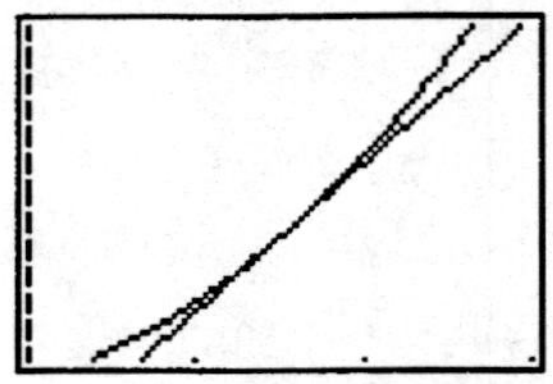

[2, 5]1 by [10, 60]5

Use the tangent drawing feature of the calculator by entering [GRAPH] [MORE] [F2] <DRAW> [MORE] [MORE] [F1] <TanLn> [2nd] [ALPHA] [Y] [1] [,] [3.54] [)] [ENTER].

(It may take a little while to graph.) Press [CLEAR] to remove the menu at the bottom of the screen. Press [EXIT] to display it again. The function and its tangent will be graphed as shown to the right.

EXAMPLE 2: Is the function $f(x)$ defined in Example 1 concave upwards at $x = 3.54$?

Solution:

Enter the function and store it as y1 as shown in Example 1 of this section.

[2nd] [CALC] [F4] <der2> [2nd] [alpha] [y] [1] [,] [x-VAR] [,] [3.54] [)] [ENTER]

Use the second derivative function.

```
der2(y1,x,3.54)
```

The result is positive. Hence the function is concave upwards at $x = 3.54$.

EXAMPLE 3: Find $\int_{1.32}^{2.58} \sqrt{x^5 - \sqrt[3]{x}}\ dx$. Use two decimal place accuracy.

Solution:

[GRAPH] [F1] <y(x)=> [2nd] [√] [(] [x-VAR] [^] [5] [-] [x-VAR] [^] [(] [1] [÷] [3] [)] [)] [2nd] [QUIT]

Enter the function and store it.

[2nd] [CALC] [F5] <fnInt> [2nd] [alpha] [y] [1] [,] [x-VAR] [,] [1.32] [,] [2.58] [)] [ENTER]

Get the numeric integration function built into the calculator function:

```
fnInt(y1,x,1.32,2.58)
```

This will evaluate the definite integral. The result is 6.95. The tolerance in calculating the definite integral is .00001. This can be changed by pressing [2nd] [TOLER] and changing the setting.

C-16 Statistics

EXAMPLE 1: Given the following data, find: (A) the mean, (B) the sample standard deviation, (C) an histogram with first class beginning at 10 and width 4 (use two decimal places), and (D) repeat Parts A and B after deleting the data value 89 from the list. How did eliminating this outlier effect the mean and standard deviation?

23	56	73	12	89	52	45
38	18	21	15	37	42	51
19	40	57	38	54	60	39

Solution (A) and (B):

Keystrokes	Explanation
STAT F2 <EDIT>	Get the statistics menu.
ENTER ENTER F5 <CLRxy>	Clear the statistical registers.
23 ENTER ENTER 38 ENTER ENTER 19 ENTER ENTER Etc.	Enter the data in the x1 position. The y1 position is used for frequency if there is more than one piece of data with the same number. Enter all data. There are 21 values. If you make a mistake, use the arrow keys to highlight the value and type it in again. To remove a data value, use the arrow keys to get the *x* data value and press F1 <DELi>.
EXIT	Exit the data entry screen.
F1 <CALC> ENTER ENTER F1 <1-VAR>	The mean is 41.86 and the sample standard deviation is 19.82.

Solution (C):

Keystrokes	Explanation
EXIT EXIT GRAPH F1 <y(x)=> F5 <SELCT>	This will deselect the function the cursor is on. The = sign will no longer be highlighted. Deselect all functions stored or delete them using CLEAR.
EXIT F2 <RANGE> 0 ENTER 100 ENTER 4 ENTER (-) 5 ENTER 5 ENTER 1 EXIT	The smallest value is 12 so let xMin=0. The greatest value is 89. Let us set xMax=100. Let's use xScl=4. This sets the histogram class width. Let yMin=-5 to get some room below the histogram. We do not know the frequencies yet so let yMax=5 for a first try. Let yScl=1.
STAT F3 <DRAW> F5 <CLDRW>	Clear all drawings from the display screen.

F1 <HIST>

Draw the histogram.

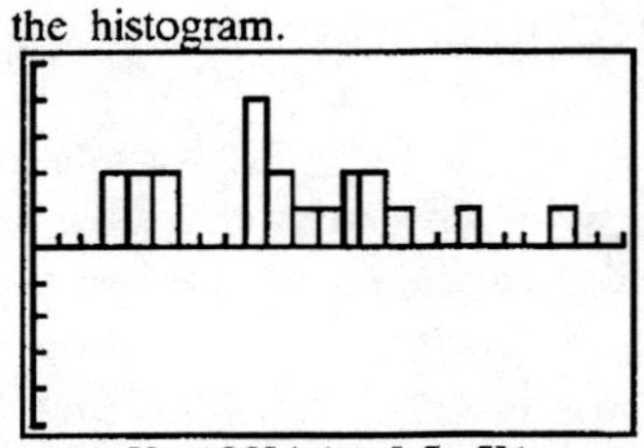

[0, 100]4 by [-5, 5]1

Solution (D):

EXIT F2 <EDIT> ENTER ENTER

Get the data list and arrow down to the 89.

▼ ... ▼ F2 <DELi>

Delete the item.

EXIT F1 <CALC> ENTER ENTER

F1 <1-VAR>

Get the statistics menu and the one variable sample statistics. The mean is now 39.5 and the sample standard deviation is 17.05. Notice how much influence an outlier has!

EXAMPLE 2: Given the following bivariate data: (A) find the correlation coefficient, (B) find the least squares line, (C) graph the least squares line, (D) plot the scatter diagram, and (E) sort the data on x and join the points using straight line segments.

x	5	8	10	29	9	24
y	3	4	8	20	3	19

Solution (A) & (B):

STAT F2 <EDIT>

Get the statistics menu.

ENTER ENTER F5 <CLRxy> 5 ENTER

Clear the statistical registers.

3 ENTER 8 ENTER 4 ENTER

Enter the data.

10 ENTER 8 ENTER etc. EXIT

F1 <CALC> ENTER ENTER

F2 <LINR>

Linear regression. r = .98
a=-1.72 and b=.79
So the least squares line is $\hat{y}$ = .79x - 1.72.

Solution (C):

EXIT EXIT

GRAPH F2 <RANGE> 0

ENTER 30 ENTER 1

(-) 10 ENTER 30 ENTER 1

Set the graph RANGE. Observe the data to determine the RANGE values. Set the y value lower so if you trace, the coordinates at the bottom of the screen do not cover up the graph.

F1 <y(x)=>

F5 <SELCT> EXIT EXIT — Deselect (or clear) all functions on the function list. See Example 1 above.

STAT F3 <DRAW> F4 <DRREG> — Draw the regression line that you just calculated.

Solution (D):

F2 <SCAT> — The scatter diagram will draw on the same graph as the least squares line.

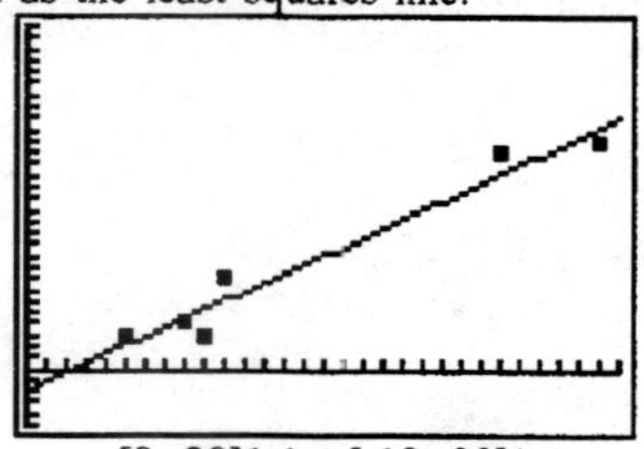

[0, 30]1 by [-10, 30]1

Solution (E):

EXIT F2 <EDIT> — Sort the data on x variable.

ENTER ENTER F3 <SORTX>

EXIT F3 <DRAW> F5 <CLDRW> — Clear the drawing. Draw the scatter diagram.

F2 <SCAT> F3 <xyLINE> — Draw the line joining the points. They will join in the order of the x data in the data list.

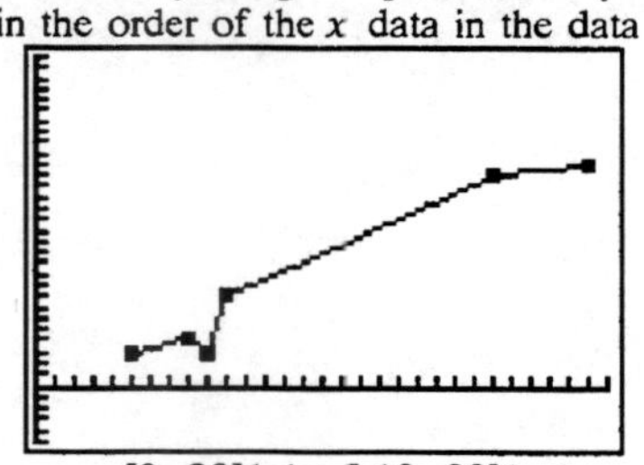

[0, 30]1 by [-10, 30]1

C-17 Setting the Display Mode for Numbers

Numbers can be entered into the calculator in scientific notation, a set number of significant digits, or a fixed number of decimal places.

EXAMPLE 1: Calculate $(9.35417\times 10^4)(-3.64389\times10^{-6})$ using:
(A) scientific notation, (B) two significant digits, and (C) Four decimal place accuracy.

Solution (A):

[2nd] [MODE] [▶] [ENTER]
[2nd] [QUIT] [9.35417] [EE] [4] [x]
[(-)] [3.64389] [EE] [(-)] [6] [ENTER]

The calculator is set to scientific mode with floating decimals. The result displayed on the calculator is -3.40855665213E-1.

Solution (B):

[2nd] [MODE] [▶] [ENTER]
[▼] [▶] [▶] [ENTER]
[2nd] [QUIT]

Set the calculator to scientific notation with 1 digit instead of floating point. Return to the Home Screen.

[2nd] [ENTRY] [ENTER]

Use the replay key to replay the calculation just done in Solution (A). The result to two significant digits is -3.4E-1.

Solution (C):

[2nd] [MODE] [ENTER]
[▼] [▶] [▶] [▶] [▶] [▶] [ENTER]
[2nd] [QUIT]

Set the calculator to normal mode with 4 digits.
Return to the Home Screen.

[2nd] [ENTRY] [ENTER]

Use the replay key to replay the last calculation. The result to four decimal places is -.3409.

Change the calculator back to floating point. The examples in this manual are done using this setting unless otherwise noted.

NOTES

Appendix D

TI-86 Operations

D-1 Operations

Press ON to turn on the calculator.

Press CLEAR to clear the screen.

Press EXIT to clear any menus displayed at the bottom of the screen.

Press 2nd ▲ to make the display darker.

Press 2nd ▼ to make the display lighter.

Press 2nd 3 to get the RESET screen.

Then press F3 <RESET> F1 <ALL> to clear all the memories, and press F4 <YES> to clear. You must now adjust the display so you can see it. Enter a few numbers and adjust the display contrast as described above.

RAM	DELET	RESET	TOL	ClrEnt
ALL	MEM	DFLTS		

Mem Cleared
Defaults set

Press 2nd OFF to turn off the calculator.

Home Screen

The screen where calculations are done is called the Home Screen. You can always get to the Home Screen by pressing 2nd QUIT. Note: QUIT is above the EXIT key.

> The item above the key will be referenced as if it is on the key. For example, QUIT, which is above the EXIT key, will be referenced from now on as 2nd QUIT.

2nd

This key must be used to access the calculator functions above and to the left of a key cap. A new cursor is displayed, ↑. These functions are in yellow on the calculator face.

ALPHA and 2nd ALPHA

This key is used to access the calculator functions and memory locations above and to the right of a key cap. These functions are in light blue on the calculator face.

The cursor changes to an A when ALPHA is pressed. Pressing ALPHA and then a letter will display an upper case letter.

The cursor changes to a a when 2nd ALPHA is pressed. Pressing 2nd ALPHA will display a lower case letter.

NOTE: There is a separate memory location for the upper case letter and for the lower case letter.

MODE

This is found above the MORE key and is used to set the operating mode of the calculator. Press 2nd MODE to get the mode screen. When the highlighted items are on the left, the calculator is in default mode. To choose a different mode: Use the arrow keys to choose the item you want to change to, then press the ENTER key. The new selection will now be highlighted. Press 2nd QUIT to return to the Home Screen.

Menus

Menus are listed across the bottom of the screen. To choose a menu item press the function key immediately below the menu item. In this manual, these functions will be referred to as the key to press in a box followed by the function in angle brackets. For example, the trace function of the calculator is found on the GRAPH menu above the F4 key is denoted by F4 <TRACE>.

Error Correction

Use the 2nd INS key to insert characters. (INS is above the DEL key.)

Use the DEL key to delete characters.

Use the arrow keys ◄ or ► to move to the left or right.

Use the up arrow key ▲ to move up one line.

Use the down arrow key ▼ to move down one line.

2nd ENTRY is used to replay the last executed operation. (ENTRY is above the ENTER key.)

D-2 Calculating

Many calculator functions are found above key caps or in menus. For example the square root function is above the x^2 key but the absolute value is found in the [MATH] menu in the [NUM] submenu.

EXAMPLE 1: Calculate $|4^2 - \sqrt{34}\,|$.

Solution: [2nd] [MATH] [F1] <NUM> [F5] <abs> [(] [4] [x^2] [-] [2nd] [√] [34] [)] [ENTER].
[ENTER] tells the calculator to calculate. The result displayed is 10.1690481052.

D-3 Plotting Points and Drawing Line Segments

EXAMPLE 1: (A) Plot the points (-3, 6) and (1, 5). (B) Draw the line segment having these points as the endpoints.

Solution (A):

Keystrokes	Explanation
[GRAPH] [$y(x)$=] [CLEAR] [EXIT] [▲] [▶] ... [▶] [ENTER] [EXIT]	Clear all functions. If one of the PLOTS is highlighted, use the up arrow and the right arrow keys to select it. Press [ENTER] to remove the highlighting.
[F3] <ZOOM> [F4] <ZSTD> [EXIT]	Now set the window variables. Using these key strokes automatically sets the variables to xMin=-10, xMax=10, xScl=1, yMin=-10, yMax=10, yScl=1 and xRes=1. This will be denoted as [-10, 10]1 by [-10, 10]1 in this manual.
[GRAPH] [MORE] [F2] <DRAW> [MORE] [MORE] [F1] <CLDRW>	Get the [DRAW] menu and clear any existing drawings from the calculator screen.
[2nd] [QUIT]	Return to the Home Screen.
[2nd] [CATLG/VARS] [F1] [▼] ... [▼]	Get the [CATALOG] listing all of the functions. Press [F1] and then hold the down arrow down, or use [F1] <Page↓>, until the pointer points to [PtOn(].
[ENTER]	Press [ENTER].
[(-)] [3] [,] [6] [)] [ENTER] [EXIT]	Enter the coordinates of the first point.

[2nd] [ENTRY]

Use the replay feature of the calculator to get the last entry.

[◄]...[◄] [DEL] [1] [,] [5] [)] [ENTER]

Use the left arrow and change the coordinates to those of the second point. Delete any unwanted characters using [DEL].

Solution (B):

[2nd] [QUIT]

Return to the Home Screen.

[2nd] [CATLG/VARS] [F1] [▼]...[▼] [ENTER]

Get the [CATALOG] listing all of the functions. Press [F1] and then the down or up arrow, or [F1] <Page↓>, or [F2] <Page↑>, until the pointer points to [Line(]. Press [ENTER].

[(-)] [3] [,] [6] [,] [1] [,] [5] [)] [ENTER]

Enter the coordinates of the first point and the second point separated by commas. Press [ENTER] to draw the line segment.

D-4 Storing and Evaluating an Expression, and Using Lists and Tables

EXAMPLE 1: Store the expression $\frac{-b+\sqrt{b^2-4ac}}{2a}$ and evaluate it to find one solution of $3x^2+4x-5=0$.

Solution:

[GRAPH] [F1] <y(x)=> [CLEAR]

We can store up to 99 functions or algebraic expressions in the function list. These can then be recalled, evaluated and/or graphed.

[(] [(-)] [ALPHA] [B] [+] [2nd] [√] [(] [ALPHA] [B] [x^2] [-] [4] [×] [ALPHA] [A] [×] [ALPHA] [C] [)] [)] [÷] [(] [2] [ALPHA] [A] [)]

Press these keys to enter the expression. It is necessary to use the multiplication symbol between letters since the calculator interprets AB as a single variable instead of A×B.

Note it is not necessary to use a multiplication symbol between a number and a letter. The calculator interprets 3A as 3×A.

[2nd] [QUIT]

Return to the Home Screen.

[3] [STO▸] [A] [ENTER] [4] [STO▸] [B] [ENTER] [(-)] [5] [STO▸] [C] [ENTER]

Store 3 as A, 4 as B and -5 as C.

<u>METHOD 1</u>

Keystrokes	Explanation
[2nd] [CATLG/VARS] [F2] <ALL> [F1] <PAGE↓>... [F1] <PAGE↓> [▼] [ENTER] [ENTER]	Get the variables list. Press [F1] <PAGE↓> and/or the arrow keys until the cursor is pointing at y1. Press [ENTER] [ENTER] The result is .786299647847.

<u>METHOD 2</u>

Keystrokes	Explanation
[2nd] [ALPHA] [Y] [1] [ENTER]	Enter y1 on the screen and press [ENTER]. The result is .786299647847.

EXAMPLE *2:* Given $f(x) = \sqrt[3]{x^4+2x}$ and $g(x) = 4^{2x-90}$. Find $f(x) + 3\sqrt{g(x)}$ at $x = 45.2$.

Solution:

Keystrokes	Explanation
[GRAPH] [F1] <y(x)=> [CLEAR] [(] [x-VAR] [^] [4] [+] [2] [x-VAR] [)] [^] [(] [1] [÷] [3] [)] [ENTER] [CLEAR] [4] [^] [(] [2] [x-VAR] [–] [90] [)] [2nd] [QUIT]	Enter the functions as y1 and y2. Press [CLEAR] to delete the old y1 function. Note that lower case x is used in functions. Upper case and lower case letters are different. Values can be stored for each of these.
[45.2] [STO▸] [x-VAR] [ENTER]	Store 45.2 as x.
[2nd] [ALPHA] [Y] [1] [+] [3] [×] [2nd] [√] [(] [2nd] [ALPHA] [Y] [2] [)] [ENTER]	Form $f(x) + 3\sqrt{g(x)}$ as y1 + 3√ y2. The expression will be evaluated when you press [ENTER]. The result is 164.97 to two decimal places.

EXAMPLE *3:* Evaluate the expression $\dfrac{3x^2 + 5x - 8}{x^2 - 1}$ to three decimal places for x = .900, .990, .999, 1.001, 1.010, and 1.100 by: (A) evaluating the expression for each value of x individually; and (B) using a list.

Solution (A): First store the expression in the Y= list. Then store the value for x. Recall the expression and evaluate it.

Keystrokes	Explanation
[GRAPH] [F1] <y(x)=> [CLEAR]	Clear the old expression for y1 by pressing [CLEAR].
[(] [3] [x-VAR] [^] [2] [+] [5] [x-VAR] [–] [8] [)] [÷] [(] [x-VAR] [^] [2] [–] [1] [)] [2nd] [QUIT]	Store the expression as y1 and return to the Home Screen.

Keystrokes	
[.9] [STO►] [x-VAR] [ENTER]	Store the values of x.
[2nd] [ALPHA] [Y] [1] [ENTER]	Recall the expression and evaluate.

Store the next value for x and reevaluate the expression. Repeat for the other values of x. The corresponding results are: 5.632, 5.513, 5.501, 5.499, 5.488, and 5.381, rounded to three decimal places.

Solution (B): Store the values of x in a list and evaluate the expression for each value of x simultaneously.

Keystrokes	
[GRAPH] [F1] <y(x)=> [CLEAR]	Clear the old expression for y1 by pressing [CLEAR].
[(] [3] [x-VAR] [^] [2] [+] [5] [x-VAR] [–] [8] [)] [÷] [(] [x-VAR] [^] [2] [–] [1] [)] [2nd] [QUIT]	Store the expression as y1 and return to the Home Screen.
[2nd] [LIST] [F1] <{> [.900] [,] [.990] [,] [.999] [,] [1.001] [,] [1.010] [,] [1.100] [F2] <}> [STO►] [x-VAR] [ENTER]	Store the values of x in the list x.
[2nd] [alpha] [Y] [1] [ENTER]	Simultaneously calculate the value of the expression stored as y1 for the values of the list x. The results are: 5.632, 5.513, 5.501, 5.499, 5.488, and 5.381, rounded to three decimal places. Use the right arrow key to view the other values in the list.

EXAMPLE 4: Evaluate the expression $\frac{3x^2 + 5x - 8}{x^2 - 1}$ to three decimal places for values of x at each tenth from 0 to 1 using a table.

Solution: First store the expression in the Y= list. Set the table parameters to begin at $x = 0$ and to have an increment of .1. Get the table.

Keystrokes	
[GRAPH] [F1] <y(x)=> [CLEAR]	Clear the old expression for y1 by pressing [CLEAR].
[(] [3] [x-VAR] [^] [2] [+] [5] [x-VAR] [–] [8] [)] [÷] [(] [x-VAR] [^] [2] [–] [1] [)] [2nd] [QUIT]	Store the expression as y1 and return to the Home Screen.

Keystrokes	Description
[TABLE] [F2] <TBLST> [0] [ENTER] [.1] [ENTER]	Set the table to begin evaluating the expression at x = 0 with a step size of .1.
[ENTER]	Set the calculator to automatically display values of x and y1.
[F1] <TABLE> [TABLE] [▼] ... [▼]	Get the table. Arrow down to see the rest of the table. The highlighted value will appear at the bottom of the table.

x	y1	
0	8	
.1	7.545455	
.2	7.166667	
.3	6.846154	
.4	6.571429	
.5	6.333333	

X=0

D-5 Evaluating Inequalities

EXAMPLE 1: Given the inequality $x^2 - 3t < 5x + 9t^2$. Test this inequality for x = 1.89 and t = -4.52.

Solution:

Keystrokes	Description
[2nd] [QUIT] [CLEAR]	
[1.89] [STO▸] [x-VAR] [ENTER]	Store the value of the variables. Lower case letters are used.
[(-)] [4.52] [STO▸] [2nd] [alpha] [T] [ENTER]	
[x-VAR] [^] [2] [-] [3]	Enter the expression and evaluate.
[2nd] [alpha] [T] [2nd] [TEST] [F2] << > [5]	
[x-VAR] [+] [9] [2nd]	
[alpha] [T] [^] [2] [ENTER]	The result is 1. This means the inequality is true for these values of the variables. If the inequality was false, the result would have been 0.

D-6 Graphing and Style of Graph

EXAMPLE 1: Graph $y = 3x^2 - 6$ and $y = 3x^2 + 1$ on the same graphing screen. Use a solid line for the first graph and a dotted line for the second.

Solution:

GRAPH MORE F2 <DRAW> — Clear drawings.

MORE MORE F1 <CLDRW>

EXIT EXIT — Turn statistical plots off by pressing the up arrow, selecting the Plot, and then pressing ENTER.

Clear or deselect all other functions by placing the cursor at the expression for the function and pressing F5 <SELCT>. Now this deselected function will not graph. Note that the = sign is no longer highlighted.

GRAPH F1 <y(x)=> CLEAR 3 x-VAR ^ 2 − 6 MORE F3 <STYLE>... F3 <STYLE>

Enter the first functions.

Select the style of graph by using the STYLE option. Press F3 <STYLE> repeatedly until the solid line shows to the left of y1.

ENTER CLEAR 3 x-VAR ^ 2 + 1 F3 <STYLE>... F3 <STYLE>

Enter the second function and select the dotted style of graph by using F3 <STYLE>.

EXIT F3 <ZOOM> F4 <ZSTD>

Exit the function menu by pressing EXIT. Press F3 and then F4 to get the standard viewing window. This will automatically set the window variables to have $-10 < x < 10$ and $-10 < y < 10$ and graph the functions.

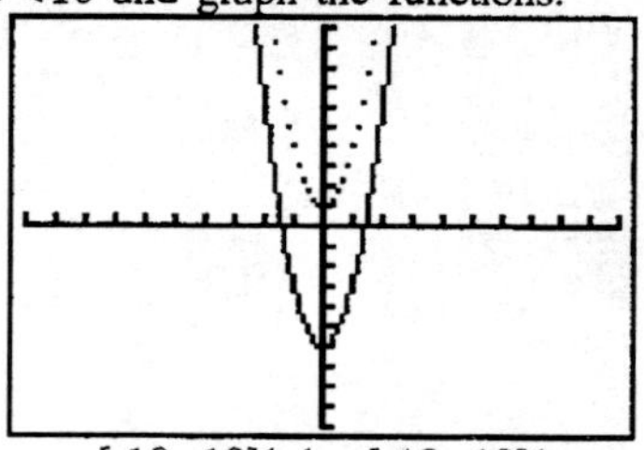

[-10, 10]1 by [-10, 10]1

CLEAR

2nd QUIT

Note that pressing CLEAR will remove the menus from the bottom of the screen.

> **ZSTD** automatically sets the viewing window variables to $-10 < x < 10$ and $-10 < y < 10$. The viewing window variables will be denoted in this manual by the notation [-10, 10]1 by [-10, 10]1. The window variables can be set by pressing GRAPH F2 <WIND> and entering the desired values.

F2 **<WIND>** means to choose the WIND option from the menu on the screen above the F2 key.

The function will <u>not</u> graph if you choose a function on the F1 <y(x)=> menu and then press F5 <SELCT>. You will notice that the highlighting disappears from the = sign. You can still evaluate the expression. To select the function, press F5 <SELCT> again.

EXAMPLE 2: Graph $f(x) = \begin{cases} .5x + 1 & x \le 2 \\ x^2 + 2 & x > 2 \end{cases}$.

Solution:
The calculator will only graph functions for values of x for which the function is defined, so we write this piecewise-defined function in a form where the pieces are defined only for the values of x that we want. Enter two functions as y1=$(.5x+1)\div(x\le2)$ and y2= $(x^2+2)\div(x>2)$. The first function is defined with $x \le 2$ and the second function is defined when $x > 2$.

GRAPH F1 <y(x)=> CLEAR	Clear the existing function.
(.5 x-VAR + 1) ÷ (x-VAR 2nd TEST F4 <≤> 2) ENTER	Enter the first piece of the piecewise-defined function.
CLEAR (x-VAR ^ 2 + 2) ÷ (x-VAR 2nd TEST F3 <>> 2) ENTER	Clear the existing function. Enter the second piece of the piecewise-defined function.
EXIT EXIT F3 <ZOOM>	
F4 <ZSTD>	Graph.

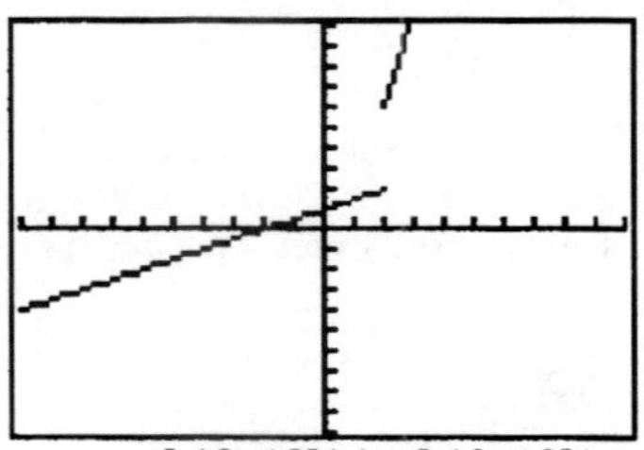

[-10, 10]1 by [-10, 10]1

D-7 TRACE, ZOOM, WINDOW, SOLVER, ROOT Finder and INTERSECT

EXAMPLE 1: Find the positive x intercept(s) for $y = 3x^2 - 6$. Use three decimal place accuracy.

There are six methods we can use to approximate the x intercepts.

1. We can trace the function and zoom in on the point by changing the window variable values.
2. We can trace the function and zoom in on the point by setting the Zoom Factors and using ZOOM.
3. We can zoom in on the point by using the Zoom Box feature and then tracing the function.
4. We can use the solver feature of the calculator.
5. We can use the root finder feature of the calculator.
6. We can use the intersect feature of the calculator.

Solution: Store the function as 3 X ^ 2 - 6 as y1. Graph using the [-10, 10]1 by [-10, 10]1. (see Example 1 from Section D-6).

F4 <TRACE>

◀ or ▶

You see a new cursor near the center of the graph screen and the coordinates of that point on the graph displayed at the bottom of the screen.

Use right and left arrow keys and move the cursor along the graph. If we had the graph of two or more functions displayed, we could use the up and down arrow keys to move between the functions. The 1 in the upper right-hand corner indicates that the cursor is on the graph of y1.

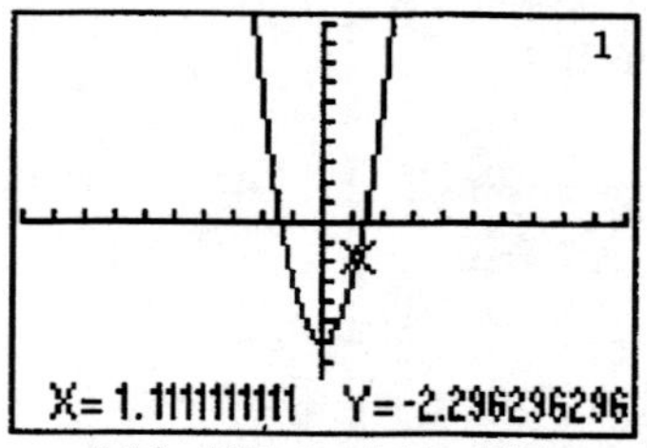

[-10, 10]1 by [-10, 10]1

METHOD 1 Set the window variables

Using the trace feature of the calculator, trace the function graphed above. We see the x intercept is near 1.5. So we shall set the window variables closer to that value.

GRAPH F2 <WIND> 1
ENTER 2 ENTER .1 ENTER
(-) .5 ENTER .5 ENTER
.1 F5 <GRAPH>
F4 <TRACE>

Set the window to [1, 2].1 by [-.5, .5].1.

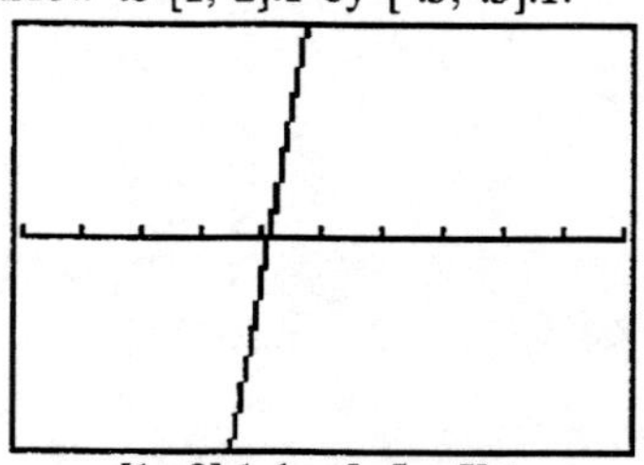
[1, 2].1 by [-.5, .5].1

Press EXIT to restore the menu at the bottom of the screen.

Use TRACE again to get a better approximation. The trace cursor always starts at the x value that is the middle of the window. It may not be on the display. Watch the values for x at the bottom of the screen.

We see that the x intercept is between 1.41 and 1.42. Set the window to [1.41, 1.42].01 by [-.01, .01].001 and graph again. We now find the x intercept to 1.414 accurate to three decimal places.

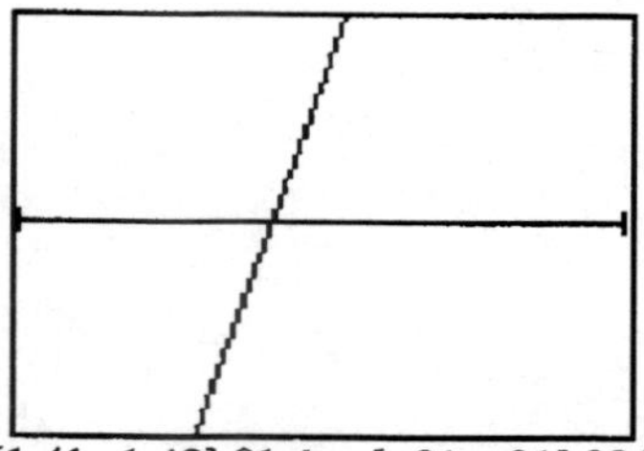

[1.41, 1.42].01 by [-.01, .01].001

METHOD 2 Zoom In

GRAPH F3 <ZOOM>
F4 <ZSTD>

Get the [-10, 10]1 by [-10, 10]1 display as shown at the beginning of this section.

MORE MORE F2 <ZFACT>
5 ENTER 5

Set the Zoom Factors to 5 and 5. The graph will be 5 times as large in both the x and y directions.

F3 <ZOOM> F2 <ZIN>

◀ or ▶ or ▲ or ▼

Use the ◀ or ▶ or ▲ or ▼ arrow keys to get the cursor to where you want the center of the new display, say at (1.42..., 0).

ENTER

The new display will be centered where you placed the cursor.

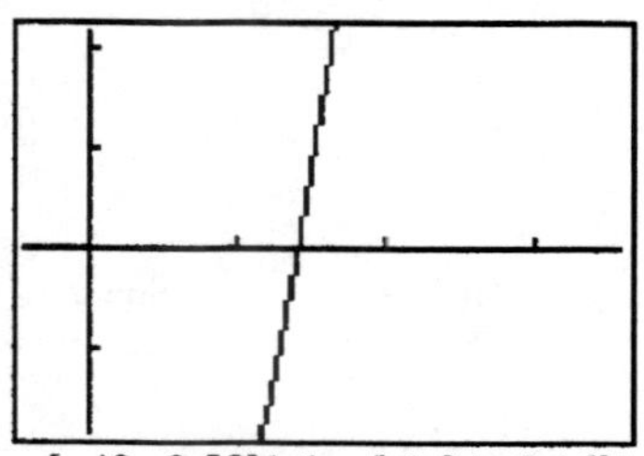

[-.42, 3.58]1 by [-1.84, 2.16]1

EXIT

Press EXIT to get the menu at the bottom of the screen. Use TRACE to get a new approximation.

Repeat the zooming in process as many times as necessary to get accuracy to three decimal places. The x intercept is 1.414 rounded to three decimal places.

METHOD 3 Zoom Box

GRAPH F3 <ZOOM>

F4 <ZSTD>

Get the [-10, 10]1 by [-10, 10]1 display as shown at the beginning of this section.

F1 <BOX>

Get the Zoom Box option.

◄ or ► or ▲ or ▼

ENTER

◄ or ► or ▲ or ▼

ENTER

Use the arrow keys repeatedly to get the cursor close to but a little above and to the left of the x intercept, say at X = 1.1111111111, Y = .9677419355. Press ENTER to place the upper corner of the Zoom Box at this point. Now move the cursor down and to the right of the x intercept, say at X = 2.0634920635 Y = -.6451612903. Press ENTER again.

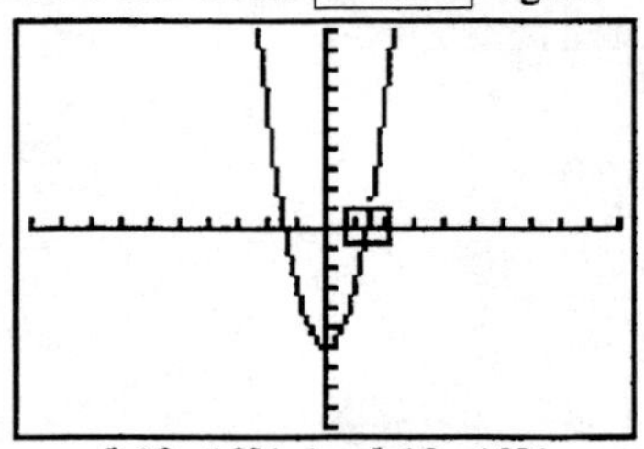

[-10, 10]1 by [-10, 10]1

The graph will be drawn again with the box as the outside of the new display.

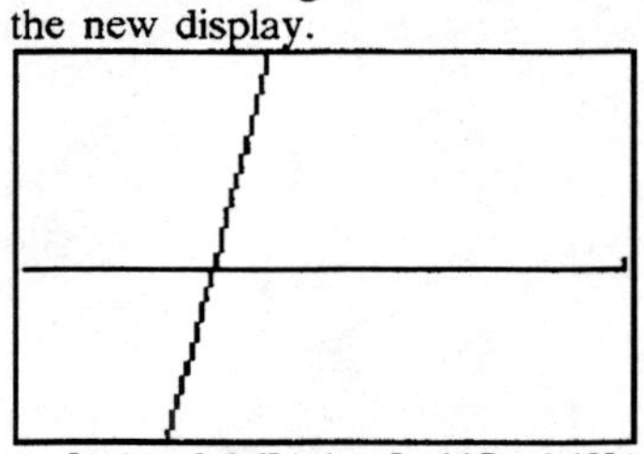

[1.11, 2.06]1 by [-.645, .968]1

Use Trace again to get another estimate of the x intercept. Use Zoom Box a second time to get a better approximation of the x intercept. The x intercept is 1.414 rounded to three decimal places.

METHOD 4 Solver

2nd SOLVER

Get the Solver feature.

F1 <y1> alpha =

0 ENTER 1.5 F5 <SOLVE>

Enter an estimate of the x intercept and solve. The x intercept is 1.414, rounded to three decimal places.

METHOD 5 Root Finder

Keystrokes	Explanation
GRAPH F3 <ZOOM>	Reset the standard viewing window variables.
F4 <ZSTD> MORE F1 <MATH> F1 <ROOT> ◄ or ► ENTER ◄ or ► ENTER ENTER	Get the MATH menu from GRAPH menu at the bottom of the screen after you press GRAPH. Choose ROOT and move the cursor to a point to the left of x intercept. Press ENTER. Move the cursor to a point to the right of the x intercept and press ENTER again. Now press ENTER a final time to get the root.

METHOD 6 Intersect

Keystrokes	Explanation
GRAPH F3 <ZOOM> F4 <ZSTD> EXIT	Reset the standard viewing window variables.
F1 <y(x)=> ▼ CLEAR 0 ENTER EXIT	Enter 0 as y2. This is the equation of the x axis.
MORE F1 <MATH> MORE F3 <ISECT>	Get the intersection option from the graphing math menu.
	Select the first curve using the arrow keys. When y1 is selected there will be a 1 in the upper right corner of the screen .
ENTER ENTER	When you press ENTER the cursor will jump to the second curve, y2.
2 ENTER	Now you'll need to enter a value of x near the x coordinate of one intersection point so the calculator knows which intersection point you are solving for. You can move the cursor and press enter, or you can press 2 and then press enter.
	The result is $x = 1.414$, $y = 0$, rounded to three decimal places.

D-8 Choosing the WINDOW Variables Values

There is no one right way to determine the window variables that should be used to display the graph of a function. Three methods to try are:

1. Graph using the default window variable values and then zoom out.
2. Analyze the features of the function and then choose the window variable values.
3. Evaluate the function at several values and then choose the values for the window variables.

EXAMPLE 1 Graph $y = .03x^3 + .02x^2 - 5x + 10$.

Solution:

METHOD 1 Graph using the default window variable values and then zoom out

Enter the function as .03 z ^ 3 + .02 z ^ 2 - 5 z + 10.

Press EXIT F3 <ZOOM> F4 <ZSTD> to get the [-10, 10]1 by [-10, 10]1 display.

Press F3 <ZOOM> MORE MORE F2 <ZFACT> and set the Zoom Factors to xFact = 4 and yFact=4. Press F3 <ZOOM> F3 <ZOUT>.

Move the cursor so it is where you want the center of the new screen to be, and press ENTER. The result below used the origin as the center of the new screen. The screen has been zoomed out to give us a better view of the graph. Note the double lines as the axes. This is because the Xscl and Yscl values are too close together. They do not change when zooming out.

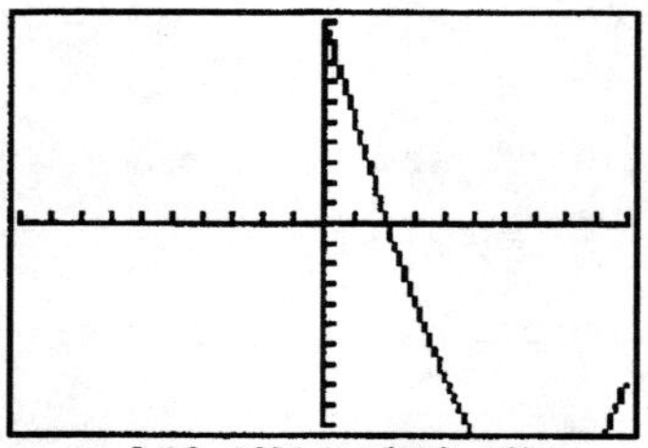

[-10, 10]1 by [-10, 10]1

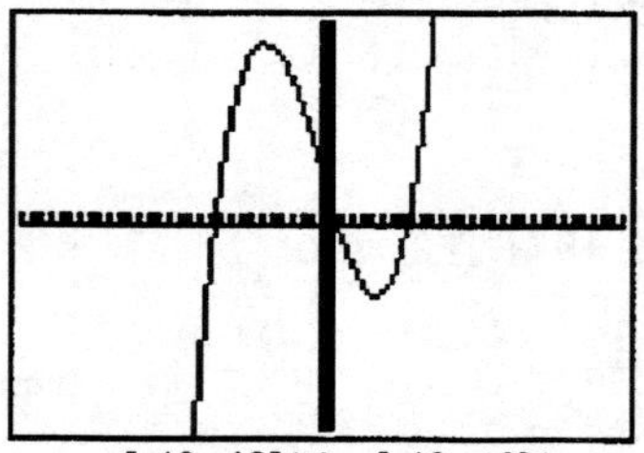

[-40, 40]1 by [-40, 40]1

METHOD 2 Analyze the function and then choose the window variable values

We will analyze the term with highest power on x since this is the dominant term of the function.

We see that the leading coefficient is .03. This means that the vertical change in the graph will be approximately .03 unit for every time x^3 increases 1 unit. An equation that represents this is $.03x^3 = y$ or $x^3 = \frac{1}{.03}y$ or $x^3 \approx 30y$. So if y increases by 1 unit, x will increase by about $\sqrt[3]{30}$ or about 3 units. This indicates that we want the x axis three times as wide as the y axis. Hence, a good first choice for the horizontal axis is -30 to 30, and a good first choice for the vertical axis is -10 to 10. Set Xmin = -30, Xmax = 30 and Xscl = 3. We choose 3 as the Xscl because this will put 20 tick marks on the axis.

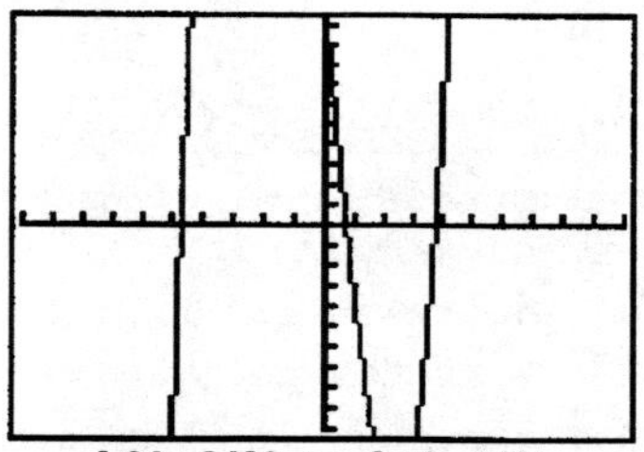

[-30, 30]3 by [-10, 10]1

The values for the vertical variables needs some adjusting since the turning points of the graph are not shown on the screen. Window variable values of [-30, 30]3 by [-40, 40]5 would show the turning points of the graph.

<u>METHOD 3 Evaluate the function at several values of *x* and then choose the window variable values</u>

Evaluate the function at several values of *x* and then choose the window variable values.The choice of *x* values to use when evaluating the function is arbitrary. Store a value of as x-VAR in the calculator and then evaluate the function. Repeat for all values of *x*.

x	$f(x)$
-25	-321.25
-10	32.00
0	10.00
10	-8.00
25	366.25

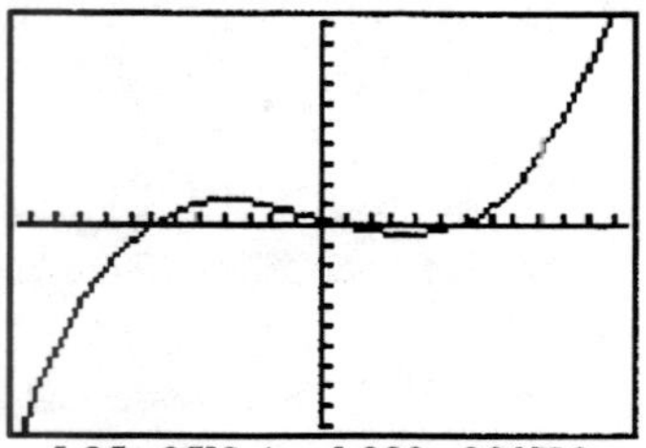

[-25, 25]2 by [-300, 300]30

Choose the Xscl and Yscl so there are about 20 marks on each axis (see METHOD 1 above).

So a good first choice for the window variables could be [-25, 25]2 by [-300, 300]30.

Another way to evaluate the function is using the TABLE feature of the calculator (see Appendix Section D-4 of this manual). Set TblStart at -25, ΔTbl at 5, and the independent variable (Indpnt:) as Auto. We see the values range from -321.25 to 366.25 as we view the list for *x* values from -25 to 25.

D-9 Equations in One Variable

An equation in one variable can be solved graphically in two ways:

1. Algebraically move all terms to one side of the equation so that you have an expression equal to zero. Create a function $f(x)$ = (the expression). Graph the function.

   There are three ways to solve an equation in the form $f(x) = 0$ or to find the x intercepts of the graph of $f(x)$:

   a. Trace and zoom

   b. Use the solver

   c. Use the root finding feature

   d. Use the intersection point finder

   The set of x intercepts is the solution to the equation.

2. Create two functions: $f(x)$ = (the left side of the equation) and $g(x)$ = (the right side of the equation). Graph both functions. Trace and zoom, use the solver, or use ISECT to find the x value of the points of intersection of the two graphs. The set of x coordinates is the solution to the equation.

EXAMPLE: Solve $\sqrt[3]{x+2} = 4x^2 - 3\sqrt{x+4}$. Use two decimal places.

Solution:

<u>METHOD 1</u> Write the equation as $\sqrt[3]{x+2} - (4x^2 - 3\sqrt{x+4}) = 0$. Create the function $f(x) = \sqrt[3]{x+2} - (4x^2 - 3\sqrt{x+4})$. Enter this function into the calculator and graph.

GRAPH F1 <y(x)=> CLEAR

(x-VAR + 2) ^ (1 ÷ 3)

− (4 x-VAR ^ 2 − 3

2nd √ (x-VAR + 4))

EXIT F3 <ZOOM> F4 <ZSTD>

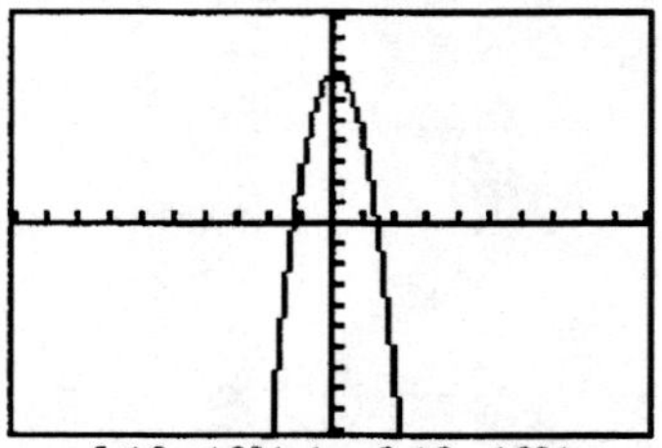

[-10, 10]1 by [-10, 10]1

We see the function crosses the x axis in two places. So we need to find two x intercepts.

Five methods for finding the x intercetps were described in Appendix Section D-7 of this manual. Following is the method using the Root Finder.

GRAPH MORE F1 <MATH>

F1 <ROOT>

◄ or ► ENTER

◄ or ► ENTER ENTER

Get the MATH menu from GRAPH menu at the bottom of the screen after you press GRAPH . Choose ROOT and move the cursor to a point to the left of x intercept. Press ENTER . Move the cursor to a point to the right of the x intercept and press ENTER again. Now press ENTER a final time to get the root.

The two intercepts are $x = -1.22$ and $x = 1.46$. Hence the solutions to the equation are -1.22 and 1.46.

NOTE: To check for extraneous solutions, store the value in the calculator as x-VAR, recall the function, and evaluate it. (See Appendix Section D-4 of this manual).

<u>METHOD 2</u> Create the two functions $f(x) = \sqrt[3]{x+2}$ and $g(x) = 4x^2 - 3\sqrt{x+4}$. Enter these into the function list as y1 and y2. Graph the functions and find the x coordinates of the intersection points. This can be done using trace and zoom, using the solver, or using the intersection point finder (see Appendix Section D-7 of this manual). The solver was used in the description below.

[GRAPH] [F1] <y(x)=> [CLEAR]
[(] [x-VAR]
[+] [2] [)] [^] [(] [1] [÷] [3] [)] [ENTER]
[CLEAR] [4] [x-VAR] [^] [2]
[-] [3] [2nd] [√] [(] [x-VAR] [+] [4] [)]
[EXIT] [F3] <ZOOM> [F4] <ZSTD>

Clear the existing functions and enter the new ones.

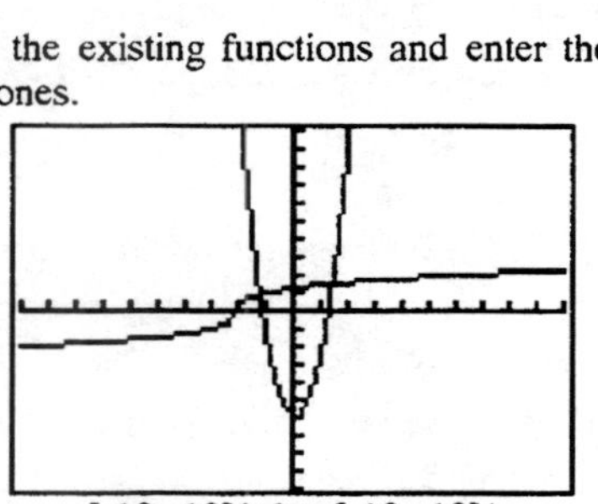

[-10, 10]1 by [-10, 10]1

[2nd] [SOLVER] [CLEAR] [F1] <y1>
[ALPHA] [=] [F2] <y2)> [ENTER] [2]
[F5] <SOLVE>

You'll need to enter a value of x near the x coordinate of one intersection point so the calculator knows which intersection point you are solving for. We chose 2. The result is $x = 1.46$, $y = 1.51$.

[(-)] [2] [F5] <SOLVE>

Repeat for the other intersection point. We used -2 as the approximate value of this x coordinate. The result is $x = -1.22$, $y = .92$.

So the solution to this equation is $x = -1.22$ and $x = 1.46$.

D-10 Inequalities in One Variable

An inequality in one variable can be solved graphically in two ways:

1. Algebraically move all terms to one side of the equation so that you have an expression equal to zero. Create a function $f(x)$ = (the expression). Graph the function.

   There are several ways to solve an equation in the form $f(x) = 0$ (or to find the x intercepts of the graph of $f(x)$). See Appendix Section D-7 of this manual.

   The set of x intercepts are the cutoff values for the solution. Now determine the intervals over which the inequality is true and write the solution.

2. Create two functions: $f(x)$ = (the left side of the equation) and $g(x)$ = (the right side of the equation). Graph both functions. Find the x coordinate of the points of intersection of the two graphs (see Appendix Section D-8 Methods 2 of this manual). The set of x coordinates are the cutoff values for the solution. Determine the intervals over which the inequality is true and write the solution.

EXAMPLE: Solve $\sqrt[3]{x + 2} < 4x^2 - 3\sqrt{x+4}$. Use two decimal places.

Solution:

<u>METHOD 1</u> Write the inequality as: $\sqrt[3]{x + 2} - (4x^2 - 3\sqrt{x+4}) < 0$. Change the sign to an equal sign: $\sqrt[3]{x + 2} - (4x^2 - 3\sqrt{x+4}) = 0$. Create the function $f(x) = \sqrt[3]{x + 2} - (4x^2 - 3\sqrt{x+4})$, graph, and find the x intercepts. This was done above in Appendix Section D-9 Method 1 of this

manual. The cutoff values are $x = -1.22$ and $x = 1.46$. Observing the graph, we see it is below the x axis for $x < -1.22$ and $x > 1.46$. Hence the solution to this inequality can be written as $(-\infty, -1.22) \cup (1.46, \infty)$.

METHOD 2 Create the two functions $f(x) = \sqrt[3]{x + 2}$ and $g(x) = 4x^2 - 3\sqrt{x+4}$. Graph these and find the x values of the intersection points. This was done above in Appendix Section D-9 Method 2 of this manual. The intersection points are (-1.22, .92) and (1.46, 1.51). Inspecting the graph we see that $f(x) < g(x)$ for $x < -1.22$ and $x > 1.46$. Hence the solution to this inequality can be written as $(-\infty, -1.22) \cup (1.46, \infty)$.

D-11 Permutations, Combinations, and Random Numbers

EXAMPLE 1: Find: (A) $P_{19,13}$ and (B) $C_{19,13}$.

Solution (A): This is a direct calculation.

Keystrokes	Description
19 2nd MATH F2 <PROB> F2 <nPr> 13 ENTER	Enter the numbers and get the built-in function from the MATH menu. The result is displayed in scientific notation as 1.68951528346E14.
EXIT EXIT	To return to the Home Screen.

Solution (B): This is a direct calculation.

Keystrokes	Description
19 2nd MATH F2 <PROB> F3 <nCr> 13 ENTER	Enter the numbers and get the built-in function from the MATH menu. The result is 27132.
EXIT EXIT	To return to the Home Screen.

EXAMPLE 2: Generate ten unique random numbers at least 0 and less than 50.

Solution:

Keystrokes	Description
5 STO▸ 2nd MATH F2 <PROB> F4 <rand> ENTER	We need to store a "seed" number in the storage location RAND. This tells the random number generator where to start.
EXIT F1 <NUM> F2 <iPart> (50 x EXIT F2 <PROB> F4 <rand>) ENTER ENTER	RAND generates a number greater than 0 and less than 1. If we multiply by 50 and take the integer part, the result is a number greater than or equal to 0 and less than 50.
ENTER ENTER ENTER ENTER ENTER ENTER ENTER ENTER ENTER	The 10 random numbers are: 36 13 6 17 21 ~~21~~ 20 5 31 16 48 Note that repetitions can occur but 21 cannot be used twice if unique random numbers are desired.

D-12 Matrices

Given the matrices $A = \begin{bmatrix} 1 & -4 & 6 \\ 2 & -1 & 0 \\ 5 & -3 & 6 \end{bmatrix}$ and $B = \begin{bmatrix} 4 & 15 \\ -2 & 6 \\ 2 & 5 \end{bmatrix}$.

EXAMPLE 1: Find -2.5AB.

Solution:

Keystrokes	Explanation
[2nd] [MATRX] [F2] <EDIT> [A] [ENTER] [3] [ENTER] [3] [ENTER]	Choose matrix A and enter the matrix dimensions.
[1] [ENTER] [(-)] [4] [ENTER] [6] [ENTER] [2] [ENTER] [(-)] [1] [ENTER] [0] [ENTER] [5] [ENTER] [(-)] [3] [ENTER] [6] [EXIT]	Enter the matrix elements row by row.
[2nd] [MATRX] [F2] <EDIT> [B] [ENTER] [3] [ENTER] [2] [ENTER] [4] [ENTER] [15] [ENTER]	Choose matrix B and enter the matrix dimensions.
[(-)] [2] [ENTER] [6] [ENTER] [2] [ENTER] [5]	Enter the matrix elements.
[EXIT]	Return to the Home Screen.
[(-)] [2.5] [2nd] [MATRX] [F1] <NAMES> [F1] <A> [×] [F2] <B> [ENTER]	Enter the matrix operations. The result is $\begin{bmatrix} -60 & -52.5 \\ -25 & -60 \\ -95 & -217.5 \end{bmatrix}$.

EXAMPLE 2: Find A^{-1}.

Solution:

Keystrokes	Explanation
[2nd] [MATRX] [F1] <NAMES> [F1] <A> [2nd] [x^{-1}] [ENTER]	The operations are done from the Home Screen. The result rounded to three decimal places is: $A = \begin{bmatrix} -.167 & .167 & .167 \\ -.333 & -.667 & .333 \\ -.028 & -.472 & .194 \end{bmatrix}$.

When performing operations on matrices, some numbers may look like decimals when in fact they are zero. This happens because the calculator rounds numbers to a certain number of decimal places. Examining the numbers using the arrow keys you will find they are decimals that are very close to zero. This situation can be eliminated by setting the number of decimal places to a specified number, say 3 (see Appendix Section D-17 of this manual).

EXAMPLE 3: Solve the system of equations using the inverse of the coefficient matrix. Round answers to two decimal places.

$$x - 4y + 6z = 2.5$$
$$2x - y = 3.2$$
$$5x - 3y + 6z = -3.4$$

Solution: The coefficient matrix was entered in Appendix Section D-12 Example 2 of this manual. Enter the constants as matrix B and find $A^{-1}B$.

[2nd] [MATRX]

[F2] <EDIT> [B] [ENTER]

[3] [ENTER] [1] [ENTER] [2.5] [ENTER]

[3.2] [ENTER] [(-)] [3.4] [EXIT]

Enter the matrix dimensions and enter the matrix elements.

[2nd] [MATRX] [F1] <NAMES>

[F1] <A> [2nd] [x^{-1}] [×] [F2] <B> [ENTER]

Return to the Home Screen.
Perform the matrix operations.
The result, rounded to two decmal places, is $\begin{bmatrix} -.45 \\ -4.10 \\ -2.24 \end{bmatrix}$. Hence, the solution to the system of equations is $x = -.45$, $y = -4.10$, $z = -2.24$, rounded to two decimal places.

EXAMPLE 4: Find the solution to the system of Example 3 above using Gauss-Jordan reduction. In other words, find the reduced form of the augmented coefficient matrix $\begin{bmatrix} 1 & -4 & 6 & 2.5 \\ 2 & -1 & 0 & 3.2 \\ 5 & -3 & 6 & -3.4 \end{bmatrix}$. Use two decimal places.

Solution:

<u>METHOD 1</u> Use row operations on the matrix to get the reduced form.

[2nd] [MATRX] [F2] <EDIT>

[A] [ENTER] [3] [ENTER] [4] [ENTER]

Enter the matrix dimensions.

[1] [ENTER] [(-)] [4] [ENTER] [6]

[ENTER] [2.5] [ENTER] [2] [ENTER]

[(-)] [1] [ENTER] [0] [ENTER] [3.2]

[ENTER] [5] [ENTER] [(-)] [3] [ENTER]

[6] [ENTER] [(-)] [3.4] [EXIT] [EXIT]

Enter the matrix elements. Notice that if you are continuing from Example 1 - 3 of this section, you need only enter the elements in column 4. Use the down arrow to move from one row to the next.

2nd MATRX F4 <OPS>
MORE F5 <mRAdd>
(-) 2 , EXIT F1 <NAMES>
F1 <A> , 1 , 2) ENTER

We wish to multiply by -2, matrix A, row 1 and add it to row 2. The result will be put in row 2.

`mRadd(-2,A,1,2)`

The result is $\begin{bmatrix} 1 & -4 & 6 & 2.5 \\ 0 & 7 & -12 & -1.8 \\ 5 & -3 & 6 & -3.4 \end{bmatrix}$.

STO▸ F1 <A>

Store the result in matrix A.

ENTER 2nd MATRX
F4 <OPS> MORE F5 <mRAdd>
(-) 5 , EXIT F1 <NAMES>
F1 <A> , 1 , 3) ENTER

Now multiply row 1 by -5 and add it to row 3 putting the result in row 3.

`mRadd(-5,A,1,3)`

The result is $\begin{bmatrix} 1 & -4 & 6 & 2.5 \\ 0 & 7 & -12 & -1.8 \\ 0 & 17 & -24 & -15.9 \end{bmatrix}$.

STO▸ A ENTER

Store the result in matrix A.

2nd MATRX F4 <OPS> MORE
F4 <multR> 1 ÷ 7 , ALPHA A
, 2) ENTER

Multiply row 2 by 1/7 .

The result is $\begin{bmatrix} 1 & -4 & 6 & 2.5 \\ 0 & 1 & -1.71 & -.26 \\ 0 & 17 & -24 & -15.9 \end{bmatrix}$

rounded to two decimal places.

STO▸ A

Store the result in matrix A.

Continue in this manner until you get

$$\begin{bmatrix} 1 & 0 & 0 & -.45 & \ldots \\ 0 & 1 & 0 & -4.1 & \ldots \\ 0 & 0 & 1 & -2.241666666 & \ldots \end{bmatrix}.$$

The solution to the system of equations, rounded to two decimal places, is

$x = -.45, y = -4.10; z = -2.24$.

> Note: To swap rows of a matrix use F4 <OPS> MORE F2 <rSwap> from the MATRX menu. The command rSwap(A,2,3) will swap rows 2 and 3 in matrix A.

METHOD 2 Use the built-in capabilities of the calculator to get the reduced form.

2nd MODE ▼ ▶ ▶ ▶ ENTER

Set the mode to two decimal places.

2nd MATRX F2 <EDIT>

Enter the matrix dimensions.

A ENTER 3 ENTER 4 ENTER
1 ENTER (-) 4 ENTER 6
ENTER 2.5 ENTER 2 ENTER

Enter the matrix elements. Notice that if you are continuing from Example 1 - 3 of this section, you need only enter the elements in column 4. Use the down arrow to move from one row to the next.

[(-)] [1] [ENTER] [0] [ENTER] [3.2] [ENTER]
[5] [ENTER] [(-)] [3] [ENTER]
[6] [ENTER] [(-)] [3.4] [EXIT]
[2nd] [MATRX] [F4] <OPS> [F5] <rref>
[EXIT] [F1] <NAMES>
[F1] <A> [ENTER]

We wish to get the reduced form of this matrix. Use the matrix operation rref to get this on the calculator.

The result is $\begin{bmatrix} 1.00 & 0.00 & 0.00 & -.45 \\ 0.00 & 1.00 & 0.00 & -4.10 \\ 0.00 & 0.00 & 1.00 & -2.24 \end{bmatrix}$.

Use the right and left arrow keys to view the entire matrix.

Hence, the solution to the system of equations, rounded to two decimal places, is $x = -.45$, $y = -4.10$, $z = -2.24$.

D-13 Logarithmic, Exponential and Hyperbolic Functions

EXAMPLE 1: Graph $f(x) = 10^{3.2x}$ for $-.5 < x < .5$.

Solution: Set the window variables to [-.5,.5].1 by [-10,10]1.

[GRAPH] [F1] <y(x)=> [CLEAR] — Clear all functions from the function list.
[10] [^] [(] [3.2] [x-VAR] [)] [EXIT]
[F2] <WIND> [(-)] [.5] [ENTER]
[.5] [ENTER] [.1] [ENTER] [(-)] [10] [ENTER]
[10] [ENTER]
[1] [F5] <GRAPH>

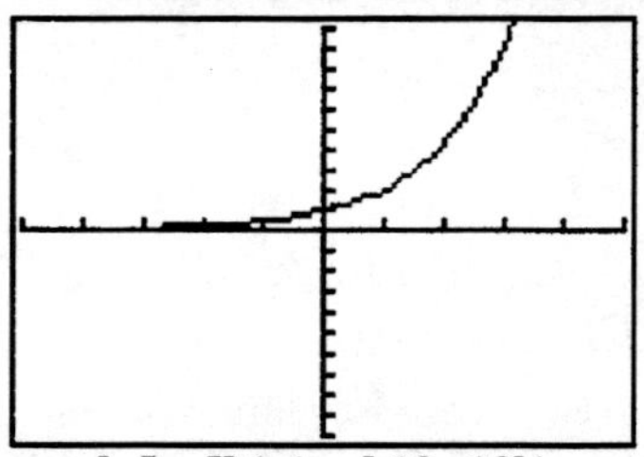

[-.5, .5].1 by [-10, 10]1

EXAMPLE 2: Graph $y = \dfrac{e^{x} + e^{-x}}{2}$.

Solution: This could be entered using [(] [2nd] [e^x] [x-VAR] [+] [2nd] [e^x] [(-)] [x-VAR] [)] [÷] [2]. However the built-in hyperbolic cosine function can also be used.

GRAPH F1 <y(x)=> CLEAR
2nd MATH F4 <HYP> F2 <cosh>
x-VAR EXIT EXIT F3 <ZOOM>
F4 <ZSTD>

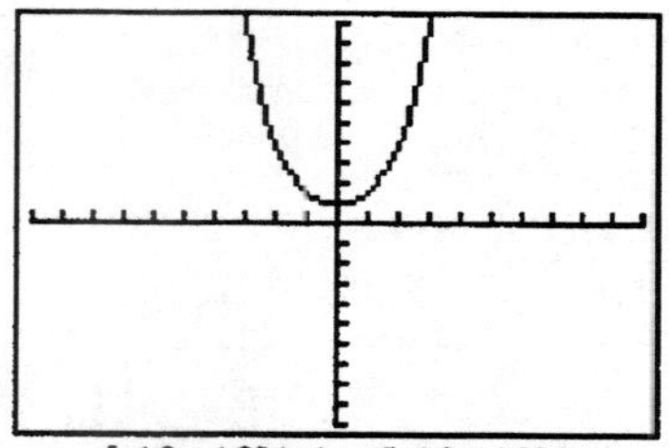

[-10, 10]1 by [-10, 10]1

EXAMPLE 3: Solve $\ln(x^2 + x) = 3x + 2$

Solution:

GRAPH F1 <y(x)=> CLEAR
LN (x-VAR ^ 2 + x-VAR

Enter the function $f(x) = \ln(x + x^2)$ as y1 in the function list.

) ENTER CLEAR 3
x-VAR + 2 EXIT

Enter the function $g(x) = 3x + 2$ as y2 in the function list.

F3 <ZOOM>
F4 <ZSTD>

Graph them on the same graphing screen. Find the x value of the point of intersection (see Appendix Section D-7 Method 6 of this manual). This will be the solution to the equation.

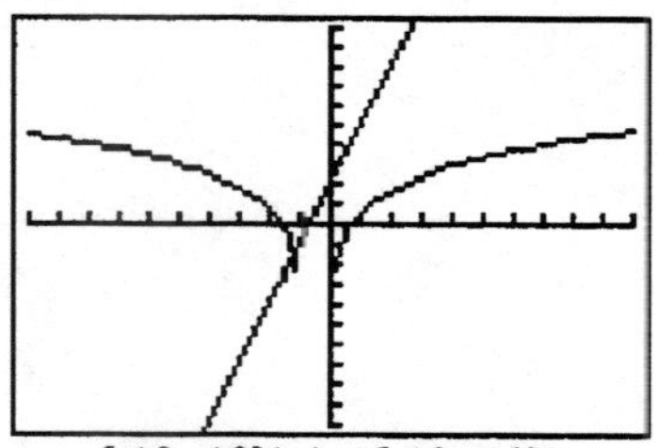

[-10, 10]1 by [-10, 10]1

The solution is -1.18, rounded to two decimal places.

D-14 Trigonometric Functions

EXAMPLE 1: Evaluate $f(x) = \cos x + \csc x$ at (A) 53°45'18" and (B) 78.52 radians.

Solution: (A)

GRAPH F1 <y(x)=> CLEAR
COS x-VAR + 1 ÷ SIN
x-VAR EXIT EXIT

Clear or deselect all existing functions. Store the function as y1.

2nd MODE ▼ ▼ ▶ ENTER

Change the calculator to degree angle measure.

Keystrokes	Explanation
CLEAR 53 + 45 ÷ 60 + 18 ÷ 3600 STO▸ x-VAR	Change the angle to decimal degrees while storing it as the variable x. Store 53.755° as x.
ENTER	
2nd ALPHA y 1 ENTER	Evaluate the function. The result to two decimal places is 1.83.

Solution: (B)

Keystrokes	Explanation
2nd MODE ▼ ▼ ENTER	Change the calculator to radian angle measure.
2nd QUIT	Since the function was entered in part (A) above, we need only to store the angle measure and evaluate the stored function.
78.52 STO▸	Store 78.52 as x.
x-VAR ENTER	Recall the function. The result to two decimal places is 49.47.
2nd ALPHA y 1 ENTER	

EXAMPLE 2: Graph $f(x) = \cos x + \csc x$.

Solution: The function was entered in the calculator in Example 1 above.

Keystrokes	Explanation
GRAPH ZOOM MORE	Change the graph mode to Dot. (See Appendix Section D-6 of this manual.) This will eliminate the "vertical lines."
F3 <ZTRIG>	This option automatically sets the window variables to [-8.24,8.24]1.57 by [-4,4]1.

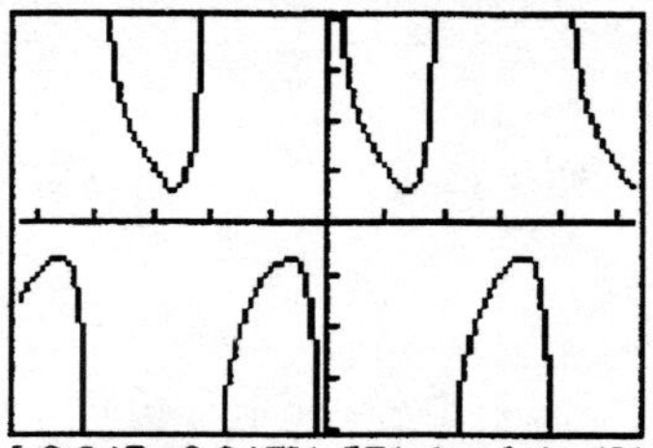

[-8.247, 8.247]1.571 by [-4, 4]1

D-15 Numeric Differentiation and Integration, Tangent Lines, and Differential Equations

EXAMPLE 1: (A) Evaluate $f(x) = \sqrt{x^5 - \sqrt[3]{x}}$ at $x = 3.54$. (B) Find the equation of the tangent line to $f(x)$ at 3.54. Use two decimal place accuracy. (C) Sketch the function and its tangent line using [2, 5]1 by [0, 60]5.

Solution (A): The function and it's derivative can be evaluated at this value of x .

Keystrokes	Comment
GRAPH F1 <y(x)=> CLEAR	Clear or deselect all existing functions.
2nd √ (x-VAR ^ 5 -	Enter the function and store it as y1.
x-VAR ^ (1 ÷ 3))	
2nd QUIT	Return to the Home Screen and store the value of x.
3.54 STO▸ x-VAR ENTER	
2nd ALPHA	Recall the function and press ENTER to evaluate the function. $f(3.54) = 23.55$.
Y 1 ENTER	

Soltuion (B):

Keystrokes	Comment
2nd CALC F3 <derl> 2nd alpha	Get the numeric derivative built-in function: `derl(y1,x,3.54)`
Y 1 , x-VAR , 3.54)	This will evaluate the derivative with respect to x of the function stored in y1 at x= 3.54.
ENTER	The result is 16.67. (The default step size used in x increments is .001.)

The equation of the tangent line is $y - 23.55 = 16.67(x - 3.54)$.

Solution (C):

<u>Method 1</u> Enter the function as y1 in the calculator and the equation of the tangent line as y2 in the calculator and graph.

<u>Method 2</u> Enter the function as y1 in the calculator.
Set the viewing rectangle to [2, 5]1 by [10, 60]5.

Use the tangent drawing feature of the calculator by entering

GRAPH MORE F2 <DRAW> MORE

MORE MORE F2 <TanLn>

2nd ALPHA Y 1 , 3.54) ENTER .

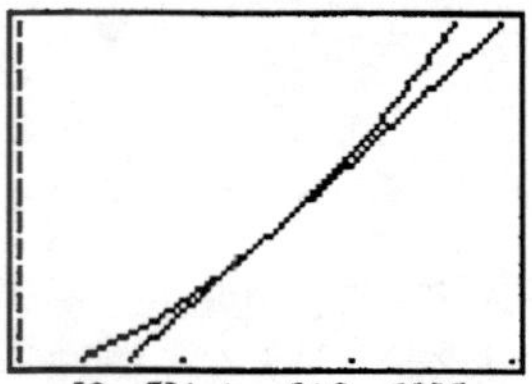

[2, 5]1 by [10, 60]5

(It may take a little while to graph.) Press CLEAR to remove the menu at the bottom of the screen. Press EXIT to display it again. The function and its tangent will be graphed as shown.

EXAMPLE 2: Is the function $f(x) = \sqrt{x^5 - \sqrt[3]{x}}$ concave upwards at $x = 3.54$?

Solution:

Enter the function and store it as y1 as shown in Example 1 of this section.

[2nd] [CALC] [F4] <der2> [2nd] [alpha] [Y] [1] [,] [x-VAR] [,] [3.54] [)] [ENTER]

Use the second derivative function.

```
der2(y1,x,3.54)
```

The result is positive. Hence the function is concave upwards at $x = 3.54$.

EXAMPLE 3: Find $\int_{1.32}^{2.58} \sqrt{x^5 - \sqrt[3]{x}}\, dx$. Use two decimal place accuracy.

Solution:

[GRAPH] [F1] <y(x)=> [CLEAR]

Clear or deselect existing functions. Enter the function and store it.

[2nd] [√] [(] [x-VAR] [^] [5]
[–] [x-VAR] [^] [(] [1] [÷] [3]
[)] [)] [2nd] [QUIT]

[2nd] [CALC] [F5] <fnInt> [2nd] [alpha]
[y] [1] [,] [x-VAR] [,] [1.32] [,] [2.58] [)]
[ENTER]

Get the numeric integration function built into the calculator function:

```
fnInt(y1,x,1.32,2.58)
```

This will evaluate the definite integral. The result is 6.95. The default tolerance in calculating the definite integral is .00001.

This can be changed by pressing

[2nd] [MEM] [F4] <TOL>

EXAMPLE 4: Given the first-order differential equation $y' = 7y + 7x - 3$ with $y(0) = 1$.

(A) Graph the approximate solution and the slope field for $0 \le x \le 1$ using Euler's method with a step size of $\Delta x = .1$ using [0, 1].1 by [0, 3].1. Use trace and fill in the table below.

(B) Graph the approximate solution and the slope field for $0 \le x \le 1$ using Runge-Kutta method with a step size of $\Delta x = .1$ on the same set of coordinate axes. Use trace and fill in the table below.

(C) The exact solution of this equation is $y = 5e^x - 7x - 4$. Fill in the table below, using three-decimal place accuracy, to compare the numerical solutions and the exact solution.

(D) Graph the exact solution on the same set of coordinate axes.

x	Euler's Method	Runge-Kutta Method	Exact Solution
0			
0.1			
0.2			
0.3			
0.4			
0.5			
0.6			
0.7			
0.8			
0.9			
1.0			

Solution (A) and (C):

Keystrokes	Explanation
2nd MODE ▼ ▼ ▼ ▼ ▶ ▶ ▶	Set the calculator in DifEq mode.
ENTER 2nd QUIT	Return to the Home Screen.
GRAPH MORE F1 <FORMT> ▼ ▼ ▼ ▼ ▶ ENTER ▼ ENTER	Set the calculator in Euler mode with the slope field turned on
2nd QUIT	Return to the Home Screen.
	Turn all the statistical plots off by pressing: 2nd STAT F3 <PLOT> F5 <PlOff>
GRAPH	Q is used in place of y , and t is used in place of x in the differential equations mode. Change y to Q and x to t in the differential equation. We now have: Q'(t)=Q+7t-3. Up to nine differential equations can be entered in the list. We will use Q1.
F1 <Q'(t)=> F2 <Q> 1 + 7 F1 <t> - 3 ENTER EXIT	Enter the differential equation into the equation editor as Q1+7t-3.
F2 <WIND> 0 ENTER 1 ENTER .1 ENTER 0 ENTER	Set the step parameters and the window variables. $0 \le t \le 1$ with a step size of .1. This means tMax = 1 and tStep = .1.
0 ENTER 1 ENTER .1 ENTER 0 ENTER 3 ENTER .5 ENTER	The window variables can be set at [0, 1].1 by [0, 3].5.
1 ENTER	Set the EStep to 1.
F3 <INITC> 1 ENTER	Set the initial conditions at tMin=0, tMax=1, and tStep=.1.

[F5] <GRAPH>

Get the graph.

[MORE] [F4] <TRACE> [▶] ... [▶]

Use the arrow keys and read the numbers of the bottom of the screen. Record them in the table in Part (C).

[GRAPH] [MORE] [F1] <FORMT>
[▼] [▼] [▼] [▼] [▼] [▶] [▶]
[ENTER] [F5] <GRAPH>

The graph shown below does not have the slope field displayed. To turn the slope field display off, select [f1] <FORMT> on the differential equation graph menu, arrow down to the bottom line, arrow to the right to select FldOff, and press enter.

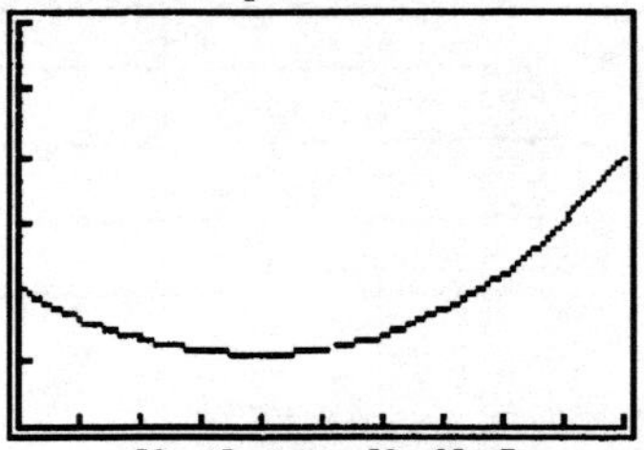

[0, 1] .1 by [0, 3] .5

Solution (B) and (C):

[GRAPH] [MORE] [F1] <FORMT>
[▼] [▼] [▼] [▼] [ENTER]
[F5] <GRAPH>

This is the same as Part (A) but with the calculator to RK method with the slope field turned off. All we need to do is to change from the Euler mode to the RK mode. Repeat tracing the graph and record the results in the table.
You cannot display two graphical solutions at one time on the calculator.

To compare graphs, plot by hand on a piece of paper.

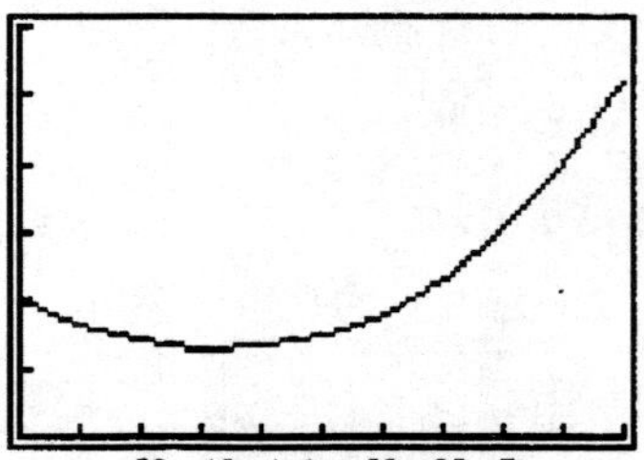

[0, 1] .1 by [0, 3] .5

Solution (C):

Keystrokes	Explanation
[2nd] [QUIT] [2nd] [MODE]	Change the back to function mode.
[▼] [▼] [▼] [▼] [ENTER]	Store the exact solution in the calculator and evaluate it for the values of x in the table.
[GRAPH] [F1] <y(x)> etc.	See Section D-4 Example 3 Solution (B).

To compare the three graph, plot by hand on a piece of paper. They should look like:

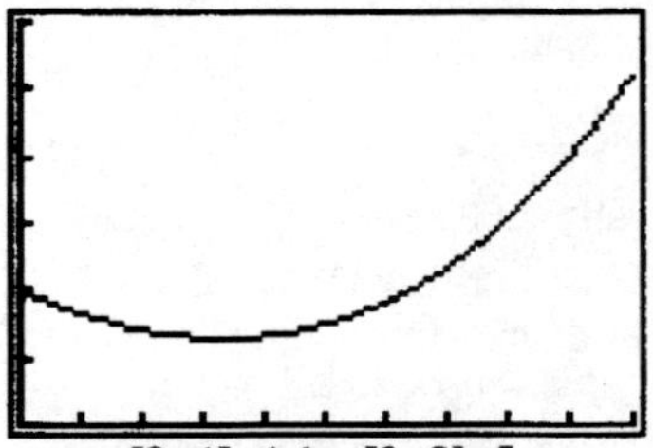

Runge-Kutta

[0, 1] .1 by [0, 3] .5

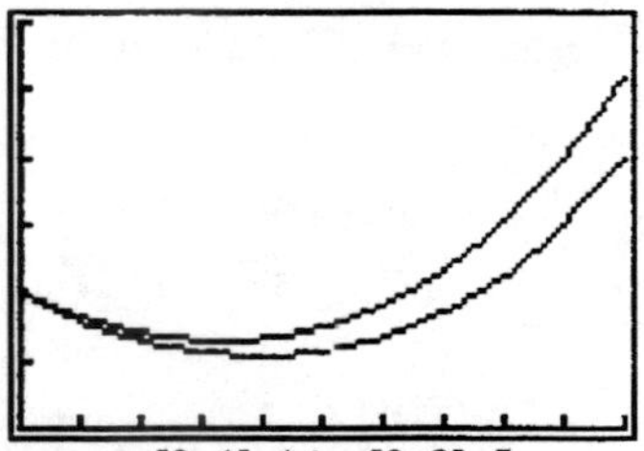

Exact

Euler

[0, 1] .1 by [0, 3] .5

Notice that the exact solution and the approximate solution found by the Runge-Kutta method are almost identical. The Runge-Kutta yields a more accurate approximation, but is not as fast to perform on the calculator.

x	Euler's Method	Runge-Kutta Method	Exact Solution
0	1	1	1
0.1	.800	.826	.826
0.2	.650	.707	.707
0.3	.555	.649	.649
0.4	.521	.659	.659
0.5	.553	.743	.744
0.6	.657	.910	.911
0.7	.844	1.168	1.169
0.8	1.118	1.527	1.528
0.9	1.490	1.997	1.998
1.0	1.969	2.590	2.591

D-16 Statistics

EXAMPLE 1: Given the following data, find: (A) the mean, (B) the sample standard deviation, (C) a histogram with first class beginning at 10 and width 4, and (D) repeat Parts A and B after deleting the data value 89 from the list. Use two decimal place accuracy. How did eliminating this outlier effect the mean and standard deviation?

23	56	73	12	89	52	45
38	18	21	15	37	42	51
19	40	57	38	54	60	39

Solution (A) and (B):

Keystrokes	Explanation
[2nd] [QUIT]	Return to the Home Screen.
[2nd] [STAT] [F2] <EDIT> [▲] [CLEAR] [▼]	Get the statistics menu. Clear the statistical registers by placing the cursor on the name of the list at the top of the screen by using the up arrow key and pressing [CLEAR]. Return to the list using the down arrow key and enter the data as *xStat*.
[23] [ENTER] [56] [ENTER] ... [39] [ENTER] Use the down arrow key [▼] or press [ENTER] to move down one line in the list.	If you make a mistake when entering data, use the arrow keys to highlight the value and type it in again.
[EXIT] to exit the edit mode. [2nd] [STAT] [F1] <CALC> [F1] <OneVa> [2nd] [ALPHA] [x] [ALPHA] [S] [2nd] [ALPHA] [T] [2nd] [ALPHA] [A] [2nd] [ALPHA] [T] [ENTER]	To remove a data value, use the arrow keys to get the data value and press [DEL]. Get the calculate menu and choose OneVa, one-variable, calculations. Enter the name of the list as xStat. This can be found on the CATLG-VARS menu or you can type it in using the [ALPHA] key and the letters.
	The mean is 41.86 and the sample standard deviation is 19.82.

Solution (C):

Keystrokes	Explanation
[2nd] [QUIT]	Return to the Home Screen.
[2nd] [STAT] [F3] <PLOT> [F1] <PLOT1> [ENTER] [▼] [F4] <HIST> [▼] [F1] <xSTAT> [EXIT] [EXIT]	Set up the statistical plot to plot a histogram.
[GRAPH] [F1] <y(x)=>	Clear or deselect functions stored in the function list.
[F5] <SELCT> [EXIT]	This will deselect the function the cursor is on. The = sign will no longer be highlighted. Deselect all functions stored or delete them using [CLEAR].
[MORE] [F2] <DRAW> [MORE] [MORE] [F1] <CLDRW> [EXIT]	Clear the drawings.

Keystrokes	Explanation
MORE MORE F2 <WIND> 0 ENTER 100 ENTER 4 ENTER (-) 5 ENTER 5 ENTER 1	Set the window variables. The smallest value is 12 so let xMin=0. The greatest value is 89. Let us set xMax=100. Let's use xScl=4. This sets the histogram class width. Let yMin=-5 to get some room below the histogram. We do not know the frequencies yet so let yMax=5 for a first try. Let yScl=1.
F5 <GRAPH>	Draw the histogram.

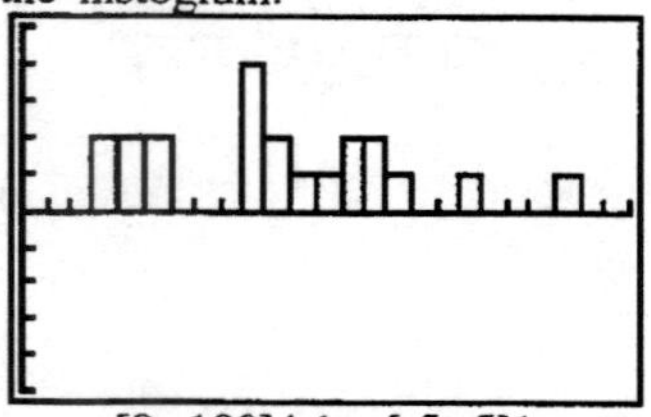

[0, 100]4 by [-5, 5]1

Solution (D):

Keystrokes	Explanation
2nd QUIT	Return to the Home Screen.
2nd STAT F2 <EDIT> ▼ ... ▼	Get the data list and arrow down to the 89.
DEL EXIT	Delete the item.
2nd STAT F1 <CALC> F1 <OneVa>	Get the statistics menu and the one variable sample statistics.
2nd ALPHA x ALPHA S 2nd ALPHA T 2nd ALPHA A 2nd ALPHA T ENTER	The mean is now 39.5 and the sample standard deviation is 17.05. Notice how much influence an outlier has!

EXAMPLE 2: Given the following bivariate data: (A) find the correlation coefficient, (B) find the least squares line, (C) graph the least squares line, (D) plot the scatter diagram, and (E) sort the data on x and join the points using straight line segments.

x	5	8	10	29	9	24
y	3	4	8	20	3	19

Solution (A) & (B):

Keystrokes	Explanation
2nd QUIT	Return to the Home Screen.
2nd STAT F2 <EDIT>	Get the statistics menu.
▲ CLEAR ▼ 5 ENTER 8 ENTER ...	Clear the statistical register for xStat and enter the data.

Keystrokes	Explanation
[24] [ENTER]	Clear the statistical register for yStat and enter the data. Move to the fStat list.
[▶]	Move to the yStat list.
[▲] [CLEAR] [▼] [3] [ENTER] [4] [ENTER]... [19] [ENTER]	Clear the statistical register for fStat and enter 1's for the frequency of the pairs of data (xStat, yStat).
[▶] [▲] [CLEAR] [▼] [1] [ENTER] [1] [ENTER]... [1] [ENTER] [EXIT]	
[2nd] [STAT] [F1] <CALC> [F3] <LinR> [ENTER]	Linear regression correlation coefficient: corr = .98. a=-1.72 and b=.79, so the least squares line is $\hat{y}$ = .79x - 1.72.

Solution (C):

Keystrokes	Explanation
[GRAPH] [F1] <y(x)=> [CLEAR]	Clear or deselect all functions stored in the function list.
[.79] [x-VAR] [-] [1.72]	Set the cursor at $y1$ and enter the equation of the regression line.
[▲] [ENTER]... [▼] [EXIT]	If any of the plots are highlighted at the top of the screen, move the cursor up using the arrow keys and press enter to deselect them (to turn them off).
[F2] <WIND> [0] [ENTER] [30] [ENTER] [1] [ENTER] [(-)] [10] [ENTER] [30] [ENTER] [1] [F5] <GRAPH>	Set the window variables. Observe the data to determine what the dimensions of the viewing window should be.

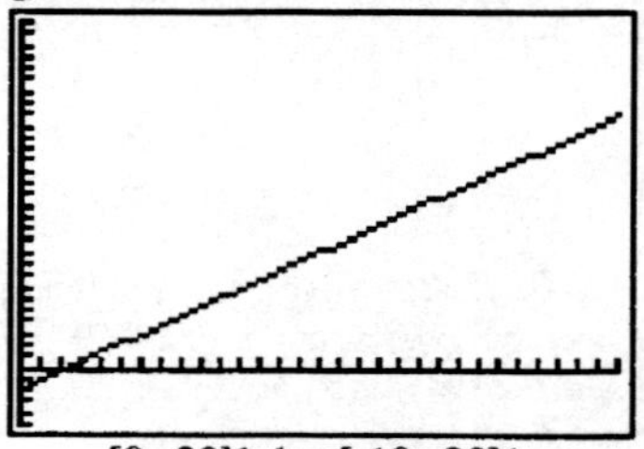

[0, 30]1 by [-10, 30]1

Solution (D):

2nd QUIT

Return to the Home Screen.

2nd STAT

Set up the statistical plot for a scatter diagram.

F3 <PLOT> F1 <PLOT1>

ENTER ▼ F1 <SCAT> ▼ F1 <xStat>

▼ F2 <yStat> ▼ F1 <□>

EXIT EXIT

GRAPH F1 <y(x)=> F5 <SELCT> EXIT

Clear or deselect functions stored in the function list except the equation of the regression line.

F2 <WIND> 0

Set the window variables. Observe the data to determine the dimensions of the viewing window should be. Set the y value less than the minimum value so if you trace, the coordinates at the bottom of the screen do not cover up the graph.

ENTER 30 ENTER 1

(-) 10 ENTER 30 ENTER 1

F5 <GRAPH>

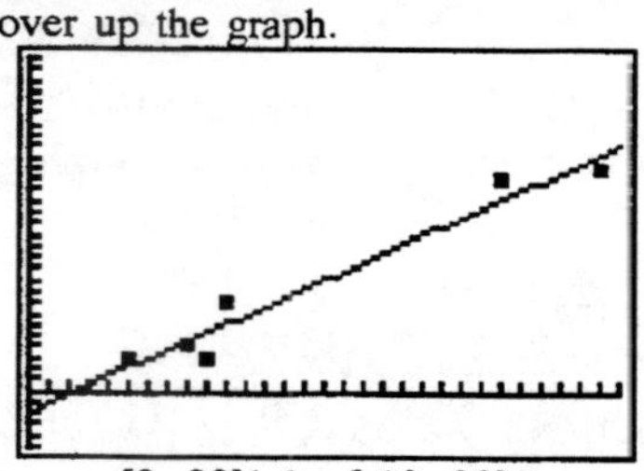

[0, 30]1 by [-10, 30]1

Solution (E):

2nd QUIT

Return to the Home Screen.

2nd QUIT 2nd LIST F5 <OPS>

Sort the data on x variable.

MORE MORE F5 <Sortx> ENTER

EXIT EXIT

GRAPH F1 <y(x)=> CLEAR

Clear or deselect all functions in the function list.

EXIT EXIT

2nd STAT F4 <DRAW>

Draw the line that joins the points in the order they are listed in the xStat list.

F3 <xyLINE>

The points will be joined in the order of the x data in the data list.

CLEAR

Remove the menu bars.

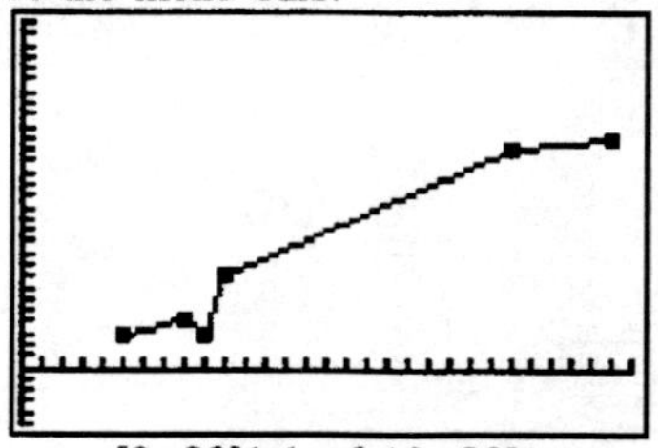

[0, 30]1 by [-10, 30]1

If the Plot is turned on, the points will also be plotted (see Part (D) above).

D-17 Setting the Display Mode for Numbers

Numbers can be entered into the calculator in scientific notation, a set number of significant digits, or a fixed number of decimal places.

EXAMPLE 1: Calculate $(9.35417\times 10^4)(-3.64389\times 10^{-6})$ using:
(A) scientific notation, (B) two significant digits, and (C) Four decimal place accuracy.

Solution (A):

2nd MODE ▶ ENTER

2nd QUIT 9.35417 EE 4 x

(-) 3.64389 EE (-) 6 ENTER

The calculator is set to scientific mode with floating decimals. The result displayed on the calculator is -3.40855665213E-1.

Solution (B):

2nd MODE ▶ ENTER

▼ ▶ ▶ ENTER

Set the calculator to scientific notation with 1 digit instead of floating point.

2nd QUIT

2nd ENTRY ENTER

Return to the Home Screen.

Use the replay key to replay the calculation just done in Solution (A). The result to two significant digits is -3.4E-1.

Solution (C):

Keystrokes	Explanation
[2nd] [MODE] [ENTER] [▼] [►] [►] [►] [►] [►] [ENTER]	Set the calculator to normal mode with 4 digits.
[2nd] [QUIT]	Return to the Home Screen.
[2nd] [ENTRY] [ENTER]	Use the replay key to replay the last calculation. The result to four decimal places is -.3409.

Change the calculator back to floating point. The examples in this manual are done using this setting unless otherwise noted.

NOTES

NOTES

Answers to Selected Problems

Chapter 1 A Beginning Library of Elementary Functions

1. The domain is $\{x \mid x \le -1, x \ge 1\}$.

3. -5.25, -5.09, -5.0099, -5.000999. The value of $Q(x)$ gets close to -5 as x gets close to 2.

5.

h	.2	.1	.01	.001	→	0
$\frac{f(1.5+h) - f(1.5)}{h}$	.9	.8	.71	.701	→	.7

7.

x	$p(x)$	$p(x)$-p
2	1118	8
5	1010	20
10	830	15
14	686	9
21	434	1

9. (A)

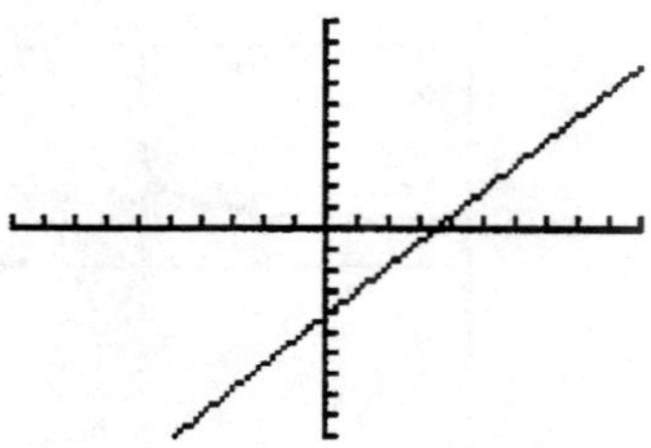

(B) & (C) x intercept is 3.5, y intercept is -4.2.

(D) $\{x \mid x > 3.5\}$

(E) The slope of the line is greater but the y intercept remains the same. Hence the x intercept of $g(x) = 2.2x - 4.2$ would be less than that of $f(x) = 1.2x - 4.2$. In fact, the x intercept of $g(x)$ is 1.9.

11. Yes.

13. Point of discontinuity at $x = 0$

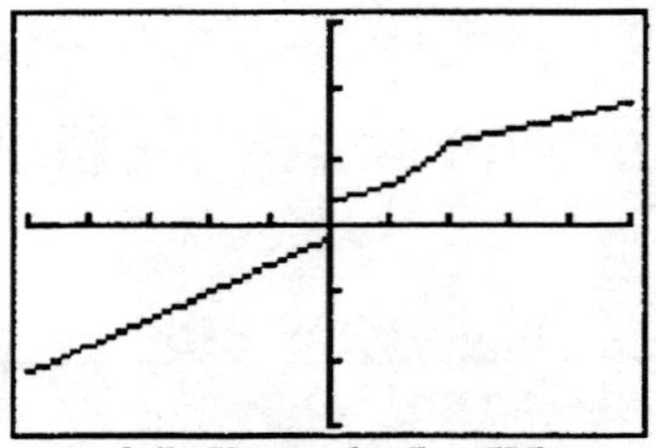

[-5, 5]1 by [-15, 15]5

15. The two lines are almost parallel. Compare the slopes 5.23/8.92 and 5.21/8.89.

17. The second graph is stretched vertically. All y values of the second graph are three times those of the first graph.
    The third graph is stretched vertically by a factor of 1.5 and reflected about the x axis. All y values of the first graph are multiplied by -1.5 to obtain the third graph.

Chapter 2 Additional Elementary Functions

1. -.84, 1.26, 11.29

3. $E(x) = .02x^3 - .10x^2 + 3x + 26.80$; $E(35) = \$1072.8$ billion

5. The vertical asymptote is $x = 1$.
   The horizontal asymptote is $y = 2$.

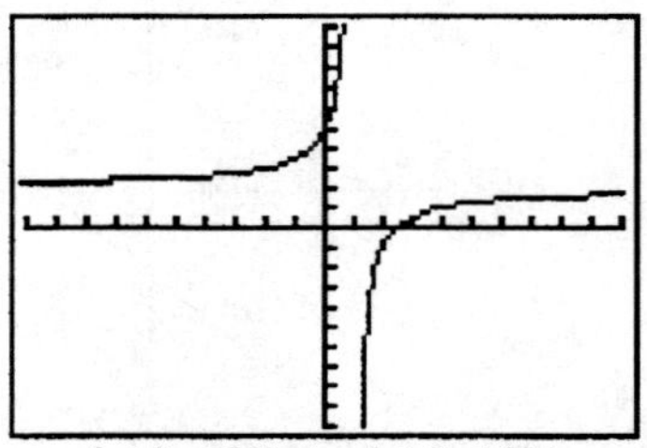

[-10, 10]1 by [-10, 10]1

7.

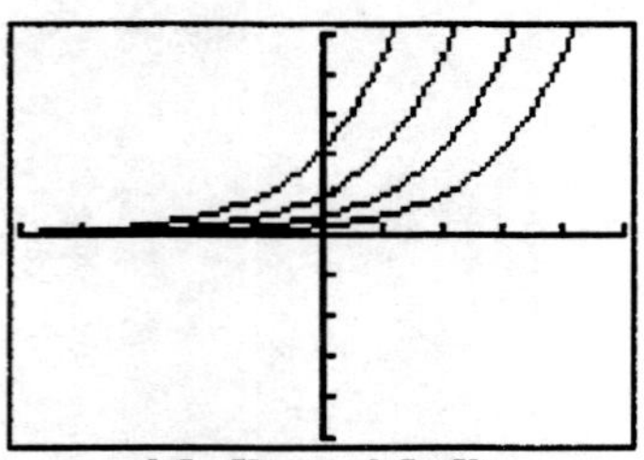

[-5, 5]1 by [-5, 5]1

Adding a constant to the x shifts the graph to the left.
Subtracting a constant to the x shifts the graph to the right.

9. 22.52825

11.

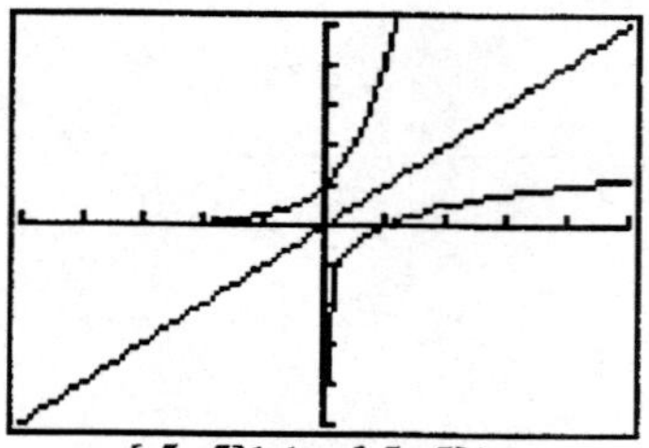

[-5, 5]1 by [-5, 5]1

Use the change-of-base formula to graph the inverse function.

Inverse of $y = 4^x$ is $y = \log_4 x = \dfrac{\ln(x)}{\ln(4)}$.

Chapter 3 Mathematics of Finance

1. 2, 175.70, 4351.41, 10878.52

3.

Quarter Number	Principal at Start of Quarter	Amount at End of Quarter	Interest Earned During the Quarter
1	6925.00	7037.53	112.53
2	7037.53	7151.89	114.36
3	7151.89	7268.11	116.22
4	7268.11	7386.22	118.11
5	7386.22	7506.24	120.03
6	7506.24	7628.22	121.98

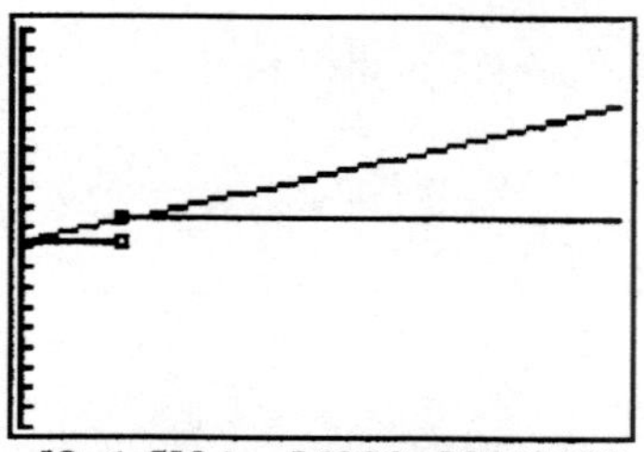

[0, 1.5]0 by [6000, 8000]100

5. S = \$1256.85, A = \$1318.52; Total interest earned = \$75.05.

7.

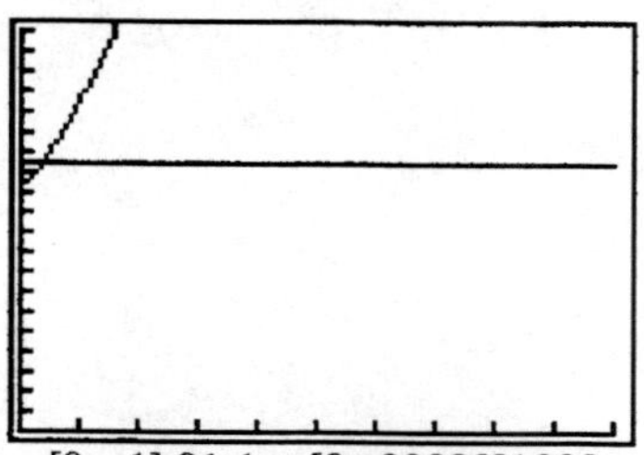

[0, .1].01 by [0, 20000]1000

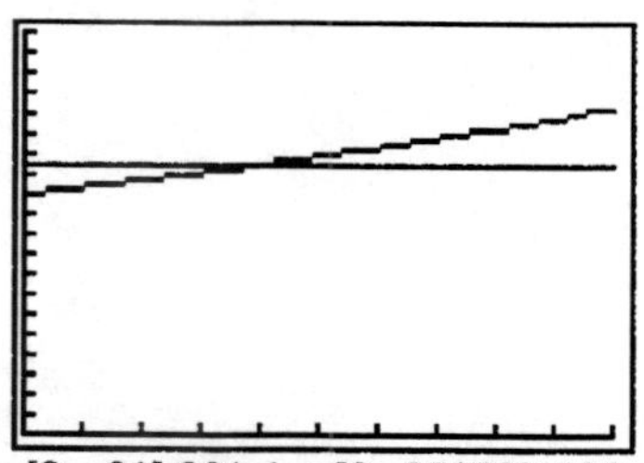

[0, .01].001 by [0, 20000]1000

$i = .004$, $r = 12i = .048 = 4.8\%$.

9. The payment is \$307.38. The total interest paid is \$188.56.

Period Number	Payment	Interest	Balance Reduction	Unpaid Balance
1	307.38	28.58	278.80	3221.20
2	307.38	26.31	281.07	2940.13
3	307.38	24.01	283.37	2656.76
4	307.38	21.70	285.68	2371.08
5	307.38	19.36	288.02	2083.06
6	307.38	17.01	290.37	1792.69
7	307.38	14.64	292.74	1499.95
8	307.38	12.25	295.13	1204.82
9	307.38	9.84	297.54	907.28
10	307.38	7.41	299.97	607.31
11	307.38	4.96	302.42	304.89
12	307.38	2.49	304.89	0.00

Chapter 4 Systems of Linear Equations; Matrices

1. The lines appear to intersect in a single point on the graph using [-10,10]1 by [-10,10]1. Graph using viewing window variables of [.4, .5].01 by [-2.5, -2.0].1 to see that the lines do not intersect in a single point. Hence there is no solution.

3. There will still be no solution. Changing the constant to 14 shifts the line downward 2 units.

5. $x = .29, y = -2.85, z = -3.65$

7. $x = 1.04, y = -.95, z = 1.07$

9. $\begin{bmatrix} \$54 & \$48 \\ \$36 & \$32 \end{bmatrix}$ and $\begin{bmatrix} \$60.53 & \$48 \\ \$40.35 & \$32 \end{bmatrix}$
The labor costs increase \$6.53 per ski for the trick ski and \$4.35 per ski for the slalom ski.

Chapter 5 Linear Inequalities and Linear Programming

1. Corner points are (0, 7.14), (0, 10.06), (6.13, 0), (26.83, 0), and (2.66, 1.45)

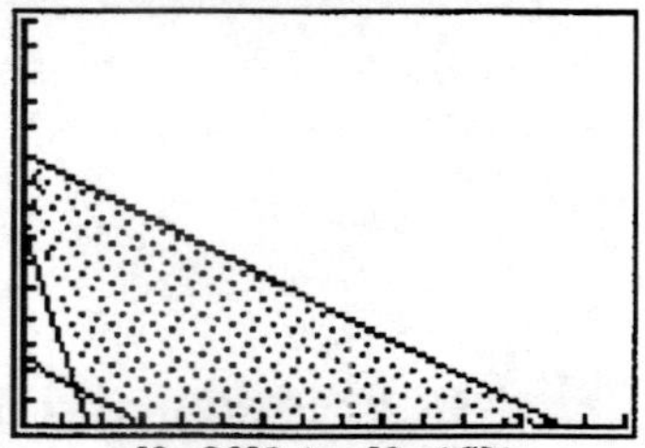

[0, 30]2 by [0, 15]1

3. Maximum at (26.83, 0), minimum at (0, 7.14).

5. A maximum profit of $2520 is achieved if 28 standard tents and 0 expedition tents are manufactured and sold. Substituting the solution into the slack variable equations shows that $s_1 = 4$ and s_2 is zero. This means that there are 4 hours in the cutting department that are not used. All available hours in the assembly department are used.

Chapter 6 Probability

1. (A) 2,496,960 (B) 6,991,488 (C) 621,075 (D) 1,366,365

3. Using a random number generator seed of 10 and entering IPart (25 x RAND) we get: 11, 13, 6, 17, and 21. Note that we use 25 as the number in the command since we want to generate numbers x such that $0 \le x \le 24$.

5. (A) $\begin{bmatrix} .294 & .130 & .015 \\ .106 & .083 & .092 \\ .076 & .065 & .139 \end{bmatrix}$ (B) .618

   (C) P(H) =.476; the events are not independent.

7. (A) (B) .25

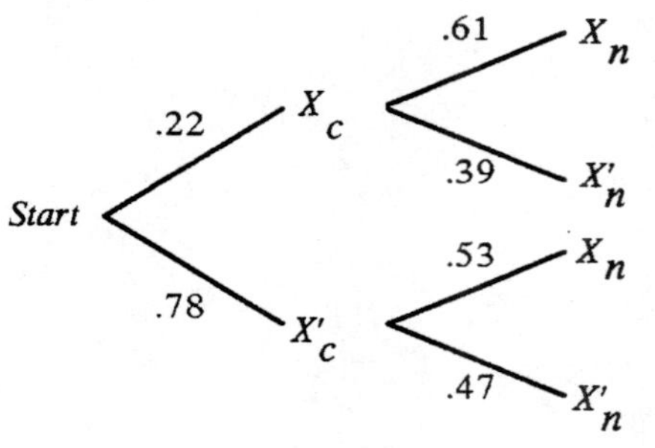

9. (A) .1

   (B) In the long run this player will win $.10 each game on the average. Numbers 5 and 6 are both the best numbers for the player to choose since they have the highest probability.

Chapter 7 Data Description and Probability Distributions

1. 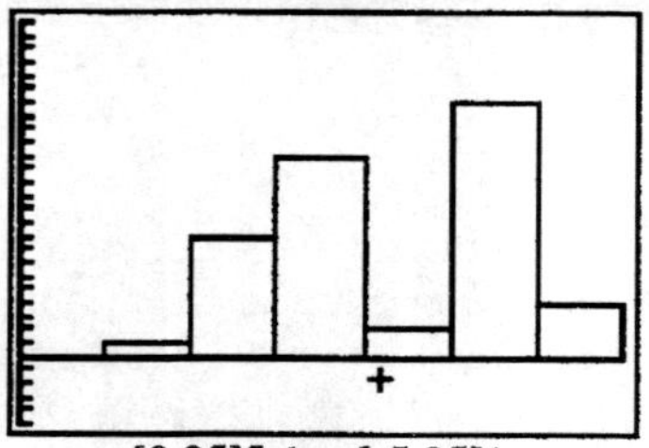

[0,35]5 by [-5,25]1

3. $\bar{x}$ = 21.12, s = 7.07.

5. P(47)+P(46)+P(45) = .422 chance that 3, 4, or 5 will not recover.

Chapter 8 Games and Decisions

1. 1.17 This means that R is expected to win $1.17 per game in the long run.

3. $P^* = [.604 \quad .396]$ $Q^* = \begin{bmatrix} .597 \\ .403 \end{bmatrix}$ The game is not fair since $P^*MQ^* = .824 \neq 0$.

Chapter 9 Markov Chains

1. (A)

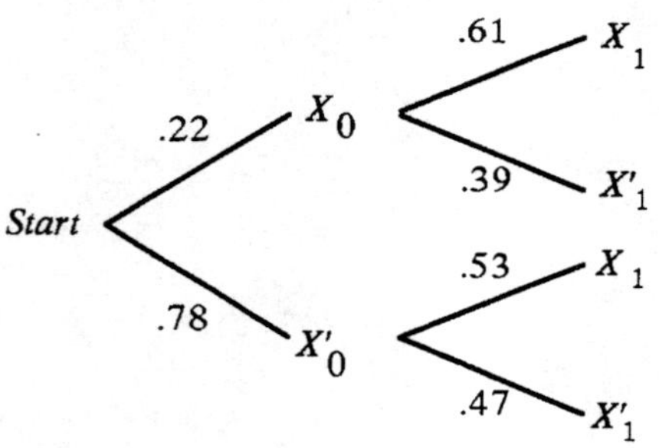

(B) Use Bayes formula.

$$
\begin{aligned}
P(X_0 \mid X_1) &= \frac{P(X_0 \text{ and } X_1)}{P(X_1)} \\
&= \frac{P(X_0 \text{ and } X_1)}{P(X_0 \text{ and } X_1) + P(X_0' \text{ and } X_1)} \\
&= \frac{.22\text{x}.61}{.22\text{x}.61 + .78\text{x}.53} = .245
\end{aligned}
$$

3. Transition matrix is
$$
\begin{array}{c} \\ X_0 \\ X_0' \end{array}
\begin{array}{c} X_1 \quad X_1' \\ \begin{bmatrix} .61 & .39 \\ .53 & .47 \end{bmatrix} \end{array}
$$
. The initial state is [.22 .78].

   (A) After five purchases, the company will have 58% of the market.

   (B) In the long run, the company will have 58% of the market.

5. (A) After five purchases, Product X will have 35% of the market and Product Y will have 29% of the market.

   (B) In the long run, Product X will have 35% of the market and Product Y will have 30% of the market.

NOTES

NOTES

EXPONENTIAL AND LOGARITHMIC FUNCTIONS

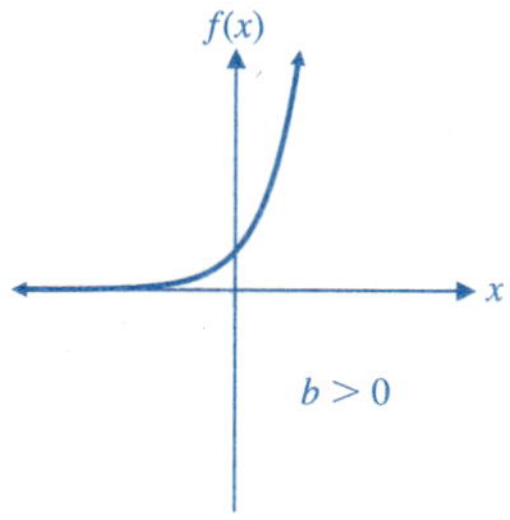

Exponential function
$f(x) = b^x$

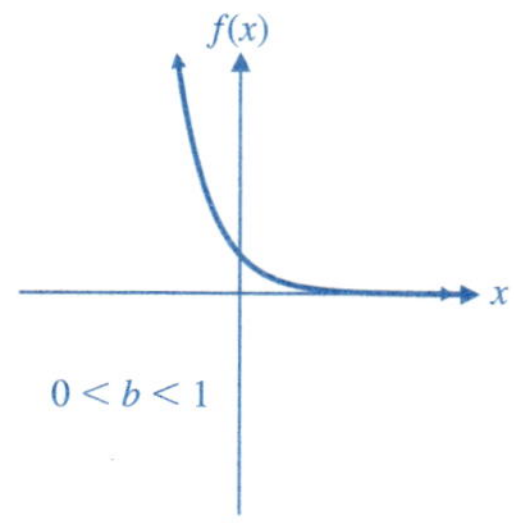

Exponential function
$f(x) = b^x$

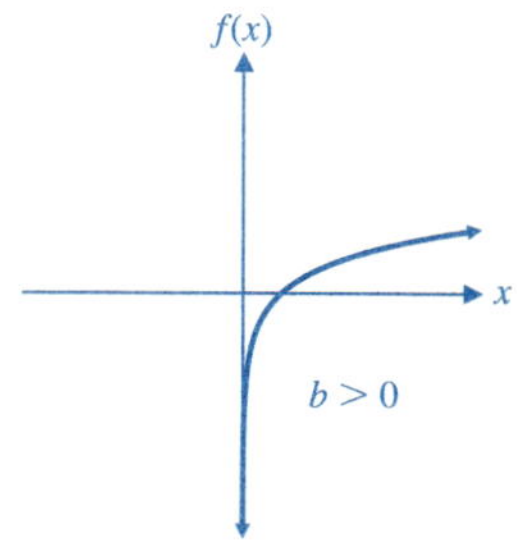

Logarithmic function
$f(x) = \log_b x$

REPRESENTATIVE POLYNOMIAL FUNCTIONS (DEGREE > 2)

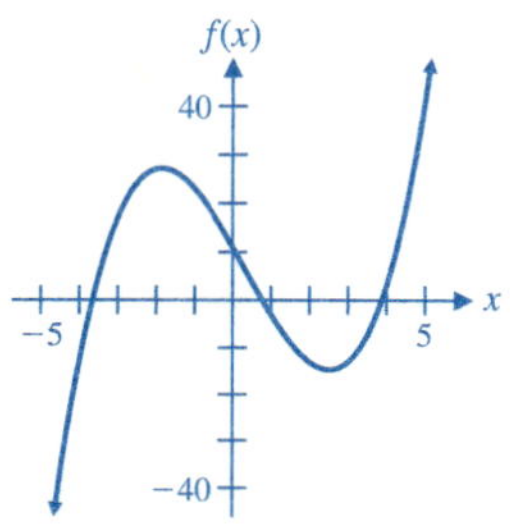

Third-degree polynomial
$f(x) = x^3 - x^2 - 14x + 11$

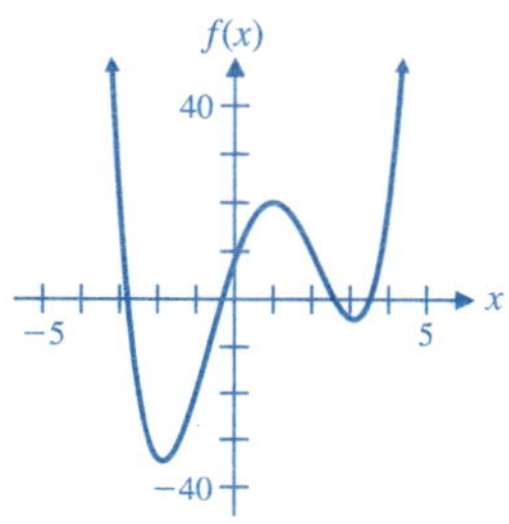

Fourth-degree polynomial
$f(x) = x^4 - 3x^3 - 9x^2 + 23x + 8$

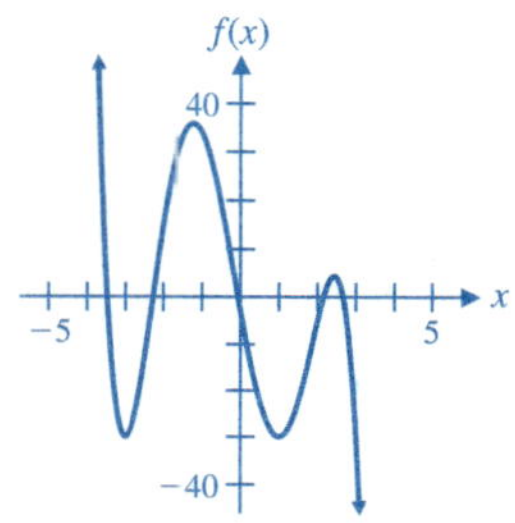

Fifth-degree polynomial
$f(x) = -x^5 - x^4 + 14x^3 + 6x^2 - 45x - 3$

REPRESENTATIVE RATIONAL FUNCTIONS

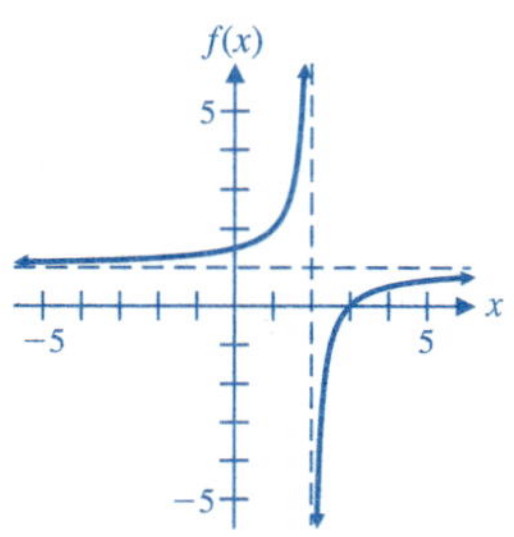

$$f(x) = \frac{x - 3}{x - 2}$$

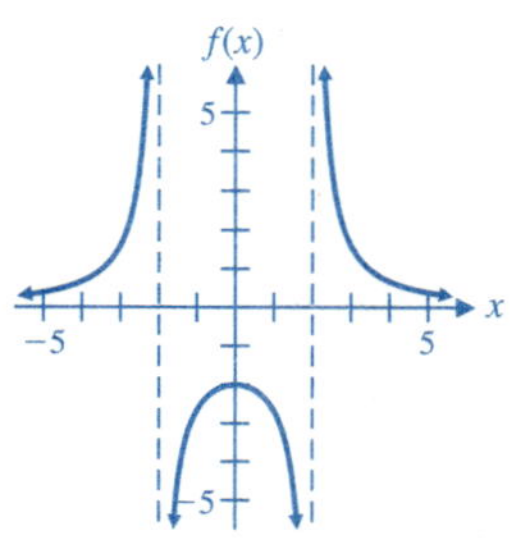

$$f(x) = \frac{8}{x^2 - 4}$$

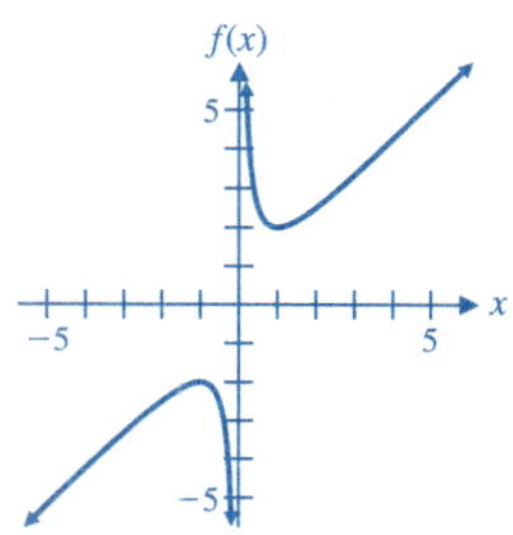

$$f(x) = x + \frac{1}{x}$$

GRAPH TRANSFORMATIONS

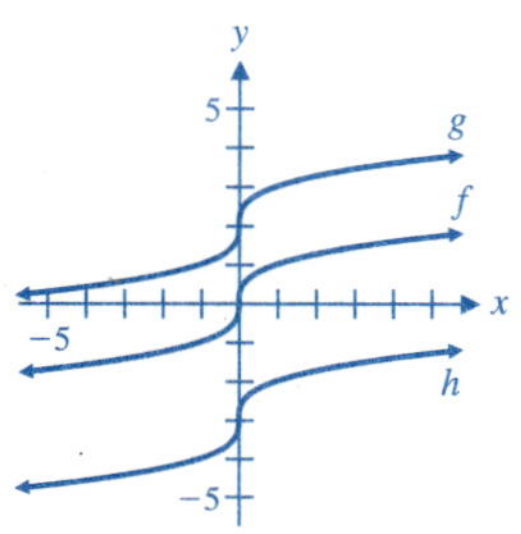

Vertical translation
$g(x) = f(x) + 2$
$h(x) = f(x) - 3$

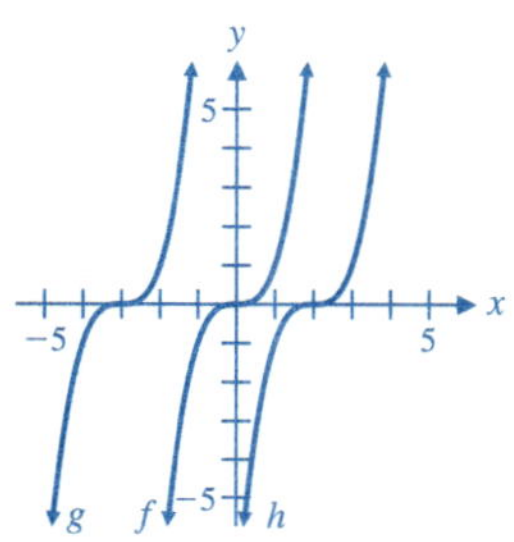

Horizontal translation
$g(x) = f(x + 3)$
$h(x) = f(x - 2)$

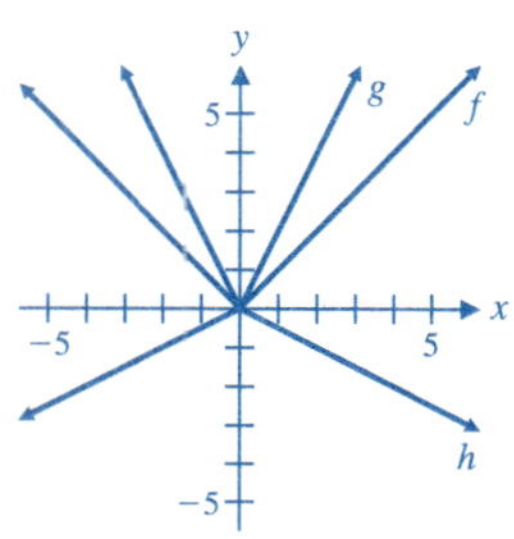

Expansion, contraction, and reflection
$g(x) = 2f(x)$
$h(x) = -0.5f(x)$

Finite Mathematics

For Business, Economics, Life Sciences, and Social Sciences

Finite Mathematics

For Business, Economics, Life Sciences, and Social Sciences

Eighth Edition

Raymond A. Barnett
Merritt College

Michael R. Ziegler
Marquette University

Karl E. Byleen
Marquette University

Prentice Hall
Upper Saddle River, New Jersey 07458

LIBRARY OF CONGRESS CATALOGING-IN-PUBLICATION DATA

Barnett, Raymond A.
Finite mathematics for business, economics, life sciences, and social sciences / Raymond A. Barnett, Michael R. Ziegler, Karl E. Byleen.—8th ed.
p. cm.
Includes indexes.
ISBN 0-13-913179-5 (hc.)
1. Mathematics. 2. Social sciences—Mathematics. 3. Biomathematics. I. Ziegler, Michael R. II. Byleen, Karl. III. Title.
QA39.2.B367 1998 98-27337
510—dc21 CIP

Executive Acquisitions Editor: Sally Simpson
Sponsoring Editor/Supplements Editor: Gina M. Huck
Marketing Manager: Patrice Lumumba Jones
Marketing Assistant: Amy Lysik
Editor-in-Chief: Jerome Grant
Editorial Director: Tim Bozik
Production Editor: Phyllis Niklas
Senior Managing Editor: Linda Mihatov Behrens
Executive Managing Editor: Kathleen Schiaparelli
Assistant Vice President of Production and Manufacturing: David W. Riccardi
Manufacturing Buyer: Alan Fischer
Manufacturing Manager: Trudy Pisciotti
Editorial Assistant: Joanne Wendelken
New Media Editor: Audra J. Walsh
Art Director: Maureen Eide
Associate Creative Director: Amy Rosen
Director of Creative Services: Paula Maylahn
Assistant to Art Director: John Christiana
Art Manager: Gus Vibal
Art Editor: Grace Hazeldine
Interior/Cover Designer: Jill Little
Cover Photo: W. Yanagida/IMA USA/Photonica Division
Art Studio: Scientific Illustrators

© 1999, 1996, 1993, 1990, 1987, 1984, 1981, 1979 Prentice-Hall, Inc.
Simon & Schuster/A Viacom Company
Upper Saddle River, NJ 07458

All rights reserved. No part of this book may be reproduced, in any form or by any means, without permission in writing from the publisher.

Printed in the United States of America
10 9 8 7 6 5 4 3 2 1

ISBN 0-13-913179-5

Prentice-Hall International (UK) Limited, *London*
Prentice-Hall of Australia Pty. Limited, *Sydney*
Prentice-Hall Canada, Inc., *Toronto*
Prentice-Hall Hispanoamericano, S. A., *Mexico*
Prentice-Hall of India Private Limited, *New Delhi*
Prentice-Hall of Japan, Inc., *Tokyo*
Simon & Schuster Asia Pte. Ltd., *Singapore*
Editora Prentice-Hall do Brasil, Ltda., *Rio de Janeiro*

Contents

Chapter Dependencies

PART ONE A LIBRARY OF ELEMENTARY FUNCTIONS*

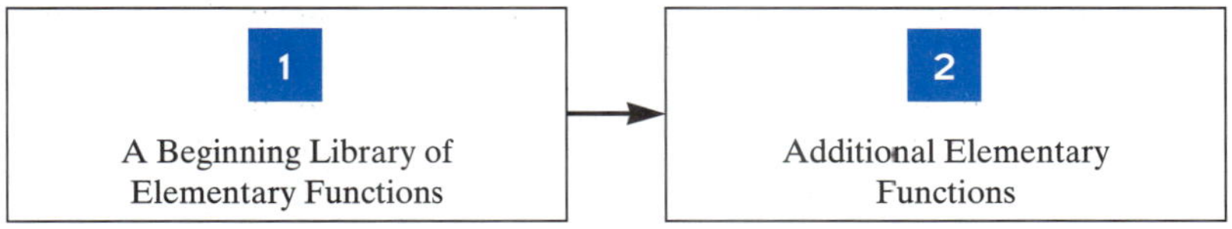

PART TWO FINITE MATHEMATICS

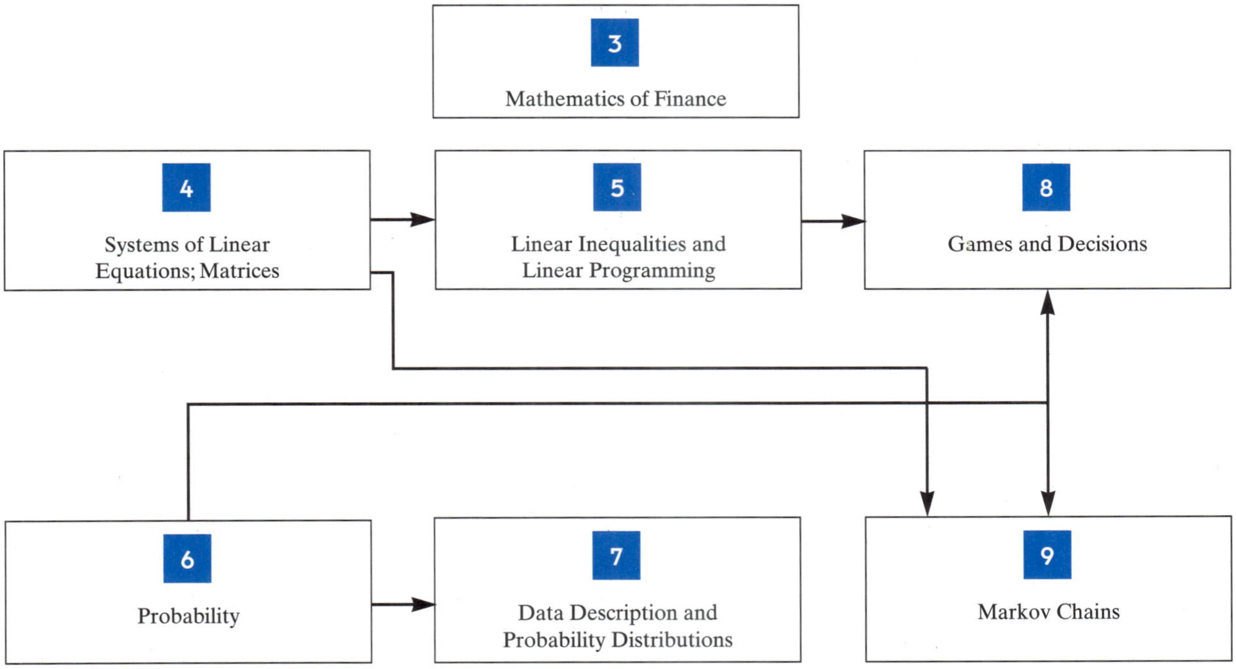

APPENDIXES

A Self-Test Basic Algebra Review

B Special Topics

*Selected topics from Part One may be referred to as needed in Part Two or reviewed systematically before starting Part Two.

Preface

The eighth edition of *Finite Mathematics for Business, Economics, Life Sciences, and Social Sciences* is designed for a one-term course in finite mathematics and for students who have had $1\frac{1}{2}$–2 years of high school algebra or the equivalent. The choice and independence of topics make the text readily adaptable to a variety of courses (see the Chapter Dependency Chart on page ix). It is one of eight books in the authors' college mathematics series.

Improvements in this edition evolved out of the generous response from a large number of users of the last and previous editions as well as survey results from instructors, mathematics departments, course outlines, and college catalogs. Fundamental to a book's growth and effectiveness is classroom use and feedback. Now in its eighth edition, *Finite Mathematics for Business, Economics, Life Sciences, and Social Sciences* has had the benefit of having a substantial amount of both.

Emphasis and Style

The text is **written for student comprehension.** Great care has been taken to write a book that is mathematically correct and accessible to students. Emphasis is on computational skills, ideas, and problem solving rather than mathematical theory. Most derivations and proofs are omitted except where their inclusion adds significant insight into a particular concept. General concepts and results are usually presented only after particular cases have been discussed.

Examples and Matched Problems

Over 260 completely worked examples are used to introduce concepts and to demonstrate problem-solving techniques. Many examples have multiple parts, significantly increasing the total number of worked examples. Each example is followed by a similar **matched problem for the student to work** while reading the material. This actively involves the student in the learning process. The answers to these matched problems are included at the end of each section for easy reference.

Exploration and Discussion

Every section contains **Explore–Discuss** problems interspersed at appropriate places to encourage the student to think about a relationship or process before a result is stated, or to investigate additional consequences of a development in the text. **Verbalization** of mathematical concepts, results, and processes is encouraged in these Explore–Discuss problems, as well as in some matched problems, and in some problems in almost every exercise set. The Explore–Discuss material also can be used as in-class or out-of-class **group activities.** In addition, at the end of every chapter, we have included two special **chapter group**

activities that involve several of the concepts discussed in the chapter. Problems in the exercise sets that require verbalization are indicated by color problem numbers.

Exercise Sets

The book contains over 3,300 problems. Many problems have multiple parts, significantly increasing the total number of problems. Each exercise set is designed so that an average or below-average student will experience success and a very capable student will be challenged. Exercise sets are mostly divided into A (routine, easy mechanics), B (more difficult mechanics), and C (difficult mechanics and some theory) levels.

Applications

A major objective of this book is to give the student substantial experience in **modeling and solving real-world problems.** Enough applications are included to convince even the most skeptical student that mathematics is really useful (see the Applications Index inside the back cover). Worked examples involving applications are identified by . **Almost every exercise set contains application problems,** usually divided into business and economics, life science, and social science groupings. An instructor with students from all three disciplines can let them choose applications from their own field of interest; if most students are from one of the three areas, then special emphasis can be placed there. Most of the applications are simplified versions of actual real-world problems taken from professional journals and books. No specialized experience is required to solve any of the applications.

Technology

The generic term **graphing utility** is used to refer to any of the various graphing calculators or computer software packages that might be available to a student using this book. (See the description of the software accompanying this book later in this Preface.) Although **access to a graphing utility is not assumed,** it is likely that many students will want to make use of one of these devices. To assist these students, **optional graphing utility activities** are included in appropriate places in the book. These include brief discussions in the text, examples or portions of examples solved on a graphing utility, problems for the student to solve, and a **group activity that involves the use of technology** at the end of each chapter. In the group activity at the end of Chapter 1, and continuing through Chapter 2, **linear regression** on a graphing utility is used at appropriate points to illustrate **mathematical modeling with real data.** All the optional graphing utility material is clearly identified by either or C and can be omitted without loss of continuity, if desired.

Graphs

All graphs are computer-generated to ensure mathematical accuracy. Graphing utility screens displayed in the text are actual output from a graphing calculator.

Additional Pedagogical Features

Annotation of examples and developments, in small color type, is found throughout the text to help students through critical stages (see Sections 1-1 and 4-2). **Think boxes** (dashed boxes) are used to enclose steps that are usually performed mentally (see Sections 1-1 and 4-1). **Boxes** are used to highlight important definitions, results, and step-by-step processes (see Sections 1-1 and 1-4). **Caution** statements appear throughout the text where student errors often occur (see Sections 4-3 and 4-5). **Functional use of color** improves the clarity of many illustrations, graphs, and developments, and guides students through certain critical steps (Sections 1-1 and 4-2). **Boldface type** is used to introduce new terms and highlight important comments. **Chapter review** sections include a review of all important terms and symbols and a comprehensive review exercise. **Answers to most review exercises,** keyed to appropriate sections, are included in the back of the book. Answers to all other odd-numbered problems are also in the back of the book. Answers to application problems in linear programming include both the mathematical model and the numeric answer.

Content

The text begins with the development of a library of elementary functions in Chapters 1 and 2, including their properties and uses. We encourage students to investigate mathematical ideas and processes **graphically** and **numerically,** as well as **algebraically.** This development lays a firm foundation for studying mathematics both in this book and in future endeavors. Depending on the syllabus for the course and the background of the students, some or all of this material can be covered at the beginning of a course, or selected portions can be referred to as needed later in the course.

The material in Part Two (Finite Mathematics) can be thought of as four units: **mathematics of finance** (Chapter 3); linear algebra, including **matrices, linear systems, and linear programming** (Chapters 4 and 5); **probability and statistics** (Chapters 6 and 7); and applications of linear algebra and probability to **game theory and Markov chains** (Chapters 8 and 9). The first three units are independent of each other, while the last two chapters are dependent on some of the earlier chapters (see the Chapter Dependency Chart preceding this Preface).

Chapter 3 presents a thorough treatment of simple and compound interest and present and future value of ordinary annuities. Appendix B contains a section on arithmetic and geometric sequences that can be covered in conjunction with this chapter, if desired.

Chapter 4 covers linear systems and matrices with an **emphasis on using row operations and Gauss–Jordan elimination** to solve systems and to find matrix inverses. This chapter also contains numerous applications of **mathematical modeling** utilizing systems and matrices. To assist students in formulating solutions, **all the answers in the back of the book to application problems** in Exercises 4-3, 4-5, and the chapter Review Exercise **contain both the mathematical model and its solution.** The row operations discussed in Sections 4-2 and 4-3 are required for the simplex method in Chapter 5. Matrix multiplication, matrix inverses, and systems of equations are required for Markov chains in Chapter 9.

Chapter 5 provides **broad and flexible coverage of linear programming.** The first two sections cover two-variable graphing techniques. Instructors who wish to emphasize techniques can cover the basic simplex method in Sections

5-3 and 5-4 and then discuss any or all of the following: the dual method (Section 5-5), the big M method (Section 5-6), or the two-phase simplex method (Group Activity 1). Those who want to emphasize modeling can discuss the formation of the mathematical model for any of the application examples in Sections 5-4, 5-5, and 5-6, and either omit the solution or use software to find the solution (see the description of the software that accompanies this text later in this Preface). To facilitate this approach, **all the answers in the back of the book to application problems** in Exercises 5-4, 5-5, 5-6, and the chapter Review Exercise **contain both the mathematical model and its solution.** Geometric, simplex, and dual solution methods are required for portions of Chapter 8.

Chapter 6 covers **counting techniques and basic probability,** including Bayes' formula and random variables. Appendix A contains a review of basic set theory and notation to support the use of sets in probability. Some of the topics discussed in Chapter 6 are required for Chapter 7.

Chapter 7 deals with basic **descriptive statistics** and more advanced probability distributions, including the important **normal distribution.** Appendix B contains a short discussion of the binomial theorem that can be used in conjunction with the development of the binomial distribution in Section 7-5.

Each of the last two chapters ties together concepts developed in earlier chapters and applies them to two interesting topics: **game theory** (Chapter 8) and **Markov chains** (Chapter 9). Either chapter provides an excellent unifying conclusion to a finite mathematics course.

Appendix A contains a **self-test** and a **concise review of basic algebra** that also may be covered as part of the course or referred to as needed. As mentioned above, Appendix B contains additional topics that can be covered in conjunction with certain sections in the text, if desired.

Supplements for the Student

1. A **Student Solutions Manual and Explorations in Finite Mathematics** by Garret J. Etgen and David Schneider is available through your book store. The manual includes detailed solutions to all odd-numbered problems and all review exercises. *Explorations in Finite Mathematics* by David Schneider contains over twenty routines that provide additional insight into the topics discussed in the text. Although this software has much of the computing power of standard mathematical software packages, it is primarily a teaching tool that focuses on understanding mathematical concepts, rather than on computing. Included are routines for Gaussian elimination, matrix inversion, solution of linear programming problems by both the geometric method and the simplex method, Markov chains, probability and statistics, and mathematics of finance. All the routines in this software package are menu-driven and are very easy to use. The matrix routines use and display rational numbers, and matrices may be saved and printed. The software will run on DOS or Windows platforms.
2. A **Graphing Calculator Manual** by Carolyn L. Meitler contains examples illustrating the use of graphics calculators to solve problems similar to those discussed in the text. The manual follows the chapter organization of the text, making it easy to find examples in the manual illustrating appropriate calculator solution methods for problems in the text. The manual includes

keystrokes for the TI-82, TI-83, TI-85, and TI-86 calculators. However, the examples and techniques can be used with any graphing utility.

3. The **PH Companion Website,** designed to complement and expand upon the text, offers a variety of teaching and learning tools, including links to related websites, practice work for students, and the ability for instructors to monitor and evaluate students' work on the website. For more information, contact your local Prentice Hall representative.
www.prenhall.com/barnett

Supplements for the Instructor

For a summary of all available supplementary materials and detailed information regarding examination copy requests and orders, see page xviii.

1. **PH Custom Test, a menu-driven random test system** for either Windows or Macintosh is available to instructors. The test system has been greatly expanded and now offers **on-line testing.** Carefully constructed algorithms use random-number generators to produce different, yet equivalent, versions of each of these problems. In addition, the system incorporates a unique **editing function** that allows the instructor to create additional problems, or alter any of the existing problems in the test, using a full set of mathematical notation. The test system offers **free-response, multiple-choice, and mixed exams.** An almost unlimited number of quizzes, review exercises, chapter tests, midterms, and final examinations, each different from the other, can be generated quickly and easily. At the same time, the system will produce answer keys, student worksheets, and a gradebook for the instructor, if desired.
2. A **Test Item File,** prepared by Laurel Technical Services, provides a hard copy of the test items available in PH Custom Test.
3. An **Instructor's Resource Manual** provides over 100 transparency masters and all the answers not included in the text. This manual is available to instructors without charge.
4. A **Student Solutions Manual and Explorations in Finite Mathematics** by Garret J. Etgen and David Schneider (see Supplements for the Student) is available to instructors.
5. A **Graphing Calculator Manual** by Carolyn L. Meitler (see Supplements for the Student) is available to instructors. The manual contains all the necessary information for a student with no previous experience with a graphing calculator, eliminating the need for the instructor to prepare materials related to calculator usage. In particular, separate appendixes for the TI-82, TI-83, TI-85, and TI-86 graphing calculators contain detailed instructions, including calculator-specific keystrokes, for performing the various operations required to effectively use each of these calculators to solve problems in the text. Furthermore, the methods illustrated for these calculators are easily adapted to other graphing utilities. The manual is very effective both for a class where all students purchase the same calculator and in a setting where students are using a variety of different calculators—an important consideration as more and more students arrive at college having already purchased a graphing calculator.

6. The **PH Companion Website,** designed to complement and expand upon the text, offers a variety of interactive teaching and learning tools, including links to related websites, practice work for students, and the ability for instructors to monitor and evaluate students' work on the website. For more information, contact your local Prentice Hall representative. *www.prenhall.com/barnett*

Error Check

Because of the careful checking and proofing by a number of mathematics instructors (acting independently), the authors and publisher believe this book to be substantially error-free. For any errors remaining, the authors would be grateful if they were sent to: Michael R. Ziegler, 509 W. Dean Court, Fox Point, WI 53217; or, by e-mail, to: michaelziegler@execpc.com

Acknowledgments

In addition to the authors, many others are involved in the successful publication of a book.

We wish to thank the following reviewers of the seventh edition:

Lou D'Alotto, St. Johns University
Paul Hutchins, Florissant Community College
Inessa Levi, University of Louisville
Wayne Miller, Lee College
Dix Petty, University of Missouri
John Ryan, Texas A & M University

We also wish to thank our colleagues who have provided input on previous editions:

Chris Boldt, Bob Bradshaw, Bruce Chaffee, Robert Chaney, Dianne Clark, Charles E. Cleaver, Barbara Cohen, Richard L. Conlon, Catherine Cron, Madhu Deshpande, Kenneth A. Dodaro, Michael W. Ecker, Jerry R. Ehman, Lucina Gallagher, Martha M. Harvey, Sue Henderson, Lloyd R. Hicks, Louis F. Hoelzle, Paul Hutchins, K. Wayne James, Robert H. Johnston, Robert Krystock, James T. Loats, Frank Lopez, Roy H. Luke, Mel Mitchell, Ronald Persky, Kenneth A. Peters, Jr., Tom Plavchak, Bob Prielipp, Stephen Rodi, Arthur Rosenthal, Sheldon Rothman, Elaine Russell, Daniel E. Scanlon, George R. Schriro, Arnold L. Schroeder, Hari Shanker, Joan Smith, Steven Terry, Delores A. Williams, Caroline Woods, Charles W. Zimmerman, and Pat Zrolka.

We also express our thanks to:

Carolyn Meitler, Stephen Merrill, Robert Mullins, and Caroline Woods for providing a careful and thorough check of all the mathematical calculations in the book, the *Student Solutions Manual*, and the *Instructor's Resource Manual* (a tedious but extremely important job).

Garret Etgen, Carolyn Meitler, and David Schneider for developing the supplemental manuals that are so important to the success of a text.

Jeanne Wallace for accurately and efficiently producing most of the manuals that supplement the text.

George Morris and his staff at Scientific Illustrators for their effective illustrations and accurate graphs.

Phyllis Niklas for guiding the book smoothly through all publication details.

All the people at Prentice Hall who contributed their efforts to the production of this book, especially Sally Simpson, our executive acquisitions editor, and Gina Huck, our sponsoring editor.

Producing this new edition with the help of all these extremely competent people has been a most satisfying experience.

R. A. Barnett
M. R. Ziegler
K. E. Byleen

■ Ordering Information

When requesting examination copies or placing orders for this text or any of the related supplementary material listed below, please refer to the corresponding ISBN numbers.

Title	ISBN Number
Finite Mathematics for Business, Economics, Life Sciences, and Social Sciences, Eighth Edition	0-13-913179-5
Test Item File to accompany Finite Mathematics, Eighth Edition	0-13-010702-6
Computer-generated random test system for Finite Mathematics, Eighth Edition:	
PH Custom Test Windows	0-13-931792-9
PH Custom Test Mac	0-13-931933-6
Instructor's Resource Manual to accompany Finite Mathematics, Eighth Edition	0-13-931750-3
Student Solutions Manual and Explorations in Finite Mathematics	0-13-932568-9
Graphing Calculator Manual to accompany Finite Mathematics, Eighth Edition	0-13-932195-0

PART ONE

A LIBRARY OF ELEMENTARY FUNCTIONS

Matched Problem 5 ➡ Find the domains of functions F, G, and H:

$$F(x) = x^2 - 3x + 1 \qquad G(x) = \frac{5}{x + 3} \qquad H(x) = \sqrt{2 - x}$$

In addition to evaluating functions at specific numbers, it is important to be able to evaluate functions at expressions that involve one or more variables. For example, the **difference quotient**

$$\frac{f(x + h) - f(x)}{h} \quad x \text{ and } x + h \text{ in the domain of } f,\ h \neq 0$$

is studied extensively in calculus.

Explore–Discuss 4

Let x and h be real numbers.

(A) If $f(x) = 4x + 3$, which of the following is true?

(1) $f(x + h) = 4x + 3 + h$
(2) $f(x + h) = 4x + 4h + 3$
(3) $f(x + h) = 4x + 4h + 6$

(B) If $g(x) = x^2$, which of the following is true?

(1) $g(x + h) = x^2 + h$
(2) $g(x + h) = x^2 + h^2$
(3) $g(x + h) = x^2 + 2hx + h^2$

(C) If $M(x) = x^2 + 4x + 3$, describe the operations that must be performed to evaluate $M(x + h)$.

Example 6 ➡ **Using Function Notation** For $f(x) = x^2 - 2x + 7$, find:

(A) $f(a)$ (B) $f(a + h)$ (C) $\dfrac{f(a + h) - f(a)}{h}$

SOLUTION

(A) $f(a) = a^2 - 2a + 7$

(B)
$$\begin{aligned} f(a + h) &= (a + h)^2 - 2(a + h) + 7 \\ &= a^2 + 2ah + h^2 - 2a - 2h + 7 \end{aligned}$$

(C)
$$\begin{aligned} \frac{f(a + h) - f(a)}{h} &= \frac{(a^2 + 2ah + h^2 - 2a - 2h + 7) - (a^2 - 2a + 7)}{h} \\ &= \frac{2ah + h^2 - 2h}{h} = \frac{h(2a + h - 2)}{h} = 2a + h - 2 \end{aligned}$$

Matched Problem 6 ➡ Repeat Example 6 for $f(x) = x^2 - 4x + 9$.

Applications

We now turn to the important concepts of **break-even** and **profit–loss** analysis, which we will return to a number of times in this text. Any manufacturing company has **costs,** C, and **revenues,** R. The company will have a **loss** if $R < C$, will **break even** if $R = C$, and will have a **profit** if $R > C$. Costs include **fixed costs** such as plant overhead, product design, setup, and promotion; and **variable costs,** which are dependent on the number of items produced at a certain cost per item. In addition, **price–demand** functions, usually established by financial departments using historical data or sampling techniques, play an important part in profit–loss analysis. We will let x, the number of units manufactured and sold, represent the independent variable. Cost functions, revenue functions, profit functions, and price–demand functions are often stated in the following forms, where a, b, m, and n are constants determined from the context of a particular problem:

Cost Function

$$\begin{aligned} C &= (\text{Fixed costs}) + (\text{Variable costs}) \\ &= a + bx \end{aligned}$$

Price–Demand Function

$$p = m - nx$$ *x is the number of items that can be sold at \$$p$ per item.*

Revenue Function

$$\begin{aligned} R &= (\text{Number of items sold}) \times (\text{Price per item}) \\ &= xp = x(m - nx) \end{aligned}$$

Profit Function

$$\begin{aligned} P &= R - C \\ &= x(m - nx) - (a + bx) \end{aligned}$$

Example 7 and Matched Problem 7 explore the relationships among the algebraic definition of a function, the numerical values of the function, and the graphical representation of the function. The interplay among algebraic, numeric, and graphic viewpoints is an important aspect of our treatment of functions and their use. In Example 7, we also see how a function can be used to describe data from the real world, a process that is often referred to as *mathematical modeling*. The material in this example will be returned to in subsequent sections so that we can analyze it in greater detail and from different points of view.

Example 7 **Price–Demand and Revenue Modeling** A manufacturer of a popular automatic camera wholesales the camera to retail outlets throughout the United States. Using statistical methods, the financial department in the company produced the price–demand data in Table 4 where p is the wholesale price per camera at which x million cameras are sold. Notice that as the price goes down, the number sold goes up.

Table 4

PRICE–DEMAND

x (MILLION)	p (\$)
2	87
5	68
8	53
12	37

Using special analytical techniques (regression analysis), an analyst arrived at the following price–demand function that models the Table 4 data:

$$p(x) = 94.8 - 5x \qquad 1 \leqslant x \leqslant 15 \tag{5}$$

Table 5
REVENUE

x (MILLION)	$R(x)$ (MILLION \$)
1	90
3	
6	
9	
12	
15	

(A) Plot the data in Table 4. Then sketch a graph of the price–demand function in the same coordinate system.
(B) What is the company's revenue function for this camera, and what is the domain of this function?
(C) Complete Table 5, computing revenues to the nearest million dollars.
(D) Plot the data in Table 5. Then sketch a graph of the revenue function using these points.
C (E) Plot the revenue function on a graphing utility.

SOLUTION (A)

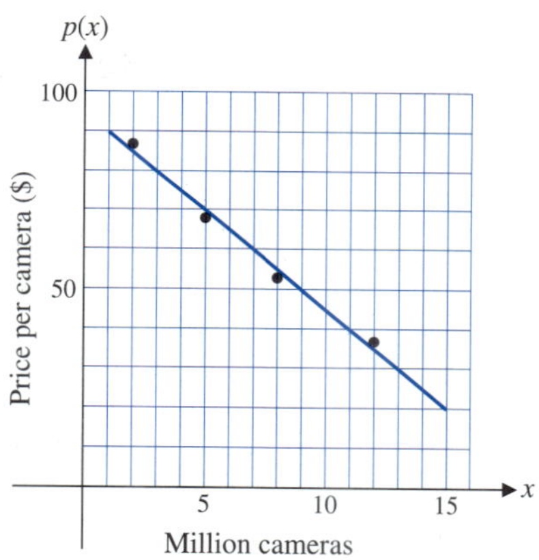

FIGURE 7
Price–demand

In Figure 7, notice that the model approximates the actual data in Table 4, and it is assumed that it gives realistic and useful results for all other values of x between 1 million and 15 million.

(B) $R(x) = xp(x) = x(94.8 - 5x)$ million dollars
Domain: $1 \leqslant x \leqslant 15$
[Same domain as the price–demand function, equation (5).]

(C) **Table 5**
REVENUE

x (MILLION)	$R(x)$ (MILLION \$)
1	90
3	239
6	389
9	448
12	418
15	297

(D)

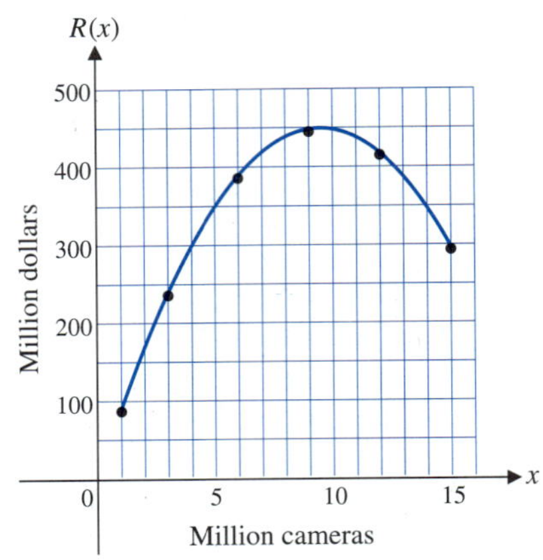

(E)

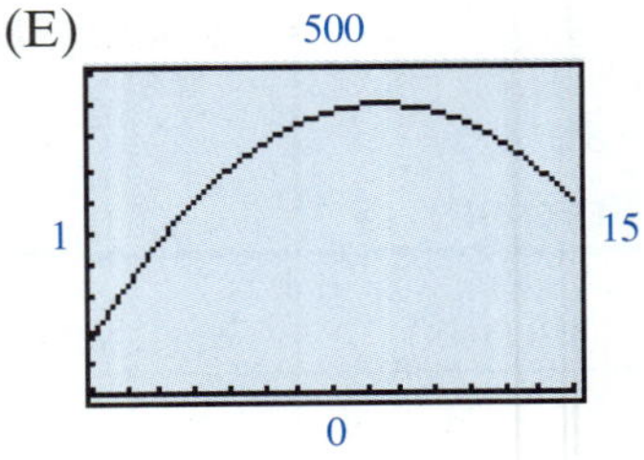

Matched Problem 7 The financial department in Example 7, using statistical techniques, produced the data in Table 6, where $C(x)$ is the cost in millions of dollars for manufacturing and selling x million cameras.

Table 6

COST DATA

x (MILLION)	$C(x)$ (MILLION $)
1	175
5	260
8	305
12	395

Using special analytical techniques (regression analysis), an analyst produced the following cost function to model the data:

$$C(x) = 156 + 19.7x \qquad 1 \le x \le 15 \tag{6}$$

(A) Plot the data in Table 6. Then sketch a graph of equation (6) in the same coordinate system.

(B) What is the company's profit function for this camera, and what is its domain?

(C) Complete Table 7, computing profits to the nearest million dollars.

Table 7

PROFIT

x (MILLION)	$P(x)$ (MILLION $)
1	−86
3	
6	
9	
12	
15	

(D) Plot the points from part (C). Then sketch a graph of the profit function through these points.

(E) Plot the profit function on a graphing utility.

Answers to Matched Problems

1.

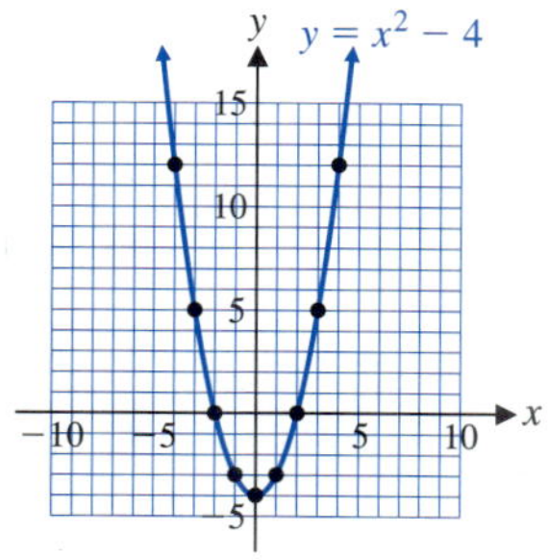

2. (A) Does not specify a function
(B) Specifies a function

3. $x \geq 2$ (inequality notation) or $[2, \infty)$ (interval notation)

4. (A) -3 (B) 0 (C) Does not exist (D) 6

5. Domain of F: R; Domain of G: All real numbers except -3;
Domain of H: $x \leq 2$ (inequality notation) or $(-\infty, 2]$ (interval notation)

6. (A) $a^2 - 4a + 9$ (B) $a^2 + 2ah + h^2 - 4a - 4h + 9$
(C) $2a + h - 4$

7. (A)

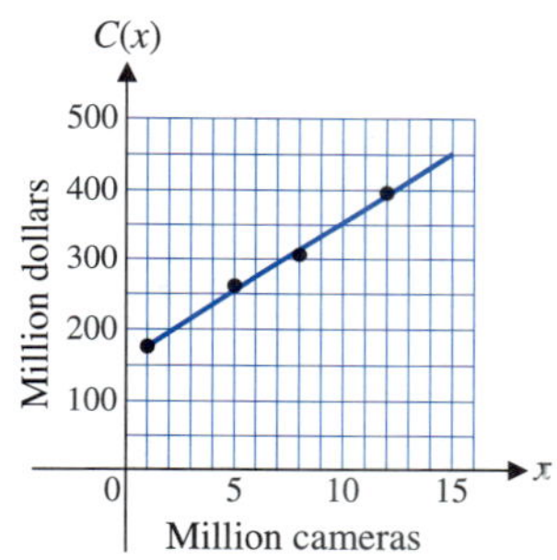

(B) $P(x) = R(x) - C(x) = x(94.8 - 5x) - (156 + 19.7x)$;
Domain: $1 \leq x \leq 15$

(C) Table 7

PROFIT

x (MILLION)	$P(x)$ (MILLION \$)
1	−86
3	24
6	115
9	115
12	25
15	−155

(D)

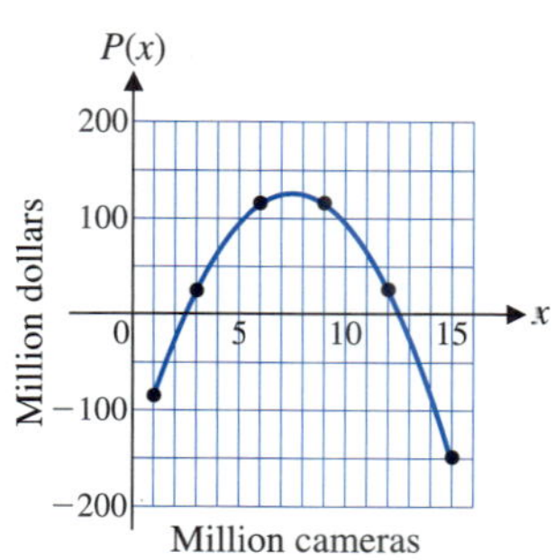

(E)

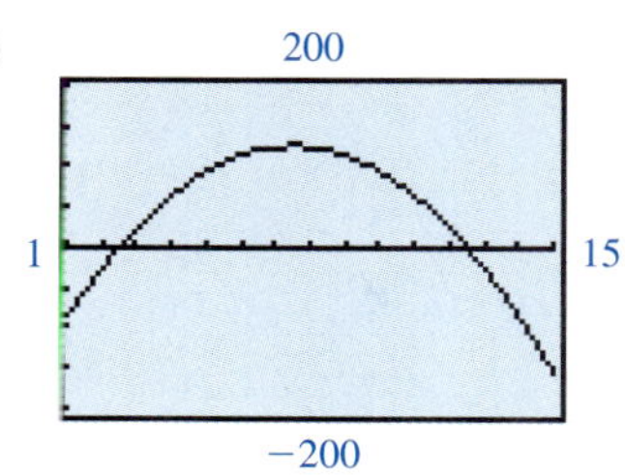

EXERCISE 1-1

A *Indicate whether each table specifies a function.*

1.

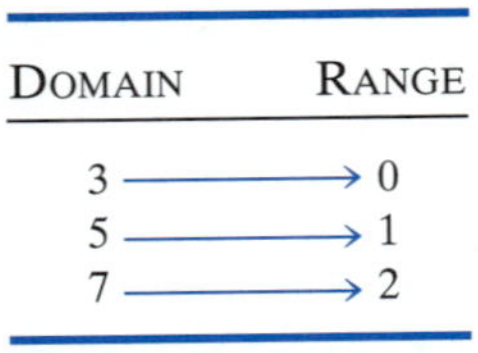

2.

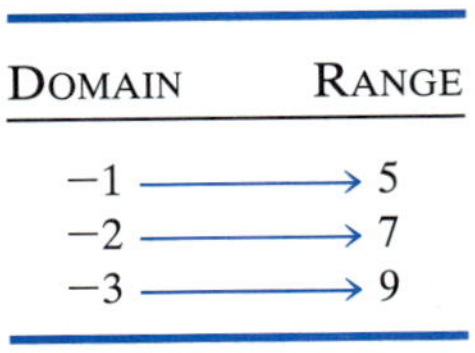

3.

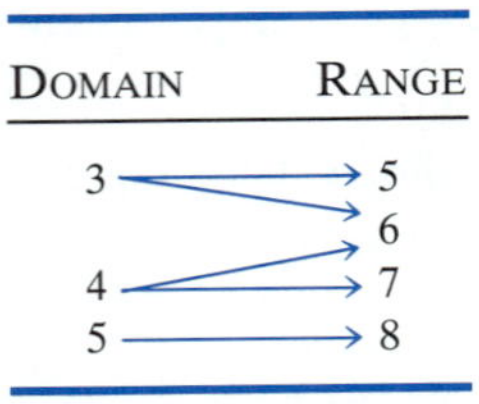

4.

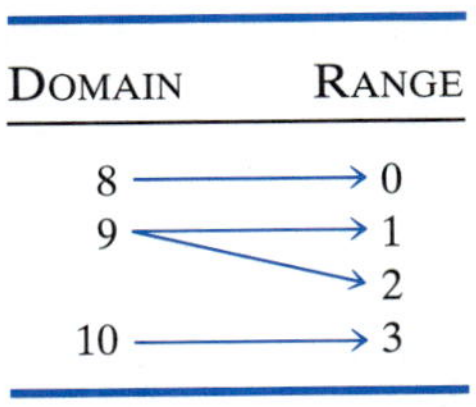

5.

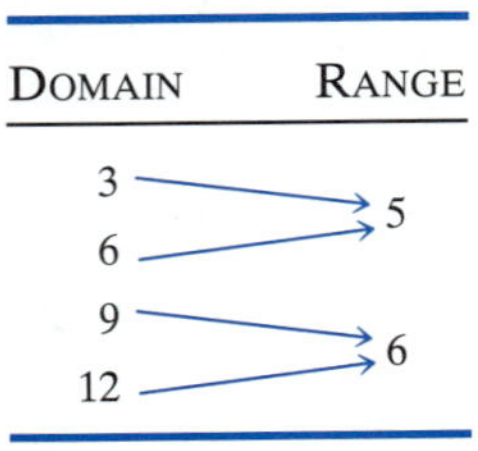

6.

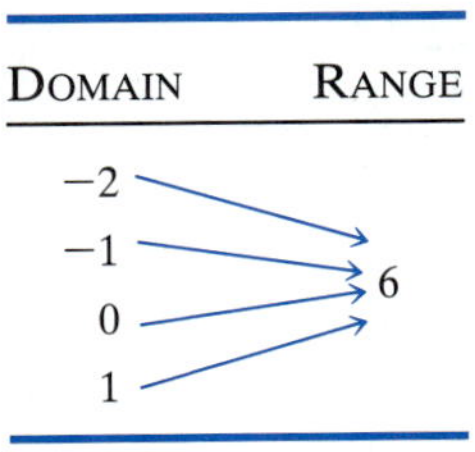

Indicate whether each graph in Problems 7–12 specifies a function

7.

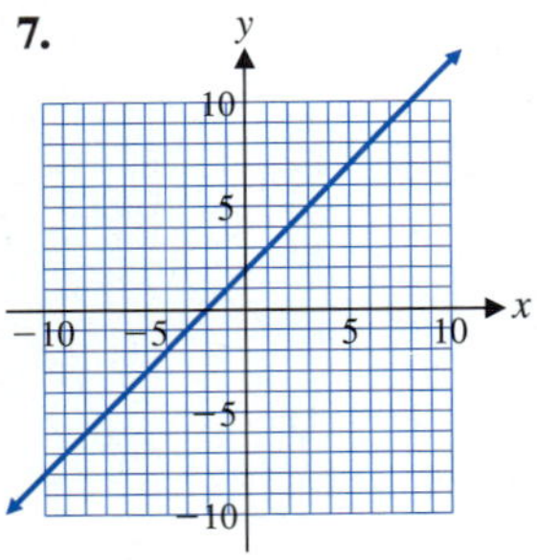

8.

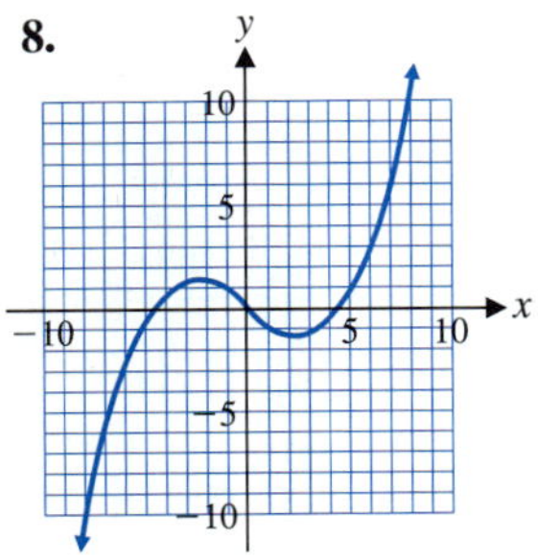

9.

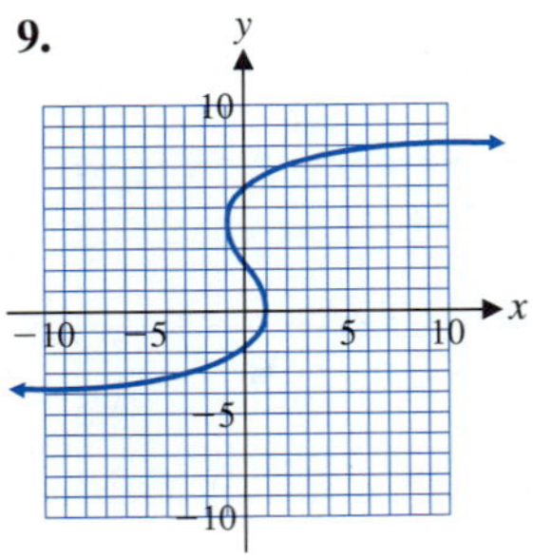

10.

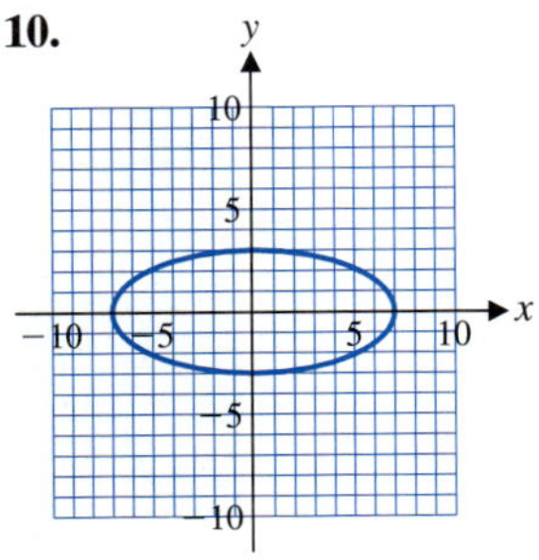

11.

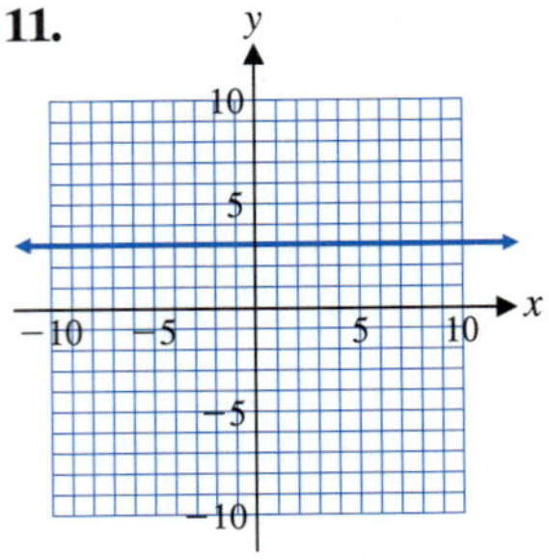

12.

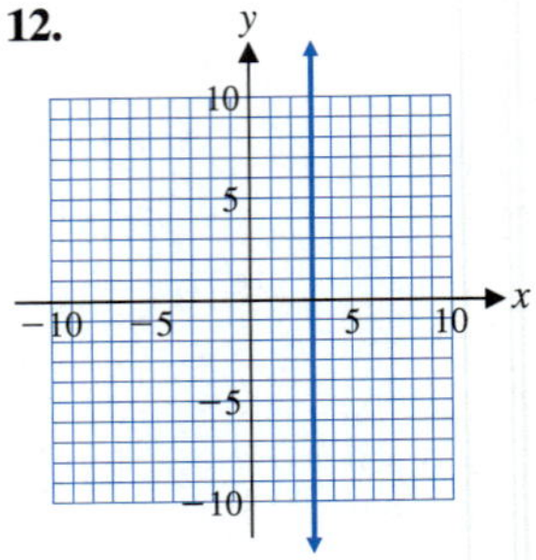

If $f(x) = 3x - 2$ and $g(x) = x - x^2$, find each of the expressions in Problems 13–30.

13. $f(2)$ **14.** $f(1)$ **15.** $f(-1)$

16. $f(-2)$ **17.** $g(3)$ **18.** $g(1)$

19. $f(0)$ **20.** $f(\frac{1}{3})$ **21.** $g(-3)$

22. $g(-2)$ **23.** $f(1) + g(2)$ **24.** $g(1) + f(2)$

25. $g(2) - f(2)$ **26.** $f(3) - g(3)$ **27.** $g(3) \cdot f(0)$

28. $g(0) \cdot f(-2)$ **29.** $\dfrac{g(-2)}{f(-2)}$ **30.** $\dfrac{g(-3)}{f(2)}$

In Problems 31–38, use the following graph of a function f to determine x or y to the nearest integer, as indicated. Some problems may have more than one answer.

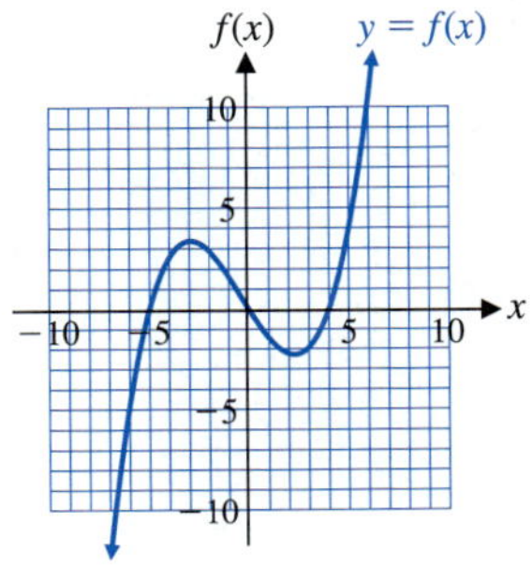

31. $y = f(-5)$ **32.** $y = f(4)$ **33.** $y = f(5)$

34. $y = f(-2)$ **35.** $0 = f(x)$ **36.** $3 = f(x), x < 0$

37. $-4 = f(x)$ **38.** $4 = f(x)$

B *In Problems 39–48, find the domain of each function.*

39. $F(x) = 2x^3 - x^2 + 3$ **40.** $H(x) = 7 - 2x^2 - x^4$

41. $f(x) = \dfrac{x-2}{x+4}$ **42.** $g(x) = \dfrac{x+1}{x-2}$

43. $F(x) = \dfrac{x+2}{x^2+3x-4}$ **44.** $G(x) = \dfrac{x-7}{x^2+x-6}$

45. $g(x) = \sqrt{7-x}$ **46.** $f(x) = \sqrt{5+x}$

47. $G(x) = \dfrac{1}{\sqrt{7-x}}$ **48.** $F(x) = \dfrac{1}{\sqrt{5+x}}$

49. Two people are discussing the function

$$f(x) = \frac{x^2-4}{x^2-9}$$

and one says to the other, "$f(2)$ exists but $f(3)$ does not." Explain what they are talking about.

50. Referring to the function in Problem 49, do $f(-2)$ and $f(-3)$ exist? Explain.

The verbal statement "function f multiplies the square of the domain element by 3 and then subtracts 7 from the result" and the algebraic statement "$f(x) = 3x^2 - 7$" define the same function. In Problems 51–54, translate each verbal definition of a function into an algebraic definition.

51. Function g subtracts 5 from twice the cube of the domain element.

52. Function f multiplies the domain element by -3 and adds 4 to the result.

53. Function G multiplies the square root of the domain element by 2 and subtracts the square of the domain element from the result.

54. Function F multiplies the cube of the domain element by -8 and adds 3 times the square root of 3 to the result.

In Problems 55–58, translate each algebraic definition of the function into a verbal definition.

55. $f(x) = 2x - 3$ **56.** $g(x) = -2x + 7$

57. $F(x) = 3x^3 - 2\sqrt{x}$ **58.** $G(x) = 4\sqrt{x} - x^2$

Determine which of the equations in Problems 59–68 specify functions with independent variable x. For those that do, find the domain. For those that do not, find a value of x to which there corresponds more than one value of y.

59. $4x - 5y = 20$ **60.** $3y - 7x = 15$

61. $x^2 - y = 1$ **62.** $x - y^2 = 1$

63. $x + y^2 = 10$ **64.** $x^2 + y = 10$

65. $xy - 4y = 1$ **66.** $xy + y - x = 5$

67. $x^2 + y^2 = 25$ **68.** $x^2 - y^2 = 16$

69. If $F(t) = 4t + 7$, find: $\dfrac{F(3+h) - F(3)}{h}$

70. If $G(r) = 3 - 5r$, find: $\dfrac{G(2+h) - G(2)}{h}$

71. If $Q(x) = x^2 - 5x + 1$, find: $\dfrac{Q(2+h) - Q(2)}{h}$

72. If $P(x) = 2x^2 - 3x - 7$, find: $\dfrac{P(3+h) - P(3)}{h}$

C *In Problems 73–80, find and simplify:*

$$\frac{f(a+h) - f(a)}{h}$$

73. $f(x) = 4x - 3$ **74.** $f(x) = -3x + 9$

75. $f(x) = 4x^2 - 7x + 6$ **76.** $f(x) = 3x^2 + 5x - 8$

77. $f(x) = x^3$ **78.** $f(x) = x^3 - x$

79. $f(x) = \sqrt{x}$ **80.** $f(x) = \dfrac{1}{x}$

Problems 81–84 refer to the area A and perimeter P of a rectangle with length l and width w (see the figure).

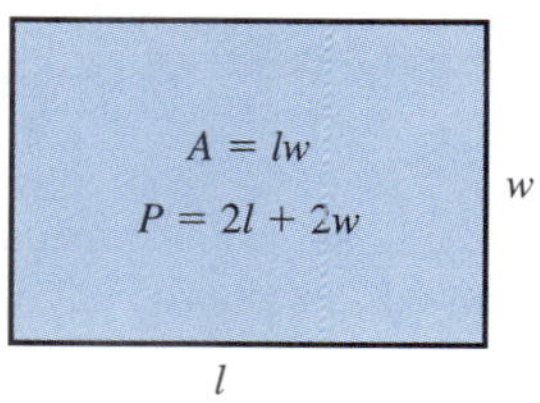

81. The area of a rectangle is 25 square inches. Express the perimeter $P(w)$ as a function of the width w, and state the domain of this function.

82. The area of a rectangle is 81 square inches. Express the perimeter $P(l)$ as a function of the length l, and state the domain of this function.

83. The perimeter of a rectangle is 100 meters. Express the area $A(l)$ as a function of the length l, and state the domain of this function.

84. The perimeter of a rectangle is 160 meters. Express the area $A(w)$ as a function of the width w, and state the domain of this function.

Applications

Business & Economics

85. *Price–demand.* A company manufactures memory chips for microcomputers. Its marketing research department, using statistical techniques, collected the data shown in Table 8, where p is the wholesale price per chip at which x million chips can be sold. Using special analytical techniques (regression analysis), an analyst produced the following price–demand function to model the data:

$$p(x) = 119 - 6x \qquad 1 \leqslant x \leqslant 15$$

Table 8

Price–Demand

x (MILLION)	p ($)
1	115
6	80
10	65
15	31

Plot the data points in Table 8, and sketch a graph of the price–demand function in the same coordinate system. What would be the estimated price per chip for a demand of 8 million chips? For a demand of 11 million chips?

86. *Price–demand.* A company manufactures "Notebook" computers. Its marketing research department, using statistical techniques, collected the data shown in Table 9, where p is the wholesale price per computer at which x thousand computers can be sold. Using special analytical techniques (regression analysis), an analyst produced the following price–demand function to model the data:

$$p(x) = 1{,}190 - 36x \qquad 1 \leqslant x \leqslant 25$$

Table 9

Price–Demand

x (THOUSAND)	p ($)
2	1,110
5	1,030
10	815
14	695
21	435

Plot the data points in Table 9, and sketch a graph of the price–demand function in the same coordinate system. What would be the estimated price per computer for a demand of 9 thousand computers? For a demand of 18 thousand computers?

87. *Revenue.*

(A) Using the price–demand function

$$p(x) = 119 - 6x \qquad 1 \leqslant x \leqslant 15$$

from Problem 85, write the company's revenue function and indicate its domain.

(B) Complete Table 10, computing revenues to the nearest million dollars.

Table 10

Revenue

x (MILLION)	$R(x)$ (MILLION $)
1	113
3	
6	
9	
12	
15	

(C) Plot the points from part (B) and sketch a graph of the revenue function through these points. Choose millions for the units on the horizontal and vertical axes. (C A graphing utility can be used as an aid.)

88. *Revenue.*

(A) Using the price–demand function

$$p(x) = 1{,}190 - 36x \qquad 1 \leqslant x \leqslant 25$$

from Problem 86, write the company's revenue function and indicate its domain.

(B) Complete Table 11, computing revenues to the nearest thousand dollars.

Table 11

Revenue

x (THOUSAND)	$R(x)$ (THOUSAND $)
2	2,236
5	
10	
15	
20	
25	

(C) Plot the points from part (B) and sketch a graph of the revenue function through these points. Choose thousands for the units on the horizontal and vertical axes. (A graphing utility can be used as an aid.)

89. *Profit.* The financial department for the company in Problems 85 and 87 established the following cost function for producing and selling x million memory chips:

$$C(x) = 234 + 23x \text{ million dollars}$$

(A) Write a profit function for producing and selling x million memory chips, and indicate its domain.

(B) Complete Table 12, computing profits to the nearest million dollars.

Table 12

PROFIT

x (MILLION)	$P(x)$ (MILLION \$)
1	−144
3	
6	
9	
12	
15	

(C) Plot the points in part (B) and sketch a graph of the profit function through these points. (A graphing utility can be used as an aid.)

90. *Profit.* The financial department for the company in Problems 86 and 88 established the following cost function for producing and selling x thousand "Notebook" computers:

$$C(x) = 4{,}320 + 146x \text{ thousand dollars}$$

(A) Write a profit function for producing and selling x thousand "Notebook" computers, and indicate the domain of this function.

(B) Complete Table 13, computing profits to the nearest thousand dollars.

Table 13

PROFIT

x (THOUSAND)	$P(x)$ (THOUSAND \$)
2	−2,376
5	
10	
15	
20	
25	

(C) Plot the points in part (B) and sketch a graph of the profit function through these points. (A graphing utility can be used as an aid.)

91. *Packaging.* A candy box is to be made out of a piece of cardboard that measures 8 by 12 inches. Equal-sized squares x inches on a side will be cut out of each corner, and then the ends and sides will be folded up to form a rectangular box.

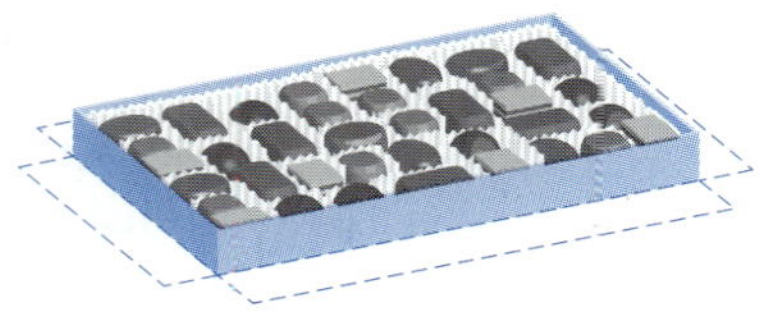

(A) Express the volume of the box $V(x)$ in terms of x.

(B) What is the domain of the function V (determined by the physical restrictions)?

(C) Complete Table 14.

Table 14

VOLUME

x	$V(x)$
1	
2	
3	

(D) Plot the points in part (C) and sketch a graph of the volume function through these points. (A graphing utility can be used as an aid.)

92. *Packaging.* Refer to Problem 91.

(A) Table 15 shows the volume of the box for some values of x between 1 and 2. Use these values to estimate to one decimal place the value of x between 1 and 2 that would produce a box with a volume of 65 cubic inches.

Table 15

VOLUME

x	$V(x)$
1.1	62.524
1.2	64.512
1.3	65.988
1.4	66.976
1.5	67.5
1.6	67.584
1.7	67.252

(B) Describe how you could refine this table to estimate x to two decimal places.

C (C) Carry out the refinement you described in part (B) and approximate x to two decimal places.

93. *Packaging.* Refer to Problems 91 and 92.

(A) Examine the graph of $V(x)$ from Problem 91D and discuss the possible locations of other values of x that would produce a box with a volume of 65 cubic inches. Construct a table like Table 15 to estimate any such value to one decimal place.

C (B) Refine the table you constructed in part (A) to provide an approximation to two decimal places.

94. *Packaging.* A parcel delivery service will only deliver packages with length plus girth (distance around) not exceeding 108 inches. A rectangular shipping box with square ends x inches on a side is to be used.

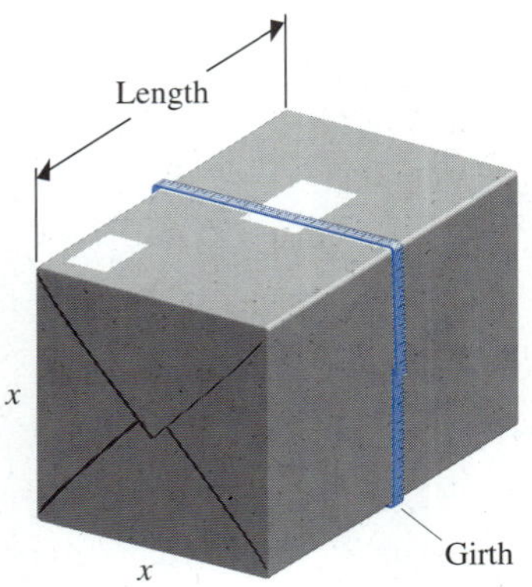

(A) If the full 108 inches is to be used, express the volume of the box $V(x)$ in terms of x.

(B) What is the domain of the function V (determined by the physical restrictions)?

(C) Complete Table 16.

Table 16

VOLUME

x	$V(x)$
5	
10	
15	
20	
25	

(D) Plot the points in part (C) and sketch a graph of the volume function through these points. (C A graphing utility can be used as an aid.)

Life Sciences

95. *Muscle contraction.* In a study of the speed of muscle contraction in frogs under various loads, noted British biophysicist and Nobel prize winner A. W. Hill determined that the weight w (in grams) placed on the muscle and the speed of contraction v (in centimeters per second) are approximately related by an equation of the form

$$(w + a)(v + b) = c$$

where a, b, and c are constants. Suppose that for a certain muscle, $a = 15$, $b = 1$, and $c = 90$. Express v as a function of w. Find the speed of contraction if a weight of 16 grams is placed on the muscle.

Social Sciences

96. *Politics.* The percentage s of seats in the House of Representatives won by Democrats and the percentage v of votes cast for Democrats (when expressed as decimal fractions) are related by the equation

$$5v - 2s = 1.4 \qquad 0 < s < 1, \quad 0.28 < v < 0.68$$

(A) Express v as a function of s, and find the percentage of votes required for the Democrats to win 51% of the seats.

(B) Express s as a function of v, and find the percentage of seats won if Democrats receive 51% of the votes.

SECTION 1-2

Elementary Functions: Graphs and Transformations

- A Beginning Library of Elementary Functions
- Vertical and Horizontal Shifts
- Reflections, Expansions, and Contractions
- Piecewise-Defined Functions

The functions

$$g(x) = x^2 - 4 \qquad h(x) = (x - 4)^2 \qquad k(x) = -4x^2$$

all can be expressed in terms of the function $f(x) = x^2$ as follows:

$$g(x) = f(x) - 4 \qquad h(x) = f(x - 4) \qquad k(x) = -4f(x)$$

In this section we will see that the graphs of functions g, h, and k are closely related to the graph of function f. Insight gained by understanding these relationships will help us analyze and interpret the graphs of many different functions.

A Beginning Library of Elementary Functions

As you progress through this book, and most any other mathematics course beyond this one, you will repeatedly encounter a relatively small list of elementary functions. We will identify these functions, study their basic properties, and include them in a library of elementary functions (see the inside front cover). This library will become an important addition to your mathematical toolbox and can be used in any course or activity where mathematics is applied.

Figure 1 shows six basic functions that you will encounter frequently. You should know the definition, domain, and range of each, and be able to recognize their graphs. For Figure 1B, recall the definition of *absolute value:*

$$|x| = \begin{cases} -x & \text{if } x < 0 \\ x & \text{if } x \geqslant 0 \end{cases}$$

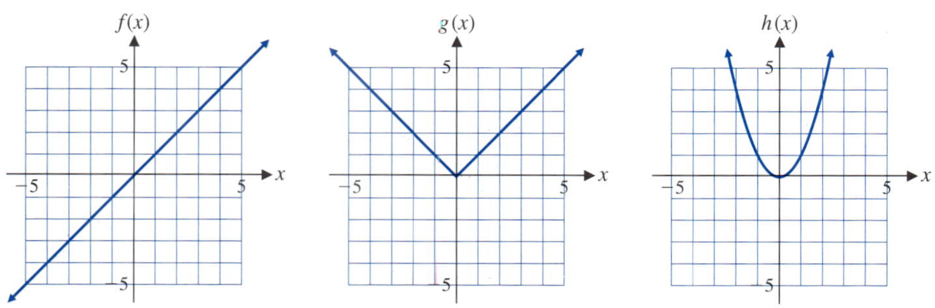

(A) **Identity function**
$f(x) = x$
Domain: R
Range: R

(B) **Absolute value function**
$g(x) = |x|$
Domain: R
Range: $[0, \infty)$

(C) **Square function**
$h(x) = x^2$
Domain: R
Range: $[0, \infty)$

FIGURE 1
Some basic functions and their graphs.
Note: Letters used to designate these functions may vary from context to context; R is the set of all real numbers.

(continued)

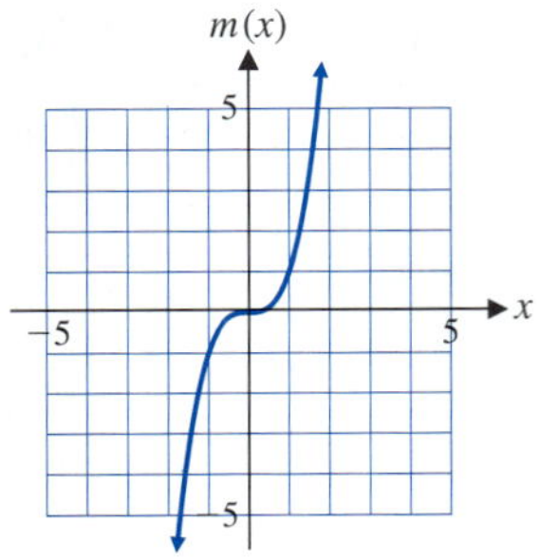

(D) **Cube function**
$m(x) = x^3$
Domain: R
Range: R

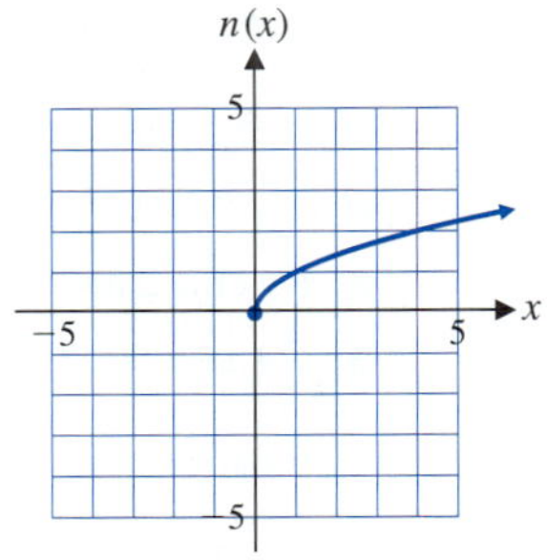

(E) **Square root function**
$n(x) = \sqrt{x}$
Domain: $[0, \infty)$
Range: $[0, \infty)$

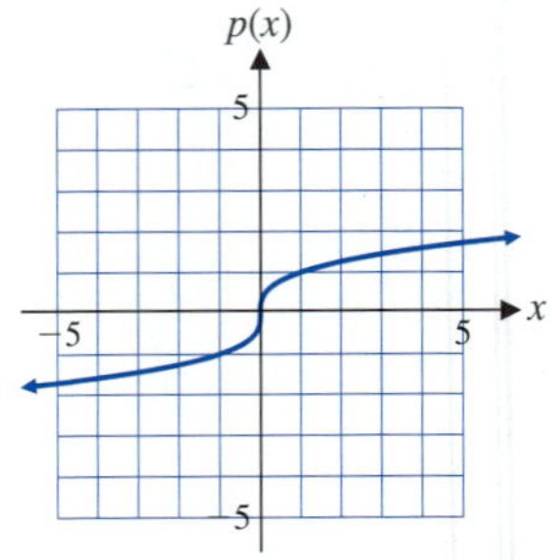

(F) **Cube root function**
$p(x) = \sqrt[3]{x}$
Domain: R
Range: R

FIGURE 1 (*continued*)

Vertical and Horizontal Shifts

If a new function is formed by performing an operation on a given function, then the graph of the new function is called a **transformation** of the graph of the original function. For example, graphs of both $y = f(x) + k$ and $y = f(x + h)$ are transformations of the graph of $y = f(x)$.

Explore–Discuss 1

Let $f(x) = x^2$.

(A) Graph $y = f(x) + k$ for $k = -4, 0,$ and 2 simultaneously in the same coordinate system. Describe the relationship between the graph of $y = f(x)$ and the graph of $y = f(x) + k$ for k any real number.

(B) Graph $y = f(x + h)$ for $h = -4, 0,$ and 2 simultaneously in the same coordinate system. Describe the relationship between the graph of $y = f(x)$ and the graph of $y = f(x + h)$ for h any real number.

Example 1 **Vertical and Horizontal Shifts**

(A) How are the graphs of $y = |x| + 4$ and $y = |x| - 5$ related to the graph of $y = |x|$? Confirm your answer by graphing all three functions simultaneously in the same coordinate system.

(B) How are the graphs of $y = |x + 4|$ and $y = |x - 5|$ related to the graph of $y = |x|$? Confirm your answer by graphing all three functions simultaneously in the same coordinate system.

SOLUTION (A) The graph of $y = |x| + 4$ is the same as the graph of $y = |x|$ shifted upward 4 units, and the graph of $y = |x| - 5$ is the same as the graph of $y = |x|$ shifted downward 5 units. Figure 2 confirms these conclusions. [It appears that the graph of $y = f(x) + k$ is the graph of $y = f(x)$ shifted up if k is positive and down if k is negative.]

(B) The graph of $y = |x + 4|$ is the same as the graph of $y = |x|$ shifted to the left 4 units, and the graph of $y = |x - 5|$ is the same as the graph of $y = |x|$ shifted to the right 5 units. Figure 3 confirms these conclusions. [It appears that the graph of $y = f(x + h)$ is the graph of $y = f(x)$ shifted right if h is negative and left if h is positive—the opposite of what you might expect.]

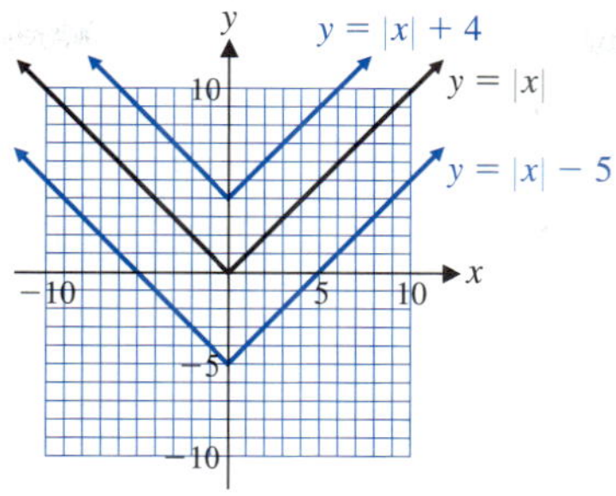

FIGURE 2
Vertical shifts

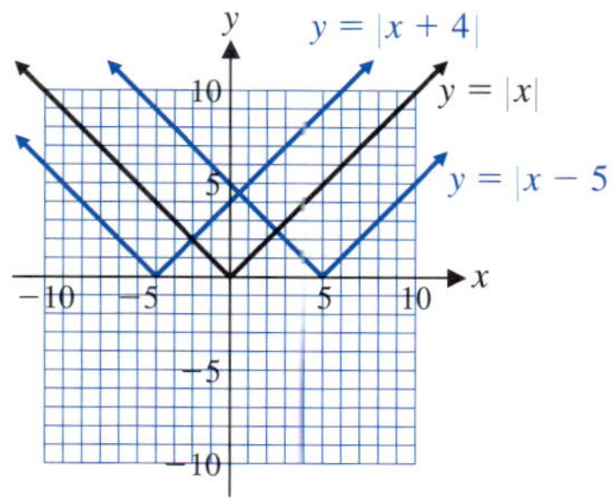

FIGURE 3
Horizontal shifts

Matched Problem 1 (A) How are the graphs of $y = \sqrt{x} + 5$ and $y = \sqrt{x} - 4$ related to the graph of $y = \sqrt{x}$? Confirm your answer by graphing all three functions simultaneously in the same coordinate system.

(B) How are the graphs of $y = \sqrt{x + 5}$ and $y = \sqrt{x - 4}$ related to the graph of $y = \sqrt{x}$? Confirm your answer by graphing all three functions simultaneously in the same coordinate system.

Comparing the graphs of $y = f(x) + k$ with the graph of $y = f(x)$, we see that the graph of $y = f(x) + k$ can be obtained from the graph of $y = f(x)$ by **vertically translating** (shifting) the graph of the latter upward k units if k is positive and downward $|k|$ units if k is negative. Comparing the graphs of $y = f(x + h)$ with the graph of $y = f(x)$, we see that the graph of $y = f(x + h)$ can be obtained from the graph of $y = f(x)$ by **horizontally translating** (shifting) the graph of the latter h units to the left if h is positive and $|h|$ units to the right if h is negative.

Example 2 **Vertical and Horizontal Translations (Shifts)** The graphs in Figure 4 are either horizontal or vertical shifts of the graph of $f(x) = x^2$. Write appropriate equations for functions H, G, M, and N in terms of f.

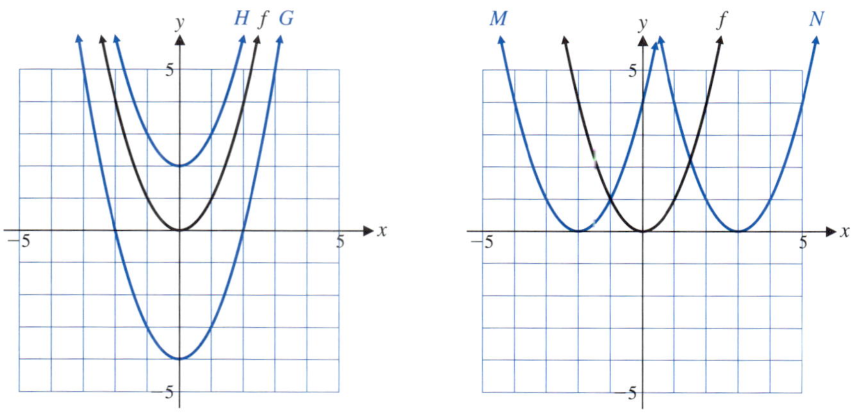

(A) (B)

FIGURE 4
Vertical and horizontal shifts

SOLUTION Functions H and G are vertical shifts given by

$$H(x) = x^2 + 2 \qquad G(x) = x^2 - 4$$

Functions M and N are horizontal shifts given by

$$M(x) = (x + 2)^2 \qquad N(x) = (x - 3)^2$$

Matched Problem 2 The graphs in Figure 5 are either horizontal or vertical shifts of the graph of $f(x) = \sqrt[3]{x}$. Write appropriate equations for functions H, G, M, and N in terms of f.

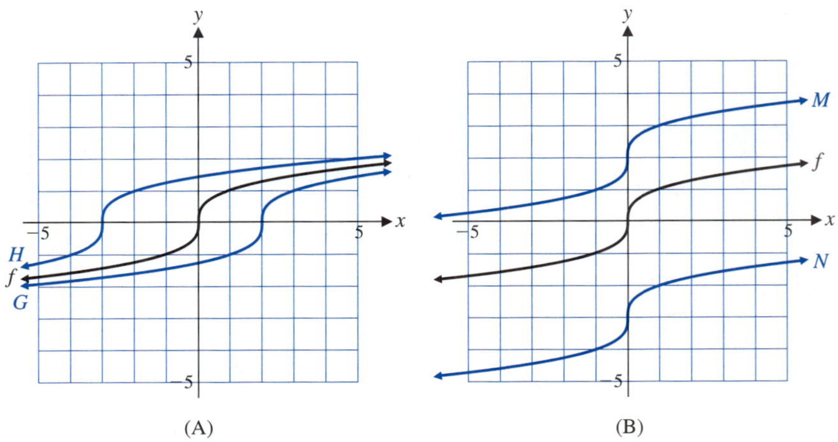

FIGURE 5
Vertical and horizontal shifts

■ REFLECTIONS, EXPANSIONS, AND CONTRACTIONS

We now investigate how the graph of $y = Af(x)$ is related to the graph of $y = f(x)$ for different real numbers A.

Explore–Discuss 2

(A) Graph $y = Ax^2$ for $A = 1, 4,$ and $\frac{1}{4}$ simultaneously in the same coordinate system.
(B) Graph $y = Ax^2$ for $A = -1, -4,$ and $-\frac{1}{4}$ simultaneously in the same coordinate system.
(C) Describe the relationship between the graph of $h(x) = x^2$ and the graph of $G(x) = Ax^2$ for A any real number.

Comparing $y = Af(x)$ to $y = f(x)$, we see that the graph of $y = Af(x)$ can be obtained from the graph of $y = f(x)$ by multiplying each ordinate value of the latter by A. The result is a **vertical expansion** of the graph of $y = f(x)$ if $A > 1$, a **vertical contraction** of the graph of $y = f(x)$ if $0 < A < 1$, and a **reflection in the x axis** if $A = -1$. If A is a negative number other than -1, then the result is a combination of a reflection in the x axis and either a vertical expansion or a vertical contraction.

Example 3 **Reflections, Expansions, and Contractions**

(A) How are the graphs of $y = 2|x|$ and $y = 0.5|x|$ related to the graph of $y = |x|$? Confirm your answer by graphing all three functions simultaneously in the same coordinate system.

(B) How is the graph of $y = -2|x|$ related to the graph of $y = |x|$? Confirm your answer by graphing both functions simultaneously in the same coordinate system.

Solution (A) The graph of $y = 2|x|$ is a vertical expansion of the graph of $y = |x|$ by a factor of 2, and the graph of $y = 0.5|x|$ is a vertical contraction of the graph of $y = |x|$ by a factor of 0.5. Figure 6 confirms this conclusion.

(B) The graph of $y = -2|x|$ is a reflection in the x axis and a vertical expansion of the graph of $y = |x|$. Figure 7 confirms this conclusion.

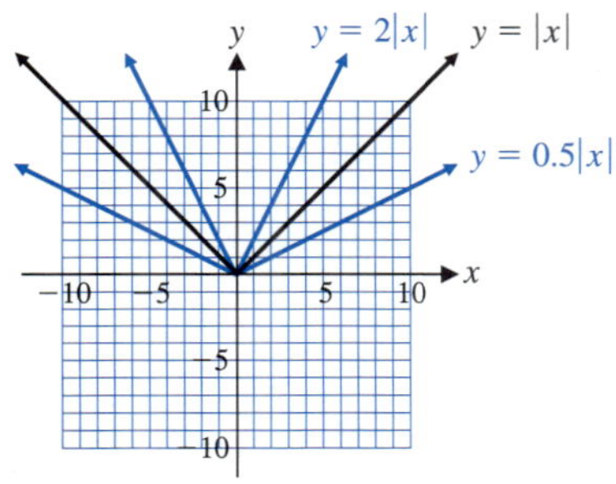

FIGURE 6
Vertical expansion and contraction

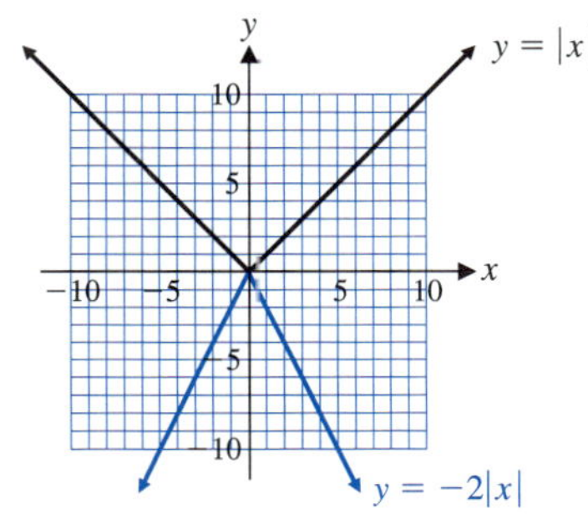

FIGURE 7
Reflection and vertical expansion

Matched Problem 3

(A) How are the graphs of $y = 2x$ and $y = 0.5x$ related to the graph of $y = x$? Confirm your answer by graphing all three functions simultaneously in the same coordinate system.

(B) How is the graph of $y = -0.5x$ related to the graph of $y = x$? Confirm your answer by graphing both functions in the same coordinate system.

The various transformations considered above are summarized in the following box for easy reference:

Graph Transformations (Summary)

Vertical Translation:

$$y = f(x) + k \quad \begin{cases} k > 0 & \text{Shift graph of } y = f(x) \text{ up } k \text{ units.} \\ k < 0 & \text{Shift graph of } y = f(x) \text{ down } |k| \text{ units.} \end{cases}$$

Horizontal Translation:

$$y = f(x + h) \quad \begin{cases} h > 0 & \text{Shift graph of } y = f(x) \text{ left } h \text{ units.} \\ h < 0 & \text{Shift graph of } y = f(x) \text{ right } |h| \text{ units.} \end{cases}$$

(continued)

Graph Transformations (Summary) (*continued*)

Reflection:

$y = -f(x)$ Reflect the graph of $y = f(x)$ in the x axis.

Vertical Expansion and Contraction:

$$y = Af(x) \quad \begin{cases} A > 1 & \text{Vertically expand graph of } y = f(x) \\ & \text{by multiplying each ordinate value by } A. \\ 0 < A < 1 & \text{Vertically contract graph of } y = f(x) \\ & \text{by multiplying each ordinate value by } A. \end{cases}$$

Explore–Discuss 3

Use a graphing utility, if available, to explore the graph of $y = A(x + h)^2 + k$ for various values of the constants A, h, and k. Discuss how the graph of $y = A(x + h)^2 + k$ is related to the graph of $y = x^2$.

Example 4 **Combining Graph Transformations** Discuss how the graph of $y = -|x - 3| + 1$ is related to the graph of $y = |x|$. Confirm your answer by graphing both functions simultaneously in the same coordinate system.

Solution The graph of $y = -|x - 3| + 1$ is a reflection in the x axis, a horizontal translation of 3 units to the right, and a vertical translation of 1 unit upward of the graph of $y = |x|$. Figure 8 confirms this description.

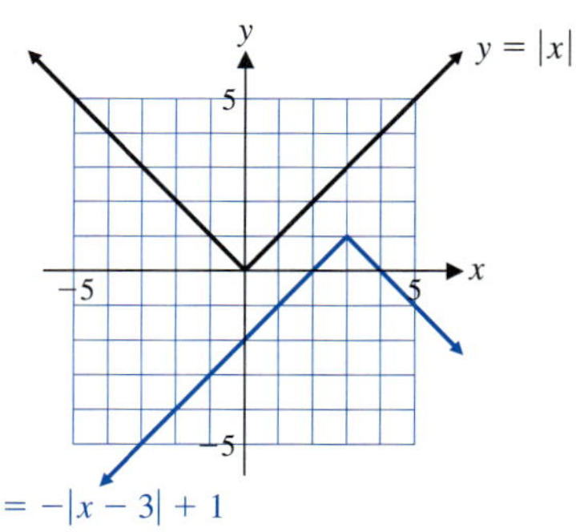

FIGURE 8
Combined transformations

Matched Problem 4 The graph of $y = G(x)$ in Figure 9 involves a reflection and a translation of the graph of $y = x^3$. Describe how the graph of function G is related to the graph of $y = x^3$ and find an equation of the function G.

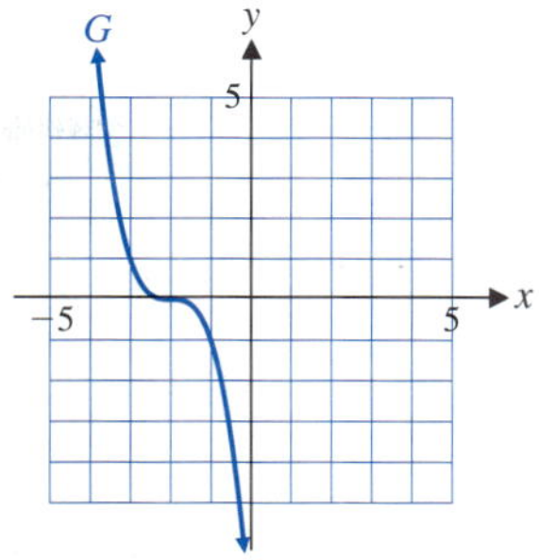

FIGURE 9
Combined transformations

Piecewise-Defined Functions

Earlier we noted that the absolute value of a real number x is defined as

$$|x| = \begin{cases} -x & \text{if } x < 0 \\ x & \text{if } x \geq 0 \end{cases}$$

Notice that this function is defined by different rules for different parts of its domain. Functions whose definitions involve more than one rule are called **piecewise-defined functions.** As the next example illustrates, piecewise-defined functions occur naturally in many applications.

Example 5 **Discount Pricing** Nordic Office Products uses volume discounting to promote sales. Table 1 shows the price per box for $\frac{1}{4}$ inch staples. The discounted prices for volume purchases apply to all boxes purchased.

Table 1
Price per Box

Number of Boxes	1	10	20
Price ($)	2.99	2.28	1.96

(A) Write a piecewise definition for the cost $C(x)$ of ordering x boxes of staples.
(B) Graph C.
(C) Explain why a customer would never want to order 8 or 9 boxes of staples.

Solution (A) The purchase cost of these staples is the price per box times the number of boxes purchased. Since the price per box depends on the volume purchased, the cost function cannot be given by a single rule and must be written as a piecewise-defined function:

$$C(x) = \begin{cases} 2.99x & \text{if } 1 \leq x < 10 \\ 2.28x & \text{if } 10 \leq x < 20 \\ 1.96x & \text{if } 20 \leq x \end{cases}$$

(B) Note that this cost function is only defined for nonnegative integer values of x (an order for a partial box of staples is not practical). However, to facilitate graphing of this cost function, it is convenient to assume that $C(x)$ is defined for all $x \geq 1$. To graph C, first note that each rule in the definition of C represents a transformation of the identity function $f(x) = x$. Graphing each transformation over the indicated interval produces the graph of C shown in Figure 10. We use this graph to represent the cost function, with the understanding that noninteger values of x may require special interpretation.

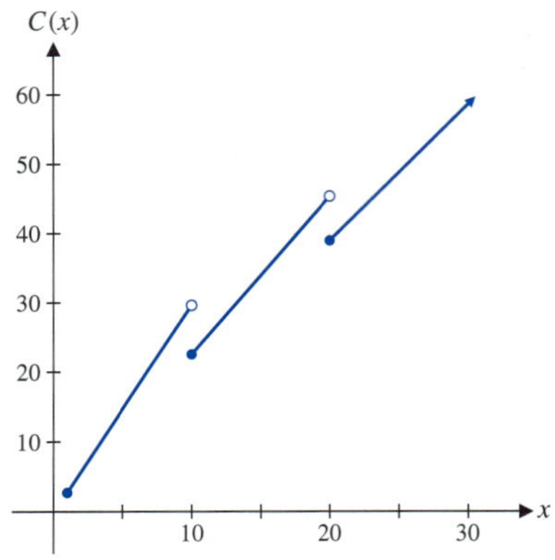

Note: A solid dot on the graph indicates that the point is part of the graph; an open dot indicates that the point is not part of the graph.

FIGURE 10
Cost of purchasing x boxes of staples

Table 2

x	$C(x)$ (\$)
8	23.92
9	26.91
10	22.80

(C) The drops in price at 10 boxes and at 20 boxes produce breaks in the graph of C. Examining the values of $C(x)$ near $x = 10$ (see Table 2), we see that the cost of 10 boxes is less than the cost of either 8 or 9 boxes. So it would be to a customer's advantage to order 10 boxes, rather than 8 or 9 boxes.

Matched Problem 5

Table 3 shows the price per pack for packs of 3×5 inch index cards at Nordic Office Products.

Table 3
Price per Pack

Number of Packs	1	5	10
Price (\$)	2.19	1.99	1.78

(A) Write a piecewise definition for the cost $C(x)$ of ordering x packs of index cards.

(B) Graph C.

(C) Explain why a customer would never want to order 9 packs of index cards.

Functions like the cost function C in Figure 10 are said to be **discontinuous** at the points where breaks occur. Thus, C is discontinuous at $x = 10$ and at $x = 20$. (The study of calculus includes a more formal presentation of this topic.)

Explore–Discuss 4

How can we graph the piecewise-defined function C in Example 5 on a graphing utility (see Fig. 10)? With your graphing utility set in the connected mode, enter y_1, y_2, and y_3, as shown in Figure 11A and graph for $0 \le x \le 30, 0 \le y \le 70$. Then modify the definitions of y_1, y_2, and y_3, as shown in Figure 11B, and graph $y_4 = y_1 + y_2 + y_3$. Now try graphing y_4 in the dot mode. Which graph gives the most useful representation of Figure 10? Which method would be most useful for identifying points of discontinuity?

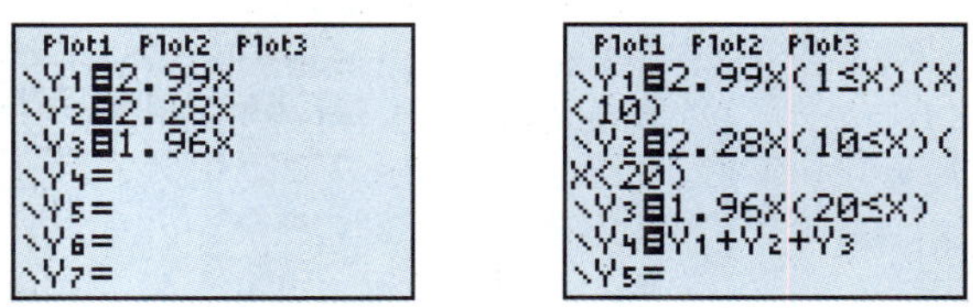

(A) (B)

FIGURE 11

Answers to Matched Problems

1. (A) The graph of $y = \sqrt{x} + 5$ is the same as the graph of $y = \sqrt{x}$ shifted upward 5 units, and the graph of $y = \sqrt{x} - 4$ is the same as the graph of $y = \sqrt{x}$ shifted downward 4 units. The figure confirms these conclusions.

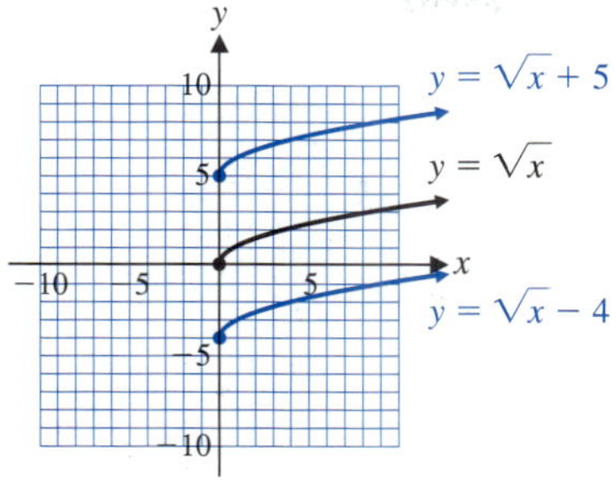

(B) The graph of $y = \sqrt{x + 5}$ is the same as the graph of $y = \sqrt{x}$ shifted to the left 5 units, and the graph of $y = \sqrt{x - 4}$ is the same as the graph of $y = \sqrt{x}$ shifted to the right 4 units. The figure confirms these conclusions.

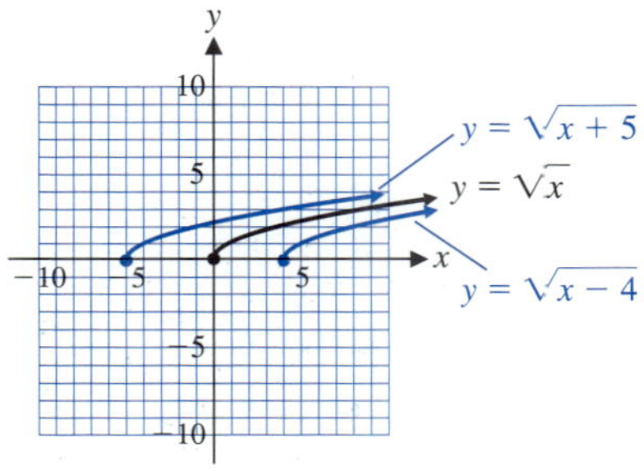

2. $H(x) = \sqrt[3]{x + 3}$, $G(x) = \sqrt[3]{x - 2}$, $M(x) = \sqrt[3]{x} + 2$, $N(x) = \sqrt[3]{x} - 3$

3. (A) The graph of $y = 2x$ is a vertical expansion of the graph of $y = x$, and the graph of $y = 0.5x$ is a vertical contraction of the graph of $y = x$. The figure confirms these conclusions.

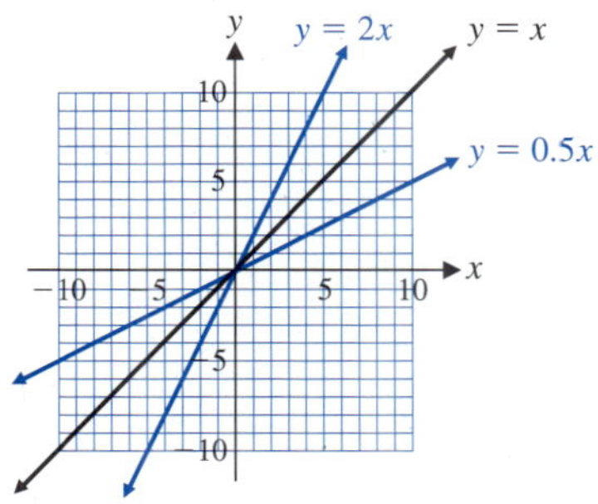

(B) The graph of $y = -0.5x$ is a vertical contraction and a reflection in the x axis of the graph of $y = x$. The figure confirms this conclusion.

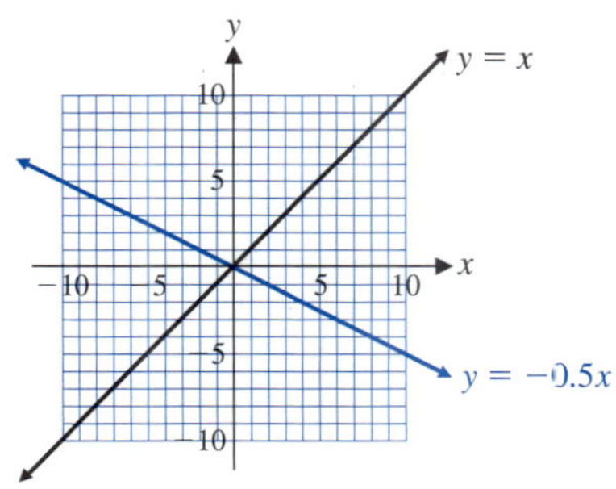

4. The graph of function G is a reflection in the x axis and a horizontal translation of 2 units to the left of the graph of $y = x^3$. An equation for G is $G(x) = -(x + 2)^3$.

5. (A) $C(x) = \begin{cases} 2.19x & \text{if } 1 \leq x < 5 \\ 1.99x & \text{if } 5 \leq x < 10 \\ 1.78x & \text{if } 10 \leq x \end{cases}$

(B)

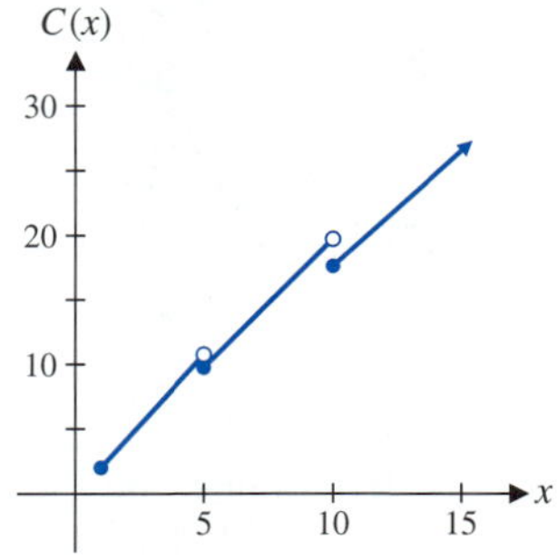

(C) The cost of 9 packs is \$17.91 and the cost of 10 packs is \$17.80. So it is always to the customer's advantage to order 10 packs instead of 9.

EXERCISE 1-2

A *Without looking back in the text, indicate the domain and range of each of the following functions. (Making rough sketches on scratch paper may help.)*

1. $f(x) = 0.4x$ **2.** $g(x) = 3x$

3. $h(x) = -x^2$ **4.** $m(x) = -|x|$

5. $g(x) = -2\sqrt{x}$ **6.** $f(x) = -0.5\sqrt[3]{x}$

7. $F(x) = -0.1x^3$ **8.** $G(x) = 5x^3$

Graph each of the functions in Problems 9–20 using the graphs of functions f and g below:

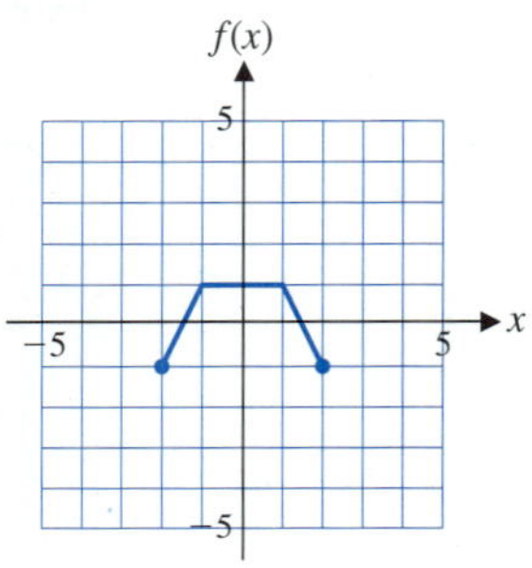

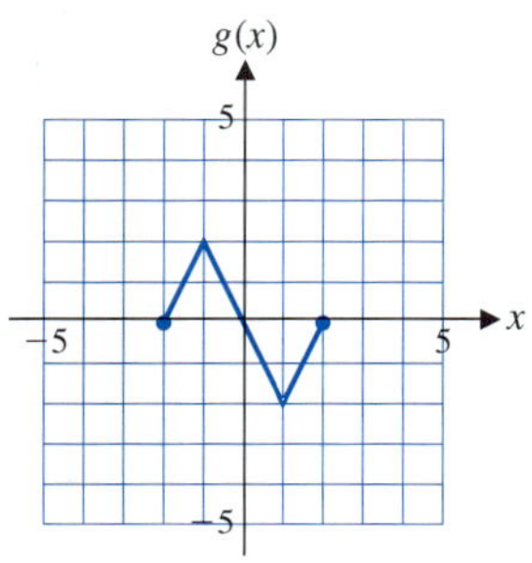

9. $y = f(x) + 2$ **10.** $y = g(x) - 1$

11. $y = f(x + 2)$ **12.** $y = g(x - 1)$

13. $y = g(x - 3)$ **14.** $y = f(x + 3)$

15. $y = g(x) - 3$ **16.** $y = f(x) + 3$

17. $y = -f(x)$ **18.** $y = -g(x)$

19. $y = 0.5g(x)$ **20.** $y = 2f(x)$

B *In Problems 21–28, indicate how the graph of each function is related to the graph of one of the six basic functions in Figure 1 (at the beginning of this section). Sketch a graph of each function.*

21. $g(x) = -|x + 3|$ **22.** $h(x) = -|x - 5|$

23. $f(x) = (x - 4)^2 - 3$ **24.** $m(x) = (x + 3)^2 + 4$

25. $f(x) = 7 - \sqrt{x}$ **26.** $g(x) = -6 + \sqrt[3]{x}$

27. $h(x) = -3|x|$ **28.** $m(x) = -0.4x^2$

Check your descriptions and graphs in Problems 21–28 by graphing each function on a graphing utility.

Each graph in Problems 29–36 is the result of applying a sequence of transformations to the graph of one of the six basic functions in Figure 1 (at the beginning of this section). Identify the basic function and describe the transformation verbally. Write an equation for the given graph.

29.

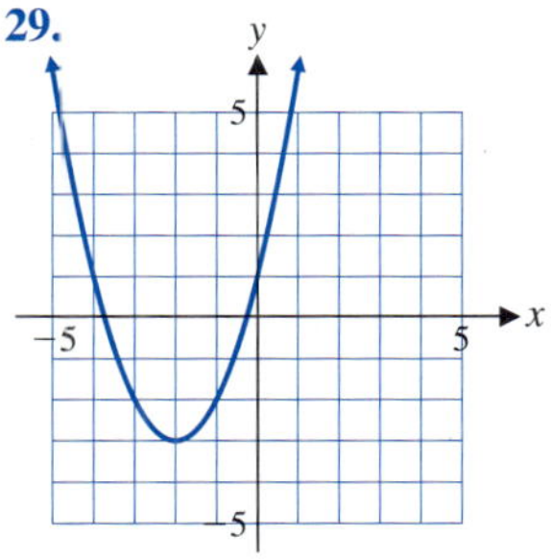

30.

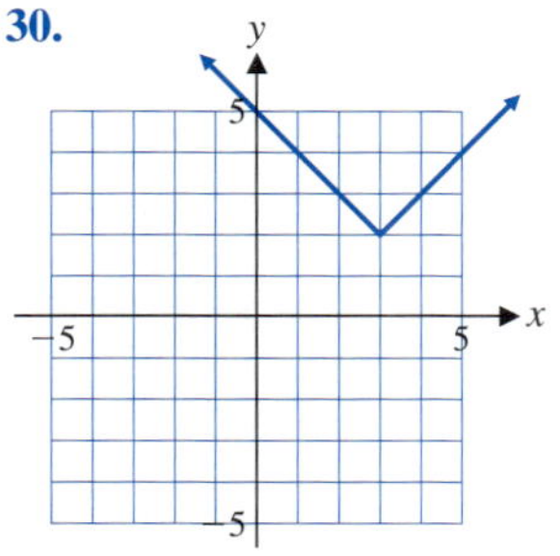

31.

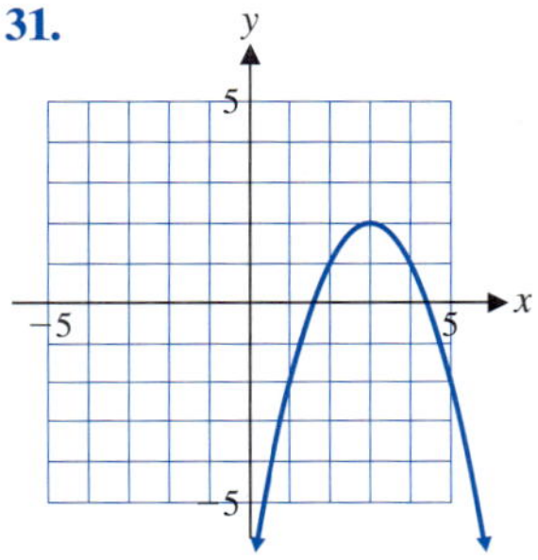

32.

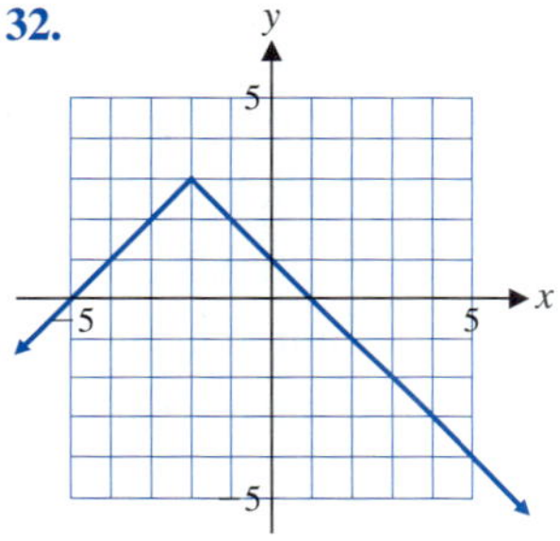

33.

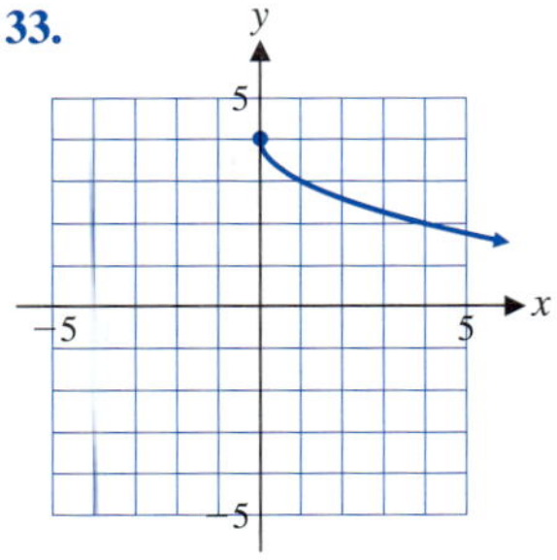

34.

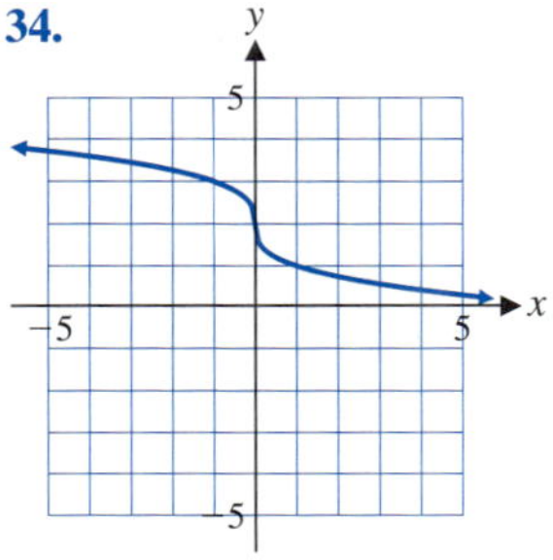

35.

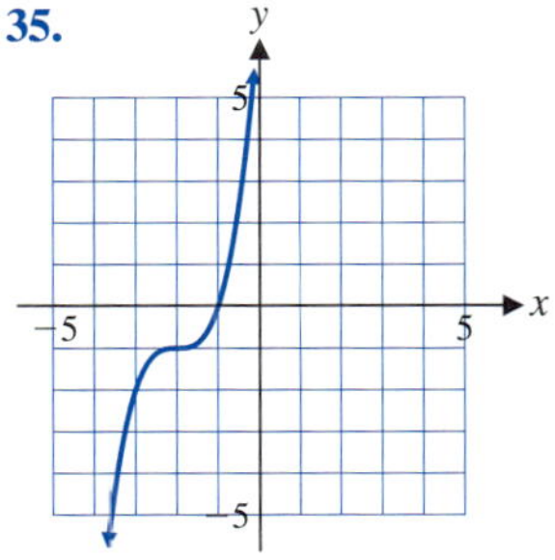

36.

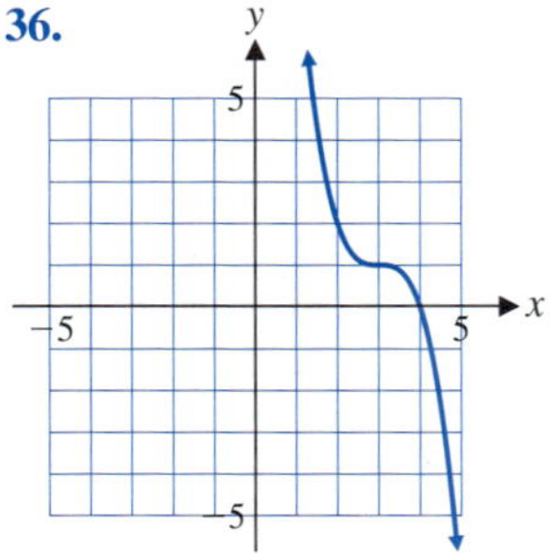

Check your equations in Problems 29–36 by graphing each on a graphing utility.

In Problems 37–42, the graph of the function g is formed by applying the indicated sequence of transformations to the given function f. Find an equation for the function g and graph g using $-5 \leq x \leq 5$ and $-5 \leq y \leq 5$.

37. The graph of $f(x) = \sqrt{x}$ is shifted 2 units to the right and 3 units down.

38. The graph of $f(x) = \sqrt[3]{x}$ is shifted 3 units to the left and 2 units up.

39. The graph of $f(x) = |x|$ is reflected in the x axis and shifted to the left 3 units.

40. The graph of $f(x) = |x|$ is reflected in the x axis and shifted to the right 1 unit.

41. The graph of $f(x) = x^3$ is reflected in the x axis and shifted 2 units to the right and down 1 unit.

42. The graph of $f(x) = x^2$ is reflected in the x axis and shifted to the left 2 units and up 4 units.

Graph each function in Problems 43–48 and find any points of discontinuity.

43. $f(x) = \begin{cases} x + 1 & \text{if } x < 0 \\ x - 1 & \text{if } x \geq 0 \end{cases}$

44. $g(x) = \begin{cases} x^2 - 1 & \text{if } x < 0 \\ 2 - x^2 & \text{if } x \geq 0 \end{cases}$

45. $h(x) = \begin{cases} -x & \text{if } x \leq 0 \\ \sqrt{x} & \text{if } x > 0 \end{cases}$

46. $k(x) = \begin{cases} \sqrt[3]{x} & \text{if } x \leq 0 \\ x^3 & \text{if } x > 0 \end{cases}$

47. $p(x) = \begin{cases} -2x & \text{if } x \leq -1 \\ x^2 & \text{if } -1 < x < 1 \\ \sqrt{x - 1} & \text{if } 1 \leq x \end{cases}$

48. $q(x) = \begin{cases} \sqrt[3]{x + 2} & \text{if } x < -2 \\ \frac{1}{2}x & \text{if } -2 \leq x \leq 2 \\ \sqrt[3]{x - 2} & \text{if } 2 < x \end{cases}$

Check Problems 43–48 by graphing on a graphing utility.

C *Each of the graphs in Problems 49–54 involves a reflection in the x axis and/or a vertical expansion or contraction of one of the basic functions in Figure 1 (at the beginning of this section). Identify the basic function, and describe the transformation verbally. Write an equation for the given graph.*

49.

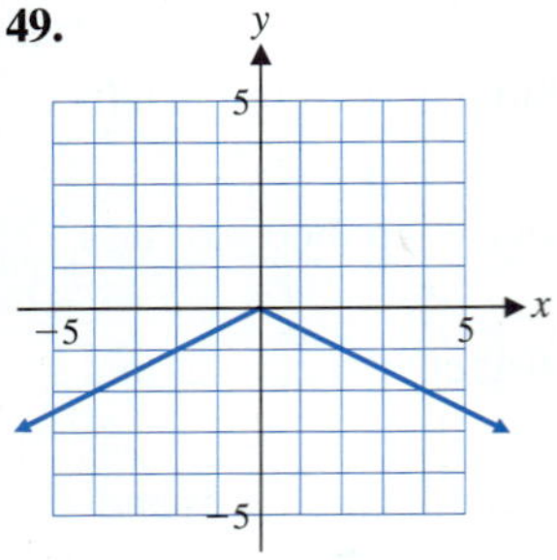

50.

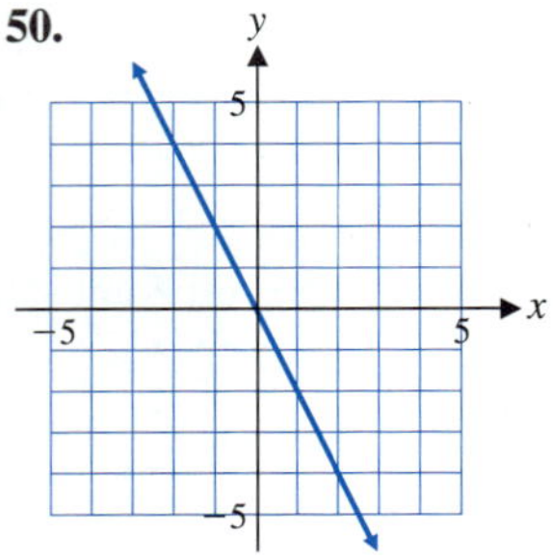

51.

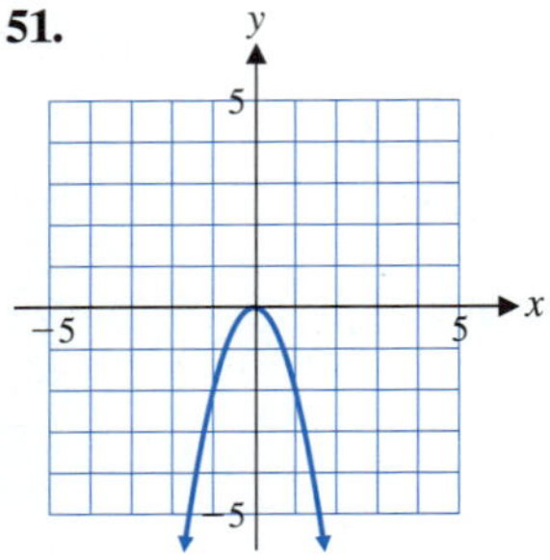

52.

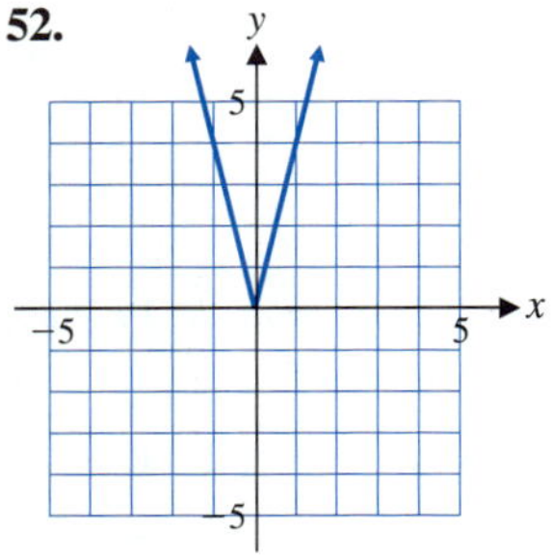

53.

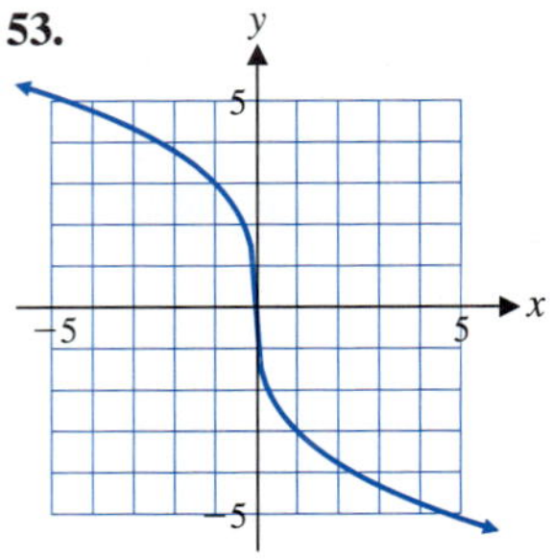

54.

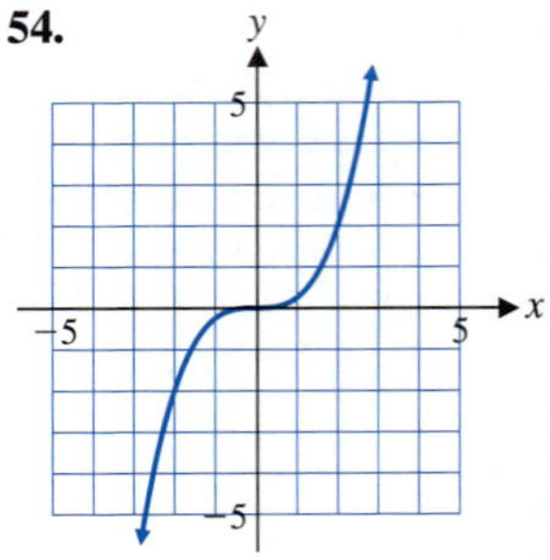

Check your equations in Problems 49–54 by graphing each on a graphing utility.

Changing the order in a sequence of transformations may change the final result. Investigate each pair of transformations in Problems 55–60 to determine if reversing their order can produce a different result. Support your conclusions with specific examples and/or mathematical arguments.

55. Vertical shift; horizontal shift

56. Vertical shift; reflection in y axis

57. Vertical shift; reflection in x axis

58. Vertical shift; expansion

59. Horizontal shift; reflection in y axis

60. Horizontal shift; contraction

Applications

Business & Economics

61. *Price–demand.* A retail chain sells CD players. The retail price $p(x)$ (in dollars) and the weekly demand x for a particular model are related by

$$p(x) = 115 - 4\sqrt{x} \qquad 9 \leq x \leq 289$$

(A) Describe how the graph of function p can be obtained from the graph of one of the basic functions in Figure 1 (at the beginning of this section).

(B) Sketch a graph of function p using part (A) as an aid.

62. *Price–supply.* The manufacturers of the CD players in Problem 61 are willing to supply x players at a price of $p(x)$ as given by the equation

$$p(x) = 4\sqrt{x} \qquad 9 \leq x \leq 289$$

(A) Describe how the graph of function p can be obtained from the graph of one of the basic functions in Figure 1 (at the beginning of this section).

(B) Sketch a graph of function p using part (A) as an aid.

63. *Hospital costs.* Using statistical methods, the financial department of a hospital arrived at the cost equation

$$C(x) = 0.00048(x - 500)^3 + 60{,}000 \qquad 100 \leq x \leq 1{,}000$$

where $C(x)$ is the cost in dollars for handling x cases per month.

(A) Describe how the graph of function C can be obtained from the graph of one of the basic functions in Figure 1 (at the beginning of this section).

C (B) Sketch a graph of function C using part (A) and a graphing utility as aids.

64. *Price–demand.* A company manufactures and sells in-line skates. Their financial department has established the price–demand function

$$p(x) = 190 - 0.013(x - 10)^2 \qquad 10 \leq x \leq 100$$

where $p(x)$ is the price at which x thousand pairs of skates can be sold.

(A) Describe how the graph of function p can be obtained from the graph of one of the basic functions in Figure 1 (at the beginning of this section).

C (B) Sketch a graph of function p using part (A) and a graphing utility as aids.

65. *Discount pricing.* The prices for boxes of double-pocket portfolios at Nordic Office Products are given in Table 4.
(A) Write a piecewise definition for the cost $C(x)$ of ordering x boxes of portfolios.
(B) Graph C.
(C) Would a customer ever want to purchase 3 boxes of portfolios? 9 boxes of portfolios? Explain.

Table 4

PRICE PER BOX

NUMBER OF BOXES	1	4	10
PRICE ($)	10.06	8.52	7.31

66. *Discount pricing.* The prices for purchasing rolls of transparent tape at Nordic Office Products are given in Table 5.
(A) Write a piecewise definition for the cost $C(x)$ of ordering x rolls of tape.
(B) Graph C.
(C) Would a customer ever want to purchase 11 rolls of tape? 22 or 23 rolls of tape? Explain.

Table 5

PRICE PER ROLL

NUMBER OF ROLLS	1	12	24
PRICE ($)	2.39	2.19	1.99

67. *Discount pricing.* A car rental company charges $0.50 per mile for the first 50 miles driven, $0.35 per mile for the next 50 miles, and $0.20 per mile for any additional miles.
(A) Write a piecewise definition for the total charges $C(x)$ for driving x miles.
(B) Graph C.
(C) How does this pricing method differ from the discount pricing used by Nordic Office Products in Problems 65 and 66? What effect does this have on the graph of C?

68. *Discount pricing.* A telephone company charges $0.06 per call for the first 60 calls during a 1 month billing period, $0.04 for the next 60 calls, and $0.02 for any additional calls.
(A) Write a piecewise definition for the total charges $C(x)$ for making x calls.
(B) Graph C.
(C) How does this pricing method differ from the discount pricing used by Nordic Office Products in Problems 65 and 66? What effect does this have on the graph of C?

Life Sciences

69. *Physiology.* A good approximation of the normal weight of a person 60 inches or taller but not taller than 80 inches is given by $w(x) = 5.5x - 220$, where x is height in inches and $w(x)$ is weight in pounds.

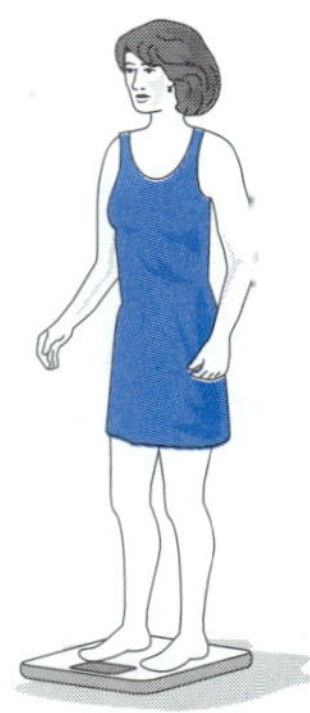

(A) Describe how the graph of function w can be obtained from the graph of one of the basic functions in Figure 1 (at the beginning of this section).
(B) Sketch a graph of function w using part (A) as an aid.

70. *Physiology.* The average weight of a particular species of snake is given by $w(x) = 463x^3$, $0.2 \le x \le 0.8$, where x is length in meters and $w(x)$ is weight in grams.
(A) Describe how the graph of function w can be obtained from the graph of one of the basic functions in Figure 1 (at the beginning of this section).
(B) Sketch a graph of function w using part (A) as an aid.

Social Sciences

71. *Safety research.* Under ideal conditions, if a person driving a vehicle slams on the brakes and skids to a stop, the speed of the vehicle $v(x)$ (in miles per hour) is given approximately by $v(x) = C\sqrt{x}$, where x is the length of skid marks (in feet) and C is a constant that depends on the road conditions and the weight of the vehicle. For a particular vehicle, $v(x) = 7.08\sqrt{x}$ and $4 \le x \le 144$.
(A) Describe how the graph of function v can be obtained from the graph of one of the basic functions in Figure 1 (at the beginning of this section).
(B) Sketch a graph of function v using part (A) as an aid.

72. *Learning.* A production analyst has found that on the average it takes a new person $T(x)$ minutes to perform a particular assembly operation after x performances of the operation, where $T(x) = 10 - \sqrt[3]{x}$, $0 \le x \le 155$.
(A) Describe how the graph of function T can be obtained from the graph of one of the basic functions in Figure 1 (at the beginning of this section).
(B) Sketch a graph of function T using part (A) as an aid.

SECTION 1-3

Linear Functions and Straight Lines

- INTERCEPTS
- LINEAR FUNCTIONS, EQUATIONS, AND INEQUALITIES
- GRAPHS OF $Ax + By = C$
- SLOPE OF A LINE
- EQUATIONS OF LINES—SPECIAL FORMS
- APPLICATIONS

In this section we will add another important class of functions to our basic list of elementary functions. These functions are called *linear functions* and include the identity function $f(x) = x$ as a special case. We will investigate the relationship between linear functions and the solutions to linear equations and inequalities. (A detailed treatment of algebraic solutions to linear equations and inequalities can be found in Appendix A-8. Finally, we will review the concept of slope and some of the standard equations of straight lines. These new tools will be applied to a variety of significant applications, including cost and price–demand functions.

INTERCEPTS

Figure 1 illustrates the graphs of three functions *f*, *g*, and *h*.

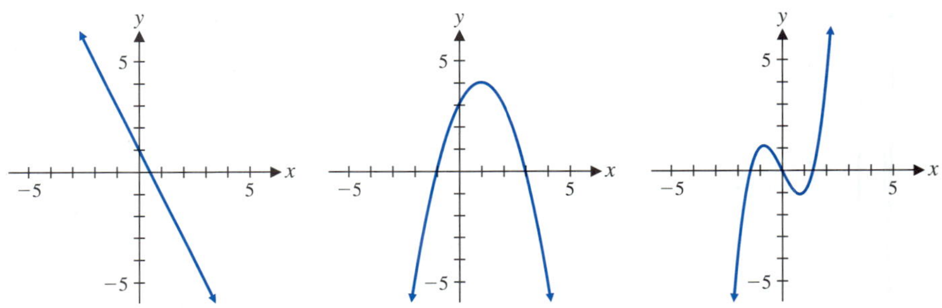

(A) $f(x) = -2x + 1$ (B) $g(x) = 3 + 2x - x^2$ (C) $h(x) = x^3 - 2x$

FIGURE 1
Graphs of several functions

If the graph of a function f crosses the x axis at a point with x coordinate a, then a is called an **x intercept** of f. If the graph of f crosses the y axis at a point with y coordinate b, then b is called the **y intercept.** It is common practice to refer to both the numbers a and b and the points $(a, 0)$ and $(0, b)$ as the x and y intercepts. If the y intercept exists, then 0 must be in the domain of f and the y intercept is simply $f(0)$. Thus, the graph of a function can have at most one y intercept. The x intercepts are all real solutions or roots of $f(x) = 0$, which may vary from none to an unlimited number. In Figure 1, function f has one y intercept and one x intercept; function g has one y intercept and two x intercepts; and function h has one y intercept and three x intercepts.

Linear Functions, Equations, and Inequalities

In Figure 1, the graph of $f(x) = -2x + 1$ is a straight line, and because of this, we choose to call this type of function a *linear function.* In general:

Linear and Constant Functions

A function f is a **linear function** if

$$f(x) = mx + b \qquad m \neq 0$$

where m and b are real numbers. The **domain** is the set of all real numbers, and the **range** is the set of all real numbers. If $m = 0$, then f is called a **constant function,**

$$f(x) = b$$

which has the set of all real numbers as its **domain** and the constant b as its **range.**

Since $mx + b$, $m \neq 0$, is a first-degree polynomial, linear functions are also called **first-degree functions.** Figure 2 shows the graphs of two linear functions f and g, and a constant function h.

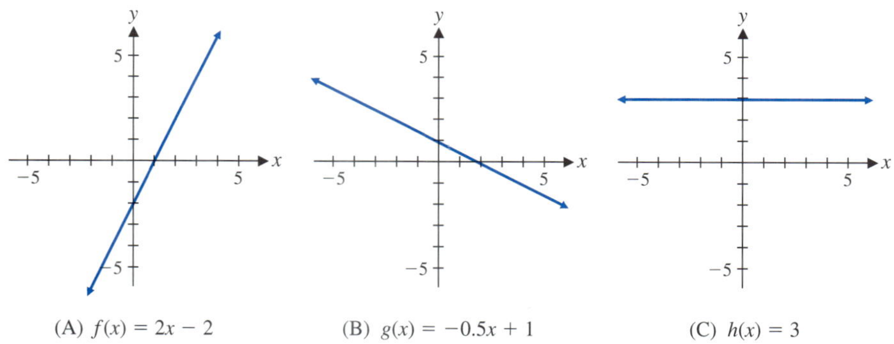

FIGURE 2
Two linear functions and a constant function

In Section 1-2 we saw that the graph of the identity function $f(x) = x$ is a straight line. The graph of a linear function $g(x) = mx + b$, $m \neq 0$, is the same as the graph of $f(x) = x$ reflected in the x axis if m is negative, vertically expanded or contracted by a factor of $|m|$, and shifted up or down $|b|$ units. In general, it can be shown that:

The graph of a linear function is a straight line that is neither horizontal nor vertical.

The graph of a constant function is a horizontal straight line.

What about vertical lines? Recall from Section 1-1 that the graph of a function cannot contain two points with the same x coordinate and different

y coordinates. Since *all* points on a vertical line have the same x coordinate, the graph of a function can never be a vertical line. Later in this section we will discuss equations of vertical lines, but these equations never define functions.

Explore–Discuss 1

(A) Is it possible for a linear function to have two x intercepts? No x intercepts? If either of your answers is yes, give an example.
(B) Is it possible for a linear function to have two y intercepts? No y intercept? If either of your answers is yes, give an example.
(C) Discuss the possible number of x and y intercepts for a constant function.

Example 1 **Intercepts, Equations, and Inequalities**

(A) Graph $f(x) = \frac{3}{2}x - 4$ in a rectangular coordinate system.
(B) Find the x and y intercepts algebraically to two decimal places.
C (C) Graph $f(x) = \frac{3}{2}x - 4$ in a standard viewing window.
C (D) Find the x and y intercepts to two decimal places using trace and zoom or an appropriate built-in routine in your graphing utility.
(E) Solve $\frac{3}{2}x - 4 \leq 0$ graphically to two decimal places using parts (A) and (B) or (C) and (D).

SOLUTION (A) Graph in a rectangular coordinate system:

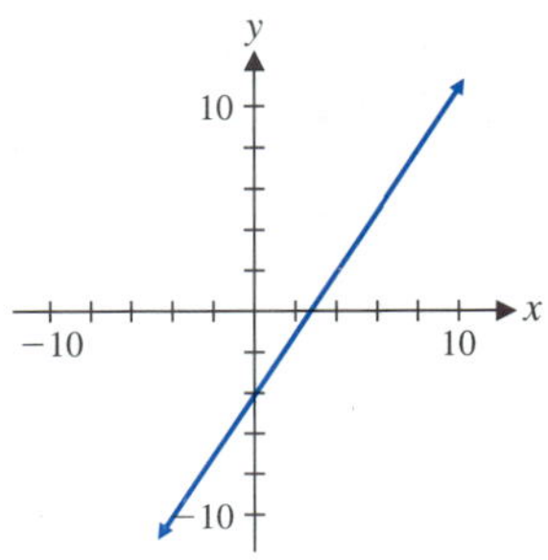

(B) Finding intercepts algebraically:

$$y \text{ intercept:} \quad f(0) = \tfrac{3}{2}(0) - 4 = -4$$
$$x \text{ intercept:} \quad f(x) = 0$$
$$\tfrac{3}{2}x - 4 = 0$$
$$\tfrac{3}{2}x = 4$$
$$x = \tfrac{8}{3} \approx 2.67$$

C (C) Graph in a graphing utility:

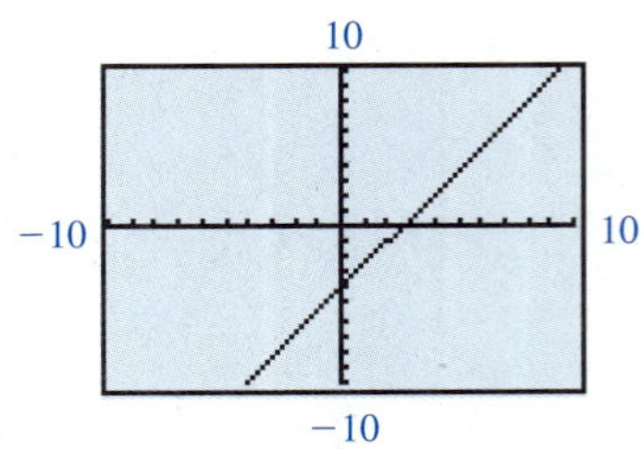

C (D) Finding intercepts graphically in a graphing utility:

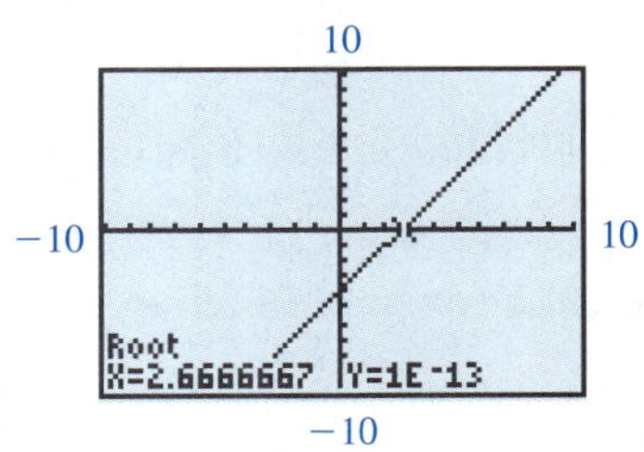

x intercept: 2.67

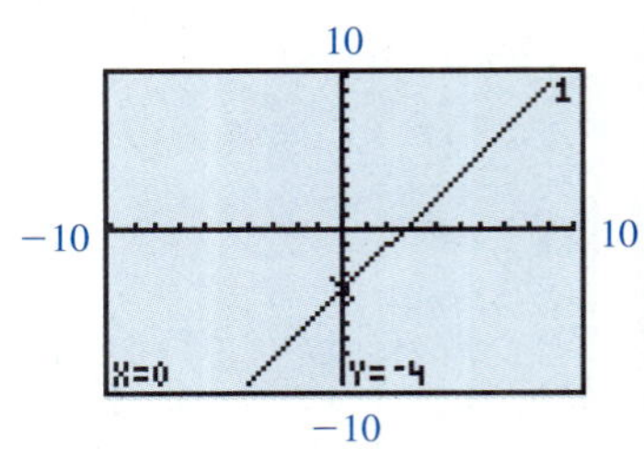

y intercept: -4

(E) Solving $\frac{3}{2}x - 4 \leq 0$ graphically using parts (A) and (B) or (C) and (D): The linear inequality $\frac{3}{2}x - 4 \leq 0$ holds for those values of x for which the graph of $f(x) = \frac{3}{2}x - 4$ in the figure in part (A) or (C) is at or below the x axis. This happens for x less than or equal to the x intercept found in parts (B) or (D). Thus, the solution set for the linear inequality is $x \leq 2.67$ or $(-\infty, 2.67]$.

Matched Problem 1

(A) Graph $f(x) = -\frac{4}{3}x + 5$ in a rectangular coordinate system.
(B) Find the x and y intercepts algebraically to two decimal places.
C (C) Graph $f(x) = -\frac{4}{3}x + 5$ in a standard viewing window.
C (D) Find the x and y intercepts to two decimal places using trace and zoom or an appropriate built-in routine in your graphing utility.
(E) Solve $-\frac{4}{3}x + 5 \geq 0$ graphically to two decimal places using parts (A) and (B) or (C) and (D).

Graphs of $Ax + By = C$

We now investigate graphs of linear, or first-degree, equations in two variables:

$$Ax + By = C \tag{1}$$

where A and B are not both 0. Depending on the values of A and B, this equation defines a linear function, a constant function, or no function at all. If $A \neq 0$ and $B \neq 0$, then equation (1) can be written as

$$y = -\frac{A}{B}x + \frac{C}{B} \quad \text{Linear function (slanted line)} \tag{2}$$

which is in the form $f(x) = mx + b$, $m \neq 0$, and hence is a linear function. If $A = 0$ and $B \neq 0$, then equation (1) can be written as

$$0x + By = C$$
$$y = \frac{C}{B} \quad \text{Constant function (horizontal line)} \tag{3}$$

which is in the form $g(x) = b$, and hence is a constant function. If $A \neq 0$ and $B = 0$, then equation (1) can be written as

$$Ax + 0y = C$$
$$x = \frac{C}{A} \quad \text{Not a function (vertical line)} \tag{4}$$

We can see that the graph of (4) is a vertical line, since the equation is satisfied for any value of y as long as x is the constant C/A. Hence, this form does not define a function.

The following theorem is a generalization of the above discussion:

Theorem 1 Graph of a Linear Equation in Two Variables

In a Cartesian plane, the graph of any equation of the form

Standard Form $\quad Ax + By = C \quad (5)$

where A, B, and C are real constants (A and B not both 0), is a straight line. Every straight line in a Cartesian plane coordinate system is the graph of an equation of this type.

(continued)

Theorem 1 *(continued)* Vertical and horizontal lines have particularly simple equations, which are special cases of equation (5):

Horizontal line with y intercept $C/B = b$: $y = b$
Vertical line with x intercept $C/A = a$: $x = a$

Explore–Discuss 2

Graph the following three special cases of $Ax + By = C$ in the same coordinate system:

(A) $3x + 2y = 6$ (B) $0x - 3y = 12$ (C) $2x + 0y = 10$

Which cases define functions? Explain why, or why not.

C Graph each case in the same viewing window using a graphing utility. (Check your manual on how to graph vertical lines.)

Sketching the graphs of equations of either form

$$Ax + By = C \qquad \text{or} \qquad y = mx + b$$

is very easy, since the graph of each equation is a straight line. All that is necessary is to plot any two points from the solution set and use a straightedge to draw a line through these two points. The x and y intercepts are usually the easiest to find.

Example 2 **Sketching Graphs of Lines**

(A) Graph $x = -4$ and $y = 6$ simultaneously in the same rectangular coordinate system. C Also, graph in a graphing utility.
(B) Write the equations of the vertical and horizontal lines that pass through the point $(7, -5)$.
(C) Graph the equation $2x - 3y = 12$ by hand. C Also, graph in a graphing utility.

SOLUTION (A) Graphing $x = -4$ and $y = 6$:

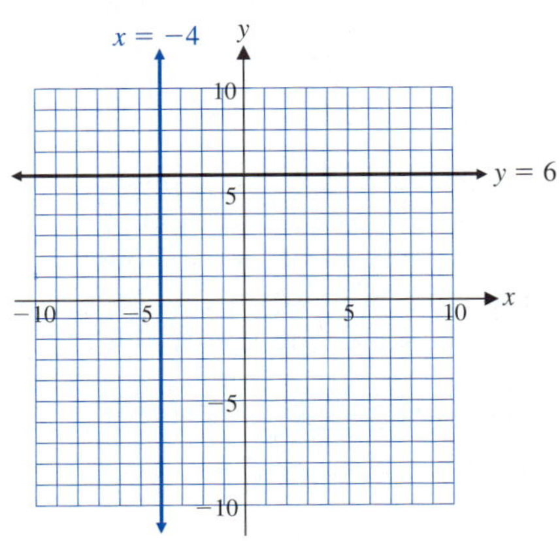

By hand

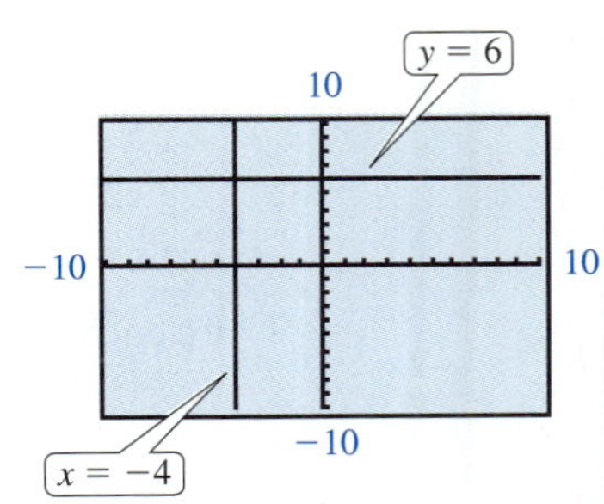

C In a graphing utility

(B) Horizontal line through $(7, -5)$: $y = -5$
Vertical line through $(7, -5)$: $x = 7$

(C) Graphing $2x - 3y = 12$: For the hand-drawn graph, find the intercepts by first letting $x = 0$ and solving for y and then letting $y = 0$ and solving for x. Then draw a line through the intercepts. **C** To graph in a graphing utility, solve the equation for y in terms of x and enter the result.

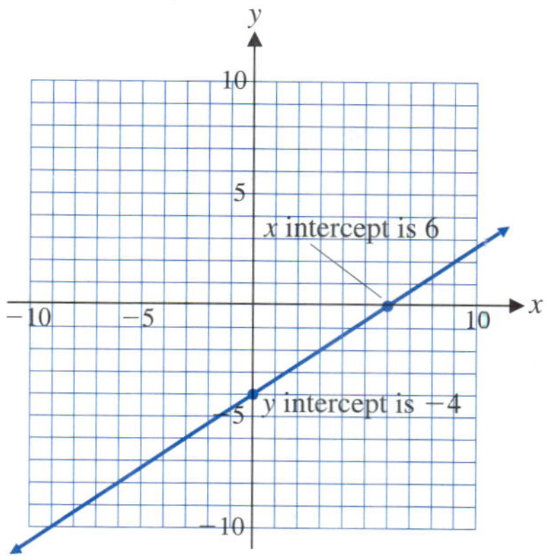

By hand

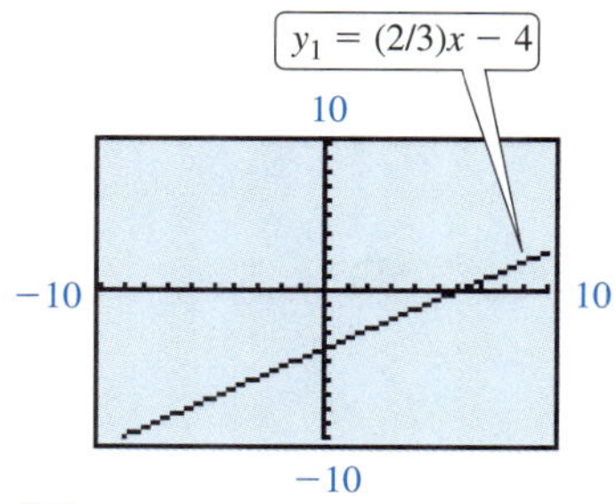

C In a graphing utility

Matched Problem 2

(A) Graph $x = 5$ and $y = -3$ simultaneously in the same rectangular coordinate system. **C** Also, graph in a graphing utility.
(B) Write the equations of the vertical and horizontal lines that pass through the point $(-8, 2)$.
(C) Graph the equation $3x + 4y = 12$ by hand. **C** Also, graph in a graphing utility.

Slope of a Line

If we take two points $P_1(x_1, y_1)$ and $P_2(x_2, y_2)$, on a line, then the ratio of the change in y to the change in x as the point moves from point P_1 to point P_2 is called the **slope** of the line. In a sense, slope provides a measure of the "steepness" of a line relative to the x axis. The change in x is often called the **run** and the change in y the **rise**.

Slope of a Line

If a line passes through two distinct points $P_1(x_1, y_1)$ and $P_2(x_2, y_2)$, then its slope is given by the formula

$$m = \frac{y_2 - y_1}{x_2 - x_1} \qquad x_1 \neq x_2$$

$$= \frac{\text{Vertical change (Rise)}}{\text{Horizontal change (Run)}}$$

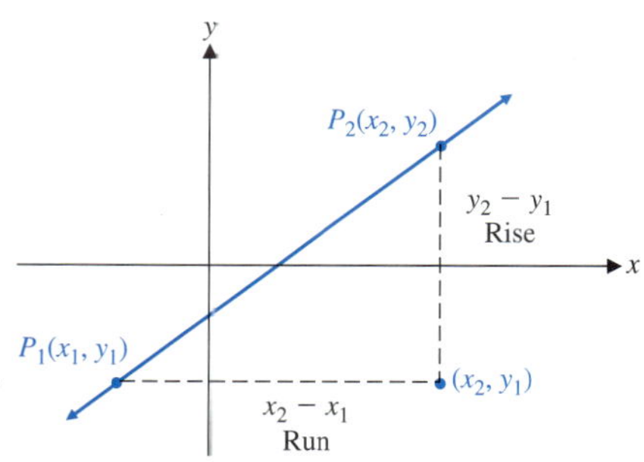

For a horizontal line, y does not change; hence, its slope is 0. For a vertical line, x does not change; hence, $x_1 = x_2$ and its slope is not defined. In general, the slope of a line may be positive, negative, 0, or not defined. Each case is illustrated geometrically in Table 1.

Table 1
Geometric Interpretation of Slope

Line	Slope	Example
Rising as x moves from left to right	Positive	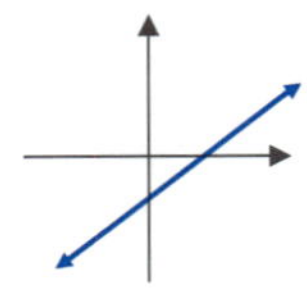
Falling as x moves from left to right	Negative	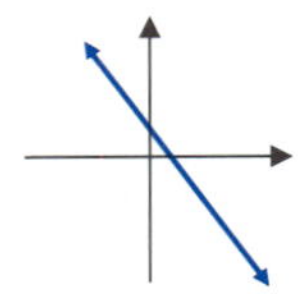
Horizontal	0	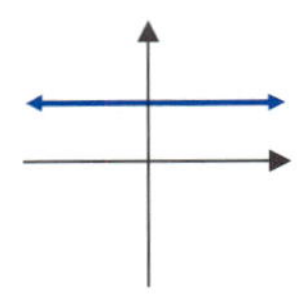
Vertical	Not defined	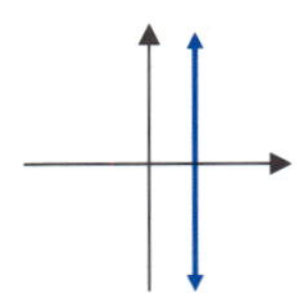

In using the formula to find the slope of the line through two points, it does not matter which point is labeled P_1 or P_2, since changing the labeling will change the sign in both the numerator and denominator of the slope formula, resulting in equivalent expressions. In addition, it is important to note that the definition of slope does not depend on the two points chosen on the line as long as they are distinct. This follows from the fact that the ratios of corresponding sides of similar triangles are equal.

Example 3 **Finding Slopes** Sketch a line through each pair of points, and find the slope of each line.

(A) $(-3, -2), (3, 4)$ (B) $(-1, 3), (2, -3)$
(C) $(-2, -3), (3, -3)$ (D) $(-2, 4), (-2, -2)$

SOLUTION

(A)

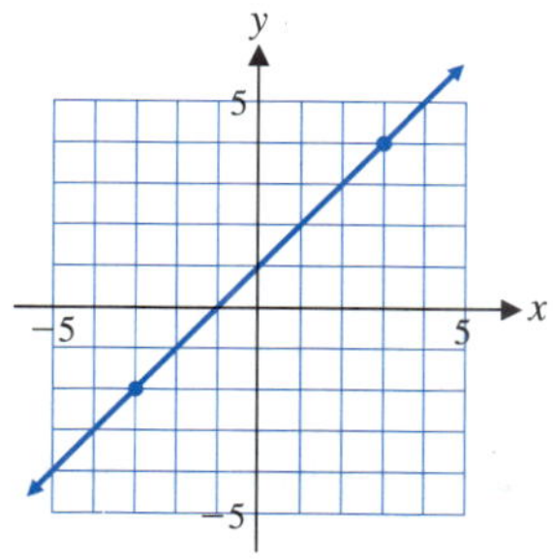

$$m = \frac{4 - (-2)}{3 - (-3)} = \frac{6}{6} = 1$$

(B)

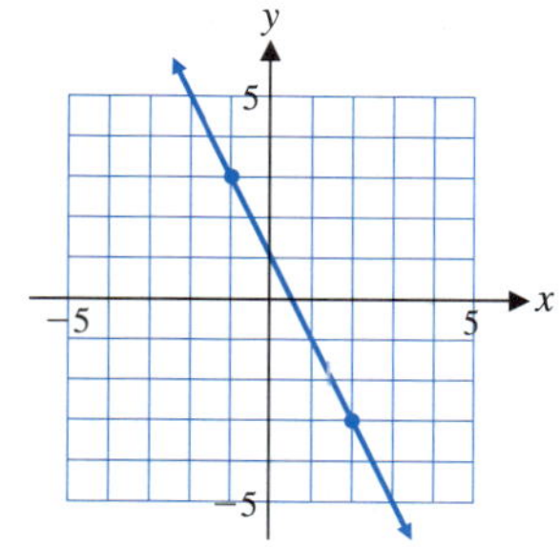

$$m = \frac{-3 - 3}{2 - (-1)} = \frac{-6}{3} = -2$$

(C)

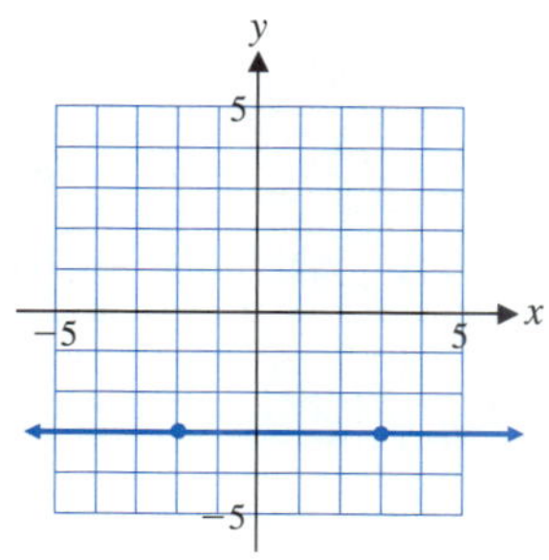

$$m = \frac{-3 - (-3)}{3 - (-2)} = \frac{0}{5} = 0$$

(D)

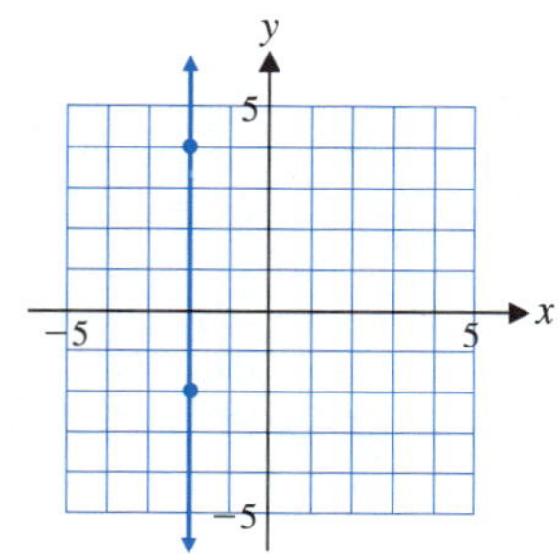

$$m = \frac{-2 - 4}{-2 - (-2)} = \frac{-6}{0}$$

Slope is not defined

Matched Problem 3 Find the slope of the line through each pair of points.

(A) $(-2, 4), (3, 4)$ (B) $(-2, 4), (0, -4)$
(C) $(-1, 5), (-1, -2)$ (D) $(-1, -2), (2, 1)$

■ EQUATIONS OF LINES—SPECIAL FORMS

Let us start by investigating why $y = mx + b$ is called the *slope–intercept form* for a line.

Explore–Discuss 3

(A) Graph $y = x + b$ for $b = -5, -3, 0, 3$, and 5 simultaneously in the same coordinate system. Verbally describe the geometric significance of b.
(B) Graph $y = mx - 1$ for $m = -2, -1, 0, 1$, and 2 simultaneously in the same coordinate system. Verbally describe the geometric significance of m.
(C) Using a graphing utility, explore the graph of $y = mx + b$ for different values of m and b.

As you can see from the above exploration, constants m and b in $y = mx + b$ have special geometric significance, which we now explicitly state.

If we let $x = 0$, then $y = b$, and we observe that the graph of $y = mx + b$ crosses the y axis at $(0, b)$. The constant b is the *y intercept*. For example, the y intercept of the graph of $y = -4x - 1$ is -1.

To determine the geometric significance of m, we proceed as follows: If $y = mx + b$, then by setting $x = 0$ and $x = 1$, we conclude that $(0, b)$ and $(1, m + b)$ lie on its graph (a line). Hence, the slope of this graph (line) is given by:

$$\text{Slope} = \frac{y_2 - y_1}{x_2 - x_1} = \frac{(m + b) - b}{1 - 0} = m$$

Thus, m is the slope of the line given by $y = mx + b$.

Slope–Intercept Form

The equation

$$y = mx + b \qquad m = \text{Slope}, b = y \text{ intercept} \tag{6}$$

is called the **slope–intercept form** of an equation of a line.

Example 4 **Using the Slope–Intercept Form**

(A) Find the slope and y intercept, and graph $y = -\frac{2}{3}x - 3$.
(B) Write the equation of the line with slope $\frac{2}{3}$ and y intercept -2.

SOLUTION (A) Slope $= m = -\frac{2}{3}$
y intercept $= b = -3$

(B) $m = \frac{2}{3}$ and $b = -2$;
thus, $y = \frac{2}{3}x - 2$

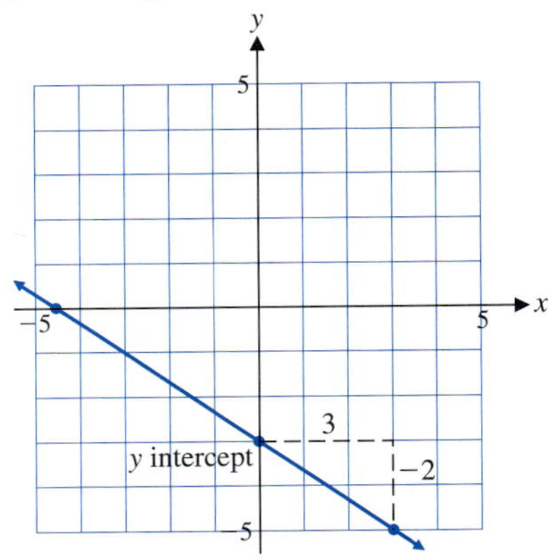

Matched Problem 4 Write the equation of the line with slope $\frac{1}{2}$ and y intercept -1. Graph.

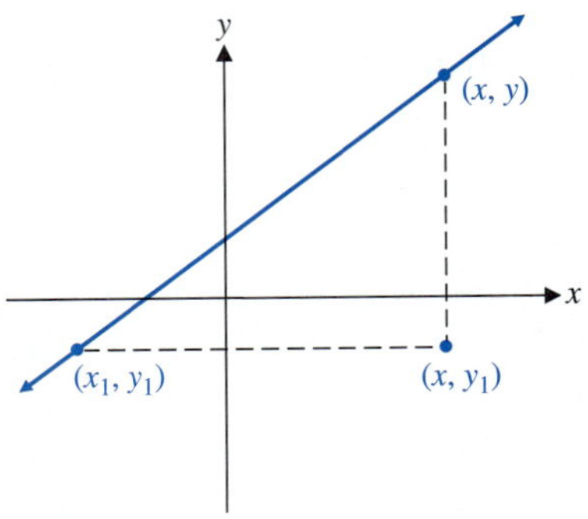

FIGURE 3

Suppose a line has slope m and passes through a fixed point (x_1, y_1). If the point (x, y) is any other point on the line (Fig. 3), then

$$\frac{y - y_1}{x - x_1} = m$$

That is,

$$y - y_1 = m(x - x_1) \tag{7}$$

We now observe that (x_1, y_1) also satisfies equation (7) and conclude that equation (7) is an equation of a line with slope m that passes through (x_1, y_1).

Point–Slope Form

An equation of a line with slope m that passes through (x_1, y_1) is

$$y - y_1 = m(x - x_1) \tag{7}$$

which is called the **point–slope form** of an equation of a line.

The point–slope form is extremely useful, since it enables us to find an equation for a line if we know its slope and the coordinates of a point on the line or if we know the coordinates of two points on the line.

Example 5 **Using the Point–Slope Form**

(A) Find an equation for the line that has slope $\frac{1}{2}$ and passes through $(-4, 3)$. Write the final answer in the form $Ax + By = C$.
(B) Find an equation for the line that passes through the two points $(-3, 2)$ and $(-4, 5)$. Write the resulting equation in the form $y = mx + b$.

SOLUTION (A) Use $y - y_1 = m(x - x_1)$. Let $m = \frac{1}{2}$ and $(x_1, y_1) = (-4, 3)$. Then

$$\begin{aligned} y - 3 &= \tfrac{1}{2}[x - (-4)] \\ y - 3 &= \tfrac{1}{2}(x + 4) \qquad \text{Multiply by 2.} \\ 2y - 6 &= x + 4 \\ -x + 2y &= 10 \quad \text{or} \quad x - 2y = -10 \end{aligned}$$

(B) First, find the slope of the line by using the slope formula:

$$m = \frac{y_2 - y_1}{x_2 - x_1} = \frac{5 - 2}{-4 - (-3)} = \frac{3}{-1} = -3$$

Now use $y - y_1 = m(x - x_1)$ with $m = -3$ and $(x_1, y_1) = (-3, 2)$:

$$\begin{aligned} y - 2 &= -3[x - (-3)] \\ y - 2 &= -3(x + 3) \\ y - 2 &= -3x - 9 \\ y &= -3x - 7 \end{aligned}$$

Matched Problem 5

(A) Find an equation for the line that has slope $\frac{2}{3}$ and passes through $(6, -2)$. Write the resulting equation in the form $Ax + By = C$, $A > 0$.
(B) Find an equation for the line that passes through $(2, -3)$ and $(4, 3)$. Write the resulting equation in the form $y = mx + b$.

The various forms of the equation of a line that we have discussed are summarized in Table 2 for convenient reference.

Table 2

EQUATIONS OF A LINE

Standard form	$Ax + By = C$	A and B not both 0
Slope–intercept form	$y = mx + b$	Slope: m; y intercept: b
Point–slope form	$y - y_1 = m(x - x_1)$	Slope: m; Point: (x_1, y_1)
Horizontal line	$y = b$	Slope: 0
Vertical line	$x = a$	Slope: Undefined

Applications

We will now see how equations of lines occur in certain applications.

Example 6 **Cost Equation** The management of a company that manufactures roller skates has fixed costs (costs at 0 output) of \$300 per day and total costs of \$4,300 per day at an output of 100 pairs of skates per day. Assume that cost C is linearly related to output x.

(A) Find the slope of the line joining the points associated with outputs of 0 and 100; that is, the line passing through $(0, 300)$ and $(100, 4{,}300)$.

(B) Find an equation of the line relating output to cost. Write the final answer in the form $C = mx + b$.

(C) Graph the cost equation from part (B) for $0 \leq x \leq 200$.

SOLUTION (A) $m = \frac{y_2 - y_1}{x_2 - x_1} = \frac{4{,}300 - 300}{100 - 0} = \frac{4{,}000}{100} = 40$

(B) We must find an equation of the line that passes through $(0, 300)$ with slope 40. We use the slope–intercept form:

$$C = mx + b$$
$$C = 40x + 300$$

(C)

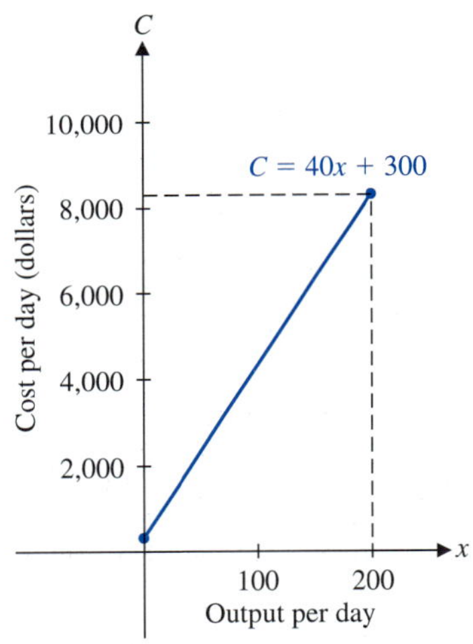

In Example 6, the **fixed cost** of \$300 per day covers plant cost, insurance, and so on. This cost is incurred whether or not there is any production. The **variable cost** is $40x$, which depends on the day's output. Note that the slope 40 is the cost of producing one pair of skates; that is, the cost per unit output.

Matched Problem 6 Answer parts (A) and (B) in Example 6 for fixed costs of \$250 per day and total costs of \$3,450 per day at an output of 80 pairs of skates per day.

Example 7 **Price–Demand** A company manufactures and sells a specialty watch. The financial research department, using statistical and analytical methods, determined that at a price of \$88 each, the demand would be 2 thousand watches, and at \$38 each, 12 thousand watches. Assuming a linear relationship between

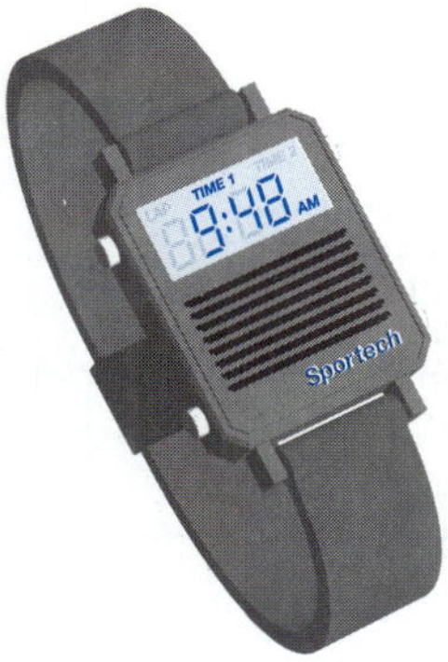

price and demand, find a linear function that models the price–demand relationship in the form $p(x) = mx + b$. What would be the price at a demand of 8 thousand watches? 15 thousand watches?

SOLUTION Find the equation of the line that passes through (2, 88) and (12, 38). We first find the slope of the line:

$$m = \frac{38 - 88}{12 - 2} = \frac{-50}{10} = -5$$

Use the point–slope form to find the equation of the line:

$$\begin{aligned} y - y_1 &= m(x - x_1) \\ y - 88 &= -5(x - 2) \\ y - 88 &= -5x + 10 \\ y &= -5x + 98 \end{aligned}$$

or

$$\begin{aligned} p(x) &= -5x + 98 \qquad \text{Price–demand equation} \\ p(8) &= -5(8) + 98 = \$58 \\ p(15) &= -5(15) + 98 = \$23 \end{aligned}$$

Thus, the price is \$58 when the demand is 8,000 and S23 when the demand is 15,000.

Matched Problem 7 The company in Example 7 also manufactures and sells a watch designed for sailboat racing. The financial analyst found that the company could sell 3 hundred of these watches at a wholesale price of \$140 each, and 11 hundred watches at a wholesale price of \$92 each. Assuming a linear relationship between price and demand, find a linear function that models the price–demand relationship in the form $p(x) = mx - b$. What would be the price at a demand of 7 hundred watches? 12 hundred watches?

Answers to Matched Problems

1. (A)

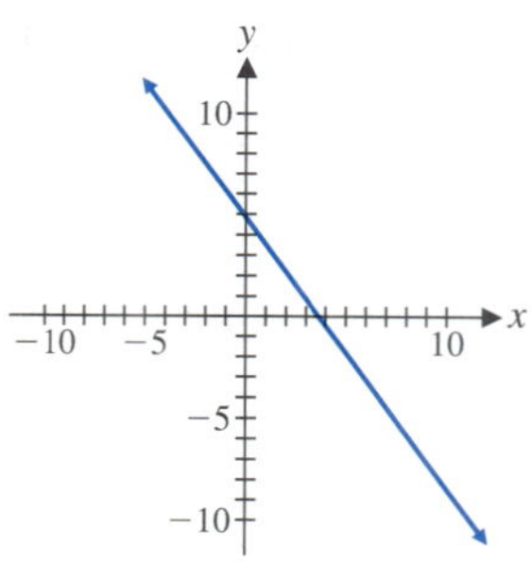

(B) x intercept: 3.75; y intercept: 5

(C)

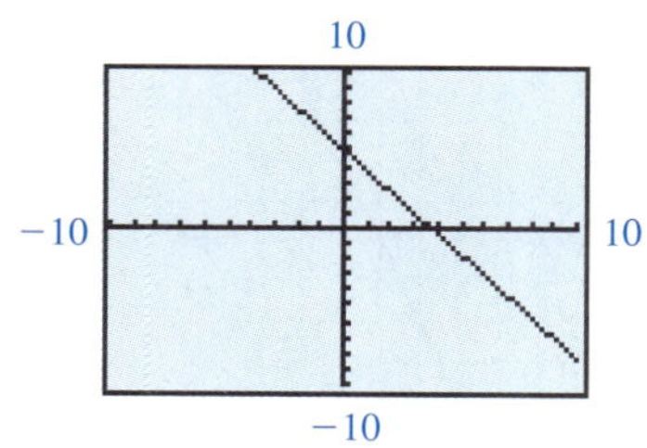

(D) x intercept: 3.75; y intercept: 5

(E) $x \le 3.75$ or $(-\infty, 3.75]$

2. (A)

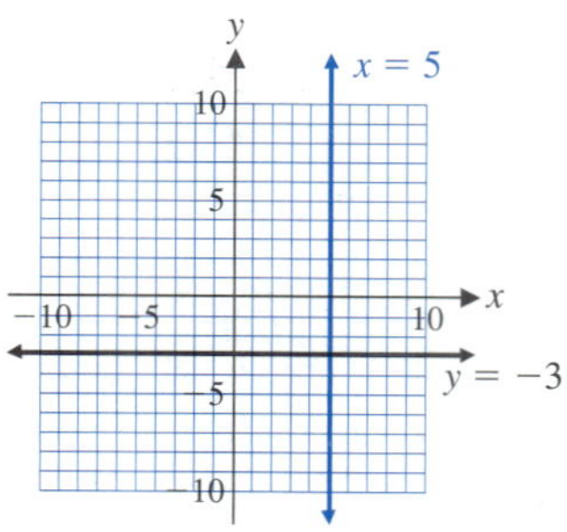

By hand

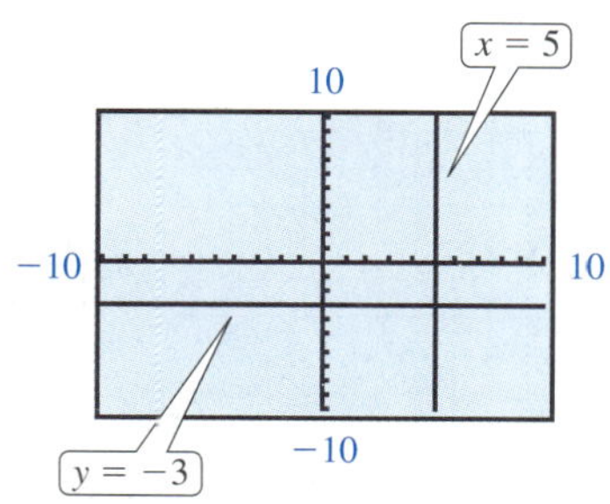

In graphing utility

(B) Horizontal line: $y = 2$; Vertical line: $x = -8$

(C)

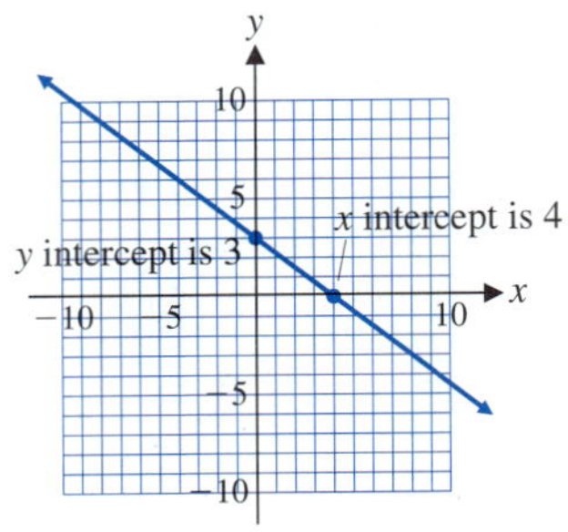

By hand

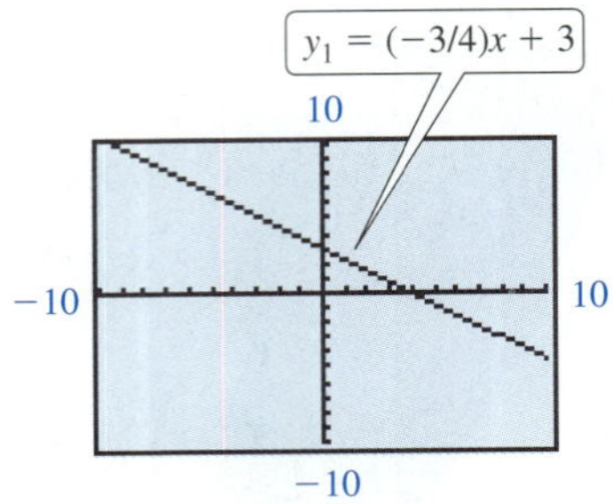

In graphing utility

3. (A) 0 (B) −4 (C) Not defined (D) 1

4. $y = \frac{1}{2}x - 1$

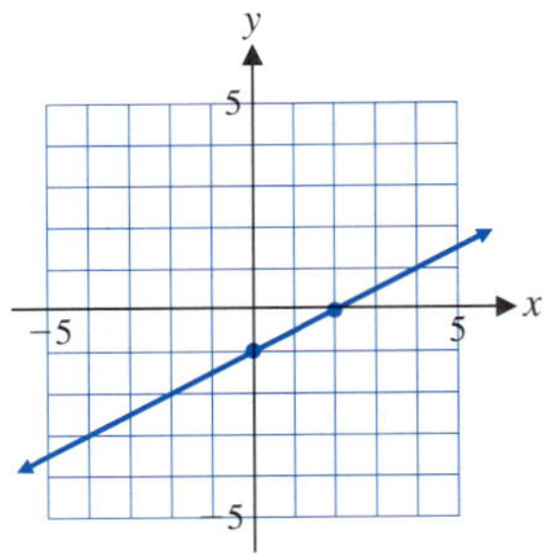

5. (A) $2x - 3y = 18$ (B) $y = 3x - 9$

6. (A) $m = 40$ (B) $C = 40x + 250$

7. $p(x) = -6x + 158$; $p(7) = \$116$; $p(12) = \$86$

EXERCISE 1-3

A *Problems 1–4 refer to graphs (A)–(D).*

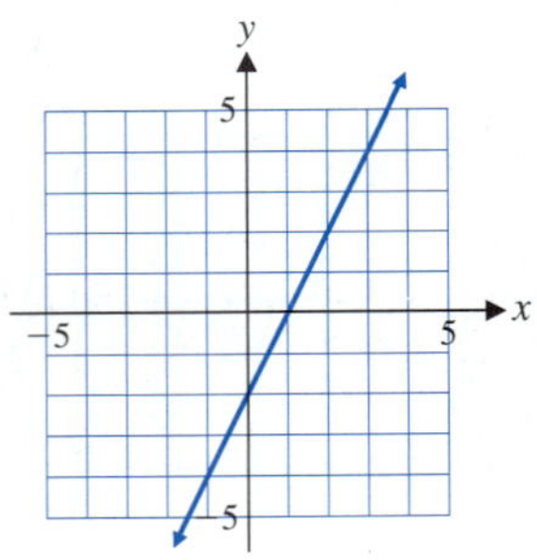

(A)

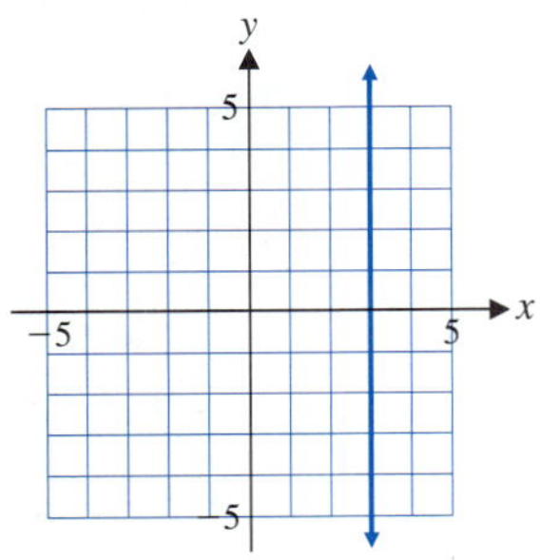

(B)

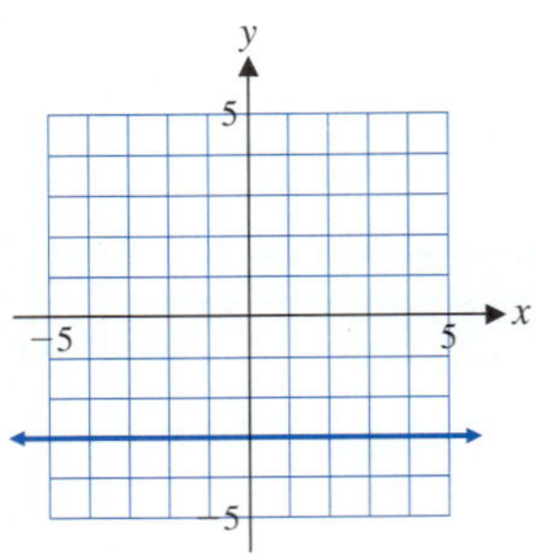

(C)

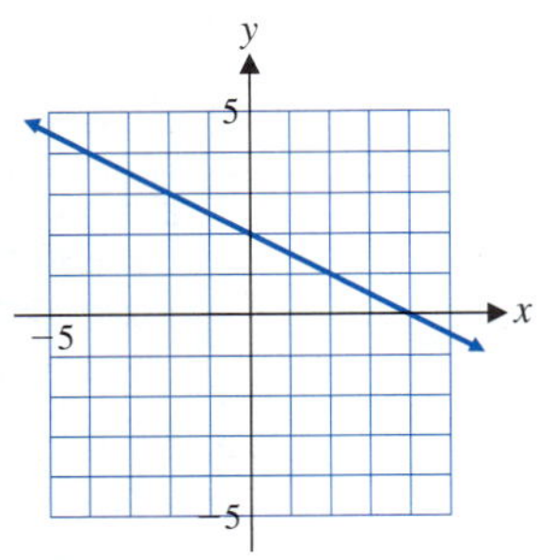

(D)

1. Identify the graph(s) of linear functions with a negative slope.

2. Identify the graph(s) of linear functions with a positive slope.

3. Identify the graph(s) of any constant functions. What is the slope of the graph?

4. Identify any graphs that are not the graphs of functions. What can you say about their slopes?

Sketch a graph of each equation in a rectangular coordinate system.

5. $y = 2x - 3$

6. $y = \frac{x}{2} + 1$

7. $2x + 3y = 12$

8. $8x - 3y = 24$

Find the slope and y intercept of the graph of each equation.

9. $y = 2x - 3$

10. $y = \frac{x}{2} + 1$

11. $y = -\frac{2}{3}x + 2$

12. $y = \frac{3}{4}x - 2$

Write an equation of the line with the indicated slope and y intercept.

13. Slope $= -2$
y intercept $= 4$

14. Slope $= -\frac{2}{3}$
y intercept $= -2$

15. Slope $= -\frac{3}{5}$
y intercept $= 3$

16. Slope $= 1$
y intercept $= -2$

B *Sketch a graph of each equation or pair of equations in a rectangular coordinate system.*

17. $y = -\frac{2}{3}x - 2$

18. $y = -\frac{3}{2}x + 1$

19. $3x - 2y = 10$

20. $5x - 6y = 15$

21. $x = 3;\ y = -2$

22. $x = -3;\ y = 2$

Check your graphs for Problems 17–22 by graphing each in a graphing utility.

In Problems 23–26, find the slope of the graph of each equation. (First write the equation in the form $y = mx + b$.)

23. $3x + y = 5$

24. $2x - y = -3$

25. $2x + 3y = 12$

26. $3x - 2y = 10$

27. (A) Graph $f(x) = 1.2x - 4.2$ in a rectangular coordinate system.
(B) Find the x and y intercepts algebraically to one decimal place.
C (C) Graph $f(x) = 1.2x - 4.2$ in a graphing utility.
C (D) Find the x and y intercepts to one decimal place using trace and zoom or an appropriate built-in routine in your graphing utility.
(E) Using the results of parts (A) and (B) or (C) and (D), find the solution set for the linear inequality

$$1.2x - 4.2 > 0$$

28. (A) Graph $f(x) = -0.8x + 5.2$ in a rectangular coordinate system.
(B) Find the x and y intercepts algebraically to one decimal place.
C (C) Graph $f(x) = -0.8x + 5.2$ in a graphing utility.
C (D) Find the x and y intercepts to one decimal place using trace and zoom or an appropriate built-in routine in your graphing utility.
(E) Using the results of parts (A) and (B) or (C) and (D), find the solution set for the linear inequality

$$-0.8x + 5.2 < 0$$

Write the equations of the vertical and horizontal lines through each point.

29. $(3, -5)$

30. $(-2, 7)$

31. $(-1, -3)$

32. $(96, -4)$

Write the equation of the line through each indicated point with the indicated slope. Write the final answer in the form $y = mx + b$.

33. $m = -3;\ (4, -1)$

34. $m = -2;\ (-3, 2)$

35. $m = \frac{2}{3};\ (-6, -5)$

36. $m = \frac{1}{2};\ (-4, 3)$

37. $m = 0;\ (3, -5)$

38. $m = 0;\ (-4, 7)$

Find the slope of the line that passes through the given points.

39. $(1, 3)$ and $(7, 5)$

40. $(2, 1)$ and $(10, 5)$

41. $(-5, -2)$ and $(5, -4)$

42. $(3, 7)$ and $(-6, 4)$

43. $(2, 7)$ and $(2, -3)$

44. $(-2, 3)$ and $(-2, -1)$

45. $(2, 3)$ and $(-5, 3)$

46. $(-3, -3)$ and $(0, -3)$

Write an equation of the line through each indicated pair of points. Write the final answer in the form $Ax + By = C$.

47. $(1, 3)$ and $(7, 5)$

48. $(2, 1)$ and $(10, 5)$

49. $(-5, -2)$ and $(5, -4)$

50. $(3, 7)$ and $(-6, 4)$

51. $(2, 7)$ and $(2, -3)$

52. $(-2, 3)$ and $(-2, -1)$

53. $(2, 3)$ and $(-5, 3)$

54. $(-3, -3)$ and $(0, -3)$

In Problems 55–60, graph the equations obtained from Problems 49–54 and indicate which define a linear function, a constant function, or no function at all.

55. Problem 49

56. Problem 50

57. Problem 51

58. Problem 52

59. Problem 53

60. Problem 54

61. Discuss the relationship among the graphs of the lines with equation $y = mx + 2$, where m is any real number.

62. Discuss the relationship among the graphs of the lines with equation $y = -0.5x + b$, where b is any real number.

C

63. (A) Graph the following equations in the same coordinate system:

$$3x + 2y = 6 \qquad 3x + 2y = 3$$
$$3x + 2y = -6 \qquad 3x + 2y = -3$$

(B) From your observations in part (A), describe the family of lines obtained by varying C in $Ax + By = C$ while holding A and B fixed.

64. (A) Graph the following two equations in the same coordinate system:

$$3x + 4y = 12 \qquad 4x - 3y = 12$$

(B) Graph the following two equations in the same coordinate system:

$$2x + 3y = 12 \qquad 3x - 2y = 12$$

(C) From your observations in parts (A) and (B), describe the apparent relationship of the graphs of

$$Ax + By = C \quad \text{and} \quad Bx - Ay = C$$

65. Describe the relationship between the graphs of $f(x) = mx + b$ and $g(x) = |mx + b|$, $m \neq 0$, and illustrate with examples. Is $g(x)$ always, sometimes, or never a linear function?

66. Describe the relationship between the graphs of $f(x) = mx + b$ and $g(x) = m|x| + b$, $m \neq 0$, and illustrate with examples. Is $g(x)$ always, sometimes, or never a linear function?

Applications

Business & Economics

67. *Simple interest.* If \$$P$ (the principal) is invested at an interest rate of r, then the amount A that is due after t years is given by

$$A = Prt + P$$

If \$100 is invested at 6% ($r = 0.06$), then $A = 6t + 100$, $t \geq 0$.

(A) What will \$100 amount to after 5 years? After 20 years?
(B) Sketch a graph of $A = 6t + 100$ for $0 \leq t \leq 20$.
(C) Find the slope of the graph and interpret verbally.

68. *Simple interest.* Use the simple interest formula from Problem 67. If \$1,000 is invested at 7.5% ($r = 0.075$), then $A = 75t + 1{,}000$, $t \geq 0$.

(A) What will \$1,000 amount to after 5 years? After 20 years?
(B) Sketch a graph of $A = 75t + 1{,}000$ for $0 \leq t \leq 1{,}000$.
(C) Find the slope of the graph and interpret verbally.

69. *Cost function.* The management of a company that manufactures surfboards has fixed costs (at 0 output) of \$200 per day and total costs of \$3,800 per day at a daily output of 20 boards.

(A) Assuming the total cost per day, $C(x)$, is linearly related to the total output per day, x, write an equation for the cost function.
(B) What are the total costs for an output of 12 boards per day?
(C) Graph the cost function for $0 \leq x \leq 20$.

70. *Cost function.* Repeat Problem 69 if the company has fixed costs of \$300 per day and total costs per day at an output of 20 boards of \$5,100.

71. *Price–demand function.* A manufacturing company is interested in introducing a new power mower. Its market research department gave the management the price–demand forecast listed in Table 3.

Table 3

Price–Demand

Demand x	Wholesale price (\$) $p(x)$
0	200
2,400	160
4,800	120
7,800	70

(A) Plot these points, letting $p(x)$ represent the price at which x number of mowers can be sold (demand). Label the horizontal axis x.
(B) Note that the points in part (A) lie along a straight line. Find an equation for the price–demand function.
(C) What would be the price for a demand of 3,000 units?
(D) Write a brief verbal interpretation of the slope of the line found in part (B).

72. *Depreciation.* Office equipment was purchased for \$20,000 and is assumed to have a scrap value of \$2,000 after 10 years. If its value is depreciated linearly (for tax purposes) from \$20,000 to \$2,000:

(A) Find the linear equation that relates value (V) in dollars to time (t) in years.
(B) What would be the value of the equipment after 6 years?
(C) Graph the equation for $0 \leq t \leq 10$.
(D) Write a brief verbal interpretation of the slope of the line found in part (A).

Merck & Co., Inc. is the world's largest pharmaceutical company. Problems 73 and 74 refer to the data in Table 4 taken from the company's 1993 annual report.

Table 4

Selected Financial Data for Merck & Co., Inc.

(BILLION $)	1988	1989	1990	1991	1992
SALES	5.9	6.5	7.7	8.6	9.7
NET INCOME	1.2	1.5	1.8	2.1	2.4

73. *Sales analysis.* A mathematical model for Merck's sales is given by

$$f(x) = 5.74 + 0.97x$$

where $x = 0$ corresponds to 1988.

(A) Complete the following table. Round values of $f(x)$ to one decimal place.

x	0	1	2	3	4
SALES	5.9	6.5	7.7	8.6	9.7
$f(x)$					

(B) Sketch the graph of f and the sales data on the same axes.

(C) Use the modeling equation to estimate the sales in 1993. In 2000.

(D) Write a brief verbal description of the company's sales from 1988 to 1992.

74. *Income analysis.* A mathematical model for Merck's income is given by

$$f(x) = 1.2 + 0.3x$$

where $x = 0$ corresponds to 1988.

(A) Complete the following table. Round values of $f(x)$ to one decimal place.

x	0	1	2	3	4
NET INCOME	1.2	1.5	1.8	2.1	2.4
$f(x)$					

(B) Sketch the graph of f and the income data on the same axes.

(C) Use the modeling equation to estimate the income in 1993. In 2000.

(D) Write a brief verbal description of the company's income from 1988 to 1992.

Life Sciences

75. *Nutrition.* In a nutrition experiment, a biologist wants to prepare a special diet for the experimental animals. Two food mixes, A and B, are available. If mix A contains 20% protein and mix B contains 10% protein, what combination of each mix will provide exactly 20 grams of protein? Let x be the amount of A used and let y be the amount of B used. Then write a linear equation relating x, y, and 20. Graph this equation for $x \geq 0$ and $y \geq 0$.

76. *Ecology.* As one descends into the ocean, pressure increases linearly. The pressure is 15 pounds per square inch on the surface and 30 pounds per square inch 33 feet below the surface.

(A) If p is the pressure in pounds and d is the depth below the surface in feet, write an equation that expresses p in terms of d. [*Hint:* Find an equation of the line that passes through (0, 15) and (33, 30).]

(B) What is the pressure at 12,540 feet (the average depth of the ocean)?

(C) Graph the equation for $0 \leq d \leq 12{,}540$.

(D) Write a brief verbal interpretation of the slope of the line found in part (A).

Social Sciences

77. *Psychology.* In an experiment on motivation, J. S. Brown trained a group of rats to run down a narrow passage in a cage to obtain food in a goal box. Using a harness, he then connected the rats to an overhead wire that was attached to a spring scale. A rat was placed at different distances d (in centimeters) from the goal box, and the pull p (in grams) of the rat toward the food was measured. Brown found that the relationship between these two variables was very close to being linear and could be approximated by the equation

$$p = -\tfrac{1}{5}d + 70 \qquad 30 \leq d \leq 175$$

(See J. S. Brown, *Journal of Comparative and Physiological Psychology,* 1948, 41:450–465.)

(A) What was the pull when $d = 30$? When $d = 175$?

(B) Graph the equation.

(C) What is the slope of the line?

SECTION 1-4

Quadratic Functions

- QUADRATIC FUNCTIONS, EQUATIONS, AND INEQUALITIES
- PROPERTIES OF QUADRATIC FUNCTIONS AND THEIR GRAPHS
- APPLICATIONS

If the degree of a linear function is increased by one, we obtain a *second-degree function,* usually called a *quadratic function,* another basic function that we will need in our library of elementary functions. We will investigate relationships between quadratic functions and the solutions to quadratic equations and inequalities. (A detailed treatment of algebraic solutions to quadratic equations can be found in Appendix A-9.) Other important properties of quadratic functions will also be investigated, including maximum and minimum properties. We will then be in a position to solve important practical problems such as finding production levels that will produce maximum revenue or maximum profit.

QUADRATIC FUNCTIONS, EQUATIONS, AND INEQUALITIES

A quadratic function is defined as follows:

Quadratic Function

A function f is a **quadratic function** if

$$f(x) = ax^2 + bx + c \qquad a \neq 0$$

where a, b, and c are real numbers. The domain of a quadratic function is the set of all real numbers.

The graphs of three quadratic functions are shown in Figure 1.

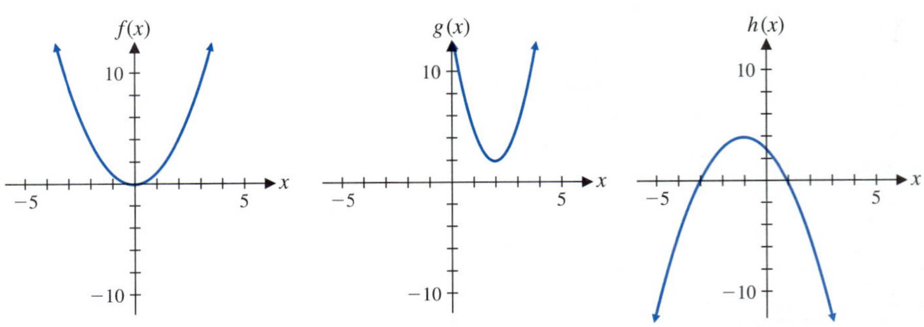

(A) $f(x) = x^2$ (B) $g(x) = 3x^2 - 12x + 14$ (C) $h(x) = 3 - 2x - x^2$

FIGURE 1
Graphs of quadratic functions

The graph of a quadratic function is called a **parabola.** We will discuss this in more detail later in this section.

Example 1 **Intercepts, Equations, and Inequalities**

(A) Sketch a graph of $f(x) = -x^2 + 5x + 3$ in a rectangular coordinate system.

(B) Find x and y intercepts algebraically to two decimal places.

C (C) Graph $f(x) = -x^2 + 5x + 3$ in a standard viewing window.

C (D) Find the x and y intercepts to two decimal places using trace and zoom or an appropriate built-in routine in your graphing utility.

(E) Solve the quadratic inequality $-x^2 + 5x + 3 \geqslant 0$ graphically to two decimal places using the results of parts (A) and (B) or (C) and (D).

C (F) Solve the equation $-x^2 + 5x + 3 = 4$ graphically to two decimal places using trace and zoom or an appropriate built-in routine in your graphing utility.

SOLUTION (A) Hand-sketching a graph of f:

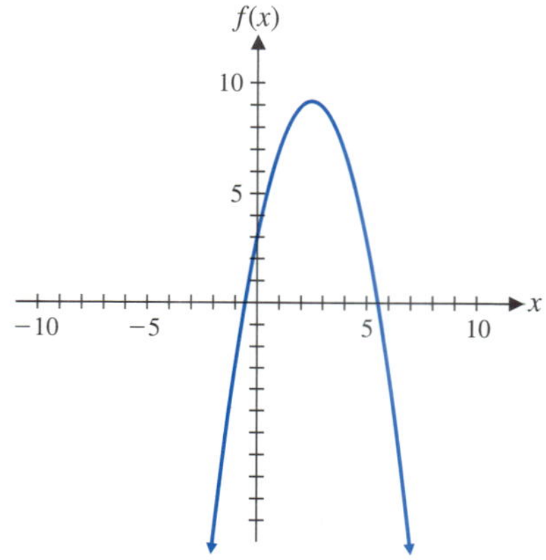

(B) Finding intercepts algebraically:

y intercept: $f(0) = -(0)^2 + 5(0) + 3 = 3$

x intercepts: $f(x) = 0$

$$-x^2 + 5x + 3 = 0 \qquad \text{Quadratic equation}$$

$$x = \frac{-b \pm \sqrt{b^2 - 4ac}}{2a} \qquad \text{Quadratic formula (see Appendix A-9)}$$

$$x = \frac{-(5) \pm \sqrt{5^2 - 4(-1)(3)}}{2(-1)}$$

$$= \frac{-5 \pm \sqrt{37}}{-2} = -0.54 \quad \text{or} \quad 5.54$$

C (C) Graphing in a graphing utility:

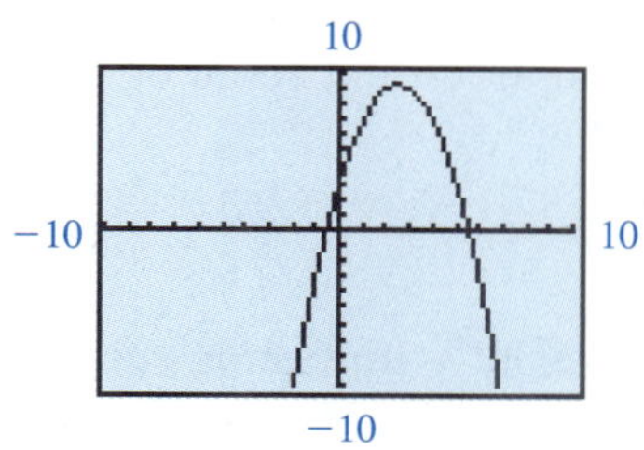

(D) Finding intercepts graphically using a graphing utility:

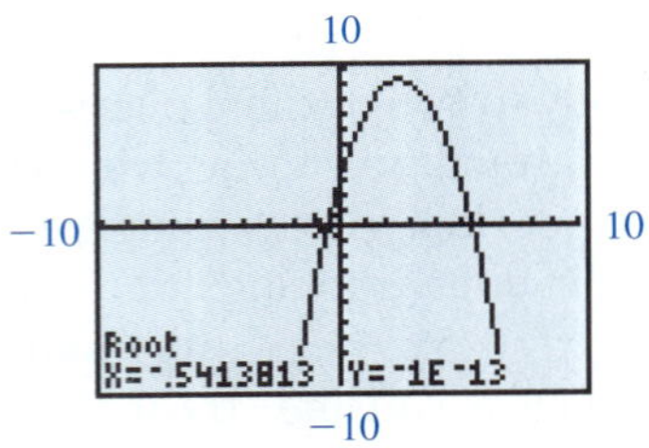

x intercept: -0.54

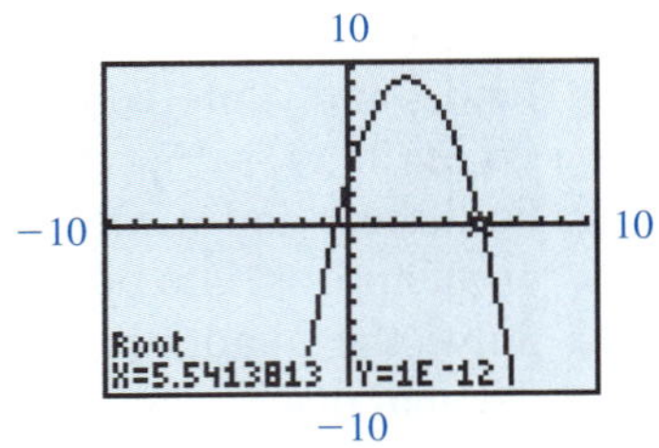

x intercept: 5.54

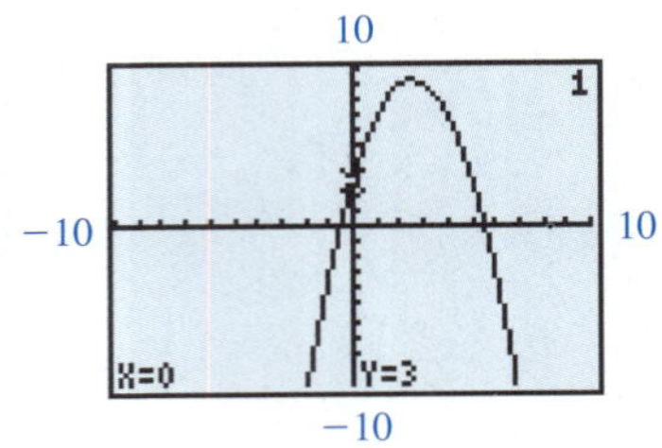

y intercept: 3

(E) Solving $-x^2 + 5x + 3 \geq 0$ graphically: The quadratic inequality

$$-x^2 + 5x + 3 \geq 0$$

holds for those values of x for which the graph of $f(x) = -x^2 + 5x + 3$ in the figures in parts (A) and (C) is at or above the x axis. This happens for x between the two x intercepts [found in parts (B) or (D)], including the two x intercepts. Thus, the solution set for the quadratic inequality is $-0.54 \leq x \leq 5.54$ or $[-0.54, 5.54]$.

(F) Solving the equation $-x^2 + 5x + 3 = 4$ using a graphing utility:

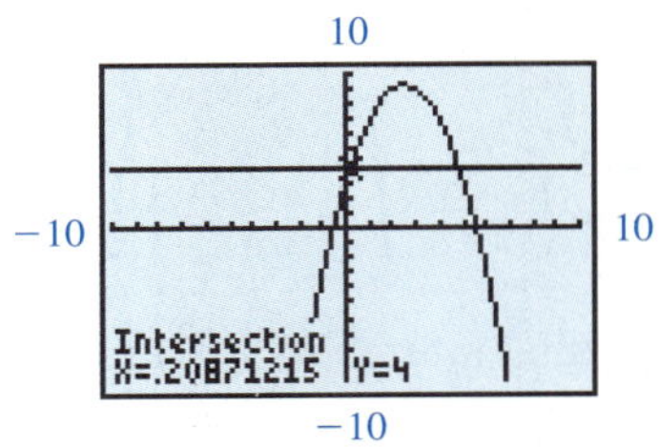

$-x^2 + 5x + 3 = 4$ at $x = 0.21$

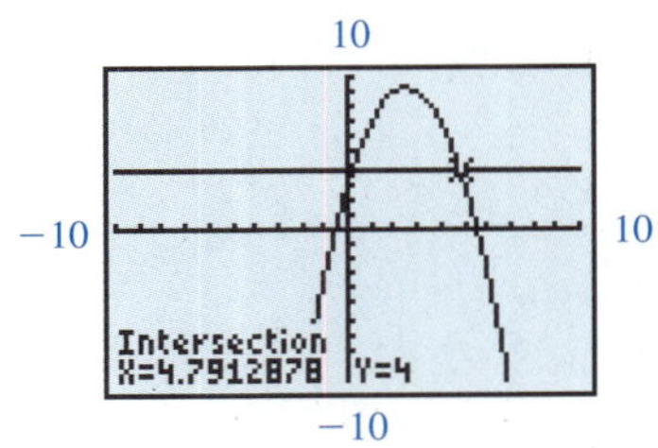

$-x^2 + 5x + 3 = 4$ at $x = 4.79$

Matched Problem 1

(A) Sketch a graph of $g(x) = 2x^2 - 5x - 5$ in a rectangular coordinate system.
(B) Find x and y intercepts algebraically to two decimal places.
(C) Graph $g(x) = 2x^2 - 5x - 5$ in a standard viewing window.
(D) Find the x and y intercepts to two decimal places using trace and zoom or an appropriate built-in routine in your graphing utility.
(E) Solve $2x^2 - 5x - 5 \geq 0$ graphically to two decimal places using the results of parts (A) and (B) or (C) and (D).
(F) Solve the equation $2x^2 - 5x - 5 = -3$ graphically to two decimal places using trace and zoom or an appropriate built-in routine in your graphing utility.

Explore–Discuss 1

How many x intercepts can the graph of a quadratic function have? How many y intercepts? Explain your reasoning.

■ PROPERTIES OF QUADRATIC FUNCTIONS AND THEIR GRAPHS

Many useful properties of the quadratic function can be uncovered by transforming

$$f(x) = ax^2 + bx + c \qquad a \neq 0 \tag{1}$$

into the form

$$f(x) = a(x - h)^2 + k \tag{2}$$

The process of *completing the square* (see Appendix A-9) is central to the transformation. We illustrate the process through a specific example and then generalize the results.

Consider the quadratic function given by

$$f(x) = -2x^2 + 16x - 24 \tag{3}$$

We start by transforming equation (3) into the form (2) by completing the square:

$$\begin{aligned} f(x) &= -2x^2 + 16x - 24 \\ &= -2(x^2 - 8x) - 24 \\ &= -2(x^2 - 8x + \mathbf{?}) - 24 \\ &= -2(x^2 - 8x + \mathbf{16}) - 24 + \mathbf{32} \\ &= -2(x - 4)^2 + 8 \end{aligned}$$

Factor the coefficient of x^2 out of the first two terms.

Add 16 to complete the square inside the parentheses. Because of the -2 outside the parentheses, we have actually added -32, so we must add 32 to the outside.

The transformation is complete and can be checked by multiplying out.

Thus,

$$f(x) = -2(x - 4)^2 + 8 \tag{4}$$

If $x = 4$, then $-2(x - 4)^2 = 0$ and $f(4) = 8$. For any other value of x, the negative number $-2(x - 4)^2$ is added to 8, making it smaller. (Think about this.) Therefore,

$$f(4) = 8$$

is the *maximum value* of $f(x)$ for all x—a very important result! Furthermore, if we choose any two x values that are the same distance from 4, we will obtain the same function value. For example, $x = 3$ and $x = 5$ are each one unit from $x = 4$ and their function values are

$$\begin{aligned} f(3) &= -2(3 - 4)^2 + 8 = 6 \\ f(5) &= -2(5 - 4)^2 + 8 = 6 \end{aligned}$$

Thus, the vertical line $x = 4$ is a line of symmetry. That is, if the graph of equation (3) is drawn on a piece of paper and the paper is folded along the line $x = 4$, then the two sides of the parabola will match exactly. All these results are illustrated by graphing equations (3) and (4) and the line $x = 4$ simultaneously in the same coordinate system (Fig. 2).

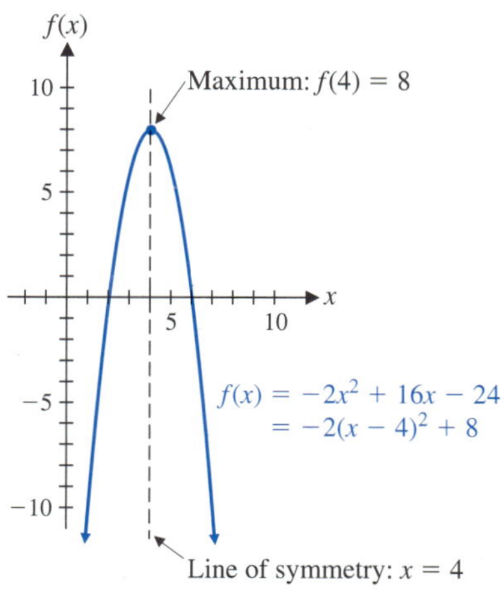

FIGURE 2
Graph of a quadratic function

From the preceding discussion, we see that as x moves from left to right, $f(x)$ is increasing on $(-\infty, 4]$, and decreasing on $[4, \infty)$, and that $f(x)$ can assume no value greater than 8. Thus,

Range of f: $y \leq 8$ or $(-\infty, 8]$

In general, the graph of a quadratic function is a parabola with line of symmetry parallel to the vertical axis. The lowest or highest point on the parabola, whichever exists, is called the **vertex.** The maximum or minimum value of a quadratic function always occurs at the vertex of the parabola. The line of symmetry through the vertex is called the **axis** of the parabola. In the above example, $x = 4$ is the axis of the parabola and $(4, 8)$ is its vertex.

Applying the graph transformation properties discussed in Section 1-2 to the transformed equation,

$$\begin{aligned} f(x) &= -2x^2 + 16x - 24 \\ &= -2(x - 4)^2 + 8 \end{aligned}$$

we see that the graph of $f(x) = -2x^2 + 16x - 24$ is the graph of $g(x) = x^2$ vertically expanded by a factor of 2, reflected in the x axis, and shifted to the right 4 units and up 8 units, as shown in Figure 3.

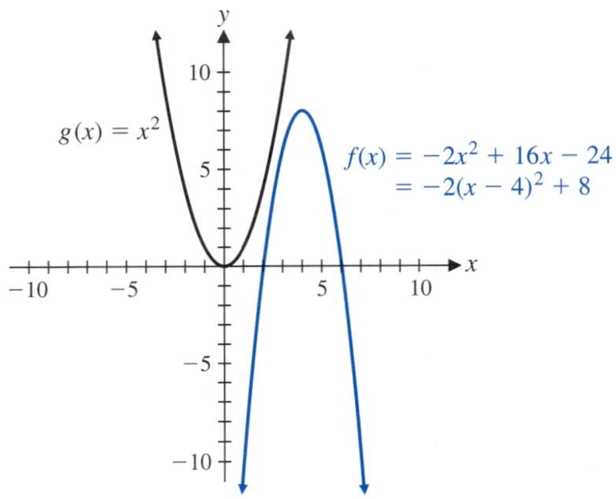

FIGURE 3
Graph of f is the graph of g transformed

Note the important results we have obtained by transforming equation (3) into equation (4):

- The vertex of the parabola
- The axis of the parabola
- The maximum value of $f(x)$
- The range of the function f
- The relationship between the graph of $g(x) = x^2$ and the graph of $f(x) = -2x^2 + 16x - 24$

Now, let us explore the effects of changing the constants a, h, and k on the graph of $f(x) = a(x - h)^2 + k$.

Explore–Discuss 2

(A) Let $a = 1$ and $h = 5$. Graph $f(x) = a(x - h)^2 + k$ for $k = -4, 0$, and 3 simultaneously in the same coordinate system. Explain the effect of changing k on the graph of f.

(B) Let $a = 1$ and $k = 2$. Graph $f(x) = a(x - h)^2 + k$ for $h = -4$, 0, and 5 simultaneously in the same coordinate system. Explain the effect of changing h on the graph of f.

(C) Let $h = 5$ and $k = -2$. Graph $f(x) = a(x - h)^2 + k$ for $a = 0.25$, 1, and 3 simultaneously in the same coordinate system. Graph function f for $a = 1, -1$, and -0.25 simultaneously in the same coordinate system. Explain the effect of changing a on the graph of f.

C (D) Discuss parts (A)–(C) using a graphing utility and a standard viewing window.

The preceding discussion is generalized for all quadratic functions in the following box:

Properties of a Quadratic Function and Its Graph

Given a quadratic function

$$f(x) = ax^2 + bx + c \qquad a \neq 0$$

and the form obtained by completing the square,

$$f(x) = a(x - h)^2 + k$$

we summarize general properties as follows:

1. The graph of f is a parabola:

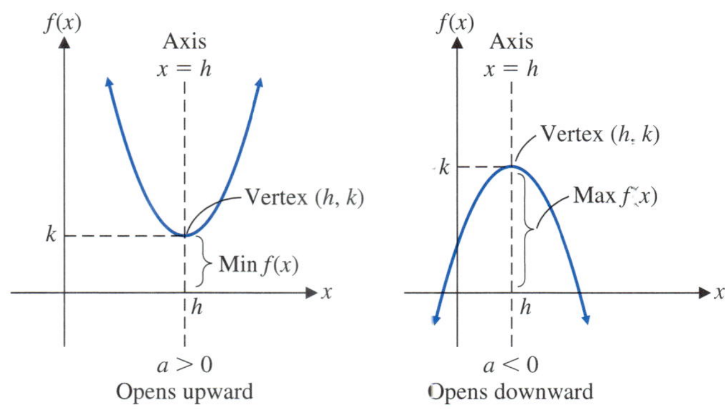

2. Vertex: (h, k) (Parabola increases on one side of the vertex and decreases on the other)
3. Axis (of symmetry): $x = h$ (Parallel to y axis)
4. $f(h) = k$ is the minimum if $a > 0$ and the maximum if $a < 0$
5. Domain: All real numbers
   Range: $(-\infty, k]$ if $a < 0$ or $[k, \infty)$ if $a > 0$
6. The graph of f is the graph of $g(x) = ax^2$ translated horizontally h units and vertically k units.

Example 2 **Analyzing a Quadratic Function** Given the quadratic function

$$f(x) = 0.5x^2 - 6x + 21$$

(A) Find the vertex and the maximum or minimum (to the nearest integer) algebraically by completing the square. State the range of f.

(B) Referring to the completed square form in part (A), describe how the graph of function f can be obtained from the graph of $g(x) = x^2$ using transformations discussed in Section 1-2.

(C) Using parts (A) and/or (B), sketch a graph of function f in a rectangular coordinate system.

C (D) Graph function f using a suitable viewing window.

C (E) Find the vertex and the maximum or minimum (to the nearest integer) graphically using trace and zoom or an appropriate built-in routine. State the range of f.

SOLUTION (A) Finding the vertex, minimum, and range algebraically: Complete the square:

$$\begin{aligned} f(x) &= 0.5x^2 - 6x + 21 \\ &= 0.5(x^2 - 12x + ?) + 21 \\ &= 0.5(x^2 - 12x + 36) + 21 - 18 \\ &= 0.5(x - 6)^2 + 3 \end{aligned}$$

From the last form, we see that $h = 6$ and $k = 3$. Thus, vertex: (6, 3); minimum: $f(6) = 3$; range: $y \geq 3$ or $[3, \infty)$.

(B) The graph of $f(x) = 0.5(x - 6)^2 + 3$ is the same as the graph of $g(x) = x^2$ vertically contracted by a factor of 0.5, and shifted to the right 6 units and up 3 units.

(C) Graph in a rectangular coordinate system:

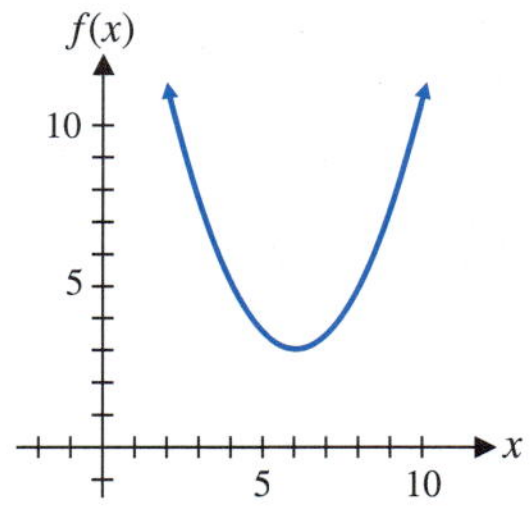

C (D) Graph in a graphing utility:

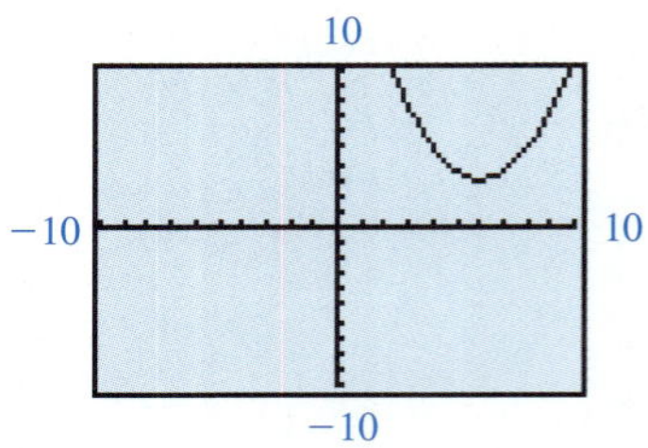

C (E) Finding the vertex, minimum, and range graphically using a graphing utility:

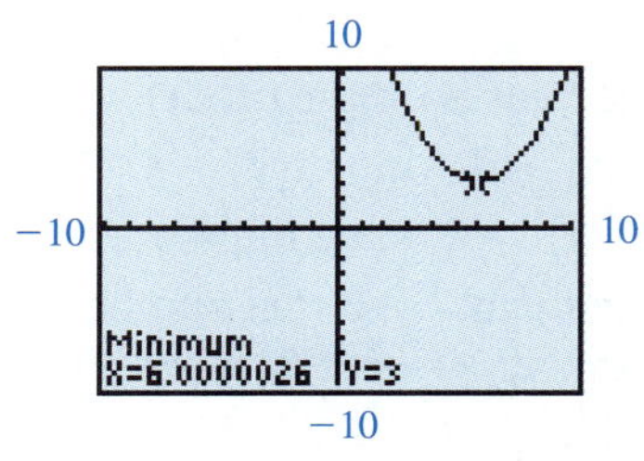

Vertex: (6, 3); minimum: $f(6) = 3$; range: $y \geq 3$ or $[3, \infty)$.

Matched Problem 2 Given the quadratic function: $f(x) = -0.25x^2 - 2x + 2$

(A) Find the vertex and the maximum or minimum (to the nearest integer) algebraically by completing the square. State the range of f.

(B) Referring to the completed square form in part (A), describe how the graph of function f can be obtained from the graph of $g(x) = x^2$ using transformations discussed in Section 1-2.

(C) Using parts (A) and/or (B), sketch a graph of function f in a rectangular coordinate system.

C (D) Graph function f using a suitable viewing window.

C (E) Find the vertex and the maximum or minimum (to the nearest integer) graphically using trace and zoom or an appropriate built-in routine. State the range of f.

Applications

Example 3 **Maximum Revenue** This is a continuation of Example 7 in Section 1-1. Recall that the financial department in the company that produces an automatic camera arrived at the following price–demand function and the corresponding revenue function:

$$p(x) = 94.8 - 5x \quad \text{Price–demand function}$$
$$R(x) = xp(x) = x(94.8 - 5x) \quad \text{Revenue function}$$

where $p(x)$ is the wholesale price per camera at which x million cameras can be sold and $R(x)$ is the corresponding revenue (in million dollars). Both functions have domain $1 \leq x \leq 15$.

(A) Find the output to the nearest thousand cameras that will produce the maximum revenue. What is the maximum revenue to the nearest thousand dollars? Solve the problem algebraically by completing the square.

(B) What is the wholesale price per camera (to the nearest dollar) that produces the maximum revenue?

C (C) Graph the revenue function using an appropriate viewing window.

C (D) Find the output to the nearest thousand cameras that will produce the maximum revenue. What is the maximum revenue to the nearest thousand dollars? Solve the problem graphically using trace and zoom or an appropriate built-in routine.

Solution (A) Algebraic solution:

$$\begin{aligned} R(x) &= x(94.8 - 5x) \\ &= -5x^2 + 94.8x \\ &= -5(x^2 - 18.96x + ?) \\ &= -5(x^2 - 18.96x + 89.8704) + 449.352 \\ &= -5(x - 9.48)^2 + 449.352 \end{aligned}$$

The maximum revenue of 449.352 million dollars ($449,352,000) occurs when $x = 9.480$ million cameras (9,480,000 cameras).

(B) Finding the wholesale price per camera: Use the price–demand function for an output of 9.480 million cameras:

$$\begin{aligned} p(x) &= 94.8 - 5x \\ p(9.480) &= 94.8 - 5(9.480) \\ &= \$47 \end{aligned}$$

(C) Graph in a graphing utility:

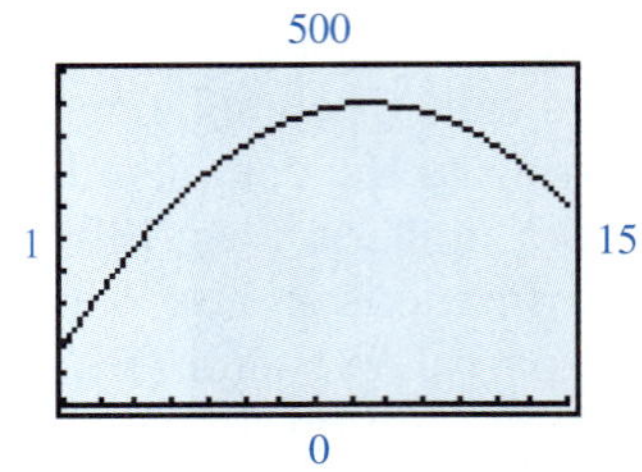

(D) Graphical solution using a graphing utility:

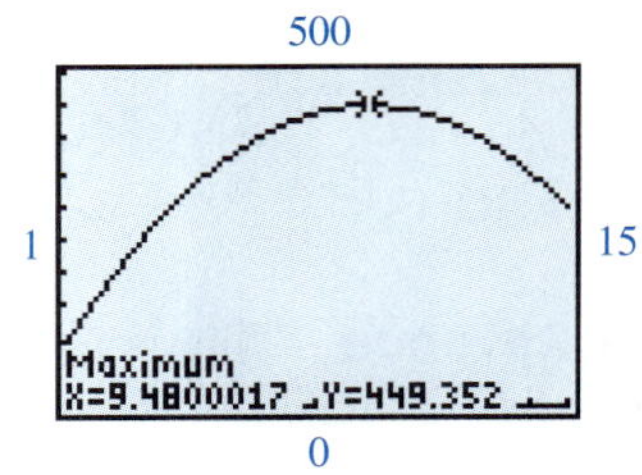

An output of 9.480 million cameras (9,480,000 cameras) will produce a maximum revenue of 449.352 million dollars ($449,352,000).

Matched Problem 3 ➡ The financial department in Example 3, using statistical and analytical techniques (see Matched Problem 7 in Section 1-1), arrived at the cost function

$$C(x) = 156 + 19.7x \quad \text{Cost function}$$

where $C(x)$ is the cost (in million dollars) for manufacturing and selling x million cameras.

(A) Using the revenue function from Example 3 and the cost function above, write an equation for the profit function.
(B) Find the output to the nearest thousand cameras that will produce the maximum profit. What is the maximum profit to the nearest thousand dollars? Solve the problem algebraically by completing the square.
(C) What is the wholesale price per camera (to the nearest dollar) that produces the maximum profit?
(D) Graph the profit function using an appropriate viewing window.
(E) Find the output to the nearest thousand cameras that will produce the maximum profit. What is the maximum profit to the nearest thousand dollars? Solve the problem graphically using trace and zoom or an appropriate built-in routine.

Example 4 ➡ **Break-Even Analysis** Use the revenue function from Example 3 and the cost function from Matched Problem 3:

$$R(x) = x(94.8 - 5x) \quad \text{Revenue function}$$
$$C(x) = 156 + 19.7x \quad \text{Cost function}$$

Both have domain $1 \leq x \leq 15$.

(A) Sketch the graphs of both functions in the same coordinate system.
(B) **Break-even points** are the production levels at which $R(x) = C(x)$. Find the break-even points algebraically to the nearest thousand cameras.

(C) Plot both functions simultaneously in the same viewing window.

(D) Find the break-even points graphically to the nearest thousand cameras using trace and zoom or an appropriate built-in routine.

(E) Recall that a loss occurs if $R(x) < C(x)$ and a profit occurs if $R(x) > C(x)$. For what outputs (to the nearest thousand cameras) will a loss occur? A profit?

Solution (A) Sketch of functions:

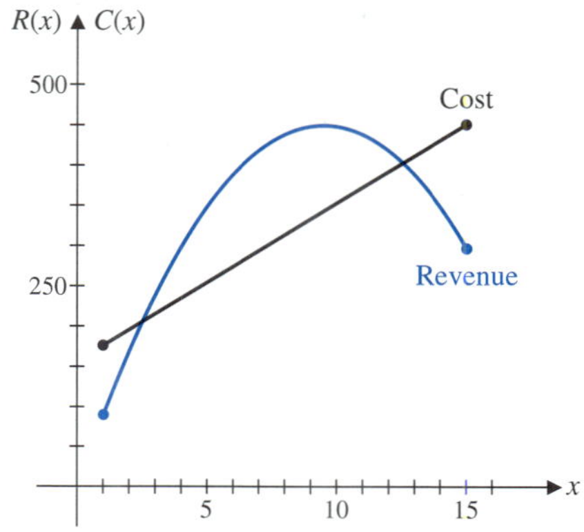

(B) Find x such that $R(x) = C(x)$:

$$x(94.8 - 5x) = 156 + 19.7x$$
$$-5x^2 + 75.1x - 156 = 0$$
$$x = \frac{-75.1 \pm \sqrt{75.1^2 - 4(-5)(-156)}}{2(-5)} \qquad \textit{Quadratic formula}$$
$$= \frac{-75.1 \pm \sqrt{2{,}520.01}}{-10}$$
$$x = 2.490 \quad \text{or} \quad 12.530$$

The company breaks even at $x = 2.490$ and 12.530 million cameras.

(C) Graph in a graphing utility:

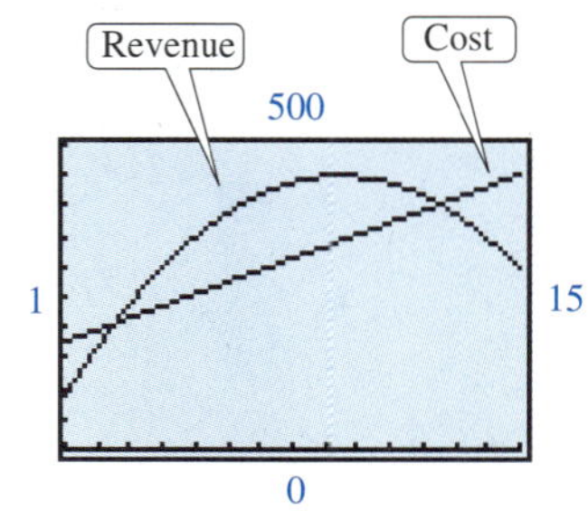

(D) Graphical solution:

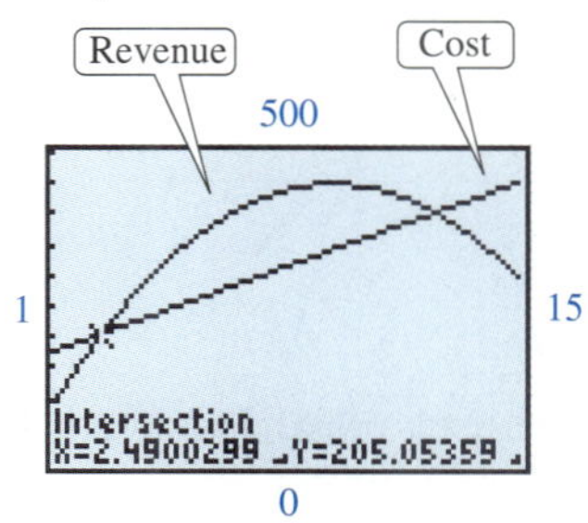

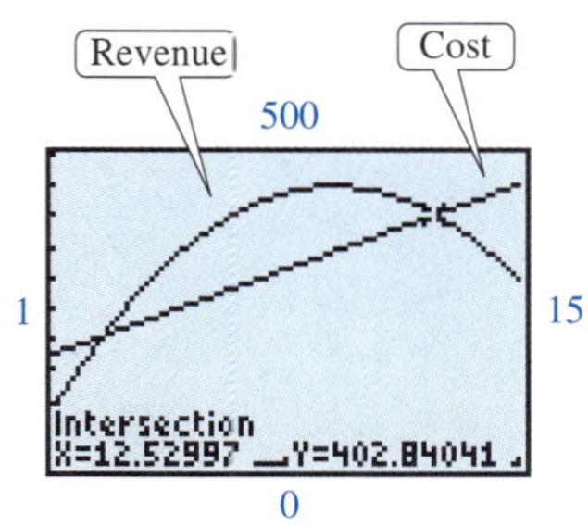

The company breaks even at $x = 2.490$ and 12.530 million cameras.

(E) Use the results from parts (A) and (B) or (C) and (D):

$$\text{Loss:} \quad 1 \leqslant x < 2.490 \quad \text{or} \quad 12.530 < x \leqslant 15$$
$$\text{Profit:} \quad 2.490 < x < 12.530$$

Matched Problem 4 Use the profit equation from Matched Problem 3:

$$P(x) = R(x) - C(x)$$
$$= -5x^2 + 75.1x - 156 \quad \text{Profit function}$$

Domain: $1 \leqslant x \leqslant 15$

(A) Sketch a graph of the profit function in a rectangular coordinate system.

(B) Break-even points occur when $P(x) = 0$. Find the break-even points algebraically to the nearest thousand cameras.

(C) Plot the profit function in an appropriate viewing window.

(D) Find the break-even points graphically to the nearest thousand cameras using trace and zoom or an appropriate built-in routine.

(E) A loss occurs if $P(x) < 0$, and a profit occurs if $P(x) > 0$. For what outputs (to the nearest thousand cameras) will a loss occur? A profit?

Answers to Matched Problems

1. (A)

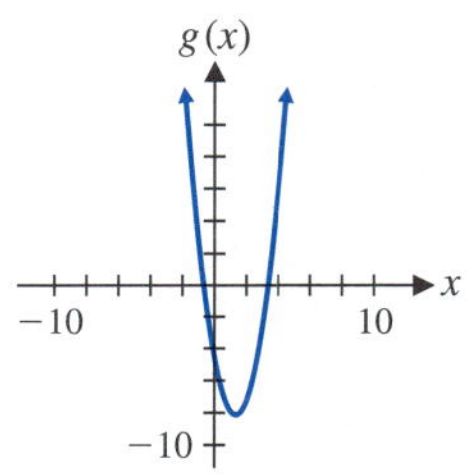

(B) x intercepts: $-0.77, 3.27$; y intercept: -5

(C)

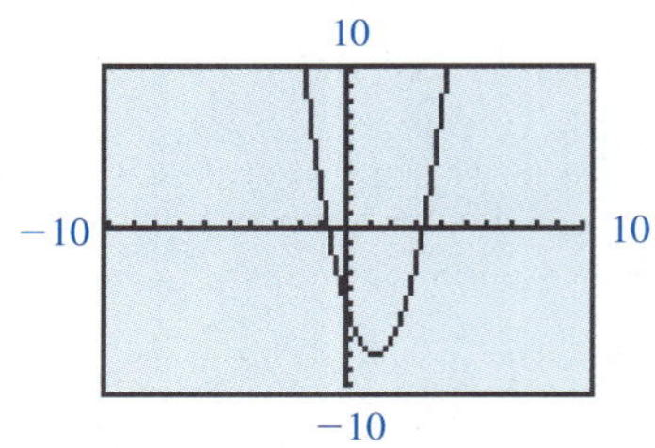

(D) x intercepts: $-0.77, 3.27$; y intercept: -5

(E) $x \leqslant -0.77$ or $x \geqslant 3.27$; or $(-\infty, -0.77]$ or $[3.27, \infty)$

(F) $x = -0.35, 2.85$

2. (A) $f(x) = -0.25(x + 4)^2 + 6$; Vertex: $(-4, 6)$; Maximum: $f(-4) = 6$; Range: $y \leqslant 6$ or $(-\infty, 6]$

(B) The graph of $f(x) = -0.25(x + 4)^2 + 6$ is the same as the graph of $g(x) = x^2$ vertically contracted by a factor of 0.25, reflected in the x axis, and shifted 4 units to the left and 6 units up.

(C)

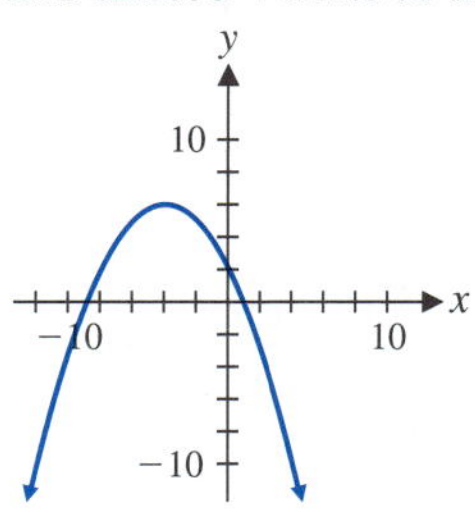

(D)

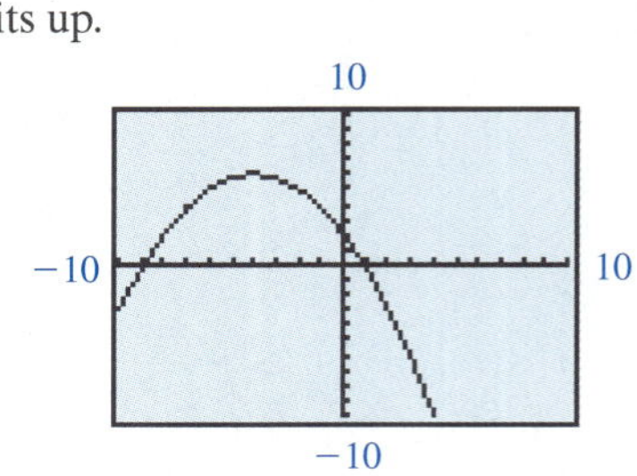

(E) Vertex: $(-4, 6)$; Maximum: $f(-4) = 6$; Range: $y \leqslant 6$ or $(-\infty, 6]$

3. (A) $P(x) = R(x) - C(x) = -5x^2 + 75.1x - 156$

(B) $P(x) = R(x) - C(x) = -5(x - 7.51)^2 + 126.0005$; output of 7.510 million cameras will produce a maximum profit of 126.001 million dollars.

(C) $p(7.510) = \$57$

(D)

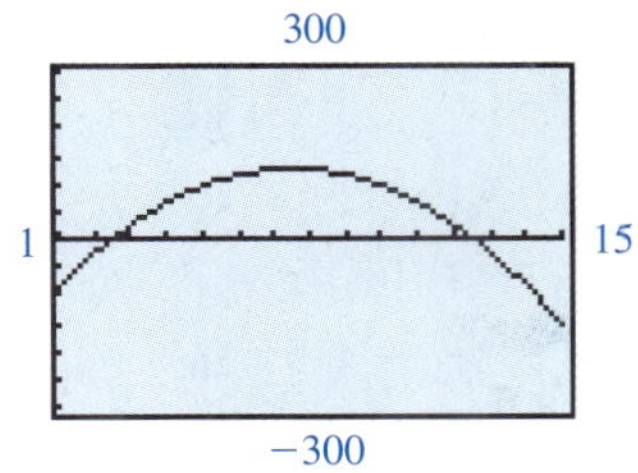

(E)

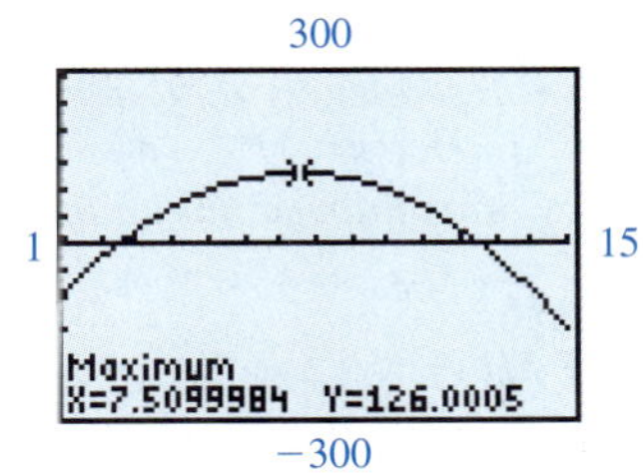

An output of 7.51 million cameras will produce a maximum profit of 126.001 million dollars. (Notice that maximum profit does not occur at the same output where maximum revenue occurs.)

4. (A)

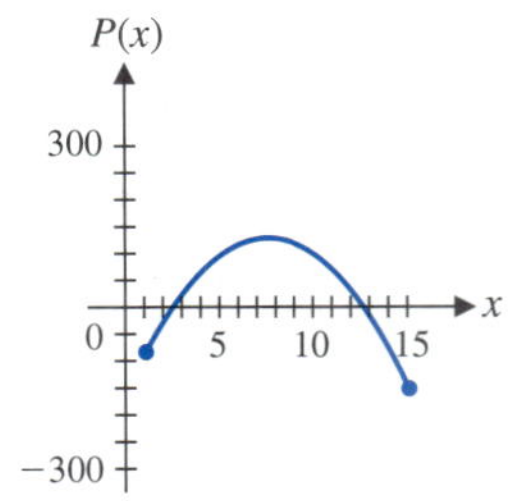

(B) $x = 2.490$ or 12.530 million cameras

(C)

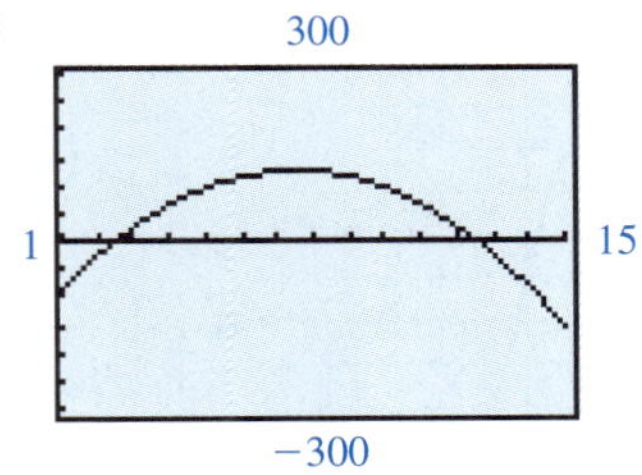

(D) $x = 2.490$ or 12.530 million cameras

(E) Loss: $1 \leqslant x < 2.490$ or $12.530 < x \leqslant 15$; Profit: $2.490 < x < 12.530$

EXERCISE 1-4

A

1. Indicate which equations define a quadratic function with x the independent variable.

(A) $y = 3 - x^2$ (B) $y^2 + x^2 = 4$

(C) $y = (2 - 3x)^2$ (D) $2x - y = 3$

(E) $y = 2x(3 - x)$ (F) $y = -2x^2 + 5x - 1$

2. Indicate which equations define a quadratic function with x the independent variable.

(A) $x^2 - y^2 = 9$ (B) $y = x(3x - 5)$

(C) $y = 1 - \sqrt{x}$ (D) $y = 4 - 3x - 2x^2$

(E) $y = 2(x - 3)^2$ (F) $y = 8.3 - 0.3x^2$

3. Match each equation with a graph of one of the functions f, g, m, or n in the figure.

(A) $y = -(x + 2)^2 + 1$ (B) $y = (x - 2)^2 - 1$

(C) $y = (x + 2)^2 - 1$ (D) $y = -(x - 2)^2 + 1$

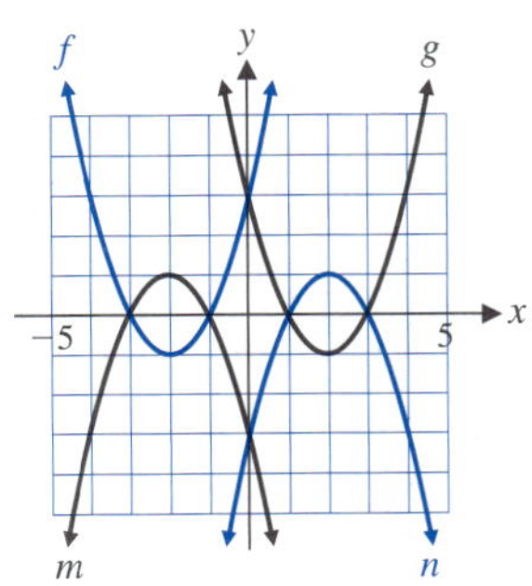

Figure for 3

4. Match each equation with a graph of one of the functions f, g, m, or n in the figure.

(A) $y = (x - 3)^2 - 4$ (B) $y = -(x + 3)^2 + 4$

(C) $y = -(x - 3)^2 + 4$ (D) $y = (x + 3)^2 - 4$

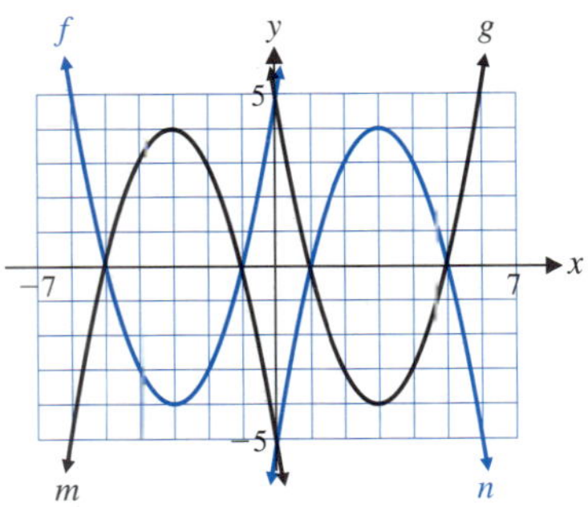

Figure for 4

For the functions indicated in Problems 5–8, find each of the following to the nearest integer by referring to the graphs for Problems 3 and Problem 4.

(A) Intercepts *(B) Vertex*

(C) Maximum or minimum *(D) Range*

(E) Increasing interval *(F) Decreasing interval*

5. Function n in the figure for Problem 3

6. Function m in the figure for Problem 4

7. Function f in the figure for Problem 3

8. Function g in the figure for Problem 4

In Problems 9–12, find each of the following algebraically (to the nearest integer) without referring to any graphs.

(A) Intercepts *(B) Vertex*
(C) Maximum or minimum *(D) Range*

9. $f(x) = -(x-2)^2 + 1$ **10.** $g(x) = -(x+3)^2 + 4$

11. $M(x) = (x+2)^2 - 1$ **12.** $N(x) = (x-3)^2 - 4$

B *In Problems 13–16, write an equation for each graph in the form $y = a(x-h)^2 + k$, where a is either 1 or −1 and h and k are integers.*

13.

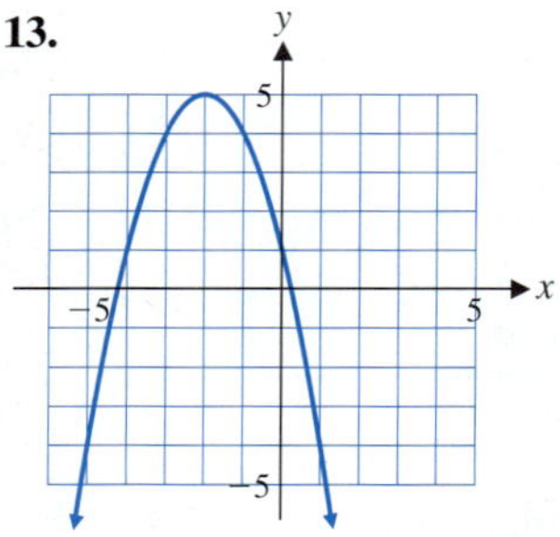

14.

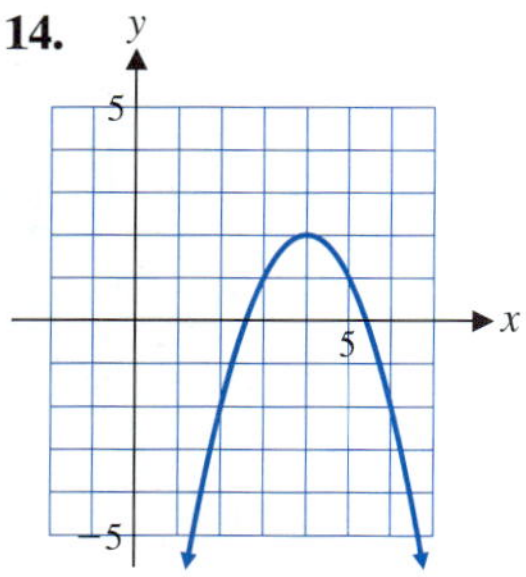

15.

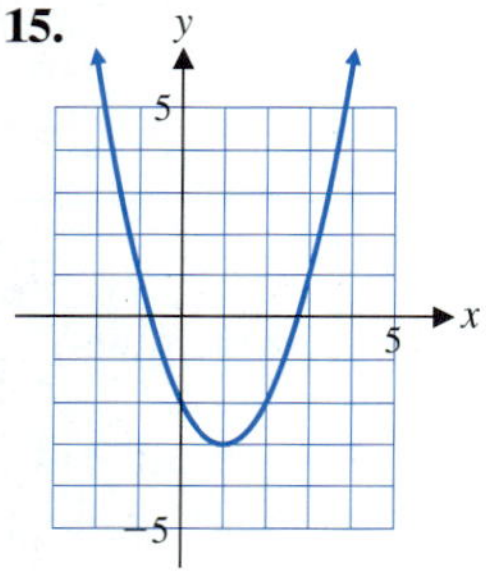

16.

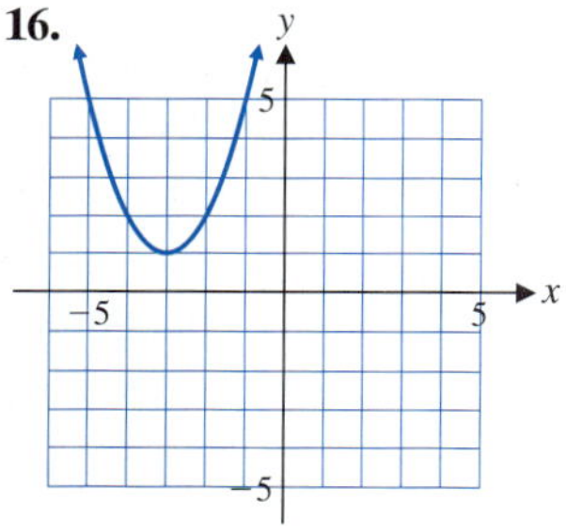

In Problems 17–22, first write each function in the form $f(x) = a(x-h)^2 + k$; then find each of the following algebraically (to one decimal place) without using any graphs:

(A) Intercepts *(B) Vertex*
(C) Maximum or minimum *(D) Range*

17. $f(x) = x^2 - 8x + 13$ **18.** $g(x) = x^2 + 10x + 20$

19. $M(x) = 1 - 6x - x^2$ **20.** $N(x) = -10 + 8x - x^2$

21. $G(x) = 0.5x^2 - 4x + 10$

22. $H(x) = -0.5x^2 - 2x - 3$

Solve Problems 17–22 using a graphing utility without completing the square. Solve graphically using trace and zoom or an appropriate built-in routine.

23. C Let $f(x) = 0.3x^2 - x - 8$. Solve each equation graphically to two decimal places.
(A) $f(x) = 4$ (B) $f(x) = -1$ (C) $f(x) = -9$

24. C Let $g(x) = -0.6x^2 + 3x + 4$. Solve each equation graphically to two decimal places.
(A) $g(x) = -2$ (B) $g(x) = 5$ (C) $g(x) = 8$

25. Explain under what graphical conditions a quadratic function has exactly one real zero.

26. Explain under what graphical conditions a quadratic function has no real zeros.

C *In Problems 27–30, first write each function in the form $f(x) = a(x-h)^2 + k$; then find each of the following algebraically (to two decimal places) without using any graphs:*

(A) Intercepts *(B) Vertex*
(C) Maximum or minimum *(D) Range*

27. $g(x) = 0.25x^2 - 1.5x - 7$

28. $m(x) = 0.20x^2 - 1.6x - 1$

29. $f(x) = -0.12x^2 + 0.96x + 1.2$

30. $n(x) = -0.15x^2 - 0.90x + 3.3$

Solve Problems 31–36 graphically to two decimal places using a graphing utility. Do not change the form of any equation or inequality.

31. $2 - 5x - x^2 = 0$ **32.** $7 + 3x - 2x^2 = 0$

33. $1.9x^2 - 1.5x - 5.6 < 0$ **34.** $3.4 + 2.9x - 1.1x^2 \geq 0$

35. $2.8 + 3.1x - 0.9x^2 \leq 0$ **36.** $1.8x^2 - 3.1x - 4.9 > 0$

37. Given that f is a quadratic function with minimum $f(x) = f(2) = 4$, find the axis, vertex, range, and x intercepts.

38. Given that f is a quadratic function with maximum $f(x) = f(-3) = -5$, find the axis, vertex, range, and x intercepts.

In Problems 39–42:

(A) Graph f and g in the same coordinate system.
(B) Solve $f(x) = g(x)$ algebraically to two decimal places.
(C) Solve $f(x) > g(x)$ using parts (A) and (B).
(D) Solve $f(x) < g(x)$ using parts (A) and (B).

39. $f(x) = -0.4x(x - 10)$
$g(x) = 0.3x + 5$
$0 \leq x \leq 10$

40. $f(x) = -0.7x(x - 7)$
$g(x) = 0.5x + 3.5$
$0 \leq x \leq 7$

41. $f(x) = -0.9x^2 + 7.2x$
$g(x) = 1.2x + 5.5$
$0 \leq x \leq 8$

42. $f(x) = -0.7x^2 + 6.3x$
$g(x) = 1.1x + 4.8$
$0 \leq x \leq 9$

43. Give a simple example of a quadratic function that has no real zeros. Explain how its graph is related to the x axis.

44. Give a simple example of a quadratic function that has exactly one real zero. Explain how its graph is related to the x axis.

Applications

Business & Economics

45. *Tire mileage.* An automobile tire manufacturer collected the data in the table relating tire pressure x (in pounds per square inch) and mileage (in thousands of miles):

x	28	30	32	34	36
MILEAGE	45	52	55	51	47

A mathematical model for the data is given by

$$f(x) = -0.518x^2 + 33.3x - 481$$

(A) Complete the following table. Round values of $f(x)$ to one decimal place.

x	28	30	32	34	36
MILEAGE	45	52	55	51	47
$f(x)$					

(B) Sketch the graph of f and the mileage data in the same coordinate system.

(C) Use values of the modeling function rounded to two decimal places to estimate the mileage for a tire pressure of 31 pounds per square inch. For 35 pounds per square inch.

(D) Write a brief description of the relationship between tire pressure and mileage.

46. *Automobile production.* The table lists General Motors' total U.S. vehicle production in millions of units from 1989 to 1993:

YEAR	1989	1990	1991	1992	1993
PRODUCTION	4.7	4.1	3.5	3.7	5.0

A mathematical model for GM's production data is given by

$$f(x) = 0.33x^2 - 1.3x + 4.8$$

where $x = 0$ corresponds to 1989.

(A) Complete the following table. Round values of $f(x)$ to one decimal place.

x	0	1	2	3	4
PRODUCTION	4.7	4.1	3.5	3.7	5.0
$f(x)$					

(B) Sketch the graph of f and the production data in the same coordinate system.

(C) Use values of the modeling function f rounded to two decimal places to estimate the production in 1994. In 1995.

(D) Write a brief verbal description of GM's production from 1989 to 1993.

47. *Revenue.* Refer to Problems 85 and 87, Exercise 1-1. We found that the marketing research department for the company that manufactures and sells memory chips for microcomputers established the following price–demand and revenue functions:

$p(x) = 119 - 6x$ Price–demand function

$R(x) = xp(x) = x(119 - 6x)$ Revenue function

where $p(x)$ is the wholesale price in dollars at which x million chips can be sold, and $R(x)$ is in millions of dollars. Both functions have domain $1 \le x \le 15$.

(A) Sketch a graph of the revenue function in a rectangular coordinate system.

(B) Find the output (to the nearest thousand chips) that will produce the maximum revenue. What is the maximum revenue to the nearest thousand dollars? Solve the problem algebraically by completing the square.

(C) Graph the revenue function using an appropriate viewing window.

(D) Find the output (to the nearest thousand chips) that will produce the maximum revenue. What is the maximum revenue to the nearest thousand dollars? Solve the problem graphically using trace and zoom or an appropriate built-in routine.

(E) What is the wholesale price per chip (to the nearest dollar) that produces the maximum revenue?

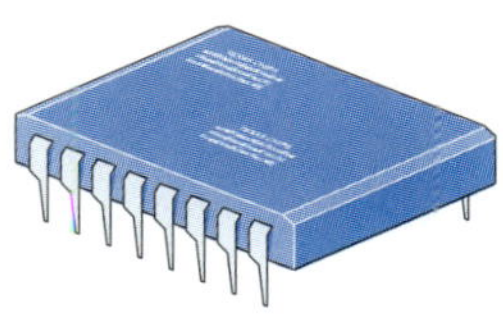

48. *Revenue.* Refer to Problems 86 and 88, Exercise 1-1. We found that the marketing research department for the company that manufactures and sells "Notebook" computers established the following price–demand and revenue functions:

$$p(x) = 1{,}190 - 36x \quad \text{Price–demand function}$$
$$R(x) = xp(x) \quad \text{Revenue function}$$
$$= x(1{,}190 - 36x)$$

where $p(x)$ is the wholesale price in dollars at which x thousand computers can be sold, and $R(x)$ is in thousands of dollars. Both functions have domain $1 \leq x \leq 25$.

(A) Sketch a graph of the revenue function in a rectangular coordinate system.

(B) Find the output (to the nearest hundred computers) that will produce the maximum revenue. What is the maximum revenue to the nearest thousand dollars? Solve the problem algebraically by completing the square.

C (C) Graph the revenue function using an appropriate viewing window.

C (D) Find the output (to the nearest hundred computers) that will produce the maximum revenue. What is the maximum revenue to the nearest thousand dollars? Solve the problem graphically using trace and zoom or an appropriate built-in routine.

(E) What is the wholesale price per computer (to the nearest dollar) that produces the maximum revenue?

49. *Break-even analysis.* Use the revenue function from Problem 47 in this exercise and the cost function from Problem 89, Exercise 1-1:

$$R(x) = x(119 - 6x) \quad \text{Revenue function}$$
$$C(x) = 234 + 23x \quad \text{Cost function}$$

where x is in millions of chips, and $R(x)$ and $C(x)$ are in millions of dollars. Both functions have domain $1 \leq x \leq 15$.

(A) Sketch a graph of both functions in the same rectangular coordinate system.

(B) Find the break-even points algebraically to the nearest thousand chips.

C (C) Graph both functions simultaneously in the same viewing window.

C (D) Find the break-even points graphically to the nearest thousand chips using trace and zoom or an appropriate built-in routine.

(E) For what outputs will a loss occur? A profit?

50. *Break-even analysis.* Use the revenue function from Problem 48 in this exercise and the cost function from Problem 90, Exercise 1-1:

$$R(x) = x(1{,}190 - 36x) \quad \text{Revenue function}$$
$$C(x) = 4{,}320 + 146x \quad \text{Cost function}$$

where x is thousands of computers, and $R(x)$ and $C(x)$ are in thousands of dollars. Both functions have domain $1 \leq x \leq 25$.

(A) Sketch a graph of both functions in the same rectangular coordinate system.

(B) Find the break-even points algebraically to the nearest hundred computers.

C (C) Graph both functions simultaneously in the same viewing window.

C (D) Find the break-even points graphically to the nearest hundred computers using trace and zoom or an appropriate built-in routine.

(E) For what outputs will a loss occur? Will a profit occur?

51. *Profit–loss analysis.* Use the revenue function from Problem 47 in this exercise and the cost function from Problem 89, Exercise 1-1:

$$R(x) = x(119 - 6x) \quad \text{Revenue function}$$
$$C(x) = 234 + 23x \quad \text{Cost function}$$

where x is in millions of chips, and $R(x)$ and $C(x)$ are in millions of dollars. Both functions have domain $1 \leq x \leq 15$.

(A) Form a profit function P, and graph R, C, and P in the same rectangular coordinate system.

C (B) Graph R, C, and P simultaneously in an appropriate viewing window.

(C) Discuss the relationship between the intersection points of the graphs of R and C and the x intercepts of P.

(D) Find the x intercepts of P algebraically to the nearest thousand chips. Find the break-even points to the nearest thousand chips.

C (E) Solve part (D) graphically.

(F) Refer to the graph drawn in part (A) or (B). Does the maximum profit appear to occur at the same output level as the maximum revenue? Are the maximum profit and the maximum revenue equal? Explain.

(G) Verify your conclusion in part (F) by completing the square for the profit function to find the output (to the nearest thousand chips) that produces the maximum profit. Find the maximum profit (to the nearest thousand dollars), and compare with Problem 47D.

C (H) Solve part (G) graphically.

52. *Profit–loss analysis.* Use the revenue function from Problem 48 in this exercise and the cost function from Problem 90, Exercise 1-1:

$$R(x) = x(1{,}190 - 36x) \quad \text{Revenue function}$$
$$C(x) = 4{,}320 + 146x \quad \text{Cost function}$$

where x is thousands of computers, and $R(x)$ and $C(x)$ are in thousands of dollars. Both functions have domain $1 \leq x \leq 25$.

(A) Form a profit function P, and graph R, C, and P in the same rectangular coordinate system.

C (B) Graph R, C, and P simultaneously in an appropriate viewing window.

(C) Discuss the relationship between the intersection points of the graphs of R and C and the x intercepts of P.

(D) Find the x intercepts of P algebraically to the nearest hundred computers. Find the break-even points to the nearest hundred computers.

C (E) Solve part (D) graphically.

(F) Refer to the graph drawn in part (A) or (B). Does the maximum profit appear to occur at the same output level as the maximum revenue? Are the maximum profit and the maximum revenue equal? Explain.

(G) Verify your conclusion in part (F) by completing the square for the profit function to find the output (to the nearest hundred computers) that produces the maximum profit. Find the maximum profit (to the nearest thousand dollars), and compare with Problem 48D.

C (H) Solve part (G) graphically.

Life Sciences

53. *Medicine.* The French physician Poiseuille was the first to discover that blood flows faster near the center of an artery than near the edge. Experimental evidence has shown that the rate of flow v (in centimeters per second) at a point x centimeters from the center of an artery (see the figure) is given by

$$v = f(x) = 1{,}000(0.04 - x^2) \qquad 0 \leqslant x \leqslant 0.2$$

(A) Find the distance from the center that the rate of flow is 20 centimeters per second. Solve algebraically to two decimal places.

C (B) Use a graphing utility to solve part (A).

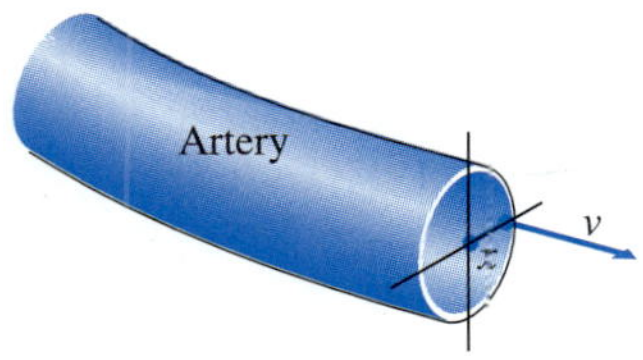

Figure for 53 and 54

54. *Medicine.* Refer to Problem 53.

(A) Find the distance from the center that the rate of flow is 30 centimeters per second. Solve algebraically to two decimal places.

C (B) Use a graphing utility to solve part (A).

IMPORTANT TERMS AND SYMBOLS

1-1 *Functions.* Cartesian or rectangular coordinate system; coordinate axes; quadrants; ordered pairs; coordinates; abscissa; ordinate; fundamental theorem of analytic geometry; solution and solution set for an equation in two variables; graph of an equation in two variables; point-by-point plotting; function; domain; range; functions specified by equations; input; output; independent variable; dependent variable; vertical-line test; function notation; difference quotient; cost function; price–demand function; revenue function; profit function

$f(x)$; $C = a + bx$; $p = m - nx$;
$R = xp$; $P = R - C$

1-2 *Elementary Functions: Graphs and Transformations.* Six basic functions: identity, absolute value, square, cube, square root, cube root; transformations; horizontal shift or translation; vertical shift or translation; reflection in the x axis; vertical expansion; vertical contraction; piecewise-defined function; discontinuity

$f(x) = x$; $g(x) = |x|$; $h(x) = x^2$;
$m(x) = x^3$; $n(x) = \sqrt{x}$; $p(x) = \sqrt[3]{x}$

See the inside front cover for the graphs of these functions.

1-3 *Linear Functions and Straight Lines.* x intercepts; y intercept; linear function; constant function; first-degree function; graph of a linear function; graph of a constant function; graphs of linear equations in two variables; standard form for the equation of a line; equations of vertical lines and horizontal lines; slope; slope–intercept form; point–slope form

$f(x) = mx + b, m \neq 0$; $f(x) = b$; $x = a$;
$Ax + By = C$; $y = mx + b$; $y - y_1 = m(x - x_1)$;
$m = \dfrac{y_2 - y_1}{x_2 - x_1}, x_1 \neq x_2$

1-4 *Quadratic Functions.* Quadratic function; parabola; finding intercepts; vertex; axis; maximum or minimum; range; break-even points

$f(x) = ax^2 + bx + c, a \neq 0$;
$f(x) = a(x - h)^2 + k, a \neq 0$

REVIEW EXERCISE

Work through all the problems in this chapter review and check your answers in the back of the book. Answers to all review problems are there along with section numbers in italics to indicate where each type of problem is discussed. Where weaknesses show up, review appropriate sections in the text.

A

1. Use point-by-point plotting to sketch a graph of $y = 5 - x^2$. Use integer values for x from -3 to 3.

2. Indicate whether each graph specifies a function:

(A)

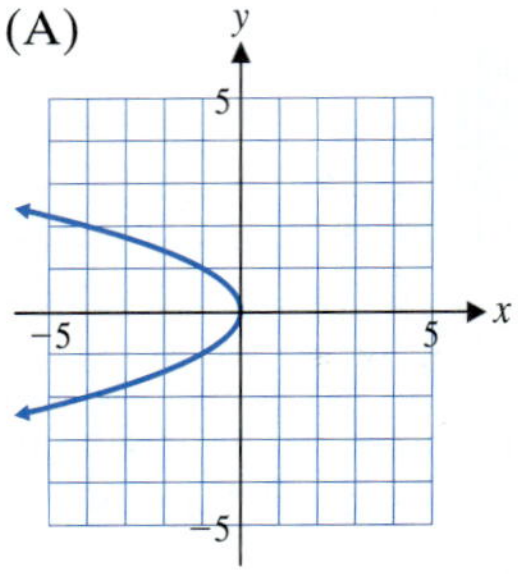

(B)

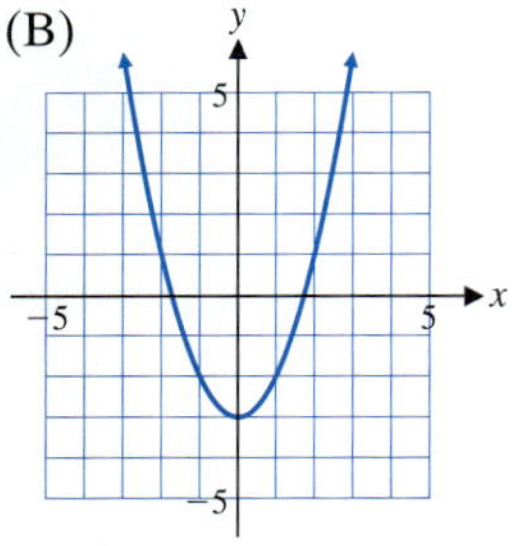

(C)

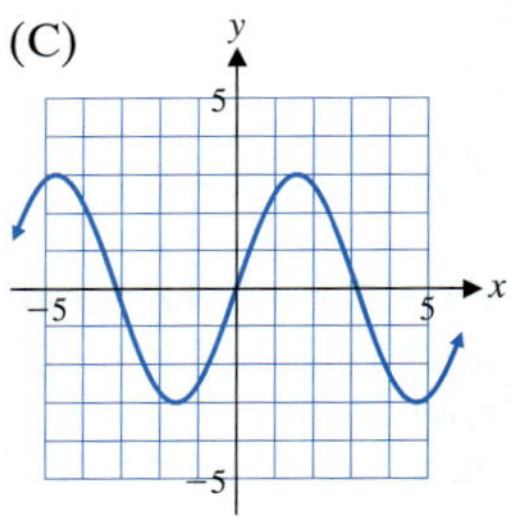

(D)

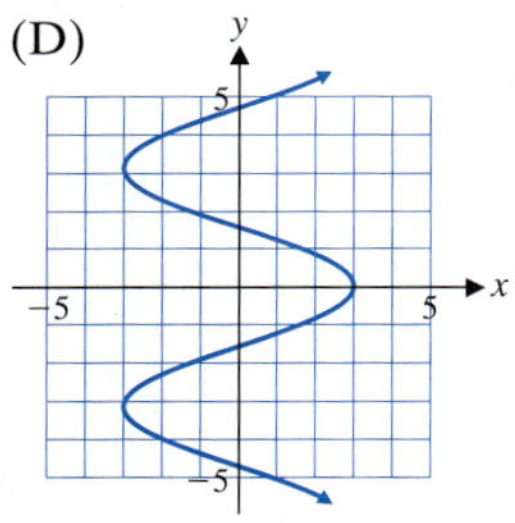

3. For $f(x) = 2x - 1$ and $g(x) = x^2 - 2x$, find:

(A) $f(-2) + g(-1)$ (B) $f(0) \cdot g(4)$

(C) $\dfrac{g(2)}{f(3)}$ (D) $\dfrac{f(3)}{g(2)}$

4. Use the graph of function f in the figure to determine (to the nearest integer) x or y as indicated.

(A) $y = f(0)$ (B) $4 = f(x)$

(C) $y = f(3)$ (D) $3 = f(x)$

(E) $y = f(-6)$ (F) $-1 = f(x)$

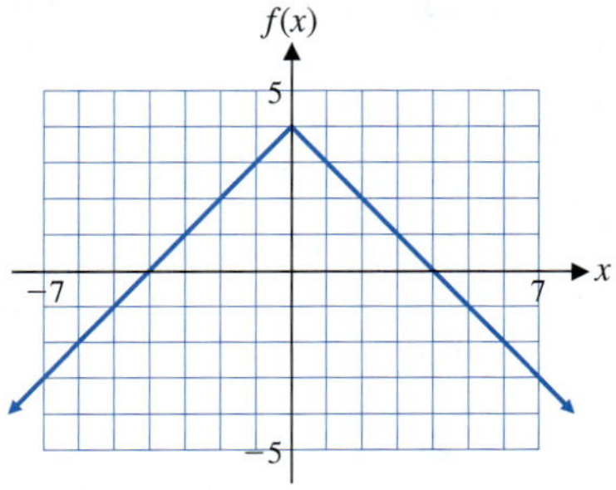

Figure for 4

5. Sketch a graph of each of the functions in parts (A)–(D) using the graph of function f in the figure below.

(A) $y = -f(x)$ (B) $y = f(x) + 4$

(C) $y = f(x - 2)$ (D) $y = -f(x + 3) - 3$

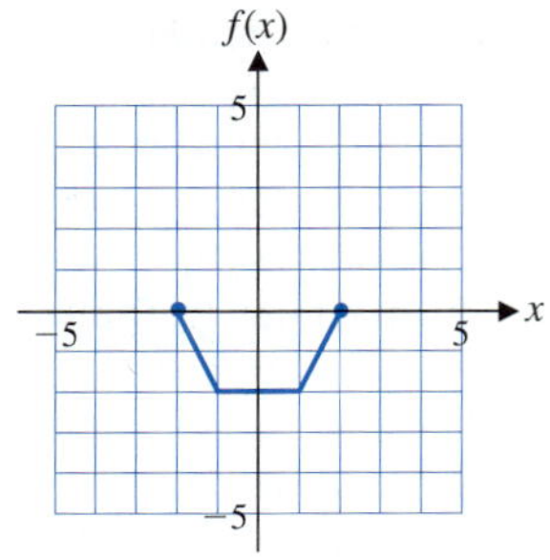

Figure for 5

6. Refer to the figure below.

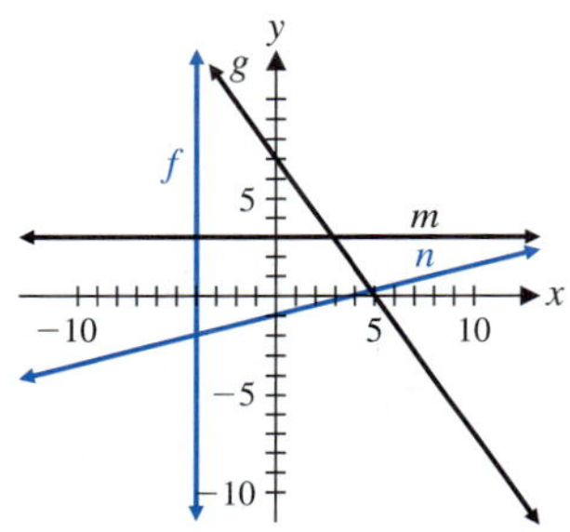

Figure for 6

(A) Identify the graphs of any linear functions with positive slopes.

(B) Identify the graphs of any linear functions with negative slopes.

(C) Identify the graphs of any constant functions. What are their slopes?

(D) Identify any graphs that are not graphs of functions. What can you say about their slopes?

7. Write an equation in the form $y = mx + b$ for a line with slope $-\frac{2}{3}$ and y intercept 6.

8. Write the equations of the vertical line and the horizontal line that pass through $(-6, 5)$.

9. Sketch a graph of $2x - 3y = 18$. What are the intercepts and slope of the line?

10. Indicate which equations define a quadratic function with x the independent variable.

(A) $y^2 = 4 - x^2$ (B) $x^2 - 3 = y$

(C) $y = -2(x - 3)^2 + 1$ (D) $y = x(x + 7)$

(E) $y = 3x + 9$ (F) $y = 0.2x^2 - 5.7$

11. Match each equation with a graph of one of the functions *f, g, m,* or *n* in the figure.

(A) $y = (x - 2)^2 - 4$ (B) $y = -(x + 2)^2 + 4$

(C) $y = -(x - 2)^2 + 4$ (D) $y = (x + 2)^2 - 4$

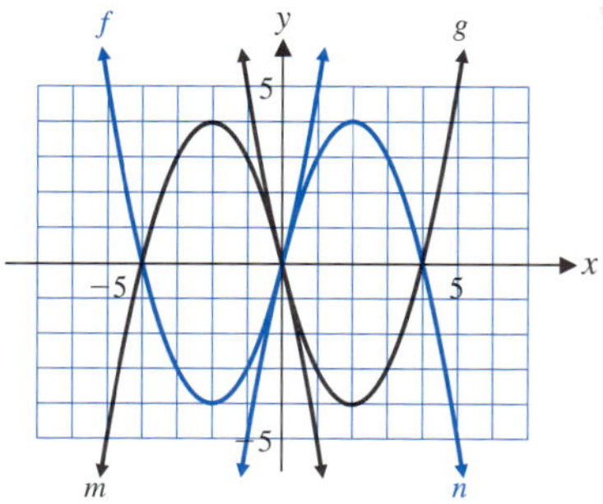

Figure for 11

12. Referring to the graph of function *f* in the figure for Problem 11 and using known properties of quadratic functions, find each of the following to the nearest integer:

(A) Intercepts (B) Vertex

(C) Maximum or minimum (D) Range

(E) Increasing interval (F) Decreasing interval

B

13. Indicate which of the following equations define a linear function or a constant function:

(A) $2x - 3y = 5$ (B) $x = -2$

(C) $y = 4 - 3x$ (D) $y = -5$

(E) $x = 3y + 5$ (F) $\frac{x}{2} - \frac{y}{3} = 1$

14. Find the domain of each function:

(A) $f(x) = \frac{2x - 5}{x^2 - x - 6}$ (B) $g(x) = \frac{3x}{\sqrt{5 - x}}$

15. The function g is defined by $g(x) = 2x - 3\sqrt{x}$. Translate into a verbal definition.

16. Describe the graphs of $x = -3$ and $y = 2$. Graph both simultaneously in the same rectangular coordinate system.

17. Let $f(x) = 0.4x(x + 4)(2 - x)$.

C (A) Sketch a graph of f on graph paper by first plotting points using odd integer values of x from -3 to 3. Then complete the graph using a graphing utility.

(B) Discuss the number of solutions of each of the following equations:

$$f(x) = 3 \qquad f(x) = 2 \qquad f(x) = 1$$

(C) Approximate the solutions of each equation in part (B) to two decimal places.

18. Find $\frac{f(2 + h) - f(2)}{h}$ for $f(x) = 3 - 2x$.

19. Find $\frac{f(a + h) - f(a)}{h}$ for $f(x) = x^2 - 3x + 1$.

20. Explain how the graph of $m(x) = -|x - 4|$ is related to the graph of $y = |x|$.

21. Explain how the graph of $g(x) = 0.3x^3 + 3$ is related to the graph of $y = x^3$.

22. The following graph is the result of applying a sequence of transformations to the graph of $y = x^2$. Describe the transformations verbally and write an equation for the given graph.

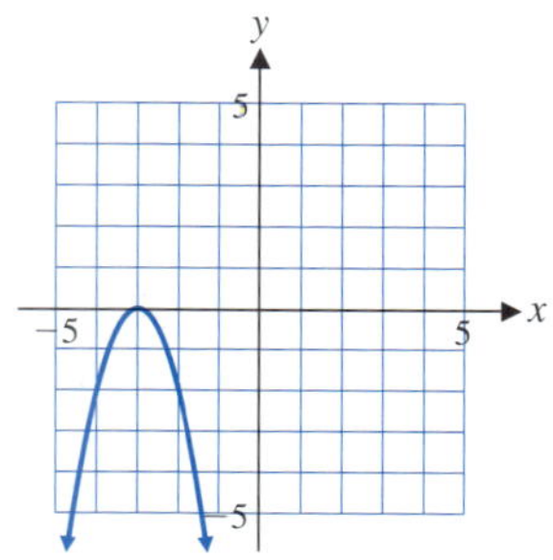

Figure for 22

23. The graph of a function f is formed by vertically expanding the graph of $y = \sqrt{x}$ by a factor of 2, and shifting it to the left 3 units and down 1 unit. Find an equation for function f and graph it for $-5 \leq x \leq 5$ and $-5 \leq y \leq 5$.

24. Let

$$f(x) = \begin{cases} -x - 2 & \text{for } x < 0 \\ 0.2x^2 & \text{for } x \geq 0 \end{cases}$$

(A) Sketch the graph of f.

(B) Find any points of discontinuity.

C (C) Check by graphing on a graphing utility.

25. Write the equation of a line through each indicated point with the indicated slope. Write the final answer in the form $y = mx + b$.

(A) $m = -\frac{2}{3}$; $(-3, 2)$ (B) $m = 0$; $(3, 3)$

26. Write the equation of the line through the two indicated points. Write the final answer in the form $Ax + By = C$.

(A) $(-3, 5), (1, -1)$ (B) $(-1, 5), (4, 5)$

(C) $(-2, 7), (-2, -2)$

27. Write an equation for the graph shown in the form $y = a(x - h)^2 + k$, where a is either -1 or $+1$ and h and k are integers.

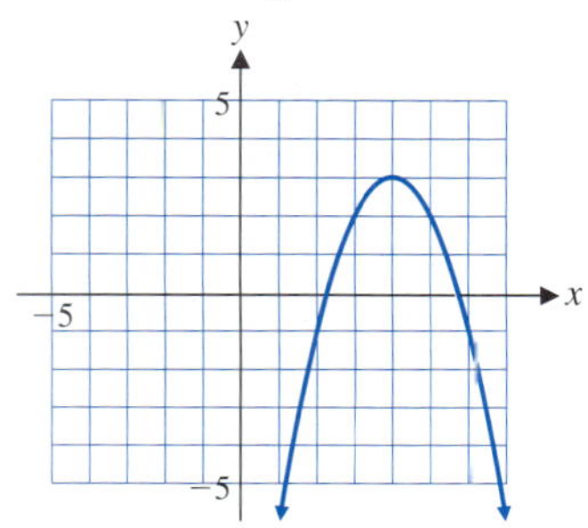

Figure for 27

28. Given $f(x) = -0.4x^2 + 3.2x + 1.2$, find the following algebraically (to one decimal place) without referring to a graph:
(A) Intercepts (B) Vertex
(C) Maximum or minimum (D) Range

29. C Graph $f(x) = -0.4x^2 + 3.2x + 1.2$ in a graphing utility and find the following (to one decimal place) using trace and zoom or an appropriate graphing utility:
(A) Intercepts (B) Vertex
(C) Maximum or minimum (D) Range

C

30. The following graph is the result of applying a sequence of transformations to the graph of $y = \sqrt[3]{x}$. Describe the transformations verbally, and write an equation for the graph.

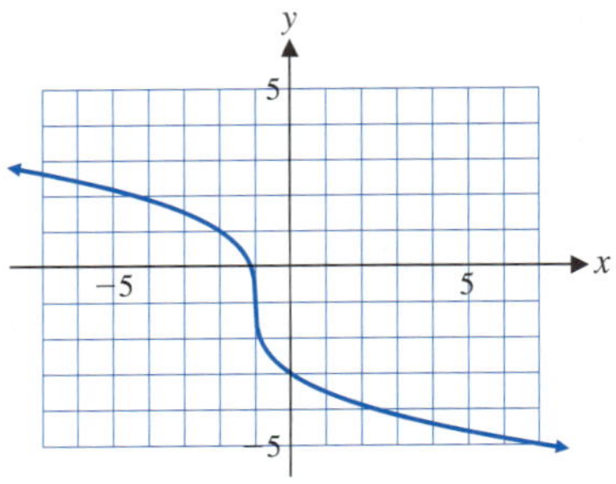

Figure for 30

31. Graph

$$y = mx + b \quad \text{and} \quad y = -\frac{1}{m}x + b$$

simultaneously in the same coordinate system for b fixed and several different values of m, $m \neq 0$. Describe the apparent relationship between the graphs of the two equations.

32. Find and simplify $\frac{f(x + h) - f(x)}{h}$ for each of the following functions:
(A) $f(x) = \sqrt{x}$ (B) $f(x) = \frac{1}{x}$

33. Given $G(x) = 0.3x^2 + 1.2x - 6.9$, find the following algebraically (to one decimal place) without the use of a graph:
(A) Intercepts (B) Vertex
(C) Maximum or minimum (D) Range
(E) Increasing and decreasing intervals

34. C Graph $G(x) = 0.3x^2 + 1.2x - 6.9$ in a standard viewing window. Then find each of the following (to one decimal place) using trace or zoom or an appropriate built-in routine:
(A) Intercepts (B) Vertex
(C) Maximum or minimum (D) Range
(E) Increasing and decreasing intervals

APPLICATIONS

Business & Economics

35. *Linear depreciation.* A computer system was purchased by a small business for $12,000 and, for tax purposes, is assumed to have a salvage value of $2,000 after 8 years. If its value is depreciated linearly from $12,000 to $2,000:
(A) Find the linear equation that relates the value V in dollars to the time t in years. Then graph the equation in a rectangular coordinate system.
(B) What would be the value of the system after 5 years?

36. C *Compound interest.* If $1,000 is invested at $100r\%$ compounded annually, at the end of 3 years it will grow to $A = 1{,}000(1 + r)^3$.
(A) Graph the equation in a graphing utility for $0 \leq r \leq 0.25$; that is, for money invested at between 0% and 25% compounded annually.
(B) At what rate of interest would $1,000 have to be invested to amount to $1,500 in 3 years? Solve for r graphically (to four decimal places) using trace and zoom or an appropriate built-in routine.

37. *Markup.* A sporting goods store sells a tennis racket that cost $130 for $208 and court shoes that cost $50 for $80.

(A) If the markup policy of the store for items that cost over $10 is assumed to be linear and is reflected in the pricing of these two items, write an equation that relates retail price R to cost C.
(B) What would be the retail price of a pair of in-line skates that cost $120?
(C) What would be the cost of a pair of cross-country skis that had a retail price of $176?
(D) What is the slope of the graph of the equation found in part (A)? Interpret the slope relative to the problem.

38. *Demand.* Egg consumption has been decreasing for some time, presumably because of increasing awareness of the high cholesterol content in egg yolks. The table lists the annual per capita consumption of eggs in the United States.

YEAR	1970	1975	1980	1985	1990
NUMBER OF EGGS	309	276	271	255	233

Source: U.S. Department of Agriculture.

A mathematical model for the data is given by

$$f(x) = 303.4 - 3.46x$$

where $x = 0$ corresponds to 1970.

(A) Complete the following table. Round values of $f(x)$ to the nearest integer.

x	0	5	10	15	20
CONSUMPTION	309	276	271	255	233
$f(x)$					

(B) Graph $y = f(x)$ and the data in the table in the same coordinate system.

(C) Use the modeling function f to estimate the per capita egg consumption in 1995. In 2000.

(D) Based on the information in the table, write a brief verbal description of egg consumption from 1970 to 1990.

39. *Pricing.* An office supply store sells ballpoint pens for \$0.49 each. For an order of 3 dozen or more pens, the price per pen for all pens ordered is reduced to \$0.44, and for an order of 6 dozen or more, the price per pen for all pens ordered is reduced to \$0.39.

(A) If $C(x)$ is the total cost in dollars for an order of x pens, write a piecewise definition for C.

(B) Graph $y = C(x)$ for $0 \leqslant x \leqslant 108$ and identify any points of discontinuity.

40. *Break-even analysis.* A video production company is planning to produce an instructional videotape. The producer estimates that it will cost \$84,000 to shoot the video and \$15 per unit to copy and distribute the tape. The wholesale price of the tape is \$50 per unit.

(A) Write cost and revenue equations, and graph both simultaneously in a rectangular coordinate system.

(B) Algebraically determine when $R = C$. Then, with the aid of part (A), determine when $R < C$ and $R > C$.

(C) Using a graphing utility, determine when $R = C$, $R < C$, and $R > C$.

41. *Break-even analysis.* The research department in a company that manufacturers AM/FM clock radios established the following price–demand, cost, and revenue functions:

$$\begin{aligned} p(x) &= 50 - 1.25x && \text{Price–demand function} \\ C(x) &= 160 + 10x && \text{Cost function} \\ R(x) &= xp(x) \\ &= x(50 - 1.25x) && \text{Revenue function} \end{aligned}$$

where x is in thousands of units, and $C(x)$ and $R(x)$ are in thousands of dollars. All three functions have domain $1 \leqslant x \leqslant 40$.

(A) Graph the cost function and the revenue function simultaneously in the same coordinate system.

(B) Algebraically determine when $R = C$. Then, with the aid of part (A), determine when $R < C$ and $R > C$ to the nearest unit.

(C) Algebraically determine the maximum revenue (to the nearest thousand dollars) and the output (to the nearest unit) that produces the maximum revenue. What is the wholesale price of the radio (to the nearest dollar) at this output?

42. *Profit–loss analysis.* Use the cost and revenue functions from Problem 41.

(A) Write a profit function and graph it in a graphing utility.

(B) Graphically determine when $P = 0$, $P < 0$, and $P > 0$ to the nearest unit.

(C) Graphically determine the maximum profit (to the nearest thousand dollars) and the output (to the nearest unit) that produces the maximum profit. What is the wholesale price of the radio (to the nearest dollar) at this output? [Compare with Problem 41C.]

43. *Construction.* A construction company has 840 feet of chain-link fence that is used to enclose storage areas for equipment and materials at construction sites. The supervisor wants to set up two identical rectangular storage areas sharing a common fence (see the figure):

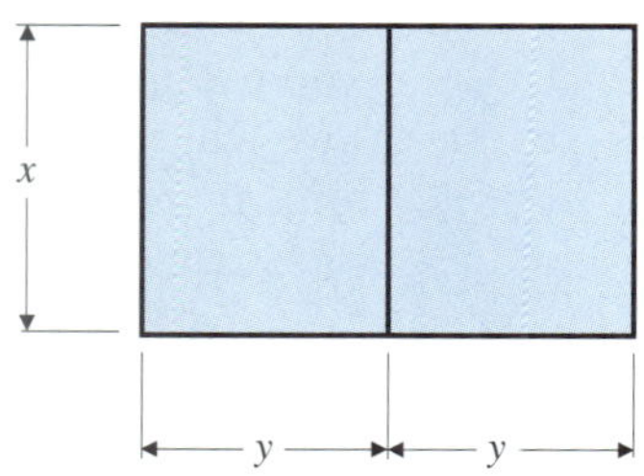

Figure for 43

Assuming all fencing is used:

(A) Express the total area $A(x)$ enclosed by both pens as a function of x.

(B) From physical considerations, what is the domain of the function A?

(C) Graph function A in a rectangular coordinate system.
(D) Use the graph to discuss the number and approximate locations of values of x that would produce storage areas with a combined area of 25,000 square feet.
C (E) Approximate graphically (to the nearest foot) the values of x that would produce storage areas with a combined area of 25,000 square feet.
(F) Algebraically determine the dimensions of the storage areas that have the maximum total combined area. What is the maximum area?

Life Sciences

44. *Air pollution.* On an average summer day in a large city, the pollution index at 8:00 AM is 20 parts per million, and it increases linearly by 15 parts per million each hour until 3:00 PM. Let $P(x)$ be the amount of pollutants in the air x hours after 8:00 AM.
(A) Express $P(x)$ as a linear function of x.
(B) What is the air pollution index at 1:00 PM?
(C) Graph the function P for $0 \leqslant x \leqslant 7$.
(D) What is the slope of the graph? (The slope is the amount of increase in pollution for each additional hour of time.)

Social Sciences

45. *Psychology—sensory perception.* One of the oldest studies in psychology concerns the following question: Given a certain level of stimulation (light, sound, weight lifting, electric shock, and so on), how much should the stimulation be increased for a person to notice the difference? In the middle of the nineteenth century, E. H. Weber (a German physiologist) formulated a law that still carries his name: If Δs is the change in stimulus that will just be noticeable at a stimulus level s, then the ratio of Δs to s is a constant:

$$\frac{\Delta s}{s} = k$$

Hence, the amount of change that will be noticed is a linear function of the stimulus level, and we note that the greater the stimulus, the more it takes to notice a difference. In an experiment on weight lifting, the constant k for a given individual was found to be $\frac{1}{30}$.
(A) Find Δs (the difference that is just noticeable) at the 30 pound level. At the 90 pound level.
(B) Graph $\Delta s = s/30$ for $0 \leqslant s \leqslant 120$.
(C) What is the slope of the graph?

Group Activity 1 Introduction to Regression Analysis

In real-world applications collected data may be assembled in table form and then examined to find a function to model the data. A very powerful mathematical tool called *regression analysis* is frequently used for this purpose. Several of the modeling functions stated in examples and exercises in Chapter 1 were constructed using *regression techniques* and a graphing utility.

Regression analysis is the process of fitting a function to a set of data points. This process is also referred to as **curve fitting.** In this group activity, we will restrict our attention to **linear regression**—that is, to fitting data with linear functions. Other types of regression analysis will be discussed in the next chapter.

At this time, we will not be interested in discussing the underlying mathematical methods used to construct a particular regression equation. Instead, we will concentrate on the mechanics of using a graphing utility to apply regression techniques to data sets.

In Example 7, Section 1-1, we were given the data in Table 1 and the corresponding modeling function

$$p(x) = 94.8 - 5x \tag{1}$$

where p is the wholesale price of a camera and x is the demand in millions.

Table 1
PRICE–DEMAND

x (MILLION)	p (\$)
2	87
5	68
8	53
12	37

Now we want to see how the function p was determined using linear regression on a graphing utility. On most graphing utilities, this process can be broken down into three steps:

Step 1. Enter the data set.

Step 2. Compute the desired regression equation.

Step 3. Graph the data set and the regression equation in the same viewing window.

The details for carrying out each step vary greatly from one graphing utility to another. Consult your manual. If you are using a Texas Instruments graphing calculator or a spreadsheet on a computer, you may also want to consult the supplements for this text (see Preface).

Figure 1 shows the results of applying this process to the data in Table 1 on a graphing calculator, and Figure 2 shows the same process on a spreadsheet.

FIGURE 1
Linear regression on a graphing calculator

Price-Demand	
x (millions)	p ($)
2	87
5	68
8	53
12	37
Slope	y intercept
-4.9634703	94.7534247

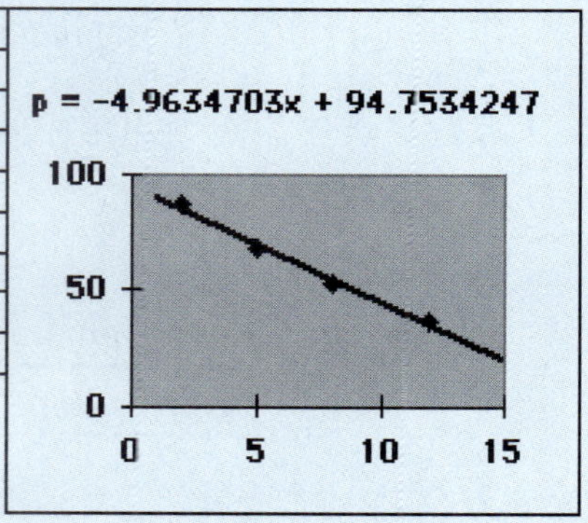

FIGURE 2
Linear regression on a spreadsheet

Notice how well the line graphed in Figures 1 and 2 appears to fit this set of data. At this time, we will not discuss any mathematical techniques for determining how well a curve fits a given set of data. Instead, we will simply assume that, in some sense, the linear regression line for a given data set is the "best fit" for that data.

Most graphing utilities compute coefficients of regression equations to many more decimal places than we will need for our purposes. Normally, when a regression equation is written, the coefficients are rounded. For example, if we round each coefficient in Figures 1 or 2 to one decimal place, we will obtain the modeling function f we used in Example 7, Section 1-1 [see equation (1)].

(A) If you have not already done so, carry out steps 1–3 for the data in Table 1. Write a brief summary of the details for performing these steps on your graphing utility and keep it for future reference. (Some of the optional exercises in subsequent chapters will require you to compute regression equations.)

(B) Each of the following examples and exercises from Chapter 1 contains a data set and an equation that models the data. In each case, use a graphing utility to compute a linear regression equation for the data. Plot the data and graph the equation in the same viewing rectangle. Discuss the difference between the graphing utility's regression equation and the modeling function stated in the exercise.

- Section 1-1, Matched Problem 7
- Exercise 1-1, Problems 85 and 86
- Exercise 1-3, Problems 73 and 74

Group Activity 2 Mathematical Modeling in Business

A company manufactures and sells mountain bikes. The management would like to have price–demand and cost functions for break-even and profit–loss analysis. Price–demand and cost functions are often established by collecting appropriate data at different levels of output, and then finding a model in the form of a basic elementary function (from our library of elementary functions) that "closely fits" the collected data. The financial department, using statistical techniques, arrived at the price–demand and cost data in Tables 1 and 2, where p is the wholesale price of a bike for a demand of x hundred bikes, $0 \leq x \leq 220$, and C is the cost (in hundreds of dollars) of producing and selling x bikes, $0 \leq x \leq 220$.

Table 1
PRICE–DEMAND

x (HUNDREDS)	p (\$)
0	525
64	370
125	270
185	130

Table 2
COST

x (HUNDREDS)	C (HUNDRED \$)
0	8,470
30	13,510
120	19,140
180	22,580
220	28,490

(A) *Building a Mathematical Model for Price–Demand.* Plot the data in Table 1 and observe that the relationship between p and x appears to be linear.

C

1. If you have completed Group Activity 1, use your graphing utility to find the linear regression line for the data in Table 1. Graph the line and the data in the same viewing window.
2. The linear regression line found in part 1 is a mathematical model for price–demand and is given by

$$p(x = -2.09x + 519 \quad \text{Price–demand equation}$$

Graph the data points from Table 1 and the price–demand equation in the same rectangular coordinate system.

3. The linear regression line defines a linear function. Interpret the slope of the line. Discuss its domain and range. Using the mathematical model, determine the price for a demand of 10,000 bikes. For a demand of 20,000 bikes.

(B) *Building a Mathematical Model for Cost.* Plot the data in Table 2 in a rectangular coordinate system. Which type of function appears to best fit the data?

1. If you have completed Group Activity 1, use your graphing utility to find the linear regression line for the data in Table 2. Graph the line and the data in the same viewing window.
2. The linear regression line found in part 1 is a mathematical model for cost and is given by

$$C(x) = 81x + 9{,}498 \quad \text{Cost equation}$$

Graph the data points from Table 2 and the cost equation in the same rectangular coordinate system.

3. Interpret the slope of the cost equation function. Discuss its domain and range. Using the mathematical model, determine the cost for an output and sales of 10,000 bikes. For an output and sales of 20,000 bikes.

(C) *Break-Even and Profit–Loss Analysis.* Refer to the price–demand equation in part (A) and the cost equation in part (B). Write an equation for the revenue function and state its domain. Write an equation for the profit function and state its domain.

1. Graph the revenue function and the cost function simultaneously in the same rectangular coordinate system. Algebraically determine at what outputs (to the nearest unit) the break-even points occur. Determine the outputs where costs exceed revenues and where revenues exceed costs.
2. Graph the revenue function and the cost function simultaneously in the same viewing window. Graphically determine at what outputs (to the nearest unit) break-even occurs, costs exceed revenues, and revenues exceed costs.
3. Graph the profit function in a rectangular coordinate system. Algebraically determine at what outputs (to the nearest unit) the break-even points occur. Determine where profits occur and where losses occur. At what output and price will a maximum profit occur? Does the maximum revenue and maximum profit occur for the same output? Discuss.
4. Graph the profit function in a graphing utility. Graphically determine at what outputs (to the nearest unit) break-even occurs, losses occur, and profits occur. At what output and price will a maximum profit occur? Does the maximum revenue and maximum profit occur for the same output? Discuss.

CHAPTER 2

ADDITIONAL ELEMENTARY FUNCTIONS

INTRODUCTION

In this chapter we add the following four general classes of functions to the beginning library of elementary functions started in Chapter 1: polynomial, rational, exponential, and logarithmic. This expanded library of elementary functions should take care of most of your function needs in this and many other courses. The linear and quadratic functions studied in Chapter 1 are special cases of the more general class of functions called *polynomials.*

SECTION 2-1 Polynomial and Rational Functions

- POLYNOMIAL FUNCTIONS
- POLYNOMIAL ROOT APPROXIMATION
- REGRESSION POLYNOMIALS
- RATIONAL FUNCTIONS
- APPLICATION

POLYNOMIAL FUNCTIONS

In Chapter 1 you were introduced to the basic functions

$f(x) = b$ — Constant function

$f(x) = ax + b \qquad a \neq 0$ — Linear function

$f(x) = ax^2 + bx + c \qquad a \neq 0$ — Quadratic function

as well as some special cases of

$f(x) = ax^3 + bx^2 + cx + d \qquad a \neq 0$ — Cubic function

Most of the earlier applications we considered, including cost, revenue, profit, loss, and packaging applications, made use of these functions. Notice the evolving pattern going from the constant function to the cubic function—the terms in each equation are of the form ax^n, where n is a nonnegative integer and a is a real number. All these functions are special cases of the general class of functions called *polynomial functions:*

Polynomial Function

A **polynomial function** is a function of the form

$$f(x) = a_n x^n + a_{n-1}x^{n-1} + \cdots + a_1 x + a_0$$

for n a nonnegative integer, called the **degree** of the polynomial. The coefficients $a_0, a_1, \ldots, a_n$ are real numbers with $a_n \neq 0$. The **domain** of a polynomial function is the set of all real numbers.

The shape of the graph of a polynomial function is connected to the degree of the polynomial. The shapes of odd-degree polynomial functions have something in common, and the shapes of even-degree polynomial functions have something in common. Figure 1 shows graphs of representative

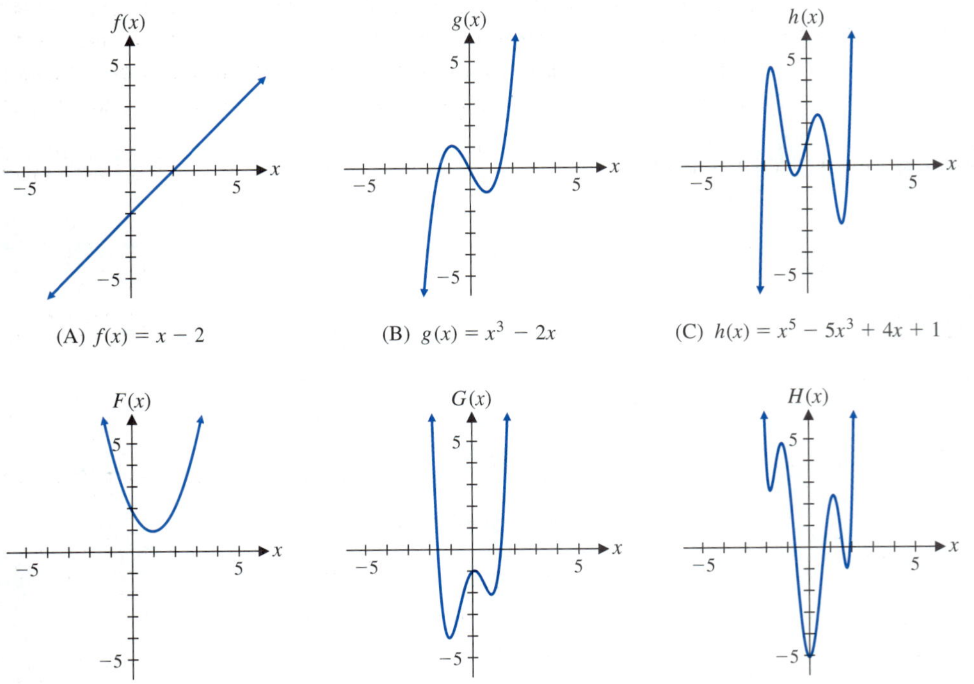

(A) $f(x) = x - 2$ (B) $g(x) = x^3 - 2x$ (C) $h(x) = x^5 - 5x^3 + 4x + 1$

(D) $F(x) = x^2 - 2x + 2$ (E) $G(x) = 2x^4 - 4x^2 + x - 1$ (F) $H(x) = x^6 - 7x^4 + 14x^2 - x - 5$

FIGURE 1
Graphs of polynomial functions

polynomial functions from degrees 1 to 6, and suggests some general properties of graphs of polynomial functions.

Notice that the odd-degree polynomial graphs start negative, end positive, and cross the x axis at least once. The even-degree polynomial graphs start positive, end positive, and may not cross the x axis at all. In all cases in Figure 1, the coefficient of the highest-degree term was chosen positive. If any leading coefficient had been chosen negative, then we would have a similar graph but flipped over.

The graph of a polynomial function is **continuous,** with no holes or breaks. That is, the graph can be drawn without removing a pen from the paper. Also, the graph of a polynomial has no sharp corners. Figure 2 shows the graphs of two functions, one that is not continuous, and the other that is continuous, but with a sharp corner. Neither function is a polynomial.

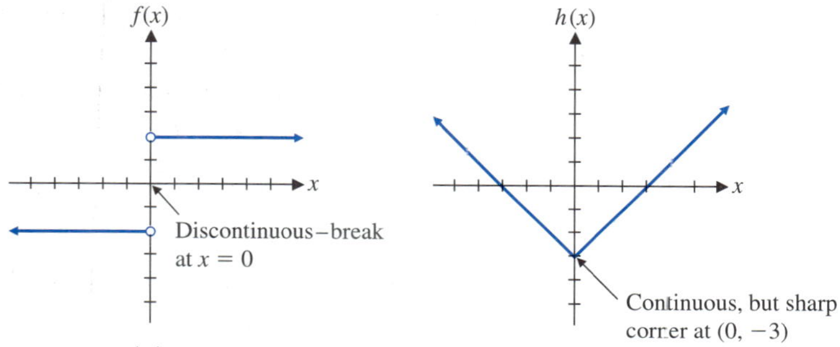

(A) $f(x) = \frac{2|x|}{x}$ (B) $h(x) = |x| - 3$

FIGURE 2
Discontinuous and sharp-corner functions

Figure 1 gives examples of polynomial functions with graphs containing the maximum number of *turning points* possible for a polynomial of that degree. A **turning point** on a continuous graph is a point that separates an increasing portion from a decreasing portion or vice versa. In general, it can be shown that:

The graph of a polynomial function of positive degree n can have at most $(n - 1)$ turning points and can cross the x axis at most n times.

Explore–Discuss 1

(A) What is the least number of turning points an odd-degree polynomial function can have? An even-degree polynomial function?
(B) What is the maximum number of x intercepts the graph of a polynomial function of degree n can have?
(C) What is the maximum number of real solutions an nth-degree polynomial equation can have?
(D) What is the least number of x intercepts the graph of a polynomial function of odd degree can have? Of even degree?
(E) What is the least number of real solutions a polynomial function of odd degree can have? Of even degree?

We now compare the graphs of two polynomial functions relative to points close to the origin and then "zoom out" to compare points distant from the origin. Compare the graphs in Figure 3.

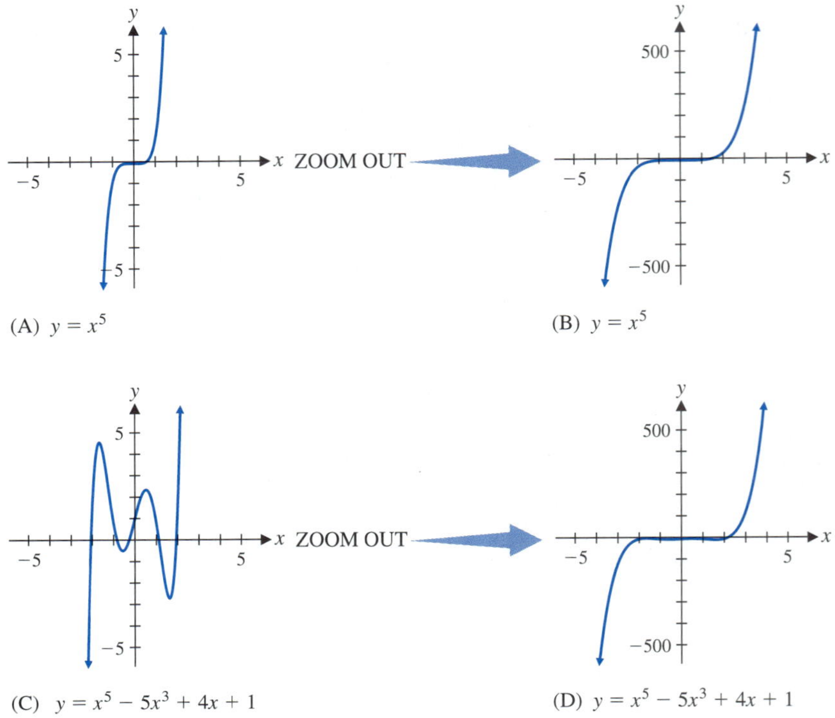

FIGURE 3
Close and distant comparisons

Figure 3 clearly shows that the highest-degree term in the polynomial dominates all other terms combined in the polynomial. As we "zoom out," the graph of $y = x^5 - 5x^3 + 4x + 1$ looks more and more like the graph of $y = x^5$. This is a general property of polynomial functions.

Explore–Discuss 2

Compare the graphs of $y = x^6$ and $y = x^6 - 7x^4 + 14x^2 - x - 5$ in the following two viewing windows:

(A) $-5 \le x \le 5, -5 \le y \le 5$
(B) $-5 \le x \le 5, -500 \le y \le 500$

Polynomial Root Approximation

An x intercept of a function f is also called a **zero*** of the function and a **root** of the equation $f(x) = 0$. Approximating the zeros of a function with a graphing utility is a simple matter, provided that we can find a window that shows where the graph of the function touches or crosses the x axis. But how can we be sure that a particular window shows all the zeros? For polynomial

*Only real numbers can be x intercepts. Functions may have roots or zeros that are not real numbers, but these will not appear as x intercepts. Since we have agreed to discuss only functions whose domains are sets of real numbers, we will not consider any complex zeros or roots that ar not real numbers.

functions, Theorem 1 can be used to locate a window that contains all the zeros of the polynomial. Theorem 1 is due to A. L. Cauchy (1789–1857), a French mathematician who has many theorems and concepts associated with his name, and illustrates that classical mathematical results are often useful tools in modern applications.

Theorem 1 ■■ LOCATING ZEROS OF A POLYNOMIAL

If r is a zero of the polynomial

$$P(x) = x^n + a_{n-1}x^{n-1} + a_{n-2}x^{n-2} + \cdots + a_1x + a_0$$

then*

$$|r| < 1 + \max\{|a_{n-1}|, |a_{n-2}|, \ldots, |a_1|, |a_0|\}$$ ■■

In a polynomial, the coefficient of the term containing the highest power of x is often referred to as the **leading coefficient.** Notice that Theorem 1 requires that this leading coefficient must be a 1.

Example 1 **Approximating the Zeros of a Polynomial** Approximate (to two decimal places) the real zeros of

$$P(x) = 2x^4 - 5x^3 - 4x^2 + 3x + 6$$

SOLUTION Since the leading coefficient of $P(x)$ is 2, Theorem 1 cannot be applied to $P(x)$. But it can be applied to

$$\begin{aligned} Q(x) = \tfrac{1}{2}P(x) &= \tfrac{1}{2}(2x^4 - 5x^3 - 4x^2 + 3x + 6) \\ &= x^4 - \tfrac{5}{2}x^3 - 2x^2 + \tfrac{3}{2}x + 3 \end{aligned}$$

Since multiplying a function by a positive constant expands or contracts the graph (see Section 1-2), but does not change the x intercepts, $P(x)$ and $Q(x)$ have the same zeros. Thus, Theorem 1 implies that any zero, r, of $P(x)$ must satisfy

$$|r| < 1 + \max\{|-\tfrac{5}{2}|, |-2|, |\tfrac{3}{2}|, |3|\} = 1 + 3 = 4$$

and all zeros of $P(x)$ must be between -4 and 4. Graphing $P(x)$ (Fig. 4A), we see two real zeros of $P(x)$ and we can be certain that there are no zeros outside this window. Using a built-in approximation routine, the real zeros of $P(x)$ (to two decimal places) are 1.15 (Fig. 4B) and 2.89 (Fig. 4C). Note also that $P(x)$ has three turning points in this window, the maximum allowable for a fourth-degree polynomial. Thus, there can't be any turning points outside this window.

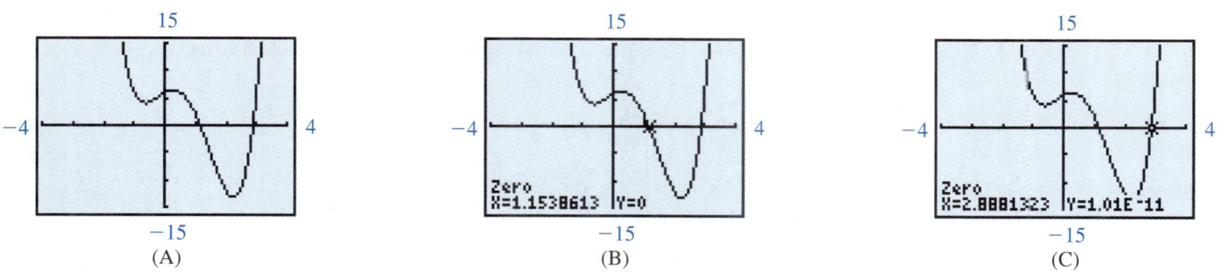

FIGURE 4 ■

*Note: $\max\{|a_{n-1}|, |a_{n-2}|, \ldots, |a_1|, |a_0|\}$ is the largest number in the list $|a_{n-1}|, |a_{n-2}|, \ldots, |a_0|$

Matched Problem 1 Approximate (to two decimal places) the real zeros of

$$P(x) = 3x^3 + 12x^2 + 9x + 4$$

Regression Polynomials

In Group Activity 1 at the end of Chapter 1, we saw that regression techniques can be used to fit a straight line to a set of data. Linear functions are not the only ones that can be applied in this manner. Most graphing utilities have the ability to fit a variety of curves to a given set of data. We will discuss polynomial regression models in this section and other types of regression models in later sections.

Example 2 **Estimating the Weight of a Fish** Using the length of a fish to estimate its weight is of interest to both scientists and sport anglers. The data in Table 1 give the average weights of lake trout for certain lengths. Use the data and regression techniques to find a polynomial model that can be used to estimate the weight of a lake trout for any length. Estimate (to the nearest ounce) the weights of lake trout of lengths 39, 40, 41, 42, and 43 inches, respectively.

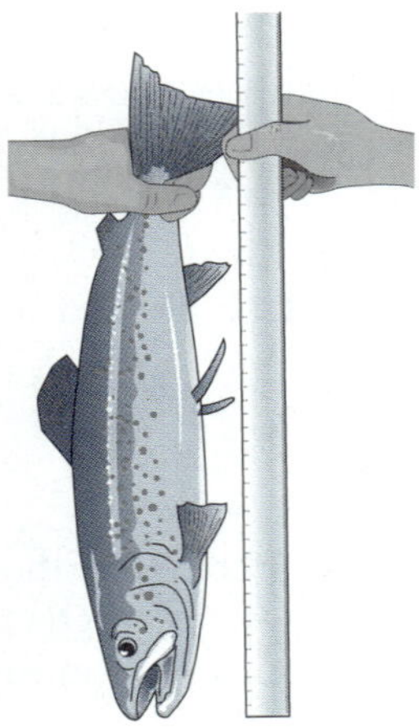

Table 1

Lake Trout

Length (in.)	Weight (oz)	Length (in.)	Weight (oz)
x	y	x	y
10	5	30	152
14	12	34	226
18	26	38	326
22	56	44	536
26	96		

Source: www.thefishernet.com

Solution The graph of the data in Table 1 (Fig. 5A) indicates that a linear regression model would not be appropriate in this case. And, in fact, we would not expect a linear relationship between length and weight. Instead, it is more likely that the weight would be related to the cube of the length. We use a cubic regression polynomial to model the data (Fig. 5B). (Consult your manual for the details of calculating regression polynomials on your graphing utility.) Figure 5C adds the graph of the polynomial model to the graph of the data. The graph in Figure 5C shows that this cubic polynomial does provide a good fit for the data. (We will have more to say about the choice of functions and the accuracy of the fit provided by regression analysis later in the text.) Figure 5D shows the estimated weights for the requested lengths.

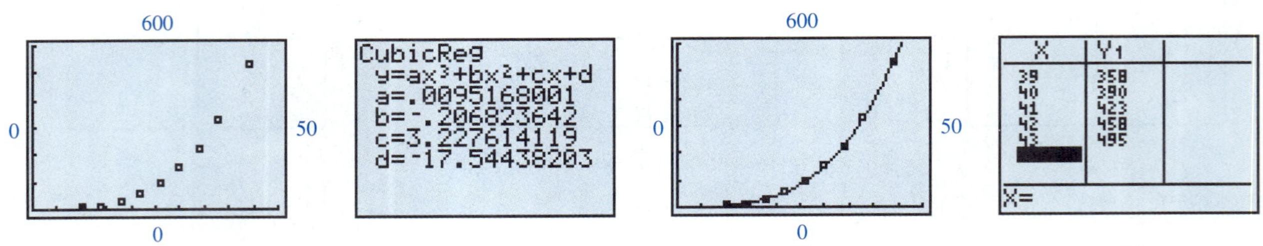

(A) (B) (C) (D)

FIGURE 5

Matched Problem 2 The data in Table 2 give the average weights of pike for certain lengths. Use a cubic regression polynomial to model the data. Estimate (to the nearest ounce) the weights of pike of lengths 39, 40, 41, 42, and 43 inches, respectively.

Table 2

PIKE

LENGTH (IN.)	WEIGHT (OZ)	LENGTH (IN.)	WEIGHT (OZ)
x	y	x	y
10	5	30	108
14	12	34	154
18	26	38	210
22	44	44	326
26	72	52	522

Source: www.thefishernet.com

RATIONAL FUNCTIONS

Just as rational numbers are defined in terms of quotients of integers, *rational functions* are defined in terms of quotients of polynomials. The following equations define rational functions:

$$f(x) = \frac{1}{x} \qquad g(x) = \frac{x-2}{x^2 - x - 6} \qquad h(x) = \frac{x^3 - 8}{x}$$

$$p(x) = 3x^2 - 5x \qquad q(x) = 7 \qquad r(x) = 0$$

Rational Function

A **rational function** is any function of the form

$$f(x) = \frac{n(x)}{d(x)} \qquad d(x) \neq 0$$

where $n(x)$ and $d(x)$ are polynomials. The **domain** is the set of all real numbers such that $d(x) \neq 0$. We assume $n(x)/d(x)$ is reduced to lowest terms.

Example 3 **Domain and Intercepts** Find the domain and intercepts for the rational function

$$f(x) = \frac{x-2}{x+1}$$

SOLUTION *Domain:* The denominator is 0 at $x = -1$. Therefore, the domain is the set of all real numbers except -1. The graph of f cannot cross the vertical line $x = -1$.

x intercepts: Find x such that $f(x) = 0$. This happens only if $x - 2 = 0$, that is, at $x = 2$. Thus, 2 is the only x intercept.

y-intercept: The y intercept is

$$f(0) = \frac{0-2}{0+1} = -2$$

Matched Problem 3 Find the domain and intercepts for the rational function: $g(x) = \dfrac{2x}{x - 2}$

In the next example we investigate the graph of $f(x) = (x - 2)/(x + 1)$ near the *point of discontinuity,* $x = -1$, and the behavior of the graph as x increases or decreases without bound. Using this information, we can complete a sketch of the graph of function f with little additional trouble. The investigation uncovers some characteristic features of graphs of rational functions.

Example 4 **Graph of a Rational Function** Given function f from Example 3:

$$f(x) = \frac{x - 2}{x + 1}$$

(A) Investigate the graph of f near the point of discontinuity, $x = -1$.
(B) Investigate the graph of f as x increases or decreases without bound.
(C) Sketch a graph of function f.

SOLUTION (A) *Let x approach -1 from the left:*

x	-2	-1.1	-1.01	-1.001	-1.0001	-1.00001
$f(x)$	4	31	301	3,001	30,001	300,001

We see that $f(x)$ increases without bound as x approaches -1 from the left. Symbolically,

$$f(x) \to \infty \quad \text{as} \quad x \to -1^-$$

Let x approach -1 from the right:

x	0	-0.9	-0.99	-0.999	-0.9999	-0.99999
$f(x)$	-2	-29	-299	$-2,999$	$-29,999$	$-299,999$

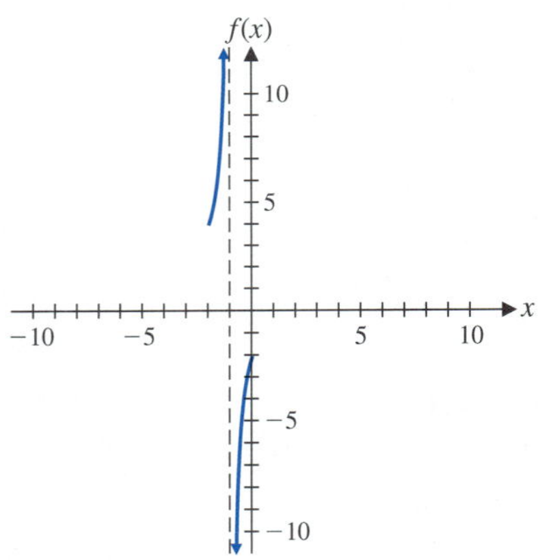

FIGURE 6
Graph near vertical asymptote $x = -1$

We see that $f(x)$ decreases without bound as x approaches -1 from the right. Symbolically,

$$f(x) \to -\infty \quad \text{as} \quad x \to -1^+$$

The vertical line $x = -1$ is called a *vertical asymptote.* The graph of f gets closer to this line as x gets closer to -1. Sketching the vertical asymptote provides an aid to drawing the graph of f near the asymptote (Fig. 6).

(B) Divide each term in the numerator and denominator of $f(x)$ by x, the highest power of x to occur in the numerator and denominator:

$$f(x) = \frac{x - 2}{x + 1} = \frac{1 - \dfrac{2}{x}}{1 + \dfrac{1}{x}}$$

As x increases or decreases without bound, $2/x$ and $1/x$ approach 0 and $f(x)$ gets closer and closer to 1. The horizontal line $y = 1$ is called a *horizontal asymptote.* The graph of $y = f(x)$ gets closer to this line as

x decreases or increases without bound. But how does the graph of $y = f(x)$ approach the horizontal line $y = 1$? Does it approach from above? From below? Or from both? To answer these questions we investigate table values as follows:

Let x approach ∞:

x	10	100	1,000	10,000	100,000
$f(x)$	0.72727	0.97030	0.99700	0.99970	0.99997

The graph of $y = f(x)$ approaches the line $y = 1$ from below as x increases without bound.

Let x approach $-\infty$:

x	-10	-100	$-1,000$	$-10,000$	$-100,000$
$f(x)$	1.33333	1.03030	1.00300	1.00030	1.00003

The graph of $y = f(x)$ approaches the line $y = 1$ from above as x decreases without bound.

Sketching the horizontal asymptote first provides an aid to drawing the graph of f as x moves away from the origin (Fig. 7).

(C) It is now easy to complete the sketch of the graph of f using the intercepts from Example 3 and filling in any points of uncertainty (Fig. 8).

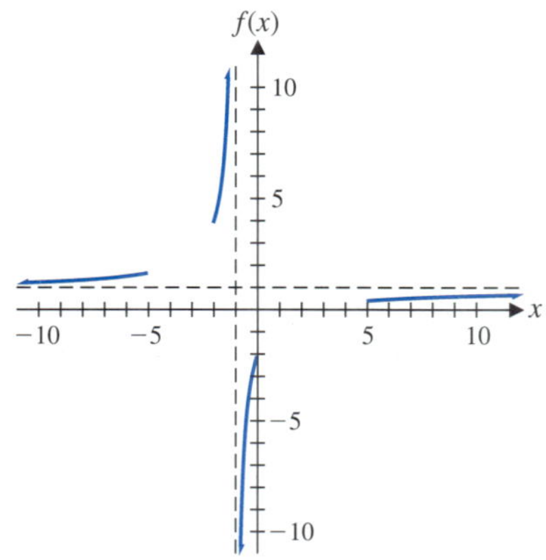

FIGURE 7
Graph of $y = f(x)$ near horizontal asymptote $y = 1$

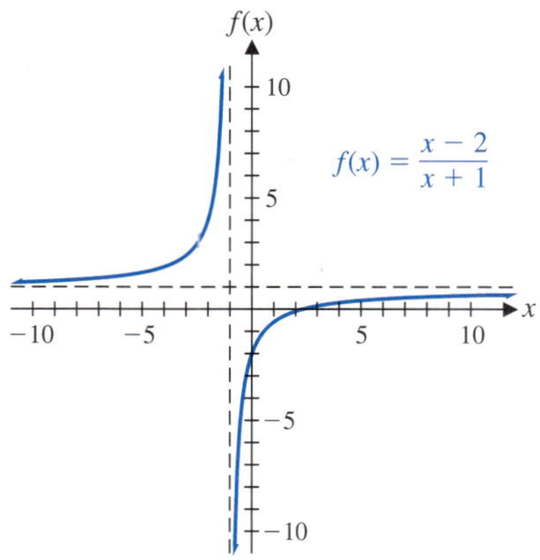

FIGURE 8
Sketch of the rational function f

Matched Problem 4 Given function g from Matched Problem 3: $g(x) = \dfrac{2x}{x-2}$

(A) Investigate the graph of g near the point of discontinuity, $x = 2$.
(B) Investigate the graph of g as x increases or decreases without bound.
(C) Sketch a graph of function g.

Graphing rational functions is considerably aided by locating vertical and horizontal asymptotes first, if they exist. The following general procedures are suggested by Example 4:

Vertical and Horizontal Asymptotes and Rational Functions

Given the rational function

$$f(x) = \frac{n(x)}{d(x)}$$

where $n(x)$ and $d(x)$ are polynomials without common factors:

1. If a is a real number such that $d(a) = 0$, then the line $x = a$ is a **vertical asymptote** of the graph of $y = f(x)$.
2. **Horizontal asymptotes,** if any exist, can be found by dividing each term of the numerator $n(x)$ and denominator $d(x)$ by the highest power of x that appears in the numerator and denominator, and then proceeding as in Example 4.

Example 5 **Graphing Rational Functions** Given the rational function:

$$f(x) = \frac{3x}{x^2 - 4}$$

(A) Find intercepts and equations for any vertical and horizontal asymptotes.
(B) Using the information from part (A) and additional points as necessary, sketch a graph of f for $-7 \leq x \leq 7$ and $-7 \leq y \leq 7$.

SOLUTION (A) *x intercepts:* $f(x) = 0$ only if $3x = 0$, or $x = 0$. Thus, the only x intercept is 0.

y intercept:

$$f(0) = \frac{3 \cdot 0}{0^2 - 4} = \frac{0}{-4} = 0$$

Thus, the y intercept is 0.

Vertical asymptotes:

$$f(x) = \frac{3x}{x^2 - 4} = \frac{3x}{(x - 2)(x + 2)}$$

The denominator is 0 at $x = -2$ and $x = 2$; hence, $x = -2$ and $x = 2$ are vertical asymptotes.

Horizontal asymptotes: Divide each term in the numerator and denominator by x^2, the highest power of x in the numerator and denominator:

$$f(x) = \frac{3x}{x^2 - 4} = \frac{\frac{3x}{x^2}}{\frac{x^2}{x^2} - \frac{4}{x^2}} = \frac{\frac{3}{x}}{1 - \frac{4}{x^2}}$$

As x increases or decreases without bound, the numerator tends to 0 and the denominator tends to 1; thus, $f(x)$ tends to 0. The line $y = 0$ is a horizontal asymptote.

(B) Use the information from part (A) and plot additional points as necessary to complete the graph, as shown in Figure 9.

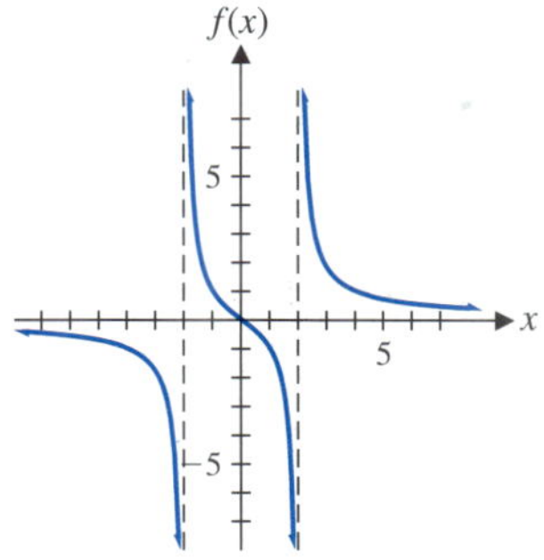

FIGURE 9

Matched Problem 5 Given the rational function: $g(x) = \frac{3x + 3}{x^2 - 9}$

(A) Find all intercepts and equations for any vertical and horizontal asymptotes.

(B) Using the information from part (A) and additional points as necessary, sketch a graph of g for $-10 \leqslant x \leqslant 10$ and $-10 \leqslant y \leqslant 10$.

APPLICATION

Rational functions occur naturally in many types of applications.

Example 6 **Employee Training** A company that manufactures computers has established that, on the average, a new employee can assemble $N(t)$ components per day after t days of on-the-job training, as given by

$$N(t) = \frac{50t}{t + 4} \qquad t \geqslant 0$$

Sketch a graph of N, $0 \leqslant t \leqslant 100$, including any vertical or horizontal asymptotes. What does $N(t)$ approach as t increases without bound?

SOLUTION *Vertical asymptotes:* None for $t \geqslant 0$

Horizontal asymptote:

$$N(t) = \frac{50t}{t + 4} = \frac{50}{1 + \frac{4}{t}}$$

$N(t)$ approaches 50 as t increases without bound. Thus, $y = 50$ is a horizontal asymptote.

Sketch of graph: The graph is shown in the margin.

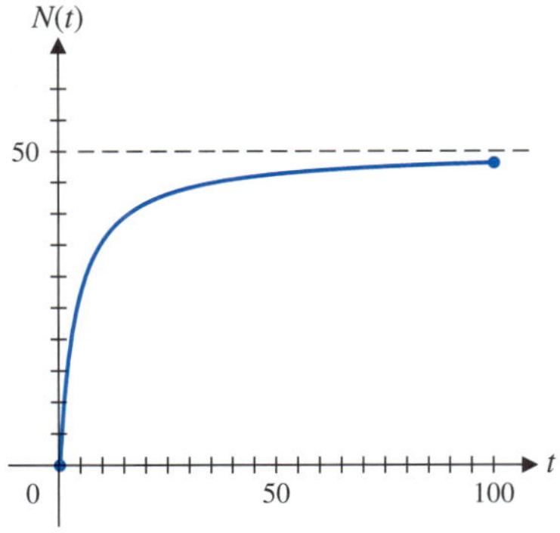

$N(t)$ approaches 50 as t increases without bound. It appears that 50 components per day would be the upper limit that an employee would be expected to assemble.

Matched Problem 6 ➡ Repeat Example 6 for: $N(t) = \dfrac{25t + 5}{t + 5} \qquad t \geq 0$

Answers to Matched Problems

1. -3.19

2.

3. Domain: All real numbers except 2; x intercept: 0; y intercept: 0

4. (A) Vertical asymptote: $x = 2$ (B) Horizontal asymptote: $y = 2$

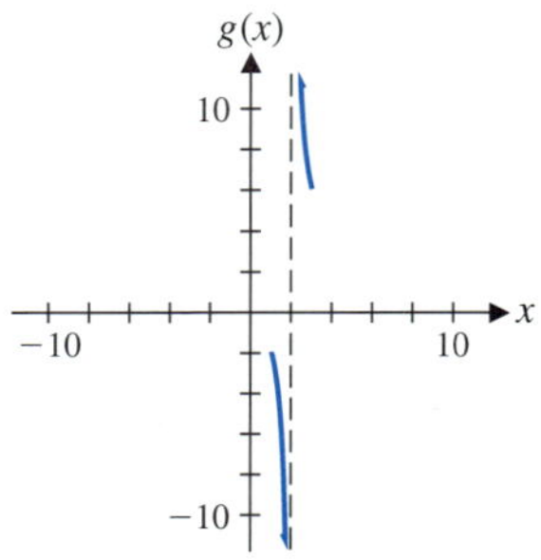

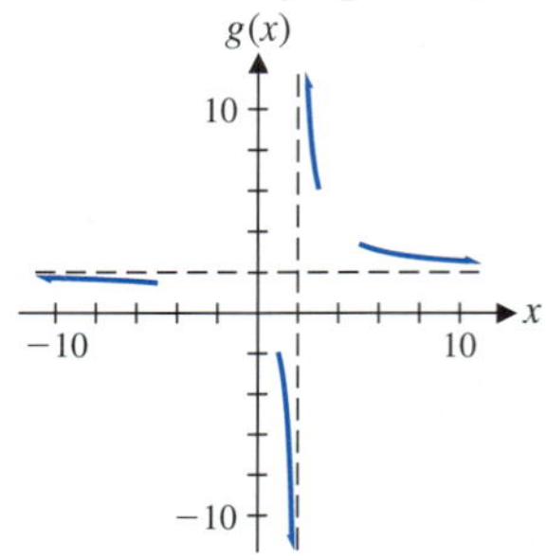

(C)

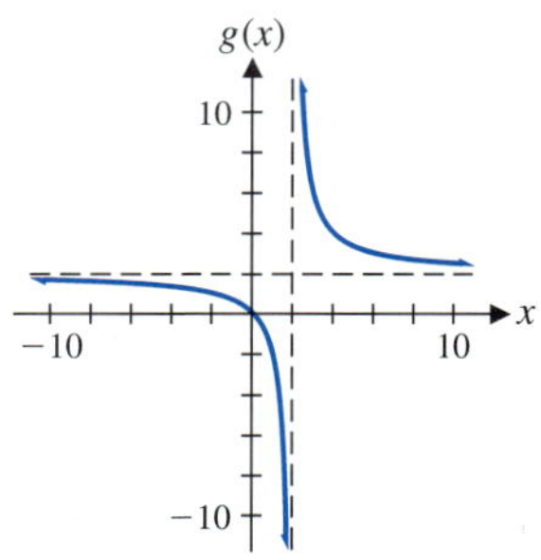

5. (A) x intercept: -1; y intercept; $-\frac{1}{3}$;
Vertical asymptotes: $x = -3$ and $x = 3$;
Horizontal asymptote: $y = 0$

(B)

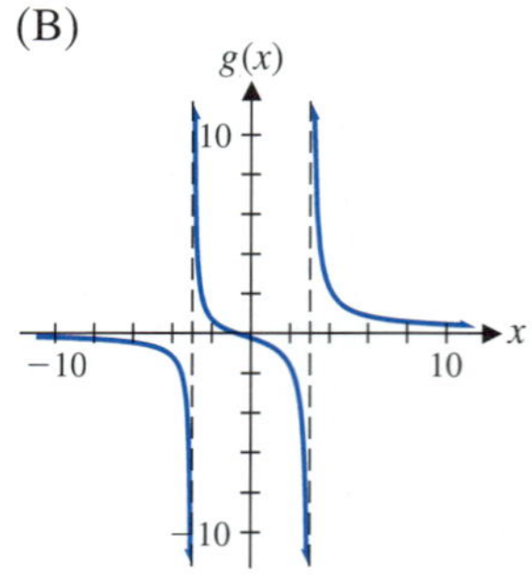

6. No vertical asymptotes for $t \geq 0$; $y = 25$ is a horizontal asymptote. $N(t)$ approaches 25 as t increases without bound. It appears that 25 components per day would be the upper limit that an employee would be expected to assemble.

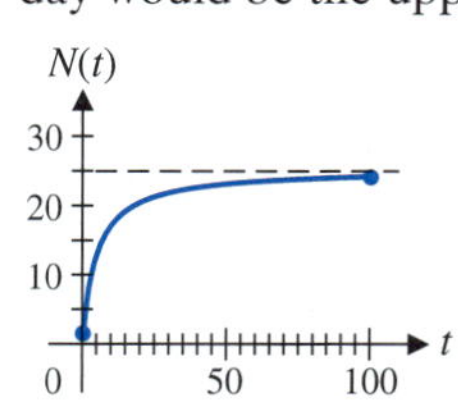

Exercise 2-1

A *For each polynomial function in Problems 1–6, find the following:*

(A) Degree of the polynomial
(B) Maximum number of turning points of the graph
(C) Maximum number of x intercepts of the graph
(D) Minimum number of x intercepts of the graph
(E) Maximum number of y intercepts of the graph
(F) Minimum number of y intercepts of the graph

1. $f(x) = ax^2 + bx + c, a \neq 0$

2. $f(x) = ax + b, a \neq 0$

3. $f(x) = ax^5 + bx^4 + cx^3 + dx^2 + ex + f, a \neq 0$

4. $f(x) = ax^4 + bx^3 + cx^2 + dx + e, a \neq 0$

5. $f(x) = ax^6 + bx^5 + cx^4 + dx^3 + ex^2 + fx + g,$ $a \neq 0$

6. $f(x) = ax^3 + bx^2 + cx + d, a \neq 0$

Each graph in Problems 7–14 is the graph of a polynomial function. Answer the following questions for each graph:

(A) How many turning points are on the graph?
(B) What is the minimum degree of a polynomial function that could have the graph?
(C) Is the leading coefficient of the polynomial negative or positive?

7.

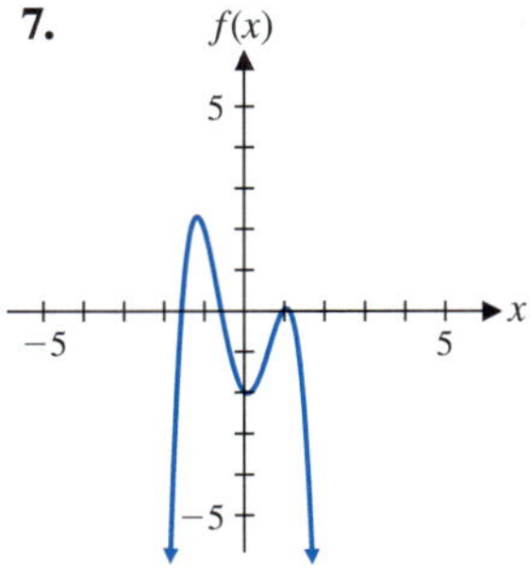

8.

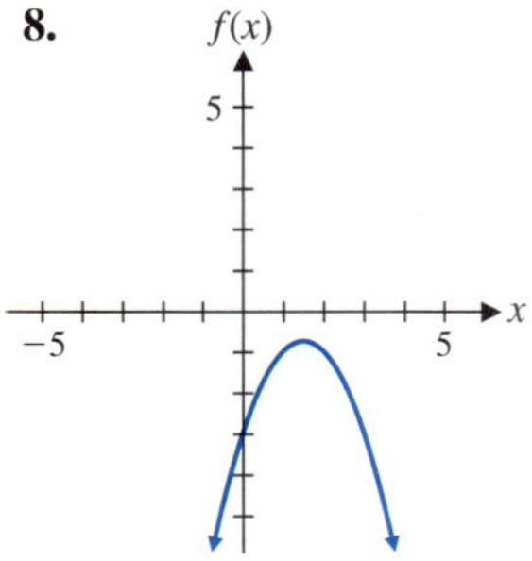

9.

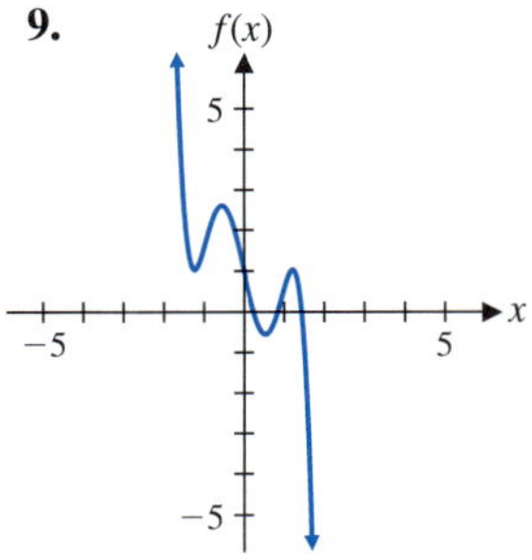

10.

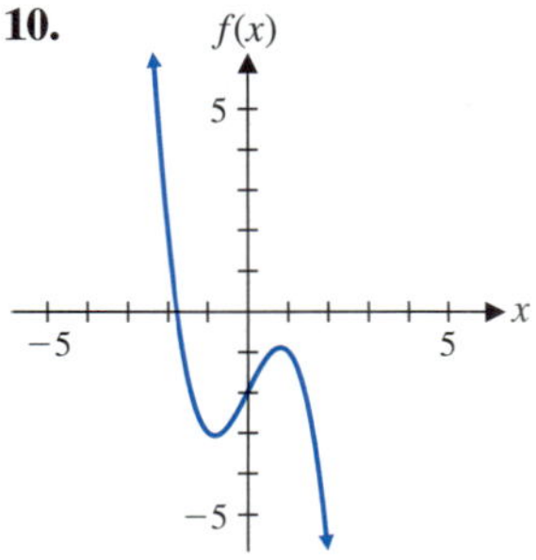

11.

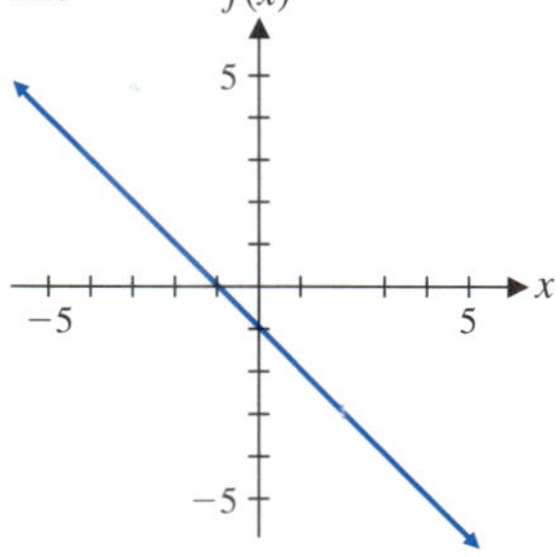

12.

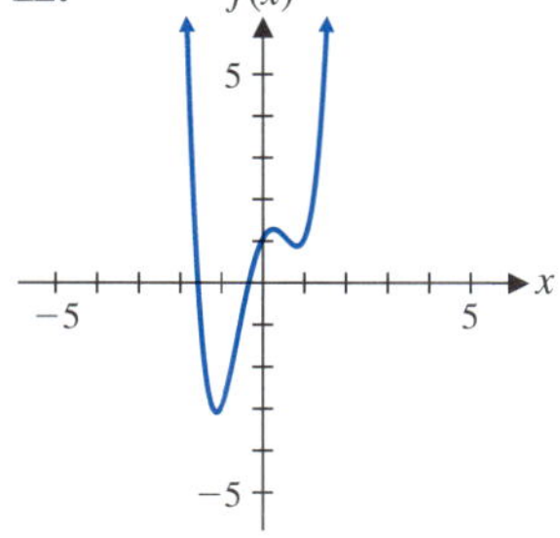

13.

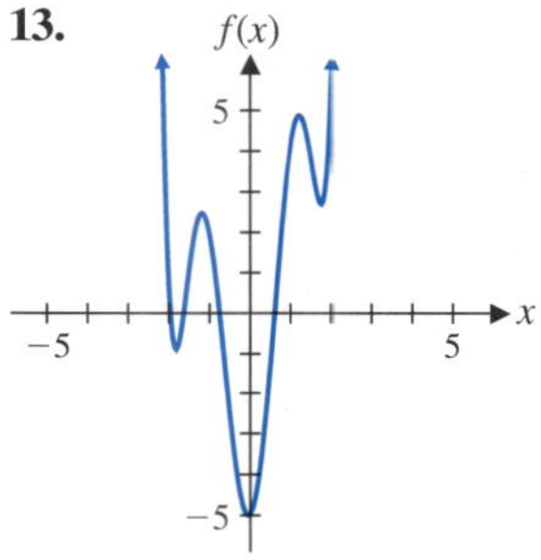

14.

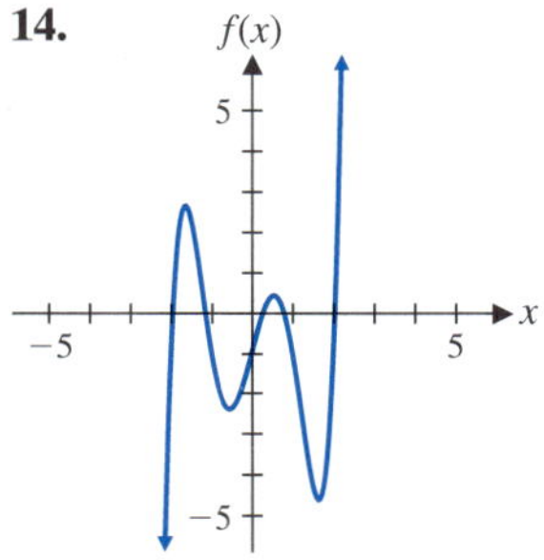

B *For each rational function in Problems 15–20:*

(A) Find the intercepts for the graph.
(B) Determine the domain.
(C) Find any vertical or horizontal asymptotes for the graph.
(D) Sketch any asymptotes as dashed lines. Then sketch a graph of $y = f(x)$ for $-10 \leqslant x \leqslant 10$ and $-10 \leqslant y \leqslant 10$.
(E) Graph $y = f(x)$ in a standard viewing window using a **C** *graphing utility.*

15. $f(x) = \dfrac{x + 2}{x - 2}$

16. $f(x) = \dfrac{x - 3}{x + 3}$

17. $f(x) = \dfrac{3x}{x + 2}$

18. $f(x) = \dfrac{2x}{x - 3}$

19. $f(x) = \dfrac{4 - 2x}{x - 4}$

20. $f(x) = \dfrac{3 - 3x}{x - 2}$

21. How does the graph of $f(x) = 2x^4 - 5x^2 + x + 2$ compare to the graph of $y = 2x^4$ as we "zoom out" (see Fig. 3)?

22. How does the graph of $f(x) = x^3 - 2x + 2$ compare to the graph of $y = x^3$ as we "zoom out"?

23. How does the graph of $f(x) = -x^5 + 4x^3 - 4x + 1$ compare to the graph of $y = -x^5$ as we "zoom out"?

24. How does the graph of $f(x) = -x^5 + 5x^3 + 4x - 1$ compare to the graph of $y = -x^5$ as we "zoom out"?

25. Compare the graph of $y = 2x^4$ to the graph of $y = 2x^4 - 5x^2 + x + 2$ in the following two viewing windows:
(A) $-5 \le x \le 5, -5 \le y \le 5$
(B) $-5 \le x \le 5, -500 \le y \le 500$

26. Compare the graph of $y = x^3$ to the graph of $y = x^3 - 2x + 2$ in the following two viewing windows:
(A) $-5 \le x \le 5, -5 \le y \le 5$
(B) $-5 \le x \le 5, -500 \le y \le 500$

27. Compare the graph of $y = -x^5$ to the graph of $y = -x^5 + 4x^3 - 4x + 1$ in the following two viewing windows:
(A) $-5 \le x \le 5, -5 \le y \le 5$
(B) $-5 \le x \le 5, -500 \le y \le 500$

28. Compare the graph of $y = -x^5$ to the graph of $y = -x^5 + 5x^3 - 5x + 2$ in the following two viewing windows:
(A) $-5 \le x \le 5, -5 \le y \le 5$
(B) $-5 \le x \le 5, -500 \le y \le 500$

In Problems 29–34, approximate the real zeros of $P(x)$ to two decimal places.

29. $P(x) = 2x^3 - x^2 - 8x - 7$

30. $P(x) = 3x^3 + 12x^2 + 8x + 5$

31. $P(x) = x^4 + 2x^3 - 3x^2 - 2x + 3$

32. $P(x) = x^4 - 3x^3 - x^2 + 2x + 2$

33. $P(x) = x^5 - 12x^4 + 8x^3 + 11$

34. $P(x) = x^5 + 14x^4 - 12x^2 - 13$

35. Graph the line $y = 0.5x + 3$. Choose any two distinct points on this line and find the linear regression model for the data set consisting of the two points you chose. Experiment with other lines of your choice. Discuss the relationship between a linear regression model for two points and the line that goes through the two points.

36. Graph the parabola $y = x^2 - 5x$. Choose any three distinct points on this parabola and find the quadratic regression model for the data set consisting of the three points you chose. Experiment with other parabolas of your choice. Discuss the relationship between a quadratic regression model for three noncollinear points and the parabola that goes through the three points.

C *For each rational function in Problems 37–42:*
(A) Find any intercepts for the graph.
(B) Find any vertical and horizontal asymptotes for the graph.
(C) Sketch any asymptotes as dashed lines. Then sketch a graph of f for $-10 \le x \le 10$ and $-10 \le y \le 10$.
(D) Graph the function in a standard viewing window using a graphing utility.

37. $f(x) = \dfrac{2x^2}{x^2 - x - 6}$

38. $f(x) = \dfrac{3x^2}{x^2 + x - 6}$

39. $f(x) = \dfrac{6 - 2x^2}{x^2 - 9}$

40. $f(x) = \dfrac{3 - 3x^2}{x^2 - 4}$

41. $f(x) = \dfrac{-4x}{x^2 + x - 6}$

42. $f(x) = \dfrac{5x}{x^2 + x - 12}$

43. Write an equation for the lowest-degree polynomial function with the graph and intercepts shown in the figure.

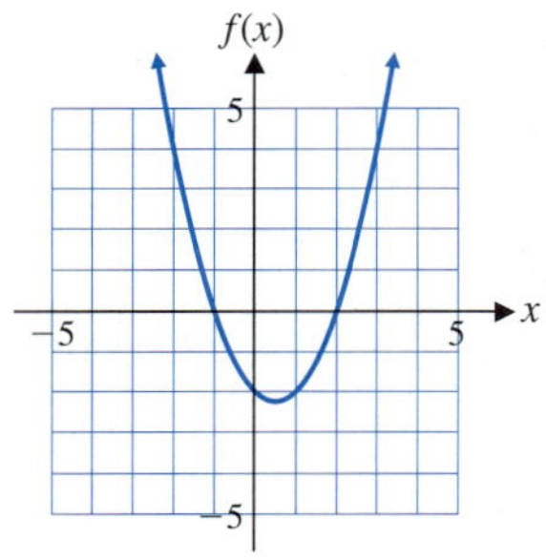

Figure for 43

44. Write an equation for the lowest-degree polynomial function with the graph and intercepts shown in the figure.

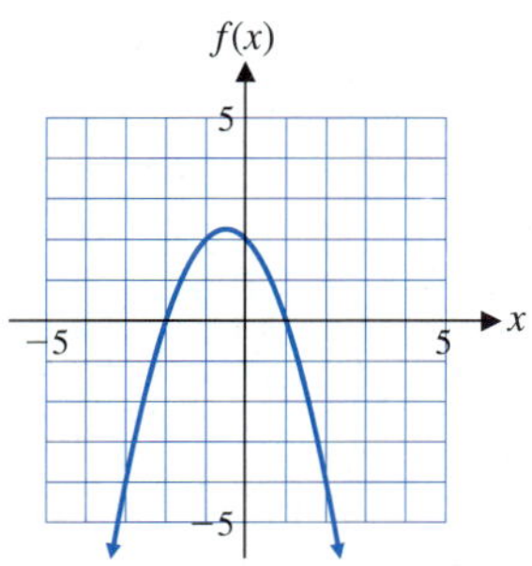

Figure for 44

45. Write an equation for the lowest-degree polynomial function with the graph and intercepts shown in the figure.

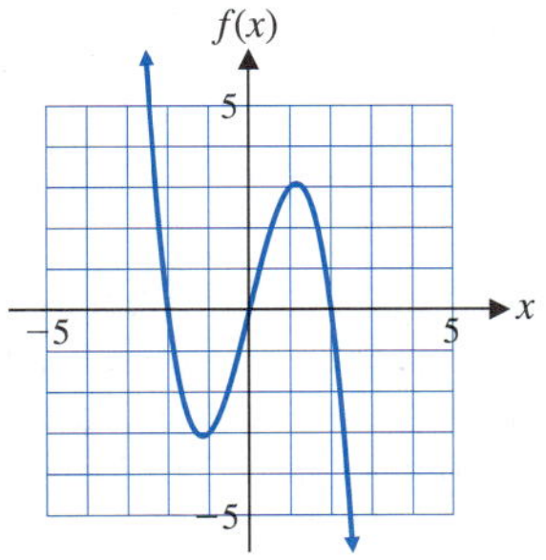

46. Write an equation for the lowest-degree polynomial function with the graph and intercepts shown in the figure.

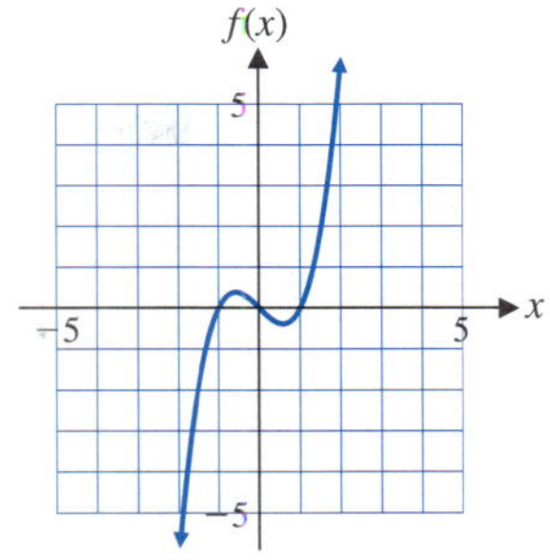

Applications

Business & Economics

47. *Average cost.* A company manufacturing snowboards has fixed costs of $200 per day and total costs of $3,800 per day at a daily output of 20 boards.

(A) Assuming the total cost per day, $C(x)$, is linearly related to the total output per day, x, write an equation for the cost function.

(B) The average cost per board for an output of x boards is given by $\overline{C}(x) = C(x)/x$. Find the average cost function.

(C) Sketch a graph of the average cost function, including any asymptotes, for $1 \le x \le 30$.

(D) What does the average cost per board tend to as production increases?

48. *Average cost.* A company manufacturing surfboards has fixed costs of $300 per day and total costs of $5,100 per day at a daily output of 20 boards.

(A) Assuming the total cost per day, $C(x)$, is linearly related to the total output per day, x, write an equation for the cost function.

(B) The average cost per board for an output of x boards is given by $\overline{C}(x) = C(x)/x$. Find the average cost function.

(C) Sketch a graph of the average cost function, including any asymptotes, for $1 \le x \le 30$.

(D) What does the average cost per board tend to as production increases?

49. *Replacement time.* An office copier has an initial price of $2,500. A service contract costs $200 for the first year and increases $50 per year thereafter. It can be shown that the total cost of the copier after n years is given by

$$C(n) = 2{,}500 + 175n + 25n^2$$

The average cost per year for n years is $\overline{C}(n) = C(n)/n$.

(A) Find the rational function $\overline{C}$.

(B) Sketch a graph of $\overline{C}$ for $2 \le n \le 20$.

(C) When is the average cost per year at a minimum, and what is the minimum average annual cost? [*Hint:* Refer to the sketch in part (B) and evaluate $\overline{C}(n)$ at appropriate integer values until a minimum value is found.] The time when the average cost is minimum is frequently referred to as the **replacement time** for the piece of equipment.

(D) Graph the average cost function $\overline{C}$ in a graphing utility and use trace and zoom or an appropriate built-in routine to find when the average annual cost is at a minimum.

50. *Minimum average cost.* Financial analysts in a company that manufactures audio CD players arrived at the following daily cost equation for manufacturing x CD players per day:

$$C(x) = x^2 + 2x + 2{,}000$$

The average cost per unit at a production level of x players per day is $\overline{C}(x) = C(x)/x$.

(A) Find the rational function $\overline{C}$.

(B) Sketch a graph of $\overline{C}$ for $5 \le x \le 150$.

(C) For what daily production level (to the nearest integer) is the average cost per unit at a minimum, and what is the minimum average cost per player (to the nearest cent)? [*Hint:* Refer to the sketch in part (B) and evaluate $\overline{C}(x)$ at appropriate integer values until a minimum value is found.]

(D) Graph the average cost function $\overline{C}$ in a graphing utility and use trace and zoom or an appropriate built-in routine to find the daily production level (to the nearest integer) at which the average cost per player is at a minimum. What is the minimum average cost to the nearest cent?

51. *Minimum average cost.* A consulting firm, using statistical methods, provided a veterinary clinic with the cost equation

$$C(x) = 0.00048(x - 500)^3 + 60{,}000 \qquad 100 \le x \le 1{,}000$$

where $C(x)$ is the cost in dollars for handling x cases per month. The average cost per case is given by $\overline{C}(x) = C(x)/x$.

(A) Write the equation for the average cost function $\overline{C}$.
(B) Graph $\overline{C}$ on a graphing utility.
(C) Use trace and zoom or an appropriate built-in routine to find the monthly caseload for the minimum average cost per case. What is the minimum average cost per case?

52. *Minimum average cost.* The financial department of a hospital, using statistical methods, arrived at the cost equation

$$C(x) = 20x^3 - 360x^2 + 2{,}300x - 1{,}000 \qquad 1 \le x \le 12$$

where $C(x)$ is the cost in thousands of dollars for handling x thousand cases per month. The average cost per case is given by $\overline{C}(x) = C(x)/x$.

(A) Write the equation for the average cost function $\overline{C}$.
(B) Graph $\overline{C}$ on a graphing utility.
(C) Use trace and zoom or an appropriate built-in routine to find the monthly caseload for the minimum average cost per case. What is the minimum average cost per case to the nearest dollar?

*Problems 53 and 54 involve the following concepts: In a free competitive market, the price of a product is determined by the relationship between supply and demand. If $p = D(x)$ and $p = S(x)$ are the price–demand and price–supply equations, respectively, for a product and if $(\bar{x}, \bar{p})$ is the point of intersection of these equations, then $(\bar{x}, \bar{p})$ is the **equilibrium point,** $\bar{p}$ is the **equilibrium price,** and $\bar{x}$ is the **equilibrium quantity.** The equilibrium price is the price at which the supply and the demand are the same.*

53. *Equilibrium point.* A particular CD is sold through a chain of stores in a city. A marketing company has established price–demand and price–supply tables for selected prices for this CD (Tables 3 and 4), where x is the daily number of CD's people are willing to buy and the store is willing to sell at a price of p dollars per CD.

Table 3
PRICE–DEMAND

x	$p = D(x)$ ($)
25	19.50
100	14.25
175	10.00
250	8.25

Table 4
PRICE–SUPPLY

x	$p = S(x)$ ($)
25	2.10
100	3.80
175	8.50
250	15.70

(A) Use a linear regression equation to model the data in Table 3 and a quadratic regression equation to model the data in Table 4.
(B) Find the point of intersection of the two equations from part (A). Write the equilibrium price to the nearest cent and the equilibrium quantity to the nearest unit.

54. *Equilibrium point.* Repeat Problem 53 with the price–demand and price–supply tables below, except use a quadratic regression model for the data in Table 5 and a linear regression model for the data in Table 6.

Table 5
PRICE–DEMAND

x	$p = D(x)$ ($)
0	24
40	23
65	20
115	11

Table 6
PRICE–SUPPLY

x	$p = S(x)$ ($)
0	5
40	10
65	12
115	16

Life Sciences

55. *Health care.* Table 7 shows the total national expenditures (in billion dollars) and the per capita expenditures (in dollars) for selected years since 1960.

(A) Let x represent the number of years since 1960 and find a cubic regression polynomial for the total national expenditures.
(B) Use the polynomial model from part (A) to estimate the total national expenditures (to the nearest tenth of a billion) for 1995.

Table 7
NATIONAL HEALTH EXPENDITURES

DATE	TOTAL EXPENDITURES (BILLION $)	PER CAPITA EXPENDITURES ($)
1960	27.1	143
1965	41.6	204
1970	74.4	346
1975	132.9	592
1980	247.2	1,002
1985	428.2	1,666
1990	697.5	2,588

Source: U.S. Bureau of the Census

56. *Health care.* Refer to Table 7.

(A) Let x represent the number of years since 1960 and find a cubic regression polynomial for the per capita expenditures.
(B) Use the polynomial model from part (A) to estimate the per capita expenditures (to the nearest dollar) for 1995.

57. *Physiology.* In a study on the speed of muscle contraction in frogs under various loads, researchers W. O. Fems and J. Marsh found that the speed of contraction decreases with increasing loads. In particular, they found that the relationship between speed of contraction v (in centimeters per second) and load x (in grams) is given approximately by

$$v(x) = \frac{26 + 0.06x}{x} \qquad x \geq 5$$

(A) What does $v(x)$ approach as x increases?
(B) Sketch a graph of function v.

Social Sciences

58. *Learning theory.* In 1917, L. L. Thurstone, a pioneer in quantitative learning theory, proposed the rational function

$$f(x) = \frac{a(x + c)}{(x + c) + b}$$

to model the number of successful acts per unit time that a person could accomplish after x practice sessions. Suppose that for a particular person enrolled in a typing class,

$$f(x) = \frac{55(x + 1)}{(x + 8)} \qquad x \geq 0$$

where $f(x)$ is the number of words per minute the person is able to type after x weeks of lessons.

(A) What does $f(x)$ approach as x increases?
(B) Sketch a graph of function f, including any vertical or horizontal asymptotes.

59. C *Marriage.* Table 8 shows the marriage and divorce rates per 1,000 population for selected years since 1950.
(A) Let x represent the number of years since 1950 and find a cubic regression polynomial for the marriage rate.
(B) Use the polynomial model from part (A) to estimate the marriage rate (to one decimal place) for 1995.

Table 8

MARRIAGES AND DIVORCES (PER 1,000 POPULATION)

DATE	MARRIAGES	DIVORCES
1950	11.1	2.6
1955	9.3	2.3
1960	8.5	2.2
1965	9.3	2.5
1970	10.6	3.5
1975	10.0	4.8
1980	10.6	5.2
1985	10.1	5.0
1990	9.8	4.7

Source: U.S. Bureau of the Census

60. C *Divorce.* Refer to Table 8.
(A) Let x represent the number of years since 1950 and find a cubic regression polynomial for the divorce rate.
(B) Use the polynomial model from part (A) to estimate the divorce rate (to one decimal place) for 1995.

SECTION 2-2

Exponential Functions

- EXPONENTIAL FUNCTIONS
- BASE e EXPONENTIAL FUNCTIONS
- GROWTH AND DECAY APPLICATIONS
- COMPOUND INTEREST
- CONTINUOUS COMPOUND INTEREST

This section introduces the important class of functions called *exponential functions.* These functions are used extensively in modeling and solving a wide variety of real-world problems, including growth of money at compound interest; growth of populations of people, animals, and bacteria; radioactive decay; and learning associated with the mastery of such devices as a new computer or an assembly process in a manufacturing plant.

Exponential Functions

We start by noting that

$$f(x) = 2^x \qquad \text{and} \qquad g(x) = x^2$$

are not the same function. Whether a variable appears as an exponent with a constant base or as a base with a constant exponent, makes a big difference. The function g is a quadratic function, which we have already discussed. The function f is a new type of function called an *exponential function.* In general:

Exponential Function

The equation

$$f(x) = b^x \qquad b > 0, b \neq 1$$

defines an **exponential function** for each different constant b, called the **base.** The **domain** of f is the set of all real numbers, and the **range** of f is the set of all positive real numbers.

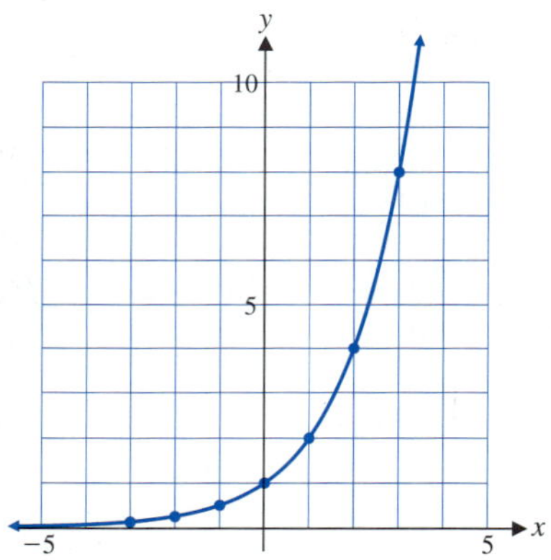

FIGURE 1
$y = 2^x$

We require the base b to be positive to avoid imaginary numbers such as $(-2)^{1/2} = \sqrt{-2} = i\sqrt{2}$. We exclude $b = 1$ as a base, since $f(x) = 1^x = 1$ is a constant function, which we have already considered.

Many students, if asked to hand sketch graphs of equations such as $y = 2^x$ or $y = 2^{-x}$, would not hesitate at all. [*Note:* $2^{-x} = 1/2^x = (1/2)^x$.] They would likely make up tables by assigning integers to x, plot the resulting points, and then join these points with a smooth curve as in Figure 1. The only catch is that we have not defined 2^x for all real numbers. From Appendix A-7, we know what 2^5, 2^{-3}, $2^{2/3}$, $2^{-3/5}$, $2^{1.4}$, and $2^{-3.14}$ mean (that is, 2^p, where p is a rational number), but what does

$$2^{\sqrt{2}}$$

mean? The question is not easy to answer at this time. In fact, a precise definition of $2^{\sqrt{2}}$ must wait for more advanced courses, where it is shown that

$$2^x$$

names a positive real number for x any real number, and that the graph of $y = 2^x$ is indeed as indicated in Figure 1.

It is useful to compare the graphs of $y = 2^x$ and $y = 2^{-x}$ by plotting both on the same set of coordinate axes, as shown in Figure 2A. The graph of

$$f(x) = b^x \qquad b > 1 \text{ (Fig. 2B)}$$

looks very much like the graph of $y = 2^x$, and the graph of

$$f(x) = b^x \qquad 0 < b < 1 \text{ (Fig. 2B)}$$

looks very much like the graph of $y = 2^{-x}$. Note that in both cases the x axis is a horizontal asymptote for the graphs.

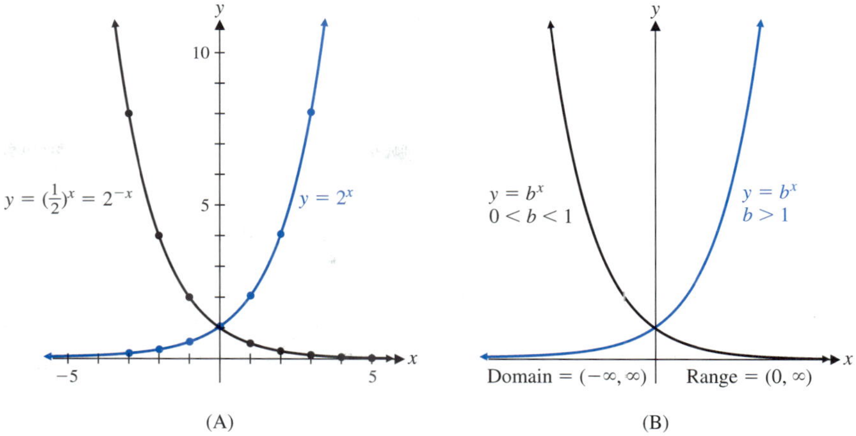

FIGURE 2
Exponential functions

The graphs in Figure 2 suggest the following important general properties of exponential functions, which we state without proof:

Basic Properties of the Graph of $f(x) = b^x$, $b > 0$, $b \neq 1$

1. All graphs will pass through the point (0, 1). *$b^0 = 1$ for any permissible base b.*
2. All graphs are continuous curves, with no holes or jumps.
3. The x axis is a horizontal asymptote.
4. If $b > 1$, then b^x increases as x increases.
5. If $0 < b < 1$, then b^x decreases as x increases.

The use of a calculator with the key $\boxed{y^x}$, or its equivalent, makes the graphing of exponential functions almost routine. Example 1 illustrates the process.

Example 1 **Graphing Exponential Functions** Sketch a graph of $y = (\frac{1}{2})4^x$, $-2 \leqslant x \leqslant 2$.

SOLUTION Use a calculator to create the table of values shown. Plot these points, and then join them with a smooth curve as in Figure 3.

x	y
−2	0.031
−1	0.125
0	0.50
1	2.00
2	8.00

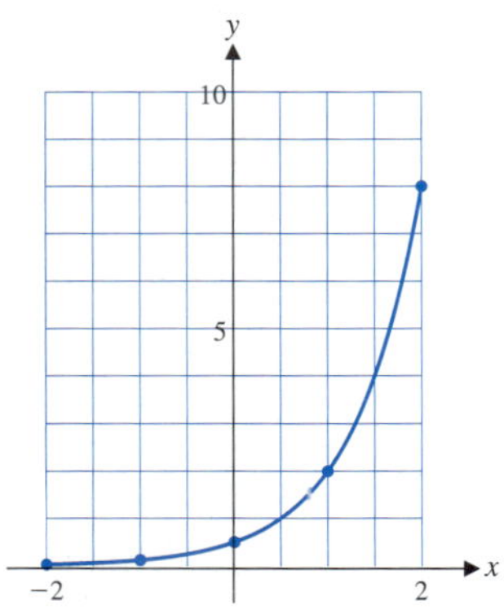

FIGURE 3
Graph of $y = (\frac{1}{2})4^x$

Matched Problem 1 Sketch a graph of $y = (\frac{1}{2})4^{-x}, -2 \leq x \leq 2$.

Explore–Discuss 1

Graph the functions $f(x) = 2^x$ and $g(x) = 3^x$ on the same set of coordinate axes. At which values of x do the graphs intersect? For which values of x is the graph of f above the graph of g? Below the graph of g? Are the graphs close together as x increases without bound? Are the graphs close together as x decreases without bound? Discuss.

Exponential functions, whose domains include irrational numbers, obey the familiar laws of exponents discussed in Appendix A-7 for rational exponents. We summarize these exponent laws here and add two other important and useful properties.

Exponential Function Properties

For a and b positive, $a \neq 1, b \neq 1$, and x and y real:

1. Exponent laws:

$$a^x a^y = a^{x+y} \qquad \frac{a^x}{a^y} = a^{x-y} \qquad \frac{4^{2y}}{4^{5y}} = 4^{2y-5y} = 4^{-3y}$$

$$(a^x)^y = a^{xy} \qquad (ab)^x = a^x b^x \qquad \left(\frac{a}{b}\right)^x = \frac{a^x}{b^x}$$

2. $a^x = a^y$ if and only if $x = y$ If $7^{5t+1} = 7^{3t-3}$, then $5t + 1 = 3t - 3$, and $t = -2$.

3. For $x \neq 0$,

   $a^x = b^x$ if and only if $a = b$ If $a^5 = 2^5$, then $a = 2$.

■ Base e Exponential Functions

Of all the possible bases b we can use for the exponential function $y = b^x$, which ones are the most useful? If you look at the keys on a calculator, you will likely see $\boxed{10^x}$ and $\boxed{e^x}$. It is clear why base 10 would be important, because our number system is a base 10 system. But what is e, and why is it included as a base? It turns out that base e is used more frequently than all other bases combined. The reason for this is that certain formulas and the results of certain processes found in calculus and more advanced mathematics take on their simplest form if this base is used. This is why you will see e used extensively in expressions and formulas that model real-world phenomena. In fact, its use is so prevalent that you will often hear people refer to $y = e^x$ as *the* exponential function.

The base e is an irrational number, and like π, it cannot be represented exactly by any finite decimal fraction. However, e can be approximated as closely as we like by evaluating the expression

$$\left(1 + \frac{1}{x}\right)^x \tag{1}$$

Table 1

x	$\left(1 + \frac{1}{x}\right)^x$
1	2
10	2.593 74...
100	2.704 81...
1,000	2.716 92...
10,000	2.718 14...
100,000	2.718 27...
1,000,000	2.718 28...
⋮	⋮

for sufficiently large x. What happens to the value of expression (1) as x increases without bound? Think about this for a moment before proceeding. Maybe you guessed that the value approaches 1, because

$$1 + \frac{1}{x}$$

approaches 1, and 1 raised to any power is 1. Let us see if this reasoning is correct by actually calculating the value of the expression for larger and larger values of x. Table 1 summarizes the results.

Interestingly, the value of expression (1) is never close to 1, but seems to be approaching a number close to 2.7183. In fact, as x increases without bound, the value of expression (1) approaches an irrational number that we call ***e***. The irrational number e to twelve decimal places is

$$e = 2.718\ 281\ 828\ 459$$

Compare this value of e with the value of e^1 from a calculator. Exactly who discovered the constant e is still being debated. It is named after the great Swiss mathematician Leonhard Euler (1707–1783).

Exponential Function with Base *e*

Exponential functions with base e and base $1/e$, respectively, are defined by

$y = e^x$ and $y = e^{-x}$
Domain: $(-\infty, \infty)$
Range: $(0, \infty)$

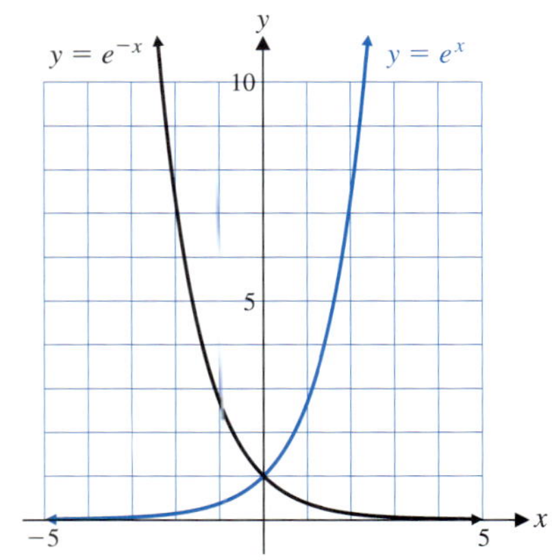

Explore–Discuss 2

Graph the functions $f(x) = e^x$, $g(x) = 2^x$, and $h(x) = 3^x$ on the same set of coordinate axes. At which values of x do the graphs intersect? For positive values of x, which of the three graphs lies above the other two? Below the other two? How does your answer change for negative values of x?

Growth and Decay Applications

Most exponential growth and decay problems are modeled using base e exponential functions. We present two applications here and many more in Exercise 2-2.

Example 2 **Exponential Growth** Cholera, an intestinal disease, is caused by a cholera bacterium that multiplies exponentially by cell division as given approximately by

$$N = N_0e^{1.386t}$$

where N is the number of bacteria present after t hours and N_0 is the number of bacteria present at the start ($t = 0$). If we start with 25 bacteria, how many bacteria (to the nearest unit) will be present:

(A) In 0.6 hour? (B) In 3.5 hours?

SOLUTION Substituting $N_0 = 25$ into the above equation, we obtain

$$N = 25e^{1.386t}$$ The graph is shown in Figure 4.

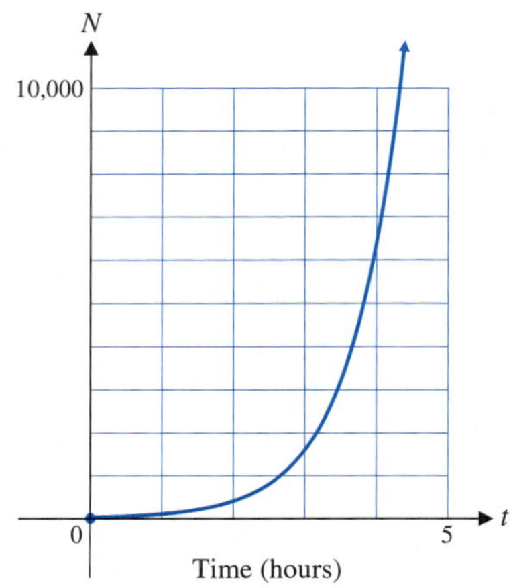

FIGURE 4

(A) Solve for N when $t = 0.6$:

$$\begin{aligned} N &= 25e^{1.386(0.6)} && \text{Use a calculator.} \\ &= 57 \text{ bacteria} \end{aligned}$$

(B) Solve for N when $t = 3.5$:

$$\begin{aligned} N &= 25e^{1.386(3.5)} && \text{Use a calculator.} \\ &= 3{,}197 \text{ bacteria} \end{aligned}$$

Matched Problem 2 Refer to the exponential growth model for cholera in Example 2. If we start with 55 bacteria, how many bacteria (to the nearest unit) will be present:

(A) In 0.85 hour? (B) In 7.25 hours?

Example 3 **Exponential Decay** Cosmic-ray bombardment of the atmosphere produces neutrons, which in turn react with nitrogen to produce radioactive carbon-14 (^{14}C). Radioactive ^{14}C enters all living tissues through carbon dioxide, which is first absorbed by plants. As long as a plant or animal is alive, ^{14}C is maintained in the living organism at a constant level. Once the organism dies, however, ^{14}C decays according to the equation

$$A = A_0e^{-0.000124t}$$

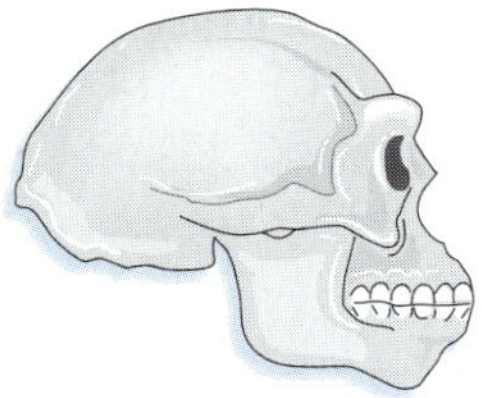

where A is the amount present after t years and A_0 is the amount present at time $t = 0$. If 500 milligrams of ^{14}C are present in a sample from a skull at the time of death, how many milligrams will be present in the sample in:

(A) 15,000 years? (B) 45,000 years?

Compute answers to two decimal places.

Solution Substituting $A_0 = 500$ in the decay equation, we have

$$A = 500e^{-0.000124t}$$ *See the graph in Figure 5.*

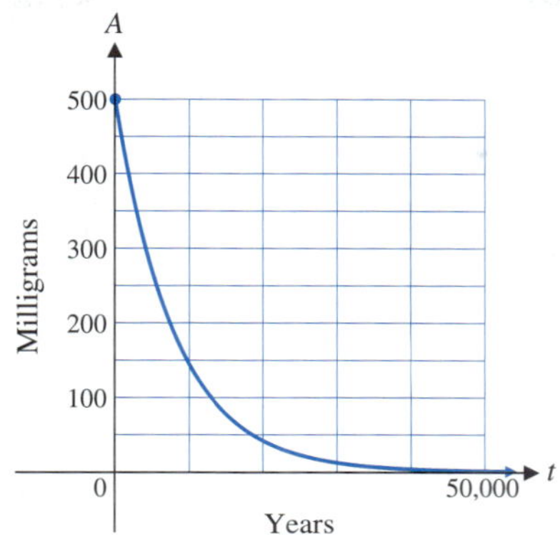

FIGURE 5

(A) Solve for A when $t = 15{,}000$:

$$\begin{aligned} A &= 500e^{-0.000124(15{,}000)} \\ &= 77.84 \text{ milligrams} \end{aligned}$$ *Use a calculator.*

(B) Solve for A when $t = 45{,}000$:

$$\begin{aligned} A &= 500e^{-0.000124(45{,}000)} \\ &= 1.89 \text{ milligrams} \end{aligned}$$ *Use a calculator.*

Matched Problem 3 Refer to the exponential decay model in Example 3. How many milligrams of ^{14}C would have to be present at the beginning in order to have 25 milligrams present after 18,000 years? Compute the answer to the nearest milligram.

Explore–Discuss 3

(A) On the same set of coordinate axes, graph the three decay equations $A = A_0e^{-0.35t}$, $t \geq 0$, for $A_0 = 10, 20$, and 30.
(B) Identify any asymptotes for the three graphs in part (A).
(C) Discuss the long-term behavior for the equations in part (A).

Example 4 **Depreciation** Table 2 gives the market value of a minivan (in dollars) x years after its purchase. Find an exponential regression model of the form $y = ab^x$ for this data set. Estimate the purchase price of the van. Estimate the value of the van 10 years after its purchase. Round answers to the nearest dollar.

Table 2

x	1	2	3	4	5	6
VALUE ($)	12,575	9,455	8,115	6,845	5,225	4,485

SOLUTION Enter the data into a graphing utility (Fig. 6A) and find the exponential regression equation (Fig. 6B). The estimated purchase price is $y_1(0) = \$14{,}910$. The data set and the regression equation are graphed in Figure 6C. Using the trace feature, we see that the estimated value after 10 years is $1,959.

(A) (B) (C)

FIGURE 6

Matched Problem 4 ➡ Table 3 gives the market value of a luxury sedan (in dollars) x years after its purchase. Find an exponential regression model of the form $y = ab^x$ for this data set. Estimate the purchase price of the sedan. Estimate the value of the sedan 10 years after its purchase. Round answers to the nearest dollar.

Table 3

x	1	2	3	4	5	6
VALUE ($)	23,125	19,050	15,625	11,875	9,450	7,125

COMPOUND INTEREST

We now turn to the growth of money at compound interest. The fee paid to use another's money is called **interest.** It is usually computed as a percent (called **interest rate**) of the principal over a given period of time. If, at the end of a payment period, the interest due is reinvested at the same rate, then the interest earned as well as the principal will earn interest during the next payment period. Interest paid on interest reinvested is called **compound interest,** and may be calculated using the following compound interest formula:

Compound Interest

If a **principal *P* (present value)** is invested at an annual **rate *r*** (expressed as a decimal) compounded m times a year, then the **amount *A* (future value)** in the account at the end of t years is given by

$$A = P\left(1 + \frac{r}{m}\right)^{mt}$$

[*Note:* P could be replaced by A_0, but convention dictates otherwise.]

Example 5 ➡ **Compound Growth** If $1,000 is invested in an account paying 10% compounded monthly, how much will be in the account at the end of 10 years? Compute the answer to the nearest cent.

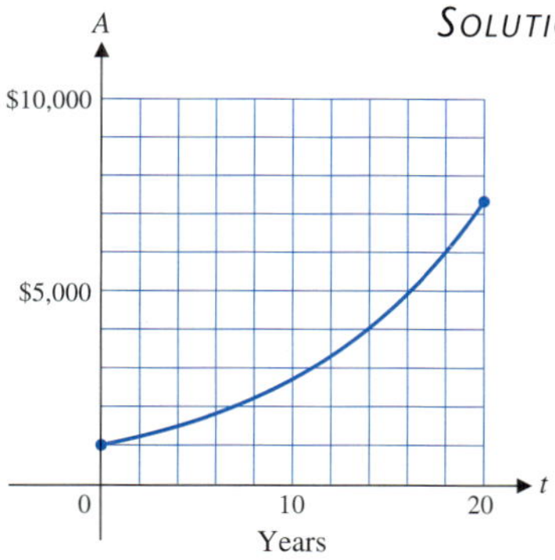

FIGURE 7

SOLUTION We use the compound interest formula as follows:

$$\begin{aligned} A &= P\left(1 + \frac{r}{m}\right)^{mt} \\ &= 1{,}000\left(1 + \frac{0.10}{12}\right)^{(12)(10)} \qquad \text{Use a calculator.} \\ &= \$2{,}707.04 \end{aligned}$$

The graph of

$$A = 1{,}000\left(1 + \frac{0.10}{12}\right)^{12t}$$

for $0 \leq t \leq 20$ is shown in Figure 7.

Matched Problem 5 If you deposit $5,000 in an account paying 9% compounded daily, how much will you have in the account in 5 years? Compute the answer to the nearest cent.

Explore–Discuss 4

Suppose that $1,000 is deposited in a savings account at an annual rate of 5%. Guess the amount in the account at the end of 1 year if interest is compounded (1) quarterly, (2) monthly, (3) daily, (4) hourly. Use the compound interest formula to compute the amounts at the end of 1 year to the nearest cent. Discuss the accuracy of your initial guesses.

Continuous Compound Interest

Returning to the compound interest formula,

$$A = P\left(1 + \frac{r}{m}\right)^{mt}$$

suppose the principal P, the annual rate r, and the time t are held fixed, and the number of compounding periods per year m is increased without bound. Will the amount A increase without bound, or will it tend to some limiting value?

Starting with $P = \$100$, $r = 0.08$, and $t = 2$ years, we construct Table 4 for several values of m with the aid of a calculator. Notice that the largest gain appears in going from annual to semiannual compounding. Then, the gains slow down as m increases. It appears that A gets closer and closer to $117.35 as m gets larger and larger.

Table 4

Compounding Frequency	m	$A = 100\left(1 + \frac{0.08}{m}\right)^{2m}$
Annually	1	$116.6400
Semiannually	2	116.9859
Quarterly	4	117.1659
Weekly	52	117.3367
Daily	365	117.3490
Hourly	8,760	117.3510

It can be shown that

$$P\left(1 + \frac{r}{m}\right)^{mt}$$

gets closer and closer to Pe^{rt} as the number of compounding periods m gets larger and larger. The latter is referred to as the **continuous compound interest formula,** a formula that is widely used in business, banking, and economics.

Continuous Compound Interest Formula

If a principal P is invested at an annual rate r (expressed as a decimal) compounded continuously, then the amount A in the account at the end of t years is given by

$$A = Pe^{rt}$$

Example 6 **Compounding Daily and Continuously** What amount will an account have after 2 years if \$5,000 is invested at an annual rate of 8%:

(A) Compounded daily? (B) Compounded continuously?

Compute answers to the nearest cent.

SOLUTION (A) Use the compound interest formula

$$A = P\left(1 + \frac{r}{m}\right)^{mt}$$

with $P = 5{,}000$, $r = 0.08$, $m = 365$, and $t = 2$:

$$A = 5{,}000\left(1 + \frac{0.08}{365}\right)^{(365)(2)} \quad \text{Use a calculator.}$$
$$= \$5{,}867.45$$

(B) Use the continuous compound interest formula

$$A = Pe^{rt}$$

with $P = 5{,}000$, $r = 0.08$, and $t = 2$:

$$A = 5{,}000e^{(0.08)(2)} \quad \text{Use a calculator.}$$
$$= \$5{,}867.55$$

Matched Problem 6 What amount will an account have after 1.5 years if \$8,000 is invested at an annual rate of 9%:

(A) Compounded weekly? (B) Compounded continuously?

Compute answers to the nearest cent.

The formulas for simple interest, compound interest, and continuous compound interest are summarized in the box at the top of the next page for convenient reference.

> **Interest Formulas**
>
> | | |
> |---|---|
> | **Simple interest** | $A = P(1 + rt)$ |
> | **Compound interest** | $A = P\left(1 + \frac{r}{m}\right)^{mt}$ |
> | **Continuous compound interest** | $A = Pe^{rt}$ |

Answers to Matched Problems

1.

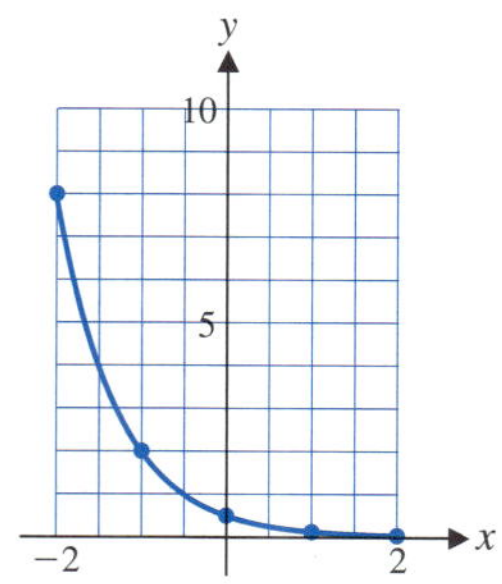

2. (A) 179 bacteria (B) 1,271,659 bacteria

3. 233 mg **4.** Purchase price: $30,363; value after 10 yr: $2,864

5. $7,841.13 **6.** (A) $9,155.23 (B) $9,156.29

Exercise 2-2

A *Graph each function in Problems 1–12 over the indicated interval.*

1. $y = 5^x; [-2, 2]$

2. $y = 3^x; [-3, 3]$

3. $y = (\frac{1}{5})^x = 5^{-x}; [-2, 2]$

4. $y = (\frac{1}{3})^x = 3^{-x}; [-3, 3]$

5. $f(x) = -5^x; [-2, 2]$

6. $g(x) = -3^{-x}; [-3, 3]$

7. $y = -e^{-x}; [-3, 3]$

8. $y = -e^x; [-3, 3]$

9. $y = 100e^{0.1x}; [-5, 5]$

10. $y = 10e^{0.2x}; [-10, 10]$

11. $g(t) = 10e^{-0.2t}; [-5, 5]$

12. $f(t) = 100e^{-0.1t}; [-5, 5]$

Simplify each expression in Problems 13–18.

13. $(4^{3x})^{2y}$

14. $10^{3x-1}10^{4-x}$

15. $\dfrac{e^{x-3}}{e^{x-4}}$

16. $\dfrac{e^x}{e^{1-x}}$

17. $(2e^{1.2t})^3$

18. $(3e^{-1.4x})^2$

B *In Problems 19–26, describe the transformations that can be used to obtain the graph of g from the graph of f (see Section 1–2).*

19. $g(x) = -2^x; f(x) = 2^x$

20. $g(x) = 2^{x-2}; f(x) = 2^x$

21. $g(x) = 3^{x+1}; f(x) = 3^x$

22. $g(x) = -3^x; f(x) = 3^x$

23. $g(x) = e^x + 1; f(x) = e^x$

24. $g(x) = e^x - 2; f(x) = e^x$

25. $g(x) = 2e^{-(x+2)}; f(x) = e^{-x}$

26. $g(x) = 0.5e^{-(x-1)}; f(x) = e^{-x}$

Check the answers to Problems 19–26 by graphing each pair of functions in the same viewing window of a graphing utility.

In Problems 27–36, graph each function over the indicated interval.

27. $f(t) = 2^{t/10}; [-30, 30]$

28. $G(t) = 3^{t/100}; [-200, 200]$

29. $y = -3 + e^{1+x}; [-4, 2]$

30. $y = 2 + e^{x-2}; [-1, 5]$

31. $y = e^{|x|}; [-3, 3]$

32. $y = e^{-|x|}; [-3, 3]$

33. $C(x) = \dfrac{e^x + e^{-x}}{2}; [-5, 5]$

34. $M(x) = e^{x/2} + e^{-x/2}; [-5, 5]$

35. $y = e^{-x^2}; [-3, 3]$

36. $y = 2^{-x^2}; [-3, 3]$

37. Find all real numbers a such that $a^2 = a^{-2}$. Explain why this does not violate the second exponential function property in the box on page 96.

38. Find real numbers a and b such that $a \neq b$ but $a^4 = b^4$. Explain why this does not violate the third exponential function property in the box on page 96.

39. A graphing utility was used to graph the exponential functions $f(x) = 2^x$ and $g(x) = e^x$ in the same viewing window, as shown in the figure. Which curve belongs to which function? Explain.

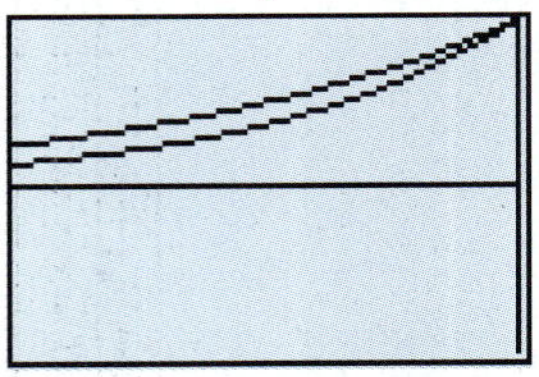

Figure for 39

40. A graphing utility was used to graph the exponential functions $f(x) = 2^x$ and $g(x) = e^x$ in the same viewing window, as shown in the figure. Which curve belongs to which function? Explain.

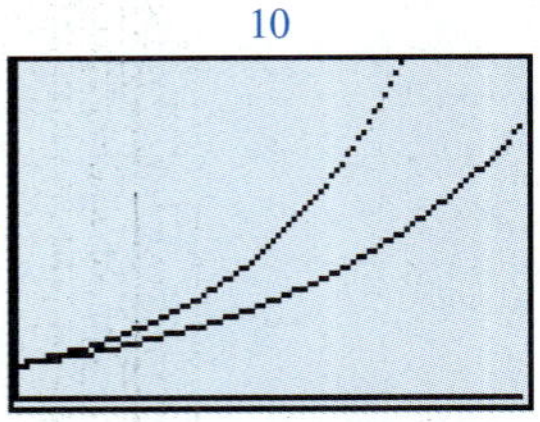

Figure for 40

41. A graphing utility was used to graph the exponential functions $f(x) = 2^{-x}$ and $g(x) = e^{-x}$ in the same viewing window, as shown in the figure. Which curve belongs to which function? Explain.

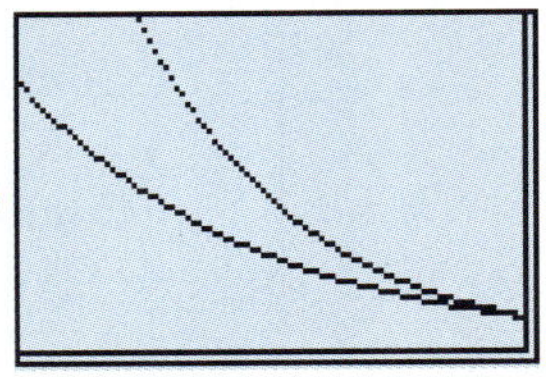

Figure for 41

42. A graphing utility was used to graph the exponential functions $f(x) = 2^{-x}$ and $g(x) = e^{-x}$ in the same viewing window, as shown in the figure. Which curve belongs to which function? Explain.

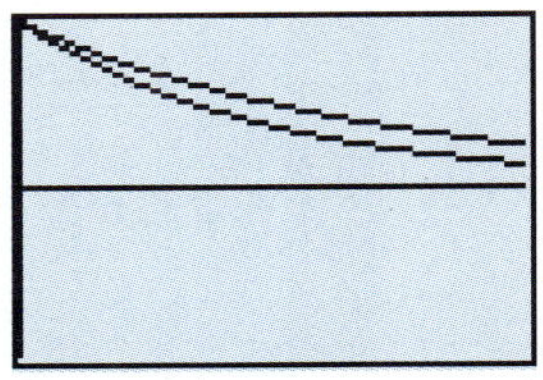

Figure for 42

Solve each equation in Problems 43–48 for x.

43. $10^{2-3x} = 10^{5x-6}$

44. $5^{3x} = 5^{4x-2}$

45. $4^{5x-x^2} = 4^{-6}$

46. $7^{x^2} = 7^{2x+3}$

47. $5^3 = (x + 2)^3$

48. $(1 - x)^5 = (2x - 1)^5$

C *Solve each equation in Problems 49–52 for x. (Remember, $e^x \neq 0$ and $e^{-x} \neq 0$.)*

49. $(x - 3)e^x = 0$

50. $2xe^{-x} = 0$

51. $3xe^{-x} + x^2e^{-x} = 0$

52. $x^2e^x - 5xe^x = 0$

Graph each function in Problems 53–56 over the indicated interval.

53. $h(x) = x(2^x); [-5, 0]$

54. $m(x) = x(3^{-x}); [0, 3]$

55. $N = \dfrac{100}{1 + e^{-t}}; [0, 5]$

56. $N = \dfrac{200}{1 + 3e^{-t}}; [0, 5]$

In Problems 57–60, approximate the real zeros of each function to two decimal places.

57. $f(x) = 4^x - 7$

58. $f(x) = 5 - 3^{-x}$

59. $f(x) = 2 + 3x + 10^x$

60. $f(x) = 7 - 2x^2 + 2^{-x}$

APPLICATIONS

Business & Economics

61. *Finance.* Suppose \$2,500 is invested at 7% compounded quarterly. How much money will be in the account in:
(A) $\frac{3}{4}$ year? (B) 15 years?

Compute answers to the nearest cent.

62. *Finance.* Suppose \$4,000 is invested at 11% compounded weekly. How much money will be in the account in:
(A) $\frac{1}{2}$ year? (B) 10 years?

Compute answers to the nearest cent.

63. *Money growth.* If you invest \$7,500 in an account paying 8.35% compounded continuously, how much money will be in the account at the end of:
(A) 5.5 years? (B) 12 years?

64. *Money growth.* If you invest \$5,250 in an account paying 11.38% compounded continuously, how much money will be in the account at the end of:
(A) 6.25 years? (B) 17 years?

65. *Finance.* A person wishes to have \$15,000 cash for a new car 5 years from now. How much should be placed in an account now, if the account pays 9.75% compounded weekly? Compute the answer to the nearest dollar.

66. *Finance.* A couple just had a baby. How much should they invest now at 8.25% compounded daily in order to have \$40,000 for the child's education 17 years from now? Compute the answer to the nearest dollar.

67. *Money growth.* *Barron's* (a national business and financial weekly) published the following "Top Savings Deposit Yields" for 1 year certificate of deposit accounts:

(A) Alamo Savings, 8.25% compounded quarterly
(B) Lamar Savings, 8.05% compounded continuously

Compute the value of \$10,000 invested in each account at the end of 1 year.

68. *Money growth.* Refer to Problem 67. In another issue of *Barron's*, $2\frac{1}{2}$ year certificate of deposit accounts included the following:
(A) Gill Saving, 8.30% compounded continuously
(B) Richardson Savings and Loan, 8.40% compounded quarterly
(C) USA Savings, 8.25% compounded daily

Compute the value of \$1,000 invested in each account at the end of $2\frac{1}{2}$ years.

69. *Present value.* A promissory note will pay \$50,000 at maturity $5\frac{1}{2}$ years from now. How much should you be willing to pay for the note now if money is worth 10% compounded continuously?

70. *Present value.* A promissory note will pay \$30,000 at maturity 10 years from now. How much should you be willing to pay for the note now if money is worth 9% compounded continuously?

71. *Advertising.* A company is trying to introduce a new product to as many people as possible through television advertising in a large metropolitan area with 2 million possible viewers. A model for the number of people N (in millions) who are aware of the product after t days of advertising was found to be

$$N = 2(1 - e^{-0.037t})$$

Graph this function for $0 \leq t \leq 50$. What value does N tend to as t increases without bound?

72. *Learning curve.* People assigned to assemble circuit boards for a computer manufacturing company undergo on-the-job training. From past experience it was found that the learning curve for the average employee is given by

$$N = 40(1 - e^{-0.12t})$$

where N is the number of boards assembled per day after t days of training. Graph this function for $0 \leq t \leq 30$. What is the maximum number of boards an average employee can be expected to produce in 1 day?

73. *Sports salaries.* Table 5 gives the average salary (in thousands of dollars) for players in the National Football League (NFL) and the National Basketball Association (NBA) in selected years since 1980.

(A) Let x represent the number of years since 1980 and find an exponential regression model ($y = ab^x$) for the NFL's average salary data. Estimate the average salary (to the nearest thousand dollars) for 1995 and 2010.

(B) The average salary in the NFL for 1995 was $637,000. How does this compare with the estimated value from the model in part (A)? How does this 1995 salary information affect the estimated salary in 2010? Explain.

Table 5

AVERAGE SALARIES (THOUSAND $)

YEAR	1980	1984	1985	1986	1987	1988	1989	1990
NFL	79	178	194	198	203	239	295	352
NBA	170	275	325	375	440	510	603	817

Source: U.S. Bureau of the Census

74. *Sports salaries.* Refer to Table 5.

(A) Let x represent the number of years since 1980 and find an exponential regression model ($y = ab^x$) for the NBA's average salary data. Estimate the average salary (to the nearest thousand dollars) for 1994 and 2010.

(B) The average salary in the NBA for 1994 was $1,700,000. How does this compare with the estimated value from the model in part (A)? How does this 1994 salary information affect the estimated salary in 2010? Explain.

Life Sciences

75. *Marine biology.* Marine life is dependent upon the microscopic plant life that exists in the *photic zone,* a zone that goes to a depth where about 1% of the surface light still remains. In some waters with a great deal of sediment, the photic zone may go down only 15–20 feet. In some murky harbors, the intensity of light d feet below the surface is given approximately by

$$I = I_0e^{-0.23d}$$

What percentage of the surface light will reach a depth of:

(A) 10 feet? (B) 20 feet?

76. *Marine biology.* Refer to Problem 75. Light intensity I relative to depth d (in feet) for one of the clearest bodies of water in the world, the Sargasso Sea in the West Indies, can be approximated by

$$I = I_0e^{-0.00942d}$$

where I_0 is the intensity of light at the surface. What percent of the surface light will reach a depth of:

(A) 50 feet? (B) 100 feet?

77. *AIDS epidemic.* The U.S. Department of Health and Human Services reports that prior to 1993 there were about 40,000 cases of AIDS among intravenous drug users in the United States. It was estimated that the disease was spreading in this group at an annual rate of 21% compounded continuously. Let 1992 be year 0 and assume this rate does not change.

(A) Write an equation that models the growth of AIDS in this group, starting in 1992.

(B) Based on the model, how many cases (to the nearest thousand) should we expect by the end of the year 2000? The year 2005?

(C) Sketch a graph of the equation found in part (A). Cover the years from 1992 through 2005.

78. *AIDS epidemic.* The U.S. Department of Health and Human Services reports that prior to 1993 there were about 245,000 cases of AIDS among the general population in the United States. It was estimated that the disease was spreading at an annual rate of 18% compounded continuously. Let 1992 be year 0 and assume this rate does not change.

(A) Write an equation that models the growth of AIDS, starting in 1992.

(B) Based on the model, how many cases (to the nearest hundred thousand) should we expect by the end of the year 2000? The year 2005?

(C) Sketch a graph of the equation found in part (A). Cover the years from 1992 through 2005.

Social Sciences

79. *World population growth.* It took from the dawn of humanity to 1830 for the population to grow to the first billion people, just 100 more years (by 1930) for the second billion, and a billion more were added in only 60 more years (by 1990). In 1995, the estimated world population was 5.7 billion. In 1994, the World Bank estimated the world population would be growing at an annual rate of 1.14% compounded continuously until 2030.

(A) Write an equation that models the world population growth, letting 1995 be year 0.

(B) Based on the model, what is the expected world population (to the nearest hundred million) in 2010? In 2030?

(C) Sketch a graph of the equation found in part (A). Cover the years from 1995 through 2030.

80. *Population growth in Ethiopia.* In 1995, the estimated population in Ethiopia was 88 million people. In 1994, the World Bank estimated the population would grow at an annual rate of 1.67% compounded continuously until 2030.

(A) Write an equation that models the population growth in Ethiopia, letting 1995 be year 0.

(B) Based on the model, what is the expected population in Ethiopia (to the nearest million) in 2010? In 2030?

(C) Sketch a graph of the equation found in part (A). Cover the years from 1995 through 2030.

81. C *Internet growth.* An Internet host is a computer that can be reached directly by other computers on the Internet. The number of Internet hosts almost doubled annually from 1990 to 1994 (Table 6).

(A) Let x represent the number of years since 1990. Find an exponential regression model ($y = ab^x$) for this data set and estimate the number of hosts in 1995 and 1996 (to the nearest thousand).

(B) Discuss the implications of this model if the number of Internet hosts continues to grow at this rate.

Table 6

INTERNET HOSTS (THOUSANDS)

YEAR	1990	1991	1992	1993	1994
HOSTS	313	617	1,136	2,056	3,864

Source: Network Wizards, www.nw.com

82. C *National debt growth.* The national debt has been growing rapidly for decades. Table 7 lists the debt (in billions of dollars) for selected years since 1970. Let x represent the number of years since 1970. Find an exponential regression model for this data set and estimate the national debt in 2000 and 2010 (to the nearest billion).

Table 7

NATIONAL DEBT (BILLION $)

YEAR	1970	1975	1980	1985	1990	1995
DEBT	389	577	930	1,946	3,233	4,974

Source: Bureau of the Public Debt, www.savingsbonds.gov

SECTION 2-3 Logarithmic Functions

- INVERSE FUNCTIONS
- LOGARITHMIC FUNCTIONS
- PROPERTIES OF LOGARITHMIC FUNCTIONS
- CALCULATOR EVALUATION OF LOGARITHMS
- APPLICATION

Find the exponential function keys $\boxed{10^x}$ and $\boxed{e^x}$ on your calculator. Close to these keys you will find $\boxed{\text{LOG}}$ and $\boxed{\text{LN}}$ keys. The latter represent *logarithmic functions,* and each is closely related to the exponential function it is near. In fact, the exponential function and the corresponding logarithmic function are said to be *inverses* of each other. In this section we will develop the concept of inverse functions and use it to define a logarithmic function as the inverse of an exponential function. We will then investigate basic properties of logarithmic functions, use a calculator to evaluate them for particular values of x, and apply them to real-world problems.

Logarithmic functions are used in modeling and solving many types of problems. For example, the decibel scale is a logarithmic scale used to measure sound intensity, and the Richter scale is a logarithmic scale used to measure the strength of the force of an earthquake. An important business application has to do with finding the time it takes money to double if it is invested at a certain rate compounded a given number of times a year or compounded continuously. This requires the solution of an exponential equation, and logarithms play a central role in the process.

Inverse Functions

Look at the graphs of $f(x) = \frac{x}{2}$ and $g(x) = \frac{|x|}{2}$ in Figure 1:

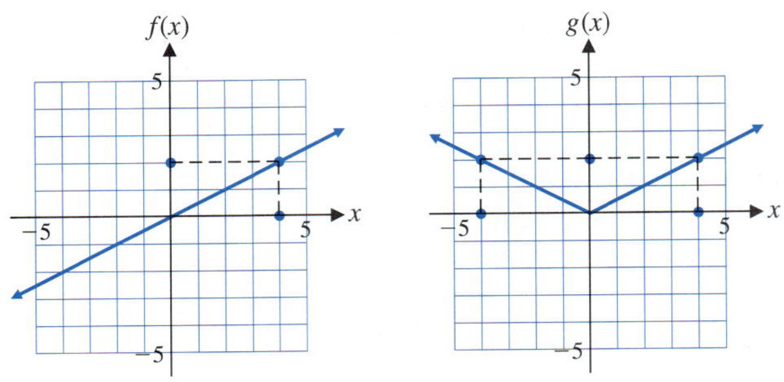

(A) $f(x) = \frac{x}{2}$ (B) $g(x) = \frac{|x|}{2}$

FIGURE 1

Because both f and g are functions, each domain value corresponds to exactly one range value. For which function does each range value correspond to exactly one domain value? This is the case only for function f. Note that for the range value 2, the corresponding domain value is 4. For function g the range value 2 corresponds to both -4 and 4. Function f is said to be *one-to-one.* In general:

One-to-One Functions

A function f is said to be **one-to-one** if each range value corresponds to exactly one domain value.

It can be shown that any continuous function that is either increasing or decreasing for all domain values is one-to-one. If a continuous function increases for some domain values and decreases for others, it cannot be one-to-one. Figure 1 shows an example of each case.

Explore–Discuss 1

Graph $f(x) = 2^x$ and $g(x) = x^2$. For a range value of 4, what are the corresponding domain values for each function? Which of the two functions is one-to-one? Explain why.

Starting with a one-to-one function f we can obtain a new function called the *inverse* of f as follows:

Inverse of a Function

If f is a one-to-one function, then the **inverse** of f is the function formed by interchanging the independent and dependent variables for f. Thus, if (a, b) is a point on the graph of f, then (b, a) is a point on the graph of the inverse of f.

[*Note:* If f is not one-to-one, then f **does not have an inverse.**]

A number of important functions in any library of elementary functions are the inverses of other basic functions in the library. In this course, we are interested in the inverses of exponential functions, called *logarithmic functions.*

Logarithmic Functions

If we start with the exponential function f defined by

$$y = 2^x \tag{1}$$

and interchange the variables, we obtain the inverse of f:

$$x = 2^y \tag{2}$$

We call the inverse the **logarithmic function with base 2,** and write

$$y = \log_2 x \quad \text{if and only if} \quad x = 2^y$$

We can graph $y = \log_2 x$ by graphing $x = 2^y$, since they are equivalent. Any ordered pair of numbers on the graph of the exponential function will be on the graph of the logarithmic function if we interchange the order of the components. For example, (3, 8) satisfies equation (1) and (8, 3) satisfies equation (2). The graphs of $y = 2^x$ and $y = \log_2 x$ are shown in Figure 2. Note that if we fold the paper along the dashed line $y = x$ in Figure 2, the two graphs match exactly. The line $y = x$ is a line of symmetry for the two graphs.

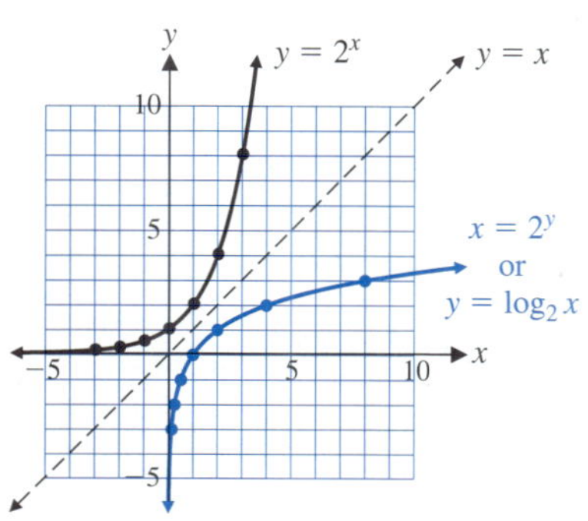

FIGURE 2

Exponential Function		Logarithmic Function	
x	$y = 2^x$	$x = 2^y$	y
−3	$\frac{1}{8}$	$\frac{1}{8}$	−3
−2	$\frac{1}{4}$	$\frac{1}{4}$	−2
−1	$\frac{1}{2}$	$\frac{1}{2}$	−1
0	1	1	0
1	2	2	1
2	4	4	2
3	8	8	3

Ordered pairs reversed

In general, since the graphs of all exponential functions of the form $f(x) = b^x$, $b \neq 1$, $b > 0$, are either increasing or decreasing (see Section 2-2), exponential functions have inverses.

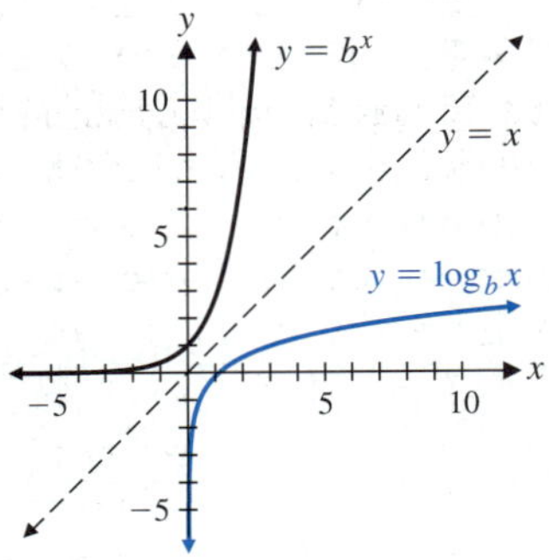

Base $b > 1$

Logarithmic Functions

The inverse of an exponential function is called a **logarithmic function.** For $b > 0$ and $b \neq 1$,

Logarithmic form		Exponential form
$y = \log_b x$	is equivalent to	$x = b^y$

The log to the base b of x is the exponent to which b must be raised to obtain x. [*Remember:* A logarithm is an exponent.] The **domain** of the logarithmic function is the set of all positive real numbers, which is also the range of the corresponding exponential function; and the **range** of the logarithmic function is the set of all real numbers, which is also the domain of the corresponding exponential function. Typical graphs of an exponential function and its inverse, a logarithmic function, are shown in the figure in the margin.

The following examples involve converting logarithmic forms to equivalent exponential forms and vice versa.

Example 1 **Logarithmic–Exponential Conversions** Change each logarithmic form to an equivalent exponential form:

(A) $\log_5 25 = 2$ (B) $\log_9 3 = \frac{1}{2}$ (C) $\log_2(\frac{1}{4}) = -2$

SOLUTION
(A) $\log_5 25 = 2$ is equivalent to $25 = 5^2$
(B) $\log_9 3 = \frac{1}{2}$ is equivalent to $3 = 9^{1/2}$
(C) $\log_2(\frac{1}{4}) = -2$ is equivalent to $\frac{1}{4} = 2^{-2}$

Matched Problem 1 Change each logarithmic form to an equivalent exponential form:

(A) $\log_3 9 = 2$ (B) $\log_4 2 = \frac{1}{2}$ (C) $\log_3(\frac{1}{9}) = -2$

Example 2 **Exponential–Logarithmic Conversions** Change each exponential form to an equivalent logarithmic form:

(A) $64 = 4^3$ (B) $6 = \sqrt{36}$ (C) $\frac{1}{8} = 2^{-3}$

SOLUTION
(A) $64 = 4^3$ is equivalent to $\log_4 64 = 3$
(B) $6 = \sqrt{36}$ is equivalent to $\log_{36} 6 = \frac{1}{2}$
(C) $\frac{1}{8} = 2^{-3}$ is equivalent to $\log_2(\frac{1}{8}) = -3$

Matched Problem 2 Change each exponential form to an equivalent logarithmic form.

(A) $49 = 7^2$ (B) $3 = \sqrt{9}$ (C) $\frac{1}{3} = 3^{-1}$

To gain a little deeper understanding of logarithmic functions and their relationship to the exponential functions, we consider a few problems where we want to find x, b, or y in $y = \log_b x$, given the other two values. All values are chosen so that the problems can be solved exactly without a calculator.

Example 3 **Solutions of the Equation $y = \log_b x$** Find y, b, or x, as indicated.

(A) Find y: $y = \log_4 16$ (B) Find x: $\log_2 x = -3$
(C) Find y: $y = \log_8 4$ (D) Find b: $\log_b 100 = 2$

SOLUTION (A) $y = \log_4 16$ is equivalent to $16 = 4^y$. Thus,

$$y = 2$$

(B) $\log_2 x = -3$ is equivalent to $x = 2^{-3}$. Thus,

$$x = \frac{1}{2^3} = \frac{1}{8}$$

(C) $y = \log_8 4$ is equivalent to

$$4 = 8^y \qquad \text{or} \qquad 2^2 = 2^{3y}$$

Thus,

$$3y = 2$$
$$y = \tfrac{2}{3}$$

(D) $\log_b 100 = 2$ is equivalent to $100 = b^2$. Thus,

$$b = 10$$ Recall that b cannot be negative.

Matched Problem 3 Find y, b, or x, as indicated.

(A) Find y: $y = \log_9 27$ (B) Find x: $\log_3 x = -1$
(C) Find b: $\log_b 1{,}000 = 3$

Properties of Logarithmic Functions

Logarithmic functions have many powerful and useful properties. We list eight basic properties in Theorem 1.

Theorem 1 PROPERTIES OF LOGARITHMIC FUNCTIONS

If b, M, and N are positive real numbers, $b \neq 1$, and p and x are real numbers, then:

1. $\log_b 1 = 0$
2. $\log_b b = 1$
3. $\log_b b^x = x$
4. $b^{\log_b x} = x, \quad x > 0$
5. $\log_b MN = \log_b M + \log_b N$
6. $\log_b \dfrac{M}{N} = \log_b M - \log_b N$
7. $\log_b M^p = p \log_b M$
8. $\log_b M = \log_b N$ if and only if $M = N$

The first four properties in Theorem 1 follow directly from the definition of a logarithmic function. Here we will sketch a proof of property 5. The other properties are established in a similar way. Let

$$u = \log_b M \qquad \text{and} \qquad v = \log_b N$$

Or, in equivalent exponential form,

$$M = b^u \qquad \text{and} \qquad N = b^v$$

Now, see if you can provide reasons for each of the following steps:

$$\log_b MN = \log_b b^u b^v = \log_b b^{u+v} = u + v = \log_b M + \log_b N$$

Example 4 **Using Logarithmic Properties**

(A) $\log_b \dfrac{wx}{yz} = \log_b wx - \log_b yz$

$= \log_b w + \log_b x - (\log_b y + \log_b z)$

$= \log_b w + \log_b x - \log_b y - \log_b z$

(B) $\log_b(wx)^{3/5} = \frac{3}{5}\log_b wx = \frac{3}{5}(\log_b w + \log_b x)$

Matched Problem 4 Write in simpler logarithmic forms, as in Example 4.

(A) $\log_b \dfrac{R}{ST}$ (B) $\log_b\left(\dfrac{R}{S}\right)^{2/3}$

The following examples and problems, though somewhat artificial, will give you additional practice in using basic logarithmic properties.

Example 5 **Solving Logarithmic Equations** Find x so that:

$$\tfrac{3}{2}\log_b 4 - \tfrac{2}{3}\log_b 8 + \log_b 2 = \log_b x$$

SOLUTION

$$\begin{aligned}
\tfrac{3}{2}\log_b 4 - \tfrac{2}{3}\log_b 8 + \log_b 2 &= \log_b x \\
\log_b 4^{3/2} - \log_b 8^{2/3} + \log_b 2 &= \log_b x && \text{Property 7} \\
\log_b 8 - \log_b 4 + \log_b 2 &= \log_b x \\
\log_b \frac{8 \cdot 2}{4} &= \log_b x && \text{Properties 5 and 6} \\
\log_b 4 &= \log_b x \\
x &= 4 && \text{Property 8}
\end{aligned}$$

Matched Problem 5 Find x so that: $3\log_b 2 + \frac{1}{2}\log_b 25 - \log_b 20 = \log_b x$

Example 6 **Solving Logarithmic Equations** Solve: $\log_{10} x + \log_{10}(x + 1) = \log_{10} 6$

SOLUTION

$$\begin{aligned}
\log_{10} x + \log_{10}(x + 1) &= \log_{10} 6 \\
\log_{10}[x(x + 1)] &= \log_{10} 6 && \text{Property 5} \\
x(x + 1) &= 6 && \text{Property 8} \\
x^2 + x - 6 &= 0 && \text{Solve by factoring.} \\
(x + 3)(x - 2) &= 0 \\
x &= -3, 2
\end{aligned}$$

We must exclude $x = -3$, since the domain of the function $\log_{10}(x + 1)$ is $x > -1$ or $(-1, \infty)$; hence, $x = 2$ is the only solution.

Matched Problem 6 Solve: $\log_3 x + \log_3(x - 3) = \log_3 10$

Explore–Discuss 2

Discuss the relationship between each of the following pairs of expressions. If the two expressions are equivalent, explain why. If they are not, give an example.

(A) $\log_b M - \log_b N$; $\dfrac{\log_b M}{\log_b N}$

(B) $\log_b M - \log_b N$; $\log_b \dfrac{M}{N}$

(C) $\log_b M + \log_b N$; $\log_b MN$

(D) $\log_b M + \log_b N$; $\log_b(M + N)$

Calculator Evaluation of Logarithms

Of all possible logarithmic bases, the base e and the base 10 are used almost exclusively. Before we can use logarithms in certain practical problems, we need to be able to approximate the logarithm of any positive number either to base 10 or to base e. And conversely, if we are given the logarithm of a number to base 10 or base e, we need to be able to approximate the number. Historically, tables were used for this purpose, but now calculators make computations faster and far more accurate.

Common logarithms (also called **Briggsian logarithms**) are logarithms with base 10. **Natural logarithms** (also called **Napierian logarithms**) are logarithms with base e. Most calculators have a key labeled "log" (or "LOG") and a key labeled "ln" (or "LN"). The former represents a common (base 10) logarithm and the latter a natural (base e) logarithm. In fact, "log" and "ln" are both used extensively in mathematical literature, and whenever you see either used in this book without a base indicated they will be interpreted as follows:

Logarithmic Notation

Common logarithm: $\log x = \log_{10} x$
Natural logarithm: $\ln x = \log_e x$

Finding the common or natural logarithm using a calculator is very easy. On some calculators, you simply enter a number from the domain of the function and press [LOG] or [LN]. On other calculators, you press either [LOG] or [LN], enter a number from the domain, and then press [ENTER]. Check the user's manual for your calculator.

Example 7 **Calculator Evaluation of Logarithms** Use a calculator to evaluate each to six decimal places:

(A) $\log 3{,}184$ (B) $\ln 0.000\ 349$ (C) $\log(-3.24)$

SOLUTION (A) $\log 3{,}184 = 3.502\ 973$ (B) $\ln 0.000\ 349 = -7.960\ 439$
(C) $\log(-3.24) = \text{Error}^*$ -3.24 is not in the domain of the log function.

*Some calculators use a more advanced definition of logarithms involving complex numbers and will display an ordered pair of real numbers as the value of $\log(-3.24)$. You should interpret such a result as an indication that the number entered is not in the domain of the logarithm function as we have defined it.

Matched Problem 7 ➡ Use a calculator to evaluate each to six decimal places:

(A) $\log 0.013\ 529$ (B) $\ln 28.693\ 28$ (C) $\ln(-0.438)$

We now turn to the second problem mentioned above: Given the logarithm of a number, find the number. We make direct use of the logarithmic–exponential relationships, which follow from the definition of logarithmic function given at the beginning of this section.

Logarithmic–Exponential Relationships

$\log x = y$	is equivalent to	$x = 10^y$
$\ln x = y$	is equivalent to	$x = e^y$

Example 8 ➡ **Solving $\log_b x = y$ for x** Find x to four decimal places, given the indicated logarithm:

(A) $\log x = -2.315$ (B) $\ln x = 2.386$

SOLUTION (A)

$$\begin{aligned} \log x &= -2.315 && \text{Change to equivalent exponential form.} \\ x &= 10^{-2.315} && \text{Evaluate with a calculator.} \\ x &= 0.0048 \end{aligned}$$

(B)

$$\begin{aligned} \ln x &= 2.386 && \text{Change to equivalent exponential form.} \\ x &= e^{2.386} && \text{Evaluate with a calculator.} \\ x &= 10.8699 \end{aligned}$$

Matched Problem 8 ➡ Find x to four decimal places, given the indicated logarithm:

(A) $\ln x = -5.062$ (B) $\log x = 2.0821$

Example 9 ➡ **Solving Exponential Equations** Solve for x to four decimal places:

(A) $10^x = 2$ (B) $e^x = 3$ (C) $3^x = 4$

SOLUTION (A)

$$\begin{aligned} 10^x &= 2 && \text{Take common logarithms of both sides.} \\ \log 10^x &= \log 2 && \text{Property 3} \\ x &= \log 2 && \text{Use a calculator.} \\ x &= 0.3010 \end{aligned}$$

(B)

$$\begin{aligned} e^x &= 3 && \text{Take natural logarithms of both sides.} \\ \ln e^x &= \ln 3 && \text{Property 3} \\ x &= \ln 3 && \text{Use a calculator.} \\ x &= 1.0986 \end{aligned}$$

(C)

$$\begin{aligned} 3^x &= 4 && \text{Take either natural or common logarithms of both sides. (We choose common logarithms.)} \\ \log 3^x &= \log 4 && \text{Property 7} \\ x \log 3 &= \log 4 && \text{Solve for } x. \\ x &= \frac{\log 4}{\log 3} && \text{Use a calculator.} \\ x &= 1.2619 \end{aligned}$$

Exponential equations can also be solved graphically by graphing both sides of an equation and finding the points of intersection. Figure 3 illustrates this approach for the equations in Example 9.

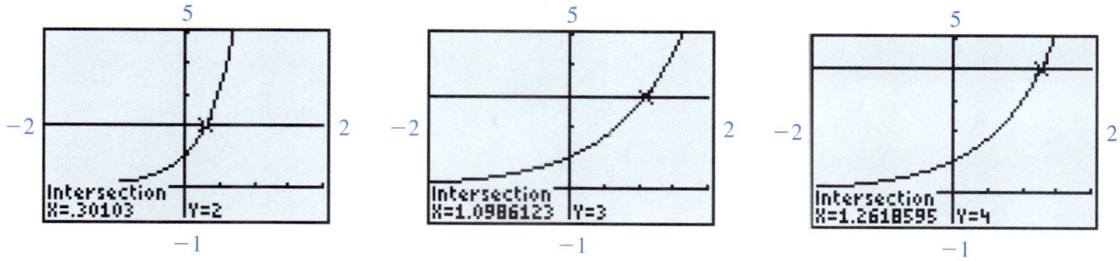

(A) $y_1 = 10^x$, $y_2 = 2$ (B) $y_1 = e^x$, $y_2 = 3$ (C) $y_1 = 3^x$, $y_2 = 4$

FIGURE 3
Graphical solution of exponential equations

Matched Problem 9 Solve for x algebraically and graphically (to four decimal places):

(A) $10^x = 7$ (B) $e^x = 6$ (C) $4^x = 5$

Explore–Discuss 3

Discuss how you could find $y = \log_5 38.25$ using either natural or common logarithms on a calculator. [*Hint:* Start by rewriting the equation in exponential form.]

Application

A convenient and easily understood way of comparing different investments is to use their **doubling times**—the length of time it takes the value of an investment to double. Logarithm properties, as you will see in Example 10, provide us with just the right tool for solving some doubling-time problems.

Example 10 **Doubling Time for an Investment** How long (to the next whole year) will it take money to double if it is invested at 20% compounded annually?

SOLUTION We use the compound interest formula discussed in Section 2-2:

$$A = P\left(1 + \frac{r}{m}\right)^{mt} \quad \text{Compound interest}$$

The problem is to find t, given $r = 0.20$, $m = 1$, and $A = 2p$; that is,

$$\begin{aligned} 2P &= P(1 + 0.2)^t \\ 2 &= 1.2^t \\ 1.2^t &= 2 && \text{Solve for } t \text{ by taking the natural or common logarithm of both sides (we choose the natural logarithm).} \\ \ln 1.2^t &= \ln 2 \\ t \ln 1.2 &= \ln 2 && \text{Property 7} \\ t &= \frac{\ln 2}{\ln 1.2} && \text{Use a calculator.} \\ &= 3.8 \text{ years} && [\textit{Note:} \ (\ln 2)/(\ln 1.2) \neq \ln 2 - \ln 1.2] \\ &\approx 4 \text{ years} && \text{To the next whole year} \end{aligned}$$

When interest is paid at the end of 3 years, the money will not be doubled; when paid at the end of 4 years, the money will be slightly more than doubled.

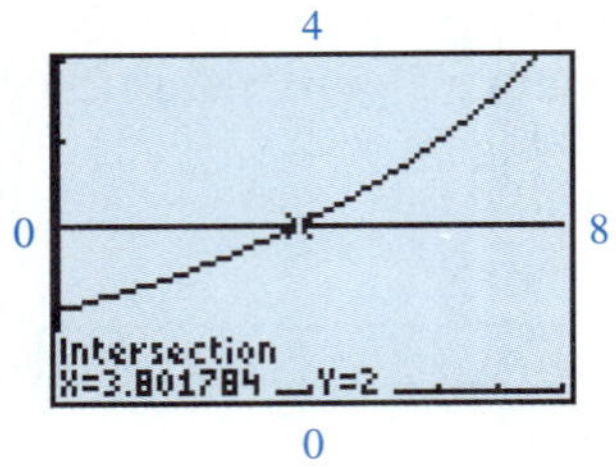

FIGURE 4
$y_1 = 1.2^x, y_2 = 2$

Example 10 can also be solved graphically by graphing both sides of the equation $2 = 1.2^t$, and finding the intersection point (Fig. 4).

Matched Problem 10 How long (to the next whole year) will it take money to triple if it is invested at 13% compounded annually?

It is interesting and instructive to graph the doubling times for various rates compounded annually. We proceed as follows:

$$\begin{aligned} A &= P(1 + r)^t \\ 2P &= P(1 + r)^t \\ 2 &= (1 + r)^t \\ (1 + r)^t &= 2 \\ \ln(1 + r)^t &= \ln 2 \\ t \ln(1 + r) &= \ln 2 \\ t &= \frac{\ln 2}{\ln(1 + r)} \end{aligned}$$

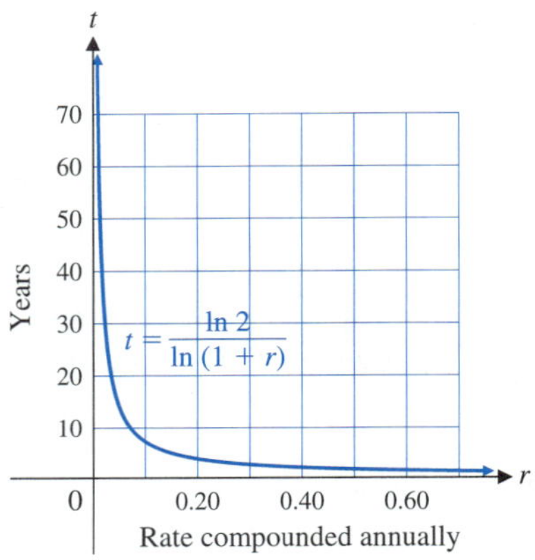

FIGURE 5
Doubling time (in years) at various rates of interest compounded annually.

Figure 5 shows the graph of this equation (doubling time in years) for interest rates compounded annually from 1% to 70% (expressed as decimals). Note the dramatic change in doubling time as rates change from 1% to 20% (from 0.01 to 0.20).

Answers to Matched Problems

1. (A) $9 = 3^2$ (B) $2 = 4^{1/2}$ (C) $\frac{1}{9} = 3^{-2}$
2. (A) $\log_7 49 = 2$ (B) $\log_9 3 = \frac{1}{2}$ (C) $\log_3(\frac{1}{3}) = -1$
3. (A) $y = \frac{3}{2}$ (B) $x = \frac{1}{3}$ (C) $b = 10$
4. (A) $\log_b R - \log_b S - \log_b T$ (B) $\frac{2}{3}(\log_b R - \log_b S)$ **5.** $x = 2$
6. $x = 5$ **7.** (A) $-1.868\ 734$ (B) $3.356\ 663$ (C) Not defined
8. (A) 0.0063 (B) 120.8092
9. (A) 0.8451 (B) 1.7918 (C) 1.1610 **10.** 6 yr

EXERCISE 2-3

A *Rewrite in equivalent exponential form:*

1. $\log_3 27 = 3$ **2.** $\log_2 32 = 5$ **3.** $\log_{10} 1 = 0$
4. $\log_e 1 = 0$ **5.** $\log_4 8 = \frac{3}{2}$ **6.** $\log_9 27 = \frac{3}{2}$

Rewrite in equivalent logarithmic form:

7. $49 = 7^2$ **8.** $36 = 6^2$ **9.** $8 = 4^{3/2}$
10. $9 = 27^{2/3}$ **11.** $A = b^u$ **12.** $M = b^x$

Evaluate each of the following without a calculator:

13. $\log_{10} 1$ **14.** $\log_e 1$ **15.** $\log_e e$
16. $\log_{10} 10$ **17.** $\log_{0.2} 0.2$ **18.** $\log_{13} 13$
19. $\log_{10} 10^3$ **20.** $\log_{10} 10^{-5}$ **21.** $\log_2 2^{-3}$
22. $\log_3 3^5$ **23.** $\log_{10} 1{,}000$ **24.** $\log_6 36$

Write in terms of simpler logarithmic forms, as in Example 4.

25. $\log_b \frac{P}{Q}$ **26.** $\log_b FG$ **27.** $\log_b L^5$

28. $\log_b w^{15}$ **29.** $\log_b \frac{p}{qrs}$ **30.** $\log_b PQR$

B *Find x, y, or b without a calculator.*

31. $\log_3 x = 2$ **32.** $\log_2 x = 2$

33. $\log_7 49 = y$ **34.** $\log_3 27 = y$

35. $\log_b 10^{-4} = -4$ **36.** $\log_b e^{-2} = -2$

37. $\log_4 x = \frac{1}{2}$ **38.** $\log_{25} x = \frac{1}{2}$

39. $\log_{1/3} 9 = y$ **40.** $\log_{49}(\frac{1}{7}) = y$

41. $\log_b 1{,}000 = \frac{3}{2}$ **42.** $\log_b 4 = \frac{2}{3}$

Write in terms of simpler logarithmic forms, going as far as you can with logarithmic properties (see Example 4).

43. $\log_b \frac{x^5}{y^3}$ **44.** $\log_b(x^2y^3)$

45. $\log_b \sqrt[3]{N}$ **46.** $\log_b \sqrt[5]{Q}$

47. $\log_b(x^2 \sqrt[3]{y})$ **48.** $\log_b \sqrt[3]{\frac{x^2}{y}}$

49. $\log_b(50 \cdot 2^{-0.2t})$ **50.** $\log_b(100 \cdot 1.06^t)$

51. $\log_b[P(1 + r)^t]$ **52.** $\log_e Ae^{-0.3t}$

53. $\log_e 100e^{-0.01t}$ **54.** $\log_{10}(67 \cdot 10^{-0.12x})$

Find x.

55. $\log_b x = \frac{2}{3} \log_b 8 + \frac{1}{2} \log_b 9 - \log_b 6$

56. $\log_b x = \frac{2}{3} \log_b 27 + 2 \log_b 2 - \log_b 3$

57. $\log_b x = \frac{3}{2} \log_b 4 - \frac{2}{3} \log_b 8 + 2 \log_b 2$

58. $\log_b x = 3 \log_b 2 + \frac{1}{2} \log_b 25 - \log_b 20$

59. $\log_b x + \log_b(x - 4) = \log_b 21$

60. $\log_b(x + 2) + \log_b x = \log_b 24$

61. $\log_{10}(x - 1) - \log_{10}(x + 1) = 1$

62. $\log_{10}(x + 6) - \log_{10}(x - 3) = 1$

Graph Problems 63 and 64 by converting to exponential form first.

63. $y = \log_2(x - 2)$ **64.** $y = \log_3(x + 2)$

65. Explain how the graph of the equation in Problem 63 can be obtained from the graph of $y = \log_2 x$ using a simple transformation (see Section 1-2).

66. Explain how the graph of the equation in Problem 64 can be obtained from the graph of $y = \log_3 x$ using a simple transformation (see Section 1-2).

67. What are the domain and range of the function defined by $y = 1 + \ln(x + 1)$?

68. What are the domain and range of the function defined by $y = \log(x - 1) - 1$?

Evaluate to five decimal places using a calculator.

69. (A) $\log 3{,}527.2$ (B) $\log 0.006\ 913\ 2$
(C) $\ln 277.63$ (D) $\ln 0.040\ 883$

70. (A) $\log 72.604$ (B) $\log 0.033\ 041$
(C) $\ln 40{,}257$ (D) $\ln 0.005\ 926\ 3$

Find x to four decimal places.

71. (A) $\log x = 1.1285$ (B) $\log x = -2.0497$
(C) $\ln x = 2.7763$ (D) $\ln x = -1.8879$

72. (A) $\log x = 2.0832$ (B) $\log x = -1.1577$
(C) $\ln x = 3.1336$ (D) $\ln x = -4.3281$

Solve each equation to four decimal places.

73. $10^x = 12$ **74.** $10^x = 153$

75. $e^x = 4.304$ **76.** $e^x = 0.3059$

77. $1.03^x = 2.475$ **78.** $1.075^x = 1.837$

79. $1.005^{12t} = 3$ **80.** $1.02^{4t} = 2$

Check Problems 73–80 by solving graphically.

Graph Problems 81–88 using a calculator and point-by-point plotting. Indicate increasing and decreasing intervals.

81. $y = \ln x$ **82.** $y = -\ln x$

83. $y = |\ln x|$ **84.** $y = \ln|x|$

85. $y = 2\ln(x + 2)$ **86.** $y = 2\ln x + 2$

87. $y = 4\ln x - 3$ **88.** $y = 4\ln(x - 3)$

Check Problems 81–88 by graphing on a graphing utility.

C

89. Explain why the calculator display shown does not contradict the logarithmic property

$$\log \frac{M}{N} = \log M - \log N$$

Figure for 89

90. Explain why the calculator display shown does not contradict the logarithmic property

$$\log MN = \log M + \log N$$

```
log 6*9
        7.003361253
log 6+log 9
         1.73239376
```

Figure for 90

91. Explain why the logarithm of 1 for any permissible base is 0.

92. Explain why 1 is not a suitable logarithmic base.

93. Write $\log_{10} y - \log_{10} c = 0.8x$ in an exponential form that is free of logarithms.

94. Write $\log_e x - \log_e 25 = 0.2t$ in an exponential form that is free of logarithms.

95. Let $p(x) = \ln x$, $q(x) = \sqrt{x}$, and $r(x) = x$. Use a graphing utility to draw graphs of all three functions in the same viewing window for $1 \leqslant x \leqslant 16$. Discuss what it means for one function to be larger than another on an interval, and then order the three functions from largest to smallest for $1 < x \leqslant 16$.

96. Let $p(x) = \log x$, $q(x) = \sqrt[3]{x}$, and $r(x) = x$. Use a graphing utility to draw graphs of all three functions in the same viewing window for $1 \leqslant x \leqslant 16$. Discuss what it means for one function to be smaller than another on an interval, and then order the three functions from smallest to largest for $1 < x \leqslant 16$.

Applications

Business & Economics

97. *Doubling time.* How long (to the next whole year) will it take money to double if it is invested at 6% interest compounded annually?

98. *Doubling time.* How long (to the next whole year) will it take money to double if it is invested at 3% interest compounded annually?

99. *Investing.* How many years (to two decimal places) will it take \$1,000 to grow to \$1,800 if it is invested at 6% compounded quarterly? Compounded continuously?

100. *Investing.* How many years (to two decimal places) will it take \$5,000 to grow to \$7,500 if it is invested at 8% compounded semiannually? Compounded continuously?

101. *Investment.* A newly married couple wishes to have \$20,000 in 8 years for the down payment on a house. At what rate of interest compounded continuously (to three decimal places) must \$10,000 be invested now to accomplish this goal?

102. *Investment.* The parents of a newborn child want to have \$45,000 for the child's college education 17 years from now. At what rate of interest compounded continuously (to three decimal places) must a grandparent's gift of \$10,000 be invested now to achieve this goal?

103. *Supply and demand.* A cordless screwdriver is sold through a national chain of discount stores. A marketing company established price–demand and price–supply tables (Tables 1 and 2), where x is the number of screwdrivers people are willing to buy and the store is willing to sell each month at a price of p dollars per screwdriver.

Table 1

Price–Demand

x	$p = D(x)$ (\$)
1,000	91
2,000	73
3,000	64
4,000	56
5,000	53

Table 2

Price–Supply

x	$p = S(x)$ (\$)
1,000	9
2,000	26
3,000	34
4,000	38
5,000	41

(A) Find a logarithmic regression model ($y = a + b \ln x$) for the data in Table 1. Estimate the demand (to the nearest unit) at a price level of \$50.

(B) Find a logarithmic regression model ($y = a + b \ln x$) for the data in Table 2. Estimate the supply (to the nearest unit) at a price level of \$50.

(C) Does a price level of \$50 represent a stable condition, or is the price likely to increase or decrease? Explain.

104. *Equilibrium point.* Use the models constructed in Problem 103 to find the equilibrium point. Write the equilibrium price to the nearest cent and the equilibrium quantity to the nearest unit.

Life Sciences

105. *Sound intensity—decibels.* Because of the extraordinary range of sensitivity of the human ear (a range of over 1,000 million millions to 1), it is helpful to use a logarithmic scale, rather than an absolute scale, to measure sound intensity over this range. The unit of measure is called the *decibel,* after the inventor of the telephone, Alexander Graham Bell. If we let N be the number of decibels, I the power of the sound in question (in watts per square centimeter), and I_0 the power of sound just below the threshold of hearing (approximately 10^{-16} watt per square centimeter), then

$$I = I_0 10^{N/10}$$

Show that this formula can be written in the form

$$N = 10 \log \frac{I}{I_0}$$

106. *Sound intensity—decibels.* Use the formula in Problem 105 (with $I_0 = 10^{-16}$ watt/cm^2) to find the decibel ratings of the following sounds:

(A) Whisper: 10^{-13} watt/cm^2
(B) Normal conversation: 3.16×10^{-10} watt/cm^2
(C) Heavy traffic: 10^{-8} watt/cm^2
(D) Jet plane with afterburner: 10^{-1} watt/cm^2

107. C *Agriculture.* Table 3 shows the yield (in bushels per acre) and the total production (in millions of bushels) for corn in the United States for selected years since 1950. Let x represent years since 1900.

Table 3

UNITED STATES CORN PRODUCTION

YEAR	x	YIELD (BUSHELS PER ACRE)	TOTAL PRODUCTION (MILLION BUSHELS)
1950	50	37.6	2,782
1960	60	55.6	3,479
1970	70	81.4	4,802
1980	80	97.7	6,867
1990	90	115.6	7,802

Source: U.S. Department of Agriculture

(A) Find a logarithmic regression model ($y = a + b \ln x$) for the yield. Estimate (to one decimal place) the yield in 1996 and in 2010.

(B) The actual yield in 1996 was 127.1 bushels per acre. How does this compare with the estimated yield in part (A)? What effect will this additional 1996 information have on the estimate for 2010? Explain.

108. C *Agriculture.* Refer to Table 3.

(A) Find a logarithmic regression model ($y = a + b \ln x$) for the total production. Estimate (to the nearest million) the production in 1996 and in 2010.

(B) The actual production in 1996 was 7,949 billion bushels. How does this compare with the estimated production in part (A)? What effect will this 1996 production information have on the estimate for 2010? Explain.

Social Sciences

109. *World population.* If the world population is now 5.8 billion people and if it continues to grow at an annual rate of 1.14% compounded continuously, how long (to the nearest year) will it take before there is only 1 square yard of land per person? (The Earth contains approximately 1.68×10^{14} square yards of land.)

110. *Archaeology—carbon-14 dating.* The radioactive carbon-14 (^{14}C) in an organism at the time of its death decays according to the equation

$$A = A_0 e^{-0.000124t}$$

where t is time in years and A_0 is the amount of ^{14}C present at time $t = 0$. (See Example 3 in Section 2-2.) Estimate the age of a skull uncovered in an archaeological site if 10% of the original amount of ^{14}C is still present. [*Hint:* Find t such that $A = 0.1A_0$.]

Important Terms and Symbols

2-1 *Polynomial and Rational Functions.* Polynomial function; degree; continuity; turning point; root; zero; leading coefficient; rational function; points o discontinuity; vertical and horizontal asymptotes

$$f(x) = a_n x^n + a_{n-1}x^{n-1} + \cdots + a_1 x + a_0, a_n \neq 0;$$

$$f(x) = \frac{n(x)}{d(x)}, d(x) \neq 0$$

2-2 *Exponential Functions.* Exponential function; base; basic graphs; horizontal asymptote; basic properties; irrational number *e;* exponential function with base *e;* exponential growth; exponential decay; compound interest; principal (present value); amount (future value); continuous compound interest

$$f(x) = b^x, b > 0, b \neq 1; \quad y = e^x; \quad N = N_0 e^{kt};$$

$$A = A_0 e^{-kt}; \quad A = P\left(1 + \frac{r}{m}\right)^{mt}; \quad A = Pe^{rt}$$

2-3 *Logarithmic Functions.* Inverse functions; one-to-one functions; logarithmic function; base; equivalent exponential form; properties; common logarithm; natural logarithm; calculator evaluation; solving logarithmic and exponential equations; doubling time

$y = \log_b x$ is equivalent to $x = b^y$;

$\log_b x, b > 0, b \neq 1; \quad \log x; \quad \ln x$

Review Exercise

Work through all the problems in this chapter review and check your answers in the back of the book. Answers to all review problems are there along with section numbers in italics to indicate where each type of problem is discussed. Where weaknesses show up, review appropriate sections in the text.

A

1. Write in logarithmic form using base *e*: $u = e^v$

2. Write in logarithmic form using base 10: $x = 10^y$

3. Write in exponential form using base *e*: $\ln M = N$

4. Write in exponential form using base 10: $\log u = v$

Simplify:

5. $\dfrac{5^{x+4}}{5^{4-x}}$ **6.** $\left(\dfrac{e^u}{e^{-u}}\right)^u$

Solve for x exactly without using a calculator.

7. $\log_3 x = 2$ **8.** $\log_x 36 = 2$

9. $\log_2 16 = x$

Solve for x to three decimal places.

10. $10^x = 143.7$ **11.** $e^x = 503{,}000$

12. $\log x = 3.105$ **13.** $\ln x = -1.147$

For each polynomial function in Problems 14 and 15, find the following:

(A) The degree of the polynomial
(B) The maximum number of turning points of the graph
(C) The maximum number of x intercepts of the graph
(D) The minimum number of x intercepts of the graph
(E) The maximum number of y intercepts of the graph
(F) The minimum number of y intercepts of the graph

14. $p(x) = ax^3 + bx^2 + cx + d, a \neq 0$

15. $p(x) = ax^4 + bx^3 + cx^2 + dx + e, a \neq 0$

Each graph in Problems 16 and 17 is the graph of a polynomial function. Answer the following questions for each graph:

(A) How many turning points are on the graph?
(B) What is the minimum degree of a polynomial function that could have the graph?
(C) Is the leading coefficient of the polynomial negative or positive?

16.

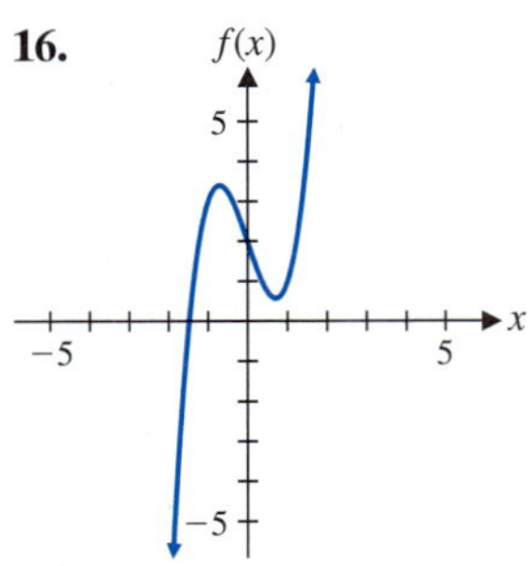

17.

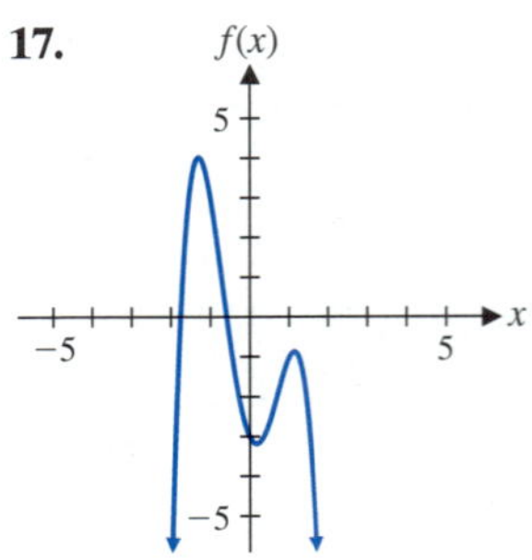

B *For each rational function in Problems 18 and 19:*

(A) *Find the intercepts for the graph.*
(B) *Determine the domain.*
(C) *Find any vertical or horizonal asymptotes for the graph.*
(D) *Sketch any asymptotes as dashed lines. Then sketch a graph of f for* $-10 \leq x \leq 10$ *and* $-10 \leq y \leq 10$.
(E) *Graph* $y = f(x)$ *in a standard viewing window using a* **C** *graphing utility.*

18. $f(x) = \dfrac{x + 4}{x - 2}$ **19.** $f(x) = \dfrac{3x - 4}{2 + x}$

Solve for x exactly without using a calculator.

20. $\log(x + 5) = \log(2x - 3)$
21. $2 \ln(x - 1) = \ln(x^2 - 5)$
22. $9^{x-1} = 3^{1+x}$ **23.** $e^{2x} = e^{x^2-3}$
24. $2x^2e^x = 3xe^x$ **25.** $\log_{1/3} 9 = x$
26. $\log_x 8 = -3$ **27.** $\log_9 x = \frac{3}{2}$

Solve Problems 28–37 for x to four decimal places.

28. $x = 3(e^{1.49})$ **29.** $x = 230(10^{-0.161})$
30. $\log x = -2.0144$ **31.** $\ln x = 0.3618$
32. $35 = 7(3^x)$ **33.** $0.01 = e^{-0.05x}$
34. $8{,}000 = 4{,}000(1.08^x)$ **35.** $5^{2x-3} = 7.08$
36. $x = \log_2 7$ **37.** $x = \log_{0.2} 5.321$

38. How does the graph of $f(x) = x^4 - 4x^2 + 1$ compare to the graph of $y = x^4$ as we "zoom out"?

39. **C** Compare the graphs of $y = x^4$ and $y = x^4 - 4x^2 + 1$ in the following two viewing windows:
(A) $-5 \leq x \leq 5, -5 \leq y \leq 5$
(B) $-5 \leq x \leq 5, -500 \leq y \leq 500$

40. **C** Let $p(x) = 2x^4 - 11x^3 - 15x^2 - 14x - 16$. Approximate the real zeros of $p(x)$ to two decimal places.

41. **C** Let $f(x) = e^x - 1$ and $g(x) = \ln(x + 2)$. Find all points of intersection for the graphs of f and g. Round answers to two decimal places.

Simplify.

42. $e^x(e^{-x} + 1) - (e^x + 1)(e^{-x} - 1)$
43. $(e^x - e^{-x})^2 - (e^x + e^{-x})(e^x - e^{-x})$

Graph Problems 44–46 over the indicated interval. Indicate increasing and decreasing intervals.

44. $y = 2^{x-1}; [-2, 4]$ **45.** $f(t) = 10e^{-0.08t}; t \geq 0$
46. $y = \ln(x + 1); (-1, 10]$

C

47. **C** Noting that $\pi = 3.141\ 592\ 654\ldots$ and $\sqrt{2} = 1.414\ 213\ 562\ldots$, explain why the calculator results shown here are obvious. Discuss similar connections between the natural logarithmic function and the exponential function with base e.

Figure for 47

Solve Problems 48–51 exactly without using a calculator.

48. $\log x - \log 3 = \log 4 - \log(x + 4)$
49. $\ln(2x - 2) - \ln(x - 1) = \ln x$
50. $\ln(x + 3) - \ln x = 2 \ln 2$
51. $\log 3x^2 = 2 + \log 9x$

52. Write $\ln y = -5t + \ln c$ in an exponential form free of logarithms. Then solve for y in terms of the remaining variables.

53. Explain why 1 cannot be used as a logarithmic base.

Applications

Business & Economics

The two formulas below will be of use in some of the problems that follow:

$$A = P\left(1 + \frac{r}{m}\right)^{mt} \quad \text{Compound interest}$$

$$A = Pe^{rt} \quad \text{Continuous compound interest}$$

54. *Money growth.* If \$5,000 is invested at 12% compounded weekly, how much (to the nearest cent) will be in the account 6 years from now?

55. *Money growth.* If \$5,000 is invested at 12% compounded continuously, how much (to the nearest cent) will be in the account 6 years from now?

56. *Finance.* Find the tripling time (to the next whole year) for money invested at 15% compounded annually.

57. *Finance.* Find the doubling time (to two decimal places) for money invested at 10% compounded continuously.

58. *Minimum average cost.* The financial department of a company that manufactures in-line skates has fixed costs of \$300 per day and total costs of \$4,300 per day at an output of 100 pairs of skates per day. Assume the cost $C(x)$ is linearly related to output x.

(A) Find an expression for the cost function $C(x)$ and the average cost function $\overline{C}(x) = C(x)/x$.
(B) Sketch a graph of the average cost function for $5 \le x \le 200$.
(C) Identify any asymptotes.
(D) What does the average cost approach as production increases?

59. C *Minimum average cost.* The cost $C(x)$ in thousands of dollars for operating a hospital for a year is given by

$$C(x) = 20x^3 - 360x^2 + 2{,}300x - 1{,}000$$

where x is the number of cases per year (in thousands). The average cost function $\overline{C}$ is given by $\overline{C}(x) = C(x)/x$.
(A) Write an equation for the average cost function.
(B) Graph the average cost function for $1 \le x \le 12$.
(C) Use trace and zoom or a built-in routine to find the number of cases per year the hospital should handle to have the minimum average cost. What is the minimum average cost?

60. C *Equilibrium point.* A company is planning to introduce a ten-piece set of nonstick cookware. A marketing company established price–demand and price–supply tables for selected prices (Tables 1 and 2), where x is the number of cookware sets people are willing to buy and the company is willing to sell each month at a price of p dollars per set.

Table 1
PRICE–DEMAND

x	$p = D(x)$ (\$)
985	330
2,145	225
2,950	170
4,225	105
5,100	50

Table 2
PRICE–SUPPLY

x	$p = S(x)$ (\$)
985	30
2,145	75
2,950	110
4,225	155
5,100	190

(A) Find a quadratic regression model for the data in Table 1. Estimate the demand at a price level of \$180.
(B) Find a linear regression model for the data in Table 2. Estimate the supply at a price level of \$180.
(C) Does a price level of \$180 represent a stable condition, or is the price likely to increase or decrease? Explain.
(D) Use the models in parts (A) and (B) to find the equilibrium point. Write the equilibrium price to the nearest cent and the equilibrium quantity to the nearest unit.

Life Sciences

61. *Medicine.* One leukemic cell injected into a healthy mouse will divide into 2 cells in about $\frac{1}{2}$ day. At the end of the day these 2 cells will divide into 4. This doubling continues until 1 billion cells are formed; then the animal dies with leukemic cells in every part of the body.
(A) Write an equation that will give the number N of leukemic cells at the end of t days.
(B) When, to the nearest day, will the mouse die?

62. *Marine biology.* The intensity of light entering water is reduced according to the exponential equation

$$I = I_0 e^{-kd}$$

where I is the intensity d feet below the surface, I_0 is the intensity at the surface, and k is the coefficient of extinction. Measurements in the Sargasso Sea in the West Indies have indicated that half of the surface light reaches a depth of 73.6 feet. Find k (to five decimal places), and find the depth (to the nearest foot) at which 1% of the surface light remains.

63. C *Agriculture.* The total United States corn consumption (in millions of bushels) is shown in Table 3 for selected years since 1975. Let x represent years since 1900.

Table 3
CORN CONSUMPTION

YEAR	x	TOTAL CONSUMPTION (MILLION BUSHELS)
1975	75	522
1980	80	659
1985	85	1,152
1990	90	1,373
1995	95	1,690

Source: U.S. Department of Agriculture

(A) Find a logarithmic regression model ($y = a + b \ln x$) for the data. Estimate (to the nearest million bushels) the total consumption in 1996 and in 2010.

(B) The actual consumption in 1996 was 1,583 million bushels. How does this compare with the estimated consumption in part (A)? What effect will this additional 1996 information have on the estimate for 2010? Explain.

Social Sciences

64. *Population growth.* Many countries have a population growth rate of 3% (or more) per year. At this rate, how many years (to the nearest tenth of a year) will it take a population to double? Use the annual compounding growth model $P = P_0(1 + r)^t$.

65. *Population growth.* Repeat Problem 64 using the continuous compounding growth model $P = P_0e^{rt}$.

66. C *Medicare.* The annual expenditures for Medicare (in billions of dollars) by the United States government for selected years since 1980 are shown in Table 4. Let x represent years since 1980.

Table 4

MEDICARE EXPENDITURES

YEAR	BILLION $
1980	37
1985	72
1990	111
1995	181

Source: U.S. Bureau of the Census

(A) Find an exponential regression model ($y = ab^x$) for the data. Estimate (to the nearest billion) the total expenditures in 1996 and in 2010.

(B) When will the total expenditures reach 500 billion dollars?

Group Activity 1 Comparing the Growth of Exponential and Polynomial Functions, and Logarithmic and Root Functions

(A) An exponential function such as $f(x) = 2^x$ increases extremely rapidly for large values of x, more rapidly than any polynomial function. Show that the graphs of $f(x) = 2^x$ and $g(x) = x^2$ intersect three times. The intersection points divide the x axis into four regions. Describe which function is greater than the other relative to each region.

(B) A logarithmic function such as $r(x) = \ln x$ increases extremely slowly for large values of x, more slowly than a function like $s(x) = \sqrt[3]{x}$. Sketch graphs of both functions in the same coordinate system for $x > 0$, and determine how many times the two graphs intersect. Describe which function is greater than the other relative to the regions determined by the intersection points.

Group Activity 2 *Comparing Regression Models*

We have used polynomial, exponential, and logarithmic regression models to fit curves to data sets. And there are other equations that can be used for curve fitting. (The TI-83 graphing calculator has twelve different equations on its STAT-CALC menu.) How can we determine which equation provides the best fit for a given set of data? There are two principal ways to select models. The first is to use information about the type of data to help make a choice. For example, we expect the weight of a fish to be related to the cube of its length. And we expect most populations to grow exponentially, at least over the short term. The second method for choosing between equations involves developing a measure of how close an equation fits a given data set. This is best introduced through an example. Consider the data set in Figure 1, where L1 represents the x coordinates and L2 represents the y coordinates. The graph of this data set is shown in Figure 2. Suppose we arbitrarily choose the equation $y_1 = 0.6x + 2$ to model the data (Fig. 3).

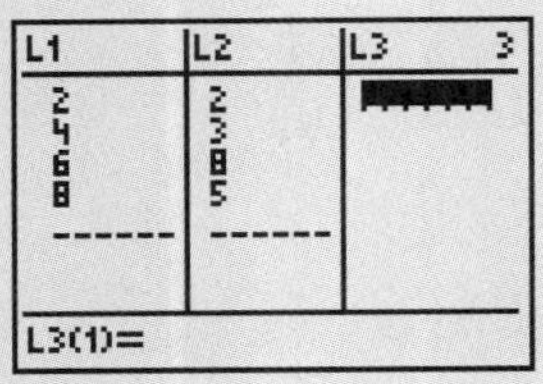

FIGURE 1

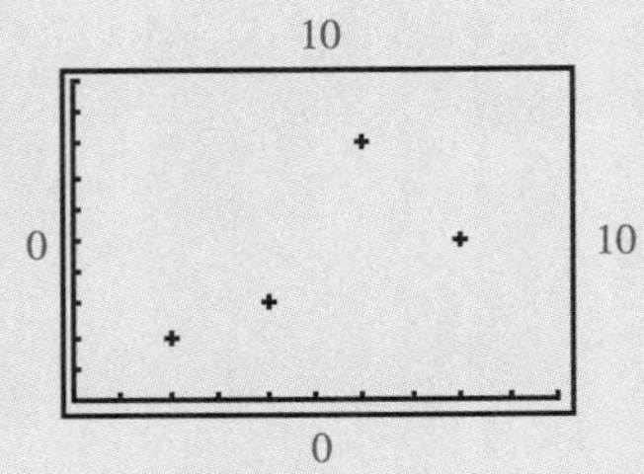

FIGURE 2

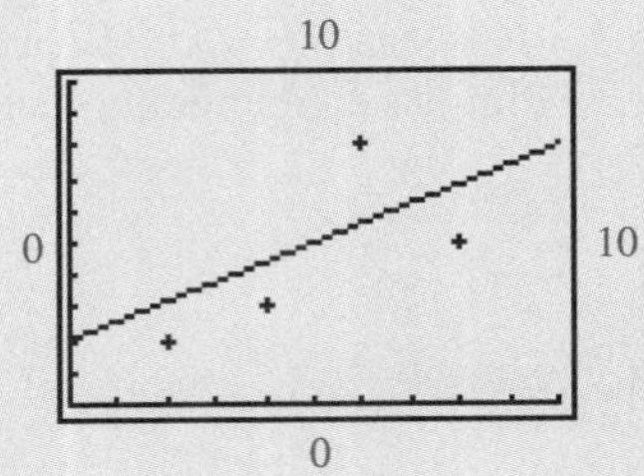

FIGURE 3
$y_1 = 0.6x + 2$

To measure how well the graph of y_1 fits the data, we examine the difference between the y coordinates in the data set and the corresponding y coordinates on the graph of y_1 (L3 in Figs. 4 and 5). Each of these differences is called a **residual.** The most commonly accepted measure of the fit provided by a given model is the **sum of the squares of the residuals (SSR).** Computing this quantity is a simple matter on a graphing calculator (Fig. 6) or a spreadsheet (Fig. 7).

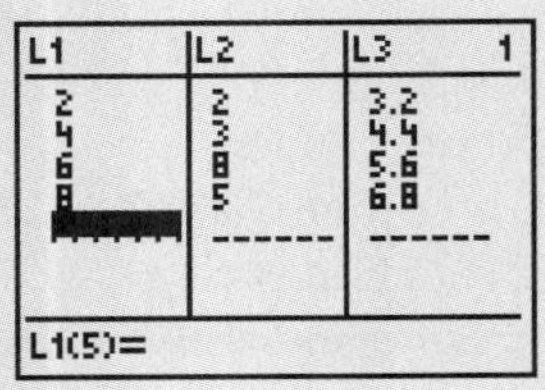

FIGURE 4

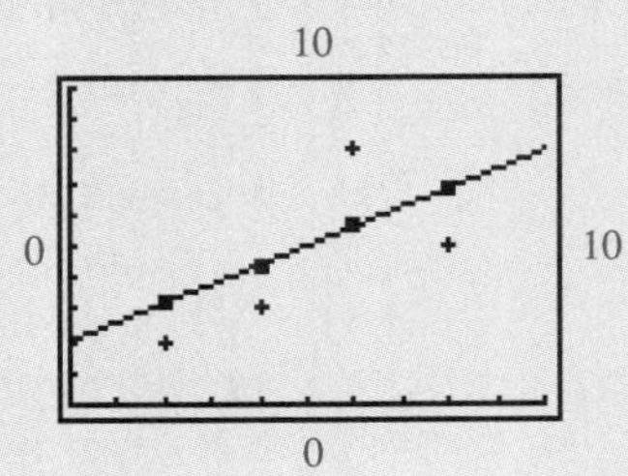

FIGURE 5
+ L2; ■ L3

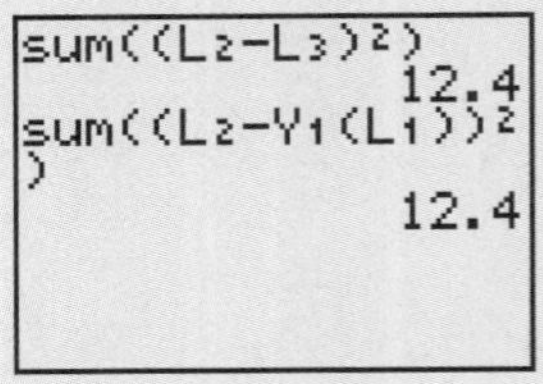

FIGURE 6
Two ways to calculate SSR

	A	B	C	D	E
1	Data Set				
2	x	y	y1=0.6x + 2	Residual	Residual^2
3	2	2	3.2	-1.2	1.44
4	4	3	4.4	-1.4	1.96
5	6	8	5.6	2.4	5.76
6	8	5	6.8	-1.8	3.24
7				SSR	12.4

FIGURE 7

(A) Find the linear regression model for the data in Figure 1, compute the SSR for this equation, and compare it with the one we computed for y_1.

It turns out that among all possible linear polynomials, **the linear regression model minimizes the sum of the squares of the residuals.** For this reason, the linear regression model is often called the **least squares line.** A similar statement can be made for polynomials of any fixed degree. That is, the quadratic regression model minimizes the SSR over all quadratic polynomials, the cubic regression model minimizes the SSR over all cubic polynomials, and so on. The same statement cannot be made for exponential or logarithmic regression models. Nevertheless, the SSR can still be used to compare exponential, logarithmic, and polynomial models.

(B) Find the exponential and logarithmic regression models for the data in Figure 1, compute their SSR's, and compare with the linear model.

(C) National annual advertising expenditures for selected years since 1950 are shown in Table 1, where x is years since 1950 and y is total expenditures in billions of dollars. Which regression model would fit the data best: a quadratic model, a cubic model, or an exponential model? Use the SSR's to support your choice.

Table 1

ANNUAL ADVERTISING EXPENDITURES, 1950–1995

x (YEARS)	0	5	10	15	20	25	30	35	40	45
y (BILLION $)	5.7	9.2	12.0	15.3	19.6	27.9	53.6	94.8	128.6	160.9

Source: U.S. Bureau of the Census

PART TWO

FINITE MATHEMATICS

CHAPTER 3

MATHEMATICS OF FINANCE

INTRODUCTION

This chapter is independent of the others; you can study it at any time. In particular, we do not assume that you have studied Chapter 2, where a few of the topics in this chapter were briefly discussed as applications of exponential and logarithmic functions.

The low cost and convenience of the calculators that are currently available make them excellent tools for solving problems on compound interest, annuities, amortization, and so on. Any calculator with logarithmic and exponentiation keys is sufficient for solving the problems in this chapter. A graphing utility offers the additional advantage of enabling us to visualize the rate at which an investment grows, or the rate at which the principal on a loan is amortized.

If time permits, you may wish to cover arithmetic and geometric sequences, discussed in Appendix B-2, before beginning this chapter. Though not necessary, these topics will provide additional insight into some of the topics covered.

To avoid repeating the statement many times, we now point out:

Throughout the chapter, interest rates are to be converted to decimal form before they are used in a formula.

SECTION 3-1

Simple Interest

Simple interest is generally used only on short-term notes—often of duration less than 1 year. The concept of simple interest, however, forms the basis of much of the rest of the material developed in this chapter, for which time periods may be much longer than a year.

If you deposit a sum of money P in a savings account or if you borrow a sum of money P from a lending agent, then P is referred to as the **principal.** When money is borrowed—whether it is a savings institution borrowing from you when you deposit money in your account or you borrowing from a lending agent—a fee is charged for the money borrowed. This fee is rent paid for the use of another's money, just as rent is paid for the use of another's house. The fee is called **interest.** It is usually computed as a percentage (called the **interest rate**)* of the principal over a given period of time. The interest rate, unless otherwise stated, is an annual rate. **Simple interest** is given by the following formula:

Simple Interest

$$I = Prt \tag{1}$$

where

P = Principal
r = Annual simple interest rate (written as a decimal)
t = Time in years

For example, the interest on a loan of \$100 at 12% for 9 months would be

$$\begin{aligned} I &= Prt \\ &= (100)(0.12)(0.75) \\ &= \$9 \end{aligned}$$

Convert 12% to a decimal (0.12) and 9 months to years ($\frac{9}{12} = 0.75$).

At the end of 9 months, the borrower would repay the principal (\$100) plus the interest (\$9), or a total of \$109.

In general, if a principal P is borrowed at a rate r, then after t years the borrower will owe the lender an amount A that will include the principal P (the **face value** of the note) plus the interest I (the rent paid for the use of the money). Since P is the amount that is borrowed now and A is the amount that must be paid back in the future, P is often referred to as the **present value** and A as the **future value.** The formula relating A and P is as follows:

Amount—Simple Interest

$$\begin{aligned} A &= P + Prt \\ &= P(1 + rt) \end{aligned} \tag{2}$$

where

P = Principal, or present value
r = Annual simple interest rate (written as a decimal)
t = Time in years
A = Amount, or future value

*If r is the interest rate written as a decimal, then $100r\%$ is the rate using %. For example, if $r = 0.12$, then using the percent symbol, %, we have $100r\% = 100(0.12)\% = 12\%$. The expressions 0.12 and 12% are equivalent.

Given any three of the four variables A, P, r, and t in (2), we can solve for the fourth. The following examples illustrate several types of common problems that can be solved by using formula (2).

Example 1 **Total Amount Due on a Loan** Find the total amount due on a loan of \$800 at 18% simple interest at the end of 4 months.

SOLUTION To find the amount A (future value) due in 4 months, we use formula (2) with $P = 800$, $r = 0.18$, and $t = \frac{4}{12} = \frac{1}{3}$ year. Thus,

$$\begin{aligned} A &= P(1 + rt) \\ &= 800[1 + 0.18(\tfrac{1}{3})] \\ &= 800(1.06) \\ &= \$848 \end{aligned}$$

Matched Problem 1 Find the total amount due on a loan of \$500 at 12% simple interest at the end of 30 months.

Explore–Discuss 1

(A) Your dear sister has loaned you \$1,000 with the understanding that the principal plus 4% simple interest are to be repaid when you are able. How much would you owe her if you repaid the loan after 1 year? After 2 years? After 5 years? After 10 years?

(B) How is the interest after 10 years related to the interest after 1 year? After 2 years? After 5 years?

(C) Explain why your answers are consistent with the fact that for simple interest the graph of future value as a function of time is a straight line (see Fig. 1).

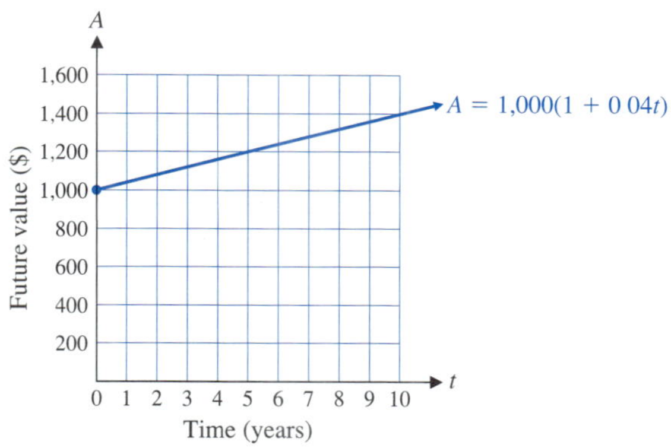

FIGURE 1

Example 2 **Present Value of an Investment** If you want to earn an annual rate of 10% on your investments, how much (to the nearest cent) should you pay for a note that will be worth \$5,000 in 9 months?

SOLUTION We again use formula (2), but now we are interested in finding the principal P (present value), given $A = \$5{,}000$, $r = 0.1$, and $t = \frac{9}{12} = 0.75$ year. Thus,

$$\begin{aligned} A &= P(1 + rt) \\ 5{,}000 &= P[1 + 0.1(0.75)] \\ 5{,}000 &= (1.075)P \\ P &= \$4{,}651.16 \end{aligned}$$

Replace A, r, and t with the given values, and solve for P.

Matched Problem 2 Repeat Example 2 with a time period of 6 months.

Example 3 **Interest Rate Earned on a Note** If you must pay \$960 for a note that will be worth \$1,000 in 6 months, what annual simple interest rate will you earn? (Express the answer as a percentage, correct to two decimal places.)

SOLUTION Again we use formula (2), but this time we are interested in finding r, given $P = \$960, A = \$1{,}000$, and $t = \frac{6}{12} = 0.5$ year. Thus,

$$\begin{aligned} A &= P(1 + rt) \\ 1{,}000 &= 960[1 + r(0.5)] \\ 1{,}000 &= 960 + 960r(0.5) \\ 40 &= 480r \\ r &= \tfrac{40}{480} \approx 0.0833 \quad \text{or} \quad 8.33\% \end{aligned}$$

Replace P, A, and t with the given values, and solve for r.

Matched Problem 3 Repeat Example 3 assuming you have paid \$952 for the note.

REMARK

It certainly isn't necessary to use a graphing utility to solve Example 3. However, in later sections we will encounter more complicated formulas where using a graphing utility may be desirable and, in some cases, absolutely necessary. There are a number of different ways to solve equations like $1{,}000 = 960(1 + 0.5r)$, depending on the type of graphing utility you are using. Figure 2A shows the standard graphical approach of graphing both sides of the equation and finding the intersection point. Figure 2B illustrates the output from using an equation solver on a graphing calculator, and Figure 2C shows similar output from a spreadsheet. Since the details for using an equation solver vary greatly from one graphing utility to another, you should consult the manual for your graphing utility or one of the manuals that accompany this book (see Preface) for specific instructions. Use some of the problems in this section to practice solution techniques on your graphing utility before encountering more complicated problems later in this book.

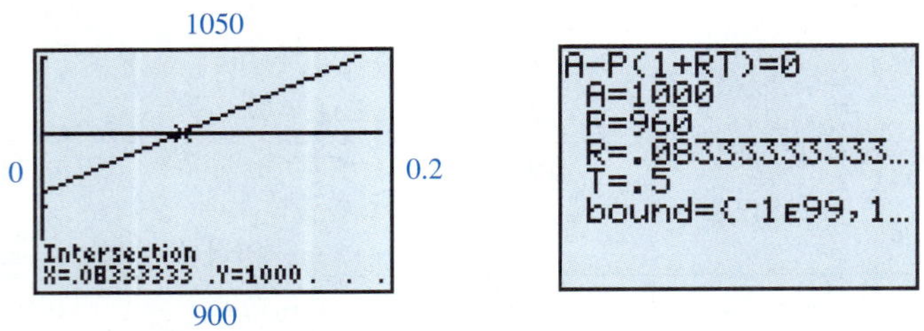

(A) Graphical approximation (B) Equation solver

	A	B
1	Amount:	\$1,000.00
2	Principal:	\$960.00
3	Rate:	0.08333333
4	Time:	0.5

(C) Spreadsheet

FIGURE 2
Various methods for solving equations on a graphing utility

Example 4 **Interest Rate Earned on an Investment** Suppose after buying a new car you decide to sell your old car to a friend. You accept a 270 day note for \$3,500 at 10% simple interest as payment. (Both principal and interest will be paid at the end of 270 days.) Sixty days later you find that you need the money and sell the note to a third party for \$3,550. What annual interest rate will the third party receive for the investment? (Express the answer as a percentage, correct to three decimal places.)

Solution **Step 1.** Find the amount that will be paid at the end of 270 days to the holder of the note. Some financial institutions use a 365 day year and others a 360 day year. In all the problems in this section involving days, we assume a 360 day year.*

$$\begin{aligned} A &= P(1 + rt) \\ &= \$3{,}500[1 + (0.1)(\tfrac{270}{360})] \\ &= \$3{,}762.50 \end{aligned}$$

Step 2. For the third party we are to find the annual rate of interest r required to make \$3,550 grow to \$3,762.50 in 210 days (270 − 60); that is, we are to find r (which is to be converted to $100r\%$), given $A = \$3{,}762.50$, $P = \$3{,}550$, and $t = \frac{210}{360}$.

$$A = P + Prt \qquad \text{Solve for } r.$$

$$r = \frac{A - P}{Pt}$$

$$r = \frac{3{,}762.50 - 3{,}550}{(3{,}550)(\frac{210}{360})} = 0.102\ 62 \quad \text{or} \quad 10.262\%$$

Matched Problem 4 Repeat Example 4 assuming that 90 days after it was initially signed, the note was sold to a third party for \$3,500.

Explore–Discuss 2

(A) Starting with formula (2), derive each of the following formulas:

$$P = \frac{A}{1 + rt} \qquad r = \frac{A - P}{Pt} \qquad t = \frac{A - P}{Pr}$$

(B) Explain why it is unnecessary to memorize the above formulas for P, r, and t if you know formula (2).

Answers to Matched Problems **1.** \$650 **2.** \$4,761.90 **3.** 10.08% **4.** 15.0%

Exercise 3-1

A *In Problems 1–4, make the indicated conversions assuming a 360 day year.*

1. 9.5% = ? (decimal); 60 days = ? year
2. 8.75% = ? (decimal); 3 quarters = ? year
3. 0.18 = ? (percentage); 5 months = ? year
4. 0.0525 = ? (percentage); 240 days = ? year

Using formula (1) for simple interest, find each of the indicated quantities in Problems 5–8.

5. $P = \$500$; $r = 8\%$; $t = 6$ months; $I = ?$
6. $P = \$900$; $r = 10\%$; $t = 9$ months; $I = ?$
7. $I = \$80$; $P = \$500$; $t = 2$ years; $r = ?$
8. $I = \$40$; $P = \$400$; $t = 4$ years; $r = ?$

B *Use formula (2) in an appropriate form to find the indicated quantities in Problems 9–12.*

9. $P = \$100$; $r = 8\%$; $t = 18$ months; $A = ?$
10. $P = \$6{,}000$; $r = 6\%$; $t = 8$ months; $A = ?$
11. $A = \$1{,}000$; $r = 10\%$; $t = 15$ months; $P = ?$
12. $A = \$8{,}000$; $r = 12\%$; $t = 7$ months; $P = ?$

Check Problems 11 and 12 by solving on a graphing utility.

*In other sections we will use a 365 day year. The choice will always be clearly stated.

C *In Problems 13–16, solve each formula for the indicated variable.*

13. $I = Prt$; for r

14. $I = Prt$; for P

15. $A = P + Prt$; for P

16. $A = P + Prt$; for r

17. Discuss the similarities and differences in the graphs of future value A as a function of time t if $1,000 is invested at simple interest at rates of 4%, 8%, and 12%, respectively (see the figure).

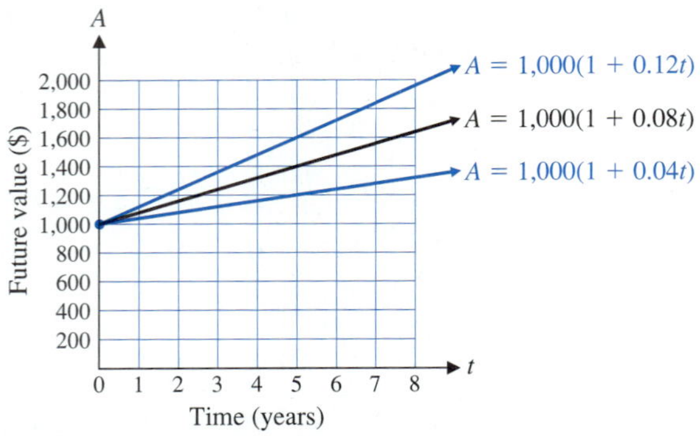

Figure for 17

18. Discuss the similarities and differences in the graphs of future value A as a function of time t for loans of $400, $800, and $1,200, respectively, each at 7.5% simple interest (see the figure).

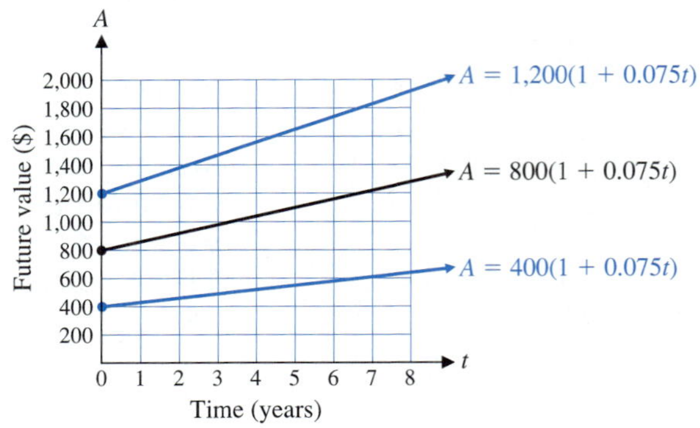

Figure for 18

APPLICATIONS*

Business & Economics

In all problems involving days, a 360 day year is assumed. When annual rates are requested as an answer, express the rate as a percentage, correct to three decimal places.

19. If $3,000 is loaned for 4 months at a 14% annual rate, how much interest is earned?

20. If $5,000 is loaned for 10 months at a 10% annual rate, how much interest is earned?

21. How much interest will you have to pay for a credit card balance of $554 that is 1 month overdue, if a 20% annual rate is charged?

22. A department store charges an 18% annual rate for overdue accounts. How much interest will be owed on an $835 account that is 2 months overdue?

23. A loan of $7,250 was repaid at the end of 8 months. What size repayment check (principal and interest) was written, if a 9% annual rate of interest was charged?

24. A loan of $10,000 was repaid at the end of 14 months. What amount (principal and interest) was repaid, if a 12% annual rate of interest was charged?

25. A loan of $4,000 was repaid at the end of 8 months with a check for $4,270. What annual rate of interest was charged?

26. A check for $3,262.50 was used to retire a 15 month $3,000 loan. What annual rate of interest was charged?

27. If you paid $30 to a loan company for the use of $1,000 for 60 days, what annual rate of interest did they charge?

28. If you paid $120 to a loan company for the use of $2,000 for 90 days, what annual rate of interest did they charge?

29. A radio commercial for a loan company states: "You only pay 50¢ a day for each $500 borrowed." If you borrow $1,500 for 120 days, what amount will you repay, and what annual interest rate is the company actually charging?

30. George finds a company that charges 70¢ per day for each $1,000 borrowed. If he borrows $3,000 for 60 days, what amount will he repay, and what annual interest rate will he be paying the company?

31. You are interested in buying a 13 week T-bill (treasury bill) from the U.S. Treasury Department. If you buy a T-bill with a maturity value of $10,000 for $9,776.94, what annual interest rate will you earn?

*The authors wish to thank Professor Roy Luke of Pierce College for his many useful suggestions of applications in this chapter.

32. If you buy a 26 week T-bill with a maturity value of $10,000 for $9,562.56 from the U.S. Treasury Department, what annual interest rate will you earn?

33. If an investor wants to earn an annual interest rate of 12.63% on a 13 week T-bill with a maturity value of $10,000, how much should the investor pay for the T-bill?

34. If an investor wants to earn an annual interest rate of 10.58% on a 26 week T-bill with a maturity value of $10,000, how much should the investor pay for the T-bill?

35. For services rendered, an attorney accepts a 90 day note for $5,500 at 12% simple interest from a client. (Both interest and principal will be repaid at the end of 90 days.) Wishing to be able to use her money sooner, the attorney sells the note to a third party for $5,540 after 30 days. What annual interest rate will the third party receive for the investment?

36. To complete the sale of a house, the seller accepts a 180 day note for $10,000 at 10% simple interest. (Both interest and principal will be repaid at the end of 180 days.) Wishing to be able to use the money sooner for the purchase of another house, the seller sells the note to a third party for $10,100 after 60 days. What annual interest rate will the third party receive for the investment?

The following buying and selling commission schedule is from a well-known discount brokerage house:

DOLLAR RANGE PER TRANSACTION	COMMISSION
$0–2,500	$19 + 1.6% of principal amount
$2,501–6,000	$44 + 0.6% of principal amount
$6,001–22,000	$62 + 0.3% of principal amount
$22,001–50,000	$84 + 0.2% of principal amount
$50,001–500,000	$134 + 0.1% of principal amount
$500,000 +	$234 + 0.08% of principal amount

Example: The commission on 500 shares at $15 per share is $84.50.

Taking into consideration the buying and selling commissions in the schedule, find the annual rate of interest earned by each investment in Problems 37–40.

37. An investor purchases 500 shares at $14.20 a share, holds the stock for 39 weeks, and then sells the stock for $16.84 a share.

38. An investor purchases 450 shares at $64.84 a share, holds the stock for 26 weeks, and then sells the stock for $72.08 a share.

39. An investor purchases 2,000 shares at $23.75 a share, holds the stock for 300 days, and then sells the stock for $26.15 a share.

40. An investor purchases 75 shares at $31.50 a share, holds the stock for 150 days, and then sells the stock for $35.40 a share.

SECTION 3-2

Compound Interest

- Compound Interest
- Effective Rate
- Growth and Time

COMPOUND INTEREST

If at the end of a payment period the interest due is reinvested at the same rate, then the interest as well as the original principal will earn interest during the next payment period. Interest paid on interest reinvested is called **compound interest.**

For example, suppose you deposit $1,000 in a bank that pays 8% compounded quarterly. How much will the bank owe you at the end of a year?

Compounding quarterly means that earned interest is paid to your account at the end of each 3 month period and that interest as well as the principal earns interest for the next quarter. Using the simple interest formula (2) from the previous section, we compute the amount in the account at the end of the first quarter after interest has been paid:

$$\begin{aligned} A &= P(1 + rt) \\ &= 1{,}000[1 + 0.08(\tfrac{1}{4})] \\ &= 1{,}000(1.02) = \$1{,}020 \end{aligned}$$

Now, \$1,020 is your new principal for the second quarter. At the end of the second quarter, after interest is paid, the account will have

$$\begin{aligned} A &= \$1{,}020[1 + 0.08(\tfrac{1}{4})] \\ &= \$1{,}020(1.02) = \$1{,}040.40 \end{aligned}$$

Similarly, at the end of the third quarter, you will have

$$\begin{aligned} A &= \$1{,}040.40[1 + 0.08(\tfrac{1}{4})] \\ &= \$1{,}040.40(1.02) = \$1{,}061.21 \end{aligned}$$

Finally, at the end of the fourth quarter, the account will have

$$\begin{aligned} A &= \$1{,}061.21([1 + 0.08(\tfrac{1}{4})] \\ &= \$1{,}061.21(1.02) = \$1{,}082.43 \end{aligned}$$

How does this compound amount compare with simple interest? The amount with simple interest would be

$$\begin{aligned} A &= P(1 + rt) \\ &= \$1{,}000[1 + 0.08(1)] \\ &= \$1{,}000(1.08) = \$1{,}080 \end{aligned}$$

We see that compounding quarterly yields \$2.43 more than simple interest would provide.

Let us look over the above calculations for compound interest to see if we can uncover a pattern that might lead to a general formula for computing compound interest for arbitrary cases:

$$\begin{aligned} A &= 1{,}000(1.02) && \text{End of first quarter} \\ A &= [1{,}000(1.02)](1.02) = 1{,}000(1.02)^2 && \text{End of second quarter} \\ A &= [1{,}000(1.02)^2](1.02) = 1{,}000(1.02)^3 && \text{End of third quarter} \\ A &= [1{,}000(1.02)^3](1.02) = 1{,}000(1.02)^4 && \text{End of fourth quarter} \end{aligned}$$

It appears that at the end of n quarters, we would have

$$A = 1{,}000(1.02)^n \qquad \text{End of } n\text{th quarter}$$

or

$$\begin{aligned} A &= 1{,}000[1 + 0.08(\tfrac{1}{4})]^n \\ &= 1{,}000[1 + \tfrac{0.08}{4}]^n \end{aligned}$$

where $\frac{0.08}{4} = 0.02$ is the interest rate per quarter. Since interest rates are generally quoted as *annual nominal rates,* the **rate per compounding period** is found by dividing the annual nominal rate by the number of compounding periods per year.

In general, if P is the principal earning interest compounded m times a year at an annual rate of r, then (by repeated use of the simple interest formula, using $i = r/m$, the rate per period) the amount A at the end of each period is

$$A = P(1 + i) \quad \text{End of the first period}$$
$$A = [P(1 + i)](1 + i) = P(1 + i)^2 \quad \text{End of second period}$$
$$A = [P(1 + i)^2](1 + i) = P(1 + i)^3 \quad \text{End of third period}$$
$$\vdots$$
$$A = [P(1 + i)^{n-1}](1 + i) = P(1 + i)^n \quad \text{End of nth period}$$

We summarize this important result in the following box:

Amount—Compound Interest

$$A = P(1 + i)^n \tag{1}$$

where $i = r/m$ and

r = Annual nominal rate*
m = Number of compounding periods per year
i = Rate per compounding period
n = Total number of compounding periods
P = Principal (present value)
A = Amount (future value) at the end of n periods

*This is often shortened to "annual rate" or just "rate."

Several examples will illustrate different uses of formula (1). If any three of the four variables in (1) are given, we can solve for the fourth using some algebra and a calculator. In particular, if A, P, and i are given, we can solve for the exponent n using properties of the logarithm and a calculator.

Example 1 **Comparing Interest for Various Compounding Periods** If \$1,000 is invested at 8% compounded

(A) annually (B) semiannually (C) quarterly (D) monthly

what is the amount after 5 years? Write answers to the nearest cent.

SOLUTION (A) Compounding annually means that there is one interest payment period per year. Thus, $n = 5$ and $i = r = 0.08$.

$$\begin{aligned} A &= P(1 + i)^n \\ &= 1{,}000(1 + 0.08)^5 \quad \text{Use a calculator.} \\ &= 1{,}000(1.469\ 328) \\ &= \$1{,}469.33 \quad \text{Interest earned} = A - P = \$469.33 \end{aligned}$$

(B) Compounding semiannually means that there are two interest payment periods per year. Thus, the number of payment periods in 5 years is $n = 2(5) = 10$, and the interest rate per period is

$$i = \frac{r}{m} = \frac{0.08}{2} = 0.04$$

So,

$$\begin{aligned} A &= P(1 + i)^n \\ &= 1{,}000(1 + 0.04)^{10} \quad \text{Use a calculator.} \\ &= 1{,}000(1.480\ 244) \\ &= \$1{,}480.24 \quad \text{Interest earned} = A - P = \$480.24 \end{aligned}$$

(C) Compounding quarterly means that there are four interest payments per year. Thus, $n = 4(5) = 20$ and $i = \frac{0.08}{4} = 0.02$. So,

$$\begin{aligned} A &= P(1 + i)^n \\ &= 1{,}000(1 + 0.02)^{20} \quad \text{Use a calculator.} \\ &= 1{,}000(1.485\ 947) \\ &= \$1{,}485.95 \quad \text{Interest earned} = A - P = \$485.95 \end{aligned}$$

(D) Compounding monthly means that there are twelve interest payments per year. Thus, $n = 12(5) = 60$ and $i = \frac{0.08}{12} = 0.006\ 66\overline{6}$.* So,

$$\begin{aligned} A &= P(1 + i)^n \\ &= 1{,}000\left(1 + \frac{0.08}{12}\right)^{60} \quad \text{Use a calculator.} \\ &= 1{,}000(1.489\ 846) \\ &= \$1{,}489.85 \quad \text{Interest earned} = A - P = \$489.85 \end{aligned}$$

Matched Problem 1 Repeat Example 1 with an annual interest rate of 6% over an 8 year period.

Notice the rather significant increase in interest earned in going from annual compounding to monthly compounding. One might wonder what happens if we compound daily, or every minute, or every second, and so on. Figure 1 shows the effect of increasing the number of compounding periods per year when \$1,000 is invested at 8% compound interest for 5 years.

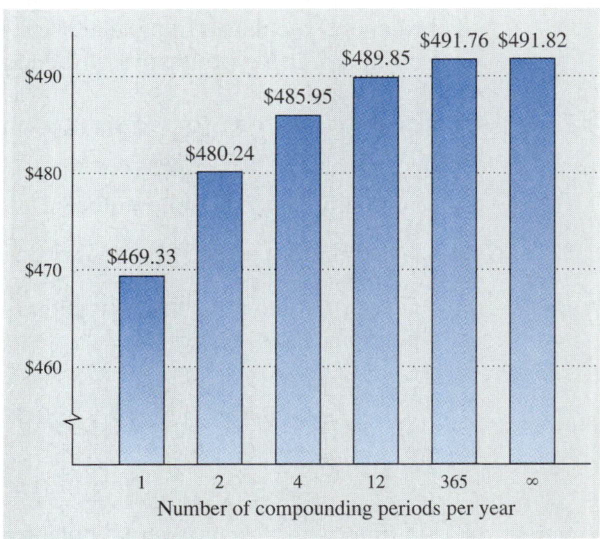

FIGURE 1

Interest on \$1,000 for 5 years at 8% with various compounding periods

---

*Recall that the bar over the 6 indicates a repeating decimal expansion. Rounding i to a small number of decimal places, such as 0.007 or 0.0067, can result in round-off errors. To avoid this, use as many decimal places for i as your calculator is capable of displaying.

Note from Figure 1 that the difference in interest earned between annual and semiannual compounding is \$480.24 − \$469.33 = \$10.91, but the difference between semiannual and quarterly compounding is only \$5.71. Furthermore, the difference between quarterly and monthly compounding is only \$3.90, even though the number of periods is tripled, and the difference between monthly and daily compounding is only \$1.91, even though the number of compounding periods has increased dramatically. These facts suggest that the interest earned approaches a limit. The limit is reached at compounding *continuously,* which yields \$491.82—only 6¢ more than compounding daily. (If interest is compounded continuously, then $A = Pe^{rt}$, as discussed in Section 2-2.) Compare the results in Figure 1 with simple interest earned over the same time period:

$$I = Prt = 1{,}000(0.08)5 = \$400$$

Example 2 ➡ **Visualizing Investments with a Graphing Utility**

(A) Use a graphing utility to graph the growth of investments of \$1,000 at 8% simple interest and of \$1,000 at 8% compounded monthly, both over a period of 5 years.

(B) Use a graphing utility to graph the growth of both investments in part (A) during the fifth year.

SOLUTION (A) To graph the first investment, the formula for simple interest, $A = P(1 + rt)$, is entered as $y_1 = 1000*(1 + .08*x)$, where the variable x represents time in years and y_1 represents the amount after x years. Similarly, to graph the second investment, the formula for compound interest, $A = P(1 + i)^n$, is entered as $y_2 = 1000*(1 + .08/12)$^$(12*x)$. Note that n, the number of periods, is entered as $12*x$ (12 months per year times x years), making the variable x have the same meaning in both equations. This allows us to compare y_1 and y_2 at various values of x using the trace feature of the graphing utility, for example. Figure 2 shows the equations that represent each investment, the window variables, and the resulting graphs. (An alternative approach would be to let the variable x represent months rather than years in both equations, and to use $y_1 = 1000*(1 + .08*(x/12))$, $y_2 = 1000*(1 + .08/12)$^x, Xmin=0, Xmax=60.)

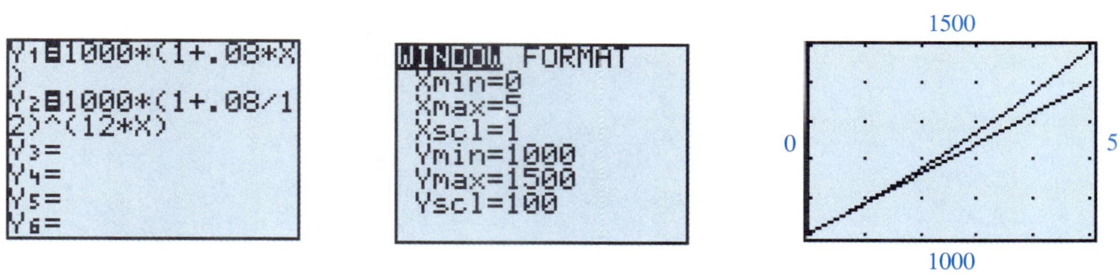

(A) (B) (C)

FIGURE 2

(B) Only the window variables need to be modified. From Figure 2 we see that the investment at simple interest is worth something more than \$1,300 when $x = 4$. By setting Xmin = 4, Xmax = 5, Ymin = 1300, Ymax = 1500, our viewing window will show the growth of both investments during the fifth year. We set Xscl $= \frac{1}{12}$ to obtain a scale for the x axis in months, and set Yscl = 20 to obtain a scale for the y axis in increments of \$20. Figure 3 shows the equations that represent each investment, the window variables, and the resulting graphs.

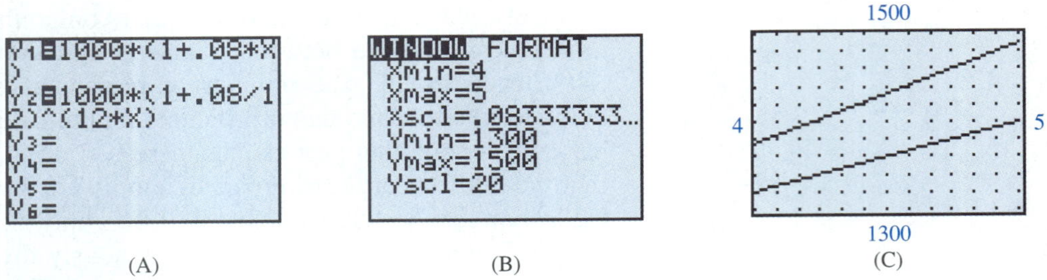

FIGURE 3

Matched Problem 2 Use a graphing utility to graph the growth of an investment of \$5,000 at 10% compounded semiannually, and of a second investment of \$5,000 at 10% compounded monthly, both over a period of 30 years.

REMARK

Refer to Figure 3C. Since compound interest is paid at the end of each compounding period, the graph of y_2 in Figure 3C is an approximation of the growth of the investment. A more accurate graph can be produced by considering only the integer part of $12x$ and placing the graphing utility in the dot mode, as shown in Figure 4. However, for graphical comparisons, it is common practice to use the approximate graph in Figure 3C instead of the exact graph in Figure 4B.

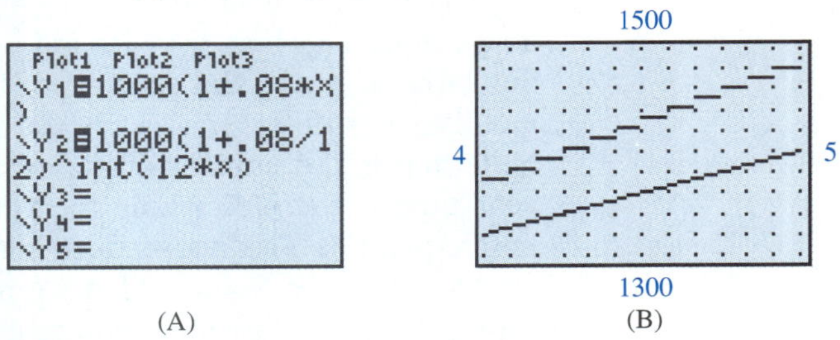

FIGURE 4

Explore–Discuss 1

(A) Which would be the better way to invest \$1,000: at 9% simple interest for 10 years, or at 7% compounded monthly for 10 years?

(B) Explain why the graph of future value as a function of time is a straight line for simple interest, but for compound interest the graph curves upward (see Fig. 5).

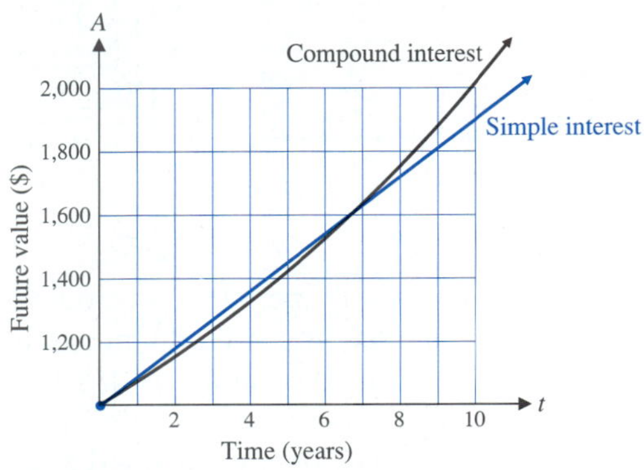

FIGURE 5

Another use of the compound interest formula is in determining how much you should invest now to have a given amount at a future date.

Example 3 **Finding Present Value** How much should you invest now at 10% compounded quarterly to have \$8,000 toward the purchase of a car in 5 years?

SOLUTION We are given a future value $A = \$8{,}000$ for a compound interest investment, and we need to find the present value (principal P) given $i = \frac{0.10}{4} = 0.025$ and $n = 4(5) = 20$.

$$A = P(1 + i)^n$$

$$8{,}000 = P(1 + 0.025)^{20}$$

$$P = \frac{8{,}000}{(1 + 0.025)^{20}} \qquad \text{Use a calculator.}$$

$$= \frac{8{,}000}{1.638\ 616} = \$4{,}882.17$$

Thus, your initial investment of \$4,882.17 will grow to \$8,000 in 5 years.

Matched Problem 3 How much should new parents invest now at 8% compounded semiannually to have \$80,000 toward their child's college education in 17 years?

A graphing utility is a useful tool for studying compound interest. In Figure 6, we use a spreadsheet to illustrate the growth of the investment in Example 3 both numerically and graphically. Similar results can be obtained from most graphing calculators. The concepts and formulas discussed in this chapter are used extensively in spreadsheets to produce tables and graphs for a wide variety of business applications.

	A	B	C
1	Period	Interest	Amount
2	0		\$4,882.17
3	1	\$122.05	\$5,004.22
4	2	\$125.11	\$5,129.33
5	3	\$128.23	\$5,257.56
6	4	\$131.44	\$5,389.00
7	5	\$134.73	\$5,523.73
8	6	\$138.09	\$5,661.82
9	7	\$141.55	\$5,803.37
10	8	\$145.08	\$5,948.45
11	9	\$148.71	\$6,097.16
12	10	\$152.43	\$6,249.59
13	11	\$156.24	\$6,405.83
14	12	\$160.15	\$6,565.98
15	13	\$164.15	\$6,730.13
16	14	\$168.25	\$6,898.38
17	15	\$172.46	\$7,070.84
18	16	\$176.77	\$7,247.61
19	17	\$181.19	\$7,428.80
20	18	\$185.72	\$7,614.52
21	19	\$190.36	\$7,804.88
22	20	\$195.12	\$8,000.00

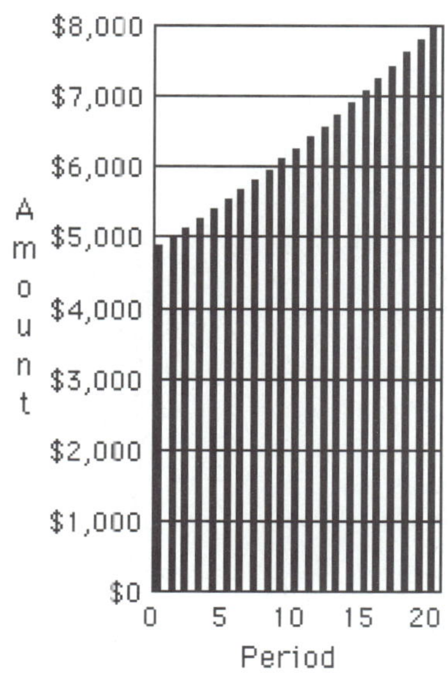

FIGURE 6
Growth of \$4,882.17 at 10% compounded quarterly for 5 years

Explore–Discuss 2

(A) To become a millionaire, you intend to deposit an amount P at rate r compounded quarterly on your 25th birthday, and withdraw 1 million dollars on your 75th birthday. How much would you have to deposit if $r = 4\%$? If $r = 8\%$? If $r = 12\%$?

(B) Suppose you deposit $2,500 on your 25th birthday. What rate r compounded quarterly must your deposit earn in order to grow to 1 million dollars by your 75th birthday?

Effective Rate

Suppose you read in the newspaper that one investment pays 15% compounded monthly and another pays 15.2% compounded semiannually. Which has the better return? A good way to compare investments is to determine their **effective rates**—the simple interest rates that would produce the same returns in *1 year* if the same principal had been invested at simple interest without compounding. (Effective rates are also called **annual yields** or **true interest rates.**)

If principal P is invested at an annual (nominal) rate r compounded m times a year, then in 1 year,

$$A = P\left(1 + \frac{r}{m}\right)^m$$

What simple interest rate will produce the same amount A in 1 year? We call this simple interest rate the effective rate, and denote it by r_e. To find r_e we proceed as follows:

$$\begin{pmatrix}\text{Amount at}\\ \text{simple interest}\\ \text{after 1 year}\end{pmatrix} = \begin{pmatrix}\text{Amount at}\\ \text{compound interest}\\ \text{after 1 year}\end{pmatrix}$$

$$P(1 + r_e) = P\left(1 + \frac{r}{m}\right)^m \qquad \text{Divide both sides by } P.$$

$$1 + r_e = \left(1 + \frac{r}{m}\right)^m \qquad \text{Isolate } r_e \text{ on the left side.}$$

$$r_e = \left(1 + \frac{r}{m}\right)^m - 1$$

Effective Rate

If principal P is invested at the annual (nominal) rate r compounded m times a year, then the effective rate r_e is given by

$$r_e = \left(1 + \frac{r}{m}\right)^m - 1$$

If interest is compounded continuously at the annual rate r, then the effective rate is given in terms of the exponential function with base e (introduced in Section 2-2) by $r_e = e^r - 1$.

Example 4 **Computing Effective Rate** A savings and loan pays 8% compounded quarterly. What is the effective rate? (Express the answer as a percentage, correct to three decimal places.)

SOLUTION

$$\begin{aligned} r_e &= \left(1 + \frac{r}{m}\right)^m - 1 \\ &= \left(1 + \frac{0.08}{4}\right)^4 - 1 \\ &= (1.02)^4 - 1 \qquad \text{Use a calculator.} \\ &= 1.082\ 432 - 1 \\ &= 0.082\ 432 \quad \text{or} \quad 8.243\% \end{aligned}$$

This shows that money invested at 8.243% simple interest earns the same amount of interest in *1 year* as money invested at 8% compounded quarterly. Thus, the effective rate of 8% compounded quarterly is 8.243%.

Matched Problem 4 What is the effective rate of money invested at 6% compounded quarterly?

Example 5 **Computing the Annual Nominal Rate Given the Effective Rate** A savings and loan wants to offer a CD (certificate of deposit) with a monthly compounding rate that has an effective rate of 7.5%. What annual nominal rate compounded monthly should they use? **C** Check with a graphing utility.

SOLUTION

$$\begin{aligned} r_e &= \left(1 + \frac{r}{m}\right)^m - 1 \\ 0.075 &= \left(1 + \frac{r}{12}\right)^{12} - 1 \\ 1.075 &= \left(1 + \frac{r}{12}\right)^{12} \\ \sqrt[12]{1.075} &= 1 + \frac{r}{12} \\ \sqrt[12]{1.075} - 1 &= \frac{r}{12} \\ r &= 12(\sqrt[12]{1.075} - 1) \qquad \text{Use a calculator.} \\ &= 0.072\ 539 \quad \text{or} \quad 7.254\% \end{aligned}$$

Thus, an annual nominal rate of 7.254% compounded monthly is equivalent to an effective rate of 7.5%.

C CHECK We use an equation solver on a graphing calculator to check this result (Fig. 7).

FIGURE 7

Matched Problem 5 What is the annual nominal rate compounded quarterly for a bond that has an effective rate of 8.8%? **C** Check with a graphing utility.

Example 6 **Comparing Two Investments** An investor has an opportunity to purchase two different notes: Note A pays 15% compounded monthly, and note B pays 15.2% compounded semiannually. Which is the better investment, assuming all else is equal?

SOLUTION Nominal rates with different compounding periods cannot be compared directly. We must first find the effective rate of each nominal rate and then compare the effective rates to determine which investment will yield the larger return.

Effective Rate for Note A:

$$\begin{aligned} r_e &= \left(1 + \frac{r}{m}\right)^m - 1 \\ &= \left(1 + \frac{0.15}{12}\right)^{12} - 1 \\ &= (1.0125)^{12} - 1 \\ &= 1.160\ 755 - 1 \\ &= 0.160\ 755 \quad \text{or} \quad 16.076\% \end{aligned}$$

Use a calculator.

Effective Rate for Note B:

$$\begin{aligned} r_e &= \left(1 + \frac{r}{m}\right)^m - 1 \\ &= \left(1 + \frac{0.152}{2}\right)^{2} - 1 \\ &= (1.076)^{2} - 1 \\ &= 1.157\ 776 - 1 \\ &= 0.157\ 776 \quad \text{or} \quad 15.778\% \end{aligned}$$

Since the effective rate for note A is greater than the effective rate for note B, note A is the preferred investment.

Matched Problem 6 Repeat Example 6 if note A pays 9% compounded monthly and note B pays 9.2% compounded semiannually.

GROWTH AND TIME

Investments are also compared by computing their **growth time**—the time it takes a given principal to grow to a particular value (the shorter the time, the greater the return on the investment). Example 7 illustrates two methods for making this calculation.

Example 7 **Computing Growth Time** How long will it take \$10,000 to grow to \$12,000 if it is invested at 9% compounded monthly?

SOLUTION **Method 1.** Use logarithms and a calculator:

$$\begin{aligned} A &= P(1 + i)^n \\ 12{,}000 &= 10{,}000\left(1 + \frac{0.09}{12}\right)^n \\ 1.2 &= 1.0075^n \end{aligned}$$

Now, solve for n by taking logarithms of both sides:

$$\ln 1.2 = \ln 1.0075^n$$ Logarithms to any base can be used; we choose the natural logarithm (base e) and use the property $\log_b M^p = p \log_b M$.

$$\ln 1.2 = n \ln 1.0075$$

$$n = \frac{\ln 1.2}{\ln 1.0075}$$

$$= 24.40 \approx 25 \text{ months} \quad \text{or} \quad 2 \text{ years and } 1 \text{ month}$$

[*Note:* 24.40 is rounded up to 25 to guarantee reaching $12,000, since interest is paid at the end of each month.]

Method 2. Use a graphing utility: To solve this problem using graphical approximation techniques, we graph both sides of the equation $12{,}000 = 10{,}000(1.0075)^n$ and find that the graphs intersect at $x = n = 24.40$ months (Fig. 8A). Thus, the growth time is 25 months. We also come to the same conclusion by using an equation solver (Fig. 8B).

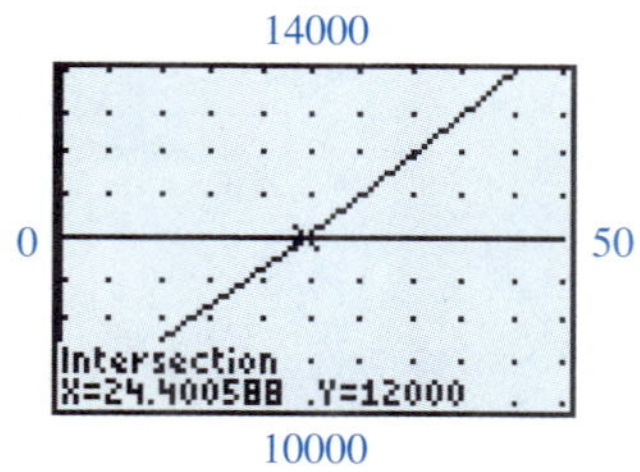

(A) $y_1 = 10000(1.0075)^x$
$y_2 = 12000$

(B)

FIGURE 8

Matched Problem 7 How long will it take $10,000 to grow to $25,000 if it is invested at 18% compounded quarterly?

CAUTION

Each compound interest problem involves two interest rates. Referring to Example 7, $r = 0.09$ or 9% is the annual nominal compounding rate, and $i = r/12 = 0.0075$ or 0.75% is the interest rate per month. Do not confuse these two rates by using r in place of i in the compound interest formula. If interest is compounded annually, then $i = r/1 = r$. In all other cases, r and i are not the same.

Answers to Matched Problems

1. (A) $1,593.85 (B) $1,604.71 (C) $1,610.32 (D) $1,614.14

2.

(A)

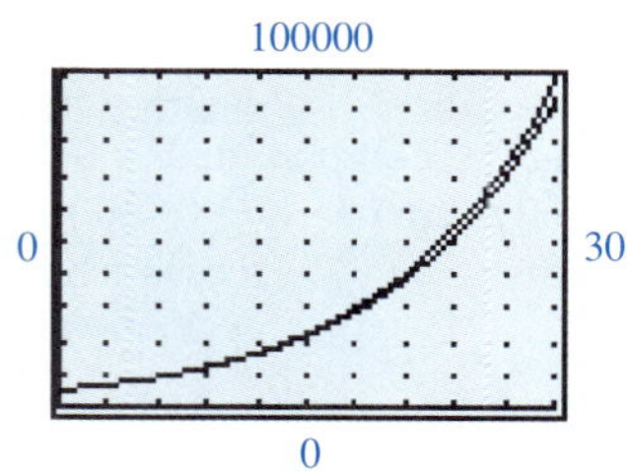

(B)

3. $21,084.17 **4.** 6.136% **5.** 8.524%

6. Note B is better (effective rate of note A is 9.381% and of note B is 9.412%).

7. $20.82 \approx 21$ quarters, or 5 years and 3 months

EXERCISE 3-2

Find all dollar amounts to the nearest cent. When an interest rate is requested as an answer, express the rate as a percentage correct to two decimal places, unless directed otherwise. In all problems involving days, use a 365 day year.

A *In Problems 1–8, use compound interest formula (1) to find each of the indicated values.*

1. $P = \$100; i = 0.01; n = 12; A = ?$

2. $P = \$1{,}000; i = 0.015; n = 20; A = ?$

3. $P = \$800; i = 0.06; n = 25; A = ?$

4. $P = \$10{,}000; i = 0.08; n = 30; A = ?$

5. $A = \$10{,}000; i = 0.03; n = 48; P = ?$

6. $A = \$1{,}000; i = 0.015; n = 60; P = ?$

7. $A = \$18{,}000; i = 0.01; n = 90; P = ?$

8. $A = \$50{,}000; i = 0.005; n = 70; P = ?$

Given the annual rate and the compounding period in Problems 9–12, find i, the interest rate per compounding period.

9. 9% compounded monthly

10. 15% compounded annually

11. 7% compounded quarterly

12. 11% compounded semiannually

Given the rate per compounding period in Problems 13–16, find r, the annual rate.

13. 0.8% per month

14. 5% per year

15. 4.5% per half-year

16. 2.3% per quarter

B

17. If $100 is invested at 6% compounded
(A) annually (B) quarterly (C) monthly
what is the amount after 4 years? How much interest is earned?

18. If $2,000 is invested at 7% compounded
(A) annually (B) quarterly (C) monthly
what is the amount after 5 years? How much interest is earned?

19. If $5,000 is invested at 18% compounded monthly, what is the amount after
(A) 2 years? (B) 4 years?

20. If $20,000 is invested at 6% compounded monthly, what is the amount after
(A) 5 years? (B) 8 years?

21. Discuss the similarities and the differences in the graphs of future value A as a function of time t if $1,000 is invested for 8 years and interest is compounded monthly at annual rates of 4%, 8%, and 12%, respectively (see the figure).

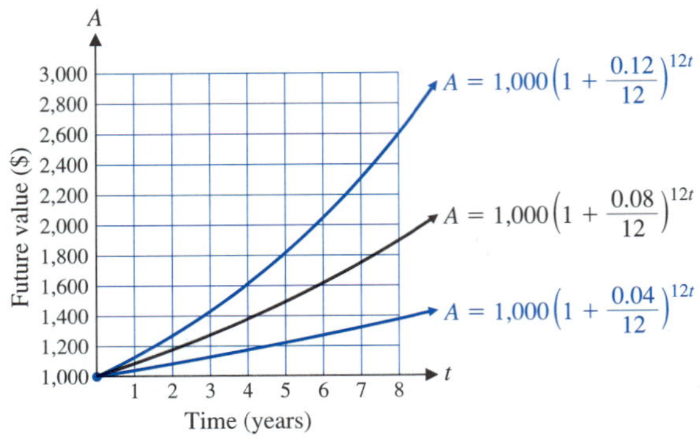

Figure for 21

22. Discuss the similarities and differences in the graphs of future value A as a function of time t for loans of $4,000, $8,000, and $12,000, respectively, each at 7.5% compounded monthly for 8 years (see the figure).

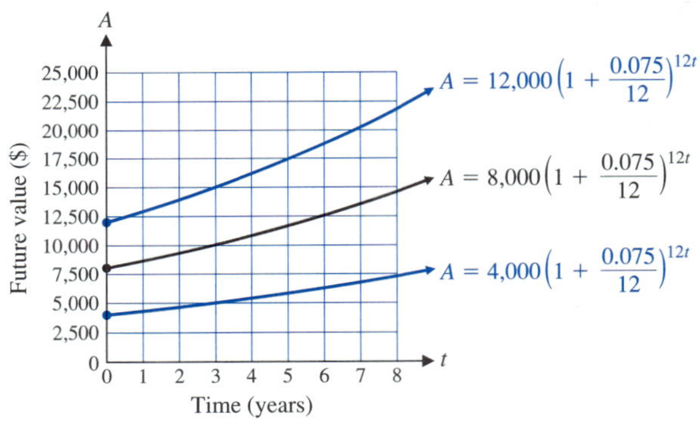

Figure for 22

23. If $1,000 is invested in an account that earns 9.75% compounded annually for 6 years, find the interest earned during each year and the amount in the account at the end of each year. Organize your results in a table.

24. If $2,000 is invested in an account that earns 8.25% compounded annually for 5 years, find the interest earned during each year and the amount in the account at the end of each year. Organize your results in a table.

Check Problems 23 and 24 by constructing a table on a graphing utility.

25. If an investment company pays 8% compounded semiannually, how much should you deposit now to have $10,000
(A) 5 years from now? (B) 10 years from now?

26. If an investment company pays 10% compounded quarterly, how much should you deposit now to have \$6,000 (A) 3 years from now? (B) 6 years from now?

27. What is the effective rate of interest for money invested at
   (A) 10% compounded quarterly?
   (B) 12% compounded monthly?

28. What is the effective rate of interest for money invested at
   (A) 6% compounded monthly?
   (B) 14% compounded semiannually?

29. How long will it take \$4,000 to grow to \$9,000 if it is invested at 15% compounded monthly?

30. How long will it take \$5,000 to grow to \$7,000 if it is invested at 8% compounded quarterly?

C *In Problems 31 and 32, use the compound interest formula (1) to find n to the nearest larger integer value.*

31. $A = 2P; i = 0.06; n = ?$

32. $A = 2P; i = 0.05; n = ?$

33. How long will it take money to double if it is invested at
   (A) 10% compounded quarterly?
   (B) 12% compounded quarterly?

34. How long will it take money to double if it is invested at
   (A) 14% compounded semiannually?
   (B) 10% compounded semiannually?

APPLICATIONS

Business & Economics

35. A newborn child receives a \$5,000 gift toward a college education from her grandparents. How much will the \$5,000 be worth in 17 years if it is invested at 9% compounded quarterly?

36. A person with \$8,000 is trying to decide whether to purchase a car now, or to invest the money at 12% compounded semiannually and then buy a more expensive car. How much will be available for the purchase of a car at the end of 3 years?

37. What will a \$110,000 house cost 10 years from now if the inflation rate over that period averages 6% compounded annually?

38. If the inflation rate averages 8% per year compounded annually for the next 5 years, what will a car costing \$10,000 now cost 5 years from now?

39. Rental costs for office space have been going up at 7% per year compounded annually for the past 5 years. If office space rent is now \$20 per square foot per month, what were the rental rates 5 years ago?

40. In a suburb of a city, housing costs have been increasing at 8% per year compounded annually for the past 8 years. A house with a \$160,000 value now would have had what value 8 years ago?

41. If the population in a particular third-world country is growing at 4% compounded annually, how long will it take the population to double? (Round up to the next higher year if not exact.)

42. If the world population is now about 5 billion people and is growing at 2% compounded annually, how long will it take the population to grow to 8 billion people? (Round up to the next higher year if not exact.)

43. Which is the better investment and why: 9% compounded monthly or 9.3% compounded annually?

44. Which is the better investment and why: 8% compounded quarterly or 8.3% compounded annually?

45. (A) If an investment of \$100 were made in the year the Declaration of Independence was signed, and if it earned 3% compounded quarterly, how much would it be worth in 1998?
   (B) Discuss the effect of compounding interest monthly, daily, and continuously (rather than quarterly) on the \$100 investment.
   C (C) Use a graphing utility to graph the growth of the investment of part (A).

46. (A) Starting with formula (1), derive each of the following formulas:

$$P = \frac{A}{(1+i)^n} \qquad i = \left(\frac{A}{P}\right)^{1/n} - 1 \qquad n = \frac{\ln A - \ln P}{\ln(1+i)}$$

   (B) Explain why it is unnecessary to memorize the above formulas for *P*, *i*, and *n* if you know formula (1).

47. You have saved \$7,000 toward the purchase of a car costing \$9,000. How long will the \$7,000 have to be invested at 9% compounded monthly to grow to \$9,000? (Round up to the next higher month if not exact.)

48. A newly married couple has \$15,000 toward the purchase of a house. For the type of house they are interested in buying, they estimate that a \$20,000 down payment will be necessary. How long will the money have to be invested at 10% compounded quarterly to grow to \$20,000? (Round up to the next higher quarter if not exact.)

49. An Individual Retirement Account (IRA) has \$20,000 in it, and the owner decides not to add any more money to the account other than interest earned at 8% compounded daily. How much will be in the account 35 years from now when the owner reaches retirement age?

50. If \$1 had been placed in a bank account at the birth of Christ and forgotten until now, how much would be in the account at the end of 2010 if the money earned 2% interest compounded annually? 2% simple interest? (Now you can see the power of compounding and see why inactive accounts are closed after a relatively short period of time.)

51. How long will it take money to double if it is invested at 14% compounded daily? 15% compounded annually? (Compute answers in years to three decimal places.)

52. How long will it take money to triple if it is invested at 10% compounded daily? 11% compounded annually? (Compute answers in years to three decimal places.)

53. In a conversation with a friend, you mention that you have two real estate investments, one that has doubled in value in the past 9 years and another that has doubled in value in the past 12 years. Your friend replies immediately that the first investment has been growing at approximately 8% compounded annually and the second at 6% compounded annually. How did your friend make these estimates? The **rule of 72** states that the annual compound rate of growth r of an investment that doubles in n years can be approximated by $r = 72/n$. Construct a table comparing the exact rate of growth and the approximate rate provided by the rule of 72 for doubling times of $n = 6, 7, \dots, 12$ years. Round both rates to one decimal place.

54. Refer to Problem 53. Show that the exact annual compound rate of growth of an investment that doubles in n years is given by $r = 100(2^{1/n} - 1)$. C Graph this equation and the rule of 72 on a graphing utility for $5 \le n \le 20$.

Solve Problems 55–58 using graphical approximation techniques on a graphing utility.

55. How long does it take for a \$2,400 investment at 13% compounded quarterly to be worth more than a \$3,000 investment at 9% compounded quarterly?

56. How long does it take for a \$4,800 investment at 10% compounded monthly to be worth more than a \$5,000 investment at 7% compounded monthly?

57. One investment pays 10% simple interest and another pays 7% compounded annually. Which investment would you choose? Why?

58. One investment pays 9% simple interest and another pays 6% compounded monthly. Which investment would you choose? Why?

59. What is the annual nominal rate compounded daily for a bond that has an effective rate of 7.4%?

60. What is the annual nominal rate compounded monthly for a CD that has an effective rate of 8.2%?

Problems 61–64 refer to zero coupon bonds. A ***zero coupon bond*** *is a bond that is sold now at a discount and will pay its face value at some time in the future when it matures—no interest payments are made.*

61. Parents wishing to have enough money for their child's college education 17 years from now decide to buy a \$30,000 face value zero coupon bond. If money is worth 10% compounded annually, what should they pay for the bond?

62. How much should a \$20,000 face value zero coupon bond, maturing in 10 years, be sold for now if its rate of return is to be 8% compounded annually?

63. If the parents in Problem 61 pay \$6,844.79 for the \$30,000 face value zero coupon bond, what annual compound rate of return will they be receiving?

64. If you pay \$5,893.24 for a \$12,000 face value zero coupon bond that matures in 7 years, what is your annual compound rate of return?

65. *Barron's* (a national business and financial weekly) published the following "Top Savings Deposit Yields" for money market deposit accounts:

(A) Virginia Beach Federal S&L
8.28% compounded monthly

(B) Franklin Savings
8.25% compounded daily

(C) Guaranty Federal Savings
8.25% compounded monthly

What is the effective yield for each?

66. *Barron's* also published the following "Top Savings Deposit Yields" for 1 year CD accounts:

(A) Spindletop Savings
9.00% compounded daily

(B) Alamo Savings of Texas
9.10% compounded quarterly

(C) Atlas Savings and Loan
9.00% compounded quarterly

What is the effective yield for each?

ATLAS
Savings & Loan
Association

67. If you just sold a stock for \$32,456.32 (net) that cost you \$24,766.81 (net) 2 years ago, what annual compound rate of return did you make on your investment?

68. If you just sold a stock for \$27,339.79 (net) that cost you \$14,664.76 (net) 4 years ago, what annual compound rate of return did you make on your investment?

69. What annual nominal rate compounded monthly has the same effective rate as 8% compounded quarterly?

70. What annual nominal rate compounded daily has the same effective rate as 9.6% compounded monthly?

SECTION 3-3

Future Value of an Annuity; Sinking Funds

- Future Value of an Annuity
- Sinking Funds
- Approximating Interest Rates

FUTURE VALUE OF AN ANNUITY

An **annuity** is any sequence of equal periodic payments. If payments are made at the end of each time interval, then the annuity is called an **ordinary annuity.** We will consider only ordinary annuities in this book. The amount, or **future value,** of an annuity is the sum of all payments plus all interest earned.

Suppose you decide to deposit \$100 every 6 months into an account that pays 6% compounded semiannually. If you make six deposits, one at the end of each interest payment period, over 3 years, how much money will be in the account after the last deposit is made? To solve this problem, let us look at it in terms of a time line. Using the compound amount formula $A = P(1 + i)^n$, we can find the value of each deposit after it has earned compound interest up through the sixth deposit, as shown in Figure 1.

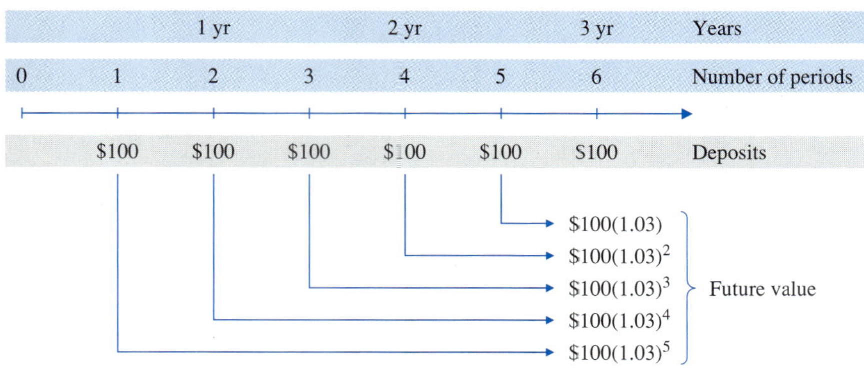

FIGURE 1

We could, of course, evaluate each of the future values in Figure 1 using a calculator and then add the results to find the amount in the account at the time of the sixth deposit—a tedious project at best. Instead, we take another approach that leads directly to a formula that will produce the same result in a few steps (even when the number of deposits is very large). We start by writing the total amount in the account after the sixth deposit in the form

$$S = 100 + 100(1.03) + 100(1.03)^2 + 100(1.03)^3 + 100(1.03)^4 + 100(1.03)^5 \quad (1)$$

We would like a simple way to sum these terms. Let us multiply each side of (1) by 1.03 to obtain

$$1.03S = 100(1.03) + 100(1.03)^2 + 100(1.03)^3 + 100(1.03)^4 + 100(1.03)^5 + 100(1.03)^6 \quad (2)$$

Subtracting (1) from (2), left side from left side and right side from right side, we obtain

$$1.03S - S = 100(1.03)^6 - 100$$

Notice how many terms drop out.

$$0.03S = 100[(1.03)^6 - 1]$$

$$S = 100\frac{(1 + 0.03)^6 - 1}{0.03} \quad (3)$$

We write S in this form to observe a general pattern.

In general, if R is the periodic deposit, i the rate per period, and n the number of periods, then the future value is given by

$$S = R + R(1 + i) + R(1 + i)^2 + \cdots + R(1 + i)^{n-1}$$

Note how this compares to (1).

and proceeding as in the above example, we obtain the general formula for the future value of an ordinary annuity:

$$S = R\frac{(1 + i)^n - 1}{i} \quad (4)$$

Note how this compares to (3).

Explore–Discuss 1

Verify formula (4) by applying to the equation

$$S = R + R(1 + i) + R(1 + i)^2 + \cdots + R(1 + i)^{n-1}$$

the technique we used above to sum equation (1).

Returning to the example above, we use a calculator to complete the problem:

$$S = 100\frac{(1.03)^6 - 1}{0.03}$$
$$= \$646.84$$

For improved accuracy, keep all values in the calculator until the end; then round to the required number of decimal places.

An alternative to the above computation of S is provided by the use of tables of values of certain functions of the mathematics of finance. One such function is the fractional factor of (4), denoted by the symbol $s_{\overline{n}|i}$ (read "s angle n at i"):

$$s_{\overline{n}|i} = \frac{(1 + i)^n - 1}{i}$$

Tables found in books on finance and mathematical handbooks list values of $s_{\overline{n}|i}$ for various values of n and i. To complete the computation of S, R is multiplied by $s_{\overline{n}|i}$ as indicated by (4). (A main advantage of using a calculator rather than tables is that a calculator can handle many more situations than a table, no matter how large the table.)

It is common to use FV (future value) for S and PMT (payment) for R in formula (4). Making these changes, we have the formula in the box at the top of the next page.

Future Value of an Ordinary Annuity

$$FV = PMT\frac{(1 + i)^n - 1}{i} = PMT\, s_{\overline{n}|i} \qquad (5)$$

where

PMT = Periodic payment
i = Rate per period
n = Number of payments (periods)
FV = Future value (amount)

[*Note*: Payments are made at the end of each period.]

Example 1 **Future Value of an Ordinary Annuity** What is the value of an annuity at the end of 20 years if $2,000 is deposited each year into an account earning 8.5% compounded annually? How much of this value is interest?

SOLUTION To find the value of the annuity, use formula (5) with $PMT = \$2{,}000$, $i = r = 0.085$, and $n = 20$.

$$FV = PMT\frac{(1 + i)^n - 1}{i}$$

$$= 2{,}000\frac{(1.085)^{20} - 1}{0.085} = \$96{,}754.03 \qquad \text{Use a calculator.}$$

To find the amount of interest earned, subtract the total amount deposited in the annuity (20 payments of $2,000) from the total value of the annuity after the 20th payment.

$$\text{Deposits} = 20(2{,}000) = \$40{,}000$$
$$\text{Interest} = \text{Value} - \text{Deposits} = 96{,}754.03 - 40{,}000 = \$56{,}754.03$$

Figure 2, which was generated using a spreadsheet, illustrates the growth of this account over 20 years.

	A	B	C	D
1	Period	Payment	Interest	Balance
2	1	$2,000.00	$0.00	$2,000.00
3	2	$2,000.00	$170.00	$4,170.00
4	3	$2,000.00	$354.45	$6,524.45
5	4	$2,000.00	$554.58	$9,079.03
6	5	$2,000.00	$771.72	$11,850.75
7	6	$2,000.00	$1,007.31	$14,858.06
8	7	$2,000.00	$1,262.94	$18,120.99
9	8	$2,000.00	$1,540.28	$21,661.28
10	9	$2,000.00	$1,841.21	$25,502.49
11	10	$2,000.00	$2,167.71	$29,670.20
12	11	$2,000.00	$2,521.97	$34,192.17
13	12	$2,000.00	$2,906.33	$39,098.50
14	13	$2,000.00	$3,323.37	$44,421.87
15	14	$2,000.00	$3,775.86	$50,197.73
16	15	$2,000.00	$4,266.81	$56,464.54
17	16	$2,000.00	$4,799.49	$63,264.02
18	17	$2,000.00	$5,377.44	$70,641.47
19	18	$2,000.00	$6,004.52	$78,645.99
20	19	$2,000.00	$6,684.91	$87,330.90
21	20	$2,000.00	$7,423.13	$96,754.03
22	Totals	$40,000.00	$56,754.03	

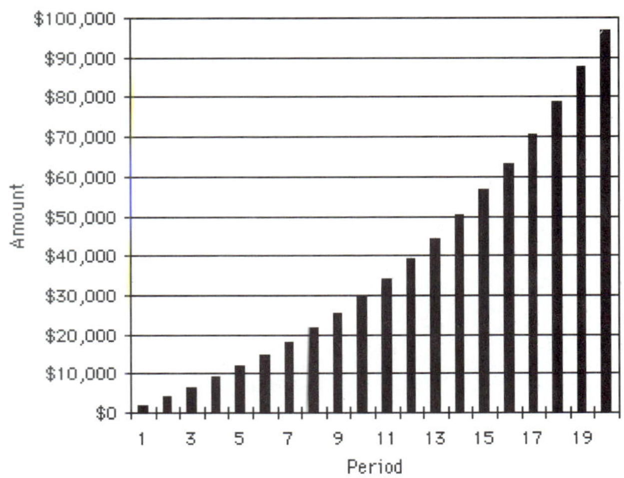

FIGURE 2
Ordinary annuity at 8.5% compounded annually for 20 years

Matched Problem 1 What is the value of an annuity at the end of 10 years if \$1,000 is deposited every 6 months into an account earning 8% compounded semiannually? How much of this value is interest?

The table in Figure 2 is called a **balance sheet.** Let's take a closer look at the construction of this table. The first line is a special case because the payment is made at the end of the period and no interest is earned. Each subsequent line of the table is computed as follows:

Payment	+ Interest	+ Old balance	= New balance	
2,000	+ 0.085(2,000)	+ 2,000	= 4,170	Period 2
2,000	+ 0.085(4,170)	+ 4,170	= 6,524.45	Period 3

And so on. The amounts at the bottom of each column in the balance sheet agree with the results we obtained by using formula (5), as you would expect. Although balance sheets are appropriate for certain situations, we will concentrate on applications of formula (5). There are many important problems that can be solved only by using this formula.

Explore–Discuss 2

(A) Discuss the similarities and differences in the graphs of future value FV as a function of time t for ordinary annuities in which \$100 is deposited each month for 8 years and interest is compounded monthly at annual rates of 4%, 8%, and 12%, respectively (Fig. 3).

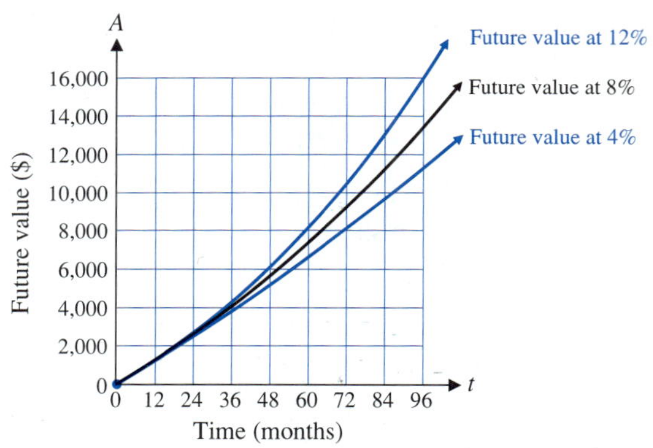

FIGURE 3

(B) Discuss the connections between the graph of the equation $y = 100t$, where t is time in months, and the graphs of part (A).

Sinking Funds

The formula for the future value of an ordinary annuity has another important application. Suppose the parents of a newborn child decide that on each of the child's birthdays up to the 17th year, they will deposit \$$PMT$ in an account that pays 6% compounded annually. The money is to be used for college expenses. What should the annual deposit \$$PMT$ be in order for the amount in the account to be \$80,000 after the 17th deposit?

We are given *FV, i,* and *n* in formula (5), and our problem is to find *PMT.* Thus,

$$FV = PMT\frac{(1+i)^n - 1}{i}$$

$$80{,}000 = PMT\frac{(1.06)^{17} - 1}{0.06} \qquad \text{Solve for } PMT.$$

$$PMT = 80{,}000\frac{0.06}{(1.06)^{17} - 1} \qquad \text{Use a calculator.}$$

$$= \$2{,}835.58 \text{ per year}$$

An annuity of 17 annual deposits of $2,835.58 at 6% compounded annually will amount to $80,000 in 17 years.

This is one of many examples of a similar type that are referred to as *sinking fund problems.* In general, any account that is established for accumulating funds to meet future obligations or debts is called a **sinking fund.** If the payments are to be made in the form of an ordinary annuity, then we have only to solve formula (5) for the **sinking fund payment** *PMT:*

$$PMT = FV\frac{i}{(1+i)^n - 1} \tag{6}$$

It is important to understand that formula (6), which is convenient to use, is simply a variation of formula (5). You can always find the sinking fund payment by first substituting the appropriate values into formula (5) and then solving for *PMT,* as we did in the college fund example discussed above. Or you can substitute directly into formula (6), as we do in the next example. Use whichever method is easier for you.

Example 2 **Computing the Payment for a Sinking Fund** A company estimates that it will have to replace a piece of equipment at a cost of $10,000 in 5 years. To have this money available in 5 years, a sinking fund is established by making fixed monthly payments into an account paying 6% compounded monthly. How much should each payment be?

Solution To find *PMT,* we can use either formula (5) or (6). We choose formula (6) with $FV = \$10{,}000$, $i = \frac{0.06}{12} = 0.005$, and $n = 5(12) = 60$:

$$PMT = FV\frac{i}{(1+i)^n - 1}$$

$$= 10{,}000\frac{0.005}{(1.005)^{60} - 1} \qquad \text{Use a calculator.}$$

$$= \$143.33 \text{ per month}$$

Matched Problem 2 A bond issue is approved for building a marina in a city. The city is required to make regular payments every 6 months into a sinking fund paying 6% compounded semiannually. At the end of 10 years, the bond obligation will be retired at a cost of $5,000,000. What should each payment be?

Explore–Discuss 3

Suppose you intend to establish an ordinary annuity earning 7.5% compounded monthly that will be worth 1 million dollars on your 70th birthday. How much would your monthly payments be if you started making payments at age 20? At age 35? At age 50?

APPROXIMATING INTEREST RATES

Algebra can be used to solve the future value formula (5) for *PMT* or *n*, but not for *i*. However, graphical techniques or equation solvers can be used to approximate *i* to as many decimal places as desired.

Example 3 **Approximating an Interest Rate** An individual makes monthly deposits of \$100 into an ordinary annuity. After 30 years, the annuity is worth \$160,000. What annual rate compounded monthly has this annuity earned during this 30 year period? Express the answer as a percentage, correct to two decimal places.

SOLUTION Substituting $FV = \$160{,}000$, $PMT = \$100$, and $n = 30(12) = 360$ in (5) produces the following equation:

$$160{,}000 = 100\frac{(1 + i)^{360} - 1}{i}$$

We can approximate the solution to this equation by using graphical techniques (Figs. 4A and 4B) or an equation solver (Fig. 4C). From Figure 4B or 4C, we see that $i = 0.006\ 956\ 7$ and $12(i) = 0.083\ 480\ 4$. Thus, the annual rate (to two decimal places) is $r = 8.35\%$.

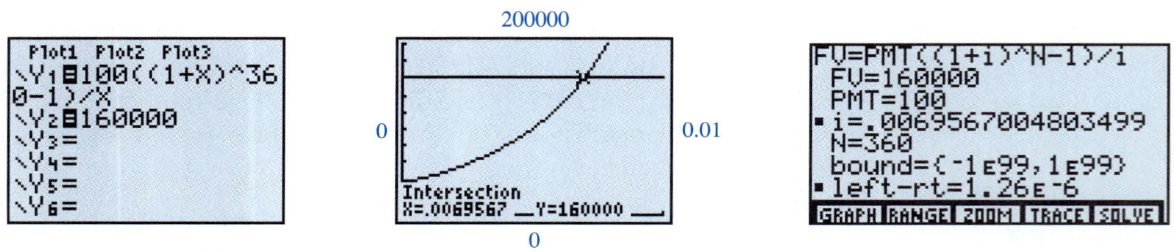

(A) (B) (C)

FIGURE 4

Matched Problem 3 An individual makes annual deposits of \$1,000 into an ordinary annuity. After 20 years, the annuity is worth \$55,000. What annual compound rate has this annuity earned during this 20 year period? Express the answer as a percentage, correct to two decimal places.

Answers to Matched Problems

1. Value: \$29,778.08; Interest: \$9,778.08 **2.** \$186,078.54 every 6 mo for 10 yr
3. 9.64%

EXERCISE 3-3

In Problems 1–12, use future value formula (5) to find each of the indicated values.

A

1. $n = 20; i = 0.03; PMT = \$500; FV = ?$
2. $n = 25; i = 0.04; PMT = \$100; FV = ?$
3. $n = 40; i = 0.02; PMT = \$1{,}000; FV = ?$
4. $n = 30; i = 0.01; PMT = \$50; FV = ?$

B

5. $FV = \$3{,}000; n = 20; i = 0.02; PMT = ?$
6. $FV = \$8{,}000; n = 30; i = 0.03; PMT = ?$
7. $FV = \$5{,}000; n = 15; i = 0.01; PMT = ?$
8. $FV = \$2{,}500; n = 10; i = 0.08; PMT = ?$

C

9. $FV = \$4{,}000; i = 0.02; PMT = 200; n = ?$

10. $FV = \$8{,}000; i = 0.04; PMT = 500; n = ?$

11. $FV = \$7{,}600; PMT = \$500; n = 10; i = ?$
C (Round answer to two decimal places.)

12. $FV = \$4{,}100; PMT = \$100; n = 20; i = ?$
C (Round answer to two decimal places.)

APPLICATIONS

Business & Economics

13. What is the value of an ordinary annuity at the end of 10 years if \$500 per quarter is deposited into an account earning 8% compounded quarterly? How much of this value is interest?

14. What is the value of an ordinary annuity at the end of 20 years if \$1,000 per year is deposited into an account earning 7% compounded annually? How much of this value is interest?

15. In order to accumulate enough money for a down payment on a house, a couple deposits \$300 per month into an account paying 6% compounded monthly. If payments are made at the end of each period, how much money will be in the account in 5 years?

16. A self-employed person has a Keogh retirement plan. (This type of plan is free of taxes until money is withdrawn.) If deposits of \$7,500 are made each year into an account paying 8% compounded annually, how much will be in the account after 20 years?

17. In 5 years a couple would like to have \$25,000 for a down payment on a house. What fixed amount should be deposited each month into an account paying 9% compounded monthly?

18. A person wishes to have \$200,000 in an account for retirement 15 years from now. How much should be deposited quarterly in an account paying 8% compounded quarterly?

19. A company estimates it will need \$100,000 in 8 years to replace a computer. If it establishes a sinking fund by making fixed monthly payments into an account paying 12% compounded monthly, how much should each payment be?

20. Parents have set up a sinking fund in order to have \$120,000 in 15 years for their children's college education. How much should be paid semiannually into an account paying 10% compounded semiannually?

21. If \$1,000 is deposited at the end of each year for 5 years into an ordinary annuity earning 8.32% compounded annually, construct a balance sheet showing the interest earned during each year and the balance at the end of each year.

22. If \$2,000 is deposited at the end of each quarter for 2 years into an ordinary annuity earning 7.9% compounded quarterly, construct a balance sheet showing the interest earned during each quarter and the balance at the end of each quarter.

Check Problems 21 and 22 by constructing a balance sheet on a graphing utility.

23. Beginning in January, a person plans to deposit \$100 at the end of each month into an account earning 9% compounded monthly. Each year taxes must be paid on the interest earned during that year. Find the interest earned during each year for the first 3 years.

24. If \$500 is deposited each quarter into an account paying 12% compounded quarterly for 3 years, find the interest earned during each of the 3 years.

25. Why does it make sense to open an Individual Retirement Account (IRA) early in one's life? (For people in certain income brackets, money deposited into an IRA and earnings from an IRA are tax-deferred until withdrawal.) Compare the following:
   (A) Jane deposits \$2,000 a year into an IRA earning 9% compounded annually. She makes her first deposit on her 24th birthday and her last deposit on her 31st birthday (8 deposits in all). Making no additional deposits, she leaves the accumulated amount from the 8 deposits in the account, earning interest at 9% compounded annually, until her 65th birthday. How much (to the nearest dollar) will be in her account on her 65th birthday?
   (B) John procrastinates and does not make his first \$2,000 deposit into an IRA until he is 32, but then he continues to deposit \$2,000 on every birthday until he is 65 (34 deposits in all). If his account also earns 9% compounded annually, how much (to the nearest dollar) will he have in his account when he makes his last deposit on his 65th birthday?
   (*Surprise*—Jane will have more money than John!)

26. Starting on his 24th birthday, and continuing on every birthday up to and including his 65th, a person deposits \$2,000 a year into an IRA. How much (to the nearest dollar) will be in the account on the 65th birthday, if the account earns:
   (A) 6% compounded annually?
   (B) 8% compounded annually?
   (C) 10% compounded annually?
   (D) 12% compounded annually?

27. You wish to have \$10,000 in 4 years to buy a car. How much should you deposit each month into an account paying 8% compounded monthly? How much interest will the account earn in the 4 years?

28. A company establishes a sinking fund to upgrade a plant in 5 years at an estimated cost of \$1,500,000. How much should be invested each quarter into an account paying 9.15% compounded quarterly? How much interest will the account earn in the 5 years?

29. You can afford monthly deposits of only \$150 into an account that pays 8.5% compounded monthly. How long will it be until you have \$7,000 to buy a boat? (Round to the next higher month if not exact.)

30. A company establishes a sinking fund for upgrading office equipment with monthly payments of \$1,000 into an account paying 10% compounded monthly. How long will it be before the account has \$100,000? (Round up to the next higher month if not exact.)

In Problems 31–34, use graphical approximation techniques or an equation solver to approximate the desired interest rate. Express each answer as a percentage, correct to two decimal places.

31. An individual makes annual payments of \$1,000 into an ordinary annuity. At the end of 5 years, the amount in the annuity is \$6,300. What annual nominal compounding rate has this annuity earned?

32. An individual invests \$2,000 annually in an IRA. At the end of 6 years, the amount in the fund is \$15,000. What annual nominal compounding rate has this fund earned?

33. At the end of each month, an employee deposits \$50 into a Christmas club fund. At the end of the year, the fund contains \$620. What annual nominal rate compounded monthly has this fund earned?

34. At the end of each month, an employee deposits \$80 into a credit union account. At the end of 2 years, the account contains \$2,100. What annual nominal rate compounded monthly has this account earned?

In Problems 35 and 36, use graphical approximation techniques to answer the questions.

35. When would an ordinary annuity consisting of quarterly payments of \$500 at 6% compounded quarterly be worth more than a principal of \$5,000 invested at 4% simple interest?

36. When would an ordinary annuity consisting of monthly payments of \$200 at 5% compounded monthly be worth more than a principal of \$10,000 invested at 7.5% compounded monthly?

SECTION 3-4 Present Value of an Annuity; Amortization

- Present Value of an Annuity
- Amortization
- Amortization Schedules
- A General Problem-Solving Strategy

PRESENT VALUE OF AN ANNUITY

How much should you deposit in an account paying 6% compounded semiannually in order to be able to withdraw \$1,000 every 6 months for the next 3 years? (After the last payment is made, no money is to be left in the account.)

Actually, we are interested in finding the **present value** of each \$1,000 that is paid out during the 3 years. We can do this by solving for P in the compound interest formula:

$$A = P(1 + i)^n$$
$$P = \frac{A}{(1 + i)^n} = A(1 + i)^{-n}$$

The rate per period is $i = \frac{0.06}{2} = 0.03$. The present value P of the first payment is $1{,}000(1.03)^{-1}$, the present value of the second payment is $1{,}000(1.03)^{-2}$, and so on. Figure 1 shows this in terms of a time line.

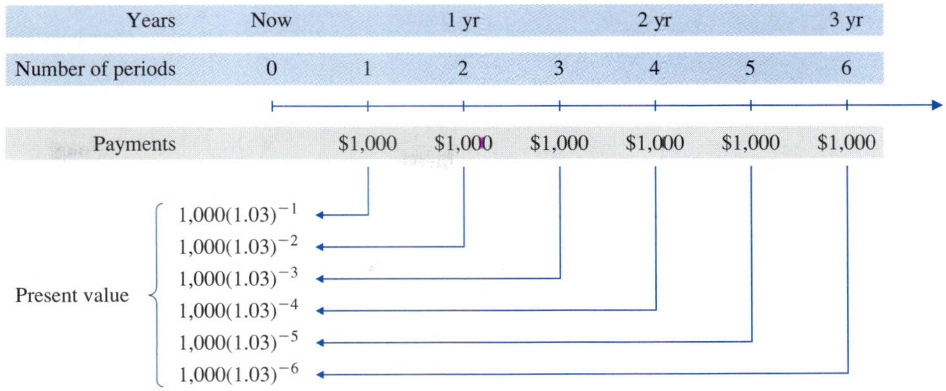

FIGURE 1

We could evaluate each of the present values in Figure 1 using a calculator and add the results to find the total present values of all the payments (which will be the amount that is needed now to buy the annuity). Since this is generally a tedious process, particularly when the number of payments is large, we will use the same device we used in the last section to produce a formula that will accomplish the same result in a couple of steps. We start by writing the sum of the present values in the form

$$P = 1{,}000(1.03)^{-1} + 1{,}000(1.03)^{-2} + \cdots + 1{,}000(1.03)^{-6} \quad (1)$$

Multiplying both sides of (1) by 1.03, we obtain

$$1.03P = 1{,}000 + 1{,}000(1.03)^{-1} + \cdots + 1{,}000(1.03)^{-5} \quad (2)$$

Now subtract (1) from (2):

$$1.03P - P = 1{,}000 - 1{,}000(1.03)^{-6}$$ *Notice how many terms drop out.*

$$0.03P = 1{,}000[1 - (1 + 0.03)^{-6}]$$

$$P = 1{,}000\frac{1 - (1 + 0.03)^{-6}}{0.03} \quad (3)$$ *We write P in this form to observe a general pattern.*

In general, if R is the periodic payment, i the rate per period, and n the number of periods, then the present value of all payments is given by

$$P = R(1 + i)^{-1} + R(1 + i)^{-2} + \cdots + R(1 + i)^{-n}$$ *Note how this compares to (1).*

Proceeding as in the above example, we obtain the general formula for the present value of an ordinary annuity:

$$P = R\frac{1 - (1 + i)^{-n}}{i} \quad (4)$$ *Note how this compares to (3).*

Explore–Discuss 1

Verify formula (4) by applying the technique we used above to sum equation (1) to

$$P = R(1 + i)^{-1} + R(1 + i)^{-2} + \cdots + R(1 + i)^{-n}$$

Returning to the example above, we use a calculator to complete the problem:

$$P = 1{,}000\frac{1 - (1.03)^{-6}}{0.03}$$
$$= \$5{,}417.19$$

The fractional factor of (4) may be denoted by the symbol $a_{\overline{n}|i}$, read as "*a* angle *n* at *i*":

$$a_{\overline{n}|i} = \frac{1 - (1 + i)^{-n}}{i}$$

An alternative approach to the above computation of P is to use a table that gives values of $a_{\overline{n}|i}$ for various values of n and i. The computation of P is then completed by multiplying R by $a_{\overline{n}|i}$ as indicated by (4). (As we noted earlier, a main advantage of using a calculator is that it can handle many more situations than any table.)

It is common to use PV (present value) for P and PMT (payment) for R in formula (4). Making these changes, we have the following:

Present Value of an Ordinary Annuity

$$PV = PMT\frac{1 - (1 + i)^{-n}}{i} = PMT a_{\overline{n}|i} \qquad (5)$$

where

PMT = Periodic payment
i = Rate per period
n = Number of periods
PV = Present value of all payments

[*Note:* Payments are made at the end of each period.]

Example 1 ➡ **Present Value of an Annuity** What is the present value of an annuity that pays \$200 per month for 5 years if money is worth 6% compounded monthly?

SOLUTION To solve this problem, use formula (5) with $PMT = \$200$, $i = \frac{0.06}{12} = 0.005$, and $n = 12(5) = 60$:

$$PV = PMT\frac{1 - (1 + i)^{-n}}{i}$$
$$= 200\frac{1 - (1.005)^{-60}}{0.005} \qquad \text{Use a calculator.}$$
$$= \$10{,}345.11$$

Matched Problem 1 ➡ How much should you deposit in an account paying 8% compounded quarterly in order to receive quarterly payments of \$1,000 for the next 4 years?

Example 2 **Retirement Planning** A person wants to make annual deposits for 25 years into an ordinary annuity that earns 9.4% compounded annually in order to then make 20 equal annual withdrawals of \$25,000, reducing the balance in the account to zero. How much must be deposited annually to accumulate sufficient funds to provide for these payments? How much total interest is earned during this entire 45 year process?

SOLUTION This problem involves both future and present values. Figure 2 illustrates the flow of money into and out of the annuity.

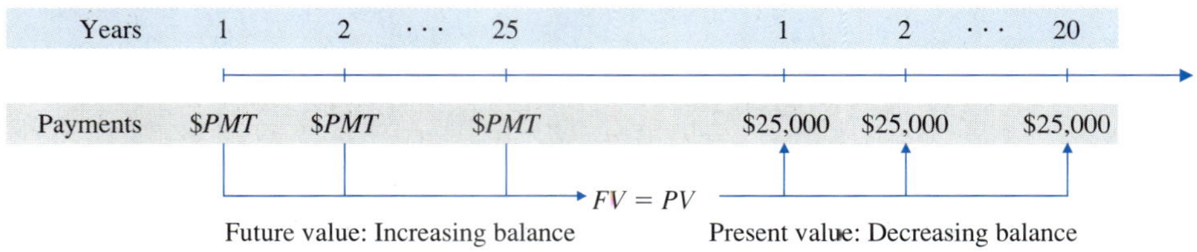

FIGURE 2

Since we are given the required withdrawals, we begin by finding the present value necessary to provide for these withdrawals. Using formula (5) with $PMT = \$25{,}000$, $i = 0.094$, and $n = 20$, we have

$$\begin{aligned} PV &= PMT\frac{1 - (1 + i)^{-n}}{i} \\ &= 25{,}000\frac{1 - (1.094)^{-20}}{0.094} \qquad \text{Use a calculator.} \\ &= \$221{,}854.69 \end{aligned}$$

Now we find the deposits that will produce a future value of \$221,854.69 in 25 years. Using formula (6) from Section 3-3 with $FV = \$221{,}854.69$, $i = 0.094$, and $n = 25$, we have

$$\begin{aligned} PMT &= FV\frac{i}{(1 + i)^n - 1} \\ &= 221{,}854.69\frac{0.094}{(1.094)^{25} - 1} \qquad \text{Use a calculator.} \\ &= \$2{,}467.96 \end{aligned}$$

Thus, depositing \$2,467.96 annually for 25 years will provide for 20 annual withdrawals of \$25,000.

The interest earned during the entire 45 year process is

$$\begin{aligned} \text{Interest} &= (\text{Total withdrawals}) - (\text{Total deposits}) \\ &= 20(25{,}000) - 25(2{,}467.96) \\ &= \$438{,}301 \end{aligned}$$

Matched Problem 2 Refer to Example 2. If \$4,000 is deposited annually for the first 25 years, how much can be withdrawn annually for the next 20 years?

■ AMORTIZATION

The present value formula for an ordinary annuity (5) has another important use. Suppose you borrow \$5,000 from a bank to buy a car and agree to repay the loan in 36 equal monthly payments, including all interest due. If the bank charges 1% per month on the unpaid balance (12% per year compounded monthly), how much should each payment be to retire the total debt including interest in 36 months?

Actually, the bank has bought an annuity from you. The question is: If the bank pays you $5,000 (present value) for an annuity paying them $\$PMT$ per month for 36 months at 12% interest compounded monthly, what are the monthly payments (PMT)? (Note that the value of the annuity at the end of 36 months is zero.) To find PMT, we have only to use formula (5) with $PV = \$5{,}000$, $i = 0.01$, and $n = 36$:

$$\begin{aligned} PV &= PMT\frac{1-(1+i)^{-n}}{i} \\ 5{,}000 &= PMT\frac{1-(1.01)^{-36}}{0.01} \qquad \text{Solve for } PMT \text{ and use a calculator.} \\ &= \$166.07 \text{ per month} \end{aligned}$$

At $166.07 per month, the car will be yours after 36 months. That is, you have *amortized* the debt in 36 equal monthly payments. (*Mort* means "death"; you have "killed" the loan in 36 months.) In general, **amortizing a debt** means that the debt is retired in a given length of time by equal periodic payments that include compound interest. We are usually interested in computing the equal periodic payments. Solving the present value formula (5) for PMT in terms of the other variables, we obtain the following **amortization formula:**

$$PMT = PV\frac{i}{1-(1+i)^{-n}} \tag{6}$$

Formula (6) is simply a variation of formula (5), and either formula can be used to find the periodic payment PMT.

Example 3 **Monthly Payment and Total Interest on an Amortized Debt** Assume that you buy a television set for $800 and agree to pay for it in 18 equal monthly payments at $1\frac{1}{2}\%$ interest per month on the unpaid balance.

(A) How much are your payments? (B) How much interest will you pay?

Solution (A) Use formula (6) with $PV = \$800$, $i = 0.015$, and $n = 18$:

$$\begin{aligned} PMT &= PV\frac{i}{1-(1+i)^{-n}} \\ &= 800\frac{0.015}{1-(1.015)^{-18}} \qquad \text{Use a calculator.} \\ &= \$51.04 \text{ per month} \end{aligned}$$

(B) $$\begin{aligned} \text{Total interest paid} &= (\text{Amount of all payments}) - (\text{Initial loan}) \\ &= 18(\$51.04) - \$800 \\ &= \$118.72 \end{aligned}$$

Matched Problem 3 If you sell your car to someone for $2,400 and agree to finance it at 1% per month on the unpaid balance, how much should you receive each month to amortize the loan in 24 months? How much interest will you receive?

Explore–Discuss 2

To purchase a home, a family plans to sign a mortgage of $70,000 at 8% on the unpaid balance. Discuss the advantages and disadvantages of a 20 year mortgage as opposed to a 30 year mortgage. Include a comparison of monthly payments and total interest paid.

■ Amortization Schedules

What happens if you are amortizing a debt with equal periodic payments and at some point decide to pay off the remainder of the debt in one lump-sum payment? This occurs each time a home with an outstanding mortgage is sold. In order to understand what happens in this situation, we must take a closer look at the amortization process. We begin with an example that is simple enough to allow us to examine the effect each payment has on the debt.

Example 4 **Constructing an Amortization Schedule** If you borrow $500 that you agree to repay in 6 equal monthly payments at 1% interest per month on the unpaid balance, how much of each monthly payment is used for interest and how much is used to reduce the unpaid balance?

Solution First, we compute the required monthly payment using formula (5) or (6). We choose formula (6) with $PV = \$500$, $i = 0.01$, and $n = 6$:

$$
\begin{aligned}
PMT &= PV\frac{i}{1 - (1 + i)^{-n}} \\
&= 500\frac{0.01}{1 - (1.01)^{-6}} \qquad \text{Use a calculator.} \\
&= \$86.27 \text{ per month}
\end{aligned}
$$

At the end of the first month, the interest due is

$$\$500(0.01) = \$5.00$$

The amortization payment is divided into two parts, payment of the interest due and reduction of the unpaid balance (repayment of principal):

Monthly payment		Interest due		Unpaid balance: reduction
$86.27	=	$5.00	+	$81.27

The unpaid balance for the next month is

Previous unpaid balance		Unpaid balance reduction		New unpaid balance
$500.00	−	$81.27	=	$418.73

At the end of the second month, the interest due on the unpaid balance of $418.73 is

$$\$418.73(0.01) = \$4.19$$

Thus, at the end of the second month, the monthly payment of $86.27 covers interest and unpaid balance reduction as follows:

$$\$86.27 = \$4.19 + \$82.08$$

and the unpaid balance for the third month is

$$\$418.73 - \$82.08 = \$336.65$$

This process continues until all payments have been made and the unpaid balance is reduced to zero. The calculations for each month are listed in Table 1, which is referred to as an **amortization schedule.**

Table 1

AMORTIZATION SCHEDULE

PAYMENT NUMBER	PAYMENT	INTEREST	UNPAID BALANCE REDUCTION	UNPAID BALANCE
0				$500.00
1	$ 86.27	$ 5.00	$ 81.27	418.73
2	86.27	4.19	82.08	336.65
3	86.27	3.37	82.90	253.75
4	86.27	2.54	83.73	170.02
5	86.27	1.70	84.57	85.45
6	86.30	0.85	85.45	0.00
Totals	$517.65	$17.65	$500.00	

Notice that the last payment had to be increased by $0.03 in order to reduce the unpaid balance to zero. This small discrepancy is due to round-off errors that occur in the computations. In almost all cases, the last payment must be adjusted slightly in order to obtain a final unpaid balance of exactly zero.

Matched Problem 4 Construct the amortization schedule for a $1,000 debt that is to be amortized in 6 equal monthly payments at 1.25% interest per month on the unpaid balance.

Example 5 **Equity in a Home** A family purchased a home 10 years ago for $80,000. The home was financed by paying 20% down and signing a 30 year mortgage at 9% on the unpaid balance. The net market value of the house (amount received after subtracting all costs involved in selling the house) is now $120,000, and the family wishes to sell the house. How much equity (to the nearest dollar) does the family have in the house now after making 120 monthly payments? [**Equity** = (Current net market value) − (Unpaid loan balance).]*

SOLUTION How can we find the unpaid loan balance after 10 years or 120 monthly payments? One way to proceed would be to construct an amortization schedule, but this would require a table with 120 lines. Fortunately, there is an easier way. The unpaid balance after 120 payments is the amount of the loan that can be paid off with the remaining 240 monthly payments (20 remaining years on the loan). Since the lending institution views a loan as an annuity

*We use the word *equity* in keeping with common usage. If a family wants to sell a house and buy another more expensive house, then the price of a new house that they can afford to buy will often depend on their equity in the first house, where equity is defined by the equation given here. In refinancing a house or taking out an "equity loan," the new mortgage (or second mortgage) often will be based on the equity in the house. Other, more technical definitions of equity do not concern us here.

that they bought from the family, **the unpaid balance of a loan with n remaining payments is the present value of that annuity and can be computed by using formula (5).** Since formula (5) requires knowledge of the monthly payment, we compute PMT first using formula (6).

Step 1. Find the monthly payment:

$$PMT = PV\frac{i}{1-(1+i)^{-n}}$$

$PV = (0.80)(\$80{,}000) = \$64{,}000$
$i = \frac{0.09}{12} = 0.0075$
$n = 12(30) = 360$

$$PMT = 64{,}000\frac{0.0075}{1-(1.0075)^{-360}}$$

Use a calculator.

$$= \$514.96 \text{ per month}$$

Step 2. Find the present value of a \$514.96 per month 20 year annuity:

$$PV = PMT\frac{1-(1+i)^{-n}}{i}$$

$PMT = \$514.96$
$n = 12(20) = 240$
$i = \frac{0.09}{12} = 0.0075$

$$PV = 514.96\frac{1-(1.0075)^{-240}}{0.0075}$$

Use a calculator.

$$= \$57{,}235$$

Unpaid loan balance

Step 3. Find the equity:

$$\begin{aligned}\text{Equity} &= (\text{Current net market value}) - (\text{Unpaid loan balance})\\ &= \$120{,}000 - \$57{,}235\\ &= \$62{,}765\end{aligned}$$

Thus, if the family sells the house for \$120,000 net, they will have \$62,765 after paying off the unpaid loan balance of \$57,235.

Matched Problem 5 A couple purchased a home 20 years ago for \$65,000. The home was financed by paying 20% down and signing a 30 year mortgage at 8% on the unpaid balance. The net market value of the house is now \$130,000, and the couple wishes to sell the house. How much equity (to the nearest dollar) does the couple have in the house now after making 240 monthly payments?

The unpaid loan balance in Example 5 may seem a surprisingly large amount to owe after having made payments for 10 years, but long-term amortizations start out with very small reductions in the unpaid balance. For example, the interest due at the end of the very first period of the loan in Example 5 was

$$\$64{,}000(0.0075) = \$480.00$$

The first monthly payment was divided as follows:

Monthly payment		Interest due		Unpaid balance reduction
\$514.96	−	\$480.00	=	\$34.96

Thus, only \$34.96 was applied to the unpaid balance.

Explore–Discuss 3

(A) A family has an \$85,000, 30 year mortgage at 9.6% compounded monthly. Show that the monthly payments are \$720.94.

(B) Explain why the equation

$$y = 720.94\,\frac{1 - (1.008)^{-12(30-x)}}{0.008}$$

gives the unpaid balance of the loan after x years.

(C) Find the unpaid balance after 5 years, after 10 years, and after 15 years.

(D) When is the unpaid balance exactly half of the original \$85,000?

C (E) Solve part (D) using graphical approximation techniques on a graphing utility (see Fig. 3).

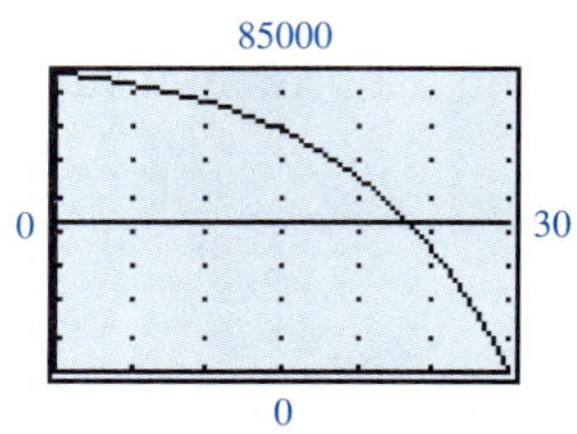

FIGURE 3

General Problem-Solving Strategy

After working the problems in Exercise 3-4, it is very important that you work the problems in the Review Exercise. This will give you valuable experience in distinguishing among the various types of problems we have considered in this chapter. It is impossible to completely categorize all the problems you will encounter, but you may find the following guidelines helpful in determining which of the four basic formulas is involved in a particular problem. Be aware that some problems may involve more than one of these formulas and others may not involve any of them.

A Strategy for Solving Mathematics of Finance Problems

Step 1. Determine whether the problem involves a single payment or a sequence of equal periodic payments. Simple and compound interest problems involve a single present value and a single future value. Ordinary annuities may be concerned with a present value or a future value but always involve a sequence of equal periodic payments.

Step 2. If a single payment is involved, determine whether simple or compound interest is used. Simple interest is usually used for durations of a year or less and compound interest for longer periods.

Step 3. If a sequence of periodic payments is involved, determine whether the payments are being made into an account that is increasing in value—a future value problem—or the payments are being made out of an account that is decreasing in value—a present value problem. Remember that amortization problems always involve the present value of an ordinary annuity.

Steps 1–3 will help you choose the correct formula for a problem, as indicated in Figure 4. Then you must determine the values of the quantities in the formula that are given in the problem and those that must be computed, and solve the problem.

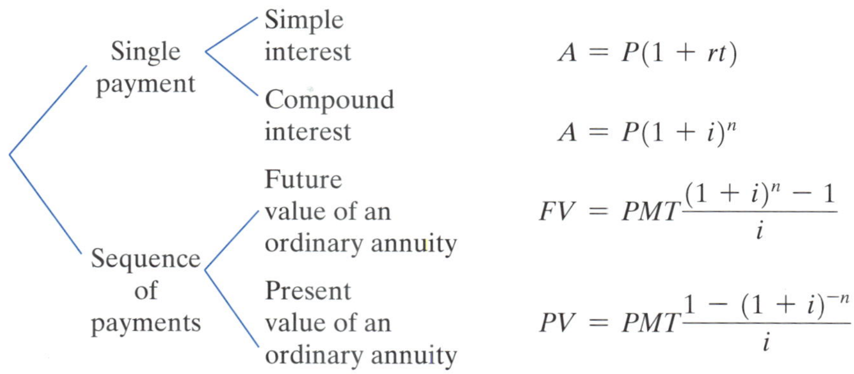

FIGURE 4

Selecting the correct formula for a problem

Answers to Matched Problems

1. \$13,577.71 **2.** \$40,519.27 **3.** PMT = \$112.98/mo; Total interest = \$311.52

4.

Payment Number	Payment	Interest	Unpaid Balance Reduction	Unpaid Balance
0				\$1,000.00
1	\$ 174.03	\$12.50	\$ 161.53	838.47
2	174.03	10.48	163.55	674.92
3	174.03	8.44	165.59	509.33
4	174.03	6.37	167.66	341.67
5	174.03	4.27	169.76	171.91
6	174.06	2.15	171.91	0.00
Totals	\$1,044.21	\$44.21	\$1,000.00	

5. \$98,551

Exercise 3-4

Use formula (5) or (6) to solve each problem.

A

1. $n = 30; i = 0.04; PMT = \$200; PV = ?$

2. $n = 40; i = 0.01; PMT = \$400; PV = ?$

3. $n = 25; i = 0.025; PMT = \$250; PV = ?$

4. $n = 60; i = 0.0075; PMT = \$500; PV = ?$

B

5. $PV = \$6,000; n = 36; i = 0.01; PMT = ?$

6. $PV = \$1,200; n = 40; i = 0.025; PMT = ?$

7. $PV = \$40,000; n = 96; i = 0.0075; PMT = ?$

8. $PV = \$14,000; n = 72; i = 0.005; PMT = ?$

C

9. $PV = \$5,000; i = 0.01; PMT = \$200; n = ?$

10. $PV = \$20,000; i = 0.0175; PMT = \$500; n = ?$

11. $PV = \$9,000; PMT = \$600; n = 20; i = ?$
C (Round answer to three decimal places.)

12. $PV = \$12,000; PMT = \$400; n = 40; i = ?$
C (Round answer to three decimal places.)

APPLICATIONS

Business & Economics

13. A relative wills you an annuity paying $4,000 per quarter for the next 10 years. If money is worth 8% compounded quarterly, what is the present value of this annuity?

14. How much should you deposit in an account paying 12% compounded monthly in order to receive $1,000 per month for the next 2 years?

15. Parents of a college student wish to set up an annuity that will pay $350 per month to the student for 4 years. How much should they deposit now at 9% interest compounded monthly to establish this annuity? How much will the student receive in the 4 years?

16. A person pays $120 per month for 48 months for a car, making no down payment. If the loan costs 1.5% interest per month on the unpaid balance, what was the original cost of the car? How much total interest will be paid?

17. (A) If you buy a stereo set for $600 and agree to pay for it in 18 equal installments at 1% interest per month on the unpaid balance, how much are your monthly payments? How much interest will you pay?
    (B) Repeat part (A) for 1.5% interest per month on the unpaid balance.

18. (A) A company buys a large copy machine for $12,000 and finances it at 12% interest compounded monthly. If the loan is to be amortized in 6 years in equal monthly payments, how much is each payment? How much interest will be paid?
    (B) Repeat part (A) with 18% interest compounded monthly.

19. A sailboat costs $16,000. You pay 25% down and amortize the rest with equal monthly payments over a 6 year period. If you must pay 1.5% interest per month on the unpaid balance (18% compounded monthly), what is your monthly payment? How much interest will you pay over the 6 years?

20. A law firm buys a computerized word-processing system costing $10,000. If the firm pays 20% down and amortizes the rest with equal monthly payments over 5 years at 9% compounded monthly, what will be the monthly payment? How much interest will the firm pay?

21. Construct the amortization schedule for a $5,000 debt that is to be amortized in 8 equal quarterly payments at 4.5% interest per quarter on the unpaid balance.

22. Construct the amortization schedule for a $10,000 debt that is to be amortized in 6 equal quarterly payments at 3.5% interest per quarter on the unpaid balance.

23. A woman borrows $6,000 at 12% compounded monthly, which is to be amortized over 3 years in equal monthly payments. For tax purposes, she needs to know the amount of interest paid during each year of the loan. Find the interest paid during the first year, the second year, and the third year of the loan. [*Hint:* Find the unpaid balance after 12 payments and after 24 payments.]

24. A man establishes an annuity for retirement by depositing $50,000 into an account that pays 9% compounded monthly. Equal monthly withdrawals will be made each month for 5 years, at which time the account will have a zero balance. Each year taxes must be paid on the interest earned by the account during that year. How much interest was earned during the first year? [*Hint:* The amount in the account at the end of the first year is the present value of a 4 year annuity.]

25. Some friends tell you that they paid $25,000 down on a new house and are to pay $525 per month for 30 years. If interest is 9.8% compounded monthly, what was the selling price of the house? How much interest will they pay in 30 years?

26. A family is thinking about buying a new house costing $120,000. They must pay 20% down, and the rest is to be amortized over 30 years in equal monthly payments. If money costs 9.6% compounded monthly, what will their monthly payment be? How much total interest will be paid over the 30 years?

27. A student receives a federally backed student loan of $6,000 at 3.5% interest compounded monthly. After finishing college in 2 years, the student must amortize the loan in the next 4 years by making equal monthly payments. What will the payments be and what total interest will the student pay? [*Hint:* This is a two-part problem. First find the amount of the debt at the end of the first 2 years; then amortize this amount over the next 4 years.]

28. A person establishes a sinking fund for retirement by contributing $7,500 per year at the end of each year for 20 years. For the next 20 years, equal yearly payments are withdrawn, at the end of which time the account will have a zero balance. If money is worth 9% compounded annually, what yearly payments will the person receive for the last 20 years?

29. A family has a $75,000, 30 year mortgage at 13.2% compounded monthly. Find the monthly payment. Also find the unpaid balance after:
    (A) 10 years (B) 20 years (C) 25 years

30. A family has a \$50,000, 20 year mortgage at 10.8% compounded monthly. Find the monthly payment. Also find the unpaid balance after:
(A) 5 years (B) 10 years (C) 15 years

31. A family has a \$30,000, 20 year mortgage at 15% compounded monthly.
(A) Find the monthly payment and the total interest paid.
(B) Suppose the family decides to add an extra \$100 to its mortgage payment each month starting with the very first payment. How long will it take the family to pay off the mortgage? How much interest will the family save?

32. At the time they retire, a couple has \$200,000 in an account that pays 8.4% compounded monthly.
(A) If they decide to withdraw equal monthly payments for 10 years, at the end of which time the account will have a zero balance, how much should they withdraw each month?
(B) If they decide to withdraw \$3,000 a month until the balance in the account is zero, how many withdrawals can they make?

33. An ordinary annuity that earns 7.5% compounded monthly has a current balance of \$500,000. The owner of the account is about to retire and has to decide how much to withdraw from the account each month. Find the number of withdrawals under each of the following options:
(A) \$5,000 monthly (B) \$4,000 monthly
(C) \$3,000 monthly

34. Refer to Problem 33. If the account owner decides to withdraw \$3,000 monthly, how much is in the account after 10 years? After 20 years? After 30 years?

35. An ordinary annuity pays 7.44% compounded monthly.
(A) An individual deposits \$100 monthly for 30 years and then makes equal monthly withdrawals for the next 15 years, reducing the balance to zero. What are the monthly withdrawals? How much interest is earned during the entire 45 year process?
(B) If the individual wants to make withdrawals of \$2,000 per month for the last 15 years, how much must be deposited monthly for the first 30 years?

36. An ordinary annuity pays 6.48% compounded monthly.
(A) An individual wants to make equal monthly deposits into the account for 15 years in order to then make equal monthly withdrawals of \$1,500 for the next 20 years, reducing the balance to zero. How much should be deposited each month for the first 15 years? What is the total interest earned during this 35 year process?
(B) If the individual makes monthly deposits of \$1,000 for the first 15 years, how much can be withdrawn monthly for the next 20 years?

37. A couple wishes to borrow money using the equity in their home for collateral. A loan company will loan them up to 70% of their equity. They purchased their home 12 years ago for \$79,000. The home was financed by paying 20% down and signing a 30 year mortgage at 12% on the unpaid balance. Equal monthly payments were made to amortize the loan over the 30 year period. The net market value of the house is now \$100,000. After making their 144th payment, they applied to the loan company for the maximum loan. How much (to the nearest dollar) will they receive?

38. A person purchased a house 10 years ago for \$100,000. The house was financed by paying 20% down and signing a 30 year mortgage at 9.6% on the unpaid balance. Equal monthly payments were made to amortize the loan over a 30 year period. The owner now (after the 120th payment) wishes to refinance the house because of a need for additional cash. If the loan company agrees to a new 30 year mortgage of 80% of the new appraised value of the house, which is \$136,000, how much cash (to the nearest dollar) will the owner receive after repaying the balance of the original mortgage?

39. Discuss the similarities and differences in the graphs of unpaid balance as a function of time for 30 year mortgages of \$50,000, \$75,000, and \$100,000, respectively, each at 9% compounded monthly (see the figure). Include computations of the monthly payment and total interest paid in each case.

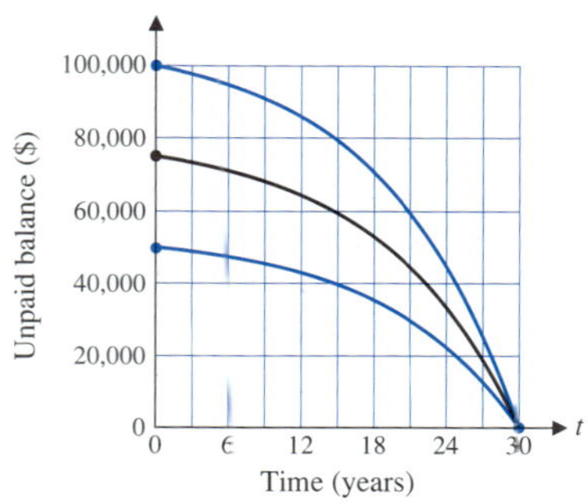

Figure for 39

40. Discuss the similarities and differences in the graphs of unpaid balance as a function of time for 30 year mortgages of \$60,000 at rates of 7%, 10%, and 13%, respectively (see the figure). Include computations of the monthly payment and total interest paid in each case.

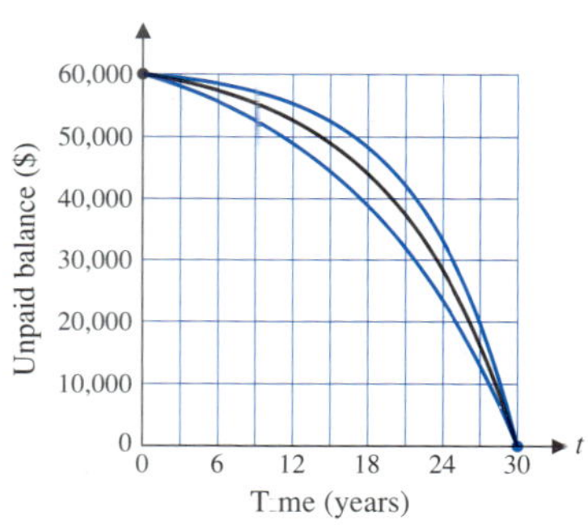

Figure for 40

In Problems 41–44, use graphical approximation techniques or an equation solver to approximate the desired interest rate. Express each answer as a percentage, correct to two decimal places.

41. A discount electronics store offers to let you pay for a \$1,000 stereo in 12 equal \$90 installments. The store claims that since you repay \$1,080 in 1 year, the \$80 finance charge represents an 8% annual rate. This would be true if you repaid the loan in a single payment at the end of the year. But since you start repayment after 1 month, this is an amortized loan, and 8% is not the correct rate. What is the annual nominal compounding rate for this loan? [Did you expect the rate to be this high? Loans of this type are called **add-on interest loans** and were very common before Congress enacted the **Truth in Lending Act.** Now credit agreements must fully disclose all credit costs and must express interest rates in terms of the rates used in the amortization process.]

42. A \$2,000 computer can be financed by paying \$100 per month for 2 years. What is the annual nominal compounding rate for this loan?

43. The owner of a small business has received two offers of purchase. The first prospective buyer offers to pay the owner \$100,000 in cash now. The second offers to pay the owner \$10,000 now and monthly payments of \$1,200 for 10 years. In effect, the second buyer is asking the owner for a \$90,000 loan. If the owner accepts the second offer, what annual nominal compounding rate will the owner receive for financing this purchase?

44. At the time they retire, a couple has \$200,000 invested in an annuity. They can take the entire amount in a single payment, or they can receive monthly payments of \$2,000 for 15 years. If they elect to receive the monthly payments, what annual nominal compounding rate will they earn on the money invested in the annuity?

Important Terms and Symbols

3-1 *Simple Interest.* Principal; interest; interest rate; simple interest; face value; present value; future value

$$I = Prt; \quad A = P(1 + rt)$$

3-2 *Compound Interest.* Compound interest; rate per compounding period; principal (present value); amount (future value); annual nominal rate; effective rate (or annual yield); growth time; rule of 72; zero coupon bond

$$A = P(1 + i)^n; \quad i = \frac{r}{m}; \quad r_e = \left(1 + \frac{r}{m}\right)^m - 1$$

3-3 *Future Value of an Annuity; Sinking Funds.* Annuity; ordinary annuity; future value; balance sheet; sinking fund; sinking fund payment

Future value: $FV = PMT\dfrac{(1 + i)^n - 1}{i}$

Sinking fund: $PMT = FV\dfrac{i}{(1 + i)^n - 1}$

3-4 *Present Value of an Annuity; Amortization.* Present value; amortizing a debt; amortization schedule; equity; problem-solving strategy

Present value: $PV = PMT\dfrac{1 - (1 + i)^{-n}}{i}$

Amortization: $PMT = PV\dfrac{i}{1 - (1 + i)^{-n}}$

Review Exercise

Work through all the problems in this chapter review and check your answers in the back of the book. Answers to all review problems are there along with section numbers in italics to indicate where each type of problem is discussed. Where weaknesses show up, review appropriate sections in the text.

A *In Problems 1–4, find the indicated quantity, given $A = P(1 + rt)$.*

1. $A = ?$; $P = \$100$; $r = 9\%$; $t = 6$ months

2. $A = \$808$; $P = ?$; $r = 12\%$; $t = 1$ month

3. $A = \$212$; $P = \$200$; $r = 8\%$; $t = ?$

4. $A = \$4{,}120$; $P = \$4{,}000$; $r = ?$; $t = 6$ months

B *In Problems 5 and 6, find the indicated quantity, given* $A = P(1 + i)^n$.

5. $A = ?$; $P = \$1{,}200$; $i = 0.005$; $n = 30$

6. $A = \$5{,}000$; $P = ?$; $i = 0.0075$; $n = 60$

In Problems 7 and 8, find the indicated quantity, given

$$FV = PMT\frac{(1 + i)^n - 1}{i}$$

7. $FV = ?$; $PMT = \$1{,}000$; $i = 0.005$; $n = 60$

8. $FV = \$8{,}000$; $PMT = ?$; $i = 0.015$; $n = 48$

In Problems 9 and 10, find the indicated quantity, given

$$PV = PMT\frac{1 - (1 + i)^{-n}}{i}$$

9. $PV = ?$; $PMT = \$2{,}500$; $i = 0.02$; $n = 16$

10. $PV = \$8{,}000$; $PMT = ?$; $i = 0.0075$; $n = 60$

C

11. Solve the equation $2{,}500 = 1{,}000(1.06)^n$ for n to the nearest integer using:
    (A) Logarithms
    C (B) Graphical approximation techniques or an equation solver on a graphing utility

12. Solve the equation

$$5{,}000 = 100\frac{(1.01)^n - 1}{0.01}$$

for n to the nearest integer using:
    (A) Logarithms
    C (B) Graphical approximation techniques or an equation solver on a graphing utility

Applications

Business & Economics

Find all dollar amounts correct to the nearest cent. When an interest rate is requested as an answer, express the rate as a percentage, correct to two decimal places.

13. If you borrow \$3,000 at 14% simple interest for 10 months, how much will you owe in 10 months? How much interest will you pay?

14. Grandparents deposited \$6,000 into a grandchild's account toward a college education. How much money (to the nearest dollar) will be in the account 17 years from now if the account earns 9% compounded monthly?

15. How much should you deposit initially in an account paying 10% compounded semiannually in order to have \$25,000 in 10 years?

16. A savings account pays 5.4% compounded annually. Construct a balance sheet showing the interest earned during each year and the balance at the end of each year for 4 years if:
    (A) A single deposit of \$400 is made at the beginning of the first year.
    (B) Four deposits of \$100 are made at the end of each year.

17. C One investment pays 13% simple interest and another 9% compounded annually. Which investment would you choose? Why?

18. A \$10,000 retirement account is left to earn interest at 7% compounded daily. How much money will be in the account 40 years from now when the owner reaches 65? (Use a 365 day year and round answer to the nearest dollar.)

19. Which is the better investment and why: 9% compounded quarterly or 9.25% compounded annually?

20. What is the value of an ordinary annuity at the end of 8 years if \$200 per month is deposited into an account earning 9% compounded monthly? How much of this value is interest?

21. A credit card company charges a 22% annual rate for overdue accounts. How much interest will be owed on a \$635 account 1 month overdue?

22. What will an \$8,000 car cost (to the nearest dollar) 5 years from now if the inflation rate over that period averages 5% compounded annually?

23. What would the \$8,000 car in Problem 22 have cost (to the nearest dollar) 5 years ago if the inflation rate over that period had averaged 5% compounded annually?

24. A loan of \$2,500 was repaid at the end of 10 months with a check for \$2,812.50. What annual rate of interest was charged?

25. A car salesperson tells you that you can buy the car you are looking at for \$3,000 down and \$200 a month for 48 months. If interest is 14% compounded monthly, what is the selling price of the car and how much interest will you pay during the 48 months?

26. You have \$2,500 toward the purchase of a boat that will cost \$3,000. How long will it take the \$2,500 to grow to \$3,000 if it is invested at 9% compounded quarterly? (Round up to the next higher quarter if not exact.)

27. How long will it take money to double if it is invested at 12% compounded monthly? 18% compounded monthly? (Round up to the next higher month if not exact.)

28. Starting on his 21st birthday, and continuing on every birthday up to and including his 65th, John deposits \$2,000 a year into an IRA. How much (to the nearest dollar) will be in the account on John's 65th birthday, if the account earns:

(A) 7% compounded annually?
(B) 11% compounded annually?

29. If you just sold a stock for \$17,388.17 (net) that cost you \$12,903.28 (net) 3 years ago, what annual compound rate of return did you make on your investment?

30. If you paid \$100 to a loan company for the use of \$1,500 for 120 days, what annual rate of interest did they charge? (Use a 360 day year.)

31. A scholarship committee wishes to establish a sinking fund that will pay \$1,500 per quarter to a student for 2 years. How much should they deposit now into an account that pays 8% compounded quarterly in order to make the scholarship payments starting 3 months from now?

32. An individual wants to establish an annuity for retirement purposes. He wants to make quarterly deposits for 20 years so that he can then make quarterly withdrawals of \$5,000 for 10 years. The annuity earns 12% interest compounded quarterly.

(A) How much will have to be in the account at the time he retires?
(B) How much should be deposited each quarter for 20 years in order to accumulate the required amount?
(C) What is the total amount of interest earned during the 30 year period?

33. A state-of-the-art compact disk stereo system costs \$3,000. You pay one-third down and amortize the rest with equal monthly payments over a 2 year period. If you are charged 1.5% interest per month on the unpaid balance, what is your monthly payment? How much interest will you pay over the 2 years?

34. A company decides to establish a sinking fund to replace a piece of equipment in 6 years at an estimated cost of \$50,000. To accomplish this, they decide to make fixed monthly payments into an account that pays 9% compounded monthly. How much should each payment be?

35. How long will it take money to double if it is invested at 10% compounded daily? 10% compounded annually? (Express answers in years to two decimal places and assume a 365 day year.)

36. A student receives a student loan for \$8,000 at 5.5% interest compounded monthly to help her finish the last 1.5 years of college. Starting 1 year after finishing college, the student must amortize the loan in the next 5 years by making equal monthly payments. What will the payments be and what total interest will the student pay?

37. C A company makes a payment of \$1,200 each month into a sinking fund that earns 6% compounded monthly. Use graphical approximation techniques on a graphing utility to determine when the fund will be worth \$100,000.

38. C A couple has a \$50,000, 20 year mortgage at 9% compounded monthly. Use graphical approximation techniques on a graphing utility to determine when the unpaid balance will drop below \$10,000.

39. A loan company advertises in the paper that you will pay only 8¢ a day for each \$100 borrowed. What annual rate of interest are they charging? (Use a 360 day year.)

40. Construct the amortization schedule for a \$1,000 debt that is to be amortized in 4 equal quarterly payments at 2.5% interest per quarter on the unpaid balance.

41. You can afford monthly deposits of only \$200 into an account that pays 7.98% compounded monthly. How long will it be until you will have \$2,500 to purchase a used car? (Round to the next higher month if not exact.)

42. A company establishes a sinking fund for plant retooling in 6 years at an estimated cost of \$850,000. How much should be invested semiannually into an account paying 8.76% compounded semiannually? How much interest will the account earn in the 6 years?

43. Security Savings & Loan pays 9.38% compounded monthly and West Lake Savings & Loan pays 9.35% compounded daily. Which is the better investment? For each investment, express the effective rate as a percentage, correct to three decimal places. Use a 365 day year.

44. If you buy a 13 week T-bill with a maturity value of \$5,000 for \$4,899.08 from the U.S. Treasury Department, what annual interest rate will you earn?

45. In order to save enough money for the down payment on a condominium, a young couple deposits $200 each month into an account that pays 9% interest compounded monthly. If they need $10,000 for a down payment, how many deposits will they have to make?

46. A business borrows $80,000 at 15% interest compounded monthly for 8 years.
(A) What is the monthly payment?
(B) What is the unpaid balance at the end of the first year?
(C) How much interest was paid during the first year?

47. You unexpectedly inherit $10,000 just after you have made the 72nd monthly payment on a 30 year mortgage of $60,000 at 10.25% compounded monthly. Discuss the relative merits of using the inheritance to reduce the principal of the loan, or to buy a certificate of deposit paying 8.75% compounded monthly.

48. Your parents are considering a $75,000, 30 year mortgage to purchase a new home. The bank at which they have done business for many years offers a rate of 12% compounded monthly. A competitor is offering 11.25% compounded monthly. Would it be worthwhile for your parents to switch banks?

49. How much should a $5,000 face value zero coupon bond, maturing in 5 years, be sold for now, if its rate of return is to be 9.5% compounded annually?

50. If you pay $4,476.20 for a $10,000 face value zero coupon bond that matures in 10 years, what is your annual compound rate of return?

51. If an investor wants to earn an annual interest rate of 10.76% on a 26 week T-bill with a maturity value of $5,000, how much should the investor pay for the T-bill?

52. Two years ago you borrowed $10,000 at 12% interest compounded monthly, which was to be amortized over 5 years. Now you have acquired some additional funds and decide that you want to pay off this loan. What is the unpaid balance after making equal monthly payments for 2 years?

53. A savings and loan company pays 9% compounded monthly. What is the effective rate?

54. (A) A man deposits $2,000 in an IRA on his 21st birthday and on each subsequent birthday up to, and including, his 29th (9 deposits in all). The account earns 8% compounded annually. If he then leaves the money in the account without making any more deposits, how much will he have on his 65th birthday, assuming the account continues to earn the same rate of interest?
(B) How much would be in the account (to the nearest dollar) on his 65th birthday if he had started the deposits on his 30th birthday and continued making deposits on each birthday until (and including) his 65th birthday?

55. In a new housing development, the houses are selling for $100,000 and require a 20% down payment. The buyer is given a choice of 30 year or 15 year financing, both at 10.75% compounded monthly.
(A) What is the monthly payment for the 30 year choice? For the 15 year choice?
(B) What is the unpaid balance after 10 years for the 30 year choice? For the 15 year choice?

56. A loan company will loan up to 60% of the equity in a home. A family purchased their home 8 years ago for $83,000. The home was financed by paying 20% down and signing a 30 year mortgage at 11.25% for the balance. Equal monthly payments were made to amortize the loan over the 30 year period. The market value of the house is now $95,000. After making their 96th payment, the family applied to the loan company for the maximum loan. How much (to the nearest dollar) will they receive?

57. C A $600 stereo is financed for 6 months by making monthly payments of $110. What is the annual nominal compounding rate for this loan?

58. C An individual deposits $2,000 each year for 25 years into an IRA. When she retires immediately after making the 25th deposit, the IRA is worth $220,000.
(A) Find the interest rate earned by the IRA over the 25 year period leading up to retirement.
(B) Assume that the IRA continues to earn the interest rate found in part (A). How long can the retiree withdraw $30,000 per year? How long can she withdraw $24,000 per year?

Group Activity 1 *Reducing Interest Payments on a Home Mortgage*

In appreciation for her parents' financial support of her college education, a recent graduate intends to save her parents some money on the financing of their home. The parents have just made the 180th payment on a 30 year loan of $80,000 at 11%.

(A) How much would be saved in interest payments if the graduate were able to find a lender willing to refinance the unpaid balance over 15 years at 9%? At 8%?

C (B) At what interest rate would the refinancing reduce the total interest payments by $10,000?

(C) Instead of refinancing, the graduate decides to contribute $100 each month to reduce the principal until the loan is paid off. How much will the graduate save her parents in interest payments? How long would she contribute? How much would she contribute?

(D) Refer to part (C). How much would the graduate need to contribute each month in order to save her parents a total of $10,000 in interest payments?

Ask a parent, relative, or friend about the terms of their mortgage. Then do the calculations necessary to develop specific strategies for reducing the interest payments on the loan by refinancing and/or early payment of the principal.

Group Activity 2 *Yield to Maturity and Internal Rate of Return*

A salesman receives annual bonuses for 4 years (see Table 1), which he immediately invests in a mutual fund. The interest earned by this fund varies from year to year.

Table 1

Year	Bonus
1	$3,000
2	$7,000
3	$4,000
4	$8,000

One year after making the last investment, the salesman closes out the fund and receives a payment of $30,000. He would like to find a compound interest rate that represents the interest his investments earned over the entire 4 year period. This rate is called the **yield to maturity.** Since he deposited different amounts each year, none of the annuity formulas discussed earlier can be used in this situation. However, the compound interest formula discussed in Section 3-2 can be used. If r is the yield to maturity, then the compound interest formula expresses the value of each investment at the end of the 4th year in terms of r (Fig. 1).

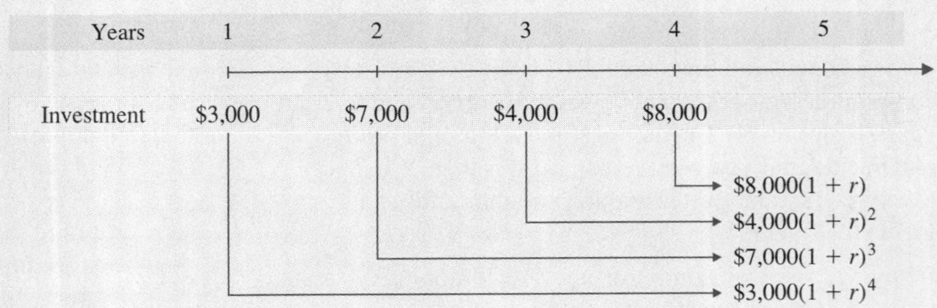

FIGURE 1

The sum of all these values must equal $30,000, the amount in the fund at the end of the 4th year. Thus, the yield to maturity must satisfy the following equation:

$$8{,}000(1 + r) + 4{,}000(1 + r)^2 + 7{,}000(1 + r)^3 + 3{,}000(1 + r)^4 = 30{,}000$$

Using graphical approximation techniques to solve this equation (Fig. 2), the yield to maturity (to two decimal places) is $r = 14.39\%$.

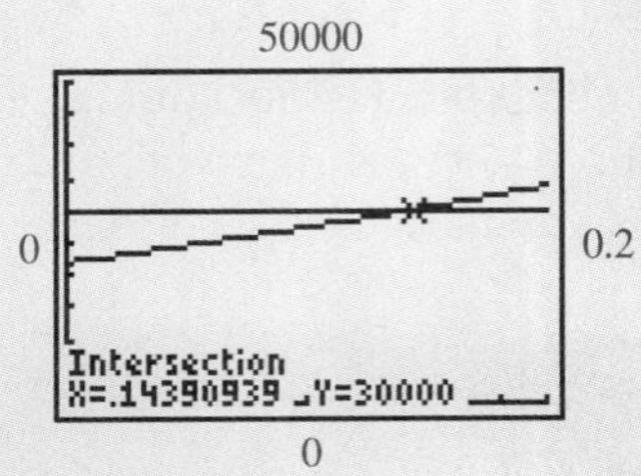

FIGURE 2

(A) In the preceding discussion, we used graphical approximation techniques to find the yield to maturity. If you have access to a graphing utility with an equation solver, use it to find the yield to maturity.

(B) Suppose your grandparents give you $1,000 in stock certificates on your 16th birthday, $1,500 on your 17th birthday, and $2,000 on your 18th birthday. On your 21st birthday, you sell all the stock for $7,000. What is the yield to maturity for this sequence of investments?

(C) A single investment of $10,000 in a small business returns payments of $3,000 at the end of the first year, $4,000 at the end of the second year, and a final payment of $5,000 at the end of the third year. If this investment is to return earnings of $100r\%$, then the sum of the present values of the three annual returns at $100r\%$ must equal the original investment. Show that this leads to the equation

$$10{,}000 = \frac{3{,}000}{1 + r} + \frac{4{,}000}{(1 + r)^2} + \frac{5{,}000}{(1 + r)^3}$$

The solution to this equation is called the **internal rate of return.** Use graphical approximation techniques or an equation solver to find the solution.

(D) An individual pays $160,000 for a two-family housing unit that generates annual profits of $15,000. After 4 years, the property is sold for $190,000. Find the internal rate of return for this investment.

CHAPTER 4

SYSTEMS OF LINEAR EQUATIONS; MATRICES

INTRODUCTION

In this chapter we first review how systems of linear equations involving two variables are solved using techniques learned in elementary algebra. (Basic properties of linear equations and lines are reviewed in Section 1-3 and Appendix A-8.) Because these techniques are not suitable for linear systems involving larger numbers of equations and variables, we then turn to a different method of solution involving the concept of an *augmented matrix,* which arises quite naturally when dealing with larger linear systems. We then study *matrices* and *matrix operations* in their own right as a new mathematical form. With these new operations added to our mathematical toolbox, we return to systems of equations from a fresh point of view. Finally, we discuss Wassily Leontief's Nobel prize-winning application of matrices to an important economics problem.

Many of the matrix methods introduced in this chapter can be performed on a variety of graphing utilities, including graphing calculators, spreadsheets, and software packages such as MatLab or *Explorations in Finite Mathematics* (see Preface). We will see that certain problems can be solved using either graphical approximation methods or matrix methods, whereas others can be solved by only one of these methods. To avoid confusion between the two methods, we will always clearly state which is to be used. Of course, all graphing utility activities continue to be optional and will be so marked, as we have done throughout the rest of the book.

SECTION 4-1

Review: Systems of Linear Equations in Two Variables

- SYSTEMS IN TWO VARIABLES
- GRAPHING
- SUBSTITUTION
- ELIMINATION BY ADDITION
- APPLICATIONS

SYSTEMS IN TWO VARIABLES

To establish basic concepts, consider the following simple example: If 2 adult tickets and 1 child ticket cost \$8, and if 1 adult ticket and 3 child tickets cost \$9, what is the price of each?

Let: x = Price of adult ticket
y = Price of child ticket

Then:
$$\begin{aligned} 2x + y &= 8 \\ x + 3y &= 9 \end{aligned}$$

We now have a system of two linear equations in two variables. It is easy to find ordered pairs (x, y) that satisfy one or the other of these equations. For example, the ordered pair $(4, 0)$ satisfies the first equation, but not the second, and the ordered pair $(6, 1)$ satisfies the second, but not the first. To solve this system, we must find all ordered pairs of real numbers that satisfy both equations at the same time. In general, we have the following definition:

Systems of Two Equations in Two Variables

Given the **linear system**

$$\begin{aligned} ax + by &= h \\ cx + dy &= k \end{aligned}$$

where $a, b, c, d, h,$ and k are real constants, a pair of numbers $x = x_0$ and $y = y_0$ [also written as an ordered pair (x_0, y_0)] is a **solution** of this system if each equation is satisfied by the pair. The set of all such ordered pairs is called the **solution set** for the system. To **solve** a system is to find its solution set.

We will consider three methods of solving such systems: *graphing, substitution,* and *elimination by addition.* Each method has certain advantages, depending on the situation.

Graphing

Recall that the graph of a line is a graph of all the ordered pairs that satisfy the equation of the line. To solve the ticket problem by graphing, we graph both equations in the same coordinate system. The coordinates of any points that the graphs have in common must be solutions to the system, since they must satisfy both equations.

Example 1 **Solving a System by Graphing** Solve the ticket problem by graphing:

$$\begin{aligned} 2x + y &= 8 \\ x + 3y &= 9 \end{aligned}$$

Solution Graph both equations in the same coordinate system (both graphs are straight lines). Then estimate the coordinates of any common points on the two lines.

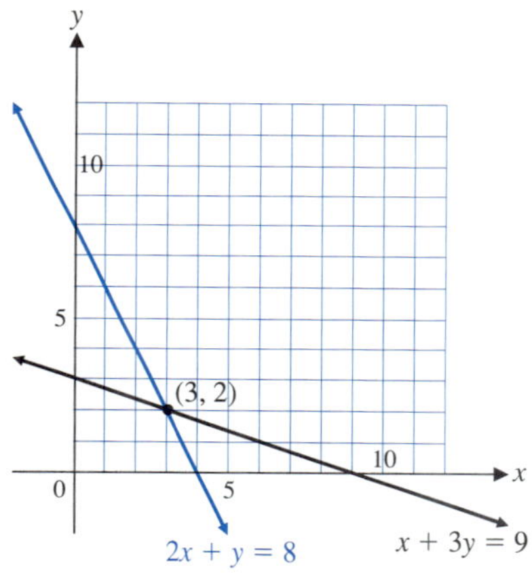

$x = \$3$ Adult ticket
$y = \$2$ Child ticket

Check

$$\begin{aligned} 2x + y &= 8 \\ 2(3) + 2 &\overset{?}{=} 8 \\ 8 &\overset{\checkmark}{=} 8 \end{aligned} \qquad \begin{aligned} x + 3y &= 9 \\ 3 + 3(2) &\overset{?}{=} 9 \\ 9 &\overset{\checkmark}{=} 9 \end{aligned}$$

(3, 2) must satisfy each original equation for a complete check.

Matched Problem 1 Solve by graphing and check:

$$\begin{aligned} 2x - y &= -3 \\ x + 2y &= -4 \end{aligned}$$

It is clear that Example 1 has exactly one solution, since the lines have exactly one point of intersection. In general, lines in a rectangular coordinate system are related to each other in one of the three ways illustrated in the next example.

Example 2 **Solving Systems by Graphing** Solve each of the following systems by graphing:

(A) $\begin{aligned} x - 2y &= 2 \\ x + y &= 5 \end{aligned}$ (B) $\begin{aligned} x + 2y &= -4 \\ 2x + 4y &= 8 \end{aligned}$ (C) $\begin{aligned} 2x + 4y &= 8 \\ x + 2y &= 4 \end{aligned}$

Solution

(A)

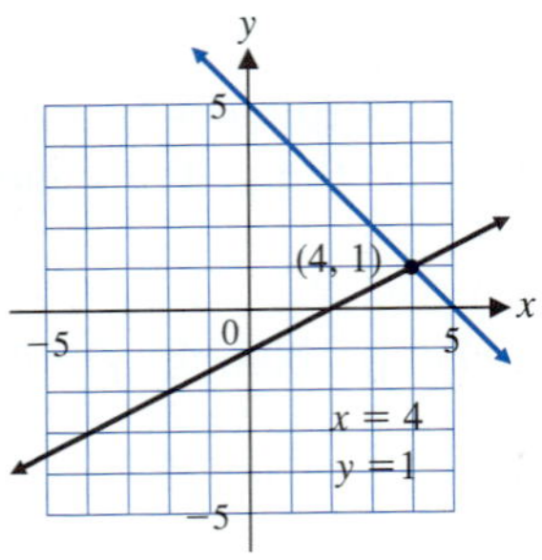

Intersection at one point only—exactly one solution

(B)

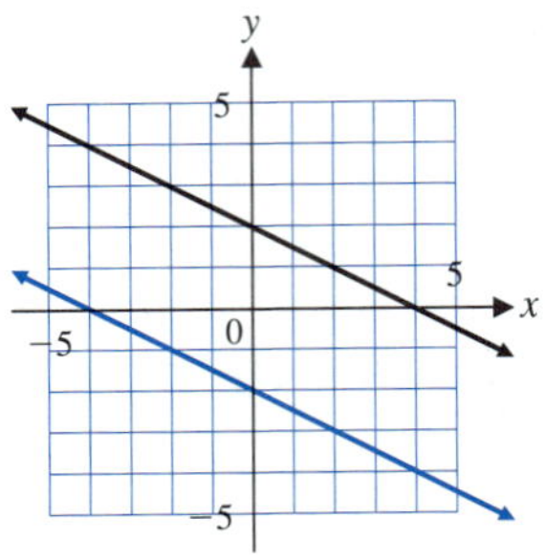

Lines are parallel (each has slope $-\frac{1}{2}$)—no solutions

(C)

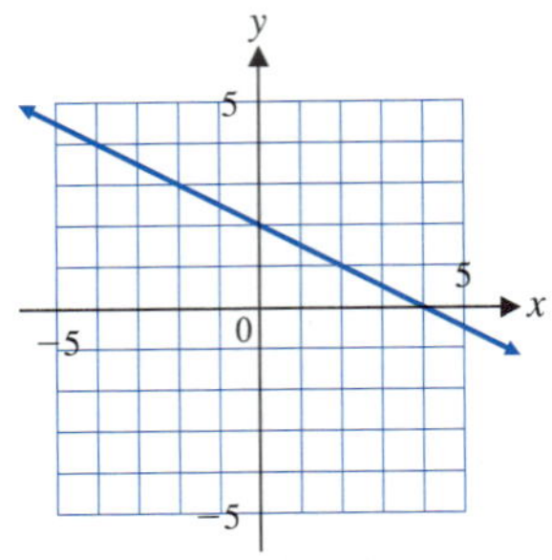

Lines coincide—infinite number of solutions

Matched Problem 2 Solve each of the following systems by graphing:

(A) $\begin{aligned} x + y &= 4 \\ 2x - y &= 2 \end{aligned}$ (B) $\begin{aligned} 6x - 3y &= 9 \\ 2x - y &= 3 \end{aligned}$ (C) $\begin{aligned} 2x - y &= 4 \\ 6x - 3y &= -18 \end{aligned}$

We now define some terms that we can use to describe the different types of solutions to systems of equations that we will encounter.

Systems of Linear Equations: Basic Terms

A system of linear equations is **consistent** if it has one or more solutions and **inconsistent** if no solutions exist. Furthermore, a consistent system is said to be **independent** if it has exactly one solution (often referred to as the **unique solution**) and **dependent** if it has more than one solution.

Referring to the three systems in Example 2, the system in part (A) is a consistent and independent system with the unique solution $x = 4$, $y = 1$. The system in part (B) is inconsistent. And the system in part (C) is a consistent and dependent system with an infinite number of solutions (all the points on the two coinciding lines).

Explore–Discuss 1

Can a consistent and dependent system have exactly two solutions? Exactly three solutions? Explain.

By geometrically interpreting a system of two linear equations in two variables, we gain useful information about what to expect in the way of solutions to the system. In general, any two lines in a coordinate plane must intersect in exactly one point, be parallel, or coincide (have identical graphs).

Thus, the systems in Example 2 illustrate the only three possible types of solutions for systems of two linear equations in two variables. These ideas are summarized in Theorem 1.

Theorem 1 — Possible Solutions to a Linear System

The linear system

$$\begin{aligned} ax + by &= h \\ cx + dy &= k \end{aligned}$$

must have:

(A) Exactly one solution — **Consistent and independent**

Or:

(B) No solution — **Inconsistent**

Or:

(C) Infinitely many solutions — **Consistent and dependent**

There are no other possibilities.

In the past, one drawback of the graphical solution method was the inaccuracy of hand-drawn graphs. The recent advent of graphing utilities has changed that. Graphical solutions performed on a graphing utility provide both a useful geometrical interpretation and an accurate approximation of the solution to a system of linear equations in two variables. Example 3 will demonstrate such a solution.

Example 3 **Solving a System Using a Graphing Utility** Solve to two decimal places using graphical approximation techniques on a graphing utility:

$$\begin{aligned} 5x + 2y &= 15 \\ 2x - 3y &= 16 \end{aligned}$$

(A) Equation definitions

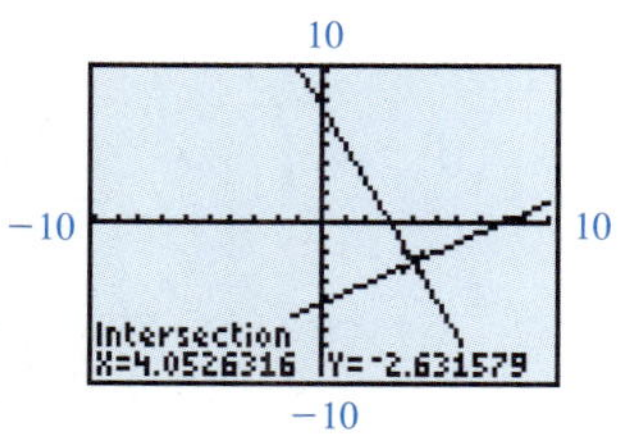

(B) Intersection point

FIGURE 1

SOLUTION First, solve each equation for y:

$$\begin{aligned} 5x + 2y &= 15 \\ 2y &= -5x + 15 \\ y &= -2.5x + 7.5 \end{aligned} \qquad \begin{aligned} 2x - 3y &= 16 \\ -3y &= -2x + 16 \\ y &= \tfrac{2}{3}x - \tfrac{16}{3} \end{aligned}$$

Next, enter each equation in the graphing utility (Fig. 1A), graph in an appropriate viewing window, and approximate the intersection point (Fig. 1B).

Rounding the values in Figure 1B to two decimal places, we see that the solution is $x = 4.05$ and $y = -2.63$, or $(4.05, -2.63)$.

CHECK

$$\begin{aligned} 5x + 2y &= 15 \\ 5(4.05) + 2(-2.63) &\stackrel{?}{=} 15 \\ 14.99 &\approx 15 \end{aligned} \qquad \begin{aligned} 2x - 3y &= 16 \\ 2(4.05) - 3(-2.63) &\stackrel{?}{=} 16 \\ 15.99 &\approx 16 \end{aligned}$$

The checks are not exact because the values of x and y are approximations.

Matched Problem 3 ➡ Solve to two decimal places using graphical approximation techniques on a graphing utility:

$$\begin{aligned} 2x - 5y &= -25 \\ 4x + 3y &= 5 \end{aligned}$$

Graphic methods help us visualize a system and its solutions, frequently reveal relationships that might otherwise be hidden, and, with the assistance of a graphing utility, provide very accurate approximations to solutions.

Substitution

Now we will review an algebraic method that is easy to use and provides exact solutions to a system of two equations in two variables, provided solutions exist. In this method, first we choose one of two equations in a system and solve for one variable in terms of the other. (We make a choice that avoids fractions, if possible.) Then we **substitute** the result into the other equation and solve the resulting linear equation in one variable. Finally, we substitute this result back into the results of the first step to find the second variable. An example should make the process clear.

Example 4 ➡ **Solving a System by Substitution** Solve by substitution:

$$\begin{aligned} 5x + y &= 4 \\ 2x - 3y &= 5 \end{aligned}$$

Solution Solve either equation for one variable in terms of the other; then substitute into the remaining equation. In this problem we can avoid fractions by choosing the first equation and solving for y in terms of x:

$$\begin{aligned} 5x + y &= 4 && \text{Solve the first equation for } y \text{ in terms of } x. \\ y &= 4 - 5x && \text{Substitute into second equation.} \\ 2x - 3y &= 5 && \text{Second equation} \\ 2x - 3(4 - 5x) &= 5 && \text{Solve for } x. \\ 2x - 12 + 15x &= 5 \\ 17x &= 17 \\ x &= 1 \end{aligned}$$

Now, replace x with 1 in $y = 4 - 5x$ to find y:

$$\begin{aligned} y &= 4 - 5x \\ y &= 4 - 5(1) \\ y &= -1 \end{aligned}$$

Thus, the solution is $x = 1$, $y = -1$ or $(1, -1)$.

Check

$$\begin{aligned} 5x + y &= 4 \\ 5(1) + (-1) &\stackrel{?}{=} 4 \\ 4 &\stackrel{\checkmark}{=} 4 \end{aligned} \qquad \begin{aligned} 2x - 3y &= 5 \\ 2(1) - 3(-1) &\stackrel{?}{=} 5 \\ 5 &\stackrel{\checkmark}{=} 5 \end{aligned}$$

Matched Problem 4 Solve by substitution:

$$\begin{aligned} 3x + 2y &= -2 \\ 2x - y &= -6 \end{aligned}$$

Explore–Discuss 2

Return to Example 2 and solve each system by substitution. Based on your results, describe how you can recognize a dependent system or an inconsistent system when using substitution.

Elimination by Addition

The methods of graphing and substitution both work well for systems involving two variables. However, neither is easily extended to larger systems. Now we turn to **elimination by addition.** This is probably the most important method of solution. It readily generalizes to larger systems and forms the basis for computer-based solution methods.

You are already familiar with operations that can be performed on a single equation without changing its solution set (see Appendix A-8). Similarly, elimination by addition involves performing appropriate operations on a system of equations to produce new and simpler *equivalent* systems with the same solution set. In general, we say that two systems of equations are **equivalent** if they have the same solution set. Theorem 2 lists three useful operations that produce equivalent systems.

Theorem 2 Operations That Produce Equivalent Systems

A system of linear equations is transformed into an equivalent system if:

(A) Two equations are interchanged.
(B) An equation is multiplied by a nonzero constant.
(C) A constant multiple of one equation is added to another equation.

Any one of the three operations in Theorem 2 can be used to produce an equivalent system, but the operations in parts (B) and (C) will be of most use to us now. Part (A) becomes useful when we apply the theorem to larger systems. The use of Theorem 2 is best illustrated by examples.

Example 5 **Solving a System Using Elimination by Addition** Solve the following system using elimination by addition:

$$\begin{aligned} 3x - 2y &= 8 \\ 2x + 5y &= -1 \end{aligned}$$

Solution We use Theorem 2 to eliminate one of the variables, thus obtaining a system with an obvious solution:

$$\begin{aligned} 3x - 2y &= 8 \\ 2x + 5y &= -1 \end{aligned}$$ Multiply the top equation by 5 and the bottom equation by 2 (Theorem 2B).

$$\begin{aligned} \mathbf{5}(3x - 2y) &= \mathbf{5}(8) \\ \mathbf{2}(2x + 5y) &= \mathbf{2}(-1) \end{aligned}$$

$$\begin{aligned} 15x - 10y &= 40 \\ 4x + 10y &= -2 \\ \hline 19x \qquad &= 38 \end{aligned}$$

Add the top equation to the bottom equation (Theorem 2C), eliminating the y terms.

Divide both sides by 19, which is the same as multiplying the equation by $\frac{1}{19}$ (Theorem 2B).

$$\mathbf{x = 2}$$

This equation paired with either of the two original equations produces a system equivalent to the original system.

Knowing that $x = 2$, we substitute this number back into either of the two original equations (we choose the second) to solve for y:

$$\begin{aligned} 2(\mathbf{2}) + 5y &= -1 \\ 5y &= -5 \\ \mathbf{y} &= \mathbf{-1} \end{aligned}$$

Thus, the solution is $x = 2$, $y = -1$ or $(2, -1)$.

CHECK

$$\begin{aligned} 3x - \quad 2y &\stackrel{?}{=} 8 \\ 3(2) - 2(-1) &\stackrel{?}{=} 8 \\ 8 &\stackrel{\checkmark}{=} 8 \end{aligned} \qquad \begin{aligned} 2x + \quad 5y &= -1 \\ 2(2) + 5(-1) &\stackrel{?}{=} -1 \\ -1 &\stackrel{\checkmark}{=} -1 \end{aligned}$$

Matched Problem 5 ➡ Solve the following system using elimination by addition:

$$\begin{aligned} 5x - 2y &= 12 \\ 2x + 3y &= 1 \end{aligned}$$

Let us see what happens in the elimination process when a system has either no solution or infinitely many solutions. Consider the following system:

$$\begin{aligned} 2x + 6y &= -3 \\ x + 3y &= 2 \end{aligned}$$

Multiplying the second equation by -2 and adding, we obtain

$$\begin{aligned} 2x + 6y &= -3 \\ -2x - 6y &= -4 \\ \hline 0 &= -7 \end{aligned}$$

Not possible

We have obtained a contradiction. The assumption that the original system has solutions must be false (otherwise, we have proved that $0 = -7$!). Thus, the system has no solutions, and its solution set is the empty set. The graphs of the equations are parallel and the system is inconsistent.

Now consider the system

$$\begin{aligned} x - \tfrac{1}{2}y &= 4 \\ -2x + y &= -8 \end{aligned}$$

If we multiply the top equation by 2 and add the result to the bottom equation, we obtain

$$\begin{array}{r} 2x - y = 8 \\ -2x + y = -8 \\ \hline 0 = 0 \end{array}$$

Obtaining $0 = 0$ by addition implies that the equations are equivalent; that is, their graphs coincide and the system is dependent. If we let $x = k$, where k is any real number, and solve either equation for y, we obtain $y = 2k - 8$. Thus, $(k, 2k - 8)$ is a solution for any real number k. The variable k is called a **parameter,** and replacing k with a real number produces a **particular solution** to the system. For example, some particular solutions to this system are

$k = -1$	$k = 2$	$k = 5$	$k = 9.4$
$(-1, -10)$	$(2, -4)$	$(5, 2)$	$(9.4, 10.8)$

Applications

Many real-world problems are readily solved by constructing a mathematical model consisting of two linear equations in two variables and applying the solution methods we have discussed. We shall examine two applications in detail.

Example 6 **Diet** A woman wants to use milk and orange juice to increase the amount of calcium and vitamin A in her daily diet. An ounce of milk contains 37 milligrams of calcium and 57 micrograms* of vitamin A. An ounce of orange juice contains 5 milligrams of calcium and 65 micrograms of vitamin A. How many ounces of milk and orange juice should the woman drink each day to provide exactly 500 milligrams of calcium and 1,200 micrograms of vitamin A?

Solution First, we define the relevant variables:

x = Number of ounces of milk
y = Number of ounces of orange juice

	Milk	Orange Juice	Total Needed
Calcium	37	5	500
Vitamin A	57	65	1,200

Next, we summarize the given information in the table in the margin. It is convenient to organize the table so that the quantities represented by the variables correspond to columns in the table (rather than to rows), as shown.

Now we use the information in the table to form equations involving x and y:

$$\begin{pmatrix}\text{Calcium in } x \text{ oz}\\ \text{of milk}\end{pmatrix} + \begin{pmatrix}\text{Calcium in } y \text{ oz}\\ \text{of orange juice}\end{pmatrix} = \begin{pmatrix}\text{Total calcium}\\ \text{needed (mg)}\end{pmatrix}$$

$$37x + 5y = 500$$

$$\begin{pmatrix}\text{Vitamin A in } x \text{ oz}\\ \text{of milk}\end{pmatrix} + \begin{pmatrix}\text{Vitamin A in } y \text{ oz}\\ \text{of orange juice}\end{pmatrix} = \begin{pmatrix}\text{Total vitamin A}\\ \text{needed } (\mu\text{g})\end{pmatrix}$$

$$57x + 65y = 1{,}200$$

*A microgram (μg) is one-millionth (10^{-6}) of a gram.

Thus, we have the following model to solve:

$$\begin{aligned} 37x + 5y &= 500 \\ 57x + 65y &= 1{,}200 \end{aligned}$$

We multiply the first equation by -13 and use elimination by addition:

$$\begin{array}{rcl} -481x - 65y &=& -6{,}500 \\ 57x + 65y &=& 1{,}200 \\ \hline -424x &=& -5{,}300 \\ x &=& \mathbf{12.5} \end{array} \qquad \begin{array}{rcl} 37(\mathbf{12.5}) + 5y &=& 500 \\ 5y &=& 37.5 \\ y &=& \mathbf{7.5} \end{array}$$

Drinking 12.5 ounces of milk and 7.5 ounces of orange juice each day will provide the required amounts of calcium and vitamin A.

Check

$$\begin{array}{rcl} 37x + 5y &=& 500 \\ 37(12.5) + 5(7.5) &\stackrel{?}{=}& 500 \\ 500 &\stackrel{\checkmark}{=}& 500 \end{array} \qquad \begin{array}{rcl} 57x + 65y &=& 1{,}200 \\ 57(12.5) + 65(7.5) &\stackrel{?}{=}& 1{,}200 \\ 1{,}200 &\stackrel{\checkmark}{=}& 1{,}200 \end{array}$$

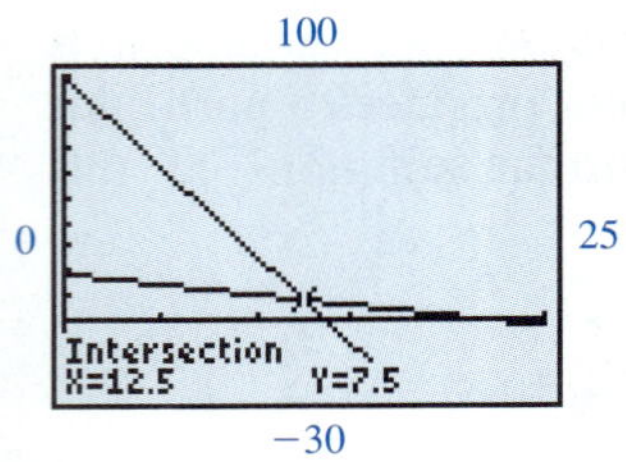

FIGURE 2
$y_1 = (500 - 37x)/5$
$y_2 = (1{,}200 - 57x)/65$

Figure 2 illustrates a solution to Example 6 using graphical approximation techniques.

Matched Problem 6 A man wants to use cottage cheese and yogurt to increase the amount of protein and calcium in his daily diet. An ounce of cottage cheese contains 3 grams of protein and 15 milligrams of calcium. An ounce of yogurt contains 1 gram of protein and 41 milligrams of calcium. How many ounces of cottage cheese and yogurt should he eat each day to provide exactly 62 grams of protein and 760 milligrams of calcium?

Example 7 **Supply and Demand** The quantity of a product that people are willing to buy during some period of time depends on its price. Generally, the higher the price, the less the demand; the lower the price, the greater the demand. Similarly, the quantity of a product that a supplier is willing to sell during some period of time also depends on the price. Generally, a supplier will be willing to supply more of a product at higher prices and less of a product at lower prices. The simplest supply and demand model is a linear model where the graphs of a demand equation and a supply equation are straight lines.

Suppose we are interested in analyzing the sale of cherries each day in a particular city. Using special analytical techniques (regression analysis) and collected data, an analyst arrives at the following price–demand and price–supply models:

$$p = -0.2q + 4 \qquad \text{Price–demand equation (consumer)}$$
$$p = 0.07q + 0.76 \qquad \text{Price–supply equation (supplier)}$$

where q represents the quantity of cherries in thousands of pounds and p represents the price in dollars. For example, we see that consumers will purchase 10 thousand pounds ($q = 10$) when the price is $p = -0.2(10) + 4 = \$2$ per pound. On the other hand, suppliers will be willing to supply 17.714 thousand pounds of cherries at \$2 per pound (solve $2 = 0.07q + 0.76$). Thus, at

\$2 per pound the suppliers are willing to supply more cherries than consumers are willing to purchase. The supply exceeds the demand at that price and the price will come down. At what price will cherries stabilize for the day? That is, at what price will supply equal demand? This price, if it exists, is called the **equilibrium price,** and the quantity sold at that price is called the **equilibrium quantity.** The point where the two curves for the price–demand equation and the price–supply equation intersect is called the **equilibrium point.** How do we find these quantities? We solve the linear system

$$\begin{aligned} p &= -0.2q + 4 &&\text{Demand equation}\\ p &= 0.07q + 0.76 &&\text{Supply equation}\end{aligned}$$

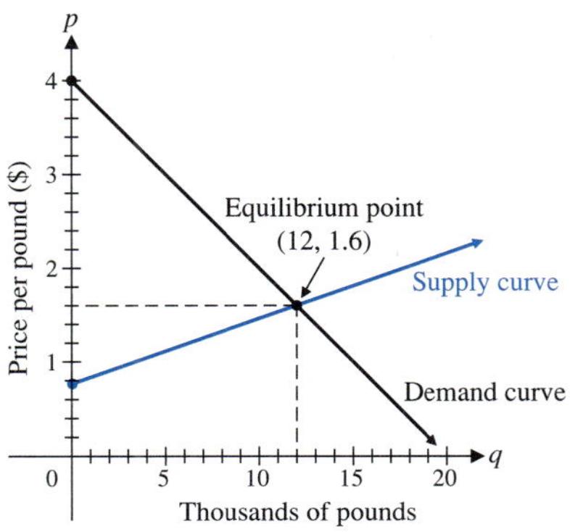

FIGURE 3

by using substitution (substituting $p = -0.2q + 4$ into the second equation):

$$\begin{aligned} -0.2q + 4 &= 0.07q + 0.76\\ -0.27q &= -3.24\\ q &= \mathbf{12\ thousand\ pounds} &&\text{Equilibrium quantity}\end{aligned}$$

Now substitute $q = 12$ back into either of the original equations in the system and solve for p (we choose the first equation):

$$\begin{aligned} p &= -0.2(12) + 4\\ p &= \mathbf{\$1.60\ per\ pound} &&\text{Equilibrium price}\end{aligned}$$

These results are interpreted geometrically in Figure 3.

If the price is above the equilibrium price of \$1.60 per pound, the supply will exceed the demand and the price will come down. If the price is below the equilibrium price of \$1.60 per pound, the demand will exceed the supply and the price will rise. Thus, the price will reach equilibrium at \$1.60. At this price, suppliers will supply 12 thousand pounds of cherries and consumers will purchase 12 thousand pounds.

Matched Problem 7 ➡ Repeat Example 7 (including drawing the graph) given:

$$\begin{aligned} p &= -0.1q + 3 &&\text{Demand equation}\\ p &= 0.08q + 0.66 &&\text{Supply equation}\end{aligned}$$

Answers to Matched Problems

1. $x = -2, y = -1$

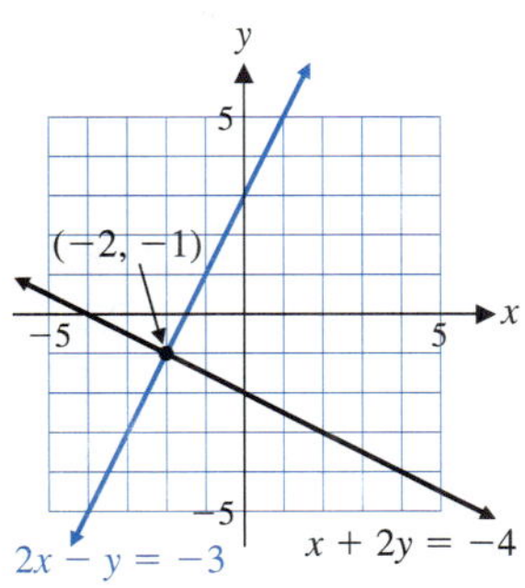

Check:

$$\begin{aligned} 2x - y &= -3\\ 2(-2) - (-1) &\overset{?}{=} -3\\ -3 &\overset{\checkmark}{=} -3\end{aligned}$$

$$\begin{aligned} x + 2y &= -4\\ (-2) + 2(-1) &\overset{?}{=} -4\\ -4 &\overset{\checkmark}{=} -4\end{aligned}$$

2. (A) $x = 2, y = 2$ (B) Infinitely many solutions (C) No solution

3. $x = -1.92, y = 4.23$ **4.** $x = -2, y = 2$ **5.** $x = 2, y = -1$

6. 16.5 oz of cottage cheese, 12.5 oz of yogurt

7. Equilibrium quantity = 13 thousand pounds; Equilibrium price = \$1.70 per pound

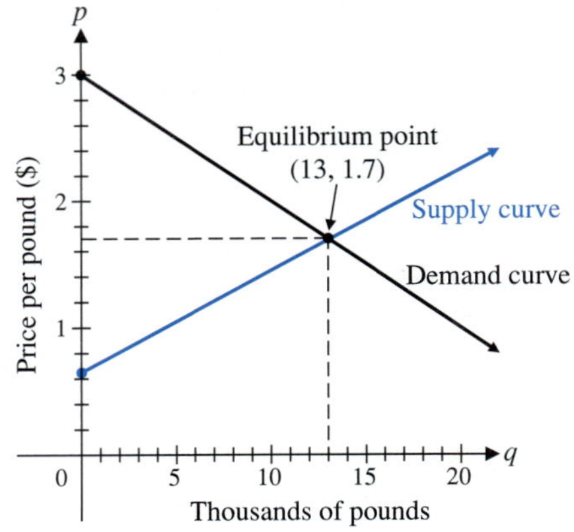

Exercise 4-1

A *Match each system in Problems 1–4 with one of the following graphs, and use the graph to solve the system.*

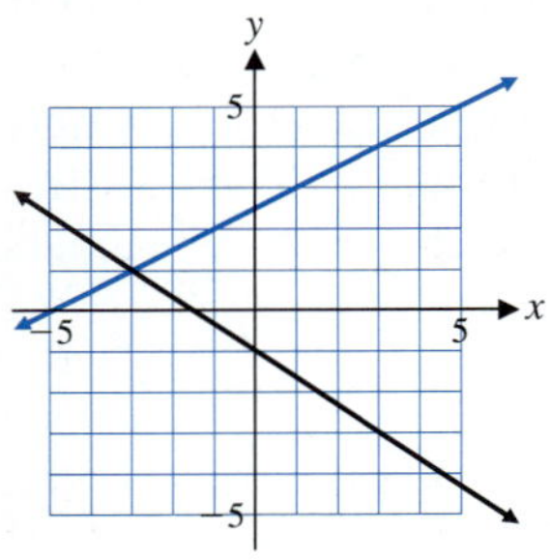

(A)

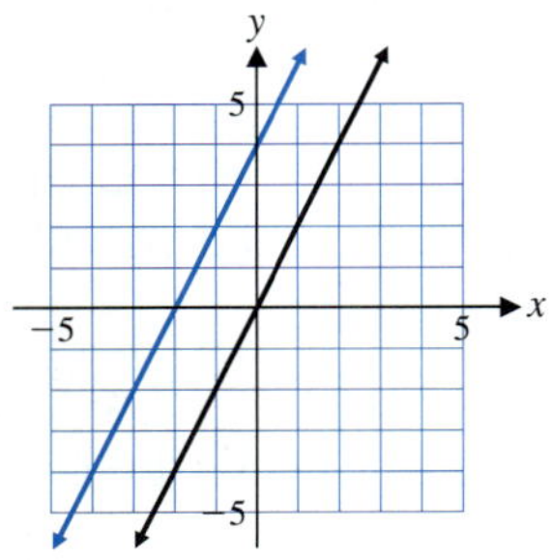

(B)

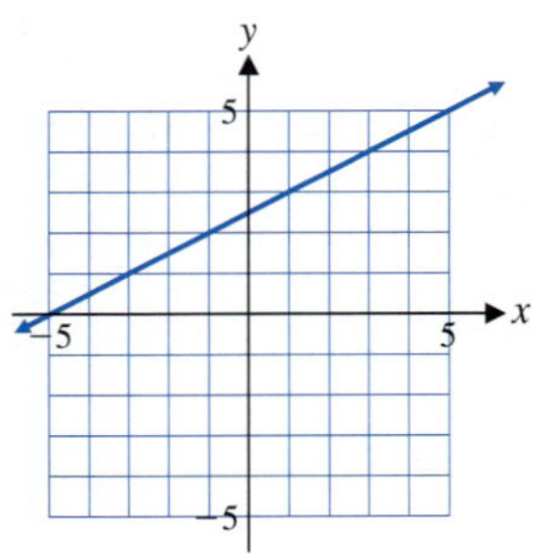

(C)

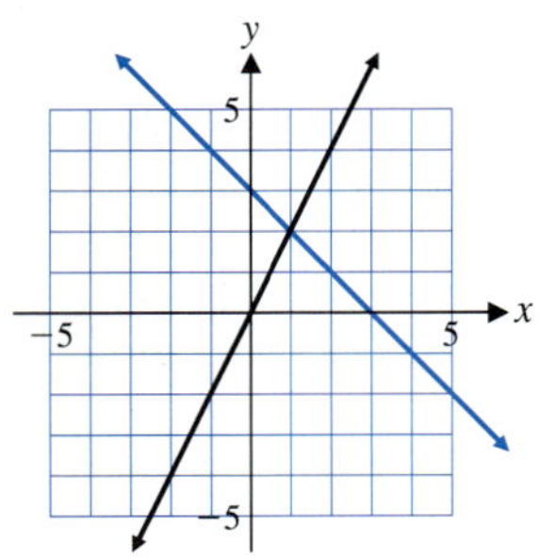

(D)

1. $\begin{aligned} -4x + 2y &= 8 \\ 2x - y &= 0 \end{aligned}$

2. $\begin{aligned} x + y &= 3 \\ 2x - y &= 0 \end{aligned}$

3. $\begin{aligned} -x + 2y &= 5 \\ 2x + 3y &= -3 \end{aligned}$

4. $\begin{aligned} 2x - 4y &= -10 \\ -x + 2y &= 5 \end{aligned}$

Solve Problems 5–8 by graphing.

5. $\begin{aligned} 3x - y &= 2 \\ x + 2y &= 10 \end{aligned}$

6. $\begin{aligned} 3x - 2y &= 12 \\ 7x + 2y &= 8 \end{aligned}$

7. $\begin{aligned} m + 2n &= 4 \\ 2m + 4n &= -8 \end{aligned}$

8. $\begin{aligned} 3u + 5v &= 15 \\ 6u + 10v &= -30 \end{aligned}$

Solve Problems 9–12 using substitution.

9. $\begin{aligned} y &= 2x - 3 \\ x + 2y &= 14 \end{aligned}$

10. $\begin{aligned} y &= x - 4 \\ x + 3y &= 12 \end{aligned}$

11. $\begin{aligned} 2x + y &= 6 \\ x - y &= -3 \end{aligned}$

12. $\begin{aligned} 3x - y &= 7 \\ 2x + 3y &= 1 \end{aligned}$

Solve Problems 13–16 using elimination by addition.

13. $\begin{aligned} 3u - 2v &= 12 \\ 7u + 2v &= 8 \end{aligned}$

14. $\begin{aligned} 2x - 3y &= -8 \\ 5x + 3y &= 1 \end{aligned}$

15. $\begin{aligned} 2m - n &= 10 \\ m - 2n &= -4 \end{aligned}$

16. $\begin{aligned} 2x + 3y &= 1 \\ 3x - y &= 7 \end{aligned}$

B *Solve Problems 17–30 using substitution or elimination by addition.*

17. $\begin{aligned} 9x - 3y &= 24 \\ 11x + 2y &= 1 \end{aligned}$

18. $\begin{aligned} 4x + 3y &= 26 \\ 3x - 11y &= -7 \end{aligned}$

19. $\begin{aligned} 2x - 3y &= -2 \\ -4x + 6y &= 7 \end{aligned}$

20. $\begin{aligned} 3x - 6y &= -9 \\ -2x + 4y &= 12 \end{aligned}$

21. $\begin{aligned} 3x + 8y &= 4 \\ 15x + 10y &= -10 \end{aligned}$

22. $\begin{aligned} 7m + 12n &= -1 \\ 5m - 3n &= 7 \end{aligned}$

23. $\begin{aligned} -6x + 10y &= -30 \\ 3x - 5y &= 15 \end{aligned}$

24. $\begin{aligned} 2x + 4y &= -8 \\ x + 2y &= 4 \end{aligned}$

25. $\begin{aligned} x + y &= 1 \\ 0.3x - 0.4y &= 0 \end{aligned}$

26. $\begin{aligned} x + y &= 1 \\ 0.5x - 0.4y &= 0 \end{aligned}$

27. $\begin{aligned} 0.2x - 0.5y &= 0.07 \\ 0.8x - 0.3y &= 0.79 \end{aligned}$

28. $\begin{aligned} 0.3u - 0.6v &= 0.18 \\ 0.5u + 0.2v &= 0.54 \end{aligned}$

29. $\begin{aligned} \tfrac{2}{5}x + \tfrac{3}{2}y &= 2 \\ \tfrac{7}{3}x - \tfrac{5}{4}y &= -5 \end{aligned}$

30. $\begin{aligned} \tfrac{7}{2}x - \tfrac{5}{6}y &= 10 \\ \tfrac{2}{5}x + \tfrac{4}{3}y &= 6 \end{aligned}$

In Problems 31–34, use a graphing calculator to approximate the solution of each system to two decimal places.

31. $\begin{aligned} 3x - 2y &= 5 \\ 4x + 3y &= 13 \end{aligned}$

32. $\begin{aligned} 3x - 7y &= -20 \\ 2x + 5y &= 8 \end{aligned}$

33. $\begin{aligned} -2.4x + 3.5y &= 0.1 \\ -1.7x + 2.6y &= -0.2 \end{aligned}$

34. $\begin{aligned} 4.2x + 5.4y &= -12.9 \\ 6.4x + 3.7y &= -4.5 \end{aligned}$

C *In Problems 35–40, graph all three equations in the same coordinate system, and then find the coordinates of any points where two or more lines intersect.*

35. $\begin{aligned} x - 2y &= -6 \\ 2x + y &= 8 \\ x + 2y &= -2 \end{aligned}$

36. $\begin{aligned} x + y &= 3 \\ x + 3y &= 15 \\ 3x - y &= 5 \end{aligned}$

37. $\begin{aligned} x + y &= 1 \\ x - 2y &= -8 \\ 3x + y &= -3 \end{aligned}$

38. $\begin{aligned} x - y &= 6 \\ x - 2y &= 8 \\ x + 4y &= -4 \end{aligned}$

39. $\begin{aligned} 4x - 3y &= -24 \\ 2x + 3y &= 12 \\ 8x - 6y &= 24 \end{aligned}$

40. $\begin{aligned} 2x + 3y &= 18 \\ 2x - 6y &= -6 \\ 4x + 6y &= -24 \end{aligned}$

41. The coefficients of the three systems given below are very similar. One might guess that the solution sets to the three systems would also be nearly identical. Develop evidence for or against this guess by considering graphs of the systems and solutions obtained using substitution or elimination by addition.

(A) $\begin{aligned} 5x + 4y &= 4 \\ 11x + 9y &= 4 \end{aligned}$ (B) $\begin{aligned} 5x + 4y &= 4 \\ 11x + 8y &= 4 \end{aligned}$

(C) $\begin{aligned} 5x + 4y &= 4 \\ 10x + 8y &= 4 \end{aligned}$

42. Repeat Problem 41 for the following systems:

(A) $\begin{aligned} 6x - 5y &= 10 \\ -13x + 11y &= -20 \end{aligned}$ (B) $\begin{aligned} 6x - 5y &= 10 \\ -13x + 10y &= -20 \end{aligned}$

(C) $\begin{aligned} 6x - 5y &= 10 \\ -12x + 10y &= -20 \end{aligned}$

APPLICATIONS

Business & Economics

43. *Supply and demand.* Suppose the supply and demand equations for printed T-shirts in a resort town for a particular week are

$$\begin{aligned} p &= 0.7q + 3 && \text{Supply equation} \\ p &= -1.7q + 15 && \text{Demand equation} \end{aligned}$$

where p is the price in dollars and q is the quantity in hundreds.

(A) Find the supply and the demand (to the nearest unit) if T-shirts are priced at $4 each. Discuss the stability of the T-shirt market at this price level.

(B) Find the supply and the demand (to the nearest unit) if T-shirts are priced at $9 each. Discuss the stability of the T-shirt market at this price level.

(C) Find the equilibrium price and quantity.

(D) Graph the two equations in the same coordinate system and identify the equilibrium point, supply curve, and demand curve.

44. *Supply and demand.* Suppose the supply and demand for printed baseball caps in a resort town for a particular week are

$$\begin{aligned} p &= 0.4q + 3.2 && \text{Supply equation} \\ p &= -1.9q + 17 && \text{Demand equation} \end{aligned}$$

where p is the price in dollars and q is the quantity in hundreds.

(A) Find the supply and the demand (to the nearest unit) if baseball caps are priced at $4 each. Discuss the stability of the baseball cap market at this price level.

(B) Find the supply and the demand (to the nearest unit) if baseball caps are priced at $9 each. Discuss the stability of the baseball cap market at this price level.

(C) Find the equilibrium price and quantity.

(D) Graph the two equations in the same coordinate system and identify the equilibrium point, supply curve, and demand curve.

45. *Supply and demand.* At \$0.60 per bushel, the daily supply for wheat is 450 bushels, and the daily demand is 570 bushels. When the price is raised to \$0.75 per bushel, the daily supply increases to 600 bushels, and the daily demand decreases to 495 bushels. Assume that the supply and demand equations are linear.

(A) Find the supply equation. [*Hint:* Write the supply equation in the form $p = aq + b$ and solve for a and b.]

(B) Find the demand equation.

(C) Find the equilibrium price and quantity.

(D) Graph the two equations in the same coordinate system and identify the equilibrium point, supply curve, and demand curve.

46. *Supply and demand.* At \$1.40 per bushel, the daily supply for oats is 850 bushels, and the daily demand is 580 bushels. When the price falls to \$1.20 per bushel, the daily supply decreases to 350 bushels, and the daily demand increases to 980 bushels. Assume that the supply and demand equations are linear.

(A) Find the supply equation.

(B) Find the demand equation.

(C) Find the equilibrium price and quantity.

(D) Graph the two equations in the same coordinate system and identify the equilibrium point, supply curve, and demand curve.

47. *Break-even analysis.* A small company manufactures portable home computers. The plant has fixed costs (leases, insurance, and so on) of \$48,000 per month and variable costs (labor, materials, and so on) of \$1,400 per unit produced. The computers are sold for \$1,800 each. Thus, the cost and revenue equations are

$$y = 48{,}000 + 1{,}400x \quad \text{Cost equation}$$
$$y = 1{,}800x \quad \text{Revenue equation}$$

where x is the total number of computers produced and sold each month, and the monthly costs and revenue are in dollars.

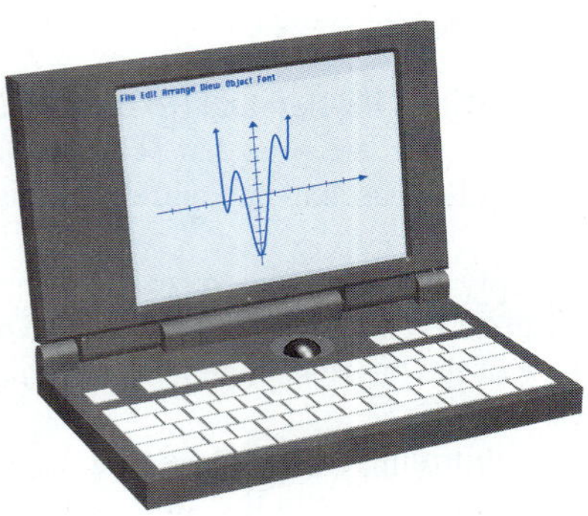

(A) How many units must be manufactured and sold each month for the company to break even?

(B) Graph both equations in the same coordinate system and show the break-even point. Interpret the regions between the lines to the left and to the right of the break-even point.

48. *Break-even analysis.* Repeat Problem 47 with the cost and revenue equations

$$y = 65{,}000 + 1{,}100x \quad \text{Cost equation}$$
$$y = 1{,}600x \quad \text{Revenue equation}$$

49. *Break-even analysis.* A mail-order company markets videotapes that sell for \$19.95, including shipping and handling. The monthly fixed costs (advertising, rent, and so on) are \$24,000, and the variable costs (materials, shipping, and so on) are \$7.45 per tape.

(A) How many tapes must be sold each month for the company to break even?

(B) Graph the cost and revenue equations in the same coordinate system and show the break-even point. Interpret the regions between the lines to the left and to the right of the break-even point.

50. *Break-even analysis.* Repeat Problem 49 if the monthly fixed costs increase to \$27,200, the variable costs increase to \$9.15, and the company raises the selling price of the tapes to \$21.95.

51. *Delivery charges.* United Express, a nationwide package delivery service, charges a base price for overnight delivery of packages weighing 1 pound or less and a surcharge for each additional pound (or fraction thereof). A customer is billed \$27.75 for shipping a 5 pound package and \$64.50 for shipping a 20 pound package. Find the base price and the surcharge for each additional pound.

52. *Delivery charges.* Refer to Problem 51. Federated Shipping, a competing overnight delivery service, informs the customer in Problem 51 that they would ship the 5 pound package for \$29.95 and the 20 pound package for \$59.20.

(A) If Federated Shipping computes its cost in the same manner as United Express, find the base price and the surcharge for Federated Shipping.

(B) Devise a simple rule that the customer can use to choose the cheaper of the two services for each package shipped. Justify your answer.

53. *Resource allocation.* A coffee manufacturer uses Colombian and Brazilian coffee beans to produce two blends, robust and mild. A pound of the robust blend requires 12 ounces of Colombian beans and 4 ounces of Brazilian beans. A pound of the mild blend requires 6 ounces of Colombian beans and 10 ounces of Brazilian beans. Coffee is shipped in 132 pound burlap bags. The company has 50 bags of Colombian beans and 40 bags of Brazilian beans on hand. How many pounds of each blend should they produce in order to use all the available beans?

54. *Resource allocation.* Refer to Problem 53.

(A) If the company decides to discontinue production of the robust blend and produce only the mild blend, how many pounds of the mild blend can they produce and how many beans of each type will they use? Are there any beans that are not used?

(B) Repeat part (A) if the company decides to discontinue production of the mild blend and produce only the robust blend.

Life Sciences

55. *Nutrition.* Animals in an experiment are to be kept under a strict diet. Each animal is to receive, among other things, 20 grams of protein and 6 grams of fat. The laboratory technician is able to purchase two food mixes of the following compositions: Mix A has 10% protein and 6% fat; mix B has 20% protein and 2% fat. How many grams of each mix should be used to obtain the right diet for a single animal?

56. *Nutrition—plants.* A fruit grower can use two types of fertilizer in an orange grove, brand A and brand B. Each bag of brand A contains 8 pounds of nitrogen and 4 pounds of phosphoric acid. Each bag of brand B contains 7 pounds of nitrogen and 6 pounds of phosphoric acid. Tests indicate that the grove needs 720 pounds of nitrogen and 500 pounds of phosphoric acid. How many bags of each brand should be used to provide the required amounts of nitrogen and phosphoric acid?

Social Sciences

57. *Psychology—approach and avoidance.* People often approach certain situations with "mixed emotions." For example, public speaking often brings forth the positive response of recognition and the negative response of failure. Which dominates? J. S. Brown, in an experiment on approach and avoidance, trained rats by feeding them from a goal box. Then the rats received mild electric shocks from the same goal box. This established an approach–avoidance conflict relative to the goal box. Using appropriate apparatus, Brown arrived at the following relationships:

$$p = -\tfrac{1}{5}d + 70 \qquad \text{Approach equation}$$

$$p = -\tfrac{4}{3}d + 230 \qquad \text{Avoidance equation}$$

where $30 \le d \le 172.5$. The approach equation gives the pull (in grams) toward the food goal box when the rat is placed d centimeters away from it. The avoidance equation gives the pull (in grams) away from the shock goal box when the rat is placed d centimeters from it.

(A) Graph the approach equation and the avoidance equation in the same coordinate system.

(B) Find the value of d for the point of intersection of these two equations.

(C) What do you think the rat would do when placed the distance d from the box found in part (B)?

(For additional discussion of this phenomenon, see J. S. Brown, "Gradients of Approach and Avoidance Responses and Their Relation to Motivation," *Journal of Comparative and Physiological Psychology,* 1948, 41:450–465.)

SECTION 4-2

Systems of Linear Equations and Augmented Matrices

- MATRICES
- SOLVING LINEAR SYSTEMS USING AUGMENTED MATRICES
- SUMMARY

Most linear systems of any consequence involve large numbers of equations and variables. It is impractical to try to solve such systems by hand. In the past, these complex systems could be solved only on large computers. Now a wide array of graphing utilities can be used to solve linear systems.

These range from graphing calculators (such as the Texas Instruments TI-83), to software packages [such as *Explorations in Finite Mathematics* (see Preface) or MatLab], to spreadsheets (such as Excel). All these graphing utilities have one thing in common: **The user is expected to be familiar with the techniques used to solve large linear systems.** In the remainder of this chapter we will develop several *matrix methods* for solving systems, with the understanding that these methods are usually used in conjunction with a graphing utility. It is important to keep in mind that we are not presenting these techniques as more efficient methods for solving linear systems manually—there are none. So we will not stress computational shortcuts for hand calculations. Instead, we will emphasize formulation of mathematical models and interpretation of the results—two activities that graphing utilities cannot perform for you.

As we mentioned earlier, when referring to the optional problems that should be solved with a graphing utility, we will continue to clearly state whether matrix methods or graphical approximation methods are to be used.

MATRICES

In solving systems of equations using elimination by addition, the coefficients of the variables and the constant terms played a central role. The process can be made more efficient for generalization and computer work by the introduction of a mathematical form called a *matrix.* A **matrix** is a rectangular array of numbers written within brackets. Two examples are

$$A = \begin{bmatrix} 1 & -4 & 5 \\ 7 & 0 & -2 \end{bmatrix} \qquad B = \begin{bmatrix} -4 & 5 & 12 \\ 0 & 1 & 8 \\ -3 & 10 & 9 \\ -6 & 0 & -1 \end{bmatrix} \tag{1}$$

Each number in a matrix is called an **element** of the matrix. Matrix A has 6 elements arranged in 2 rows and 3 columns. Matrix B has 12 elements arranged in 4 rows and 3 columns. If a matrix has m rows and n columns, it is called an $\boldsymbol{m \times n}$ **matrix** (read "m by n matrix"). The expression $m \times n$ is called the **size** of the matrix, and the numbers m and n are called the **dimensions** of the matrix. It is important to note that the number of rows is always given first. Referring to (1) above, A is a 2×3 matrix and B is a 4×3 matrix. A matrix with n rows and n columns is called a **square matrix of order $\boldsymbol{n}$.** A matrix with only 1 column is called a **column matrix,** and a matrix with only 1 row is called a **row matrix.** These definitions are illustrated by the following:

3×3

$$\begin{bmatrix} 0.5 & 0.2 & 1.0 \\ 0.0 & 0.3 & 0.5 \\ 0.7 & 0.0 & 0.2 \end{bmatrix}$$

Square matrix of order 3

4×1

$$\begin{bmatrix} 3 \\ -2 \\ 1 \\ 0 \end{bmatrix}$$

Column matrix

1×4

$$\begin{bmatrix} 2 & \frac{1}{2} & 0 & -\frac{2}{3} \end{bmatrix}$$

Row matrix

The **position** of an element in a matrix is given by the row and column containing the element. This is usually denoted using **double subscript notation a_{ij}**, where i is the row and j is the column containing the element a_{ij}, as illustrated below:

$$A = \begin{bmatrix} 1 & -4 & 5 \\ 7 & 0 & -2 \end{bmatrix} \qquad \begin{matrix} a_{11} = 1, & a_{12} = -4, & a_{13} = 5 \\ a_{21} = 7, & a_{22} = 0, & a_{23} = -2 \end{matrix}$$

Note that a_{12} is read "a sub one two" (*not* "a sub twelve"). The elements $a_{11} = 1$ and $a_{22} = 0$ make up the *principal diagonal* of A. In general, the **principal diagonal** of a matrix A consists of the elements $a_{11}, a_{22}, a_{33}, \ldots$.

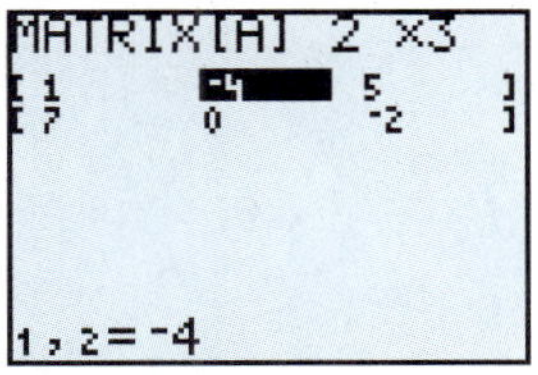

FIGURE 1
Matrix notation on a graphing calculator

REMARK

Most graphing utilities are capable of storing and manipulating matrices. Figure 1 shows matrix A displayed in the editing screen of a graphing calculator. The size of the matrix is given at the top of the screen. The position and value of the currently selected element is given at the bottom. Notice that a comma is used in the notation for the position. This is common practice on many graphing utilities, but not in mathematical literature. In a spreadsheet, matrices are referred to by their location in the spreadsheet (upper left corner to lower right corner), using either row and column numbers (Fig. 2A) or row numbers and column letters (Fig. 2B).

	1	2	3
1	1	-4	5
2	7	0	-2

(A) Location of matrix A: R1C1:R2C3

	A	B	C	D	E	F
1						
2						
3						
4						
5				1	-4	5
6				7	0	-2

(B) Location of matrix A: D5:F6

FIGURE 2
Matrix notation in a spreadsheet

The coefficients and constant terms in a system of linear equations can be used to form several matrices of interest. Related to the system

$$\begin{aligned} 2x - 3y &= 5 \\ x + 2y &= -3 \end{aligned} \tag{2}$$

are the following matrices:

Coefficient matrix $\begin{bmatrix} 2 & -3 \\ 1 & 2 \end{bmatrix}$ Constant matrix $\begin{bmatrix} 5 \\ -3 \end{bmatrix}$ Augmented coefficient matrix $\left[\begin{array}{cc|c} 2 & -3 & 5 \\ 1 & 2 & -3 \end{array}\right]$

The augmented coefficient matrix contains all the essential parts of the system—both the coefficients and the constants. The vertical bar is included only as a visual aid to help us separate the coefficients from the constant terms. (Matrices entered and displayed on a graphing calculator or computer will not display this line.) Later in this chapter we will make use of the coefficient and constant matrices. For now, we will find the augmented coefficient matrix sufficient for our needs.

For ease of generalization to the larger systems in the following sections, we are now going to change the notation for the variables in (2) to a subscript form. (We would soon run out of letters, but we will not run out of subscripts.) That is, in place of x and y, we use x_1 and x_2, respectively, and (2) is rewritten as

$$\begin{aligned} 2x_1 - 3x_2 &= 5 \\ x_1 + 2x_2 &= -3 \end{aligned}$$

In general, associated with each linear system of the form

$$\begin{aligned} a_{11}x_1 + a_{12}x_2 &= k_1 \\ a_{21}x_1 + a_{22}x_2 &= k_2 \end{aligned} \tag{3}$$

where x_1 and x_2 are variables, is the **augmented matrix** of the system:

Column 1 (C_1), Column 2 (C_2), Column 3 (C_3)

$$\left[\begin{array}{cc|c} a_{11} & a_{12} & k_1 \\ a_{21} & a_{22} & k_2 \end{array}\right] \begin{array}{l} \leftarrow \text{Row 1 } (R_1) \\ \leftarrow \text{Row 2 } (R_2) \end{array}$$

This matrix contains the essential parts of system (3). Our objective is to learn how to manipulate augmented matrices in order to solve system (3), if a solution exists. The manipulative process is a direct outgrowth of the elimination process discussed in Section 4-1.

Recall that two linear systems are said to be **equivalent** if they have exactly the same solution set. How did we transform linear systems into equivalent linear systems? We used the operations listed below (Theorem 2, Section 4-1):

Operations That Produce Equivalent Systems

A system of linear equations is transformed into an equivalent system if:

(A) Two equations are interchanged.
(B) An equation is multiplied by a nonzero constant.
(C) A constant multiple of one equation is added to another equation.

Paralleling the earlier discussion, we say that two augmented matrices are **row-equivalent,** denoted by the symbol ~ placed between the two matrices, if they are augmented matrices of equivalent systems of equations. (Think about this.) How do we transform augmented matrices into row-equivalent matrices? We use Theorem 1, which is a direct consequence of the operations listed above.

Theorem 1 ■■ OPERATIONS THAT PRODUCE ROW-EQUIVALENT MATRICES

An augmented matrix is transformed into a row-equivalent matrix by performing any of the following **row operations:**

(A) Two rows are interchanged ($R_i \leftrightarrow R_j$).
(B) A row is multiplied by a nonzero constant ($kR_i \rightarrow R_i$).
(C) A constant multiple of one row is added to another row ($kR_j + R_i \rightarrow R_i$).

[*Note:* The arrow $\rightarrow$ means "replaces."] ■■

Solving Linear Systems Using Augmented Matrices

The use of Theorem 1 in solving systems in the form of (3) is best illustrated by examples.

Example 1 **Solving a System Using Augmented Matrix Methods** Solve using augmented matrix methods:

$$\begin{aligned} 3x_1 + 4x_2 &= 1 \\ x_1 - 2x_2 &= 7 \end{aligned} \qquad (4)$$

Solution We start by writing the augmented matrix corresponding to (4):

$$\left[\begin{array}{cc|c} 3 & 4 & 1 \\ 1 & -2 & 7 \end{array}\right] \qquad (5)$$

Our objective is to use row operations from Theorem 1 to try to transform (5) into the form

$$\left[\begin{array}{cc|c} 1 & 0 & m \\ 0 & 1 & n \end{array}\right] \qquad (6)$$

where m and n are real numbers. The solution to system (4) will then be obvious, since matrix (6) will be the augmented matrix of the following system (a row in an augmented matrix always corresponds to an equation in a linear system):

$$\begin{aligned} x_1 &= m \qquad & x_1 + 0x_2 &= m \\ x_2 &= n \qquad & 0x_1 + x_2 &= n \end{aligned}$$

We now proceed to use row operations to transform (5) into form (6).

Step 1. To get a 1 in the upper left corner, we interchange R_1 and R_2 (Theorem 1A):

$$\left[\begin{array}{cc|c} 3 & 4 & 1 \\ 1 & -2 & 7 \end{array}\right] \underset{\sim}{R_1 \leftrightarrow R_2} \left[\begin{array}{cc|c} 1 & -2 & 7 \\ 3 & 4 & 1 \end{array}\right]$$

Step 2. To get a 0 in the lower left corner, we multiply R_1 by (-3) and add to R_2 (Theorem 1C)—this changes R_2 but not R_1. Some people find it useful to write $(-3R_1)$ outside the matrix to help reduce errors in arithmetic, as shown:

$$\left[\begin{array}{cc|c} 1 & -2 & 7 \\ 3 & 4 & 1 \end{array}\right] \underset{(-3)R_1 + R_2 \to R_2}{\sim} \left[\begin{array}{cc|c} 1 & -2 & 7 \\ 0 & 10 & -20 \end{array}\right]$$

$$-3 \quad 6 \quad -21$$

Step 3. To get a 1 in the second row, second column, we multiply R_2 by $\frac{1}{10}$ (Theorem 1B):

$$\left[\begin{array}{cc|c} 1 & -2 & 7 \\ 0 & 10 & -20 \end{array}\right] \underset{\frac{1}{10}R_2 \to R_2}{\sim} \left[\begin{array}{cc|c} 1 & -2 & 7 \\ 0 & 1 & -2 \end{array}\right]$$

Step 4. To get a 0 in the first row, second column, we multiply R_2 by 2 and add the result to R_1 (Theorem 1C)—this changes R_1 but not R_2:

$$0 \quad 2 \quad -4$$

$$\left[\begin{array}{cc|c} 1 & -2 & 7 \\ 0 & 1 & -2 \end{array}\right] \underset{\sim}{2R_2 + R_1 \to R_1} \left[\begin{array}{cc|c} 1 & 0 & 3 \\ 0 & 1 & -2 \end{array}\right]$$

We have accomplished our objective! The last matrix is the augmented matrix for the system

$$\begin{aligned} x_1 &= 3 \qquad & x_1 + 0x_2 &= 3 \\ x_2 &= -2 \qquad & 0x_1 + x_2 &= -2 \end{aligned} \tag{7}$$

Since system (7) is equivalent to system (4), our starting system, we have solved (4); that is, $x_1 = 3$ and $x_2 = -2$.

CHECK

$$\begin{aligned} 3x_1 + 4x_2 &= 1 \\ 3(3) + 4(-2) &\overset{?}{=} 1 \\ 1 &\overset{\checkmark}{=} 1 \end{aligned} \qquad \begin{aligned} x_1 - 2x_2 &= 7 \\ 3 - 2(-2) &\overset{?}{=} 7 \\ 7 &\overset{\checkmark}{=} 7 \end{aligned}$$

The above process may be written more compactly as follows:

Step 1: Need a 1 here.

$$\left[\begin{array}{rr|r} 3 & 4 & 1 \\ 1 & -2 & 7 \end{array}\right] \quad R_1 \leftrightarrow R_2$$

Step 2: Need a 0 here.

$$\sim \left[\begin{array}{rr|r} 1 & -2 & 7 \\ 3 & 4 & 1 \end{array}\right] \quad (-3)R_1 + R_2 \rightarrow R_2$$

$$\begin{array}{rr|r} -3 & 6 & -21 \end{array}$$

Step 3: Need a 1 here.

$$\sim \left[\begin{array}{rr|r} 1 & -2 & 7 \\ 0 & 10 & -20 \end{array}\right] \quad \tfrac{1}{10}R_2 \rightarrow R_2$$

$$\begin{array}{rr|r} 0 & 2 & -4 \end{array}$$

Step 4: Need a 0 here.

$$\sim \left[\begin{array}{rr|r} 1 & -2 & 7 \\ 0 & 1 & -2 \end{array}\right] \quad 2R_2 + R_1 \rightarrow R_1$$

$$\sim \left[\begin{array}{rr|r} 1 & 0 & 3 \\ 0 & 1 & -2 \end{array}\right]$$

Therefore, $x_1 = 3$ and $x_2 = -2$.

Matched Problem 1 Solve using augmented matrix methods:

$$\begin{aligned} 2x_1 - x_2 &= -7 \\ x_1 + 2x_2 &= 4 \end{aligned}$$

Many graphing utilities can perform row operations. Figure 3 shows the results of performing the row operations used in the solution of Example 1. Consult your manual for the details of performing row operations on your graphing utility.

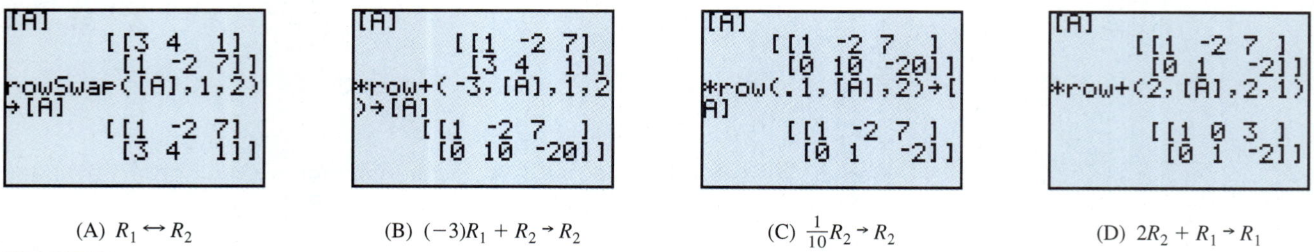

(A) $R_1 \leftrightarrow R_2$ (B) $(-3)R_1 + R_2 \rightarrow R_2$ (C) $\frac{1}{10}R_2 \rightarrow R_2$ (D) $2R_2 + R_1 \rightarrow R_1$

FIGURE 3
Row operations on a graphing utility

Explore–Discuss 1

The summary following Example 1 shows five augmented coefficient matrices. Write the linear system that each matrix represents, solve each system graphically, and discuss the relationships among these solutions.

Example 2 **Solving a System Using Augmented Matrix Methods** Solve using augmented matrix methods:

$$\begin{aligned} 2x_1 - 3x_2 &= 6 \\ 3x_1 + 4x_2 &= \tfrac{1}{2} \end{aligned}$$

Step 1: Need a 1 here.

$$\left[\begin{array}{cc|c} 2 & -3 & 6 \\ 3 & 4 & \frac{1}{2} \end{array}\right] \quad \tfrac{1}{2}R_1 \to R_1$$

Step 2: Need a 0 here.

$$\sim \left[\begin{array}{cc|c} 1 & -\frac{3}{2} & 3 \\ 3 & 4 & \frac{1}{2} \end{array}\right] \quad (-3)R_1 + R_2 \to R_2$$

$$-3 \quad \tfrac{9}{2} \quad -9$$

Step 3: Need a 1 here.

$$\sim \left[\begin{array}{cc|c} 1 & -\frac{3}{2} & 3 \\ 0 & \frac{17}{2} & -\frac{17}{2} \end{array}\right] \quad \tfrac{2}{17}R_2 \to R_2$$

$$0 \quad \tfrac{3}{2} \quad -\tfrac{3}{2}$$

Step 4: Need a 0 here.

$$\sim \left[\begin{array}{cc|c} 1 & -\frac{3}{2} & 3 \\ 0 & 1 & -1 \end{array}\right] \quad \tfrac{3}{2}R_2 + R_1 \to R_1$$

$$\sim \left[\begin{array}{cc|c} 1 & 0 & \frac{3}{2} \\ 0 & 1 & -1 \end{array}\right]$$

Thus, $x_1 = \frac{3}{2}$ and $x_2 = -1$. The check is left to the reader.

Matched Problem 2 Solve using augmented matrix methods:

$$\begin{aligned} 5x_1 - 2x_2 &= 11 \\ 2x_1 + 3x_2 &= \tfrac{5}{2} \end{aligned}$$

Example 3 **Solving a System Using Augmented Matrix Methods** Solve using augmented matrix methods:

$$\begin{aligned} 2x_1 - x_2 &= 4 \\ -6x_1 + 3x_2 &= -12 \end{aligned} \qquad (8)$$

SOLUTION

$$\left[\begin{array}{cc|c} 2 & -1 & 4 \\ -6 & 3 & -12 \end{array}\right] \quad \begin{array}{l} \tfrac{1}{2}R_1 \to R_1 \text{ (to get a 1 in the upper left corner)} \\ \tfrac{1}{3}R_2 \to R_2 \text{ (this simplifies } R_2\text{)} \end{array}$$

$$\sim \left[\begin{array}{cc|c} 1 & -\frac{1}{2} & 2 \\ -2 & 1 & -4 \end{array}\right] \quad 2R_1 + R_2 \to R_2 \text{ (to get a 0 in the lower left corner)}$$

$$2 \quad -1 \quad 4$$

$$\sim \left[\begin{array}{cc|c} 1 & -\frac{1}{2} & 2 \\ 0 & 0 & 0 \end{array}\right]$$

The last matrix corresponds to the system

$$\begin{aligned} x_1 - \tfrac{1}{2}x_2 &= 2 \qquad x_1 - \tfrac{1}{2}x_2 = 2 \\ 0 &= 0 \qquad 0x_1 + 0x_2 = 0 \end{aligned} \tag{9}$$

This system is equivalent to the original system. Geometrically, the graphs of the two original equations coincide and there are infinitely many solutions. In general, if we end up with a row of zeros in an augmented matrix for a two-equation–two-variable system, the system is dependent and there are infinitely many solutions.

We represent the infinitely many solutions using the same method that was used in Section 4-1; that is, by introducing a parameter. We start by solving $x_1 - \frac{1}{2}x_2 = 2$, the first equation in (9), for either variable in terms of the other. We choose to solve for x_1 in terms of x_2 because it is easier:

$$x_1 = \tfrac{1}{2}x_2 + 2 \tag{10}$$

Now we introduce a parameter t (we can use other letters, such as k, s, p, q, and so on, to represent a parameter just as well). If we let $x_2 = t$, then for t any real number,

$$\begin{aligned} x_1 &= \tfrac{1}{2}t + 2 \\ x_2 &= t \end{aligned} \tag{11}$$

represents a solution of system (8). Using ordered pair notation, we may also write:

For any real number t,

$$\left(\tfrac{1}{2}t + 2, t\right) \tag{12}$$

is a solution of system (8). And more formally, we may write:

$$\text{Solution set} = \left\{\left(\tfrac{1}{2}t + 2, t\right) \mid t \in R\right\} \tag{13}$$

We will generally use the less formal forms (11) and (12) to represent the solution set for problems of this type.

CHECK The following is a check that (11) provides a solution for system (8) for any real number t:

$$\begin{aligned} 2x_1 - x_2 &= 4 \\ 2\left(\tfrac{1}{2}t + 2\right) - t &\stackrel{?}{=} 4 \\ t + 4 - t &\stackrel{?}{=} 4 \\ 4 &\stackrel{\checkmark}{=} 4 \end{aligned} \qquad \begin{aligned} -6x_1 + 3x_2 &= -12 \\ -6\left(\tfrac{1}{2}t + 2\right) + 3t &\stackrel{?}{=} -12 \\ -3t - 12 + 3t &\stackrel{?}{=} -12 \\ -12 &\stackrel{\checkmark}{=} -12 \end{aligned}$$

Matched Problem 3 ➡ Solve using augmented matrix methods:

$$\begin{aligned} -2x_1 + 6x_2 &= 6 \\ 3x_1 - 9x_2 &= -9 \end{aligned}$$

Explore–Discuss 2

The solution of Example 3 involved three augmented coefficient matrices. Write the linear system that each matrix represents, solve each system graphically, and discuss the relationships among these solutions.

Example 4 **Solving a System Using Augmented Matrix Methods** Solve using augmented matrix methods:

$$\begin{aligned} 2x_1 + 6x_2 &= -3 \\ x_1 + 3x_2 &= 2 \end{aligned}$$

SOLUTION

$$\left[\begin{array}{cc|c} 2 & 6 & -3 \\ 1 & 3 & 2 \end{array}\right] \quad R_1 \leftrightarrow R_2$$

$$\sim \left[\begin{array}{cc|c} 1 & 3 & 2 \\ 2 & 6 & -3 \end{array}\right] \quad (-2)R_1 + R_2 \rightarrow R_2$$

$$\begin{array}{ccc} -2 & -6 & -4 \end{array}$$

$$\sim \left[\begin{array}{cc|c} 1 & 3 & 2 \\ 0 & 0 & -7 \end{array}\right] \quad R_2 \text{ implies the contradiction } 0 = -7.$$

This is the augmented matrix of the system

$$\begin{aligned} x_1 + 3x_2 &= 2 \qquad & x_1 + 3x_2 &= 2 \\ 0 &= -7 \qquad & 0x_1 + 0x_2 &= -7 \end{aligned}$$

The second equation is not satisfied by any ordered pair of real numbers. Hence, as we saw in Section 4-1, the original system is inconsistent and has no solution—otherwise, we have once again proved that $0 = -7$! Thus, if in a row of an augmented matrix we obtain all zeros to the left of the vertical bar and a nonzero number to the right, then the system is inconsistent and there are no solutions.

Matched Problem 4 Solve using augmented matrix methods:

$$\begin{aligned} 2x_1 - x_2 &= 3 \\ 4x_1 - 2x_2 &= -1 \end{aligned}$$

SUMMARY

Summary

Form 1: A Unique Solution (Consistent and Independent)	**Form 2: Infinitely Many Solutions (Consistent and Dependent)**	**Form 3: No Solution (Inconsistent)**
$\left[\begin{array}{cc\|c} 1 & 0 & m \\ 0 & 1 & n \end{array}\right]$	$\left[\begin{array}{cc\|c} 1 & m & n \\ 0 & 0 & 0 \end{array}\right]$	$\left[\begin{array}{cc\|c} 1 & m & n \\ 0 & 0 & p \end{array}\right]$

m, n, p real numbers; $p \neq 0$

The process of solving systems of equations described in this section is referred to as **Gauss–Jordan elimination.** We will formalize this method in the next section so that it will apply to systems of any size, including systems where the number of equations and the number of variables are not the same.

Answers to Matched Problems

1. $x_1 = -2, x_2 = 3$ **2.** $x_1 = 2, x_2 = -\frac{1}{2}$
3. The system is dependent. For t any real number, a solution is $x_1 = 3t - 3, x_2 = t$.
4. Inconsistent—no solution

EXERCISE 4-2

A *Problems 1–10 refer to the following matrices:*

$$A = \begin{bmatrix} 2 & -4 & 0 \\ 6 & 1 & -5 \end{bmatrix} \qquad B = \begin{bmatrix} -1 & 9 & 0 \\ -4 & 8 & 7 \\ 2 & 4 & 0 \end{bmatrix}$$

$$C = \begin{bmatrix} 2 & -3 & 0 \end{bmatrix} \qquad D = \begin{bmatrix} -5 \\ 8 \end{bmatrix}$$

1. What is the size of A? Of C?

2. What is the size of B? Of D?

3. Identify all row matrices.

4. Identify all column matrices.

5. Identify all square matrices.

6. For matrix B, find b_{21} and b_{13}.

7. For matrix A, find a_{12} and a_{23}.

8. For matrices C and D, find c_{13} and d_{21}.

9. Find the elements on the principal diagonal of matrix B.

10. Find the elements on the principal diagonal of matrix A.

Problems 11 and 12 refer to the matrices shown below.

$$E = \begin{bmatrix} 1 & -2 & 3 & 9 \\ -5 & 0 & 7 & -8 \end{bmatrix} \qquad F = \begin{bmatrix} 4 & -6 \\ 2 & 3 \\ -5 & 7 \end{bmatrix}$$

11. (A) What is the size of E?
(B) How many additional columns would F have to have to be a square matrix?
(C) Find e_{23} and f_{12}.

12. (A) What is the size of F?
(B) How many additional rows would E have to have to be a square matrix?
(C) Find e_{14} and f_{31}.

Perform the row operations indicated in Problems 13–24 on the following matrix:

$$\left[\begin{array}{cc|c} 1 & -3 & 2 \\ 4 & -6 & -8 \end{array}\right]$$

13. $R_1 \leftrightarrow R_2$ **14.** $\frac{1}{2}R_2 \to R_2$

15. $-4R_1 \to R_1$ **16.** $-2R_1 \to R_1$

17. $2R_2 \to R_2$ **18.** $-1R_2 \to R_2$

19. $(-4)R_1 + R_2 \to R_2$ **20.** $(-\frac{1}{2})R_2 + R_1 \to R_1$

21. $(-2)R_1 + R_2 \to R_2$ **22.** $(-3)R_1 + R_2 \to R_2$

23. $(-1)R_1 + R_2 \to R_2$ **24.** $R_1 + R_2 \to R_2$

Each of the matrices in Problems 25–30 is the result of performing a single row operation on the matrix A shown below. Identify the row operation.

$$A = \left[\begin{array}{cc|c} -1 & 2 & -3 \\ 6 & -3 & 12 \end{array}\right]$$

25. $\left[\begin{array}{cc|c} -1 & 2 & -3 \\ 2 & -1 & 4 \end{array}\right]$ **26.** $\left[\begin{array}{cc|c} -2 & 4 & -6 \\ 6 & -3 & 12 \end{array}\right]$

27. $\left[\begin{array}{cc|c} -1 & 2 & -3 \\ 0 & 9 & -6 \end{array}\right]$ **28.** $\left[\begin{array}{cc|c} 3 & 0 & 5 \\ 6 & -3 & 12 \end{array}\right]$

29. $\left[\begin{array}{cc|c} 1 & 1 & 1 \\ 6 & -3 & 12 \end{array}\right]$ **30.** $\left[\begin{array}{cc|c} -1 & 2 & -3 \\ 2 & 5 & 0 \end{array}\right]$

Check Problems 25–30 by performing the row operation you identified on a graphing utility.

Solve Problems 31 and 32 using augmented matrix methods. Write the linear system represented by each augmented matrix in your solution, and solve each of these systems graphically. Discuss the relationships among the solutions of these systems.

31. $x_1 + x_2 = 5$
$x_1 - x_2 = 1$

32. $x_1 - x_2 = 2$
$x_1 + x_2 = 6$

B *Solve Problems 33–52 using augmented matrix methods.*

33. $x_1 - 2x_2 = 1$
$2x_1 - x_2 = 5$

34. $x_1 + 3x_2 = 1$
$3x_1 - 2x_2 = 14$

35. $x_1 - 4x_2 = -2$
$-2x_1 + x_2 = -3$

36. $x_1 - 3x_2 = -5$
$-3x_1 - x_2 = 5$

37. $3x_1 - x_2 = 2$
$x_1 + 2x_2 = 10$

38. $2x_1 + x_2 = 0$
$x_1 - 2x_2 = -5$

39. $x_1 + 2x_2 = 4$
$2x_1 + 4x_2 = -8$

40. $2x_1 - 3x_2 = -2$
$-4x_1 + 6x_2 = 7$

41. $\begin{aligned} 2x_1 + x_2 &= 6 \\ x_1 - x_2 &= -3 \end{aligned}$

42. $\begin{aligned} 3x_1 - x_2 &= -5 \\ x_1 + 3x_2 &= 5 \end{aligned}$

43. $\begin{aligned} 3x_1 - 6x_2 &= -9 \\ -2x_1 + 4x_2 &= 6 \end{aligned}$

44. $\begin{aligned} 2x_1 - 4x_2 &= -2 \\ -3x_1 + 6x_2 &= 3 \end{aligned}$

45. $\begin{aligned} 4x_1 - 2x_2 &= 2 \\ -6x_1 + 3x_2 &= -3 \end{aligned}$

46. $\begin{aligned} -6x_1 + 2x_2 &= 4 \\ 3x_1 - x_2 &= -2 \end{aligned}$

47. $\begin{aligned} 2x_1 + x_2 &= 1 \\ 4x_1 - x_2 &= -7 \end{aligned}$

48. $\begin{aligned} 2x_1 - x_2 &= -8 \\ 2x_1 + x_2 &= 8 \end{aligned}$

49. $\begin{aligned} 4x_1 - 6x_2 &= 8 \\ -6x_1 + 9x_2 &= -10 \end{aligned}$

50. $\begin{aligned} 2x_1 - 4x_2 &= -4 \\ -3x_1 + 6x_2 &= 4 \end{aligned}$

51. $\begin{aligned} -4x_1 + 6x_2 &= -8 \\ 6x_1 - 9x_2 &= 12 \end{aligned}$

52. $\begin{aligned} -2x_1 + 4x_2 &= 4 \\ 3x_1 - 6x_2 &= -6 \end{aligned}$

C *Solve Problems 53–58 using augmented matrix methods.*

53. $\begin{aligned} 3x_1 - x_2 &= 7 \\ 2x_1 + 3x_2 &= 1 \end{aligned}$

54. $\begin{aligned} 2x_1 - 3x_2 &= -8 \\ 5x_1 + 3x_2 &= 1 \end{aligned}$

55. $\begin{aligned} 3x_1 + 2x_2 &= 4 \\ 2x_1 - x_2 &= 5 \end{aligned}$

56. $\begin{aligned} 4x_1 + 3x_2 &= 26 \\ 3x_1 - 11x_2 &= -7 \end{aligned}$

57. $\begin{aligned} 0.2x_1 - 0.5x_2 &= 0.07 \\ 0.8x_1 - 0.3x_2 &= 0.79 \end{aligned}$

58. $\begin{aligned} 0.3x_1 - 0.6x_2 &= 0.18 \\ 0.5x_1 - 0.2x_2 &= 0.54 \end{aligned}$

Solve Problems 59–62 using augmented matrix methods. Use a graphing utility to perform the row operations.

59. $\begin{aligned} 0.8x_1 + 2.88x_2 &= 4 \\ 1.25x_1 + 4.34x_2 &= 5 \end{aligned}$

60. $\begin{aligned} 2.7x_1 - 15.12x_2 &= 27 \\ 3.25x_1 - 18.52x_2 &= 33 \end{aligned}$

61. $\begin{aligned} 4.8x_1 - 40.32x_2 &= 295.2 \\ -3.75x_1 + 28.7x_2 &= -211.2 \end{aligned}$

62. $\begin{aligned} 5.7x_1 - 8.55x_2 &= -35.91 \\ 4.5x_1 + 5.73x_2 &= 76.17 \end{aligned}$

SECTION 4-3 Gauss–Jordan Elimination

- REDUCED MATRICES
- SOLVING SYSTEMS BY GAUSS–JORDAN ELIMINATION
- APPLICATION

Now that you have had some experience with row operations on simple augmented matrices, we will consider systems involving more than two variables. In addition, we will not require that a system have the same number of equations as variables. It turns out that the results for two-variable–two-equation linear systems stated in Theorem 1, Section 4-1, actually hold for linear systems of any size.

> **Possible Solutions to a Linear System**
>
> It can be shown that any linear system must have exactly one solution, no solution, or an infinite number of solutions, regardless of the number of equations or number of variables in the system. The terms *unique solution, consistent, inconsistent, dependent,* and *independent* are used to describe these solutions, just as in the two-variable case.

REDUCED MATRICES

In the last section we used row operations to transform the augmented coefficient matrix for a system of two equations in two variables,

$$\left[\begin{array}{cc|c} a_{11} & a_{12} & k_1 \\ a_{21} & a_{22} & k_2 \end{array}\right] \qquad \begin{aligned} a_{11}x_1 + a_{12}x_2 &= k_1 \\ a_{21}x_1 + a_{22}x_2 &= k_2 \end{aligned}$$

into one of the following simplified forms:

$$\underset{\text{Form 1}}{\left[\begin{array}{cc|c} 1 & 0 & m \\ 0 & 1 & n \end{array}\right]} \qquad \underset{\text{Form 2}}{\left[\begin{array}{cc|c} 1 & m & n \\ 0 & 0 & 0 \end{array}\right]} \qquad \underset{\text{Form 3}}{\left[\begin{array}{cc|c} 1 & m & n \\ 0 & 0 & p \end{array}\right]} \tag{1}$$

where m, n, and p are real numbers, $p \neq 0$. Each of these reduced forms represents a system that has a different type of solution set, and no two of these forms are row-equivalent. Thus, we consider each of these to be a different simplified form. Now we want to consider larger systems with more variables and more equations.

Explore–Discuss 1

Forms 1, 2, and 3 in (1) represent systems that have a unique solution, an infinite number of solutions, and no solution, respectively. Discuss the number of solutions for the systems of three equations in three variables represented by the following augmented coefficient matrices:

(A) $\left[\begin{array}{ccc|c} 1 & 2 & 3 & 5 \\ 0 & 0 & 0 & 6 \\ 0 & 0 & 0 & 0 \end{array}\right]$ (B) $\left[\begin{array}{ccc|c} 1 & 2 & 3 & 5 \\ 0 & 0 & 0 & 0 \\ 0 & 0 & 0 & 0 \end{array}\right]$

(C) $\left[\begin{array}{ccc|c} 1 & 0 & 0 & 5 \\ 0 & 1 & 0 & 6 \\ 0 & 0 & 1 & 7 \end{array}\right]$

Since there is no upper limit on the number of variables or the number of equations in a linear system, it is not feasible to explicitly list all possible "simplified forms" for larger systems, as we did for systems of two equations in two variables. Instead, we state a general definition of a simplified form called a *reduced matrix* that can be applied to all matrices and systems, regardless of size.

Reduced Matrix

A matrix is in **reduced form** if:

1. Each row consisting entirely of zeros is below any row having at least one nonzero element.
2. The leftmost nonzero element in each row is 1.
3. All other elements in the column containing the leftmost 1 of a given row are zeros.
4. The leftmost 1 in any row is to the right of the leftmost 1 in the row above.

The following matrices are in reduced form. Check each one carefully to convince yourself that the conditions in the definition are met.

$$\left[\begin{array}{cc|c} 1 & 0 & 2 \\ 0 & 1 & -3 \end{array}\right] \qquad \left[\begin{array}{ccc|c} 1 & 0 & 0 & 2 \\ 0 & 1 & 0 & -1 \\ 0 & 0 & 1 & 3 \end{array}\right] \qquad \left[\begin{array}{cc|c} 1 & 0 & 3 \\ 0 & 1 & -1 \\ 0 & 0 & 0 \end{array}\right]$$

$$\left[\begin{array}{cccc|c} 1 & 4 & 0 & 0 & -3 \\ 0 & 0 & 1 & 0 & 2 \\ 0 & 0 & 0 & 1 & 6 \end{array}\right] \qquad \left[\begin{array}{ccc|c} 1 & 0 & 4 & 0 \\ 0 & 1 & 3 & 0 \\ 0 & 0 & 0 & 1 \end{array}\right]$$

Example 1 **Reduced Forms** The matrices below are not in reduced form. Indicate which condition in the definition is violated for each matrix. State the row operation(s) required to transform the matrix into reduced form and find the reduced form.

(A) $\left[\begin{array}{cc|c} 0 & 1 & -2 \\ 1 & 0 & 3 \end{array}\right]$ (B) $\left[\begin{array}{ccc|c} 1 & 2 & -2 & 3 \\ 0 & 0 & 1 & -1 \end{array}\right]$

(C) $\left[\begin{array}{cc|c} 1 & 0 & -3 \\ 0 & 0 & 0 \\ 0 & 1 & -2 \end{array}\right]$ (D) $\left[\begin{array}{ccc|c} 1 & 0 & 0 & -1 \\ 0 & 2 & 0 & 3 \\ 0 & 0 & 1 & -5 \end{array}\right]$

SOLUTION (A) Condition 4 is violated: The leftmost 1 in Row 2 is not to the right of the leftmost 1 in Row 1. Perform the row operation $R_1 \leftrightarrow R_2$ to obtain

$$\left[\begin{array}{cc|c} 1 & 0 & 3 \\ 0 & 1 & -2 \end{array}\right]$$

(B) Condition 3 is violated: The column containing the leftmost 1 in Row 2 has a nonzero element above the 1. Perform the row operation $2R_2 + R_1 \rightarrow R_1$ to obtain

$$\left[\begin{array}{ccc|c} 1 & 2 & 0 & 1 \\ 0 & 0 & 1 & -1 \end{array}\right]$$

(C) Condition 1 is violated: The second row contains all zeros and it is not below any row having at least one nonzero element. Perform the row operation $R_2 \leftrightarrow R_3$ to obtain

$$\left[\begin{array}{cc|c} 1 & 0 & -3 \\ 0 & 1 & -2 \\ 0 & 0 & 0 \end{array}\right]$$

(D) Condition 2 is violated: The leftmost nonzero element in Row 2 is not a 1. Perform the row operation $\frac{1}{2}R_2 \rightarrow R_2$ to obtain

$$\left[\begin{array}{ccc|c} 1 & 0 & 0 & -1 \\ 0 & 1 & 0 & \frac{3}{2} \\ 0 & 0 & 1 & -5 \end{array}\right]$$

Matched Problem 1 The matrices below are not in reduced form. Indicate which condition in the definition is violated for each matrix. State the row operation(s) required to transform the matrix into reduced form and find the reduced form.

(A) $\left[\begin{array}{cc|c} 1 & 0 & 2 \\ 0 & 3 & -6 \end{array}\right]$ (B) $\left[\begin{array}{ccc|c} 1 & 5 & 4 & 3 \\ 0 & 1 & 2 & -1 \\ 0 & 0 & 0 & 0 \end{array}\right]$

(C) $\left[\begin{array}{ccc|c} 0 & 1 & 0 & -3 \\ 1 & 0 & 0 & 0 \\ 0 & 0 & 1 & 2 \end{array}\right]$ (D) $\left[\begin{array}{ccc|c} 1 & 2 & 0 & 3 \\ 0 & 0 & 0 & 0 \\ 0 & 0 & 1 & 4 \end{array}\right]$

Solving Systems by Gauss–Jordan Elimination

We are now ready to outline the Gauss–Jordan method for solving systems of linear equations. The method systematically transforms an augmented matrix into a reduced form. The system corresponding to a reduced augmented coefficient matrix is called a **reduced system.** As we shall see, reduced systems are easy to solve.

The Gauss–Jordan elimination method is named after the German mathematician Carl Friedrich Gauss (1777–1885) and the German geodesist Wilhelm Jordan (1842–1899). Gauss, one of the greatest mathematicians of all time, used a method of solving systems of equations in his astronomical work that was later generalized by Jordan to solve problems in large-scale surveying.

Example 2 **Solving a System Using Gauss–Jordan Elimination** Solve by Gauss–Jordan elimination:

$$\begin{aligned} 2x_1 - 2x_2 + x_3 &= 3 \\ 3x_1 + x_2 - x_3 &= 7 \\ x_1 - 3x_2 + 2x_3 &= 0 \end{aligned}$$

Solution Write the augmented matrix and follow the steps indicated at the right.

Need a 1 here.

$$\left[\begin{array}{ccc|c} 2 & -2 & 1 & 3 \\ 3 & 1 & -1 & 7 \\ 1 & -3 & 2 & 0 \end{array}\right] \quad R_1 \leftrightarrow R_3$$

Step 1. Choose the leftmost nonzero column and get a 1 at the top.

Need 0's here.

$$\sim \left[\begin{array}{ccc|c} 1 & -3 & 2 & 0 \\ 3 & 1 & -1 & 7 \\ 2 & -2 & 1 & 3 \end{array}\right] \quad \begin{array}{l} (-3)R_1 + R_2 \to R_2 \\ (-2)R_1 + R_3 \to R_3 \end{array}$$

Step 2. Use multiples of the row containing the 1 from step 1 to get zeros in all remaining places in the column containing this 1.

Need a 1 here.

$$\sim \left[\begin{array}{ccc|c} 1 & -3 & 2 & 0 \\ 0 & 10 & -7 & 7 \\ 0 & 4 & -3 & 3 \end{array}\right] \quad 0.1R_2 \to R_2$$

Step 3. Repeat step 1 with the *submatrix* formed by (mentally) deleting the top row.

Need 0's here.

$$\sim \left[\begin{array}{ccc|c} 1 & -3 & 2 & 0 \\ 0 & 1 & -0.7 & 0.7 \\ 0 & 4 & -3 & 3 \end{array}\right] \quad \begin{array}{l} 3R_2 + R_1 \to R_1 \\ \\ (-4)R_2 + R_3 \to R_3 \end{array}$$

Step 4. Repeat step 2 with the *entire matrix.*

Need a 1 here.

$$\sim\left[\begin{array}{ccc|c} 1 & 0 & -0.1 & 2.1 \\ 0 & 1 & -0.7 & 0.7 \\ 0 & 0 & -0.2 & 0.2 \end{array}\right] \quad (-5)R_3 \to R_3$$

Step 3. Repeat step 1 with the *submatrix* formed by (mentally) deleting the top two rows.

Need 0's here.

$$\sim\left[\begin{array}{ccc|c} 1 & 0 & -0.1 & 2.1 \\ 0 & 1 & -0.7 & 0.7 \\ 0 & 0 & 1 & -1 \end{array}\right] \quad \begin{array}{l} 0.1R_3 + R_1 \to R_1 \\ 0.7R_3 + R_2 \to R_2 \end{array}$$

Step 4. Repeat step 2 with the *entire matrix*.

$$\sim\left[\begin{array}{ccc|c} 1 & 0 & 0 & 2 \\ 0 & 1 & 0 & 0 \\ 0 & 0 & 1 & -1 \end{array}\right]$$

The matrix is now in reduced form, and we can proceed to solve the corresponding reduced system.

$$\begin{aligned} x_1 \qquad\qquad &= 2 \\ x_2 \qquad &= 0 \\ x_3 &= -1 \end{aligned}$$

The solution to this system is $x_1 = 2, x_2 = 0, x_3 = -1$. You should check this solution in the original system.

Gauss–Jordan Elimination

Step 1. Choose the leftmost nonzero column and use appropriate row operations to get a 1 at the top.

Step 2. Use multiples of the row containing the 1 from step 1 to get zeros in all remaining places in the column containing this 1.

Step 3. Repeat step 1 with the **submatrix** formed by (mentally) deleting the row used in step 2 and all rows above this row.

Step 4. Repeat step 2 with the **entire matrix,** including the mentally deleted rows. Continue this process until it is impossible to go further.

[*Note:* If at any point in this process we obtain a row with all zeros to the left of the vertical line and a nonzero number to the right, we can stop, since we will have a contradiction: $0 = n, n \neq 0$. We can then conclude that the system has no solution.]

Remarks

1. Even though each matrix has a unique reduced form, the sequence of steps (algorithm) presented here for transforming a matrix into a reduced form is not unique. That is, other sequences of steps (using row operations) can produce a reduced matrix. (For example, it is possible to use row operations in such a way that computations involving fractions are minimized.) But we emphasize again that we are not interested in the most efficient hand methods for transforming small matrices into

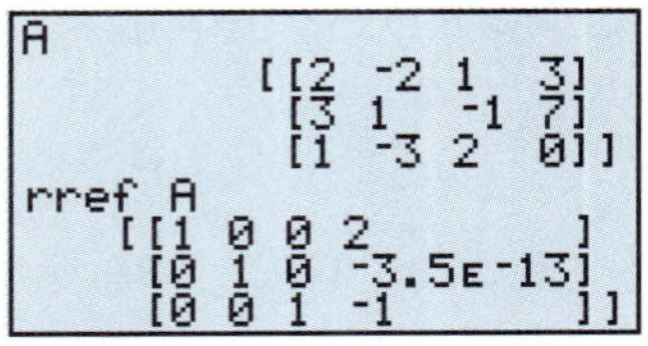

FIGURE 1
Gauss–Jordan elimination on a graphing calculator.

reduced forms. Our main interest is in giving you a little experience with a method that is suitable for solving large-scale systems on a graphing utility.

2. Most graphing utilities have the ability to find reduced forms, either directly or with some programming. Figure 1 illustrates the solution of Example 2 on a graphing calculator that has a built-in routine for finding reduced forms. Notice that in Row 2 and Column 4 of the reduced form the graphing calculator has displayed the very small number $-3.5\text{E}-13$, instead of the exact value 0. This is a common occurrence on a graphing calculator and causes no problems. Just replace any very small numbers displayed in scientific notation with 0.

Matched Problem 2 ➠ Solve by Gauss–Jordan elimination:

$$\begin{aligned} 3x_1 + x_2 - 2x_3 &= 2 \\ x_1 - 2x_2 + x_3 &= 3 \\ 2x_1 - x_2 - 3x_3 &= 3 \end{aligned}$$

Example 3 ➠ **Solving a System Using Gauss–Jordan Elimination** Solve by Gauss–Jordan elimination:

$$\begin{aligned} 2x_1 - 4x_2 + x_3 &= -4 \\ 4x_1 - 8x_2 + 7x_3 &= 2 \\ -2x_1 + 4x_2 - 3x_3 &= 5 \end{aligned}$$

SOLUTION

$$\left[\begin{array}{rrr|r} 2 & -4 & 1 & -4 \\ 4 & -8 & 7 & 2 \\ -2 & 4 & -3 & 5 \end{array}\right] \quad 0.5R_1 \to R_1$$

$$\sim \left[\begin{array}{rrr|r} 1 & -2 & 0.5 & -2 \\ 4 & -8 & 7 & 2 \\ -2 & 4 & -3 & 5 \end{array}\right] \quad \begin{array}{l} (-4)R_1 + R_2 \to R_2 \\ 2R_1 + R_3 \to R_3 \end{array}$$

$$\sim \left[\begin{array}{rrr|r} 1 & -2 & 0.5 & -2 \\ 0 & 0 & 5 & 10 \\ 0 & 0 & -2 & 1 \end{array}\right] \quad 0.2R_2 \to R_2$$

Note that Column 3 is the leftmost nonzero column in this submatrix

$$\sim \left[\begin{array}{rrr|r} 1 & -2 & 0.5 & -2 \\ 0 & 0 & 1 & 2 \\ 0 & 0 & -2 & 1 \end{array}\right] \quad \begin{array}{l} (-0.5)R_2 + R_1 \to R_1 \\ 2R_2 + R_3 \to R_3 \end{array}$$

$$\sim \left[\begin{array}{rrr|r} 1 & -2 & 0 & -3 \\ 0 & 0 & 1 & 2 \\ 0 & 0 & 0 & 5 \end{array}\right]$$

We stop the Gauss–Jordan elimination, even though the matrix is not in reduced form, since the last row produces a contradiction.

Matched Problem 3 ➠ Solve by Gauss–Jordan elimination:

$$\begin{aligned} 2x_1 - 4x_2 - x_3 &= -8 \\ 4x_1 - 8x_2 + 3x_3 &= 4 \\ -2x_1 + 4x_2 + x_3 &= 11 \end{aligned}$$

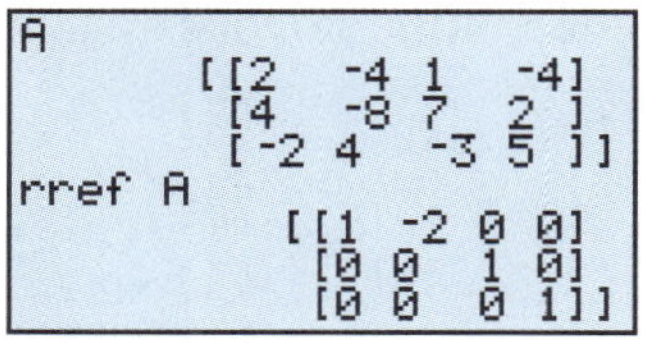

FIGURE 2
Recognizing contradictions on a graphing calculator

CAUTION

Figure 2 shows the solution to Example 3 on a graphing calculator with a built-in reduced form routine. Notice that the graphing calculator does not stop when a contradiction first occurs, but continues on to find the reduced form. Nevertheless, the last row in the reduced form still produces a contradiction. Do not confuse this type of reduced form with one that represents a consistent system (see Fig. 1).

Example 4 **Solving a System Using Gauss–Jordan Elimination** Solve by Gauss–Jordan elimination:

$$\begin{aligned} 3x_1 + 6x_2 - 9x_3 &= 15 \\ 2x_1 + 4x_2 - 6x_3 &= 10 \\ -2x_1 - 3x_2 + 4x_3 &= -6 \end{aligned}$$

SOLUTION

$$\left[\begin{array}{rrr|r} 3 & 6 & -9 & 15 \\ 2 & 4 & -6 & 10 \\ -2 & -3 & 4 & -6 \end{array}\right] \quad \tfrac{1}{3}R_1 \to R_1$$

$$\sim \left[\begin{array}{rrr|r} 1 & 2 & -3 & 5 \\ 2 & 4 & -6 & 10 \\ -2 & -3 & 4 & -6 \end{array}\right] \quad \begin{array}{l} (-2)R_1 + R_2 \to R_2 \\ 2R_1 + R_3 \to R_3 \end{array}$$

$$\sim \left[\begin{array}{rrr|r} 1 & 2 & -3 & 5 \\ 0 & 0 & 0 & 0 \\ 0 & 1 & -2 & 4 \end{array}\right] \quad R_2 \leftrightarrow R_3$$

Note that we must interchange Rows 2 and 3 to obtain a nonzero entry at the top of the second column of this submatrix.

$$\sim \left[\begin{array}{rrr|r} 1 & 2 & -3 & 5 \\ 0 & 1 & -2 & 4 \\ 0 & 0 & 0 & 0 \end{array}\right] \quad (-2)R_2 + R_1 \to R_1$$

$$\sim \left[\begin{array}{rrr|r} 1 & 0 & 1 & -3 \\ 0 & 1 & -2 & 4 \\ 0 & 0 & 0 & 0 \end{array}\right]$$

The matrix is now in reduced form. Write the corresponding reduced system and solve.

$$\begin{aligned} x_1 \qquad + x_3 &= -3 \\ x_2 - 2x_3 &= 4 \end{aligned}$$

We discard the equation corresponding to the third (all zero) row in the reduced form, since it is satisfied by all values of x_1, x_2, and x_3.

Note that the leftmost variable in each equation appears in one and only one equation. We solve for the leftmost variables x_1 and x_2 in terms of the remaining variable, x_3:

$$\begin{aligned} x_1 &= -x_3 - 3 \\ x_2 &= 2x_3 + 4 \end{aligned}$$

If we let $x_3 = t$, then for any real number t,

$$\begin{aligned} x_1 &= -t - 3 \\ x_2 &= 2t + 4 \\ x_3 &= t \end{aligned}$$

You should check that $(-t - 3, 2t + 4, t)$ is a solution of the original system for any real number t. Some particular solutions are

$$\begin{array}{ccc} t = 0 & t = -2 & t = 3.5 \\ (-3, 4, 0) & (-1, 0, -2) & (-6.5, 11, 3.5) \end{array}$$

In general:

If the number of leftmost 1's in a reduced augmented coefficient matrix is less than the number of variables in the system and there are no contradictions, then the system is dependent and has infinitely many solutions.

Describing the solution set to this type of system is not difficult. In a reduced system, the *leftmost variables* correspond to the leftmost 1's in the corresponding reduced augmented matrix. The definition of reduced form for an augmented matrix ensures that each leftmost variable in the corresponding reduced system appears in one and only one equation of the system. Solving for each leftmost variable in terms of the remaining variables and writing a general solution to the system is usually easy. (Example 5 illustrates a slightly more involved case.)

Matched Problem 4 Solve by Gauss–Jordan elimination:

$$\begin{aligned} 2x_1 - 2x_2 - 4x_3 &= -2 \\ 3x_1 - 3x_2 - 6x_3 &= -3 \\ -2x_1 + 3x_2 + x_3 &= 7 \end{aligned}$$

Explore–Discuss 2

Explain why the definition of reduced form ensures that each leftmost variable in a reduced system appears in one and only one equation and no equation contains more than one leftmost variable. Discuss methods for determining whether a consistent system is independent or dependent by examining the reduced form.

Example 5 **Solving a System Using Gauss–Jordan Elimination** Solve by Gauss–Jordan elimination:

$$\begin{aligned} x_1 + 2x_2 + 4x_3 + x_4 - x_5 &= 1 \\ 2x_1 + 4x_2 + 8x_3 + 3x_4 - 4x_5 &= 2 \\ x_1 + 3x_2 + 7x_3 + 3x_5 &= -2 \end{aligned}$$

SOLUTION

$$\left[\begin{array}{ccccc|c} 1 & 2 & 4 & 1 & -1 & 1 \\ 2 & 4 & 8 & 3 & -4 & 2 \\ 1 & 3 & 7 & 0 & 3 & -2 \end{array}\right] \begin{array}{l} \\ (-2)R_1 + R_2 \to R_2 \\ (-1)R_1 + R_3 \to R_3 \end{array}$$

$$\sim \left[\begin{array}{ccccc|c} 1 & 2 & 4 & 1 & -1 & 1 \\ 0 & 0 & 0 & 1 & -2 & 0 \\ 0 & 1 & 3 & -1 & 4 & -3 \end{array}\right] \quad R_2 \leftrightarrow R_3$$

$$\sim \left[\begin{array}{ccccc|c} 1 & 2 & 4 & 1 & -1 & 1 \\ 0 & 1 & 3 & -1 & 4 & -3 \\ 0 & 0 & 0 & 1 & -2 & 0 \end{array}\right] \begin{array}{l} (-2)R_2 + R_1 \to R_1 \\ \\ \\ \end{array}$$

$$\sim\left[\begin{array}{ccccc|c} 1 & 0 & -2 & 3 & -9 & 7 \\ 0 & 1 & 3 & -1 & 4 & -3 \\ 0 & 0 & 0 & 1 & -2 & 0 \end{array}\right] \quad \begin{array}{l} (-3)R_3 + R_1 \to R_1 \\ R_3 + R_2 \to R_2 \end{array}$$

$$\sim\left[\begin{array}{ccccc|c} 1 & 0 & -2 & 0 & -3 & 7 \\ 0 & 1 & 3 & 0 & 2 & -3 \\ 0 & 0 & 0 & 1 & -2 & 0 \end{array}\right] \quad \text{Matrix is in reduced form.}$$

$$\begin{aligned} x_1 \quad - 2x_3 \quad - 3x_5 &= 7 \\ x_2 + 3x_3 \quad + 2x_5 &= -3 \\ x_4 - 2x_5 &= 0 \end{aligned}$$

Solve for the leftmost variables x_1, x_2, and x_4 in terms of the remaining variables x_3 and x_5:

$$\begin{aligned} x_1 &= 2x_3 + 3x_5 + 7 \\ x_2 &= -3x_3 - 2x_5 - 3 \\ x_4 &= 2x_5 \end{aligned}$$

If we let $x_3 = s$ and $x_5 = t$, then for any real numbers s and t,

$$\begin{aligned} x_1 &= 2s + 3t + 7 \\ x_2 &= -3s - 2t - 3 \\ x_3 &= s \\ x_4 &= 2t \\ x_5 &= t \end{aligned}$$

is a solution. The check is left for you to perform.

Matched Problem 5 Solve by Gauss–Jordan elimination:

$$\begin{aligned} x_1 - x_2 + 2x_3 \quad - 2x_5 &= 3 \\ -2x_1 + 2x_2 - 4x_3 - x_4 + x_5 &= -5 \\ 3x_1 - 3x_2 + 7x_3 + x_4 - 4x_5 &= 6 \end{aligned}$$

Application

Dependent systems of linear equations provide an excellent opportunity to discuss mathematical modeling in a little more detail. The process of using mathematics to solve real-world problems can be broken down into three steps (Fig. 3):

Step 1. *Construct* a mathematical model whose solution will provide information about the real-world problem.

Step 2. *Solve* the mathematical model.

Step 3. *Interpret* the solution to the mathematical model in terms of the original real-world problem.

In more complex problems, this cycle may have to be repeated several times to obtain the required information about the real-world problem.

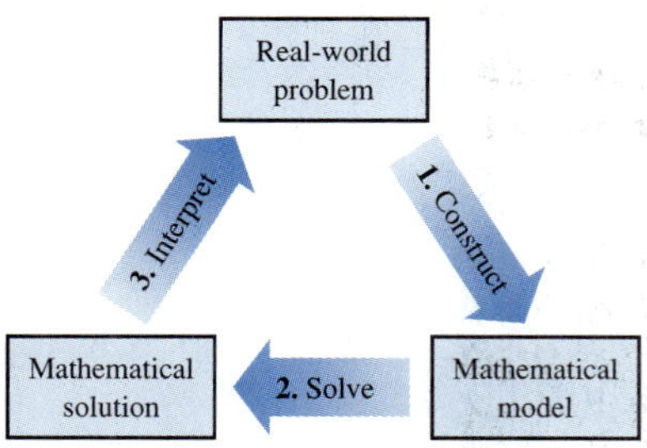

FIGURE 3

Example 6 **Production Scheduling** A casting company produces three different bronze sculptures. The casting department has available a maximum of 140 labor-hours per week, and the finishing department has a maximum of 180 labor-hours available per week. Sculpture A requires 30 hours for casting and 10 hours for finishing; sculpture B requires 10 hours for casting and 10 hours for finishing; and sculpture C requires 10 hours for casting and 30 hours for finishing. If the plant is to operate at maximum capacity, how many of each sculpture should be produced each week?

SOLUTION First, we summarize the relevant manufacturing data in a table:

	LABOR-HOURS PER SCULPTURE A	B	C	MAXIMUM LABOR-HOURS AVAILABLE PER WEEK
CASTING DEPARTMENT	30	10	10	140
FINISHING DEPARTMENT	10	10	30	180

Let: x_1 = Number of sculpture A produced per week
x_2 = Number of sculpture B produced per week
x_3 = Number of sculpture C produced per week

Then an appropriate mathematical model for this problem is

$$\begin{aligned} 30x_1 + 10x_2 + 10x_3 &= 140 \quad \text{Casting department} \\ 10x_1 + 10x_2 + 30x_3 &= 180 \quad \text{Finishing department} \end{aligned} \tag{2}$$

Now we can form the augmented matrix of the system and solve by using Gauss–Jordan elimination:

$$\left[\begin{array}{ccc|c} 30 & 10 & 10 & 140 \\ 10 & 10 & 30 & 180 \end{array}\right] \quad \begin{array}{l} \frac{1}{10}R_1 \to R_1 \quad \text{Simplify each row.} \\ \frac{1}{10}R_2 \to R_2 \end{array}$$

$$\sim \left[\begin{array}{ccc|c} 3 & 1 & 1 & 14 \\ 1 & 1 & 3 & 18 \end{array}\right] \quad R_1 \leftrightarrow R_2$$

$$\sim \left[\begin{array}{ccc|c} 1 & 1 & 3 & 18 \\ 3 & 1 & 1 & 14 \end{array}\right] \quad (-3)R_1 + R_2 \to R_2$$

$$\sim \left[\begin{array}{ccc|c} 1 & 1 & 3 & 18 \\ 0 & -2 & -8 & -40 \end{array}\right] \quad (-\tfrac{1}{2})R_2 \to R_2$$

$$\sim \left[\begin{array}{ccc|c} 1 & 1 & 3 & 18 \\ 0 & 1 & 4 & 20 \end{array}\right] \quad (-1)R_2 + R_1 \to R_1$$

$$\sim \left[\begin{array}{ccc|c} 1 & 0 & -1 & -2 \\ 0 & 1 & 4 & 20 \end{array}\right] \quad \text{Matrix is in reduced form.}$$

$$\begin{aligned} x_1 \qquad - x_3 &= -2 \\ x_2 + 4x_3 &= 20 \end{aligned} \qquad \text{or} \qquad \begin{aligned} x_1 &= x_3 - 2 \\ x_2 &= -4x_3 + 20 \end{aligned}$$

Let $x_3 = t$. Then for t any real number,

$$\begin{aligned} x_1 &= t - 2 \\ x_2 &= -4t + 20 \\ x_3 &= t \end{aligned} \tag{3}$$

is a solution to the mathematical model in (2).

Now we must interpret this solution set in terms of the original problem. Since the variables x_1, x_2, and x_3 represent numbers of sculptures, they must be nonnegative numbers. And if we assume that we cannot produce a fractional number of sculptures, then each must be a nonnegative whole number. Since $t = x_3$, it follows that t must also be a nonnegative whole number. The first and second equations in (3) place further restrictions on the values that t can assume:

$$\begin{array}{lll} x_1 = t - 2 \geqslant 0 & \text{implies that} & t \geqslant 2 \\ x_2 = 20 - 4t \geqslant 0 & \text{implies that} & t \leqslant 5 \end{array}$$

Thus, the only possible values of t that will produce meaningful solutions to the original problem are 2, 3, 4, and 5. That is, the only possible production schedules that utilize the full capacity of the plant are $x_1 = t - 2$ type A, $x_2 = 20 - 4t$ type B, and $x_3 = t$ type C sculptures, where $t = 2, 3, 4,$ or 5. A table is a convenient way to display these solutions:

	Sculpture A	Sculpture B	Sculpture C
t	x_1	x_2	x_3
2	0	12	2
3	1	8	3
4	2	4	4
5	3	0	5

Matched Problem 6 Repeat Example 6 given a casting capacity of 120 labor-hours per week and a finishing capacity of 200 labor-hours per week.

Answers to Matched Problems

1. (A) Condition 2 is violated: The 3 in Row 2 and Column 2 should be a 1. Perform the operation $\frac{1}{3}R_2 \rightarrow R_2$ to obtain

$$\left[\begin{array}{cc|c} 1 & 0 & 2 \\ 0 & 1 & -2 \end{array}\right]$$

(B) Condition 3 is violated: The 5 in Row 1 and Column 2 should be a 0. Perform the operation $(-5)R_2 + R_1 \rightarrow R_1$ to obtain

$$\left[\begin{array}{ccc|c} 1 & 0 & -6 & 8 \\ 0 & 1 & 2 & -1 \\ 0 & 0 & 0 & 0 \end{array}\right]$$

(C) Condition 4 is violated. The leftmost 1 in the second row is not to the right of the leftmost 1 in the first row. Perform the operation $R_1 \leftrightarrow R_2$ to obtain

$$\left[\begin{array}{ccc|c} 1 & 0 & 0 & 0 \\ 0 & 1 & 0 & -3 \\ 0 & 0 & 1 & 2 \end{array}\right]$$

(D) Condition 1 is violated: The all-zero second row should be at the bottom. Perform the operation $R_2 \leftrightarrow R_3$ to obtain

$$\left[\begin{array}{ccc|c} 1 & 2 & 0 & 3 \\ 0 & 0 & 1 & 4 \\ 0 & 0 & 0 & 0 \end{array}\right]$$

2. $x_1 = 1, x_2 = -1, x_3 = 0$ **3.** Inconsistent; no solution
4. $x_1 = 5t + 4, x_2 = 3t + 5, x_3 = t$, t any real number
5. $x_1 = s + 7, x_2 = s, x_3 = t - 2, x_4 = -3t - 1, x_5 = t$, s and t any real numbers
6. $x_1 = t - 4$ type A, $x_2 = 24 - 4t$ type B, and $x_3 = t$ type C sculptures, where $t = 4, 5,$ or 6

EXERCISE 4-3

A *In Problems 1–10, if a matrix is in reduced form, say so. If not, explain why and indicate the row operation(s) necessary to transform the matrix into reduced form.*

1. $\left[\begin{array}{cc|c} 1 & 0 & 2 \\ 0 & 1 & -1 \end{array}\right]$

2. $\left[\begin{array}{cc|c} 0 & 1 & 2 \\ 1 & 0 & -1 \end{array}\right]$

3. $\left[\begin{array}{ccc|c} 1 & 0 & 2 & 3 \\ 0 & 0 & 0 & 0 \\ 0 & 1 & -1 & 4 \end{array}\right]$

4. $\left[\begin{array}{ccc|c} 1 & 0 & 0 & -2 \\ 0 & 1 & 0 & 0 \\ 0 & 0 & 1 & 1 \end{array}\right]$

5. $\left[\begin{array}{ccc|c} 0 & 1 & 0 & 2 \\ 0 & 0 & 3 & -1 \\ 0 & 0 & 0 & 0 \end{array}\right]$

6. $\left[\begin{array}{ccc|c} 1 & 2 & -3 & 1 \\ 0 & 0 & 1 & 4 \\ 0 & 0 & 0 & 0 \end{array}\right]$

7. $\left[\begin{array}{ccc|c} 1 & 1 & 0 & 1 \\ 0 & 0 & 1 & 1 \\ 0 & 0 & 0 & 0 \end{array}\right]$

8. $\left[\begin{array}{ccc|c} 1 & 0 & -1 & 3 \\ 0 & 2 & 1 & 1 \\ 0 & 0 & 0 & 0 \end{array}\right]$

9. $\left[\begin{array}{cccc|c} 1 & 0 & -2 & 0 & 1 \\ 0 & 0 & 1 & 1 & 0 \end{array}\right]$

10. $\left[\begin{array}{cccc|c} 1 & -2 & 0 & 0 & 1 \\ 0 & 0 & 1 & 1 & 0 \end{array}\right]$

Write the linear system corresponding to each reduced augmented matrix in Problems 11–18 and solve.

11. $\left[\begin{array}{ccc|c} 1 & 0 & 0 & -2 \\ 0 & 1 & 0 & 3 \\ 0 & 0 & 1 & 0 \end{array}\right]$

12. $\left[\begin{array}{cccc|c} 1 & 0 & 0 & 0 & -2 \\ 0 & 1 & 0 & 0 & 0 \\ 0 & 0 & 1 & 0 & 1 \\ 0 & 0 & 0 & 1 & 3 \end{array}\right]$

13. $\left[\begin{array}{ccc|c} 1 & 0 & -2 & 3 \\ 0 & 1 & 1 & -5 \\ 0 & 0 & 0 & 0 \end{array}\right]$

14. $\left[\begin{array}{ccc|c} 1 & -2 & 0 & -3 \\ 0 & 0 & 1 & 5 \\ 0 & 0 & 0 & 0 \end{array}\right]$

15. $\left[\begin{array}{cc|c} 1 & 0 & 0 \\ 0 & 1 & 0 \\ 0 & 0 & 1 \end{array}\right]$

16. $\left[\begin{array}{cc|c} 1 & 0 & 5 \\ 0 & 1 & -3 \\ 0 & 0 & 0 \end{array}\right]$

17. $\left[\begin{array}{cccc|c} 1 & -2 & 0 & -3 & -5 \\ 0 & 0 & 1 & 3 & 2 \end{array}\right]$

18. $\left[\begin{array}{cccc|c} 1 & 0 & -2 & 3 & 4 \\ 0 & 1 & -1 & 2 & -1 \end{array}\right]$

B *Use row operations to change each matrix in Problems 19–24 to reduced form.*

19. $\left[\begin{array}{cc|c} 1 & 2 & -1 \\ 0 & 1 & 3 \end{array}\right]$

20. $\left[\begin{array}{cc|c} 1 & 3 & 1 \\ 0 & 2 & -4 \end{array}\right]$

21. $\left[\begin{array}{ccc|c} 1 & 0 & -3 & 1 \\ 0 & 1 & 2 & 0 \\ 0 & 0 & 3 & -6 \end{array}\right]$

22. $\left[\begin{array}{ccc|c} 1 & 0 & 4 & 0 \\ 0 & 1 & -3 & -1 \\ 0 & 0 & -2 & 2 \end{array}\right]$

23. $\left[\begin{array}{ccc|c} 1 & 2 & -2 & -1 \\ 0 & 3 & -6 & 1 \\ 0 & -1 & 2 & -\frac{1}{3} \end{array}\right]$

24. $\left[\begin{array}{ccc|c} 0 & -2 & 8 & 1 \\ 2 & -2 & 6 & -4 \\ 0 & -1 & 4 & \frac{1}{2} \end{array}\right]$

Solve Problems 25–44 using Gauss–Jordan elimination.

25. $\begin{aligned} 2x_1 + 4x_2 - 10x_3 &= -2 \\ 3x_1 + 9x_2 - 21x_3 &= 0 \\ x_1 + 5x_2 - 12x_3 &= 1 \end{aligned}$

26. $\begin{aligned} 3x_1 + 5x_2 - x_3 &= -7 \\ x_1 + x_2 + x_3 &= -1 \\ 2x_1 + 11x_3 &= 7 \end{aligned}$

27. $\begin{aligned} 3x_1 + 8x_2 - x_3 &= -18 \\ 2x_1 + x_2 + 5x_3 &= 8 \\ 2x_1 + 4x_2 + 2x_3 &= -4 \end{aligned}$

28. $\begin{aligned} 2x_1 + 6x_2 + 15x_3 &= -12 \\ 4x_1 + 7x_2 + 13x_3 &= -10 \\ 3x_1 + 6x_2 + 12x_3 &= -9 \end{aligned}$

29. $\begin{aligned} 2x_1 - x_2 - 3x_3 &= 8 \\ x_1 - 2x_2 &= 7 \end{aligned}$

30. $\begin{aligned} 2x_1 + 4x_2 - 6x_3 &= 10 \\ 3x_1 + 3x_2 - 3x_3 &= 6 \end{aligned}$

31. $\begin{aligned} 2x_1 - x_2 &= 0 \\ 3x_1 + 2x_2 &= 7 \\ x_1 - x_2 &= -1 \end{aligned}$

32. $\begin{aligned} 2x_1 - x_2 &= 0 \\ 3x_1 + 2x_2 &= 7 \\ x_1 - x_2 &= -2 \end{aligned}$

33. $\begin{aligned} 3x_1 - 4x_2 - x_3 &= 1 \\ 2x_1 - 3x_2 + x_3 &= 1 \\ x_1 - 2x_2 + 3x_3 &= 2 \end{aligned}$

34. $\begin{aligned} 3x_1 + 7x_2 - x_3 &= 11 \\ x_1 + 2x_2 - x_3 &= 3 \\ 2x_1 + 4x_2 - 2x_3 &= 10 \end{aligned}$

35. $\begin{aligned} 3x_1 - 2x_2 + x_3 &= -7 \\ 2x_1 + x_2 - 4x_3 &= 0 \\ x_1 + x_2 - 3x_3 &= 1 \end{aligned}$

36. $\begin{aligned} 2x_1 + 3x_2 + 5x_3 &= 21 \\ x_1 - x_2 - 5x_3 &= -2 \\ 2x_1 + x_2 - x_3 &= 11 \end{aligned}$

37. $\begin{aligned} 2x_1 + 4x_2 - 2x_3 &= 2 \\ -3x_1 - 6x_2 + 3x_3 &= -3 \end{aligned}$

38. $\begin{aligned} 3x_1 - 9x_2 + 12x_3 &= 6 \\ -2x_1 + 6x_2 - 8x_3 &= -4 \end{aligned}$

39. $\begin{aligned} 4x_1 - x_2 + 2x_3 &= 3 \\ -4x_1 + x_2 - 3x_3 &= -10 \\ 8x_1 - 2x_2 + 9x_3 &= -1 \end{aligned}$

40. $\begin{aligned} 4x_1 - 2x_2 + 2x_3 &= 5 \\ -6x_1 + 3x_2 - 3x_3 &= -2 \\ 10x_1 - 5x_2 + 9x_3 &= 4 \end{aligned}$

41. $\begin{aligned} 2x_1 - 5x_2 - 3x_3 &= 7 \\ -4x_1 + 10x_2 + 2x_3 &= 6 \\ 6x_1 - 15x_2 - x_3 &= -19 \end{aligned}$

42. $\begin{aligned} -4x_1 + 8x_2 + 10x_3 &= -6 \\ 6x_1 - 12x_2 - 15x_3 &= 9 \\ -8x_1 + 14x_2 + 19x_3 &= -8 \end{aligned}$

43. $\begin{aligned} 5x_1 - 3x_2 + 2x_3 &= 13 \\ 2x_1 - x_2 - 3x_3 &= 1 \\ 4x_1 - 2x_2 + 4x_3 &= 12 \end{aligned}$

44. $\begin{aligned} 4x_1 - 2x_2 + 3x_3 &= 3 \\ 3x_1 - x_2 - 2x_3 &= -10 \\ 2x_1 + 4x_2 - x_3 &= -1 \end{aligned}$

45. Consider a consistent system of three linear equations in three variables. Discuss the nature of the system and its solution set if the reduced form of the augmented coefficient matrix has:
(A) One leftmost 1 (B) Two leftmost 1's
(C) Three leftmost 1's (D) Four leftmost 1's

46. Consider a system of three linear equations in three variables. Give examples of two reduced forms that are not row-equivalent if the system is:
(A) Consistent and dependent
(B) Inconsistent

In Problems 47–50, discuss the relationship between the number of solutions of the system and the constant k.

47. $\begin{aligned} x_1 - x_2 &= 4 \\ 3x_1 + kx_2 &= 7 \end{aligned}$

48. $\begin{aligned} x_1 + 2x_2 &= 4 \\ -2x_1 + kx_2 &= -8 \end{aligned}$

49. $\begin{aligned} x_1 + kx_2 &= 3 \\ 2x_1 + 6x_2 &= 6 \end{aligned}$

50. $\begin{aligned} x_1 + kx_2 &= 3 \\ 2x_1 + 4x_2 &= 8 \end{aligned}$

C *Solve Problems 51–56 using Gauss–Jordan elimination.*

51. $\begin{aligned} x_1 + 2x_2 - 4x_3 - x_4 &= 7 \\ 2x_1 + 5x_2 - 9x_3 - 4x_4 &= 16 \\ x_1 + 5x_2 - 7x_3 - 7x_4 &= 13 \end{aligned}$

52. $\begin{aligned} 2x_1 + 4x_2 + 5x_3 + 4x_4 &= 8 \\ x_1 + 2x_2 + 2x_3 + x_4 &= 3 \end{aligned}$

53. $\begin{aligned} x_1 - x_2 + 3x_3 - 2x_4 &= 1 \\ -2x_1 + 4x_2 - 3x_3 + x_4 &= 0.5 \\ 3x_1 - x_2 + 10x_3 - 4x_4 &= 2.9 \\ 4x_1 - 3x_2 + 8x_3 - 2x_4 &= 0.6 \end{aligned}$

54. $\begin{aligned} x_1 + x_2 + 4x_3 + x_4 &= 1.3 \\ -x_1 + x_2 - x_3 &= 1.1 \\ 2x_1 + x_3 + 3x_4 &= -4.4 \\ 2x_1 + 5x_2 + 11x_3 + 3x_4 &= 5.6 \end{aligned}$

55. $\begin{aligned} x_1 - 2x_2 + x_3 + x_4 + 2x_5 &= 2 \\ -2x_1 + 4x_2 + 2x_3 + 2x_4 - 2x_5 &= 0 \\ 3x_1 - 6x_2 + x_3 + x_4 + 5x_5 &= 4 \\ -x_1 + 2x_2 + 3x_3 + x_4 + x_5 &= 3 \end{aligned}$

56. $\begin{aligned} x_1 - 3x_2 + x_3 + x_4 + 2x_5 &= 2 \\ -x_1 + 5x_2 + 2x_3 + 2x_4 - 2x_5 &= 0 \\ 2x_1 - 6x_2 + 2x_3 + 2x_4 + 4x_5 &= 4 \\ -x_1 + 3x_2 - x_3 - x_5 &= -3 \end{aligned}$

Applications

Construct a mathematical model for each of the following problems. (The answers in the back of the book include both the mathematical model and the interpretation of its solution.) Use Gauss–Jordan elimination to solve the model and then interpret the solution.

Business & Economics

57. *Production scheduling.* A small manufacturing plant makes three types of inflatable boats: one-person, two-person, and four-person models. Each boat requires the services of three departments, as listed in the table. The cutting, assembly, and packaging departments have available a maximum of 380, 330, and 120 labor-hours per week, respectively.

Department	One-Person Boat	Two-Person Boat	Four-Person Boat
Cutting	0.5 hr	1.0 hr	1.5 hr
Assembly	0.6 hr	0.9 hr	1.2 hr
Packaging	0.2 hr	0.3 hr	0.5 hr

(A) How many boats of each type must be produced each week for the plant to operate at full capacity?
(B) How is the production schedule in part (A) affected if the packaging department is no longer used?
(C) How is the production schedule in part (A) affected if the four-person boat is no longer produced?

58. *Production scheduling.* Repeat Problem 57 assuming the cutting, assembly, and packaging departments have available a maximum of 350, 330, and 115 labor-hours per week, respectively.

59. *Purchasing.* A chemical manufacturer wants to purchase a fleet of 24 railroad tank cars with a combined carrying capacity of 250,000 gallons. Tank cars with three different carrying capacities are available: 6,000 gallons, 8,000 gallons, and 18,000 gallons. How many of each type of tank car should be purchased?

60. *Purchasing.* A commuter airline wants to purchase a fleet of 30 airplanes with a combined carrying capacity of 960 passengers. The three available types of planes carry 18, 24, and 42 passengers, respectively. How many of each type of plane should be purchased?

61. *Income tax.* A corporation has a taxable income of $7,650,000. At this income level, the federal income tax rate is 50%, the state tax rate is 20%, and the local tax rate is 10%. If each tax rate is applied to the total taxable income, the resulting tax liability for the corporation would be 80% of taxable income. However, it is customary to deduct taxes paid to one agency before computing taxes for the other agencies. Assume that the federal taxes are based on the income that remains after the state and local taxes are deducted, and that state and local taxes are computed in a similar manner. What is the tax liability of the corporation (as a percentage of taxable income) if these deductions are taken into consideration?

62. *Income tax.* Repeat Problem 61 if local taxes are not allowed as a deduction for federal and state taxes.

63. C *Taxable income.* As a result of several mergers and acquisitions, stock in four companies has been distributed among the companies. Each row of the following table gives the percentage of stock in the four companies that a particular company owns and the annual net income of each company (in millions of dollars):

	Percentage of Stock Owned in Company				Annual Net Income
Company	*A*	*B*	*C*	*D*	*Million $*
A	71	8	3	7	3.2
B	12	81	11	13	2.6
C	11	9	72	8	3.8
D	6	2	14	72	4.4

Thus, company *A* holds 71% of its own stock, 8% of the stock in company *B*, 3% of the stock in company *C*, and so on. For the purpose of assessing a state tax on corporate income, the taxable income of each company is defined to be its share of its own annual net income plus its share of the taxable income of each of the other companies, as determined by the percentages in the table. What is the taxable income of each company (to the nearest thousand dollars)?

64. C *Taxable income.* Repeat Problem 63 if tax law is changed so that the taxable income of a company is defined to be all of its own annual net income plus its share of the taxable income of each of the other companies.

Life Sciences

65. *Nutrition.* A dietitian in a hospital is to arrange a special diet composed of three basic foods. The diet is to include exactly 340 units of calcium, 180 units of iron, and 220 units of vitamin A. The number of units per ounce of each special ingredient for each of the foods is indicated in the table.

	UNITS PER OUNCE		
	Food A	*Food B*	*Food C*
CALCIUM	30	10	20
IRON	10	10	20
VITAMIN A	10	30	20

(A) How many ounces of each food must be used to meet the diet requirements?
(B) How is the diet in part (A) affected if food *C* is not used?
(C) How is the diet in part (A) affected if the vitamin A requirement is dropped?

66. *Nutrition.* Repeat Problem 65 if the diet is to include exactly 400 units of calcium, 160 units of iron, and 240 units of vitamin A.

67. *Nutrition—plants.* A farmer can buy four types of plant food. Each barrel of mix *A* contains 30 pounds of phosphoric acid, 50 pounds of nitrogen, and 30 pounds of potash; each barrel of mix *B* contains 30 pounds of phosphoric acid, 75 pounds of nitrogen, and 20 pounds of potash; each barrel of mix *C* contains 30 pounds of phosphoric acid, 25 pounds of nitrogen, and 20 pounds of potash; and each barrel of mix *D* contains 60 pounds of phosphoric acid, 25 pounds of nitrogen, and 50 pounds of potash. Soil tests indicate that a particular field needs 900 pounds of phosphoric acid, 750 pounds of nitrogen, and 700 pounds of potash. How many barrels of each type of food should the farmer mix together to supply the necessary nutrients for the field?

68. *Nutrition—animals.* In a laboratory experiment, rats are to be fed 5 packets of food containing a total of 80 units of vitamin E. There are four different brands of food packets that can be used. A packet of brand *A* contains 5 units of vitamin E, a packet of brand *B* contains 10 units of vitamin E, a packet of brand *C* contains 15 units of vitamin E, and a packet of brand *D* contains 20 units of vitamin E. How many packets of each brand should be mixed and fed to the rats?

Social Sciences

69. *Sociology.* Two sociologists have grant money to study school busing in a particular city. They wish to conduct an opinion survey using 600 telephone contacts and 400 house contacts. Survey company *A* has personnel to do 30 telephone and 10 house contacts per hour; survey company *B* can handle 20 telephone and 20 house contacts per hour. How many hours should be scheduled for each firm to produce exactly the number of contacts needed?

70. *Sociology.* Repeat Problem 69 if 650 telephone contacts and 350 house contacts are needed.

71. *Traffic flow.* The rush-hour traffic flow for a network of four one-way streets in a city is shown in the figure. The numbers next to each street indicate the number of vehicles per hour that enter and leave the network on that street. The variables x_1, x_2, x_3, and x_4 represent the flow of traffic between the four intersections in the network.

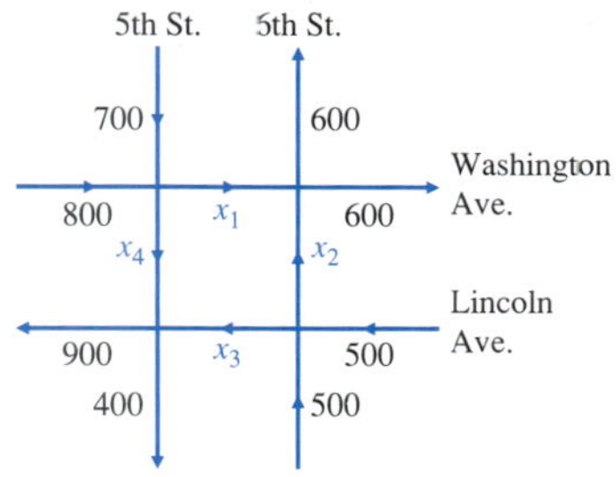

Figure for 71

(A) For a smooth traffic flow, the number of vehicles entering each intersection should always equal the number leaving. For example, since 1,500 vehicles enter the intersection of 5th Street and Washington Avenue each hour and $x_1 + x_4$ vehicles leave this intersection, we see that $x_1 + x_4 = 1{,}500$. Find the equations determined by the traffic flow at each of the other three intersections.
(B) Find the solution to the system in part (A).
(C) What is the maximum number of vehicles that can travel from Washington Avenue to Lincoln Avenue on 5th Street? What is the minimum number?
(D) If traffic lights are adjusted so that 1,000 vehicles per hour travel from Washington Avenue to Lincoln Avenue on 5th Street, determine the flow around the rest of the network.

72. *Traffic flow.* Refer to Problem 71. Closing Washington Avenue east of 6th Street for construction changes the traffic flow for the network as indicated in the figure. Repeat parts (A)–(D) of Problem 71 for this traffic flow.

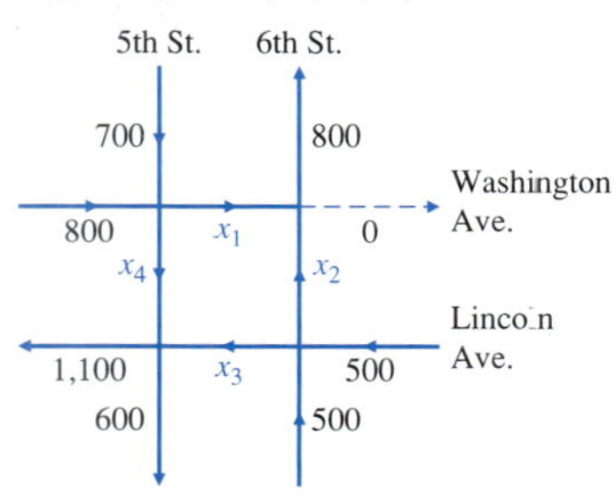

Figure for 72

SECTION 4-4

Matrices: Basic Operations

- ADDITION AND SUBTRACTION
- PRODUCT OF A NUMBER k AND A MATRIX M
- MATRIX PRODUCT

In the last two sections we introduced the important new idea of matrices. In this and the following sections, we will develop this concept further. Matrices are both a very ancient and a very current mathematical concept. References to matrices and systems of equations can be found in Chinese manuscripts dating back to around 200 B.C. More recently, the advent of computers has made matrices a very useful tool for a wide variety of applications. Most graphing calculators and computers are capable of performing calculations with matrices.

As we will see, matrix addition and multiplication are similar to real number addition and multiplication in many respects, but there are some important differences. A brief review of Appendix A-2, where the basic properties of real number operations are discussed, will help you understand the similarities and the differences.

ADDITION AND SUBTRACTION

Before we can discuss arithmetic operations for matrices, we have to define equality for matrices. Two matrices are **equal** if they have the same size and their corresponding elements are equal. For example,

$$\underset{2\times 3}{\begin{bmatrix} a & b & c \\ d & e & f \end{bmatrix}} = \underset{2\times 3}{\begin{bmatrix} u & v & w \\ x & y & z \end{bmatrix}} \quad \text{if and only if} \quad \begin{matrix} a = u & b = v & c = w \\ d = x & e = y & f = z \end{matrix}$$

The **sum of two matrices of the same size** is the matrix with elements that are the sum of the corresponding elements of the two given matrices.

Addition is not defined for matrices of different sizes.

Example 1 **Matrix Addition**

(A) $\begin{bmatrix} a & b \\ c & d \end{bmatrix} + \begin{bmatrix} w & x \\ y & z \end{bmatrix} = \begin{bmatrix} (a+w) & (b+x) \\ (c+y) & (d+z) \end{bmatrix}$

(B) $\begin{bmatrix} 2 & -3 & 0 \\ 1 & 2 & -5 \end{bmatrix} + \begin{bmatrix} 3 & 1 & 2 \\ -3 & 2 & 5 \end{bmatrix} = \begin{bmatrix} 5 & -2 & 2 \\ -2 & 4 & 0 \end{bmatrix}$

(C) $\begin{bmatrix} 5 & 0 & -2 \\ 1 & -3 & 8 \end{bmatrix} + \begin{bmatrix} -1 & 7 \\ 0 & 6 \\ -2 & 8 \end{bmatrix}$ Not defined

Matched Problem 1 Add: $\begin{bmatrix} 3 & 2 \\ -1 & -1 \\ 0 & 3 \end{bmatrix} + \begin{bmatrix} -2 & 3 \\ 1 & -1 \\ 2 & -2 \end{bmatrix}$

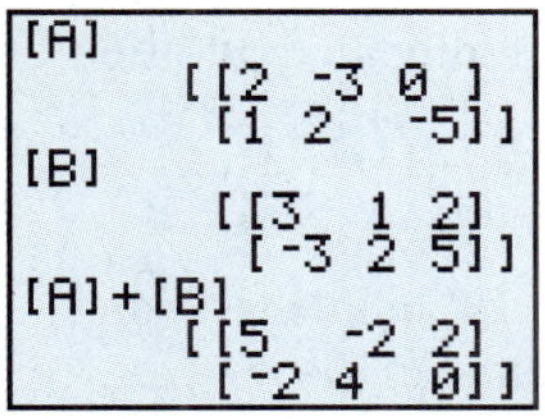

FIGURE 1
Addition on a graphing calculator

Graphing utilities can be used to solve problems involving matrix operations. Figure 1 illustrates the solution to Example 1B on a graphing calculator.

Because we add two matrices by adding their corresponding elements, it follows from the properties of real numbers that matrices of the same size are commutative and associative relative to addition. That is, if A, B, and C are matrices of the same size, then

Commutative: $A + B = B + A$
Associative: $(A + B) + C = A + (B + C)$

A matrix with elements that are all zeros is called a **zero matrix.** For example,

$$\begin{bmatrix} 0 & 0 & 0 \end{bmatrix} \qquad \begin{bmatrix} 0 & 0 \\ 0 & 0 \end{bmatrix} \qquad \begin{bmatrix} 0 \\ 0 \\ 0 \\ 0 \end{bmatrix} \qquad \begin{bmatrix} 0 & 0 & 0 & 0 \\ 0 & 0 & 0 & 0 \\ 0 & 0 & 0 & 0 \end{bmatrix}$$

are zero matrices of different sizes. [*Note:* The simpler notation "0" is often used to denote the zero matrix of an arbitrary size.] The **negative of a matrix** M, denoted by $-M$, is a matrix with elements that are the negatives of the elements in M. Thus, if

$$M = \begin{bmatrix} a & b \\ c & d \end{bmatrix} \qquad \text{then} \qquad -M = \begin{bmatrix} -a & -b \\ -c & -d \end{bmatrix}$$

Note that $M + (-M) = 0$ (a zero matrix).

If A and B are matrices of the same size, then we define **subtraction** as follows:

$$\mathbf{A - B = A + (-B)}$$

Thus, to subtract matrix B from matrix A, we simply add the negative of B to A.

Example 2 **Matrix Subtraction**

$$\begin{bmatrix} 3 & -2 \\ 5 & 0 \end{bmatrix} - \begin{bmatrix} -2 & 2 \\ 3 & 4 \end{bmatrix} = \begin{bmatrix} 3 & -2 \\ 5 & 0 \end{bmatrix} + \begin{bmatrix} 2 & -2 \\ -3 & -4 \end{bmatrix} = \begin{bmatrix} 5 & -4 \\ 2 & -4 \end{bmatrix}$$

Matched Problem 2 Subtract: $\begin{bmatrix} 2 & -3 & 5 \end{bmatrix} - \begin{bmatrix} 3 & -2 & 1 \end{bmatrix}$

Product of a Number k and a Matrix M

The **product of a number k and a matrix M,** denoted by kM, is a matrix formed by multiplying each element of M by k.

Example 3 **Multiplication of a Matrix by a Number**

$$-2\begin{bmatrix} 3 & -1 & 0 \\ -2 & 1 & 3 \\ 0 & -1 & -2 \end{bmatrix} = \begin{bmatrix} -6 & 2 & 0 \\ 4 & -2 & -6 \\ 0 & 2 & 4 \end{bmatrix}$$

Matched Problem 3 Find: $10\begin{bmatrix} 1.3 \\ 0.2 \\ 3.5 \end{bmatrix}$

Explore–Discuss 1

Multiplication of two numbers can be interpreted as repeated addition if one of the numbers is a positive integer. That is,

$$2a = a + a \qquad 3a = a + a + a \qquad 4a = a + a + a + a$$

and so on. Discuss this interpretation for the product of an integer k and matrix M. Use specific examples to illustrate your remarks.

The next example illustrates the use of matrix operations in an applied setting.

Example 4 **Sales Commissions** Ms. Smith and Mr. Jones are salespeople in a new-car agency that sells only two models. August was the last month for this year's models, and next year's models were introduced in September. Gross dollar sales for each month are given in the following matrices:

$$\begin{array}{l} \\ \text{Ms. Smith} \\ \text{Mr. Jones} \end{array} \overset{\text{August sales}}{\begin{array}{cc} \text{Compact} & \text{Luxury} \\ \end{array}} \begin{bmatrix} \$54{,}000 & \$88{,}000 \\ \$126{,}000 & 0 \end{bmatrix} = A \qquad \overset{\text{September sales}}{\begin{array}{cc} \text{Compact} & \text{Luxury} \end{array}} \begin{bmatrix} \$228{,}000 & \$368{,}000 \\ \$304{,}000 & \$322{,}000 \end{bmatrix} = B$$

(For example, Ms. Smith had \$54,000 in compact sales in August, and Mr. Jones had \$322,000 in luxury car sales in September.)

(A) What were the combined dollar sales in August and September for each salesperson and each model?
(B) What was the increase in dollar sales from August to September?
(C) If both salespeople receive 5% commissions on gross dollar sales, compute the commission for each person for each model sold in September.

SOLUTION (A)

$$A + B = \overset{\begin{array}{cc} \text{Compact} & \text{Luxury} \end{array}}{\begin{bmatrix} \$282{,}000 & \$456{,}000 \\ \$430{,}000 & \$322{,}000 \end{bmatrix}} \begin{array}{l} \text{Ms. Smith} \\ \text{Mr. Jones} \end{array}$$

(B)

$$B - A = \overset{\begin{array}{cc} \text{Compact} & \text{Luxury} \end{array}}{\begin{bmatrix} \$174{,}000 & \$280{,}000 \\ \$178{,}000 & \$322{,}000 \end{bmatrix}} \begin{array}{l} \text{Ms. Smith} \\ \text{Mr. Jones} \end{array}$$

(C)

$$0.05B = \begin{bmatrix} (0.05)(\$228{,}000) & (0.05)(\$368{,}000) \\ (0.05)(\$304{,}000) & (0.05)(\$322{,}000) \end{bmatrix}$$

$$= \begin{bmatrix} \$11{,}400 & \$18{,}400 \\ \$15{,}200 & \$16{,}100 \end{bmatrix} \begin{array}{l} \text{Ms. Smith} \\ \text{Mr. Jones} \end{array}$$

Matched Problem 4 Repeat Example 4 with

$$A = \begin{bmatrix} \$45{,}000 & \$77{,}000 \\ \$106{,}000 & \$22{,}000 \end{bmatrix} \quad \text{and} \quad B = \begin{bmatrix} \$190{,}000 & \$345{,}000 \\ \$266{,}000 & \$276{,}000 \end{bmatrix}$$

Figure 2 illustrates a solution for Example 4 on a spreadsheet.

	1	2	3	4	5	6	7
1		August Sales		September Sales		September Commissions	
2		Compact	Luxury	Compact	Luxury	Compact	Luxury
3	Smith	$54,000	$88,000	$228,000	$368,000	$11,400	$18,400
4	Jones	$126,000	$0	$304,000	$322,000	$15,200	$16,100
5		Combined Sales		Sales Increase			
6	Smith	$282,000	$456,000	$174,000	$280,000		
7	Jones	$430,000	$322,000	$178,000	$322,000		

FIGURE 2

Matrix Product

Now we are going to introduce a matrix multiplication that will seem rather strange at first. Despite this apparent strangeness, this operation is well-founded in the general theory of matrices and, as we will see, is extremely useful in many practical problems.

Historically, matrix multiplication was introduced by the English mathematician Arthur Cayley (1821–1895) in studies of systems of linear equations and linear transformations. In Section 4-6, you will see that matrix multiplication is central to the process of expressing systems of linear equations as matrix equations and to the process of solving matrix equations. Matrix equations and their solutions provide us with an alternate method of solving linear systems with the same number of variables as equations.

We start by defining the product of two special matrices, a row matrix and a column matrix.

Product of a Row Matrix and a Column Matrix

The **product** of a $1 \times n$ row matrix and an $n \times 1$ column matrix is a 1×1 matrix given by

$$\underset{1 \times n}{[a_1\ a_2\ \dots\ a_n]} \underset{n \times 1}{\begin{bmatrix} b_1 \\ b_2 \\ \vdots \\ b_n \end{bmatrix}} = [a_1b_1 + a_2b_2 + \cdots + a_nb_n]$$

Note that the number of elements in the row matrix and in the column matrix must be the same for the product to be defined.

Example 5 **Product of a Row Matrix and a Column Matrix**

$$[2\ \ -3\ \ 0]\begin{bmatrix} -5 \\ 2 \\ -2 \end{bmatrix} = [(2)(-5) + (-3)(2) + (0)(-2)] = [-10 - 6 + 0] = [-16]$$

Matched Problem 5 ➡ $\begin{bmatrix} -1 & 0 & 3 & 2 \end{bmatrix} \begin{bmatrix} 2 \\ 3 \\ 4 \\ -1 \end{bmatrix} = ?$

Refer to Example 5. The distinction between the real number −16 and the 1×1 matrix $[-16]$ is a technical one, and it is common to see 1×1 matrices written as real numbers without brackets. In the work that follows, we will frequently refer to 1×1 matrices as real numbers and omit the brackets whenever it is convenient to do so.

Example 6 ➡ **Labor Costs** A factory produces a slalom water ski that requires 3 labor-hours in the assembly department and 1 labor-hour in the finishing department. Assembly personnel receive \$9 per hour and finishing personnel receive \$6 per hour. Total labor cost per ski is given by the product:

$$\begin{bmatrix} 3 & 1 \end{bmatrix} \begin{bmatrix} 9 \\ 6 \end{bmatrix} = [(3)(9) + (1)(6)] = [27 + 6] = [33] \quad \text{or} \quad \$33 \text{ per ski}$$

Matched Problem 6 ➡ If the factory in Example 6 also produces a trick water ski that requires 5 labor-hours in the assembly department and 1.5 labor-hours in the finishing department, write a product between appropriate row and column matrices that will give the total labor cost for this ski. Compute the cost.

We now use the product of a $1 \times n$ row matrix and an $n \times 1$ column matrix to extend the definition of matrix product to more general matrices.

Matrix Product

If A is an $m \times p$ matrix and B is a $p \times n$ matrix, then the **matrix product** of A and B, denoted AB, is an $m \times n$ matrix whose element in the ith row and jth column is the real number obtained from the product of the ith row of A and the jth column of B. If the number of columns in A does not equal the number of rows in B, then the matrix product AB is **not defined.**

It is important to check sizes before starting the multiplication process. If A is an $a \times b$ matrix and B is a $c \times d$ matrix, then if $b = c$, the product AB will exist and will be an $a \times d$ matrix (see Fig. 3.). If $b \neq c$, then the product AB does not exist. The definition is not as complicated as it might first seem. An example should help clarify the process.

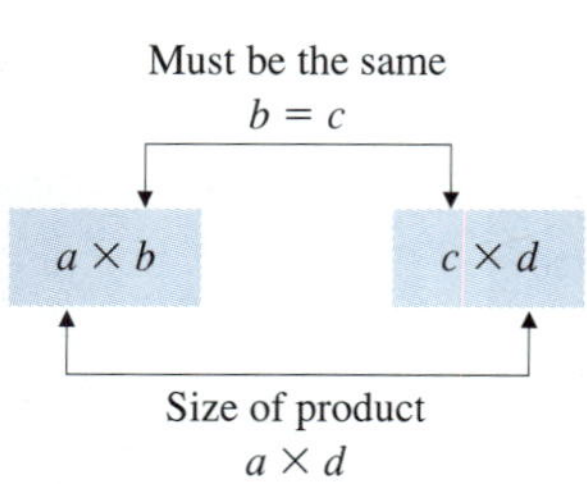

FIGURE 3

For

$$A = \begin{bmatrix} 2 & 3 & -1 \\ -2 & 1 & 2 \end{bmatrix} \quad \text{and} \quad B = \begin{bmatrix} 1 & 3 \\ 2 & 0 \\ -1 & 2 \end{bmatrix}$$

A is 2×3 and B is 3×2, so AB is 2×2. To find the first row of AB, we take the product of the first row of A with every column of B and write each result as a real number, not as a 1×1 matrix. The second row of AB is computed in the same manner. The four products of row and column matrices used to produce the four elements in AB are shown in the following dashed box. These products are usually calculated mentally or with the aid of a calculator, and need not be written out. The shaded portions highlight the steps involved in computing the element in the first row and second column of AB.

$$\underset{2\times 3}{\begin{bmatrix} 2 & 3 & -1 \\ -2 & 1 & 2 \end{bmatrix}} \underset{3\times 2}{\begin{bmatrix} 1 & 3 \\ 2 & 0 \\ -1 & 2 \end{bmatrix}} = \begin{bmatrix} [2 \ \ 3 \ \ -1]\begin{bmatrix} 1 \\ 2 \\ -1 \end{bmatrix} & [2 \ \ 3 \ \ -1]\begin{bmatrix} 3 \\ 0 \\ 2 \end{bmatrix} \\ [-2 \ \ 1 \ \ 2]\begin{bmatrix} 1 \\ 2 \\ -1 \end{bmatrix} & [-2 \ \ 1 \ \ 2]\begin{bmatrix} 3 \\ 0 \\ 2 \end{bmatrix} \end{bmatrix} = \underset{2\times 2}{\begin{bmatrix} 9 & 4 \\ -2 & -2 \end{bmatrix}}$$

Example 7 **Matrix Multiplication**

(A) $$\underset{3\times 2}{\begin{bmatrix} 2 & 1 \\ 1 & 0 \\ -1 & 2 \end{bmatrix}} \underset{2\times 4}{\begin{bmatrix} 1 & -1 & 0 & 1 \\ 2 & 1 & 2 & 0 \end{bmatrix}} = \underset{3\times 4}{\begin{bmatrix} 4 & -1 & 2 & 2 \\ 1 & -1 & 0 & 1 \\ 3 & 3 & 4 & -1 \end{bmatrix}}$$

(B) $$\underset{2\times 4}{\begin{bmatrix} 1 & -1 & 0 & 1 \\ 2 & 1 & 2 & 0 \end{bmatrix}} \underset{3\times 2}{\begin{bmatrix} 2 & 1 \\ 1 & 0 \\ -1 & 2 \end{bmatrix}}$$

Not defined.

(C) $$\begin{bmatrix} 2 & 6 \\ -1 & -3 \end{bmatrix}\begin{bmatrix} 1 & 2 \\ 3 & 6 \end{bmatrix} = \begin{bmatrix} 20 & 40 \\ -10 & -20 \end{bmatrix}$$

(D) $$\begin{bmatrix} 1 & 2 \\ 3 & 6 \end{bmatrix}\begin{bmatrix} 2 & 6 \\ -1 & -3 \end{bmatrix} = \begin{bmatrix} 0 & 0 \\ 0 & 0 \end{bmatrix}$$

(E) $$[2 \ \ -3 \ \ 0]\begin{bmatrix} -5 \\ 2 \\ -2 \end{bmatrix} = [-16]$$

(F) $$\begin{bmatrix} -5 \\ 2 \\ -2 \end{bmatrix}[2 \ \ -3 \ \ 0] = \begin{bmatrix} -10 & 15 & 0 \\ 4 & -6 & 0 \\ -4 & 6 & 0 \end{bmatrix}$$

Matched Problem 7 Find each product, if it is defined:

(A) $$\begin{bmatrix} -1 & 0 & 3 & -2 \\ 1 & 2 & 2 & 0 \end{bmatrix}\begin{bmatrix} -1 & 1 \\ 2 & 3 \\ 1 & 0 \end{bmatrix}$$

(B) $$\begin{bmatrix} -1 & 1 \\ 2 & 3 \\ 1 & 0 \end{bmatrix}\begin{bmatrix} -1 & 0 & 3 & -2 \\ 1 & 2 & 2 & 0 \end{bmatrix}$$

(C) $\begin{bmatrix} 1 & 2 \\ -1 & -2 \end{bmatrix}\begin{bmatrix} -2 & 4 \\ 1 & -2 \end{bmatrix}$ (D) $\begin{bmatrix} -2 & 4 \\ 1 & -2 \end{bmatrix}\begin{bmatrix} 1 & 2 \\ -1 & -2 \end{bmatrix}$

(E) $[3 \quad -2 \quad 1]\begin{bmatrix} 4 \\ 2 \\ 3 \end{bmatrix}$ (F) $\begin{bmatrix} 4 \\ 2 \\ 3 \end{bmatrix}[3 \quad -2 \quad 1]$

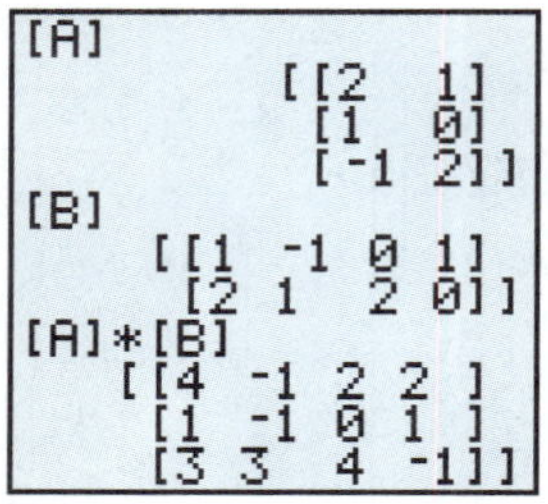

FIGURE 4
Multiplication on a graphing calculator

Figure 4 illustrates a graphing utility solution to Example 7A. What would you expect to happen if you tried to solve Example 7B on a graphing utility?

In the arithmetic of real numbers it does not matter in which order we multiply; for example, $5 \times 7 = 7 \times 5$. In matrix multiplication, however, it does make a difference. That is, AB does not always equal BA, even if both multiplications are defined and both products are the same size (see Examples 7C and 7D). Thus:

Matrix multiplication is not commutative.

Also, AB may be zero with neither A nor B equal to zero (see Example 7D). Thus:

The zero property does not hold for matrix multiplication.

(See Appendix A-2 for a discussion of the zero property for real numbers.)

Explore–Discuss 2

In addition to the commutative and zero properties, there are other significant differences between real number multiplication and matrix multiplication.

(A) In real number multiplication, the only real number whose square is 0 is the real number 0 ($0^2 = 0$). Find at least one 2×2 matrix A with all elements nonzero such that $A^2 = 0$,* where 0 is the 2×2 zero matrix.

(B) In real number multiplication the only nonzero real number that is equal to its square is the real number 1 ($1^2 = 1$). Find at least one 2×2 matrix A with all elements nonzero such that $A^2 = A$.

We will continue our discussion of properties of matrix multiplication later in this chapter. Now we consider an application of matrix multiplication.

Example 8 **Labor Costs** Let us combine the time requirements for slalom and trick water skis discussed in Example 6 and Matched Problem 6 into one matrix:

Labor-hours per ski

	Assembly department	Finishing department
Trick ski	5 hr	1.5 hr
Slalom ski	3 hr	1 hr

$$\begin{bmatrix} 5\text{ hr} & 1.5\text{ hr} \\ 3\text{ hr} & 1\text{ hr} \end{bmatrix} = L$$

*Following standard algebraic notation, we write $A^2 = AA$, $A^3 = AAA$, and so on.

Now suppose the company has two manufacturing plants, one in California and the other in Wisconsin, and that their hourly rates for each department are given in the following matrix:

$$\begin{array}{r} \\ \text{Assembly department} \\ \text{Finishing department} \end{array} \begin{array}{c} \text{Hourly wages} \\ \begin{array}{cc} \text{California} & \text{Wisconsin} \end{array} \\ \begin{bmatrix} \$12 & \$9 \\ \$8 & \$6 \end{bmatrix} \end{array} = H$$

Since H and L are both 2×2 matrices, we can take the product of H and L in either order and the result will be a 2×2 matrix:

$$HL = \begin{bmatrix} 12 & 9 \\ 8 & 6 \end{bmatrix} \begin{bmatrix} 5 & 1.5 \\ 3 & 1 \end{bmatrix} = \begin{bmatrix} 87 & 27 \\ 58 & 18 \end{bmatrix}$$

$$LH = \begin{bmatrix} 5 & 1.5 \\ 3 & 1 \end{bmatrix} \begin{bmatrix} 12 & 9 \\ 8 & 6 \end{bmatrix} = \begin{bmatrix} 72 & 54 \\ 44 & 33 \end{bmatrix}$$

How can we interpret the elements in these products? Let's begin with the product HL. The element 87 in the first row and first column of HL is the product of the first row matrix of H and the first column matrix of L:

$$\begin{array}{c} \begin{array}{cc} \text{CA} & \text{WI} \end{array} \\ [12 \quad 9] \end{array} \begin{bmatrix} 5 \\ 3 \end{bmatrix} \begin{array}{l} \text{Trick} \\ \text{Slalom} \end{array} = 12(5) + 9(3) = 60 - 27 = 87$$

Notice that \$60 is the labor cost for assembling a trick ski at the California plant and \$27 is the labor cost for assembling a slalom ski at the Wisconsin plant. Although both numbers represent labor costs, it makes no sense to add them together. They do not pertain to the same type of ski or to the same plant. Thus, even though the product HL happens to be defined mathematically, it has no useful interpretation in this problem.

Now let's consider the product LH. The element 72 in the first row and first column of LH is given by the following product:

$$\begin{array}{c} \begin{array}{cc} \text{Assembly} & \text{Finishing} \end{array} \\ [5 \qquad 1.5] \end{array} \begin{bmatrix} 12 \\ 8 \end{bmatrix} \begin{array}{l} \text{Assembly} \\ \text{Finishing} \end{array} = 5(12) + 1.5(8) = 60 + 12 = 72$$

This time, \$60 is the labor cost for assembling a trick ski at the California plant and \$12 is the labor cost for finishing a trick ski at the California plant. Thus, the sum is the total labor cost for producing a trick ski at the California plant. The other elements in LH also represent total labor costs, as indicated by the row and column labels shown below:

$$LH = \begin{array}{c} \text{Labor costs per ski} \\ \begin{array}{cc} \text{CA} & \text{WI} \end{array} \\ \begin{bmatrix} \$72 & \$54 \\ \$44 & \$33 \end{bmatrix} \end{array} \begin{array}{l} \\ \\ \text{Trick} \\ \text{Slalom} \end{array}$$

Figure 5 shows a solution to Example 8 on a spreadsheet.

	A	B	C	D	E	F
1		Labor-hours per ski			Hourly wages	
2		Assembly	Finishing		California	Wisconsin
3	Trick ski	5	1.5	Assembly	$12	$9
4	Slalom ski	3	1	Finishing	$8	$6
5		Labor costs per ski				
6		California	Wisconsin			
7	Trick ski	$72	$54			
8	Slalom ski	$44	$33			

FIGURE 5
Matrix multiplication in a spreadsheet:
The command MMULT(B3: C4, E3: F4) produces the matrix in B7:C8

Matched Problem 8 Refer to Example 8. The company wants to know how many hours to schedule in each department in order to produce 2,000 trick skis and 1,000 slalom skis. These production requirements can be represented by either of the following matrices:

$$P = \overset{\substack{\text{Trick} \\ \text{skis}}}{[2{,}000} \quad \overset{\substack{\text{Slalom} \\ \text{skis}}}{1{,}000]} \qquad Q = \begin{bmatrix} 2{,}000 \\ 1{,}000 \end{bmatrix} \begin{matrix} \text{Trick skis} \\ \text{Slalom skis} \end{matrix}$$

Using the labor-hour matrix L from Example 8, find PL or LQ, whichever has a meaningful interpretation for this problem, and label the rows and columns accordingly.

CAUTION

Example 8 and Matched Problem 8 illustrate an important point about matrix multiplication. Even if you are using a graphing utility to perform the calculations in a matrix product, it is still necessary for you to know the definition of matrix multiplication so that you can interpret the results correctly.

Answers to Matched Problems

1. $\begin{bmatrix} 1 & 5 \\ 0 & -2 \\ 2 & 1 \end{bmatrix}$ **2.** $[-1 \quad -1 \quad 4]$ **3.** $\begin{bmatrix} 13 \\ 2 \\ 35 \end{bmatrix}$

4. (A) $\begin{bmatrix} \$235{,}000 & \$422{,}000 \\ \$372{,}000 & \$298{,}000 \end{bmatrix}$ (B) $\begin{bmatrix} \$145{,}000 & \$268{,}000 \\ \$160{,}000 & \$254{,}000 \end{bmatrix}$

(C) $\begin{bmatrix} \$9{,}500 & \$17{,}250 \\ \$13{,}300 & \$13{,}800 \end{bmatrix}$

5. $[8]$ **6.** $[5 \quad 1.5]\begin{bmatrix} 9 \\ 6 \end{bmatrix} = [54]$, or \$54

7. (A) Not defined (B) $\begin{bmatrix} 2 & 2 & -1 & 2 \\ 1 & 6 & 12 & -4 \\ -1 & 0 & 3 & -2 \end{bmatrix}$ (C) $\begin{bmatrix} 0 & 0 \\ 0 & 0 \end{bmatrix}$

(D) $\begin{bmatrix} -6 & -12 \\ 3 & 6 \end{bmatrix}$ (E) $[11]$ (F) $\begin{bmatrix} 12 & -8 & 4 \\ 6 & -4 & 2 \\ 9 & -6 & 3 \end{bmatrix}$

8. $PL = \overset{\text{Assembly}}{[13{,}000} \quad \overset{\text{Finishing}}{4{,}000]}$ Labor-hours

EXERCISE 4-4

A *Perform the indicated operations in Problems 1–14, if possible.*

1. $\begin{bmatrix} 2 & -1 \\ 3 & 0 \end{bmatrix} + \begin{bmatrix} -3 & 1 \\ 2 & -3 \end{bmatrix}$

2. $\begin{bmatrix} -3 & 5 \\ 2 & 0 \\ 1 & 4 \end{bmatrix} + \begin{bmatrix} 2 & 1 \\ -6 & 3 \\ 0 & -5 \end{bmatrix}$

3. $\begin{bmatrix} 4 & -1 & 0 \\ 2 & 1 & 3 \end{bmatrix} + \begin{bmatrix} 2 & 1 \\ -6 & 3 \\ 0 & -5 \end{bmatrix}$

4. $\begin{bmatrix} -3 & 5 \\ 2 & 0 \\ 1 & 4 \end{bmatrix} + \begin{bmatrix} -2 & 1 & 3 \\ 5 & 6 & -8 \end{bmatrix}$

5. $\begin{bmatrix} 4 & -5 \\ 1 & 0 \\ 1 & -3 \end{bmatrix} - \begin{bmatrix} -1 & 2 \\ 6 & -2 \\ 1 & -7 \end{bmatrix}$

6. $\begin{bmatrix} 6 & 2 & -3 \\ 0 & -4 & 5 \end{bmatrix} - \begin{bmatrix} 4 & -1 & 2 \\ -5 & 1 & -2 \end{bmatrix}$

7. $5\begin{bmatrix} 1 & -2 & 0 & 4 \\ -3 & 2 & -1 & 6 \end{bmatrix}$

8. $10\begin{bmatrix} 2 & -1 & 3 \\ 0 & -4 & 5 \end{bmatrix}$

9. $\begin{bmatrix} 3 & 4 \\ -1 & -2 \end{bmatrix}\begin{bmatrix} -1 \\ 2 \end{bmatrix}$

10. $\begin{bmatrix} -1 & 1 \\ 2 & -3 \end{bmatrix}\begin{bmatrix} 4 \\ -2 \end{bmatrix}$

11. $\begin{bmatrix} 2 & -3 \\ 1 & 2 \end{bmatrix}\begin{bmatrix} 1 & -1 \\ 0 & -2 \end{bmatrix}$

12. $\begin{bmatrix} -3 & 2 \\ 4 & -1 \end{bmatrix}\begin{bmatrix} -2 & 5 \\ -1 & 3 \end{bmatrix}$

13. $\begin{bmatrix} 1 & -1 \\ 0 & -2 \end{bmatrix}\begin{bmatrix} 2 & -3 \\ 1 & 2 \end{bmatrix}$

14. $\begin{bmatrix} -2 & 5 \\ -1 & 3 \end{bmatrix}\begin{bmatrix} -3 & 2 \\ 4 & -1 \end{bmatrix}$

B *Find the products in Problems 15–22.*

15. $\begin{bmatrix} 5 & -2 \end{bmatrix}\begin{bmatrix} -3 \\ -4 \end{bmatrix}$

16. $\begin{bmatrix} -4 & 3 \end{bmatrix}\begin{bmatrix} -2 \\ 1 \end{bmatrix}$

17. $\begin{bmatrix} -3 \\ -4 \end{bmatrix}\begin{bmatrix} 5 & -2 \end{bmatrix}$

18. $\begin{bmatrix} -2 \\ 1 \end{bmatrix}\begin{bmatrix} -4 & 3 \end{bmatrix}$

19. $\begin{bmatrix} 3 & -2 & -4 \end{bmatrix}\begin{bmatrix} 1 \\ 2 \\ -3 \end{bmatrix}$

20. $\begin{bmatrix} 1 & -2 & 2 \end{bmatrix}\begin{bmatrix} 2 \\ -1 \\ 1 \end{bmatrix}$

21. $\begin{bmatrix} 1 \\ 2 \\ -3 \end{bmatrix}\begin{bmatrix} 3 & -2 & -4 \end{bmatrix}$

22. $\begin{bmatrix} 2 \\ -1 \\ 1 \end{bmatrix}\begin{bmatrix} 1 & -2 & 2 \end{bmatrix}$

Problems 23–40 refer to the following matrices:

$$A = \begin{bmatrix} 2 & -1 & 3 \\ 0 & 4 & -2 \end{bmatrix} \qquad B = \begin{bmatrix} -3 & 1 \\ 2 & 5 \end{bmatrix}$$

$$C = \begin{bmatrix} -1 & 0 & 2 \\ 4 & -3 & 1 \\ -2 & 3 & 5 \end{bmatrix} \qquad D = \begin{bmatrix} 3 & -2 \\ 0 & -1 \\ 1 & 2 \end{bmatrix}$$

Perform the indicated operations, if possible.

23. AC **24.** CA **25.** AB

26. BA **27.** B^2 **28.** C^2

29. $B + AD$ **30.** $C + DA$ **31.** $(0.1)DB$

32. $(0.2)CD$ **33.** $(3)BA + (4)AC$

34. $(2)DB + (5)CD$ **35.** $(-2)BA + (6)CD$

36. $(-1)AC + (3)DB$ **37.** ACD **38.** CDA

39. DBA **40.** BAD

In Problems 41 and 42, use a graphing utility to calculate $B, B^2, B^3, \ldots$ and $AB, AB^2, AB^3, \ldots$. Describe any patterns you observe in each sequence of matrices.

41. $A = \begin{bmatrix} 0.3 & 0.7 \end{bmatrix}$ and $B = \begin{bmatrix} 0.4 & 0.6 \\ 0.2 & 0.8 \end{bmatrix}$

42. $A = \begin{bmatrix} 0.4 & 0.6 \end{bmatrix}$ and $B = \begin{bmatrix} 0.9 & 0.1 \\ 0.3 & 0.7 \end{bmatrix}$

43. Find a, b, c, and d so that

$$\begin{bmatrix} a & b \\ c & d \end{bmatrix} + \begin{bmatrix} 2 & -3 \\ 0 & 1 \end{bmatrix} = \begin{bmatrix} 1 & -2 \\ 3 & -4 \end{bmatrix}$$

44. Find w, x, y, and z so that

$$\begin{bmatrix} 4 & -2 \\ -3 & 0 \end{bmatrix} + \begin{bmatrix} w & x \\ y & z \end{bmatrix} = \begin{bmatrix} 2 & -3 \\ 0 & 5 \end{bmatrix}$$

45. Find x and y so that

$$\begin{bmatrix} 2x & 4 \\ -3 & 5x \end{bmatrix} + \begin{bmatrix} 3y & -2 \\ -2 & -y \end{bmatrix} = \begin{bmatrix} -5 & 2 \\ -5 & 13 \end{bmatrix}$$

46. Find x and y so that

$$\begin{bmatrix} 5 & 3x \\ 2x & -4 \end{bmatrix} + \begin{bmatrix} 1 & -4y \\ 7y & 4 \end{bmatrix} = \begin{bmatrix} 6 & -7 \\ 5 & 0 \end{bmatrix}$$

C

47. Find x and y so that

$$\begin{bmatrix} x & -1 \\ 1 & 0 \end{bmatrix}\begin{bmatrix} 2 & 1 \\ 4 & 1 \end{bmatrix} = \begin{bmatrix} y & y \\ 2 & 1 \end{bmatrix}$$

48. Find x and y so that

$$\begin{bmatrix} 1 & 3 \\ -2 & -2 \end{bmatrix}\begin{bmatrix} x & 1 \\ 3 & 2 \end{bmatrix} = \begin{bmatrix} y & 7 \\ y & -6 \end{bmatrix}$$

49. Find a, b, c, and d so that

$$\begin{bmatrix} 1 & -2 \\ 2 & -3 \end{bmatrix}\begin{bmatrix} a & b \\ c & d \end{bmatrix} = \begin{bmatrix} 1 & 0 \\ 3 & 2 \end{bmatrix}$$

50. Find a, b, c, and d so that

$$\begin{bmatrix} 1 & 3 \\ 1 & 4 \end{bmatrix}\begin{bmatrix} a & b \\ c & d \end{bmatrix} = \begin{bmatrix} 6 & -5 \\ 7 & -7 \end{bmatrix}$$

51. A square matrix is a **diagonal matrix** if all elements not on the principal diagonal are zero. Thus, a 2×2 diagonal matrix has the form

$$A = \begin{bmatrix} a & 0 \\ 0 & d \end{bmatrix}$$

where a and d are any real numbers. Discuss the validity of each of the following statements. If the statement is always true, explain why. If not, give examples.

(A) If A and B are 2×2 diagonal matrices, then $A + B$ is a 2×2 diagonal matrix.

(B) If A and B are 2×2 diagonal matrices, then $A + B = B + A$.

(C) If A and B are 2×2 diagonal matrices, then AB is a 2×2 diagonal matrix.

(D) If A and B are 2×2 diagonal matrices, then $AB = BA$.

52. A square matrix is an **upper triangular matrix** if all elements below the principal diagonal are zero. Thus, a 2×2 upper triangular matrix has the form

$$A = \begin{bmatrix} a & b \\ 0 & d \end{bmatrix}$$

where a, b, and d are any real numbers. Discuss the validity of each of the following statements. If the statement is always true, explain why. If not, give examples.

(A) If A and B are 2×2 upper triangular matrices, then $A + B$ is a 2×2 upper triangular matrix.

(B) If A and B are 2×2 upper triangular matrices, then $A + B = B + A$.

(C) If A and B are 2×2 upper triangular matrices, then AB is a 2×2 upper triangular matrix.

(D) If A and B are 2×2 upper triangular matrices, then $AB = BA$.

Applications

Business & Economics

53. *Cost analysis.* A company with two different plants manufactures guitars and banjos. Its production costs for each instrument are given in the following matrices:

$$\begin{array}{r} \\ \\ \text{Materials} \\ \text{Labor} \end{array}\begin{array}{c} \text{Plant } X \\ \text{Guitar Banjo} \\ \begin{bmatrix} \$30 & \$25 \\ \$60 & \$80 \end{bmatrix} = A \end{array} \qquad \begin{array}{c} \text{Plant } Y \\ \text{Guitar Banjo} \\ \begin{bmatrix} \$36 & \$27 \\ \$54 & \$74 \end{bmatrix} = B \end{array}$$

Find $\frac{1}{2}(A + B)$, the average cost of production for the two plants.

54. *Cost analysis.* If both labor and materials at plant X in Problem 53 are increased by 20%, find $\frac{1}{2}(1.2A + B)$, the new average cost of production for the two plants.

55. *Markup.* An import car dealer sells three models of a car. The retail prices and the current dealer invoice prices (costs) for the basic models and indicated options are given in the following two matrices (where "Air" means air-conditioning):

Retail price

$$\begin{array}{r} \\ \text{Model } A \\ \text{Model } B \\ \text{Model } C \end{array}\begin{array}{c} \begin{array}{cccc} \text{Basic car} & \text{Air} & \text{AM/FM radio} & \text{Cruise control} \end{array} \\ \begin{bmatrix} \$10{,}900 & \$683 & \$253 & \$195 \\ \$13{,}000 & \$738 & \$382 & \$206 \\ \$16{,}300 & \$867 & \$537 & \$225 \end{bmatrix} = M \end{array}$$

Dealer invoice price

$$\begin{array}{r} \\ \text{Model } A \\ \text{Model } B \\ \text{Model } C \end{array}\begin{array}{c} \begin{array}{cccc} \text{Basic car} & \text{Air} & \text{AM/FM radio} & \text{Cruise control} \end{array} \\ \begin{bmatrix} \$9{,}400 & \$582 & \$195 & \$160 \\ \$11{,}500 & \$621 & \$295 & \$171 \\ \$14{,}100 & \$737 & \$420 & \$184 \end{bmatrix} = N \end{array}$$

We define the markup matrix to be $M - N$ (**markup** is the difference between the retail price and the dealer invoice price). Suppose the value of the dollar has had a sharp decline and the dealer invoice price is to have an across-the-board 15% increase next year. In order to stay competitive with domestic cars, the dealer increases the retail prices only 10%. Calculate a markup matrix for next year's models and the indicated options. (Compute results to the nearest dollar.)

56. *Markup.* Referring to Problem 55, what is the markup matrix resulting from a 20% increase in dealer invoice prices and an increase in retail prices of 15%? (Compute results to the nearest dollar.)

57. *Labor costs.* A company with manufacturing plants located in different parts of the country has labor-hour and wage requirements for the manufacture of three types of inflatable boats as given in the following two matrices:

Labor-hours per boat

	Cutting department	Assembly department	Packaging department	
$M =$	0.6 hr	0.6 hr	0.2 hr	One-person boat
	1.0 hr	0.9 hr	0.3 hr	Two-person boat
	1.5 hr	1.2 hr	0.4 hr	Four-person boat

Hourly wages

	Plant I	Plant II	
$N =$	\$8	\$9	Cutting department
	\$10	\$12	Assembly department
	\$5	\$6	Packaging department

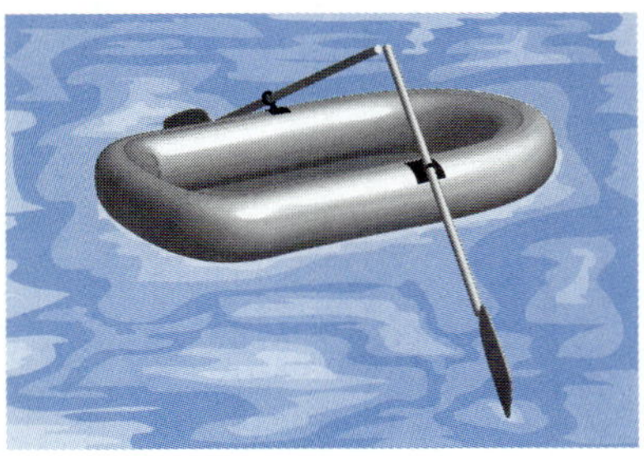

(A) Find the labor costs for a one-person boat manufactured at plant I.

(B) Find the labor costs for a four-person boat manufactured at plant II.

(C) Discuss possible interpretations of the elements in the matrix products MN and NM.

(D) If either of the products MN or NM has a meaningful interpretation, find the product and label its rows and columns.

58. *Inventory value.* A personal computer retail company sells five different computer models through three stores located in a large metropolitan area. The inventory of each model on hand in each store is summarized in matrix M. Wholesale (W) and retail (R) values of each model computer are summarized in matrix N.

Model

	A	B	C	D	E	
$M =$	4	2	3	7	1	Store 1
	2	3	5	0	6	Store 2
	10	4	3	4	3	Store 3

	W	R		
$N =$	\$700	\$840	A	Model
	\$1,400	\$1,800	B	
	\$1,800	\$2,400	C	
	\$2,700	\$3,300	D	
	\$3,500	\$4,900	E	

(A) What is the retail value of the inventory at store 2?

(B) What is the wholesale value of the inventory at store 3?

(C) Discuss possible interpretations of the elements in the matrix products MN and NM.

(D) If either of the products MN or NM has a meaningful interpretation, find the product and label its rows and columns.

(E) Discuss methods of matrix multiplication that can be used to find the total inventory of each model on hand at all three stores. State the matrices that can be used and perform the necessary operations.

(F) Discuss methods of matrix multiplication that can be used to find the total inventory of all five models at each store. State the matrices that can be used and perform the necessary operations.

59. *Air freight.* A nationwide air freight service has connecting flights between five cities as illustrated in the diagram. To represent this schedule in matrix form, we construct a 5×5 **incidence matrix** A, where the rows represent the origin of each flight and the columns represent the destinations. We place a 1 in the ith row and jth column of this matrix if there is a connecting flight from the ith city to the jth city; otherwise, insert a 0. We also place 0's on the principal diagonal, because a connecting flight with the same origin and destination makes no sense. With the schedule represented in the mathematical form of a matrix, we can perform operations on this matrix to obtain information about the schedule.

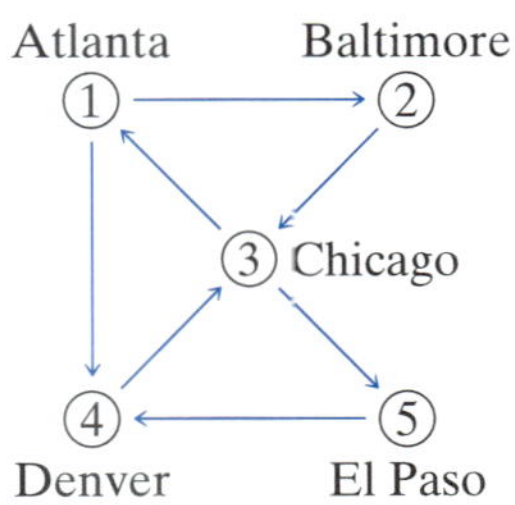

Destination

Origin	1	2	3	4	5
1	0	1	0	1	0
2	0	0	1	0	0
3	1	0	0	0	1
4	0	0	1	0	0
5	0	0	0	1	0

$= A$

(A) Find A^2. What does the 1 in Row 2 and Column 1 of A^2 indicate about the schedule? What does the 2 in Row 1 and Column 3 indicate about the schedule? In general, how would you interpret each element not on the principal diagonal of A^2? [*Hint*: Examine the diagram for possible connections between the ith city and the jth city.]

(B) Find A^3. What does the 1 in Row 4 and Column 2 of A^3 indicate about the schedule? What does the

2 in Row 1 and Column 5 indicate about the schedule? In general, how would you interpret each element not on the principal diagonal of A^3?

(C) Compute $A, A + A^2, A + A^2 + A^3, \ldots,$ until you obtain a matrix with no 0 elements (except possibly on the principal diagonal), and interpret.

60. *Air freight.* Find the incidence matrix A for the flight schedule illustrated in the diagram. Compute $A, A + A^2, A + A^2 + A^3, \ldots,$ until you obtain a matrix with no 0 elements (except possibly on the principal diagonal), and interpret.

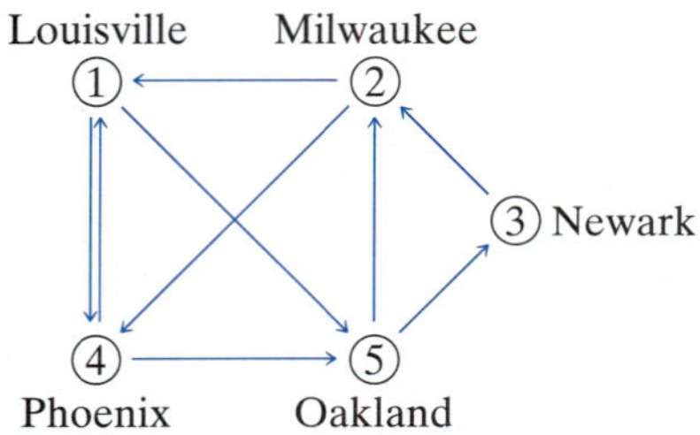

Life Sciences

61. *Nutrition.* A nutritionist for a cereal company blends two cereals in three different mixes. The amounts of protein, carbohydrate, and fat (in grams per ounce) in each cereal are given by matrix M. The amounts of each cereal used in the three mixes are given by matrix N.

$$M = \begin{matrix} & \text{Cereal A} & \text{Cereal B} & \\ & \left[\begin{matrix} 4\text{ g/oz} \\ 20\text{ g/oz} \\ 3\text{ g/oz} \end{matrix}\right. & \left.\begin{matrix} 2\text{ g/oz} \\ 16\text{ g/oz} \\ 1\text{ g/oz} \end{matrix}\right] & \begin{matrix} \text{Protein} \\ \text{Carbohydrate} \\ \text{Fat} \end{matrix} \end{matrix}$$

$$N = \begin{matrix} & \text{Mix X} & \text{Mix Y} & \text{Mix Z} & \\ & \left[\begin{matrix} 15\text{ oz} \\ 5\text{ oz} \end{matrix}\right. & \begin{matrix} 10\text{ oz} \\ 10\text{ oz} \end{matrix} & \left.\begin{matrix} 5\text{ oz} \\ 15\text{ oz} \end{matrix}\right] & \begin{matrix} \text{Cereal A} \\ \text{Cereal B} \end{matrix} \end{matrix}$$

(A) Find the amount of protein in mix X.
(B) Find the amount of fat in mix Z.
(C) Discuss possible interpretations of the elements in the matrix products MN and NM.
(D) If either of the products MN or NM has a meaningful interpretation, find the product and label its rows and columns.

62. *Heredity.* Gregor Mendel (1822–1884), an Austrian monk and botanist, made discoveries that revolutionized the science of genetics. In one experiment, he crossed dihybrid yellow round peas (yellow and round are dominant characteristics; the peas also contained genes for the recessive characteristics green and wrinkled) and obtained 560 peas of the types indicated in the matrix:

$$\begin{matrix} & \text{Round} & \text{Wrinkled} \\ \text{Yellow} & \left[\begin{matrix} 319 \\ 108 \end{matrix}\right. & \left.\begin{matrix} 101 \\ 32 \end{matrix}\right] \\ \text{Green} & & \end{matrix} = M$$

Suppose he carried out a second experiment of the same type and obtained 640 peas of the types indicated in this matrix:

$$\begin{matrix} & \text{Round} & \text{Wrinkled} \\ \text{Yellow} & \left[\begin{matrix} 370 \\ 110 \end{matrix}\right. & \left.\begin{matrix} 124 \\ 36 \end{matrix}\right] \\ \text{Green} & & \end{matrix} = N$$

If the results of the two experiments are combined, write the resulting matrix $M + N$. Compute the percentage of the total number of peas (1,200) in each category of the combined results.

Social Sciences

63. *Politics.* In a local California election, a group hired a public relations firm to promote its candidate in three ways: telephone calls, house calls, and letters. The cost per contact is given in matrix M, and the number of contacts of each type made in two adjacent cities is given in matrix N.

$$M = \begin{matrix} \text{Cost per contact} & \\ \begin{bmatrix} \$0.40 \\ \$1.00 \\ \$0.35 \end{bmatrix} & \begin{matrix} \text{Telephone call} \\ \text{House call} \\ \text{Letter} \end{matrix} \end{matrix}$$

$$N = \begin{matrix} & \text{Telephone call} & \text{House call} & \text{Letter} & \\ & \left[\begin{matrix} 1{,}000 \\ 2{,}000 \end{matrix}\right. & \begin{matrix} 500 \\ 800 \end{matrix} & \left.\begin{matrix} 5{,}000 \\ 8{,}000 \end{matrix}\right] & \begin{matrix} \text{Berkeley} \\ \text{Oakland} \end{matrix} \end{matrix}$$

(A) Find the total amount spent in Berkeley.
(B) Find the total amount spent in Oakland.
(C) Discuss possible interpretations of the elements in the matrix products MN and NM.
(D) If either of the products MN or NM has a meaningful interpretation, find the product and label its rows and columns.
(E) Discuss methods of matrix multiplication that can be used to find the total number of telephone calls, house calls, and letters. State the matrices that can be used and perform the necessary operations.
(F) Discuss methods of matrix multiplication that can be used to find the total number of contacts in Berkeley and in Oakland. State the matrices that can be used and perform the necessary operations.

64. *Averaging tests.* A teacher has given four tests to a class of five students and stored the results in the following matrix:

$$\begin{array}{c} \\ \\ \text{Ann} \\ \text{Bob} \\ \text{Carol} \\ \text{Dan} \\ \text{Eric} \end{array} \begin{array}{c} \text{Tests} \\ \begin{array}{cccc} 1 & 2 & 3 & 4 \end{array} \\ \begin{bmatrix} 78 & 84 & 81 & 86 \\ 91 & 65 & 84 & 92 \\ 95 & 90 & 92 & 91 \\ 75 & 82 & 87 & 91 \\ 83 & 88 & 81 & 76 \end{bmatrix} \end{array} = M$$

Discuss methods of matrix multiplication that the teacher can use to obtain the information indicated below. In each case, state the matrices to be used and then perform the necessary operations.

(A) The average on all four tests for each student, assuming that all four tests are given equal weight

(B) The average on all four tests for each student, assuming that the first three tests are given equal weight and the fourth is given twice this weight

(C) The class average on each of the four tests

65. *Dominance relation.* In order to rank players for an upcoming tennis tournament, a club decides to have each player play one set with every other player. The results are given in the table.

PLAYER	DEFEATED
1. Aaron	Charles, Dan, Elvis
2. Bart	Aaron, Dan, Elvis
3. Charles	Bart, Dan
4. Dan	Frank
5. Elvis	Charles, Dan, Frank
6. Frank	Aaron, Bart, Charles

(A) Express the outcomes as an incidence matrix A by placing a 1 in the ith row and jth column of A if player i defeated player j and a 0 otherwise (see Problem 59).

(B) Compute the matrix $B = A + A^2$.

(C) Discuss matrix multiplication methods that can be used to find the sum of the rows in B. State the matrices that can be used and perform the necessary operations.

(D) Rank the players from strongest to weakest. Explain the reasoning behind your ranking.

66. *Dominance relation.* Each member of a chess team plays one match with every other player. The results are given in the table.

PLAYER	DEFEATED
1. Anne	Diane
2. Bridget	Anne, Carol, Diane
3. Carol	Anne
4. Diane	Carol, Erlene
5. Erlene	Anne, Bridget, Carol

(A) Express the outcomes as an incidence matrix A by placing a 1 in the ith row and jth column of A if player i defeated player j and a 0 otherwise (see Problem 59).

(B) Compute the matrix $B = A + A^2$.

(C) Discuss matrix multiplication methods that can be used to find the sum of the rows in B. State the matrices that can be used and perform the necessary operations.

(D) Rank the players from strongest to weakest. Explain the reasoning behind your ranking.

SECTION 4-5

Inverse of a Square Matrix

- IDENTITY MATRIX FOR MULTIPLICATION
- INVERSE OF A SQUARE MATRIX
- APPLICATION: CRYPTOGRAPHY

IDENTITY MATRIX FOR MULTIPLICATION

We know that

$$1a = a1 = a \qquad \text{for all real numbers } a$$

The number 1 is called the **identity** for real number multiplication. Does the set of all matrices of a given size have an identity element for multiplication? That is, if M is an arbitrary $m \times n$ matrix, does M have an identity element

I such that $IM = MI = M$? The answer, in general, is no. However, the set of all **square matrices of order n** (matrices with n rows and n columns) does have an identity, and it is given as follows:

> The **identity element for multiplication** for the set of all square matrices of order n is the square matrix of order n, denoted by I, with 1's along the principal diagonal (from the upper left corner to the lower right) and 0's elsewhere.

For example,

$$\begin{bmatrix} 1 & 0 \\ 0 & 1 \end{bmatrix} \quad \text{and} \quad \begin{bmatrix} 1 & 0 & 0 \\ 0 & 1 & 0 \\ 0 & 0 & 1 \end{bmatrix}$$

are the identity matrices for all square matrices of order 2 and 3, respectively.

Most graphing utilities have a built-in command for generating the identity matrix of a given order (see Fig. 1)

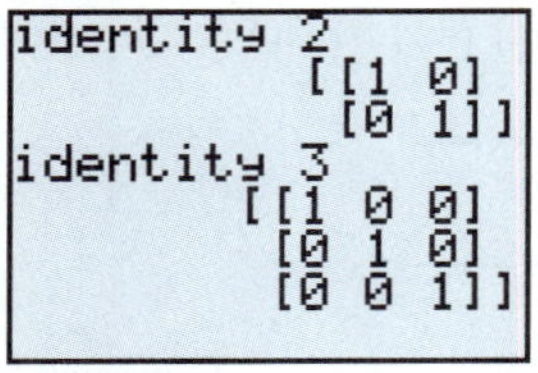

FIGURE 1
Identity matrices

Example 1 **Identity Matrix Multiplication**

(A) $\begin{bmatrix} 1 & 0 & 0 \\ 0 & 1 & 0 \\ 0 & 0 & 1 \end{bmatrix}\begin{bmatrix} a & b & c \\ d & e & f \\ g & h & i \end{bmatrix} = \begin{bmatrix} a & b & c \\ d & e & f \\ g & h & i \end{bmatrix}$

(B) $\begin{bmatrix} a & b & c \\ d & e & f \\ g & h & i \end{bmatrix}\begin{bmatrix} 1 & 0 & 0 \\ 0 & 1 & 0 \\ 0 & 0 & 1 \end{bmatrix} = \begin{bmatrix} a & b & c \\ d & e & f \\ g & h & i \end{bmatrix}$

(C) $\begin{bmatrix} 1 & 0 \\ 0 & 1 \end{bmatrix}\begin{bmatrix} a & b & c \\ d & e & f \end{bmatrix} = \begin{bmatrix} a & b & c \\ d & e & f \end{bmatrix}$

(D) $\begin{bmatrix} a & b & c \\ d & e & \text{f} \end{bmatrix}\begin{bmatrix} 1 & 0 & 0 \\ 0 & 1 & 0 \\ 0 & 0 & 1 \end{bmatrix} = \begin{bmatrix} a & b & c \\ d & e & f \end{bmatrix}$

Matched Problem 1 Multiply:

(A) $\begin{bmatrix} 1 & 0 \\ 0 & 1 \end{bmatrix}\begin{bmatrix} 2 & -3 \\ 5 & 7 \end{bmatrix}$ and $\begin{bmatrix} 2 & -3 \\ 5 & 7 \end{bmatrix}\begin{bmatrix} 1 & 0 \\ 0 & 1 \end{bmatrix}$

(B) $\begin{bmatrix} 1 & 0 & 0 \\ 0 & 1 & 0 \\ 0 & 0 & 1 \end{bmatrix}\begin{bmatrix} 4 & 2 \\ 3 & -5 \\ 6 & 8 \end{bmatrix}$ and $\begin{bmatrix} 4 & 2 \\ 3 & -5 \\ 6 & 8 \end{bmatrix}\begin{bmatrix} 1 & 0 \\ 0 & 1 \end{bmatrix}$

In general, we can show that if M is a square matrix of order n and I is the identity matrix of order n, then

$$IM = MI = M$$

If M is an $m \times n$ matrix that is not square ($m \neq n$), then it is still possible to multiply M on the left and on the right by an identity matrix, but not with the same size identity matrix (see Examples 1C and 1D). In order to avoid the complications involved with associating two different identity matrices with each nonsquare matrix, we restrict our attention in this section to square matrices.

Explore–Discuss 1

The only real number solutions to the equation $x^2 = 1$ are $x = 1$ and $x = -1$.

(A) Show that $A = \begin{bmatrix} 0 & 1 \\ 1 & 0 \end{bmatrix}$ satisfies $A^2 = I$, where I is the 2×2 identity.

(B) Show that $B = \begin{bmatrix} 0 & -1 \\ -1 & 0 \end{bmatrix}$ satisfies $B^2 = I$.

(C) Find a 2×2 matrix with all elements nonzero whose square is the 2×2 identity matrix.

Inverse of a Square Matrix

In the set of real numbers, we know that for each real number a (except 0) there exists a real number a^{-1} such that

$$a^{-1}a = 1$$

The number a^{-1} is called the **inverse** of the number a relative to multiplication, or the **multiplicative inverse** of a. For example, 2^{-1} is the multiplicative inverse of 2, since $2^{-1} \cdot 2 = 1$. We use this idea to define the inverse of a square matrix.

Inverse of a Square Matrix

Let M be a square matrix of order n and I be the identity matrix of order n. If there exists a matrix M^{-1} (read "M inverse") such that

$$M^{-1}M = MM^{-1} = I$$

then M^{-1} is called the **multiplicative inverse of M** or, more simply, the **inverse of M.**

The multiplicative inverse of a nonzero real number a also can be written as $1/a$. This notation is not used for matrix inverses.

Let us use the above definition to find M^{-1} for

$$M = \begin{bmatrix} 2 & 3 \\ 1 & 2 \end{bmatrix}$$

We are looking for

$$M^{-1} = \begin{bmatrix} a & c \\ b & d \end{bmatrix}$$

such that

$$MM^{-1} = M^{-1}M = I$$

Thus, we write

$$\overset{M}{\begin{bmatrix} 2 & 3 \\ 1 & 2 \end{bmatrix}} \overset{M^{-1}}{\begin{bmatrix} a & c \\ b & d \end{bmatrix}} = \overset{I}{\begin{bmatrix} 1 & 0 \\ 0 & 1 \end{bmatrix}}$$

and try to find a, b, c, and d so that the product of M and M^{-1} is the identity matrix I. Multiplying M and M^{-1} on the left side, we obtain

$$\begin{bmatrix} (2a + 3b) & (2c + 3d) \\ (a + 2b) & (c + 2d) \end{bmatrix} = \begin{bmatrix} 1 & 0 \\ 0 & 1 \end{bmatrix}$$

which is true only if

$$\begin{aligned} 2a + 3b &= 1 \\ a + 2b &= 0 \end{aligned} \qquad \begin{aligned} 2c + 3d &= 0 \\ c + 2d &= 1 \end{aligned}$$

Solving these two systems, we find that $a = 2$, $b = -1$, $c = -3$, and $d = 2$. Thus,

$$M^{-1} = \begin{bmatrix} 2 & -3 \\ -1 & 2 \end{bmatrix}$$

as is easily checked:

$$\overset{M}{\begin{bmatrix} 2 & 3 \\ 1 & 2 \end{bmatrix}} \overset{M^{-1}}{\begin{bmatrix} 2 & -3 \\ -1 & 2 \end{bmatrix}} = \overset{I}{\begin{bmatrix} 1 & 0 \\ 0 & 1 \end{bmatrix}} = \overset{M^{-1}}{\begin{bmatrix} 2 & -3 \\ -1 & 2 \end{bmatrix}} \overset{M}{\begin{bmatrix} 2 & 3 \\ 1 & 2 \end{bmatrix}}$$

Unlike nonzero real numbers, inverses do not always exist for square matrices. For example, if

$$N = \begin{bmatrix} 2 & 1 \\ 4 & 2 \end{bmatrix}$$

then, proceeding as above, we are led to the systems

$$\begin{aligned} 2a + b &= 1 \\ 4a + 2b &= 0 \end{aligned} \qquad \begin{aligned} 2c + d &= 0 \\ 4c + 2d &= 1 \end{aligned}$$

These are both inconsistent and have no solution. Hence, N^{-1} does not exist.

Being able to find inverses, when they exist, leads to direct and simple solutions to many practical problems. In the next section, we shall show how inverses can be used to solve systems of linear equations.

The method outlined above for finding M^{-1}, if it exists, gets very involved for matrices of order larger than 2. Now that we know what we are looking for, we can introduce the idea of the augmented matrix (considered in Sections 4-2 and 4-3) to make the process more efficient.

Example 2 **Finding the Inverse of a Matrix** Find the inverse, if it exists, of the matrix

$$M = \begin{bmatrix} 1 & -1 & 1 \\ 0 & 2 & -1 \\ 2 & 3 & 0 \end{bmatrix}$$

SOLUTION We start as before and write

$$\overset{M}{\begin{bmatrix} 1 & -1 & 1 \\ 0 & 2 & -1 \\ 2 & 3 & 0 \end{bmatrix}} \overset{M^{-1}}{\begin{bmatrix} a & d & g \\ b & e & h \\ c & f & i \end{bmatrix}} = \overset{I}{\begin{bmatrix} 1 & 0 & 0 \\ 0 & 1 & 0 \\ 0 & 0 & 1 \end{bmatrix}}$$

which is true only if

$$\begin{aligned} a - b + c &= 1 \\ 2b - c &= 0 \\ 2a + 3b &= 0 \end{aligned} \qquad \begin{aligned} d - e + f &= 0 \\ 2e - f &= 1 \\ 2d + 3e &= 0 \end{aligned} \qquad \begin{aligned} g - h + i &= 0 \\ 2h - i &= 0 \\ 2g + 3h &= 1 \end{aligned}$$

Now we write augmented matrices for each of the three systems:

$$\overset{\text{First}}{\left[\begin{array}{ccc|c} 1 & -1 & 1 & 1 \\ 0 & 2 & -1 & 0 \\ 2 & 3 & 0 & 0 \end{array}\right]} \qquad \overset{\text{Second}}{\left[\begin{array}{ccc|c} 1 & -1 & 1 & 0 \\ 0 & 2 & -1 & 1 \\ 2 & 3 & 0 & 0 \end{array}\right]} \qquad \overset{\text{Third}}{\left[\begin{array}{ccc|c} 1 & -1 & 1 & 0 \\ 0 & 2 & -1 & 0 \\ 2 & 3 & 0 & 1 \end{array}\right]}$$

Since each matrix to the left of the vertical bar is the same, exactly the same row operations can be used on each augmented matrix to transform it into a reduced form. We can speed up the process substantially by combining all three augmented matrices into the single augmented matrix form below:

$$\left[\begin{array}{ccc|ccc} 1 & -1 & 1 & 1 & 0 & 0 \\ 0 & 2 & -1 & 0 & 1 & 0 \\ 2 & 3 & 0 & 0 & 0 & 1 \end{array}\right] = [M|I] \tag{1}$$

We now try to perform row operations on matrix (1) until we obtain a row-equivalent matrix that looks like matrix (2):

$$\left[\begin{array}{ccc|ccc} \overset{}{1} & 0 & 0 & a & d & g \\ 0 & \underset{I}{1} & 0 & b & \underset{B}{e} & h \\ 0 & 0 & 1 & c & f & i \end{array}\right] = [I|B] \tag{2}$$

If this can be done, then the new matrix B to the right of the vertical bar will be M^{-1}! Now let us try to transform (1) into a form like (2). We will follow the same sequence of steps as we did in the solution of linear systems by Gauss–Jordan elimination (see Section 4-3).

$$
\begin{array}{c}
\quad M \qquad\qquad\qquad I \\
\left[\begin{array}{rrr|rrr} 1 & -1 & 1 & 1 & 0 & 0 \\ 0 & 2 & -1 & 0 & 1 & 0 \\ 2 & 3 & 0 & 0 & 0 & 1 \end{array}\right] \quad (-2)R_1 + R_3 \to R_3 \\
\sim \left[\begin{array}{rrr|rrr} 1 & -1 & 1 & 1 & 0 & 0 \\ 0 & 2 & -1 & 0 & 1 & 0 \\ 0 & 5 & -2 & -2 & 0 & 1 \end{array}\right] \quad \tfrac{1}{2}R_2 \to R_2 \\
\sim \left[\begin{array}{rrr|rrr} 1 & -1 & 1 & 1 & 0 & 0 \\ 0 & 1 & -\frac{1}{2} & 0 & \frac{1}{2} & 0 \\ 0 & 5 & -2 & -2 & 0 & 1 \end{array}\right] \quad \begin{array}{l} R_2 + R_1 \to R_1 \\ \\ (-5)R_2 + R_3 \to R_3 \end{array} \\
\sim \left[\begin{array}{rrr|rrr} 1 & 0 & \frac{1}{2} & 1 & \frac{1}{2} & 0 \\ 0 & 1 & -\frac{1}{2} & 0 & \frac{1}{2} & 0 \\ 0 & 0 & \frac{1}{2} & -2 & -\frac{5}{2} & 1 \end{array}\right] \quad 2R_3 \to R_3 \\
\sim \left[\begin{array}{rrr|rrr} 1 & 0 & \frac{1}{2} & 1 & \frac{1}{2} & 0 \\ 0 & 1 & -\frac{1}{2} & 0 & \frac{1}{2} & 0 \\ 0 & 0 & 1 & -4 & -5 & 2 \end{array}\right] \quad \begin{array}{l} (-\frac{1}{2})R_3 + R_1 \to R_1 \\ \frac{1}{2}R_3 + R_2 \to R_2 \\ \end{array} \\
\sim \left[\begin{array}{rrr|rrr} 1 & 0 & 0 & 3 & 3 & -1 \\ 0 & 1 & 0 & -2 & -2 & 1 \\ 0 & 0 & 1 & -4 & -5 & 2 \end{array}\right] = [I|B]
\end{array}
$$

Converting back to systems of equations equivalent to our three original systems (we will not have to do this step in practice), we have

$$
\begin{array}{lll}
a = 3 & d = 3 & g = -1 \\
b = -2 & e = -2 & h = 1 \\
c = -4 & f = -5 & i = 2
\end{array}
$$

And these are just the elements of M^{-1} that we are looking for! Hence,

$$
M^{-1} = \begin{bmatrix} 3 & 3 & -1 \\ -2 & -2 & 1 \\ -4 & -5 & 2 \end{bmatrix}
$$

Note that this is the matrix to the right of the vertical line in the last augmented matrix. That is, $M^{-1} = B$.

Since the definition of matrix inverse requires that

$$
M^{-1}M = I \quad \text{and} \quad MM^{-1} = I \tag{3}
$$

it appears that we must compute both $M^{-1}M$ and MM^{-1} to check our work. However, it can be shown that if one of the equations in (3) is satisfied, then the other is also satisfied. Thus, for checking purposes, it is sufficient to compute either $M^{-1}M$ or MM^{-1}; we do not need to do both.

CHECK

$$
M^{-1}M = \begin{bmatrix} 3 & 3 & -1 \\ -2 & -2 & 1 \\ -4 & -5 & 2 \end{bmatrix} \begin{bmatrix} 1 & -1 & 1 \\ 0 & 2 & -1 \\ 2 & 3 & 0 \end{bmatrix} = \begin{bmatrix} 1 & 0 & 0 \\ 0 & 1 & 0 \\ 0 & 0 & 1 \end{bmatrix} = I
$$

Matched Problem 2 ➡ Let: $M = \begin{bmatrix} 3 & -1 & 1 \\ -1 & 1 & 0 \\ 1 & 0 & 1 \end{bmatrix}$

(A) Form the augmented matrix $[M|I]$.
(B) Use row operations to transform $[M|I]$ into $[I|B]$.
(C) Verify by multiplication that $B = M^{-1}$ (that is, show that $BM = I$).

The procedure shown in Example 2 can be used to find the inverse of any square matrix, if the inverse exists, and will also indicate when the inverse does not exist. These ideas are summarized in Theorem 1.

Theorem 1 — Inverse of a Square Matrix M

If $[M|I]$ is transformed by row operations into $[I|B]$, then the resulting matrix B is M^{-1}. However, if we obtain 0's in one or more rows to the left of the vertical line, then M^{-1} does not exist.

Explore–Discuss 2

(A) Suppose that the square matrix M has a row of all zeros. Explain why M has no inverse.
(B) Suppose that the square matrix M has a column of all zeros. Explain why M has no inverse.

Example 3 ➡ **Finding a Matrix Inverse** Find M^{-1}, given: $M = \begin{bmatrix} 4 & -1 \\ -6 & 2 \end{bmatrix}$

SOLUTION

$$\left[\begin{array}{rr|rr} 4 & -1 & 1 & 0 \\ -6 & 2 & 0 & 1 \end{array}\right] \quad \tfrac{1}{4}R_1 \to R_1$$

$$\sim \left[\begin{array}{rr|rr} 1 & -\frac{1}{4} & \frac{1}{4} & 0 \\ -6 & 2 & 0 & 1 \end{array}\right] \quad 6R_1 + R_2 \to R_2$$

$$\sim \left[\begin{array}{rr|rr} 1 & -\frac{1}{4} & \frac{1}{4} & 0 \\ 0 & \frac{1}{2} & \frac{3}{2} & 1 \end{array}\right] \quad 2R_2 \to R_2$$

$$\sim \left[\begin{array}{rr|rr} 1 & -\frac{1}{4} & \frac{1}{4} & 0 \\ 0 & 1 & 3 & 2 \end{array}\right] \quad \tfrac{1}{4}R_2 + R_1 \to R_1$$

$$\sim \left[\begin{array}{rr|rr} 1 & 0 & 1 & \frac{1}{2} \\ 0 & 1 & 3 & 2 \end{array}\right]$$

Thus,

$$M^{-1} = \begin{bmatrix} 1 & \frac{1}{2} \\ 3 & 2 \end{bmatrix}$$

Check by showing $M^{-1}M = I$.

Matched Problem 3 ➡ Find M^{-1}, given: $M = \begin{bmatrix} 2 & -6 \\ 1 & -2 \end{bmatrix}$

Most graphing utilities can compute matrix inverses, as illustrated in Figure 2 for the solution to Example 3.

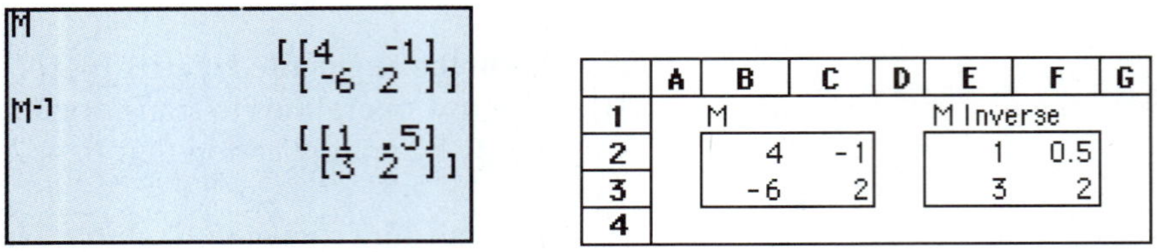

(A) The command M^{-1} produces the inverse on this graphing calculator

(B) The command MINVERSE (B2:C3) produces the inverse in this spreadsheet

FIGURE 2
Finding a matrix inverse on a graphing utility

Explore–Discuss 3

The inverse of

$$A = \begin{bmatrix} a & b \\ c & d \end{bmatrix}$$

is

$$A^{-1} = \begin{bmatrix} \dfrac{d}{ad - bc} & \dfrac{-b}{ad - bc} \\ \dfrac{-c}{ad - bc} & \dfrac{a}{ad - bc} \end{bmatrix} = \frac{1}{D}\begin{bmatrix} d & -b \\ -c & a \end{bmatrix} \qquad D = ad - bc$$

provided $D \neq 0$.

(A) Use matrix multiplication to verify this formula. What can you conclude about A^{-1} if $D = 0$?

(B) Use this formula to find the inverse of matrix A in Example 3.

Example 4 **Finding a Matrix Inverse** Find M^{-1}, given: $M = \begin{bmatrix} 2 & -4 \\ -3 & 6 \end{bmatrix}$

SOLUTION

$$\left[\begin{array}{rr|rr} 2 & -4 & 1 & 0 \\ -3 & 6 & 0 & 1 \end{array}\right] \quad \tfrac{1}{2}R_1 \rightarrow R_1$$

$$\sim \left[\begin{array}{rr|rr} 1 & -2 & \frac{1}{2} & 0 \\ -3 & 6 & 0 & 1 \end{array}\right] \quad 3R_1 + R_2 \rightarrow R_2$$

$$\sim \left[\begin{array}{rr|rr} 1 & -2 & \frac{1}{2} & 0 \\ 0 & 0 & \frac{3}{2} & 1 \end{array}\right]$$

We have all 0's in the second row to the left of the vertical bar; therefore, the inverse does not exist.

Matched Problem 4 Find N^{-1}, given: $N = \begin{bmatrix} 3 & 1 \\ 6 & 2 \end{bmatrix}$

CAUTION

Matrices that do not have inverses are called **singular matrices.** Graphing calculators and computers usually recognize singular matrices and display an appropriate error message. However, in some cases, roundoff error may cause a graphing calculator or computer to display an incorrect inverse when the matrix is singular or to display an error message when the inverse does, in fact, exist. Figure 3 illustrates the results of solving Example 4 and Matched Problem 4 on a particular graphing calculator. Notice that the calculator gives the correct answer to Example 4, but displays an inverse for the singular matrix N in Matched Problem 4. The extremely large numbers in N^{-1} suggest that an error has occurred, and a quick check on the graphing calculator shows that NN^{-1} is not the identity. Whenever the results of finding an inverse on a graphing calculator or computer are in doubt, use Theorem 1 to find the inverse.

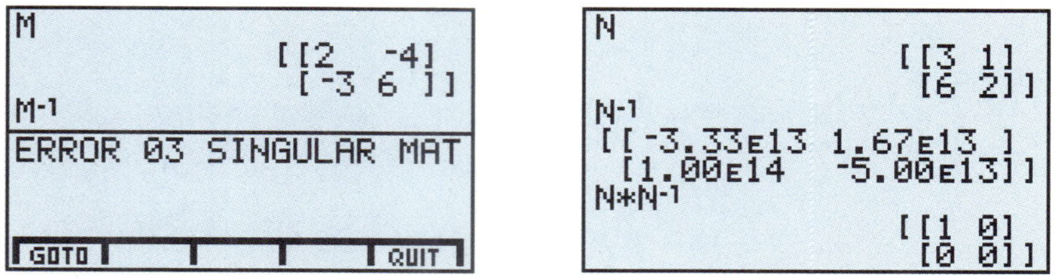

(A) Solution to Example 4 (B) Solution to Matched Problem 4

FIGURE 3

■ APPLICATION: CRYPTOGRAPHY

Matrix inverses can provide a simple and effective procedure for encoding and decoding messages. To begin, assign the numbers 1–26 to the letters in the alphabet, as shown below. Also assign the number 27 to a blank to provide for space between words. (A more sophisticated code could include both capital and lowercase letters and punctuation symbols.)

A	B	C	D	E	F	G	H	I	J	K	L	M	N	O	P	Q	R	S	T	U	V	W	X	Y	Z	Blank
1	2	3	4	5	6	7	8	9	10	11	12	13	14	15	16	17	18	19	20	21	22	23	24	25	26	27

Thus, the message "SECRET CODE" corresponds to the sequence

19 5 3 18 5 20 27 3 15 4 5

Any matrix whose elements are positive integers and whose inverse exists can be used as an **encoding matrix.** For example, to use the 2×2 matrix

$$A = \begin{bmatrix} 4 & 3 \\ 1 & 1 \end{bmatrix}$$

to encode the above message, first we divide the numbers in the sequence into groups of 2 and use these groups as the columns of a matrix B with 2 rows:

$$B = \begin{bmatrix} 19 & 3 & 5 & 27 & 15 & 5 \\ 5 & 18 & 20 & 3 & 4 & 27 \end{bmatrix}$$ Proceed down the columns, not across the rows.

(Notice that we added an extra blank at the end of the message to make the columns come out even.) Then we multiply this matrix on the left by A:

$$AB = \begin{bmatrix} 4 & 3 \\ 1 & 1 \end{bmatrix} \begin{bmatrix} 19 & 3 & 5 & 27 & 15 & 5 \\ 5 & 18 & 20 & 3 & 4 & 27 \end{bmatrix}$$
$$= \begin{bmatrix} 91 & 66 & 80 & 117 & 72 & 101 \\ 24 & 21 & 25 & 30 & 19 & 32 \end{bmatrix}$$

The coded message is

91 24 66 21 80 25 117 30 72 19 101 32

This message can be decoded simply by putting it back into matrix form and multiplying on the left by the **decoding matrix** A^{-1}. Since A^{-1} is easily determined if A is known, the encoding matrix A is the only key needed to decode messages encoded in this manner. Although simple in concept, codes of this type can be very difficult to crack.

Example 5 **Cryptography** The message

46 84 85 55 101 100 31 59 64 57 102 99 29 57 38 65 111 122

was encoded with the matrix A shown below. Decode this message.

$$A = \begin{bmatrix} 1 & 1 & 1 \\ 2 & 1 & 2 \\ 2 & 3 & 1 \end{bmatrix}$$

SOLUTION Since the encoding matrix A is 3×3, we begin by entering the coded message in the columns of a matrix C with three rows:

$$C = \begin{bmatrix} 46 & 55 & 31 & 57 & 29 & 65 \\ 84 & 101 & 59 & 102 & 57 & 111 \\ 85 & 100 & 64 & 99 & 38 & 122 \end{bmatrix}$$

If B is the matrix containing the uncoded message, then B and C are related by $C = AB$. To recover B, we find A^{-1} (details omitted) and multiply both sides of the equation $C = AB$ by A^{-1}:

$$B = A^{-1}C$$
$$= \begin{bmatrix} -5 & 2 & 1 \\ 2 & -1 & 0 \\ 4 & -1 & -1 \end{bmatrix} \begin{bmatrix} 46 & 55 & 31 & 57 & 29 & 65 \\ 84 & 101 & 59 & 102 & 57 & 111 \\ 85 & 100 & 64 & 99 & 38 & 122 \end{bmatrix}$$
$$= \begin{bmatrix} 23 & 27 & 27 & 18 & 7 & 19 \\ 8 & 9 & 3 & 12 & 1 & 19 \\ 15 & 19 & 1 & 27 & 21 & 27 \end{bmatrix}$$

Writing the numbers in the columns of this matrix in sequence and using the correspondence between numbers and letters noted earlier produces the decoded message:

23 8 15 27	9 19 27	3 1 18 12 27	7 1 21 19 19 27
W H O	I S	C A R L	G A U S S

(The answer to this question can be found earlier in this chapter.)

Matched Problem 5 The message below was also encoded with the matrix A in Example 5. Decode this message:

46 84 85 55 101 100 59 95 132 25 42 53 52 91 90 43 71 83 19 37 25

Answers to Matched Problems

1. (A) $\begin{bmatrix} 2 & -3 \\ 5 & 7 \end{bmatrix}$ (B) $\begin{bmatrix} 4 & 2 \\ 3 & -5 \\ 6 & 8 \end{bmatrix}$

2. (A) $\left[\begin{array}{rrr|rrr} 3 & -1 & 1 & 1 & 0 & 0 \\ -1 & 1 & 0 & 0 & 1 & 0 \\ 1 & 0 & 1 & 0 & 0 & 1 \end{array}\right]$

(B) $\left[\begin{array}{rrr|rrr} 1 & 0 & 0 & 1 & 1 & -1 \\ 0 & 1 & 0 & 1 & 2 & -1 \\ 0 & 0 & 1 & -1 & -1 & 2 \end{array}\right]$

(C) $\begin{bmatrix} 1 & 1 & -1 \\ 1 & 2 & -1 \\ -1 & -1 & 2 \end{bmatrix}\begin{bmatrix} 3 & -1 & 1 \\ -1 & 1 & 0 \\ 1 & 0 & 1 \end{bmatrix} = \begin{bmatrix} 1 & 0 & 0 \\ 0 & 1 & 0 \\ 0 & 0 & 1 \end{bmatrix}$

3. $\begin{bmatrix} -1 & 3 \\ -\frac{1}{2} & 1 \end{bmatrix}$ **4.** Does not exist **5.** WHO IS WILHELM JORDAN

EXERCISE 4-5

A *Perform the indicated operations in Problems 1–8.*

1. $\begin{bmatrix} 1 & 0 \\ 0 & 1 \end{bmatrix}\begin{bmatrix} 2 & -3 \\ 4 & 5 \end{bmatrix}$

2. $\begin{bmatrix} 1 & 0 \\ 0 & 1 \end{bmatrix}\begin{bmatrix} -1 & 6 \\ 0 & 2 \end{bmatrix}$

3. $\begin{bmatrix} 2 & -3 \\ 4 & 5 \end{bmatrix}\begin{bmatrix} 1 & 0 \\ 0 & 1 \end{bmatrix}$

4. $\begin{bmatrix} -1 & 6 \\ 0 & 2 \end{bmatrix}\begin{bmatrix} 1 & 0 \\ 0 & 1 \end{bmatrix}$

5. $\begin{bmatrix} 1 & 0 & 0 \\ 0 & 1 & 0 \\ 0 & 0 & 1 \end{bmatrix}\begin{bmatrix} -2 & 1 & 3 \\ 2 & 4 & -2 \\ 5 & 1 & 0 \end{bmatrix}$

6. $\begin{bmatrix} 1 & 0 & 0 \\ 0 & 1 & 0 \\ 0 & 0 & 1 \end{bmatrix}\begin{bmatrix} 3 & -4 & 0 \\ 1 & 2 & -5 \\ 6 & -3 & -1 \end{bmatrix}$

7. $\begin{bmatrix} -2 & 1 & 3 \\ 2 & 4 & -2 \\ 5 & 1 & 0 \end{bmatrix}\begin{bmatrix} 1 & 0 & 0 \\ 0 & 1 & 0 \\ 0 & 0 & 1 \end{bmatrix}$

8. $\begin{bmatrix} 3 & -4 & 0 \\ 1 & 2 & -5 \\ 6 & -3 & -1 \end{bmatrix}\begin{bmatrix} 1 & 0 & 0 \\ 0 & 1 & 0 \\ 0 & 0 & 1 \end{bmatrix}$

In Problems 9–18, examine the product of the two matrices to determine if each is the inverse of the other.

9. $\begin{bmatrix} 3 & -4 \\ -2 & 3 \end{bmatrix}; \begin{bmatrix} 3 & 4 \\ 2 & 3 \end{bmatrix}$

10. $\begin{bmatrix} -2 & -1 \\ -4 & 2 \end{bmatrix}; \begin{bmatrix} 1 & -1 \\ 2 & -2 \end{bmatrix}$

11. $\begin{bmatrix} 2 & 2 \\ -1 & -1 \end{bmatrix}; \begin{bmatrix} 1 & 1 \\ -1 & -1 \end{bmatrix}$

12. $\begin{bmatrix} 5 & -7 \\ -2 & 3 \end{bmatrix}; \begin{bmatrix} 3 & 7 \\ 2 & 5 \end{bmatrix}$

13. $\begin{bmatrix} -5 & 2 \\ -8 & 3 \end{bmatrix}; \begin{bmatrix} 3 & -2 \\ 8 & -5 \end{bmatrix}$

14. $\begin{bmatrix} 7 & 4 \\ -5 & -3 \end{bmatrix}; \begin{bmatrix} 3 & 4 \\ -5 & -7 \end{bmatrix}$

15. $\begin{bmatrix} 1 & 2 & 0 \\ 0 & 1 & 0 \\ -1 & -1 & 1 \end{bmatrix}; \begin{bmatrix} 1 & -2 & 0 \\ 0 & 1 & 0 \\ 1 & -1 & 0 \end{bmatrix}$

16. $\begin{bmatrix} 1 & 0 & 1 \\ -3 & 1 & -2 \\ 0 & 0 & 1 \end{bmatrix}; \begin{bmatrix} 1 & 0 & -1 \\ 3 & 1 & -1 \\ 0 & 0 & 1 \end{bmatrix}$

17. $\begin{bmatrix} 1 & -1 & 1 \\ 0 & 2 & -1 \\ 2 & 3 & 0 \end{bmatrix}; \begin{bmatrix} 3 & 3 & -1 \\ -2 & -2 & 1 \\ -4 & -5 & 2 \end{bmatrix}$

18. $\begin{bmatrix} 1 & 0 & -1 \\ 3 & 1 & -1 \\ 0 & 0 & 0 \end{bmatrix}; \begin{bmatrix} 1 & 0 & -1 \\ -3 & 1 & -2 \\ 0 & 0 & 1 \end{bmatrix}$

B *Given M in Problems 19–28, find M^{-1} and show that $M^{-1}M = I$.*

19. $\begin{bmatrix} -1 & 0 \\ -3 & 1 \end{bmatrix}$

20. $\begin{bmatrix} 1 & -5 \\ 0 & -1 \end{bmatrix}$

21. $\begin{bmatrix} 1 & 2 \\ 1 & 3 \end{bmatrix}$

22. $\begin{bmatrix} 2 & 1 \\ 5 & 3 \end{bmatrix}$

23. $\begin{bmatrix} 1 & 3 \\ 2 & 7 \end{bmatrix}$

24. $\begin{bmatrix} 2 & 1 \\ 1 & 1 \end{bmatrix}$

25. $\begin{bmatrix} 1 & -3 & 0 \\ 0 & 1 & 1 \\ 2 & -1 & 4 \end{bmatrix}$

26. $\begin{bmatrix} 2 & 3 & 0 \\ 1 & 2 & 3 \\ 0 & -1 & -5 \end{bmatrix}$

27. $\begin{bmatrix} 1 & 1 & 0 \\ 2 & 3 & -1 \\ 1 & 0 & 2 \end{bmatrix}$

28. $\begin{bmatrix} 1 & 0 & -1 \\ 2 & -1 & 0 \\ 1 & 1 & -2 \end{bmatrix}$

Find the inverse of each matrix in Problems 29–34, if it exists.

29. $\begin{bmatrix} 4 & 3 \\ -3 & -2 \end{bmatrix}$

30. $\begin{bmatrix} -4 & 3 \\ -5 & 4 \end{bmatrix}$

31. $\begin{bmatrix} 2 & 6 \\ 3 & 9 \end{bmatrix}$

32. $\begin{bmatrix} 2 & -4 \\ -3 & 6 \end{bmatrix}$

33. $\begin{bmatrix} 2 & 1 \\ 4 & 3 \end{bmatrix}$

34. $\begin{bmatrix} -5 & 3 \\ 2 & -2 \end{bmatrix}$

C *Find the inverse of each matrix in Problems 35–42, if it exists.*

35. $\begin{bmatrix} -5 & -2 & -2 \\ 2 & 1 & 0 \\ 1 & 0 & 1 \end{bmatrix}$

36. $\begin{bmatrix} 2 & -2 & 4 \\ 1 & 1 & 1 \\ 1 & 0 & 1 \end{bmatrix}$

37. $\begin{bmatrix} 2 & 1 & 1 \\ 1 & 1 & 0 \\ -1 & -1 & 0 \end{bmatrix}$

38. $\begin{bmatrix} 1 & -1 & 0 \\ 2 & -1 & 1 \\ 0 & 1 & 1 \end{bmatrix}$

39. $\begin{bmatrix} -1 & -2 & 2 \\ 4 & 3 & 0 \\ 4 & 0 & 4 \end{bmatrix}$

40. $\begin{bmatrix} 4 & 2 & 2 \\ 4 & 2 & 0 \\ 5 & 0 & 5 \end{bmatrix}$

41. $\begin{bmatrix} 2 & -1 & -2 \\ -4 & 2 & 8 \\ 6 & -2 & -1 \end{bmatrix}$

42. $\begin{bmatrix} -1 & -1 & 4 \\ 3 & 3 & -22 \\ -2 & -1 & 19 \end{bmatrix}$

43. Show that $(A^{-1})^{-1} = A$ for: $A = \begin{bmatrix} 4 & 3 \\ 3 & 2 \end{bmatrix}$

44. Show that $(AB)^{-1} = B^{-1}A^{-1}$ for:

$$A = \begin{bmatrix} 4 & 3 \\ 3 & 2 \end{bmatrix} \quad \text{and} \quad B = \begin{bmatrix} 2 & 5 \\ 3 & 7 \end{bmatrix}$$

45. Discuss the existence of M^{-1} for 2×2 diagonal matrices of the form

$$M = \begin{bmatrix} a & 0 \\ 0 & d \end{bmatrix}$$

Generalize your conclusions to $n \times n$ diagonal matrices.

46. Discuss the existence of M^{-1} for 2×2 upper triangular matrices of the form

$$M = \begin{bmatrix} a & b \\ 0 & d \end{bmatrix}$$

Generalize your conclusions to $n \times n$ upper triangular matrices.

In Problems 47–49, find A^{-1} and A^2.

47. $A = \begin{bmatrix} 3 & 2 \\ -4 & -3 \end{bmatrix}$

48. $A = \begin{bmatrix} -2 & -1 \\ 3 & 2 \end{bmatrix}$

49. $A = \begin{bmatrix} 4 & 3 \\ -5 & -4 \end{bmatrix}$

50. Based on your observations in Problems 47–49, if $A = A^{-1}$ for a square matrix A, what is A^2? Give a mathematical argument to support your conclusion.

In Problems 51–54, use a graphing utility to find the inverse of each matrix, if it exists.

51. $\begin{bmatrix} 6 & 2 & 0 & 4 \\ 5 & 3 & 2 & 1 \\ 0 & -1 & 1 & -2 \\ 2 & -3 & 1 & 0 \end{bmatrix}$

52. $\begin{bmatrix} -2 & 4 & 0 & -1 \\ 2 & -1 & 2 & 5 \\ 0 & 2 & -1 & 7 \\ 2 & -3 & 0 & 5 \end{bmatrix}$

53. $\begin{bmatrix} 3 & 2 & 3 & 4 & 4 \\ 5 & 4 & 3 & 2 & 1 \\ -1 & -1 & 2 & -2 & 3 \\ 3 & -3 & 1 & 0 & 1 \\ 1 & 1 & 2 & 0 & 2 \end{bmatrix}$

54. $\begin{bmatrix} 1 & 2 & 3 & 4 & 5 \\ 2 & 6 & 4 & 5 & 6 \\ -1 & -2 & -1 & 2 & 3 \\ 1 & 6 & 1 & 6 & 4 \\ 1 & -4 & 3 & -7 & -4 \end{bmatrix}$

APPLICATIONS

Social Sciences

Problems 55–58 refer to the encoding matrix:

$$A = \begin{bmatrix} 1 & 2 \\ 1 & 3 \end{bmatrix}$$

55. *Cryptography.* Encode the message "THE SUN ALSO RISES" using matrix A.

56. *Cryptography.* Encode the message "THE GRAPES OF WRATH" using matrix A.

57. *Cryptography.* The following message was encoded with matrix A. Decode this message:

37 52 24 29 73 96 49 69 62 89 36 44 59 86 41 50 22 26

58. *Cryptography.* The following message was encoded with matrix A. Decode this message:

9 13 40 49 29 34 29 30 22 26 33 36 43 57 29 34 54 74

Problems 59–62 require the use of a graphing calculator or a computer. To use the 5 × 5 encoding matrix B given below, form a matrix with 5 rows and as many columns as necessary to accommodate each message.

$$B = \begin{bmatrix} 1 & 0 & 1 & 0 & 1 \\ 0 & 1 & 1 & 0 & 3 \\ 2 & 1 & 1 & 1 & 1 \\ 0 & 0 & 1 & 0 & 2 \\ 1 & 1 & 1 & 2 & 1 \end{bmatrix}$$

59. *Cryptography.* Encode the message "THE BEST YEARS OF OUR LIVES" using matrix B.

60. *Cryptography.* Encode the message "THE BRIDGE ON THE RIVER KWAI" using matrix B.

61. *Cryptography.* The following message was encoded with matrix B. Decode this message:

32 34 87 19 94 24 21 67 11 69 54 71 112 43 112 56 92 109 55 109 48 66 98 41 89 62 135 124 81 143

62. *Cryptography.* The following message was encoded with matrix B. Decode this message:

28 22 83 11 90 57 108 102 69 105 57 78 91 57 89 66 111 112 79 125

SECTION 4-6 Matrix Equations and Systems of Linear Equations

- MATRIX EQUATIONS
- MATRIX EQUATIONS AND SYSTEMS OF LINEAR EQUATIONS
- APPLICATION

The identity matrix and inverse matrix discussed in the last section can be put to immediate use in the solution of certain simple matrix equations. Being able to solve a matrix equation gives us another important method of solving systems of equations, provided the system is independent and has the same number of variables as equations. If the system is dependent or if it has either fewer or more variables than equations, we must return to the Gauss–Jordan method of elimination.

MATRIX EQUATIONS

Before we discuss the solution of matrix equations, you will probably find it helpful to briefly review the basic properties of real numbers discussed in Appendix A-2 and the discussion of linear equations in Appendix A-8.

Explore–Discuss 1

Let a, b, and c be real numbers, with $a \neq 0$. Solve each equation for x.

(A) $ax = b$ (B) $ax + b = c$

Solving simple matrix equations follows very much the same procedures used in solving real number equations. We have, however, less freedom with matrix equations, because matrix multiplication is not commutative. In solving matrix equations, we will be guided by the properties of matrices summarized in Theorem 1.

Theorem 1 BASIC PROPERTIES OF MATRICES

Assuming all products and sums are defined for the indicated matrices A, B, C, I, and 0, then:

Addition Properties

Associative: $(A + B) + C = A + (B + C)$
Commutative: $A + B = B + A$
Additive Identity: $A + 0 = 0 + A = A$
Additive Inverse: $A + (-A) = (-A) + A = 0$

Multiplication Properties

Associative Property: $A(BC) = (AB)C$
Multiplicative Identity: $AI = IA = A$
Multiplicative Inverse: If A is a square matrix and A^{-1} exists, then $AA^{-1} = A^{-1}A = I$.

Combined Properties

Left Distributive: $A(B + C) = AB + AC$
Right Distributive: $(B + C)A = BA + CA$

Equality

Addition: If $A = B$, then $A + C = B + C$.
Left Multiplication: If $A = B$, then $CA = CB$.
Right Mulplication: If $A = B$, then $AC = BC$.

The use of these properties in the solution of matrix equations is best illustrated by an example.

Example 1 **Solving a Matrix Equation** Given an $n \times n$ matrix A and $n \times 1$ column matrices B and X, solve $AX = B$ for X. Assume all necessary inverses exist.

SOLUTION We are interested in finding a column matrix X that satisfies the matrix equation $AX = B$. To solve this equation, we multiply both sides on the left by A^{-1}, assuming it exists, to isolate X on the left side.

$$AX = B \quad \text{Use the left multiplication property.}$$
$$A^{-1}(AX) = A^{-1}B \quad \text{Use the associative property.}$$
$$(A^{-1}A)X = A^{-1}B \quad A^{-1}A = I$$
$$IX = A^{-1}B \quad IX = X$$
$$X = A^{-1}B$$

Matched Problem 1 ➡ Given an $n \times n$ matrix A and $n \times 1$ column matrices B, C, and X, solve $AX + C = B$ for X. Assume all necessary inverses exist.

CAUTION

Do not mix the left multiplication property and the right multiplication property. If $AX = B$, then

$$A^{-1}(AX) \neq BA^{-1}$$

■ Matrix Equations and Systems of Linear Equations

We now show how independent systems of linear equations with the same number of variables as equations can be solved. First, convert the system into a matrix equation of the form $AX = B$, and then use $X = A^{-1}B$ as obtained in Example 1.

Example 2 ➡ **Using Inverses to Solve Systems of Equations** Use matrix inverse methods to solve the system:

$$\begin{aligned} x_1 - x_2 + x_3 &= 1 \\ 2x_2 - x_3 &= 1 \\ 2x_1 + 3x_2 \phantom{{}-x_3} &= 1 \end{aligned} \qquad (1)$$

SOLUTION The inverse of the coefficient matrix

$$A = \begin{bmatrix} 1 & -1 & 1 \\ 0 & 2 & -1 \\ 2 & 3 & 0 \end{bmatrix}$$

provides an efficient method for solving this system. To see how, we convert system (1) into a matrix equation:

$$\overset{A}{\begin{bmatrix} 1 & -1 & 1 \\ 0 & 2 & -1 \\ 2 & 3 & 0 \end{bmatrix}} \overset{X}{\begin{bmatrix} x_1 \\ x_2 \\ x_3 \end{bmatrix}} = \overset{B}{\begin{bmatrix} 1 \\ 1 \\ 1 \end{bmatrix}}$$

Check that matrix equation (2) is equivalent to system (1) by finding the product of the left side and then equating corresponding elements on the left with those on the right. Now you see another important reason for defining matrix multiplication as we did.

We are interested in finding a column matrix X that satisfies the matrix equation $AX = B$. In Example 1 we found that if A^{-1} exists, then

$$X = A^{-1}B$$

The inverse of A was found in Example 2, Section 4-5, to be

$$A^{-1} = \begin{bmatrix} 3 & 3 & -1 \\ -2 & -2 & 1 \\ -4 & -5 & 2 \end{bmatrix}$$

Thus,

$$\overset{X}{\begin{bmatrix} x_1 \\ x_2 \\ x_3 \end{bmatrix}} = \overset{A^{-1}}{\begin{bmatrix} 3 & 3 & -1 \\ -2 & -2 & 1 \\ -4 & -5 & 2 \end{bmatrix}} \overset{B}{\begin{bmatrix} 1 \\ 1 \\ 1 \end{bmatrix}} = \begin{bmatrix} 5 \\ -3 \\ -7 \end{bmatrix}$$

and we can conclude that $x_1 = 5, x_2 = -3$, and $x_3 = -7$. Check this result in system (1).

Matched Problem 2 ➡ Use matrix inverse methods to solve the system:

$$\begin{aligned} 3x_1 - x_2 + x_3 &= 1 \\ -x_1 + x_2 &= 3 \\ x_1 + x_3 &= 2 \end{aligned}$$

[*Note:* The inverse of the coefficient matrix was found in Matched Problem 2, Section 4-5.]

At first glance, using matrix inverse methods seems to require the same amount of effort as using Gauss–Jordan elimination. In either case, row operations must be applied to an augmented matrix involving the coefficients of the system. The advantage of the inverse matrix method becomes readily apparent when solving a number of systems with a common coefficient matrix and different constant terms.

Example 3 ➡ **Using Inverses to Solve Systems of Equations** Use matrix inverse methods to solve each of the following systems:

(A)
$$\begin{aligned} x_1 - x_2 + x_3 &= 3 \\ 2x_2 - x_3 &= 1 \\ 2x_1 + 3x_2 &= 4 \end{aligned}$$

(B)
$$\begin{aligned} x_1 - x_2 + x_3 &= -5 \\ 2x_2 - x_3 &= 2 \\ 2x_1 + 3x_2 &= -3 \end{aligned}$$

SOLUTION Notice that both systems have the same coefficient matrix A as system (1) in Example 2. Only the constant terms have changed. Thus, we can use A^{-1} to solve these systems just as we did in Example 2.

(A)
$$\overset{X}{\begin{bmatrix} x_1 \\ x_2 \\ x_3 \end{bmatrix}} = \overset{A^{-1}}{\begin{bmatrix} 3 & 3 & -1 \\ -2 & -2 & 1 \\ -4 & -5 & 2 \end{bmatrix}} \overset{B}{\begin{bmatrix} 3 \\ 1 \\ 4 \end{bmatrix}} = \begin{bmatrix} 8 \\ -4 \\ -9 \end{bmatrix}$$

Thus, $x_1 = 8, x_2 = -4$, and $x_3 = -9$.

(B)
$$\overset{X}{\begin{bmatrix} x_1 \\ x_2 \\ x_3 \end{bmatrix}} = \overset{A^{-1}}{\begin{bmatrix} 3 & 3 & -1 \\ -2 & -2 & 1 \\ -4 & -5 & 2 \end{bmatrix}} \overset{B}{\begin{bmatrix} -5 \\ 2 \\ -3 \end{bmatrix}} = \begin{bmatrix} -6 \\ 3 \\ 4 \end{bmatrix}$$

Thus, $x_1 = -6, x_2 = 3$, and $x_3 = 4$.

Matched Problem 3 Use matrix inverse methods to solve each of the following systems (see Matched Problem 2):

(A) $$\begin{aligned} 3x_1 - x_2 + x_3 &= 3 \\ -x_1 + x_2 &= -3 \\ x_1 + x_3 &= 2 \end{aligned}$$

(B) $$\begin{aligned} 3x_1 - x_2 + x_3 &= -5 \\ -x_1 + x_2 &= 1 \\ x_1 + x_3 &= -4 \end{aligned}$$

As Examples 2 and 3 illustrate, inverse methods are very convenient for hand calculations because once the inverse is found, it can be used to solve any new system formed by changing only the constant terms. Since most graphing utilities can compute the inverse of a matrix, this method also adapts readily to graphing utility solutions (see Fig. 1). However, if your graphing utility also has a built-in procedure for finding the reduced form of an augmented coefficient matrix, then it is just as convenient to use Gauss–Jordan elimination. Furthermore, Gauss–Jordan elimination can be used in all cases and, as noted below, matrix inverse methods cannot always be used.

	A	B	C	D	E	F	G	H	I	J
1			A			B	X		B	X
2		1	-1	1		3	8		-5	-6
3		0	2	-1		1	-4		2	3
4		2	3	0		4	-9		-3	4

FIGURE 1
Using inverse methods on a spreadsheet:
The values in G2:G4 are produced by the command MMULT(MINVERSE(B2:D4),F2:F4)

Using Inverse Methods to Solve Systems of Equations

If the number of equations in a system equals the number of variables and the coefficient matrix has an inverse, then the system will always have a unique solution that can be found by using the inverse of the coefficient matrix to solve the corresponding matrix equation.

Matrix equation	Solution
$AX = B$	$X = A^{-1}B$

REMARK

What happens if the coefficient matrix does not have an inverse? In this case, it can be shown that the system does not have a unique solution and is either dependent or inconsistent. Gauss–Jordan elimination must be used to determine which is the case. Also, as we mentioned earlier, Gauss–Jordan elimination always must be used if the number of variables is not the same as the number of equations.

Application

The following application will illustrate the usefulness of the inverse matrix method for solving systems of equations.

Example 4 **Investment Analysis** An investment advisor currently has two types of investments available for clients: a conservative investment A that pays 10% per year and an investment B of higher risk that pays 20% per year. Clients may divide their investments between the two to achieve any total return desired between 10% and 20%. However, the higher the desired return, the higher the risk. How should each client listed in the table invest to achieve the indicated return?

	Client			
	1	2	3	k
Total investment	\$20,000	\$50,000	\$10,000	k_1
Annual return desired	\$ 2,400 (12%)	\$ 7,500 (15%)	\$ 1,300 (13%)	k_2

Solution We will solve the problem for an arbitrary client k by finding an inverse matrix. Then we will apply the result to the three specific clients.

Let: x_1 = Amount invested in A
x_2 = Amount invested in B

Then we have the following mathematical model:

$$\begin{aligned} x_1 + \quad x_2 &= k_1 && \text{Total invested} \\ 0.1x_1 + 0.2x_2 &= k_2 && \text{Total annual return desired} \end{aligned}$$

Write as a matrix equation:

$$\overset{A}{\begin{bmatrix} 1 & 1 \\ 0.1 & 0.2 \end{bmatrix}} \overset{X}{\begin{bmatrix} x_1 \\ x_2 \end{bmatrix}} = \overset{B}{\begin{bmatrix} k_1 \\ k_2 \end{bmatrix}}$$

If A^{-1} exists, then

$$X = A^{-1}B$$

We now find A^{-1} by starting with the augmented matrix $[A|I]$ and proceeding as discussed in Section 4-5:

$$\left[\begin{array}{cc|cc} 1 & 1 & 1 & 0 \\ 0.1 & 0.2 & 0 & 1 \end{array}\right] \quad 10R_2 \to R_2$$

$$\sim \left[\begin{array}{cc|cc} 1 & 1 & 1 & 0 \\ 1 & 2 & 0 & 10 \end{array}\right] \quad (-1)R_1 + R_2 \to R_2$$

$$\sim \left[\begin{array}{cc|cc} 1 & 1 & 1 & 0 \\ 0 & 1 & -1 & 10 \end{array}\right] \quad (-1)R_2 + R_1 \to R_1$$

$$\sim \left[\begin{array}{cc|cc} 1 & 0 & 2 & -10 \\ 0 & 1 & -1 & 10 \end{array}\right]$$

Thus,

$$A^{-1} = \begin{bmatrix} 2 & -10 \\ -1 & 10 \end{bmatrix} \qquad \textit{Check: } \overset{A^{-1}}{\begin{bmatrix} 2 & -10 \\ -1 & 10 \end{bmatrix}} \overset{A}{\begin{bmatrix} 1 & 1 \\ 0.1 & 0.2 \end{bmatrix}} = \overset{I}{\begin{bmatrix} 1 & 0 \\ 0 & 1 \end{bmatrix}}$$

and

$$\overset{X}{\begin{bmatrix} x_1 \\ x_2 \end{bmatrix}} = \overset{A^{-1}}{\begin{bmatrix} 2 & -10 \\ -1 & 10 \end{bmatrix}} \overset{B}{\begin{bmatrix} k_1 \\ k_2 \end{bmatrix}}$$

To solve each client's investment problem, we replace k_1 and k_2 with appropriate values from the table and multiply by A^{-1}:

Client 1

$$\begin{bmatrix} x_1 \\ x_2 \end{bmatrix} = \begin{bmatrix} 2 & -10 \\ -1 & 10 \end{bmatrix} \begin{bmatrix} 20{,}000 \\ 2{,}400 \end{bmatrix} = \begin{bmatrix} 16{,}000 \\ 4{,}000 \end{bmatrix}$$

Solution: $x_1 =$ \$16,000 in investment A, $x_2 =$ \$4,000 in investment B

Client 2

$$\begin{bmatrix} x_1 \\ x_2 \end{bmatrix} = \begin{bmatrix} 2 & -10 \\ -1 & 10 \end{bmatrix} \begin{bmatrix} 50{,}000 \\ 7{,}500 \end{bmatrix} = \begin{bmatrix} 25{,}000 \\ 25{,}000 \end{bmatrix}$$

Solution: $x_1 =$ \$25,000 in investment A, $x_2 =$ \$25,000 in investment B

Client 3

$$\begin{bmatrix} x_1 \\ x_2 \end{bmatrix} = \begin{bmatrix} 2 & -10 \\ -1 & 10 \end{bmatrix} \begin{bmatrix} 10{,}000 \\ 1{,}300 \end{bmatrix} = \begin{bmatrix} 7{,}000 \\ 3{,}000 \end{bmatrix}$$

Solution: $x_1 =$ \$7,000 in investment A, $x_2 =$ \$3,000 in investment B

Matched Problem 4 Repeat Example 4 with investment A paying 8% and investment B paying 24%.

Figure 2 illustrates a solution to Example 4 on a spreadsheet.

	A	B	C	D	E	F	G
1			Clients				
2		1	2	3		A	
3	Total Investment	$20,000	$50,000	$10,000		1	1
4	Annual Return	$2,400	$7,500	$1,300		0.1	0.2
5	Amount Invested in A	$16,000	$25,000	$7,000			
6	Amount Invested in B	$4,000	$25,000	$3,000			

FIGURE 2

Explore–Discuss 2

Refer to the mathematical model in Example 4:

$$\overset{A}{\begin{bmatrix} 1 & 1 \\ 0.1 & 0.2 \end{bmatrix}} \overset{X}{\begin{bmatrix} x_1 \\ x_2 \end{bmatrix}} = \overset{B}{\begin{bmatrix} k_1 \\ k_2 \end{bmatrix}} \qquad (3)$$

(A) Does equation (3) always have a solution for any constant matrix B?

(B) Do all these solutions make sense for the original problem? If not, give examples.

(C) If the total investment is $k_1 = \$10{,}000$, describe all possible annual returns k_2.

Answers to Matched Problems

1.
$$\begin{aligned} AX + C &= B \\ (AX + C) - C &= B - C \\ AX + (C - C) &= B - C \\ AX + 0 &= B - C \\ AX &= B - C \\ A^{-1}(AX) &= A^{-1}(B - C) \\ (A^{-1}A)X &= A^{-1}(B - C) \\ IX &= A^{-1}(B - C) \\ X &= A^{-1}(B - C) \end{aligned}$$

2. $x_1 = 2, x_2 = 5, x_3 = 0$

3. (A) $x_1 = -2, x_2 = -5, x_3 = 4$ (B) $x_1 = 0, x_2 = 1, x_3 = -4$

4. $A^{-1} = \begin{bmatrix} 1.5 & -6.25 \\ -0.5 & 6.25 \end{bmatrix}$; Client 1: \$15,000 in A and \$5,000 in B; Client 2: \$28,125 in A and \$21,875 in B; Client 3: \$6,875 in A and \$3,125 in B

EXERCISE 4-6

A *Write Problems 1–4 as systems of linear equations without matrices.*

1. $\begin{bmatrix} 3 & 1 \\ 2 & -1 \end{bmatrix} \begin{bmatrix} x_1 \\ x_2 \end{bmatrix} = \begin{bmatrix} 5 \\ -4 \end{bmatrix}$

2. $\begin{bmatrix} -2 & 1 \\ -3 & 4 \end{bmatrix} \begin{bmatrix} x_1 \\ x_2 \end{bmatrix} = \begin{bmatrix} -5 \\ 7 \end{bmatrix}$

3. $\begin{bmatrix} -3 & 1 & 0 \\ 2 & 0 & 1 \\ -1 & 3 & -2 \end{bmatrix} \begin{bmatrix} x_1 \\ x_2 \\ x_3 \end{bmatrix} = \begin{bmatrix} 3 \\ -4 \\ 2 \end{bmatrix}$

4. $\begin{bmatrix} 2 & -1 & 0 \\ -2 & 3 & -1 \\ 4 & 0 & 3 \end{bmatrix} \begin{bmatrix} x_1 \\ x_2 \\ x_3 \end{bmatrix} = \begin{bmatrix} 6 \\ -4 \\ 7 \end{bmatrix}$

Write each system in Problems 5–8 as a matrix equation of the form $AX = B$.

5. $\begin{aligned} 3x_1 - 4x_2 &= 1 \\ 2x_1 + x_2 &= 5 \end{aligned}$

6. $\begin{aligned} 2x_1 + x_2 &= 8 \\ -5x_1 + 3x_2 &= -4 \end{aligned}$

7. $\begin{aligned} x_1 - 3x_2 + 2x_3 &= -3 \\ -2x_1 + 3x_2 &= 1 \\ x_1 + x_2 + 4x_3 &= -2 \end{aligned}$

8. $\begin{aligned} 3x_1 + 2x_3 &= 9 \\ -x_1 + 4x_2 + x_3 &= -7 \\ -2x_1 + 3x_2 &= 6 \end{aligned}$

Find x_1 and x_2 in Problems 9–12.

9. $\begin{bmatrix} x_1 \\ x_2 \end{bmatrix} = \begin{bmatrix} 3 & -2 \\ 1 & 4 \end{bmatrix} \begin{bmatrix} -2 \\ 1 \end{bmatrix}$

10. $\begin{bmatrix} x_1 \\ x_2 \end{bmatrix} = \begin{bmatrix} -2 & 1 \\ -1 & 2 \end{bmatrix} \begin{bmatrix} 3 \\ -2 \end{bmatrix}$

11. $\begin{bmatrix} x_1 \\ x_2 \end{bmatrix} = \begin{bmatrix} -2 & 3 \\ 2 & -1 \end{bmatrix} \begin{bmatrix} 3 \\ 2 \end{bmatrix}$

12. $\begin{bmatrix} x_1 \\ x_2 \end{bmatrix} = \begin{bmatrix} 3 & -1 \\ 0 & 2 \end{bmatrix} \begin{bmatrix} -2 \\ 1 \end{bmatrix}$

In Problems 13–16, find x_1 and x_2.

13. $\begin{bmatrix} 1 & -1 \\ 1 & -2 \end{bmatrix} \begin{bmatrix} x_1 \\ x_2 \end{bmatrix} = \begin{bmatrix} 5 \\ 7 \end{bmatrix}$

14. $\begin{bmatrix} 1 & 3 \\ 1 & 4 \end{bmatrix} \begin{bmatrix} x_1 \\ x_2 \end{bmatrix} = \begin{bmatrix} 9 \\ 6 \end{bmatrix}$

15. $\begin{bmatrix} 1 & 1 \\ 2 & -3 \end{bmatrix} \begin{bmatrix} x_1 \\ x_2 \end{bmatrix} = \begin{bmatrix} 15 \\ 10 \end{bmatrix}$

16. $\begin{bmatrix} 1 & 1 \\ 3 & -2 \end{bmatrix} \begin{bmatrix} x_1 \\ x_2 \end{bmatrix} = \begin{bmatrix} 10 \\ 20 \end{bmatrix}$

B *In Problems 17–24, write each system as a matrix equation and solve by using inverses. [Note: The inverses were found in Problems 21–28 in Exercise 4-5.]*

17. $x_1 + 2x_2 = k_1$
$x_1 + 3x_2 = k_2$
(A) $k_1 = 1, k_2 = 3$
(B) $k_1 = 3, k_2 = 5$
(C) $k_1 = -2, k_2 = 1$

18. $2x_1 + x_2 = k_1$
$5x_1 + 3x_2 = k_2$
(A) $k_1 = 2, k_2 = 13$
(B) $k_1 = 2, k_2 = 4$
(C) $k_1 = 1, k_2 = -3$

19. $x_1 + 3x_2 = k_1$
$2x_1 + 7x_2 = k_2$
(A) $k_1 = 2, k_2 = -1$
(B) $k_1 = 1, k_2 = 0$
(C) $k_1 = 3, k_2 = -1$

20. $2x_1 + x_2 = k_1$
$x_1 + x_2 = k_2$
(A) $k_1 = -1, k_2 = -2$
(B) $k_1 = 2, k_2 = 3$
(C) $k_1 = 2, k_2 = 0$

21. $x_1 - 3x_2 = k_1$
$x_2 + x_3 = k_2$
$2x_1 - x_2 + 4x_3 = k_3$
(A) $k_1 = 1, k_2 = 0, k_3 = 2$
(B) $k_1 = -1, k_2 = 1, k_3 = 0$
(C) $k_1 = 2, k_2 = -2, k_3 = 1$

22. $2x_1 + 3x_2 = k_1$
$x_1 + 2x_2 + 3x_3 = k_2$
$-x_2 - 5x_3 = k_3$
(A) $k_1 = 0, k_2 = 2, k_3 = 1$
(B) $k_1 = -2, k_2 = 0, k_3 = 1$
(C) $k_1 = 3, k_2 = 1, k_3 = 0$

23. $x_1 + x_2 = k_1$
$2x_1 + 3x_2 - x_3 = k_2$
$x_1 + 2x_3 = k_3$
(A) $k_1 = 2, k_2 = 0, k_3 = 4$
(B) $k_1 = 0, k_2 = 4, k_3 = -2$
(C) $k_1 = 4, k_2 = 2, k_3 = 0$

24. $x_1 - x_3 = k_1$
$2x_1 - x_2 = k_2$
$x_1 + x_2 - 2x_3 = k_3$
(A) $k_1 = 4, k_2 = 8, k_3 = 0$
(B) $k_1 = 4, k_2 = 0, k_3 = -4$
(C) $k_1 = 0, k_2 = 8, k_3 = -8$

In Problems 25–30, explain why the system cannot be solved by matrix inverse methods. Discuss methods that could be used and then solve the system.

25. $-2x_1 + 4x_2 = -5$
$6x_1 - 12x_2 = 15$

26. $-2x_1 + 4x_2 = 5$
$6x_1 - 12x_2 = 15$

27. $x_1 - 3x_2 - 2x_3 = -1$
$-2x_1 + 6x_2 + 4x_3 = 3$

28. $x_1 - 3x_2 - 2x_3 = -1$
$-2x_1 + 7x_2 + 3x_3 = 3$

29. $x_1 - 2x_2 + 3x_3 = 1$
$2x_1 - 3x_2 - 2x_3 = 3$
$x_1 - x_2 - 5x_3 = 2$

30. $x_1 - 2x_2 + 3x_3 = 1$
$2x_1 - 3x_2 - 2x_3 = 3$
$x_1 - x_2 - 5x_3 = 4$

C *For $n \times n$ matrices A and B, and $n \times 1$ column matrices C, D, and X, solve each matrix equation in Problems 31–36 for X. Assume all necessary inverses exist.*

31. $AX - BX = C$

32. $AX + BX = C$

33. $AX + X = C$

34. $AX - X = C$

35. $AX - C = D - BX$

36. $AX + C = BX + D$

37. Use matrix inverse methods to solve the following system for the indicated values of k_1 and k_2.

$$\begin{aligned} x_1 + 2.001x_2 &= k_1 \\ x_1 + \quad 2x_2 &= k_2 \end{aligned}$$

(A) $k_1 = 1, k_2 = 1$
(B) $k_1 = 1, k_2 = 0$
(C) $k_1 = 0, k_2 = 1$

Discuss the effect of small changes in the constant terms on the solution set of this system.

38. Repeat Problem 37 for the following system:

$$\begin{aligned} x_1 - 3.001x_2 &= k_1 \\ x_1 - \quad 3x_2 &= k_2 \end{aligned}$$

In Problems 39–42, write each system as a matrix equation and solve by using the inverse coefficient matrix. Use a graphing utility to perform the necessary calculations.

39.
$$\begin{aligned} x_1 + 8x_2 + 7x_3 &= 135 \\ 6x_1 + 6x_2 + 8x_3 &= 155 \\ 3x_1 + 4x_2 + 6x_3 &= 75 \end{aligned}$$

40.
$$\begin{aligned} 5x_1 + 3x_2 - 2x_3 &= 112 \\ 7x_1 + 5x_2 \phantom{{}- 2x_3} &= 70 \\ 3x_1 + x_2 - 9x_3 &= 96 \end{aligned}$$

41.
$$\begin{aligned} 6x_1 + 9x_2 + 7x_3 + 5x_4 &= 250 \\ 6x_1 + 4x_2 + 7x_3 + 3x_4 &= 195 \\ 4x_1 + 5x_2 + 3x_3 + 2x_4 &= 145 \\ 4x_1 + 3x_2 + 8x_3 + 2x_4 &= 125 \end{aligned}$$

42.
$$\begin{aligned} 3x_1 + 3x_2 + 6x_3 + 5x_4 &= 10 \\ 4x_1 + 5x_2 + 8x_3 + 2x_4 &= 15 \\ 3x_1 + 6x_2 + 7x_3 + 4x_4 &= 30 \\ 4x_1 + x_2 + 6x_3 + 3x_4 &= 25 \end{aligned}$$

Applications

Construct a mathematical model for each of the following problems. (The answers in the back of the book include both the mathematical model and the interpretation of its solution.) Use matrix inverse methods to solve the model and then interpret the solution.

Business & Economics

43. *Resource allocation.* A concert hall has 10,000 seats and two categories of ticket prices, \$4 and \$8. Assume all seats in each category can be sold.

	Concert		
	1	*2*	*3*
Tickets sold	10,000	10,000	10,000
Return required	\$56,000	\$60,000	\$68,000

(A) How many tickets of each category should be sold to bring in each of the returns indicated in the table?

(B) Is it possible to bring in a return of \$90,000? Of \$30,000? Explain.

(C) Describe all the possible returns.

44. *Parking receipts.* Parking fees at a municipal zoo are \$5.00 for local residents and \$7.50 for all others. At the end of each day, the total number of vehicles parked that day and the gross receipts for the day are recorded, but the number of vehicles in each category is not. The following table contains the relevant information for a recent 4 day period:

	Day			
	1	*2*	*3*	*4*
Vehicles parked	1,200	1,550	1,740	1,400
Gross receipts	\$7,125	\$9,825	\$11,100	\$8,650

(A) How many vehicles in each category used the zoo's parking facilities each day?

(B) If 1,200 vehicles are parked in one day, is it possible to take in gross receipts of \$5,000? Of \$10,000? Explain.

(C) Describe all possible gross receipts on a day when 1,200 vehicles are parked.

45. *Production scheduling.* A supplier for the automobile industry manufactures car and truck frames at two different plants. The production rates (in frames per hour) for each plant are given in the table:

Plant	Car frames	Truck frames
A	10	5
B	8	8

How many hours should each plant be scheduled to operate to exactly fill each of the orders in the following table?

	Orders		
	1	*2*	*3*
Car frames	3,000	2,800	2,600
Truck frames	1,600	2,000	2,200

46. *Production scheduling.* Labor and material costs for manufacturing two guitar models are given in the table:

Guitar model	Labor cost	Material cost
A	\$30	\$20
B	\$40	\$30

(A) If a total of $3,000 a week is allowed for labor and material, how many of each model should be produced each week to use exactly each of the allocations of the $3,000 indicated in the following table?

	Weekly Allocation		
	1	*2*	*3*
Labor	$1,800	$1,750	$1,720
Material	$1,200	$1,250	$1,280

(B) Is it possible to use an allocation of $1,600 for labor and $1,400 for material? Of $2,000 for labor and $1,000 for material? Explain.

47. *Incentive plan.* A small company provides an incentive plan for its top executives. Each executive receives as a bonus a percentage of the portion of the annual profit that remains after the bonuses for the other executives have been deducted (see the table). If the company has an annual profit of $2 million, find the bonus for each executive. Round each bonus to the nearest hundred dollars.

Officer	Bonus
President	3%
Executive vice-president	2.5%
Associate vice-president	2%
Assistant vice-president	1.5%

48. *Incentive plan.* Repeat Problem 47 if the company decides to include a 1% bonus for the sales manager in the incentive plan.

Life Sciences

49. *Diets.* A biologist has available two commercial food mixes containing the percentages of protein and fat given in the table:

Mix	Protein (%)	Fat (%)
A	20	2
B	10	6

(A) How many ounces of each mix should be used to prepare each of the diets listed in the following table?

	Diet		
	1	*2*	*3*
Protein	20 oz	10 oz	10 oz
Fat	6 oz	4 oz	6 oz

(B) Is it possible to prepare a diet consisting of 20 ounces of protein and 14 ounces of fat? Of 20 ounces of protein and 1 ounce of fat? Explain.

Social Sciences

50. *Education—resource allocation.* A state university system is planning to hire new faculty at the rank of lecturer or instructor for several of its 2 year community colleges. The number of sections taught and the annual salary (in thousands of dollars) for each rank are given in the table:

	Rank	
	Lecturer	*Instructor*
Sections taught	3	4
Annual salary (thousand $)	20	25

The number of sections that must be taught by the new faculty and the amount budgeted for salaries (in thousands of dollars) at each of the colleges are given in the following table. How many faculty of each rank should be hired at each college to exactly meet the demand for sections and completely exhaust the salary budget?

	Community College		
	1	*2*	*3*
Demand for sections	30	33	35
Salary budget (thousand $)	200	210	220

SECTION 4-7

Leontief Input–Output Analysis

- Two-Industry Model
- Three-Industry Model

A very important application of matrices and their inverses is found in the branch of applied mathematics called **input–output analysis.** Wassily Leontief, the primary force behind these new developments, was awarded the Nobel prize in economics in 1973 because of the significant impact his work had on economic planning for industrialized countries. Among other things, he conducted a comprehensive study of how 500 sectors of the American economy interacted with each other. Of course, large-scale computers played an important role in this analysis.

Our investigation will be more modest. In fact, we will start with an economy comprised of only two industries. From these humble beginnings, ideas and definitions will evolve that can be readily generalized for more realistic economies. Input–output analysis attempts to establish equilibrium conditions under which industries in an economy have just enough output to satisfy each other's demands in addition to final (outside) demands. Given the internal demands within the industries for each other's output, the problem is to determine output levels that will meet various levels of final (outside) demands.

Two-Industry Model

To make the problem concrete, let us start with a hypothetical economy with only two industries—electric company E and water company W. Output for both companies is measured in dollars. The electric company uses both electricity and water (input) in the production of electricity (output), and the water company uses both electricity and water (input) in the production of water (output). Suppose that the production of each dollar's worth of electricity requires \$0.30 worth of electricity and \$0.10 worth of water, and the production of each dollar's worth of water requires \$0.20 worth of electricity and \$0.40 worth of water. If the final demand from the outside sector of the economy (the demand from all other users of electricity and water) is

$$d_1 = \$12 \text{ million for electricity}$$
$$d_2 = \$8 \text{ million for water}$$

how much electricity and water should be produced to meet this final demand?

To begin, suppose the electric company produces \$12 million worth of electricity and the water company produces \$8 million worth of water (the final demand). Then the production processes of the companies would require

Electricity required to produce electricity		Electricity required to produce water	
$0.3(12)$	$+$	$0.2(8)$	$= \$5.2$ million of electricity

and

Water required to produce electricity		Water required to produce water		
$0.1(12)$	$+$	$0.4(8)$	$=$	\$4.4 million of water

leaving only \$6.8 million of electricity and \$3.6 million of water to satisfy the final demand of the outside sector. Thus, in order to meet the internal demands of both companies and to end up with enough electricity for the final outside demand, both companies must produce more than just the amount demanded by the outside sector. In fact, they must produce exactly enough to meet their own internal demands plus that demanded by the outside sector.

Explore–Discuss 1

Suppose the electric company doubles its production to \$24 million of electricity and the water company doubles its production to \$16 million of water. How much electricity and water is consumed in the production process? How much is left for the final demand of the outside sector? Try other values to see if you can determine by trial and error the production levels that will result in a final demand of \$12 million for electricity and \$8 million for water.

We now state the main problem of input–output analysis.

Basic Input–Output Problem

Given the internal demands for each industry's output, determine output levels for the various industries that will meet a given final (outside) level of demand as well as the internal demand.

If

$x_1 =$ Total output from electric company
$x_2 =$ Total output from water company

then reasoning as above, the internal demands are

$0.3x_1 + 0.2x_2$ Internal demand for electricity
$0.1x_1 + 0.4x_2$ Internal demand for water

Combining the internal demand with the final demand produces the following system of equations:

Total output — Internal demand — Final demand

$$\begin{aligned} x_1 &= 0.3x_1 + 0.2x_2 + d_1 \\ x_2 &= 0.1x_1 + 0.4x_2 + d_2 \end{aligned} \quad (1)$$

or, in matrix form,

$$\begin{bmatrix} x_1 \\ x_2 \end{bmatrix} = \begin{bmatrix} 0.3 & 0.2 \\ 0.1 & 0.4 \end{bmatrix} \begin{bmatrix} x_1 \\ x_2 \end{bmatrix} + \begin{bmatrix} d_1 \\ d_2 \end{bmatrix}$$

or

$$X = MX + D \qquad (2)$$

where

$$D = \begin{bmatrix} d_1 \\ d_2 \end{bmatrix} \qquad \text{Final demand matrix}$$

$$X = \begin{bmatrix} x_1 \\ x_2 \end{bmatrix} \qquad \text{Output matrix}$$

$$M = \begin{matrix} \\ E \\ W \end{matrix} \begin{matrix} \begin{matrix} E & \quad W \end{matrix} \\ \begin{bmatrix} 0.3 & 0.2 \\ 0.1 & 0.4 \end{bmatrix} \end{matrix} \qquad \text{Technology matrix}$$

The **technology matrix** is the heart of input–output analysis. The elements in the technology matrix are determined as follows (read from left to right and then up):

$$\begin{matrix} & & \text{Output} & \\ & & E & W \\ & & \uparrow & \\ \text{Input} & E \longrightarrow & \left(\begin{matrix}\text{Input from } E\\ \text{to produce \$1}\\ \text{of electricity}\end{matrix}\right) & \left(\begin{matrix}\text{Input from } E\\ \text{to produce \$1}\\ \text{of water}\end{matrix}\right) \\ & W & \left(\begin{matrix}\text{Input from } W\\ \text{to produce \$1}\\ \text{of electricity}\end{matrix}\right) & \left(\begin{matrix}\text{Input from } W\\ \text{to produce \$1}\\ \text{of water}\end{matrix}\right) \end{matrix} = M$$

Now our problem is to solve equation (2) for X. We proceed as in Section 4-6:

$$\begin{aligned} X &= MX + D \\ X - MX &= D \\ IX - MX &= D & I = \begin{bmatrix} 1 & 0 \\ 0 & 1 \end{bmatrix} \\ (I - M)X &= D \\ X &= (I - M)^{-1}D & \text{Assuming } I - M \text{ has an inverse} \qquad (3) \end{aligned}$$

We find

$$I - M = \begin{bmatrix} 0.7 & -0.2 \\ -0.1 & 0.6 \end{bmatrix} \quad \text{and} \quad (I - M)^{-1} = \begin{bmatrix} 1.5 & 0.5 \\ 0.25 & 1.75 \end{bmatrix}$$

Then we have

$$\begin{aligned} \begin{bmatrix} x_1 \\ x_2 \end{bmatrix} &= \begin{bmatrix} 1.5 & 0.5 \\ 0.25 & 1.75 \end{bmatrix} \begin{bmatrix} d_1 \\ d_2 \end{bmatrix} \qquad (4) \\ &= \begin{bmatrix} 1.5 & 0.5 \\ 0.25 & 1.75 \end{bmatrix} \begin{bmatrix} 12 \\ 8 \end{bmatrix} = \begin{bmatrix} 22 \\ 17 \end{bmatrix} \end{aligned}$$

Therefore, the electric company must have an output of $22 million and the water company must have an output of $17 million so that each company can meet both internal and final demands.

CHECK We use equation (2) to check our work:

$$X = MX + D$$

$$\begin{bmatrix} 22 \\ 17 \end{bmatrix} \stackrel{?}{=} \begin{bmatrix} 0.3 & 0.2 \\ 0.1 & 0.4 \end{bmatrix} \begin{bmatrix} 22 \\ 17 \end{bmatrix} + \begin{bmatrix} 12 \\ 8 \end{bmatrix}$$

$$\begin{bmatrix} 22 \\ 17 \end{bmatrix} \stackrel{?}{=} \begin{bmatrix} 10 \\ 9 \end{bmatrix} + \begin{bmatrix} 12 \\ 8 \end{bmatrix}$$

$$\begin{bmatrix} 22 \\ 17 \end{bmatrix} \stackrel{\checkmark}{=} \begin{bmatrix} 22 \\ 17 \end{bmatrix}$$

Actually, (4) solves the original problem for arbitrary final demands d_1 and d_2. This is very useful, since (4) gives a quick solution not only for the final demands stated in the original problem, but also for various other projected final demands. If we had solved system (1) by Gauss–Jordan elimination, then we would have to start over for each new set of final demands.

Suppose in the original problem that the projected final demands 5 years from now are $d_1 = 24$ and $d_2 = 16$. To determine each company's output for this projection, we simply substitute these values into (4) and multiply:

$$\begin{bmatrix} x_1 \\ x_2 \end{bmatrix} = \begin{bmatrix} 1.5 & 0.5 \\ 0.25 & 1.75 \end{bmatrix} \begin{bmatrix} 24 \\ 16 \end{bmatrix} = \begin{bmatrix} 44 \\ 34 \end{bmatrix}$$

We summarize these results for convenient reference.

Solution to a Two-Industry Input–Output Problem

Given two industries, C_1 and C_2, with

Technology matrix (C_1, C_2); Output matrix; Final demand matrix

$$M = \begin{matrix} C_1 \\ C_2 \end{matrix} \begin{bmatrix} a_{11} & a_{12} \\ a_{21} & a_{22} \end{bmatrix} \qquad X = \begin{bmatrix} x_1 \\ x_2 \end{bmatrix} \qquad D = \begin{bmatrix} d_1 \\ d_2 \end{bmatrix}$$

where a_{ij} is the input required from C_i to produce a dollar's worth of output for C_j, the solution to the input–output matrix equation

Total output, Internal demand, Final demand

$$X = MX + D \tag{2}$$

is

$$X = (I - M)^{-1}D \tag{3}$$

assuming $I - M$ has an inverse.

To solve an input–output problem on a graphing utility, simply store matrices M, D, and I in memory; then use equation (3) to find X and equation (2) to check your results. Figure 1 illustrates this process on a graphing calculator.

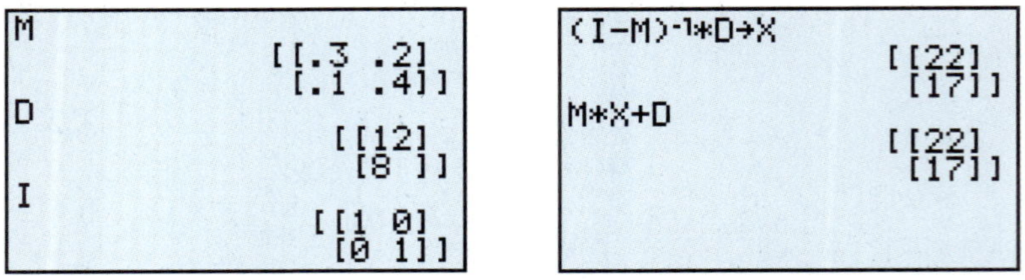

(A) Store M, D, and I in the graphing calculator's memory

(B) Compute X and check in equation (2)

FIGURE 1

Three-Industry Model

Equations (2) and (3) in the solution to a two-industry input–output problem are the same for a three-industry economy, a four-industry economy, or an economy with n industries (where n is any natural number). The steps we took going from (2) to (3) hold for arbitrary matrices as long as the matrices have the correct sizes and $(I - M)^{-1}$ exists.

Explore–Discuss 2

If equations (2) and (3) are valid for an economy with n industries, discuss the size of all the matrices in each equation.

The next example illustrates the application of equations (2) and (3) to a three-industry economy.

Example 1 **Input–Output Analysis** An economy is based on three sectors, agriculture (A), energy (E), and manufacturing (M). Production of a dollar's worth of agriculture requires an input of \$0.20 from the agriculture sector and \$0.40 from the energy sector. Production of a dollar's worth of energy requires an input of \$0.20 from the energy sector and \$0.40 from the manufacturing sector. Production of a dollar's worth of manufacturing requires an input of \$0.10 from the agriculture sector, \$0.10 from the energy sector, and \$0.30 from the manufacturing sector. Find the output from each sector that is needed to satisfy a final demand of \$20 billion for agriculture, \$10 billion for energy, and \$30 billion for manufacturing.

Solution Since this is a three-industry problem, the technology matrix will be a 3×3 matrix, and the output and final demand matrices will be 3×1 column matrices. Using the information given in the problem, we can write

Technology matrix (columns A E M; rows A, E, M) — Final demand matrix — Output matrix

$$M = \begin{matrix} A \\ E \\ M \end{matrix}\begin{bmatrix} 0.2 & 0 & 0.1 \\ 0.4 & 0.2 & 0.1 \\ 0 & 0.4 & 0.3 \end{bmatrix} \qquad D = \begin{bmatrix} 20 \\ 10 \\ 30 \end{bmatrix} \qquad X = \begin{bmatrix} x_1 \\ x_2 \\ x_3 \end{bmatrix}$$

where M, X, and D satisfy the input–output equation $X = MX + D$. Since the solution to this equation is $X = (I - M)^{-1}D$, we must first find $I - M$ and then $(I - M)^{-1}$. Omitting the details of the calculations, we have

$$I - M = \begin{bmatrix} 0.8 & 0 & -0.1 \\ -0.4 & 0.8 & -0.1 \\ 0 & -0.4 & 0.7 \end{bmatrix}$$

and

$$(I - M)^{-1} = \begin{bmatrix} 1.3 & 0.1 & 0.2 \\ 0.7 & 1.4 & 0.3 \\ 0.4 & 0.8 & 1.6 \end{bmatrix}$$

Thus, the output matrix X is given by

$$\overset{X}{\begin{bmatrix} x_1 \\ x_2 \\ x_3 \end{bmatrix}} = \overset{(I - M)^{-1}}{\begin{bmatrix} 1.3 & 0.1 & 0.2 \\ 0.7 & 1.4 & 0.3 \\ 0.4 & 0.8 & 1.6 \end{bmatrix}} \overset{D}{\begin{bmatrix} 20 \\ 10 \\ 30 \end{bmatrix}} = \begin{bmatrix} 33 \\ 37 \\ 64 \end{bmatrix}$$

An output of $33 billion for agriculture, $37 billion for energy, and $64 billion for manufacturing will meet the given final demands. You should check this result in equation (2).

Figure 2 illustrates a spreadsheet solution for Example 1.

	A	B	C	D	E	F	G	H	I	J	K	L	M
1		Technology Matrix M								Final		Output	
2		A	E	M			I - M			Demand			
3	A	0.2	0	0.1		0.8	0	-0.1		20		33	
4	E	0.4	0.2	0.1		-0.4	0.8	-0.1		10		37	
5	M	0	0.4	0.3		0	-0.4	0.7		30		64	

FIGURE 2

The command MMULT(MINVERSE(F3:H5), J3:J5) produces the output in L3:L5

Matched Problem 1 An economy is based on three sectors, coal, oil, and transportation. Production of a dollar's worth of coal requires an input of $0.20 from the coal sector and $0.40 from the transportation sector. Production of a dollar's worth of oil requires an input of $0.10 from the oil sector and $0.20 from the transportation sector. Production of a dollar's worth of transportation requires an input of $0.40 from the coal sector, $0.20 from the oil sector, and $0.20 from the transportation sector.

(A) Find the technology matrix M.

(B) Find $(I - M)^{-1}$.

(C) Find the output from each sector that is needed to satisfy a final demand of $30 billion for coal, $10 billion for oil, and $20 billion for transportation.

Answers to Matched Problem

1. (A) $\begin{bmatrix} 0.2 & 0 & 0.4 \\ 0 & 0.1 & 0.2 \\ 0.4 & 0.2 & 0.2 \end{bmatrix}$ (B) $\begin{bmatrix} 1.7 & 0.2 & 0.9 \\ 0.2 & 1.2 & 0.4 \\ 0.9 & 0.4 & 1.8 \end{bmatrix}$

(C) \$71 billion for coal, \$26 billion for oil, and \$67 billion for transportation

EXERCISE 4-7

A *Problems 1–6 pertain to the following input–output model: Assume an economy is based on two industrial sectors, agriculture (A) and energy (E). The technology matrix M and final demand matrices (in billions of dollars) are*

$$\begin{matrix} & A & E \\ A & 0.4 & 0.2 \\ E & 0.2 & 0.1 \end{matrix} = M$$

$$D_1 = \begin{bmatrix} 6 \\ 4 \end{bmatrix} \quad D_2 = \begin{bmatrix} 8 \\ 5 \end{bmatrix} \quad D_3 = \begin{bmatrix} 12 \\ 9 \end{bmatrix}$$

1. How much input from A and E are required to produce a dollar's worth of output for A?
2. How much input from A and E are required to produce a dollar's worth of output for E?
3. Find $I - M$ and $(I - M)^{-1}$.
4. Find the output for each sector that is needed to satisfy the final demand D_1.
5. Repeat Problem 4 for D_2.
6. Repeat Problem 4 for D_3.

B *Problems 7–12 pertain to the following input–output model: Assume an economy is based on three industrial sectors, agriculture (A), building (B), and energy (E). The technology matrix M and final demand matrices (in billions of dollars) are*

$$\begin{matrix} & A & B & E \\ A & 0.3 & 0.2 & 0.2 \\ B & 0.1 & 0.1 & 0.1 \\ E & 0.2 & 0.1 & 0.1 \end{matrix} = M$$

$$D_1 = \begin{bmatrix} 5 \\ 10 \\ 15 \end{bmatrix} \quad D_2 = \begin{bmatrix} 20 \\ 15 \\ 10 \end{bmatrix}$$

7. How much input from A, B, and E are required to produce a dollar's worth of output for B?
8. How much of each of B's output dollars is required as input for each of the three sectors?
9. Show that

$$I - M = \begin{bmatrix} 0.7 & -0.2 & -0.2 \\ -0.1 & 0.9 & -0.1 \\ -0.2 & -0.1 & 0.9 \end{bmatrix}$$

10. Given

$$(I - M)^{-1} = \begin{bmatrix} 1.6 & 0.4 & 0.4 \\ 0.22 & 1.18 & 0.18 \\ 0.38 & 0.22 & 1.22 \end{bmatrix}$$

show that $(I - M)^{-1}(I - M) = I$.

11. Use $(I - M)^{-1}$ in Problem 10 to find the output for each sector that is needed to satisfy the final demand D_1.
12. Repeat Problem 11 for D_2.

In Problems 13–16, find $(I - M)^{-1}$ and X.

13. $M = \begin{bmatrix} 0.2 & 0.2 \\ 0.3 & 0.3 \end{bmatrix}$; $D = \begin{bmatrix} 10 \\ 25 \end{bmatrix}$
14. $M = \begin{bmatrix} 0.4 & 0.1 \\ 0.2 & 0.3 \end{bmatrix}$; $D = \begin{bmatrix} 15 \\ 20 \end{bmatrix}$

C

15. $M = \begin{bmatrix} 0.3 & 0.1 & 0.3 \\ 0.2 & 0.1 & 0.2 \\ 0.1 & 0.1 & 0.1 \end{bmatrix}$; $D = \begin{bmatrix} 20 \\ 5 \\ 10 \end{bmatrix}$
16. $M = \begin{bmatrix} 0.3 & 0.2 & 0.3 \\ 0.1 & 0.1 & 0.1 \\ 0.1 & 0.2 & 0.1 \end{bmatrix}$; $D = \begin{bmatrix} 10 \\ 25 \\ 15 \end{bmatrix}$
17. The technology matrix for an economy based on agriculture (A) and manufacturing (M) is

$$M = \begin{matrix} & A & M \\ A & 0.3 & 0.25 \\ M & 0.1 & 0.25 \end{matrix}$$

(A) Find the output for each sector that is needed to satisfy a final demand of \$40 million for agriculture and \$40 million for manufacturing.

(B) Discuss the effect on the final demand if the agriculture output in part (A) is increased by \$20 million and manufacturing output remains unchanged.

18. The technology matrix for an economy based on energy (E) and transportation (T) is

$$M = \begin{matrix} & \begin{matrix} E & T \end{matrix} \\ \begin{matrix} E \\ T \end{matrix} & \begin{bmatrix} 0.25 & 0.25 \\ 0.4 & 0.2 \end{bmatrix} \end{matrix}$$

(A) Find the output for each sector that is needed to satisfy a final demand of \$50 million for energy and \$50 million for transportation.

(B) Discuss the effect on the final demand if the transportation output in part (A) is increased by \$40 million and the energy output remains unchanged.

19. The technology matrix for an economy based on energy (E) and mining (M) is

$$M = \begin{matrix} & \begin{matrix} E & M \end{matrix} \\ \begin{matrix} E \\ M \end{matrix} & \begin{bmatrix} 0.2 & 0.3 \\ 0.4 & 0.3 \end{bmatrix} \end{matrix}$$

The management of these two sectors would like to set the total output level so that the final demand is always 40% of the total output. Discuss methods that could be used to accomplish this objective.

20. The technology matrix for an economy based on automobiles (A) and construction (C) is

$$M = \begin{matrix} & \begin{matrix} A & C \end{matrix} \\ \begin{matrix} A \\ C \end{matrix} & \begin{bmatrix} 0.1 & 0.4 \\ 0.1 & 0.1 \end{bmatrix} \end{matrix}$$

The management of these two sectors would like to set the total output level so that the final demand is always 70% of the total output. Discuss methods that could be used to accomplish this objective.

21. All the technology matrices in the text have elements between 0 and 1. Why is this the case? Would you ever expect to find an element in a technology matrix that is negative? That is equal to 0? That is equal to 1? That is greater than 1?

22. The sum of the elements in a column of any of the technology matrices in the text is less than 1. Why is this the case? Would you ever expect to find a column with a sum equal to 1? Greater than 1? How would you describe an economic system where the sum of the elements in every column of the technology matrix is 1?

Applications

Business & Economics

23. An economy is based on two industrial sectors, coal and steel. Production of a dollar's worth of coal requires an input of \$0.10 from the coal sector and \$0.20 from the steel sector. Production of a dollar's worth of steel requires an input of \$0.20 from the coal sector and \$0.40 from the steel sector. Find the output for each sector that is needed to satisfy a final demand of \$20 billion for coal and \$10 billion for steel.

24. An economy is based on two sectors, transportation and manufacturing. Production of a dollar's worth of transportation requires \$0.10 of input from each sector and production of a dollar's worth of manufacturing requires an input of \$0.40 from each sector. Find the output for each sector that is needed to satisfy a final demand of \$5 billion for transportation and \$20 billion for manufacturing.

25. The economy of a small island nation is based on two sectors, agriculture and tourism. Production of a dollar's worth of agriculture requires an input of \$0.20 from agriculture and \$0.15 from tourism. Production of a dollar's worth of tourism requires an input of \$0.40 from agriculture and \$0.30 from tourism. Find the output from each sector that is needed to satisfy a final demand of \$60 million for agriculture and \$80 million for tourism.

26. The economy of a country is based on two sectors, agriculture and oil. Production of a dollar's worth of agriculture requires an input of \$0.40 from agriculture and \$0.35 from oil. Production of a dollar's worth of oil requires an input of \$0.20 from agriculture and \$0.05 from oil. Find the output from each sector that is needed to satisfy a final demand of \$40 million for agriculture and \$250 million for oil.

27. An economy is based on three sectors, agriculture, manufacturing, and energy. Production of a dollar's worth of agriculture requires inputs of \$0.20 from agriculture, \$0.20 from manufacturing, and \$0.20 from energy. Production of a dollar's worth of manufacturing requires inputs of \$0.40 from agriculture, \$0.10 from manufacturing, and \$0.10 from energy. Production of a dollar's worth of energy requires inputs of \$0.30 from agriculture, \$0.10 from manufacturing, and \$0.10 from energy. Find the output for each sector that is needed to satisfy a final demand of \$10 billion for agriculture, \$15 billion for manufacturing, and \$20 billion for energy.

28. A large energy company produces electricity, natural gas, and oil. The production of a dollar's worth of electricity requires inputs of \$0.30 from electricity, \$0.10 from natural gas, and \$0.20 from oil. Production of a dollar's worth of natural gas requires inputs of \$0.30 from electricity, \$0.10 from natural gas, and \$0.20 from oil. Production of a dollar's worth of oil requires inputs of \$0.10 from each sector. Find the output for each sector that is needed to satisfy a final demand of \$25 billion for electricity, \$15 billion for natural gas, and \$20 billion for oil.

Use a graphing utility to solve Problems 29 and 30.

29. An economy is based on four sectors, agriculture (A), energy (E), labor (L), and manufacturing (M). The table gives the input requirements for a dollar's worth of output for each sector, along with the projected final demand (in billions of dollars) for a 3 year period. Find the output for each sector that is needed to satisfy each of these final demands. Round answers to the nearest billion dollars.

		Output				Final Demand		
		A	*E*	*L*	*M*	*1*	*2*	*3*
Input	*A*	0.05	0.17	0.23	0.09	23	32	55
	E	0.07	0.12	0.15	0.19	41	48	62
	L	0.25	0.08	0.03	0.32	18	21	25
	M	0.11	0.19	0.28	0.16	31	33	35

30. Repeat Problem 29 with the following table:

		Output				Final Demand		
		A	*E*	*L*	*M*	*1*	*2*	*3*
Input	*A*	0.07	0.09	0.27	0.12	18	22	37
	E	0.14	0.07	0.21	0.24	26	31	42
	L	0.17	0.06	0.02	0.21	12	19	28
	M	0.15	0.13	0.31	0.19	41	45	49

Important Terms and Symbols

4-1 *Review: Systems of Linear Equations in Two Variables.* Linear system; solution of a system; solution set; graphing method; consistent; inconsistent; independent; unique solution; dependent; substitution method; elimination by addition; equivalent systems; parameter; particular solution; equilibrium price; equilibrium quantity; equilibrium point

4-2 *Systems of Linear Equations and Augmented Matrices.* Matrix; element; size; $m \times n$ matrix; dimensions; square matrix of order n; column matrix; row matrix; position of an element; double subscript notation, a_{ij}; principal diagonal; coefficient matrix; constant matrix; augmented matrix; equivalent systems; row-equivalent matrices; row operations; Gauss–Jordan elimination

$$R_i \leftrightarrow R_j; \quad kR_i \rightarrow R_i; \quad kR_j + R_i \rightarrow R_i$$

4-3 *Gauss–Jordan Elimination.* Reduced form; reduced system; Gauss–Jordan elimination; submatrix

4-4 *Matrices: Basic Operations.* Equal matrices; sum of two matrices; zero matrix; negative of a matrix; subtraction of matrices; product of a number and a matrix; product of a row matrix and a column matrix; product of two matrices; diagonal matrix; upper triangular matrix; markup; incidence matrix

4-5 *Inverse of a Square Matrix.* Identity element for multiplication; multiplicative inverse; singular matrix; encoding matrix; decoding matrix

4-6 *Matrix Equations and Systems of Linear Equations.* Matrix equation; basic properties of matrices

4-7 *Leontief Input–Output Analysis.* Input–output analysis; technology matrix; final demand matrix; output matrix

REVIEW EXERCISE

Work through all the problems in this chapter review and check your answers in the back of the book. Answers to all review problems are there along with section numbers in italics to indicate where each type of problem is discussed. Where weaknesses show up, review appropriate sections in the text.

A

1. Solve the following system by graphing:

$$\begin{aligned} 2x - y &= 4 \\ x - 2y &= -4 \end{aligned}$$

2. Solve the system in Problem 1 by substitution.

3. If a matrix is in reduced form, say so. If not, explain why and state the row operation(s) necessary to transform the matrix into reduced form.

(A) $\left[\begin{array}{cc|c} 0 & 1 & 2 \\ 1 & 0 & 3 \end{array}\right]$ (B) $\left[\begin{array}{cc|c} 1 & 0 & 2 \\ 0 & 3 & 3 \end{array}\right]$

(C) $\left[\begin{array}{ccc|c} 1 & 0 & 1 & 2 \\ 0 & 1 & 1 & 3 \end{array}\right]$ (D) $\left[\begin{array}{ccc|c} 1 & 1 & 0 & 2 \\ 0 & 1 & 1 & 3 \end{array}\right]$

4. Given matrices A and B:

$$A = \begin{bmatrix} 5 & 3 & -1 & 0 & 2 \\ -4 & 8 & 1 & 3 & 0 \end{bmatrix} \qquad B = \begin{bmatrix} -3 & 2 \\ 0 & 4 \\ -1 & 7 \end{bmatrix}$$

(A) What is the size of A? Of B?
(B) Find a_{24}, a_{15}, b_{31}, and b_{22}.
(C) Is AB defined? Is BA defined?

5. Find x_1 and x_2:

$$\begin{bmatrix} 1 & -2 \\ 1 & -3 \end{bmatrix}\begin{bmatrix} x_1 \\ x_2 \end{bmatrix} = \begin{bmatrix} 4 \\ 2 \end{bmatrix}$$

In Problems 6–14, perform the operations that are defined, given the following matrices:

$$A = \begin{bmatrix} 1 & 2 \\ 3 & 1 \end{bmatrix} \qquad B = \begin{bmatrix} 2 & 1 \\ 1 & 1 \end{bmatrix}$$

$$C = \begin{bmatrix} 2 & 3 \end{bmatrix} \qquad D = \begin{bmatrix} 1 \\ 2 \end{bmatrix}$$

6. $A + B$ **7.** $B + D$ **8.** $A - 2B$
9. AB **10.** AC **11.** AD
12. DC **13.** CD **14.** $C + D$

15. Find the inverse of the matrix A given below by appropriate row operations on $[A|I]$. Show that $A^{-1}A = I$.

$$A = \begin{bmatrix} 4 & 3 \\ 3 & 2 \end{bmatrix}$$

16. Solve the following system using elimination by addition:

$$\begin{aligned} 4x_1 + 3x_2 &= 3 \\ 3x_1 + 2x_2 &= 5 \end{aligned}$$

17. Solve the system in Problem 16 by performing appropriate row operations on the augmented matrix of the system.

18. Solve the system in Problem 16 by writing the system as a matrix equation and using the inverse of the coefficient matrix (see Problem 15). Also, solve the system if the constants 3 and 5 are replaced by 7 and 10, respectively. By 4 and 2, respectively.

B

In Problems 19–24, perform the operations that are defined, given the following matrices:

$$A = \begin{bmatrix} 2 & -2 \\ 1 & 0 \\ 3 & 2 \end{bmatrix} \qquad B = \begin{bmatrix} -1 \\ 2 \\ 3 \end{bmatrix} \qquad C = \begin{bmatrix} 2 & 1 & 3 \end{bmatrix}$$

$$D = \begin{bmatrix} 3 & -2 & 1 \\ -1 & 1 & 2 \end{bmatrix} \qquad E = \begin{bmatrix} 3 & -4 \\ -1 & 0 \end{bmatrix}$$

19. $A + D$ **20.** $E + DA$ **21.** $DA - 3E$
22. BC **23.** CB **24.** $AD - BC$

25. Find the inverse of the matrix A given below by appropriate row operations on $[A|I]$. Show that $A^{-1}A = I$.

$$A = \begin{bmatrix} 1 & 2 & 3 \\ 2 & 3 & 4 \\ 1 & 2 & 1 \end{bmatrix}$$

26. Solve by Gauss–Jordan elimination:

(A)
$$\begin{aligned} x_1 + 2x_2 + 3x_3 &= 1 \\ 2x_1 + 3x_2 + 4x_3 &= 3 \\ x_1 + 2x_2 + x_3 &= 3 \end{aligned}$$

(B)
$$\begin{aligned} x_1 + 2x_2 - x_3 &= 2 \\ 2x_1 + 3x_2 + x_3 &= -3 \\ 3x_1 + 5x_2 &= -1 \end{aligned}$$

27. Solve the system in Problem 26A by writing the system as a matrix equation and using the inverse of the coefficient matrix (see Problem 25). Also, solve the system if the constants 1, 3, and 3 are replaced by 0, 0, and −2, respectively. By −3, −4, and 1, respectively.

28. Discuss the relationship between the number of solutions of the following system and the constant k.

$$\begin{aligned} 2x_1 - 6x_2 &= 4 \\ -x_1 + kx_2 &= -2 \end{aligned}$$

29. Given the technology matrix M and the final demand matrix D (in billions of dollars), find $(I - M)^{-1}$ and the output matrix X:

$$M = \begin{bmatrix} 0.2 & 0.15 \\ 0.4 & 0.3 \end{bmatrix} \qquad D = \begin{bmatrix} 30 \\ 20 \end{bmatrix}$$

30. C Use zoom and trace or other graphical approximation techniques on a graphing utility to find the solution of the following system to two decimal places:

$$\begin{aligned} x - 5y &= -5 \\ 2x + 3y &= 12 \end{aligned}$$

C

31. Find the inverse of the matrix A given below. Show that $A^{-1}A = I$.

$$A = \begin{bmatrix} 4 & 5 & 6 \\ 4 & 5 & -4 \\ 1 & 1 & 1 \end{bmatrix}$$

32. Solve the system

$$\begin{aligned} 0.04x_1 + 0.05x_2 + 0.06x_3 &= 360 \\ 0.04x_1 + 0.05x_2 - 0.04x_3 &= 120 \\ x_1 + x_2 + x_3 &= 7{,}000 \end{aligned}$$

by writing it as a matrix equation and using the inverse of the coefficient matrix. (Before starting, multiply the first two equations by 100 to eliminate decimals. Also, see Problem 31.)

33. Solve Problem 32 by Gauss–Jordan elimination.

34. Given the technology matrix M and the final demand matrix D (in billions of dollars), find $(I - M)^{-1}$ and the output matrix X:

$$M = \begin{bmatrix} 0.2 & 0 & 0.4 \\ 0.1 & 0.3 & 0.1 \\ 0 & 0.4 & 0.2 \end{bmatrix} \qquad D = \begin{bmatrix} 40 \\ 20 \\ 30 \end{bmatrix}$$

35. Discuss the number of solutions for a system of n equations in n variables if the coefficient matrix:
(A) Has an inverse (B) Does not have an inverse

36. Discuss the number of solutions for the system corresponding to the reduced form shown below if:
(A) $m \neq 0$ (B) $m = 0$ and $n \neq 0$
(C) $m = 0$ and $n = 0$

$$\left[\begin{array}{ccc|c} 1 & 0 & -2 & 5 \\ 0 & 1 & 3 & 3 \\ 0 & 0 & m & n \end{array}\right]$$

37. One solution to the input–output equation $X = MX + D$ is given by $X = (I - M)^{-1}D$. Discuss the validity of each step in the following solutions of this equation. (Assume all necessary inverses exist.) Are both solutions correct?

(A)
$$\begin{aligned} X &= MX + D \\ X - MX &= D \\ X(I - M) &= D \\ X &= D(I - M)^{-1} \end{aligned}$$

(B)
$$\begin{aligned} X &= MX + D \\ -D &= MX - X \\ -D &= (M - I)X \\ X &= (M - I)^{-1}(-D) \end{aligned}$$

Applications

Business & Economics

38. *Break-even analysis.* A cookware manufacturer is preparing to market a new pasta machine. The company's fixed costs for research, development, tooling, etc., are \$243,000 and the variable costs are \$22.45 per machine. The company sells the pasta machine for \$59.95.
(A) Find the cost and revenue equations.
(B) Find the break-even point.
(C) Graph both equations in the same coordinate system and show the break-even point. Use the graph to determine the production levels that will result in a profit and in a loss.

39. *Resource allocation.* A mining company has two mines with ore composition as given in the table. How many tons of each ore should be used to obtain exactly 4.5 tons of nickel and 10 tons of copper? Solve using Gauss–Jordan elimination.

Ore	Nickel (%)	Copper (%)
A	1	2
B	2	5

40. *Resource allocation.*
(A) Set up Problem 39 as a matrix equation and solve using the inverse of the coefficient matrix.
(B) Solve Problem 39 as in part (A) if 2.3 tons of nickel and 5 tons of copper are needed.

41. *Purchasing.* A soft-drink distributor has budgeted \$300,000 for the purchase of 12 new delivery trucks. If a model *A* truck costs \$18,000, a model *B* truck costs \$22,000, and a model *C* truck costs \$30,000, how many trucks of each model should the distributor purchase to use exactly all the budgeted funds?

42. *Material costs.* A manufacturer wishes to make two different bronze alloys in a metal foundry. The quantities of copper, tin, and zinc needed are indicated in matrix *M*. The costs for these materials (in dollars per pound) from two suppliers are summarized in matrix *N*. The company must choose one supplier or the other.

$$M = \begin{bmatrix} 4{,}800 \text{ lb} & 600 \text{ lb} & 300 \text{ lb} \\ 6{,}000 \text{ lb} & 1{,}400 \text{ lb} & 700 \text{ lb} \end{bmatrix} \begin{matrix} \text{Alloy 1} \\ \text{Alloy 2} \end{matrix}$$
(columns: Copper, Tin, Zinc)

$$N = \begin{bmatrix} \$0.75 & \$0.70 \\ \$6.50 & \$6.70 \\ \$0.40 & \$0.50 \end{bmatrix} \begin{matrix} \text{Copper} \\ \text{Tin} \\ \text{Zinc} \end{matrix}$$
(columns: Supplier A, Supplier B)

(A) Discuss possible interpretations of the elements in the matrix products *MN* and *NM*.
(B) If either of the products *MN* or *NM* has a meaningful interpretation, find the product and label its rows and columns.
(C) Discuss methods of matrix multiplication that can be used to determine the supplier that will provide the necessary materials at the lowest cost.

43. *Labor costs.* A company with manufacturing plants in California and Texas has labor-hour and wage requirements for the manufacture of two inexpensive calculators as given in matrices *M* and *N* below:

Labor-hours per calculator

$$M = \begin{bmatrix} 0.15 \text{ hr} & 0.10 \text{ hr} & 0.05 \text{ hr} \\ 0.25 \text{ hr} & 0.20 \text{ hr} & 0.05 \text{ hr} \end{bmatrix} \begin{matrix} \text{Model A} \\ \text{Model B} \end{matrix}$$
(columns: Fabricating department, Assembly department, Packaging department)

Hourly wages

$$N = \begin{bmatrix} \$15 & \$12 \\ \$12 & \$10 \\ \$\ 4 & \$\ 4 \end{bmatrix} \begin{matrix} \text{Fabricating department} \\ \text{Assembly department} \\ \text{Packaging department} \end{matrix}$$
(columns: California plant, Texas plant)

(A) Find the labor cost for producing one model *B* calculator at the California plant.
(B) Discuss possible interpretations of the elements in the matrix products *MN* and *NM*.
(C) If either of the products *MN* or *NM* has a meaningful interpretation, find the product and label its rows and columns.

44. *Investment analysis.* A person has \$5,000 to invest, part at 5% and the rest at 10%. How much should be invested at each rate to yield \$400 per year? Solve using augmented matrix methods.

45. *Investment analysis.* Solve Problem 44 by using a matrix equation and the inverse of the coefficient matrix.

46. *Investment analysis.* In Problem 44, is it possible to have an annual yield of \$200? Of \$500? Describe all possible annual yields.

47. *Resource allocation.* An outdoor amphitheater has 25,000 seats. Ticket prices are \$8, \$12, and \$20, and the number of tickets priced at \$8 must equal the number priced at \$20. How many tickets of each type should be sold (assuming all seats can be sold) to bring in each of the returns indicated in the table? Solve using the inverse of the coefficient matrix.

	Concert 1	Concert 2	Concert 3
Tickets sold	25,000	25,000	25,000
Return required	\$320,000	\$330,000	\$340,000

48. *Resource allocation.* Discuss the effect on the solutions to Problem 47 if it is no longer required to have an equal number of \$8 tickets and \$20 tickets.

49. *Input–output analysis.* An economy is based on two industrial sectors, agriculture and fabrication. Production of a dollar's worth of agriculture requires an input of \$0.30 from the agriculture sector and \$0.20 from the fabricating sector. Production of a dollar's worth of fabrication requires \$0.10 from the agriculture sector and \$0.40 from the fabrication sector.
(A) Find the output for each sector that is needed to satisfy a final demand of \$50 billion for agriculture and \$20 billion for fabrication.
(B) Find the output for each sector that is needed to satisfy a final demand of \$80 billion for agriculture and \$60 billion for fabrication.

Social Sciences

50. *Cryptography.* The following message was encoded with the matrix *B* shown below. Decode the message:

25 8 26 24 25 33 21 41 48 41 30 50 21 32 41 52 52 79

$$B = \begin{bmatrix} 1 & 1 & 0 \\ 1 & 0 & 1 \\ 1 & 1 & 1 \end{bmatrix}$$

51. *Traffic flow.* The rush-hour traffic flow (in vehicles per hour) for a network of four one-way streets is shown in the figure.

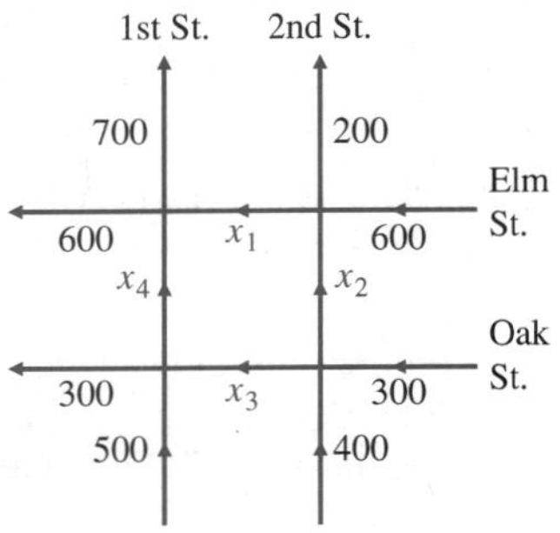

Figure for 51

(A) Write the system of equations determined by the flow of traffic through the four intersections.

(B) Find the solution of the system in part (A).

(C) What is the maximum number of vehicles per hour that can travel from Oak Street to Elm Street on 1st Street? What is the minimum number?

(D) If traffic lights are adjusted so that 500 vehicles per hour travel from Oak Street to Elm Street on 1st Street, determine the flow around the rest of the network.

52. *Dominance relation.* In a tournament between four football teams, each team plays one game with every other team with the following results: Team *A* defeats team *B*, team *B* defeats teams *C* and *D*, team *C* defeats teams *A* and *D*, and team *D* defeats team *A*.

(A) Represent the results of the tournament as an incidence matrix *M*.

(B) Compute $M + M^2$ and use this matrix to rank the four teams from high to low. Explain the reasoning behind your ranking.

Group Activity 1 *Using Matrices to Find Cost, Revenue, and Profit*

A toy distributor purchases model train components from various suppliers and packages these components in three different ready-to-run train sets, the Limited, the Empire, and the Comet. The components used in each set are listed in Table 1. For convenience, the total labor time required to prepare a set for shipping is included as a component.

Table 1

PRODUCT COMPONENTS

	TRAIN SET		
COMPONENT	*Limited*	*Empire*	*Comet*
Locomotive	1	1	2
Car	5	6	8
Track piece	20	24	32
Track switch	1	2	4
Power pack	1	1	1
Labor (minutes)	15	18	24

The current costs of the components are given in Table 2, and the distributor's selling prices for the sets are given in Table 3.

Table 2

COMPONENT COSTS

COMPONENT	COST PER UNIT ($)
Locomotive	12.52
Car	1.43
Track piece	0.25
Track switch	2.29
Power pack	12.54
Labor (per minute)	0.15

Table 3

SELLING PRICES

SET	PRICE ($)
Limited	54.60
Empire	62.28
Comet	81.15

Table 4

CUSTOMER ORDER

SET	QUANTITY
Limited	48
Empire	24
Comet	12

The distributor has just received the order shown in Table 4 from a retail toy store.

The distributor wants to store the information in each table in a matrix and use matrix operations to find the following information:

1. The inventory (parts and labor) required to fill the order
2. The cost (parts and labor) of filling the order
3. The revenue (sales) received from the customer
4. The profit realized on the order

(A) Use a single letter to designate the matrix representing each table and write matrix expressions in terms of these letters that will provide the required information. Discuss the size of the matrix you must use to represent each table so that all the pertinent matrix operations are defined.

(B) Evaluate the matrix expressions in part (A).

Shortly after filling the order in Table 4, a supplier informs the distributor that the cars and locomotives used in these train sets are no longer available. The distributor currently has 30 locomotives and 134 cars in stock.

(C) How many train sets of each type can the distributor produce using all the available locomotives and cars? Assume that the distributor has unlimited quantities of the other components used in these sets.

(D) How much profit will the distributor make if all these sets are sold? If there is more than one way to use all the available locomotives and cars, which one will produce the largest profit?

Group Activity 2 *Direct and Indirect Operating Costs*

An electronics firm has three production departments that produce three different finished products: copiers, printers, and fax machines. The company also has four other departments, payroll, advertising, research, and maintenance, that provide services for the production departments. Some of the service departments also provide services for the other service departments and themselves. The monthly direct cost of operating each department is given in Table 1, along with the percentage of service time each department receives from each service department (including themselves). The total cost of operating each department consists of its direct costs plus the cost of services received, which are determined by the percentages in Table 1.

Table 1

		PERCENTAGE OF SERVICE TIME RECEIVED FROM:			
DEPARTMENT	DIRECT COSTS (MONTHLY, $)	*Payroll*	*Advertising*	*Research*	*Maintenance*
Payroll	100,000	15	0	0	10
Advertising	150,000	15	0	0	10
Research	120,000	15	0	25	20
Maintenance	60,000	10	0	0	0
Copiers	270,000	15	20	40	20
Printers	140,000	15	20	20	20
Fax machines	190,000	15	60	15	20

Let x_1, x_2, x_3, and x_4 denote the total cost of operating the payroll, advertising, research, and maintenance departments, respectively, and y_1, y_2, and y_3 denote the total cost of operating the copier, printer, and fax machine departments, respectively.

(A) The total costs for the payroll department, x_1, must satisfy

$$x_1 = 100{,}000 + 0.15x_1 + 0.1x_4$$

Write similar equations for the other three service departments.

(B) Express the system in part (A) as a matrix equation of the form $X = D + AX$.

(C) The total costs for the copier department, y_1, must satisfy

$$y_1 = 270{,}000 + 0.15x_1 + 0.2x_2 + 0.4x_3 + 0.2x_4$$

Write similar equations for the other two production departments.

(D) Express the system in part (C) as a matrix equation of the form $Y = C + BX$.

(E) Use matrix inverse methods to solve the equation in part (B) and use this solution to solve the equation in part (D).

(F) Compare the sum of the direct costs of all seven departments as listed in Table 1 with the sum of the total costs of the three production departments found in part (E), and interpret.

CHAPTER 5

LINEAR INEQUALITIES AND LINEAR PROGRAMMING

INTRODUCTION

In this chapter we will discuss linear inequalities in two and more variables. In addition, we will introduce a relatively new and powerful mathematical tool called *linear programming,* which will be used to solve a variety of interesting practical problems. The row operations on matrices introduced in Chapter 4 will be particularly useful in Sections 5-4, 5-5, and 5-6.

SECTION 5-1 Systems of Linear Inequalities in Two Variables

- GRAPHING LINEAR INEQUALITIES IN TWO VARIABLES
- SOLVING SYSTEMS OF LINEAR INEQUALITIES GRAPHICALLY
- APPLICATIONS

Many applications of mathematics involve systems of inequalities rather than systems of equations. A graph is often the most convenient way to represent the solutions of a system of linear inequalities in two variables. In this section we discuss techniques for graphing both a single linear inequality in two variables and a system of linear inequalities in two variables.

Graphing Linear Inequalities in Two Variables

We know how to graph first-degree equations such as

$$y = 2x - 3 \qquad \text{and} \qquad 2x - 3y = 5$$

but how do we graph first-degree inequalities such as the following?

$$y \leq 2x - 3 \qquad \text{and} \qquad 2x - 3y > 5$$

We will find that graphing these inequalities is almost as easy as graphing the equations, but first we must discuss some important subsets of a plane in a rectangular coordinate system.

A line divides the plane into two halves called **half-planes.** A vertical line divides it into **left** and **right half-planes;** a nonvertical line divides it into **upper** and **lower half-planes** (Fig. 1).

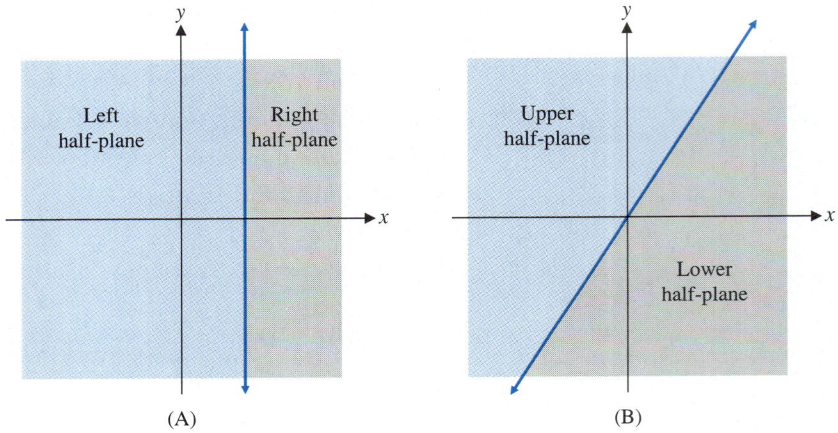

FIGURE 1

Explore–Discuss 1

Consider the following linear equation and related linear inequalities:

$$(1)\ 3x - 4y = 24 \qquad (2)\ 3x - 4y < 24 \qquad (3)\ 3x - 4y > 24$$

(A) Graph the line with equation (1).
(B) Find the point on this line with x coordinate 4 and draw a vertical line through this point. Discuss the relationship between the y coordinates of the points on this line and statements (1), (2), and (3).
(C) Repeat part (B) for $x = -4$. For $x = 12$.
(D) Based on your observations in parts (B) and (C), write a verbal description of all the points in the plane that satisfy equation (1), those that satisfy inequality (2), and those that satisfy inequality (3).

To investigate the half-planes determined by a linear equation such as $y - x = -2$, we rewrite the equation as $y = x - 2$. For any given value of x, there is exactly one value for y such that (x, y) lies on the line. For example, for $x = 4$, we have $y = 4 - 2 = 2$. For the same x and smaller values of y, the point (x, y) will lie below the line, since $y < x - 2$. Thus, the lower half-plane corresponds to the solution of the inequality $y < x - 2$. Similarly, the upper half-plane corresponds to $y > x - 2$, as shown in Figure 2.

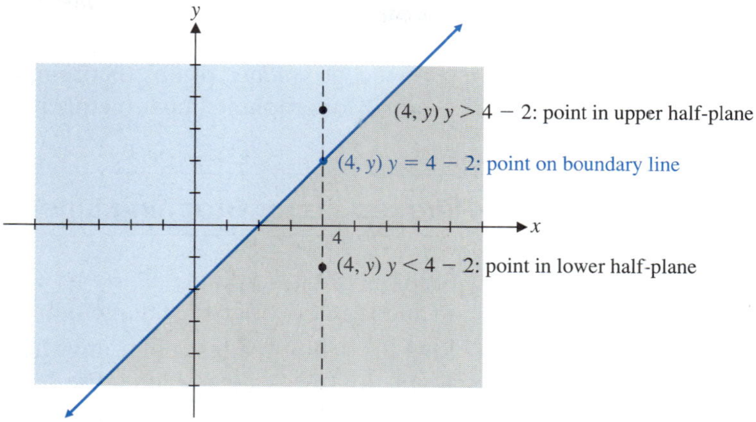

FIGURE 2

The four inequalities formed from $y = x - 2$ by replacing the $=$ sign by $>$, $\geqslant$, $<$, and $\leqslant$, respectively, are

$$y > x - 2 \qquad y \geqslant x - 2 \qquad y < x - 2 \qquad y \leqslant x - 2$$

The graph of each is a half-plane, excluding the boundary line for $<$ and $>$, and including the boundary line for $\leqslant$ and $\geqslant$. In Figure 3, the half-planes are indicated with small arrows on the graph of $y = x - 2$ and then graphed as shaded regions. Excluded boundary lines are shown as dashed lines, and included boundary lines are shown as solid lines.

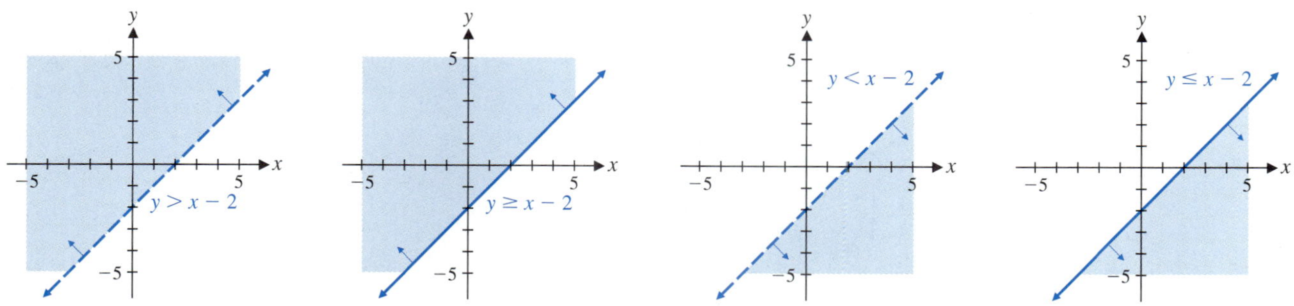

FIGURE 3

The preceding discussion suggests the following theorem, which is stated without proof:

Theorem 1 ■■ GRAPHS OF LINEAR INEQUALITIES

The graph of the linear inequality

$$Ax + By < C \qquad \text{or} \qquad Ax + By > C$$

with $B \neq 0$, is either the upper half-plane or the lower half-plane (but not both) determined by the line $Ax + By = C$.

(continued)

If $B = 0$ and $A \neq 0$, the graph of

$$Ax < C \qquad \text{or} \qquad Ax > C$$

is either the left half-plane or the right half-plane (but not both) determined by the line $Ax = C$. ■■

As a consequence of this theorem, we state a simple and fast mechanical procedure for graphing linear inequalities.

Procedure for Graphing Linear Inequalities

Step 1. First graph $Ax + By = C$ as a dashed line if equality is not included in the original statement or as a solid line if equality is included.

Step 2. Choose a test point anywhere in the plane not on the line [the origin (0, 0) usually requires the least computation] and substitute the coordinates into the inequality.

Step 3. The graph of the original inequality includes the half-plane containing the test point if the inequality is satisfied by that point or the half-plane not containing the test point if the inequality is not satisfied by that point.

Example 1 **Graphing a Linear Inequality** Graph $2x - 3y \leqslant 6$.

C Check on a graphing utility.

SOLUTION **Step 1.** Graph $2x - 3y = 6$ as a solid line, since equality is included in the original statement:

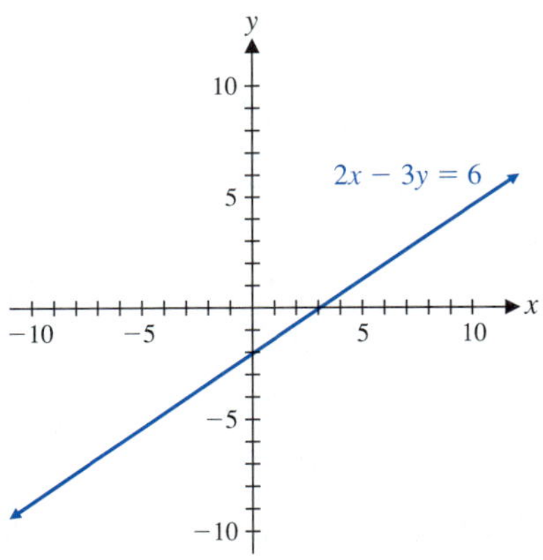

Step 2. Pick a convenient test point above or below the line. The origin (0, 0) requires the least computation, so substituting (0, 0) into the inequality, we get

$$\begin{aligned} 2x - 3y &\leqslant 6 \\ 2(0) - 3(0) = 0 &\leqslant 6 \end{aligned}$$

This is a true statement; therefore, the point (0, 0) is in the solution set.

Step 3. The line $2x - 3y = 6$ and the half-plane containing the origin form the graph of $2x - 3y \leqslant 6$, as shown in Figure 4.

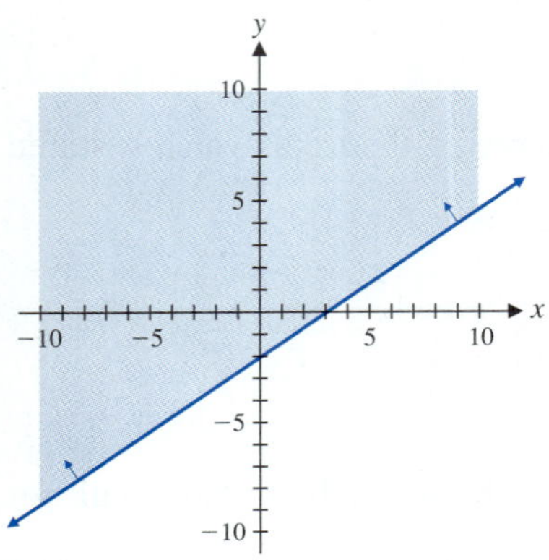

FIGURE 4

CHECK Figure 5 shows a check of this solution on a graphing utility. In Figure 5A, the small triangle to the left of y_1 indicates that the option to shade above the graph was selected. Consult the manual to see how to shade graphs on your graphing utility.

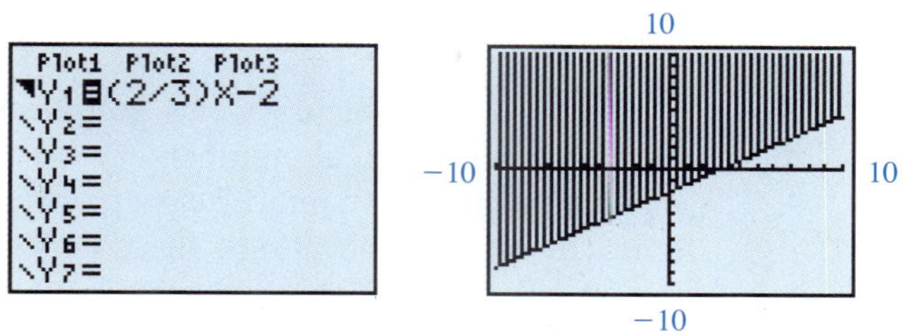

(A) (B)

FIGURE 5

Matched Problem 1 Graph $6x - 3y > 18$.

Check on a graphing utility.

Example 2 **Graphing Inequalities** Graph:

(A) $y > -3$ (B) $2x \leqslant 5$ (C) $x \leqslant 3y$

SOLUTION (A)

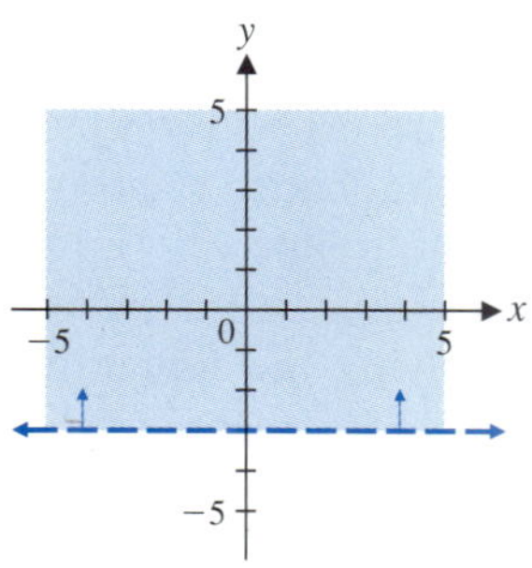

(B)

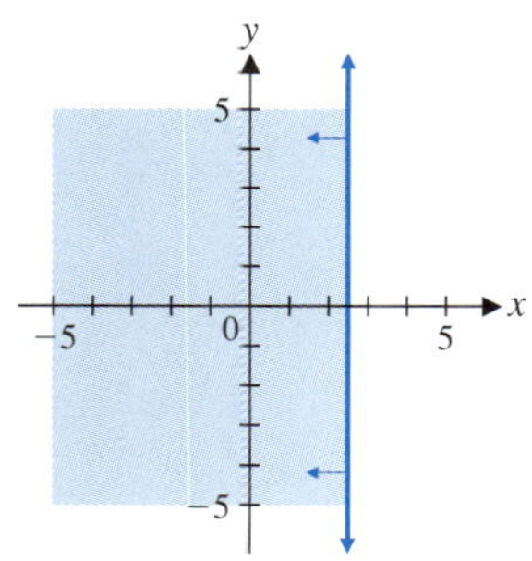

(C)

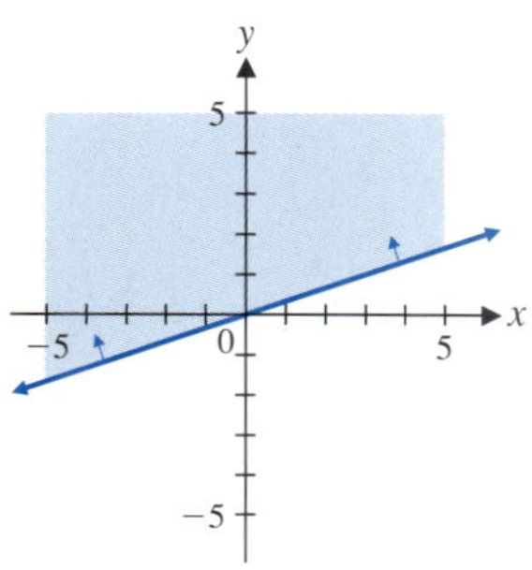

Matched Problem 2 Graph:

(A) $y < 4$ (B) $4x \geqslant -9$ (C) $3x \geqslant 2y$

Solving Systems of Linear Inequalities Graphically

We now consider systems of linear inequalities such as

$$\begin{aligned} x + y &\geq 6 \\ 2x - y &\geq 0 \end{aligned} \qquad \text{and} \qquad \begin{aligned} 2x + y &\leq 22 \\ x + y &\leq 13 \\ 2x + 5y &\leq 50 \\ x &\geq 0 \\ y &\geq 0 \end{aligned}$$

We wish to **solve** such systems **graphically;** that is, to find the graph of all ordered pairs of real numbers (x, y) that simultaneously satisfy all the inequalities in the system. The graph is called the **solution region** for the system. (In many applications, the solution region is also called the **feasible region.**) To find the solution region, we graph each inequality in the system and then take the intersection of all the graphs. To simplify the discussion that follows, **we will consider only systems of linear inequalities where equality is included in each statement in the system.**

Example 3 **Solving a System of Linear Inequalities Graphically** Solve the following system of linear inequalities graphically:

$$\begin{aligned} x + y &\geq 6 \\ 2x - y &\geq 0 \end{aligned}$$

Solution Graph the line $x + y = 6$ and shade the region that satisfies the linear inequality $x + y \geq 6$. This region is shaded with gray lines in Figure 6A. Next, graph the line $2x - y = 0$ and shade the region that satisfies the inequality $2x - y \geq 0$. This region is shaded with blue lines in Figure 6A. The solution region for the system of inequalities is the intersection of these two regions. This is the region shaded in both gray and blue in Figure 6A and redrawn in Figure 6B with only the solution region shaded. The coordinates of any point in the shaded region of Figure 6B specify a solution to the system. For example, the points (2, 4), (6, 3), and (7.43, 8.56) are three of infinitely many solutions, as can be easily checked. The intersection point (2, 4) is obtained by solving the equations $x + y = 6$ and $2x - y = 0$ simultaneously using any of the techniques discussed in Chapter 4.

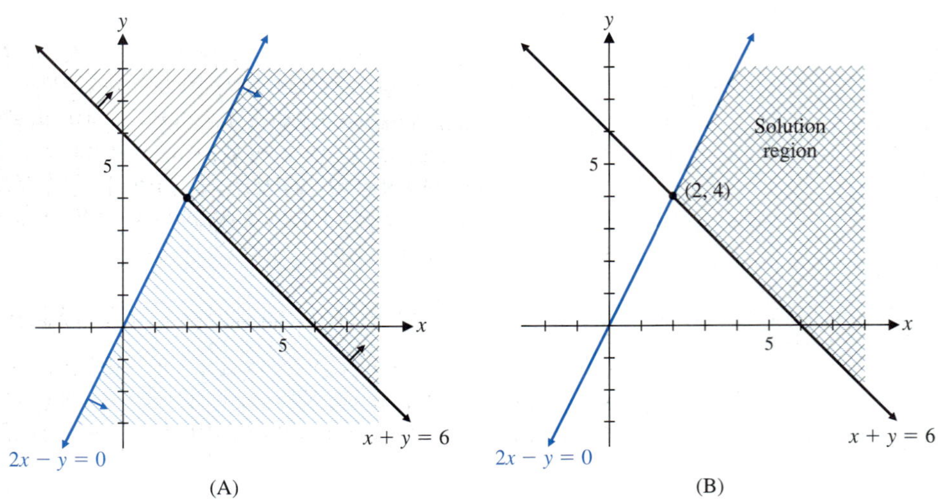

FIGURE 6

Matched Problem 3 ➡ Solve the following system of linear inequalities graphically:

$$\begin{aligned} 3x + y &\leq 21 \\ x - 2y &\leq 0 \end{aligned}$$

Explore–Discuss 2

Refer to Example 3. Graph each boundary line and shade the region containing the points that do *not* satisfy each inequality. That is, shade the region of the plane that corresponds to the inequality $x + y < 6$ and then shade the region that corresponds to the inequality $2x - y < 0$. What portion of the plane is left unshaded? Compare this method with the one used in the solution to Example 3.

The method of solving inequalities investigated in Explore–Discuss 2 works very well on a graphing utility that allows the user to shade above and below a graph. Referring to Example 3, the unshaded region in Figure 7B corresponds to the solution region in Figure 6B.

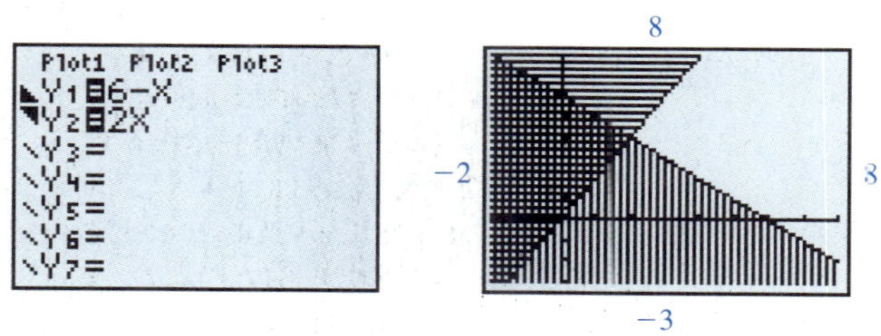

(A) (B)

FIGURE 7

The points of intersection of the lines that form the boundary of a solution region will play a fundamental role in the solution of linear programming problems, which are discussed in the next section.

Corner Point

A **corner point** of a solution region is a point in the solution region that is the intersection of two boundary lines.

For example, the point (2, 4) is the only corner point of the solution region in Example 3 (Fig. 6).

Example 4 ➡ **Solving a System of Linear Inequalities Graphically** Solve the following system of linear inequalities graphically, and find the corner points:

$$\begin{aligned} 2x + y &\leq 22 \\ x + y &\leq 13 \\ 2x + 5y &\leq 50 \\ x &\geq 0 \\ y &\geq 0 \end{aligned}$$

SOLUTION The inequalities $x \geq 0$ and $y \geq 0$ indicate that the solution region will lie in the first quadrant.* Thus, we can restrict our attention to that portion of the plane. First, we graph the lines

$$\begin{aligned} 2x + y &= 22 \\ x + y &= 13 \\ 2x + 5y &= 50 \end{aligned}$$

Find the x and y intercepts of each line; then sketch the line through these points.

Next, choosing (0, 0) as a test point, we see that the graph of each of the first three inequalities in the system consists of its corresponding line and the half-plane lying below the line, as indicated by the small arrows in Figure 8. Thus, the solution region of the system consists of the points in the first quadrant that simultaneously lie on or below all three of these lines (see the shaded region in Figure 8).

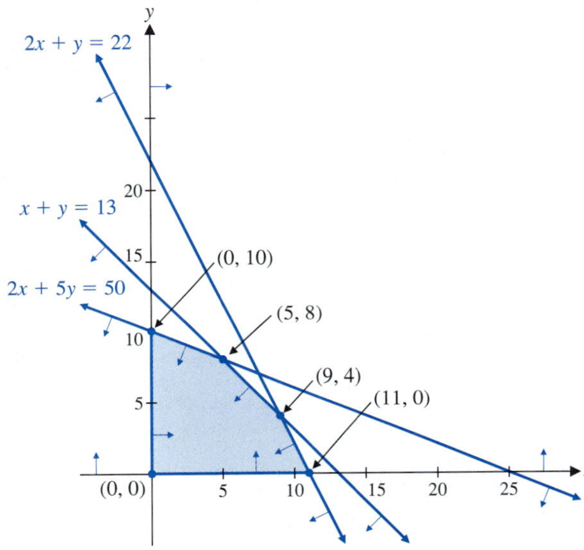

FIGURE 8

The corner points (0, 0), (0, 10), and (11, 0) can be determined from the graph. The other two corner points are determined as follows:

Solve the system	Solve the system
$2x + 5y = 50$ $x + y = 13$	$2x + y = 22$ $x + y = 13$
to obtain (5, 8).	to obtain (9, 4).

Note that the lines $2x + 5y = 50$ and $2x + y = 22$ also intersect, but the intersection point is not part of the solution region, and hence, is not a corner point.

*The inequalities $x \geq 0$ and $y \geq 0$ occur frequently in applications involving systems of inequalities, since x and y often represent quantities that cannot be negative (number of units produced, number of hours worked, etc.).

Matched Problem 4 ➡ Solve the following system of linear inequalities graphically, and find the corner points:

$$\begin{aligned} 5x + y &\geq 20 \\ x + y &\geq 12 \\ x + 3y &\geq 18 \\ x &\geq 0 \\ y &\geq 0 \end{aligned}$$

If we compare the solution regions of Example 3 and Example 4, we see that there is a fundamental difference between these two regions. We can draw a circle around the solution region in Example 4; however, it is impossible to include all the points in the solution region in Example 3 in any circle, no matter how large we draw it. This leads to the following definition:

> **Bounded and Unbounded Solution Regions**
>
> A solution region of a system of linear inequalities is **bounded** if it can be enclosed within a circle. If it cannot be enclosed within a circle, then it is **unbounded.**

Thus, the solution region for Example 4 is bounded, and the solution region for Example 3 is unbounded. This definition will be important in the next section.

APPLICATIONS

Example 5 ➡ **Medicine** A patient in a hospital is required to have at least 84 units of drug *A* and 120 units of drug *B* each day (assume that an overdosage of either drug is harmless). Each gram of substance *M* contains 10 units of drug *A* and 8 units of drug *B*, and each gram of substance *N* contains 2 units of drug *A* and 4 units of drug *B*. How many grams of substances *M* and *N* can be mixed to meet the minimum daily requirements?

SOLUTION To clarify relationships, we summarize the information in the following table:

	AMOUNT OF DRUG PER GRAM		MINIMUM DAILY
	Substance M	*Substance N*	REQUIREMENT
DRUG *A*	10 units	2 units	84 units
DRUG *B*	8 units	4 units	120 units

Let: x = Number of grams of substance *M* used
y = Number of grams of substance *N* used
Then: $10x$ = Number of units of drug *A* in x grams of substance *M*
$2y$ = Number of units of drug *A* in y grams of substance *N*
$8x$ = Number of units of drug *B* in x grams of substance *M*
$4y$ = Number of units of drug *B* in y grams of substance *N*

The following conditions must be satisfied to meet daily requirements:

$$\begin{pmatrix}\text{Number of units of} \\ \text{drug } A \\ \text{in } x \text{ grams of substance } M\end{pmatrix} + \begin{pmatrix}\text{Number of units of} \\ \text{drug } A \\ \text{in } y \text{ grams of substance } N\end{pmatrix} \geq 84$$

$$\begin{pmatrix}\text{Number of units of} \\ \text{drug } B \\ \text{in } x \text{ grams of substance } M\end{pmatrix} + \begin{pmatrix}\text{Number of units of} \\ \text{drug } B \\ \text{in } y \text{ grams of substance } N\end{pmatrix} \geq 120$$

$$(\text{Number of grams of substance } M \text{ used}) \geq 0$$
$$(\text{Number of grams of substance } N \text{ used}) \geq 0$$

Converting these verbal statements into symbolic statements by using the variables x and y introduced above, we obtain the following model consisting of a system of linear inequalities:

$$\begin{aligned} 10x + 2y &\geq 84 && \text{Drug A restriction} \\ 8x + 4y &\geq 120 && \text{Drug B restriction} \\ x &\geq 0 && \text{Cannot use a negative amount of M} \\ y &\geq 0 && \text{Cannot use a negative amount of N} \end{aligned}$$

Graphing this system of linear inequalities, we obtain the set of feasible solutions, or the feasible region (solution region), as shown in Figure 9. Thus, any point in the shaded area (including the straight-line boundaries) will meet the daily requirements; any point outside the shaded area will not. For example, 4 units of drug M and 23 units of drug N will meet the daily requirements, but 4 units of drug M and 21 units of drug N will not. (Note that the feasible region is unbounded.)

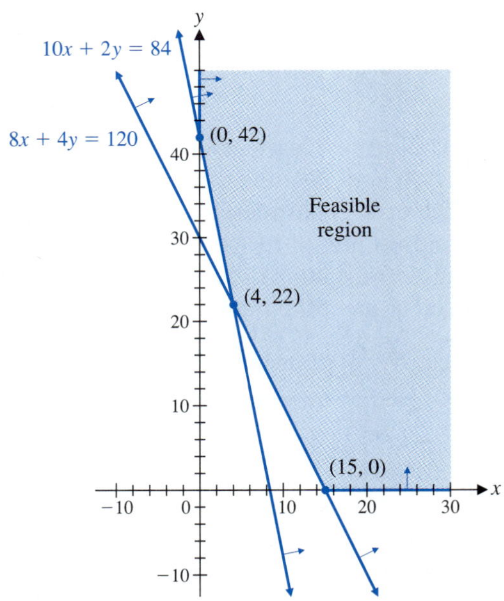

FIGURE 9

Matched Problem 5 **Resource Allocation** A manufacturing plant makes two types of inflatable boats, a two-person boat and a four-person boat. Each two-person boat requires 0.9 labor-hour in the cutting department and 0.8 labor-hour in the assembly department. Each four-person boat requires 1.8 labor-hours in the cutting department and 1.2 labor-hours in the assembly department. The maximum labor-hours available each month in the cutting and assembly departments are 864 and 672, respectively.

(A) Summarize this information in a table.
(B) If x two-person boats and y four-person boats are manufactured each month, write a system of linear inequalities that reflect the conditions indicated. Find the set of feasible solutions graphically.

REMARK

Refer to Example 5 and Matched Problem 5. In Example 5, it is certainly reasonable to consider noninteger values for x and y. The situation is not as clear in Matched Problem 5. How do we interpret a production schedule of 214.5 two-person boats and 347.75 four-person boats? It is not possible to manufacture a fraction of a boat. But it is possible to *average* 214.5 two-person and 347.75 four-person boats per month. In general, we will assume that all points in the feasible region represent acceptable solutions, even though noninteger solutions might require special interpretation.

Answers to Matched Problems

1. Graph $6x - 3y = 18$ as a dashed line (since equality is not included). Choosing the origin (0, 0) as a test point, we see that $6(0) - 3(0) > 18$ is a false statement; thus, the lower half-plane determined by $6x - 3y = 18$ is the graph of $6x - 3y > 18$.

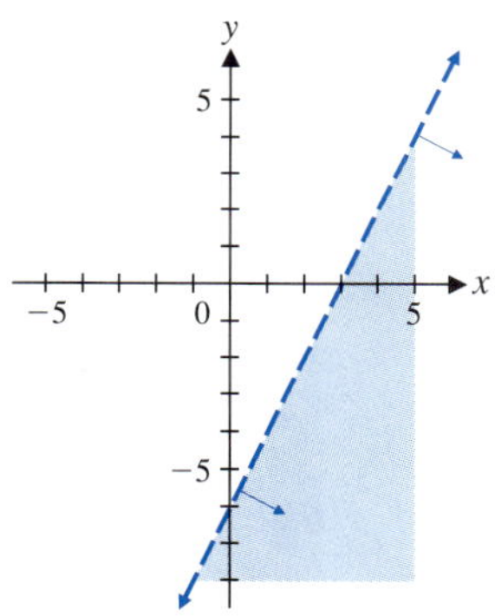

2. (A)

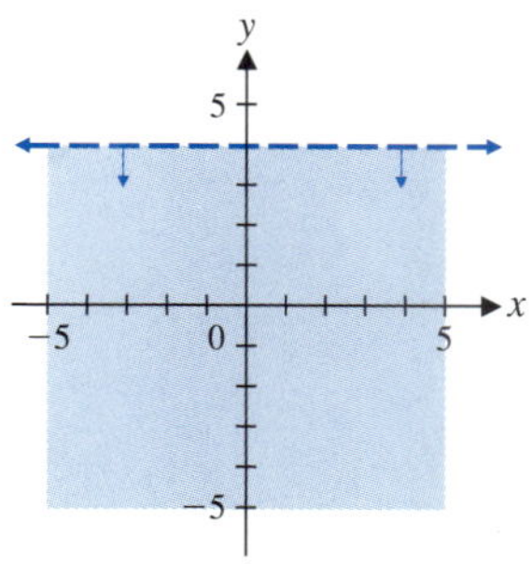

(B)

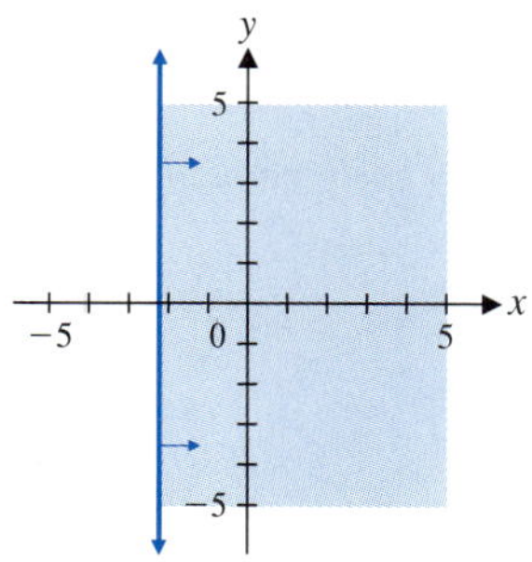

(C)

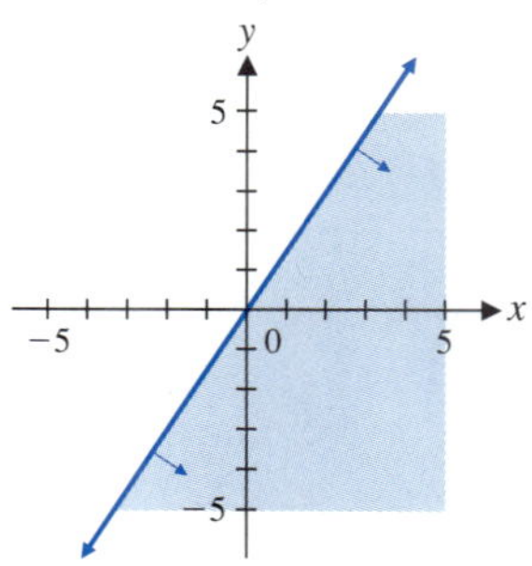

3.

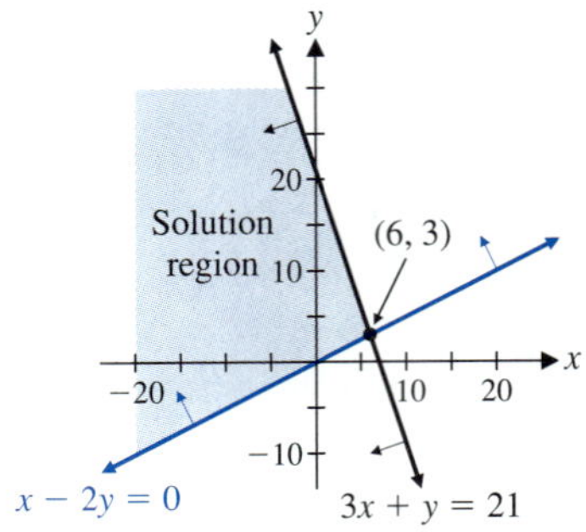

4.

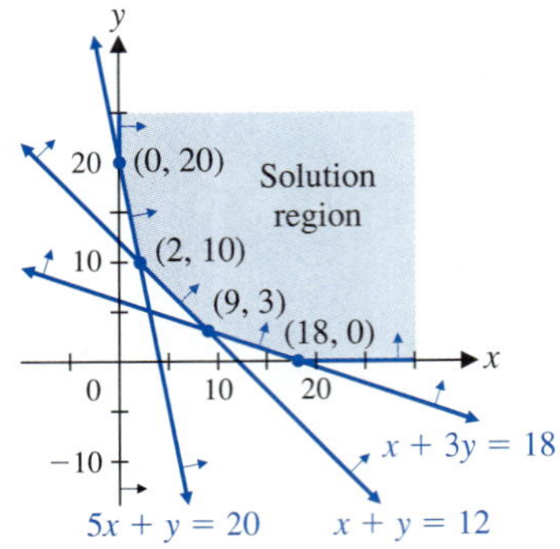

5. (A)

	Labor-hours required: *Two-person boat*	Labor-hours required: *Four-person boat*	Maximum labor-hours available per month
Cutting department	0.9	1.8	864
Assembly department	0.8	1.2	672

(B)
$$\begin{aligned} 0.9x + 1.8y &\leqslant 864 \\ 0.8x + 1.2y &\leqslant 672 \\ x &\geqslant 0 \\ y &\geqslant 0 \end{aligned}$$

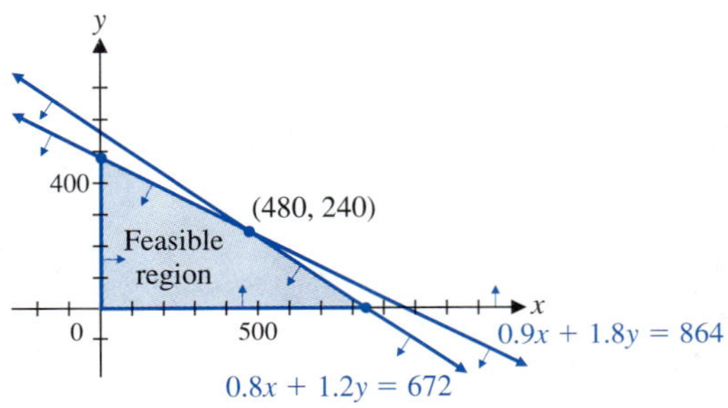

Exercise 5-1

A *Graph each inequality in Problems 1–10.*

1. $y \leqslant x - 1$

2. $y > x + 1$

3. $3x - 2y > 6$

4. $2x - 5y \leqslant 10$

5. $x \geqslant -4$

6. $y < 5$

7. $6x + 4y \geqslant 24$

8. $4x + 8y \geqslant 32$

9. $5x \leqslant -2y$

10. $6x \geqslant 4y$

In Problems 11–14, match the solution region of each system of linear inequalities with one of the four regions shown in the figure.

11. $x + 2y \leqslant 8$
$3x - 2y \geqslant 0$

12. $x + 2y \geqslant 8$
$3x - 2y \leqslant 0$

13. $x + 2y \geqslant 8$
$3x - 2y \geqslant 0$

14. $x + 2y \leqslant 8$
$3x - 2y \leqslant 0$

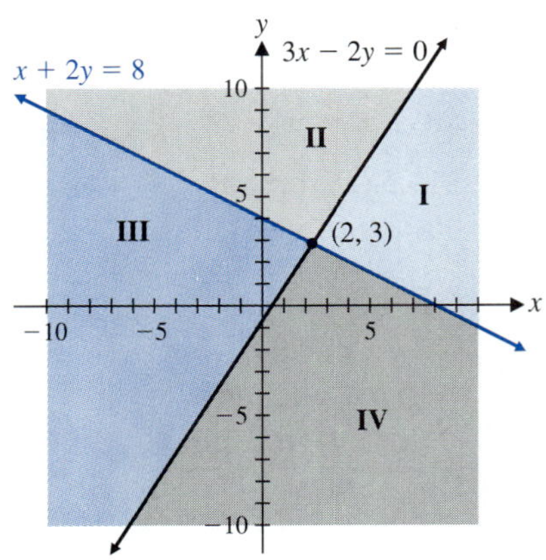

Figure for 11–14

In Problems 15–18, solve each system of linear inequalities graphically.

15. $\begin{aligned} 3x + y &\geq 6 \\ x &\leq 4 \end{aligned}$

16. $\begin{aligned} 3x + 4y &\leq 12 \\ y &\geq -3 \end{aligned}$

17. $\begin{aligned} x - 2y &\leq 12 \\ 2x + y &\geq 4 \end{aligned}$

18. $\begin{aligned} 2x + 5y &\leq 20 \\ x - 5y &\geq -5 \end{aligned}$

Problems 19–22 require a graphing utility that gives the user the option of shading above or below a graph.

(A) Graph the boundary lines in a standard viewing window and shade the region that contains the points that satisfy each inequality.

(B) Repeat part (A), but this time shade the region that contains the points that do not satisfy each inequality. (See Explore–Discuss 2 and Fig. 7.)

Explain how you can recognize the solution region in each graph.

19. $\begin{aligned} x + y &\leq 5 \\ 2x - y &\leq 1 \end{aligned}$

20. $\begin{aligned} x - 2y &\leq 1 \\ x + 3y &\geq 12 \end{aligned}$

21. $\begin{aligned} 2x + y &\geq 4 \\ 3x - y &\leq 7 \end{aligned}$

22. $\begin{aligned} 3x + y &\geq -2 \\ x - 2y &\geq -6 \end{aligned}$

B *In Problems 23–26, match the solution region of each system of linear inequalities with one of the four regions shown in the figure. Identify the corner points of each solution region.*

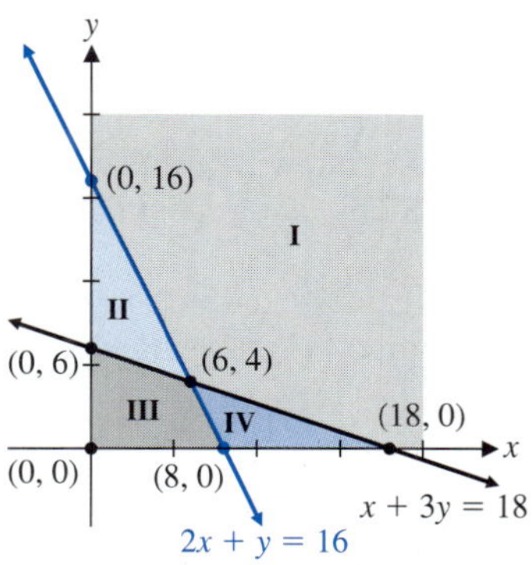

Figure for 23–26

23. $\begin{aligned} x + 3y &\leq 18 \\ 2x + y &\geq 16 \\ x &\geq 0 \\ y &\geq 0 \end{aligned}$

24. $\begin{aligned} x + 3y &\leq 18 \\ 2x + y &\leq 16 \\ x &\geq 0 \\ y &\geq 0 \end{aligned}$

25. $\begin{aligned} x + 3y &\geq 18 \\ 2x + y &\geq 16 \\ x &\geq 0 \\ y &\geq 0 \end{aligned}$

26. $\begin{aligned} x + 3y &\geq 18 \\ 2x + y &\leq 16 \\ x &\geq 0 \\ y &\geq 0 \end{aligned}$

Solve the systems in Problems 27–36 graphically, and indicate whether each solution region is bounded or unbounded. Find the coordinates of each corner point.

27. $\begin{aligned} 2x + 3y &\leq 12 \\ x &\geq 0 \\ y &\geq 0 \end{aligned}$

28. $\begin{aligned} 3x + 4y &\leq 24 \\ x &\geq 0 \\ y &\geq 0 \end{aligned}$

29. $\begin{aligned} 2x + y &\leq 10 \\ x + 2y &\leq 8 \\ x &\geq 0 \\ y &\geq 0 \end{aligned}$

30. $\begin{aligned} 6x + 3y &\leq 24 \\ 3x + 6y &\leq 30 \\ x &\geq 0 \\ y &\geq 0 \end{aligned}$

31. $\begin{aligned} 2x + y &\geq 10 \\ x + 2y &\geq 8 \\ x &\geq 0 \\ y &\geq 0 \end{aligned}$

32. $\begin{aligned} 4x + 3y &\geq 24 \\ 3x + 4y &\geq 8 \\ x &\geq 0 \\ y &\geq 0 \end{aligned}$

33. $\begin{aligned} 2x + y &\leq 10 \\ x + y &\leq 7 \\ x + 2y &\leq 12 \\ x &\geq 0 \\ y &\geq 0 \end{aligned}$

34. $\begin{aligned} 3x + y &\leq 21 \\ x + y &\leq 9 \\ x + 3y &\leq 21 \\ x &\geq 0 \\ y &\geq 0 \end{aligned}$

35. $\begin{aligned} 2x + y &\geq 16 \\ x + y &\geq 12 \\ x + 2y &\geq 14 \\ x &\geq 0 \\ y &\geq 0 \end{aligned}$

36. $\begin{aligned} 3x + y &\geq 24 \\ x + y &\geq 16 \\ x + 3y &\geq 30 \\ x &\geq 0 \\ y &\geq 0 \end{aligned}$

Check Problems 27–36 on a graphing utility.

C *Solve the systems in Problems 37–46 graphically, and indicate whether each solution region is bounded or unbounded. Find the coordinates of each corner point.*

37. $\begin{aligned} x + 4y &\leq 32 \\ 3x + y &\leq 30 \\ 4x + 5y &\geq 51 \end{aligned}$

38. $\begin{aligned} x + y &\leq 11 \\ x + 5y &\geq 15 \\ 2x + y &\geq 12 \end{aligned}$

39. $\begin{aligned} 4x + 3y &\leq 48 \\ 2x + y &\geq 24 \\ x &\leq 9 \end{aligned}$

40. $\begin{aligned} 2x + 3y &\geq 24 \\ x + 3y &\leq 15 \\ y &\geq 4 \end{aligned}$

41. $\begin{aligned} x - y &\leq 0 \\ 2x - y &\leq 4 \\ 0 \leq x &\leq 8 \end{aligned}$

42. $\begin{aligned} 2x + 3y &\geq 12 \\ -x + 3y &\leq 3 \\ 0 \leq y &\leq 5 \end{aligned}$

43. $\begin{aligned} -x + 3y &\geq 1 \\ 5x - y &\geq 9 \\ x + y &\leq 9 \\ x &\leq 5 \end{aligned}$

44. $\begin{aligned} x + y &\leq 10 \\ 5x + 3y &\geq 15 \\ -2x + 3y &\leq 15 \\ 2x - 5y &\leq 6 \end{aligned}$

45. C $\begin{aligned} 16x + 13y &\leq 119 \\ 12x + 16y &\geq 101 \\ -4x + 3y &\leq 11 \end{aligned}$

46. C $\begin{aligned} 8x + 4y &\leq 41 \\ -15x + 5y &\leq 19 \\ 2x + 6y &\geq 37 \end{aligned}$

Problems 47 and 48 introduce an algebraic process for finding the corner points of a solution region without drawing a graph. We will have a great deal more to say about this process later in this chapter.

47. Consider the following system of inequalities and corresponding boundary lines:

$$\begin{aligned} 3x + 4y &\le 36 \\ 3x + 2y &\le 30 \\ x &\ge 0 \\ y &\ge 0 \end{aligned} \qquad \begin{aligned} 3x + 4y &= 36 \\ 3x + 2y &= 30 \\ x &= 0 \\ y &= 0 \end{aligned}$$

(A) Use algebraic methods to find the intersection points (if any exist) for each possible pair of boundary lines. (There are six different possible pairs.)

(B) Test each intersection point in all four inequalities to determine which are corner points.

48. Repeat Problem 47 for

$$\begin{aligned} 2x + y &\le 16 \\ 2x + 3y &\le 36 \\ x &\ge 0 \\ y &\ge 0 \end{aligned} \qquad \begin{aligned} 2x + y &= 16 \\ 2x + 3y &= 36 \\ x &= 0 \\ y &= 0 \end{aligned}$$

APPLICATIONS

Business & Economics

49. *Manufacturing—resource allocation.* A manufacturing company makes two types of water skis, a trick ski and a slalom ski. The trick ski requires 6 labor-hours for fabricating and 1 labor-hour for finishing. The slalom ski requires 4 labor-hours for fabricating and 1 labor-hour for finishing. The maximum labor-hours available per day for fabricating and finishing are 108 and 24, respectively. If x is the number of trick skis and y is the number of slalom skis produced per day, write a system of linear inequalities that indicates appropriate restraints on x and y. Find the set of feasible solutions graphically for the number of each type of ski that can be produced.

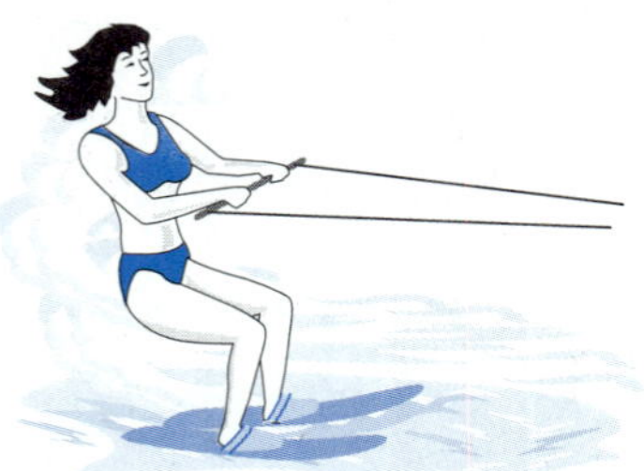

50. *Manufacturing—resource allocation.* A furniture manufacturing company manufactures dining room tables and chairs. A table requires 8 labor-hours for assembling and 2 labor-hours for finishing. A chair requires 2 labor-hours for assembling and 1 labor-hour for finishing. The maximum labor-hours available per day for assembly and finishing are 400 and 120, respectively. If x is the number of tables and y is the number of chairs produced per day, write a system of linear inequalities that indicates appropriate restraints on x and y. Find the set of feasible solutions graphically for the number of tables and chairs that can be produced.

51. *Manufacturing—resource allocation.* Refer to Problem 49. The company makes a profit of \$50 on each trick ski and a profit of \$60 on each slalom ski.

(A) If the company makes 10 trick skis and 10 slalom skis per day, the daily profit will be \$1,100. Are there other production schedules that will result in a daily profit of \$1,100? How are these schedules related to the graph of the line $50x + 60y = 1{,}100$?

(B) Find a production schedule that will produce a daily profit greater than \$1,100 and repeat part (A) for this schedule.

(C) Discuss methods for using lines like those in parts (A) and (B) to find the largest possible daily profit.

52. *Manufacturing—resource allocation.* Refer to Problem 50. The company makes a profit of \$50 on each table and a profit of \$15 on each chair.

(A) If the company makes 20 tables and 20 chairs per day, the daily profit will be \$1,300. Are there other production schedules that will result in a daily profit of \$1,300? How are these schedules related to the graph of the line $50x + 15y = 1{,}300$?

(B) Find a production schedule that will produce a daily profit greater than \$1,300 and repeat part (A) for this schedule.

(C) Discuss methods for using lines like those in parts (A) and (B) to find the largest possible daily profit.

Life Sciences

53. *Nutrition—plants.* A farmer can buy two types of plant food, mix *A* and mix *B*. Each cubic yard of mix *A* contains 20 pounds of phosphoric acid, 30 pounds of nitrogen, and 5 pounds of potash. Each cubic yard of mix *B* contains 10 pounds of phosphoric acid, 30 pounds of nitrogen, and 10 pounds of potash. The minimum monthly requirements are 460 pounds of phosphoric acid, 960 pounds of nitrogen, and 220 pounds of potash. If x is the number of cubic yards of mix *A* used and y is the number of cubic yards of mix *B* used, write a system of linear inequalities that indicates appropriate restraints on x and y. Find the set of feasible solutions graphically for the amounts of mix *A* and mix *B* that can be used.

54. *Nutrition—people.* A dietitian in a hospital is to arrange a special diet using two foods. Each ounce of food *M* contains 30 units of calcium, 10 units of iron, and 10 units of vitamin A. Each ounce of food *N* contains 10 units of calcium, 10 units of iron, and 30 units of vitamin A. The minimum requirements in the diet are 360 units of calcium, 160 units of iron, and 240 units of vitamin A. If x is the number of ounces of food *M* used and y is the number of ounces of food *N* used, write a system of linear inequalities that reflects the conditions indicated. Find the set of feasible solutions graphically for the amount of each kind of food that can be used.

Social Sciences

55. *Psychology.* In an experiment on conditioning, a psychologist uses two types of Skinner (conditioning) boxes with mice and rats. Each mouse spends 10 minutes per day in box *A* and 20 minutes per day in box *B*. Each rat spends 20 minutes per day in box *A* and 10 minutes per day in box *B*. The total maximum time available per day is 800 minutes for box *A* and 640 minutes for box *B*. We are interested in the various numbers of mice and rats that can be used in the experiment under the conditions stated. If we let x be the number of mice used and y the number of rats used, write a system of linear inequalities that indicates appropriate restrictions on x and y. Find the set of feasible solutions graphically.

SECTION 5-2

Linear Programming in Two Dimensions—A Geometric Approach

- A LINEAR PROGRAMMING PROBLEM
- LINEAR PROGRAMMING—A GENERAL DESCRIPTION
- GEOMETRIC SOLUTION OF LINEAR PROGRAMMING PROBLEMS
- APPLICATIONS

Several problems discussed in the last section are related to a more general type of problem called a *linear programming problem.* Linear programming is a mathematical process that has been developed to help management in decision-making, and it has become one of the most widely used and best-known tools of management science. We will introduce this topic by considering an example in detail, using an intuitive geometric approach. Insight gained from this approach will prove invaluable when we later consider an algebraic approach that is less intuitive but necessary in solving most real-world problems.

NOTATION CHANGE

For ease of generalization to the larger problems in later sections, we will now change variable notation from letters such as x and y to subscript forms such as x_1 and x_2.

A Linear Programming Problem

We begin our discussion with a concrete example. The geometric method of solution will suggest two important theorems and a simple general geometric procedure for solving linear programming problems in two variables.

Example 1 **Production Scheduling** A manufacturer of lightweight mountain tents makes a standard model and an expedition model for national distribution. Each standard tent requires 1 labor-hour from the cutting department and 3 labor-hours from the assembly department. Each expedition tent requires 2 labor-hours from the cutting department and 4 labor-hours from the assembly department. The maximum labor-hours available per day in the cutting department and the assembly department are 32 and 84, respectively. If the company makes a profit of \$50 on each standard tent and \$80 on each expedition tent, how many tents of each type should be manufactured each day to maximize the total daily profit (assuming all tents can be sold)?

Solution This is an example of a linear programming problem. To see relationships more clearly, we summarize the manufacturing requirements, objectives, and restrictions in Table 1.

Table 1

	Labor-hours per tent		Maximum
	Standard model	*Expedition model*	labor-hours available per day
Cutting department	1	2	32
Assembly department	3	4	84
Profit per tent	\$50	\$80	

We now proceed to formulate a mathematical model for the problem and then to solve it by using geometric methods.

Objective Function The *objective* of management is to *decide* how many of each tent model should be produced each day in order to maximize profit.

Let x_1 = Number of standard tents produced per day
x_2 = Number of expedition tents produced per day } Decision variables

The following equation gives the total profit for x_1 standard tents and x_2 expedition tents manufactured each day, assuming all tents manufactured are sold:

$$P = 50x_1 + 80x_2$$ Objective function

Mathematically, the manufacturer needs to decide on values for the **decision variables** (x_1, x_2) that achieve its objective—that is, maximize the **objective function** (profit) $P = 50x_1 + 80x_2$. (Note that P is a function of two independent variables.) It appears that the profit can be made as large as we like by manufacturing more and more tents—or can it?

CONSTRAINTS Any manufacturing company, no matter how large or small, has manufacturing limits imposed by available resources, plant capacity, demand, and so forth. These limits are referred to as **problem constraints.**

CUTTING DEPARTMENT CONSTRAINT

$$\begin{pmatrix}\text{Daily cutting}\\ \text{time for } x_1\\ \text{standard tents}\end{pmatrix} + \begin{pmatrix}\text{Daily cutting}\\ \text{time for } x_2\\ \text{expedition tents}\end{pmatrix} \le \begin{pmatrix}\text{Maximum labor-}\\ \text{hours available}\\ \text{per day}\end{pmatrix}$$

$$1x_1 + 2x_2 \le 32$$

ASSEMBLY DEPARTMENT CONSTRAINT

$$\begin{pmatrix}\text{Daily assembly}\\ \text{time for } x_1\\ \text{standard tents}\end{pmatrix} + \begin{pmatrix}\text{Daily assembly}\\ \text{time for } x_2\\ \text{expedition tents}\end{pmatrix} \le \begin{pmatrix}\text{Maximum labor-}\\ \text{hours available}\\ \text{per day}\end{pmatrix}$$

$$3x_1 + 4x_2 \le 84$$

NONNEGATIVE CONSTRAINTS It is not possible to manufacture a negative number of tents; thus, we have the **nonnegative constraints**

$$x_1 \ge 0$$
$$x_2 \ge 0$$

which we usually write in the form

$$x_1, x_2 \ge 0$$

MATHEMATICAL MODEL We now have a **mathematical model** for the problem under consideration:

$$\begin{aligned} \text{Maximize} \quad & P = 50x_1 + 80x_2 && \text{Objective function} \\ \text{Subject to} \quad & x_1 + 2x_2 \le 32 && \text{Problem constraints} \\ & 3x_1 + 4x_2 \le 84 && \\ & x_1, x_2 \ge 0 && \text{Nonnegative constraints} \end{aligned}$$

GRAPHIC SOLUTION **Solving** the set of linear inequality constraints **graphically** (see the last section), we obtain the feasible region for production schedules (Figure 1).

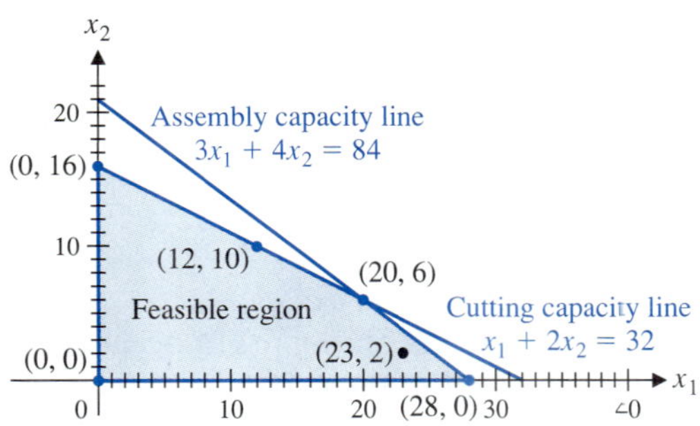

FIGURE 1

By choosing a production schedule (x_1, x_2) from the feasible region, a profit can be determined using the objective function

$$P = 50x_1 + 80x_2$$

For example, if $x_1 = 12$ and $x_2 = 10$, then the profit for the day would be

$$\begin{aligned} P &= 50(12) + 80(10) \\ &= \$1{,}400 \end{aligned}$$

Or if $x_1 = 23$ and $x_2 = 2$, then the profit for the day would be

$$\begin{aligned} P &= 50(23) + 80(2) \\ &= \$1{,}310 \end{aligned}$$

But the question is, out of all possible production schedules (x_1, x_2) from the feasible region, which schedule(s) produces the *maximum* profit? Thus, we have a **maximization problem.** Since point-by-point checking is impossible (there are infinitely many points to check), we must find another way.

By assigning P in $P = 50x_1 + 80x_2$ a particular value and plotting the resulting equation in the coordinate system shown in Figure 1, we obtain a **constant-profit line (isoprofit line).** Every point in the feasible region on this line represents a production schedule that will produce the same profit. By doing this for a number of values for P, we obtain a family of constant-profit lines (Fig. 2) that are parallel to each other, since they all have the same slope. To see this, we write $P = 50x_1 + 80x_2$ in the slope–intercept form

$$x_2 = -\frac{5}{8}x_1 + \frac{P}{80}$$

and note that for any profit P, the constant-profit line has slope $-\frac{5}{8}$. We also observe that as the profit P increases, the x_2 intercept $(P/80)$ increases, and the line moves away from the origin.

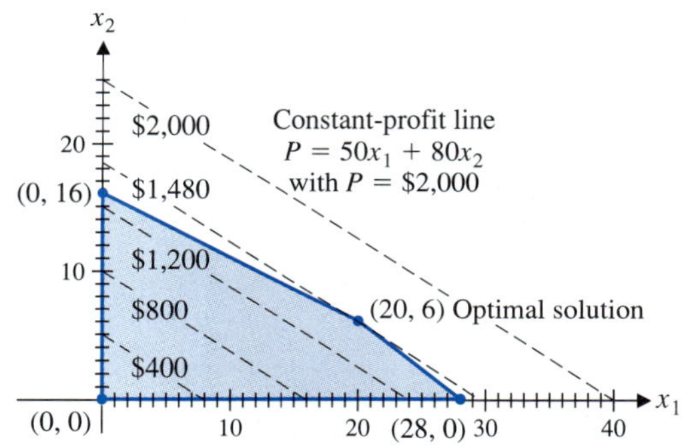

FIGURE 2
Constant-profit lines

Thus, the maximum profit occurs at a point where a constant-profit line is the farthest from the origin but still in contact with the feasible region. In this example, this occurs at (20, 6), as is seen in Figure 2. Thus, if the manufacturer makes 20 standard tents and 6 expedition tents per day, the profit will be maximized at

$$\begin{aligned} P &= 50(20) + 80(6) \\ &= \$1{,}480 \end{aligned}$$

The point (20, 6) is called an **optimal solution** to the problem, because it maximizes the objective (profit) function and is in the feasible region. In general, it appears that a maximum profit occurs at one of the corner points. We also note that the minimum profit $(P = 0)$ occurs at the corner point $(0, 0)$.

Matched Problem 1

A manufacturing plant makes two types of inflatable boats, a two-person boat and a four-person boat. Each two-person boat requires 0.9 labor-hour from the cutting department and 0.8 labor-hour from the assembly department. Each four-person boat requires 1.8 labor-hours from the cutting department and 1.2 labor-hours from the assembly department. The maximum labor-hours available per month in the cutting department and the assembly department are 864 and 672, respectively. The company makes a profit of \$25 on each two-person boat and \$40 on each four-person boat.

(A) Summarize the relevant material in a table similar to Table 1 in Example 1.
(B) Identify the decision variables.
(C) Write the objective function P.
(D) Write the problem constraints and the nonnegative constraints.
(E) Graph the feasible region. Include graphs of the objective function for $P = \$5{,}000$, $P = \$10{,}000$, $P = \$15{,}000$, and $P = \$21{,}600$.
(F) From the graph and constant-profit lines, determine how many boats should be manufactured each month to maximize the profit. What is the maximum profit?

Explore–Discuss 1

Refer to the feasible region S shown in the figure.

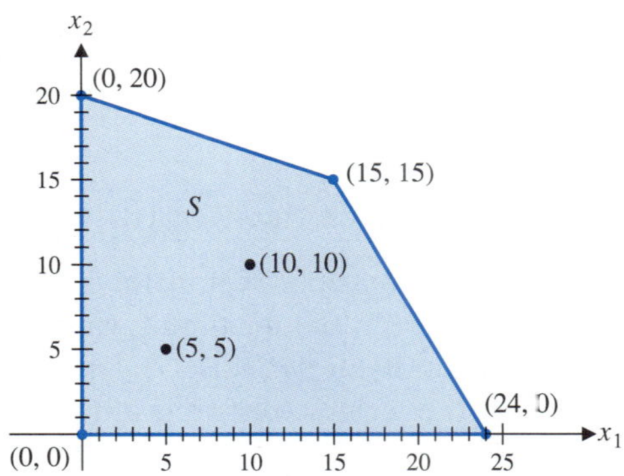

(A) Let $P = x_1 + x_2$. Graph the isoprofit lines through the points $(5, 5)$ and $(10, 10)$. Place a straightedge along the line with the smaller profit and slide it in the direction of increasing profit, without changing its slope. What is the maximum value of P? Where does this maximum value occur?
(B) Repeat part (A) for $P = x_1 + 10x_2$.
(C) Repeat part (A) for $P = 10x_1 + x_2$.

Linear Programming—A General Description

In Example 1 and Matched Problem 1, the optimal solution occurs at a corner point of the feasible region. Is this always the case? The answer is a qualified yes, as will be seen in Theorem 1. First, we give a few general definitions.

A **linear programming problem** is one that is concerned with finding the **optimal value** (maximum or minimum value) of a linear **objective function** of the form

$$z = c_1x_1 + c_2x_2 + \cdots + c_nx_n$$

where the **decision variables** $x_1, x_2, \ldots, x_n$ are subject to **problem constraints** in the form of linear inequalities and equations. In addition, the decision variables must satisfy the **nonnegative constraints** $x_i \geq 0, i = 1, 2, \ldots, n$. The set of points satisfying both the problem constraints and the nonnegative constraints is called the **feasible region** for the problem. Any point in the feasible region that produces the optimal value of the objective function over the feasible region is called an **optimal solution.**

Theorem 1 ■■ FUNDAMENTAL THEOREM OF LINEAR PROGRAMMING—VERSION 1

If the optimal value of the objective function in a linear programming problem exists, then that value must occur at one (or more) of the corner points of the feasible region. ■■

Theorem 1 provides a simple procedure for solving a linear programming problem, *provided the problem has an optimal solution—not all do.* In order to use Theorem 1, we must know that the problem under consideration has an optimal solution. Theorem 2 provides some conditions that will ensure that a linear programming problem has an optimal solution.

Theorem 2 ■■ EXISTENCE OF OPTIMAL SOLUTIONS

(A) If the feasible region for a linear programming problem is bounded, then both the maximum value and the minimum value of the objective function always exist.
(B) If the feasible region is unbounded and the coefficients of the objective function are positive, then the minimum value of the objective function exists, but the maximum value does not.
(C) If the feasible region is empty (that is, there are no points that satisfy all the constraints), then both the maximum value and the minimum value of the objective function do not exist. ■■

Theorem 2 does not cover all possibilities. For example, what happens if the feasible region is unbounded and one (or both) of the coefficients of the objective function are negative? Problems of this type must be solved by carefully examining the graph of the objective function for various feasible solutions, as we did in Example 1.

■ GEOMETRIC SOLUTION OF LINEAR PROGRAMMING PROBLEMS

The discussion above leads to the following procedure for the geometric solution of linear programming problems with two decision variables:

Geometric Solution of a Linear Programming Problem with Two Decision Variables

Step 1. For an applied problem, summarize relevant material in table form (see Table 1 in Example 1).

Step 2. Form a mathematical model for the problem:

(A) Introduce decision variables, and write a linear objective function.
(B) Write problem constraints using linear inequalities and/or equations.
(C) Write nonnegative constraints.

Step 3. Graph the feasible region. Then, if an optimal solution exists according to Theorem 2, find the coordinates of each corner point.

Step 4. Make a table listing the value of the objective function at each corner point.

Step 5. Determine the optimal solution(s) from the table in step 4.

Step 6. For an applied problem, interpret the optimal solution(s) in terms of the original problem.

Before we consider additional applications, let us use this procedure to solve some linear programming problems where the model has already been determined.

Example 2 **Solving a Linear Programming Problem**

(A) Minimize and maximize

$$z = 3x_1 + x_2$$

Subject to

$$\begin{aligned} 2x_1 + x_2 &\le 20 \\ 10x_1 + x_2 &\ge 36 \\ 2x_1 + 5x_2 &\ge 36 \\ x_1, x_2 &\ge 0 \end{aligned}$$

(B) Minimize and maximize

$$z = 10x_1 + 20x_2$$

Subject to

$$\begin{aligned} 6x_1 + 2x_2 &\ge 36 \\ 2x_1 + 4x_2 &\ge 32 \\ x_2 &\le 20 \\ x_1, x_2 &\ge 0 \end{aligned}$$

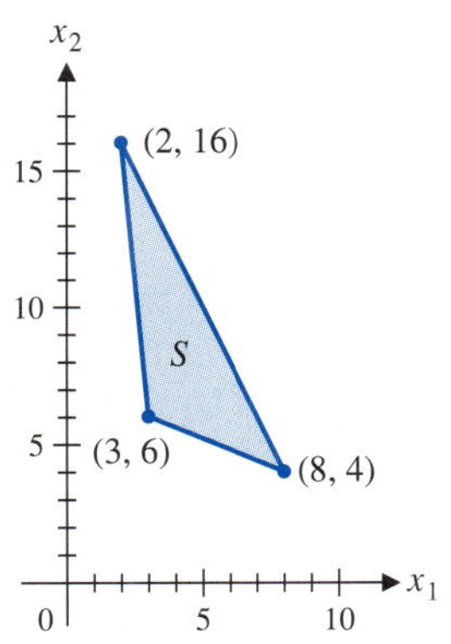

CORNER POINT (x_1, x_2)	$z = 3x_1 + x_2$
(3, 6)	15
(2, 16)	22
(8, 4)	28

SOLUTION (A) We begin with step 3, since step 1 does not apply and step 2 has already been done for us.

Step 3. Graph the feasible region S. Then, after checking Theorem 2 to determine that an optimal solution exists, find the coordinates of each corner point. Since S is bounded, z will have both a maximum and a minimum on S (Theorem 2A) and these will both occur at corner points (Theorem 1).

Step 4. Evaluate the objective function at each corner point, as shown in the table in the margin.

Step 5. Determine the optimal solutions from step 4. Examining the values in the table, we see that the minimum value of z is 15 at (3, 6) and the maximum value of z is 28 at (8, 4).

(B) Again, we can begin with step 3.

Step 3. Graph the feasible region S. Then, after checking Theorem 2 to determine that an optimal solution exists, find the coordinates of each corner point. Since S is unbounded and the coefficients of the objective function are positive, z has a minimum value on S but no maximum value (Theorem 2B).

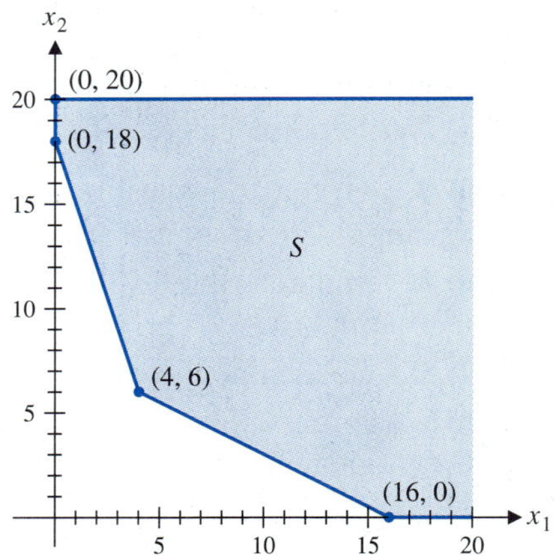

CORNER POINT (x_1, x_2)	$z = 10x_1 + 20x_2$
(0, 20)	400
(0, 18)	360
(4, 6)	160
(16, 0)	160

Step 4. Evaluate the objective function at each corner point, as shown in the table in the margin.

Step 5. Determine the optimal solution from step 4. The minimum value of z is 160 at $(4, 6)$ and at $(16, 0)$.

The solution to Example 2 is a **multiple optimal solution.**

In general, if two corner points are both optimal solutions to a linear programming problem, then any point on the line segment joining them is also an optimal solution.

This is the only time that optimal solutions also occur at noncorner points.

Matched Problem 2

(A) Maximize and minimize $z = 4x_1 + 2x_2$ subject to the constraints given in Example 2A.

(B) Maximize and minimize $z = 20x_1 + 5x_2$ subject to the constraints given in Example 2B.

Explore–Discuss 2

In Example 2B we saw that there was no optimal solution for the problem of maximizing the objective function P over the feasible region S. We want to add an additional constraint to modify the feasible region so that an optimal solution for the maximization problem does exist. Which of the following constraints will accomplish this objective?

(A) $x_1 \leq 20$ (B) $x_2 \geq 4$ (C) $x_1 \leq x_2$ (D) $x_2 \leq x_1$

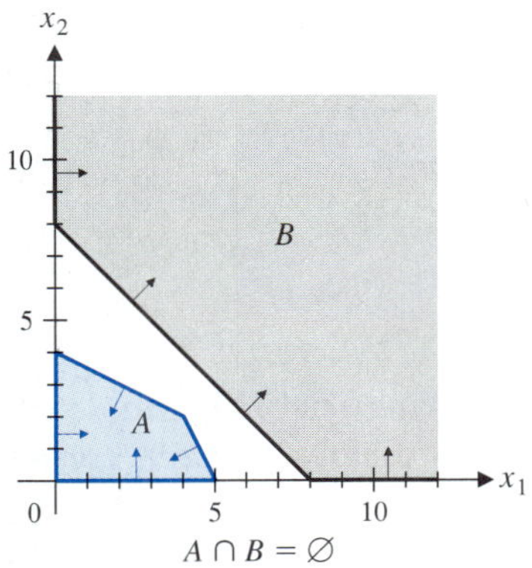

FIGURE 3

For an illustration of Theorem 2C, consider the following:

$$\begin{aligned} \text{Maximize} \quad & P = 2x_1 + 3x_2 \\ \text{Subject to} \quad & x_1 + x_2 \geq 8 \\ & x_1 + 2x_2 \leq 8 \\ & 2x_1 + x_2 \leq 10 \\ & x_1, x_2 \geq 0 \end{aligned}$$

The intersection of the graphs of the constraint inequalities is the empty set (Fig. 3); hence, the *feasible region is empty* (see Appendix A-1). If this happens, then the problem should be reexamined to see if it has been formulated properly. If it has, then the management may have to reconsider items such as labor-hours, overtime, budget, and supplies allocated to the project in order to obtain a nonempty feasible region and a solution to the original problem.

■ APPLICATIONS

Example 3 ➠ **Medicine** We now convert Example 5 from the preceding section into a linear programming problem. A patient in a hospital is required to have at least 84 units of drug D_1 and 120 units of drug D_2 each day (assume that an overdosage of either drug is harmless). Each gram of substance M contains 10 units of drug D_1 and 8 units of drug D_2, and each gram of substance N contains 2 units of drug D_1 and 4 units of drug D_2. Now, suppose both M and N contain an undesirable drug D_3, 3 units per gram in M and 1 unit per gram in N. How many grams of each of substances M and N should be mixed to meet the minimum daily requirements and at the same time minimize the intake of drug D_3? How many units of the undesirable drug D_3 will be in this mixture?

SOLUTION **Step 1.** Summarize relevant material in table form:

	AMOUNT OF DRUG PER GRAM		MINIMUM DAILY
	Substance M	*Substance N*	REQUIREMENT
DRUG D_1	10 units	2 units	84 units
DRUG D_2	8 units	4 units	120 units
DRUG D_3	3 units	1 unit	

Step 2. Form a mathematical model for the problem.

Let: x_1 = Number of grams of substance M used
x_2 = Number of grams of substance N used } Decision variables

We form the linear objective function

$$C = 3x_1 + x_2$$

which gives the amount of the undesirable drug D_3 in x_1 grams of M and x_2 grams of N. Proceeding as before, we formulate the following mathematical model for the problem:

$$\begin{aligned} \text{Minimize} \quad & C = 3x_1 + x_2 && \text{Objective function} \\ \text{Subject to} \quad & 10x_1 + 2x_2 \geq 84 && \text{Drug } D_1 \text{ constraint} \\ & 8x_1 + 4x_2 \geq 120 && \text{Drug } D_2 \text{ constraint} \\ & x_1, x_2 \geq 0 && \text{Nonnegative constraints} \end{aligned}$$

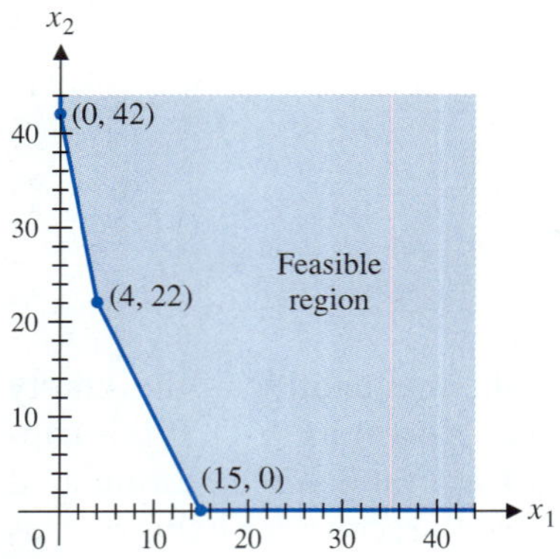

FIGURE 4

Step 3. Graph the feasible region. Then, after checking Theorem 2 to determine that an optimal solution exists, find the coordinates of each corner point. Solving the system of constraint inequalities graphically, we obtain the feasible region shown in Figure 4. Since the feasible region is unbounded and the coefficients of the objective function are positive, this minimization problem has a solution.

Step 4. Evaluate the objective function at each corner point, as shown in the table.

CORNER POINT (x_1, x_2)	$C = 3x_1 + x_2$
(0, 42)	42
(4, 22)	34
(15, 0)	45

Step 5. Determine the optimal solution from step 4. The optimal solution is $C = 34$ at the corner point (4, 22).

Step 6. Interpret the optimal solution in terms of the original problem. If we use 4 grams of substance M and 22 grams of substance N, we will supply the minimum daily requirements for drugs D_1 and D_2 and minimize the intake of the undesirable drug D_3 at 34 units. (Any other combination of M and N from the feasible region will result in a larger amount of the undesirable drug D_3.)

Matched Problem 3 **Agriculture** A chicken farmer can buy a special food mix A at 20¢ per pound and a special food mix B at 40¢ per pound. Each pound of mix A contains 3,000 units of nutrient N_1 and 1,000 units of nutrient N_2; each pound of mix B contains 4,000 units of nutrient N_1 and 4,000 units of nutrient N_2. If the minimum daily requirements for the chickens collectively are 36,000 units of nutrient N_1 and 20,000 units of nutrient N_2, how many pounds of each food mix should be used each day to minimize daily food costs while meeting (or exceeding) the minimum daily nutrient requirements? What is the minimum daily cost?

REMARK

Refer to Example 3. If we change the minimum requirement for drug D_2 from 120 to 125, the optimal solution changes to 3.6 grams of substance M and 24.1 grams of substance N, correct to one decimal place.

Now refer to Example 1. If we change the maximum labor-hours available per day in the assembly department from 84 to 79, the solution changes to 15 standard tents and 8.5 expedition tents.

We can measure 3.6 grams of substance M and 24.1 grams of substance N, but how can we make 8.5 tents? Should we make 8 tents? Or 9 tents? If the solutions to a problem must be integers and the optimal solution found graphically involves decimals, rounding the decimal value to the nearest integer does not always produce the *optimal integer solution* (see Problem 36 in Exercise 5-2). Finding optimal integer solutions to a linear programming problem is called *integer programming* and requires special techniques that are beyond the scope of this book. As mentioned earlier, if we encounter a solution like 8.5 tents per day, we will interpret this as an *average* value over many days of production.

Answers to Matched Problems

1. (A)

	LABOR-HOURS REQUIRED *Two-person boat*	*Four-person boat*	MAXIMUM LABOR-HOURS AVAILABLE PER MONTH
CUTTING DEPARTMENT	0.9	1.8	864
ASSEMBLY DEPARTMENT	0.8	1.2	672
PROFIT PER BOAT	\$25	\$40	

(B) x_1 = Number of two-person boats produced each month
x_2 = Number of four-person boats produced each month

(C) $P = 25x_1 + 40x_2$ (D) $0.9x_1 + 1.8x_2 \le 864$
$0.8x_1 + 1.2x_2 \le 672$
$x_1, x_2 \ge 0$

(E)

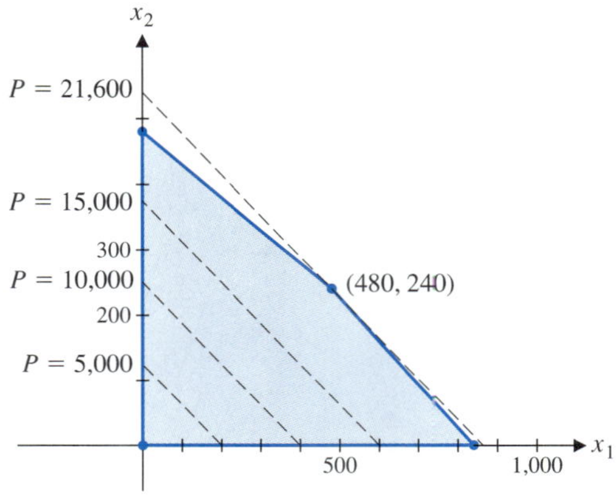

(F) 480 two-person boats, 240 four-person boats; Max P = \$21,600 per month

2. (A) Min $z = 24$ at $(3, 6)$; Max $z = 40$ at $(2, 16)$ and $(8, 4)$ (multiple optimal solution)
(B) Min $z = 90$ at $(0, 18)$; no maximum value

3. 8 lb of mix *A*, 3 lb of mix *B*; Min C = \$2.80 per day

EXERCISE 5-2

A *Find the maximum value of each objective function in Problems 1–4 over the feasible region S shown in the figure.*

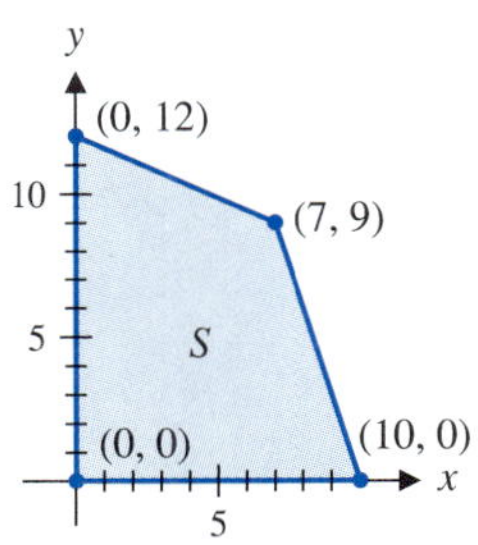

1. $z = x + y$ **2.** $z = 4x + y$

3. $z = 3x + 7y$ **4.** $z = 9x + 3y$

Find the minimum value of each objective function in Problems 5–8 over the feasible region T shown in the figure.

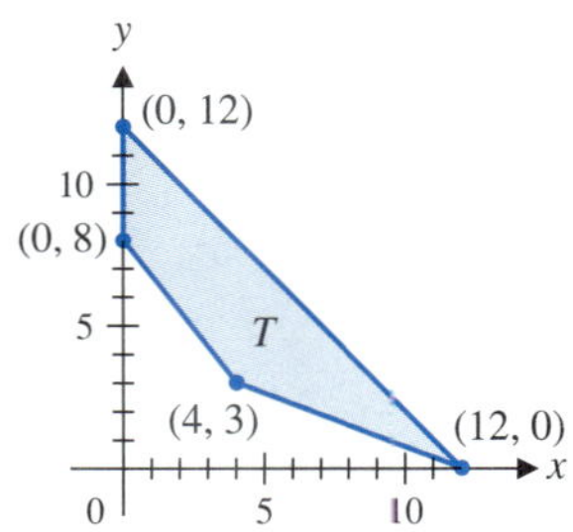

5. $z = 7x + 4y$ **6.** $z = 7x + 9y$

7. $z = 3x + 8y$ **8.** $z = 5x + 4y$

B *Solve the linear programming problems stated in Problems 9–26.*

9. Maximize $P = 5x_1 + 5x_2$
Subject to
$$\begin{aligned} 2x_1 + x_2 &\leqslant 10 \\ x_1 + 2x_2 &\leqslant 8 \\ x_1, x_2 &\geqslant 0 \end{aligned}$$

10. Maximize $P = 3x_1 + 2x_2$
Subject to
$$\begin{aligned} 6x_1 + 3x_2 &\leqslant 24 \\ 3x_1 + 6x_2 &\leqslant 30 \\ x_1, x_2 &\geqslant 0 \end{aligned}$$

11. Minimize and maximize
$$z = 2x_1 + 3x_2$$
Subject to
$$\begin{aligned} 2x_1 + x_2 &\geqslant 10 \\ x_1 + 2x_2 &\geqslant 8 \\ x_1, x_2 &\geqslant 0 \end{aligned}$$

12. Minimize and maximize
$$z = 8x_1 + 7x_2$$
Subject to
$$\begin{aligned} 4x_1 + 3x_2 &\geqslant 24 \\ 3x_1 + 4x_2 &\geqslant 8 \\ x_1, x_2 &\geqslant 0 \end{aligned}$$

13. Maximize $P = 30x_1 + 40x_2$
Subject to
$$\begin{aligned} 2x_1 + x_2 &\leqslant 10 \\ x_1 + x_2 &\leqslant 7 \\ x_1 + 2x_2 &\leqslant 12 \\ x_1, x_2 &\geqslant 0 \end{aligned}$$

14. Maximize $P = 20x_1 + 10x_2$
Subject to
$$\begin{aligned} 3x_1 + x_2 &\leqslant 21 \\ x_1 + x_2 &\leqslant 9 \\ x_1 + 3x_2 &\leqslant 21 \\ x_1, x_2 &\geqslant 0 \end{aligned}$$

15. Minimize and maximize
$$z = 10x_1 + 30x_2$$
Subject to
$$\begin{aligned} 2x_1 + x_2 &\geqslant 16 \\ x_1 + x_2 &\geqslant 12 \\ x_1 + 2x_2 &\geqslant 14 \\ x_1, x_2 &\geqslant 0 \end{aligned}$$

16. Minimize and maximize
$$z = 400x_1 + 100x_2$$
Subject to
$$\begin{aligned} 3x_1 + x_2 &\geqslant 24 \\ x_1 + x_2 &\geqslant 16 \\ x_1 + 3x_2 &\geqslant 30 \\ x_1, x_2 &\geqslant 0 \end{aligned}$$

17. Minimize and maximize
$$P = 30x_1 + 10x_2$$
Subject to
$$\begin{aligned} 2x_1 + 2x_2 &\geqslant 4 \\ 6x_1 + 4x_2 &\leqslant 36 \\ 2x_1 + x_2 &\leqslant 10 \\ x_1, x_2 &\geqslant 0 \end{aligned}$$

18. Minimize and maximize
$$P = 2x_1 + x_2$$
Subject to
$$\begin{aligned} x_1 + x_2 &\geqslant 2 \\ 6x_1 + 4x_2 &\leqslant 36 \\ 4x_1 + 2x_2 &\leqslant 20 \\ x_1, x_2 &\geqslant 0 \end{aligned}$$

19. Minimize and maximize
$$P = 3x_1 + 5x_2$$
Subject to
$$\begin{aligned} x_1 + 2x_2 &\leqslant 6 \\ x_1 + x_2 &\leqslant 4 \\ 2x_1 + 3x_2 &\geqslant 12 \\ x_1, x_2 &\geqslant 0 \end{aligned}$$

20. Minimize and maximize
$$P = -x_1 + 3x_2$$
Subject to
$$\begin{aligned} 2x_1 - x_2 &\geqslant 4 \\ -x_1 + 2x_2 &\leqslant 4 \\ x_2 &\leqslant 6 \\ x_1, x_2 &\geqslant 0 \end{aligned}$$

21. Minimize and maximize
$$P = 20x_1 + 10x_2$$
Subject to
$$\begin{aligned} 2x_1 + 3x_2 &\geqslant 30 \\ 2x_1 + x_2 &\leqslant 26 \\ -2x_1 + 5x_2 &\leqslant 34 \\ x_1, x_2 &\geqslant 0 \end{aligned}$$

22. Minimize and maximize
$$P = 12x_1 + 14x_2$$
Subject to
$$\begin{aligned} -2x_1 + x_2 &\geqslant 6 \\ x_1 + x_2 &\leqslant 15 \\ 3x_1 - x_2 &\geqslant 0 \\ x_1, x_2 &\geqslant 0 \end{aligned}$$

23. Maximize $P = 20x_1 + 30x_2$
Subject to
$$\begin{aligned} 0.6x_1 + 1.2x_2 &\leqslant 960 \\ 0.03x_1 + 0.04x_2 &\leqslant 36 \\ 0.3x_1 + 0.2x_2 &\leqslant 270 \\ x_1, x_2 &\geqslant 0 \end{aligned}$$

24. Minimize $C = 30x_1 + 10x_2$
Subject to
$$\begin{aligned} 1.8x_1 + 0.9x_2 &\geq 270 \\ 0.3x_1 + 0.2x_2 &\geq 54 \\ 0.01x_1 + 0.03x_2 &\geq 3.9 \\ x_1, x_2 &\geq 0 \end{aligned}$$

25. Maximize $P = 525x_1 + 478x_2$
C Subject to
$$\begin{aligned} 275x_1 + 322x_2 &\leq 3{,}381 \\ 350x_1 + 340x_2 &\leq 3{,}762 \\ 425x_1 + 306x_2 &\leq 4{,}114 \\ x_1, x_2 &\geq 0 \end{aligned}$$

26. Maximize $P = 300x_1 + 460x_2$
C Subject to
$$\begin{aligned} 245x_1 + 452x_2 &\leq 4{,}181 \\ 290x_1 + 379x_2 &\leq 3{,}888 \\ 390x_1 + 299x_2 &\leq 4{,}407 \\ x_1, x_2 &\geq 0 \end{aligned}$$

In Problems 27 and 28, explain why Theorem 2 cannot be used to conclude that a maximum or minimum value exists. Graph the feasible regions and use graphs of the objective function $z = x_1 - x_2$ for various values of z to discuss the existence of a maximum value and a minimum value.

27. Minimize and maximize

$$z = x_1 - x_2$$

Subject to
$$\begin{aligned} x_1 - 2x_2 &\leq 0 \\ 2x_1 - x_2 &\leq 6 \\ x_1, x_2 &\geq 0 \end{aligned}$$

28. Minimize and maximize

$$z = x_1 - x_2$$

Subject to
$$\begin{aligned} x_1 - 2x_2 &\geq -6 \\ 2x_1 - x_2 &\geq 0 \\ x_1, x_2 &\geq 0 \end{aligned}$$

C

29. The corner points for the bounded feasible region determined by the system of linear inequalities

$$\begin{aligned} x_1 + 2x_2 &\leq 10 \\ 3x_1 + x_2 &\leq 15 \\ x_1, x_2 &\geq 0 \end{aligned}$$

are $O = (0, 0)$, $A = (0, 5)$, $B = (4, 3)$, and $C = (5, 0)$. If $P = ax_1 + bx_2$ and $a, b > 0$, determine conditions on a and b that will ensure that the maximum value of P occurs:

(A) Only at A
(B) Only at B
(C) Only at C
(D) At both A and B
(E) At both B and C

30. The corner points for the feasible region determined by the system of linear inequalities

$$\begin{aligned} x_1 + 4x_2 &\geq 30 \\ 3x_1 + x_2 &\geq 24 \\ x_1, x_2 &\geq 0 \end{aligned}$$

are $A = (0, 24)$, $B = (6, 6)$, and $D = (30, 0)$. If $C = ax_1 + bx_2$ and $a, b > 0$, determine conditions on a and b that will ensure that the minimum value of C occurs:

(A) Only at A
(B) Only at B
(C) Only at D
(D) At both A and B
(E) At both B and D

Applications

Business & Economics

31. *Manufacturing—resource allocation.* A manufacturing company makes two types of water skis, a trick ski and a slalom ski. The relevant manufacturing data are given in the table.

Department	Labor-hours per ski: Trick ski	Labor-hours per ski: Slalom ski	Maximum labor-hours available per day
Fabricating	6	4	108
Finishing	1	1	24

(A) If the profit on a trick ski is \$40 and the profit on a slalom ski is \$30, how many of each type of ski should be manufactured each day to realize a maximum profit? What is the maximum profit?
(B) Discuss the effect on the production schedule and the maximum profit if the profit on a slalom ski decreases to \$25.
(C) Discuss the effect on the production schedule and the maximum profit if the profit on a slalom ski increases to \$45.

32. *Manufacturing—resource allocation.* A furniture manufacturing company manufactures dining room tables and chairs. The relevant manufacturing data are given in the table.

Department	Labor-Hours per Unit: Table	Labor-Hours per Unit: Chair	Maximum Labor-Hours Available per Day
Assembly	8	2	400
Finishing	2	1	120
Profit per Unit	$90	$25	

(A) How many tables and chairs should be manufactured each day to realize a maximum profit? What is the maximum profit?

(B) Discuss the effect on the production schedule and the maximum profit if the marketing department of the company decides that the number of chairs produced should be at least four times the number of tables produced.

33. *Manufacturing—production scheduling.* A furniture company has two plants that produce the lumber used in manufacturing tables and chairs. In 1 day of operation, plant A can produce the lumber required to manufacture 20 tables and 60 chairs, and plant B can produce the lumber required to manufacture 25 tables and 50 chairs. The company needs enough lumber to manufacture at least 200 tables and 500 chairs.

(A) If it costs $1,000 to operate plant A for 1 day and $900 to operate plant B for 1 day, how many days should each plant be operated in order to produce a sufficient amount of lumber at a minimum cost? What is the minimum cost?

(B) Discuss the effect on the operating schedule and the minimum cost if the daily cost of operating plant A is reduced to $600 and all other data in part (A) remain the same.

(C) Discuss the effect on the operating schedule and the minimum cost if the daily cost of operating plant B is reduced to $800 and all other data in part (A) remain the same.

34. *Manufacturing—resource allocation.* An electronics firm manufactures two types of personal computers, a standard model and a portable model. The production of a standard computer requires a capital expenditure of $400 and 40 hours of labor. The production of a portable computer requires a capital expenditure of $250 and 30 hours of labor. The firm has $20,000 capital and 2,160 labor-hours available for production of standard and portable computers.

(A) What is the maximum number of computers the company is capable of producing?

(B) If each standard computer contributes a profit of $320 and each portable model contributes a profit of $220, how much profit will the company make by producing the maximum number of computers determined in part (A)? Is this the maximum profit? If not, what is the maximum profit?

35. *Transportation.* The officers of a high school senior class are planning to rent buses and vans for a class trip. Each bus can transport 40 students, requires 3 chaperones, and costs $1,200 to rent. Each van can transport 8 students, requires 1 chaperone, and costs $100 to rent. Since there are 400 students in the senior class that may be eligible to go on the trip, the officers must plan to accommodate at least 400 students. Since only 36 parents have volunteered to serve as chaperones, the officers must plan to use at most 36 chaperones. How many vehicles of each type should the officers rent in order to minimize the transportation costs? What are the minimal transportation costs?

36. *Transportation.* Refer to Problem 35. If each van can transport 7 people and there are 35 available chaperones, show that the optimal solution found graphically involves decimals. Find all feasible solutions with integer coordinates and identify the one that minimizes the transportation costs. Can this optimal integer solution be obtained by rounding the optimal decimal solution? Explain.

37. *Investment.* An investor has $60,000 to invest in a CD and a mutual fund. The CD yields 5% and the mutual fund yields on the average 9%. The mutual fund requires a minimum investment of $10,000 and the investor requires that twice as much should be invested in CD's as in the mutual fund. How much should be invested in CD's and how much in the mutual fund to maximize the return? What is the maximum return?

38. *Investment.* An investor has $24,000 to invest in bonds of AAA and B qualities. The AAA bonds yield on the average 6% and the B bonds yield 10%. The investor requires that at least three times as much money should be invested in AAA bonds as in B bonds. How much should be invested in each type of bond to maximize the return? What is the maximum return?

39. *Pollution control.* Because of new federal regulations on pollution, a chemical plant introduced a new, more expensive process to supplement or replace an older process used in the production of a particular chemical. The older process emitted 20 grams of sulfur dioxide and 40 grams of particulate matter into the atmosphere for each gallon of chemical produced. The new process emits 5 grams of sulfur dioxide and 20 grams of particulate matter for each gallon produced. The company makes a profit of 60¢ per gallon and 20¢ per gallon on the old and new processes, respectively.

(A) If the government allows the plant to emit no more than 16,000 grams of sulfur dioxide and 30,000 grams of particulate matter daily, how many gallons of the chemical should be produced by each process to maximize daily profit? What is the maximum daily profit?

(B) Discuss the effect on the production schedule and the maximum profit if the government decides to restrict emissions of sulfur dioxide to 11,500 grams daily and all other data remain unchanged.

(C) Discuss the effect on the production schedule and the maximum profit if the government decides to restrict emissions of sulfur dioxide to 7,200 grams daily and all other data remain unchanged.

40. *Capital expansion.* A fast-food chain plans to expand by opening several new restaurants. The chain operates two types of restaurants, drive-through and full-service. A drive-through restaurant costs $100,000 to construct, requires 5 employees, and has an expected annual revenue of $200,000. A full-service restaurant costs $150,000 to construct, requires 15 employees, and has an expected annual revenue of $500,000. The chain has $2,400,000 in capital available for expansion. Labor contracts require that they hire no more than 210 employees, and licensing restrictions require that they open no more than 20 new restaurants. How many restaurants of each type should the chain open in order to maximize the expected revenue? What is the maximum expected revenue? How much of their capital will they use and how many employees will they hire?

Life Sciences

41. *Nutrition—plants.* A fruit grower can use two types of fertilizer in his orange grove, brand *A* and brand *B*. The amounts (in pounds) of nitrogen, phosphoric acid, and chlorine in a bag of each brand are given in the table. Tests indicate that the grove needs at least 1,000 pounds of phosphoric acid and at most 400 pounds of chlorine.

	Pounds per bag	
	Brand A	*Brand B*
Nitrogen	8	3
Phosphoric acid	4	4
Chlorine	2	1

(A) If the grower wants to maximize the amount of nitrogen added to the grove, how many bags of each mix should be used? How much nitrogen will be added?

(B) If the grower wants to minimize the amount of nitrogen added to the grove, how many bags of each mix should be used? How much nitrogen will be added?

42. *Nutrition—people.* A dietitian in a hospital is to arrange a special diet composed of two foods, *M* and *N*. Each ounce of food *M* contains 30 units of calcium, 10 units of iron, 10 units of vitamin A, and 8 units of cholesterol. Each ounce of food *N* contains 10 units of calcium, 10 units of iron, 30 units of vitamin A, and 4 units of cholesterol. If the minimum daily requirements are 360 units of calcium, 160 units of iron, and 240 units of vitamin A, how many ounces of each food should be used to meet the minimum requirements and at the same time minimize the cholesterol intake? What is the minimum cholesterol intake?

43. *Nutrition—plants.* A farmer can buy two types of plant food, mix *A* and mix *B*. Each cubic yard of mix *A* contains 20 pounds of phosphoric acid, 30 pounds of nitrogen, and 5 pounds of potash. Each cubic yard of mix *B* contains 10 pounds of phosphoric acid, 30 pounds of nitrogen, and 10 pounds of potash. The minimum monthly requirements are 460 pounds of phosphoric acid, 960 pounds of nitrogen, and 220 pounds of potash. If mix *A* costs $30 per cubic yard and mix *B* costs $35 per cubic yard, how many cubic yards of each mix should the farmer blend to meet the minimum monthly requirements at a minimal cost? What is this cost?

44. *Nutrition—animals.* A laboratory technician in a medical research center is asked to formulate a diet from two commercially packaged foods, food *A* and food *B*, for a group of animals. Each ounce of food *A* contains 8 units of fat, 16 units of carbohydrate, and 2 units of protein. Each ounce of food *B* contains 4 units of fat, 32 units of carbohydrate, and 8 units of protein. The minimum daily requirements are 176 units of fat, 1,024 units of carbohydrate, and 384 units of protein. If food *A* costs 5¢ per ounce and food *B* costs 5¢ per ounce, how many ounces of each food should be used to meet the minimum daily requirements at the least cost? What is the cost for this amount of food?

Social Sciences

45. *Psychology.* In an experiment on conditioning, a psychologist uses two types of Skinner boxes with mice and rats. The amount of time (in minutes) each mouse and each rat spends in each box per day is given in the table. What is the maximum number of mice and rats that can be used in this experiment? How many mice and how many rats produce this maximum?

	Time		Maximum time available per day
	Mice	*Rats*	
Skinner box *A*	10 min	20 min	800 min
Skinner box *B*	20 min	10 min	640 min

46. *Sociology.* A city council voted to conduct a study on inner-city community problems. A nearby university was contacted to provide sociologists and research assistants. Allocation of time and costs per week are given in the table. How many sociologists and how many research assistants should be hired to minimize the cost and meet the weekly labor-hour requirements? What is the minimum weekly cost?

	Labor-Hours		Minimum Labor-Hours Needed per Week
	Sociologist	*Research assistant*	
Fieldwork	10	30	180
Research Center	30	10	140
Costs per week	\$500	\$300	

SECTION 5-3

A Geometric Introduction to the Simplex Method

- Standard Maximization Problems in Standard Form
- Slack Variables
- Basic and Nonbasic Variables; Basic Solutions and Basic Feasible Solutions
- Basic Feasible Solutions and the Simplex Method

The geometric method of solving linear programming problems provides us with an overview of the subject and some useful terminology. But, practically speaking, the method is useful only for problems involving two decision variables and relatively few problem constraints. What happens when we need more decision variables and more problem constraints? We use an algebraic method called the *simplex method,* which was developed by George B. Dantzig in 1947 while on assignment to the U.S. Department of the Air Force. Ideally suited to computer use, the method is used routinely on applied problems involving hundreds and even thousands of variables and problem constraints.

The algebraic procedures utilized in the simplex method require the problem constraints to be written as equations rather than inequalities. This new form of the linear programming problem also prompts the use of some new terminology. We introduce this new form and associated terminology through a simple example and an appropriate geometric interpretation. From this example we can illustrate what the simplex method does geometrically before we immerse ourselves in the algebraic details of the process.

Standard Maximization Problems in Standard Form

We now return to the tent production problem in Example 1 from the last section. Recall the mathematical model for the problem:

$$\begin{aligned} &\text{Maximize} & P &= 50x_1 + 80x_2 & &\text{Objective function} \\ &\text{Subject to} & x_1 + 2x_2 &\le 32 & &\text{Cutting department constraint} \\ & & 3x_1 + 4x_2 &\le 84 & &\text{Assembly department constraint} \\ & & x_1, x_2 &\ge 0 & &\text{Nonnegative constraints} \end{aligned} \qquad (1)$$

where the decision variables x_1 and x_2 are the number of standard and expedition tents, respectively, produced each day.

Notice that the problem constraints involve $\leq$ inequalities with positive constants to the right of the inequality. Maximization problems that satisfy this condition are called *standard maximization problems.* In this and the next section we will restrict our attention to standard maximization problems.

A Standard Maximization Problem in Standard Form

A linear programming problem is said to be a **standard maximization problem in standard form** if its mathematical model is of the following form:

Maximize the objective function

$$P = c_1x_1 + c_2x_2 + \cdots + c_nx_n$$

Subject to problem constraints of the form

$$a_1x_1 + a_2x_2 + \cdots + a_nx_n \leq b \qquad b \geq 0$$

With nonnegative constraints

$$x_1, x_2, \ldots, x_n \geq 0$$

[*Note:* Mathematical model (1) above is a standard maximization problem in standard form. Also note that the coefficients of the objective function can be any real numbers.]

Explore–Discuss 1

Find an example of a standard maximization problem in standard form involving two variables and one problem constraint such that:

(A) The feasible region is bounded.
(B) The feasible region is unbounded.

Is it possible for a standard maximization problem to have no solution? Explain.

Slack Variables

To adapt a linear programming problem to the matrix methods used in the simplex process (as discussed in the next section), we convert the problem constraint inequalities into a system of linear equations by using a simple device called a *slack variable.* In particular, to convert the system of problem constraint inequalities from (1),

$$\begin{aligned} x_1 + 2x_2 &\leq 32 \quad \text{Cutting department constraint} \\ 3x_1 + 4x_2 &\leq 84 \quad \text{Assembly department constraint} \end{aligned} \tag{2}$$

into a system of equations, we add variables s_1 and s_2 to the left sides of (2) to obtain

$$\begin{aligned} x_1 + 2x_2 + s_1 &= 32 \\ 3x_1 + 4x_2 + s_2 &= 84 \end{aligned} \tag{3}$$

The variables s_1 and s_2 are called **slack variables** because each makes up the difference (takes up the slack) between the left and right sides of the inequalities in (2). For example, if we produced 20 standard tents ($x_1 = 20$) and 5 expedition tents ($x_2 = 5$), then the number of labor-hours used in the cutting department would be $20 + 2(5) = 30$, leaving a slack of 2 unused labor-hours out of the 32 available. Thus, s_1 would have the value of 2.

Notice that if the decision variables x_1 and x_2 satisfy the system of constraint inequalities (2), then the slack variables s_1 and s_2 are nonnegative. We will have more to say about this later in this discussion.

Basic and Nonbasic Variables; Basic Solutions and Basic Feasible Solutions

Observe that system (3) has infinitely many solutions—just solve for s_1 and s_2 in terms of x_1 and x_2, and then assign x_1 and x_2 arbitrary values. Certain solutions of system (3), called *basic solutions,* are related to the intersection points of the (extended) boundary lines of the feasible region in Figure 1.

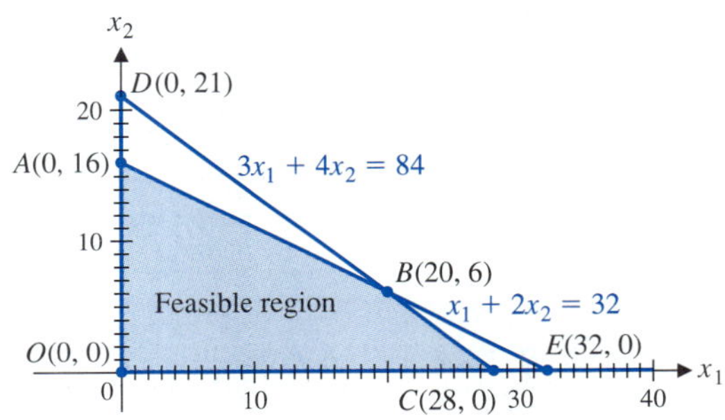

FIGURE 1

How are basic solutions to system (3) determined? System (3) involves four variables and two equations. We divide the four variables into two groups, called *basic variables* and *nonbasic variables,* as follows: Basic variables are selected arbitrarily with the restriction that there be as many basic variables as there are equations. The remaining variables are nonbasic variables.

Since system (3) has two equations, we can select any two of the four variables as basic variables. The remaining two variables are then nonbasic variables. A solution found by setting the two nonbasic variables equal to 0 and solving for the two basic variables is a basic solution. [Note that setting two variables equal to 0 in system (3) results in a system of two equations with two variables, which has (from Chapter 4) exactly one solution, infinitely many solutions, or no solution.]

Example 1 **Basic Solutions and the Feasible Region**

(A) Find two basic solutions for system (3) by first selecting s_1 and s_2 as basic variables, and then by selecting x_2 and s_1 as basic variables.

(B) Associate each basic solution found in part (A) with an intersection point of the (extended) boundary lines of the feasible region in Figure 1, and indicate which boundary lines produce each intersection point.

(C) Indicate which of the intersection points found in part (B) are in the feasible region.

SOLUTION (A) With s_1 and s_2 selected as basic variables, x_1 and x_2 are nonbasic variables. A basic solution is found by setting the nonbasic variables equal to 0 and solving for the basic variables. If $x_1 = 0$ and $x_2 = 0$, then system (3) becomes

$$\begin{aligned} s_1 &= 32 \\ s_2 &= 84 \end{aligned} \qquad \begin{aligned} \overset{0}{x_1} + \overset{0}{2x_2} + s_1 \quad &= 32 \\ 3x_1 + 4x_2 \quad + s_2 &= 84 \end{aligned}$$

and the basic solution is

$$x_1 = 0, \quad x_2 = 0, \quad s_1 = 32, \quad s_2 = 84 \tag{4}$$

If we select x_2 and s_1 as basic variables, then x_1 and s_2 are nonbasic variables. Setting the nonbasic variables equal to 0, system (3) becomes

$$\begin{aligned} 2x_2 + s_1 &= 32 \\ 4x_2 &= 84 \end{aligned} \qquad \begin{aligned} \overset{0}{x_1} + 2x_2 + s_1 \quad &= 32 \\ 3x_1 + 4x_2 \quad + \overset{0}{s_2} &= 84 \end{aligned}$$

Solving, we see that $x_2 = 21$ and $s_1 = -10$, and the basic solution is

$$x_1 = 0, \quad x_2 = 21, \quad s_1 = -10, \quad s_2 = 0 \tag{5}$$

(B) Basic solution (4)—since $x_1 = 0$ and $x_2 = 0$—corresponds to the origin $O(0, 0)$ in Figure 1, which is the intersection of the boundary lines $x_1 = 0$ and $x_2 = 0$. Basic solution (5)—since $x_1 = 0$ and $x_2 = 21$—corresponds to the intersection point $D(0, 21)$, which is the intersection of the boundary lines $x_1 = 0$ and $3x_1 + 4x_2 = 84$.

(C) The intersection point $O(0, 0)$ is in the feasible region; hence, it is natural to call the corresponding basic solution a *basic feasible solution.* The intersection point $D(0, 21)$ is not in the feasible region; hence, the corresponding basic solution is not feasible.

Matched Problem 1 (A) Find two basic solutions for system (3) by first selecting x_1 and s_1 as basic variables, and then by selecting x_1 and s_2 as basic variables.

(B) Associate each basic solution found in part (A) with an intersection point of the (extended) boundary lines of the feasible region in Figure 1, and indicate which boundary lines produce each intersection point.

(C) Indicate which of the intersection points found in part (B) are in the feasible region.

Proceeding systematically as in Example 1 and Matched Problem 1, we can obtain all basic solutions to system (3). The results are summarized in Table 1, which also includes geometric interpretations of the basic solutions relative to Figure 1. Figure 2 summarizes these interpretations. A careful study of Table 1 and Figure 2 is very worthwhile. (Note that to be sure we have listed all possible basic solutions in Table 1, it is convenient to organize the table in terms of the zero values of the nonbasic variables.)

Table 1
BASIC SOLUTIONS

Basic Solutions				Intersection	Intersecting	
x_1	x_2	s_1	s_2	Point	Boundary Lines	Feasible
0	0	32	84	$O(0,0)$	$x_1 = 0$ $x_2 = 0$	Yes
0	16	0	20	$A(0,16)$	$x_1 = 0$ $x_1 + 2x_2 = 32$	Yes
0	21	−10	0	$D(0,21)$	$x_1 = 0$ $3x_1 + 4x_2 = 84$	No
32	0	0	−12	$E(32,0)$	$x_2 = 0$ $x_1 + 2x_2 = 32$	No
28	0	4	0	$C(28,0)$	$x_2 = 0$ $3x_1 + 4x_2 = 84$	Yes
20	6	0	0	$B(20,6)$	$x_1 + 2x_2 = 32$ $3x_1 + 4x_2 = 84$	Yes

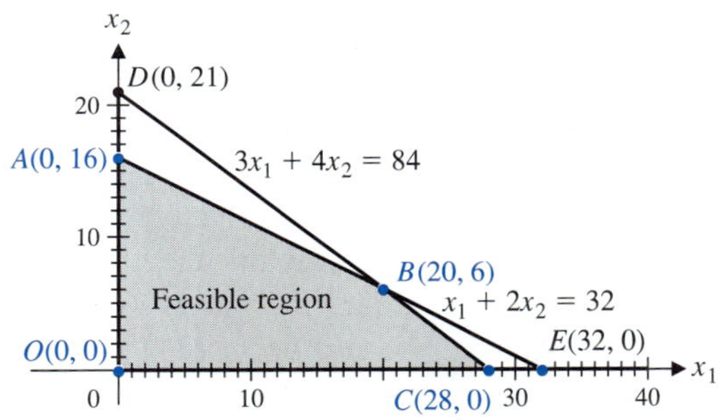

FIGURE 2
Basic feasible solutions: O, A, B, C
Basic solutions that are not feasible: D, E

OBSERVATIONS FROM TABLE 1 AND FIGURE 2 IMPORTANT TO THE DEVELOPMENT OF THE SIMPLEX METHOD

1. In Table 1, observe that a basic solution that is not feasible includes at least one negative value and that a basic feasible solution does not include any negative values. That is, we can determine the feasibility of a basic solution simply by examining the signs of all the variables in the solution.
2. In Table 1 and Figure 2, observe that basic feasible solutions are associated with the corner points of the feasible region, which include the optimal solution to the original linear programming problem.

Before proceeding further, let us formalize the definitions alluded to in the discussion above so that they apply to standard maximization problems in general, without any reference to geometric forms.

Basic Variables and Nonbasic Variables; Basic Solutions and Basic Feasible Solutions

Given a system of linear equations associated with a linear programming problem (such a system will always have more variables than equations):

> The variables are divided into two (mutually exclusive) groups, as follows: **Basic variables** are selected arbitrarily with the one restriction that there be as many basic variables as there are equations. The remaining variables are called **nonbasic variables.**
>
> A solution found by setting the nonbasic variables equal to 0 and solving for the basic variables is called a **basic solution.** If a basic solution has no negative values, it is a **basic feasible solution.**

Example 2 **Basic Variables and Basic Solutions** Suppose there is a system of three problem constraint equations with eight (slack and decision) variables associated with a standard maximization problem.

(A) How many basic variables and how many nonbasic variables are associated with the system?
(B) Setting the nonbasic variables equal to 0 will result in a system of how many linear equations with how many variables?
(C) If a basic solution has all nonnegative elements, is it feasible or not feasible?

Solution
(A) Since there are three equations in the system, there should be three basic variables and five nonbasic variables.
(B) Three equations with three variables
(C) Feasible

Matched Problem 2 Suppose there is a system of five problem constraint equations with eleven (slack and decision) variables associated with a standard maximization problem.

(A) How many basic variables and how many nonbasic variables are associated with the system?
(B) Setting the nonbasic variables equal to 0 will result in a system of how many linear equations with how many variables?
(C) If a basic solution has one or more negative elements, is it feasible or not feasible?

Basic Feasible Solutions and the Simplex Method

The following important theorem, which is equivalent to the fundamental theorem (Theorem 1 in the preceding section), is stated without proof:

Theorem 1 FUNDAMENTAL THEOREM OF LINEAR PROGRAMMING—VERSION 2

If the optimal value of the objective function in a linear programming problem exists, then that value must occur at one (or more) of the basic feasible solutions.

Now you can understand why the concepts of basic and nonbasic variables and basic solutions and basic feasible solutions are so important—these concepts are central to the process of finding optimal solutions to linear programming problems.

Explore–Discuss 2

If we know that a standard maximization problem has an optimal solution, how could we use a table of basic solutions like Table 1 to find the optimal solution?

We have taken the first step toward finding a general method of solving linear programming problems involving any number of variables and problem constraints. That is, **we have found a method of identifying all the corner points (basic feasible solutions) of a feasible region without drawing its graph.** This is a critical step if we want to consider problems with more than two decision variables. Unfortunately, the number of corner points increases dramatically as the number of variables and constraints increases. In real-world problems, it is not practical to find all the corner points in order to find the optimal solution. Thus, the next step is to find a method of locating the optimal solution without finding every corner point. The procedure for doing this is the simplex method mentioned at the beginning of this section.

The simplex method, using a special matrix and row operations, automatically moves from one basic feasible solution to another—that is, from one corner point of the feasible region to another—each time getting closer to an optimal solution (if one exists), until an optimal solution is reached. Then the process stops. A remarkable property of the simplex method is that in large linear programming problems it usually arrives at an optimal solution (if one exists) by testing relatively few of the large number of basic feasible solutions (corner points) available.

With this background, we are now ready to discuss the algebraic details of the simplex method in the next section.

Answers to Matched Problems

1. (A) Basic solution corresponding to basic variables x_1 and s_1: $x_1 = 28, x_2 = 0, s_1 = 4, s_2 = 0$. Basic solution corresponding to basic variables x_1 and s_2: $x_1 = 32, x_2 = 0, s_1 = 0, s_2 = -12$.
(B) The first basic solution corresponds to $C(28, 0)$, which is the intersection of the boundary lines $x_2 = 0$ and $3x_1 + 4x_2 = 84$. The second basic solution corresponds to $E(32, 0)$, which is the intersection of the boundary lines $x_2 = 0$ and $x_1 + 2x_2 = 32$.
(C) $C(28, 0)$ is in the feasible region (hence, the corresponding basic solution is a basic feasible solution); $E(32, 0)$ is not in the feasible region (hence, the corresponding basic solution is not feasible).

2. (A) Five basic variables and six nonbasic variables
(B) Five equations and five variables (C) Not feasible

EXERCISE 5-3

A

1. Discuss the relationship between a standard maximization problem with two problem constraints and three decision variables, and the associated system of problem constraint equations. In particular, find each of the following quantities and explain how each was determined:
(A) The number of slack variables that must be introduced to form the system of problem constraint equations
(B) The number of basic variables and the number of nonbasic variables associated with the system
(C) The number of linear equations and the number of variables in the system formed by setting the nonbasic variables equal to 0

2. Repeat Problem 1 for a standard maximization problem with three problem constraints and four decision variables.

3. Discuss the relationship between a standard maximization problem and the associated system of problem constraint equations, if the system of problem constraint equations has nine variables, including five slack variables. In particular, find each of the following quantities and explain how each was determined:
(A) The number of constraint equations in the system
(B) The number of decision variables in the system
(C) The number of basic variables and the number of nonbasic variables associated with the system
(D) The number of linear equations and the number of variables in the system formed by setting the nonbasic variables equal to 0

4. Repeat Problem 3 if the system of problem constraint equations has ten variables, including four slack variables.

5. Listed in the table below are all the basic solutions for the system

$$\begin{aligned} 2x_1 + 3x_2 + s_1 \quad &= 24 \\ 4x_1 + 3x_2 \quad + s_2 &= 36 \end{aligned}$$

For each basic solution, identify the nonbasic variables and the basic variables. Then classify each basic solution as feasible or not feasible.

	x_1	x_2	s_1	s_2
(A)	0	0	24	36
(B)	0	8	0	12
(C)	0	12	−12	0
(D)	12	0	0	−12
(E)	9	0	6	0
(F)	6	4	0	0

6. Repeat Problem 5 for the system

$$\begin{aligned} 2x_1 + x_2 + s_1 \quad &= 30 \\ x_1 + 5x_2 \quad + s_2 &= 60 \end{aligned}$$

whose basic solutions are given in the following table:

	x_1	x_2	s_1	s_2
(A)	0	0	30	60
(B)	0	30	0	−90
(C)	0	12	18	0
(D)	15	0	0	45
(E)	60	0	−90	0
(F)	10	10	0	0

7. Listed in the table below are all the possible choices of nonbasic variables for the system

$$\begin{aligned} 2x_1 + x_2 + s_1 \quad &= 50 \\ x_1 + 2x_2 \quad + s_2 &= 40 \end{aligned}$$

In each case, find the values of the basic variables and determine whether the basic solution is feasible.

	x_1	x_2	s_1	s_2
(A)	0	0	?	?
(B)	0	?	0	?
(C)	0	?	?	0
(D)	?	0	0	?
(E)	?	0	?	0
(F)	?	?	0	0

8. Repeat Problem 7 for the system

$$\begin{aligned} x_1 + 2x_2 + s_1 \quad &= 12 \\ 3x_1 + 2x_2 \quad + s_2 &= 24 \end{aligned}$$

B *Graph the systems of inequalities in Problems 9–12. Introduce slack variables to convert each system of inequalities to a system of equations, and find all the basic solutions of the system. Construct a table (similar to Table 1) listing each basic solution, the corresponding point on the graph, and whether the basic solution is feasible. (You do not need to list the intersecting lines.)*

9. $$\begin{aligned} x_1 + x_2 &\le 16 \\ 2x_1 + x_2 &\le 20 \\ x_1, x_2 &\ge 0 \end{aligned}$$

10. $$\begin{aligned} 5x_1 + x_2 &\le 35 \\ 4x_1 + x_2 &\le 32 \\ x_1, x_2 &\ge 0 \end{aligned}$$

11. $$\begin{aligned} 2x_1 + x_2 &\le 22 \\ x_1 + x_2 &\le 12 \\ x_1 + 2x_2 &\le 20 \\ x_1, x_2 &\ge 0 \end{aligned}$$

12. $$\begin{aligned} 4x_1 + x_2 &\le 28 \\ 2x_1 + x_2 &\le 16 \\ x_1 + x_2 &\le 13 \\ x_1, x_2 &\ge 0 \end{aligned}$$

SECTION 5-4

The Simplex Method: Maximization with Problem Constraints of the Form $\leq$

- INITIAL SYSTEM
- THE SIMPLEX TABLEAU
- THE PIVOT OPERATION
- INTERPRETING THE SIMPLEX PROCESS GEOMETRICALLY
- THE SIMPLEX METHOD SUMMARIZED
- APPLICATION

We are now ready to develop the simplex method for a standard maximization problem. Specific details in the presentation of the method generally vary from from one book to another, even though the underlying process is the same. The presentation developed here emphasizes concept development and understanding.

As pointed out in the last section, the simplex method is most useful when used with computers. Consequently, it is not intended that you become expert in manually solving linear programming problems using the simplex method. But it is important that you become proficient in constructing the models for linear programming problems so that they can be solved using a computer, and it is also important that you develop skill in interpreting the results. One way to gain this proficiency and interpretive skill is to set up and manually solve a number of fairly simple linear programming problems using the simplex method. This is the main goal here and in Sections 5-5 and 5-6. To assist you in learning to develop the models, the answer sections for Exercises 5-4, 5-5, and 5-6 contain both the model and its solution. The software that accompanies *Explorations in Finite Mathematics* (see Preface) can be used to solve the linear programming problems in this chapter, as can spreadsheets, such as Excel.

INITIAL SYSTEM

We will introduce the concepts and procedures involved in the simplex method through an example—the tent production example we have discussed earlier. We restate the problem here in standard form for convenient reference:

$$\begin{array}{lrl} \text{Maximize} & P = 50x_1 + 80x_2 & \text{Objective function} \\ \text{Subject to} & \left.\begin{aligned} x_1 + 2x_2 &\leq 32 \\ 3x_1 + 4x_2 &\leq 84 \end{aligned}\right\} & \text{Problem constraints} \\ & x_1, x_2 \geq 0 & \text{Nonnegative constraints} \end{array} \qquad (1)$$

Introducing slack variables s_1 and s_2, we convert the problem constraint inequalities in (1) into the following system of problem constraint equations:

$$\begin{aligned} x_1 + 2x_2 + s_1 &= 32 \\ 3x_1 + 4x_2 + s_2 &= 84 \\ x_1, x_2, s_1, s_2 &\geq 0 \end{aligned} \qquad (2)$$

Since a basic solution of (2) is not feasible if it contains any negative values, we have also included the nonnegative constraints for both the decision variables x_1 and x_2 and the slack variables s_1 and s_2. From our discussion in the last section, we know that out of the infinitely many solutions to system (2), an optimal solution is among the basic feasible solutions, which correspond to the corner points of the feasible region.

As part of the simplex method we add the objective function equation $P = 50x_1 + 80x_2$ in the form $-50x_1 - 80x_2 + P = 0$ to system (2) to create what is called the **initial system:**

$$\begin{aligned} x_1 + 2x_2 + s_1 \qquad &= 32 \\ 3x_1 + 4x_2 \qquad + s_2 &= 84 \\ -50x_1 - 80x_2 \qquad\qquad + P &= 0 \\ x_1, x_2, s_1, s_2 &\geq 0 \end{aligned} \tag{3}$$

When we add the objective function equation to system (2), we must slightly modify the earlier definitions of basic solution and basic feasible solution so that they apply to the initial system (3).

Basic Solutions and Basic Feasible Solutions for Initial Systems

1. The objective function variable P is always selected as a basic variable and is never selected as a nonbasic variable.
2. Note that a basic solution of system (3) is also a basic solution of system (2) after P is deleted.
3. If a basic solution of system (3) is a basic feasible solution of system (2) after deleting P, then the basic solution of (3) is called a **basic feasible solution** of (3).
4. A basic feasible solution of system (3) can contain a negative number, but only if it is the value of P, the objective function variable.

These changes lead to a small change in the second version of the fundamental theorem (see Theorem 1 in Section 5-3).

Theorem 1 ■■ FUNDAMENTAL THEOREM OF LINEAR PROGRAMMING—VERSION 3

If the optimal value of the objective function in a linear programming problem exists, then that value must occur at one (or more) of the basic feasible solutions of the initial system. ■■

With these adjustments understood, we start the simplex process with a basic feasible solution of the initial system (3), which we will refer to as an **initial basic feasible solution.** An initial basic feasible solution that is easy to find is the one associated with the origin.

Since system (3) has three equations and five variables, it has three basic variables and two nonbasic variables. Looking at the system, we see that x_1 and x_2 appear in all equations and s_1, s_2, and P each appears only once and each in a different equation. A basic solution can be found by inspection by selecting s_1, s_2, and P as the basic variables (remember, P is always selected as a basic variable) and x_1 and x_2 as the nonbasic variables to be set equal to 0. Setting x_1 and x_2 equal to 0 and solving for the basic variables, we obtain the basic solution:

$$x_1 = 0, \quad x_2 = 0, \quad s_1 = 32, \quad s_2 = 84, \quad P = 0$$

This basic solution is feasible since none of the variables (excluding P) are negative. Thus, this is the initial basic feasible solution we seek.

Now you can see why we wanted to add the objective function equation to system (2): A basic feasible solution of (3) not only includes a basic feasible solution of (2), but, in addition, it includes the value of P for that basic feasible solution of (2).

The initial basic feasible solution we just found is associated with the origin. Of course, if we do not produce any tents, we do not expect a profit, so $P = \$0$. Starting with this easily obtained initial basic feasible solution, the simplex process moves through each iteration (each repetition) to another basic feasible solution, each time improving the profit, and the process continues until the maximum profit is reached. Then the process stops.

The Simplex Tableau

To facilitate the search for the optimal solution, we now turn to matrix methods discussed in Chapter 4. Our first step is to write the augmented matrix for the initial system (3). This matrix is called the **initial simplex tableau,** and it is simply a tabulation of the coefficients in system (3).

$$\begin{array}{c} \\ s_1 \\ s_2 \\ P \end{array} \begin{array}{c} \begin{array}{ccccc} x_1 & x_2 & s_1 & s_2 & P \end{array} \\ \left[\begin{array}{ccccc|c} 1 & 2 & 1 & 0 & 0 & 32 \\ 3 & 4 & 0 & 1 & 0 & 84 \\ \hdashline -50 & -80 & 0 & 0 & 1 & 0 \end{array}\right] \end{array} \quad \text{Initial simplex tableau} \qquad (4)$$

In tableau (4), the row below the dashed line always corresponds to the objective function. Each of the basic variables we selected above, s_1, s_2, and P, is also placed on the left of the tableau so that the intersection element in its row and column is not 0. For example, we place the basic variable s_1 on the left so that the intersection element of the s_1 row and the s_1 column is 1 and not 0. The basic variable s_2 is similarly placed. The objective function variable P is always placed at the bottom. The reason for writing the basic variables on the left in this way is that this placement makes it possible to read certain basic feasible solutions directly from the tableau. If $x_1 = 0$ and $x_2 = 0$, then the basic variables on the left of tableau (4) are lined up with their corresponding values, 32, 84, and 0, to the right of the vertical line.

Looking at tableau (4) relative to the choice of s_1, s_2, and P as basic variables, we see that each basic variable is above a column that has all

0 elements except for a single 1 and that no two such columns contain 1's in the same row. These observations lead to a formalization of the process of selecting basic and nonbasic variables that is an important part of the simplex method:

Selecting Basic and Nonbasic Variables for the Simplex Method

Given a simplex tableau:

Step 1. Determine the number of basic variables and the number of nonbasic variables. These numbers do not change during the simplex process.

Step 2. *Selecting Basic Variables.* A variable can be selected as a basic variable only if it corresponds to a column in the tableau that has exactly one nonzero element (usually 1) and the nonzero element in the column is not in the same row as the nonzero element in the column of another basic variable. (This procedure always selects P as a basic variable, since the P column never changes during the simplex process.)

Step 3. *Selecting Nonbasic Variables.* After the basic variables are selected in step 2, the remaining variables are selected as the nonbasic variables. (The tableau columns under the nonbasic variables usually contain more than one nonzero element.)

The earlier selection of s_1, s_2, and P as basic variables and x_1 and x_2 as nonbasic variables conforms to this prescribed convention of selecting basic and nonbasic variables for the simplex process.

The Pivot Operation

The simplex method will now switch one of the nonbasic variables, x_1 or x_2, for one of the basic variables, s_1 or s_2 (but not P), as a step toward improving the profit. For a nonbasic variable to be classified as a basic variable we need to perform appropriate row operations on the tableau so that the newly selected basic variable will end up with exactly one nonzero element in its column. In this process, the old basic variable will usually gain additional nonzero elements in its column as it becomes nonbasic.

Which nonbasic variable should we select to become basic? It makes sense to select the nonbasic variable that will increase the profit the most per unit change in that variable. Looking at the objective function

$$P = 50x_1 + 80x_2$$

we see that if x_1 stays a nonbasic variable (set equal to 0) and if x_2 becomes a new basic variable, then

$$P = 50(0) + 80x_2 = 80x_2$$

and for each unit increase in x_2, P will increase \$80. If x_2 stays a nonbasic variable and x_1 becomes a new basic variable, then (reasoning in the same way) for each unit increase in x_1, P will increase only \$50. So, we select the nonbasic variable x_2 to enter the set of basic variables, and call it the **entering variable.** (The basic variable leaving the set of basic variables to become a nonbasic variable is called the **exiting variable.** Exiting variables will be discussed shortly.)

call the column corresponding to the entering variable the **pivot colu** Looking at the bottom row in tableau (4)—the objective function row ow the dashed line—we see that the pivot column is associated with column to the left of the P column that has the most negative bottom elen . In general, the most negative element in the bottom row to the left of the P column *indicates* the variable above it that will produce the greatest increase in P for a unit increase in that variable. For this reason, we call the elements in the bottom row of the tableau to the left of the P column **indicators.**

We illustrate the indicators, the pivot column, the entering variable, and the initial basic feasible solution below:

Entering variable ↓

$$\begin{array}{c} \\ s_1 \\ s_2 \\ P \end{array} \begin{array}{c} \begin{array}{cccccc} x_1 & x_2 & s_1 & s_2 & P & \end{array} \\ \left[\begin{array}{ccccc|c} 1 & 2 & 1 & 0 & 0 & 32 \\ 3 & 4 & 0 & 1 & 0 & 84 \\ \hdashline -50 & -80 & 0 & 0 & 1 & 0 \end{array}\right] \end{array} \qquad (5)$$

Initial simplex tableau

Indicators are shown in color.

↑ Pivot column

$$x_1 = 0, \quad x_2 = 0, \quad s_1 = 32, \quad s_2 = 84, \quad P = 0$$ Initial basic feasible solution

Now that we have chosen the nonbasic variable x_2 as the entering variable (the nonbasic variable to become basic), which of the two basic variables, s_1 or s_2, should we choose as the exiting variable (the basic variable to become nonbasic)? We saw above that for $x_1 = 0$, each unit increase in the entering variable x_2 results in an increase of \$80 for P. Can we increase x_2 without limit? No! A limit is imposed by the nonnegative requirements for s_1 and s_2. (Remember that if any of the basic variables except P become negative, we no longer have a feasible solution.) So we rephrase the question and ask: How much can x_2 be increased when $x_1 = 0$ without causing s_1 or s_2 to become negative? To see how much x_2 can be increased, we refer to tableau (5) or system (3) and write the two problem constraint equations with $x_1 = 0$:

$$\begin{aligned} 2x_2 + s_1 &= 32 \\ 4x_2 + s_2 &= 84 \end{aligned}$$

Solving for s_1 and s_2, we have

$$\begin{aligned} s_1 &= 32 - 2x_2 \\ s_2 &= 84 - 4x_2 \end{aligned}$$

For s_1 and s_2 to be nonnegative, x_2 must be chosen so that both $32 - 2x_2$ and $84 - 4x_2$ are nonnegative. That is, so that

$$\begin{aligned} 32 - 2x_2 &\geqslant 0 \\ -2x_2 &\geqslant -32 \\ x_2 &\leqslant \tfrac{32}{2} = 16 \end{aligned} \qquad \text{and} \qquad \begin{aligned} 84 - 4x_2 &\geqslant 0 \\ -4x_2 &\geqslant -84 \\ x_2 &\leqslant \tfrac{84}{4} = 21 \end{aligned}$$

For both inequalities to be satisfied, x_2 must be less than or equal to the smaller of the values, which is 16. Thus, x_2 can increase to 16 without either s_1 or s_2 becoming negative. Now, observe how each value (16 and 21) can be obtained directly from tableau (6) below:

Entering variable ↓

$$\begin{array}{c} \\ s_1 \\ s_2 \\ P \end{array}\begin{array}{c} \begin{array}{cccccc} x_1 & x_2 & s_1 & s_2 & P & \end{array} \\ \left[\begin{array}{rrrrr|r} 1 & 2 & 1 & 0 & 0 & 32 \\ 3 & 4 & 0 & 1 & 0 & 84 \\ \hdashline -50 & -80 & 0 & 0 & 1 & 0 \end{array}\right] \end{array} \begin{array}{l} \\ \frac{32}{2} = 16 \text{ (smallest)} \\ \frac{84}{4} = 21 \\ \\ \end{array} \qquad (6)$$

↑ Pivot column

From (6) we can determine the amount the entering variable can increase by choosing the smallest of the quotients obtained by dividing each element in the last column above the dashed line by the corresponding *positive* element in the pivot column. The row with the smallest quotient is called the **pivot row,** and the variable to the left of the pivot row is the exiting variable. In this case, s_1 will be the exiting variable, and the roles of x_2 and s_1 will be interchanged. The element at the intersection of the pivot column and the pivot row is called the **pivot element,** and we circle this element for ease of recognition. Since a negative or 0 element in the pivot column places no restriction on the amount an entering variable can increase, it is not necessary to compute the quotient for negative or 0 values in the pivot column.

A negative or 0 element is never selected for the pivot element.

Tableau (7) illustrates this process, which is summarized in the next box.

Entering variable ↓

Pivot element (circled)

$$\begin{array}{r} \\ \text{Exiting variable} \rightarrow s_1 \\ s_2 \\ P \end{array}\begin{array}{c} \begin{array}{cccccc} x_1 & x_2 & s_1 & s_2 & P & \end{array} \\ \left[\begin{array}{rrrrr|r} 1 & \textcircled{2} & 1 & 0 & 0 & 32 \\ 3 & 4 & 0 & 1 & 0 & 84 \\ \hdashline -50 & -80 & 0 & 0 & 1 & 0 \end{array}\right] \end{array} \begin{array}{l} \\ \leftarrow \text{Pivot row} \\ \\ \\ \end{array} \qquad (7)$$

↑ Pivot column

Selecting the Pivot Element

Step 1. Locate the most negative indicator in the bottom row of the tableau to the left of the P column (the negative number with the largest absolute value). The column containing this element is the *pivot column.* If there is a tie for the most negative, choose either.

Step 2. Divide each *positive* element in the pivot column above the dashed line into the corresponding element in the last column. The *pivot row* is the row corresponding to the smallest quotient obtained. If there is a tie for the smallest quotient, choose either. If the pivot column above the dashed line has no positive elements, then there is no solution, and we stop.

Step 3. The *pivot* (or *pivot element*) is the element in the intersection of the pivot column and pivot row. [*Note:* The pivot element is always positive and is never in the bottom row.]

[*Remember:* The entering variable is at the top of the pivot column, and the exiting variable is at the left of the pivot row.]

In order for x_2 to be classified as a basic variable, we perform row operations on tableau (7) so that the pivot element is transformed into 1 and all other elements in the column into 0's. This procedure for transforming a nonbasic variable into a basic variable is called a *pivot operation,* or *pivoting,* and is summarized in the box.

Performing a Pivot Operation

A **pivot operation,** or **pivoting,** consists of performing row operations as follows:

Step 1. Multiply the pivot row by the reciprocal of the pivot element to transform the pivot element into a 1. (If the pivot element is already a 1, omit this step.)

Step 2. Add multiples of the pivot row to other rows in the tableau to transform all other nonzero elements in the pivot column into 0's.

[*Note:* Rows are not to be interchanged while performing a pivot operation. The only way the (positive) pivot element can be transformed into 1 (if it is not a 1 already) is for the pivot row to be multiplied by the reciprocal of the pivot element.]

Performing a pivot operation has the following effects:

1. The (entering) nonbasic variable becomes a basic variable.
2. The (exiting) basic variable becomes a nonbasic variable.
3. The value of the objective function is increased, or, in some cases, remains the same.

We now carry out the pivot operation on tableau (7). (To facilitate the process, we do not repeat the variables after the first tableau, and we use "Enter" and "Exit" for "Entering variable" and "Exiting variable," respectively.)

$$\begin{array}{c} \\ \text{Exit} \rightarrow \end{array}\quad
\begin{matrix} & x_1 & \overset{\text{Enter}}{\overset{\downarrow}{x_2}} & s_1 & s_2 & P & \\
s_1 & 1 & \textcircled{2} & 1 & 0 & 0 & 32 \\
s_2 & 3 & 4 & 0 & 1 & 0 & 84 \\
P & -50 & -80 & 0 & 0 & 1 & 0 \end{matrix} \qquad \tfrac{1}{2}R_1 \rightarrow R_1$$

$$\sim \left[\begin{array}{ccccc|c} \frac{1}{2} & \textcircled{1} & \frac{1}{2} & 0 & 0 & 16 \\ 3 & 4 & 0 & 1 & 0 & 84 \\ \hdashline -50 & -80 & 0 & 0 & 1 & 0 \end{array}\right] \qquad \begin{array}{l} (-4)R_1 + R_2 \rightarrow R_2 \\ 80R_1 + R_3 \rightarrow R_3 \end{array}$$

$$\sim \left[\begin{array}{ccccc|c} \frac{1}{2} & 1 & \frac{1}{2} & 0 & 0 & 16 \\ 1 & 0 & -2 & 1 & 0 & 20 \\ \hdashline -10 & 0 & 40 & 0 & 1 & 1{,}280 \end{array}\right]$$

We have completed the pivot operation, and now we must insert appropriate variables for this new tableau. Since x_2 replaced s_1, the basic variables are now x_2, s_2, and P, as indicated by the labels on the left side of the new tableau. Note that this selection of basic variables agrees with the procedure outlined on page 305 for selecting basic variables. We write the new basic feasible solution by setting the nonbasic variables x_1 and s_1 equal to 0 and solving for the basic variables by inspection. (Remember, the values of the basic variables listed on the left are the corresponding numbers to the right of the vertical line. To see this, substitute $x_1 = 0$ and $s_1 = 0$ in the corresponding system shown next to the simplex tableau.)

$$\begin{matrix} & x_1 & x_2 & s_1 & s_2 & P & \\
x_2 & \frac{1}{2} & 1 & \frac{1}{2} & 0 & 0 & 16 \\
s_2 & 1 & 0 & -2 & 1 & 0 & 20 \\
P & -10 & 0 & 40 & 0 & 1 & 1{,}280 \end{matrix} \qquad
\begin{aligned} \tfrac{1}{2}x_1 + x_2 + \tfrac{1}{2}s_1 \qquad &= 16 \\ x_1 \qquad - 2s_1 + s_2 \qquad &= 20 \\ -10x_1 + \qquad 40s_1 \qquad + P &= 1{,}280 \end{aligned}$$

$$x_1 = 0, \quad x_2 = 16, \quad s_1 = 0, \quad s_2 = 20, \quad P = \$1{,}280$$

A profit of $1,280 is a marked improvement over the $0 profit produced by the initial basic feasible solution. But we can improve P still further, since a negative indicator still remains in the bottom row. To see why, we write out the objective function:

$$-10x_1 + 40s_1 + P = 1{,}280$$

or

$$P = 10x_1 - 40s_1 + 1{,}280$$

If s_1 stays a nonbasic variable (set equal to 0) and x_1 becomes a new basic variable, then

$$P = 10x_1 - 40(0) + 1{,}280 = 10x_1 + 1{,}280$$

and for each unit increase in x_1, P will increase $10.

We now go through another iteration of the simplex process (that is, we repeat the above sequence of steps) using another pivot element. The pivot element and the entering and exiting variables are shown in the following tableau:

$$\begin{array}{r c} & \begin{array}{ccccc|c} \overset{\text{Enter}}{\downarrow} & & & & & \\ x_1 & x_2 & s_1 & s_2 & P & \end{array} \\ \begin{array}{r} x_2 \\ \text{Exit} \rightarrow s_2 \\ P \end{array} & \left[\begin{array}{ccccc|c} \frac{1}{2} & 1 & \frac{1}{2} & 0 & 0 & 16 \\ \textcircled{1} & 0 & -2 & 1 & 0 & 20 \\ \hdashline -10 & 0 & 40 & 0 & 1 & 1{,}280 \end{array}\right] \begin{array}{l} \frac{16}{1/2} = 32 \\ \frac{20}{1} = 20 \\ \end{array} \end{array}$$

We now pivot on (the circled) 1. That is, we perform a pivot operation using this 1 as the pivot element. Since the pivot element is 1, we do not need to perform the first step in the pivot operation, so we proceed to the second step to get 0's above and below the pivot element 1. As before, to facilitate the process, we omit writing the variables, except for the first tableau.

$$\begin{array}{r c} & \begin{array}{ccccc|c} \overset{\text{Enter}}{\downarrow} & & & & & \\ x_1 & x_2 & s_1 & s_2 & P & \end{array} \\ \begin{array}{r} x_2 \\ \text{Exit} \rightarrow s_2 \\ P \end{array} & \left[\begin{array}{ccccc|c} \frac{1}{2} & 1 & \frac{1}{2} & 0 & 0 & 16 \\ \textcircled{1} & 0 & -2 & 1 & 0 & 20 \\ \hdashline -10 & 0 & 40 & 0 & 1 & 1{,}280 \end{array}\right] \begin{array}{l} (-\frac{1}{2})R_2 + R_1 \rightarrow R_1 \\ \\ 10R_2 + R_3 \rightarrow R_3 \end{array} \\ \sim & \left[\begin{array}{ccccc|c} 0 & 1 & \frac{3}{2} & -\frac{1}{2} & 0 & 6 \\ 1 & 0 & -2 & 1 & 0 & 20 \\ \hdashline 0 & 0 & 20 & 10 & 1 & 1{,}480 \end{array}\right] \end{array}$$

Since there are no more negative indicators in the bottom row, we are through. Let us insert the appropriate variables for this last tableau and write the corresponding basic feasible solution. The basic variables are now x_1, x_2, and P, so to get the corresponding basic feasible solution, we set the nonbasic variables s_1 and s_2 equal to 0 and solve for the basic variables by inspection.

$$\begin{array}{r c} & \begin{array}{ccccc|c} x_1 & x_2 & s_1 & s_2 & P & \end{array} \\ \begin{array}{r} x_2 \\ x_1 \\ P \end{array} & \left[\begin{array}{ccccc|c} 0 & 1 & \frac{3}{2} & -\frac{1}{2} & 0 & 6 \\ 1 & 0 & -2 & 1 & 0 & 20 \\ \hdashline 0 & 0 & 20 & 10 & 1 & 1{,}480 \end{array}\right] \end{array}$$

$$x_1 = 20, \quad x_2 = 6, \quad s_1 = 0, \quad s_2 = 0, \quad P = 1{,}480$$

To see why this is the maximum, we rewrite the objective function from the bottom row:

$$\begin{aligned} 20s_1 + 10s_2 + P &= 1{,}480 \\ P &= 1{,}480 - 20s_1 - 10s_2 \end{aligned}$$

Since s_1 and s_2 cannot be negative, any increase of either from 0 will make the profit smaller.

Finally, returning to our original problem, we conclude that a production schedule of 20 standard tents and 6 expedition tents will produce a maximum profit of $1,480 per day, which is the same as the geometric solution obtained in Section 5-2. The fact that the slack variables are both 0 means that for this production schedule, the plant will operate at full capacity—there is no slack in either the cutting department or the assembly department.

Interpreting the Simplex Process Geometrically

We can now interpret the simplex process just completed geometrically in terms of the feasible region graphed in the preceding section. Table 1 lists the three basic feasible solutions we just found using the simplex method (in the order they were found). The table also includes the corresponding corner points of the feasible region illustrated in Figure 1.

Table 1

Basic feasible solutions (obtained above)

x_1	x_2	s_1	s_2	P ($)	Corner point
0	0	32	84	0	$O(0, 0)$
0	16	0	20	1,280	$A(0, 16)$
20	6	0	0	1,480	$B(20, 6)$

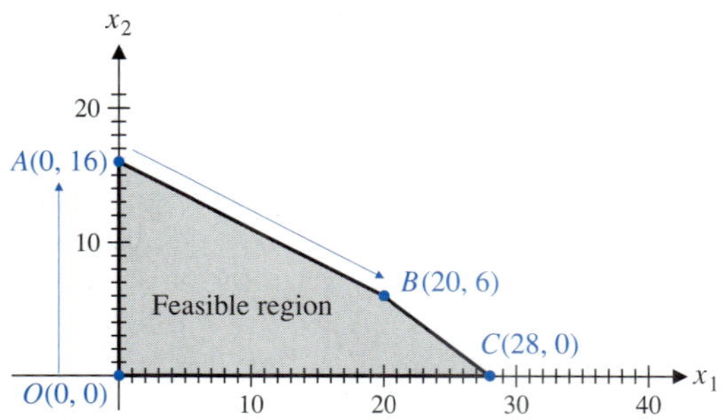

FIGURE 1

Looking at Table 1 and Figure 1, we see that the simplex process started at the origin, moved to the adjacent corner point $A(0, 16)$, and then to the optimal solution $B(20, 6)$ at the next adjacent corner point. This is typical of the simplex process.

■ The Simplex Method Summarized

Before commencing with further examples, we summarize the important parts of the simplex method schematically in Figure 2.

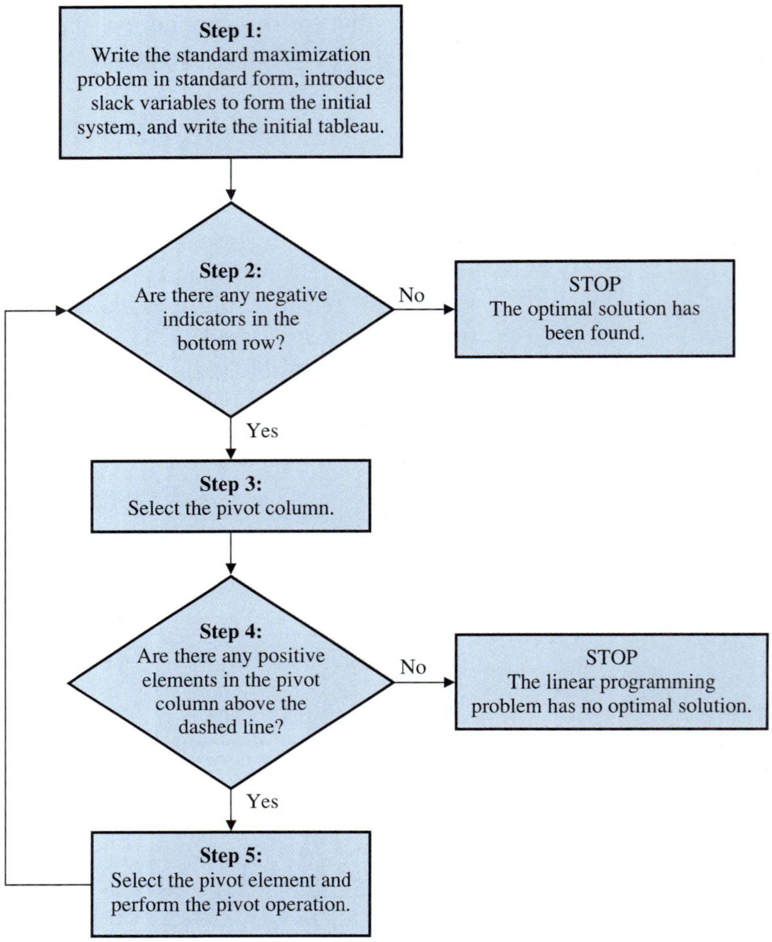

FIGURE 2

Simplex algorithm for standard maximization problems
(Problem constraints are of the $\leqslant$ form with nonnegative constants on the right. The coefficients of the objective function can be any real numbers.)

Example 1 ➡ **Using the Simplex Method** Solve the following linear programming problem using the simplex method:

$$\begin{aligned} \text{Maximize} \quad & P = 10x_1 + 5x_2 \\ \text{Subject to} \quad & 4x_1 + x_2 \leqslant 28 \\ & 2x_1 + 3x_2 \leqslant 24 \\ & x_1, x_2 \geqslant 0 \end{aligned}$$

SOLUTION Introduce slack variables s_1 and s_2, and write the initial system:

$$\begin{aligned} 4x_1 + x_2 + s_1 \qquad &= 28 \\ 2x_1 + 3x_2 \qquad + s_2 \qquad &= 24 \\ -10x_1 - 5x_2 \qquad\qquad + P &= 0 \\ x_1, x_2, s_1, s_2 &\geq 0 \end{aligned}$$

Write the simplex tableau, and identify the first pivot element and the entering and exiting variables:

$$\begin{array}{r} \\ \text{Exit} \rightarrow s_1 \\ s_2 \\ P \end{array} \begin{array}{c} \overset{\text{Enter}}{\downarrow} \\ \end{array} \begin{array}{c} \begin{matrix} x_1 & x_2 & s_1 & s_2 & P \end{matrix} \\ \left[\begin{array}{ccccc|c} \textcircled{4} & 1 & 1 & 0 & 0 & 28 \\ 2 & 3 & 0 & 1 & 0 & 24 \\ \hline -10 & -5 & 0 & 0 & 1 & 0 \end{array}\right] \end{array} \begin{array}{l} \\ \frac{28}{4} = 7 \\ \frac{24}{2} = 12 \\ \\ \end{array}$$

Perform the pivot operation:

$$\begin{array}{r} \\ \text{Exit} \rightarrow s_1 \\ s_2 \\ P \end{array} \begin{array}{c} \overset{\text{Enter}}{\downarrow} \\ \begin{matrix} x_1 & x_2 & s_1 & s_2 & P \end{matrix} \\ \left[\begin{array}{ccccc|c} \textcircled{4} & 1 & 1 & 0 & 0 & 28 \\ 2 & 3 & 0 & 1 & 0 & 24 \\ \hline -10 & -5 & 0 & 0 & 1 & 0 \end{array}\right] \end{array} \begin{array}{l} \frac{1}{4}R_1 \rightarrow R_1 \\ \\ \\ \end{array}$$

$$\sim \left[\begin{array}{ccccc|c} \textcircled{1} & 0.25 & 0.25 & 0 & 0 & 7 \\ 2 & 3 & 0 & 1 & 0 & 24 \\ \hline -10 & -5 & 0 & 0 & 1 & 0 \end{array}\right] \begin{array}{l} \\ (-2)R_1 + R_2 \rightarrow R_2 \\ 10R_1 + R_3 \rightarrow R_3 \end{array}$$

$$\begin{array}{r} x_1 \\ \sim s_2 \\ P \end{array} \left[\begin{array}{ccccc|c} 1 & 0.25 & 0.25 & 0 & 0 & 7 \\ 0 & 2.5 & -0.5 & 1 & 0 & 10 \\ \hline 0 & -2.5 & 2.5 & 0 & 1 & 70 \end{array}\right]$$

Since there is still a negative indicator in the last row, we repeat the process by finding a new pivot element:

$$\begin{array}{r} \\ x_1 \\ \text{Exit} \rightarrow s_2 \\ P \end{array} \begin{array}{c} \begin{matrix} x_1 & \overset{\text{Enter}}{\underset{\downarrow}{x_2}} & s_1 & s_2 & P \end{matrix} \\ \left[\begin{array}{ccccc|c} 1 & 0.25 & 0.25 & 0 & 0 & 7 \\ 0 & \textcircled{2.5} & -0.5 & 1 & 0 & 10 \\ \hline 0 & -2.5 & 2.5 & 0 & 1 & 70 \end{array}\right] \end{array} \begin{array}{l} \\ \frac{7}{0.25} = 28 \\ \frac{10}{2.5} = 4 \\ \\ \end{array}$$

Performing the pivot operation, we obtain

$$
\begin{array}{c}
 \\ \\ x_1 \\ \text{Exit} \rightarrow s_2 \\ P
\end{array}
\begin{array}{c}
\quad \text{Enter} \downarrow \\
\begin{array}{ccccc} x_1 & x_2 & s_1 & s_2 & P \end{array} \\
\left[\begin{array}{ccccc|c}
1 & 0.25 & 0.25 & 0 & 0 & 7 \\
0 & \textcircled{2.5} & -0.5 & 1 & 0 & 10 \\
\hdashline
0 & -2.5 & 2.5 & 0 & 1 & 70
\end{array}\right]
\end{array}
\quad \tfrac{1}{2.5}R_2 \rightarrow R_2
$$

$$
\sim \left[\begin{array}{ccccc|c}
1 & 0.25 & 0.25 & 0 & 0 & 7 \\
0 & \textcircled{1} & -0.2 & 0.4 & 0 & 4 \\
\hdashline
0 & -2.5 & 2.5 & 0 & 1 & 70
\end{array}\right]
\quad
\begin{array}{l}
(-0.25)R_2 + R_1 \rightarrow R_1 \\
\\
2.5R_2 + R_3 \rightarrow R_3
\end{array}
$$

$$
\begin{array}{c} x_1 \\ \sim x_2 \\ P \end{array}
\left[\begin{array}{ccccc|c}
1 & 0 & 0.3 & -0.1 & 0 & 6 \\
0 & 1 & -0.2 & 0.4 & 0 & 4 \\
\hdashline
0 & 0 & 2 & 1 & 1 & 80
\end{array}\right]
$$

Since all the indicators in the last row are nonnegative, we stop and read the optimal solution:

$$\text{Max } P = 80 \qquad \text{at} \qquad x_1 = 6, \quad x_2 = 4, \quad s_1 = 0, \quad s_2 = 0$$

(To see why this makes sense, write the objective function corresponding to the last row to see what happens to P when you try to increase s_1 or s_2.)

Matched Problem 1 Solve the following linear programming problem using the simplex method:

$$
\begin{array}{ll}
\text{Maximize} & P = 2x_1 + x_2 \\
\text{Subject to} & 5x_1 + x_2 \leq 9 \\
 & \quad x_1 + x_2 \leq 5 \\
 & \quad\; x_1, x_2 \geq 0
\end{array}
$$

Explore–Discuss 1

Graph the feasible region for the linear programming problem in Example 1 and trace the path to the optimal solution determined by the simplex method.

Example 2 **Using the Simplex Method** Solve using the simplex method:

$$
\begin{array}{ll}
\text{Maximize} & P = 6x_1 + 3x_2 \\
\text{Subject to} & -2x_1 + 3x_2 \leq 9 \\
 & \;\; -x_1 + 3x_2 \leq 12 \\
 & \qquad x_1, x_2 \geq 0
\end{array}
$$

SOLUTION Write the initial system using the slack variables s_1 and s_2:

$$
\begin{array}{rcl}
-2x_1 + 3x_2 + s_1 \qquad\qquad & = & 9 \\
-x_1 + 3x_2 \qquad + s_2 \qquad & = & 12 \\
-6x_1 - 3x_2 \qquad\qquad + P & = & 0
\end{array}
$$

Write the simplex tableau and identify the first pivot element:

$$\begin{array}{c} \\ s_1 \\ s_2 \\ P \end{array}\begin{array}{c} \begin{array}{ccccc} x_1 & x_2 & s_1 & s_2 & P \end{array} \\ \left[\begin{array}{rrrrr|r} -2 & 3 & 1 & 0 & 0 & 9 \\ -1 & 3 & 0 & 1 & 0 & 12 \\ \hdashline -6 & -3 & 0 & 0 & 1 & 0 \end{array}\right] \end{array}$$

↑ Pivot column

Since both elements in the pivot column above the dashed line are negative, we are unable to select a pivot row. We stop and conclude that there is no optimal solution.

Matched Problem 2 Solve using the simplex method:

$$\begin{aligned} \text{Maximize} \quad & P = 2x_1 + 3x_2 \\ \text{Subject to} \quad & -3x_1 + 4x_2 \le 12 \\ & x_2 \le 2 \\ & x_1, x_2 \ge 0 \end{aligned}$$

Explore–Discuss 2

In Example 2 we encountered a tableau with a pivot column and no pivot row, indicating a problem with no optimal solution.

(A) What happens if you violate the rule for selecting the pivot row, select the negative element -1 for the pivot, and then continue with the simplex method?

(B) There was another negative indicator in the tableau in Example 2. What happens if you violate the rule for selecting the pivot column, select the second column for a pivot column, and then continue with the simplex method?

(C) Graph the feasible region in Example 2, graph the lines corresponding to $P = 36$ and $P = 66$, and explain why this problem has no optimal solution.

Refer to Examples 1 and 2. In Example 1 we concluded that we had found the optimal solution because we could not select a pivot column. In Example 2 we concluded that the problem had no optimal solution because we selected a pivot column and then could not select a pivot row. Notice that we do not try to continue with the simplex method by selecting a negative pivot element or using a different column for the pivot column. Remember:

If it is not possible to select a pivot column, the simplex method stops and we conclude that the optimal solution has been found. If the pivot column has been selected and it is not possible to select a pivot row, the simplex method stops and we conclude that there is no optimal solution.

Application

Example 3 **Agriculture** A farmer owns a 100 acre farm and plans to plant at most three crops. The seed for crops *A*, *B*, and *C* costs \$40, \$20, and \$30 per acre, respectively. A maximum of \$3,200 can be spent on seed. Crops *A*, *B*, and *C* require 1, 2, and 1 workdays per acre, respectively, and there are a maximum of 160 workdays available. If the farmer can make a profit of \$100 per acre on crop *A*, \$300 per acre on crop *B*, and \$200 per acre on crop *C*, how many acres of each crop should be planted to maximize profit?

SOLUTION

Let: x_1 = Number of acres of crop A
x_2 = Number of acres of crop B
x_3 = Number of acres of crop C
P = Total profit

Then we have the following linear programming problem:

$$\begin{aligned}
\text{Maximize} \quad & P = 100x_1 + 300x_2 + 200x_3 && \text{Objective function}\\
\text{Subject to} \quad & \left.\begin{aligned} x_1 + x_2 + x_3 &\le 100\\ 40x_1 + 20x_2 + 30x_3 &\le 3{,}200\\ x_1 + 2x_2 + x_3 &\le 160 \end{aligned}\right\} && \text{Problem constraints}\\
& x_1, x_2, x_3 \ge 0 && \text{Nonnegative constraints}
\end{aligned}$$

Next, we introduce slack variables and form the initial system:

$$\begin{aligned}
x_1 + x_2 + x_3 + s_1 \qquad\qquad &= 100\\
40x_1 + 20x_2 + 30x_3 \quad + s_2 \qquad &= 3{,}200\\
x_1 + 2x_2 + x_3 \qquad + s_3 \quad &= 160\\
-100x_1 - 300x_2 - 200x_3 \qquad\qquad + P &= 0\\
x_1, x_2, x_3, s_1, s_2, s_3 &\ge 0
\end{aligned}$$

Notice that the initial system has $7 - 4 = 3$ nonbasic variables and 4 basic variables. Now we form the simplex tableau and solve by the simplex method:

Enter ↓ (x_2)

$$\begin{matrix} \\ s_1\\ s_2\\ \text{Exit}\rightarrow s_3\\ P \end{matrix}
\begin{matrix} x_1 & x_2 & x_3 & s_1 & s_2 & s_3 & P & \\
\left[\begin{matrix} 1 & 1 & 1 & 1 & 0 & 0 & 0\\ 40 & 20 & 30 & 0 & 1 & 0 & 0\\ 1 & \textcircled{2} & 1 & 0 & 0 & 1 & 0\\ \hdashline -100 & -300 & -200 & 0 & 0 & 0 & 1 \end{matrix}\right. & \left|\begin{matrix} 100\\ 3{,}200\\ 160\\ \hdashline 0 \end{matrix}\right] \end{matrix}
\quad \begin{matrix} \\ \\ 0.5R_3 \rightarrow R_3\\ \\ \end{matrix}$$

$$\sim \left[\begin{array}{rrrrrrr|r} 1 & 1 & 1 & 1 & 0 & 0 & 0 & 100\\ 40 & 20 & 30 & 0 & 1 & 0 & 0 & 3{,}200\\ 0.5 & \textcircled{1} & 0.5 & 0 & 0 & 0.5 & 0 & 80\\ \hdashline -100 & -300 & -200 & 0 & 0 & 0 & 1 & 0 \end{array}\right]
\quad \begin{matrix} (-1)R_3 + R_1 \rightarrow R_1\\ (-20)R_3 + R_2 \rightarrow R_2\\ \\ 300R_3 + R_4 \rightarrow R_4 \end{matrix}$$

$$
\begin{array}{c}
\\
\\
\text{Exit} \rightarrow \\
\\
\\
\\
\end{array}
\begin{array}{c}
\\
s_1 \\ s_2 \\ x_2 \\ P
\end{array}
\begin{array}{c}
 & & \text{Enter} & & & & \\
 & & \downarrow & & & & \\
x_1 & x_2 & x_3 & s_1 & s_2 & s_3 & P
\end{array}
$$

$$
\begin{array}{c} s_1 \\ s_2 \\ x_2 \\ P \end{array}
\left[\begin{array}{ccccccc|c}
0.5 & 0 & \textcircled{0.5} & 1 & 0 & -0.5 & 0 & 20 \\
30 & 0 & 20 & 0 & 1 & -10 & 0 & 1{,}600 \\
0.5 & 1 & 0.5 & 0 & 0 & 0.5 & 0 & 80 \\
\hdashline
50 & 0 & -50 & 0 & 0 & 150 & 1 & 24{,}000
\end{array}\right]
\quad 2R_1 \rightarrow R_1
$$

$$
\sim
\left[\begin{array}{ccccccc|c}
1 & 0 & \textcircled{1} & 2 & 0 & -1 & 0 & 40 \\
30 & 0 & 20 & 0 & 1 & -10 & 0 & 1{,}600 \\
0.5 & 1 & 0.5 & 0 & 0 & 0.5 & 0 & 80 \\
\hdashline
50 & 0 & -50 & 0 & 0 & 150 & 1 & 24{,}000
\end{array}\right]
\quad
\begin{array}{l}
\\
(-20)R_1 + R_2 \rightarrow R_2 \\
(-0.5)R_1 + R_3 \rightarrow R_3 \\
50R_1 + R_4 \rightarrow R_4
\end{array}
$$

$$
\sim
\begin{array}{c} x_3 \\ s_2 \\ x_2 \\ P \end{array}
\left[\begin{array}{ccccccc|c}
1 & 0 & 1 & 2 & 0 & -1 & 0 & 40 \\
10 & 0 & 0 & -40 & 1 & 10 & 0 & 800 \\
0 & 1 & 0 & -1 & 0 & 1 & 0 & 60 \\
\hdashline
100 & 0 & 0 & 100 & 0 & 100 & 1 & 26{,}000
\end{array}\right]
$$

All indicators in the bottom row are nonnegative, and we can now read the optimal solution:

$$x_1 = 0, \quad x_2 = 60, \quad x_3 = 40, \quad s_1 = 0, \quad s_2 = 800, \quad s_3 = 0, \quad P = \$26{,}000$$

Thus, if the farmer plants 60 acres in crop *B*, 40 acres in crop *C*, and no crop *A*, the maximum profit of \$26,000 will be realized. The fact that $s_2 = 800$ tells us (look at the second row in the equations at the start) that this maximum profit is reached by using only \$2,400 of the \$3,200 available for seed; that is, we have a slack of \$800 that can be used for some other purpose. ■

Figure 3 illustrates a solution to Example 3 in Excel, a popular spreadsheet for personal computers. The software in *Explorations in Finite Mathematics* (see Preface) can also be used to solve this problem.

	A	B	C	D	E	F
1	Resources	Crop A	Crop B	Crop C	Available	Used
2	Acres	1	1	1	100	100
3	Seed($)	40	20	30	3,200	2,400
4	Workdays	1	2	1	160	160
5	Profit Per Acre	100	300	200	26,000	<-Total
6	Acres to plant	0	60	40		Profit

FIGURE 3

Matched Problem 3 ➡ Repeat Example 3 modified as follows:

	Investment per acre			Maximum
	Crop A	*Crop B*	*Crop C*	available
Seed cost	\$24	\$40	\$30	\$3,600
Workdays	1	2	2	160
Profit	\$140	\$200	\$160	

■

REMARKS

1. It can be shown that the feasible region for the linear programming problem in Example 3 has eight corner points, yet the simplex method found the solution in only two steps. Now you begin to see the power of the simplex method. In larger problems, the difference between the total number of corner points and the number of steps required by the simplex method is even more dramatic. A feasible region may have hundreds or even thousands of corner points, yet the simplex method will often find the optimal solution in 10 or 15 steps.
2. Refer to the second problem constraint in the model for Example 3:

   $$40x_1 + 20x_2 + 30x_3 \leq 3{,}200$$

   Multiplying both sides of this inequality by $\frac{1}{10}$ before introducing a slack variable simplifies subsequent calculations. However, performing this operation has a side effect—it changes the units of the slack variable from dollars to tens of dollars. To see why this happens, compare the equations

   $$40x_1 + 20x_2 + 30x_3 + s_2 = 3{,}200 \qquad s_2 \text{ represents dollars}$$

   and

   $$4x_1 + 2x_2 + 3x_3 + s_2' = 320 \qquad s_2' \text{ represents tens of dollars}$$

   In general, if you multiply a problem constraint by a number, remember to take this into account when you interpret the value of the slack variable for that constraint.
3. It is important to realize that in order to keep this introduction as simple as possible, we have purposely avoided certain degenerate cases that lead to difficulties. Discussion and resolution of these problems is left to a more advanced treatment of the subject.

Answers to Matched Problems

1. Max $P = 6$ when $x_1 = 1$ and $x_2 = 4$
2. No optimal solution
3. 40 acres of crop A, 60 acres of crop B, no crop C; Max $P = \$17{,}600$ (since $s_2 = 240$, \$240 out of the \$3,600 will not be spent)

EXERCISE 5-4

A *For the simplex tableaux in Problems 1–4:*

(A) Identify the basic and nonbasic variables.

(B) Find the corresponding basic feasible solution.

(C) Determine whether the optimal solution has been found, an additional pivot is required, or the problem has no optimal solution.

1.

$$\begin{array}{ccccc} x_1 & x_2 & s_1 & s_2 & P & \\ \end{array}$$
$$\left[\begin{array}{rrrrr|r} 2 & 1 & 0 & 3 & 0 & 12 \\ 3 & 0 & 1 & -2 & 0 & 15 \\ \hline -4 & 0 & 0 & 4 & 1 & 20 \end{array}\right]$$

2.

$$\begin{array}{ccccc} x_1 & x_2 & s_1 & s_2 & P \end{array}$$
$$\left[\begin{array}{rrrrr|r} 1 & 4 & -2 & 0 & 0 & 10 \\ 0 & 2 & 3 & 1 & 0 & 25 \\ \hline 0 & 5 & 6 & 0 & 1 & 35 \end{array}\right]$$

3.

$$\begin{array}{ccccccc} x_1 & x_2 & x_3 & s_1 & s_2 & s_3 & P \end{array}$$
$$\left[\begin{array}{rrrrrrr|r} -2 & 0 & 1 & 3 & 1 & 0 & 0 & 5 \\ 0 & 1 & 0 & -2 & 0 & 0 & 0 & 15 \\ -1 & 0 & 0 & 4 & 1 & 1 & 0 & 12 \\ \hline -4 & 0 & 0 & 2 & 4 & 0 & 1 & 45 \end{array}\right]$$

4.

$$\begin{array}{ccccccc} x_1 & x_2 & x_3 & s_1 & s_2 & s_3 & P \end{array}$$
$$\left[\begin{array}{rrrrrrr|r} 0 & 2 & -1 & 1 & 4 & 0 & 0 & 5 \\ 0 & 1 & 2 & 0 & -2 & 1 & 0 & 2 \\ 1 & 3 & 0 & 0 & 5 & 0 & 0 & 11 \\ \hline 0 & -5 & 4 & 0 & -3 & 0 & 1 & 27 \end{array}\right]$$

In Problems 5–8, find the pivot element, identify the entering and exiting variables, and perform one pivot operation.

5.
$$\begin{array}{c} \\ \\ \\ \end{array}\begin{array}{ccccc|c} x_1 & x_2 & s_1 & s_2 & P & \\ \hline 1 & 4 & 1 & 0 & 0 & 4 \\ 3 & 5 & 0 & 1 & 0 & 24 \\ \hline -8 & -5 & 0 & 0 & 1 & 0 \end{array}$$

6.
$$\begin{array}{ccccc|c} x_1 & x_2 & s_1 & s_2 & P & \\ \hline 1 & 6 & 1 & 0 & 0 & 36 \\ 3 & 1 & 0 & 1 & 0 & 5 \\ \hline -1 & -2 & 0 & 0 & 1 & 0 \end{array}$$

7.
$$\begin{array}{cccccc|c} x_1 & x_2 & s_1 & s_2 & s_3 & P & \\ \hline 2 & 1 & 1 & 0 & 0 & 0 & 4 \\ 3 & 0 & 1 & 1 & 0 & 0 & 8 \\ 0 & 0 & 2 & 0 & 1 & 0 & 2 \\ \hline -4 & 0 & -3 & 0 & 0 & 1 & 5 \end{array}$$

8.
$$\begin{array}{cccccc|c} x_1 & x_2 & s_1 & s_2 & s_3 & P & \\ \hline 0 & 0 & 2 & 1 & 1 & 0 & 2 \\ 1 & 0 & -4 & 0 & 1 & 0 & 3 \\ 0 & 1 & 5 & 0 & 2 & 0 & 11 \\ \hline 0 & 0 & -6 & 0 & -5 & 1 & 18 \end{array}$$

In Problems 9–12:
(A) Using slack variables, write the initial system for each linear programming problem.
(B) Write the simplex tableau, circle the first pivot, and identify the entering and exiting variables.
(C) Use the simplex method to solve the problem.

9. Maximize $P = 15x_1 + 10x_2$
Subject to $2x_1 + x_2 \leq 10$
$x_1 + 3x_2 \leq 10$
$x_1, x_2 \geq 0$

10. Maximize $P = 3x_1 + 2x_2$
Subject to $5x_1 + 2x_2 \leq 20$
$3x_1 + 2x_2 \leq 16$
$x_1, x_2 \geq 0$

11. Repeat Problem 9 with the objective function changed to $P = 30x_1 + x_2$.

12. Repeat Problem 10 with the objective function changed to $P = x_1 + 3x_2$.

B *Solve the linear programming problems in Problems 13–28 using the simplex method.*

13. Maximize $P = 30x_1 + 40x_2$
Subject to $2x_1 + x_2 \leq 10$
$x_1 + x_2 \leq 7$
$x_1 + 2x_2 \leq 12$
$x_1, x_2 \geq 0$

14. Maximize $P = 15x_1 + 20x_2$
Subject to $2x_1 + x_2 \leq 9$
$x_1 + x_2 \leq 6$
$x_1 + 2x_2 \leq 10$
$x_1, x_2 \geq 0$

15. Maximize $P = 2x_1 + 3x_2$
Subject to $-2x_1 + x_2 \leq 2$
$-x_1 + x_2 \leq 5$
$x_2 \leq 6$
$x_1, x_2 \geq 0$

16. Repeat Problem 15 with $P = -x_1 + 3x_2$.

17. Maximize $P = -x_1 + 2x_2$
Subject to $-x_1 + x_2 \leq 2$
$-x_1 + 3x_2 \leq 12$
$x_1 - 4x_2 \leq 4$
$x_1, x_2 \geq 0$

18. Repeat Problem 17 with $P = x_1 + 2x_2$.

19. Maximize $P = 5x_1 + 2x_2 - x_3$
Subject to $x_1 + x_2 - x_3 \leq 10$
$2x_1 + 4x_2 + 3x_3 \leq 30$
$x_1, x_2, x_3 \geq 0$

20. Maximize $P = 4x_1 - 3x_2 + 2x_3$
Subject to $x_1 + 2x_2 - x_3 \leq 5$
$3x_1 + 2x_2 + 2x_3 \leq 22$
$x_1, x_2, x_3 \geq 0$

21. Maximize $P = 2x_1 + 3x_2 + 4x_3$
Subject to $x_1 + x_3 \leq 4$
$x_2 + x_3 \leq 3$
$x_1, x_2, x_3 \geq 0$

22. Maximize $P = x_1 + x_2 + 2x_3$
Subject to $x_1 - 2x_2 + x_3 \leq 9$
$2x_1 + x_2 + 2x_3 \leq 28$
$x_1, x_2, x_3 \geq 0$

23. Maximize $P = 4x_1 + 3x_2 + 2x_3$
Subject to $3x_1 + 2x_2 + 5x_3 \leq 23$
$2x_1 + x_2 + x_3 \leq 8$
$x_1 + x_2 + 2x_3 \leq 7$
$x_1, x_2, x_3 \geq 0$

24. Maximize $P = 4x_1 + 2x_2 + 3x_3$
Subject to $x_1 + x_2 + x_3 \leq 11$
$2x_1 + 3x_2 + x_3 \leq 20$
$x_1 + 3x_2 + 2x_3 \leq 20$
$x_1, x_2, x_3 \geq 0$

C

25. Maximize $P = 20x_1 + 30x_2$
Subject to
$$\begin{aligned} 0.6x_1 + 1.2x_2 &\le 960 \\ 0.03x_1 + 0.04x_2 &\le 36 \\ 0.3x_1 + 0.2x_2 &\le 270 \\ x_1, x_2 &\ge 0 \end{aligned}$$

26. Repeat Problem 25 with $P = 20x_1 + 20x_2$.

27. Maximize $P = x_1 + 2x_2 + 3x_3$
Subject to
$$\begin{aligned} 2x_1 + 2x_2 + 8x_3 &\le 600 \\ x_1 + 3x_2 + 2x_3 &\le 600 \\ 3x_1 + 2x_2 + x_3 &\le 400 \\ x_1, x_2, x_3 &\ge 0 \end{aligned}$$

28. Maximize $P = 10x_1 + 50x_2 + 10x_3$
Subject to
$$\begin{aligned} 3x_1 + 3x_2 + 3x_3 &\le 66 \\ 6x_1 - 2x_2 + 4x_3 &\le 48 \\ 3x_1 + 6x_2 + 9x_3 &\le 108 \\ x_1, x_2, x_3 &\ge 0 \end{aligned}$$

In Problems 29 and 30, first solve the linear programming problem by the simplex method, keeping track of the basic feasible solutions at each step. Then graph the feasible region and illustrate the path to the optimal solution determined by the simplex method.

29. Maximize $P = 2x_1 + 5x_2$
Subject to
$$\begin{aligned} x_1 + 2x_2 &\le 40 \\ x_1 + 3x_2 &\le 48 \\ x_1 + 4x_2 &\le 60 \\ x_2 &\le 14 \\ x_1, x_2 &\ge 0 \end{aligned}$$

30. Maximize $P = 5x_1 + 3x_2$
Subject to
$$\begin{aligned} 5x_1 + 4x_2 &\le 100 \\ 2x_1 + x_2 &\le 28 \\ 4x_1 + x_2 &\le 42 \\ x_1 &\le 10 \\ x_1, x_2 &\ge 0 \end{aligned}$$

In Problems 31–34, there is a tie for the choice of the first pivot column. Use the simplex method to solve each problem two different ways: first by choosing Column 1 as the first pivot column, and then by choosing Column 2 as the first pivot column. Discuss the relationship between these two solutions.

31. Maximize $P = x_1 + x_2$
Subject to
$$\begin{aligned} 2x_1 + x_2 &\le 16 \\ x_1 &\le 6 \\ x_2 &\le 10 \\ x_1, x_2 &\ge 0 \end{aligned}$$

32. Maximize $P = x_1 + x_2$
Subject to
$$\begin{aligned} x_1 + 2x_2 &\le 10 \\ x_1 &\le 6 \\ x_2 &\le 4 \\ x_1, x_2 &\ge 0 \end{aligned}$$

33. Maximize $P = 3x_1 + 3x_2 + 2x_3$
Subject to
$$\begin{aligned} x_1 + x_2 + 2x_3 &\le 20 \\ 2x_1 + x_2 + 4x_3 &\le 32 \\ x_1, x_2, x_3 &\ge 0 \end{aligned}$$

34. Maximize $P = 2x_1 + 2x_2 + x_3$
Subject to
$$\begin{aligned} x_1 + x_2 + 3x_3 &\le 10 \\ 2x_1 + 4x_2 + 5x_3 &\le 24 \\ x_1, x_2, x_3 &\ge 0 \end{aligned}$$

APPLICATIONS

In Problems 35–48, construct a mathematical model in the form of a linear programming problem. (The answers in the back of the book for these application problems include the model.) Then solve the problem using the simplex method. Include an interpretation of any nonzero slack variables in the optimal solution.

Business & Economics

35. *Manufacturing—resource allocation.* A small company manufactures three different electronic components for computers. Component *A* requires 2 hours of fabrication and 1 hour of assembly; component *B* requires 3 hours of fabrication and 1 hour of assembly; and component *C* requires 2 hours of fabrication and 2 hours of assembly. The company has up to 1,000 labor-hours of fabrication time and 800 labor-hours of assembly time available per week. The profit on each component, *A, B,* and *C,* is \$7, \$8, and \$10, respectively. How many components of each type should the company manufacture each week in order to maximize its profit (assuming all components that it manufactures can be sold)? What is the maximum profit?

36. *Manufacturing—resource allocation.* Solve Problem 35 with the additional restriction that the combined total number of components produced each week cannot exceed 420. Discuss the effect of this restriction on the solution to Problem 35.

37. *Investment.* An investor has at most \$100,000 to invest in government bonds, mutual funds, and money market funds. The average yields for government bonds, mutual funds, and money market funds are 8%, 13%, and 15%, respectively. The investor's policy requires that the total amount invested in mutual and money market funds not exceed the amount invested in government bonds. How much should be invested in each type of investment in order to maximize the return? What is the maximum return?

38. *Investment.* Repeat Problem 37 under the additional assumption that no more than $30,000 can be invested in money market funds.

39. *Advertising.* A department store chain has up to $20,000 to spend on television advertising for a sale. All ads will be placed with one television station, where a 30 second ad costs $1,000 on daytime TV and is viewed by 14,000 potential customers, $2,000 on prime-time TV and is viewed by 24,000 potential customers, and $1,500 on late-night TV and is viewed by 18,000 potential customers. The television station will not accept a total of more than 15 ads in all three time periods. How many ads should be placed in each time period in order to maximize the number of potential customers who will see the ads? How many potential customers will see the ads? (Ignore repeated viewings of the ad by the same potential customer.)

40. *Advertising.* Repeat Problem 39 if the department store increases its budget to $24,000 and requires that at least half of the ads be placed in prime-time shows.

41. *Construction—resource allocation.* A contractor is planning to build a new housing development consisting of colonial, split-level, and ranch-style houses. A colonial house requires $\frac{1}{2}$ acre of land, $60,000 capital, and 4,000 labor-hours to construct, and returns a profit of $20,000. A split-level house requires $\frac{1}{2}$ acre of land, $60,000 capital, and 3,000 labor-hours to construct, and returns a profit of $18,000. A ranch house requires 1 acre of land, $80,000 capital, and 4,000 labor-hours to construct, and returns a profit of $24,000. The contractor has available 30 acres of land, $3,200,000 capital, and 180,000 labor-hours.

(A) How many houses of each type should be constructed to maximize the contractor's profit? What is the maximum profit?

(B) A decrease in demand for colonial houses causes the profit on a colonial house to drop from $20,000 to $17,000. Discuss the effect of this change on the number of houses built and on the maximum profit.

(C) An increase in demand for colonial houses causes the profit on a colonial house to rise from $20,000 to $25,000. Discuss the effect of this change on the number of houses built and on the maximum profit.

42. *Manufacturing—resource allocation.* A company manufactures three-speed, five-speed, and ten-speed bicycles. Each bicycle passes through three departments, fabrication, painting & plating, and final assembly. The relevant manufacturing data are given in the table.

	Labor-Hours per Bicycle			Maximum Labor-Hours Available per Day
	Three-speed	*Five-speed*	*Ten-speed*	
Fabrication	3	4	5	120
Painting & Plating	5	3	5	130
Final Assembly	4	3	5	120
Profit per Bicycle ($)	80	70	100	

(A) How many bicycles of each type should the company manufacture per day in order to maximize its profit? What is the maximum profit?

(B) Discuss the effect on the solution to part (A) if the profit on a ten-speed bicycle increases to $110 and all other data in part (A) remain the same.

(C) Discuss the effect on the solution to part (A) if the profit on a five-speed bicycle increases to $110 and all other data in part (A) remain the same.

43. *Packaging—product mix.* A candy company makes three types of candy, solid-center, fruit-filled, and cream-filled, and packages these candies in three different assortments. A box of assortment I contains 4 solid-center, 4 fruit-filled, and 12 cream-filled candies, and sells for $9.40. A box of assortment II contains 12 solid-center, 4 fruit-filled, and 4 cream-filled candies, and sells for $7.60. A box of assortment III contains 8 solid-center, 8 fruit-filled, and 8 cream-filled candies, and sells for $11.00. The manufacturing costs per piece of candy are $0.20 for solid-center, $0.25 for fruit-filled, and $0.30 for cream-filled. The company can manufacture 4,800 solid-center, 4,000 fruit-filled, and 5,600 cream-filled candies weekly.

(A) How many boxes of each type should the company produce each week in order to maximize their profit? What is the maximum profit?

(B) Discuss the effect on the solution to part (A) if the number of fruit-filled candies manufactured weekly is increased to 5,000 and all other data in part (A) remain the same.

(C) Discuss the effect on the solution to part (A) if the number of fruit-filled candies and the number of solid-center candies manufactured weekly are each increased to 6,000 and all other data in part (A) remain the same.

44. *Scheduling—resource allocation.* A small accounting firm prepares tax returns for three types of customers: individual, commercial, and industrial. The tax preparation process begins with a 1 hour interview with the customer. The data collected during this interview are entered into a time-sharing computer system, which produces the customer's tax return. It takes 1 hour to enter the data for an individual customer, 2 hours for a commercial customer, and $1\frac{1}{2}$ hours for an industrial customer. It takes 10 minutes of computer time to process an individual return, 25 minutes to process a commercial return, and 20 minutes to process an industrial return. The firm has one employee who conducts the initial interview and two who enter the data into the computer. The interviewer can work a maximum of 50 hours a week, and each of the data-entry employees can work a maximum of 40 hours a week. The computer is available for a maximum of 1,025 minutes a week. The firm makes a profit of $50 on each individual customer, $65 on each commercial customer, and $60 on each industrial customer.

(A) How many customers of each type should the firm schedule each week in order to maximize its profit? What is the maximum profit?

(B) Discuss the effect on the solution to part (A) if the maximum number of hours the interviewer works is decreased to 35 hours per week and all other data in part (A) remain the same.

(C) Discuss the effect on the solution to part (A) if the maximum number of minutes of available computer time is decreased to 450 minutes per week and all other data in part (A) remain the same.

Life Sciences

45. *Nutrition—animals.* The natural diet of a certain animal consists of three foods, *A, B,* and *C.* The number of units of calcium, iron, and protein in 1 gram of each food and the average daily intake are given in the table. A scientist wants to investigate the effect of increasing the protein in the animal's diet while not allowing the units of calcium and iron to exceed their average daily intakes. How many grams of each food should be used to maximize the amount of protein in the diet? What is the maximum amount of protein?

	Units per Gram			Average Daily Intake
	Food A	*Food B*	*Food C*	(Units)
Calcium	1	3	2	30
Iron	2	1	2	24
Protein	3	4	5	60

46. *Nutrition—animals.* Repeat Problem 45 if the scientist wants to maximize the daily calcium intake while not allowing the intake of iron or protein to exceed the average daily intake.

Social Sciences

47. *Opinion survey.* A political scientist has received a grant to fund a research project involving voting trends. The budget of the grant includes $3,200 for conducting door-to-door interviews the day before an election. Undergraduate students, graduate students, and faculty members will be hired to conduct the interviews. Each undergraduate student will conduct 18 interviews and be paid $100. Each graduate student will conduct 25 interviews and be paid $150. Each faculty member will conduct 30 interviews and be paid $200. Due to limited transportation facilities, no more than 20 interviewers can be hired. How many undergraduate students, graduate students, and faculty members should be hired in order to maximize the number of interviews that will be conducted? What is the maximum number of interviews?

48. *Opinion survey.* Repeat Problem 47 if one of the requirements of the grant is that at least 50% of the interviewers be undergraduate students.

Section 5-5 The Dual; Minimization with Problem Constraints of the Form $\geq$

- Formation of the Dual Problem
- Solution of Minimization Problems
- Application: A Transportation Problem
- Summary of Problem Types and Solution Methods

In the last section we restricted ourselves to standard maximization problems (problem constraints of the form $\leq$, with nonnegative constants on the right and objective function coefficients any real numbers). Now we will consider minimization problems with $\geq$ problem constraints. These two types of problems turn out to be very closely related.

FORMATION OF THE DUAL PROBLEM

Associated with each minimization problem with $\geq$ constraints is a maximization problem called the **dual problem.** To illustrate the procedure for forming the dual problem, consider the following minimization problem:

$$\begin{aligned} \text{Minimize} \quad & C = 16x_1 + 45x_2 \\ \text{Subject to} \quad & 2x_1 + 5x_2 \geq 50 \\ & x_1 + 3x_2 \geq 27 \\ & x_1, x_2 \geq 0 \end{aligned} \tag{1}$$

The first step in forming the dual problem is to construct a matrix by using the problem constraints and the objective function written in the following form:

$$\begin{aligned} 2x_1 + \ \ 5x_2 &\geq 50 \\ x_1 + \ \ 3x_2 &\geq 27 \\ 16x_1 + 45x_2 &= C \end{aligned} \qquad A = \left[\begin{array}{cc|c} 2 & 5 & 50 \\ 1 & 3 & 27 \\ \hline 16 & 45 & 1 \end{array}\right]$$

CAUTION

Do not confuse matrix A with the simplex tableau. We use a solid horizontal line in matrix A to help distinguish the dual matrix from the simplex tableau. No slack variables are involved in matrix A, and the coefficient of C is in the same column as the constants from the problem constraints.

Now we will form a second matrix called the *transpose of A*. In general, the **transpose** of a given matrix A is the matrix A^{T} formed by interchanging the rows and corresponding columns of A (first row with first column, second row with second column, and so on).

$$A = \left[\begin{array}{cc|c} 2 & 5 & 50 \\ 1 & 3 & 27 \\ \hline 16 & 45 & 1 \end{array}\right]$$

R_1 in $A = C_1$ in A^{T}

R_2 in $A = C_2$ in A^{T}

R_3 in $A = C_3$ in A^{T}

$$A^{\mathrm{T}} = \left[\begin{array}{cc|c} 2 & 1 & 16 \\ 5 & 3 & 45 \\ \hline 50 & 27 & 1 \end{array}\right]$$

A^{T} is the transpose of A.

If you are using a graphing utility for matrix operations, consult your manual for information on finding the transpose of a matrix (see Fig. 1).

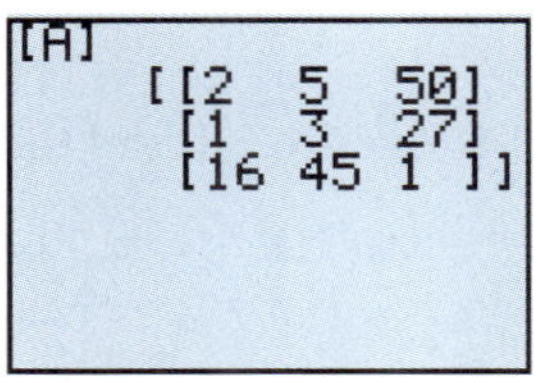

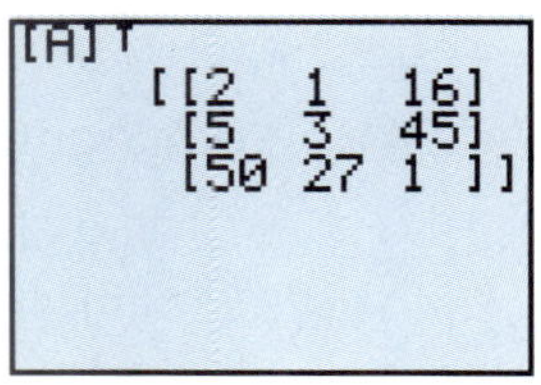

FIGURE 1

Finding A^{T} on a graphing utility

Now, we use the rows of A^T to define a new linear programming problem. This new problem will always be a maximization problem with $\leq$ problem constraints. To avoid confusion, we shall use different variables in this new problem:

$$\begin{aligned} 2y_1 + y_2 &\leq 16 \\ 5y_1 + 3y_2 &\leq 45 \\ 50y_1 + 27y_2 &= P \end{aligned} \qquad A^T = \begin{array}{c} \begin{matrix} y_1 & y_2 & \end{matrix} \\ \left[\begin{array}{cc|c} 2 & 1 & 16 \\ 5 & 3 & 45 \\ \hline 50 & 27 & 1 \end{array}\right] \end{array}$$

The dual of the minimization problem (1) is the following maximization problem:

$$\begin{aligned} \text{Maximize} \quad & P = 50y_1 + 27y_2 \\ \text{Subject to} \quad & 2y_1 + y_2 \leq 16 \\ & 5y_1 + 3y_2 \leq 45 \\ & y_1, y_2 \geq 0 \end{aligned} \qquad (2)$$

Explore–Discuss 1

Excluding the nonnegative constraints, the components of a linear programming problem can be divided into three categories: the coefficients of the objective function, the coefficients of the problem constraints, and the constants on the right side of the problem constraints. Write a verbal description of the relationship between the components of the original minimization problem (1) and the dual maximization problem (2).

The procedure for forming the dual problem is summarized in the box below:

Formation of the Dual Problem

Given a minimization problem with $\geq$ problem constraints:

Step 1. Use the coefficients and constants in the problem constraints and the objective function to form a matrix A with the coefficients of the objective function in the last row.

Step 2. Interchange the rows and columns of matrix A to form the matrix A^T, the transpose of A.

Step 3. Use the rows of A^T to form a maximization problem with $\leq$ problem constraints.

Example 1 **Forming the Dual Problem** Form the dual problem:

$$\begin{aligned} \text{Minimize} \quad & C = 40x_1 + 12x_2 + 40x_3 \\ \text{Subject to} \quad & 2x_1 + x_2 + 5x_3 \geq 20 \\ & 4x_1 + x_2 + x_3 \geq 30 \\ & x_1, x_2, x_3 \geq 0 \end{aligned}$$

SOLUTION **Step 1.** Form the matrix A:

$$A = \left[\begin{array}{ccc|c} 2 & 1 & 5 & 20 \\ 4 & 1 & 1 & 30 \\ \hline 40 & 12 & 40 & 1 \end{array}\right]$$

Step 2. Form the matrix A^{T}, the transpose of A:

$$A^{T} = \left[\begin{array}{cc|c} 2 & 4 & 40 \\ 1 & 1 & 12 \\ 5 & 1 & 40 \\ \hline 20 & 30 & 1 \end{array}\right]$$

Step 3. State the dual problem:

$$\begin{aligned} \text{Maximize} \quad & P = 20y_1 + 30y_2 \\ \text{Subject to} \quad & 2y_1 + 4y_2 \le 40 \\ & y_1 + y_2 \le 12 \\ & 5y_1 + y_2 \le 40 \\ & y_1, y_2 \ge 0 \end{aligned}$$

Matched Problem 1 Form the dual problem:

$$\begin{aligned} \text{Minimize} \quad & C = 16x_1 + 9x_2 + 21x_3 \\ \text{Subject to} \quad & x_1 + x_2 + 3x_3 \ge 12 \\ & 2x_1 + x_2 + x_3 \ge 16 \\ & x_1, x_2, x_3 \ge 0 \end{aligned}$$

SOLUTION OF MINIMIZATION PROBLEMS

The following theorem establishes the relationship between the solution of a minimization problem and the solution of its dual:

Theorem 1 THE FUNDAMENTAL PRINCIPLE OF DUALITY

A minimization problem has a solution if and only if its dual problem has a solution. If a solution exists, then the optimal value of the minimization problem is the same as the optimal value of the dual problem.

The proof of Theorem 1 is beyond the scope of this text. However, we can illustrate Theorem 1 by solving minimization problem (1) and its dual maximization problem (2) geometrically. (Note that Theorem 2 in Section 5-2 guarantees that both problems have solutions.)

Original Problem (1)

Minimize $C = 16x_1 + 45x_2$
Subject to $2x_1 + 5x_2 \geq 50$
$x_1 + 3x_2 \geq 27$
$x_1, x_2 \geq 0$

Dual Problem (2)

Maximize $P = 50y_1 + 27y_2$
Subject to $2y_1 + y_2 \leq 16$
$5y_1 + 3y_2 \leq 45$
$y_1, y_2 \geq 0$

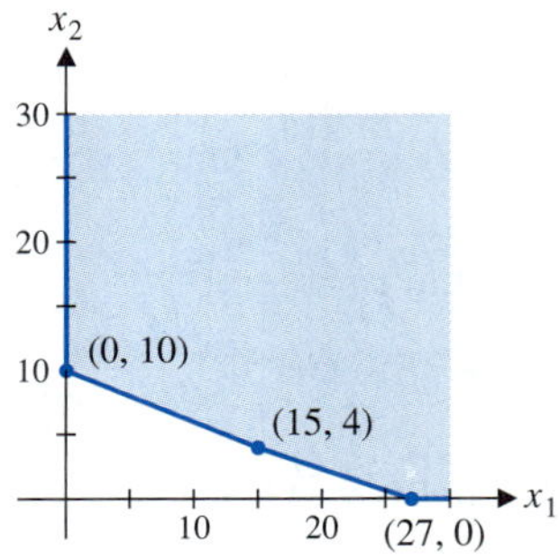

Corner point (x_1, x_2)	$C = 16x_1 + 45x_2$
(0, 10)	450
(15, 4)	420
(27, 0)	432

Min $C = 420$ at $(15, 4)$

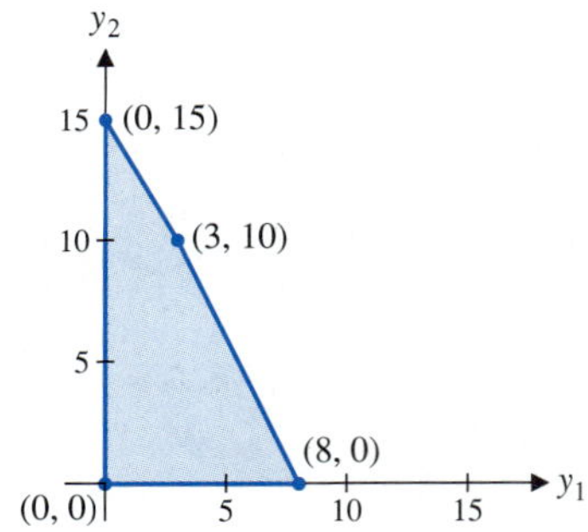

Corner point (y_1, y_2)	$P = 50y_1 + 27y_2$
(0, 0)	0
(0, 15)	405
(3, 10)	420
(8, 0)	400

Max $P = 420$ at $(3, 10)$

Thus, the minimum value of C in problem (1) is the same as the maximum value of P in problem (2). Notice that the optimal solutions that produce this optimal value are different: (15, 4) is the optimal solution for problem (1), and (3, 10) is the optimal solution for problem (2). Theorem 1 only guarantees that the optimal values of a minimization problem and its dual are equal, not that the optimal solutions are the same. In general, it is not possible to determine an optimal solution for a minimization problem by examining the feasible set for the dual problem. However, it is possible to apply the simplex method to the dual problem and find both the optimal value and an optimal solution to the original minimization problem. To see how this is done, we will now solve problem (2) by the simplex method.

For reasons that will become clear later, we will use the variables x_1 and x_2 from the original problem as the slack variables in the dual problem:

$$\begin{aligned} 2y_1 + y_2 + x_1 \qquad &= 16 \\ 5y_1 + 3y_2 \qquad + x_2 &= 45 \\ -50y_1 - 27y_2 \qquad + P &= 0 \end{aligned}$$

Initial system for the dual problem

$$
\begin{array}{c}
\\ x_1 \\ x_2 \\ P
\end{array}
\begin{array}{c}
\begin{array}{ccccc|c} y_1 & y_2 & x_1 & x_2 & P & \end{array}\\
\left[\begin{array}{ccccc|c}
\textcircled{2} & 1 & 1 & 0 & 0 & 16 \\
5 & 3 & 0 & 1 & 0 & 45 \\
\hline
-50 & -27 & 0 & 0 & 1 & 0
\end{array}\right]
\end{array}
\quad 0.5R_1 \to R_1
$$

$$
\sim \left[\begin{array}{ccccc|c}
\textcircled{1} & 0.5 & 0.5 & 0 & 0 & 8 \\
5 & 3 & 0 & 1 & 0 & 45 \\
\hline
-50 & -27 & 0 & 0 & 1 & 0
\end{array}\right]
\quad \begin{array}{l} (-5)R_1 + R_2 \to R_2 \\ 50R_1 + R_3 \to R_3 \end{array}
$$

$$
\begin{array}{r} y_1 \\ \sim x_2 \\ P \end{array}
\left[\begin{array}{ccccc|c}
1 & 0.5 & 0.5 & 0 & 0 & 8 \\
0 & \textcircled{0.5} & -2.5 & 1 & 0 & 5 \\
\hline
0 & -2 & 25 & 0 & 1 & 400
\end{array}\right]
\quad 2R_2 \to R_2
$$

$$
\sim \left[\begin{array}{ccccc|c}
1 & 0.5 & 0.5 & 0 & 0 & 8 \\
0 & \textcircled{1} & -5 & 2 & 0 & 10 \\
\hline
0 & -2 & 25 & 0 & 1 & 400
\end{array}\right]
\quad \begin{array}{l} (-0.5)R_2 + R_1 \to R_1 \\ \\ 2R_2 + R_3 \to R_3 \end{array}
$$

$$
\begin{array}{r} y_1 \\ \sim y_2 \\ P \end{array}
\left[\begin{array}{ccccc|c}
1 & 0 & 3 & -1 & 0 & 3 \\
0 & 1 & -5 & 2 & 0 & 10 \\
\hline
0 & 0 & 15 & 4 & 1 & 420
\end{array}\right]
$$

Since all indicators in the bottom row are nonnegative, the solution to the dual problem is

$$y_1 = 3, \quad y_2 = 10, \quad x_1 = 0, \quad x_2 = 0, \quad P = 420$$

which agrees with our earlier geometric solution. Furthermore, examining the bottom row of the final simplex tableau, we see the same optimal solution to the minimization problem that we obtained directly by the geometric method:

$$\text{Min } C = 420 \qquad \text{at} \qquad x_1 = 15, \quad x_2 = 4$$

This is no accident.

An optimal solution to a minimization problem always can be obtained from the bottom row of the final simplex tableau for the dual problem.

Now we can see that using x_1 and x_2 as slack variables in the dual problem makes it easy to identify the solution of the original problem.

Explore–Discuss 2

The simplex method can be used to solve any standard maximization problem. Which of the following minimization problems have dual problems that are standard maximization problems? (Do not solve the problems.)

(A) Minimize $C = 2x_1 + 3x_2$
Subject to
$$\begin{aligned} 2x_1 - 5x_2 &\geq 4 \\ x_1 - 3x_2 &\geq -6 \\ x_1, x_2 &\geq 0 \end{aligned}$$

(B) Minimize $C = 2x_1 - 3x_2$
Subject to
$$\begin{aligned} -2x_1 + 5x_2 &\geq 4 \\ -x_1 + 3x_2 &\geq 6 \\ x_1, x_2 &\geq 0 \end{aligned}$$

In general, what conditions must a minimization problem satisfy so that its dual problem is a standard maximization problem?

The procedure for solving a minimization problem by applying the simplex method to its dual problem is summarized in the following box:

Solution of a Minimization Problem

Given a minimization problem with nonnegative coefficients in the objective function:

Step 1. Write all problem constraints as $\geqslant$ inequalities. (This may introduce negative numbers on the right side of some problem constraints.)

Step 2. Form the dual problem.

Step 3. Write the initial system of the dual problem, using the variables from the minimization problem as slack variables.

Step 4. Use the simplex method to solve the dual problem.

Step 5. Read the solution of the minimization problem from the bottom row of the final simplex tableau in step 4. [*Note:* If the dual problem has no optimal solution, then the minimization problem has no optimal solution.]

Example 2 **Solving a Minimization Problem** Solve the following minimization problem by maximizing the dual:

$$\begin{aligned} \text{Minimize} \quad & C = 40x_1 + 12x_2 + 40x_3 \\ \text{Subject to} \quad & 2x_1 + x_2 + 5x_3 \geqslant 20 \\ & 4x_1 + x_2 + x_3 \geqslant 30 \\ & x_1, x_2, x_3 \geqslant 0 \end{aligned}$$

Solution From Example 1 the dual is

$$\begin{aligned} \text{Maximize} \quad & P = 20y_1 + 30y_2 \\ \text{Subject to} \quad & 2y_1 + 4y_2 \leqslant 40 \\ & y_1 + y_2 \leqslant 12 \\ & 5y_1 + y_2 \leqslant 40 \\ & y_1, y_2 \geqslant 0 \end{aligned}$$

Using x_1, x_2, and x_3 for slack variables, we obtain the initial system for the dual:

$$\begin{aligned} 2y_1 + 4y_2 + x_1 \qquad\qquad\qquad &= 40 \\ y_1 + y_2 \qquad + x_2 \qquad\qquad &= 12 \\ 5y_1 + y_2 \qquad\qquad + x_3 \qquad &= 40 \\ -20y_1 - 30y_2 \qquad\qquad\qquad + P &= 0 \end{aligned}$$

Now we form the simplex tableau and solve the dual problem:

$$
\begin{array}{c}
 \\ x_1 \\ x_2 \\ x_3 \\ P
\end{array}
\begin{array}{c}
\begin{array}{cccccc} y_1 & y_2 & x_1 & x_2 & x_3 & P \end{array} \\
\left[\begin{array}{cccccc|c}
2 & \textcircled{4} & 1 & 0 & 0 & 0 & 40 \\
1 & 1 & 0 & 1 & 0 & 0 & 12 \\
5 & 1 & 0 & 0 & 1 & 0 & 40 \\
\hdashline
-20 & -30 & 0 & 0 & 0 & 1 & 0
\end{array}\right]
\end{array}
\quad \tfrac{1}{4}R_1 \to R_1
$$

$$
\sim \left[\begin{array}{cccccc|c}
\frac{1}{2} & \textcircled{1} & \frac{1}{4} & 0 & 0 & 0 & 10 \\
1 & 1 & 0 & 1 & 0 & 0 & 12 \\
5 & 1 & 0 & 0 & 1 & 0 & 40 \\
\hdashline
-20 & -30 & 0 & 0 & 0 & 1 & 0
\end{array}\right]
\quad
\begin{array}{l}
(-1)R_1 + R_2 \to R_2 \\
(-1)R_1 + R_3 \to R_3 \\
30R_1 + R_4 \to R_4
\end{array}
$$

$$
\sim
\begin{array}{c}
y_2 \\ x_2 \\ x_3 \\ P
\end{array}
\left[\begin{array}{cccccc|c}
\frac{1}{2} & 1 & \frac{1}{4} & 0 & 0 & 0 & 10 \\
\textcircled{\frac{1}{2}} & 0 & -\frac{1}{4} & 1 & 0 & 0 & 2 \\
\frac{9}{2} & 0 & -\frac{1}{4} & 0 & 1 & 0 & 30 \\
\hdashline
-5 & 0 & \frac{15}{2} & 0 & 0 & 1 & 300
\end{array}\right]
\quad 2R_2 \to R_2
$$

$$
\sim \left[\begin{array}{cccccc|c}
\frac{1}{2} & 1 & \frac{1}{4} & 0 & 0 & 0 & 10 \\
\textcircled{1} & 0 & -\frac{1}{2} & 2 & 0 & 0 & 4 \\
\frac{9}{2} & 0 & -\frac{1}{4} & 0 & 1 & 0 & 30 \\
\hdashline
-5 & 0 & \frac{15}{2} & 0 & 0 & 1 & 300
\end{array}\right]
\quad
\begin{array}{l}
(-\frac{1}{2})R_2 + R_1 \to R_1 \\
 \\
(-\frac{9}{2})R_2 + R_3 \to R_3 \\
5R_2 + R_4 \to R_4
\end{array}
$$

$$
\sim
\begin{array}{c}
y_2 \\ y_1 \\ x_3 \\ P
\end{array}
\left[\begin{array}{cccccc|c}
0 & 1 & \frac{1}{2} & -1 & 0 & 0 & 8 \\
1 & 0 & -\frac{1}{2} & 2 & 0 & 0 & 4 \\
0 & 0 & 2 & -9 & 1 & 0 & 12 \\
\hdashline
0 & 0 & 5 & 10 & 0 & 1 & 320
\end{array}\right]
$$

From the bottom row of this tableau, we see that

$$\text{Min } C = 320 \quad \text{at} \quad x_1 = 5, \quad x_2 = 10, \quad x_3 = 0$$

Matched Problem 2 Solve the following minimization problem by maximizing the dual (see Matched Problem 1):

$$
\begin{array}{ll}
\text{Minimize} & C = 16x_1 + 9x_2 + 21x_3 \\
\text{Subject to} & x_1 + x_2 + 3x_3 \geq 12 \\
 & 2x_1 + x_2 + x_3 \geq 16 \\
 & x_1, x_2, x_3 \geq 0
\end{array}
$$

CAUTION

In the preceding section, we noted that multiplying a problem constraint by a number (usually in order to simplify calculations) changes the units of the slack variable. This requires some special interpretation of the value of the slack variable in the optimal solution, but causes no serious problems. However, when using the dual method, multiplying a problem constraint in the dual problem by a number can have some very serious consequences—the bottom row of the final simplex tableau may no longer give the correct solution to the minimization problem. To see this, refer to the first problem constraint of the dual problem in Example 2:

$$2y_1 + 4y_2 \leqslant 40$$

If we multiply this constraint by $\frac{1}{2}$ and then solve, the final tableau is (verify this):

$$\begin{array}{c} \begin{matrix} y_1 & y_2 & x_1 & x_2 & x_3 & P \end{matrix} \\ \left[\begin{array}{cccccc|c} 0 & 1 & 1 & -1 & 0 & 0 & 8 \\ 1 & 0 & -1 & 2 & 0 & 0 & 4 \\ 0 & 0 & 4 & -9 & 1 & 0 & 12 \\ \hdashline 0 & 0 & 10 & 10 & 0 & 1 & 320 \end{array}\right] \end{array}$$

The bottom row of this tableau indicates that the optimal solution to the minimization problem is $C = 320$ at $x_1 = 10$ and $x_2 = 10$. This is not the correct answer ($x_1 = 5$ is the correct answer). Thus, **you should never multiply a problem constraint in a maximization problem by a number if that maximization problem is being used to solve a minimization problem.** You may still simplify problem constraints in a minimization problem before forming the dual problem.

Example 3 **Solving a Minimization Problem** Solve the following minimization problem by maximizing the dual:

$$\begin{aligned} \text{Minimize} \quad & C = 5x_1 + 10x_2 \\ \text{Subject to} \quad & x_1 - x_2 \geqslant 1 \\ & -x_1 + x_2 \geqslant 2 \\ & x_1, x_2 \geqslant 0 \end{aligned}$$

SOLUTION $$A = \left[\begin{array}{cc|c} 1 & -1 & 1 \\ -1 & 1 & 2 \\ \hline 5 & 10 & 1 \end{array}\right] \qquad A^{T} = \left[\begin{array}{cc|c} 1 & -1 & 5 \\ -1 & 1 & 10 \\ \hline 1 & 2 & 1 \end{array}\right]$$

The dual problem is

$$\begin{aligned} \text{Maximize} \quad & P = y_1 + 2y_2 \\ \text{Subject to} \quad & y_1 - y_2 \leqslant 5 \\ & -y_1 + y_2 \leqslant 10 \\ & y_1, y_2 \geqslant 0 \end{aligned}$$

Introduce slack variables x_1 and x_2, and form the initial system for the dual:

$$\begin{aligned} y_1 - y_2 + x_1 \qquad\qquad &= 5 \\ -y_1 + y_2 \qquad + x_2 \qquad &= 10 \\ -y_1 - 2y_2 \qquad\qquad + P &= 0 \end{aligned}$$

Form the simplex tableau and solve:

$$
\begin{array}{c}
 \\ x_1 \\ x_2 \\ P
\end{array}
\begin{array}{c}
\begin{matrix} y_1 & y_2 & x_1 & x_2 & P & \end{matrix} \\
\left[\begin{array}{rrrrr|r}
1 & -1 & 1 & 0 & 0 & 5 \\
-1 & \textcircled{1} & 0 & 1 & 0 & 10 \\
\hdashline
-1 & -2 & 0 & 0 & 1 & 0
\end{array}\right]
\end{array}
\begin{array}{l}
R_2 + R_1 \to R_1 \\
\\
2R_2 + R_3 \to R_3
\end{array}
$$

$$
\sim \left[\begin{array}{rrrrr|r}
0 & 0 & 1 & 1 & 0 & 15 \\
-1 & 1 & 0 & 1 & 0 & 10 \\
\hdashline
-3 & 0 & 0 & 2 & 1 & 20
\end{array}\right]
\begin{array}{l}
\text{No positive elements} \\
\text{above dashed line in} \\
\text{pivot column}
\end{array}
$$

$\uparrow$
Pivot column

The -3 in the bottom row indicates that Column 1 is the pivot column. Since no positive elements appear in the pivot column above the dashed line, we are unable to select a pivot row. We stop the pivot operation and conclude that this maximization problem has no optimal solution (see the flowchart in Fig. 2, Section 5-4). Theorem 1 now implies that the original minimization problem has no solution. The graph of the inequalities in the minimization problem (Fig. 2) shows that the feasible region is empty; thus, it is not surprising that an optimal solution does not exist.

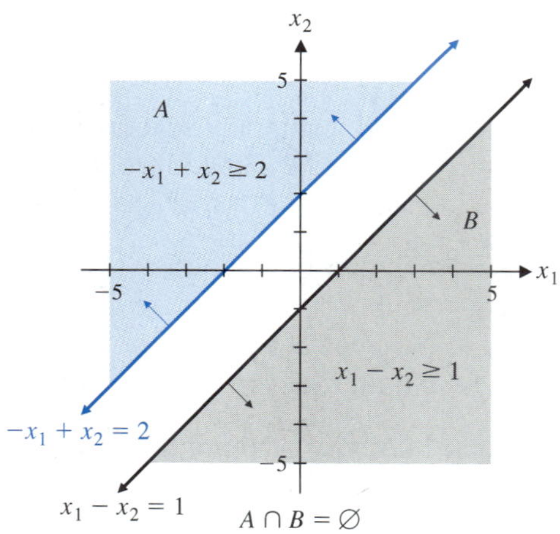

FIGURE 2

Matched Problem 3 ➡ Solve the following minimization problem by maximizing the dual:

$$
\begin{aligned}
\text{Minimize} \quad & C = 2x_1 + 3x_2 \\
\text{Subject to} \quad & x_1 - 2x_2 \geq 2 \\
& -x_1 + x_2 \geq 1 \\
& x_1, x_2 \geq 0
\end{aligned}
$$

■ Application: A Transportation Problem

One of the first applications of linear programming was to the problem of minimizing the cost of transporting materials. Problems of this type are referred to as **transportation problems.**

Example 4 **A Transportation Problem** A computer manufacturing company has two assembly plants, plant A and plant B, and two distribution outlets, outlet I and outlet II. Plant A can assemble at most 700 computers a month, and plant B can assemble at most 900 computers a month. Outlet I must have at least 500 computers a month, and outlet II must have at least 1,000 computers a month. Transportation costs for shipping one computer from each plant to each outlet are as follows: \$6 from plant A to outlet I; \$5 from plant A to outlet II; \$4 from plant B to outlet I; \$8 from plant B to outlet II. Find a shipping schedule that will minimize the total cost of shipping the computers from the assembly plants to the distribution outlets. What is this minimum cost?

SOLUTION First, we summarize the relevant data in a table:

	Distribution Outlet I	Distribution Outlet II	Assembly Capacity
Plant A	\$6	\$5	700
Plant B	\$4	\$8	900
Minimum Required	500	1,000	

In order to find a shipping schedule, we must determine the number of computers that should be shipped from each plant to each outlet. This will require the use of four decision variables:

x_1 = Number of computers shipped from plant A to outlet I
x_2 = Number of computers shipped from plant A to outlet II
x_3 = Number of computers shipped from plant B to outlet I
x_4 = Number of computers shipped from plant B to outlet II

The total number of computers shipped from plant A is $x_1 + x_2$. Since this cannot exceed the assembly capacity at A, we have

$$x_1 + x_2 \leq 700 \quad \text{Number shipped from plant } A$$

Similarly, the total number shipped from plant B must satisfy

$$x_3 + x_4 \leq 900 \quad \text{Number shipped from plant } B$$

The total number shipped to each outlet must satisfy

$$x_1 + x_3 \geq 500 \quad \text{Number shipped to outlet I}$$

and

$$x_2 + x_4 \geq 1{,}000 \quad \text{Number shipped to outlet II}$$

Using the shipping charges in the table, the total shipping charges are

$$C = 6x_1 + 5x_2 + 4x_3 + 8x_4$$

Thus, we must solve the following linear programming problem:

$$
\begin{array}{llll}
\text{Minimize} & C = 6x_1 + 5x_2 + 4x_3 + 8x_4 & & \\
\text{Subject to} & x_1 + x_2 \qquad\qquad \leqslant 700 & & \text{Available from A} \\
& \qquad\qquad x_3 + x_4 \leqslant 900 & & \text{Available from B} \\
& x_1 \qquad + x_3 \qquad \geqslant 500 & & \text{Required at I} \\
& \qquad x_2 \qquad + x_4 \geqslant 1{,}000 & & \text{Required at II} \\
& \qquad x_1, x_2, x_3, x_4 \geqslant 0 & &
\end{array}
$$

Before we can solve this problem, we must multiply the first two constraints by -1 so that all the problem constraints are of the $\geqslant$ type. This will introduce negative constants into the minimization problem but not into the dual. Since the coefficients of C are nonnegative, the constants in the dual problem will be nonnegative and the dual will be a standard maximization problem. The problem can now be stated as

$$
\begin{array}{ll}
\text{Minimize} & C = 6x_1 + 5x_2 + 4x_3 + 8x_4 \\
\text{Subject to} & -x_1 - x_2 \qquad\qquad \geqslant -700 \\
& \qquad\qquad - x_3 - x_4 \geqslant -900 \\
& x_1 \qquad + x_3 \qquad \geqslant 500 \\
& \qquad x_2 \qquad + x_4 \geqslant 1{,}000 \\
& \qquad x_1, x_2, x_3, x_4 \geqslant 0
\end{array}
$$

$$
A = \left[\begin{array}{cccc|c}
-1 & -1 & 0 & 0 & -700 \\
0 & 0 & -1 & -1 & -900 \\
1 & 0 & 1 & 0 & 500 \\
0 & 1 & 0 & 1 & 1{,}000 \\
\hline
6 & 5 & 4 & 8 & 1
\end{array}\right]
$$

$$
A^{\mathrm{T}} = \left[\begin{array}{cccc|c}
-1 & 0 & 1 & 0 & 6 \\
-1 & 0 & 0 & 1 & 5 \\
0 & -1 & 1 & 0 & 4 \\
0 & -1 & 0 & 1 & 8 \\
\hline
-700 & -900 & 500 & 1{,}000 & 1
\end{array}\right]
$$

The dual problem is

$$
\begin{array}{ll}
\text{Maximize} & P = -700y_1 - 900y_2 + 500y_3 + 1{,}000y_4 \\
\text{Subject to} & -y_1 \qquad + y_3 \qquad \leqslant 6 \\
& -y_1 \qquad\qquad + y_4 \leqslant 5 \\
& \qquad -y_2 + y_3 \qquad \leqslant 4 \\
& \qquad -y_2 \qquad + y_4 \leqslant 8 \\
& \qquad y_1, y_2, y_3, y_4 \geqslant 0
\end{array}
$$

Introduce slack variables x_1, x_2, x_3, and x_4, and form the initial system for the dual:

$$
\begin{array}{rrrrrrrrrl}
-y_1 & & + y_3 & & + x_1 & & & & & = 6 \\
-y_1 & & & + y_4 & & + x_2 & & & & = 5 \\
& -y_2 & + y_3 & & & & + x_3 & & & = 4 \\
& -y_2 & & + y_4 & & & & + x_4 & & = 8 \\
700y_1 & + 900y_2 & - 500y_3 & - 1{,}000y_4 & & & & & + P & = 0
\end{array}
$$

Form the simplex tableau and solve:

$$
\begin{array}{c}
\begin{array}{c|ccccccccc|c|l}
 & y_1 & y_2 & y_3 & y_4 & x_1 & x_2 & x_3 & x_4 & P & & \\
\hline
x_1 & -1 & 0 & 1 & 0 & 1 & 0 & 0 & 0 & 0 & 6 & \\
x_2 & -1 & 0 & 0 & \textcircled{1} & 0 & 1 & 0 & 0 & 0 & 5 & \\
x_3 & 0 & -1 & 1 & 0 & 0 & 0 & 1 & 0 & 0 & 4 & \\
x_4 & 0 & -1 & 0 & 1 & 0 & 0 & 0 & 1 & 0 & 8 & (-1)R_2 + R_4 \to R_4 \\
\hdashline
P & 700 & 900 & -500 & -1{,}000 & 0 & 0 & 0 & 0 & 1 & 0 & 1{,}000R_2 + R_5 \to R_5
\end{array} \\
\\
\sim
\begin{array}{c|ccccccccc|c|l}
x_1 & -1 & 0 & 1 & 0 & 1 & 0 & 0 & 0 & 0 & 6 & (-1)R_3 + R_1 \to R_1 \\
y_4 & -1 & 0 & 0 & 1 & 0 & 1 & 0 & 0 & 0 & 5 & \\
x_3 & 0 & -1 & \textcircled{1} & 0 & 0 & 0 & 1 & 0 & 0 & 4 & \\
x_4 & 1 & -1 & 0 & 0 & 0 & -1 & 0 & 1 & 0 & 3 & \\
\hdashline
P & -300 & 900 & -500 & 0 & 0 & 1{,}000 & 0 & 0 & 1 & 5{,}000 & 500R_3 + R_5 \to R_5
\end{array} \\
\\
\sim
\begin{array}{c|ccccccccc|c|l}
x_1 & -1 & 1 & 0 & 0 & 1 & 0 & -1 & 0 & 0 & 2 & R_4 + R_1 \to R_1 \\
y_4 & -1 & 0 & 0 & 1 & 0 & 1 & 0 & 0 & 0 & 5 & R_4 + R_2 \to R_2 \\
y_3 & 0 & -1 & 1 & 0 & 0 & 0 & 1 & 0 & 0 & 4 & \\
x_4 & \textcircled{1} & -1 & 0 & 0 & 0 & -1 & 0 & 1 & 0 & 3 & \\
\hdashline
P & -300 & 400 & 0 & 0 & 0 & 0 & 0 & 0 & 1 & 7{,}000 & 300R_4 + R_5 \to R_5
\end{array} \\
\\
\sim
\begin{array}{c|ccccccccc|c}
x_1 & 0 & 0 & 0 & 0 & 1 & -1 & -1 & 1 & 0 & 6 \\
y_4 & 0 & -1 & 0 & 1 & 0 & 0 & 0 & 1 & 0 & 5 \\
y_3 & 0 & -1 & 1 & 0 & 0 & 0 & 1 & 0 & 0 & 4 \\
y_1 & 1 & -1 & 0 & 0 & 0 & -1 & 0 & 1 & 0 & 3 \\
\hdashline
P & 0 & 100 & 0 & 0 & 0 & 700 & 500 & 300 & 1 & 7{,}900
\end{array}
\end{array}
$$

From the bottom row of this tableau, we have

$$\text{Min } C = 7{,}900 \quad \text{at} \quad x_1 = 0, \quad x_2 = 700, \quad x_3 = 500, \quad x_4 = 300$$

The shipping schedule that minimizes the shipping charges is 700 from plant *A* to outlet II, 500 from plant *B* to outlet I, and 300 from plant *B* to outlet II. The total shipping cost is \$7,900.

Figure 3 shows a solution to Example 4 in Excel, a popular spreadsheet for personal computers. Notice that Excel permits the user to organize the original data and the solution in a format that is clear and easy to read. This is one of the main advantages of using spreadsheets to solve linear programming problems. The software in *Explorations in Finite Mathematics* (see Preface) can also be used to solve this problem.

	A	B	C	D	E	F	G	H	I
1		DATA					SHIPPING SCHEDULE		
2		DISTRIBUTION					DISTRIBUTION		
3		OUTLET		ASSEMBLY			OUTLET		
4		I	II	CAPACITY			I	II	TOTAL
5	PLANT A	\$6	\$5	700		PLANT A	0	700	700
6	PLANT B	\$4	\$8	900		PLANT B	500	300	800
7	MINIMUM					TOTAL	500	1,000	
8	REQUIRED	500	1,000					TOTAL COST	\$7,900

FIGURE 3

Matched Problem 4 ➡ Repeat Example 4 if the shipping charge from plant A to outlet I is increased to \$7 and the shipping charge from plant B to outlet II is decreased to \$3.

Summary of Problem Types and Solution Methods

In this and the preceding sections, we have solved both maximization and minimization problems, but with certain restrictions on the problem constraints, constants on the right, and/or objective function coefficients. Table 1 summarizes the types of problems and methods of solution we have considered so far.

Table 1

Summary of Problem Types and Simplex Solution Methods

Problem type	Problem constraints	Right-side constants	Coefficients of objective function	Method of solution
1. Maximization	$\le$	Nonnegative	Any real numbers	Simplex method with slack variables
2. Minimization	$\ge$	Any real numbers	Nonnegative	Form dual and solve by simplex method with slack variables

The next section develops a generalized version of the simplex method that can handle both maximization and minimization problems with any combination of $\le$, $\ge$, and $=$ problem constraints.

Answers to Matched Problems

1. Maximize $P = 12y_1 + 16y_2$
Subject to
$$\begin{aligned} y_1 + 2y_2 &\le 16 \\ y_1 + y_2 &\le 9 \\ 3y_1 + y_2 &\le 21 \\ y_1, y_2 &\ge 0 \end{aligned}$$

2. Min $C = 136$ at $x_1 = 4$, $x_2 = 8$, $x_3 = 0$

3. Dual problem:
Maximize $P = 2y_1 + y_2$
Subject to
$$\begin{aligned} y_1 - y_2 &\le 2 \\ -2y_1 + y_2 &\le 3 \\ y_1, y_2 &\ge 0 \end{aligned}$$
No optimal solution

4. 600 from plant A to outlet II, 500 from plant B to outlet I, 400 from plant B to outlet II; total shipping cost is \$6,200

Exercise 5-5

A *In Problems 1–8, find the transpose of each matrix.*

1. $\begin{bmatrix} -5 & 0 & 3 & -1 & 8 \end{bmatrix}$

2. $\begin{bmatrix} 1 & 0 & -7 & 3 & -2 \end{bmatrix}$

3. $\begin{bmatrix} 1 \\ -2 \\ 0 \\ 4 \end{bmatrix}$

4. $\begin{bmatrix} 9 \\ 5 \\ -4 \\ 0 \end{bmatrix}$

5. $\begin{bmatrix} 2 & 1 & -6 & 0 & -1 \\ 5 & 2 & 0 & 1 & 3 \end{bmatrix}$

6. $\begin{bmatrix} 7 & 3 & -1 & 3 \\ -6 & 1 & 0 & -9 \end{bmatrix}$

7. $\begin{bmatrix} 1 & 2 & -1 \\ 0 & 2 & -7 \\ 8 & 0 & 1 \\ 4 & -1 & 3 \end{bmatrix}$

8. $\begin{bmatrix} 1 & -1 & 3 & 2 \\ 1 & -4 & 0 & 2 \\ 4 & -5 & 6 & 1 \\ -3 & 8 & 0 & -1 \\ 2 & 7 & -3 & 1 \end{bmatrix}$

In Problems 9 and 10:
(A) Form the dual problem.
(B) Write the initial system for the dual problem.
(C) Write the initial simplex tableau for the dual problem and label the columns of the tableau.

9. Minimize $C = 8x_1 + 9x_2$
Subject to $x_1 + 3x_2 \geq 4$
$2x_1 + x_2 \geq 5$
$x_1, x_2 \geq 0$

10. Minimize $C = 12x_1 + 5x_2$
Subject to $2x_1 + x_2 \geq 7$
$3x_1 + x_2 \geq 9$
$x_1, x_2 \geq 0$

In Problems 11 and 12, a minimization problem, the corresponding dual problem, and the final simplex tableau in the solution of the dual problem are given.
(A) Find the optimal solution of the dual problem.
(B) Find the optimal solution of the minimization problem.

11. Minimize $C = 21x_1 + 50x_2$
Subject to $2x_1 + 5x_2 \geq 12$
$3x_1 + 7x_2 \geq 17$
$x_1, x_2 \geq 0$

Maximize $P = 12y_1 + 17y_2$
Subject to $2y_1 + 3y_2 \leq 21$
$5y_1 + 7y_2 \leq 50$
$y_1, y_2 \geq 0$

$$\begin{array}{c} \begin{matrix} y_1 & y_2 & x_1 & x_2 & P & \end{matrix} \\ \left[\begin{array}{ccccc|c} 0 & 1 & 5 & -2 & 0 & 5 \\ 1 & 0 & -7 & 3 & 0 & 3 \\ \hline 0 & 0 & 1 & 2 & 1 & 121 \end{array}\right] \end{array}$$

12. Minimize $C = 16x_1 + 25x_2$
Subject to $3x_1 + 5x_2 \geq 30$
$2x_1 + 3x_2 \geq 19$
$x_1, x_2 \geq 0$

Maximize $P = 30y_1 + 19y_2$
Subject to $3y_1 + 2y_2 \leq 16$
$5y_1 + 3y_2 \leq 25$
$y_1, y_2 \geq 0$

$$\begin{array}{c} \begin{matrix} y_1 & y_2 & x_1 & x_2 & P & \end{matrix} \\ \left[\begin{array}{ccccc|c} 0 & 1 & 5 & -3 & 0 & 5 \\ 1 & 0 & -3 & 2 & 0 & 2 \\ \hline 0 & 0 & 5 & 3 & 1 & 155 \end{array}\right] \end{array}$$

In Problems 13–20:
(A) Form the dual problem.
(B) Find the solution to the original problem by applying the simplex method to the dual problem.

13. Minimize $C = 9x_1 + 2x_2$
Subject to $4x_1 + x_2 \geq 13$
$3x_1 + x_2 \geq 12$
$x_1, x_2 \geq 0$

14. Minimize $C = x_1 + 4x_2$
Subject to $x_1 + 2x_2 \geq 5$
$x_1 + 3x_2 \geq 6$
$x_1, x_2 \geq 0$

15. Minimize $C = 7x_1 + 12x_2$
Subject to $2x_1 + 3x_2 \geq 15$
$x_1 + 2x_2 \geq 8$
$x_1, x_2 \geq 0$

16. Minimize $C = 3x_1 + 5x_2$
Subject to $2x_1 + 3x_2 \geq 7$
$x_1 + 2x_2 \geq 4$
$x_1, x_2 \geq 0$

17. Minimize $C = 11x_1 + 4x_2$
Subject to $2x_1 + x_2 \geq 8$
$-2x_1 + 3x_2 \geq 4$
$x_1, x_2 \geq 0$

18. Minimize $C = 40x_1 + 10x_2$
Subject to $2x_1 + x_2 \geq 12$
$3x_1 - x_2 \geq 3$
$x_1, x_2 \geq 0$

19. Minimize $C = 7x_1 + 9x_2$
Subject to $-3x_1 + x_2 \geq 6$
$x_1 - 2x_2 \geq 4$
$x_1, x_2 \geq 0$

20. Minimize $C = 10x_1 + 15x_2$
Subject to $-4x_1 + x_2 \geq 12$
$12x_1 - 3x_2 \geq 10$
$x_1, x_2 \geq 0$

B *Solve the linear programming problems in Problems 21–32 by applying the simplex method to the dual problem.*

21. Minimize $C = 3x_1 + 9x_2$
Subject to $2x_1 + x_2 \geq 8$
$x_1 + 2x_2 \geq 8$
$x_1, x_2 \geq 0$

22. Minimize $C = 2x_1 + x_2$
Subject to $x_1 + x_2 \geq 8$
$x_1 + 2x_2 \geq 4$
$x_1, x_2 \geq 0$

23. Minimize $C = 7x_1 + 5x_2$
Subject to
$$\begin{aligned} x_1 + x_2 &\geq 4 \\ x_1 - 2x_2 &\geq -8 \\ -2x_1 + x_2 &\geq -8 \\ x_1, x_2 &\geq 0 \end{aligned}$$

24. Minimize $C = 10x_1 + 4x_2$
Subject to
$$\begin{aligned} 2x_1 + x_2 &\geq 6 \\ x_1 - 4x_2 &\geq -24 \\ -8x_1 + 5x_2 &\geq -24 \\ x_1, x_2 &\geq 0 \end{aligned}$$

25. Minimize $C = 10x_1 + 30x_2$
Subject to
$$\begin{aligned} 2x_1 + x_2 &\geq 16 \\ x_1 + x_2 &\geq 12 \\ x_1 + 2x_2 &\geq 14 \\ x_1, x_2 &\geq 0 \end{aligned}$$

26. Minimize $C = 40x_1 + 10x_2$
Subject to
$$\begin{aligned} 3x_1 + x_2 &\geq 24 \\ x_1 + x_2 &\geq 16 \\ x_1 + 4x_2 &\geq 30 \\ x_1, x_2 &\geq 0 \end{aligned}$$

27. Minimize $C = 5x_1 + 7x_2$
Subject to
$$\begin{aligned} x_1 &\geq 4 \\ x_1 + x_2 &\geq 8 \\ x_1 + 2x_2 &\geq 10 \\ x_1, x_2 &\geq 0 \end{aligned}$$

28. Minimize $C = 4x_1 + 5x_2$
Subject to
$$\begin{aligned} 2x_1 + x_2 &\geq 12 \\ x_1 + x_2 &\geq 9 \\ + x_2 &\geq 4 \\ x_1, x_2 &\geq 0 \end{aligned}$$

29. Minimize $C = 10x_1 + 7x_2 + 12x_3$
Subject to
$$\begin{aligned} x_1 + x_2 + 2x_3 &\geq 7 \\ 2x_1 + x_2 + x_3 &\geq 4 \\ x_1, x_2, x_3 &\geq 0 \end{aligned}$$

30. Minimize $C = 14x_1 + 8x_2 + 20x_3$
Subject to
$$\begin{aligned} x_1 + x_2 + 3x_3 &\geq 6 \\ 2x_1 + x_2 + x_3 &\geq 9 \\ x_1, x_2, x_3 &\geq 0 \end{aligned}$$

31. Minimize $C = 5x_1 + 2x_2 + 2x_3$
Subject to
$$\begin{aligned} x_1 - 4x_2 + x_3 &\geq 6 \\ -x_1 + x_2 - 2x_3 &\geq 4 \\ x_1, x_2, x_3 &\geq 0 \end{aligned}$$

32. Minimize $C = 6x_1 + 8x_2 + 3x_3$
Subject to
$$\begin{aligned} -3x_1 - 2x_2 + x_3 &\geq 4 \\ x_1 + x_2 - x_3 &\geq 2 \\ x_1, x_2, x_3 &\geq 0 \end{aligned}$$

33. A minimization problem has 4 variables and 2 problem constraints. How many variables and problem constraints are in the dual problem?

34. A minimization problem has 3 variables and 5 problem constraints. How many variables and problem constraints are in the dual problem?

35. If you want to solve a minimization problem by applying the geometric method to the dual problem, how many variables and problem constraints must be in the original problem?

36. If you want to solve a minimization problem by applying the geometric method to the original problem, how many variables and problem constraints must be in the original problem?

In Problems 37–40, determine whether a minimization problem with the indicated condition can be solved by applying the simplex method to the dual problem. If your answer is yes, describe any necessary modifications that must be made before forming the dual problem. If your answer is no, explain why.

37. A coefficient of the objective function is negative.

38. A coefficient of a problem constraint is negative.

39. A problem constraint is of the ≤ form.

40. A problem constraint has a negative constant on the right side.

C *Solve the linear programming problems in Problems 41–44 by applying the simplex method to the dual problem.*

41. Minimize $C = 16x_1 + 8x_2 + 4x_3$
Subject to
$$\begin{aligned} 3x_1 + 2x_2 + 2x_3 &\geq 16 \\ 4x_1 + 3x_2 + x_3 &\geq 14 \\ 5x_1 + 3x_2 + x_3 &\geq 12 \\ x_1, x_2, x_3 &\geq 0 \end{aligned}$$

42. Minimize $C = 6x_1 + 8x_2 + 12x_3$
Subject to
$$\begin{aligned} x_1 + 3x_2 + 3x_3 &\geq 6 \\ x_1 + 5x_2 + 5x_3 &\geq 4 \\ 2x_1 + 2x_2 + 3x_3 &\geq 8 \\ x_1, x_2, x_3 &\geq 0 \end{aligned}$$

43. Minimize $C = 5x_1 + 4x_2 + 5x_3 + 6x_4$
Subject to
$$\begin{aligned} x_1 + x_2 &\leq 12 \\ x_3 + x_4 &\leq 25 \\ x_1 + x_3 &\geq 20 \\ x_2 + x_4 &\geq 15 \\ x_1, x_2, x_3, x_4 &\geq 0 \end{aligned}$$

44. Repeat Problem 43 with $C = 4x_1 - 7x_2 + 5x_3 + 6x_4$.

APPLICATIONS

In Problems 45–52, construct a mathematical model in the form of a linear programming problem. (The answers in the back of the book for these application problems include the model.) Then solve the problem by applying the simplex method to the dual problem.

Business & Economics

45. *Manufacturing—production scheduling.* A food processing company produces regular and deluxe ice cream at three plants. Per hour of operation, the plant in Cedarburg produces 20 gallons of regular ice cream and 10 gallons of deluxe ice cream, the Grafton plant produces 10 gallons of regular and 20 gallons of deluxe, and the West Bend plant produces 20 gallons of regular and 20 gallons of deluxe. It costs \$70 per hour to operate the Cedarburg plant, \$75 per hour to operate the Grafton plant, and \$90 per hour to operate the West Bend plant.

(A) The company needs at least 300 gallons of regular ice cream and at least 200 gallons of deluxe ice cream each day. How many hours per day should each plant be scheduled to operate in order to produce the required amounts of ice cream and minimize the cost of production? What is the minimum production cost?

(B) Discuss the effect on the production schedule and the minimum production cost if the demand for deluxe ice cream increases to 300 gallons per day and all other data in part (A) remain the same.

(C) Repeat part (B) if the demand for deluxe ice cream increases to 400 gallons per day.

46. *Mining—production scheduling.* A mining company operates two mines, each of which produces three grades of ore. The West Summit mine can produce 2 tons of low-grade ore, 3 tons of medium-grade ore, and 1 ton of high-grade ore per hour of operation. The North Ridge mine can produce 2 tons of low-grade ore, 1 ton of medium-grade ore, and 2 tons of high-grade ore per hour of operation. To satisfy existing orders, the company needs at least 100 tons of low-grade ore, 60 tons of medium-grade ore, and 80 tons of high-grade ore. The cost of operating each mine varies, depending on the conditions encountered while extracting the ore.

(A) If it costs \$400 per hour to operate the West Summit mine and \$600 per hour to operate the North Ridge mine, how many hours should each mine be operated to supply the required amounts of ore and minimize the cost of production? What is the minimum production cost?

(B) Discuss the effect on the production schedule and the minimum production cost if it costs \$300 per hour to operate the West Summit mine, \$700 per hour to operate the North Ridge mine, and all other data in part (A) remain the same.

(C) Repeat part (B) if it costs \$800 per hour to operate the West Summit mine and \$200 per hour to operate the North Ridge mine.

47. *Purchasing.* Acme Micros markets computers with single-sided and double-sided disk drives. The disk drives are supplied by two other companies, Associated Electronics and Digital Drives. Associated Electronics charges \$250 for a single-sided disk drive and \$350 for a double-sided disk drive. Digital Drives charges \$290 for a single-sided disk drive and \$320 for a double-sided disk drive. Each month, Associated Electronics can supply at most 1,000 disk drives in any combination of single-sided and double-sided drives. The combined monthly total supplied by Digital Drives cannot exceed 2,000 disk drives. Acme Micros needs at least 1,200 single-sided drives and at least 1,600 double-sided drives each month. How many disk drives of each type should Acme Micros order from each supplier in order to meet its monthly demand and minimize the purchase cost? What is the minimum purchase cost?

48. *Transportation.* A feed company stores grain in elevators located in Ames, Iowa, and Bedford, Indiana. Each month the grain is shipped to processing plants in Columbia, Missouri, and Danville, Illinois. The monthly supply (in tons) of grain at each elevator, the monthly demand (in tons) at each processing plant, and the cost per ton for transporting the grain are given in the table. Determine a shipping schedule that will minimize the cost of transporting the grain. What is the minimum cost?

	SHIPPING COST (\$ PER TON)		SUPPLY
	Columbia	*Danville*	(TONS)
AMES	22	38	700
BEDFORD	46	24	500
DEMAND (TONS)	400	600	

Life Sciences

49. *Nutrition—people.* A dietitian in a hospital is to arrange a special diet using three foods, *L, M,* and *N.* Each ounce of food *L* contains 20 units of calcium, 10 units of iron, 10 units of vitamin A, and 20 units of cholesterol. Each ounce of food *M* contains 10 units of calcium, 10 units of iron, 15 units of vitamin A, and 24 units of cholesterol. Each ounce of food *N* contains 10 units of calcium, 10 units of iron, 10 units of vitamin A, and 18 units of cholesterol. If the minimum daily requirements are 300 units of calcium, 200 units of iron, and 240 units of vitamin A, how many ounces of each food should be used to meet the minimum requirements and at the same time minimize the cholesterol intake? What is the minimum cholesterol intake?

50. *Nutrition—plants.* A farmer can buy three types of plant food, mix *A*, mix *B*, and mix *C*. Each cubic yard of mix *A* contains 20 pounds of phosphoric acid, 10 pounds of nitrogen, and 10 pounds of potash. Each cubic yard of mix *B* contains 10 pounds of phosphoric acid, 10 pounds of nitrogen, and 15 pounds of potash. Each cubic yard of mix *C* contains 20 pounds of phosphoric acid, 20 pounds of nitrogen, and 5 pounds of potash. The minimum monthly requirements are 480 pounds of phosphoric acid, 320 pounds of nitrogen, and 225 pounds of potash. If mix *A* costs \$30 per cubic yard, mix *B* costs \$36 per cubic yard, and mix *C* costs \$39 per cubic yard, how many cubic yards of each mix should the farmer blend to meet the minimum monthly requirements at a minimal cost? What is the minimum cost?

Social Sciences

51. *Education—resource allocation.* A metropolitan school district has two high schools that are overcrowded and two that are underenrolled. In order to balance the enrollment, the school board has decided to bus students from the overcrowded schools to the underenrolled schools. North Division High School has 300 more students than it should have, and South Division High School has 500 more students than it should have. Central High School can accommodate 400 additional students, and Washington High School can accommodate 500 additional students. The weekly cost of busing a student from North Division to Central is \$5, from North Division to Washington is \$2, from South Division to Central is \$3, and from South Division to Washington is \$4. Determine the number of students that should be bused from each of the overcrowded schools to each of the underenrolled schools in order to balance the enrollment and minimize the cost of busing the students. What is the minimum cost?

52. *Education—resource allocation.* Repeat Problem 51 if the weekly cost of busing a student from North Division to Washington is \$7 and all the other information remains the same.

SECTION 5-6

Maximization and Minimization with Mixed Problem Constraints

- An Introduction to the Big *M* Method
- The Big *M* Method
- Minimization by the Big *M* Method
- Summary of Methods of Solution
- Larger Problems—A Refinery Application

In the preceding two sections, we have seen how to solve both maximization and minimization problems, but with rather severe restrictions on problem constraints, right-side constants, and/or objective function coefficients (see the summary in Table 1 of the last section). In this section we will present a generalized version of the simplex method that will solve both maximization and minimization problems with any combination of $\leq$, $\geq$, and $=$ problem constraints. The only requirement is that each problem constraint have a nonnegative constant on the right side. (This restriction is easily accommodated, as you will see.)

An Introduction to the Big *M* Method

We introduce the *big M method* through a simple maximization problem with mixed problem constraints. The key parts of the method will then be summarized and applied to more complex problems.

Consider the following problem:

$$\begin{aligned} \text{Maximize} \quad & P = 2x_1 + x_2 \\ \text{Subject to} \quad & x_1 + x_2 \leqslant 10 \\ & -x_1 + x_2 \geqslant 2 \\ & x_1, x_2 \geqslant 0 \end{aligned} \tag{1}$$

To form an equation out of the first inequality, we introduce a slack variable s_1, as before, and write

$$x_1 + x_2 + s_1 = 10$$

How can we form an equation out of the second inequality? We introduce a second variable s_2 and subtract it from the left side so that we can write

$$-x_1 + x_2 - s_2 = 2$$

The variable s_2 is called a **surplus variable,** because it is the amount (surplus) by which the left side of the inequality exceeds the right side.

We now express the linear programming problem (1) as a system of equations:

$$\begin{aligned} x_1 + x_2 + s_1 \qquad\qquad &= 10 \\ -x_1 + x_2 \qquad - s_2 \qquad &= 2 \\ -2x_1 - x_2 \qquad\qquad + P &= 0 \\ x_1, x_2, s_1, s_2 &\geqslant 0 \end{aligned} \tag{2}$$

It can be shown that a basic solution of (2) is not feasible if any of the variables (excluding P) are negative. Thus, **a surplus variable is required to satisfy the nonnegative constraint.**

The basic solution found by setting the nonbasic variables x_1 and x_2 equal to 0 is

$$x_1 = 0, \quad x_2 = 0, \quad s_1 = 10, \quad s_2 = -2, \quad P = 0$$

But this basic solution is not feasible, since the surplus variable s_2 is negative (which is a violation of the nonnegative requirements of all variables except P). The simplex method works only when the basic solution for a tableau is feasible, so we cannot solve this problem simply by writing the tableau for (2) and starting pivot operations.

Explore–Discuss 1

To see that the simplex method will not work when the basic solution is not feasible, write the tableau for (2) and perform a pivot operation. Is the new basic solution feasible? Is it possible to select another pivot element?

In order to use the simplex method on problems with mixed constraints, we turn to an ingenious device called an *artificial variable.* This variable has no physical meaning in the original problem (which explains the use of the word "artificial") and is introduced solely for the purpose of obtaining a basic feasible solution so that we can apply the simplex method. An **artificial variable** is a variable introduced into each equation that has a surplus variable. As before, to ensure that we consider only feasible basic solutions, **an artificial variable is required to satisfy the nonnegative constraint.** (As we shall see later, artificial variables are also used to augment equality problem constraints when they are present.)

Returning to the problem at hand, we introduce an artificial variable a_1 into the equation involving the surplus variable s_2:

$$-x_1 + x_2 - s_2 + a_1 = 2$$

To prevent an artificial variable from becoming part of an optimal solution to the original problem, a very large "penalty" is introduced into the objective function. This penalty is created by choosing a positive constant M so large that the artificial variable is forced to be 0 in any final optimal solution of the original problem. (Since the constant M can be made as large as we wish in computer solutions, M is often selected as the largest number the computer can hold!) We then add the term $-Ma_1$ to the objective function:

$$P = 2x_1 + x_2 - Ma_1$$

We now have a new problem, which we call the **modified problem:**

$$\begin{array}{ll} \text{Maximize} & P = 2x_1 + x_2 - Ma_1 \\ \text{Subject to} & x_1 + x_2 + s_1 \qquad\qquad = 10 \\ & -x_1 + x_2 \qquad - s_2 + a_1 = 2 \\ & \qquad x_1, x_2, s_1, s_2, a_1 \geq 0 \end{array} \tag{3}$$

The initial system for the modified problem (3) is

$$\begin{aligned} x_1 + x_2 + s_1 \qquad\qquad\qquad &= 10 \\ -x_1 + x_2 \qquad - s_2 + \quad a_1 \qquad &= 2 \\ -2x_1 - x_2 \qquad\qquad + Ma_1 + P &= 0 \\ x_1, x_2, s_1, s_2, a_1 &\geq 0 \end{aligned} \tag{4}$$

We next write the augmented coefficient matrix for (4), which we call the **preliminary simplex tableau** for the modified problem. (The reason we call it the "preliminary" simplex tableau instead of the "initial" simplex tableau will be made clear shortly.)

$$\begin{array}{cccccc|c} x_1 & x_2 & s_1 & s_2 & a_1 & P & \\ 1 & 1 & 1 & 0 & 0 & 0 & 10 \\ -1 & 1 & 0 & -1 & 1 & 0 & 2 \\ \hline -2 & -1 & 0 & 0 & M & 1 & 0 \end{array} \tag{5}$$

To start the simplex process, including any necessary pivot operations, the preliminary simplex tableau should either meet the two requirements given in the following box or be transformed by row operations into a tableau that meets these two requirements.

Initial Simplex Tableau Requirements

For a system tableau to be considered an **initial simplex tableau,** it must satisfy the following two requirements:

1. The requisite number of basic variables must be selectable by the process described in Section 5-4. That is, a variable can be selected as a basic variable only if it corresponds to a column in the tableau that has exactly one nonzero element and the nonzero element in the column is not in the same row as the nonzero element in the column of another basic variable. The remaining variables are then selected as nonbasic variables to be set equal to 0 in determining a basic solution.
2. The basic solution found by setting the nonbasic variables equal to 0 is feasible.

Tableau (5) satisfies the first initial simplex tableau requirement, since s_1, s_2, and P can be selected as basic variables according to the criterion stated. (Not all preliminary simplex tableaux satisfy the first requirement; see Example 2.) However, tableau (5) does not satisfy the second initial simplex tableau requirement, since the basic solution is not feasible ($s_2 = -2$). To use the simplex method, we must first use row operations to transform (5) into an equivalent matrix that satisfies both initial simplex tableau requirements. **Note that this transformation is not a pivot operation.**

To get an idea of how to proceed, notice in tableau (5) that -1 in the s_2 column is in the same row as 1 in the a_1 column. This is not an accident! The artificial variable a_1 was introduced so that this would happen. If we eliminate M from the bottom of the a_1 column, then the nonbasic variable a_1 will become a basic variable and the troublesome basic variable s_2 will become a nonbasic variable. Thus, we proceed to eliminate M from the a_1 column using row operations:

$$\begin{matrix} x_1 & x_2 & s_1 & s_2 & a_1 & P & \end{matrix}$$

$$\left[\begin{array}{cccccc|c} 1 & 1 & 1 & 0 & 0 & 0 & 10 \\ -1 & 1 & 0 & -1 & 1 & 0 & 2 \\ \hdashline -2 & -1 & 0 & 0 & M & 1 & 0 \end{array}\right] \quad (-M)R_2 + R_3 \to R_3$$

$$\sim \left[\begin{array}{cccccc|c} 1 & 1 & 1 & 0 & 0 & 0 & 10 \\ -1 & 1 & 0 & -1 & 1 & 0 & 2 \\ \hdashline M-2 & -M-1 & 0 & M & 0 & 1 & -2M \end{array}\right]$$

From this last matrix we see that the basic variables are s_1, a_1, and P. The basic solution found by setting the nonbasic variables x_1, x_2, and s_2 equal to 0 is

$$x_1 = 0, \quad x_2 = 0, \quad s_1 = 10, \quad s_2 = 0, \quad a_1 = 2, \quad P = -2M$$

The basic solution is feasible (remember, P can be negative), and both requirements for an initial simplex tableau are met. We can now commence with the simplex process using pivot operations.

The pivot column is determined by the most negative indicator in the bottom row of the tableau. Since M is a positive number, $-M - 1$ is certainly a negative indicator. What about the indicator $M - 2$? Remember that M is a very large positive number. We will assume that M is so large that any expression of the form $M - k$ is positive. Thus, the only negative indicator in the bottom row is $-M - 1$.

$$\text{Pivot row} \rightarrow \quad \begin{array}{c} \\ \\ \\ \\ \end{array}\begin{bmatrix} \overset{x_1}{1} & \overset{x_2}{1} & \overset{s_1}{1} & \overset{s_2}{0} & \overset{a_1}{0} & \overset{P}{0} & | & 10 \\ -1 & \textcircled{1} & 0 & -1 & 1 & 0 & | & 2 \\ \hdashline M-2 & -M-1 & 0 & M & 0 & 1 & | & -2M \end{bmatrix} \quad \begin{array}{l} \frac{10}{1} = 10 \\ \frac{2}{1} = 2 \\ \\ \end{array}$$

$$\begin{array}{c} \uparrow \\ \text{Pivot column} \end{array}$$

Having identified the pivot element, we now begin pivoting:

$$\begin{array}{c} \\ s_1 \\ a_1 \\ P \end{array}\begin{bmatrix} \overset{x_1}{1} & \overset{x_2}{1} & \overset{s_1}{1} & \overset{s_2}{0} & \overset{a_1}{0} & \overset{P}{0} & | & 10 \\ -1 & \textcircled{1} & 0 & -1 & 1 & 0 & | & 2 \\ \hdashline M-2 & -M-1 & 0 & M & 0 & 1 & | & -2M \end{bmatrix} \quad \begin{array}{l} (-1)R_2 + R_1 \rightarrow R_1 \\ \\ (M+1)R_2 + R_3 \rightarrow R_3 \end{array}$$

$$\begin{array}{c} s_1 \\ \sim x_2 \\ P \end{array}\begin{bmatrix} \textcircled{2} & 0 & 1 & 1 & -1 & 0 & | & 8 \\ -1 & 1 & 0 & -1 & 1 & 0 & | & 2 \\ \hdashline -3 & 0 & 0 & -1 & M+1 & 1 & | & 2 \end{bmatrix} \quad \begin{array}{l} \frac{1}{2}R_1 \rightarrow R_1 \\ \\ \\ \end{array}$$

$$\sim \begin{bmatrix} \textcircled{1} & 0 & \frac{1}{2} & \frac{1}{2} & -\frac{1}{2} & 0 & | & 4 \\ -1 & 1 & 0 & -1 & 1 & 0 & | & 2 \\ \hdashline -3 & 0 & 0 & -1 & M+1 & 1 & | & 2 \end{bmatrix} \quad \begin{array}{l} \\ R_1 + R_2 \rightarrow R_2 \\ 3R_1 + R_3 \rightarrow R_3 \end{array}$$

$$\begin{array}{c} x_1 \\ \sim x_2 \\ P \end{array}\begin{bmatrix} 1 & 0 & \frac{1}{2} & \frac{1}{2} & -\frac{1}{2} & 0 & | & 4 \\ 0 & 1 & \frac{1}{2} & -\frac{1}{2} & \frac{1}{2} & 0 & | & 6 \\ \hdashline 0 & 0 & \frac{3}{2} & \frac{1}{2} & M-\frac{1}{2} & 1 & | & 14 \end{bmatrix}$$

Since all the indicators in the last row are nonnegative $\left(M - \frac{1}{2}\right.$ is nonnegative because M is a very large positive number$\left.\right)$, we can stop and write the optimal solution:

$$\text{Max } P = 14 \quad \text{at} \quad x_1 = 4, \quad x_2 = 6, \quad s_1 = 0, \quad s_2 = 0, \quad a_1 = 0$$

This is an optimal solution to the modified problem (3). How is it related to the original problem (2)? Since $a_1 = 0$ in this solution,

$$x_1 = 4, \quad x_2 = 6, \quad s_1 = 0, \quad s_2 = 0, \quad P = 14 \qquad (6)$$

is certainly a feasible solution for (2). [You can verify this by direct substitution into (2).] Surprisingly, it turns out that (6) is an optimal solution to the original problem. To see that this is true, suppose we were able to find feasible values of x_1, x_2, s_1, and s_2 that satisfy the original system (2) and produce a value of $P > 14$. Then by using these same values in (3) along with $a_1 = 0$, we have found a feasible solution of (3) with $P > 14$. This contradicts the fact that $P = 14$ is the maximum value of P for the modified problem. Thus, (6) is an optimal solution for the original problem.

As this example illustrates, if $a_1 = 0$ is an optimal solution for the modified problem, then deleting a_1 produces an optimal solution for the original problem. What happens if $a_1 \neq 0$ in the optimal solution for the modified problem? In this case, it can be shown that the original problem has no optimal solution because its feasible set is empty.

Explore–Discuss 2

Graph the feasible region for problem (1), plot the (x_1, x_2) coordinates of the basic solution for each simplex tableau in the solution of this problem, and illustrate the path to the optimal solution.

In larger problems, each $\geqslant$ problem constraint will require the introduction of a surplus variable and an artificial variable. If one of the problem constraints is an equation rather than an inequality, there is no need to introduce a slack or surplus variable. However, each $=$ problem constraint will require the introduction of another artificial variable to prevent the initial basic solution from violating the equality constraint—the decision variables are often 0 in the initial basic solution (see Example 2). Finally, each artificial variable also must be included in the objective function for the modified problem. The same constant M can be used for each artificial variable. Because of the role that the constant M plays in this approach, this method is often called the **big M method.**

The Big M Method

We now summarize the key steps of the big M method and use them to solve several problems.

The Big M Method—Introducing Slack, Surplus, and Artificial Variables to Form the Modified Problem

Step 1. If any problem constraints have negative constants on the right side, multiply both sides by -1 to obtain a constraint with a nonnegative constant. (If the constraint is an inequality, this will reverse the direction of the inequality.)

Step 2. Introduce a slack variable in each $\leqslant$ constraint.

Step 3. Introduce a surplus variable and an artificial variable in each $\geqslant$ constraint.

Step 4. Introduce an artificial variable in each $=$ constraint.

Step 5. For each artificial variable a_i, add $-Ma_i$ to the objective function. Use the same constant M for all artificial variables.

Example 1 **Finding the Modified Problem** Find the modified problem for the following linear programming problem. (Do not attempt to solve the problem.)

$$\begin{array}{ll} \text{Maximize} & P = 2x_1 + 5x_2 + 3x_3 \\ \text{Subject to} & x_1 + 2x_2 - x_3 \leqslant 7 \\ & -x_1 + x_2 - 2x_3 \leqslant -5 \\ & x_1 + 4x_2 + 3x_3 \geqslant 1 \\ & 2x_1 - x_2 + 4x_3 = 6 \\ & x_1, x_2, x_3 \geqslant 0 \end{array}$$

Solution First, we multiply the second constraint by -1 to change -5 to 5:

$$(-1)(-x_1 + x_2 - 2x_3) \geq (-1)(-5)$$
$$x_1 - x_2 + 2x_3 \geq 5$$

Next, we introduce the slack, surplus, and artificial variables according to the rules stated in the box:

$$\begin{aligned}
x_1 + 2x_2 - x_3 + s_1 &= 7 \\
x_1 - x_2 + 2x_3 - s_2 + a_1 &= 5 \\
x_1 + 4x_2 + 3x_3 - s_3 + a_2 &= 1 \\
2x_1 - x_2 + 4x_3 + a_3 &= 6
\end{aligned}$$

Finally, we add $-Ma_1$, $-Ma_2$, and $-Ma_3$ to the objective function:

$$P = 2x_1 + 5x_2 + 3x_3 - Ma_1 - Ma_2 - Ma_3$$

The modified problem is

$$\begin{aligned}
\text{Maximize} \quad & P = 2x_1 + 5x_2 + 3x_3 - Ma_1 - Ma_2 - Ma_3 \\
\text{Subject to} \quad & x_1 + 2x_2 - x_3 + s_1 = 7 \\
& x_1 - x_2 + 2x_3 - s_2 + a_1 = 5 \\
& x_1 + 4x_2 + 3x_3 - s_3 + a_2 = 1 \\
& 2x_1 - x_2 + 4x_3 + a_3 = 6 \\
& x_1, x_2, x_3, s_1, s_2, s_3, a_1, a_2, a_3 \geq 0
\end{aligned}$$

Matched Problem 1 ➡ Repeat Example 1 for:

$$\begin{aligned}
\text{Maximize} \quad & P = 3x_1 - 2x_2 + x_3 \\
\text{Subject to} \quad & x_1 - 2x_2 + x_3 \geq 5 \\
& -x_1 - 3x_2 + 4x_3 \leq -10 \\
& 2x_1 + 4x_2 + 5x_3 \leq 20 \\
& 3x_1 - x_2 - x_3 = -15 \\
& x_1, x_2, x_3 \geq 0
\end{aligned}$$

We now list the key steps for solving a problem using the big M method. The various steps and remarks are based on a number of important theorems, which we assume without proof. In particular, step 2 is based on the fact that (except for some degenerate cases not considered here) if the modified linear programming problem has an optimal solution, then the preliminary simplex tableau will be transformed into an initial simplex tableau by eliminating the M's from the columns corresponding to the artificial variables in the preliminary simplex tableau. Having obtained an initial simplex tableau, we can then commence with pivot operations.

The Big M Method—Solving the Problem

Step 1. Form the preliminary simplex tableau for the modified problem.

Step 2. Use row operations to eliminate the M's in the bottom row of the preliminary simplex tableau in the columns corresponding to the artificial variables. The resulting tableau is the initial simplex tableau.

Step 3. Solve the modified problem by applying the simplex method to the initial simplex tableau found in step 2.

(continued)

The Big M Method—Solving the Problem (*continued*)

Step 4. Relate the optimal solution of the modified problem to the original problem.

(A) If the modified problem has no optimal solution, then the original problem has no optimal solution.

(B) If all artificial variables are 0 in the optimal solution to the modified problem, then delete the artificial variables to find an optimal solution to the original problem.

(C) If any artificial variables are nonzero in the optimal solution to the modified problem, then the original problem has no optimal solution.

Example 2 **Using the Big M Method** Solve the following linear programming problem using the big M method:

$$\begin{aligned} &\text{Maximize} && P = x_1 - x_2 + 3x_3 \\ &\text{Subject to} && x_1 + x_2 \leqslant 20 \\ &&& x_1 + x_3 = 5 \\ &&& x_2 + x_3 \geqslant 10 \\ &&& x_1, x_2, x_3 \geqslant 0 \end{aligned}$$

Solution State the modified problem:

$$\begin{aligned} &\text{Maximize} && P = x_1 - x_2 + 3x_3 - Ma_1 - Ma_2 \\ &\text{Subject to} && x_1 + x_2 + s_1 = 20 \\ &&& x_1 + x_3 + a_1 = 5 \\ &&& x_2 + x_3 - s_2 + a_2 = 10 \\ &&& x_1, x_2, x_3, s_1, s_2, a_1, a_2 \geqslant 0 \end{aligned}$$

Write the preliminary simplex tableau for the modified problem, and find the initial simplex tableau by eliminating the M's from the artificial variable columns:

$$\begin{array}{c} \\ \\ \end{array}\left[\begin{array}{cccccccc|c} x_1 & x_2 & x_3 & s_1 & a_1 & s_2 & a_2 & P & \\ 1 & 1 & 0 & 1 & 0 & 0 & 0 & 0 & 20 \\ 1 & 0 & 1 & 0 & 1 & 0 & 0 & 0 & 5 \\ 0 & 1 & 1 & 0 & 0 & -1 & 1 & 0 & 10 \\ \hline -1 & 1 & -3 & 0 & M & 0 & M & 1 & 0 \end{array}\right]$$

Eliminate M from the a_1 column. $(-M)R_2 + R_4 \to R_4$

$$\sim \left[\begin{array}{cccccccc|c} 1 & 1 & 0 & 1 & 0 & 0 & 0 & 0 & 20 \\ 1 & 0 & 1 & 0 & 1 & 0 & 0 & 0 & 5 \\ 0 & 1 & 1 & 0 & 0 & -1 & 1 & 0 & 10 \\ \hline -M-1 & 1 & -M-3 & 0 & 0 & 0 & M & 1 & -5M \end{array}\right]$$

Eliminate M from the a_2 column. $(-M)R_3 + R_4 \to R_4$

$$\sim \left[\begin{array}{cccccccc|c} 1 & 1 & 0 & 1 & 0 & 0 & 0 & 0 & 20 \\ 1 & 0 & 1 & 0 & 1 & 0 & 0 & 0 & 5 \\ 0 & 1 & 1 & 0 & 0 & -1 & 1 & 0 & 10 \\ \hline -M-1 & -M+1 & -2M-3 & 0 & 0 & M & 0 & 1 & -15M \end{array}\right]$$

From this last matrix we see that the basic variables are s_1, a_1, a_2, and P. The basic solution found by setting the nonbasic variables x_1, x_2, x_3, and s_2 equal to 0 is

$$x_1 = 0, \quad x_2 = 0, \quad x_3 = 0, \quad s_1 = 20, \quad a_1 = 5, \quad s_2 = 0, \quad a_2 = 10, \quad P = -15M$$

The basic solution is feasible, and both requirements for an initial simplex tableau are met. We can now commence with pivot operations to find the optimal solution.

	x_1	x_2	x_3	s_1	a_1	s_2	a_2	P		
s_1	1	1	0	1	0	0	0	0	20	
a_1	1	0	(1)	0	1	0	0	0	5	
a_2	0	1	1	0	0	−1	1	0	10	$(-1)R_2 + R_3 \to R_3$
P	$-M-1$	$-M+1$	$-2M-3$	0	0	M	0	1	$-15M$	$(2M+3)R_2 + R_4 \to R_4$

~

	x_1	x_2	x_3	s_1	a_1	s_2	a_2	P		
s_1	1	1	0	1	0	0	0	0	20	$(-1)R_3 + R_1 \to R_1$
x_3	1	0	1	0	1	0	0	0	5	
a_2	−1	(1)	0	0	−1	−1	1	0	5	
P	$M+2$	$-M+1$	0	0	$2M+3$	M	0	1	$-5M+15$	$(M-1)R_3 + R_4 \to R_4$

~

	x_1	x_2	x_3	s_1	a_1	s_2	a_2	P	
s_1	2	0	0	1	1	1	−1	0	15
x_3	1	0	1	0	1	0	0	0	5
x_2	−1	1	0	0	−1	−1	1	0	5
P	3	0	0	0	$M+4$	1	$M-1$	1	10

Since the bottom row has no negative indicators, we can stop and write the optimal solution to the modified problem:

$$x_1 = 0, \quad x_2 = 5, \quad x_3 = 5, \quad s_1 = 15, \quad a_1 = 0, \quad s_2 = 0, \quad a_2 = 0, \quad P = 10$$

Since $a_1 = 0$ and $a_2 = 0$, the solution to the original problem is

$$\text{Max } P = 10 \quad \text{at} \quad x_1 = 0, \quad x_2 = 5, \quad x_3 = 5$$

Matched Problem 2 ➡ Solve the following linear programming problem using the big M method:

$$\begin{aligned} \text{Maximize} \quad & P = x_1 + 4x_2 + 2x_3 \\ \text{Subject to} \quad & x_2 + x_3 \le 4 \\ & x_1 - x_3 = 6 \\ & x_1 - x_2 - x_3 \ge 1 \\ & x_1, x_2, x_3 \ge 0 \end{aligned}$$

Example 3 ➡ **Using the Big M Method** Solve the following linear programming problem using the big M method:

$$\begin{aligned} \text{Maximize} \quad & P = 3x_1 + 5x_2 \\ \text{Subject to} \quad & 2x_1 + x_2 \le 4 \\ & x_1 + 2x_2 \ge 10 \\ & x_1, x_2 \ge 0 \end{aligned}$$

SOLUTION Introducing slack, surplus, and artificial variables, we obtain the modified problem:

$$\begin{aligned} 2x_1 + x_2 + s_1 \qquad\qquad &= 4 \\ x_1 + 2x_2 \qquad - s_2 + a_1 \qquad &= 10 \\ -3x_1 - 5x_2 \qquad\qquad + Ma_1 + P &= 0 \end{aligned} \quad \text{Modified problem}$$

Preliminary simplex tableau

$$\left[\begin{array}{cccccc|c} x_1 & x_2 & s_1 & s_2 & a_1 & P & \\ 2 & 1 & 1 & 0 & 0 & 0 & 4 \\ 1 & 2 & 0 & -1 & 1 & 0 & 10 \\ \hdashline -3 & -5 & 0 & 0 & M & 1 & 0 \end{array}\right] \quad \begin{array}{l} \text{Eliminate } M \text{ in the } a_1 \text{ column.} \\ \\ (-M)R_2 + R_3 \rightarrow R_3 \end{array}$$

Initial simplex tableau

$$\begin{array}{c} s_1 \\ \sim a_1 \\ P \end{array}\left[\begin{array}{cccccc|c} 2 & \textcircled{1} & 1 & 0 & 0 & 0 & 4 \\ 1 & 2 & 0 & -1 & 1 & 0 & 10 \\ \hdashline -M-3 & -2M-5 & 0 & M & 0 & 1 & -10M \end{array}\right] \quad \begin{array}{l} \text{Begin pivot operations.} \\ (-2)R_1 + R_2 \rightarrow R_2 \\ (2M+5)R_1 + R_3 \rightarrow R_3 \end{array}$$

$$\begin{array}{c} x_2 \\ \sim a_1 \\ P \end{array}\left[\begin{array}{cccccc|c} 2 & 1 & 1 & 0 & 0 & 0 & 4 \\ -3 & 0 & -2 & -1 & 1 & 0 & 2 \\ \hdashline 3M+7 & 0 & 2M+5 & M & 0 & 1 & -2M+20 \end{array}\right]$$

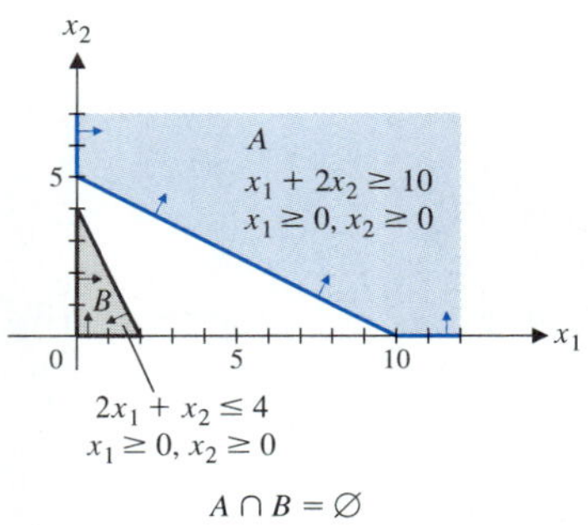

FIGURE 1

The optimal solution of the modified problem is

$$x_1 = 0, \quad x_2 = 4, \quad s_1 = 0, \quad s_2 = 0, \quad a_1 = 2, \quad P = -2M + 20$$

Since $a_1 \neq 0$, the original problem has no optimal solution. Figure 1 shows that the feasible region for the original problem is empty.

Matched Problem 3 Solve the following linear programming problem using the big M method:

$$\begin{aligned} \text{Maximize} \quad & P = 3x_1 + 2x_2 \\ \text{Subject to} \quad & x_1 + 5x_2 \leqslant 5 \\ & 2x_1 + x_2 \geqslant 12 \\ & x_1, x_2 \geqslant 0 \end{aligned}$$

MINIMIZATION BY THE BIG M METHOD

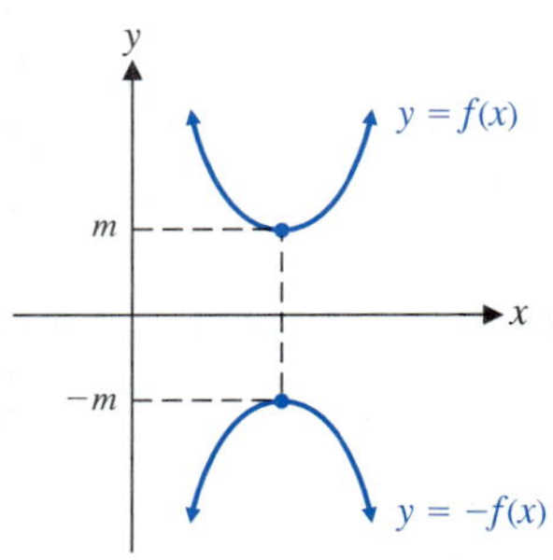

FIGURE 2

In addition to solving any maximization problem, the big M method can be used to solve minimization problems. To minimize an objective function, we have only to maximize its negative. Figure 2 illustrates the fact that the minimum value of a function f occurs at the same point as the maximum value of the function $-f$. Furthermore, if m is the minimum value of f, then $-m$ is the maximum value of $-f$, and conversely. Thus, we can find the minimum value of a function f by finding the maximum value of $-f$ and then changing the sign of the maximum value.

Example 4 **Production Scheduling—A Minimization Problem** A small jewelry manufacturing company employs a person who is a highly skilled gem cutter, and it wishes to use this person at least 6 hours per day for this purpose. On the other hand, the polishing facilities can be used in any amounts up to 10 hours per day. The company specializes in three kinds of semiprecious gemstones, *J*, *K*, and *L*. Relevant cutting, polishing, and cost requirements are listed in the table. How many gemstones of each type should be processed each day to minimize the cost of the finished stones? What is the minimum cost?

	J	*K*	*L*
CUTTING	1 hr	1 hr	1 hr
POLISHING	2 hr	1 hr	2 hr
COST PER STONE	$30	$30	$10

SOLUTION If we let x_1, x_2, and x_3 represent the number of type *J*, *K*, and *L* stones finished per day, respectively, then we have the following linear programming problem to solve, where *C* is the cost of the stones:

$$\begin{aligned} \text{Minimize} \quad & C = 30x_1 + 30x_2 + 10x_3 && \text{Objective function} \\ \text{Subject to} \quad & x_1 + x_2 + x_3 \geq 6 && \text{Problem constraints} \\ & 2x_1 + x_2 + 2x_3 \leq 10 && \\ & x_1, x_2, x_3 \geq 0 && \text{Nonnegative constraints} \end{aligned}$$

We convert this to a maximization problem by letting

$$P = -C = -30x_1 - 30x_2 - 10x_3$$

Thus, we get

$$\begin{aligned} \text{Maximize} \quad & P = -30x_1 - 30x_2 - 10x_3 \\ \text{Subject to} \quad & x_1 + x_2 + x_3 \geq 6 \\ & 2x_1 + x_2 + 2x_3 \leq 10 \\ & x_1, x_2, x_3 \geq 0 \end{aligned}$$

and Min $C = -$Max P. To solve, we first state the modified problem:

$$\begin{aligned} x_1 + x_2 + x_3 - s_1 + a_1 \qquad &= 6 \\ 2x_1 + x_2 + 2x_3 \qquad + s_2 &= 10 \\ 30x_1 + 30x_2 + 10x_3 + Ma_1 \qquad + P &= 0 \\ x_1, x_2, x_3, s_1, s_2, a_1 &\geq 0 \end{aligned}$$

$$\left[\begin{array}{ccccccc|c} x_1 & x_2 & x_3 & s_1 & a_1 & s_2 & P & \\ 1 & 1 & 1 & -1 & 1 & 0 & 0 & 6 \\ 2 & 1 & 2 & 0 & 0 & 1 & 0 & 10 \\ \hline 30 & 30 & 10 & 0 & M & 0 & 1 & 0 \end{array}\right]$$

Eliminate M in the a_1 column.

$(-M)R_1 + R_3 \to R_3$

Begin pivot operations. Assume M is so large that $-M + 30$ and $-M + 10$ are negative.

$$\begin{array}{c} \\ a_1 \\ \sim s_2 \\ P \end{array} \left[\begin{array}{ccccccc|c} x_1 & x_2 & x_3 & s_1 & a_1 & s_2 & P & \\ 1 & 1 & 1 & -1 & 1 & 0 & 0 & 6 \\ 2 & 1 & \textcircled{2} & 0 & 0 & 1 & 0 & 10 \\ \hline -M+30 & -M+30 & -M+10 & M & 0 & 0 & 1 & -6M \end{array}\right]$$

$0.5R_2 \to R_2$

$$\sim \left[\begin{array}{ccccccc|c} 1 & 1 & 1 & -1 & 1 & 0 & 0 & 6 \\ 1 & 0.5 & \textcircled{1} & 0 & 0 & 0.5 & 0 & 5 \\ \hline -M+30 & -M+30 & -M+10 & M & 0 & 0 & 1 & -6M \end{array}\right]$$

$(-1)R_2 + R_1 \to R_1$

$(M - 10)R_2 + R_3 \to R_3$

$$\begin{array}{c} \\ a_1 \\ \sim x_3 \\ P \end{array}\left[\begin{array}{ccccccc|c} x_1 & x_2 & x_3 & s_1 & a_1 & s_2 & P & \\ 0 & \boxed{0.5} & 0 & -1 & 1 & -0.5 & 0 & 1 \\ 1 & 0.5 & 1 & 0 & 0 & 0.5 & 0 & 5 \\ \hline 20 & -0.5M+25 & 0 & M & 0 & 0.5M-5 & 1 & -M-50 \end{array}\right] \quad 2R_1 \to R_1$$

$$\sim\left[\begin{array}{ccccccc|c} 0 & \boxed{1} & 0 & -2 & 2 & -1 & 0 & 2 \\ 1 & 0.5 & 1 & 0 & 0 & 0.5 & 0 & 5 \\ \hline 20 & -0.5M+25 & 0 & M & 0 & 0.5M-5 & 1 & -M-50 \end{array}\right] \quad \begin{array}{l} (-0.5)R_1 + R_2 \to R_2 \\ (0.5M-25)R_1 + R_3 \to R_3 \end{array}$$

$$\begin{array}{c} x_2 \\ \sim x_3 \\ P \end{array}\left[\begin{array}{ccccccc|c} 0 & 1 & 0 & -2 & 2 & -1 & 0 & 2 \\ 1 & 0 & 1 & 1 & -1 & 1 & 0 & 4 \\ \hline 20 & 0 & 0 & 50 & M-50 & 20 & 1 & -100 \end{array}\right]$$

The bottom row has no negative indicators, so the optimal solution for the modified problem is

$$x_1 = 0, \quad x_2 = 2, \quad x_3 = 4, \quad s_1 = 0, \quad a_1 = 0, \quad s_2 = 0, \quad P = -100$$

Since $a_1 = 0$, deleting a_1 produces the optimal solution to the original maximization problem and also to the minimization problem. Thus,

$$\text{Min } C = -\text{Max } P = -(-100) = 100 \qquad \text{at} \qquad x_1 = 0, \quad x_2 = 2, \quad x_3 = 4$$

That is, a minimum cost of \$100 for gemstones will be realized if no type J, 2 type K, and 4 type L stones are processed each day.

Matched Problem 4 ➡ Repeat Example 4 if the gem cutter is to be used for at least 8 hours a day and all other data remain the same.

Summary of Methods of Solution

The big M method can be used to solve any minimization problem, including those that can be solved by the dual method. (Note that Example 4 could have been solved, by the dual method.) Both methods of solving minimization problems are important. You will be instructed to solve most minimization problems in Exercise 5-6 by the big M method in order to gain more experience with this method. If the method of solution is not specified, then the dual method is usually easier.

Table 1 should help you select the proper method of solution for any linear programming problem.

Table 1

Summary of Problem Types and Simplex Solution Methods

Problem type	Problem constraints	Right-side constants	Coefficients of objective function	Method of solution
1. Maximization	$\le$	Nonnegative	Any real numbers	Simplex method with slack variables
2. Minimization	$\ge$	Any real numbers	Nonnegative	Form dual and solve by the preceding method
3. Maximization	Mixed $(\le, \ge, =)$	Nonnegative	Any real numbers	Form modified problem with slack, surplus, and artificial variables, and solve by the big M method
4. Minimization	Mixed $(\le, \ge, =)$	Nonnegative	Any real numbers	Maximize negative of objective function by the preceding method

■ Larger Problems—A Refinery Application

Up to this point, all the problems we have considered could be solved by hand. However, the real value of the simplex method lies in its ability to solve problems with a large number of variables and constraints, where a computer is generally used to perform the actual pivot operations. As a final application, we will consider a problem that would require the use of a computer to complete the solution.

Example 5 **Petroleum Blending** A refinery produces two grades of gasoline, regular and premium, by blending together two components, A and B. Component A has an octane rating of 90 and costs \$28 a barrel. Component B has an octane rating of 110 and costs \$32 a barrel. The octane rating for regular gasoline must be at least 95, and the octane rating for premium must be at least 105. Regular gasoline sells for \$34 a barrel and premium sells for \$40 a barrel. Currently, the company has 30,000 barrels of component A and 20,000 barrels of component B. It also has orders for 20,000 barrels of regular and 10,000 barrels of premium that must be filled. Assuming that all the gasoline produced can be sold, determine the maximum possible profit.

Solution First we organize the information given in the problem in Table 2.

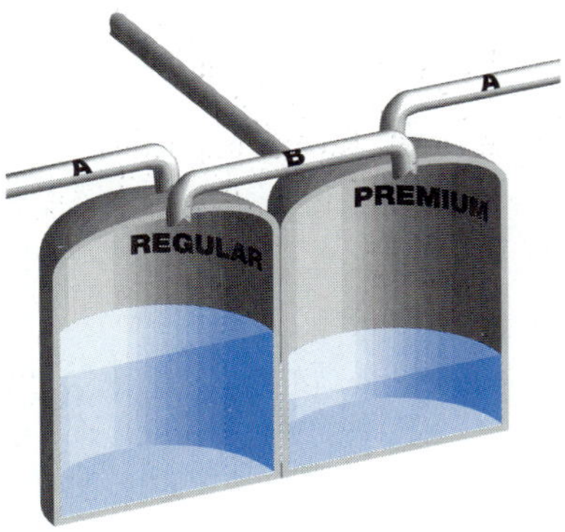

Table 2

Component	Octane rating	Cost (\$)	Available supply
A	90	28	30,000 barrels
B	110	32	20,000 barrels

Grade	Minimum octane rating	Selling price (\$)	Existing orders
Regular	95	34	20,000 barrels
Premium	105	40	10,000 barrels

Let: x_1 = Number of barrels of component A used in regular gasoline
x_2 = Number of barrels of component A used in premium gasoline
x_3 = Number of barrels of component B used in regular gasoline
x_4 = Number of barrels of component B used in premium gasoline

The total amount of component A used is $x_1 + x_2$. This cannot exceed the available supply. Thus, one constraint is

$$x_1 + x_2 \leq 30{,}000$$

The corresponding inequality for component B is

$$x_3 + x_4 \leq 20{,}000$$

The amounts of regular and premium gasoline produced must be sufficient to meet the existing orders:

$$x_1 + x_3 \geq 20{,}000 \quad \text{Regular}$$
$$x_2 + x_4 \geq 10{,}000 \quad \text{Premium}$$

Now consider the octane ratings. The octane rating of a blend is simply the proportional average of the octane ratings of the components. Thus, the octane rating for regular gasoline is

$$90\frac{x_1}{x_1 + x_3} + 110\frac{x_3}{x_1 + x_3}$$

where $x_1/(x_1 + x_3)$ is the percentage of component A used in regular gasoline and $x_3/(x_1 + x_3)$ is the percentage of component B. The final octane rating of regular gasoline must be at least 95; thus,

$$\begin{aligned} 90\frac{x_1}{x_1 + x_3} + 110\frac{x_3}{x_1 + x_3} &\geq 95 && \text{Multiply by } x_1 + x_3. \\ 90x_1 + 110x_3 &\geq 95(x_1 + x_3) && \text{Collect like terms on the left side.} \\ -5x_1 + 15x_3 &\geq 0 && \text{Octane rating for regular} \end{aligned}$$

The corresponding inequality for premium gasoline is

$$\begin{aligned} 90\frac{x_2}{x_2 + x_4} + 110\frac{x_4}{x_2 + x_4} &\geq 105 \\ 90x_2 + 110x_4 &\geq 105(x_2 + x_4) \\ -15x_2 + 5x_4 &\geq 0 && \text{Octane rating for premium} \end{aligned}$$

The cost of the components used is

$$C = 28(x_1 + x_2) + 32(x_3 + x_4)$$

The revenue from selling all the gasoline is

$$R = 34(x_1 + x_3) + 40(x_2 + x_4)$$

and the profit is

$$\begin{aligned} P &= R - C \\ &= 34(x_1 + x_3) + 40(x_2 + x_4) - 28(x_1 + x_2) - 32(x_3 + x_4) \\ &= (34 - 28)x_1 + (40 - 28)x_2 + (34 - 32)x_3 + (40 - 32)x_4 \\ &= 6x_1 + 12x_2 + 2x_3 + 8x_4 \end{aligned}$$

To find the maximum profit, we must solve the following linear programming problem:

$$\begin{aligned} \text{Maximize } \quad & P = 6x_1 + 12x_2 + 2x_3 + 8x_4 && \text{Profit} \\ \text{Subject to } \quad & x_1 + x_2 \leq 30{,}000 && \text{Available } A \\ & x_3 + x_4 \leq 20{,}000 && \text{Available } B \\ & x_1 + x_3 \geq 20{,}000 && \text{Required regular} \\ & x_2 + x_4 \geq 10{,}000 && \text{Required premium} \\ & -5x_1 + 15x_3 \geq 0 && \text{Octane for regular} \\ & -15x_2 + 5x_4 \geq 0 && \text{Octane for premium} \\ & x_1, x_2, x_3, x_4 \geq 0 \end{aligned}$$

The tableau for this problem would have seven rows and sixteen columns. Solving this problem by hand is possible but would require considerable effort. In actual practice, computers are used to solve most linear programming problems. Table 3 displays typical input and output for a computer solution of this problem. [This problem could also be solved with a spreadsheet or with the software in *Explorations in Finite Mathematics* (see Preface).]

Table 3

INPUT TO PROGRAM

```
        * - LINEAR PROGRAMMING - *
              PROBLEM DISPLAY

MAXIMIZE
P = 6X1 + 12X2 + 2X3 + 8X4
SUBJECT TO:
   X1 +     X2              <= 30000
                 X3 +  X4 <= 20000
   X1           +   X3      >= 20000
            X2        +  X4 >= 10000
-5X1           + 15X3      >=     0
        -15X2         + 5X4 >=     0
XI >= 0, I = 1 TO 4
```

OUTPUT FROM PROGRAM

```
        * - LINEAR PROGRAMMING - *
              SOLUTION DISPLAY

DECISION VARIABLES      SLACK VARIABLES
X1 = 26250              S1 = 0
X2 = 3750               S2 = 0
X3 = 8750               SURPLUS VARIABLES
X4 = 11250              S3 = 15000
                        S4 = 5000
                        S5 = 0
                        S6 = 0
MAXIMUM VALUE OF OBJECTIVE FUNCTION
310000
```

According to the output in Table 3, the refinery should blend 26,250 barrels of component A and 8,750 barrels of component B to produce 35,000 barrels of regular. They should blend 3,750 barrels of component A and 11,250 barrels of component B to produce 15,000 barrels of premium. This will result in a maximum profit of \$310,000.

Explore–Discuss 3

Interpret the values of the slack and surplus variables in the computer solution in Table 3.

Matched Problem 5 ➡ Suppose the refinery in Example 5 has 35,000 barrels of component A, which costs \$25 a barrel, and 15,000 barrels of component B, which costs \$35 a barrel. If all the other information is unchanged, formulate a linear programming problem whose solution is the maximum profit. Do not attempt to solve the problem (unless you have access to software that solves linear programming problems).

Answers to Matched Problems

1. Maximize $P = 3x_1 - 2x_2 + x_3 - Ma_1 - Ma_2 - Ma_3$

Subject to

$$\begin{aligned} x_1 - 2x_2 + x_3 - s_1 + a_1 &= 5 \\ x_1 + 3x_2 - 4x_3 - s_2 + a_2 &= 10 \\ 2x_1 + 4x_2 + 5x_3 + s_3 &= 20 \\ -3x_1 + x_2 + x_3 + a_3 &= 15 \\ x_1, x_2, x_3, s_1, a_1, s_2, a_2, s_3, a_3 &\geqslant 0 \end{aligned}$$

2. Max $P = 22$ at $x_1 = 6$, $x_2 = 4$, $x_3 = 0$ **3.** No optimal solution

4. A minimum cost of \$200 is realized when no type J, 6 type K, and 2 type L stones are processed each day.

5. Maximize $P = 9x_1 + 15x_2 - x_3 + 5x_4$

Subject to
$$\begin{aligned} x_1 + x_2 &\le 35{,}000 \\ x_3 + x_4 &\le 15{,}000 \\ x_1 + x_3 &\ge 20{,}000 \\ x_2 + x_4 &\ge 10{,}000 \\ -5x_1 + 15x_3 &\ge 0 \\ -15x_2 + 5x_4 &\ge 0 \\ x_1, x_2, x_3, x_4 &\ge 0 \end{aligned}$$

Exercise 5-6

A *In Problems 1–8:*

(A) Introduce slack, surplus, and artificial variables and form the modified problem.

(B) Write the preliminary simplex tableau for the modified problem and find the initial simplex tableau.

(C) Find the optimal solution of the modified problem by applying the simplex method to the initial simplex tableau.

(D) Find the optimal solution of the original problem, if it exists.

1. Maximize $P = 5x_1 + 2x_2$
Subject to $x_1 + 2x_2 \le 12$
$x_1 + x_2 \ge 4$
$x_1, x_2 \ge 0$

2. Maximize $P = 3x_1 + 7x_2$
Subject to $2x_1 + x_2 \le 16$
$x_1 + x_2 \ge 6$
$x_1, x_2 \ge 0$

3. Maximize $P = 3x_1 + 5x_2$
Subject to $2x_1 + x_2 \le 8$
$x_1 + x_2 = 6$
$x_1, x_2 \ge 0$

4. Maximize $P = 4x_1 + 3x_2$
Subject to $x_1 + 3x_2 \le 24$
$x_1 + x_2 = 12$
$x_1, x_2 \ge 0$

5. Maximize $P = 4x_1 + 3x_2$
Subject to $-x_1 + 2x_2 \le 2$
$x_1 + x_2 \ge 4$
$x_1, x_2 \ge 0$

6. Maximize $P = 3x_1 + 4x_2$
Subject to $x_1 - 2x_2 \le 2$
$x_1 + x_2 \ge 5$
$x_1, x_2 \ge 0$

7. Maximize $P = 5x_1 + 10x_2$
Subject to $x_1 + x_2 \le 3$
$2x_1 + 3x_2 \ge 12$
$x_1, x_2 \ge 0$

8. Maximize $P = 4x_1 + 6x_2$
Subject to $x_1 + x_2 \le 2$
$3x_1 + 5x_2 \ge 15$
$x_1, x_2 \ge 0$

B *Use the big M method to solve Problems 9–22.*

9. Minimize and maximize $P = 2x_1 - x_2$
Subject to $x_1 + x_2 \le 8$
$5x_1 + 3x_2 \ge 30$
$x_1, x_2 \ge 0$

10. Minimize and maximize $P = -4x_1 + 16x_2$
Subject to $3x_1 + x_2 \le 28$
$x_1 + 2x_2 \ge 16$
$x_1, x_2 \ge 0$

11. Maximize $P = 2x_1 + 5x_2$
Subject to $x_1 + 2x_2 \le 18$
$2x_1 + x_2 \le 21$
$x_1 + x_2 \ge 10$
$x_1, x_2 \ge 0$

12. Maximize $P = 6x_1 + 2x_2$
Subject to $x_1 + 2x_2 \le 20$
$2x_1 + x_2 \le 16$
$x_1 + x_2 \ge 9$
$x_1, x_2 \ge 0$

13. Maximize $P = 10x_1 + 12x_2 + 20x_3$
Subject to $3x_1 + x_2 + 2x_3 \ge 12$
$x_1 - x_2 + 2x_3 = 6$
$x_1, x_2, x_3 \ge 0$

14. Maximize $P = 5x_1 + 7x_2 + 9x_3$
Subject to $x_1 - x_2 + x_3 \ge 20$
$2x_1 + x_2 + 5x_3 = 35$
$x_1, x_2, x_3 \ge 0$

15. Minimize $C = -5x_1 - 12x_2 + 16x_3$
Subject to $x_1 + 2x_2 + x_3 \le 10$
$2x_1 + 3x_2 + x_3 \ge 6$
$2x_1 + x_2 - x_3 = 1$
$x_1, x_2, x_3 \ge 0$

16. Minimize $C = -3x_1 + 15x_2 - 4x_3$
Subject to
$$\begin{aligned} 2x_1 + x_2 + 3x_3 &\le 24 \\ x_1 + 2x_2 + x_3 &\ge 6 \\ x_1 - 3x_2 + x_3 &= 2 \\ x_1, x_2, x_3 &\ge 0 \end{aligned}$$

17. Maximize $P = 3x_1 + 5x_2 + 6x_3$
Subject to
$$\begin{aligned} 2x_1 + x_2 + 2x_3 &\le 8 \\ 2x_1 + x_2 - 2x_3 &= 0 \\ x_1, x_2, x_3 &\ge 0 \end{aligned}$$

18. Maximize $P = 3x_1 + 6x_2 + 2x_3$
Subject to
$$\begin{aligned} 2x_1 + 2x_2 + 3x_3 &\le 12 \\ 2x_1 - 2x_2 + x_3 &= 0 \\ x_1, x_2, x_3 &\ge 0 \end{aligned}$$

19. Maximize $P = 2x_1 + 3x_2 + 4x_3$
Subject to
$$\begin{aligned} x_1 + 2x_2 + x_3 &\le 25 \\ 2x_1 + x_2 + 2x_3 &\le 60 \\ x_1 + 2x_2 - x_3 &\ge 10 \\ x_1, x_2, x_3 &\ge 0 \end{aligned}$$

20. Maximize $P = 5x_1 + 2x_2 + 9x_3$
Subject to
$$\begin{aligned} 2x_1 + 4x_2 + x_3 &\le 150 \\ 3x_1 + 3x_2 + x_3 &\le 90 \\ -x_1 + 5x_2 + x_3 &\ge 120 \\ x_1, x_2, x_3 &\ge 0 \end{aligned}$$

21. Maximize $P = x_1 + 2x_2 + 5x_3$
Subject to
$$\begin{aligned} x_1 + 3x_2 + 2x_3 &\le 60 \\ 2x_1 + 5x_2 + 2x_3 &\ge 50 \\ x_1 - 2x_2 + x_3 &\ge 40 \\ x_1, x_2, x_3 &\ge 0 \end{aligned}$$

22. Maximize $P = 2x_1 + 4x_2 + x_3$
Subject to
$$\begin{aligned} 2x_1 + 3x_2 + 5x_3 &\le 280 \\ 2x_1 + 2x_2 + x_3 &\ge 140 \\ 2x_1 + x_2 &\ge 150 \\ x_1, x_2, x_3 &\ge 0 \end{aligned}$$

23. Solve Problems 5 and 7 by the geometric method. Compare the conditions in the big M method that indicate no optimal solution exists with the conditions stated in Theorem 2 in Section 5-2.

24. Repeat Problem 23 with Problems 6 and 8.

C *Problems 25–32 are mixed. Some can be solved by the methods presented in Sections 5-4 and 5-5, while others must be solved by the big M method.*

25. Minimize $C = 10x_1 - 40x_2 - 5x_3$
Subject to
$$\begin{aligned} x_1 + 3x_2 &\le 6 \\ 4x_2 + x_3 &\le 3 \\ x_1, x_2, x_3 &\ge 0 \end{aligned}$$

26. Maximize $P = 7x_1 - 5x_2 + 2x_3$
Subject to
$$\begin{aligned} x_1 - 2x_2 + x_3 &\ge -8 \\ x_1 - x_2 + x_3 &\le 10 \\ x_1, x_2, x_3 &\ge 0 \end{aligned}$$

27. Maximize $P = -5x_1 + 10x_2 + 15x_3$
Subject to
$$\begin{aligned} 2x_1 + 3x_2 + x_3 &\le 24 \\ x_1 - 2x_2 - 2x_3 &\ge 1 \\ x_1, x_2, x_3 &\ge 0 \end{aligned}$$

28. Minimize $C = -5x_1 + 10x_2 + 15x_3$
Subject to
$$\begin{aligned} 2x_1 + 3x_2 + x_3 &\le 24 \\ x_1 - 2x_2 - 2x_3 &\ge 1 \\ x_1, x_2, x_3 &\ge 0 \end{aligned}$$

29. Minimize $C = 10x_1 + 40x_2 + 5x_3$
Subject to
$$\begin{aligned} x_1 + 3x_2 &\ge 6 \\ 4x_2 + x_3 &\ge 3 \\ x_1, x_2, x_3 &\ge 0 \end{aligned}$$

30. Maximize $P = 8x_1 + 2x_2 - 10x_3$
Subject to
$$\begin{aligned} x_1 + x_2 - 3x_3 &\le 6 \\ 4x_1 - x_2 + 2x_3 &\le -7 \\ x_1, x_2, x_3 &\ge 0 \end{aligned}$$

31. Maximize $P = 12x_1 + 9x_2 + 5x_3$
Subject to
$$\begin{aligned} x_1 + 3x_2 + x_3 &\le 40 \\ 2x_1 + x_2 + 3x_3 &\le 60 \\ x_1, x_2, x_3 &\ge 0 \end{aligned}$$

32. Minimize $C = 10x_1 + 12x_2 + 28x_3$
Subject to
$$\begin{aligned} 4x_1 + 2x_2 + 3x_3 &\ge 20 \\ 3x_1 - x_2 - 4x_3 &\le 10 \\ x_1, x_2, x_3 &\ge 0 \end{aligned}$$

APPLICATIONS

In Problems 33–40, construct a mathematical model in the form of a linear programming problem. (The answers in the back of the book for these application problems include the model.) Then solve the problem using the big M method.

Business & Economics

33. *Manufacturing—resource allocation.* An electronics company manufactures two types of add-on memory modules for microcomputers, a 16k module and a 64k module. Each 16k module requires 10 minutes for assembly and 2 minutes for testing. Each 64k module requires 15 minutes for assembly and 4 minutes for testing. The company makes a profit of $18 on each 16k module and $30 on each 64k module. The assembly department can work a maximum of 2,200 minutes per day, and the testing department can work a maximum of 500 minutes a day. In order to satisfy current orders, the company must produce at least 50 of the 16k modules per day. How many units of each module should the company manufacture each day in order to maximize the daily profit? What is the maximum profit?

34. *Manufacturing—resource allocation.* Discuss the effect on the solution to Problem 33 if the assembly department can work only 2,100 minutes daily.

35. *Advertising.* A company planning an advertising campaign to attract new customers wants to place a total of at most 10 ads in 3 newspapers. Each ad in the *Sentinel* costs $200 and will be read by 2,000 people. Each ad in the *Journal* costs $200 and will be read by 500 people. Each ad in the *Tribune* costs $100 and will be read by 1,500 people. The company wants at least 16,000 people to read its ads. How many ads should it place in each paper in order to minimize the advertising costs? What is the minimum cost?

36. *Advertising.* Discuss the effect on the solution to Problem 35 if the *Tribune* will not accept more than 4 ads from the company.

Life Sciences

37. *Nutrition—people.* An individual on a high-protein, low-carbohydrate diet requires at least 100 units of protein and at most 24 units of carbohydrates daily. The diet will consist entirely of three special liquid diet foods, *A*, *B*, and *C*. The contents and costs of the diet foods are given in the table. How many bottles of each brand of diet food should be consumed daily in order to meet the protein and carbohydrate requirements at minimal cost? What is the minimum cost?

	Units per bottle		
	A	*B*	*C*
Protein	10	10	20
Carbohydrates	2	3	4
Cost per bottle ($)	0.60	0.40	0.90

38. *Nutrition—people.* Discuss the effect on the solution to Problem 37 if the cost of brand *C* liquid diet food increases to $1.50 per bottle.

39. *Nutrition—plants.* A farmer can use three types of plant food, mix *A*, mix *B*, and mix *C*. The amounts (in pounds) of nitrogen, phosphoric acid, and potash in a cubic yard of each mix are given in the table. Tests performed on the soil in a large field indicate that the field needs at least 800 pounds of potash. The tests also indicate that no more than 700 pounds of phosphoric acid should be added to the field. The farmer plans to plant a crop that requires a great deal of nitrogen. How many cubic yards of each mix should be added to the field in order to satisfy the potash and phosphoric acid requirements and maximize the amount of nitrogen added? What is the maximum amount of nitrogen?

	Pounds per cubic yard		
	A	*B*	*C*
Nitrogen	12	16	8
Phosphoric acid	12	8	16
Potash	16	8	16

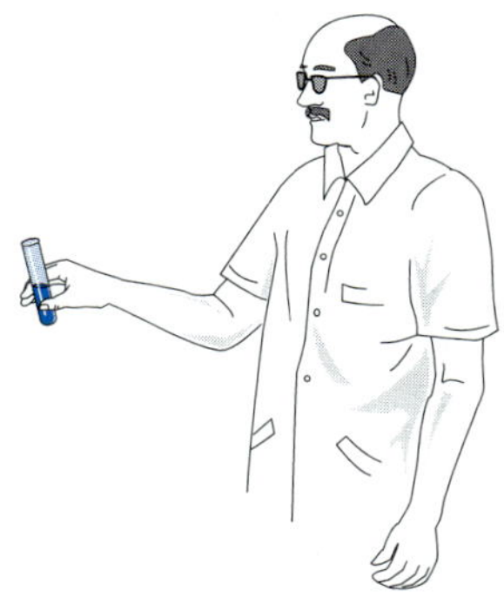

40. *Nutrition—plants.* Discuss the effect on the solution to Problem 39 if the limit on phosphoric acid is increased to 1,000 pounds.

In Problems 41–49, construct a mathematical model in the form of a linear programming problem. Do not solve.

Business & Economics

41. *Manufacturing—production scheduling.* A company manufactures car and truck frames at plants in Milwaukee and Racine. The Milwaukee plant has a daily operating budget of $50,000 and can produce at most 300 frames daily in any combination. It costs $150 to manufacture a car frame and $200 to manufacture a truck frame at the Milwaukee plant. The Racine plant has a daily operating budget of $35,000, can produce a maximum combined total of 200 frames daily, and produces a car frame at a cost of $135 and a truck frame at a cost of $180. Based on past demand, the company wants to limit production to a maximum of 250 car frames and 350 truck frames per day. If the company realizes a profit of $50 on each car frame and $70 on each truck frame, how many frames of each type should be produced at each plant to maximize the daily profit?

42. *Finances—loan distributions.* A savings and loan company has $3 million to lend. The types of loans and annual returns offered by the company are given in the table. State laws require that at least 50% of the money loaned for mortgages must be for first mortgages and that at least 30% of the total amount loaned must be for either first or second mortgages. Company policy requires that the amount of signature and automobile loans cannot exceed 25% of the total amount loaned and that signature loans cannot exceed 15% of the total amount loaned. How much money should be allocated to each type of loan in order to maximize the company's return?

TYPE OF LOAN	ANNUAL RETURN (%)
Signature	18
First mortgage	12
Second mortgage	14
Automobile	16

43. *Blending—petroleum.* A refinery produces two grades of gasoline, regular and premium, by blending together three components, *A*, *B*, and *C*. Component *A* has an octane rating of 90 and costs $28 a barrel, component *B* has an octane rating of 100 and costs $30 a barrel, and component *C* has an octane rating of 110 and costs $34 a barrel. The octane rating for regular must be at least 95 and the octane rating for premium must be at least 105. Regular gasoline sells for $38 a barrel and premium sells for $46 a barrel. The company has 40,000 barrels of component *A*, 25,000 barrels of component *B*, and 15,000 barrels of component *C*, and must produce at least 30,000 barrels of regular and 25,000 barrels of premium. How should the components be blended in order to maximize profit?

44. *Blending—food processing.* A company produces two brands of trail mix, regular and deluxe, by mixing dried fruits, nuts, and cereal. The recipes for the mixes are given in the table. The company has 1,200 pounds of dried fruits, 750 pounds of nuts, and 1,500 pounds of cereal to be used in producing the mixes. The company makes a profit of $0.40 on each pound of regular mix and $0.60 on each pound of deluxe mix. How many pounds of each ingredient should be used in each mix in order to maximize the company's profit?

TYPE OF MIX	INGREDIENTS
Regular	At least 20% nuts
	At most 40% cereal
Deluxe	At least 30% nuts
	At most 25% cereal

45. *Investment strategy.* An investor is planning to divide her investments among high-tech mutual funds, global mutual funds, corporate bonds, municipal bonds, and CD's. Each of these investments has an estimated annual yield and a risk factor (see the table). The risk level for each choice is the product of its risk factor and the percentage of the total funds invested in that choice. The total risk level is the sum of the risk levels for all the investments. The investor wants at least 20% of her investments to be in CD's and does not want the risk level to exceed 1.8. What percentage of her total investments should be invested in each choice to maximize the return?

INVESTMENT	ANNUAL RETURN (%)	RISK FACTOR
High-tech funds	0.11	2.7
Global funds	0.1	1.8
Corporate bonds	0.09	1.2
Muncipal bonds	0.08	0.5
CD's	0.05	0

46. *Investment strategy.* Refer to Problem 45. Suppose the investor decides that she would like to minimize the total risk factor, as long as her return does not fall below 9%. What percentage of her total investments should be invested in each choice to minimize the total risk level?

Life Sciences

47. *Nutrition—people.* A dietitian in a hospital is to arrange a special diet using the foods *L*, *M*, and *N*. The table gives the nutritional contents and the cost of 1 ounce of each food. The daily requirements for the diet are at least 400 units of calcium, at least 200 units of iron, at least 300 units of vitamin A, at most 150 units of cholesterol, and at most 900 calories. How many ounces of each food should be used in order to meet the requirements of the diet at minimal cost?

	UNITS PER BOTTLE		
	L	*M*	*N*
CALCIUM	30	10	30
IRON	10	10	10
VITAMIN A	10	30	20
CHOLESTEROL	8	4	6
CALORIES	60	40	50
COST PER OUNCE ($)	0.40	0.60	0.80

48. *Nutrition—feed mixtures.* A farmer grows three crops, corn, oats, and soybeans, which he mixes together to feed his cows and pigs. At least 40% of the feed mix for the cows must be corn. The feed mix for the pigs must contain at least twice as much soybeans as corn. He has harvested 1,000 bushels of corn, 500 bushels of oats, and 1,000 bushels of soybeans. He needs 1,000 bushels of each feed mix for his livestock. The unused corn, oats, and soybeans can be sold for $4, $3.50, and $3.25 a bushel, respectively (thus, these amounts also represent the cost of the crops used to feed the livestock). How many bushels of each crop should be used in each feed mix in order to produce sufficient food for the livestock at minimal cost?

Social Sciences

49. *Education—resource allocation.* Three towns are forming a consolidated school district with two high schools. Each high school has a maximum capacity of 2,000 students. Town A has 500 high school students, town B has 1,200, and town C has 1,800. The weekly costs of transporting a student from each town to each school are given in the table. In order to keep the enrollment balanced, the school board has decided that each high school must enroll at least 40% of the total student population. Furthermore, no more than 60% of the students in any town should be sent to the same high school. How many students from each town should be enrolled in each school in order to meet these requirements and minimize the cost of transporting the students?

	Weekly transportation cost per student ($)	
	School I	*School II*
Town A	4	8
Town B	6	4
Town C	3	9

Important Terms and Symbols

5-1 *Systems of Linear Inequalities in Two Variables.* Graph of a linear inequality in two variables; left half-plane; right half-plane; upper half-plane; lower half-plane; graphical solution of systems of linear inequalities; solution region; feasible region; corner point; bounded regions; unbounded regions

5-2 *Linear Programming in Two Dimensions—A Geometric Approach.* Linear programming problem; decision variables; objective function; problem constraints; nonnegative constraints; mathematical model; graphical solution; maximization problem; constant-profit line; optimal solution; feasible region; fundamental theorem of linear programming; multiple optimal solution; empty feasible region; unbounded objective function

5-3 *A Geometric Introduction to the Simplex Method.* Standard maximization problem in standard form; slack variables; basic variables; nonbasic variables; basic solution; basic feasible solution; fundamental theorem of linear programming

5-4 *The Simplex Method: Maximization with Problem Constraints of the Form* $\leqslant$. Initial system; basic solutions and basic feasible solutions for initial systems; fundamental theorem of linear programming; initial basic feasible solution; simplex tableau; initial simplex tableau; selecting basic and nonbasic variables for the simplex method; pivot operation; entering variable; exiting variable; pivot column; indicators; pivot row; pivot element; pivoting

5-5 *The Dual; Minimization with Problem Constraints of the Form* $\geqslant$. Dual problem; transpose; A^T; solution of a minimization problem; fundamental principal of duality; transportation problem

5-6 *Maximization and Minimization with Mixed Problem Constraints.* Surplus variable; artificial variable; modified problem; preliminary simplex tableau; initial simplex tableau; big M method

Summary of Problem Types and Simplex Solution Methods

Problem type	Problem constraints	Right-side constants	Coefficients of objective function	Method of solution
1. Maximization	$\leqslant$	Nonnegative	Any real numbers	Simplex method with slack variables
2. Minimization	$\geqslant$	Any real numbers	Nonnegative	Form dual and solve by the preceding method
3. Maximization	Mixed ($\leqslant, \geqslant, =$)	Nonnegative	Any real numbers	Form modified problem with slack, surplus, and artificial variables, and solve by the big M method
4. Minimization	Mixed ($\leqslant, \geqslant, =$)	Nonnegative	Any real numbers	Maximize negative of objective function by the preceding method

Review Exercise

Work through all the problems in this chapter review and check your answers in the back of the book. Answers to all review problems are there along with section numbers in italics to indicate where each type of problem is discussed. Where weaknesses show up, review appropriate sections in the text.

A *Solve the systems in Problems 1 and 2 graphically, and indicate whether each solution region is bounded or unbounded. Find the coordinates of each corner point.*

1. $$\begin{aligned} 2x_1 + x_2 &\le 8 \\ 3x_1 + 9x_2 &\le 27 \\ x_1, x_2 &\ge 0 \end{aligned}$$

2. $$\begin{aligned} 3x_1 + x_2 &\ge 9 \\ 2x_1 + 4x_2 &\ge 16 \\ x_1, x_2 &\ge 0 \end{aligned}$$

3. Solve the linear programming problem geometrically:

$$\begin{aligned} \text{Maximize} \quad & P = 6x_1 + 2x_2 \\ \text{Subject to} \quad & 2x_1 + x_2 \le 8 \\ & x_1 + 2x_2 \le 10 \\ & x_1, x_2 \ge 0 \end{aligned}$$

4. Convert the problem constraints in Problem 3 into a system of equations using slack variables.

5. How many basic variables and how many nonbasic variables are associated with the system in Problem 4?

6. Find all basic solutions for the system in Problem 4, and determine which basic solutions are feasible.

7. Write the simplex tableau for Problem 3, and circle the pivot element. Indicate the entering and exiting variables.

8. Solve Problem 3 using the simplex method.

9. For the simplex tableau below, identify the basic and nonbasic variables. Find the pivot element, the entering and exiting variables, and perform one pivot operation.

$$\begin{array}{c} \begin{matrix} x_1 & x_2 & x_3 & s_1 & s_2 & s_3 & P & \end{matrix} \\ \left[\begin{array}{rrrrrrr|r} 2 & 1 & 3 & -1 & 0 & 0 & 0 & 20 \\ 3 & 0 & 4 & 1 & 1 & 0 & 0 & 30 \\ 2 & 0 & 5 & 2 & 0 & 1 & 0 & 10 \\ \hline -8 & 0 & -5 & 3 & 0 & 0 & 1 & 50 \end{array}\right] \end{array}$$

10. Find the basic solution for each tableau. Determine whether the optimal solution has been reached, additional pivoting is required, or the problem has no optimal solution.

(A)
$$\begin{array}{c} \begin{matrix} x_1 & x_2 & s_1 & s_2 & P & \end{matrix} \\ \left[\begin{array}{rrrrr|r} 4 & 1 & 0 & 0 & 0 & 2 \\ 2 & 0 & 1 & 1 & 0 & 5 \\ \hline -2 & 0 & 3 & 0 & 1 & 12 \end{array}\right] \end{array}$$

(B)
$$\begin{array}{c} \begin{matrix} x_1 & x_2 & s_1 & s_2 & P & \end{matrix} \\ \left[\begin{array}{rrrrr|r} -1 & 3 & 0 & 1 & 0 & 7 \\ 0 & 2 & 1 & 0 & 0 & 0 \\ \hline -2 & 1 & 0 & 0 & 1 & 22 \end{array}\right] \end{array}$$

(C)
$$\left[\begin{array}{rrrrr|r} 1 & -2 & 0 & 4 & 0 & 6 \\ 0 & 2 & 1 & 6 & 0 & 15 \\ \hline 0 & 3 & 0 & 2 & 1 & 10 \end{array}\right]$$

11. Solve the linear programming problem geometrically:

$$\begin{aligned} \text{Minimize} \quad & C = 5x_1 + 2x_2 \\ \text{Subject to} \quad & x_1 + 3x_2 \ge 15 \\ & 2x_1 + x_2 \ge 20 \\ & x_1, x_2 \ge 0 \end{aligned}$$

12. Form the dual of Problem 11.

13. Write the initial system for the dual in Problem 12.

14. Write the first simplex tableau for the dual in Problem 12 and label the columns.

15. Use the simplex method to find the optimal solution of the dual in Problem 12.

16. Use the final simplex tableau from Problem 15 to find the optimal solution of Problem 11.

B

17. Solve the linear programming problem geometrically:

$$\begin{aligned} \text{Maximize} \quad & P = 3x_1 + 4x_2 \\ \text{Subject to} \quad & 2x_1 + 4x_2 \le 24 \\ & 3x_1 + 3x_2 \le 21 \\ & 4x_1 + 2x_2 \le 20 \\ & x_1, x_2 \ge 0 \end{aligned}$$

18. Solve Problem 17 using the simplex method.

19. Solve the linear programming problem geometrically:

$$\begin{aligned} \text{Minimize} \quad & C = 3x_1 + 8x_2 \\ \text{Subject to} \quad & x_1 + x_2 \ge 10 \\ & x_1 + 2x_2 \ge 15 \\ & x_2 \ge 3 \\ & x_1, x_2 \ge 0 \end{aligned}$$

20. Form the dual of Problem 19.

21. Solve Problem 19 by applying the simplex method to the dual in Problem 20.

Solve the linear programming Problems 22 and 23.

22. $$\begin{aligned} \text{Maximize} \quad & P = 5x_1 + 3x_2 - 3x_3 \\ \text{Subject to} \quad & x_1 - x_2 - 2x_3 \le 3 \\ & 2x_1 + 2x_2 - 5x_3 \le 10 \\ & x_1, x_2, x_3 \ge 0 \end{aligned}$$

23. $$\begin{aligned} \text{Maximize} \quad & P = 5x_1 + 3x_2 - 3x_3 \\ \text{Subject to} \quad & x_1 - x_2 - 2x_3 \le 3 \\ & x_1 + x_2 \le 5 \\ & x_1, x_2, x_3 \ge 0 \end{aligned}$$

In Problems 24 and 25:

(A) Introduce slack, surplus, and artificial variables and form the modified problem.

(B) Write the preliminary simplex tableau for the modified problem and find the initial simplex tableau.

(C) Find the optimal solution of the modified problem by applying the simplex method to the initial simplex tableau.

(D) Find the optimal solution of the original problem, if it exists.

24. Maximize $P = x_1 + 3x_2$
Subject to
$$\begin{aligned} x_1 + x_2 &\geq 6 \\ x_1 + 2x_2 &\leq 8 \\ x_1, x_2 &\geq 0 \end{aligned}$$

25. Maximize $P = x_1 + x_2$
Subject to
$$\begin{aligned} x_1 + x_2 &\geq 5 \\ x_1 + 2x_2 &\leq 4 \\ x_1, x_2 &\geq 0 \end{aligned}$$

26. Find the modified problem for the following linear programming problem. (Do not solve.)

Maximize $P = 2x_1 + 3x_2 + x_3$
Subject to
$$\begin{aligned} x_1 - 3x_2 + x_3 &\leq 7 \\ -x_1 - x_2 + 2x_3 &\leq -2 \\ 3x_1 + 2x_2 - x_3 &= 4 \\ x_1, x_2, x_3 &\geq 0 \end{aligned}$$

Write a brief verbal description of the type of linear programming problem that can be solved by the method indicated in Problems 27–30. Include the type of optimization, the number of variables, the type of constraints, and any restrictions on the coefficients and constants.

27. Geometric method

28. Basic simplex method with slack variables

29. Dual method

30. Big M method

C

31. Solve the following linear programming problem by the simplex method, keeping track of the obvious basic solution at each step. Then graph the feasible region and illustrate the path to the optimal solution determined by the simplex method.

Maximize $P = 2x_1 + 3x_2$
Subject to
$$\begin{aligned} x_1 + 2x_2 &\leq 22 \\ 3x_1 + x_2 &\leq 26 \\ x_1 &\leq 8 \\ x_2 &\leq 10 \\ x_1, x_2 &\geq 0 \end{aligned}$$

32. Solve by the dual method:

Minimize $C = 3x_1 + 2x_2$
Subject to
$$\begin{aligned} 2x_1 + x_2 &\leq 20 \\ 2x_1 + x_2 &\geq 9 \\ x_1 + x_2 &\geq 6 \\ x_1, x_2 &\geq 0 \end{aligned}$$

33. Solve Problem 32 by the big M method.

34. Solve by the dual method:

Minimize $C = 15x_1 + 12x_2 + 15x_3 + 18x_4$
Subject to
$$\begin{aligned} x_1 + x_2 &\leq 240 \\ x_3 + x_4 &\leq 500 \\ x_1 + x_3 &\geq 400 \\ x_2 + x_4 &\geq 300 \\ x_1, x_2, x_3, x_4 &\geq 0 \end{aligned}$$

Applications

In Problems 35–39, construct a mathematical model in the form of a linear programming problem. (The answers in the back of the book for these application problems include the model.) Then solve the problem by the indicated method.

Business & Economics

35. *Manufacturing—resource allocation.* South Shore Sail Loft manufactures regular and competition sails. Each regular sail takes 2 hours to cut and 4 hours to sew. Each competition sail takes 3 hours to cut and 10 hours to sew. There are 150 hours available in the cutting department and 380 hours available in the sewing department. Use the geometric method.

(A) If the Loft makes a profit of \$100 on each regular sail and \$200 on each competition sail, how many sails of each type should the company manufacture to maximize their profit? What is the maximum profit?

(B) An increase in the demand for competition sails causes the profit on a competition sail to rise to \$260. Discuss the effect of this change on the number of sails manufactured and on the maximum profit.

(C) A decrease in the demand for competition sails causes the profit on a competition sail to drop to \$140. Discuss the effect of this change on the number of sails manufactured and on the maximum profit.

36. *Investment.* An investor has \$150,000 to invest in oil stock, steel stock, and government bonds. The bonds are guaranteed to yield 5%, but the yield for each stock can vary. To protect against major losses, the investor decides that the amount invested in oil stock should not exceed \$50,000 and that the total amount invested in stock cannot exceed the amount invested in bonds by more than \$25,000. Use the simplex method to answer the following questions.

(A) If the oil stock yields 12% and the steel stock yields 9%, how much money should be invested in each alternative in order to maximize the return? What is the maximum return?

(B) Repeat part (A) if the oil stock yields 9% and the steel stock yields 12%.

37. *Transportation—shipping schedule.* A company produces motors for washing machines at factory *A* and factory *B*. The motors are then shipped to either plant *X* or plant *Y*, where the washing machines are assembled. The maximum number of motors that can be produced at each factory monthly, the minimum number required monthly for each plant to meet anticipated demand, and the shipping charges for one motor are given in the table. Determine a shipping schedule that will minimize the cost of transporting the motors from the factories to the assembly plants. Use the dual method.

	Plant *X*	Plant *Y*	Maximum Production
Factory *A*	\$5	\$8	1,500
Factory *B*	\$9	\$7	1,000
Minimum Requirement	900	1,200	

38. *Blending—food processing.* A company blends long-grain rice and wild rice to produce two brands of rice mixes, brand *A*, which is marketed under the company's name, and brand *B*, which is marketed as a generic brand. Brand *A* must contain at least 10% wild rice, and brand *B* must contain at least 5% wild rice. Long-grain rice costs the company \$0.70 per pound, and wild rice costs \$3.40 per pound. The company sells brand *A* for \$1.50 a pound and brand *B* for \$1.20 a pound. The company has 8,000 pounds of long-grain rice and 500 pounds of wild rice on hand. How should the company use the available rice to maximize their profit on these two brands of rice? What is the maximum profit? Use the simplex method.

Life Sciences

39. *Nutrition—animals.* A special diet for laboratory animals is to contain at least 850 units of vitamins, 800 units of minerals, and 1,150 calories. There are two feed mixes available, mix *A* and mix *B*. A gram of mix *A* contains 2 units of vitamins, 2 units of minerals, and 4 calories. A gram of mix *B* contains 5 units of vitamins, 4 units of minerals, and 5 calories. Use the geometric method.

(A) If mix *A* costs \$0.04 per gram and mix *B* costs \$0.09 per gram, how many grams of each mix should be used to satisfy the requirements of the diet at minimal cost? What is the minimum cost?

(B) If the price of mix *B* decreases to \$0.06 per gram, discuss the effect of this change on the solution in part (A).

(C) If the price of mix *B* increases to \$0.12 per gram, discuss the effect of this change on the solution in part (A).

Group Activity 1 The Two-Phase Method (An Alternative to the Big M Method)

In the big *M* method, we added artificial variables to $\geqslant$ and $=$ constraints in order to obtain a basic feasible solution. Then we added terms of the form $-Ma_i$ to the objective function and assumed that *M* was an arbitrarily large number, which forced the values of the artificial variables to be zero in any optimal solution. There is another commonly used method, called the *two-phase method,* that treats the artificial variables in a different manner. To illustrate this method, consider the following example:

$$\begin{aligned} \text{Maximize} \quad & P = 2x_1 + x_2 \\ \text{Subject to} \quad & x_1 + x_2 \leqslant 10 \\ & x_1 \geqslant 3 \\ & x_2 \geqslant 4 \\ & x_1, x_2 \geqslant 0 \end{aligned}$$

(A) Graph the feasible region and solve this problem by the geometric method for comparison with the new method that follows.

If we add slack and surplus variables, we obtain the following system:

$$\begin{aligned} x_1 + x_2 + s_1 \qquad\qquad &= 10 \\ x_1 \qquad - s_2 \qquad &= 3 \\ x_2 \qquad - s_3 &= 4 \\ -2x_1 - x_2 \qquad\qquad + P &= 0 \end{aligned} \tag{1}$$

The basic solution for this system is

$$x_1 = 0, \quad x_2 = 0, \quad s_1 = 10, \quad s_2 = -3, \quad s_3 = -4, \quad P = 0$$

This solution is not feasible, and we cannot start the simplex process. As before, we add artificial variables to obtain a basic feasible solution, but this time we do not modify the objective function:

$$\begin{aligned} x_1 + x_2 + s_1 \qquad\qquad\qquad &= 10 \\ x_1 \qquad - s_2 + a_1 \qquad\quad &= 3 \\ x_2 \qquad\qquad - s_3 + a_2 &= 4 \\ -2x_1 - x_2 \qquad\qquad\qquad + P &= 0 \end{aligned} \tag{2}$$

The basic solution for this system is

$$x_1 = 0, \quad x_2 = 0, \quad s_1 = 10, \quad s_2 = 0, \quad a_1 = 3, \quad s_3 = 0, \quad a_2 = 4, \quad P = 0$$

This solution is feasible, and we can start the simplex process.

(B) Write the simplex tableau for system (2) and apply the simplex procedure. Compare your result with the geometric solution from part (A).

In general, applying the simplex method to system (2) will produce an optimal solution of (2) with one or more of the artificial variables having nonzero values. In most cases, simply deleting the artificial variables will not produce a basic feasible solution to (1). However, if we can find a basic feasible solution of (2) with the artificial variables all equal to zero, then we can delete the artificial variables and find a basic feasible solution to (1). In the big M method, we accomplished this by adding terms to the objective function. In this new method, we consider another objective function:

$$C = a_1 + a_2$$

If we minimize C and the minimum value turns out to be zero, then, since a_1 and a_2 are nonnegative, they will both be zero. To minimize C, we maximize

$$T = -C = -a_1 - a_2$$

We cannot add T to system (2) in this form. To preserve the necessary form for a simplex tableau, we must express T in terms of the nonbasic variables. Solving the second and third equations in (2) for a_1 and a_2, respectively, and adding, we obtain

$$T = x_1 + x_2 - s_2 - s_3 - 7$$

Transposing all the variables to the left side, we add this equation to (2) to form system (3):

$$\begin{aligned}
x_1 + x_2 + s_1 \qquad\qquad\qquad\qquad &= 10 \\
x_1 \qquad - s_2 + a_1 \qquad\qquad &= 3 \\
x_2 \qquad\qquad - s_3 + a_2 \qquad &= 4 \\
-2x_1 - x_2 \qquad\qquad\qquad + P \quad &= 0 \\
-x_1 - x_2 + s_2 + s_3 \qquad\qquad + T &= -7
\end{aligned} \tag{3}$$

(C) Write the simplex tableau for system (3) and apply the simplex method, with the following restriction: *Do not choose any elements in the objective function row as pivot elements.* When you find the optimal solution, write the corresponding system, which is shown below.

$$\begin{aligned}
s_1 + s_2 - a_1 + s_3 - a_2 \qquad\qquad &= 3 \\
x_1 \qquad - s_2 + a_1 \qquad\qquad\qquad &= 3 \\
x_2 \qquad\qquad - s_3 + a_2 \qquad\qquad &= 4 \\
- 2s_2 + 2a_1 - s_3 + a_2 + P \qquad &= 10 \\
a_1 + a_2 \qquad + T &= 0
\end{aligned} \tag{4}$$

The basic feasible solution for (4) is

$$x_1 = 3, \quad x_2 = 4, \quad s_1 = 3, \quad s_2 = 0, \quad a_1 = 0,$$
$$s_3 = 0, \quad a_2 = 0, \quad P = 10, \quad T = 0$$

Note that the maximum value of T is 0 and that it occurs for $a_1 = a_2 = 0$. Thus, deleting the equation for T and the columns of the artificial variables gives us system (5), which is equivalent to system (1) but has a basic solution that is feasible:

$$\begin{aligned}
s_1 + s_2 + s_3 \qquad &= 3 \\
x_1 \qquad - s_2 \qquad\qquad &= 3 \\
x_2 \qquad\qquad - s_3 \qquad &= 4 \\
- 2s_2 - s_3 + P &= 10
\end{aligned} \tag{5}$$

(D) Write the simplex tableau for (5) and apply the simplex procedure to find the optimal solution. Compare the results with the solution in part (A).

This method is called the *two-phase method* because it involves solving two problems:

Phase 1. Minimize the sum of the artificial variables to obtain a basic feasible solution with all artificial variables equal to zero.

Phase 2. Delete the artificial variables and complete the solution of the original problem.

(E) Write a step-by-step procedure for using the two-phase method to solve linear programming problems. Discuss situations where this method will not produce an optimal solution. (Considering some simple examples might be helpful.) Use your procedure to solve Problems 13, 15, 17, and 19 in Exercise 5-6.

Group Activity 2 *Production Scheduling*

A company manufactures a line of outdoor furniture consisting of small tables, regular chairs, rocking chairs, chaise longues, and large tables. Each piece of furniture passes through three different production departments: fabrication, assembly, and finishing. Table 1 gives the time each piece must spend in each department, the labor-hours available in each department, the profit from each piece, and the existing orders for each product. The company must make a sufficient number of pieces of furniture to fill all the existing orders.

Table 1

	Production Department (Labor-hours per unit)				
Product	*Fabrication*	*Assembly*	*Finishing*	Profit per unit ($)	Orders
Small table	1	1	1	8	200
Regular chair	1	2	3	17	160
Rocking chair	2	2	3	24	180
Chaise longue	3	4	2	31	240
Large table	5	4	5	52	100
Available labor-hours	2,500	3,000	3,500		

(A) Construct a mathematical model in the form of a linear programming problem and use a graphing utility to show that the maximum profit of \$27,830 is realized when the company produces 200 small tables, 270 regular chairs, 180 rocking chairs, 240 chaise longues, and 190 large tables.

(B) Discuss the effect on the optimal solution in part (A) if the profit on a small table increases to \$12.

(C) Discuss the effect on the optimal solution in part (A) if the profit on a large table decreases to \$30.

(D) Discuss the effect on the optimal solution in part (A) if the orders for rocking chairs are canceled.

(E) Discuss the effect on the optimal solution in part (A) if the orders for chaise longues are increased to 300.

(F) Discuss the effect on the optimal solution in part (A) if the available hours in the assembly department are increased to 3,300.

(G) Discuss the effect on the optimal solution in part (A) if the available hours in the fabrication department are increased to 2,800.

CHAPTER 6

PROBABILITY

INTRODUCTION

Probability, like many branches of mathematics, evolved out of practical considerations. Girolamo Cardano (1501–1576), a gambler and physician, produced some of the best mathematics of his time, including a systematic analysis of gambling problems. In 1654, another gambler, the Chevalier de Méré, plagued with bad luck, approached the well-known French philosopher and mathematician Blaise Pascal (1623–1662) regarding certain dice problems. Pascal became interested in these problems, studied them, and discussed them with Pierre de Fermat (1601–1665), another French mathematician. Thus, out of the gaming rooms of western Europe the study of probability was born.

In spite of this lowly birth, probability has matured into a highly respected and immensely useful branch of mathematics. It is used in practically every field. Probability theory can be thought of as the science of uncertainty. If, for example, a card is drawn from a deck of 52 playing cards, it is uncertain which card will be drawn. But suppose a card is drawn and replaced in the deck and a card is again drawn and replaced, and this action is repeated a large number of times. A particular card, say the ace of spades, will be drawn over the long run with a relative frequency that is approximately predictable. Probability theory is concerned with determining the long-run frequency of the occurrence of a given event.

How do we assign probabilities to events? There are two basic approaches to this problem, one theoretical and the other empirical. An example will illustrate the difference between the two approaches.

Suppose you were asked, "What is the probability of obtaining a 2 on a single throw of a die?" Using a *theoretical approach,* we would reason as follows: Since there are 6 *equally likely* ways the die can turn up (assuming the die is fair) and there is only 1 way a 2 can turn up, then the probability of obtaining a 2 is $\frac{1}{6}$. Here, we have arrived at a probability assignment without even rolling a die once; we have used certain assumptions and a reasoning process.

What does the result have to do with reality? We would expect that in the long run (after rolling a die many times), the 2 would appear approximately $\frac{1}{6}$ of the time.

With the *empirical approach,* we make no assumption about the equally likely ways in which the die can turn up. We simply set up an experiment and roll the die a large number of times. Then we compute the percentage of times the 2 appears and use this number as an estimate of the probability of obtaining a 2 on a single roll of the die. Each approach has advantages and drawbacks; these will be discussed in the following sections.

We will first consider the theoretical approach and develop procedures that will lead to the solution of a large variety of interesting problems. These procedures require counting the number of ways certain events can happen, and this is not always easy. However, powerful mathematical tools can assist us in this counting task. The development of these tools is the subject matter of the next two sections.

SECTION 6-1 Basic Counting Principles

- Addition Principle
- Venn Diagrams
- Multiplication Principle

Addition Principle

If the enrollment in a college chemistry class consists of 13 males and 15 females, then it is easy to see that there are a total of 28 students enrolled in the class. This is a simple example of a **counting technique**—a method for determining the number of elements in a set without actually enumerating the elements one by one. Set operations play a fundamental role in many counting techniques (see Appendix A-1 for a review of set operations). For example, if M is the set of male students in the chemistry class and F is the set of female students, then the *union* of sets M and F, denoted $M \cup F$, is the set of all students in the class. Since these sets have no elements in common, the *intersection* of sets M and F, denoted $M \cap F$, is the *empty set* $\varnothing$; we then say that M and F are *disjoint sets.* The total number of students enrolled in the class is the **number of elements** in $M \cup F$, denoted by $n(M \cup F)$ and given by

$$\begin{aligned} n(M \cup F) &= n(M) + n(F) \\ &= 13 + 15 = 28 \end{aligned} \qquad (1)$$

Thus, we see that in this example the number of elements in the union of sets M and F is the sum of the number of elements in M and in F. However, this does not work for all pairs of sets. To see why, consider another example. Suppose that the enrollment in a mathematics class consists of 22 math majors and 16 physics majors, and that 7 of these students have majors in both subjects. If M represents the set of math majors and P represents the set of physics majors, then $M \cap P$ represents the set of double majors. It is tempting to proceed as before and conclude that there are $22 + 16 = 38$ students in the class, but this is incorrect. We have counted the double majors twice, once as math majors and again as physics majors. To correct for this double counting, we subtract the number of double majors from this sum. Thus, the total number of students enrolled in this class is given by

$$\begin{aligned} n(M \cup P) &= n(M) + n(P) - n(M \cap P) \\ &= 22 + 16 - 7 = 31 \end{aligned} \tag{2}$$

Equation (1) for the chemistry class and equation (2) for the math class are illustrations of the *addition principle* for counting the elements in the union of two sets.

Addition Principle (for Counting)

For any two sets A and B,

$$n(A \cup B) = n(A) + n(B) - n(A \cap B) \tag{3}$$

If A and B are disjoint, then

$$n(A \cup B) = n(A) + n(B) \tag{4}$$

Example 1 **Employee Benefits** According to a survey of business firms in a certain city, 750 firms offer their employees health insurance, 640 offer dental insurance, and 280 offer health insurance and dental insurance. How many firms offer their employees health insurance or dental insurance?

SOLUTION If H is the set of firms that offer their employees health insurance and D is the set that offer dental insurance, then

$H \cap D$ = Set of firms that offer health insurance **and** dental insurance
$H \cup D$ = Set of firms that offer health insurance **or** dental insurance

Thus,

$$n(H) = 750 \qquad n(D) = 640 \qquad n(H \cap D) = 280$$

and

$$\begin{aligned} n(H \cup D) &= n(H) + n(D) - n(H \cap D) \\ &= 750 + 640 - 280 = 1{,}110 \end{aligned}$$

Thus, 1,110 firms offer their employees health insurance or dental insurance.

Matched Problem 1 The survey in Example 1 also indicated that 345 firms offer their employees group life insurance, 285 offer long-term disability insurance, and 115 offer group life insurance and long-term disability insurance. How many firms offer their employees group life insurance or long-term disability insurance?

VENN DIAGRAMS

The next example introduces some additional set concepts that provide useful counting techniques.

Example 2 **Market Research** A city has two daily newspapers, the *Sentinel* and the *Journal*. The following information was obtained from a survey of 100 residents of the city: 35 people subscribe to the *Sentinel*, 60 subscribe to the *Journal*, and 20 subscribe to both newspapers.

(A) How many people in the survey subscribe to the *Sentinel* but not to the *Journal?*
(B) How many subscribe to the *Journal* but not to the *Sentinel?*
(C) How many do not subscribe to either paper?
(D) Organize this information in a table.

SOLUTION Let U be the group of people surveyed, let S be the set of people who subscribe to the *Sentinel*, and let J be the set of people who subscribe to the *Journal*. Since U contains all the elements under consideration, it is the *universal set* for this problem. The *complement* of S, denoted S', is the set of people in the survey group U who do not subscribe to the *Sentinel*. Similarly, J' is the set of people in the group who do not subscribe to the *Journal*. Using the sets S and J, their complements, and set intersection, we can divide the set U into the four disjoint subsets defined below and illustrated in the *Venn diagram* in Figure 1.

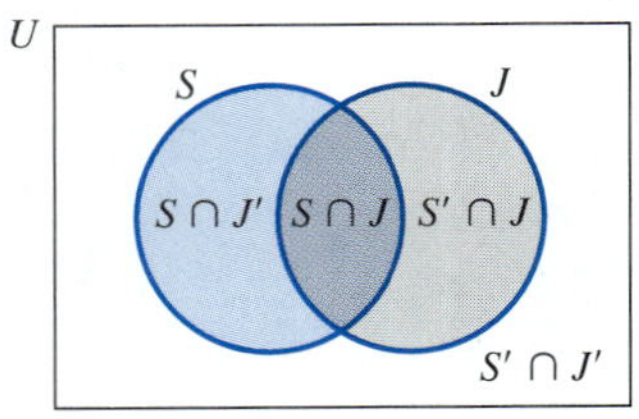

FIGURE 1
Venn diagram for the newspaper survey

$S \cap J$ = Set of people who subscribe to both papers
$S \cap J'$ = Set of people who subscribe to the *Sentinel* but not the *Journal*
$S' \cap J$ = Set of people who subscribe to the *Journal* but not the *Sentinel*
$S' \cap J'$ = Set of people who do not subscribe to either paper

The given survey information can be expressed in terms of set notation as

$$n(U) = 100 \qquad n(S) = 35 \qquad n(J) = 60 \qquad n(S \cap J) = 20$$

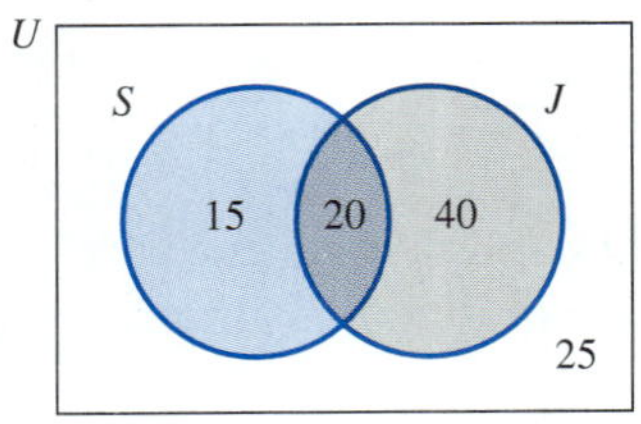

FIGURE 2
Newspaper survey results

We can use this information and a Venn diagram to answer parts (A)–(C). To begin, we place 20 in $S \cap J$ in the diagram (see Fig. 2). As we proceed through parts (A)–(C), we add each answer to the diagram.

(A) Since 35 people subscribe to the *Sentinel* and 20 subscribe to both papers, the number of people who subscribe to the *Sentinel* but not to the *Journal* is

$$n(S \cap J') = 35 - 20 = 15$$

(B) In a similar manner, the number of people who subscribe to the *Journal* but not to the *Sentinel* is

$$n(S' \cap J) = 60 - 20 = 40$$

(C) The total number of newspaper subscribers is $20 + 15 + 40 = 75$. Thus, the number of people who do not subscribe to either paper is

$$n(S' \cap J') = 100 - 75 = 25$$

(D) Venn diagrams are useful tools for determining the number of elements in the various sets in a survey, but often the results must be presented in the form of a table, rather than a diagram. The following table contains the same information as Figure 2, but organized in a different format:

		JOURNAL		
		Subscriber, J	*Nonsubscriber, J′*	*Totals*
SENTINEL	*Subscriber, S*	20	15	35
	Nonsubscriber, S′	40	25	65
	Totals	60	40	100

Matched Problem 2

A small town has two radio stations, an AM station and an FM station. A survey of 100 residents of the town produced the following results: In the last 30 days, 65 people have listened to the AM station, 45 have listened to the FM station, and 30 have listened to both stations.

(A) How many people in the survey have listened to the AM station, but not to the FM station, in this 30 day period?
(B) How many have listened to the FM station, but not to the AM station?
(C) How many have not listened to either station?
(D) Organize this information in a table.

Explore–Discuss 1

Let A, B, and C be three sets. Use a Venn diagram to explain the following equation:

$$n(A \cup B \cup C) = n(A) + n(B) + n(C) - n(A \cap B) - n(A \cap C) - n(B \cap C) + n(A \cap B \cap C)$$

Multiplication Principle

As we have just seen, if the elements of a set are determined by the union operation, then addition and subtraction are used to count the number of elements in the set. Now we want to consider sets whose elements are determined by a sequence of operations. We will see that multiplication is used to count the number of elements in sets formed in this manner. The best way to see how this works is to start with an example.

Example 3 **Product Mix** A retail store stocks windbreaker jackets in small, medium, large, and extra large, and all are available in blue or red. What are the combined choices and how many combined choices are there?

SOLUTION To solve the problem we use a tree diagram:

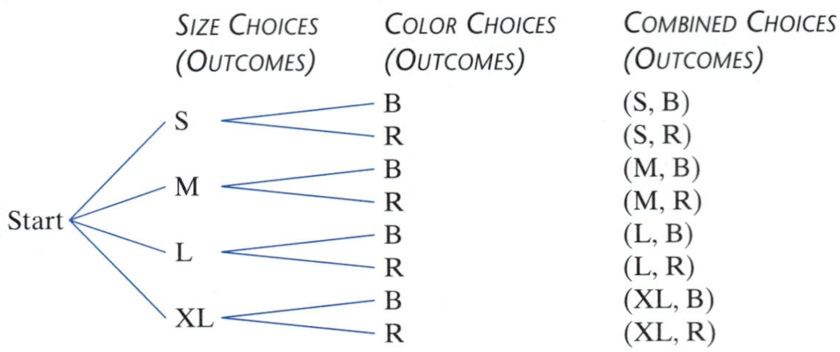

Thus, there are 8 possible combined choices (outcomes). There are 4 ways a size can be chosen and 2 ways a color can be chosen. The first element in the ordered pair represents a size choice, and the second element represents a color choice.

Matched Problem 3 A company offers its employees health plans from three different companies, R, S, and T. Each company offers two levels of coverage, A and B, with one level requiring additional employee contributions. What are the combined choices, and how many choices are there? Solve using a tree diagram.

Now suppose you asked, "From the 26 letters in the alphabet, how many ways can 3 letters appear in a row on a license plate if no letter is repeated?" To try to count the possibilities using a tree diagram would be extremely tedious, to say the least. The following **multiplication principle** will enable us to solve this problem easily; in addition, it forms the basis for several other counting devices that are developed in the next section.

Multiplication Principle (for Counting)

1. If two operations O_1 and O_2 are performed in order, with N_1 possible outcomes for the first operation and N_2 possible outcomes for the second operation, then there are

   $$N_1 \cdot N_2$$

   possible combined outcomes of the first operation followed by the second.
2. In general, if n operations $O_1, O_2, \ldots, O_n$ are performed in order, with possible number of outcomes $N_1, N_2, \ldots, N_n$, respectively, then there are

   $$N_1 \cdot N_2 \cdot \cdots \cdot N_n$$

   possible combined outcomes of the operations performed in the given order.

In Example 3, we see that there are 4 possible outcomes in choosing a size (the first operation) and 2 possible outcomes in choosing a color (the second operation); hence, by the multiplication principle, there are $4 \cdot 2 = 8$ possible combined outcomes. Use the multiplication principle to solve Matched Problem 3. [*Answer:* $3 \cdot 2 = 6$]

To answer the license plate question: There are 26 ways the first letter can be chosen; after a first letter is chosen, there are 25 ways a second letter can be chosen; and after 2 letters are chosen, there are 24 ways a third letter can be chosen. Hence, using the multiplication principle, there are $26 \cdot 25 \cdot 24 = 15{,}600$ possible ways 3 letters can be chosen from the alphabet without repeats.

Explore–Discuss 2

A state is about to adopt a 6-character format for its license plates, consisting of a block of letters followed by a block of numbers. The format must accommodate up to 20 million vehicles. What size would you recommend for each block?

Example 4 **Computer-Assisted Testing** Many colleges and universities are now using computer-assisted testing procedures. Suppose a screening test is to consist of 5 questions, and a computer stores 5 comparable questions for the first test question, 8 for the second, 6 for the third, 5 for the fourth, and 10 for the fifth. How many different 5-question tests can the computer select? (Two tests are considered different if they differ in one or more questions.)

SOLUTION

O_1:	Selecting the first question	N_1:	5 ways
O_2:	Selecting the second question	N_2:	8 ways
O_3:	Selecting the third question	N_3:	6 ways
O_4:	Selecting the fourth question	N_4:	5 ways
O_5:	Selecting the fifth question	N_5:	10 ways

Thus, the computer can generate

$$5 \cdot 8 \cdot 6 \cdot 5 \cdot 10 = 12{,}000 \text{ different tests}$$

Matched Problem 4 Each question on a multiple-choice test has 5 choices. If there are 5 such questions on a test, how many different response sheets are possible if only 1 choice is marked for each question?

Example 5 **Code Words** How many 3-letter code words are possible using the first 8 letters of the alphabet if:

(A) No letter can be repeated? (B) Letters can be repeated?
(C) Adjacent letters cannot be alike?

SOLUTION To form 3-letter code words from the 8 letters available, we select a letter for the first position, one for the second position, and one for the third position. Altogether, there are three operations.

(A) No letter can be repeated:

O_1:	Selecting the first letter	N_1:	8 ways	
O_2:	Selecting the second letter	N_2:	7 ways	Since 1 letter has been used
O_3:	Selecting the third letter	N_3:	6 ways	Since 2 letters have been used

Thus, there are

$8 \cdot 7 \cdot 6 = 336$ possible code words   *Possible combined operations*

(B) Letters can be repeated:

O_1: Selecting the first letter	N_1: 8 ways	
O_2: Selecting the second letter	N_2: 8 ways	Repeats allowed
O_3: Selecting the third letter	N_3: 8 ways	Repeats allowed

Thus, there are

$8 \cdot 8 \cdot 8 = 8^3 = 512$ possible code words

(C) Adjacent letters cannot be alike:

O_1: Selecting the first letter	N_1: 8 ways	
O_2: Selecting the second letter	N_2: 7 ways	Cannot be the same as the first
O_3: Selecting the third letter	N_3: 7 ways	Cannot be the same as the second, but can be the same as the first

Thus, there are

$8 \cdot 7 \cdot 7 = 392$ possible code words

Matched Problem 5 How many 4-letter code words are possible using the first 10 letters of the alphabet under the three different conditions stated in Example 5?

Answers to Matched Problems

1. 515

2. (A) 35 (B) 15 (C) 20

(D)

		FM Listener	FM Nonlistener	Totals
AM	*Listener*	30	35	65
AM	*Nonlistener*	15	20	35
	Totals	45	55	100

3. There are 6 combined choices:

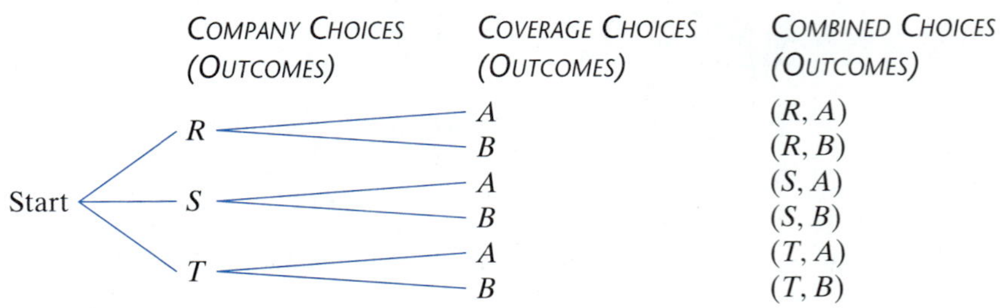

4. 5^5, or 3,125

5. (A) $10 \cdot 9 \cdot 8 \cdot 7 = 5{,}040$ (B) $10 \cdot 10 \cdot 10 \cdot 10 = 10{,}000$
(C) $10 \cdot 9 \cdot 9 \cdot 9 = 7{,}290$

EXERCISE 6-1

A *Problems 1–12 refer to the following Venn diagram. Find the number of elements in each of the indicated sets.*

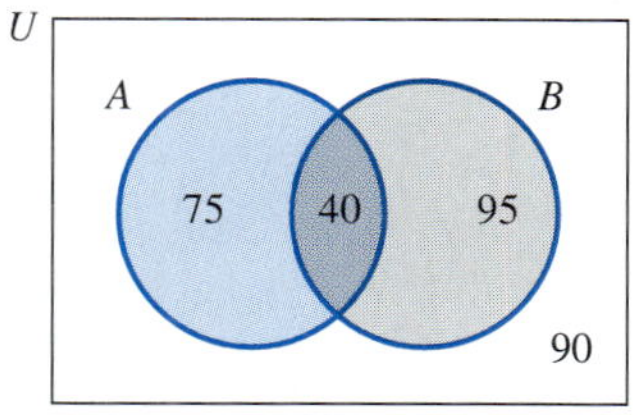

1. A **2.** B **3.** U

4. A' **5.** B' **6.** $A \cap B$

7. $A \cup B$ **8.** $A \cap B'$ **9.** $A' \cap B$

10. $A' \cap B'$ **11.** $(A \cap B)'$ **12.** $(A \cup B)'$

Solve Problems 13–16 two ways: (A) using a tree diagram and (B) using the multiplication principle.

13. How many ways can 2 coins turn up—heads, H, or tails, T—if the combined outcome (H, T) is to be distinguished from the outcome (T, H)?

14. How many 2-letter code words can be formed from the first 3 letters of the alphabet if no letter can be used more than once?

15. A coin is tossed with possible outcomes heads, H, or tails, T. Then a single die is tossed with possible outcomes 1, 2, 3, 4, 5, or 6. How many combined outcomes are there?

16. In how many ways can 3 coins turn up—heads, H, or tails, T—if combined outcomes such as (H, T, H), (H, H, T), and (T, H, H) are to be considered different?

B *In Problems 17–20, use the given information to determine the number of elements in each of the four disjoint subsets in the following Venn diagram.*

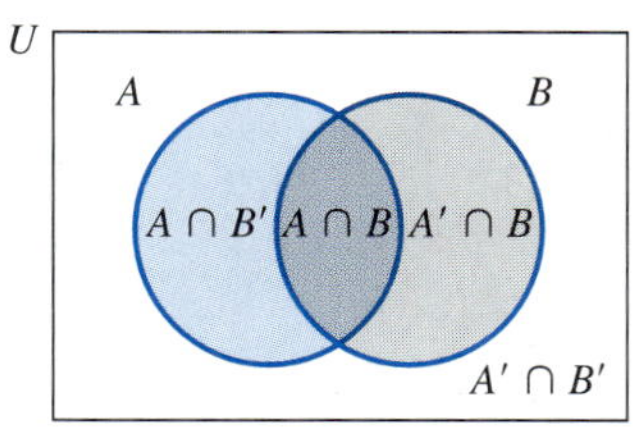

17. $n(A) = 80,\ n(B) = 50,\ n(A \cap B) = 20,\ n(U) = 200$

18. $n(A) = 45,\ n(B) = 35,\ n(A \cap B) = 15,\ n(U) = 100$

19. $n(A) = 25,\ n(B) = 55,\ n(A \cup B) = 60,\ n(U) = 100$

20. $n(A) = 70,\ n(B) = 90,\ n(A \cup B) = 120,\ n(U) = 200$

In Problems 21–24, use the given information to complete the following table.

	A	A'	*Totals*
B	?	?	?
B'	?	?	?
Totals	?	?	?

21. $n(A) = 70,\ n(B) = 90,\ n(A \cap B) = 30,\ n(U) = 200$

22. $n(A) = 55,\ n(B) = 65,\ n(A \cap B) = 35,\ n(U) = 100$

23. $n(A) = 45,\ n(B) = 55,\ n(A \cup B) = 80,\ n(U) = 100$

24. $n(A) = 80,\ n(B) = 70,\ n(A \cup B) = 110,\ n(U) = 200$

In Problems 25 and 26, discuss the validity of each statement. If the statement is always true, explain why. If not, give a counterexample.

25. (A) If A or B is the empty set, then A and B are disjoint.

(B) If A and B are disjoint, then A or B is the empty set.

26. (A) If A and B are disjoint, then $n(A \cap B) = n(A) + n(B)$.

(B) If $n(A \cup B) = n(A) + n(B)$, then A and B are disjoint.

27. A particular new car model is available with 5 choices of color, 3 choices of transmission, 4 types of interior, and 2 types of engine. How many different variations of this model car are possible?

28. A delicatessen serves meat sandwiches with the following options: 3 kinds of bread, 5 kinds of meat, and lettuce or sprouts. How many different sandwiches are possible, assuming one item is used out of each category?

29. How many 4-letter code words are possible from the first 6 letters of the alphabet if no letter is repeated? If letters are repeated? If adjacent letters must be different?

30. How many 5-letter code words are possible from the first 7 letters of the alphabet if no letter is repeated? If letters are repeated? If adjacent letters must be different?

31. A combination lock has 5 wheels, each labeled with the 10 digits from 0 to 9. How many 5-digit opening combinations are possible if no digit is repeated? If digits are repeated? If successive digits must be different?

32. A small combination lock on a suitcase has 3 wheels, each labeled with the 10 digits from 0 to 9. How many 3-digit combinations are possible if no digit is repeated? If digits are repeated? If successive digits must be different?

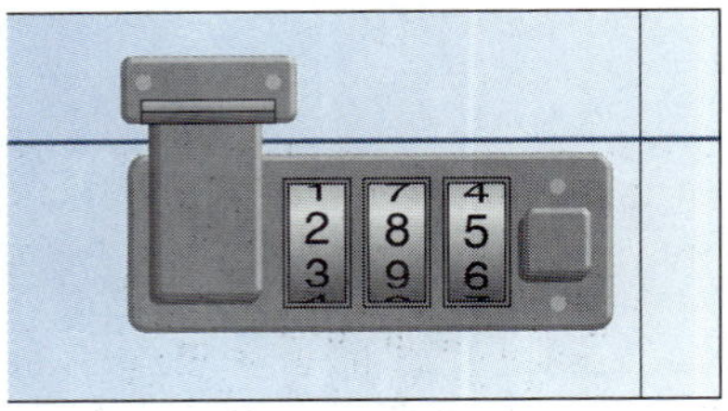

33. How many different license plates are possible if each contains 3 letters (out of the 26 letters of the alphabet) followed by 3 digits (from 0 to 9)? How many of these license plates contain no repeated letters and no repeated digits?

34. How many 5-digit ZIP code numbers are possible? How many of these numbers contain no repeated digits?

35. In Example 3 does it make any difference in which order the selection operations are performed? That is, if we select a jacket color first and then select a size, are there as many combined choices available as selecting a size first and then a color? Justify your answer using tree diagrams and the multiplication principle.

36. Explain how three sets, *A*, *B*, and *C*, can be related to each other in order for the following equation to hold (Venn diagrams may be helpful):

$$n(A \cup B \cup C) = n(A) + n(B) + n(C) - n(A \cap C) - n(B \cap C)$$

C

37. A group of 75 people includes 32 who play tennis, 37 who play golf, and 8 who play both tennis and golf. How many people in the group play neither sport?

38. A class of 30 music students includes 13 who play the piano, 16 who play the guitar, and 5 who play both the piano and the guitar. How many students in the class play neither instrument?

39. A group of 100 people touring Europe includes 42 people who speak French, 55 who speak German, and 17 who speak neither language. How many people in the group speak both French and German?

40. A high school football team with 40 players includes 16 players who played offense last year, 17 who played defense, and 12 who were not last year's team. How many players from last year played both offense and defense?

Applications

Business & Economics

41. *Management selection.* A management selection service classifies its applicants (using tests and interviews) as high-IQ, middle-IQ, or low-IQ and as aggressive or passive. How many combined classifications are possible?

(A) Solve by using a tree diagram.
(B) Solve by using the multiplication principle.

42. *Management selection.* A corporation plans to fill 2 different positions for vice-president, V_1 and V_2, from administrative officers in 2 of its manufacturing plants. Plant *A* has 6 officers and plant *B* has 8. How many ways can these 2 positions be filled if the V_1 position is to be filled from plant *A* and the V_2 position from plant *B?* How many ways can the 2 positions be filled if the selection is made without regard to plant?

43. *Transportation.* A sales representative who lives in city *A* wishes to start from home and fly to 3 different cities, *B, C,* and *D.* If there are 2 choices of local transportation (drive her own car to and from the airport or use a taxi for both trips), and if all cities are interconnected by airlines, how many travel plans can be constructed to visit each city exactly once and return home?

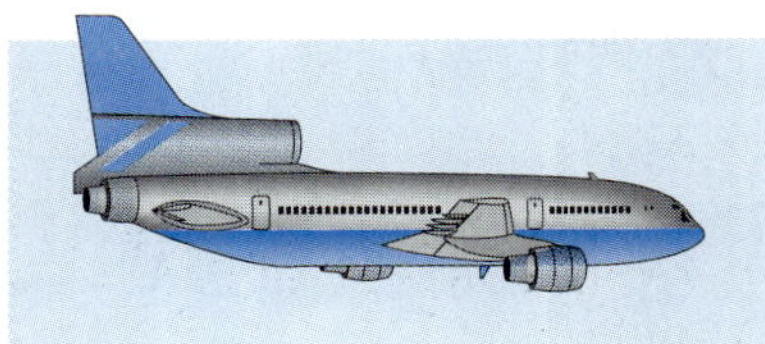

44. *Transportation.* A manufacturing company in city *A* wishes to truck its product to 4 different cities, *B, C, D,* and *E.* If the cities are all interconnected by roads, how many different route plans can be constructed so that a single truck, starting from *A,* will visit each city exactly once, then return home?

45. *Market research.* A survey of 1,200 people in a certain city indicates that 850 own microwave ovens, 740 own VCR's, and 580 own microwave ovens and VCR's.
(A) How many people in the survey own either a microwave oven or a VCR?
(B) How many own neither a microwave oven nor a VCR?
(C) How many own a microwave oven and do not own a VCR?

46. *Market research.* A survey of 800 small businesses indicates that 250 own photocopiers, 420 own fax machines, and 180 own photocopiers and fax machines.
(A) How many businesses in the survey own either a photocopier or a fax machine?
(B) How many own neither a photocopier nor a fax machine?
(C) How many own a fax machine and do not own a photocopier?

47. *Communications.* A cable television company has 8,000 subscribers in a suburban community. The company offers two premium channels, HBO and Showtime. If 2,450 subscribers receive HBO, 1,940 receive Showtime, and 5,180 do not receive any premium channel, how many subscribers receive both HBO and Showtime?

48. *Communications.* A local telephone company offers its 10,000 customers two special services: call forwarding and call waiting. If 3,770 customers use call forwarding, 3,250 use call waiting, and 4,530 do not use either of these services, how many customers use both call forwarding and call waiting?

49. *Minimum wage.* The table below gives the number of male and female workers (in thousands) earning at or below the minimum wage for several age categories.

	Workers per age group (in thousands)			
	16–19	*20–24*	*25+*	*Totals*
Males at minimum wage	343	154	237	734
Males below minimum wage	118	102	159	379
Females at minimum wage	367	186	503	1,056
Females below minimum wage	251	202	540	993
Totals	1,079	644	1,439	3,162

(A) How many males are age 20–24 and below minimum wage?
(B) How many females are age 20 or older and at minimum wage?
(C) How many workers are age 16–19 or males at minimum wage?
(D) How many workers are below minimum wage?

50. *Minimum wage.* Refer to the table in Problem 49.
(A) How many females are age 16–19 and at minimum wage?
(B) How many males are age 16–24 and below minimum wage?
(C) How many workers are age 20–24 or females below minimum wage?
(D) How many workers are at minimum wage?

Life Sciences

51. *Medicine.* A medical researcher classifies subjects according to male or female; smoker or nonsmoker; and underweight, average weight, or overweight. How many combined classifications are possible?
(A) Solve using a tree diagram.
(B) Solve using the multiplication principle.

52. *Family planning.* A couple is planning to have 3 children. How many boy–girl combinations are possible? Distinguish between combined outcomes such as (B, B, G), (B, G, B), and (G, B, B).
(A) Solve by using a tree diagram.
(B) Solve by using the multiplication principle.

Social Sciences

53. *Politics.* A politician running for a third term in office is planning to contact all contributors to her first two campaigns. If 1,475 individuals contributed to the first campaign, 2,350 individuals contributed to the second campaign, and 920 individuals contributed to the first and second campaigns, how many individuals have contributed to the first or second campaign?

54. *Politics.* If 12,457 people voted for a politician in his first election, 15,322 voted for him in his second election, and 9,345 voted for him in the first and second elections, how many people voted for this politician in the first or second election?

SECTION 6-2

Permutations and Combinations

- FACTORIAL
- PERMUTATIONS
- COMBINATIONS
- APPLICATIONS

The multiplication principle discussed in the preceding section can be used to develop two additional devices for counting that are extremely useful in more complicated counting problems. Both of these devices use a function called a *factorial function,* which we introduce first.

FACTORIAL

When using the multiplication principle, we encountered expressions of the form

$$26 \cdot 25 \cdot 24 \qquad 8 \cdot 7 \cdot 6$$

where each natural number factor is decreased by 1 as we move from left to right. The factors in the following product continue to decrease by 1 until a factor of 1 is reached:

$$5 \cdot 4 \cdot 3 \cdot 2 \cdot 1$$

Products of this type are encountered so frequently in counting problems that it is useful to be able to express them in a concise notation. The product of the first n natural numbers is called ***n* factorial** and is denoted by ***n*!**. Also, we define **zero factorial, 0!,** to be 1. Symbolically:

> **Factorial**
>
> For n a natural number,
>
> $$n! = n(n-1)(n-2) \cdot \cdots \cdot 2 \cdot 1 \qquad 4! = 4 \cdot 3 \cdot 2 \cdot 1$$
> $$0! = 1$$
> $$n! = n \cdot (n-1)!$$
>
> [*Note:* Many calculators have an [n!] key or its equivalent.]

Example 1 **Computing Factorials**

(A) $5! = 5 \cdot 4 \cdot 3 \cdot 2 \cdot 1 = 120$

(B) $\dfrac{7!}{6!} = \dfrac{7 \cdot \cancel{6!}}{\cancel{6!}} = 7$

(C) $\dfrac{8!}{5!} = \dfrac{8 \cdot 7 \cdot 6 \cdot \cancel{5!}}{\cancel{5!}} = 8 \cdot 7 \cdot 6 = 336$

(D) $\dfrac{52!}{5!47!} = \dfrac{52 \cdot 51 \cdot 50 \cdot 49 \cdot 48 \cdot \cancel{47!}}{5 \cdot 4 \cdot 3 \cdot 2 \cdot 1 \cdot \cancel{47!}} = 2{,}598{,}960$

Matched Problem 1 ➡ Find:

(A) $6!$ (B) $\frac{10!}{9!}$ (C) $\frac{10!}{7!}$ (D) $\frac{5!}{0!3!}$ (E) $\frac{20!}{3!17!}$

It is interesting and useful to note that $n!$ grows very rapidly. Compare the following:

$$5! = 120 \qquad 10! = 3{,}628{,}800 \qquad 15! = 1{,}307{,}674{,}368{,}000$$

Try 69!, 70!, and 71! on your calculator.

Permutations

A particular (horizontal) arrangement of a set of paintings on a wall is called a *permutation* of the set of paintings. In general:

A Permutation of a Set of Objects

A **permutation** of a set of distinct objects is an arrangement of the objects in a specific order without repetition.

Suppose 4 pictures are to be arranged from left to right on one wall of an art gallery. How many permutations (ordered arrangements) are possible? Using the multiplication principle, there are 4 ways of selecting the first picture; after the first picture is selected, there are 3 ways of selecting the second picture; after the first 2 pictures are selected, there are 2 ways of selecting the third picture; and after the first 3 pictures are selected, there is only 1 way to select the fourth. Thus, the number of permutations (ordered arrangements) of the set of 4 pictures is

$$4 \cdot 3 \cdot 2 \cdot 1 = 4! = 24$$

In general, how many permutations of a set of n distinct objects are possible? Reasoning as above, there are n ways in which the first object can be chosen, there are $n - 1$ ways in which the second object can be chosen, and so on. Using the multiplication principle, we have the following:

The Number of Permutations of n Objects

The number of permutations of n distinct objects without repetition, denoted by $P_{n,n}$, is

$$P_{n,n} = n(n-1) \cdot \cdots \cdot 2 \cdot 1 = n! \qquad n \text{ factors}$$

Example: The number of permutations of 7 objects is

$$P_{7,7} = 7 \cdot 6 \cdot 5 \cdot 4 \cdot 3 \cdot 2 \cdot 1 = 7! \qquad 7 \text{ factors}$$

Now suppose the director of the art gallery decides to use only 2 of the 4 available paintings, and they will be arranged on the wall from left to right. We are now talking about a particular arrangement of 2 paintings out of the 4, which is called a *permutation of 4 objects taken 2 at a time.* In general:

A Permutation of n Objects Taken r at a Time

A permutation of a set of n distinct objects taken r at a time without repetition is an arrangement of r of the n objects in a specific order.

How many ordered arrangements of 2 pictures can be formed from the 4? That is, how many permutations of 4 objects taken 2 at a time are there? There are 4 ways the first picture can be selected; after selecting the first picture, there are 3 ways the second picture can be selected. Thus, the number of permutations of a set of 4 objects taken 2 at a time, which is denoted by $P_{4,2}$, is given by

$$P_{4,2} = 4 \cdot 3$$

In terms of factorials, we have

$$P_{4,2} = 4 \cdot 3 = \frac{4 \cdot 3 \cdot \mathbf{2!}}{\mathbf{2!}} = \frac{4!}{2!} \qquad \text{Multiplying } 4 \cdot 3 \text{ by 1 in the form } 2!/2!$$

Reasoning in the same way as in the example, we find that the number of permutations of n distinct objects taken r at a time without repetition $(0 \le r \le n)$ is given by

$$\begin{aligned} P_{n,r} &= n(n-1)(n-2) \cdot \cdots \cdot (n-r+1) && r \text{ factors} \\ P_{9,6} &= 9(9-1)(9-2) \cdot \cdots \cdot (9-6+1) && 6 \text{ factors} \\ &= 9 \cdot 8 \cdot 7 \cdot 6 \cdot 5 \cdot 4 \end{aligned}$$

Multiplying the right side of the equation for $P_{n,r}$ by 1 in the form $(n-r)!/(n-r)!$, we obtain a factorial form for $P_{n,r}$:

$$P_{n,r} = n(n-1)(n-2) \cdot \cdots \cdot (n-r+1)\frac{(n-r)!}{(n-r)!}$$

But, since

$$n(n-1)(n-2) \cdot \cdots \cdot (n-r+1)(n-r)! = n!$$

The above expression simplifies to

$$P_{n,r} = \frac{n!}{(n-r)!}$$

We summarize these results in the following box:

> **Number of Permutations of n Objects Taken r at a Time**
>
> The number of permutations of n distinct objects taken r at a time without repetition is given by*
>
> $$P_{n,r} = n(n-1)(n-2) \cdot \cdots \cdot (n-r+1) \quad r \text{ factors}$$
>
> $$P_{5,2} = 5 \cdot 4 \quad 2 \text{ factors}$$
>
> or
>
> $$P_{n,r} = \frac{n!}{(n-r)!} \quad 0 \leqslant r \leqslant n \quad P_{5,2} = \frac{5!}{(5-2)!} = \frac{5!}{3!}$$
>
> [*Note*: $P_{n,n} = \frac{n!}{(n-n)!} = \frac{n!}{0!} = n!$ permutations of n objects taken n at a time. Remember, by definition, $0! = 1$.]
>
> *In place of the symbol $P_{n,r}$, the symbols P^n_r, ${}_nP_r$, and $P(n, r)$ are often used.

Example 2 **Permutations** Given the set $\{A, B, C\}$, how many permutations are there of this set of 3 objects taken 2 at a time? Answer the question:

(A) Using a tree diagram (B) Using the multiplication principle
(C) Using the two formulas for $P_{n,r}$

SOLUTION (A) Using a tree diagram:

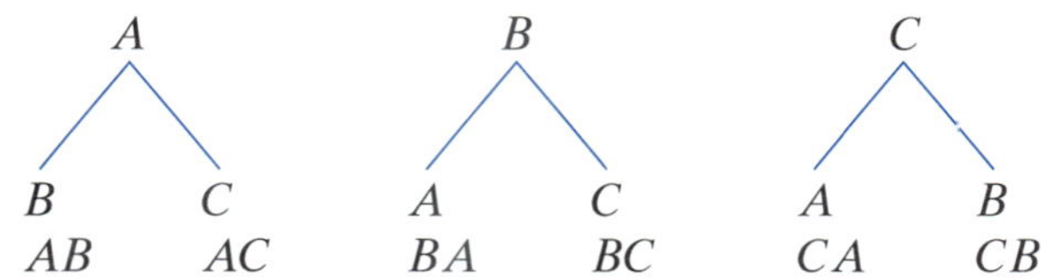

There are 6 permutations of 3 objects taken 2 at a time.

(B) Using the multiplication principle:

O_1: Fill the first position N_1: 3 ways
O_2: Fill the second position N_2: 2 ways

Thus, there are

$3 \cdot 2 = 6$ permutations of 3 objects taken 2 at a time

(C) Using the two formulas for $P_{n,r}$:

$$\overset{\text{2 factors}}{\downarrow}$$
$$P_{3,2} = 3 \cdot 2 = 6 \quad \text{or} \quad P_{3,2} = \frac{3!}{(3-2)!} = \frac{3 \cdot 2 \cdot 1}{1} = 6$$

Thus, there are 6 permutations of 3 objects taken 2 at a time.

Of course, all three methods produce the same answer.

Matched Problem 2 Given the set $\{A, B, C, D\}$, how many permutations are there of this set of 4 objects taken 2 at a time? Answer the question:

(A) Using a tree diagram (B) Using the multiplication principle
(C) Using the two formulas for $P_{n,r}$

In Example 2 you probably found the multiplication principle the easiest method to use. But for large values of n and r you will find that the factorial formula is more convenient. In fact, many calculators have functions that compute $n!$ and $P_{n,r}$ directly. See Figure 1 in Example 3 and the user's manual for your calculator.

Example 3 **Permutations** Find the number of permutations of 13 objects taken 8 at a time. Compute the answer using a calculator.

Solution We use the factorial formula for $P_{n,r}$:

$$P_{13,8} = \frac{13!}{(13-8)!} = \frac{13!}{5!} = 51{,}891{,}840$$

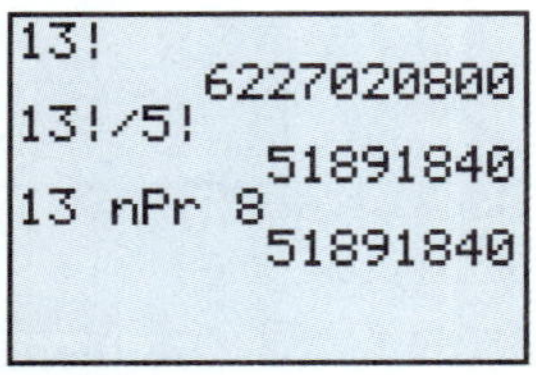

FIGURE 1

Using a tree diagram to solve this problem would involve a monumental effort. Using the multiplication principle would involve multiplying $13 \cdot 12 \cdot 11 \cdot 10 \cdot 9 \cdot 8 \cdot 7 \cdot 6$ (8 factors), which is not too bad. A calculator can provide instant results. See Figure 1.

Matched Problem 3 Find the number of permutations of 30 objects taken 4 at a time. Compute the answer exactly using a calculator.

Combinations

Now suppose that an art museum owns 8 paintings by a given artist and another art museum wishes to borrow 3 of these paintings for a special show. In selecting 3 of the 8 paintings for shipment, the order would not matter, and we would simply be selecting a 3 element subset from the set of 8 paintings. That is, we would be selecting what is called *a combination of 8 objects taken 3 at a time.* In general:

A Combination of n Objects Taken r at a Time

A **combination** of a set of n distinct objects taken r at a time without repetition is an r element subset of the set of n objects. The arrangement of the elements in the subset does not matter.

How many ways can the 3 paintings be selected for shipment out of the 8 available? That is, what is the number of combinations of 8 objects taken 3 at a time? To answer this question, and to get a better insight into the general problem, we return to Example 2.

In Example 2 we were given the set $\{A, B, C\}$ and found the number of permutations of 3 objects taken 2 at a time using a tree diagram. From this tree diagram we also can determine the number of combinations of 3 objects taken 2 at a time (the number of 2 element subsets from a 3 element set), and compare it with the number of permutations (see Fig. 2)

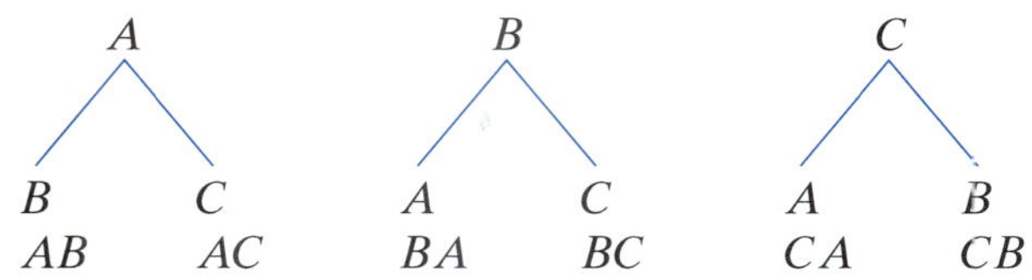

(A, B) (B, A)	(A, C) (C, A)	(B, C) (C, B)	6 permutations of 3 objects taken 2 at a time **(Order does matter)**
$\{A, B\}$	$\{A, C\}$	$\{B, C\}$	3 combinations of 3 objects taken 2 at a time **(Order does not matter)**

FIGURE 2

There are fewer combinations than permutations, as we would expect. To each subset (combination) there corresponds two ordered pairs (permutations). We denote the number of combinations in Figure 2 by

$$C_{3,2} \qquad \text{or} \qquad \binom{3}{2}$$

Our final goal is to find a factorial formula for $C_{n,r}$, the number of combinations of n objects taken r at a time. But first, we will develop a formula for $C_{3,2}$, and then we will generalize from this experience.

We know the number of permutations of 3 objects taken 2 at a time is given by $P_{3,2}$, and we have a formula for computing this number. Now, suppose we think of $P_{3,2}$ in terms of two operations:

O_1: Selecting a subset of 2 elements — N_1: $C_{3,2}$ ways
O_2: Arranging the subset in a given order — N_2: $2!$ ways

The combined operation, O_1 followed by O_2, produces a permutation of 3 objects taken 2 at a time. Thus,

$$P_{3,2} = C_{3,2} \cdot 2! \qquad \text{or} \qquad C_{3,2} = \frac{P_{3,2}}{2!}$$

To find $C_{3,2}$, the number of combinations of 3 objects taken 2 at a time, we substitute

$$P_{3,2} = \frac{3!}{(3-2)!}$$

and solve for $C_{3,2}$:

$$C_{3,2} = \frac{3!}{2!(3-2)!} = \frac{3 \cdot 2 \cdot 1}{(2 \cdot 1)(1)} = 3$$

This result agrees with the result we got using a tree diagram. Note that the number of combinations of 3 objects taken 2 at a time is the same as the number of permutations of 3 objects taken 2 at a time divided by the number of permutations of the elements in a 2 element subset. This observation also can be made in Figure 2.

Reasoning the same way as in the example, the number of combinations of n objects taken r at a time ($0 \leq r \leq n$) is given by

$$C_{n,r} = \frac{P_{n,r}}{r!}$$

$$= \frac{n!}{r!(n-r)!} \qquad \text{Since } P_{n,r} = \frac{n!}{(n-r)!}$$

In summary:

Number of Combinations of n Objects Taken r at a Time

The number of combinations of n distinct objects taken r at a time without repetition is given by*

$$C_{n,r} = \binom{n}{r} \qquad\qquad C_{52,5} = \binom{52}{5}$$

$$= \frac{P_{n,r}}{r!} \qquad\qquad = \frac{P_{52,5}}{5!}$$

$$= \frac{n!}{r!(n-r)!} \quad 0 \leq r \leq n \qquad\qquad = \frac{52!}{5!(52-5)!}$$

*In place of the symbols $C_{n,r}$ and $\binom{n}{r}$, the symbols C_r^n, ${}_nC_r$, and $C(n, r)$ are often used.

Now we can answer the question posed earlier in the museum example. There are

$$C_{8,3} = \frac{8!}{3!(8-3)!} = \frac{8!}{3!5!} = \frac{8 \cdot 7 \cdot 6 \cdot \cancel{5!}}{3 \cdot 2 \cdot 1 \cdot \cancel{5!}} = 56$$

ways the 3 paintings can be selected for shipment. That is, there are 56 combinations of 8 objects taken 3 at a time.

Example 4 **Permutations and Combinations** From a committee of 10 people:

(A) In how many ways can we choose a chairperson, a vice-chairperson, and a secretary, assuming that one person cannot hold more than one position?

(B) In how many ways can we choose a subcommittee of 3 people?

SOLUTION Note how parts (A) and (B) differ. In part (A), order of choice makes a difference in the selection of the officers. In part (B), the ordering does not matter in choosing a 3 person subcommittee. Thus, in part (A), we are interested in the number of *permutations* of 10 objects taken 3 at a time; and in part (B), we are interested in the number of *combinations* of 10 objects taken 3 at a time. These quantities are computed as follows (and since the numbers are not large, we do not need to use a calculator):

$$\text{(A)}\quad P_{10,3} = \frac{10!}{(10-3)!} = \frac{10!}{7!} = \boxed{\frac{10 \cdot 9 \cdot 8 \cdot \cancel{7!}}{\cancel{7!}}} = 720 \text{ ways}$$

$$\text{(B)}\quad C_{10,3} = \frac{10!}{3!(10-3)!} = \frac{10!}{3!7!} = \boxed{\frac{10 \cdot 9 \cdot 8 \cdot \cancel{7!}}{3 \cdot 2 \cdot 1 \cdot \cancel{7!}}} = 120 \text{ ways}$$

Matched Problem 4 From a committee of 12 people:

(A) In how many ways can we choose a chairperson, a vice-chairperson, a secretary, and a treasurer, assuming that one person cannot hold more than one position?

(B) In how many ways can we choose a subcommittee of 4 people?

If n and r are other than small numbers (as in Example 4), a calculator is a useful aid in evaluating expressions involving factorials. Many calculators have a function that computes $C_{n,r}$ directly. (See Fig. 3 below.)

Example 5 **Combinations** Find the number of combinations of 13 objects taken 8 at a time. Compute the answer exactly, using a calculator.

SOLUTION

$$C_{13,8} = \binom{13}{8} = \frac{13!}{8!(13-8)!} = \frac{13!}{8!5!} = 1{,}287$$

Compare the result in Example 5 with that obtained in Example 3, and note that $C_{13,8}$ is substantially smaller than $P_{13,8}$. See Figure 3.

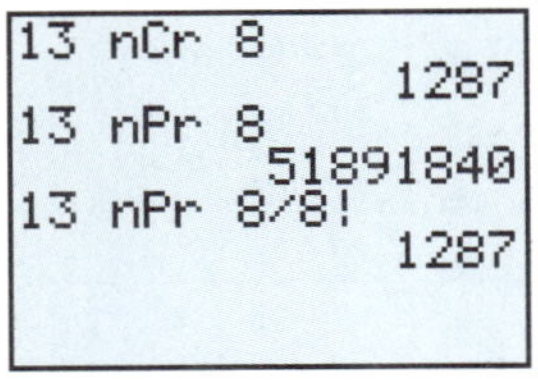

FIGURE 3

Matched Problem 5 Find the number of combinations of 30 objects taken 4 at a time. Compute the answer exactly using a calculator.

REMEMBER

In a permutation, order matters.
In a combination, order does not matter.

To determine whether permutations or combinations are involved in a problem, see if rearranging the collection or listing produces a different object. If so, use permutations; if not, use combinations.

Explore–Discuss 1

(A) List alphabetically by the first letter, all 3-letter license plate codes consisting of 3 different letters chosen from M, A, T, H. Discuss how this list relates to $P_{n,r}$.

(B) Reorganize the list from part (A) so that now all codes without M come first, then all codes without A, then all codes without T, and finally all codes without H. Discuss how this list illustrates the formula $P_{n,r} = r!C_{n,r}$.

Applications

We now consider some applications of the concepts discussed above. Several applications in this and the following sections involve a standard 52-card deck of playing cards, which is described below:

Standard 52-Card Deck of Playing Cards

A standard deck of 52 cards (see Fig. 4) has four 13-card suits: diamonds, hearts, clubs, and spades. The diamonds and hearts are red, and the clubs and spades are black. Each 13-card suit contains cards numbered from 2 to 10, a jack, a queen, a king, and an ace. The jack, queen, and king are called *face cards.* Depending on the game, the ace may be counted as the lowest and/or the highest card in the suit.

FIGURE 4

Example 6 **Counting Techniques** How many 5-card hands will have 3 aces and 2 kings?

Solution The solution involves both the multiplication principle and combinations. Think of selecting the 5-card hand in terms of the following two operations:

O_1: Choosing 3 aces out of 4 possible (order is not important) N_1: $C_{4,3}$
O_2: Choosing 2 kings out of 4 possible (order is not important) N_2: $C_{4,2}$

Using the multiplication principle, we have

$$\begin{aligned}\text{Number of hands} &= C_{4,3} \cdot C_{4,2}\\ &= \frac{4!}{3!(4-3)!} \cdot \frac{4!}{2!(4-2)!}\\ &= 4 \cdot 6 = 24\end{aligned}$$

Matched Problem 6 How many 5-card hands will have 3 hearts and 2 spades?

Example 7 **Counting Techniques** Serial numbers for a product are to be made using 2 letters followed by 3 numbers. If the letters are to be taken from the first 8 letters of the alphabet with no repeats and the numbers are to be taken from the 10 digits (0–9) with no repeats, how many serial numbers are possible?

Solution The solution involves both the multiplication principle and permutations. Think of selecting a serial number in terms of the following two operations:

O_1: Choosing 2 letters out of 8 available (order is important) N_1: $P_{8,2}$
O_2: Choosing 3 numbers out of 10 available (order is important) N_2: $P_{10,3}$

Using the multiplication principle, we have

$$\begin{aligned}\text{Number of serial numbers} &= P_{8,2} \cdot P_{10,3}\\ &= \frac{8!}{(8-2)!} \cdot \frac{10!}{(10-3)!}\\ &= 56 \cdot 720 = 40{,}320\end{aligned}$$

Matched Problem 7 Repeat Example 7 under the same conditions, except the serial numbers are now to have 3 letters followed by 2 digits (no repeats).

Example 8 **Counting Techniques** A company has 7 senior and 5 junior officers. An ad hoc legislative committee is to be formed. In how many ways can a 4-officer committee be formed so that it is composed of:

(A) Any 4 officers?
(B) 4 senior officers?
(C) 3 senior officers and 1 junior officer?
(D) 2 senior and 2 junior officers?
(E) At least 2 senior officers?

SOLUTION (A) Since there are a total of 12 officers in the company, the number of different 4-member committees is

$$C_{12,4} = \frac{12!}{4!(12-4)!} = \frac{12!}{4!8!} = 495$$

(B) If only senior officers can be on the committee, the number of different committees is

$$C_{7,4} = \frac{7!}{4!(7-4)!} = \frac{7!}{4!3!} = 35$$

(C) The 3 senior officers can be selected in $C_{7,3}$ ways, and the 1 junior officer can be selected in $C_{5,1}$ ways. Applying the multiplication principle, the number of ways that 3 senior officers and 1 junior officer can be selected is

$$C_{7,3} \cdot C_{5,1} = \frac{7!}{3!(7-3)!} \cdot \frac{5!}{1!(5-1)!} = \frac{7!5!}{3!4!1!4!} = 175$$

(D)
$$C_{7,2} \cdot C_{5,2} = \frac{7!}{2!(7-2)!} \cdot \frac{5!}{2!(5-2)!} = \frac{7!5!}{2!5!2!3!} = 210$$

(E) The committees with *at least* 2 senior officers can be divided into three disjoint collections:

1. Committees with 4 senior officers and 0 junior officers
2. Committees with 3 senior officers and 1 junior officer
3. Committees with 2 senior officers and 2 junior officers

The number of committees of types 1, 2, and 3 were computed in parts (B), (C), and (D), respectively. The total number of committees of all three types is the sum of these quantities:

$$\underset{\text{Type 1}}{C_{7,4}} + \underset{\text{Type 2}}{C_{7,3} \cdot C_{5,1}} + \underset{\text{Type 3}}{C_{7,2} \cdot C_{5,2}} = 35 + 175 + 210 = 420$$

Matched Problem 8 Given the information in Example 8, answer the following questions:

(A) How many 4-officer committees with 1 senior officer and 3 junior officers can be formed?
(B) How many 4-officer committees with 4 junior officers can be formed?
(C) How many 4-officer committees with at least 2 junior officers can be formed?

Explore–Discuss 2

(A) Compute the numbers $C_{n,0}, C_{n,1}, C_{n,2}, \ldots, C_{n,n}$ for $n = 4, 5, 6$. Discuss the patterns you observe.

(B) Compute the sum

$$C_{n,0} + C_{n,1} + C_{n,2} + \cdots + C_{n,n}$$

and the alternating sum

$$C_{n,0} - C_{n,1} + C_{n,2} - \cdots + (-1)^n C_{n,n}$$

for $n = 4, 5, 6$. Guess the sum and alternating sum for $n = 7$, and verify your guess.

Answers to Matched Problems

1. (A) 720 (B) 10 (C) 720 (D) 20 (E) 1,140

2. (A)

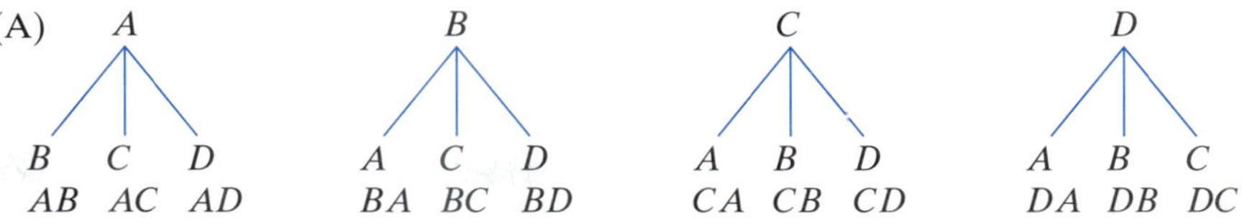

12 permutations of 4 objects taken 2 at a time

(B) O_1: Fill first position N_1: 4 ways
O_2: Fill second position N_2: 3 ways
$4 \cdot 3 = 12$

(C) $P_{4,2} = 4 \cdot 3 = 12;\ P_{4,2} = \dfrac{4!}{(4-2)!} = 12$

3. $P_{30,4} = \dfrac{30!}{(30-4)!} = 657{,}720$

4. (A) $P_{12,4} = \dfrac{12!}{(12-4)!} = 11{,}880$ ways (B) $C_{12,4} = \dfrac{12!}{4!(12-4)!} = 495$ ways

5. $C_{30,4} = \dfrac{30!}{4!(30-4)!} = 27{,}405$ **6.** $C_{13,3} \cdot C_{13,2} = 22{,}308$

7. $P_{8,3} \cdot P_{10,2} = 30{,}240$

8. (A) $C_{7,1} \cdot C_{5,3} = 70$ (B) $C_{5,4} = 5$
(C) $C_{7,2} \cdot C_{5,2} + C_{7,1} \cdot C_{5,3} + C_{5,4} = 285$

EXERCISE 6-2

A *Evaluate the expressions in Problems 1–20.*

1. $4!$ **2.** $6!$ **3.** $\dfrac{9!}{8!}$

4. $\dfrac{14!}{13!}$ **5.** $\dfrac{11!}{8!}$ **6.** $\dfrac{14!}{12!}$

7. $\dfrac{5!}{2!3!}$ **8.** $\dfrac{6!}{4!2!}$ **9.** $\dfrac{7!}{4!(7-4)!}$

10. $\dfrac{8!}{3!(8-3)!}$ **11.** $\dfrac{7!}{7!(7-7)!}$ **12.** $\dfrac{8!}{0!(8-0)!}$

13. $P_{5,3}$ **14.** $P_{4,2}$ **15.** $P_{52,4}$

16. $P_{52,2}$ **17.** $C_{5,3}$ **18.** $C_{4,2}$

19. $C_{52,4}$ **20.** $C_{52,2}$

In Problems 21 and 22, would you consider the selection to be a combination or a permutation? Explain your reasoning.

21. (A) The recently elected university president named 3 new officers: a vice-president of finance, a vice-president of academic affairs, and a vice-president of student affairs.

(B) A university president selected 2 of the vice-presidents to attend the dedication ceremony of a new branch campus.

22. (A) A student checked out 4 novels from the library to read over the holidays.

(B) A student did some holiday shopping by buying 4 books, 1 for his father, 1 for his mother, 1 for his younger sister, and 1 for his older brother.

23. In a horse race, how many different finishes among the first 3 places are possible if 10 horses are running? (Exclude ties.)

24. In a long-distance foot race, how many different finishes among the first 5 places are possible if 50 people are running? (Exclude ties.)

25. How many ways can a 3-person subcommittee be selected from a committee of 7 people? How many ways can a president, vice-president, and secretary be chosen from a committee of 7 people?

26. Nine cards are numbered with the digits from 1 to 9. A 3-card hand is dealt, 1 card at a time. How many hands are possible where:

(A) Order is taken into consideration?

(B) Order is not taken into consideration?

B

27. Discuss the relative growth rates of $x!$, 3^x, and x^3.

28. Discuss the relative growth rates of $x!$, 2^x, and x^2.

29. From a standard 52-card deck, how many 5-card hands will have all hearts?

30. From a standard 52-card deck, how many 5-card hands will have all face cards? All face cards, but no kings?

31. From a standard 52-card deck, how many 7-card hands have exactly 5 spades and 2 hearts?

32. From a standard 52-card deck, how many 5-card hands will have 2 clubs and 3 hearts?

33. A catering service offers 8 appetizers, 10 main courses, and 7 desserts. A banquet committee is to select 3 appetizers, 4 main courses, and 2 desserts. How many ways can this be done?

34. Three departments have 12, 15, and 18 members, respectively. If each department is to select a delegate and an alternate to represent the department at a conference, how many ways can this be done?

In Problems 35 and 36, refer to the table in the graphing calculator display below, which shows $y_1 = P_{n,r}$ and $y_2 = C_{n,r}$ for $n = 6$.

X	Y1	Y2
0	1	1
1	6	6
2	30	15
3	120	20
4	360	15
5	720	6
6	720	1

Y2=6 nCr X

35. Discuss and explain the symmetry of the numbers in the y_2 column of the table.

36. Explain how the table illustrates the formula

$$P_{n,r} = r!C_{n,r}$$

C

37. Eight distinct points are selected on the circumference of a circle.
(A) How many chords can be drawn by joining the points in all possible ways?
(B) How many triangles can be drawn using these 8 points as vertices?
(C) How many quadrilaterals can be drawn using these 8 points as vertices?

38. Five distinct points are selected on the circumference of a circle.
(A) How many chords can be drawn by joining the points in all possible ways?
(B) How many triangles can be drawn using these 5 points as vertices?

39. How many ways can 2 people be seated in a row of 5 chairs? 3 people? 4 people? 5 people?

40. Each of 2 countries sends 5 delegates to a negotiating conference. A rectangular table is used with 5 chairs on each long side. If each country is assigned a long side of the table (operation 1), how many seating arrangements are possible?

41. A basketball team has 5 distinct positions. Out of 8 players, how many starting teams are possible if:
(A) The distinct positions are taken into consideration?
(B) The distinct positions are not taken into consideration?
(C) The distinct positions are not taken into consideration, but either Mike or Ken (but not both) must start?

42. How many 4-person committees are possible from a group of 9 people if:
(A) There are no restrictions?
(B) Both Jim and Mary must be on the committee?
(C) Either Jim or Mary (but not both) must be on the committee?

43. Find the largest integer k such that your calculator can compute $k!$ without an overflow error.

44. Find the largest integer k such that your calculator can compute $C_{2k,k}$ without an overflow error.

45. C Note from the table in the graphing calculator display below that the largest value of $C_{n,r}$ when $n = 20$ is $C_{20,10} = 184{,}756$. Use a similar table to find the largest value of $C_{n,r}$ when $n = 24$.

X	Y1
7	77520
8	125970
9	167960
10	184756
11	167960
12	125970
13	77520

Y1=20 nCr X

46. C Note from the table in the graphing calculator display below that the largest value of $C_{n,r}$ when $n = 21$ is $C_{21,10} = C_{21,11} = 352{,}716$. Use a similar table to find the largest value of $C_{n,r}$ when $n = 17$.

X	Y1
7	116280
8	203490
9	293930
10	352716
11	352716
12	293930
13	203490

Y1=21 nCr X

Applications

Business & Economics

47. *Quality control.* A computer store receives a shipment of 24 laser printers, including 5 that are defective. Three of these printers are selected to be displayed in the store.
(A) How many selections can be made?
(B) How many of these selections will contain no defective printers?

48. *Quality control.* An electronics store receives a shipment of 30 graphing calculators, including 6 that are defective. Four of these calculators are selected to be sent to a local high school.
(A) How many selections can be made?
(B) How many of these selections will contain no defective calculators?

49. *Business closings.* A jewelry store chain with 8 stores in Georgia, 12 in Florida, and 10 in Alabama is planning to close 10 of these stores.
(A) How many ways can this be done?
(B) The company decides to close 2 stores in Georgia, 5 in Florida, and 3 in Alabama. How many ways can this be done?

50. *Employee layoffs.* A real estate company with 14 employees in their central office, 8 in their north office, and 6 in their south office is planning to lay off 12 employees.
(A) How many ways can this be done?
(B) The company decides to lay off 5 employees from the central office, 4 from the north office, and 3 from the south office. How many ways can this be done?

51. *Personnel selection.* Suppose 6 female and 5 male applicants have been successfully screened for 5 positions. In how many ways can the following compositions be selected?
(A) 3 females and 2 males
(B) 4 females and 1 male
(C) 5 females
(D) 5 people regardless of sex
(E) At least 4 females

52. *Committee selection.* A 4-person grievance committee is to be selected out of 2 departments, A and B, with 15 and 20 people, respectively. In how many ways can the following committees be selected?
(A) 3 from A and 1 from B
(B) 2 from A and 2 from B
(C) All from A
(D) 4 people regardless of department
(E) At least 3 from department A

Life Sciences

53. *Medicine.* There are 8 standard classifications of blood type. An examination for prospective laboratory technicians consists of having each candidate determine the type for 3 blood samples. How many different examinations can be given if no 2 of the samples provided for the candidate have the same type? If 2 or more samples can have the same type?

54. *Medical research.* Because of the limited funds, 5 research centers are to be chosen out of 8 suitable ones for a study on heart disease. How many choices are possible?

Social Sciences

55. *Politics.* A nominating convention is to select a president and vice-president from among 4 candidates. Campaign buttons, listing a president and a vice-president, are to be designed for each possible outcome before the convention. How many different kinds of buttons should be designed?

56. *Politics.* In how many different ways can 6 candidates for an office be listed on a ballot?

Section 6-3 Sample Spaces, Events, and Probability

- Experiments
- Sample Spaces and Events
- Probability of an Event
- Equally Likely Assumption

This section provides a relatively brief and informal introduction to probability. More detailed and formal treatments can be found in books and courses devoted entirely to the subject. Probability studies involve many subtle ideas, and care must be taken at the beginning to understand the fundamental concepts.

EXPERIMENTS

Our first step in constructing a mathematical model for probability studies is to describe the type of experiments on which probability studies are based. Some experiments do not yield the same results each time they are performed no matter how carefully they are repeated under the same conditions. These experiments are called **random experiments.** Familiar examples of random experiments are flipping coins, rolling dice, observing the frequency of defective items from an assembly line, or observing the frequency of deaths in a certain age group.

Probability theory is a branch of mathematics that has been developed to deal with outcomes of random experiments, both real and conceptual. In the work that follows, we will simply use the word **experiment** to mean a random experiment.

SAMPLE SPACES AND EVENTS

Associated with outcomes of experiments are *sample spaces* and *events.* Our second step in constructing a mathematical model for probability studies is to define these two terms. Set concepts, which are reviewed in Appendix A-1, are useful in this regard.

FIGURE 1

Consider the experiment, "A wheel with 18 numbers on the perimeter (Fig. 1) is spun and allowed to come to rest so that a pointer points within a numbered sector."

What outcomes might we observe? When the wheel stops, we might be interested in which number is next to the pointer, or whether that number is an odd number, or whether that number is divisible by 5, or whether that number is prime, or whether the pointer is in a shaded or white sector, and so on. The list of possible outcomes appears endless. In general, there is no unique method of analyzing all possible outcomes of an experiment. Therefore, before conducting an experiment, it is important to decide just what outcomes are of interest.

Suppose we limit our interest to the set of numbers on the wheel and to various subsets of these numbers, such as the set of prime numbers on the wheel or the set of odd numbers. Having decided what to observe, we make a list of outcomes of the experiment, called *simple outcomes* or *simple events,* such that in each trial of the experiment (each spin of the wheel), one and only one of the outcomes on the list will occur. For our stated interests, we choose each number on the wheel as a simple event and form the set

$$S = \{1, 2, 3, \dots, 17, 18\}$$

The set of simple events S for the experiment is called a *sample space* for the experiment.

Now consider the outcome, "When the wheel comes to rest, the number next to the pointer is divisible by 4." This outcome is not a simple outcome (or simple event), since it is not associated with one and only one element in the sample space S. The outcome will occur whenever any one of the simple events 4, 8, 12, or 16 occurs; that is, whenever an element in the subset

$$E = \{4, 8, 12, 16\}$$

occurs. Subset E is called a *compound event* (and the outcome, a *compound outcome*).

In general:

Sample Spaces and Events

If we formulate a set S of outcomes (events) of an experiment in such a way that in each trial of the experiment one and only one of the outcomes (events) in the set will occur, then we call the set S a **sample space** for the experiment. Each element in S is called a **simple outcome,** or **simple event.**

An **event E** is defined to be any subset of S (including the empty set $\varnothing$ and the sample space S). Event E is a **simple event** if it contains only one element and a **compound event** if it contains more than one element. We say that **an event E occurs** if any of the simple events in E occurs.

We use the terms *event* and *outcome of an experiment* interchangeably. Technically, an event is the mathematical counterpart of an outcome of an experiment, but we will not insist on a strict adherence to this distinction in our development of probability.

REAL WORLD	MATHEMATICAL MODEL
Experiment (real or conceptual)	Sample space (set S)
Outcome (simple or compound)	Event (subset of S; simple or compound)

Example 1 **Simple and Compound Events** Relative to the number wheel experiment (see Fig. 1) and the sample space

$$S = \{1, 2, 3, \ldots, 17, 18\}$$

what is the event E (subset of the sample space S) that corresponds to each of the following outcomes? Indicate whether the event is a simple event or a compound event.

(A) The outcome is a prime number. (B) The outcome is the square of 4.

SOLUTION (A) The outcome is a prime number if any of the simple events 2, 3, 5, 7, 11, 13, or 17 occurs.* Thus, to say the event "A prime number occurs" is the same as saying the experiment has an outcome in the set

$$E = \{2, 3, 5, 7, 11, 13, 17\}$$

Since event E has more than one element, it is a compound event.

*Technically, we should write $\{2\}$, $\{3\}$, $\{5\}$, $\{7\}$, $\{11\}$, $\{13\}$, and $\{17\}$ for the simple events, since there is a logical distinction between an element of a set and a subset consisting of only that element. But we will just keep this in mind and drop the braces for simple events to simplify the notation.

(B) The outcome is the square of 4 if 16 occurs. Thus, to say the event "The square of 4 occurs" is the same as saying the experiment has an outcome in the set

$$E = \{16\}$$

Since E has only one element, it is a simple event.

Matched Problem 1 ➡ Repeat Example 1 for:

(A) The outcome is a number divisible by 12.
(B) The outcome is an even number greater than 15.

Example 2 ➡ **Sample Spaces** A nickel and a dime are tossed. How shall we identify a sample space for this experiment? There are a number of possibilities, depending on our interest. We shall consider three.

(A) If we are interested in whether each coin falls heads (H) or tails (T), then, using a tree diagram, we can easily determine an appropriate sample space for the experiment:

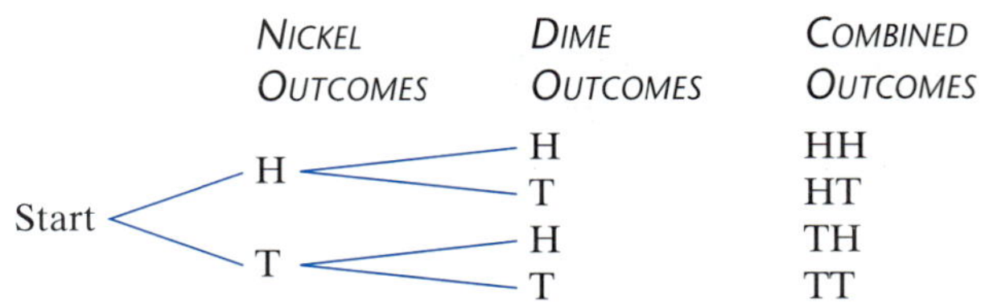

Thus,

$$S_1 = \{\text{HH, HT, TH, TT}\}$$

and there are 4 simple events in the sample space.

(B) If we are interested only in the number of heads that appear on a single toss of the two coins, then we can let

$$S_2 = \{0, 1, 2\}$$

and there are 3 simple events in the sample space.

(C) If we are interested in whether the coins match (M) or do not match (D), then we can let

$$S_3 = \{M, D\}$$

and there are only 2 simple events in the sample space.

In Example 2, which sample space would be appropriate for all three interests? Sample space S_1 contains more information than either S_2 or S_3. If we know which outcome has occurred in S_1, then we know which outcome has occurred in S_2 and S_3. However, the reverse is not true. (Note that the simple events in S_2 and S_3 are compound events in S_1.) In this sense, we say that S_1 is a more **fundamental sample space** than either S_2 or S_3. Thus, we would choose S_1 as an appropriate sample space for all three expressed interests.

Choosing Sample Spaces

There is no one correct sample space for a given experiment. When specifying a sample space for an experiment, we include as much detail as is necessary to answer *all* questions of interest regarding the outcomes of the experiment. When in doubt, choose a sample space with more elements rather than fewer.

Matched Problem 2 ➡ An experiment consists of recording the boy–girl composition of 2-child families.

(A) What is an appropriate sample space if we are interested in the sex of each child in the order of their births? Draw a tree diagram.

(B) What is an appropriate sample space if we are interested only in the number of girls in a family?

(C) What is an appropriate sample space if we are interested only in whether the sexes are alike (A) or different (D)?

(D) What is an appropriate sample space for all three interests expressed in parts (A)–(C)?

Example 3 ➡ **Sample Spaces and Events** Consider an experiment of rolling two dice. A convenient sample space that will enable us to answer many questions about interesting events is shown in Figure 2. Let S be the set of all ordered pairs in the figure. The simple event $(3, 2)$ is to be distinguished from the simple event $(2, 3)$. The former indicates that a 3 turned up on the first die and a 2 on the second, while the latter indicates that a 2 turned up on the first die and a 3 on the second.

SECOND DIE

FIRST DIE

(1, 1)	(1, 2)	(1, 3)	(1, 4)	(1, 5)	(1, 6)
(2, 1)	(2, 2)	(2, 3)	(2, 4)	(2, 5)	(2, 6)
(3, 1)	(3, 2)	(3, 3)	(3, 4)	(3, 5)	(3, 6)
(4, 1)	(4, 2)	(4, 3)	(4, 4)	(4, 5)	(4, 6)
(5, 1)	(5, 2)	(5, 3)	(5, 4)	(5, 5)	(5, 6)
(6, 1)	(6, 2)	(6, 3)	(6, 4)	(6, 5)	(6, 6)

FIGURE 2

What is the event (subset of the sample space S) that corresponds to each of the following outcomes?

(A) A sum of 7 turns up. (B) A sum of 11 turns up.
(C) A sum less than 4 turns up. (D) A sum of 12 turns up.

SOLUTION (A) By "A sum of 7 turns up," we mean that the sum of all dots on both turned-up faces is 7. This outcome corresponds to the event

$$\{(6, 1), (5, 2), (4, 3), (3, 4), (2, 5), (1, 6)\}$$

(B) "A sum of 11 turns up" corresponds to the event

$$\{(6, 5), (5, 6)\}$$

(C) "A sum less than 4 turns up" corresponds to the event

$$\{(1, 1), (2, 1), (1, 2)\}$$

(D) "A sum of 12 turns up" corresponds to the event

$$\{(6, 6)\}$$

Matched Problem 3 Refer to the sample space shown in Figure 2. What is the event that corresponds to each of the following outcomes?

(A) A sum of 5 turns up.
(B) A sum that is a prime number greater than 7 turns up.

As indicated earlier, we often use the terms *event* and *outcome of an experiment* interchangeably. Thus, in Example 3 we might say "the event, 'A sum of 11 turns up' " in place of "the outcome, 'A sum of 11 turns up,' " or even write

$$E = \text{A sum of 11 turns up} = \{(6, 5), (5, 6)\}$$

PROBABILITY OF AN EVENT

The next step in developing our mathematical model for probability studies is the introduction of a *probability function.* This is a function that assigns to an arbitrary event associated with a sample space a real number between 0 and 1, inclusive. We start by discussing ways in which probabilities are assigned to simple events in the sample space S.

Probabilities for Simple Events

Given a sample space

$$S = \{e_1, e_2, \ldots, e_n\}$$

with n simple events, to each simple event e_i we assign a real number, denoted by $\boldsymbol{P(e_i)}$, called the **probability of the event e_i.** These numbers can be assigned in an arbitrary manner as long as the following two conditions are satisfied:

(continued)

Probabilities for Simple Events (*continued*)

1. The probability of a simple event is a number between 0 and 1, inclusive. That is,

$$0 \leqslant P(e_i) \leqslant 1$$

2. The sum of the probabilities of all simple events in the sample space is 1. That is,

$$P(e_1) + P(e_2) + \cdots + P(e_r) = 1$$

Any probability assignment that meets conditions 1 and 2 is said to be an **acceptable probability assignment.**

Our mathematical theory does not explain how acceptable probabilities are assigned to simple events. These assignments are generally based on the expected or actual percentage of times a simple event occurs when an experiment is repeated a large number of times. Assignments based on this principle are called **reasonable.**

Let an experiment be the flipping of a single coin, and let us choose a sample space S to be

$$S = \{H, T\}$$

If a coin appears to be fair, we are inclined to assign probabilities to the simple events in S as follows:

$$P(H) = \tfrac{1}{2} \qquad \text{and} \qquad P(T) = \tfrac{1}{2}$$

These assignments are based on reasoning that, since there are 2 ways a coin can land, in the long run, a head will turn up half the time and a tail will turn up half the time. These probability assignments are acceptable, since both conditions for acceptable probability assignments stated in the box are satisfied:

1. $0 \leqslant P(H) \leqslant 1, \qquad 0 \leqslant P(T) \leqslant 1$
2. $P(H) + P(T) = \tfrac{1}{2} + \tfrac{1}{2} = 1$

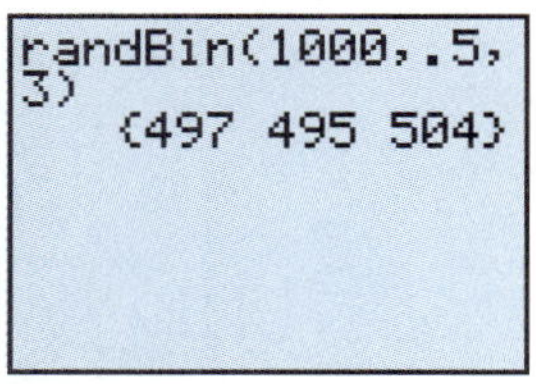

FIGURE 3

If we were to flip a coin 1,000 times, we would expect a head to turn up approximately, but not exactly, 500 times. The random number feature on a graphing utility can be used to simulate 1,000 flips of a coin. Figure 3 shows the results of 3 such simulations: 497 heads the first time, 495 heads the second, and 504 heads the third.

If, however, we get only 376 heads in 1,000 flips of a coin, we might suspect the coin is not fair. Then we might assign the simple events in the sample space S the following probabilities, based on our experimental results:

$$P(H) = .376 \qquad \text{and} \qquad P(T) = .624$$

This is also an acceptable assignment. However, the probability assignment

$$P(H) = 1 \qquad \text{and} \qquad P(T) = 0$$

though acceptable, is not reasonable (unless the coin has 2 heads). And the assignment

$$P(H) = .6 \qquad \text{and} \qquad P(T) = .8$$

is not acceptable, since $.6 + .8 = 1.4$, which violates condition 2 in the box. [*Note:* In probability studies, the 0 to the left of the decimal is usually omitted. Thus, we write .6 and .8 instead of 0.6 and 0.8.]

It is important to keep in mind that out of the infinitely many possible acceptable probability assignments to simple events in a sample space, we are generally inclined to choose one assignment over another based on reasoning or experimental results.

Given an acceptable probability assignment for simple events in a sample space S, how do we define the probability of an arbitrary event E associated with S?

Probability of an Event E

Given an acceptable probability assignment for the simple events in a sample space S, we define the **probability of an arbitrary event E**, denoted by $P(E)$, as follows:

(A) If E is the empty set, then $P(E) = 0$.
(B) If E is a simple event, then $P(E)$ has already been assigned.
(C) If E is a compound event, then $P(E)$ is the sum of the probabilities of all the simple events in E.
(D) If E is the sample space S, then $P(E) = P(S) = 1$ [this is a special case of part (C)].

Example 4 **Probabilities of Events** Let us return to Example 2, the tossing of a nickel and a dime, and the sample space

$$S = \{\text{HH, HT, TH, TT}\}$$

Since there are 4 simple outcomes and the coins are assumed to be fair, it would appear that each outcome would occur 25% of the time, in the long run. Let us assign the same probability of $\frac{1}{4}$ to each simple event in S:

SIMPLE EVENT e_i	HH	HT	TH	TT
$P(e_i)$	$\frac{1}{4}$	$\frac{1}{4}$	$\frac{1}{4}$	$\frac{1}{4}$

This is an acceptable assignment according to conditions 1 and 2, and it is a reasonable assignment for ideal (perfectly balanced) coins or coins close to ideal.

(A) What is the probability of getting 1 head (and 1 tail)?
(B) What is the probability of getting at least 1 head?
(C) What is the probability of getting at least 1 head or at least 1 tail?
(D) What is the probability of getting 3 heads?

Solution (A) E_1 = Getting 1 head = {HT, TH}

Since E_1 is a compound event, we use part (C) in the box and find $P(E_1)$ by adding the probabilities of the simple events in E_1:

$$P(E_1) = P(\text{HT}) + P(\text{TH}) = \tfrac{1}{4} + \tfrac{1}{4} = \tfrac{1}{2}$$

(B) E_2 = Getting at least 1 head = {HH, HT, TH}

$$P(E_2) = P(\text{HH}) + P(\text{HT}) + P(\text{TH}) = \tfrac{1}{4} + \tfrac{1}{4} + \tfrac{1}{4} = \tfrac{3}{4}$$

(C) $E_3 = \{\text{HH, HT, TH, TT}\} = S$

$$P(E_3) = P(S) = 1 \quad \tfrac{1}{4} + \tfrac{1}{4} + \tfrac{1}{4} + \tfrac{1}{4} = 1$$

(D) E_4 = Getting 3 heads = $\varnothing$ Empty set

$$P(\varnothing) = 0$$

Steps for Finding the Probability of an Event *E*

Step 1. Set up an appropriate sample space *S* for the experiment.
Step 2. Assign acceptable probabilities to the simple events in *S*.
Step 3. To obtain the probability of an arbitrary event *E*, add the probabilities of the simple events in *E*.

The function *P* defined in steps 2 and 3 is a **probability function** whose domain is all possible events (subsets) in the sample space *S* and whose range is a set of real numbers between 0 and 1, inclusive.

Matched Problem 4 Suppose in Example 4 that, after flipping the nickel and dime 1,000 times, we find that HH turns up 273 times, HT turns up 206 times, TH turns up 312 times, and TT turns up 209 times. On the basis of this evidence, we assign probabilities to the simple events in *S* as follows:

Simple Event e_i	HH	HT	TH	TT
$P(e_i)$	.273	.206	.312	.209

This is an acceptable and reasonable probability assignment for the simple events in *S*. What are the probabilities of the following events?

(A) E_1 = Getting at least 1 tail
(B) E_2 = Getting 2 tails
(C) E_3 = Getting at least 1 head or at least 1 tail

Example 4 and Matched Problem 4 illustrate two important ways in which acceptable and reasonable probability assignments are made for simple events in a sample space S. Each approach has its advantage in certain situations:

1. *Theoretical.* We use assumptions and a deductive reasoning process to assign probabilities to simple events. No experiments are actually conducted. This is what we did in Example 4.
2. *Empirical.* We assign probabilities to simple events based on the results of actual experiments. This is what we did in Matched Problem 4.

Empirical probability concepts are stated more precisely as follows: If we conduct an experiment n times and event E occurs with **frequency $f(E)$,** then the ratio $f(E)/n$ is called the **relative frequency** of the occurrence of event E in n trials. We define the **empirical probability** of E, denoted by **$P(E)$,** by the number (if it exists) that the relative frequency $f(E)/n$ approaches as n gets larger and larger. For any particular n, the relative frequency $f(E)/n$ is also called the **approximate empirical probability** of event E.

Empirical Probability Approximation

$$P(E) \approx \frac{\text{Frequency of occurrence of } E}{\text{Total number of trials}} = \frac{f(E)}{n}$$

(The larger n is, the better the approximation.)

For most of this section we will emphasize the theoretical approach. In the next section we will return to the empirical approach.

Equally Likely Assumption

In tossing a nickel and a dime (Example 4), we assigned the same probability, $\frac{1}{4}$, to each simple event in the sample space $S = \{\text{HH}, \text{HT}, \text{TH}, \text{TT}\}$. By assigning the same probability to each simple event in S, we are actually making the assumption that each simple event is as likely to occur as any other. We refer to this as an **equally likely assumption.** In general, we have the following:

Probability of a Simple Event Under an Equally Likely Assumption

If, in a sample space

$$S = \{e_1, e_2, \ldots, e_n\}$$

with n elements, we assume each simple event e_i is as likely to occur as any other, then we assign the probability $1/n$ to each. That is,

$$P(e_i) = \frac{1}{n}$$

Under an equally likely assumption, we can develop a very useful formula for finding probabilities of arbitrary events associated with a sample space S. Consider the following example.

If a single die is rolled and we assume each face is as likely to come up as any other, then for the sample space

$$S = \{1, 2, 3, 4, 5, 6\}$$

we assign a probability of $\frac{1}{6}$ to each simple event, since there are 6 simple events. The probability of

$$E = \text{Rolling a prime number} = \{2, 3, 5\}$$

is

Number of elements in E ↓

$$P(E) = P(2) + P(3) + P(5) = \tfrac{1}{6} + \tfrac{1}{6} + \tfrac{1}{6} = \tfrac{3}{6} = \tfrac{1}{2}$$

↑ Number of elements in S

Thus, under the assumption that each simple event is as likely to occur as any other, the computation of the probability of the occurrence of any event E in a sample space S is the number of elements in E divided by the number of elements in S.

Theorem 1 ■■ PROBABILITY OF AN ARBITRARY EVENT UNDER AN EQUALLY LIKELY ASSUMPTION

If we assume each simple event in sample space S is as likely to occur as any other, then the probability of an arbitrary event E in S is given by

$$P(E) = \frac{\text{Number of elements in } E}{\text{Number of elements in } S} = \frac{n(E)}{n(S)}$$ ■■

Example 5 **Probabilities and Equally Likely Assumptions** Let us again consider rolling two dice, and assume each simple event in the sample space shown in Figure 2 (page 393) is as likely as any other. Find the probabilities of the following events:

(A) E_1 = A sum of 7 turns up (B) E_2 = A sum of 11 turns up
(C) E_3 = A sum less than 4 turns up (D) E_4 = A sum of 12 turns up

SOLUTION Referring to Figure 2 (page 393) and the results found in Example 3, we find:

(A) $P(E_1) = \frac{n(E_1)}{n(S)} = \frac{6}{36} = \frac{1}{6}$ $E_1 = \{(6, 1), (5, 2), (4, 3), (3, 4), (2, 5), (1, 6)\}$

(B) $P(E_2) = \frac{n(E_2)}{n(S)} = \frac{2}{36} = \frac{1}{18}$ $E_2 = \{(6, 5), (5, 6)\}$

(C) $P(E_3) = \frac{n(E_3)}{n(S)} = \frac{3}{36} = \frac{1}{12}$ $E_3 = \{(1, 1), (2, 1), (1, 2)\}$

(D) $P(E_4) = \frac{n(E_4)}{n(S)} = \frac{1}{36}$ $E_4 = \{(6, 6)\}$ ■

Matched Problem 5 Under the conditions in Example 5, find the probabilities of the following events (each event refers to the sum of the dots facing up on both dice):

(A) E_5 = A sum of 5 turns up
(B) E_6 = A sum that is a prime number greater than 7 turns up

Example 6 **Simulation and Empirical Probabilities** Use output from the random number feature of a graphing utility to simulate 100 rolls of two dice. Determine the empirical probabilities of the following events, and compare with the theoretical probabilities:

(A) E_1 = A sum of 7 turns up (B) E_2 = A sum of 11 turns up

SOLUTION A graphing utility can be used to select a random integer from 1 to 6. Each of the six integers in the given range is equally likely to be selected. Therefore, by selecting a random integer from 1 to 6 and adding it to a second random integer from 1 to 6, we simulate rolling two dice and recording the sum (see the first command in Figure 4A and your user's manual). The second command in Figure 4A simulates 100 rolls of two dice; the sums are stored in list L_1. From the statistical plot of L_1 in Figure 4B we obtain the empirical probabilities:*

(A) The empirical probability of E_1 is $\frac{15}{100} = .15$; the theoretical probability of E_1 (see Example 5A) is $\frac{6}{36} = .167$.
(B) The empirical probability of E_2 is $\frac{6}{100} = .06$; the theoretical probability of E_2 (see Example 5B) is $\frac{2}{36} = .056$.

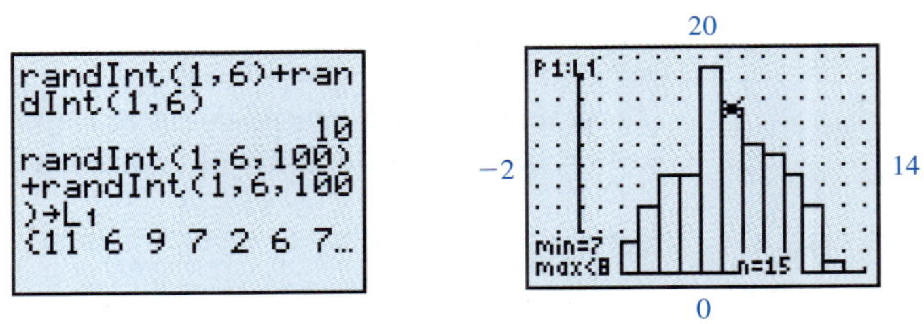

(A) (B)

FIGURE 4

Matched Problem 6 Use the graphing utility output in Figure 4B to determine the empirical probabilities of the following events, and compare with the theoretical probabilities:

(A) E_3 = A sum less than 4 turns up (B) E_4 = A sum of 12 turns up

Explore–Discuss 1

A shipment box contains 12 graphing calculators, out of which 2 are defective. A calculator is drawn at random from the box and then, without replacement, a second calculator is drawn. Discuss whether the equally likely assumption would be appropriate for the sample space $S = \{GG, GD, DG, DD\}$, where G is a good calculator and D is a defective one.

*If you simulate this experiment on your graphing utility, you should not expect to get the same empirical probabilities.

We now turn to some examples that make use of the counting techniques developed in the preceding sections.

Example 7 **Probability and Equally Likely Assumption** In drawing 5 cards from a 52-card deck without replacement, what is the probability of getting 5 spades?

SOLUTION Let the sample space S be the set of all 5-card hands from a 52-card deck. Since the order in a hand does not matter, $n(S) = C_{52,5}$. Let event E be the set of all 5-card hands from 13 spades. Again, the order does not matter and $n(E) = C_{13,5}$. Thus, assuming each 5-card hand is as likely as any other,

$$P(E) = \frac{n(E)}{n(S)} = \frac{C_{13,5}}{C_{52,5}} = \frac{1{,}287}{2{,}598{,}960} \approx .0005$$

Matched Problem 7 In drawing 7 cards from a 52-card deck without replacement, what is the probability of getting 7 hearts?

Example 8 **Probability and Equally Likely Assumption** The board of regents of a university is made up of 12 men and 16 women. If a committee of 6 is chosen at random, what is the probability that it will contain 3 men and 3 women?

SOLUTION Let S be the set of all 6-person committees out of 28 people. Then

$$n(S) = C_{28,6}$$

Let E be the set of all 6-person committees with 3 men and 3 women. To find $n(E)$, we use the multiplication principle and the following two operations:

O_1: Select 3 men out of the 12 available $\quad N_1$: $C_{12,3}$
O_2: Select 3 women out of the 16 available $\quad N_2$: $C_{16,3}$

Thus,

$$n(E) = N_1 \cdot N_2 = C_{12,3} \cdot C_{16,3}$$

and

$$P(E) = \frac{n(E)}{n(S)} = \frac{C_{12,3} \cdot C_{16,3}}{C_{28,6}} \approx .327$$

Matched Problem 8 What is the probability that the committee in Example 8 will have 4 men and 2 women?

It needs to be pointed out that there are many counting problems for which it is not possible to produce a simple formula that will yield the number of possible cases. In situations of this type, we often revert back to tree diagrams and counting branches.

Explore–Discuss 2

Five football teams play each other in the Central Division. At the beginning of the season we are interested in discussing what will be the relative standing of the five teams at the end of the season, excluding ties. Describe an appropriate sample space for this discussion. How many simple events does it contain? If each possible final standing of the five teams is as likely as any other, what would be the probability of the occurrence of a particular final standing? Discuss the reasons that a sports writer would or would not use the equally likely assumption in predicting the final standing of the five teams.

Answers to Matched Problems

1. (A) $E = \{12\}$; simple event (B) $E = \{16, 18\}$; compound event

2. (A) $S_1 = \{BB, BG, GB, GG\}$;

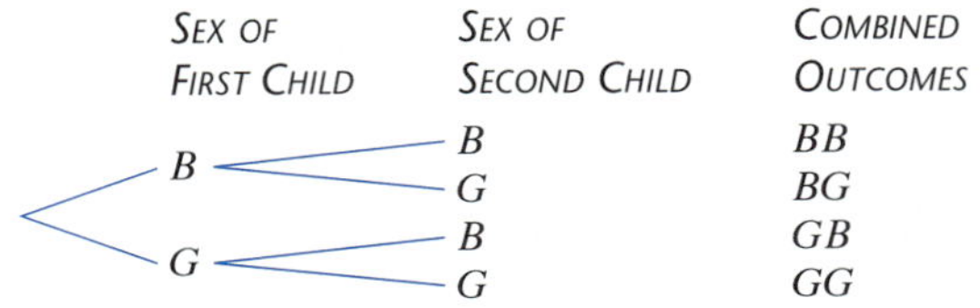

(B) $S_2 = \{0, 1, 2\}$ (C) $S_3 = \{A, D\}$ (D) S_1

3. (A) $\{(4, 1), (3, 2), (2, 3), (1, 4)\}$ (B) $\{(6, 5), (5, 6)\}$

4. (A) .727 (B) .209 (C) 1 **5.** (A) $P(E_5) = \frac{1}{9}$ (B) $P(E_6) = \frac{1}{18}$

6. (A) $\frac{9}{100} = .09$ (empirical); $\frac{3}{36} = .083$ (theoretical)

(B) $\frac{1}{100} = .01$ (empirical); $\frac{1}{36} = .028$ (theoretical)

7. $C_{13,7}/C_{52,7} \approx 1.3 \times 10^{-5}$ **8.** $C_{12,4}C_{16,2}/C_{28,6} \approx .158$

Exercise 6-3

A

1. How would you interpret $P(E) = 1$?
2. How would you interpret $P(E) = 0$?
3. In a family with 2 children, excluding multiple births, what is the probability of having 2 children of the opposite sex? Assume a girl is as likely as a boy at each birth.
4. In a family with 2 children, excluding multiple births, what is the probability of having 2 girls? Assume a girl is as likely as a boy at each birth.
5. A store carries four brands of CD players, *J*, *G*, *P*, and *S*. From past records, the manager found that the relative frequency of brand choice among customers varied. Which of the following probability assignments for a particular customer choosing a particular brand of CD player would have to be rejected? Why?

   (A) $P(J) = .15$, $P(G) = -.35$, $P(P) = .50$, $P(S) = .70$

   (B) $P(J) = .32$, $P(G) = .28$, $P(P) = .24$, $P(S) = .30$

   (C) $P(J) = .26$, $P(G) = .14$, $P(P) = .30$, $P(S) = .30$

6. Using the probability assignments in Problem 5C, what is the probability that a random customer will not choose brand *S*?
7. Using the probability assignments in Problem 5C, what is the probability that a random customer will choose brand *J* or brand *P*?
8. Using the probability assignments in Problem 5C, what is the probability that a random customer will not choose brand *J* or brand *P*?

B

9. In a family with 3 children, excluding multiple births, what is the probability of having 2 boys and 1 girl, in that order? Assume a boy is as likely as a girl at each birth.
10. In a family with 3 children, excluding multiple births, what is the probability of having 2 boys and 1 girl, in any order? Assume a boy is as likely as a girl at each birth.

11. A small combination lock on a suitcase has 3 wheels, each labeled with the 10 digits from 0 to 9. If an opening combination is a particular sequence of 3 digits with no repeats, what is the probability of a person guessing the right combination?

12. A combination lock has 5 wheels, each labeled with the 10 digits from 0 to 9. If an opening combination is a particular sequence of 5 digits with no repeats, what is the probability of a person guessing the right combination?

Refer to the description of a standard deck of 52 cards and Figure 4 on page 384. An experiment consists of dealing 5 cards from a standard 52-card deck. In Problems 13–16, what is the probability of being dealt:

13. 5 black cards? **14.** 5 hearts?

15. 5 face cards? **16.** 5 nonface cards?

17. Twenty thousand students are enrolled at a state university. A student is selected at random and his or her birthday (month and day, not year) is recorded. Describe an appropriate sample space for this experiment, and assign acceptable probabilities to the simple events. What are your assumptions in making this assignment?

18. In a hotly contested three-way race for the U.S. Senate, polls indicate the two leading candidates are running neck-and-neck while the third candidate is receiving half the support of either of the others. Registered voters are chosen at random and are asked which of the three will get their vote. Describe an appropriate sample space for this random survey experiment and assign acceptable probabilities to the simple events.

19. Suppose 5 thank-you notes are written and 5 envelopes are addressed. Accidentally, the notes are randomly inserted into the envelopes and mailed without checking the addresses. What is the probability that all the notes will be inserted into the correct envelopes?

20. Suppose 6 people check their coats in a checkroom. If all claim checks are lost and the 6 coats are randomly returned, what is the probability that all the people will get their own coats back?

An experiment consists of rolling two fair dice and adding the dots on the two sides facing up. Using the sample space shown in Figure 2 (page 393) and assuming each simple event is as likely as any other, find the probability of the sum of the dots indicated in Problems 21–36.

21. Sum is 2. **22.** Sum is 10.

23. Sum is 6. **24.** Sum is 8.

25. Sum is less than 5. **26.** Sum is greater than 8.

27. Sum is not 7 or 11. **28.** Sum is not 2, 4, or 6.

29. Sum is 1. **30.** Sum is 13.

31. Sum is divisible by 3. **32.** Sum is divisible by 4.

33. Sum is 7 or 11 (a "natural").

34. Sum is 2, 3, or 12 ("craps").

35. Sum is divisible by 2 or 3.

36. Sum is divisible by 2 and 3.

An experiment consists of tossing three fair (not weighted) coins, except one of the three coins has a head on both sides. Compute the probability of obtaining the indicated results in Problems 37–42.

37. 1 head **38.** 2 heads

39. 3 heads **40.** 0 heads

41. More than 1 head **42.** More than 1 tail

43. (A) Is it possible to get 19 heads in 20 flips of a fair coin? Explain.
(B) If you flipped a coin 40 times and got 37 heads, would you suspect that the coin was unfair? Why or why not? If you suspect an unfair coin, what empirical probabilities would you assign to the simple events of the sample space?

44. (A) Is it possible to get 7 double 6's in 10 rolls of a pair of fair dice? Explain.
(B) If you rolled a pair of dice 36 times and got 11 double 6's, would you suspect that the dice were unfair? Why or why not? If you suspect loaded dice, what empirical probability would you assign to the event of rolling a double 6?

C *An experiment consists of rolling two fair (not weighted) dice and adding the dots on the two sides facing up. Each die has the number 1 on two opposite faces, the number 2 on two opposite faces, and the number 3 on two opposite faces. Compute the probability of obtaining the indicated sums in Problems 45–52.*

45. 2 **46.** 3 **47.** 4

48. 5 **49.** 6 **50.** 7

51. An odd sum **52.** An even sum

Refer to the description of a standard deck of 52 cards on page 384. An experiment consists of dealing 5 cards from a standard 52-card deck. In Problems 53–60, what is the probability of being dealt:

53. 5 face cards or aces?

54. 5 numbered cards (2 through 10)?

55. 4 aces?

56. Four of a kind (4 queens, 4 kings, and so on)?

57. A 10, jack, queen, king, and ace, all in the same suit?

58. A 2, 3, 4, 5, and 6, all in the same suit?

59. 2 aces and 3 queens?

60. 2 kings and 3 aces?

In Problems 61–64, several experiments are simulated using the random number feature on a graphing utility. For example, the roll of a fair die can be simulated by selecting a random integer from 1 to 6, and 50 rolls of a fair die by selecting 50 random integers from 1 to 6 (see Fig. A for Problem 61 and your user's manual).

61. From the statistical plot of the outcomes of rolling a fair die 50 times (see Fig. B), we see, for example, that the number 4 was rolled exactly 11 times.

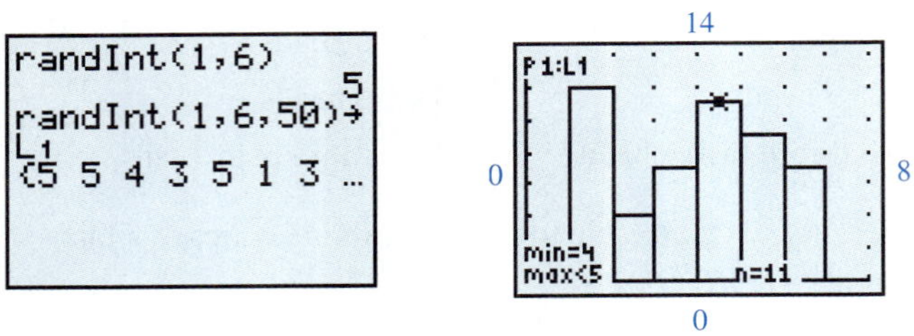

(A) (B)

Figure for 61

(A) What is the empirical probability that the number 6 was rolled?

(B) What is the probability that a 6 is rolled under the equally likely assumption?

(C) Use a graphing utility to simulate 100 rolls of a fair die and determine the empirical probabilities of the six outcomes.

62. Use a graphing utility to simulate 200 tosses of a nickel and dime, representing the outcomes HH, HT, TH, and TT by 1, 2, 3, and 4, respectively.

(A) Find the empirical probabilities of the four outcomes.

(B) What is the probability of each outcome under the equally likely assumption?

63. (A) Explain how a graphing utility can be used to simulate 500 tosses of a coin.

(B) Carry out the simulation and find the empirical probabilities of the two outcomes.

(C) What is the probability of each outcome under the equally likely assumption?

64. From a box containing a dozen balls numbered 1 through 12, one ball is drawn at random.

(A) Explain how a graphing utility can be used to simulate 400 repetitions of this experiment.

(B) Carry out the simulation and find the empirical probability of drawing the 8 ball.

(C) What is the probability of drawing the 8 ball under the equally likely assumption?

Applications

Business & Economics

65. *Consumer testing.* Twelve popular brands of beer are to be used in a blind taste study for consumer recognition.

(A) If 4 distinct brands are chosen at random from the 12 and if a consumer is not allowed to repeat any answers, what is the probability that all 4 brands could be identified by just guessing?

(B) If repeats are allowed in the 4 brands chosen at random from the 12 and if the consumer is allowed to repeat answers, what is the probability of correct identification of all 4 by just guessing?

66. *Consumer testing.* Six popular brands of cola are to be used in a blind taste study for consumer recognition.

(A) If 3 distinct brands are chosen at random from the 6 and if a consumer is not allowed to repeat any answers, what is the probability that all 3 brands could be identified by just guessing?

(B) If repeats are allowed in the 3 brands chosen at random from the 6 and if the consumer is allowed to repeat answers, what is the probability of correct identification of all 3 by just guessing?

67. *Personnel selection.* Suppose 6 female and 5 male applicants have been successfully screened for 5 positions. If the 5 positions are filled at random from the 11 finalists, what is the probability of selecting:

(A) 3 females and 2 males?

(B) 4 females and 1 male?

(C) 5 females?

(D) At least 4 females?

68. *Committee selection.* A 4-person grievance committee is to be composed of employees in 2 departments, A and B, with 15 and 20 employees, respectively. If the 4 people are selected at random from the 35 employees, what is the probability of selecting:

(A) 3 from A and 1 from B?

(B) 2 from A and 2 from B?

(C) All from A?

(D) At least 3 from A?

Life Sciences

69. *Medicine.* A prospective laboratory technician is to be tested on identifying blood types from 8 standard classifications.

(A) If 3 distinct samples are chosen at random from the 8 types and if the examinee is not allowed to repeat any answers, what is the probability that all 3 could be correctly identified by just guessing?

(B) If repeats are allowed in the 3 blood types chosen at random from the 8 and if the examinee is allowed to repeat answers, what is the probability of correct identification of all 3 by just guessing?

70. *Medical research.* Because of limited funds, 5 research centers are to be chosen out of 8 suitable ones for a study on heart disease. If the selection is made at random, what is the probability that 5 particular research centers will be chosen?

Social Sciences

71. *Membership selection.* A town council has 11 members, 6 Democrats and 5 Republicans.

(A) If the president and vice-president are selected at random, what is the probability that they are both Democrats?

(B) If a 3-person committee is selected at random, what is the probability that Republicans make up the majority?

SECTION 6-4

Union, Intersection, and Complement of Events; Odds

- UNION AND INTERSECTION
- COMPLEMENT OF AN EVENT
- ODDS
- APPLICATIONS TO EMPIRICAL PROBABILITY

Recall that in Section 6-3 we said that given a sample space

$$S = \{e_1, e_2, \ldots, e_n\}$$

any function P defined on S such that

$$0 \le P(e_i) \le 1 \qquad i = 1, 2, \ldots, n$$

and

$$P(e_1) + P(e_2) + \cdots + P(e_n) = 1$$

is called a *probability function.* In addition, we said that any subset of S is called an *event* E, and we defined the probability of E to be the sum of the probabilities of the simple events in E.

UNION AND INTERSECTION

Since events are subsets of a sample space, the union and intersection of events are simply the union and intersection of sets as defined in Appendix A-1 and in the following box. In this section we will concentrate on the union of events and consider only simple cases of intersection. The latter will be investigated in more detail in the next section.

Union and Intersection of Events*

If A and B are two events in a sample space S, then the **union** of A and B, denoted by $\boldsymbol{A \cup B}$, and the **intersection** of A and B, denoted by $\boldsymbol{A \cap B}$, are defined as follows:

$A \cup B = \{e \in S | e \in A$ **or** $e \in B\}$ $\quad A \cap B = \{e \in S | e \in A$ **and** $e \in B\}$

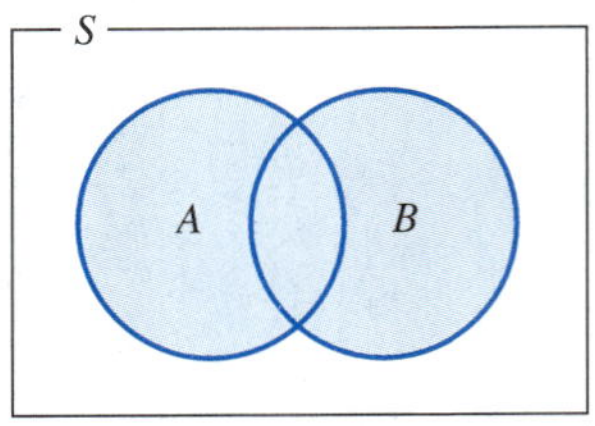

$A \cup B$

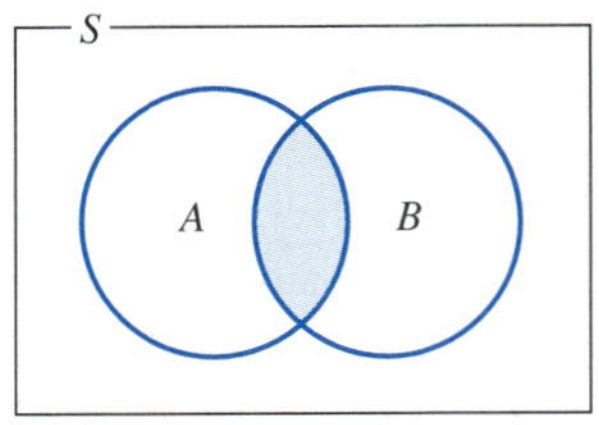

$A \cap B$

Furthermore, we define:

The **event *A* or *B*** to be $A \cup B$
The **event *A* and *B*** to be $A \cap B$

*See Appendix A-1 for a discussion of set notation.

Example 1 **Probability Involving Union and Intersection** Consider the sample space of equally likely events for the rolling of a single fair die:

$$S = \{1, 2, 3, 4, 5, 6\}$$

(A) What is the probability of rolling a number that is odd **and** exactly divisible by 3?

(B) What is the probability of rolling a number that is odd **or** exactly divisible by 3?

SOLUTION

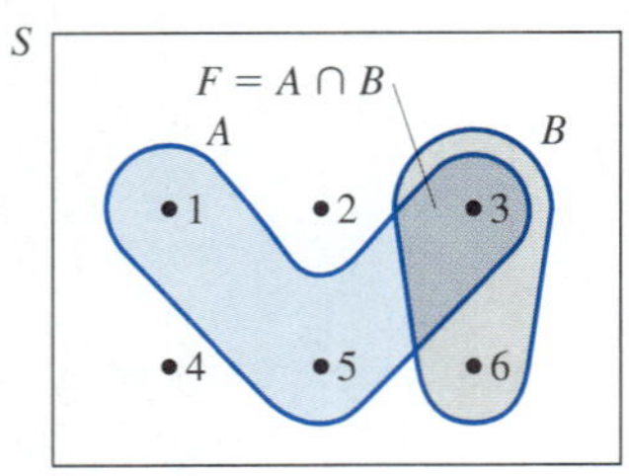

(A) Let A be the event of rolling an odd number, B the event of rolling a number divisible by 3, and F the event of rolling a number that is odd **and** divisible by 3. Then

$$A = \{1, 3, 5\} \qquad B = \{3, 6\} \qquad F = A \cap B = \{3\}$$

Thus, the probability of rolling a number that is odd **and** exactly divisible by 3 is

$$P(F) = P(A \cap B) = \frac{n(A \cap B)}{n(S)} = \frac{1}{6}$$

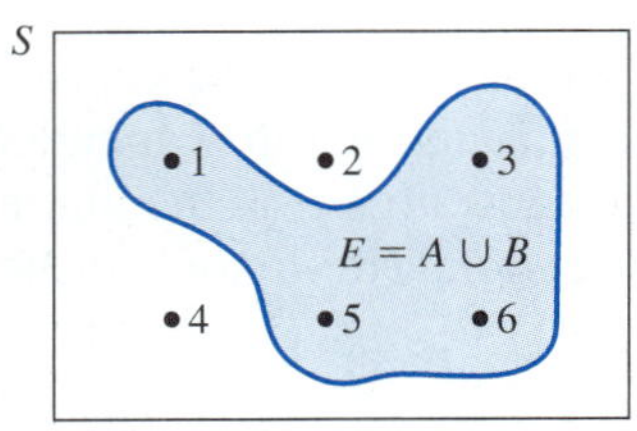

(B) Let A and B be the same events as in part (A), and let E be the event of rolling a number that is odd **or** divisible by 3. Then

$$A = \{1, 3, 5\} \qquad B = \{3, 6\} \qquad E = A \cup B = \{1, 3, 5, 6\}$$

Thus, the probability of rolling a number that is odd **or** exactly divisible by 3 is

$$P(E) = P(A \cup B) = \frac{n(A \cup B)}{n(S)} = \frac{4}{6} = \frac{2}{3}$$

Matched Problem 1 ➡ Use the sample space in Example 1 to answer the following:

(A) What is the probability of rolling an odd number **and** a prime number?
(B) What is the probability of rolling an odd number **or** a prime number?

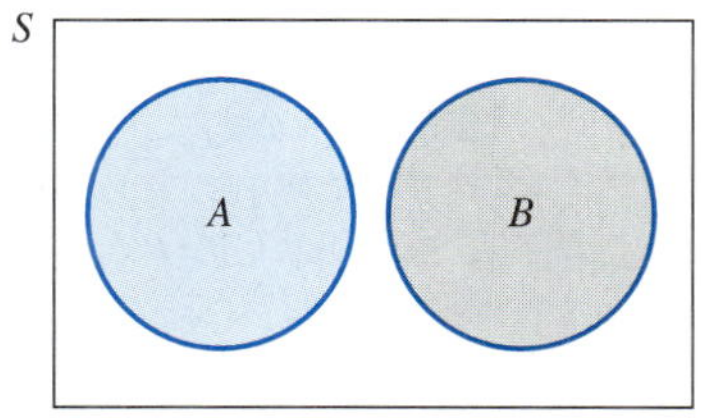

FIGURE 1
Mutually exclusive: $A \cap B = \varnothing$

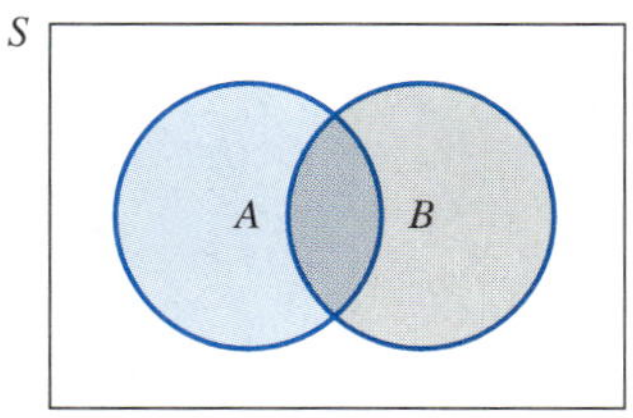

FIGURE 2
Not mutually exclusive: $A \cap B \neq \varnothing$

Suppose

$$E = A \cup B$$

Can we find $P(E)$ in terms of A and B? The answer is almost yes, but we must be careful. It would be nice if

$$P(A \cup B) = P(A) + P(B) \tag{1}$$

This turns out to be true if events A and B are **mutually exclusive (disjoint);** that is, if $A \cap B = \varnothing$ (Fig. 1). In this case, $P(A \cup B)$ is the sum of all the probabilities of simple events in A added to the sum of all the probabilities of simple events in B. But what happens if events A and B are not mutually exclusive; that is, if $A \cap B \neq \varnothing$ (see Fig. 2)? If we simply add the probabilities of the elements in A to the probabilities of the elements in B, we are including some of the probabilities twice, namely those for elements that are in both A and B. To compensate for this double counting, we subtract $P(A \cap B)$ from equation (1) to obtain

$$P(A \cup B) = P(A) + P(B) - P(A \cap B) \tag{2}$$

We notice that (2) holds for both cases, $A \cap B \neq \varnothing$ and $A \cap B = \varnothing$, since (2) reduces to (1) for the latter case $[P(A \cap B) = P(\varnothing) = 0]$. It is better to use (2) if there is any doubt that A and B are mutually exclusive. We summarize this discussion in the box for convenient reference.

Probability of a Union of Two Events

For any events A and B,

$$P(A \cup B) = P(A) + P(B) - P(A \cap B) \tag{2}$$

If A and B are mutually exclusive, then

$$P(A \cup B) = P(A) + P(B) \tag{1}$$

Example 2 ➡ **Probability Involving Union and Intersection** Suppose two fair dice are rolled:

(A) What is the probability that a sum of 7 or 11 turns up?
(B) What is the probability that both dice turn up the same or that a sum less than 5 turns up?

SOLUTION (A) If A is the event that a sum of 7 turns up and B is the event that a sum of 11 turns up, then (see Fig. 3) the event that a sum of 7 or 11 turns up is $A \cup B$, where

$$A = \{(1, 6), (2, 5), (3, 4) (4, 3), (5, 2), (6, 1)\}$$
$$B = \{(5, 6), (6, 5)\}$$

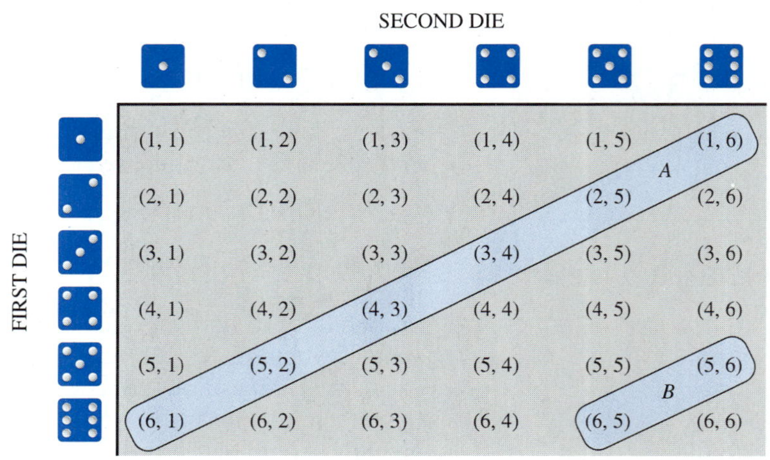

FIGURE 3

Since events A and B are mutually exclusive, we can use equation (1) to calculate $P(A \cup B)$:

$$\begin{aligned} P(A \cup B) &= P(A) + P(B) \\ &= \tfrac{6}{36} + \tfrac{2}{36} \\ &= \tfrac{8}{36} = \tfrac{2}{9} \end{aligned}$$

In this equally likely sample space, $n(A) = 6$, $n(B) = 2$, and $n(S) = 36$.

(B) If A is the event that both dice turn up the same and B is the event that the sum is less than 5, then (see Fig. 4) the event that both dice turn up the same or the sum is less than 5 is $A \cup B$, where

$$A = \{(1, 1), (2, 2), (3, 3), (4, 4), (5, 5), (6, 6)\}$$
$$B = \{(1, 1), (1, 2), (1, 3), (2, 1), (2, 2), (3, 1)\}$$

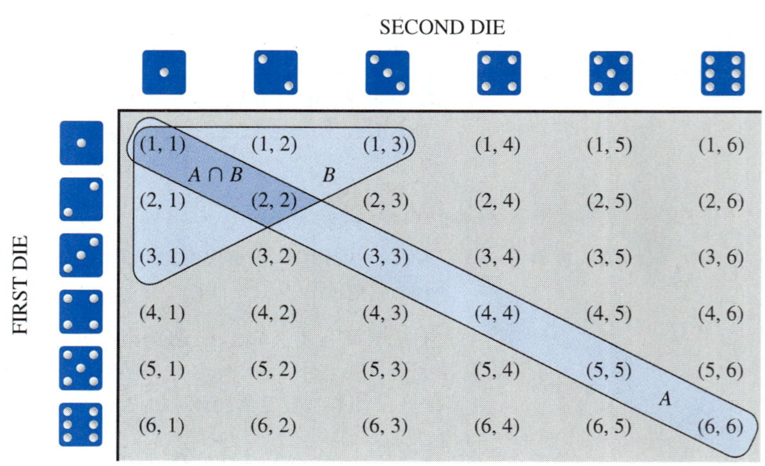

FIGURE 4

Since $A \cap B = \{(1, 1), (2, 2)\}$, A and B are not mutually exclusive and we use equation (2) to calculate $P(A \cup B)$:

$$\begin{aligned} P(A \cup B) &= P(A) + P(B) - P(A \cap B) \\ &= \tfrac{6}{36} + \tfrac{6}{36} - \tfrac{2}{36} \\ &= \tfrac{10}{36} = \tfrac{5}{18} \end{aligned}$$

Matched Problem 2 Use the sample space in Example 2 to answer the following:

(A) What is the probability that a sum of 2 or 3 turns up?
(B) What is the probability that both dice turn up the same or that a sum greater than 8 turns up?

You no doubt noticed in Example 2 that we actually did not have to use either formula (1) or (2). We could have proceeded as in Example 1 and simply counted sample points in $A \cup B$. The following example illustrates the use of formula (2) in a situation where visual representation of sample points is not practical.

Example 3 **Probability Involving Union and Intersection** What is the probability that a number selected at random from the first 500 positive integers is (exactly) divisible by 3 or 4?

SOLUTION Let A be the event that a drawn integer is divisible by 3 and B the event that a drawn integer is divisible by 4. Note that events A and B are not mutually exclusive, because multiples of 12 are divisible by both 3 and 4. Since each of the positive integers from 1 to 500 is as likely to be drawn as any other, we can use $n(A)$, $n(B)$, and $n(A \cap B)$ to determine $P(A \cup B)$, where (think about this)

$$\begin{aligned} n(A) &= \text{The largest integer less than or equal to } \tfrac{500}{3} = 166 \\ n(B) &= \text{The largest integer less than or equal to } \tfrac{500}{4} = 125 \\ n(A \cap B) &= \text{The largest integer less than or equal to } \tfrac{500}{12} = 41 \end{aligned}$$

Now we can compute $P(A \cup B)$:

$$\begin{aligned} P(A \cup B) &= P(A) + P(B) - P(A \cap B) \\ &= \frac{n(A)}{n(S)} + \frac{n(B)}{n(S)} - \frac{n(A \cap B)}{n(S)} \\ &= \frac{166}{500} + \frac{125}{500} - \frac{41}{500} = \frac{250}{500} = .5 \end{aligned}$$

Matched Problem 3 What is the probability that a number selected at random from the first 140 positive integers is (exactly) divisible by 4 or 6?

Complement of an Event

Suppose we divide a finite sample space

$$S = \{e_1, \ldots, e_n\}$$

into two subsets E and E' such that

$$E \cap E' = \varnothing$$

that is, E and E' are mutually exclusive, and

$$E \cup E' = S$$

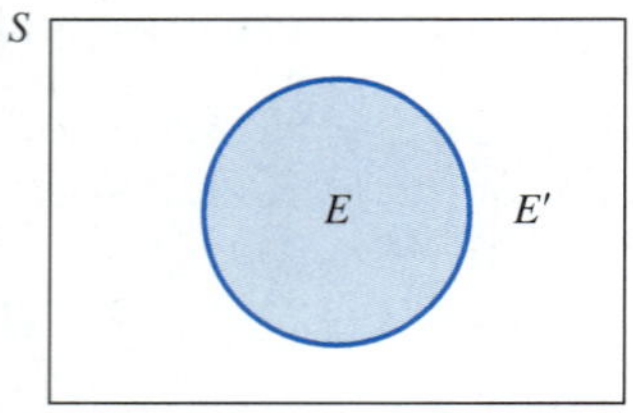

FIGURE 5

Then E' is called the **complement of E** relative to S. Thus, E' contains all the elements of S that are not in E (Fig. 5). Furthermore,

$$\begin{aligned} P(S) &= P(E \cup E') \\ &= P(E) + P(E') = 1 \end{aligned}$$

Hence, we have:

Complements

$$P(E) = 1 - P(E') \qquad P(E') = 1 - P(E) \tag{3}$$

If the probability of rain is .67, then the probability of no rain is $1 - .67 = .33$; if the probability of striking oil is .01, then the probability of not striking oil is .99. If the probability of having at least 1 boy in a 2-child family is .75, what is the probability of having 2 girls? [*Answer:* .25.]

Explore–Discuss 1

(A) Suppose that E and F are complementary events. Are E and F necessarily mutually exclusive? Explain why or why not.

(B) Suppose that E and F are mutually exclusive events. Are E and F necessarily complementary? Explain why or why not.

In looking for $P(E)$, there are situations in which it is easier to find $P(E')$ first, and then use equations (3) to find $P(E)$. The next two examples illustrate two such situations.

Example 4 **Quality Control** A shipment of 45 precision parts, including 9 that are defective, is sent to an assembly plant. The quality control division selects 10 at random for testing and rejects the whole shipment if 1 or more in the sample are found defective. What is the probability that the shipment will be rejected?

SOLUTION If E is the event that 1 or more parts in a random sample of 10 are defective, then E', the complement of E, is the event that no parts in a random sample of 10 are defective. It is easier to compute $P(E')$ than to compute $P(E)$ directly (try the latter to see why). Once $P(E')$ is found, we will use $P(E) = 1 - P(E')$ to find $P(E)$.

The sample space S for this experiment is the set of all subsets of 10 elements from the set of 45 parts shipped. Thus, since there are $45 - 9 = 36$ nondefective parts,

$$P(E') = \frac{n(E')}{n(S)} = \frac{C_{36,10}}{C_{45,10}} \approx .08$$

and

$$P(E) = 1 - P(E') \approx 1 - .08 = .92$$

Matched Problem 4 A shipment of 40 precision parts, including 8 that are defective, is sent to an assembly plant. The quality control division selects 10 at random for testing and rejects the whole shipment if 1 or more in the sample are found defective. What is the probability that the shipment will be rejected?

Example 5 **Birthday Problem** In a group of n people, what is the probability that at least 2 people have the same birthday (the same month and day, excluding February 29)? (Make a guess for a class of 40 people, and check your guess with the conclusion of this example.)

SOLUTION If we form a list of the birthdays of all the people in the group, then we have a simple event in the sample space

$$S = \text{Set of all lists of } n \text{ birthdays}$$

For any person in the group, we will assume that any birthday is as likely as any other, so that the simple events in S are equally likely. How many simple events are in the set S? Since any person could have any one of 365 birthdays (excluding February 29), the multiplication principle implies that the number of simple events in S is

$$\begin{aligned} n(S) &= \underset{\text{1st person}}{365} \cdot \underset{\text{2nd person}}{365} \cdot \underset{\text{3rd person}}{365} \cdot \cdots \cdot \underset{n\text{th person}}{365} \\ &= 365^n \end{aligned}$$

Now, let E be the event that at least 2 people in the group have the same birthday. Then E' is the event that no 2 people have the same birthday. The multiplication principle also can be used to determine the number of simple events in E':

$$\begin{aligned} n(E') &= \underset{\text{1st person}}{365} \cdot \underset{\text{2nd person}}{364} \cdot \underset{\text{3rd person}}{363} \cdot \cdots \cdot \underset{n\text{th person}}{(366 - n)} \\ &= \frac{[365 \cdot 364 \cdot 363 \cdot \cdots \cdot (366 - n)](365 - n)!}{(365 - n)!} \qquad \text{Multiply numerator and denominator by } (365 - n)! \\ &= \frac{365!}{(365 - n)!} \end{aligned}$$

Since we have assumed that S is an equally likely sample space,

$$P(E') = \frac{n(E')}{n(S)} = \frac{\dfrac{365!}{(365 - n)!}}{365^n} = \frac{365!}{365^n(365 - n)!}$$

Thus,

$$\begin{aligned} P(E) &= 1 - P(E') \\ &= 1 - \frac{365!}{365^n(365 - n)!} \end{aligned} \qquad (4)$$

Equation (4) is valid for any n satisfying $1 \leqslant n \leqslant 365$. [What is $P(E)$ if $n > 365$?] For example, in a group of 6 people,

$$\begin{aligned} P(E) &= 1 - \frac{365!}{(365)^6 359!} \\ &= 1 - \frac{\cancel{365} \cdot 364 \cdot 363 \cdot 362 \cdot 361 \cdot 360 \cdot \cancel{359!}}{\cancel{365} \cdot 365 \cdot 365 \cdot 365 \cdot 365 \cdot 365 \cdot \cancel{359!}} \\ &= .04 \end{aligned}$$

It is interesting to note that as the size of the group increases, $P(E)$ increases more rapidly than you might expect. Figure 6 shows the graph of $P(E)$ for $1 \leqslant n \leqslant 39$. Table 1 gives the value of $P(E)$ for selected values of n. Notice that for a group of only 23 people, the probability that 2 or more have the same birthday is greater than $\frac{1}{2}$. [See Problem 51 in Exercise 6-4 for a discussion of the form of equation (4) used to produce the graph in Fig. 6.]

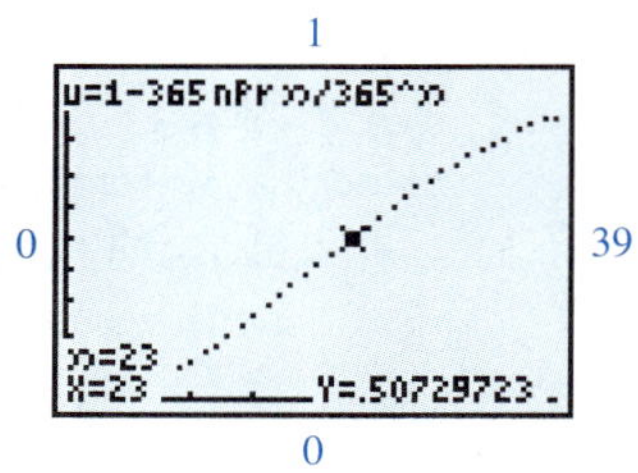

FIGURE 6

Table 1

THE BIRTHDAY PROBLEM

NUMBER OF PEOPLE IN GROUP n	PROBABILITY THAT 2 OR MORE HAVE SAME BIRTHDAY $P(E)$
5	.027
10	.117
15	.253
20	.411
23	**.507**
30	.706
40	.891
50	.970
60	.994
70	.999

Matched Problem 5 ➠ Use equation (4) to evaluate $P(E)$ for $n = 4$.

Explore–Discuss 2

Determine the smallest number n such that in a group of n people the probability that 2 or more have a birthday in the same month is greater than .5. Discuss the assumptions underlying your computation.

■ ODDS

When the probability of an event E is known, it is often customary (particularly in gaming situations) to speak of *odds* for (or against) the event E, rather than the *probability* of the event E. The relationship between these two designations is outlined in the box at the top of the next page.

From Probability to Odds

If $P(E)$ is the probability of the event E, then we define:

(A) **Odds for E** $= \dfrac{P(E)}{1 - P(E)} = \dfrac{P(E)}{P(E')}$ $\qquad P(E) \neq 1$

(B) **Odds against E** $= \dfrac{P(E')}{P(E)}$ $\qquad P(E) \neq 0$

[*Note:* When possible, odds are expressed as ratios of whole numbers.]

Example 6 **Probability and Odds** If you roll a fair die once, the probability of rolling a 4 is $\frac{1}{6}$, whereas the odds in favor of rolling a 4 are

$$\frac{P(E)}{P(E')} = \frac{\frac{1}{6}}{\frac{5}{6}} = \frac{1}{5} \qquad \text{Read as "1 to 5" and also written as "1:5"}$$

and

$$\text{Odds against rolling a 4} = \frac{5}{1}$$

In terms of a fair game, if you bet $1 on a 4 turning up, you would lose $1 to the house if any number other than 4 turns up and you would be paid $5 by the house (and in addition your bet of $1 returned) if a 4 turns up. (An experienced gambler would say that the house pays 5 to 1 on a 4 turning up on a single roll of a die.)

In general, if the odds for an event E are $a{:}b$, then **a game is fair** if your bet of $\$a$ is lost if event E does not happen, but you win $\$b$ (and in addition your bet of $\$a$ is returned) if event E does happen. (Gamblers would say that the house pays b to a on event E happening.)

Matched Problem 6

(A) What are the odds for rolling a sum of 8 in a single roll of two fair dice?
(B) If you bet $5 that a sum of 8 will turn up, what should the house pay (plus returning your $5 bet) if a sum of 8 does turn up for the game to be fair?

Now we will go in the other direction: If we are given the odds for an event, what is the probability of the event? (The verification of the following formula is left to Problem 55 in Exercise 6-4.)

From Odds to Probability

If the odds for event E are a/b, then the probability of E is

$$P(E) = \frac{a}{a + b}$$

Example 7 **Odds and Probability** If in repeated rolls of two fair dice the odds for rolling a 5 before rolling a 7 are 2 to 3, then the probability of rolling a 5 before rolling a 7 is

$$P(E) = \frac{a}{a + b} = \frac{2}{2 + 3} = \frac{2}{5}$$

Matched Problem 7 If in repeated rolls of two fair dice the odds against rolling a 6 before rolling a 7 are 6 to 5, then what is the probability of rolling a 6 before rolling a 7? (Be careful! Read the problem again.)

Applications to Empirical Probability

In the following discussions, the term *empirical probability* will mean the probability of an event determined by a sample that is used to approximate the probability of the corresponding event in the total population. How does the approximate empirical probability of an event determined from a sample relate to the actual probability of an event relative to the total population? In mathematical statistics an important theorem called the **law of large numbers** (or the **law of averages**) is proved. Informally, it states that the approximate empirical probability can be made as close to the actual probability as we please by making the sample sufficiently large.

Example 8 **Market Research** From a survey involving 1,000 people in a certain city, it was found that 500 people had tried a certain brand of diet cola, 600 had tried a certain brand of regular cola, and 200 had tried both types of cola. If a resident of the city is selected at random, what is the (empirical) probability that:

(A) The resident has tried the diet or the regular cola? What are the (empirical) odds for this event?

(B) The resident has tried one of the colas but not both? What are the (empirical) odds against this event?

Solution Let D be the event that a person has tried the diet cola and R the event that a person has tried the regular cola. The events D and R can be used to partition the residents of the city into four mutually exclusive subsets (a collection of subsets is mutually exclusive if the intersection of any two of them is the empty set):

$D \cap R$ = Set of people who have tried both colas
$D \cap R'$ = Set of people who have tried the diet cola but not the regular cola
$D' \cap R$ = Set of people who have tried the regular cola but not the diet cola
$D' \cap R'$ = Set of people who have not tried either cola

These sets are displayed in the Venn diagram in Figure 7.

The sample population of 1,000 residents is also partitioned into four mutually exclusive sets, with $n(D) = 500$, $n(R) = 600$, and $n(D \cap R) = 200$. By using a Venn diagram (Fig. 8), we can determine the number of sample points in the sets $D \cap R'$, $D' \cap R$, and $D' \cap R'$ (see Example 2 in Section 6-1).

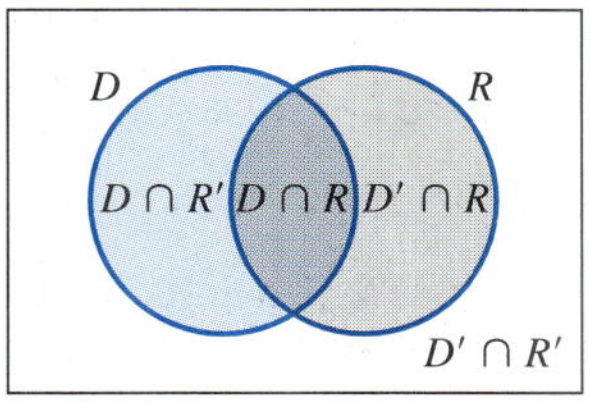

FIGURE 7
Total population

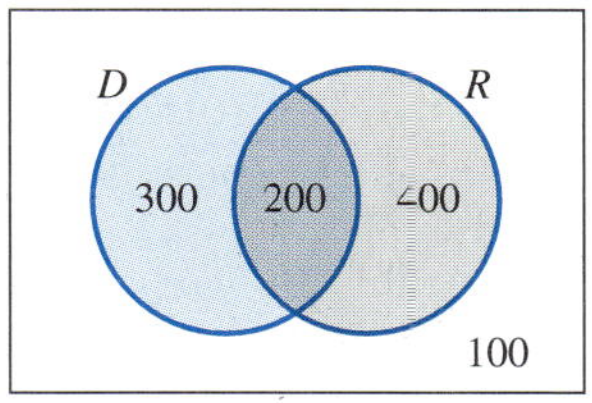

FIGURE 8
Sample population

These frequencies can be conveniently displayed in a table:

		Regular R	No regular R'	*Totals*
Diet	D	200	300	500
No diet	D'	400	100	500
	Totals	600	400	1,000

Assuming that each sample point is equally likely, we form a probability table by dividing each entry in this table by 1,000, the total number surveyed. These are empirical probabilities for the sample population, which we can use to approximate the corresponding probabilities for the total population.

		Regular R	No regular R'	*Totals*
Diet	D	.2	.3	.5
No diet	D'	.4	.1	.5
	Totals	.6	.4	1.0

Now we are ready to compute the required probabilities.

(A) The event that a person has tried the diet or the regular cola is $E = D \cup R$. We compute $P(E)$ two ways.

Method 1. Directly:

$$\begin{aligned} P(E) &= P(D \cup R) \\ &= P(D) + P(R) - P(D \cap R) \\ &= .5 + .6 - .2 = .9 \end{aligned}$$

Method 2. Using the complement of E:

$$\begin{aligned} P(E) &= 1 - P(E') \\ &= 1 - P(D' \cap R') \qquad E' = (D \cup R)' = D' \cap R' \text{ (see Fig. 7)} \\ &= 1 - .1 = .9 \end{aligned}$$

In either case,

$$\text{Odds for } E = \frac{P(E)}{P(E')} = \frac{.9}{.1} = \frac{9}{1} \quad \text{or} \quad 9:1$$

(B) The event that a person has tried one cola but not both is the event that the person has tried diet and not regular cola or has tried regular and not diet cola. In terms of sets, this is event $E = (D \cap R') \cup (D' \cap R)$. Since $D \cap R'$ and $D' \cap R$ are mutually exclusive (look at the Venn diagram in Fig. 7),

$$\begin{aligned} P(E) &= P[(D \cap R') \cup (D' \cap R)] \\ &= P(D \cap R') + P(D' \cap R) \\ &= .3 + .4 = .7 \end{aligned}$$

$$\text{Odds against } E = \frac{P(E')}{P(E)} = \frac{.3}{.7} = \frac{3}{7} \quad \text{or} \quad 3:7$$

Matched Problem 8 If a resident from the city in Example 8 is selected at random, what is the (empirical) probability that:

(A) The resident has not tried either cola? What are the (empirical) odds for this event?

(B) The resident has tried the diet cola or has not tried the regular cola? What are the (empirical) odds against this event?

Answers to Matched Problems

1. (A) $\frac{1}{3}$ (B) $\frac{2}{3}$ **2.** (A) $\frac{1}{12}$ (B) $\frac{7}{18}$ **3.** $\frac{47}{140} \approx .336$ **4.** .92
5. .016 **6.** (A) 5:31 (B) \$31 **7.** $\frac{5}{11} \approx .455$
8. (A) $P(D' \cap R') = .1$; odds for $D' \cap R' = \frac{1}{9}$ or 1:9
(B) $P(D \cup R') = .6$; odds against $D \cup R' = \frac{2}{3}$ or 2:3

EXERCISE 6-4

A

1. If a manufactured item has the probability .003 of failing within 90 days, what is the probability that the item will not fail in that time period?
2. In a particular cross of two plants the probability that the flowers will be red is .25. What is the probability that they will not be red?

A spinner is numbered from 1 through 10, and each number is as likely to come up as any other. In Problems 3–6, use equation (1) or (2), indicating which is used, to compute the probability that in a single spin the spinner will stop at:

3. A number less than 3 or larger than 7
4. A 2 or a number larger than 6
5. An even number or a number divisible by 3
6. An odd number or a number divisible by 3

Problems 7–18 refer to the Venn diagram for events A and B in an equally likely sample space S shown below. Find each of the indicated probabilities.

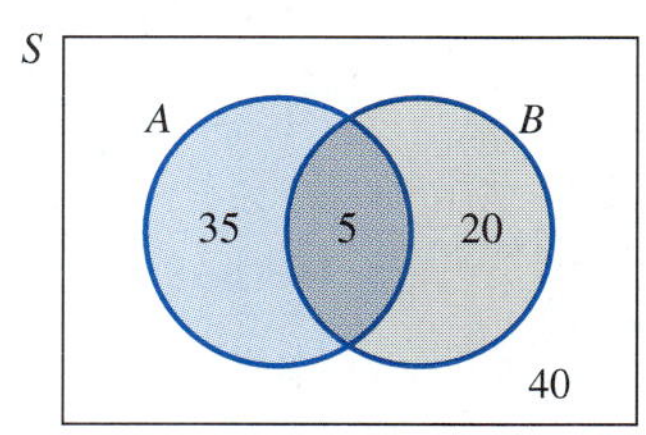

7. $P(A)$ 8. $P(A')$ 9. $P(B)$
10. $P(B')$ 11. $P(A \cap B)$ 12. $P(A \cap B')$
13. $P(A' \cap B)$ 14. $P(A' \cap B')$ 15. $P(A \cup B)$
16. $P(A \cup B')$ 17. $P(A' \cup B)$ 18. $P(A' \cup B')$

In Problems 19–22, use the equally likely sample space in Example 2 and equation (1) or (2), indicating which is used, to compute the probability of the following events:

19. A sum of 5 or 6 **20.** A sum of 9 or 10

21. The number on the first die is a 1 or the number on the second die is a 1.

22. The number on the first die is a 1 or the number on the second die is less than 3.

23. Given the following probabilities for an event E, find the odds for and against E:
(A) $\frac{3}{8}$ (B) $\frac{1}{4}$ (C) .4 (D) .55

24. Given the following probabilities for an event E, find the odds for and against E:
(A) $\frac{3}{5}$ (B) $\frac{1}{7}$ (C) .6 (D) .35

25. Compute the probability of event E if the odds in favor of E are:
(A) $\frac{3}{8}$ (B) $\frac{11}{7}$ (C) $\frac{4}{1}$ (D) $\frac{49}{51}$

26. Compute the probability of event E if the odds in favor of E are:
(A) $\frac{5}{9}$ (B) $\frac{4}{3}$ (C) $\frac{3}{7}$ (D) $\frac{23}{77}$

In Problems 27 and 28, discuss the validity of each statement. If the statement is always true, explain why. If not, give a counterexample.

27. (A) The empirical probability of an event is less than or equal to its theoretical probability.
(B) If events E and F are mutually exclusive, then

$$P(E) + P(F) = P(E \cup F) + P(E \cap F)$$

28. (A) If $P(E) = \frac{1}{2}$, then the odds for E equal the odds against E.
(B) If the odds for E are $a:b$, then the odds against E' are $b:a$.

B *In Problems 29–32, compute the odds in favor of obtaining:*

29. A head in a single toss of a coin

30. A number divisible by 3 in a single roll of a die

31. At least 1 head when a single coin is tossed 3 times

32. 1 head when a single coin is tossed twice

In Problems 33–36, compute the odds against obtaining:

33. A number greater than 4 in a single roll of a die

34. 2 heads when a single coin is tossed twice

35. A 3 or an even number in a single roll of a die

36. An odd number or a number divisible by 3 in a single roll of a die

37. (A) What are the odds for rolling a sum of 5 in a single roll of two fair dice?
(B) If you bet $1 that a sum of 5 will turn up, what should the house pay (plus returning your $1 bet) if a sum of 5 turns up for the game to be fair?

38. (A) What are the odds for rolling a sum of 10 in a single roll of two fair dice?
(B) If you bet $1 that a sum of 10 will turn up, what should the house pay (plus returning your $1 bet) if a sum of 10 turns up for the game to be fair?

A pair of dice are rolled 1,000 times with the following frequencies of outcomes:

SUM	2	3	4	5	6	7	8	9	10	11	12
FREQUENCY	10	30	50	70	110	150	170	140	120	80	70

Use these frequencies to calculate the approximate empirical probabilities and odds for the events in Problems 39 and 40.

39. (A) The sum is less than 4 or greater than 9.
(B) The sum is even or exactly divisible by 5.

40. (A) The sum is a prime number or is exactly divisible by 4.
(B) The sum is an odd number or exactly divisible by 3.

In Problems 41–44, a single card is drawn from a standard 52-card deck. Calculate the probability of and odds for each event.

41. A face card or a club is drawn.

42. A king or a heart is drawn.

43. A black card or an ace is drawn.

44. A heart or a number less than 7 (count an ace as 1) is drawn.

45. What is the probability of getting at least 1 diamond in a 5-card hand dealt from a standard 52-card deck?

46. What is the probability of getting at least 1 black card in a 7-card hand dealt from a standard 52-card deck?

47. What is the probability that a number selected at random from the first 1,000 positive integers is (exactly) divisible by 6 or 8?

48. What is the probability that a number selected at random from the first 600 positive integers is (exactly) divisible by 6 or 9?

49. Explain how the three events A, B, and C from a sample space S are related to each other in order for the following equation to hold:

$$P(A \cup B \cup C) = P(A) + P(B) + P(C) - P(A \cap B)$$

50. Explain how the three events A, B, and C from a sample space S are related to each other in order for the following equation to hold:

$$P(A \cup B \cup C) = P(A) + P(B) + P(C)$$

51. Show that the solution to the birthday problem in Example 5 can be written in the form

$$P(E) = 1 - \frac{P_{365,n}}{365^n}$$

For a calculator that has a $\boxed{P_{n,r}}$ key, explain why this form may be better for direct evaluation than the other form used in the solution to Example 5. Try direct evaluation of both forms on a calculator for $n = 25$.

52. Many (but not all) calculators experience an overflow error when computing $P_{365,n}$ for $n > 39$ and when computing 365^n. Explain how you would evaluate $P(E)$ for any $n > 39$ on such a calculator.

C

53. In a group of n people ($n \leq 12$), what is the probability that at least 2 of them have the same birth month? (Assume any birth month is as likely as any other.)

54. In a group of n people ($n \leq 100$), each person is asked to select a number between 1 and 100, write the number on a slip of paper, and place the slip in a hat. What is the probability that at least 2 of the slips in the hat have the same number written on them?

55. If the odds in favor of an event E occurring are a to b, show that

$$P(E) = \frac{a}{a + b}$$

[*Hint:* Solve the equation $P(E)/P(E') = a/b$ for $P(E)$.]

56. If $P(E) = c/d$, show that odds in favor of E occurring are c to $d - c$.

57. C The command in Figure A was used on a graphing utility to simulate 50 repetitions of rolling a pair of dice and recording their sum. A statistical plot of the results is shown in Figure B.
(A) Use Figure B to find the empirical probability of rolling a 7 or 8.

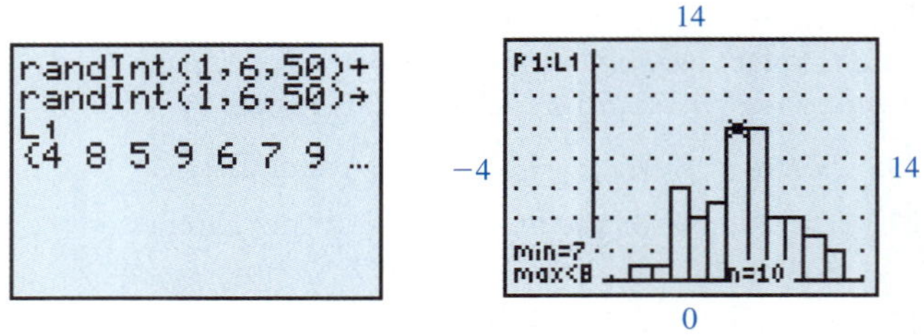

(A) (B)

Figure for 57

(B) What is the theoretical probability of rolling a 7 or 8?
(C) Use a graphing utility to simulate 200 repetitions of rolling a pair of dice and recording their sum, and find the empirical probability of rolling a 7 or 8.

58. C Consider the command in Figure A and the associated statistical plot in Figure B.

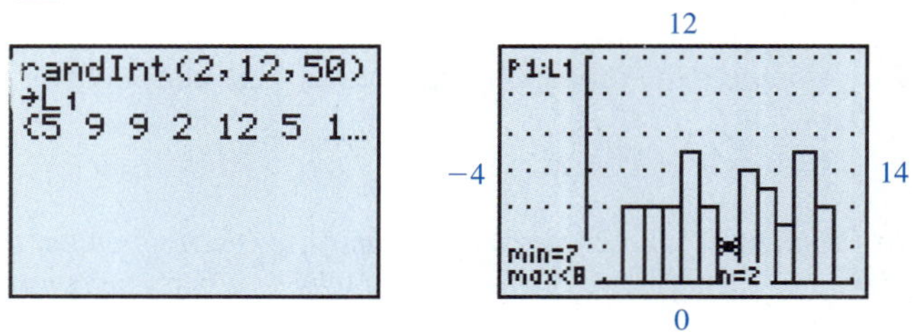

(A) (B)

Figure for 58

(A) Explain why the command does not simulate 50 repetitions of rolling a pair of dice and recording their sum.
(B) Describe an experiment that is simulated by this command.
(C) Simulate 200 repetitions of the experiment you described in part (B), and find the empirical probability of recording a 7 or 8.
(D) What is the theoretical probability of recording a 7 or 8?

APPLICATIONS

Business & Economics

59. *Market research.* From a survey involving 1,000 students at a large university, a market research company found that 750 students owned stereos, 450 owned cars, and 350 owned cars and stereos. If a student at the university is selected at random, what is the (empirical) probability that:
(A) The student owns either a car or a stereo?
(B) The student owns neither a car nor a stereo?

60. *Market research.* If a student at the university in Problem 59 is selected at random, what is the (empirical) probability that:
(A) The student does not own a car?
(B) The student owns a car but not a stereo?

61. *Insurance.* By examining the past driving records of drivers in a certain city, an insurance company has determined the (empirical) probabilities in the table shown at the top of the next page.

		Miles driven per year			
		Less than 10,000, M_1	*10,000–15,000 inclusive, M_2*	*More than 15,000, M_3*	*Totals*
Accident	A	.05	.1	.15	.3
No accident	A'	.15	.2	.35	.7
	Totals	.2	.3	.5	1.0

If a driver in this city is selected at random, what is the probability that:

(A) He or she drives less than 10,000 miles per year or has an accident?

(B) He or she drives 10,000 or more miles per year and has no accidents?

62. *Insurance.* Use the (empirical) probabilities in Problem 61 to find the probability that a driver in the city selected at random:

(A) Drives more than 15,000 miles per year or has an accident

(B) Drives 15,000 or fewer miles per year and has an accident

63. *Manufacturing.* Manufacturers of a portable computer provide a 90 day limited warranty covering only the keyboard and the disk drive. Their records indicate that during the warranty period, 6% of their computers are returned because they have defective keyboards, 5% are returned because they have defective disk drives, and 1% are returned because both the keyboard and the disk drive are defective. What is the (empirical) probability that a computer will not be returned during the warranty period?

64. *Product testing.* In order to test a new car, an automobile manufacturer wants to select 4 employees to test drive the car for 1 year. If 12 management and 8 union employees volunteer to be test drivers and the selection is made at random, what is the probability that at least 1 union employee is selected?

65. *Quality control.* A shipment of 60 inexpensive digital watches, including 9 that are defective, is sent to a department store. The receiving department selects 10 at random for testing and rejects the whole shipment if 1 or more in the sample are found defective. What is the probability that the shipment will be rejected?

66. *Quality control.* An automated manufacturing process produces 40 computer circuit boards, including 7 that are defective. The quality control department selects 10 at random (from the 40 produced) for testing and will shut down the plant for trouble shooting if 1 or more in the sample are found defective. What is the probability that the plant will be shut down?

Life Sciences

67. *Medicine.* In order to test a new drug for adverse reactions, the drug was administered to 1,000 test subjects with the following results: 60 subjects reported that their only adverse reaction was a loss of appetite, 90 subjects reported that their only adverse reaction was a loss of sleep, and 800 subjects reported no adverse reactions at all. If this drug is released for general use, what is the (empirical) probability that a person using the drug will suffer both a loss of appetite and a loss of sleep?

68. *Medicine.* Thirty animals are to be used in a medical experiment on diet deficiency: 3 male and 7 female rhesus monkeys, 6 male and 4 female chimpanzees, and 2 male and 8 female dogs. If one animal is selected at random, what is the probability of getting:

(A) A chimpanzee or a dog?

(B) A chimpanzee or a male?

(C) An animal other than a female monkey?

Social Sciences

Problems 69 and 70 refer to the data in the table below, obtained from a random survey of 1,000 residents of a state. The participants were asked their political affiliations and their preferences in an upcoming gubernatorial election. (In the table, D = Democrat, R = Republican, and U = Unaffiliated.)

		D	*R*	*U*	*Totals*
Candidate *A*	*A*	200	100	85	385
Candidate *B*	*B*	250	230	50	530
No preference	*N*	50	20	15	85
	Total	500	350	150	1,000

69. *Politics.* If a resident of the state is selected at random, what is the (empirical) probability that the resident is:

(A) Not affiliated with a political party or has no preference? What are the odds for this event?

(B) Affiliated with a political party and prefers candidate A? What are the odds against this event?

70. *Politics.* If a resident of the state is selected at random, what is the (empirical) probability that the resident is:

(A) A Democrat or prefers candidate B? What are the odds for this event?

(B) Not a Democrat and has no preference? What are the odds against this event?

71. *Sociology.* A group of 5 Blacks, 5 Asians, 5 Latinos, and 5 Whites was used in a study on racial influence in small group dynamics. If 3 people are chosen at random, what is the probability that at least 1 is Black?

SECTION 6-5

Conditional Probability, Intersection, and Independence

- CONDITIONAL PROBABILITY
- INTERSECTION OF EVENTS—PRODUCT RULE
- PROBABILITY TREES
- INDEPENDENT EVENTS
- SUMMARY

In the previous section we asked if the probability of the union of two events could be expressed in terms of the probabilities of the individual events, and we found the answer to be a qualified yes (see page 407). Now we ask the same question for the intersection of two events; that is, can the probability of the intersection of two events be represented in terms of the probabilities of the individual events? The answer again is a qualified yes. But before we find out how and under what conditions, we must investigate a related concept called *conditional probability.*

CONDITIONAL PROBABILITY

The probability of an event may change if we are told of the occurrence of another event. For example, if an adult (a person 21 years or older) is selected at random from all adults in the United States, the probability of that person having lung cancer would not be too high. However, if we are told that the person is also a heavy smoker, then we would certainly want to revise the probability upward.

In general, the probability of the occurrence of an event A, given the occurrence of another event B, is called a **conditional probability** and is denoted by $P(A|B)$.

In the above illustration, events A and B would be

$A =$ Adult has lung cancer
$B =$ Adult is a heavy smoker

And $P(A|B)$ would represent the probability of an adult having lung cancer, given that he or she is a heavy smoker.

Our objective now is to try to formulate a precise definition of $P(A|B)$. It is helpful to start with a relatively simple problem, solve it intuitively, and then generalize from this experience.

What is the probability of rolling a prime number $(2, 3, \text{or } 5)$ in a single roll of a fair die? Let

$$S = \{1, 2, 3, 4, 5, 6\}$$

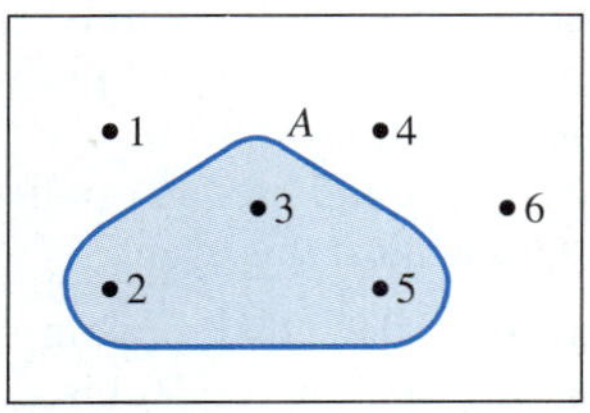

FIGURE 1

Then the event of rolling a prime number is (see Fig. 1)

$$A = \{2, 3, 5\}$$

Thus, since we assume each simple event in the sample space is equally likely,

$$P(A) = \frac{n(A)}{n(S)} = \frac{3}{6} = \frac{1}{2}$$

Now suppose you are asked, "In a single roll of a fair die, what is the probability that a prime number has turned up if we are given the additional

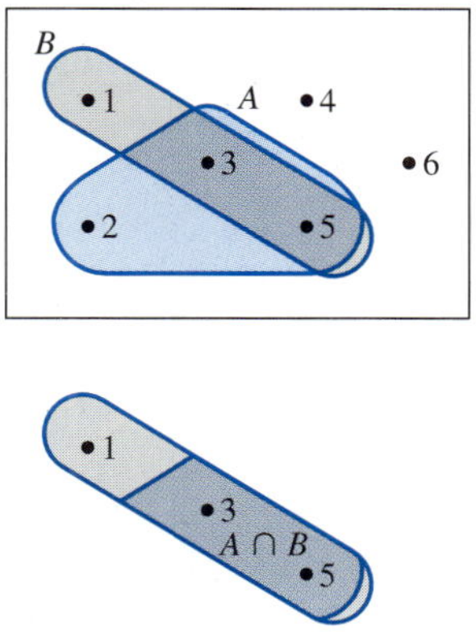

FIGURE 2
B is the new sample space

information that an odd number has turned up?" The additional knowledge that another event has occurred, namely,

$$B = \text{Odd number turns up}$$

puts the problem in a new light. We are now interested only in the part of event A (rolling a prime number) that is in event B (rolling an odd number). Event B, since we know it has occurred, becomes the new sample space. The Venn diagrams in Figure 2 illustrate the various relationships. Thus, the probability of A given B is the number of A elements in B divided by the total number of elements in B. Symbolically,

$$P(A|B) = \frac{n(A \cap B)}{n(B)} = \frac{2}{3}$$

Dividing the numerator and denominator of $n(A \cap B)/n(B)$ by $n(S)$, the number of elements in the original sample space, we can express $P(A|B)$ in terms of $P(A \cap B)$ and $P(B)$:*

$$P(A|B) = \frac{n(A \cap B)}{n(B)} = \frac{\dfrac{n(A \cap B)}{n(S)}}{\dfrac{n(B)}{n(S)}} = \frac{P(A \cap B)}{P(B)}$$

Using the right side to compute $P(A|B)$ for the example above, we obtain the same result (as we should):

$$P(A|B) = \frac{P(A \cap B)}{P(B)} = \frac{\frac{2}{6}}{\frac{3}{6}} = \frac{2}{3}$$

We use the above formula to motivate the following definition of *conditional probability,* which applies to any sample space, including those having simple events that are not equally likely (see Example 1).

> ### Conditional Probability
>
> For events A and B in an arbitrary sample space S, we define the conditional probability of A given B by
>
> $$P(A|B) = \frac{P(A \cap B)}{P(B)} \qquad P(B) \neq 0 \qquad (1)$$

Example 1 ➡ **Conditional Probability** A pointer is spun once on the circular spinner shown in Figure 3. The probability assigned to the pointer landing on a given integer (from 1 to 6) is the ratio of the area of the corresponding circular sector to the area of the whole circle, as given in the table:

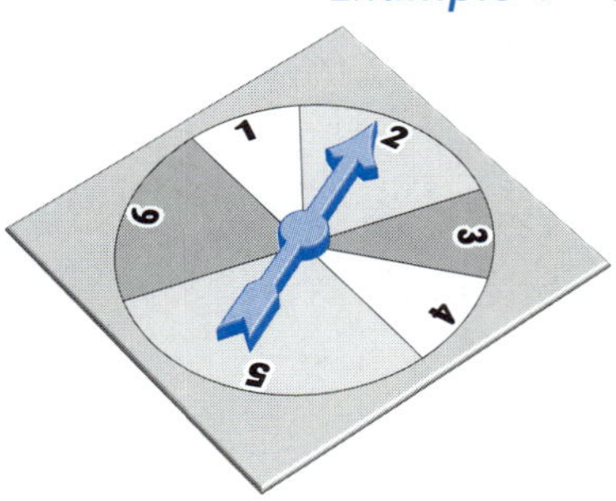

FIGURE 3

e_i	1	2	3	4	5	6
$P(e_i)$	.1	.2	.1	.1	.3	.2

$S = \{1, 2, 3, 4, 5, 6\}$

*Note that $P(A|B)$ is a probability based on the new sample space B, while $P(A \cap B)$ and $P(B)$ are both probabilities based on the original sample space S.

(A) What is the probability of the pointer landing on a prime number?
(B) What is the probability of the pointer landing on a prime number, given that it landed on an odd number?

SOLUTION Let the events E and F be defined as follows:

$$E = \text{Pointer lands on a prime number} = \{2, 3, 5\}$$
$$F = \text{Pointer lands on an odd number} = \{1, 3, 5\}$$

(A)
$$P(E) = P(2) + P(3) + P(5) = .2 + .1 + .3 = .6$$

(B) First note that $E \cap F = \{3, 5\}$.

$$P(E|F) = \frac{P(E \cap F)}{P(F)} = \frac{P(3) + P(5)}{P(1) + P(3) + P(5)} = \frac{.1 + .3}{.1 + .1 + .3} = \frac{.4}{.5} = .8$$

Matched Problem 1 Refer to the spinner and table in Example 1.

(A) What is the probability of the pointer landing on a number greater than 4?
(B) What is the probability of the pointer landing on a number greater than 4, given that it landed on an even number?

Explore–Discuss 1

Compare the spinner problem in Example 1 to the die problem at the beginning of the section. Discuss their similarities and differences.

Example 2 **Safety Research** Suppose past records in a large city produced the following probability data on a driver being in an accident on the last day of a Memorial Day weekend:

		ACCIDENT A	NO ACCIDENT A'	*Totals*
RAIN	R	.025	.335	.360
NO RAIN	R'	.015	.625	.640
	Totals	.040	.960	1.000

$S = \{RA, RA', R'A, R'A'\}$

(A) Find the probability of an accident, rain or no rain.
(B) Find the probability of rain, accident or no accident.
(C) Find the probability of an accident and rain.
(D) Find the probability of an accident, given rain.

SOLUTION (A) Let $A = \{RA, R'A\}$ Event: "accident"

$$P(A) = P(RA) + P(R'A) = .025 + .015 = .040$$

(B) Let $R = \{RA, RA'\}$ Event: "rain"

$$P(R) = P(RA) + P(RA') = .025 + .335 = .360$$

(C) $A \cap R = \{RA\}$ Event: "accident and rain"

$$P(A \cap R) = P(RA) = .025$$

(D) $P(A|R) = \dfrac{P(A \cap R)}{P(R)} = \dfrac{.025}{.360} = .069$ Event: "accident, given rain"

Compare this result with that in part (A). Note that $P(A|R) \neq P(A)$, and the probability of an accident, given rain, is higher than the probability of an accident without the knowledge of rain.

Matched Problem 2 Referring to the table in Example 2, determine the following:

(A) Probability of no rain
(B) Probability of an accident and no rain
(C) Probability of an accident, given no rain [Use formula (1) and the results of parts (A) and (B).]

Intersection of Events—Product Rule

We now return to the original problem of this section; that is, representing the probability of an intersection of two events in terms of the probabilities of the individual events. If $P(A) \neq 0$ and $P(B) \neq 0$, then using formula (1) we can write

$$P(A|B) = \frac{P(A \cap B)}{P(B)} \quad \text{and} \quad P(B|A) = \frac{P(B \cap A)}{P(A)}$$

Solving the first equation for $P(A \cap B)$ and the second equation for $P(B \cap A)$, we have

$$P(A \cap B) = P(B)P(A|B) \quad \text{and} \quad P(B \cap A) = P(A)P(B|A)$$

Since $A \cap B = B \cap A$ for any sets A and B, it follows that

$$P(A \cap B) = P(B)P(A|B) = P(A)P(B|A)$$

and we have the **product rule:**

Product Rule

For events A and B with nonzero probabilities in a sample space S,

$$P(A \cap B) = P(A)P(B|A) = P(B)P(A|B) \tag{2}$$

and we can use either $P(A)P(B|A)$ or $P(B)P(A|B)$ to compute $P(A \cap B)$.

Example 3 **Consumer Survey** If 60% of a department store's customers are female and 75% of the female customers have charge accounts at the store, what is the probability that a customer selected at random is a female and has a charge account?

SOLUTION Let: S = All store customers
F = Female customers
C = Customers with a charge account

If 60% of the customers are female, then the probability that a customer selected at random is a female is

$$P(F) = .60$$

Since 75% of the female customers have charge accounts, the probability that a customer has a charge account, given that the customer is a female, is

$$P(C|F) = .75$$

Using equation (2), the probability that a customer is a female and has a charge account is

$$P(F \cap C) = P(F)P(C|F) = (.60)(.75) = .45$$

Matched Problem 3 If 80% of the male customers of the department store in Example 3 have charge accounts, what is the probability that a customer selected at random is a male and has a charge account?

PROBABILITY TREES

We used tree diagrams in Section 6-1 to help us count the number of combined outcomes in a sequence of experiments. In much the same way we will now use probability trees to help us compute the probabilities of combined outcomes in a sequence of experiments. An example will help make the process of forming and using probability trees clear.

Example 4 **Probability Tree** Two balls are drawn in succession, without replacement, from a box containing 3 blue and 2 white balls (Fig. 4). What is the probability of drawing a white ball on the second draw?

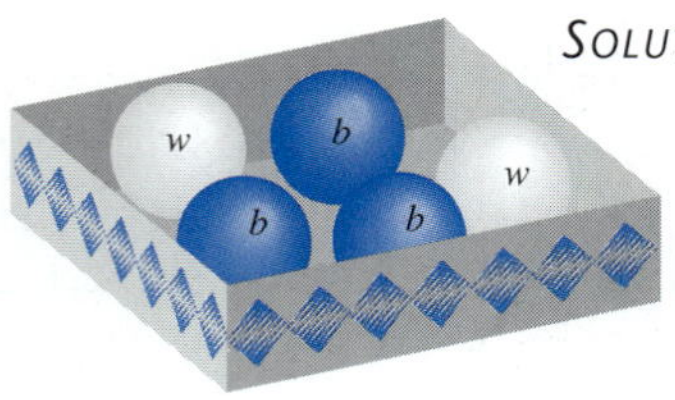

FIGURE 4

SOLUTION We start with a tree diagram (Fig. 5) showing the combined outcomes of the two experiments (first draw and second draw):

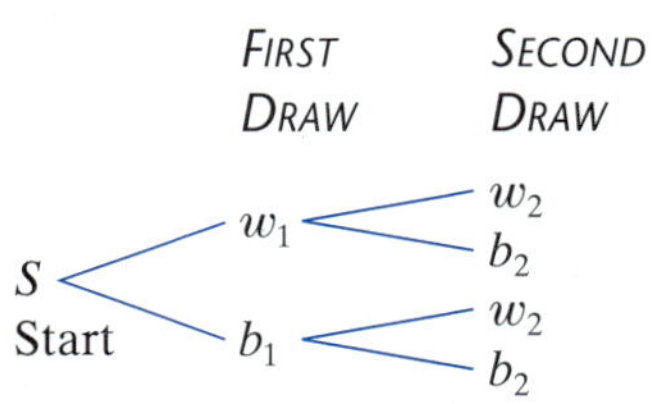

FIGURE 5

We now assign a probability to each branch on the tree (Fig. 6). For example, we assign the probability $\frac{2}{5}$ to the branch Sw_1, since this is the probability of drawing a white ball on the first draw (there are 2 white balls and 3 blue balls in the box). What probability should be assigned to the branch w_1w_2? This is the conditional probability $P(w_2|w_1)$; that is, the probability of drawing a white ball on the second draw given that a white ball was drawn on the first draw and not replaced. Since the box now contains 1 white ball and

3 blue balls, the probability is $\frac{1}{4}$. Continuing in the same way, we assign probabilities to the other branches of the tree and obtain Figure 6.

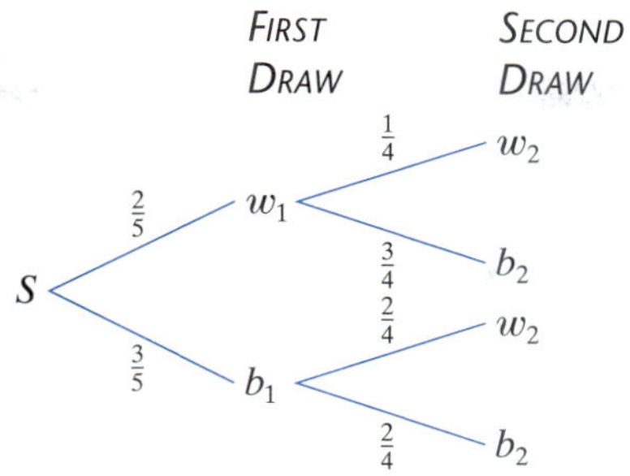

FIGURE 6

What is the probability of the combined outcome $w_1 \cap w_2$; that is, of drawing a white ball on the first draw and a white ball on the second draw?* Using the product rule (2), we have

$$\begin{aligned} P(w_1 \cap w_2) &= P(w_1)P(w_2|w_1) \\ &= \left(\tfrac{2}{5}\right)\left(\tfrac{1}{4}\right) = \tfrac{1}{10} \end{aligned}$$

The combined outcome $w_1 \cap w_2$ corresponds to the unique path Sw_1w_2 in the tree diagram, and we see that the probability of reaching w_2 along this path is just the product of the probabilities assigned to the branches on the path. Reasoning in the same way, we obtain the probability of each remaining combined outcome by multiplying the probabilities assigned to the branches on the path corresponding to the given combined outcomes. These probabilities are often written at the ends of the paths to which they correspond (Fig. 7).

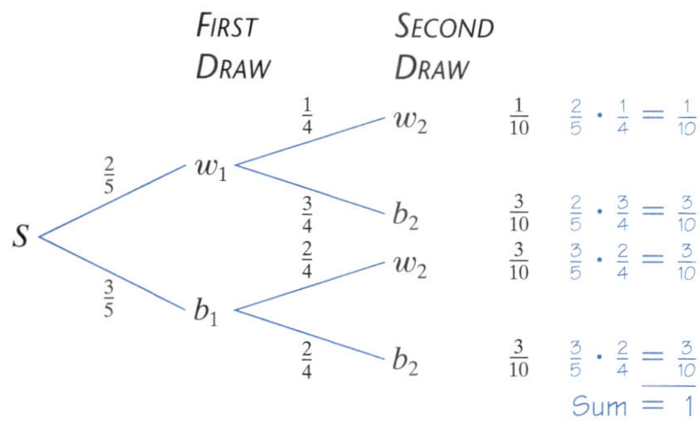

FIGURE 7

Now it is an easy matter to complete the problem. A white ball drawn on the second draw corresponds to either the combined outcome $w_1 \cap w_2$ or $b_1 \cap w_2$ occurring. Thus, since these combined outcomes are mutually exclusive,

$$\begin{aligned} P(w_2) &= P(w_1 \cap w_2) + P(b_1 \cap w_2) \\ &= \tfrac{1}{10} + \tfrac{3}{10} = \tfrac{4}{10} = \tfrac{2}{5} \end{aligned}$$

which is just the sum of the probabilities listed at the ends of the two paths terminating in w_2.

---

*The sample space for the combined outcomes is $S = \{w_1w_2, w_1b_2, b_1w_2, b_1b_2\}$. If we let $w_1 = \{w_1w_2, w_1b_2\}$ and $w_2 = \{w_1w_2, b_1w_2\}$, then $w_1 \cap w_2 = \{w_1w_2\}$.

Matched Problem 4 Two balls are drawn in succession without replacement from a box containing 4 red and 2 white balls. What is the probability of drawing a red ball on the second draw?

The sequence of two experiments in Example 4 is an example of a *stochastic process.* In general, a **stochastic process** involves a sequence of experiments where the outcome of each experiment is not certain. Our interest is in making predictions about the process as a whole. The analysis in Example 4 generalizes to stochastic processes involving any finite sequence of experiments. We summarize the procedures used in Example 4 for general application:

Constructing Probability Trees

Step 1. Draw a tree diagram corresponding to all combined outcomes of the sequence of experiments.

Step 2. Assign a probability to each tree branch. (This is the probability of the occurrence of the event on the right end of the branch subject to the occurrence of all events on the path leading to the event on the right end of the branch. The probability of the occurrence of a combined outcome that corresponds to a path through the tree is the product of all branch probabilities on the path.*)

Step 3. Use the results in steps 1 and 2 to answer various questions related to the sequence of experiments as a whole.

*If we form a sample space S such that each simple event in S corresponds to one path through the tree, and if the probability assigned to each simple event in S is the product of the branch probabilities on the corresponding path, then it can be shown that this is not only an acceptable assignment (all probabilities for the simple events in S are nonnegative and their sum is 1), but it is the only assignment consistent with the method used to assign branch probabilities within the tree.

Explore–Discuss 2

Refer to the table on rain and accidents in Example 2 and use formula (1), where appropriate, to complete the following probability tree:

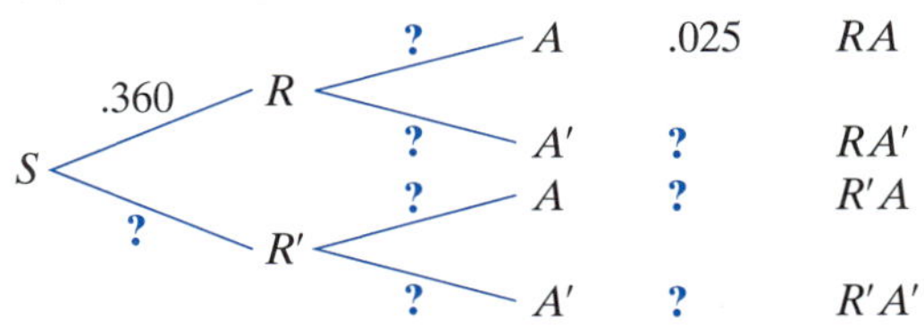

Discuss the difference between $P(R \cap A)$ and $P(A|R)$.

Example 5 **Product Defects** A large computer company A subcontracts the manufacturing of its circuit boards to two companies, 40% to company B and 60% to company C. Company B in turn subcontracts 70% of the orders it receives from company A to company D and the remaining 30% to company E, both subsidiaries of company B. When the boards are completed by companies D, E, and C, they are shipped to company A to be used in various computer models. It has been found that 1.5%, 1%, and .5% of the boards from *D*, *E*, and *C*, respectively, prove defective during the 90 day warranty period after a computer is first sold. What is the probability that a given board in a computer will be defective during the 90 day warranty period?

SOLUTION Draw a tree diagram and assign probabilities to each branch (Fig. 8):

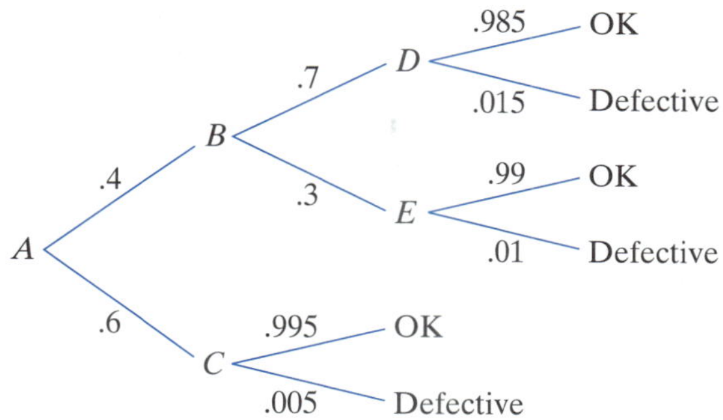

FIGURE 8

There are three paths leading to defective (the board will be defective within the 90 day warranty period). We multiply the branch probabilities on each path and add the three products:

$$\begin{aligned} P(\text{Defective}) &= (.4)(.7)(.015) + (.4)(.3)(.01) + (.6)(.005) \\ &= .0084 \end{aligned}$$

Matched Problem 5 In Example 5, what is the probability that a circuit board in a completed computer came from company *E* or *C*?

INDEPENDENT EVENTS

We return to Example 4, which involved drawing two balls in succession without replacement from a box of 3 blue and 2 white balls (Fig. 4). What difference does "without replacement" and "with replacement" make? Figure 9 shows probability trees corresponding to each case. Go over the probability assignments for the branches in part (B) to convince yourself of their correctness.

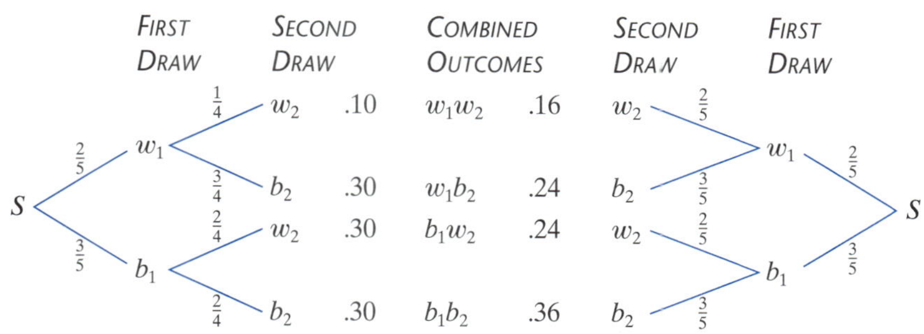

(A) Without replacement (B) With replacement

FIGURE 9

$S = \{w_1w_2, w_1b_2, b_1w_2, b_1b_2\}$

Let: $A =$ White ball on second draw $= \{w_1w_2, b_1w_2\}$
$B =$ White ball on first draw $= \{w_1w_2, w_1b_2\}$

We now compute $P(A|B)$ and $P(A)$ for each case in Figure 9.

Case 1. *Without Replacement:*

$$P(A|B) = \frac{P(A \cap B)}{P(B)} = \frac{P\{w_1w_2\}}{P\{w_1w_2, w_1b_2\}} = \frac{.10}{.10 + .30} = .25$$

(This is the assignment to branch w_1w_2 we made by looking in the box and counting.)

$$P(A) = P\{w_1w_2, b_1w_2\} = .10 + .30 = .40$$

Note that $P(A|B) \neq P(A)$, and we conclude that the probability of A is affected by the occurrence of B.

Case 2. *With Replacement:*

$$P(A|B) = \frac{P(A \cap B)}{P(B)} = \frac{P\{w_1w_2\}}{P\{w_1w_2, w_1b_2\}} = \frac{.16}{.16 + .24} = .40$$

(This is the assignment to branch w_1w_2 we made by looking in the box and counting.)

$$P(A) = P\{w_1w_2, b_1w_2\} = .16 + .24 = .40$$

Note that $P(A|B) = P(A)$, and we conclude that the probability of A is not affected by the occurrence of B.

Intuitively, if $P(A|B) = P(A)$, then it appears that event A is "independent" of B. Let us pursue this further. If events A and B are such that

$$P(A|B) = P(A)$$

then replacing the left side by its equivalent from equation (1), we obtain

$$\frac{P(A \cap B)}{P(B)} = P(A)$$

After multiplying both sides by $P(B)$, the last equation becomes

$$P(A \cap B) = P(A)P(B)$$

This result motivates the following definition of *independence:*

Independence

If A and B are any events in a sample space S, we say that **A and B are independent** if and only if

$$P(A \cap B) = P(A)P(B) \tag{3}$$

Otherwise, A and B are said to be **dependent.**

From the definition of independence one can prove (see Problems 43 and 44 in Exercise 6-5) the following theorem:

Theorem 1 ON INDEPENDENCE

If A and B are independent events with nonzero probabilities in a sample space S, then

$$P(A|B) = P(A) \quad \text{and} \quad P(B|A) = P(B) \tag{4}$$

If either equation in (4) holds, then A and B are independent.

In practice, one often has correct intuitive feelings about independence. For example, if you toss a coin twice, the second toss is independent of the first (a coin has no memory); if a card is drawn from a deck twice, with replacement, the second draw is independent of the first (the deck has no memory); if you roll a pair of dice twice, the second roll is independent of the first (dice have no memory); and so on. However, there are pairs of events that are not obviously independent or dependent, so we need equations (3) or (4) to decide these cases. Example 6 considers two events that are obviously independent, and Example 7 considers events where independence or dependence is not obvious.

Example 6 **Testing for Independence** In two tosses of a single fair coin show that the events "A head on the first toss" and "A head on the second toss" are independent.

SOLUTION Consider the sample space of equally likely outcomes for the tossing of a fair coin twice,

$$S = \{\text{HH, HT, TH, TT}\}$$

and the two events,

$$\begin{aligned} A &= \text{A head on the first toss} = \{\text{HH, HT}\} \\ B &= \text{A head on the second toss} = \{\text{HH, TH}\} \end{aligned}$$

Then

$$P(A) = \tfrac{2}{4} = \tfrac{1}{2} \qquad P(B) = \tfrac{2}{4} = \tfrac{1}{2} \qquad P(A \cap B) = \tfrac{1}{4}$$

Thus,

$$P(A \cap B) = \tfrac{1}{4} = \tfrac{1}{2} \cdot \tfrac{1}{2} = P(A)P(B)$$

and the two events are independent. (The theory agrees with our intuition—a coin has no memory.)

Matched Problem 6 In Example 6, compute $P(B|A)$ and compare with $P(B)$.

Example 7 **Testing for Independence** A single card is drawn from a standard 52-card deck. Test the following events for independence (try guessing the answer to each part before looking at the solution):

(A) E = The drawn card is a spade.
F = The drawn card is a face card.

(B) G = The drawn card is a club.
H = The drawn card is a heart.

SOLUTION (A) To test E and F for independence, we compute $P(E \cap F)$ and $P(E)P(F)$. If they are equal, then events E and F are independent; if they are not equal, then events E and F are dependent.

$$P(E \cap F) = \tfrac{3}{52} \qquad P(E)P(F) = \left(\tfrac{13}{52}\right)\left(\tfrac{12}{52}\right) = \tfrac{3}{52}$$

Events E and F are independent. (Did you guess this?)

(B) Proceeding as in part (A), we see that

$$P(G \cap H) = P(\varnothing) = 0 \qquad P(G)P(H) = \left(\tfrac{13}{52}\right)\left(\tfrac{13}{52}\right) = \tfrac{1}{16}$$

Events G and H are dependent. (Did you guess this?)

CAUTION

Students often confuse *mutually exclusive (disjoint) events* with *independent events.* One does not necessarily imply the other. In fact, it is not difficult to show (see Problem 47, Exercise 6-5) that any two mutually exclusive events A and B, with nonzero probabilities, are always dependent!

Matched Problem 7 A single card is drawn from a standard 52-card deck. Test the following events for independence:

(A) E = The drawn card is a red card.
F = The drawn card's number is divisible by 5 (face cards are not assigned values).

(B) G = The drawn card is a king.
H = The drawn card is a queen.

Explore–Discuss 3

In college basketball, would it be reasonable to assume that the following events are independent? Explain why or why not.

A = The Golden Eagles win in the first round of the NCAA tournament.
B = The Golden Eagles win in the second round of the NCAA tournament.

The notion of independence can be extended to more than two events:

Independent Set of Events

A set of events is said to be **independent** if for each finite subset $\{E_1, E_2, \dots, E_k\}$

$$P(E_1 \cap E_2 \cap \cdots \cap E_k) = P(E_1)P(E_2) \cdot \cdots \cdot P(E_k) \qquad (5)$$

The next example makes direct use of this definition.

Example 8 **Computer Control Systems** A space shuttle has four independent computer control systems. If the probability of failure (during flight) of any one system is .001, what is the probability of failure of all four systems?

SOLUTION Let: E_1 = Failure of system 1 $\quad$ E_3 = Failure of system 3
E_2 = Failure of system 2 $\quad$ E_4 = Failure of system 4

Then, since events E_1, E_2, E_3, and E_4 are given to be independent,

$$\begin{aligned} P(E_1 \cap E_2 \cap E_3 \cap E_4) &= P(E_1)P(E_2)P(E_3)P(E_4) \\ &= (.001)^4 \\ &= .000\ 000\ 000\ 001 \end{aligned}$$

Matched Problem 8 A single die is rolled 6 times. What is the probability of getting the sequence, 1, 2, 3, 4, 5, 6?

SUMMARY

The key results in this section are summarized in the following box for an overview and ease of reference:

Summary of Key Concepts

Conditional Probability

$$P(A|B) = \frac{P(A \cap B)}{P(B)} \qquad P(B|A) = \frac{P(B \cap A)}{P(A)}$$

[*Note*: $P(A|B)$ is a probability based on the new sample space B, while $P(A \cap B)$ and $P(B)$ are probabilities based on the original sample space S.]

Product Rule

$$P(A \cap B) = P(A)P(B|A) = P(B)P(A|B)$$

Independent Events

- A and B are independent if and only if

$$P(A \cap B) = P(A)P(B)$$

- If A and B are independent events with nonzero probabilities, then

$$P(A|B) = P(A) \quad \text{and} \quad P(B|A) = P(B)$$

- If A and B are events with nonzero probabilities and either $P(A|B) = P(A)$ or $P(B|A) = P(B)$, then A and B are independent.
- If $E_1, E_2, \ldots, E_n$ are independent, then

$$P(E_1 \cap E_2 \cap \cdots \cap E_n) = P(E_1)P(E_2) \cdot \cdots \cdot P(E_n)$$

Answers to Matched Problems

1. (A) .5 (B) .4

2. (A) $P(R') = .640$ (B) $P(A \cap R') = .015$

(C) $P(A|R') = \frac{P(A \cap R')}{P(R')} = .023$

3. $P(M \cap C) = P(M)P(C|M) = .32$

4. $\frac{2}{3}$ **5.** .72

6. $P(B|A) = \frac{P(A \cap B)}{P(A)} = \frac{\frac{1}{4}}{\frac{1}{2}} = \frac{1}{2} = P(B)$

7. (A) E and F are independent
(B) G and H are dependent

8. $(\frac{1}{6})^6 \approx .000\ 021\ 4$

EXERCISE 6-5

A *Given the probabilities in the table below for events in a sample space S, solve Problems 1–16 relative to these probabilities.*

	A	*B*	*C*	*Totals*
D	.10	.04	.06	.20
E	.40	.26	.14	.80
Totals	.50	.30	.20	1.00

In Problems 1–8, find each probability directly from the table.

1. $P(A)$ **2.** $P(C)$ **3.** $P(D)$

4. $P(E)$ **5.** $P(A \cap D)$ **6.** $P(A \cap E)$

7. $P(C \cap D)$ **8.** $P(C \cap E)$

In Problems 9–12 compute each probability using equation (1) and appropriate table values:

9. $P(A|D)$ **10.** $P(A|E)$ **11.** $P(C|D)$

12. $P(C|E)$

In Problems 13–16, test each pair of events for independence:

13. A and D **14.** A and E

15. C and D **16.** C and E

17. A fair coin is tossed 8 times.
(A) What is the probability of tossing a head on the 8th toss, given that the preceding 7 tosses were heads?
(B) What is the probability of getting 8 heads or 8 tails?

18. A fair die is rolled 5 times.
(A) What is the probability of getting a 6 on the 5th roll, given that a 6 turned up on the preceding 4 rolls?
(B) What is the probability that the same number turns up every time?

19. A pointer is spun once on the circular spinner shown at the top of the next column. The probability assigned to the pointer landing on a given integer (from 1 to 5) is the ratio of the area of the corresponding circular sector to the area of the whole circle, as given in the table:

e_i	1	2	3	4	5
$P(e_i)$	.3	.1	.2	.3	.1

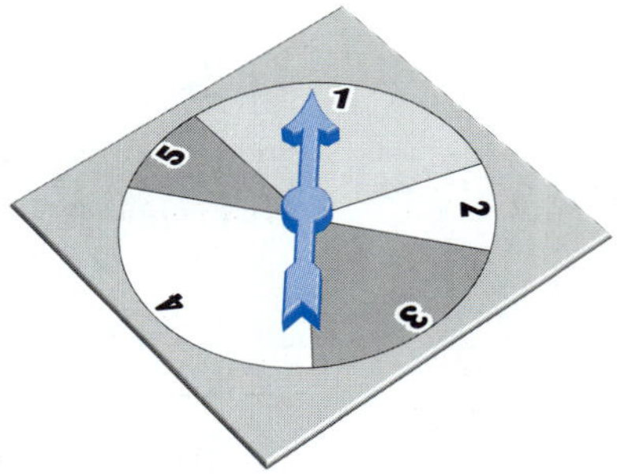

Given the events:

E = Pointer lands on an even number
F = Pointer lands on a number less than 4

(A) Find $P(F|E)$.
(B) Test events E and F for independence.

20. Repeat Problem 19 with the following events:

E = Pointer lands on an odd number
F = Pointer lands on a prime number

Compute the indicated probabilities in Problems 21 and 22 by referring to the following probability tree:

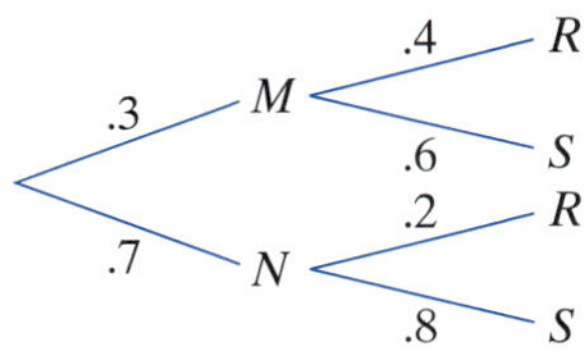

21. (A) $P(M \cap S)$ (B) $P(R)$

22. (A) $P(N \cap R)$ (B) $P(S)$

B

23. A fair coin is tossed twice. Let $S = \{HH, HT, TH, TT\}$ be the sample space of equally likely simple events. We are interested in the following events:

E_1 = A head on the first toss
E_2 = A tail on the first toss
E_3 = A tail on the second toss
E_4 = A head on the second toss

For each pair of events, discuss whether they are independent and whether they are mutually exclusive.
(A) E_1 and E_4 (B) E_1 and E_2

24. For each pair of events (see Problem 23), discuss whether they are independent and whether they are mutually exclusive.
(A) E_1 and E_3 (B) E_3 and E_4

25. In 2 throws of a fair die, what is the probability that you will get an even number on each throw? An even number on the first or second throw?

26. In 2 throws of a fair die, what is the probability that you will get at least 5 on each throw? At least 5 on the first or second throw?

27. Two cards are drawn in succession from a standard 52-card deck. What is the probability that the first card is a club and the second card is a heart:
(A) If the cards are drawn without replacement?
(B) If the cards are drawn with replacement?

28. Two cards are drawn in succession from a standard 52-card deck. What is the probability that both cards are red:
(A) If the cards are drawn without replacement?
(B) If the cards are drawn with replacement?

29. A card is drawn at random from a standard 52-card deck. Events G and H are:

G = The drawn card is black.
H = The drawn card is divisible by 3 (face cards are not valued).

(A) Find $P(H|G)$.
(B) Test H and G for independence.

30. A card is drawn at random from a standard 52-card deck. Events M and N are:

M = The drawn card is a diamond.
N = The drawn card is even (face cards are not valued).

(A) Find $P(N|M)$.
(B) Test M and N for independence.

31. Let A be the event that all of a family's children are the same sex, and let B be the event that the family has at most 1 boy. Assuming the probability of having a girl is the same as the probability of having a boy (both .5), test events A and B for independence if:
(A) The family has 2 children.
(B) The family has 3 children.

32. An experiment consists of tossing n coins. Let A be the event that at least 2 heads turn up, and let B be the event that all the coins turn up the same. Test A and B for independence if:
(A) 2 coins are tossed. (B) 3 coins are tossed.

Problems 33–36 refer to the following experiment: 2 balls are drawn in succession out of a box containing 2 red and 5 white balls. Let R_i be the event that the ith ball is red, and let W_i be the event that the ith ball is white.

33. Construct a probability tree for this experiment and find the probability of each of the events $R_1 \cap R_2$, $R_1 \cap W_2$, $W_1 \cap R_2$, $W_1 \cap W_2$, given that the first ball drawn was:
(A) Replaced before the second draw
(B) Not replaced before the second draw

34. Find the probability that the second ball was red, given that the first ball was:
(A) Replaced before the second draw
(B) Not replaced before the second draw

35. Find the probability that at least 1 ball was red, given that the first ball was:
(A) Replaced before the second draw
(B) Not replaced before the second draw

36. Find the probability that both balls were the same color, given that the first ball was:
(A) Replaced before the second draw
(B) Not replaced before the second draw

In Problems 37 and 38, discuss the validity of each statement. If the statement is always true, explain why. If not, give a counterexample.

37. (A) If A and B are independent events, then

$$P(A|B) = P(B|A)$$

(B) If $P(A \cap B) = P(A)P(B|A)$, then A and B are independent.

38. (A) If two balls are drawn in succession, with replacement, from a box containing m red and n white balls ($m \geq 1$ and $n \geq 1$), then

$$P(W_1 \cap R_2) = P(R_1 \cap W_2)$$

(B) If two balls are drawn in succession, without replacement, from a box containing m red and n white balls ($m \geq 1$ and $n \geq 1$), then

$$P(W_1 \cap R_2) = P(R_1 \cap W_2)$$

C

39. A box contains 2 red, 3 white, and 4 green balls. Two balls are drawn out of the box in succession without replacement. What is the probability that both balls are the same color?

40. For the experiment in Problem 39, what is the probability that no white balls are drawn?

41. An urn contains 2 one-dollar bills, 1 five-dollar bill, and 1 ten-dollar bill. A player draws bills one at a time without replacement from the urn until a ten-dollar bill is drawn. Then the game stops. All bills are kept by the player.
(A) What is the probability of winning \$16?
(B) What is the probability of winning all bills in the urn?
(C) What is the probability of the game stopping at the second draw?

42. Ann and Barbara are playing a tennis match. The first player to win 2 sets wins the match. For any given set, the probability that Ann wins that set is $\frac{2}{3}$. Find the probability that:
(A) Ann wins the match. (B) 3 sets are played.
(C) The player who wins the first set goes on to win the match.

43. Show that if A and B are independent events with nonzero probabilities in a sample space S, then

$$P(A|B) = P(A) \quad \text{and} \quad P(B|A) = P(B)$$

44. Show that if A and B are events with nonzero probabilities in a sample space S, and either $P(A|B) = P(A)$ or $P(B|A) = P(B)$, then events A and B are independent.

45. Show that $P(A|A) = 1$ when $P(A) \neq 0$.

46. Show that $P(A|B) + P(A'|B) = 1$.

47. Show that A and B are dependent if A and B are mutually exclusive and $P(A) \neq 0$, $P(B) \neq 0$.

48. Show that $P(A|B) = 1$ if B is a subset of A and $P(B) \neq 0$.

APPLICATIONS

Business & Economics

49. *Labor relations.* In a study to determine employee voting patterns in a recent strike election, 1,000 employees were selected at random and the following tabulation was made:

		SALARY CLASSIFICATION			
		Hourly (H)	*Salary (S)*	*Salary + Bonus (B)*	*Totals*
TO STRIKE	*Yes (Y)*	400	180	20	600
	No (N)	150	120	130	400
	Totals	550	300	150	1,000

(A) Convert this table to a probability table by dividing each entry by 1,000.
(B) What is the probability of an employee voting to strike, given the person is paid hourly?
(C) What is the probability of an employee voting to strike, given the person receives a salary plus bonus?
(D) What is the probability of an employee being on straight salary (S)? Of being on straight salary given he or she voted in favor of striking?
(E) What is the probability of an employee being paid hourly? Of being paid hourly given he or she voted in favor of striking?
(F) What is the probability of an employee being in a salary plus bonus position and voting against striking?
(G) Are events S and Y independent?
(H) Are events H and Y independent?
(I) Are events B and N independent?

50. *Quality control.* An automobile manufacturer produces 37% of its cars at plant A. If 5% of the cars manufactured at plant A have defective emission control devices, what is the probability that one of this manufacturer's cars was manufactured at plant A and has a defective emission control device?

51. *Bonus incentives.* If a salesperson has gross sales of over \$600,000 in a year, he or she is eligible to play the company's bonus game: A black box contains 1 twenty-dollar bill, 2 five-dollar bills, and 1 one-dollar bill. Bills are drawn out of the box one at a time without replacement until a twenty-dollar bill is drawn. Then the game stops. The salesperson's bonus is 1,000 times the value of the bills drawn.
(A) What is the probability of winning a \$26,000 bonus?
(B) What is the probability of winning the maximum bonus, \$31,000, by drawing out all bills in the box?
(C) What is the probability of the game stopping at the third draw?

52. *Personnel selection.* To transfer into a particular technical department, a company requires an employee to pass a screening test. A maximum of 3 attempts are allowed at 6 month intervals between trials. From past records it is found that 40% pass on the first trial; of those that fail the first trial and take the test a second time, 60% pass; and of those that fail on the second trial and take the test a third time, 20% pass. For an employee wishing to transfer:
(A) What is the probability of passing the test on the first or second try?
(B) What is the probability of failing on the first 2 trials and passing on the third?
(C) What is the probability of failing on all 3 attempts?

Life Sciences

53. *Food and Drug Administration.* An ice cream company wishes to use a red dye to enhance the color in its strawberry ice cream. The Food and Drug Administration (FDA) requires the dye to be tested for cancer-producing potential in a laboratory using laboratory rats. The results of one test on 1,000 rats are summarized in the following table:

		Developed cancer C	No cancer C'	*Totals*
Ate red dye	R	60	440	500
Did not eat red dye	R'	20	480	500
	Totals	80	920	1,000

(A) Convert the table into a probability table by dividing each entry by 1,000.
(B) Are "developing cancer" and "eating red dye" independent events?
(C) Should the FDA approve or ban the use of the dye? Explain why or why not using $P(C|R)$ and $P(C)$.
(D) Suppose the number of rats that ate red dye and developed cancer was 20, but the number that developed cancer was still 80 and the number that ate red dye was still 500. What should the FDA do, based on these results? Explain why.

54. *Genetics.* In a study to determine frequency and dependency of color-blindness relative to females and males, 1,000 people were chosen at random and the following results were recorded:

		Female F	Male F'	*Totals*
Color-blind	C	2	24	26
Normal	C'	518	456	974
	Totals	520	480	1,000

(A) Convert this table to a probability table by dividing each entry by 1,000.
(B) What is the probability that a person is a woman, given that the person is color-blind?
(C) What is the probability that a person is color-blind, given that the person is a male?
(D) Are the events color-blindness and male independent?
(E) Are the events color-blindness and female independent?

Social Sciences

55. *Psychology.* In a study to determine the frequency and dependency of IQ ranges relative to males and females, 1,000 people were chosen at random and the following results were recorded:

		IQ Below 90 (A)	IQ 90–120 (B)	IQ Above 120 (C)	*Totals*
Female	F	130	286	104	520
Male	F'	120	264	96	480
	Totals	250	550	200	1,000

(A) Convert this table to a probability table by dividing each entry by 1,000.
(B) What is the probability of a person having an IQ below 90, given that the person is a female? A male?
(C) What is the probability of a person having an IQ above 120, given that the person is a female? A male?
(D) What is the probability of a person having an IQ below 90?
(E) What is the probability of a person having an IQ between 90 and 120? Of a person having an IQ between 90 and 120, given that the person is a male?
(F) What is the probability of a person being female and having an IQ above 120?
(G) Are any of the events A, B, or C dependent relative to F or F'?

56. *Voting patterns.* A survey of the residents of a precinct in a large city revealed that 55% of the residents were members of the Democratic party and that 60% of the Democratic party members voted in the last election. What is the probability that a person selected at random from the residents of this precinct is a member of the Democratic party and voted in the last election?

SECTION 6-6

Bayes' Formula

In the preceding section we discussed the conditional probability of the occurrence of an event, given the occurrence of an earlier event. Now we are going to reverse the problem and try to find the probability of an earlier event conditioned on the occurrence of a later event. As you will see before the discussion is over, a number of practical problems are of this form. First, let us consider a relatively simple problem that will provide the basis for a generalization.

Example 1 **Probability of an Earlier Event Given a Later Event** One urn has 3 blue and 2 white balls; a second urn has 1 blue and 3 white balls (Fig. 1). A single fair die is rolled and if 1 or 2 comes up, a ball is drawn out of the first urn; otherwise, a ball is drawn out of the second urn. If the drawn ball is blue, what is the probability that it came out of the first urn? Out of the second urn?

Urn U_1 Urn U_2

FIGURE 1

SOLUTION We form a probability tree, letting U_1 represent urn 1, U_2 urn 2, B a blue ball, and W a white ball. Then, on the various outcome branches we assign appropriate probabilities. For example, $P(U_1) = \frac{1}{3}$, $P(B|U_1) = \frac{3}{5}$, and so on:

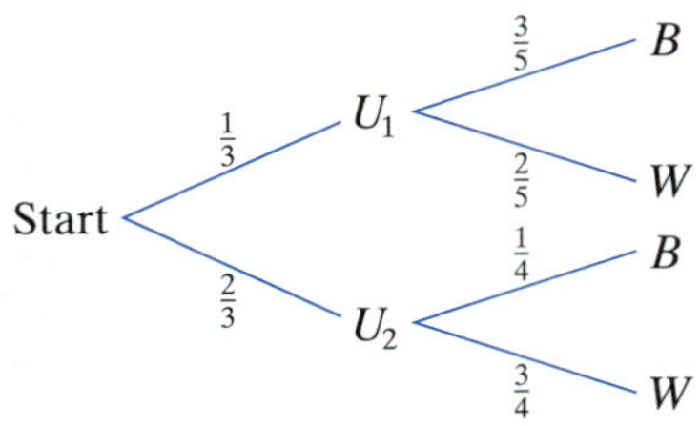

Now we are interested in finding $P(U_1|B)$; that is, the probability that the ball came out of urn 1, given the drawn ball is blue. Using equation (1) from Section 6-5, we can write

$$P(U_1|B) = \frac{P(U_1 \cap B)}{P(B)} \tag{1}$$

If we look at the tree diagram, we can see that B is at the end of two different branches; thus,

$$P(B) = P(U_1 \cap B) + P(U_2 \cap B) \tag{2}$$

After substituting equation (2) into equation (1), we get

$$\begin{aligned} P(U_1|B) &= \frac{P(U_1 \cap B)}{P(U_1 \cap B) + P(U_2 \cap B)} \qquad P(A \cap B) = P(A)P(B|A) \\ &= \frac{P(U_1)P(B|U_1)}{P(U_1)P(B|U_1) + P(U_2)P(B|U_2)} \\ &= \frac{P(B|U_1)P(U_1)}{P(B|U_1)P(U_1) + P(B|U_2)P(U_2)} \end{aligned} \tag{3}$$

Formula (3) is really a lot simpler to use than it looks. You do not need to memorize it; you simply need to understand its form relative to the probability tree above. Referring to the probability tree, we see that

$$\begin{aligned} P(B|U_1)P(U_1) &= \text{Product of branch probabilites leading to } B \text{ through } U_1 \\ &= \left(\tfrac{3}{5}\right)\left(\tfrac{1}{3}\right) \qquad \text{We usually start at } B \text{ and work back through } U_1. \\ P(B|U_2)P(U_2) &= \text{Product of branch probabilities leading to } B \text{ through } U_2 \\ &= \left(\tfrac{1}{4}\right)\left(\tfrac{2}{3}\right) \qquad \text{We usually start at } B \text{ and work back through } U_2. \end{aligned}$$

Equation (3) now can be interpreted in terms of the probability tree as follows:

$$P(U_1|B) = \frac{\text{Product of branch probabilities leading to } B \text{ through } U_1}{\text{Sum of all branch products leading to } B}$$

$$= \frac{(\frac{3}{5})(\frac{1}{3})}{(\frac{3}{5})(\frac{1}{3}) + (\frac{1}{4})(\frac{2}{3})} = \frac{6}{11} \approx .55$$

Similarly,

$$P(U_2|B) = \frac{\text{Product of branch probabilities leading to } B \text{ through } U_2}{\text{Sum of all branch products leading to } B}$$

$$= \frac{(\frac{1}{4})(\frac{2}{3})}{(\frac{3}{5})(\frac{1}{3}) + (\frac{1}{4})(\frac{2}{3})} = \frac{5}{11} \approx .45$$

[*Note*: We also could have obtained $P(U_2|B)$ by subtracting $P(U_1|B)$ from 1. Why?]

Matched Problem 1 Repeat Example 1, but find $P(U_1|W)$ and $P(U_2|W)$.

Explore–Discuss 1

Given the probability tree

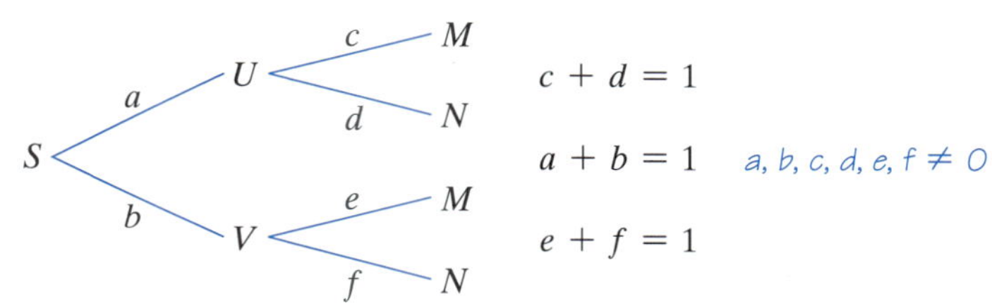

(A) Discuss the difference between $P(M|U)$ and $P(U|M)$, and between $P(N|V)$ and $P(V|N)$, in terms of *a, b, c, d, e,* and *f*.

(B) Show that $ac + ad + be + bf = 1$. What is the significance of this result?

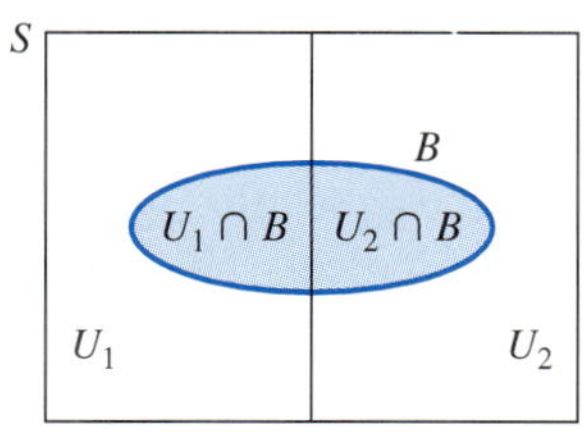

FIGURE 2

In generalizing the results in Example 1, it is helpful to look at its structure in terms of the Venn diagram shown in Figure 2. We note that U_1 and U_2 are mutually exclusive (disjoint), and their union forms *S*. The following two equations can now be interpreted in terms of this diagram:

$$P(U_1|B) = \frac{P(U_1 \cap B)}{P(B)} = \frac{P(U_1 \cap B)}{P(U_1 \cap B) + P(U_2 \cap B)}$$

$$P(U_2|B) = \frac{P(U_2 \cap B)}{P(B)} = \frac{P(U_2 \cap B)}{P(U_1 \cap B) + P(U_2 \cap B)}$$

Look over the equations and the diagram carefully.

Of course, there is no reason to stop here. Suppose U_1, U_2, and U_3 are three mutually exclusive events whose union is the whole sample space *S*. Then, for an arbitrary event *E* in *S*, with $P(E) \neq 0$, the corresponding Venn diagram looks like Figure 3, and

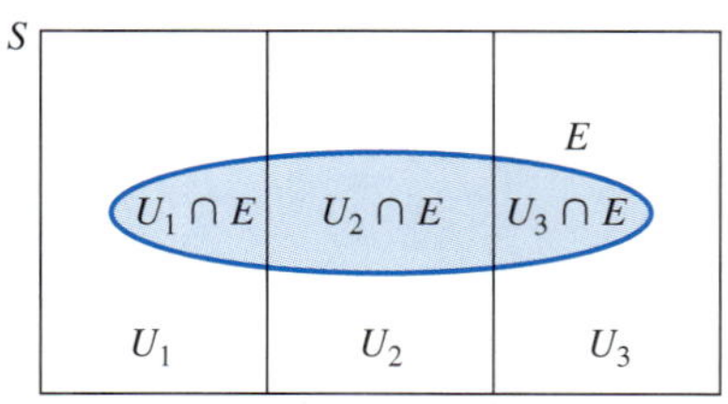

FIGURE 3

$$P(U_1|E) = \frac{P(U_1 \cap E)}{P(E)} = \frac{P(U_1 \cap E)}{P(U_1 \cap E) + P(U_2 \cap E) + P(U_3 \cap E)}$$

Similar results hold for U_2 and U_3.

Reasoning in the same way, we arrive at the following famous theorem, which was first stated by the Presbyterian minister Thomas Bayes (1702–1763):

Theorem 1 BAYES' FORMULA

Let $U_1, U_2, \ldots, U_n$ be n mutually exclusive events whose union is the sample space S. Let E be an arbitrary event in S such that $P(E) \neq 0$. Then

$$
\begin{aligned}
P(U_1|E) &= \frac{P(U_1 \cap E)}{P(E)} \\
&= \frac{P(U_1 \cap E)}{P(U_1 \cap E) + P(U_2 \cap E) + \cdots + P(U_n \cap E)} \\
&= \frac{P(E|U_1)P(U_1)}{P(E|U_1)P(U_1) + P(E|U_2)P(U_2) + \cdots + P(E|U_n)P(U_n)}
\end{aligned}
$$

Similar results hold for $U_2, U_3, \ldots, U_n$.

You do not need to memorize Bayes' formula. In practice, it is often easier to draw a probability tree and use the following:

Bayes' Formula and Probability Trees

$$P(U_1|E) = \frac{\text{Product of branch probabilities leading to } E \text{ through } U_1}{\text{Sum of all branch products leading to } E}$$

Similar results hold for $U_2, U_3, \ldots, U_n$.

Example 2 **Tuberculosis Screening** A new, inexpensive skin test is devised for detecting tuberculosis. To evaluate the test before it is put into use, a medical researcher randomly selects 1,000 people. Using precise but more expensive methods already available, it is found that 8% of the 1,000 people tested have tuberculosis. Now each of the 1,000 subjects is given the new skin test and the following results are recorded: The test indicates tuberculosis in 96% of those who have it and in 2% of those who do not. Based on these results, what is the probability of a randomly chosen person having tuberculosis given that the skin test indicates the disease? What is the probability of a person not having tuberculosis given that the skin test indicates the disease? (That is, what is the probability of the skin test giving a *false positive result?*)

SOLUTION Now we will see the power of Bayes' formula in an important application. To start, we form a tree diagram and place appropriate probabilities on each branch:

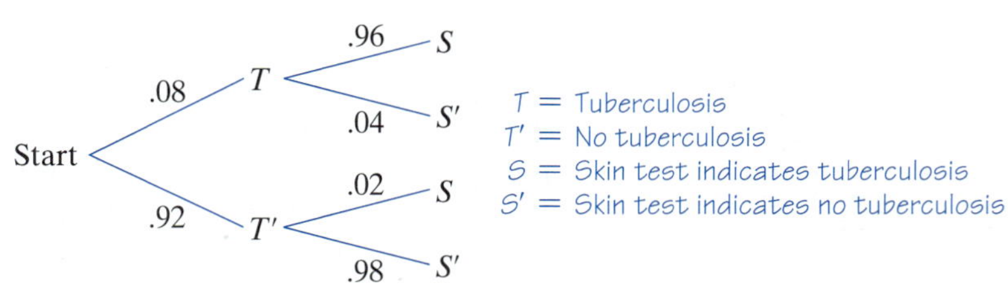

We are interested in finding $P(T|S)$; that is, the probability of a person having tuberculosis given that the skin test indicates the disease. Bayes' formula for this case is

$$P(T|S) = \frac{\text{Product of branch probabilities leading to } S \text{ through } T}{\text{Sum of all branch products leading to } S}$$

Substituting appropriate values from the probability tree, we obtain

$$P(T|S) = \frac{(.08)(.96)}{(.08)(.96) + (.92)(.02)} = .81$$

The probability of a person not having tuberculosis given that the skin test indicates the disease, denoted by $P(T'|S)$, is

$$P(T'|S) = 1 - P(T|S) = 1 - .81 = .19 \qquad P(T|S) + P(T'|S) = 1$$

Other important questions that need to be answered are indicated in Matched Problem 2.

Matched Problem 2 What is the probability that a person has tuberculosis given that the test indicates no tuberculosis is present? (That is, what is the probability of the skin test giving a *false negative result?*) What is the probability that a person does not have tuberculosis given that the test indicates no tuberculosis is present?

Example 3 **Product Defects** A company produces 1,000 refrigerators a week at three plants. Plant A produces 350 refrigerators a week, plant B produces 250 refrigerators a week, and plant C produces 400 refrigerators a week. Production records indicate that 5% of the refrigerators produced at plant A will be defective, 3% of those produced at plant B will be defective, and 7% of those produced at plant C will be defective. All the refrigerators are shipped to a central warehouse. If a refrigerator at the warehouse is found to be defective, what is the probability that it was produced at plant A?

SOLUTION We begin by constructing a tree diagram:

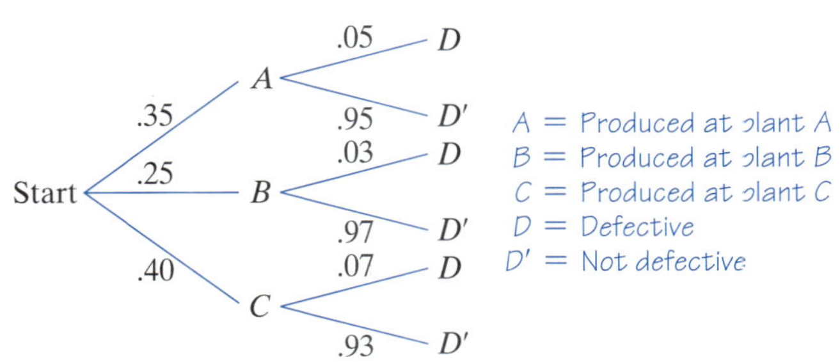

The probability that a defective refrigerator was produced at plant A is $P(A|D)$. Bayes' formula for this case is

$$P(A|D) = \frac{\text{Product of branch probabilities leading to } D \text{ through } A}{\text{Sum of all branch products leading to } D}$$

Using the values from the probability tree, we have

$$P(A|D) = \frac{(.35)(.05)}{(.35)(.05) + (.25)(.03) + (.40)(.07)} \approx .33$$

Matched Problem 3 In Example 3, what is the probability that a defective refrigerator in the warehouse was produced at plant B? At plant C?

Explore–Discuss 2

Given the probability tree:

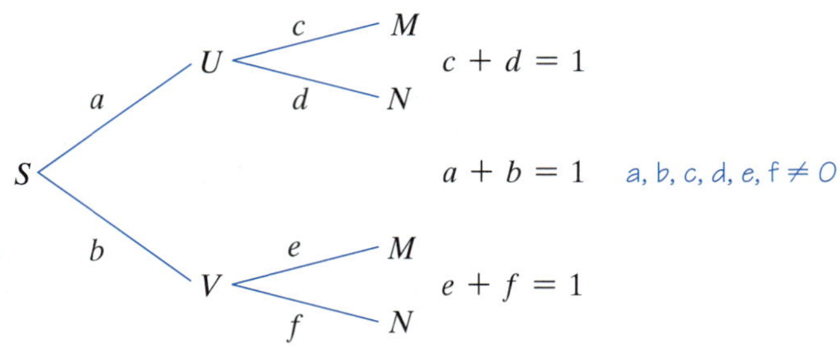

Suppose U and M are independent events. Discuss the restrictions that their independence forces on the probabilities a, b, c, d, e, and f.

Answers to Matched Problems

1. $P(U_1|W) = \frac{4}{19} \approx .21$; $P(U_2|W) = \frac{15}{19} \approx .79$
2. $P(T|S') = .004$; $P(T'|S') = .996$ **3.** $P(B|D) \approx .14$; $P(C|D) \approx .53$

Exercise 6-6

A *Find the probabilities in Problems 1–6 by referring to the tree diagram below.*

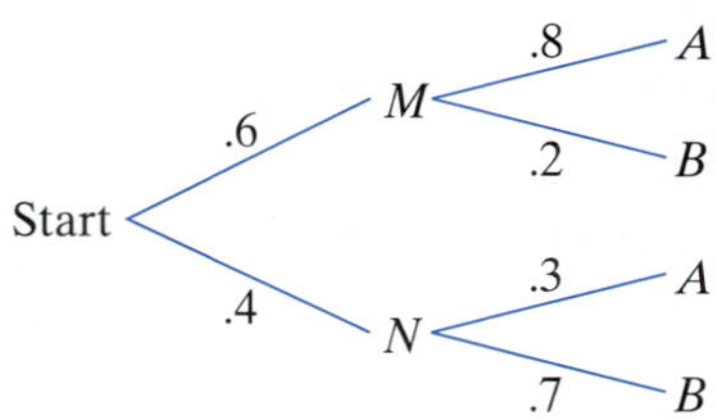

1. $P(M \cap A) = P(M)P(A|M)$

2. $P(N \cap B) = P(N)P(B|N)$

3. $P(A) = P(M \cap A) + P(N \cap A)$

4. $P(B) = P(M \cap B) + P(N \cap B)$

5. $P(M|A) = \dfrac{P(M \cap A)}{P(M \cap A) + P(N \cap A)}$

6. $P(N|B) = \dfrac{P(N \cap B)}{P(N \cap B) + P(M \cap B)}$

Find the probabilities in Problems 7–10 by referring to the following Venn diagram and using Bayes' formula (assume that the simple events in S are equally likely):

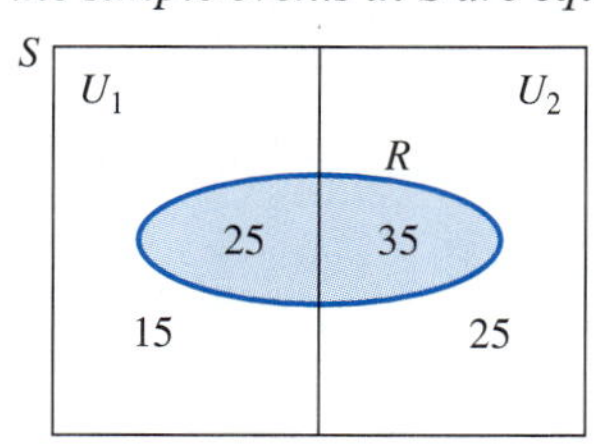

7. $P(U_1|R)$

8. $P(U_2|R)$

9. $P(U_1|R')$

10. $P(U_2|R')$

B *Find the probabilities in Problems 11–16 by referring to the following tree diagram and using Bayes' formula:*

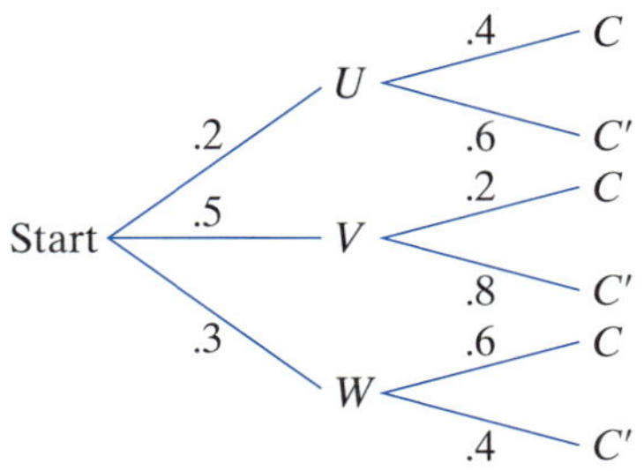

11. $P(U|C)$ **12.** $P(V|C')$ **13.** $P(W|C)$

14. $P(U|C')$ **15.** $P(V|C)$ **16.** $P(W|C')$

Find the probabilities in Problems 17–22 by referring to the following Venn diagram and using Bayes' formula (assume that the simple events in S are equally likely):

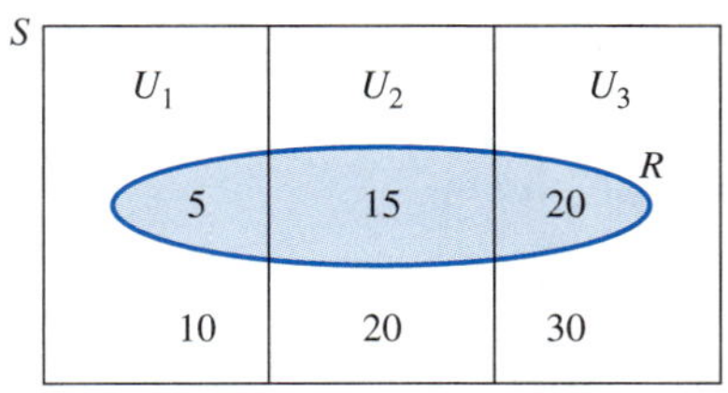

17. $P(U_1|R)$ **18.** $P(U_2|R')$ **19.** $P(U_3|R)$

20. $P(U_1|R')$ **21.** $P(U_2|R)$ **22.** $P(U_3|R')$

In Problems 23 and 24, use the probabilities in the first tree diagram to find the probability of each branch of the second tree diagram.

23.

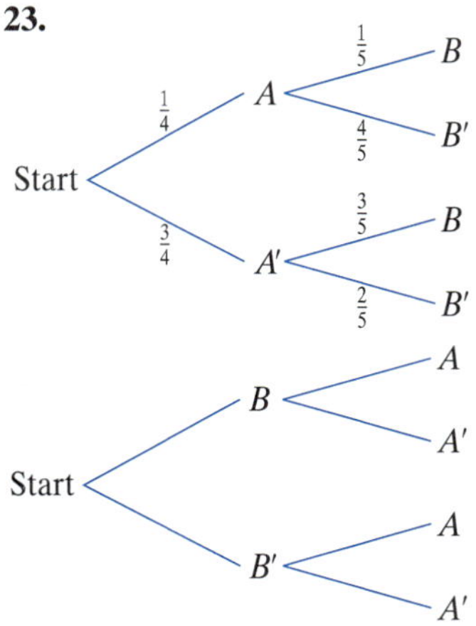

24.

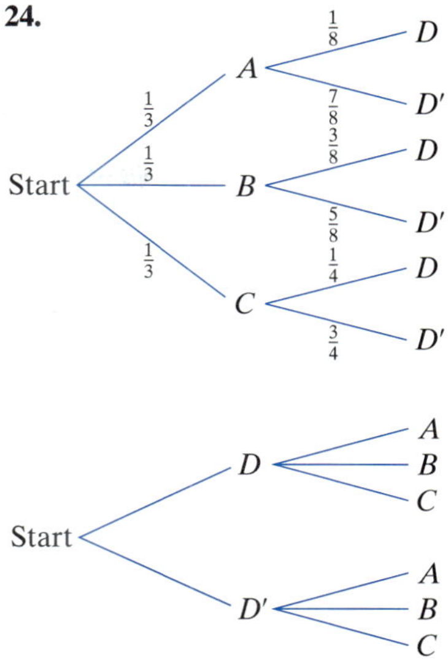

In Problems 25–28, one of two urns is chosen at random with one as likely to be chosen as the other. Then a ball is withdrawn from the chosen urn. Urn 1 contains 1 white and 4 red balls, and urn 2 has 3 white and 2 red balls.

25. If a white ball is drawn, what is the probability that it came from urn 1?

26. If a white ball is drawn, what is the probability that it came from urn 2?

27. If a red ball is drawn, what is the probability that it came from urn 2?

28. If a red ball is drawn, what is the probability that it came from urn 1?

In Problems 29 and 30, an urn contains 4 red and 5 white balls. Two balls are drawn in succession without replacement.

29. If the second ball is white, what is the probability that the first ball was white?

30. If the second ball is red, what is the probability that the first ball was red?

In Problems 31 and 32, urn 1 contains 7 red and 3 white balls. Urn 2 contains 4 red and 5 white balls. A ball is drawn from urn 1 and placed in urn 2. Then a ball is drawn from urn 2.

31. If the ball drawn from urn 2 is red, what is the probability that the ball drawn from urn 1 was red?

32. If the ball drawn from urn 2 is white, what is the probability that the ball drawn from urn 1 was white?

In Problems 33 and 34, refer to the following probability tree:

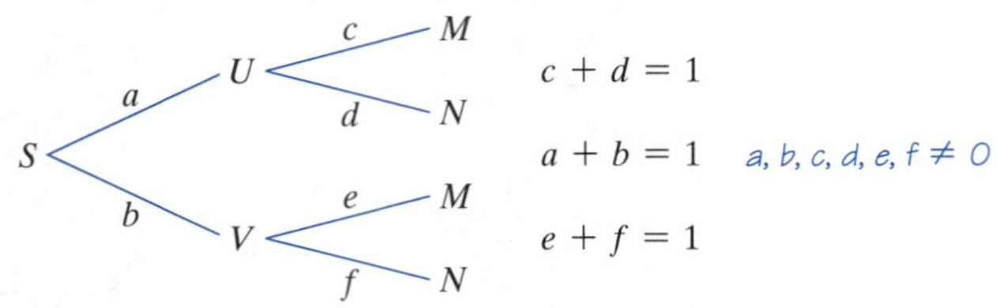

33. Suppose that $c = e$. Discuss the dependence or independence of events U and M.

34. Suppose that $c = d = e = f$. Discuss the dependence or independence of events M and N.

In Problems 35 and 36, two balls are drawn in succession from an urn containing m blue balls and n white balls ($m \geq 2$ and $n \geq 2$). Discuss the validity of each statement. If the statement is always true, explain why. If not, give a counterexample.

35. (A) If the two balls are drawn with replacement, then $P(B_1|B_2) = P(B_2|B_1)$.
(B) If the two balls are drawn without replacement, then $P(B_1|B_2) = P(B_2|B_1)$.

36. (A) If the two balls are drawn with replacement, then $P(B_1|W_2) = P(W_2|B_1)$.
(B) If the two balls are drawn without replacement, then $P(B_1|W_2) = P(W_2|B_1)$.

C

37. If 2 cards are drawn in succession from a standard 52-card deck without replacement and the second card is a heart, what is the probability that the first card is a heart?

38. A box contains 10 balls numbered 1 through 10. Two balls are drawn in succession without replacement. If the second ball drawn has the number 4 on it, what is the probability that the first ball had a smaller number on it? An even number on it?

39. Show that $P(U_1|R) + P(U_1'|R) = 1$.

40. If U_1 and U_2 are two mutually exclusive events whose union is the equally likely sample space S and if E is an arbitrary event in S such that $P(E) \neq 0$, show that

$$P(U_1|E) = \frac{n(U_1 \cap E)}{n(U_1 \cap E) + n(U_2 \cap E)}$$

Applications

In the following applications, the word "probability" is often understood to mean "approximate empirical probability."

Business & Economics

41. *Employee screening.* The management of a company finds that 30% of the secretaries hired are unsatisfactory. The personnel director is instructed to devise a test that will improve the situation. One hundred employed secretaries are chosen at random and are given a newly constructed test. Out of these, 90% of the successful secretaries pass the test and 20% of the unsuccessful secretaries pass. Based on these results, if a person applies for a secretarial job, takes the test, and passes it, what is the probability that he or she is a good secretary? If the applicant fails the test, what is the probability that he or she is a good secretary?

42. *Employee rating.* A company has rated 75% of its employees as satisfactory and 25% as unsatisfactory. Personnel records indicate that 80% of the satisfactory workers had previous work experience, while only 40% of the unsatisfactory workers had any previous work experience. If a person with previous work experience is hired, what is the probability that this person will be a satisfactory employee? If a person with no previous work experience is hired, what is the probability that this person will be a satisfactory employee?

43. *Product defects.* A manufacturer obtains clock–radios from three different subcontractors: 20% from *A*, 40% from *B*, and 40% from *C*. The defective rates for these subcontractors are 1%, 3%, and 2%, respectively. If a defective clock–radio is returned by a customer, what is the probability that it came from subcontractor *A*? From *B*? From *C*?

44. *Product defects.* A computer store sells three types of microcomputers, brand *A*, brand *B*, and brand *C*. Of the computers they sell, 60% are brand *A*, 25% are brand *B*, and 15% are brand *C*. They have found that 20% of the brand *A* computers, 15% of the brand *B* computers, and 5% of the brand *C* computers are returned for service during the warranty period. If a computer is returned for service during the warranty period, what is the probability that it is a brand *A* computer? A brand *B* computer? A brand *C* computer?

Life Sciences

45. *Cancer screening.* A new, simple test has been developed to detect a particular type of cancer. The test must be evaluated before it is put into use. A medical researcher selects a random sample of 1,000 adults and finds (by other means) that 2% have this type of cancer. Each of the 1,000 adults is given the test, and it is found that the test indicates cancer in 98% of those who have it and in 1% of those who do not. Based on these results, what is the probability of a randomly chosen person having cancer given that the test indicates cancer? Of a person having cancer given that the test does not indicate cancer?

46. *Pregnancy testing.* In a random sample of 200 women who suspect that they are pregnant, 100 turn out to be pregnant. A new pregnancy test given to these women indicated pregnancy in 92 of the 100 pregnant women and in 12 of the 100 nonpregnant women. If a woman suspects she is pregnant and this test indicates that she is pregnant, what is the probability that she is pregnant? If the test indicates that she is not pregnant, what is the probability that she is not pregnant?

47. *Medical research.* In a random sample of 1,000 people, it is found that 7% have a liver ailment. Of those who have a liver ailment, 40% are heavy drinkers, 50% are moderate drinkers, and 10% are nondrinkers. Of those who do not have a liver ailment, 10% are heavy drinkers, 70% are moderate drinkers, and 20% are nondrinkers. If a person is chosen at random and it is found that he or she is a heavy drinker, what is the probability of that person having a liver ailment? What is the probability for a nondrinker?

48. *Tuberculosis screening.* A test for tuberculosis was given to 1,000 subjects, 8% of whom were known to have tuberculosis. For the subjects who had tuberculosis, the test indicated tuberculosis in 90% of the subjects, was inconclusive for 7%, and indicated no tuberculosis in 3%. For the subjects who did not have tuberculosis, the test indicated tuberculosis in 5% of the subjects, was inconclusive for 10%, and indicated no tuberculosis in the remaining 85%. What is the probability of a randomly selected person having tuberculosis given that the test indicates tuberculosis? Of not having tuberculosis given that the test was inconclusive?

Social Sciences

49. *Police science.* A new lie-detector test has been devised and must be tested before it is put into use. One hundred people are selected at random, and each person draws and keeps a card from a box of 100 cards. Half the cards instruct the person to lie and the others instruct the person to tell the truth. The test indicates lying in 80% of those who lied and in 5% of those who did not. What is the probability that a randomly chosen subject will have lied given that the test indicates lying? That the subject will not have lied given that the test indicates lying?

50. *Politics.* In a given county, records show that of the registered voters, 45% are Democrats, 35% are Republicans, and 20% are independents. In an election, 70% of the Democrats, 40% of the Republicans, and 80% of the independents voted in favor of a parks and recreation bond proposal. If a registered voter chosen at random is found to have voted in favor of the bond, what is the probability that the voter is a Republican? An independent? A Democrat?

SECTION 6-7

Random Variable, Probability Distribution, and Expected Value

- RANDOM VARIABLE; PROBABILITY DISTRIBUTION
- EXPECTED VALUE OF A RANDOM VARIABLE
- DECISION-MAKING AND EXPECTED VALUE

■ RANDOM VARIABLE; PROBABILITY DISTRIBUTION

When performing a random experiment, a sample space S is selected in such a way that all probability problems of interest relative to the experiment can be solved. In many situations we may not be interested in each simple event in the sample space S but in some numerical value associated with the event.

For example, if 3 coins are tossed, we may be interested in the number of heads that turn up rather than in the particular pattern that turns up. Or, in selecting a random sample of students, we may be interested in the proportion that are women rather than which particular students are women. In the same way, a "craps" player is usually interested in the sum of the dots on the showing faces of the dice rather than the pattern of dots on each face.

In each of these examples, we have a rule that assigns to each simple event in S a single real number. Mathematically speaking, we are dealing with a function (see Section 1-1). Historically, this particular type of function has been called a "random variable."

Random Variable

A **random variable** is a function that assigns a numerical value to each simple event in a sample space S.

The term *random variable* is an unfortunate choice, since it is neither random nor a variable—it is a function with a numerical value, and it is defined on a sample space. But the terminology has stuck and is now standard, so we shall have to live with it. Capital letters, such as X, are used to represent random variables.

Let us return to the experiment of tossing 3 coins. A sample space S of equally likely simple events is indicated in Table 1. Suppose we are interested in the number of heads (0, 1, 2, or 3) appearing on each toss of the 3 coins and the probability of each of these events. We introduce a random variable X (a function) that indicates the number of heads for each simple event in S (see the second column in Table 1). For example, $X(e_1) = 0$, $X(e_2) = 1$, and so on. The random variable X assigns a numerical value to each simple event in the sample space S.

Table 1
Number of Heads in the Toss of 3 Coins

Sample space S	Number of heads $X(e_j)$
e_1: TTT	0
e_2: TTH	1
e_3: THT	1
e_4: HTT	1
e_5: THH	2
e_6: HTH	2
e_7: HHT	2
e_8: HHH	3

We are interested in the probability of the occurrence of each image or range value of X; that is, in the probability of the occurrence of 0 heads, 1 head, 2 heads, or 3 heads in the single toss of 3 coins. We indicate this probability by

$$p(x) \qquad \text{where} \quad x \in \{0, 1, 2, 3\}$$

The function p is called the **probability function* of the random variable X.**

What is $p(2)$, the probability of getting exactly 2 heads on the single toss of 3 coins? "Exactly 2 heads occur" is the event

$$E = \{\text{THH, HTH, HHT}\}$$

Thus,

$$p(2) = \frac{n(E)}{n(S)} = \frac{3}{8}$$

Proceeding similarly for $p(0)$, $p(1)$, and $p(3)$, we obtain the results in Table 2. This table is called a **probability distribution for the random variable X.** Probability distributions are also represented graphically, as shown in Figure 1. The graph of a probability distribution is often called a **histogram.**

*The probability function p of the random variable X is defined by $p(x) = P(\{e_i \in S | X(e_i) = x\})$, which, because of its cumbersome nature, is usually simplified to $p(x) = P(X = x)$ or, simply, $p(x)$. We will use the simplified notation.

Table 2

PROBABILITY DISTRIBUTION

NUMBER OF HEADS x	0	1	2	3
PROBABILITY $p(x)$	$\frac{1}{8}$	$\frac{3}{8}$	$\frac{3}{8}$	$\frac{1}{8}$

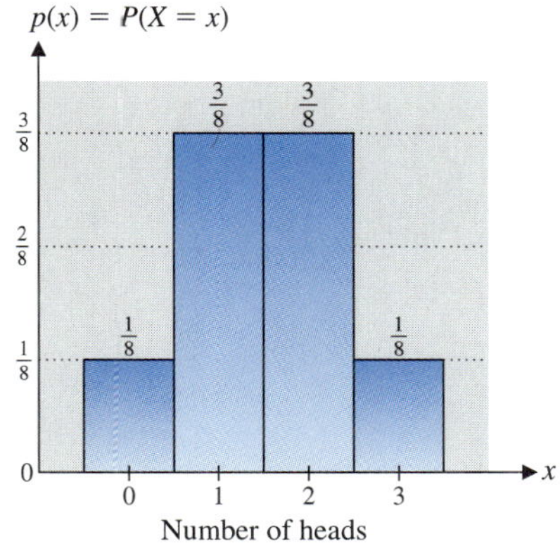

FIGURE 1

Histogram for a probability distribution

Note from Table 2 or Figure 1 that

1. $0 \le p(x) \le 1, \quad x \in \{0, 1, 2, 3\}$
2. $p(0) + p(1) + p(2) + p(3) = \frac{1}{8} + \frac{3}{8} + \frac{3}{8} + \frac{1}{8} = 1$

These are general properties that any probability distribution of a random variable X associated with a finite sample space must have.

Probability Distribution of a Random Variable X

A probability function $P(X = x) = p(x)$ is a **probability distribution of the random variable X** if

1. $0 \le p(x) \le 1, \quad x \in \{x_1, x_2, \ldots, x_n\}$
2. $p(x_1) + p(x_2) + \cdots + p(x_n) = 1$

where $\{x_1, x_2, \ldots, x_n\}$ are the (range) values of X (see Fig. 2).

Figure 2 illustrates the process of forming a probability distribution of a random variable.

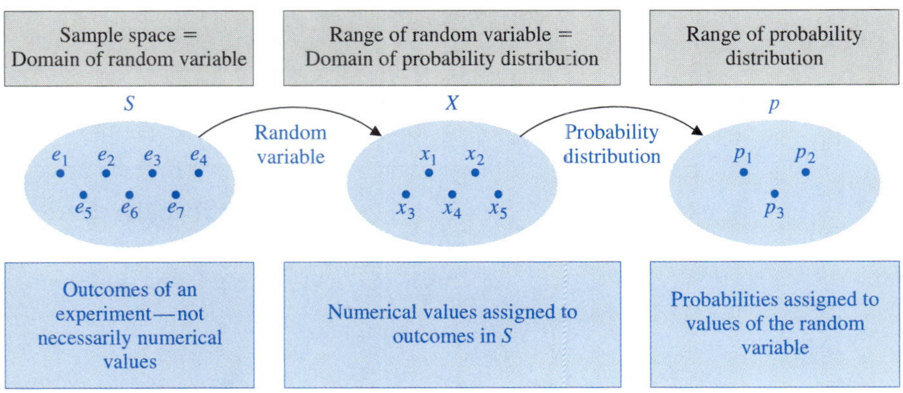

FIGURE 2

Probability distribution of a random variable for a finite sample space

Expected Value of a Random Variable

Suppose the experiment of tossing 3 coins was repeated a large number of times. What would be the average number of heads per toss (the total number of heads in all tosses divided by the total number of tosses)? Consulting the probability distribution in Table 2 or Figure 1, we see that we would expect to toss 0 heads $\frac{1}{8}$ of the time, 1 head $\frac{3}{8}$ of the time, 2 heads $\frac{3}{8}$ of the time, and 3 heads $\frac{1}{8}$ of the time. Thus, in the long run, we would expect the average number of heads per toss of the 3 coins, or the *expected value* $E(X)$, to given by

$$E(X) = 0(\tfrac{1}{8}) + 1(\tfrac{3}{8}) + 2(\tfrac{3}{8}) + 3(\tfrac{1}{8}) = \tfrac{12}{8} = 1.5$$

It is important to note that the expected value is not a value that will necessarily occur in a single experiment (1.5 heads cannot occur in the toss of 3 coins), but it is an average of what occurs over a large number of experiments. Sometimes we will toss more than 1.5 heads and sometimes less, but if the experiment is repeated many times, the average number of heads per experiment should be close to 1.5.

We now make the above discussion more precise through the following definition of expected value:

Expected Value of a Random Variable X

Given the probability distribution for the random variable X,

x_i	x_1	x_2	$\dots$	x_n
p_i	p_1	p_2	$\dots$	p_n

where $p_i = p(x_i)$, we define the **expected value of X,** denoted **$E(X)$,** by the formula

$$E(X) = x_1p_1 + x_2p_2 + \cdots + x_np_n$$

We again emphasize that the expected value is not to be expected to occur in a single experiment; it is a long-run average of repeated experiments—it is the weighted average of the possible outcomes, each weighted by its probability.

Steps for Computing the Expected Value of a Random Variable X

Step 1. Form the probability distribution of the random variable X.

Step 2. Multiply each image value of X, x_i, by its corresponding probability of occurrence p_i; then add the results.

Example 1 **Expected Value** What is the expected value (long-run average) of the number of dots facing up for the roll of a single die?

Solution If we choose

$$S = \{1, 2, 3, 4, 5, 6\}$$

as our sample space, then each simple event is a numerical outcome reflecting our interest, and each is equally likely. The random variable X in this case is just the identity function (each number is associated with itself). Thus, the probability distribution for X is

x_i	1	2	3	4	5	6
p_i	$\frac{1}{6}$	$\frac{1}{6}$	$\frac{1}{6}$	$\frac{1}{6}$	$\frac{1}{6}$	$\frac{1}{6}$

Hence,

$$E(X) = 1(\tfrac{1}{6}) + 2(\tfrac{1}{6}) + 3(\tfrac{1}{6}) + 4(\tfrac{1}{6}) + 5(\tfrac{1}{6}) + 6(\tfrac{1}{6})$$
$$= \tfrac{21}{6} = 3.5$$

Matched Problem 1 Suppose the die in Example 1 is not fair and we obtain (empirically) the following probability distribution for X:

x_i	1	2	3	4	5	6
p_i	.14	.13	.18	.20	.11	.24

[*Note:* Sum = 1.]

What is the expected value of X?

Explore–Discuss 1

In a class of 10 students, 1 student scored 95 on the first exam, 3 scored 85, 1 scored 82, 2 scored 75, 2 scored 73, and 1 scored 65. The probability distribution of an exam score for a student selected at random from the class is as follows:

x_i	65	73	75	82	85	95
p_i	.1	.2	.2	.1	.3	.1

Discuss the relationship between the expected value of the probability distribution and the class average for the exam.

Example 2 **Expected Value** A carton of 20 calculator batteries contains 2 dead ones. A random sample of 3 is selected from the 20 and tested. Let X be the random variable associated with the number of dead batteries found in a sample.

(A) Find the probability distribution of X.
(B) Find the expected number of dead batteries in a sample.

Solution (A) The number of ways of selecting a sample of 3 from 20 (order is not important) is $C_{20,3}$. This is the number of simple events in the experiment, each as likely as the other. A sample will have either 0, 1, or 2 dead batteries. These are the values of the random variable in which we are interested. The probability distribution is computed as follows:

$$p(0) = \frac{C_{18,3}}{C_{20,3}} \approx .716 \qquad p(1) = \frac{C_{2,1}C_{18,2}}{C_{20,3}} \approx .268 \qquad p(2) = \frac{C_{2,2}C_{18,1}}{C_{20,3}} \approx .016$$

We summarize the above results in a convenient table:

x_i	0	1	2
p_i	.716	.268	.016

[*Note:* $.716 + .268 + .016 = 1$.]

(B) The expected number of dead batteries in a sample is readily computed as follows:

$$E(X) = (0)(.716) + (1)(.268) + (2)(.016) = .3$$

The expected value is not one of the random variable values; rather, it is a number that the average number of dead batteries in a sample would approach as the experiment is repeated without end.

Matched Problem 2 Repeat Example 2 using a random sample of 4.

Example 3 **Expected Value of a Game** A spinner device is numbered from 0 to 5, and each of the 6 numbers is as likely to come up as any other. A player who bets \$1 on any given number wins \$4 (and gets the bet back) if the pointer comes to rest on the chosen number; otherwise, the \$1 bet is lost. What is the expected value of the game (long-run average gain or loss per game)?

Solution The sample space of equally likely events is

$$S = \{0, 1, 2, 3, 4, 5\}$$

Each sample point occurs with a probability of $\frac{1}{6}$. The random variable X assigns \$4 to the winning number and −\$1 to each of the remaining numbers. Thus, the probability distribution for X, called a **payoff table,** is as shown in the margin. The probability of winning \$4 is $\frac{1}{6}$ and of losing \$1 is $\frac{5}{6}$. We can now compute the expected value of the game:

Payoff Table (Probability Distribution for X)

x_i	\$4	−\$1
p_i	$\frac{1}{6}$	$\frac{5}{6}$

$$E(X) = \$4\left(\tfrac{1}{6}\right) + (-\$1)\left(\tfrac{5}{6}\right) = -\$\tfrac{1}{6} \approx -\$0.1667 \approx -17¢ \text{ per game}$$

Thus, in the long run the player will lose an average of about 17¢ per game.

Using the definition of a fair game in Section 6-4, it can be shown that **a game is fair if and only if $E(X) = 0$.** The game in Example 3 is not fair.

Matched Problem 3 Repeat Example 3 with the player winning \$5 instead of \$4 if the chosen number turns up. The loss is still \$1 if any other number turns up. Is this now a fair game?

Explore–Discuss 2

A coin is tossed twice and the number of heads recorded. You pay $1 to play. You keep your dollar if no heads turn up and lose your dollar if exactly 1 head turns up. What should the house pay if 2 heads turn up in order for the game to be fair? Discuss the process and reasoning used to arrive at your answer.

Example 4 **Expected Value and Insurance** Suppose you are interested in insuring a car stereo system for $500 against theft. An insurance company charges a premium of $60 for coverage for 1 year, claiming an empirically determined probability of .1 that the stereo will be stolen some time during the year. What is your expected return from the insurance company if you take out this insurance?

SOLUTION This is actually a game of chance in which your stake is $60. You have a .1 chance of receiving $440 from the insurance company ($500 minus your stake of $60) and a .9 chance of losing your stake of $60. What is the expected value of this "game"? We form a payoff table (the probability distribution for X) as shown in the margin. Then we compute the expected value as follows:

PAYOFF TABLE

x_i	\$440	−\$60
p_i	.1	.9

$$E(X) = (\$440)(.1) + (-\$60)(.9) = -\$10$$

This means that if you insure with this company over many years and the circumstances remain the same, you would have an average net loss to the insurance company of $10 per year.

Matched Problem 4 Find the expected value in Example 4 from the insurance company's point of view.

DECISION-MAKING AND EXPECTED VALUE

We conclude this section with an example in decision-making.

Example 5 **Decision Analysis** An outdoor concert featuring a very popular musical group is scheduled for a Sunday afternoon in a large open stadium. The promoter, worrying about being rained out, contacts a long-range weather forecaster who predicts the chance of rain on that Sunday to be .24. If it does not rain, the promoter is certain to net $100,000; if it does rain, the promoter estimates that the net will be only $10,000. An insurance company agrees to insure the concert for $100,000 against rain at a premium of $20,000. Should the promoter buy the insurance?

SOLUTION The promoter has a choice between two courses of action; A_1: Insure and A_2: Do not insure. As an aid in making a decision, the expected value is computed for each course of action. Probability distributions are indicated in the following payoff table (read vertically):

PAYOFF TABLE

	A_1: INSURE	A_2: DO NOT INSURE
p_i	x_i	x_i
.24 (rain)	\$90,000	\$ 10,000
.76 (no rain)	\$80,000	\$100,000

Note that the $90,000 entry comes from the insurance company's payoff ($100,000) minus the premium ($20,000) plus gate receipts ($10,000). The reasons for the other entries should be obvious. The expected value for each course of action is computed as follows:

A_1: INSURE

$$E(X) = x_1p_1 + x_2p_2 = (\$90{,}000)(.24) + (\$80{,}000)(.76) = \$82{,}400$$

A_2: DO NOT INSURE

$$E(X) = (\$10{,}000)(.24) + (\$100{,}000)(.76) = \$78{,}400$$

It appears that the promoter's best course of action is to buy the insurance at $20,000. The promoter is using a long-run average to make a decision about a single event—a common practice in making decisions in areas of uncertainty.

Matched Problem 5 In Example 5, what is the insurance company's expected value if it writes the policy?

Answers to Matched Problems

1. $E(X) = 3.73$

2. (A)

x_i	0	1	2
p_i	.632	.337	.032*

*Note: Due to roundoff error, sum = 1.001 ≈ 1.

(B) .4

3. $E(X) = \$0$; the game is fair

4. $E(X) = (-\$440)(.1) + (\$60)(.9) = \$10$ (This amount, of course, is necessary to cover expenses and profit.)

5. $E(X) = (-\$80{,}000)(.24) + (\$20{,}000)(.76) = -\$4{,}000$ (This means the insurance company had other information regarding the weather than the promoter; otherwise, the company would not have written this policy.)

EXERCISE 6-7

Where possible, construct a probability distribution or payoff table for a suitable random variable X; then complete the problem.

A

1. If the probability distribution for the random variable X is given in the table, what is the expected value of X?

x_i	−3	0	4
p_i	.3	.5	.2

2. If the probability distribution for the random variable X is given in the table, what is the expected value of X?

x_i	−2	−1	0	1	2
p_i	.1	.2	.4	.2	.1

3. In tossing 2 fair coins, what is the expected number of heads?

4. In a family with 2 children, excluding multiple births and assuming a boy is as likely as a girl at each birth, what is the expected number of boys?

5. A fair coin is flipped. If a head turns up, you win $1. If a tail turns up, you lose $1. What is the expected value of the game? Is the game fair?

6. Repeat Problem 5, assuming an unfair coin with the probability of a head being .55 and a tail being .45.

B

7. After paying $4 to play, a single fair die is rolled and you are paid back the number of dollars corresponding to the number of dots facing up. For example, if a 5 turns up, $5 is returned to you for a net gain, or payoff, of $1; if a 1 turns up, $1 is returned for a net gain of −$3; and so on. What is the expected value of the game? Is the game fair?

8. Repeat Problem 7 with the same game costing $3.50 for each play.

9. Two coins are flipped. You win $2 if either 2 heads or 2 tails turn up; you lose $3 if a head and a tail turn up. What is the expected value of the game?

10. In Problem 9, for the game to be fair, how much *should* you lose if a head and a tail turn up?

11. A friend offers the following game: She wins $1 from you if on four rolls of a single die, a 6 turns up at least once; otherwise, you win $1 from her. What is the expected value of the game to you? To her?

12. On three rolls of a single die, you will lose $10 if a 5 turns up at least once, and you will win $7 otherwise. What is the expected value of the game?

13. A single die is rolled once. You win $5 if a 1 or 2 turns up and $10 if a 3, 4, or 5 turns up. How much should you lose if a 6 turns up for the game to be fair? Describe the steps you took to arrive at your answer.

14. A single die is rolled once. You lose $12 if a number divisible by 3 turns up. How much should you win if a number not divisible by 3 turns up for the game to be fair? Describe the process and reasoning used to arrive at your answer.

15. A pair of dice is rolled once. Suppose you lose $10 if a 7 turns up and win $11 if an 11 or 12 turns up. How much should you win or lose if any other number turns up in order for the game to be fair?

16. A coin is tossed three times. Suppose you lose $3 if 3 heads appear, lose $2 if 2 heads appear, and win $3 if 0 heads appear. How much should you win or lose if 1 head appears in order for the game to be fair?

17. The payoff table for two courses of action, A_1 or A_2, is given below. Which of the two actions will produce the largest expected value? What is it?

	A_1	A_2
p_i	x_i	x_i
.1	−$200	−$100
.2	$100	$200
.4	$400	$300
.3	$100	$200

18. The payoff table for three possible courses of action is given below. Which of the three actions will produce the largest expected value? What is it?

	A_1	A_2	A_3
p_i	x_i	x_i	x_i
.2	$ 500	$ 400	$ 300
.4	$1,200	$1,100	$1,000
.3	$1,200	$1,800	$1,700
.1	$1,200	$1,800	$2,400

19. Roulette wheels in Nevada generally have 38 equally spaced slots numbered 00, 0, 1, 2, . . . , 36. A player who bets $1 on any given number wins $35 (and gets the bet back) if the ball comes to rest on the chosen number; otherwise, the $1 bet is lost. What is the expected value of this game?

20. In roulette (see Problem 19) the numbers from 1 to 36 are evenly divided between red and black. A player who bets $1 on black wins $1 (and gets the bet back) if the ball comes to rest on black; otherwise (if the ball lands on red, 0 or 00), the $1 bet is lost. What is the expected value of the game?

21. A game has an expected value to you of $100. It costs $100 to play, but if you win you receive $100,000 (including your $100 bet), for a net gain of $99,900. What is the probability of winning? Would you play this game? Discuss the factors that would influence your decision.

22. A game has an expected value to you of −$0.50. It costs $2 to play, but if you win you receive $20 (including your $2 bet), for a net gain of $18. What is the probability of winning? Would you play this game? Discuss the factors that would influence your decision.

C

23. Five thousand tickets are sold at $1 each for a charity raffle. Tickets are to be drawn at random and monetary prizes awarded as follows: 1 prize of $500; 3 prizes of $100, 5 prizes of $20, and 20 prizes of $5. What is the expected value of this raffle if you buy 1 ticket?

24. Ten thousand raffle tickets are sold at $2 each for a local library benefit. Prizes are awarded as follows: 2 prizes of $1,000, 4 prizes of $500, and 10 prizes of $100. What is the expected value of this raffle if you purchase 1 ticket?

25. A box of 10 flashbulbs contains 3 defective bulbs. A random sample of 2 is selected and tested. Let X be the random variable associated with the number of defective bulbs in the sample.
(A) Find the probability distribution of X.
(B) Find the expected number of defective bulbs in a sample.

26. A box of 8 flashbulbs contains 3 defective bulbs. A random sample of 2 is selected and tested. Let X be the random variable associated with the number of defective bulbs in a sample.
(A) Find the probability distribution of X.
(B) Find the expected number of defective bulbs in a sample.

27. One thousand raffle tickets are sold at $1 each. Three tickets will be drawn at random (without replacement) and each will pay $200. Suppose you buy 5 tickets.
(A) Create a payoff table for 0, 1, 2, and 3 winning tickets among the 5 tickets you purchased. (If you do not have any winning tickets, you lose $5; if you have 1 winning ticket, you net $195, since your initial $5 will not be returned to you; and so on.)
(B) What is the expected value of the raffle to you?

28. Repeat Problem 27 with the purchase of 10 tickets.

29. To simulate roulette on a graphing utility, a random integer between −1 and 36 is selected (−1 represents 00; see Problem 19). The command in Figure A simulates 200 games.

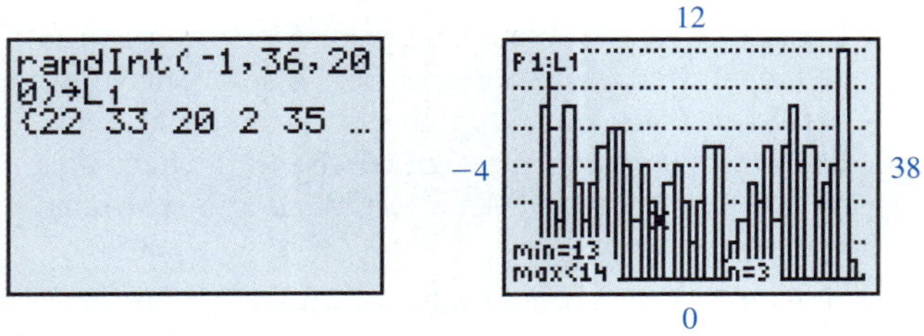

(A) (B)

Figure for 29

(A) Use the statistical plot in Figure B to determine the net gain or loss of placing a $1 bet on the number 13 in each of the 200 games.

(B) Compare the results of part (A) with the expected value of the game.

(C) Use a graphing utility to simulate betting $1 on the number 7 in each of 500 games of roulette, and compare the simulated and expected gains or losses.

30. Use a graphing utility to simulate the results of placing a $1 bet on black in each of 400 games of roulette (see Problems 20 and 29), and compare the simulated and expected gains or losses.

Applications

Business & Economics

31. *Insurance.* The annual premium for a $5,000 insurance policy against the theft of a painting is $150. If the (empirical) probability that the painting will be stolen during the year is .01, what is your expected return from the insurance company if you take out this insurance?

32. *Insurance.* Repeat Problem 31 from the point of view of the insurance company.

33. *Decision analysis.* An oil company, after careful testing and analysis, is considering drilling in two different sites. It is estimated that site A will net $30 million if successful (probability .2) and lose $3 million if not (probability .8); site B will net $70 million if successful (probability .1) and lose $4 million if not (probability .9). Which site should the company choose according to the expected return from each site?

34. *Decision analysis.* Repeat Problem 33, assuming additional analysis caused the estimated probability of success in field B to be changed from .1 to .11.

Life Sciences

35. *Genetics.* Suppose that, at each birth, having a girl is not as likely as having a boy. The probability assignments for the number of boys in a 3-child family are approximated empirically from past records and are given in the table. What is the expected number of boys in a 3-child family?

Number of boys x_i	p_i
0	.12
1	.36
2	.38
3	.14

36. *Genetics.* A pink-flowering plant is of genotype RW. If two such plants are crossed, we obtain a red plant (RR) with probability .25, a pink plant (RW or WR) with probability .50, and a white plant (WW) with probability .25, as shown in the table. What is the expected number of W genes present in a crossing of this type?

Number of W genes present x_i	p_i
0	.25
1	.50
2	.25

Social Sciences

37. *Politics.* A money drive is organized by a campaign committee for a candidate running for public office. Two approaches are considered:

A_1: A general mailing with a follow-up mailing
A_2: Door-to-door solicitation with follow-up telephone calls

From campaign records of previous committees, average donations and their corresponding probabilities are estimated to be:

A_1		A_2	
x_i (*Return per person*)	p_i	x_i (*Return per person*)	p_i
\$10	.3	\$15	.3
5	.2	3	.1
0	.5	0	.6
	1.0		1.0

What are the expected returns? Which course of action should be taken according to the expected returns?

Important Terms and Symbols

6-1 *Basic Counting Principles.* Counting technique; number of elements in a set; addition principle; tree diagram; multiplication principle

$$n(A);\quad n(A \cup B) = n(A) + n(B) - n(A \cap B);$$
$$n(A \cup B) = n(A) + n(B), \text{ if } A \cap B = \varnothing$$

6-2 *Permutations and Combinations.* n factorial; zero factorial; permutation; permutation of n objects; permutation of n objects taken r at a time; combination; combination of n objects taken r at a time

$$n! = n(n-1)(n-2)\cdot \cdots \cdot 2 \cdot 1;\quad 0! = 1;$$
$$P_{n,n} = n!;\quad P_{n,r} = \frac{n!}{(n-r)!};$$
$$C_{n,r} = \binom{n}{r} = \frac{P_{n,r}}{r!} = \frac{n!}{r!(n-r)!}, 0 \le r \le n$$

6-3 *Sample Spaces, Events, and Probability.* Random experiment; experiment; sample space; event; simple outcome; simple event; compound event; fundamental sample space; probability of an event; acceptable probability assignment; reasonable probability assignment; probability function; frequency; relative frequency; empirical probability; approximate empirical probability; equally likely assumption

$$P(E);\quad P(E) = \frac{f(E)}{n};\quad n(E)$$

6-4 *Union, Intersection, and Complement of Events; Odds.* Union; intersection; event A **or** event B; event A **and** event B; mutually exclusive; disjoint; complement of an event; odds; fair game; law of large numbers

$$A \cup B;\quad A \cap B;$$
$$P(A \cup B) = P(A) + P(B) - P(A \cap B);$$
$$P(A \cup B) = P(A) + P(B), \text{ if } A \cap B = \varnothing;$$
$$A';\quad P(A') = 1 - P(A);\quad \frac{P(E)}{P(E')} \text{ (odds for } E)$$

6-5 *Conditional Probability, Intersection, and Independence.* Conditional probability of A given B; product rule; stochastic process; independent events; dependent events

$$P(A|B) = \frac{P(A \cap B)}{P(B)}, \text{ when } P(B) \ne 0;$$
$$P(A \cap B) = P(A)P(B|A);$$
$P(A \cap B) = P(A)P(B)$, if and only if A and B are independent;
$P(A|B) = P(A)$ and $P(B|A) = P(B)$, if A and B are independent and $P(A) \ne 0$, $P(B) \ne 0$;
$P(E_1 \cap E_2 \cap \cdots \cap E_n) = P(E_1)P(E_2)\cdot \cdots \cdot P(E_n)$, if $E_1, E_2, \ldots, E_n$ are independent

6-6 *Bayes' Formula.*

$$P(U_1|E) = \frac{P(E|U_1)P(U_1)}{P(E|U_1)P(U_1) + P(E|U_2)P(U_2) + \cdots + P(E|U_n)P(U_n)}$$
$$= \frac{\left(\text{Product of branch probabilities leading to } E \text{ through } U_1\right)}{\text{Sum of all branch products leading to } E}$$

6-7 *Random Variable, Probability Distribution, and Expected Value.* Random variable; probability function of a random variable X; probability distribution for a random variable; histogram; expected value of a random variable; payoff table; fair game

$$E(X) = x_1p_1 + x_2p_2 + \cdots + x_np_n$$

Review Exercise

Work through all the problems in this chapter review and check your answers in the back of the book. Answers to all review problems are there along with section numbers in italics to indicate where each type of problem is discussed. Where weaknesses show up, review appropriate sections in the text.

A

1. A single die is rolled and a coin is flipped. How many combined outcomes are possible? Solve:
   (A) By using a tree diagram
   (B) By using the multiplication principle

2. Use the Venn diagram to find the number of elements in each of the following sets:
   (A) A (B) B
   (C) $A \cap B$ (D) $A \cup B$
   (E) U (F) A'
   (G) $(A \cap B)'$ (H) $(A \cup B)'$

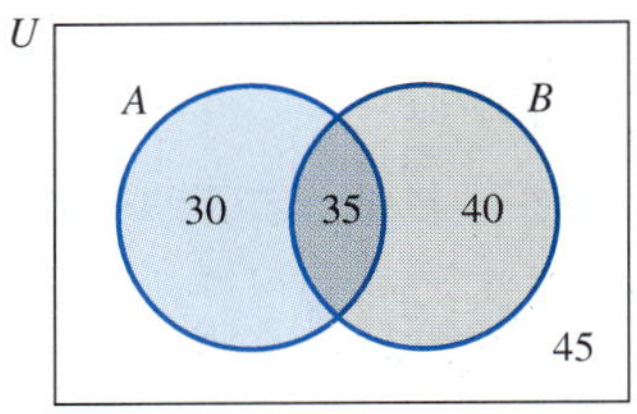

3. Evaluate $C_{6,2}$ and $P_{6,2}$.

4. How many seating arrangements are possible with 6 people and 6 chairs in a row? Solve by using the multiplication principle.

5. Solve Problem 4 using permutations or combinations, whichever is applicable.

6. In a single deal of 5 cards from a standard 52-card deck, what is the probability of being dealt 5 clubs?

7. Betty and Bill are members of a 15-person ski club. If the president and treasurer are selected by lottery, what is the probability that Betty will be president and Bill will be treasurer? (A person cannot hold more than one office.)

8. Each of the first 10 letters of the alphabet is printed on a separate card. What is the probability of drawing 3 cards and getting the code word *dig* by drawing *d* on the first draw, *i* on the second draw, and *g* on the third draw? What is the probability of being dealt a 3-card hand containing the letters *d, i,* and *g* in any order?

9. A drug has side effects for 50 out of 1,000 people in a test. What is the approximate empirical probability that a person using the drug will have side effects?

10. A spinning device has 5 numbers, 1, 2, 3, 4, and 5, each as likely to turn up as the other. A person pays $3 and then receives back the dollar amount corresponding to the number turning up on a single spin. What is the expected value of the game? Is the game fair?

11. If A and B are events in a sample space S and $P(A) = .3, P(B) = .4$, and $P(A \cap B) = .1$, find:
   (A) $P(A')$ (B) $P(A \cup B)$

12. A spinner lands on R with probability .3, on G with probability .5, and on B with probability .2. Find the probability and odds for the spinner landing on either R or G.

13. If in repeated rolls of two fair dice the odds for rolling a sum of 8 before rolling a sum of 7 are 5 to 6, then what is the probability of rolling a sum of 8 before rolling a sum of 7?

Answer Problems 14–22 using the table of probabilities shown below.

	X	*Y*	*Z*	*Totals*
S	.10	.25	.15	.50
T	.05	.20	.02	.27
R	.05	.15	.03	.23
Totals	.20	.60	.20	1.00

14. Find $P(T)$.
15. Find $P(Z)$.
16. Find $P(T \cap Z)$.
17. Find $P(R \cap Z)$.
18. Find $P(R|Z)$.
19. Find $P(Z|R)$.
20. Find $P(T|Z)$.
21. Are T and Z independent?
22. Are S and X independent?

Answer Problems 23–30 using the following probability tree:

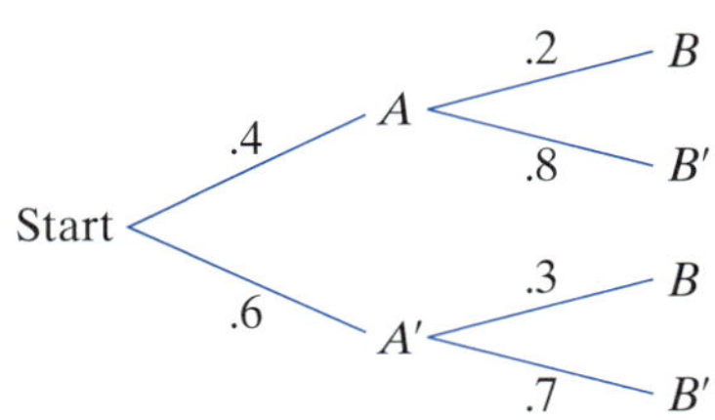

23. $P(A)$
24. $P(B|A)$
25. $P(B|A')$
26. $P(A \cap B)$
27. $P(A' \cap B)$
28. $P(B)$
29. $P(A|B)$
30. $P(A|B')$

31. (A) If 10 out of 32 students in a class were born in June, July, or August, what is the approximate empirical probability of any student being born in June, July, or August?
(B) If one is as likely to be born in any of the 12 months of a year as any other, what is the theoretical probability of being born in either June, July, or August?
(C) Discuss the discrepancy between the answers to parts (A) and (B).

In Problems 32 and 33, discuss the validity of each statement. If the statement is always true, explain why. If not, give a counterexample.

32. (A) If A or B is the empty set, then A and B are mutually exclusive.
(B) If A and B are mutually exclusive events, then

$$P(A \cap B) = P(A)P(B)$$

33. (A) If A and B are independent events, then

$$P(A \cup B) = P(A) + P(B)$$

(B) The events A and A' are independent.

B

34. A player tosses two coins and receives $5 if 2 heads turn up, loses $4 if 1 head turns up, and wins $2 if 0 heads turn up. (Would you play this game?) Compute the expected value of the game. Is the game fair?

35. A spinning device has 3 numbers, 1, 2, and 3, each as likely to turn up as the other. If the device is spun twice, what is the probability that:
(A) The same number turns up both times?
(B) The sum of the numbers turning up is 5?

36. In a single draw from a standard 52-card deck, what are the probability and odds for drawing:
(A) A jack or a queen? (B) A jack or a spade?
(C) A card other than an ace?

37. (A) What are the odds for rolling a sum of 5 on the single roll of two fair dice?
(B) If you bet $1 that a sum of 5 will turn up, what should the house pay (plus return your $1 bet) for the game to be fair?

38. Two coins are flipped 1,000 times with the following frequencies:

2 heads	210
1 head	480
0 heads	310

(A) Compute the empirical probability for each outcome.
(B) Compute the theoretical probability for each outcome.
(C) Using the theoretical probabilities computed in part (B), compute the expected frequency of each outcome, assuming fair coins.

39. A man has 5 children. Each of those children has 3 children, who in turn each have 2 children. Discuss the number of descendants that the man has.

40. A fair coin is tossed 10 times. On each of the first 9 tosses the outcome is heads. Discuss the probability of a head on the 10th toss.

41. An experiment consists of rolling a pair of fair dice. Let X be the random variable associated with the sum of the values that turn up.
(A) Find the probability distribution for X.
(B) Find the expected value of X.

42. Two dice are rolled. The sample space is chosen as the set of all ordered pairs of integers taken from $\{1, 2, 3, 4, 5, 6\}$. What is the event A that corresponds to the sum being divisible by 4? What is the event B that corresponds to the sum being divisible by 6? What are $P(A), P(B), P(A \cap B)$, and $P(A \cup B)$?

43. A person tells you that the following approximate empirical probabilities apply to the sample space $\{e_1, e_2, e_3, e_4\}$: $P(e_1) \approx .1, P(e_2) \approx -.2, P(e_3) \approx .6, P(e_4) \approx 2$. There are three reasons why P cannot be a probability function. Name them.

44. Use the following information to complete the frequency table below:

$$n(A) = 50, \quad n(B) = 45,$$
$$n(A \cup B) = 80, \quad n(U) = 100$$

	A	*A'*	*Totals*
B			
B'			
Totals			

45. A pointer is spun on a circular spinner. The probabilities of the pointer landing on the integers from 1 to 5 are given in the table below.

e_i	1	2	3	4	5
p_i	.2	.1	.3	.3	.1

(A) What is the probability of the pointer landing on an odd number?
(B) What is the probability of the pointer landing on a number less than 4 given that it landed on an odd number?

46. A card is drawn at random from a standard 52-card deck. If E is the event "The drawn card is red" and F is the event "The drawn card is an ace," then:
(A) Find $P(F|E)$.
(B) Test E and F for independence.

47. How many 3-letter code words are possible using the first 8 letters of the alphabet if no letter can be repeated? If letters can be repeated? If adjacent letters cannot be alike?

48. Solve the following problems using $P_{n,r}$ or $C_{n,r}$:
(A) How many 3-digit opening combinations are possible on a combination lock with 6 digits if the digits cannot be repeated?
(B) Five tennis players have made the finals. If each of the 5 players is to play every other player exactly once, how many games must be scheduled?

49. Use graphical techniques on a graphing utility to find the largest value of $C_{n,r}$ when $n = 25$.

In Problems 50–54, urn U_1 contains 2 white balls and 3 red balls; urn U_2 contains 2 white balls and 1 red ball.

50. Two balls are drawn out of urn U_1 in succession. What is the probability of drawing a white ball followed by a red ball if the first ball is:
(A) Replaced? (B) Not replaced?

51. Which of the two parts in Problem 50 involve dependent events?

52. In Problem 50, what is the expected number of red balls if the first ball is:
(A) Replaced? (B) Not replaced?

53. An urn is selected at random by flipping a fair coin; then a ball is drawn from the urn. Compute:
(A) $P(R|U_1)$ (B) $P(R|U_2)$
(C) $P(R)$ (D) $P(U_1|R)$
(E) $P(U_2|W)$ (F) $P(U_1 \cap R)$

54. In Problem 53, are the events "Selecting urn U_1" and "Drawing a red ball" independent?

55. From a standard deck of 52 cards, what is the probability of obtaining a 5-card hand:
(A) Of all diamonds?
(B) Of 3 diamonds and 2 spades?
Write answers in terms of $C_{n,r}$ or $P_{n,r}$; do not evaluate.

56. A group of 10 people includes one married couple. If 4 people are selected at random, what is the probability that the married couple is selected?

57. If 3 operations O_1, O_2, O_3 are performed in order, with possible number of outcomes N_1, N_2, N_3, respectively, determine the number of branches in the corresponding tree diagram.

58. A 5-card hand is drawn from a standard deck. Discuss how you can tell that the following two events are dependent without any computation.

S = Hand consists entirely of spades
H = Hand consists entirely of hearts

59. The command in Figure A was used on a graphing utility to simulate 50 repetitions of rolling a pair of dice and recording the minimum of the two numbers. A statistical plot of the results is shown in Figure B.

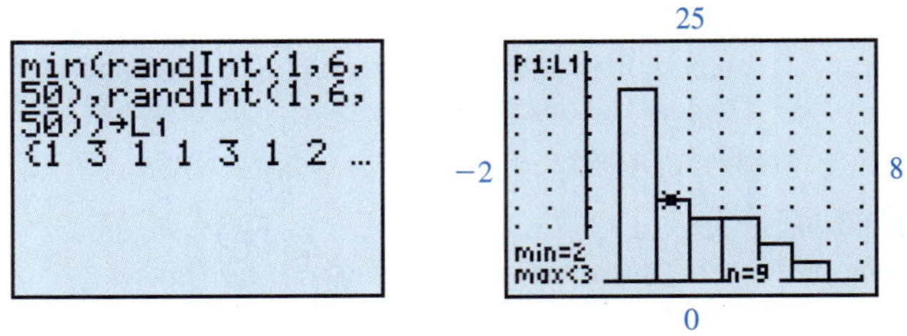

(A) (B)
Figure for 59

(A) Use Figure B to find the empirical probability that the minimum is 2.
(B) What is the theoretical probability that the minimum is 2?
(C) Use a graphing utility to simulate 200 rolls of a pair of dice, determine the empirical probability that the minimum is 4, and compare with the theoretical probability.

60. A card is drawn at random from a standard 52-card deck. Use a graphing utility to simulate 800 such draws, determine the empirical probability that the card is a black jack, and compare with the theoretical probability.

C

61. Three fair coins are tossed 1,000 times with the following frequencies of outcomes:

Number of Heads	0	1	2	3
Frequency	120	360	350	170

(A) What is the approximate empirical probability of obtaining 2 heads?
(B) What is the theoretical probability of obtaining 2 heads?
(C) What is the expected frequency of obtaining 2 heads?

62. You bet a friend \$1 that you will get 1 or more double 6's on 24 rolls of a pair of fair dice. What is your expected value for this game? What is your friend's expected value? Is the game fair?

63. A software development department consists of 6 women and 4 men.
(A) How many ways can they select a chief programmer, a backup programmer, and a programming librarian?

(B) If the positions in part (A) are selected by lottery, what is the probability that women are selected for all 3 positions?
(C) How many ways can they select a team of 3 programmers to work on a particular project?
(D) If the selections in part (C) are made by lottery, what is the probability that a majority of the team members will be women?

64. A group of 150 people includes 52 who play chess, 93 who play checkers, and 28 who play both chess and checkers. How many people in the group play neither game?

65. If 3 people are selected from a group of 7 men and 3 women, what is the probability that at least 1 woman is selected?

Two cards are drawn in succession without replacement from a standard 52-card deck. In Problems 66 and 67, compute the indicated probabilities.

66. The second card is a heart given the first card is a heart.

67. The first card is a heart given the second card is a heart.

68. Two fair (not weighted) dice are each numbered with a 3 on one side, a 2 on two sides, and a 1 on three sides. The dice are rolled and the numbers on the two up faces are added. If X is the random variable associated with the sample space $S = \{2, 3, 4, 5, 6\}$:
(A) Find the probability distribution of X.
(B) Find the expected value of X.

69. If you pay \$3.50 to play the game in Problem 68 (the dice are rolled once) and you are returned the dollar amount corresponding to the sum on the faces, what is the expected value of the game? Is the game fair? If it is not fair, how much should you pay to make the game fair?

70. How many different 5-child families are possible where the sex of each child in the order of their birth is taken into consideration [that is, birth sequences such as (B, G, G, B, B) and (G, B, G, B, B) produce different families]? How many families are possible if the order pattern is not taken into account?

71. Suppose 3 white balls and 1 black ball are placed in a box. Balls are drawn in succession without replacement until a black ball is drawn, and then the game is over. You win if the black ball is drawn on the fourth draw.
(A) What are the probability and odds for winning?
(B) If you bet \$1, what should the house pay you for winning (plus return your \$1 bet) if the game is to be fair?

72. If each of 5 people is asked to identify his or her favorite book from a list of 10 best-sellers, what is the probability that at least 2 of them identify the same book?

Applications

Business & Economics

73. *Transportation.* A distribution center A wishes to distribute its products to five different retail stores, B, C, D, E, and F, in a city. How many different route plans can be constructed so that a single truck, starting from A, will deliver to each store exactly once, and then return to the center?

74. *Market research.* A survey of 1,000 people indicates that 340 have invested in stocks, 480 have invested in bonds, and 210 have invested in stocks and bonds.
(A) How many people in the survey have invested in stocks or bonds?
(B) How many have invested in neither stocks nor bonds?
(C) How many have invested in bonds and not in stocks?

75. *Market research.* From a survey of 100 residents of a city, it was found that 40 read the daily morning paper, 70 read the daily evening paper, and 30 read both papers. What is the (empirical) probability that a resident selected at random:
(A) Reads a daily paper?
(B) Does not read a daily paper?
(C) Reads exactly one daily paper?

76. *Market research.* A market research firm has determined that 40% of the people in a certain area have seen the advertising for a new product and that 85% of those who have seen the advertising have purchased the product. What is the probability that a person in this area has seen the advertising and purchased the product?

77. *Market analysis.* A clothing company selected 1,000 persons at random and surveyed them to determine a relationship between age of purchaser and annual purchases of jeans. The results are given in the table.

	Jeans Purchased Annually				
Age	*0*	*1*	*2*	*Above 2*	*Totals*
Under 12	60	70	30	10	170
12–18	30	100	100	60	290
19–25	70	110	120	30	330
Over 25	100	50	40	20	210
Totals	260	330	290	120	1,000

Given the events:

$A =$ Person buys 2 pairs of jeans
$B =$ Person is between 12 and 18 years old
$C =$ Person does not buy more than 2 pairs of jeans
$D =$ Person buys more than 2 pairs of jeans

(A) Find $P(A), P(B), P(A \cap B), P(A|B), P(B|A)$.
(B) Are events A and B independent? Explain.
(C) Find $P(C), P(D), P(C \cap D), P(C|D), P(D|C)$.
(D) Are events C and D mutually exclusive? Independent? Explain.

78. *Decision analysis.* A company sales manager, after careful analysis, presents two sales plans. It is estimated that plan A will net \$10 million if successful (probability .8) and lose \$2 million if not (probability .2); plan B will net \$12 million if successful (probability .7) and lose \$2 million if not (probability .3). What is the expected return for each plan? Which plan should be chosen based on the expected return?

79. *Insurance.* A \$300 bicycle is insured against theft for an annual premium of \$30. If the probability that the bicycle will be stolen during the year is .08 (empirically determined), what is the expected value of the policy?

80. *Quality control.* Twelve precision parts, including 2 that are substandard, are sent to an assembly plant. The plant will select 4 at random and will return the whole shipment if 1 or more of the sample are found to be substandard. What is the probability that the shipment will be returned?

81. *Quality control.* A dozen computer circuit boards, including 2 that are defective, are sent to a computer service center. A random sample of 3 is selected and tested. Let X be the random variable associated with the number of circuit boards in a sample that is defective.
(A) Find the probability distribution of X.
(B) Find the expected number of defective boards in a sample.

Life Sciences

82. *Medicine—cardiogram test.* By testing a large number of individuals, it has been determined that 82% of the population have normal hearts, 11% have some minor heart problems, and 7% have severe heart problems. Ninety-five percent of the persons with normal hearts, 30% of those with minor problems, and 5% of those with severe problems will pass a cardiogram test. What is the probability that a person who passes the cardiogram test has a normal heart?

83. *Genetics.* Six men in 100 and 1 woman in 100 are color-blind. A person is selected at random and is found to be color-blind. What is the probability that this person is a man? (Assume the total population contains the same number of women as men.)

Social Sciences

84. *Voter preference.* In a straw poll, the 30 students in a mathematics class are asked to indicate their preference for president of the student government. Approximate empirical probabilities are assigned on the basis of the poll: candidate A should receive 53% of the vote, candidate B should receive 37%, and candidate C should receive 10%. One week later, candidate B wins the election. Discuss the factors that may account for the discrepancy between the poll and the election results.

Group Activity 1 Car and Crying Towels

The host on a television game show gives you (the contestant) a choice of three closed doors labeled *A, B,* and *C.* Behind one of the doors is a car, and behind each of the other two, a crying towel. After you have selected a door, which is not opened, the host opens one of the two doors not selected by you. The host, knowing which door hides the car, always chooses a crying towel door. The host then gives you a chance to stick to your original choice or to switch to the other unopened door. Which of the following two strategies would you choose and why?

Strategy 1. Stick to your original choice.
Strategy 2. Switch to the other unopened door.

(A) *Solve by "Common Sense" Reasoning.* Let each person in your group indicate their feeling about which strategy would be best or if it does not matter. Present the problem to friends and ask which strategy they would prefer and why. Summarize the results of these discussions.

(B) *Solve by Simulation.* To simulate this problem use a spinner with three equally likely outcomes, and suppose the car is hidden behind door *C.* Your original door choice is made by spinning the spinner once and making the choice indicated by the spinner. [*Note:* A single die can be used in place of the spinner by choosing door *A* if 1 or 2 dots turn up, door *B* for 3 or 4 dots, and door *C* for 5 or 6 dots.]

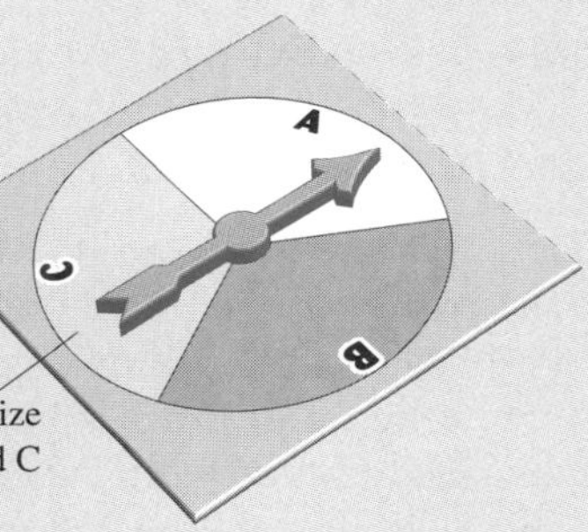

Car prize behind C

We now outline a simulation of the Stick strategy, and leave it to you to do the same for the Switch strategy.

Stick Strategy Simulation.

1. Suppose the spinner lands on door *A.* The host opens door *B,* knowing the car is behind door *C,* and reveals a crying towel. You stick to your original choice *A* and lose; getting a crying towel when door *A* is opened.
2. Suppose the spinner lands on door *B.* The host opens door *A,* knowing the car is behind *C,* and reveals a crying towel. You stick to *B* and lose; getting a crying towel when door *B* is opened.
3. Suppose the spinner lands on door *C.* The host opens either door *A* or *B,* revealing a crying towel. You stick to door *C* and win, getting a car when door *C* is opened.

Switch Strategy Simulation. Modify the simulation for the Stick strategy to apply to the Switch strategy.

Each student in your group should run at least 100 trials for each strategy simulation and record the wins and losses. Pool the results of all students in your group. Use the relative frequency of wins for each strategy to write approximate empirical probabilities for a win for that strategy. Now, which strategy do you think you would use and why?

(C) *Solve Theoretically Using Probability Trees.* The car is assumed to be behind door *C.*

Again, we outline a probability tree for the Stick strategy on the next page, and leave it to you to do the same for the Switch strategy.

Stick Strategy. Complete the following probability tree and compute the probability of a win for the Stick strategy.

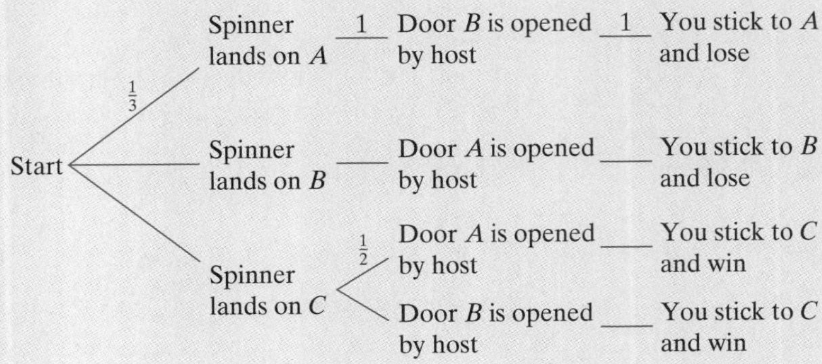

Switch Strategy. Construct a probability tree for the Switch strategy, and compute the probability of a win for this strategy.

Compare the results of parts (A), (B), and (C). Which strategy would you now choose and why? Present the problem to a friend and try to convince your friend that the strategy you have chosen is the best strategy.

Now consider the **three cards problem:** Three cards are in a box. Both sides of one card are white, both sides of a second card are black, and one side of a third card is white and the other side is black. A card is drawn from the box at random and placed on the table. A white side is showing. What is the probability that the other side is black? Describe how you would solve this problem by using a simulation approach and by using a theoretical approach. Use each approach to solve the problem.

Group Activity 2 Simulation—A Draft Lottery for Professional Sports

What is the probability that hurricane Hypatia will hit land north of Cape Hatteras? That the new county jail will be at maximum capacity by year's end? That a supermarket customer will wait more than 7 minutes in one of the checkout lanes?

It would be virtually impossible to answer such questions by assigning probabilities to the simple events of a sample space. The sample space might be too large to be manageable, or it might be extremely difficult to determine probabilities of simple events.

Simulation offers an alternative approach. In several of the exercises in this chapter we have used a graphing utility to simulate flipping a coin, rolling a pair of dice and recording their sum, playing roulette, and so on. We use such simulations and computer simulations of more complex processes to determine the approximate empirical probabilities of events.

Professional sports leagues hold annual draft lotteries. The draft lottery is used to determine the order in which teams will select players in the draft, which is held at a later date. The lottery is weighted in favor of the poorer teams but also involves an element of chance.

Suppose there are 10 teams that did not make the playoffs; call them *A*, *B*, *C*, *D*, *E*, *F*, *G*, *H*, *I*, and *J*, arranged from the team with the poorest record

(A) to the team with the best record (J). The draft order for these 10 teams is determined by drawing balls at random from a drum that initially contains 1,000 balls of 10 different colors: the numbers of balls for the 10 teams A–J are 600, 200, 100, 50, 30, 10, 4, 3, 2, and 1, respectively.

If the first ball drawn belongs to team D, then D will select first in the player draft. All other balls belonging to D are removed from the drum before the second ball is drawn. If the second ball drawn belongs to team F, then F will select second in the player draft, and so on. The drawing continues in this fashion until an order, say *DFABCHGJEI*, is established for the player draft.

(A) What is the theoretical probability that team A wins the lottery and selects first in the player draft?
(B) What is the theoretical probability that team A selects second in the player draft?

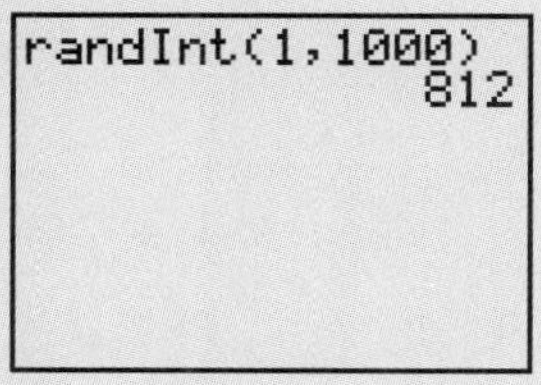

FIGURE 1

We can use a graphing utility to simulate the first draw of a ball from the drum: We use the random number feature to select a random integer from 1 to 1,000 and consider an integer in the range 1–600 as belonging to team A, 601–800 to team B, 801–900 to team C, and so on (see Fig. 1).

Figure 1 indicates that the first ball drawn belongs to team C. Now, instead of removing the balls belonging to C, it is more convenient to imagine that the drum contains all 1,000 balls for the next several draws, with the understanding that if another ball belonging to C is drawn, that draw is ignored. Now, suppose the command randInt(1,1000,5) produces {966, 432, 656, 843, 729} as the next set of 5 elements, which belong to E, A, B, C, and B, respectively. Then the order *CEAB* has been established for the first four draft picks (843 and 729 are ignored, since they belong to teams already selected).

Now imagine that the balls belonging to teams A, B, and C have been removed from the drum. Suppose the command randInt(901,1000,5) produces the next set {943, 969, 990, 928, 948}, whose elements belong to D, E, F, D, and D, respectively. Then the order *CEABDF* has been established for the first six draft picks.

Now imagine that the balls belonging to all 6 of these teams have been removed. If the command randInt(991,1000,5) produces the set {993, 995, 1000, 992, 997}, whose elements belong to G, H, J, G, and H, respectively, then the order for all 10 teams is determined, *CEABDFGHJI*, and we have completed one simulation of the lottery.

(C) Carry out at least 20 simulations of the draft lottery.
(D) Based on your simulations, what is the empirical probability that team A drafts first? That A drafts second? Compare with the theoretical probabilities.
(E) What is the empirical probability that A drafts fifth or later? That F gets one of the first three draft picks? That J gets one of the first five draft picks?
(F) Discuss the difficulty of computing the theoretical probabilities of the events of part (E).

CHAPTER 7

DATA DESCRIPTION AND PROBABILITY DISTRIBUTIONS

INTRODUCTION

In this chapter we will take a look at ways of describing data sets using graphs, tables, averages, and so on. In addition, we will develop the idea of probability distributions associated with both actual and theoretical data sets.

SECTION 7-1

Graphing Data

- BAR GRAPHS
- BROKEN-LINE GRAPHS
- PIE GRAPHS

Television, newspapers, magazines, books, and reports make substantial use of graphics to visually communicate complicated sets of data to the viewer. In this section we will look at bar graphs, broken-line graphs, and pie graphs and the techniques for producing them. It is important to remember that graphs are visual aids and should be prepared with care. The object is to provide the viewer with the maximum amount of information, while minimizing the time and effort required to "read" the information from the graph.

Bar Graphs

Bar graphs are widely used because they are easy to construct and easy to read. They are effective in presenting visual interpretations or comparisons of data. Consider Tables 1 and 2. Bar graphs are well-suited to describe these two data sets. Vertical bars are usually used for time series—that is, data that change over time, as in Table 1. The labels on the horizontal axis are then units of time (hours, days, years, and so on, whichever is appropriate), as shown in Figure 1. Horizontal bars are generally used for data that change by category, as in Table 2, because of the ease of labeling categories on the vertical axis of the bar graph (see Fig. 2). To increase clarity, a space is left between the bars. Bar graphs for the data in Tables 1 and 2 are illustrated in Figures 1 and 2.

Table 1

U.S. Public Debt

Year	Debt *Billion dollars*
1950	256.1
1960	284.1
1970	370.1
1980	907.7
1990	3,233.3

Table 2

Traffic at Busiest U.S. Airports, 1992

Airport	Arrivals and departures *Million passengers*
Chicago (O'Hare)	64
Dallas/Ft. Worth	52
Los Angeles	46
Atlanta	42
San Francisco	32

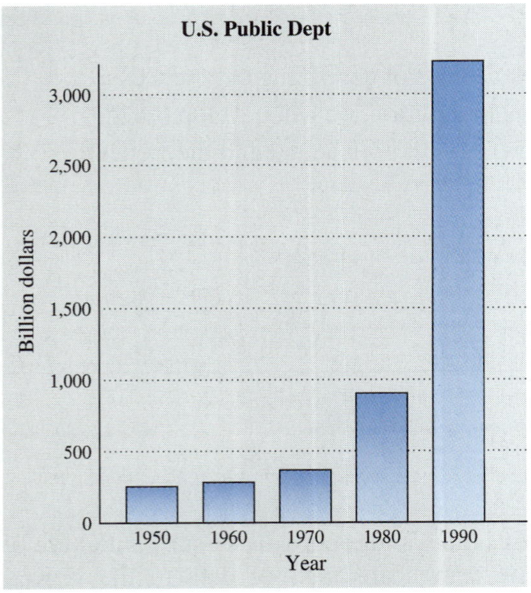

FIGURE 1
Vertical bar graph

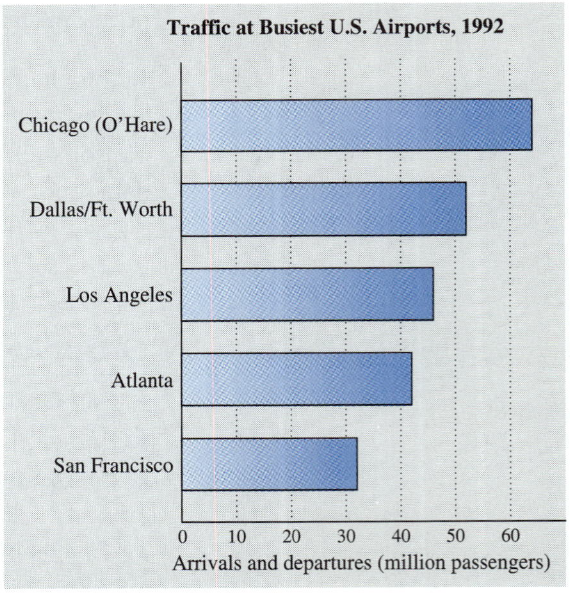

FIGURE 2
Horizontal bar graph

Two additional variations on bar graphs, the double bar graph and the divided bar graph, are illustrated in Figures 3 and 4, respectively.

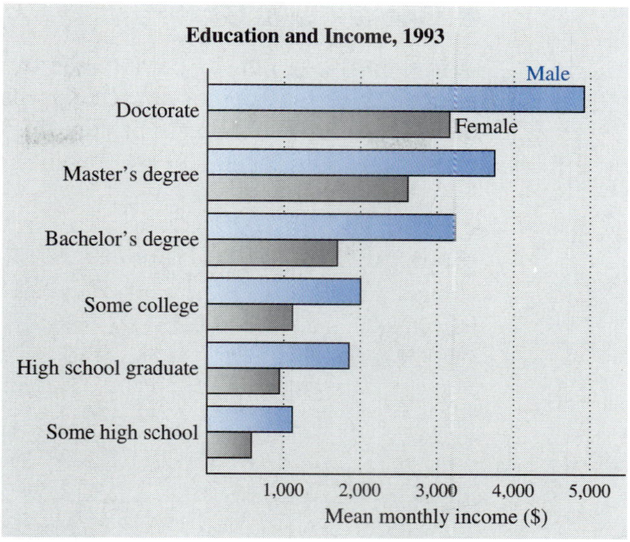

FIGURE 3
Double bar graph

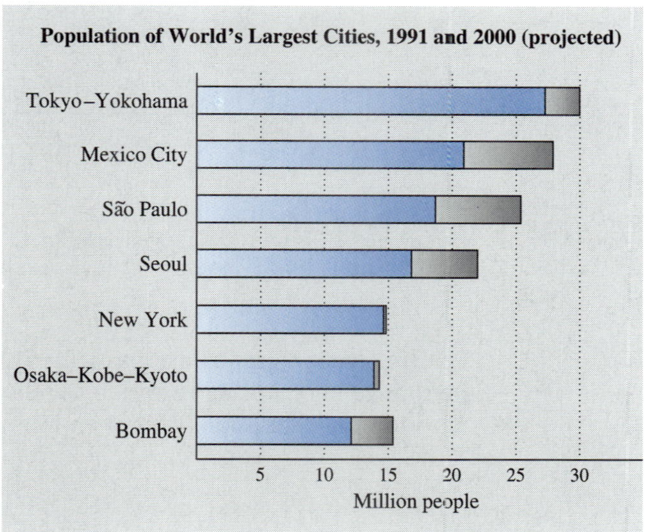

FIGURE 4
Divided bar chart

Explore–Discuss 1

(A) Using Figure 3, estimate the mean monthly income of a male with a master's degree and of a female who holds a doctorate. Which educational category has the greatest difference between male and female incomes? The least difference?

(B) Using Figure 4, estimate the population of Mexico City in the years 1991 and 2000. Which of the seven largest cities is projected to have the greatest increase in population from 1991 to 2000? The least increase? Which city would you conjecture to have the largest population in the world in 2010? Explain.

Broken-Line Graphs

A **broken-line graph** can be obtained from a vertical bar graph by joining the midpoints of the tops of consecutive bars with straight lines. For example, using Figure 1 we obtain the broken-line graph in Figure 5.

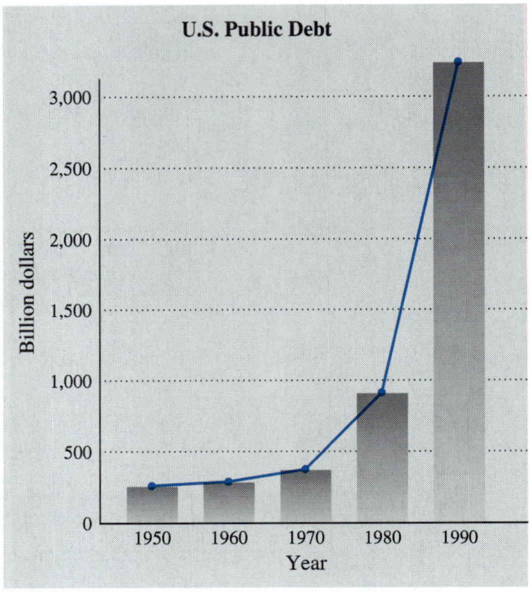

FIGURE 5
Broken-line graph

Broken-line graphs are particularly useful when we want to emphasize the change in one or more variables relative to time. Figures 6 and 7 illustrate two additional variations of broken-line graphs.

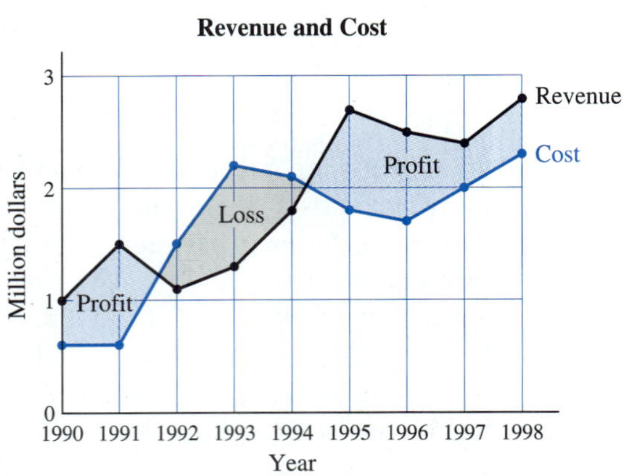

FIGURE 6
Broken-line graphs

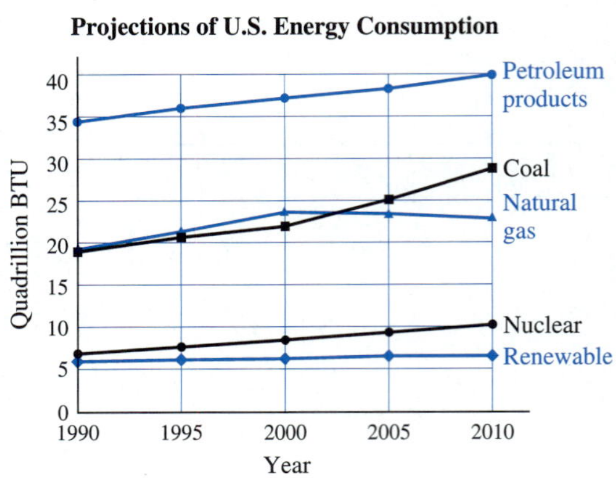

FIGURE 7
Broken-line graphs

Explore–Discuss 2

(A) Using Figure 6, estimate the revenue and costs in 1995. In which years is a profit realized? In which year is the greatest loss experienced?

(B) Using Figure 7, estimate the U.S. consumption of each of the five sources of energy in 2010. Estimate the percentage of total consumption that will come from nuclear energy in the year 2010.

PIE GRAPHS

A **pie graph** is generally used to show how a whole is divided among several categories. The amount in each category is expressed as a percentage, and then a circle is divided into segments (pieces of pie) proportional to the percentages of each category. The central angle of a segment is the percentage of 360° corresponding to the percentage of that category (see Fig. 8). In constructing pie graphs, we use relatively few categories, arrange the segments in ascending or descending order of size around the circle, and label each part.

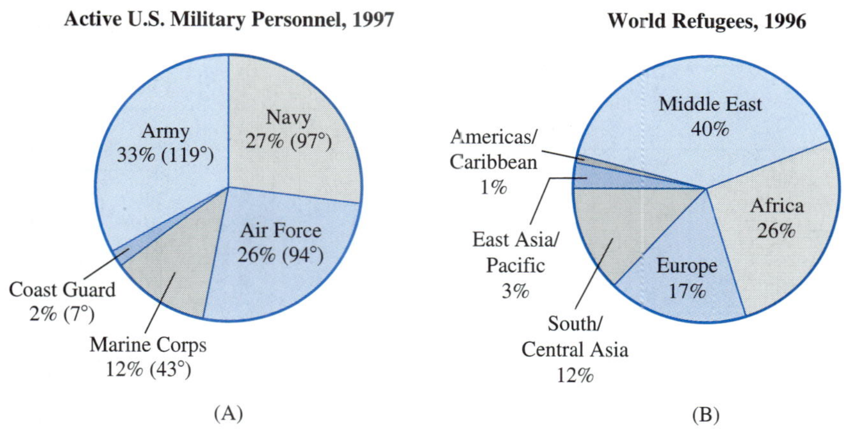

FIGURE 8
Pie graphs

Bar graphs, broken-line graphs, and pie graphs are easily constructed using a spreadsheet. After the data are entered (see Fig. 9 for the data corresponding to Fig. 8A) and the type of display (bar, broken-line, pie) is chosen, the graph is drawn automatically. Various options for axes, gridlines, patterns, and text are available to improve the clarity of the visual display.

	A	B	C	D	E	F
1		Army	Navy	Air Force	Marine Corps	Coast Guard
2		487297	397957	378681	172329	34717

FIGURE 9

EXERCISE 7-1

APPLICATIONS

Business & Economics

1. *Gross national product.* Graph the data in the following table using a bar graph:

GROSS NATIONAL PRODUCT (GNP)

YEAR	GNP *Billion dollars*
1960	515.3
1970	1,015.5
1980	2,732.0
1990	5,524.5

2. *Government debt.* Graph the data in the following table using a bar graph:

GOVERNMENT DEBT, 1993

COUNTRY	DEBT AS PERCENTAGE OF GROSS NATIONAL PRODUCT
United Kingdom	41.9
United States	59.8
Japan	64.9
Italy	108.4
Belgium	134.4

3. *Family budget.* Use the double bar graph shown to estimate the mean family expenditures on housing in 1984 and 1994. Which category of expenditures showed the greatest increase from 1984 to 1994? Which category showed the greatest percentage increase?

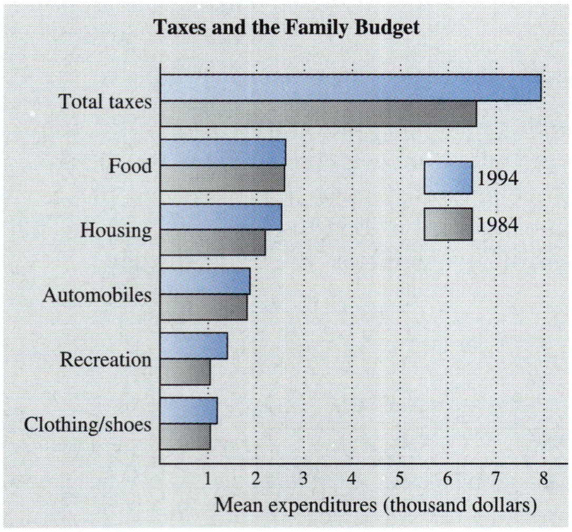

Figure for 3

4. *Steel production.* Graph the data in the following table using a double bar graph:

CRUDE STEEL PRODUCTION (THOUSAND TONS)

COUNTRY	1990	1991
Former USSR	12,868	11,660
Japan	9,194	8,992
United States	8,168	7,287
China	5,388	5,880
Germany	3,203	3,232
Italy	2,121	2,086

5. *Railroad freight.* Graph the data in the following table using a broken-line graph:

ANNUAL RAILROAD CARLOADINGS IN THE UNITED STATES

YEAR	CARLOADINGS
1940	36,358,000
1950	38,903,000
1960	27,886,950
1970	27,015,020
1980	22,223,000
1990	21,884,649

6. *Unemployment rates.* Use the broken-line graphs shown to estimate the unemployment rates in Australia, the United States, and Japan in 1990. Which country suffered the greatest increase in the unemployment rate over a single year? In which year? In which years did the unemployment rate decrease in all three countries?

Figure for 6

7. *Taxes.* Graph the data in the following table using a pie graph:

U.S. Taxes by Source, 1994

Source	Percentage of total taxes
Individual income	33
Social insurance	30
Sales and excise	14
Property	10
Corporate income	9
All other	4

8. *Automobile imports.* Graph the data in the following table using a bar graph:

New Passenger Cars Imported into the United States, 1996

Country	Thousand cars
Canada	1,590
Japan	1,013
Mexico	551
Germany	234
South Korea	141

9. *Automobile imports.* Explain how the data in Problem 8 could be displayed using a pie graph, and find the central angles of such a graph.

10. *Automobile imports.* In 1996, 4.6% of all imported new passenger cars came from Sweden, the United Kingdom, Italy, and other countries not listed in the table for Problem 8. Explain how this additional information could be incorporated into a pie graph, and find the central angles of such a graph.

Life Sciences

11. *Population growth.* Graph the data in the following table using a bar graph:

Annual World Population Growth

Year	Million people
1900	11
1925	21
1950	46
1975	71
2000	93 (projected)

12. *Population growth.* Graph the data in Problem 11 using a broken-line graph.

13. *Nutrition.* Graph the data in the following table using a double bar chart:

Recommended Daily Allowances (RDA)

Nutrient *Grams*	Males *Age 15–18*	Females *Age 15–18*
Carbohydrate	375	275
Protein	60	44
Fat	100	73

14. *Nutrition.* Graph the data in Problem 18 using two pie graphs, one for males and one for females. Describe the similarities of the two pie graphs.

15. *Nutrition*

(A) Graph the nutritional information in the following table using a double bar graph:

Fast-Food Burgers—Nutritional Information

	Calories	Calories from fat
Burger, 2 ounces, plain	268	108
Extra patty, 2 ounces beef	154	90
Cheese, 1 slice	105	81
Bacon, 3 slices	109	81
Mayonnaise, 1 tablespoon	100	99

(B) Suppose you are trying to limit the fat in your diet to at most 30% of your calories, and your calories to 2,000 per day. Should you order the double bacon cheeseburger with mayo for lunch? How would such a lunch affect your choice of breakfast and dinner? Discuss.

16. *Diseases.* Would a pie graph be more effective or less effective than the bar graph shown in presenting information on the world's deadliest diseases? Justify your answer.

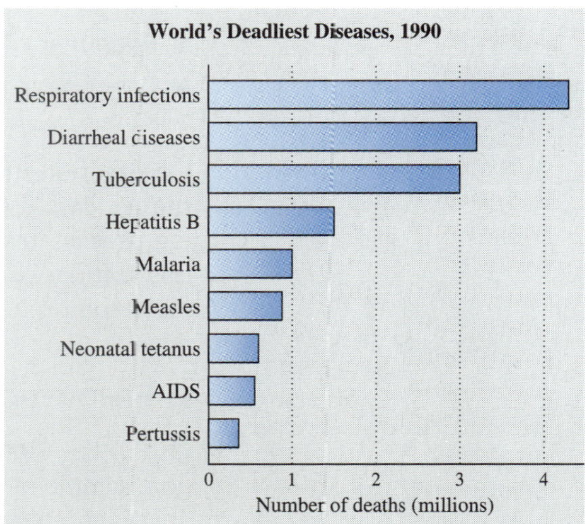

Figure for 16

Social Sciences

17. *Incarceration rates.* Graph the data in the following table using a bar graph:

PRISON INCARCERATION RATES, 1992

COUNTRY	NUMBER PER 100,000 POPULATION
United States	455
South Africa	311
Venezuela	177
Hungary	117
Canada	111

18. *Labor force.* Graph the data in the following table using a double bar graph:

CIVILIAN LABOR FORCE (THOUSANDS)

YEAR	EMPLOYED	UNEMPLOYED
1970	78,678	4,093
1975	85,846	7,929
1980	99,303	7,637
1985	107,150	8,312
1990	117,914	6,874

19. *Median age.* Use the broken-line graph shown to estimate the median age in 1900 and 1990. In which decades did the median age increase? In which did it decrease? Discuss the factors that may have contributed to the increases and decreases.

Median Age in the United States, 1900–1990

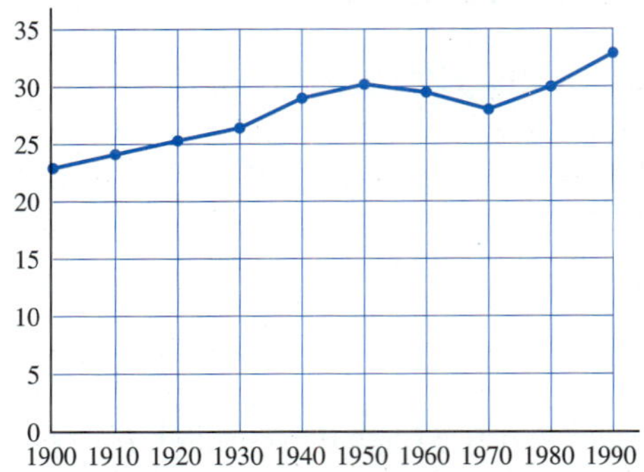

Figure for 19

20. *HIV/AIDS.* Graph the data in the following table using a pie graph:

CURRENT HIV/AIDS CASES, 1996

REGION	PERCENTAGE OF CASES
Sub-Saharan Africa	64
South/Southeast Asia	22
Latin America	6
North America	4
Europe	2
Other	2

SECTION 7-2

Graphing Quantitative Data

- FREQUENCY DISTRIBUTIONS
- COMMENTS ON STATISTICS
- HISTOGRAMS
- FREQUENCY POLYGONS
- CUMULATIVE FREQUENCY TABLES AND POLYGONS

Observations that are measured on a numerical scale are referred to as **quantitative data.** Weights, ages, bond yields, the length of a part in a manufacturing process, test scores, and so on, are all examples of quantitative data. In this section we will consider how large sets of quantitative data can be made more comprehensible through special tables and graphs.

FREQUENCY DISTRIBUTIONS

Out of the total population of entering freshmen at a large university, a random sample of 100 students is selected and their entrance examination scores are recorded (see Table 1).

Table 1

ENTRANCE EXAMINATION SCORES OF 100 ENTERING FRESHMEN

762	451	602	440	570	553	367	520	454	653
433	508	520	603	532	673	480	592	565	662
712	415	595	580	643	542	470	743	608	503
566	493	635	780	537	622	463	613	502	577
618	581	644	605	588	695	517	537	552	682
340	537	370	745	605	673	487	412	613	470
548	627	576	637	787	507	566	628	676	750
442	591	735	523	518	612	589	648	662	512
663	588	627	584	672	533	738	455	512	622
544	462	730	576	588	705	695	541	537	563

The mass of raw data in Table 1 certainly does not elicit much interest or exhibit much useful information. The data must be organized in some way so that it is comprehensible. This can be done by constructing a **frequency table.** We generally choose five to twenty **class intervals** of equal length to cover the data range—the more data, the greater the number of intervals—and tally the data relative to these intervals. The **data range** in Table 1 is 787 − 340 = 447 (found by subtracting the smallest value in the data from the largest). If we choose ten intervals, each of length 50, we will be able to cover all the scores. Table 2 shows the result of this tally.

Table 2

FREQUENCY TABLE

CLASS INTERVAL	TALLY	FREQUENCY	RELATIVE FREQUENCY
299.5–349.5	𝍷	1	.01
349.5–399.5	𝍷𝍷	2	.02
399.5–449.5	𝍸	5	.05
449.5–499.5	𝍸 𝍸	10	.10
499.5–549.5	𝍸 𝍸 𝍸 𝍸 𝍷	21	.21
549.5–599.5	𝍸 𝍸 𝍸 𝍸	20	.20
599.5–649.5	𝍸 𝍸 𝍸 𝍷𝍷𝍷𝍷	19	.19
649.5–699.5	𝍸 𝍸 𝍷	11	.11
699.5–749.5	𝍸 𝍷𝍷	7	.07
749.5–799.5	𝍷𝍷𝍷𝍷	4	.04
		100	1.00

At first it might seem appropriate to start at 300 and form the class intervals: 300–350, 350–400, 400–450, and so on. But if we do this, where will we place 350 or 400? We could, of course, adopt a convention of placing a score falling on an upper boundary of a class in the next higher class (and some people do exactly this); however, to avoid confusion, we will always use one decimal place more for class boundaries than appears in the raw data. Thus, in this case, we chose the class intervals 299.5–349.5, 349.5–399.5, and so on, so that each score could be assigned to one and only one class interval.

The number of measurements that fall within a given class interval is called the **class frequency,** and the set of all such frequencies associated with their corresponding classes is called a **frequency distribution.** Thus, Table 2 represents a frequency distribution of the set of raw scores in Table 1. If we divide each frequency by the total number of items in the original data set (in our case 100), we obtain the **relative frequency** of the data falling in each class interval—that is, the percentage of the whole that falls in each class interval (see the last column in Table 2).

The relative frequencies also can be interpreted as probabilities associated with the experiment, "A score is drawn at random out of the 100 in the sample." An appropriate sample space for this experiment would be the set of simple outcomes

$$\begin{aligned} e_1 &= \text{A score falls in the first class interval} \\ e_2 &= \text{A score falls in the second class interval} \\ &\vdots \\ e_{10} &= \text{A score falls in the tenth class interval} \end{aligned}$$

The set of relative frequencies is then referred to as the **probability distribution** for the sample space.

Example 1 **Determining Probabilities from a Frequency Table** Referring to the probability distribution just described and Table 2, determine the probability that:

(A) A randomly drawn score is between 499.5 and 549.5.
(B) A randomly drawn score is between 449.5 and 649.5.

SOLUTION (A) Since the relative frequency associated with the class interval 499.5–549.5 is .21, the probability that a randomly drawn score (from the sample of 100) falls in this interval is .21.

(B) Since a score falling in the interval 449.5–649.5 is a compound event, we simply add the probabilities for the simple events whose union is this compound event. Thus, we add the probabilities corresponding to each class interval from 449.5 to 649.5 to obtain

$$.10 + .21 + .20 + .19 = .70$$

Matched Problem 1 Repeat Example 1 for the following intervals:

(A) 649.5–699.5 (B) 299.5–499.5

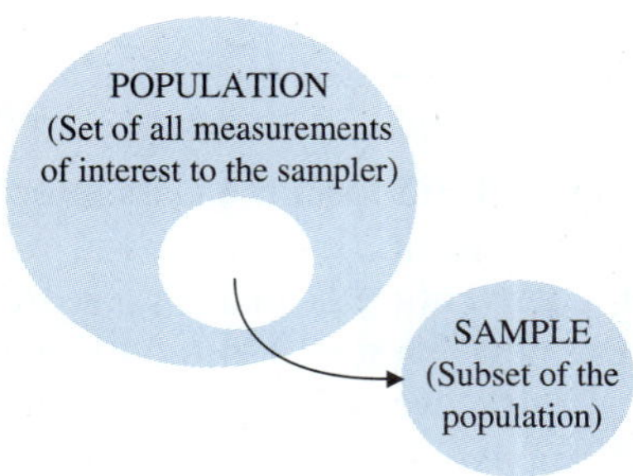

FIGURE 1
Inferential statistics: Based on information obtained from a sample, the goal of statistics is to make inferences about the population as a whole.

Comments on Statistics

Now, of course, what we are really interested in is whether the probability distribution for the sample of 100 entrance examination scores has anything to do with the total population of entering freshmen in the university in question. This is a problem for the important branch of mathematics called *statistics,* which deals with the process of making inferences about a total population based on random samples drawn from the population. Figure 1 schematically illustrates the *inferential statistical* process. We will not go too far into inferential statistics in this book, since the subject is studied in detail in any course in statistics, but our work in probability provides a good foundation for this study.

Intuitively, in the entrance examination example, we would expect that the larger the sample size, the more closely the probability distribution for the sample will approximate that for the total population. That is about all we can say at the moment.

HISTOGRAMS

A **histogram** is a special kind of vertical bar graph. In fact, if you rotate Table 2 counterclockwise 90°, the tally marks in the table take on the appearance of a bar graph. Histograms have no space between the bars, class boundaries are located on the horizontal axis, and frequencies are associated with the vertical axis. Figure 2 is a histogram for the frequency distribution in Table 2. Note that we have included both frequencies and relative frequencies on the vertical scale. You can include either one or the other, or both, depending on what needs to be emphasized. The histogram is the most common graphical representation of frequency distributions.

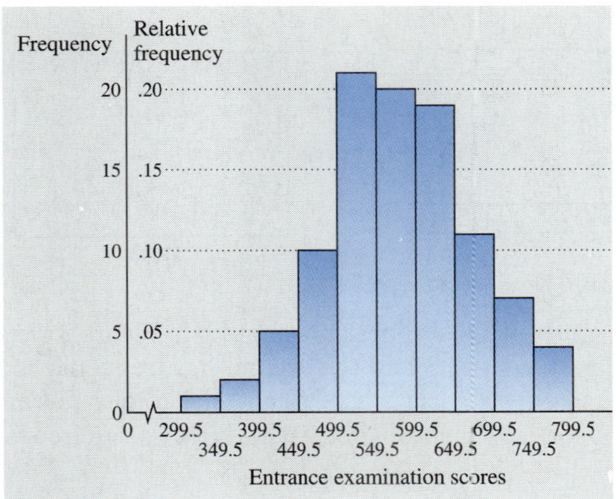

FIGURE 2
Histogram

Explore–Discuss 1

(A) Draw a histogram for the following data set using a class interval width of 0.5 starting at 0.45:

1.5	1.3	1.4	1.5	2.5	1.5	1.4	1.6
1.4	1.5	1.5	1.4	0.9	2.1	1.4	1.5

(B) Draw a histogram for the same data set using the same class interval width but starting at 0.75.

(C) The histograms of parts (A) and (B) represent the same set of data. How do they differ? Which of the two gives a better description of the data set? Explain.

Example 2 **Constructing Histograms with a Graphing Utility** Twenty vehicles were chosen at random upon arrival at a vehicle emissions inspection station, and the time elapsed (in minutes) from arrival to completion of the emissions test was recorded for each of the vehicles:

5	12	11	4	7	20	14	8	12	11
18	15	14	9	10	13	12	20	26	17

(A) Use a graphing utility to draw a histogram of the data, choosing the five class intervals 2.5–7.5, 7.5–12.5, and so on.

(B) What is the probability that for a vehicle chosen at random from the sample, the time required at the inspection station is less than 12.5 minutes? That it exceeds 22.5 minutes?

Solution (A) Various kinds of statistical plots can be drawn by most graphing utilities. To draw a histogram we enter the data as a list, specify a histogram from among the various statistical plotting options, set the window variables, and graph. Figure 3 shows the data entered as a list, the settings of the window variables, and the resulting histogram for a particular graphing calculator. Consult your manual or the graphing utility supplement to this text for details.

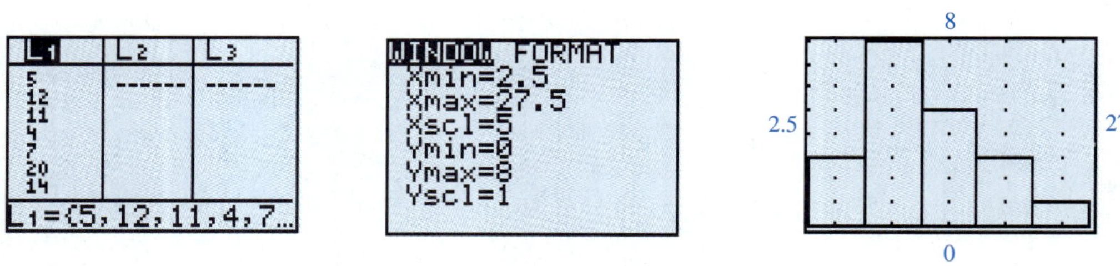

FIGURE 3

(B) From the histogram in Figure 3 we see that the first class has frequency 3 and the second has frequency 8. The upper boundary of the second class is 12.5, and the total number of data items is 20. Therefore, the probability that the time required is less than 12.5 minutes is

$$\frac{3 + 8}{20} = \frac{11}{20} = .55$$

Similarly, since the frequency of the last class is 1, the probability that the time required exceeds 22.5 minutes is

$$\frac{1}{20} = .05$$

Matched Problem 2 The weights (in pounds) were recorded for 20 kindergarten children chosen at random:

51	46	37	39	48	42	41	44	57	36
47	44	41	50	45	46	34	39	42	44

(A) Use a graphing utility to draw a histogram of the data, choosing the five class intervals 32.5–37.5, 37.5–42.5, and so on.

(B) What is the probability that a kindergarten child chosen at random from the sample weighs less than 42.5 pounds? More than 42.5 pounds?

FREQUENCY POLYGONS

A **frequency polygon** is a broken-line graph where successive midpoints of the tops of the bars in a histogram are joined by straight lines. To draw a frequency polygon for a frequency distribution, you do not need to draw a histogram first; you can just locate the midpoints and join them with straight lines. Figure 4 is a frequency polygon for the frequency distribution in Table 2. If the amount of data becomes very large and we substantially increase the number of classes, the frequency polygon will take on the appearance of a smooth curve called a **frequency curve.**

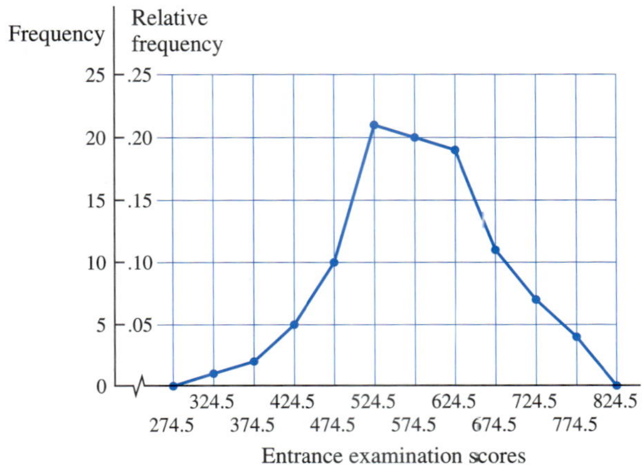

FIGURE 4
Frequency polygon

CUMULATIVE FREQUENCY TABLES AND POLYGONS

If we are interested in how many or what percentage of a total sample lies above or below a particular measurement, a **cumulative frequency table** and **polygon** are useful. Using the frequency distribution in Table 2, we accumulate the frequencies by starting with the first class and adding frequencies as we move down the column. The results are shown in Table 3. (How is the last column formed?)

Table 3
CUMULATIVE FREQUENCY TABLE

CLASS INTERVAL	FREQUENCY	CUMULATIVE FREQUENCY	RELATIVE CUMULATIVE FREQUENCY
299.5–349.5	1	1	.01
349.5–399.5	2	3	.03
399.5–449.5	5	8	.08
449.5–499.5	10	18	.18
499.5–549.5	21	39	.39
549.5–599.5	20	59	.59
599.5–649.5	19	78	.78
649.5–699.5	11	89	.89
699.5–749.5	7	96	.96
749.5–799.5	4	100	1.00

To form a cumulative frequency polygon, or **ogive** as it is also called, the cumulative frequency is plotted over the upper boundary of the corresponding class. Figure 5 is the cumulative frequency polygon for the cumulative frequency table in Table 3. Notice that we can easily see that 78% of the students scored below 649.5, while only 18% scored below 499.5. We also can conclude that the probability of a randomly selected score from the sample of 100 lying below 649.5 is .78 and above 649.5 is 1.00 − .78 = .22.

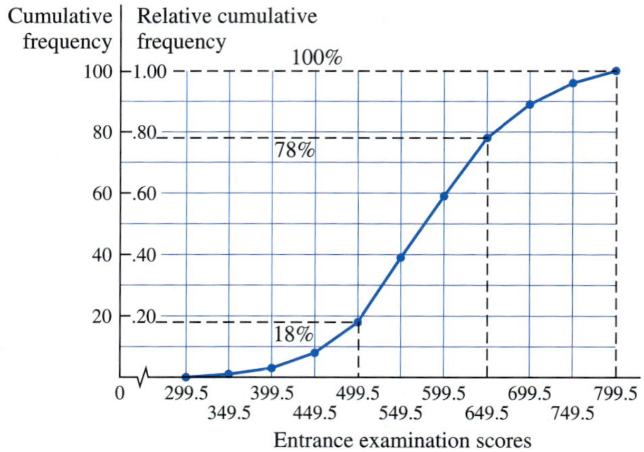

FIGURE 5
Cumulative frequency polygon (ogive)

Explore–Discuss 2

(A) Construct a histogram for the data set whose cumulative frequency polygon is shown in Figure 6.

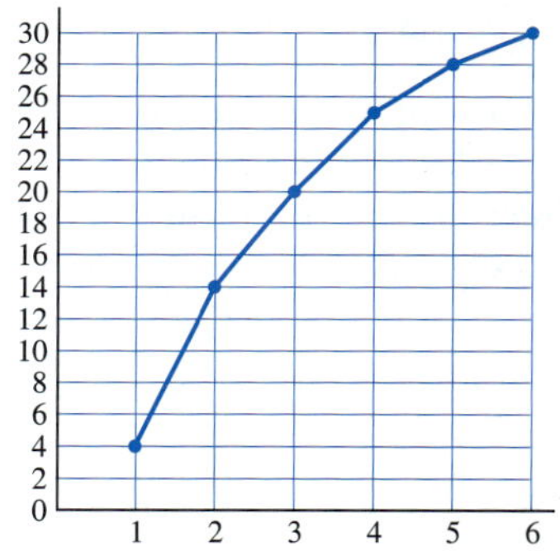

FIGURE 6

(B) Can the original data set be reconstructed from the cumulative frequency polygon? Explain.

Answers to Matched Problems **1.** (A) .11 (B) .18 **2.** (A)

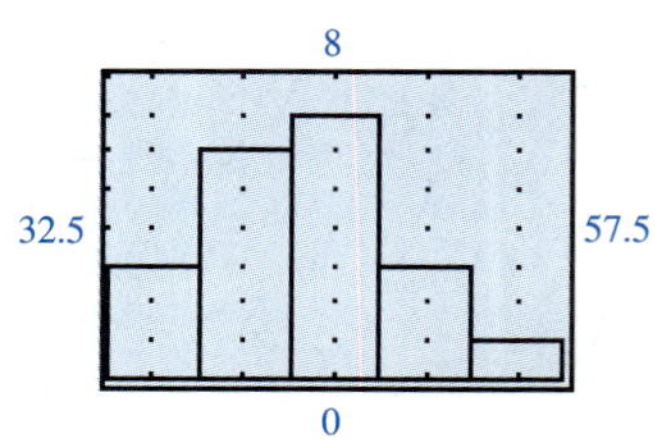

(B) .45; .55

Exercise 7-2

1. (A) Construct a frequency table and a histogram for the following data set using a class interval width of 2 starting at 0.5:

6	7	2	7	9
6	4	7	6	6

(B) Construct a frequency table and a histogram for the following data set using a class interval width of 2 starting at 0.5:

5	6	8	1	3
5	10	7	6	8

(C) How are the two histograms of parts (A) and (B) similar? How are the two data sets different?

2. (A) Construct a frequency table and a histogram for the data set of part (A) of Problem 1 using a class interval width of 1 starting at 0.5.

(B) Construct a frequency table and a histogram for the data set of part (B) of Problem 1 using a class interval width of 1 starting at 0.5.

(C) How are the histograms of parts (A) and (B) different?

3. C The graphing utility command shown in Figure A generated a set of 400 random integers from 2 to 24, stored as list L_1. The statistical plot in Figure B is a histogram of L_1, using a class interval width of 1 starting at 1.5.

(A) Explain how the window variables can be changed to display a histogram of the same data set using a class interval width of 2 starting at 1.5. A width of 4 starting at 1.5.

(B) Describe the effect of increasing the class interval width on the shape of the histogram.

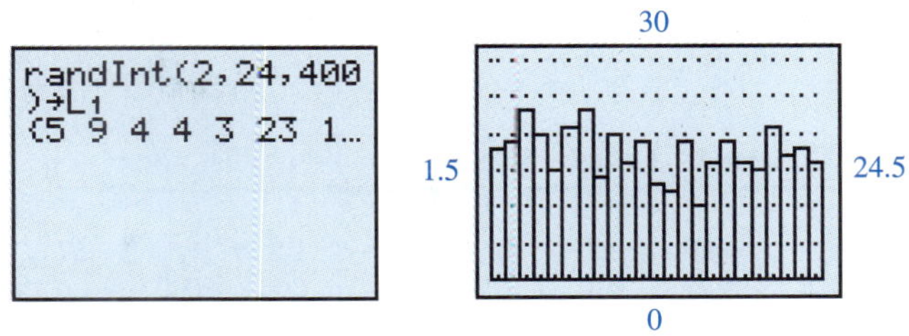

(A) (B)

Figure for 3

4. C An experiment consists of rolling a pair of dodecahedral (twelve-sided) dice and recording their sum (the sides of each die are numbered from 1 to 12). The command shown in Figure A simulated 500 rolls of the dodecahedral dice. The statistical plot in Figure B is a histogram of the 500 sums using a class interval width of 1 starting at 1.5.

(A) Explain how the window variables can be changed to display a histogram of the same data set using a class interval width of 2 starting at 1.5. A width of 3 starting at −0.5.

(B) Describe the effect of increasing the class interval width on the shape of the histogram.

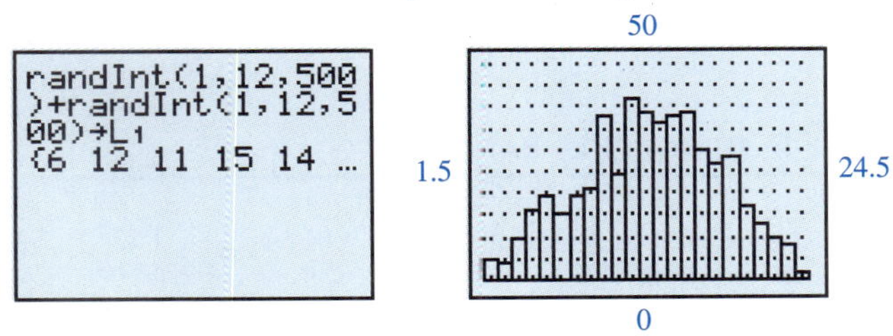

(A) (B)

Figure for 4

Applications

Business & Economics

5. *Starting salaries.* The starting salaries (in thousands of dollars) of 20 graduates, chosen at random from the graduating class of an urban university, were determined and recorded in the table:

Starting Salaries

34	29	27	39	41
28	32	37	35	36
23	31	33	34	29
27	35	29	30	32

(A) Construct a frequency and relative frequency table using a class interval width of 4 starting at 20.5.

(B) Construct a histogram.

(C) What is the probability that a graduate chosen from the sample will have a starting salary above \$32,500? Below \$28.500?

C (D) Construct a histogram using a graphing utility.

6. *Commute times.* Thirty-two persons were chosen at random from among the employees of a large corporation and their commute times (in hours) from home to work were determined and recorded in the table:

COMMUTE TIMES

0.5	0.9	0.2	0.4	0.7	1.2	1.1	0.7
0.6	0.4	0.8	1.1	0.9	0.3	0.4	1.0
0.9	1.0	0.7	0.3	0.6	1.1	0.7	1.1
0.4	1.3	0.7	0.6	1.0	0.8	0.4	0.9

(A) Construct a frequency and relative frequency table using a class interval width of 0.2 starting at 0.15.
(B) Construct a histogram.
(C) What is the probability that a person chosen at random from the sample will have a commuting time of at least an hour? Of at most half an hour?
C (D) Construct a histogram using a graphing utility.

7. *Common stocks.* The table shows price–earnings ratios of 100 common stocks chosen at random from the New York Stock Exchange.

PRICE–EARNINGS (PE) RATIOS

7	11	6	6	10	6	31	28	13	19
6	18	9	7	5	5	9	8	10	6
10	3	4	6	7	9	9	19	7	9
17	33	17	12	7	5	7	10	7	9
18	17	4	6	11	13	7	6	10	7
7	9	8	15	16	11	10	7	5	14
12	10	6	7	7	13	10	5	6	4
10	6	7	11	19	17	6	9	6	5
6	13	4	7	6	12	9	14	9	7
18	5	12	8	8	8	13	9	13	15

(A) Construct a frequency and relative frequency table using a class interval of 5 starting at −0.5.
(B) Construct a histogram.
(C) Construct a frequency polygon.
(D) Construct a cumulative frequency and relative cumulative frequency table. What is the probability of a price–earnings ratio drawn at random from the sample lying between 4.5 and 14.5?
(E) Construct a cumulative frequency polygon.

Life Sciences

8. *Mouse weights.* One hundred healthy mice were weighed at the beginning of an experiment with the following results:

MOUSE WEIGHTS (IN GRAMS)

51	54	47	53	59	46	50	50	56	46
48	50	45	49	52	55	42	57	45	51
53	55	51	47	53	53	49	51	43	48
44	48	54	46	49	51	52	50	55	51
50	53	45	49	57	54	53	49	46	48
52	48	50	52	47	50	44	46	47	49
49	51	57	49	51	42	49	53	44	52
53	55	48	52	44	46	54	54	57	55
48	50	50	55	52	48	47	52	55	50
59	52	47	46	56	54	51	56	54	55

(A) Construct a frequency and relative frequency table using a class interval of 2 starting at 41.5.
(B) Construct a histogram.
(C) Construct a frequency polygon.
(D) Construct a cumulative frequency and relative cumulative frequency table. What is the probability of a mouse weight drawn at random from the sample lying between 45.5 and 53.5?
(E) Construct a cumulative frequency polygon.

Social Sciences

9. *Grade-point averages.* One hundred seniors were chosen at random from a graduating class at a university and their grade-point averages recorded:

GRADE-POINT AVERAGES (GPA)

2.1	2.0	2.7	2.6	2.1	3.5	3.1	2.1	2.2	2.9
2.3	2.5	3.1	2.2	2.2	2.0	2.3	2.5	2.1	2.4
2.7	2.9	2.1	2.2	2.5	2.3	2.1	2.1	3.3	2.1
2.2	2.2	2.5	2.3	2.7	2.4	2.8	3.1	2.0	2.3
2.6	3.2	2.2	2.5	3.6	2.3	2.4	3.7	2.5	2.4
3.5	2.4	2.3	3.9	2.9	2.7	2.6	2.1	2.4	2.0
2.4	3.3	3.1	2.8	2.3	2.5	2.1	3.0	2.6	2.3
2.1	2.6	2.2	3.2	2.7	2.8	3.4	2.7	3.6	2.1
2.7	2.8	3.5	2.4	2.3	2.0	2.1	3.1	2.8	2.1
3.8	2.5	2.7	2.1	2.2	2.4	2.9	3.3	2.0	2.6

(A) Construct a frequency and relative frequency table using a class interval of 0.2 starting at 1.95.
(B) Construct a histogram.
(C) Construct a frequency polygon.
(D) Construct a cumulative frequency and relative cumulative frequency table. What is the probability of a GPA drawn at random from the sample being over 2.95?
(E) Construct a cumulative frequency polygon.

SECTION 7-3

Measures of Central Tendency

- MEAN
- MEDIAN
- MODE

In the preceding section, we found that graphic techniques contributed substantially to our comprehension of large masses of raw data. In this and the next section, we will discuss several important numerical measures that are used to describe sets of data. These numerical descriptions are generally of two types:

1. Measures that indicate the approximate center of a distribution, called **measures of central tendency**
2. Measures that indicate the amount of scatter about a central point, called **measures of dispersion**

In this section we will look at three widely used measures of central tendency, and in the next section we will consider measures of dispersion.

MEAN

When we speak of the average yield of 10 year municipal bonds, the average number of smog-free days per year in a certain city, or the average SAT score for students at a university, we usually interpret these averages to be **arithmetic averages,** or **means.** In general, we define the mean of a set of quantitative data as follows:

Mean (Ungrouped Data)

The **mean** of a set of quantitative data is equal to the sum of all the measurements in the data set divided by the total number of measurements in the set.

The mean is a single number that, in a sense, represents the entire data set. It involves all the measurements in the set, it is easily computed, and it enters readily into other useful formulas. Because of these and other desirable properties, the mean is the most widely used measure of central tendency.

In statistics, we are concerned with both a sample mean and the mean of the corresponding population (the sample mean is often used as an estimator for the population mean), so it is important to use different symbols to represent these two means. It is customary to use a letter with an overbar, such as $\bar{x}$, to represent a sample mean and the Greek letter μ ("mu") to represent a population mean.

Notation for the Mean

$\bar{x}$ = Sample mean $\qquad$ μ = Population mean

Before considering examples, let us formulate the concept symbolically using the **summation symbol** Σ (see Appendix B-1). If

$$x_1, x_2, \ldots, x_n$$

represents a set of n measurements, then the sum

$$x_1 + x_2 + \cdots + x_n$$

is compactly and conveniently represented by

$$\sum_{i=1}^{n} x_i \quad \text{or} \quad \Sigma_{i=1}^{n} x_i$$

We now can express the mean in symbolic form.

Mean (Ungrouped Data)

If $x_1, x_2, \ldots, x_n$ is a set of n measurements, then the **mean** of the set of measurements is given by

$$[\text{Mean}] = \frac{\sum_{i=1}^{n} x_i}{n} = \frac{x_1 + x_2 + \cdots + x_n}{n} \tag{1}$$

where

$\bar{x} = [\text{Mean}]$ if data set is a sample
$\mu = [\text{Mean}]$ if data set is the population

Example 1 **Finding the Mean** Find the mean for the sample measurements 3, 5, 1, 8, 6, 5, 4, and 6.

SOLUTION Solve using formula (1):

$$\bar{x} = \frac{\sum_{i=1}^{n} x_i}{n} = \frac{3 + 5 + 1 + 8 + 6 + 5 + 4 + 6}{8} = \frac{38}{8} = 4.75$$

Matched Problem 1 Find the mean for the sample measurements 3.2, 4.5, 2.8, 5.0, and 3.6.

If data have been grouped in a frequency table, such as Table 2 (Section 7-2), then an alternative formula for the mean is generally used, as given in the box at the top of the next page.

Mean (Grouped Data)

A data set of n measurements is grouped into k classes in a frequency table. If x_i is the midpoint of the ith class interval and f_i is the ith class frequency, then the **mean for the grouped data** is given by

$$[\text{Mean}] = \frac{\sum_{i=1}^{k} x_i f_i}{n} = \frac{x_1 f_1 + x_2 f_2 + \cdots + x_k f_k}{n} \tag{2}$$

where

$$n = \sum_{i=1}^{k} f_i = \text{Total number of measurements}$$
$$\bar{x} = [\text{Mean}] \text{ if data set is a sample}$$
$$\mu = [\text{Mean}] \text{ if data set is the population}$$

CAUTION

It is important to note that n is the total number of measurements in the entire data set—not the number of classes!

The mean computed by formula (2) is a **weighted average** of the midpoints of the class intervals. In general, this will be close to, but not exactly the same as, the mean computed by formula (1) for ungrouped data.

Example 2 **Finding the Mean for Grouped Data** Find the mean for the sample data summarized in Table 2 of Section 7-2.

SOLUTION We repeat part of Table 2 here, adding columns for the class midpoints x_i and the products $x_i f_i$ (Table 1).

Table 1

ENTRANCE EXAMINATION SCORES

CLASS INTERVAL	MIDPOINT x_i	FREQUENCY f_i	PRODUCT $x_i f_i$
299.5–349.5	324.5	1	324.5
349.5–399.5	374.5	2	749.0
399.5–449.5	424.5	5	2,122.5
449.5–499.5	474.5	10	4,745.0
499.5–549.5	524.5	21	11,014.5
549.5–599.5	574.5	20	11,490.0
599.5–649.5	624.5	19	11,865.5
649.5–699.5	674.5	11	7,419.5
699.5–749.5	724.5	7	5,071.5
749.5–799.5	774.5	4	3,098.0
		$n = \sum_{i=1}^{10} f_i = 100$	$\sum_{i=1}^{10} x_i f_i = 57{,}900.0$

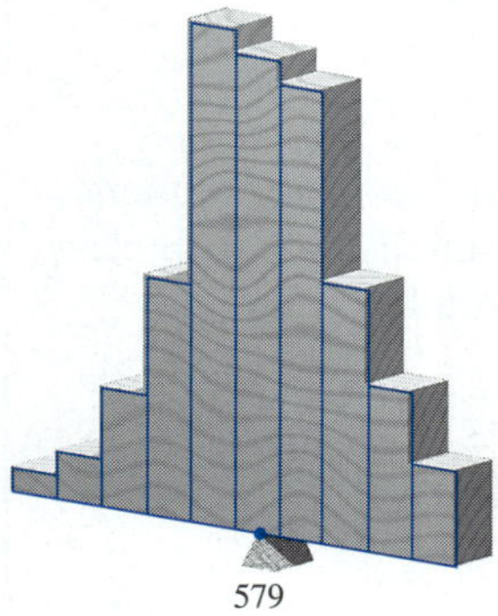

FIGURE 1
The balance point on the histogram is $\overline{x} = 579$.

Thus, the average entrance examination score for the sample of 100 entering freshmen is

$$\overline{x} = \frac{\sum_{i=1}^{k} x_i f_i}{n} = \frac{57{,}900}{100} = 579$$

If the histogram for the data in Table 1 (Fig. 2 in Section 7-2) was drawn on a piece of wood of uniform thickness and the wood cut around the outside of the figure, then the resulting object would balance exactly at the mean $\overline{x} = 579$, as shown in Figure 1.

Matched Problem 2 Compute the mean for the grouped sample data listed in Table 2.

Table 2

CLASS INTERVAL	FREQUENCY
0.5–5.5	6
5.5–10.5	20
10.5–15.5	18
15.5–20.5	4

Note that the mean also can be interpreted as the expected value of a random variable.

If the random variable X is assigned midpoint values in a table of grouped data (such as those in the second column of Table 1), then, starting with formula (2), we have

$$\begin{aligned}
[\text{Mean}] &= \frac{\sum_{i=1}^{k} x_i f_i}{n} \\
&= \frac{x_1 f_1 + x_2 f_2 + \cdots + x_k f_k}{n} \\
&= x_1\left(\frac{f_1}{n}\right) + x_2\left(\frac{f_2}{n}\right) + \cdots + x_k\left(\frac{f_k}{n}\right) \\
&= x_1 p_1 + x_2 p_2 + \cdots + x_k p_k \\
&= E(X)
\end{aligned}$$

Median

Occasionally, the mean can be misleading as a measure of central tendency. Suppose the annual salaries of seven people in a small company are \$17,000, \$20,000, \$28,000, \$18,000, \$18,000, \$120,000, and \$24,000. The mean salary is

$$\bar{x} = \frac{\sum_{i=1}^{n} x_i}{n} = \frac{\$245{,}000}{7} = \$35{,}000$$

Six of the seven salaries are below the average! The one large salary distorts the results.

A measure of central tendency that is not influenced by extreme values is the **median.** The following definition of median makes precise our intuitive notion of the "middle element" when a set of measurements is arranged in ascending or descending order. Some sets of measurements, for example, 5, 7, 8, 13, 21, have a middle element. Other sets, for example, 9, 10, 15, 20, 23, 24, have no middle element, or you might prefer to say they have two middle elements. For any number between 15 and 20, half the measurements fall above the number and half fall below.

Median

1. If the number of measurements in a set is odd, the **median** is the middle measurement when the measurements are arranged in ascending or descending order.
2. If the number of measurements in a set is even, the **median** is the mean of the two middle measurements when the measurements are arranged in ascending or descending order.

Example 3 **Finding the Median** Find the median salary in the above list of seven salaries.

SOLUTION Arrange the salaries in increasing order and choose the middle one:

SALARY	
$ 17,000	
18,000	
18,000	
20,000	← Median ($20,000)
24,000	
28,000	← Mean ($35,000)
120,000	

In this case, the median is a better measure of central tendency than the mean.

Matched Problem 3 Add the salary $100,000 to those in Example 3 and compute the median and mean for these eight salaries.

The median, as we have defined it, is easy to determine and is not influenced by extreme values. Our definition does have some minor handicaps, however. First, if the measurements we are analyzing were carried out in a laboratory and presented to us in a frequency table, we may not have access to the individual measurements. In that case we would not be able to compute the median using the above definition. Second, a set like 4, 4, 6, 7, 7, 7, 9 would have median 7 by our definition, but 7 does not possess the symmetry we expect of a "middle element" since there are three measurements below 7 but only one above.

To overcome these handicaps, we define a second concept, the *median for grouped data.* To guarantee that the median for grouped data exists and is unique, we assume that the frequency table for the grouped data has no classes of frequency 0.

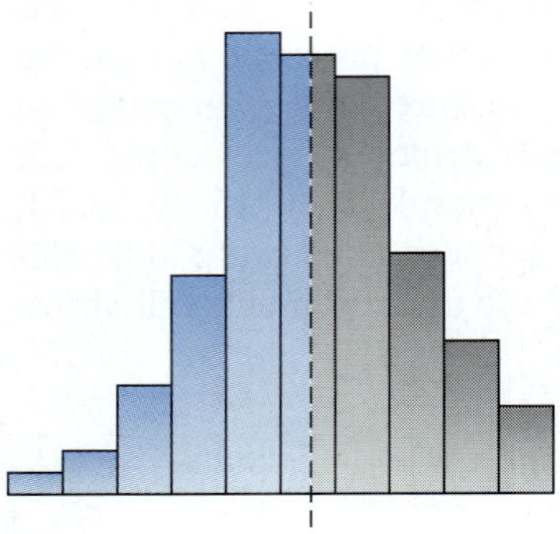

FIGURE 2
The area to the left of the median equals the area to the right.

Median for Grouped Data

The **median for grouped data** with no classes of frequency 0 is the number such that the histogram has the same area to the left of the median as to the right of the median (see Fig. 2).

Example 4 **Finding the Median for Grouped Data** Compute the median for the grouped data of Table 3.

SOLUTION We first draw the histogram of the data (Fig. 3). The total area of the histogram is 15, which is just the sum of the frequencies, since all rectangles have a base of length 1. The area to the left of the median must be half the total area—that is, $\frac{15}{2} = 7.5$. Looking at Figure 3 we see that the median M lies between 6.5 and 7.5. Thus, the area to the left of M, which is the sum of the shaded areas in Figure 3, must be 7.5:

$$(1)(3) + (1)(1) + (1)(2) + (M - 6.5)(4) = 7.5$$

Solving for M gives $M = 6.875$. That is, the median for the grouped data in Table 3 is 6.875.

Table 3

Class interval	Frequency
3.5–4.5	3
4.5–5.5	1
5.5–6.5	2
6.5–7.5	4
7.5–8.5	3
8.5–9.5	2

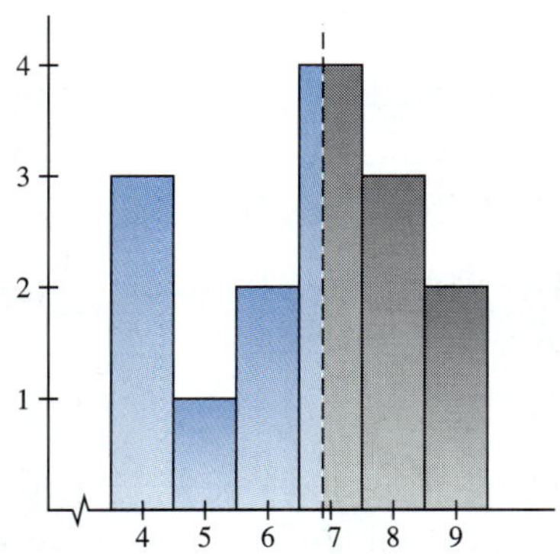

FIGURE 3

Matched Problem 4 ➡ Find the median for the grouped data in the following table:

CLASS INTERVAL	FREQUENCY
3.5–4.5	4
4.5–5.5	2
5.5–6.5	3
6.5–7.5	5
7.5–8.5	4
8.5–9.5	3

Explore–Discuss 1

We have given geometric interpretations of the mean for grouped data as the balance point of a histogram (Fig. 1), and of the median for grouped data as the point that divides a histogram into equal areas (Fig. 2).

(A) Give an example of a simple histogram in which the balance point (the mean) and the point marking equal areas (the median) are two different points. Explain how other examples in which the mean does not equal the median could be constructed.

(B) Given an example of a simple histogram in which the balance point (the mean) and the point marking equal areas (the median) are the same point. Discuss properties of a histogram that would guarantee that the mean and the median are equal.

(C) Explain why the median for grouped data is not influenced by extreme values.

MODE

A third measure of central tendency is the *mode.*

Mode

The **mode** is the most frequently occurring measurement in a data set. There may be a unique mode, several modes, or essentially no mode.

Example 5 ➡ **Finding Mode, Median, and Mean**

	DATA SET	MODE	MEDIAN	MEAN
(A)	4, 5, 5, 5, 6, 6, 7, 8, 12	5	6	6.44
(B)	1, 2, 3, 3, 3, 5, 6, 7, 7, 7, 23	3, 7	5	6.09
(C)	1, 3, 5, 6, 7, 9, 11, 15, 16	None	7	8.11

Data set (B) in Example 5 is referred to as **bimodal,** since there are two modes.

Matched Problem 5 ➡ Compute the mode(s), median, and mean for each data set:

(A) 2, 1, 2, 1, 1, 5, 1, 9, 4 (B) 2, 5, 1, 4, 9, 8, 7

(C) 8, 2, 6, 8, 3, 3, 1, 5, 1, 8, 3

The mode, median, and mean can be computed in various ways with the aid of a graphing utility. In Figure 4A the data set of Example 5B is entered as a list, and its median and mean are computed. The histogram in Figure 4B shows the two modes of the same data set.

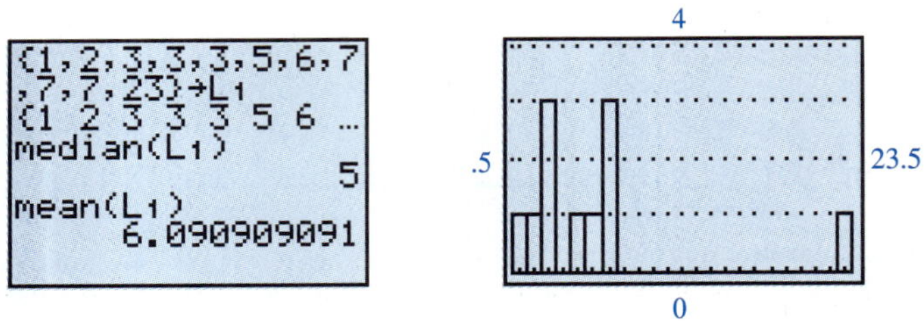

(A) (B)

FIGURE 4

As with the median, the mode is not influenced by extreme values. Suppose, in the data set of Example 5B, we replace 23 by 8. The modes remain 3 and 7 and the median is still 5, but the mean changes to 4.73. The mode is most useful for large data sets because it emphasizes data concentration. For example, a clothing retailer would be interested in the mode of sizes due to customer demand of the various items stocked in a store.

The mode also can be used for qualitative attributes—that is, attributes that are not numerical. The mean and median are not suitable in these cases. For example, the mode can be used to give an indication of a favorite brand of ice cream or the worst movie of the year. Figure 5 shows the results of a random survey of 1,000 people on entree preferences when eating dinner out. According to this survey, we would say that the modal preference is beef. Note that the mode is the only measure of central tendency (location) that can be used for this type of data; the mean and median make no sense.

In actual practice, the mean is used the most, the median next, and the mode a distant third.

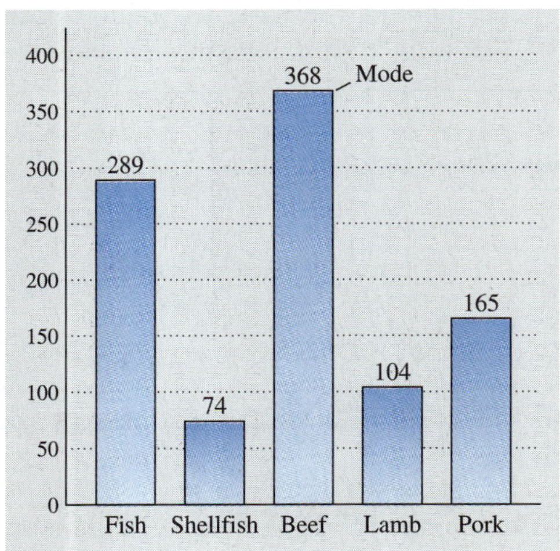

FIGURE 5
The modal preference for an entree is beef.

Explore–Discuss 2

For many sets of measurements the median lies between the mode and the mean. But this is not always so.

(A) In a class of seven students the scores on an exam were 52, 89, 89, 92, 93, 96, 99. Show that the mean is less than the mode, and that the mode is less than the median.

(B) Construct hypothetical sets of exam scores to show that all possible orders among the mean, median, and mode can occur.

Answers to Matched Problems

1. $\bar{x} \approx 3.8$ **2.** $\bar{x} \approx 10.1$ **3.** Median = \$22,000; Mean = \$43,125

4. Median for grouped data = 6.8

5. First, arrange each set of data in ascending order:

	Data Set	Mode	Median	Mean
(A)	1, 1, 1, 1, 2, 2, 4, 5, 9	1	2	2.89
(B)	1, 2, 4, 5, 7, 8, 9	None	5	5.14
(C)	1, 1, 2, 3, 3, 3, 5, 6, 8, 8, 8	3, 8	3	4.36

Exercise 7-3

A *Find the mean, median, and mode for the sets of ungrouped data given in Problems 1 and 2.*

1. 1, 2, 2, 3, 3, 3, 3, 4, 4, 5 **2.** 1, 1, 1, 1, 2, 3, 4, 5, 5, 5

Find the mean, median, and/or mode, whichever are applicable, in Problems 3 and 4.

3.

Flavor	Number Preferring
Vanilla	139
Chocolate	376
Strawberry	89
Pistachio	105
Cherry	63
Almond mocha	228

4.

Car Color	Number Preferring
Red	1,324
White	3,084
Black	1,617
Blue	2,303
Brown	2,718
Gold	1,992

Find the mean for the sets of grouped data in Problems 5 and 6.

5.

Interval	Frequency
0.5–2.5	2
2.5–4.5	5
4.5–6.5	7
6.5–8.5	1

6.

Interval	Frequency
0.5–2.5	5
2.5–4.5	1
4.5–6.5	2
6.5–8.5	7

B

7. Which single measure of central tendency—the mean, median, or mode—would you say best describes the following set of measurements? Discuss the factors that justify your preference.

8.01	7.91	8.13	6.24	7.95
8.04	7.99	8.09	6.24	81.2

8. Which single measure of central tendency—the mean, median, or mode—would you say best describes the following set of measurements? Discuss the factors that justify your preference.

47	51	80	91	85
69	91	95	81	60

9. A data set is formed by recording the results of 100 rolls of a fair die.
(A) What would you expect the mean of the data set to be? The median?
C (B) Form such a data set by using a graphing utility to simulate 100 rolls of a fair die, and find its mean and median.

10. A data set is formed by recording the sums on 200 rolls of a pair of fair dice.
(A) What would you expect the mean of the data set to be? The median?
C (B) Form such a data set by using a graphing utility to simulate 200 rolls of a pair of fair dice, and find the mean and median of the set.

C

11. (A) Construct a set of four numbers that has mean 300, median 250, and mode 175.
(B) Let $m_1 > m_2 > m_3$. Devise and discuss a procedure for constructing a set of four numbers that has mean m_1, median m_2, and mode m_3.

12. (A) Construct a set of five numbers that has mean 200, median 150, and mode 50.
(B) Let $m_1 > m_2 > m_3$. Devise and discuss a procedure for constructing a set of five numbers that has mean m_1, median m_2, and mode m_3.

Applications

Business & Economics

13. *Price–earnings ratios.* Find the mean, median, and mode for the data in the following table:

Price–Earnings Ratios for Eight Stocks in a Portfolio

5.3	10.1	18.7	35.5
12.9	8.4	16.2	10.1

14. *Gasoline tax.* Find the mean, median, and mode for the data in the following table:

State Gasoline Tax, 1993

State	Tax Rate per Gallon
Connecticut	29¢
Rhode Island	28¢
West Virginia	25¢
Nebraska	24¢
Oregon	24¢
Maryland	23¢
Wisconsin	23¢

15. *Light bulb lifetime.* Find the mean and median for the data in the following table:

Life (in Hours) of 50 Randomly Selected Light Bulbs

Interval	Frequency
799.5–899.5	3
899.5–999.5	10
999.5–1,099.5	24
1,099.5–1,199.5	12
1,199.5–1,299.5	1

16. *Price–earnings ratios.* Find the mean and median for the data in the following table:

Price–Earnings Ratios of 100 Randomly Chosen Stocks from the New York Stock Exchange

Interval	Frequency
−0.5–4.5	5
4.5–9.5	54
9.5–14.5	25
14.5–19.5	9
19.5–24.5	4
24.5–29.5	1
29.5–34.5	2

17. *Housing starts.* Find the mean, median, and mode for the data in the following table:

New Housing Starts

Year	Number
1987	1,600,000
1988	1,490,000
1989	1,380,000
1990	1,200,000
1991	1,100,000
1992	1,200,000

18. *Funding of higher education.* Find the mean, median, and mode for the data in the following table:

FUNDING OF PUBLIC HIGHER EDUCATION, 1994

STATE	DOLLARS SPENT PER STUDENT
California	4,030
Delaware	3,960
Florida	4,770
Nebraska	4,840
New York	4,840
Texas	4,440
Wisconsin	6,450

Life Sciences

19. *Mouse weights.* Find the mean and median for the data in the following table:

MOUSE WEIGHTS (IN GRAMS)

INTERVAL	FREQUENCY
41.5–43.5	3
43.5–45.5	7
45.5–47.5	13
47.5–49.5	17
49.5–51.5	19
51.5–53.5	17
53.5–55.5	15
55.5–57.5	7
57.5–59.5	2

20. *Blood cholesterol levels.* Find the mean and median for the data in the following table (given in milligrams per deciliter):

BLOOD CHOLESTEROL LEVELS

INTERVAL	FREQUENCY
149.5–169.5	4
169.5–189.5	11
189.5–209.5	15
209.5–229.5	25
229.5–249.5	13
249.5–269.5	7
269.5–289.5	3
289.5–309.5	2

Social Sciences

21. *Immigration.* Find the mean, median, and mode for the data in the following table:

IMMIGRATION TO THE UNITED STATES, 1992

COUNTRY	NUMBER
China	39,000
Great Britain	20,000
India	37,000
Ireland	12,000
Japan	11,000
Poland	25,000
Former USSR	44,000

22. *Grade-point averages.* Find the mean and median for the grouped data in the following table:

GRADUATING CLASS GRADE-POINT AVERAGES

INTERVAL	FREQUENCY
1.95–2.15	21
2.15–2.35	19
2.35–2.55	17
2.55–2.75	14
2.75–2.95	9
2.95–3.15	6
3.15–3.35	5
3.35–3.55	4
3.55–3.75	3
3.75–3.95	2

23. *Entrance examination scores.* Compute the median for the grouped data of entrance examination scores given in Table 1.

24. *Library holdings.* Find the mean and median for the grouped data in the following table:

LIBRARY HOLDINGS OF 50 MAJOR U.S. UNIVERSITIES, 1991–1992

VOLUMES *Millions*	NUMBER OF UNIVERSITIES
0–2.5	4
2.5–5.0	32
5.0–7.5	10
7.5–10.0	3
10.0–12.5	1

SECTION 7-4

Measures of Dispersion

- RANGE
- STANDARD DEVIATION—UNGROUPED DATA
- STANDARD DEVIATION—GROUPED DATA
- SIGNIFICANCE OF STANDARD DEVIATION

A measure of central tendency gives us a typical value that can be used to describe a whole set of data, but this measure does not tell us whether the data are tightly clustered or widely dispersed. We will now consider two measures of variation—*range* and *standard deviation*—that will give some indication of data scatter.

RANGE

A measure of dispersion, or scatter, that is easy to compute and is easily understood is the range. The **range for a set of ungrouped data** is the difference between the largest and the smallest values in the data set. The **range for a frequency distribution** is the difference between the upper boundary of the highest class and the lower boundary of the lowest class.

Consider the histograms in Figure 1. We see that the range adds only a little information about the amount of variation in a data set. The graphs clearly show that even though each data set has the same mean and range, all three sets differ in the amount of scatter, or variation, of the data relative to the mean. The data set in part (A) is tightly clustered about the mean; the data set in part (B) is dispersed away from the mean; and the data set in part (C) is uniformly distributed over its range.

Since the range depends only on the extreme values of the data, it does not give us any information about the dispersion of the data between these extremes. We need a measure of dispersion that will give us some idea of how the data are clustered or scattered relative to the mean. The *standard deviation* is such a measure.

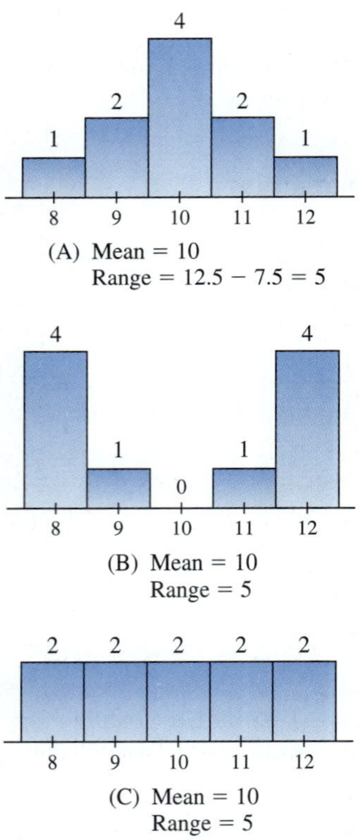

FIGURE 1

STANDARD DEVIATION—UNGROUPED DATA

We will develop the concepts of *variance* and *standard deviation*—both measures of variation—through a simple example. Suppose a random sample of five stamped parts is selected from a manufacturing process, and these parts are found to have the following lengths (in centimeters):

5.2, 5.3, 5.2, 5.5, 5.3

Computing the sample mean, we obtain

$$\bar{x} = \frac{\sum_{i=1}^{n} x_i}{n} = \frac{5.2 + 5.3 + 5.2 + 5.5 + 5.3}{5}$$
$$= 5.3 \text{ centimeters}$$

Table 1

x_i	$(x_i - \overline{x})$
5.2	−0.1
5.3	0.0
5.2	−0.1
5.5	0.2
5.3	0.0

How much variation exists between the sample mean and all measurements in the sample? As a first attempt at measuring the variation, let us represent the **deviation** of a measurement from the mean by $(x_i - \overline{x})$. Table 1 lists all the deviations for this sample.

Using these deviations, what kind of formula can we find that will give us a single measure of variation? It appears that the average of the deviations might be a good measure. But look what happens when we add the second column in Table 1. We get 0! It turns out that this will always happen for any data set. Now what? We could take the average of the absolute values of the deviations; however, this approach leads to problems relative to statistical inference. Instead, to get around the sign problem, we will take the average of the squares of the deviations and call this number the **variance** of the data set:

$$[\text{Variance}] = \frac{\sum_{i=1}^{n} (x_i - \overline{x})^2}{n} \tag{1}$$

Calculating the variance using the entries in Table 1, we have

$$[\text{Variance}] = \frac{\sum_{i=1}^{5} (x_i - 5.3)^2}{5} = 0.012 \text{ square centimeter}$$

We still have a problem, because the units in the variance are square centimeters instead of centimeters (the units of the original data set). To obtain the units of the original data set, we take the positive square root of the variance and call the result the **standard deviation** of the data set:

$$[\text{Standard deviation}] = \sqrt{\frac{\sum_{i=1}^{n} (x_i - \overline{x})^2}{n}} \tag{2}$$

$$= \sqrt{\frac{\sum_{i=1}^{5} (x_i - 5.3)^2}{5}} = 0.11 \text{ centimeter}$$

The sample variance is usually denoted by s^2 and the population variance by σ^2 (σ is the Greek letter "sigma"). The sample standard deviation is usually denoted by s and the population standard deviation by σ.

In inferential statistics the sample variance s^2 is often used as an estimator for the population variance σ^2 and the sample standard deviation s for the population standard deviation σ. It can be shown that one can obtain better estimates of the population parameters in terms of the sample parameters (particularly when using small samples) if the divisor n is replaced by $n - 1$ when computing sample variances or sample standard deviations. With these modifications, we have the following:

Variance (Ungrouped Data)*

The **sample variance** s^2 of a set of n sample measurements $x_1, x_2, \ldots, x_n$ with mean $\bar{x}$ is given by

$$s^2 = \frac{\sum_{i=1}^{n} (x_i - \bar{x})^2}{n - 1} \tag{3}$$

If $x_1, x_2, \ldots, x_n$ is the whole population with mean μ, then the **population variance** σ^2 is given by

$$\sigma^2 = \frac{\sum_{i=1}^{n} (x_i - \mu)^2}{n}$$

*In this section we will restrict our interest to the sample variance.

The standard deviation is just the positive square root of the variance. Therefore, we have the following formulas:

Standard Deviation (Ungrouped Data)*

The **sample standard deviation** s of a set of n sample measurements $x_1, x_2, \ldots, x_n$ with mean $\bar{x}$ is given by

$$s = \sqrt{\frac{\sum_{i=1}^{n} (x_i - \bar{x})^2}{n - 1}} \tag{4}$$

If $x_1, x_2, \ldots, x_n$ is the whole population with mean μ, then the **population standard deviation** σ is given by

$$\sigma = \sqrt{\frac{\sum_{i=1}^{n} (x_i - \mu)^2}{n}}$$

*In this section we will restrict our interest to the sample standard deviation.

Computing the standard deviation for the original sample measurements (Table 1), we now obtain

$$s = \sqrt{\frac{\sum_{i=1}^{5} (x_i - 5.3)^2}{5 - 1}} = 0.12 \text{ centimeter}$$

Example 1 **Finding the Standard Deviation** Find the standard deviation for the sample measurements 1, 3, 5, 4, 3.

SOLUTION To find the standard deviation for the data set, we can utilize a table or use a calculator. Most will prefer the latter. Here is what we compute:

$$\bar{x} = \frac{1 + 3 + 5 + 4 + 3}{5} = 3.2$$

$$s = \sqrt{\frac{(1 - 3.2)^2 + (3 - 3.2)^2 + (5 - 3.2)^2 + (4 - 3.2)^2 + (3 - 3.2)^2}{5 - 1}}$$

$$\approx 1.48$$

Matched Problem 1 Find the standard deviation for the sample measurements 1.2, 1.4, 1.7, 1.3, 1.5.

REMARK

Many calculators and graphing utilities can compute $\bar{x}$ and s directly after the sample measurements are entered—a helpful feature, especially when the sample is fairly large. This shortcut is illustrated in Figure 2 for a particular graphing calculator, where the data from Example 1 are entered as a list and several different one-variable statistics are immediately calculated. Included among these statistics are the mean $\bar{x}$, the sample standard deviation s (denoted by Sx in Fig. 2B), the population standard deviation σ (denoted by σx), the number n of measurements, the smallest element of the data set (denoted by minX), the largest element of the data set (denoted by maxX), the median (denoted by Med), and several statistics we have not discussed.

(A) Data (B) Statistics (C) Statistics (continued)

FIGURE 2

Explore–Discuss 1

(A) When is the sample standard deviation of a set of measurements equal to 0?

(B) Can the sample standard deviation of a set of measurements ever be greater than the range? Explain why or why not.

Formulas (4) and (2) produce very nearly the same result for large samples. The main difference in the two lies in their use for small samples. The **law of large numbers** informally states that we can make a sample standard deviation s as close to the population standard deviation σ as we like by making the sample sufficiently large.

■ STANDARD DEVIATION—GROUPED DATA

Formula (4) for sample standard deviation is extended to grouped sample data as described in the following box:

> ### Standard Deviation (Grouped Data)*
>
> Suppose a data set of n sample measurements is grouped into k classes in a frequency table, where x_i is the midpoint and f_i is the frequency of the ith class interval. Then the **sample standard deviation s** for the grouped data is
>
> $$s = \sqrt{\frac{\sum_{i=1}^{k} (x_i - \bar{x})^2 f_i}{n - 1}} \qquad (5)$$
>
> where $n = \sum_{i=1}^{k} f_i$ = Total number of measurements.
>
> If $x_1, x_2, \ldots, x_n$ is the whole population with mean μ, then the **population standard deviation σ** is given by
>
> $$\sigma = \sqrt{\frac{\sum_{i=1}^{n} (x_i - \mu)^2 f_i}{n}}$$
>
> *In this section we will restrict our interest to the sample standard deviation.

Example 2 ➠ **Finding the Standard Deviation for Grouped Data** Find the standard deviation for each set of grouped sample data.

(A)

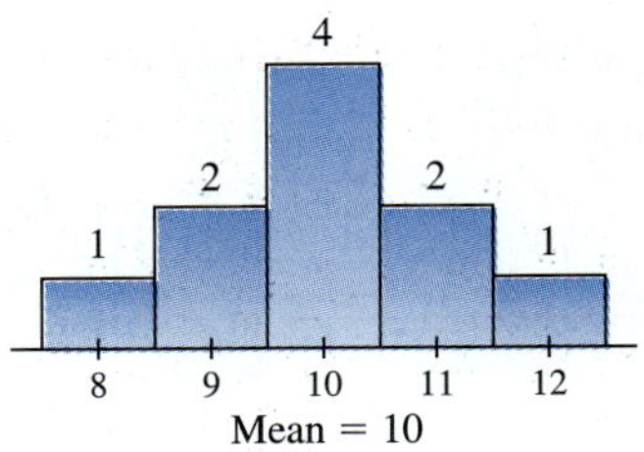

(B)

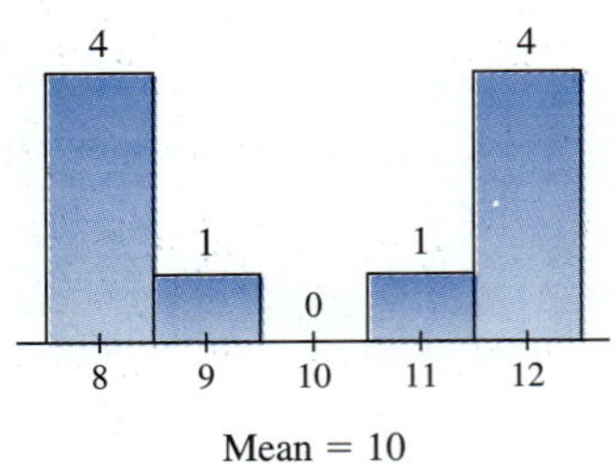

SOLUTION

(A) $$s = \sqrt{\frac{(8-10)^2(1) + (9-10)^2(2) + (10-10)^2(4) + (11-10)^2(2) + (12-10)^2(1)}{10-1}}$$

$$= \sqrt{\frac{12}{9}} \approx 1.15$$

(B) $$s = \sqrt{\frac{(8-10)^2(4) + (9-10)^2(1) + (10-10)^2(0) + (11-10)^2(1) + (12-10)^2(4)}{10-1}}$$

$$= \sqrt{\frac{34}{9}} \approx 1.94$$

Comparing the results of parts (A) and (B) in Example 2, we find that the larger standard deviation is associated with the data that deviate furthest from the mean.

Matched Problem 2 ➡ Find the standard deviation for the grouped sample data shown below:

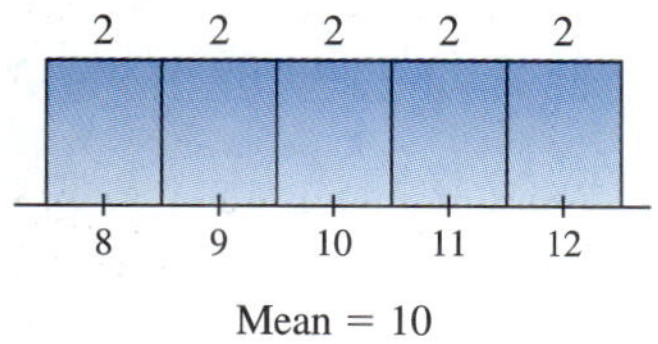

REMARK

Figure 3 illustrates the shortcut computation of the mean and standard deviation on a particular graphing calculator when the data are grouped. The sample data of Example 2A are entered. List L_1 contains the midpoints of the class intervals, and list L_2 contains the corresponding frequencies. The mean, standard deviation, and other one-variable statistics are then calculated immediately.

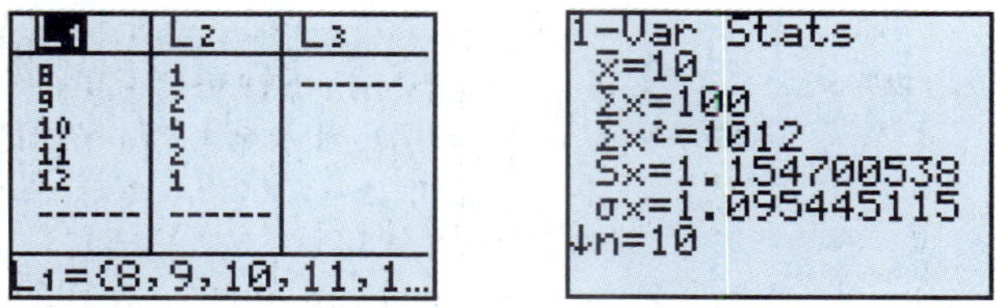

(A) (B)

FIGURE 3

■ SIGNIFICANCE OF STANDARD DEVIATION

The standard deviation can give us additional information about a frequency distribution of a set of raw data. Suppose we draw a smooth curve through the midpoints of the tops of the rectangles forming a histogram for a fairly large frequency distribution (see Fig. 4). If the resulting curve is approximately bell-shaped, then it can be shown that approximately 68% of the data will lie in the interval from $\bar{x} - s$ to $\bar{x} + s$, about 95% of the data will lie in the interval from $\bar{x} - 2s$ to $\bar{x} + 2s$, and almost all the data will lie in the interval from $\bar{x} - 3s$ to $\bar{x} + 3s$. We will have much more to say about this in Section 7-6.

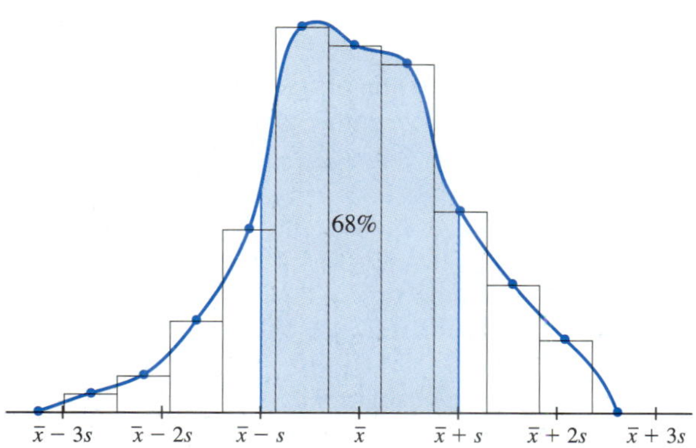

FIGURE 4

Explore–Discuss 2

(A) Verify that the following sample of 21 measurements has mean $\bar{x} = 4.95$ and sample standard deviation $s = 2.31$. Therefore, 15 of the 21 measurements, or 71% of the data, lie in the interval 2.64–7.26, that is, within 1 standard deviation of the mean. What proportion of the data lies within 2 standard deviations of the mean? Within 3 standard deviations?

4	5	9	3	4	6	1
5	0	7	2	5	8	6
3	4	7	6	5	6	8

(B) What proportion of the following data set lies within 1 standard deviation of the mean? Within 2 standard deviations? Within 3 standard deviations?

2	8	0	9	2	8	1
3	5	6	0	1	9	4
5	8	2	1	3	7	9

(C) Based on your answers to parts (A) and (B), which of the two data sets would have a histogram that is approximately bell-shaped? Confirm your conjecture by constructing a histogram with class interval width 1, starting at -0.5, for each data set.

Answers to Matched Problems

1. $s \approx 0.19$

2. $s \approx 1.49$ (a value between those found in Example 2, as expected)

Exercise 7-4

A *In Problems 1 and 2, find the standard deviation for each set of ungrouped sample data using formula (4).*

1. 1, 2, 2, 3, 3, 3, 3, 4, 4, 5

2. 1, 1, 1, 1, 2, 3, 4, 5, 5, 5

3. (A) What proportion of the following sample of ten measurements lies within 1 standard deviation of the mean? Within 2 standard deviations? Within 3 standard deviations?

4	2	3	5	3
1	6	4	2	3

(B) Based on your answers to part (A), would you conjecture that the histogram is approximately bell-shaped? Explain.

(C) To confirm your conjecture, construct a histogram with class interval width 1, starting at 0.5.

4. (A) What proportion of the following sample of ten measurements lies within 1 standard deviation of the mean? Within 2 standard deviations? Within 3 standard deviations?

3	5	1	2	1
5	4	5	1	3

(B) Based on your answers to part (A), would you conjecture that the histogram is approximately bell-shaped? Explain.

(C) To confirm your conjecture, construct a histogram with class interval width 1, starting at 0.5.

B *In Problems 5 and 6, find the standard deviation for each set of grouped sample data using formula (5).*

5.

INTERVAL	FREQUENCY
0.5–3.5	2
3.5–6.5	5
6.5–9.5	7
9.5–12.5	1

6.

INTERVAL	FREQUENCY
0.5–3.5	5
3.5–6.5	1
6.5–9.5	2
9.5–12.5	7

In Problems 7 and 8, discuss the validity of each statement. If the statement is always true, explain why. If not, give a counterexample.

7. (A) The sample variance of a set of n sample measurements is always greater than or equal to the sample standard deviation.
(B) The population variance of $x_1, x_2, \ldots, x_n$ is always greater than or equal to 0.

8. (A) The sample variance of a set of n sample measurements is always positive.
(B) For a sample x_1, x_2 of size two, the sample variance is equal to

$$\frac{(x_1 - x_2)^2}{2}$$

C

9. A data set is formed by recording the sums in 100 rolls of a pair of dice. A second data set is formed by recording the results of 100 draws of a ball from a box containing 11 balls numbered 2 through 12.
(A) Which of the two data sets would you expect to have the smaller standard deviation? Explain.
C (B) To obtain evidence for your answer to part (A), use a graphing utility to simulate both experiments, and compute the standard deviations of each of the two data sets.

10. A data set is formed by recording the results of rolling a fair die 200 times. A second data set is formed by rolling a pair of dice 200 times, each time recording the minimum of the two numbers.
(A) Which of the two data sets would you expect to have the smaller standard deviation? Explain.
C (B) To obtain evidence for your answer to part (A), use a graphing utility to simulate both experiments, and compute the standard deviations of each of the two data sets.

APPLICATIONS

Find the mean and standard deviation for each of the sample data sets given in Problems 11–18.

Use the suggestions in the remarks following Examples 1 and 2 to perform some of the computations.

Business & Economics

11. *Earnings per share.* The earnings per share (in dollars) for 12 companies selected at random from the list of *Fortune 500* companies are:

2.35	1.42	8.05	6.71
3.11	2.56	0.72	4.17
5.33	7.74	3.88	6.21

12. *Checkout times.* The checkout times (in minutes) for 12 randomly selected customers at a large supermarket during the store's busiest time are:

4.6	8.5	6.1	7.8
10.9	9.3	11.4	5.8
9.7	8.8	6.7	13.2

13. *Quality control.* The lives (in hours of continuous use) of 100 randomly selected flashlight batteries are:

INTERVAL	FREQUENCY
6.95–7.45	2
7.45–7.95	10
7.95–8.45	23
8.45–8.95	30
8.95–9.45	21
9.45–9.95	13
9.95–10.45	1

14. *Stock analysis.* The price–earning ratios of 100 randomly selected stocks from the New York Stock Exchange are:

Interval	Frequency
−0.5–4.5	5
4.5–9.5	54
9.5–14.5	25
14.5–19.5	13
19.5–24.5	0
24.5–29.5	1
29.5–34.5	2

Life Sciences

15. *Medicine.* The reaction times (in minutes) of a drug given to a random sample of 12 patients are:

4.9	5.1	3.9	4.2
6.4	3.4	5.8	6.1
5.0	5.6	5.8	4.6

16. *Nutrition—animals.* The mouse weights (in grams) of a random sample of 100 mice involved in a nutrition experiment are:

Interval	Frequency
41.5–43.5	3
43.5–45.5	7
45.5–47.5	13
47.5–49.5	17
49.5–51.5	19
51.5–53.5	17
53.5–55.5	15
55.5–57.5	7
57.5–59.5	2

Social Sciences

17. *Reading scores.* The grade-level reading scores from a reading test given to a random sample of 12 students in an urban high school graduating class are:

9	11	11	15
10	12	12	13
8	7	13	12

18. *Grade-point average.* The grade-point averages of a random sample of 100 students from the graduating class of a large university are:

Interval	Frequency
1.95–2.15	21
2.15–2.35	19
2.35–2.55	17
2.55–2.75	14
2.75–2.95	9
2.95–3.15	6
3.15–3.35	5
3.35–3.55	4
3.55–3.75	3
3.75–3.95	2

Section 7-5 Bernoulli Trials and Binomial Distributions

- Bernoulli Trials
- Binomial Formula (Brief Review)
- Binomial Distribution
- Application

In Section 7-2 we discussed frequency and relative frequency distributions, which were represented by tables and histograms (see Table 2, page 471, and Fig. 2, page 473). Frequency distributions and their corresponding probability distributions based on actual observations are *empirical* in nature. But there are many situations in which it is of interest (and possible) to determine the kind of relative frequency distribution we might expect before any data have actually been collected. What we have in mind is a **theoretical,** or **hypothetical, probability distribution**—that is, a probability distribution based on assumptions and theory rather than actual observations or measurements. Theoretical probability distributions are used to approximate properties of real-world distributions, assuming the theoretical and empirical distributions are closely matched.

There are many interesting theoretical probability distributions. One of particular interest because of its widespread use is the *binomial distribution.* The reason for the name "binomial distribution" is that the distribution is closely related to the binomial expansion of $(q + p)^n$, where n is a natural number. We start the discussion with a particular type of experiment called a *Bernoulli experiment,* or *trial.*

Bernoulli Trials

If we toss a coin, either a head occurs or it does not. If we roll a die, either a 3 shows or it fails to show. If you are vaccinated for smallpox, either you contract smallpox or you do not. What do all these situations have in common? All can be classified as experiments with two possible outcomes, each the complement of the other. An experiment for which there are only two possible outcomes, E or E', is called a **Bernoulli experiment,** or **trial,** named after Jacob Bernoulli (1654–1705), the Swiss scientist and mathematician who was one of the first to study systematically the probability problems related to a two-outcome experiment.

In a Bernoulli experiment or trial, it is customary to refer to one of the two outcomes as a **success *S*** and to the other as a **failure *F*.** If we designate the probability of success by

$$P(S) = p$$

then the probability of failure is

$$P(F) = 1 - p = q \quad \text{Note: } p + q = 1$$

Example 1 **Probability of Success in a Bernoulli Trial** Suppose we roll a fair die and ask for the probability of a 6 turning up. This can be viewed as a Bernoulli trial by identifying a success with a 6 turning up and a failure with any of the other numbers turning up. Thus,

$$p = \tfrac{1}{6} \quad \text{and} \quad q = 1 - \tfrac{1}{6} = \tfrac{5}{6}$$

Matched Problem 1 Find p and q for a single roll of a fair die, where a success is a number divisible by 3 turning up.

Now, suppose a Bernoulli trial is repeated a number of times. It becomes of interest to try to determine the probability of a given number of successes out of the given number of trials. For example, we might be interested in the probability of obtaining exactly three 5's in six rolls of a fair die or the probability that 8 people will not catch influenza out of the 10 who have been inoculated.

Suppose a Bernoulli trial is repeated five times so that each trial is completely *independent* of any other and p is the probability of success on each trial. Then the probability of the outcome $SSFFS$ would be

$$\begin{aligned} P(SSFFS) &= P(S)P(S)P(F)P(F)P(S) \quad \text{See Section 6-5.} \\ &= ppqqp \\ &= p^3q^2 \end{aligned}$$

In general, we define a *sequence of Bernoulli trials* as follows:

Bernoulli Trials

A sequence of experiments is called a **sequence of Bernoulli trials,** or a **binomial experiment,** if:

1. Only two outcomes are possible on each trial.
2. The probability of success p for each trial is a constant (probability of failure is then $q = 1 - p$).
3. All trials are independent.

The reason for calling a *sequence of Bernoulli trials* a *binomial experiment* will be made clear shortly.

Example 2 **Probability of an Outcome of a Binomial Experiment** If we roll a fair die five times and identify a success in a single roll with a 1 turning up, what is the probability of the sequence *SFFSS* occurring?

SOLUTION

$$p = \tfrac{1}{6} \qquad q = 1 - p = \tfrac{5}{6}$$

$$\begin{aligned} P(SFFSS) &= pqqpp \\ &= p^3q^2 \\ &= \left(\tfrac{1}{6}\right)^3\left(\tfrac{5}{6}\right)^2 \approx .003 \end{aligned}$$

Matched Problem 2 In Example 2, find the probability of the outcome *FSSSF*.

If we roll a fair die five times, what is the probability of obtaining exactly three 1's? Notice how this problem differs from Example 2. In that example we looked at only one way three 1's can occur. Then in Matched Problem 2 we saw another way. Thus, exactly three 1's may occur in the following two sequences (among others):

$$SFFSS \qquad FSSSF$$

We found that the probability of each sequence occurring is the same, namely,

$$\left(\tfrac{1}{6}\right)^3\left(\tfrac{5}{6}\right)^2$$

How many more sequences will produce exactly three 1's? To answer this question, think of the number of ways the following five blank positions can be filled with three *S*'s and two *F*'s:

b_1	b_2	b_3	b_4	b_5
☐	☐	☐	☐	☐

A given sequence is determined, of course, once the *S*'s are located. Thus, we are interested in the number of ways three blank positions can be selected for the S's out of the five available blank positions b_1, b_2, b_3, b_4, and b_5. This problem should sound familiar—it is just the problem of finding the number of combinations of 5 objects taken 3 at a time; that is, $C_{5,3}$. Thus, the number of different sequences of successes and failures that produce exactly three successes (exactly three 1's) is

$$C_{5,3} = \frac{5!}{3!2!} = 10$$

Since the probability of each sequence is the same,

$$p^3q^2 = (\tfrac{1}{6})^3(\tfrac{5}{6})^2$$

and there are 10 mutually exclusive sequences that produce exactly three 1's, we have

$$\begin{aligned} P(\text{Exactly three successes}) &= C_{5,3}\left(\frac{1}{6}\right)^3\left(\frac{5}{6}\right)^2 \\ &= \frac{5!}{3!2!}\left(\frac{1}{6}\right)^3\left(\frac{5}{6}\right)^2 \\ &= (10)\left(\frac{1}{6}\right)^3\left(\frac{5}{6}\right)^2 \approx .032 \end{aligned}$$

Reasoning in essentially the same way, the following important theorem can be proved:

Theorem 1 ■■ **PROBABILITY OF x SUCCESSES IN n BERNOULLI TRIALS**

The probability of exactly x successes in n independent repeated Bernoulli trials, with the probability of success of each trial p (and of failure q), is

$$P(x \text{ successes}) = C_{n,x}p^xq^{n-x} \qquad (1)$$ ■■

Example 3 ➠ **Probability of x Successes in n Bernoulli Trials** If a fair die is rolled five times, what is the probability of rolling:

(A) Exactly two 3's? (B) At least two 3's?

SOLUTION (A) Use formula (1) with $n = 5$, $x = 2$, and $p = \frac{1}{6}$:

$$\begin{aligned} P(x = 2) &= C_{5,2}\left(\frac{1}{6}\right)^2\left(\frac{5}{6}\right)^3 \\ &= \frac{5!}{2!3!}\left(\frac{1}{6}\right)^2\left(\frac{5}{6}\right)^3 \approx .161 \end{aligned}$$

(B) Notice how this problem differs from part (A). Here we have

$$P(x \geqslant 2) = P(x = 2) + P(x = 3) + P(x = 4) + P(x = 5)$$

It is actually easier to compute the probability of the complement of this event, $P(x < 2)$, and use

$$P(x \geqslant 2) = 1 - P(x < 2)$$

where

$$P(x < 2) = P(x = 0) + P(x = 1)$$

We now compute $P(x = 0)$ and $P(x = 1)$:

$$P(x = 0) = C_{5,0}\left(\frac{1}{6}\right)^0\left(\frac{5}{6}\right)^5 = \left(\frac{5}{6}\right)^5 \approx .402$$

$$P(x = 1) = C_{5,1}\left(\frac{1}{6}\right)^1\left(\frac{5}{6}\right)^4 = \frac{5!}{1!4!}\left(\frac{1}{6}\right)^1\left(\frac{5}{6}\right)^4 \approx .402$$

Thus,

$$P(x < 2) = .402 + .402 = .804$$

and

$$P(x \geqslant 2) = 1 - .804 = .196$$

Matched Problem 3 Using the same die experiment as in Example 3, what is the probability of rolling:

(A) Exactly one 3? (B) At least one 3?

Binomial Formula (Brief Review)

Before extending Bernoulli trials to binomial distributions it is worthwhile to review briefly the binomial formula. (A more detailed discussion of this formula can be found in Appendix B-3.) To start, let us calculate directly the first five natural number powers of $(a + b)^n$:

$$\begin{aligned}
(a + b)^1 &= a + b \\
(a + b)^2 &= a^2 + 2ab + b^2 \\
(a + b)^3 &= a^3 + 3a^2b + 3ab^2 + b^3 \\
(a + b)^4 &= a^4 + 4a^3b + 6a^2b^2 + 4ab^3 + b^4 \\
(a + b)^5 &= a^5 + 5a^4b + 10a^3b^2 + 10a^2b^3 + 5ab^4 + b^5
\end{aligned}$$

In general, it can be shown that a binomial expansion is given by the well-known **binomial formula:**

Binomial Formula

For n a natural number,

$$(a + b)^n = C_{n,0}a^n + C_{n,1}a^{n-1}b + C_{n,2}a^{n-2}b^2 + \cdots + C_{n,n}b^n$$

Example 4 **Finding Binomial Expansions** Use the binomial formula to expand $(q + p)^3$.

SOLUTION
$$\begin{aligned}(q + p)^3 &= C_{3,0}q^3 + C_{3,1}q^2p + C_{3,2}qp^2 + C_{3,3}p^3 \\ &= q^3 + 3q^2p + 3qp^2 + p^3\end{aligned}$$

Matched Problem 4 Use the binomial formula to expand $(q + p)^4$.

BINOMIAL DISTRIBUTION

We now generalize the discussion of Bernoulli trials to *binomial distributions.* We start by considering a sequence of three Bernoulli trials. Let the random variable X_3 (see Section 6-7) represent the number of successes in three trials, 0, 1, 2, or 3. We are interested in the probability distribution for this random variable.

Which outcomes of an experiment consisting of a sequence of three Bernoulli trials lead to the random variable values 0, 1, 2, and 3, and what are the probabilities associated with these values? Table 1 answers these questions.

Table 1

SIMPLE EVENT	PROBABILITY OF SIMPLE EVENT	X_3 *x successes in 3 trials*	$P(X_3 = x)$
FFF	$qqq = q^3$	0	q^3
FFS	$qqp = q^2p$		
FSF	$qpq = q^2p$	1	$3q^2p$
SFF	$pqq = q^2p$		
FSS	$qpp = qp^2$		
SFS	$pqp = qp^2$	2	$3qp^2$
SSF	$ppq = qp^2$		
SSS	$ppp = p^3$	3	p^3

The terms in the last column of Table 1 are the terms in the binomial expansion of $(q + p)^3$, as we saw in Example 4. The last two columns in Table 1 provide a probability distribution for the random variable X_3. Note that both conditions for a probability distribution (see Section 6-7) are met:

1. $0 \leq P(X_3 = x) \leq 1, \quad x \in \{0, 1, 2, 3\}$
2. $$\begin{aligned}1 = 1^3 &= (q + p)^3 \qquad \text{Recall that } q + p = 1. \\ &= C_{3,0}q^3 + C_{3,1}q^2p + C_{3,2}qp^2 + C_{3,3}p^3 \\ &= q^3 + 3q^2p + 3qp^2 + p^3 \\ &= P(X_3 = 0) + P(X_3 = 1) + P(X_3 = 2) + P(X_3 = 3)\end{aligned}$$

Reasoning in the same way for the general case, we see why the probability distribution of a random variable associated with the number of successes in a sequence of n Bernoulli trials is called a *binomial distribution*—the probability of each number is a term in the binomial expansion of $(q + p)^n$. For this reason, a sequence of Bernoulli trials is often referred to as a *binomial experiment.* In terms of a formula, which we already discussed from another point of view (see Theorem 1), we have:

Binomial Distribution

$$P(X_n = x) = P(x \text{ successes in } n \text{ trials}) = C_{n,x}p^x q^{n-x} \qquad x \in \{0, 1, 2, \dots, n\}$$

where p is the probability of success and q is the probability of failure on each trial.

Informally, we will write $P(x)$ in place of $P(X_n = x)$.

Example 5 **Constructing Tables and Histograms for Binomial Distributions**

Suppose a fair die is rolled three times and a success on a single roll is considered to be rolling a number divisible by 3.

(A) Write the probability function for the binomial distribution.
(B) Construct a table for this binomial distribution.
(C) Draw a histogram for this binomial distribution.

SOLUTION (A)

$p = \frac{1}{3}$ Since two numbers of out of six are divisible by 3

$q = 1 - p = \frac{2}{3}$

$n = 3$

Hence,

$$P(x) = P(x \text{ successes in 3 trials}) = C_{3,x}(\tfrac{1}{3})^x(\tfrac{2}{3})^{3-x}$$

(B)

x	$P(x)$
0	$C_{3,0}(\frac{1}{3})^0(\frac{2}{3})^3 \approx .30$
1	$C_{3,1}(\frac{1}{3})^1(\frac{2}{3})^2 \approx .44$
2	$C_{3,2}(\frac{1}{3})^2(\frac{2}{3})^1 \approx .22$
3	$C_{3,3}(\frac{1}{3})^3(\frac{2}{3})^0 \approx .04$
	1.00

(C)

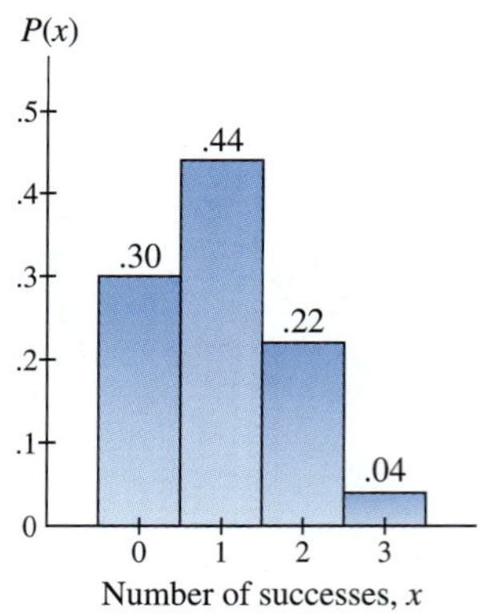

FIGURE 1

If we actually performed the binomial experiment described in Example 5 a large number of times with a fair die, we would find that we would roll no number divisible by 3 in three rolls of a die about 30% of the time, one number divisible by 3 in three rolls about 44% of the time, two numbers divisible by 3 in three rolls about 22% of the time, and three numbers divisible by 3 in three rolls only 4% of the time. Note that the sum of all the probabilities is 1, as it should be.

The graphing utility command in Figure 2A simulates 100 repetitions of the binomial experiment in Example 5. The number of successes in each trial is stored in list L_1. From Figure 2B, which shows a histogram of L_1, we note that the empirical probability of rolling one number divisible by 3 in three rolls is $\frac{40}{100} = 40\%$, close to the theoretical probability of 44%. The empirical probabilities of 0, 2, or 3 successes also would be close to the corresponding theoretical probabilities.

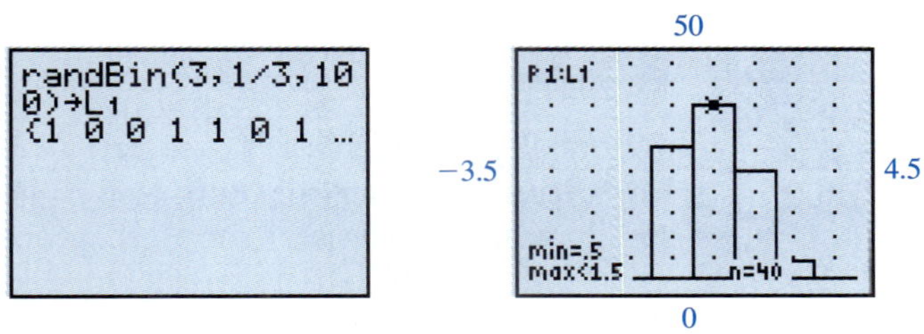

(A) (B)

FIGURE 2

Matched Problem 5 ➡ Repeat Example 5, where the binomial experiment consists of two rolls of a die instead of three rolls.

We close our discussion of binomial distributions by stating (without proof) formulas for the mean and standard deviation of the random variable associated with the distribution. These are theoretical results that apply to the total population; hence, they are denoted by the Greek letters μ and σ, respectively.

Mean and Standard Deviation (Random Variable in a Binomial Distribution)

Mean: $\mu = np$

Standard deviation: $\sigma = \sqrt{npq}$

Example 6 ➡ **Computing the Mean and Standard Deviation of a Binomial Distribution** Compute the mean and standard deviation for the random variable in Example 5.

SOLUTION

$$n = 3 \qquad p = \tfrac{1}{3} \qquad q = 1 - \tfrac{1}{3} = \tfrac{2}{3}$$

$$\mu = np = 3(\tfrac{1}{3}) = 1 \qquad \sigma = \sqrt{npq} = \sqrt{3(\tfrac{1}{3})(\tfrac{2}{3})} \approx .82$$

Matched Problem 6 ➡ Compute the mean and standard deviation for the random variable in Matched Problem 5.

Explore–Discuss 1

Let X_{100} denote the number of successes of 100 Bernoulli trials, each with probability of success p.

(A) For what values of p would the mean of X_{100} be equal to 0? 50? 100?

(B) For what values of p would the standard deviation of X_{100} be equal to 0? 5? 10?

■ Application

Binomial experiments are associated with a wide variety of practical problems: industrial sampling, drug testing, genetics, epidemics, medical diagnosis, opinion polls, analysis of social phenomena, qualifying tests, and so on. Several types of applications are included in Exercise 7-5. We will now consider one application in detail.

Example 7 ➡ **Patient Recovery** The probability of recovering after a particular type of operation is .5. Let us investigate the binomial distribution involving eight patients undergoing this operation.

(A) Write the function defining this distribution.
(B) Construct a table for the distribution.
(C) Construct a histogram for the distribution.
(D) Find the mean and standard deviation for the distribution.

Solution (A) Letting a recovery be a success, we have

$$p = .5 \qquad q = 1 - p = .5 \qquad n = 8$$

Hence,

$$P(x) = P(\text{Exactly } x \text{ successes in 8 trials}) = C_{8,x}(.5)^x(.5)^{8-x} = C_{8,x}(.5)^8$$

(B)

x	$P(x)$
0	$C_{8,0}(.5)^8 \approx .004$
1	$C_{8,1}(.5)^8 \approx .031$
2	$C_{8,2}(.5)^8 \approx .109$
3	$C_{8,3}(.5)^8 \approx .219$
4	$C_{8,4}(.5)^8 \approx .273$
5	$C_{8,5}(.5)^8 \approx .219$
6	$C_{8,6}(.5)^8 \approx .109$
7	$C_{8,7}(.5)^8 \approx .031$
8	$C_{8,8}(.5)^8 \approx .004$
	$.999 \approx 1$

The discrepancy in the sum is due to round-off errors.

(C)

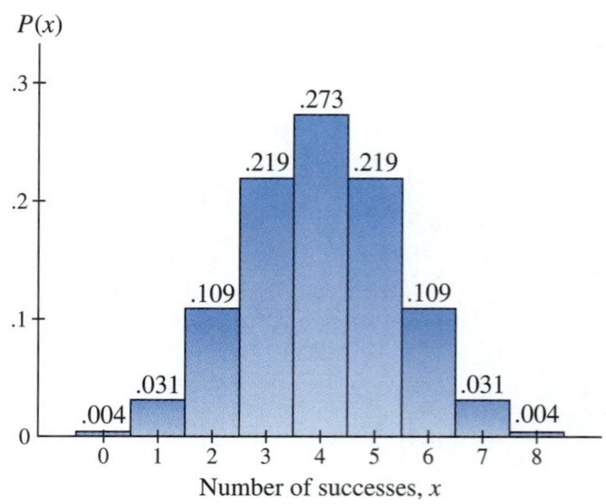

(D) $\mu = np = 8(.5) = 4 \qquad \sigma = \sqrt{npq} = \sqrt{8(.5)(.5)} \approx 1.41$

Matched Problem 7 Repeat Example 7 for four patients.

Explore–Discuss 2 The mean of a random variable is its expected value. Use the distribution tables for the random variables of Examples 5 and 7 to compute the expected values. Do your answers agree with the results obtained using the formula $\mu = np$? Explain.

Answers to Matched Problems

1. $p = \frac{1}{3}, q = \frac{2}{3}$ 2. $p^3q^2 = (\frac{1}{6})^3(\frac{5}{6})^2 \approx .003$
3. (A) .402 (B) $1 - P(x = 0) = 1 - .402 = .598$
4. $C_{4,0}q^4 + C_{4,1}q^3p + C_{4,2}q^2p^2 + C_{4,3}qp^3 + C_{4,4}p^4 = q^4 + 4q^3p + 6q^2p^2 + 4qp^3 + p^4$
5. (A) $P(x) = P(x \text{ successes in 2 trials}) = C_{2,x}(\frac{1}{3})^x(\frac{2}{3})^{2-x}, x \in \{0, 1, 2\}$

(B)

x	$P(x)$
0	$\frac{4}{9} \approx .44$
1	$\frac{4}{9} \approx .44$
2	$\frac{1}{9} \approx .11$

(C)

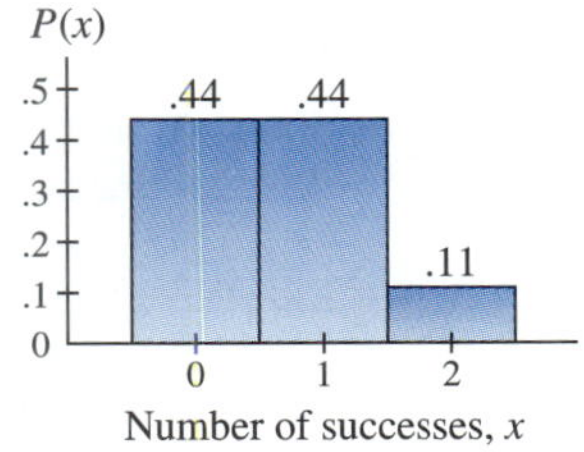

6. $\mu \approx .67; \sigma \approx .67$
7. (A) $P(x) = P(\text{Exactly } x \text{ successes in 4 trials}) = C_{4,x}(.5)^4$

(B)

x	$P(x)$
0	.06
1	.25
2	.38
3	.25
4	.06
	1.00

(C)

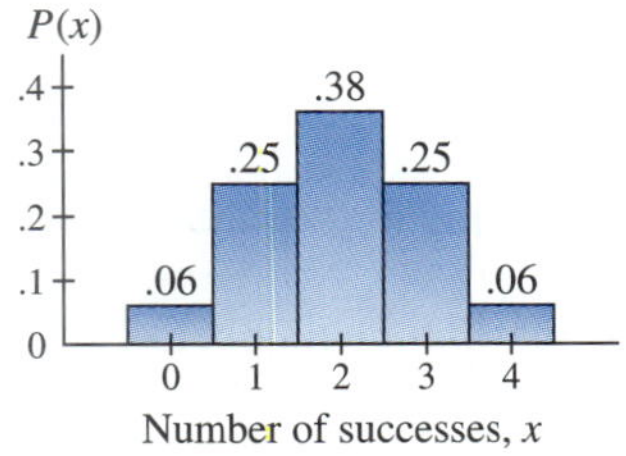

(D) $\mu = 2; \sigma = 1$

EXERCISE 7-5

A *Evaluate $C_{n,x}p^xq^{n-x}$ for the values of n, x, and p given in Problems 1–6.*

1. $n = 3, x = 2, p = \frac{1}{2}$
2. $n = 3, x = 1, p = \frac{1}{2}$
3. $n = 3, x = 0, p = \frac{1}{2}$
4. $n = 3, x = 3, p = \frac{1}{2}$
5. $n = 5, x = 3, p = .4$
6. $n = 5, x = 0, p = .4$

In Problems 7–12, a fair coin is tossed three times. What is the probability of obtaining:

7. Exactly 2 heads?
8. Exactly 1 head?
9. 0 heads?
10. 3 heads?
11. At least 2 heads?
12. At least 1 head?

Construct a histogram for each of the binomial distributions in Problems 13–16. Compute the mean and standard deviation for each distribution.

13. $P(x) = C_{2,x}(.3)^x(.7)^{2-x}$
14. $P(x) = C_{2,x}(.7)^x(.3)^{2-x}$
15. $P(x) = C_{4,x}(.5)^x(.5)^{4-x}$
16. $P(x) = C_{6,x}(.5)^x(.5)^{6-x}$

B *In Problems 17–22, a fair die is rolled four times. What is the probability of rolling:*

17. Exactly three 2's?
18. Exactly two 3's?
19. No 1's?
20. All 4's?
21. At least one 6?
22. At least one 4?

23. If a baseball player has a batting average of .350, what is the probability that the player will get the following number of hits in the next four times at bat?
(A) Exactly 2 hits (B) At least 2 hits

24. If a true–false test with 10 questions is given, what is the probability of scoring:
(A) Exactly 70% just by guessing?
(B) 70% or better just by guessing?

25. A multiple-choice test consists of 10 questions, each with choices A, B, C, D, E (of which exactly one choice is correct). Which is more likely if you simply guess at each question: that all your answers are wrong, or that at least half are right? Explain.

26. If 60% of the electorate supports the mayor, what is the probability that in a random sample of 10 voters, fewer than half support her?

Construct a histogram for each of the binomial distributions in Problems 27–30. Compute the mean and standard deviation for each distribution.

27. $P(x) = C_{6,x}(.4)^x(.6)^{6-x}$ **28.** $P(x) = C_{6,x}(.6)^x(.4)^{6-x}$

29. $P(x) = C_{8,x}(.3)^x(.7)^{8-x}$ **30.** $P(x) = C_{8,x}(.7)^x(.3)^{8-x}$

In Problems 31 and 32, use a graphing utility to construct a probability distribution table.

31. A random variable represents the number of successes in 20 Bernoulli trials, each with probability of success $p = .85$.
(A) Find the mean and standard deviation of the random variable.
(B) Find the probability that the number of successes lies within 1 standard deviation of the mean.

32. A random variable represents the number of successes in 20 Bernoulli trials, each with probability of success $p = .45$.
(A) Find the mean and standard deviation of the random variable.
(B) Find the probability that the number of successes lies within 1 standard deviation of the mean.

C *In Problems 33 and 34, a coin is loaded so that the probability of a head occurring on a single toss is $\frac{3}{4}$. In five tosses of the coin, what is the probability of getting:*

33. All heads or all tails?

34. Exactly 2 heads or exactly 2 tails?

35. Toss a coin three times or toss three coins simultaneously, and record the number of heads. Repeat the binomial experiment 100 times and compare your relative frequency distribution with the theoretical probability distribution.

36. Roll a die three times or roll three dice simultaneously, and record the number of 5's that occur. Repeat the binomial experiment 100 times and compare your relative frequency distribution with the theoretical probability distribution.

37. Find conditions on p that guarantee the histogram for a binomial distribution is symmetrical about $x = n/2$. Justify your answer.

38. Consider two binomial distributions for 1,000 repeated Bernoulli trials—the first for trials with $p = .15$, and the second for trials with $p = .85$. How are the histograms for the two distributions related? Explain.

39. A random variable represents the number of heads in ten tosses of a coin.
(A) Find the mean and standard deviation of the random variable.
C (B) Use a graphing utility to simulate 200 repetitions of the binomial experiment, and compare the mean and standard deviation of the numbers of heads from the simulation to the answers for part (A).

40. A random variable represents the number of 7's or 11's in ten rolls of a pair of dice.
(A) Find the mean and standard deviation of the random variable.
C (B) Use a graphing utility to simulate 100 repetitions of the binomial experiment, and compare the mean and standard deviation of the numbers of 7's or 11's from the simulation to the answers for part (A).

Applications

Business & Economics

41. *Management training.* Each year a company selects a number of employees for a management training program given by a nearby university. On the average, 70% of those sent complete the program. Out of 7 people sent by the company, what is the probability that:
(A) Exactly 5 complete the program?
(B) 5 or more complete the program?

42. *Employee turnover.* If the probability of a new employee in a fast-food chain still being with the company at the end of 1 year is .6, what is the probability that out of 8 newly hired people:
(A) 5 will still be with the company after 1 year?
(B) 5 or more will still be with the company after 1 year?

43. *Quality control.* A manufacturing process produces, on the average, 6 defective items out of 100. To control quality, each day a sample of 10 completed items is selected at random and inspected. If the sample produces more than 2 defective items, then the whole day's output is inspected and the manufacturing process is reviewed. What is the probability of this happening, assuming that the process is still producing 6% defective items?

44. *Guarantees.* A manufacturing process produces, on the average, 3% defective items. The company ships 10 items in each box and wishes to guarantee no more than 1 defective item per box. If this guarantee accompanies each box, what is the probability that the box will fail to satisfy the guarantee?

45. *Quality control.* A manufacturing process produces, on the average, 5 defective items out of 100. To control quality, each day a random sample of 6 completed items is selected and inspected. If a success on a single trial (inspection of 1 item) is finding the item defective, then the inspection of each of the 6 items in the sample constitutes a binomial experiment, which has a binomial distribution.
(A) Write the function defining the distribution.
(B) Construct a table for the distribution.
(C) Draw a histogram.
(D) Compute the mean and standard deviation.

46. *Management training.* Each year a company selects 5 employees for a management training program given at a nearby university. On the average, 40% of those sent complete the course in the top 10% of their class. If we consider an employee finishing in the top 10% of the class a success in a binomial experiment, then for the 5 employees entering the program there exists a binomial distribution involving $P(x$ successes out of 5).
(A) Write the function defining the distribution.
(B) Construct a table for the distribution.
(C) Draw a histogram.
(D) Compute the mean and standard deviation.

Life Sciences

47. *Medical diagnosis.* A person with tuberculosis is given a chest x ray. Four tuberculosis x-ray specialists examine each x ray independently. If each specialist can detect tuberculosis 80% of the time when it is present, what is the probability that at least 1 of the specialists will detect tuberculosis in this person?

48. *Harmful side effects of drugs.* A pharmaceutical laboratory claims that a drug it produces causes serious side effects in 20 people out of 1,000, on the average. To check this claim, a hospital administers the drug to 10 randomly chosen patients and finds that 3 suffer from serious side effects. If the laboratory's claims are correct, what is the probability of the hospital obtaining these results?

49. *Genetics.* The probability that brown-eyed parents, both with the recessive gene for blue, will have a child with brown eyes is .75. If such parents have 5 children, what is the probability that they will have:
(A) All blue-eyed children?
(B) Exactly 3 children with brown eyes?
(C) At least 3 children with brown eyes?

50. *Gene mutations.* The probability of gene mutation under a given level of radiation is 3×10^{-5}. What is the probability of the occurrence of at least 1 gene mutation if 10^5 genes are exposed to this level of radiation?

51. *Epidemics.* If the probability of a person contracting influenza on exposure is .6, consider the binomial distribution for a family of 6 that has been exposed.
(A) Write the function defining the distribution.
(B) Construct a table for the distribution.
(C) Draw a histogram.
(D) Compute the mean and standard deviation.

52. *Side effects of drugs.* The probability that a given drug will produce a serious side effect in a person using the drug is .02. In the binomial distribution for 450 people using the drug, what are the mean and standard deviation?

Social Sciences

53. *Testing.* A multiple-choice test is given with 5 choices (only one is correct) for each of 10 questions. What is the probability of passing the test with a grade of 70% or better just by guessing?

54. *Opinion polls.* An opinion poll based on a small sample can be unrepresentative of the population. To see why, let us assume that 40% of the electorate favors a certain candidate. If a random sample of 7 is asked their preference, what is the probability that a majority will favor this candidate?

55. *Testing.* A multiple-choice test is given with 5 choices (only one is correct) for each of 5 questions. Answering each of the 5 questions by guessing constitutes a binomial experiment with an associated binomial distribution.
(A) Write the function defining the distribution.
(B) Construct a table for the distribution.
(C) Draw a histogram.
(D) Compute the mean and standard deviation.

56. *Sociology.* The probability that a marriage will end in divorce within 10 years is .4. What are the mean and standard deviation for the binomial distribution involving 1,000 marriages?

57. *Sociology.* If the probability is .60 that a marriage will end in divorce within 20 years after its start, what is the probability that out of 6 couples just married, in the next 20 years:
(A) None will be divorced?
(B) All will be divorced?
(C) Exactly 2 will be divorced?
(D) At least 2 will be divorced?

SECTION 7-6

Normal Distributions

- NORMAL DISTRIBUTION
- AREAS UNDER NORMAL CURVES
- APPROXIMATING A BINOMIAL DISTRIBUTION WITH A NORMAL DISTRIBUTION

NORMAL DISTRIBUTION

If we take the histogram for a binomial distribution, say, the one we drew for Example 7 in Section 7-5 ($n = 8$, $p = .5$), and join the midpoints of the tops of the rectangles with a smooth curve, we obtain the bell-shaped curve in Figure 1.

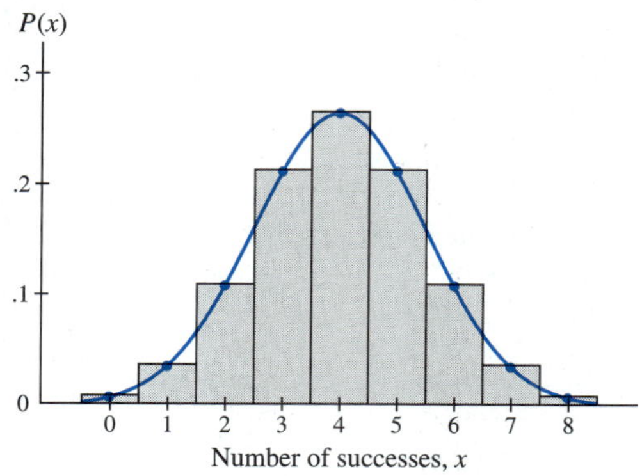

FIGURE 1
The binomial distribution and a bell-shaped curve

The mathematical foundation for this type of curve was established by Abraham De Moivre (1667–1754), Pierre Laplace (1749–1827), and Carl Gauss (1777–1855). The bell-shaped curves studied by these famous mathematicians are called **normal curves** or **normal probability distributions,** and their equations* are completely determined by the mean μ and standard deviation σ of the distribution. Figure 2 illustrates three normal curves with different means and standard deviations.

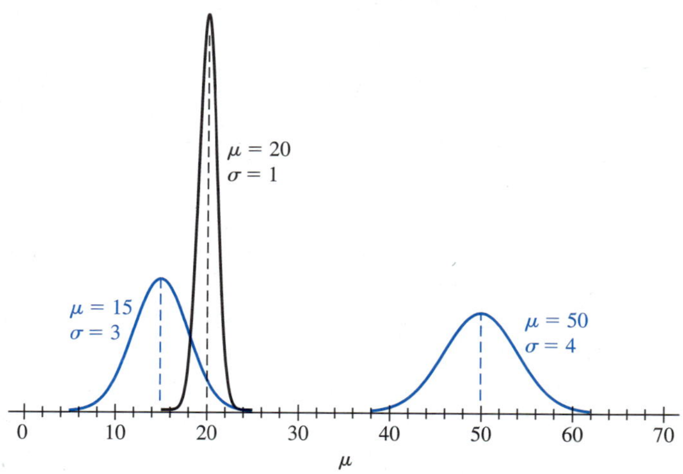

FIGURE 2
Normal probability distributions

---

*The equation for a normal curve is fairly complicated: $f(x) = \dfrac{1}{\sigma\sqrt{2\pi}} e^{-(x-\mu)^2/2\sigma^2}$ where $\pi \approx 3.1416$ and $e \approx 2.7183$.

Until now we have dealt with **discrete random variables,** that is, random variables that assume a finite or a "countably infinite" number of values (we have dealt only with the finite case). Random variables associated with normal distributions are *continuous* in nature; that is, they assume all values over an interval on a real number line. These are called **continuous random variables.** Random variables associated with people's heights, light bulb lifetimes, or the lengths of time between breakdowns of an office copier are continuous. The following is a list of some of the important properties of normal curves (normal probability distributions of a continuous random variable):

Normal Curve Properties

1. Normal curves are bell-shaped and are symmetrical with respect to a vertical line.
2. The mean is at the point where the axis of symmetry intersects the horizontal axis.
3. The shape of a normal curve is completely determined by its mean and standard deviation—a small standard deviation indicates a tight clustering about the mean and thus a tall, narrow curve; a large standard deviation indicates a large deviation from the mean and thus a broad, flat curve (see Fig. 2).
4. Irrespective of the shape, the area between the curve and the x axis is always 1.
5. Irrespective of the shape, 68.3% of the area will lie within an interval of 1 standard deviation on either side of the mean, 95.4% within 2 standard deviations on either side, and 99.7% within 3 standard deviations on either side (see Fig. 3).

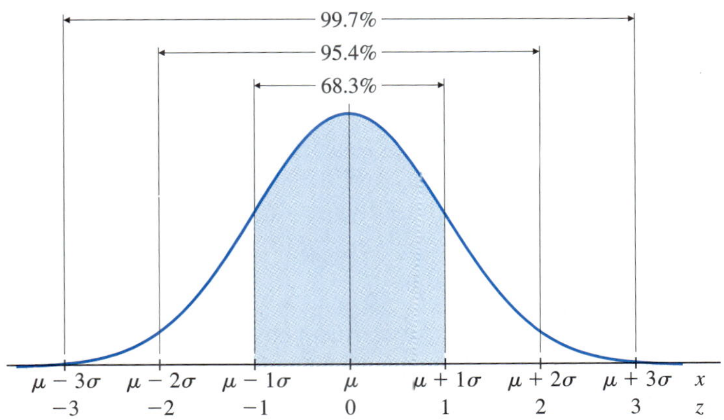

FIGURE 3
Normal curve areas

The normal probability distribution is the most important of all theoretical distributions. It is at the heart of a great deal of statistical theory, and it is also a useful tool in its own right for solving practical problems. Not only does a normal curve provide a good approximation for a binomial distribution for large n, but it also approximates many other relative frequency distributions. For example, normal curves often provide good approximations for the relative frequency distributions for heights and weights of people, measurements of manufactured parts, scores on IQ tests, college entrance examinations, civil service tests, and measurements of errors in laboratory experiments.

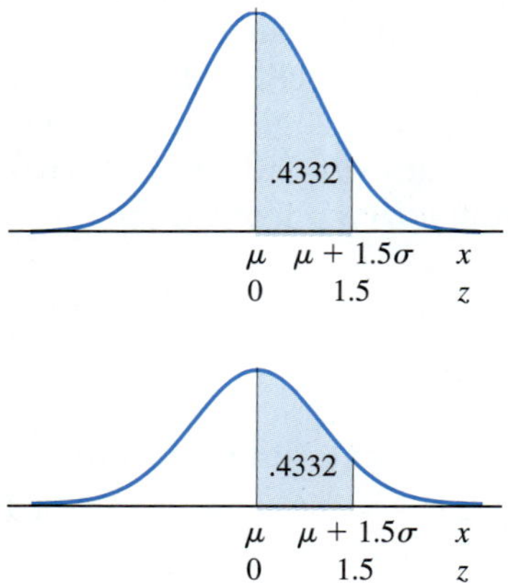

FIGURE 4

Areas Under Normal Curves

In order to use normal curves in practical problems, we must be able to determine areas under different parts of a curve. Fortunately, we can use the same approach and the same table for all normal curves, irrespective of their shapes, that is, irrespective of the values of μ and σ. It is a remarkable fact that the area under a normal curve between a mean μ and a given number of standard deviations to the right (or left) of μ is the same regardless of the shape of the normal curve (see Fig. 4). Thus, if **z represents the number of standard deviations that a measurement x is from a mean μ** (see Fig. 5), then the areas in terms of these z values can be listed in just one table. Table I in Appendix C is such a table.

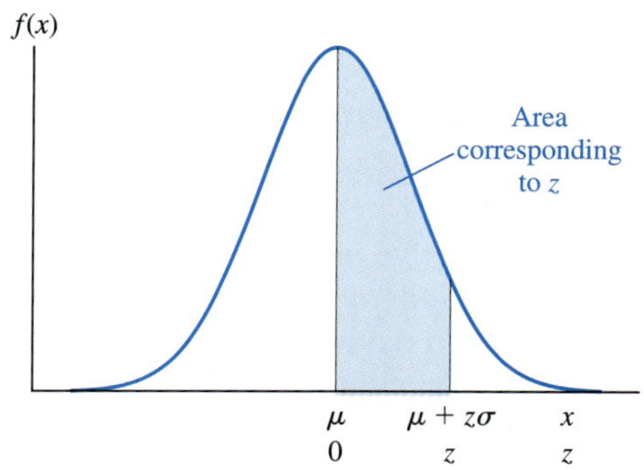

FIGURE 5
Areas and z values

Example 1 **Finding Probabilities for a Normal Distribution** A manufacturing process produces light bulbs with life expectancies that are normally distributed with a mean of 500 hours and a standard deviation of 100 hours. What percentage of the light bulbs can be expected to last between 500 and 670 hours?

Solution To answer this question, we first determine how many standard deviations 670 is from 500, the mean. This is easily done by dividing the distance between 500 and 670 by 100, the standard deviation. Thus,

$$z = \frac{670 - 500}{100} = \frac{170}{100} = 1.70$$

That is, 670 is 1.7 standard deviations from 500, the mean. Referring to Table I in Appendix C, we see that .4554 corresponds to $z = 1.70$. And since the total area under a normal curve is 1, we conclude that 45.54% of the light bulbs produced will last between 500 and 670 hours (see Fig. 6).

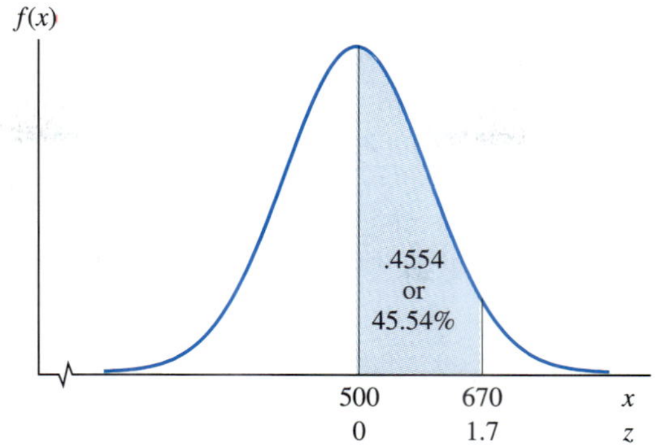

FIGURE 6
Light bulb life expectancy—positive z

Matched Problem 1 What percentage of the light bulbs can be expected to last between 500 and 750 hours?

In general, to find how many standard deviations a measurement x is from a mean μ, first determine the distance between x and μ and then divide by σ:

$$z = \frac{\text{Distance between } x \text{ and } \mu}{\text{Standard deviation}} = \frac{x - \mu}{\sigma}$$

Example 2 **Finding Probabilities for a Normal Distribution** From all light bulbs produced (see Example 1), what is the probability of a light bulb chosen at random lasting between 380 and 500 hours?

Solution To answer this, we first find z:

$$z = \frac{x - \mu}{\sigma} = \frac{380 - 500}{100} = -1.20$$

It is usually a good idea to draw a rough sketch of a normal curve and insert relevant data (see Fig. 7).

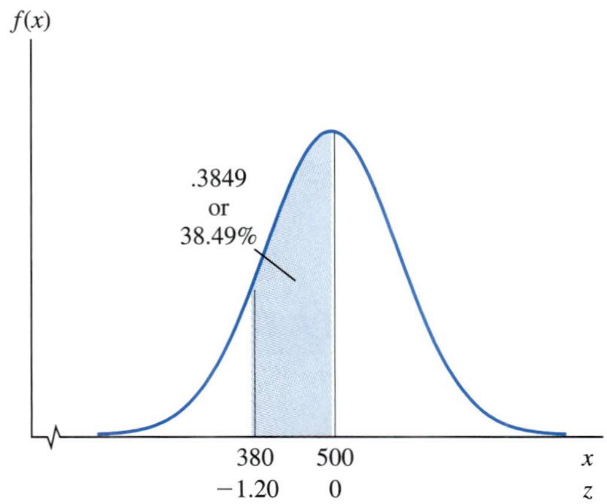

FIGURE 7
Light bulb life expectancy—negative z

Table I in Appendix C does not include negative values for z, but because normal curves are symmetrical with respect to a vertical line through the mean, we simply use the absolute value (positive value) of z for the table. Thus, the area corresponding to $z = -1.20$ is the same as the area corresponding to $z = 1.20$, which is .3849. And since the area under the whole normal curve is 1, we conclude that the probability of a light bulb chosen at random lasting between 380 and 500 hours is .3849.

Matched Problem 2 What is the probability of a light bulb chosen at random lasting between 400 and 500 hours?

The first graphing utility command in Figure 8A simulates the life expectancies of 100 light bulbs by generating 100 random numbers from the normal distribution with $\mu = 500, \sigma = 100$ of Example 1. The numbers are stored in list L_1. Note from Figure 8A that the mean and standard deviation of L_1 are close to the mean and standard deviation of the normal distribution. From Figure 8B, which shows a histogram of L_1, we note that the empirical probability that a light bulb lasts between 380 and 500 hours is

$$\frac{11 + 13 + 14}{100} = .37$$

which is close to the theoretical probability of .3849 computed in Example 2.

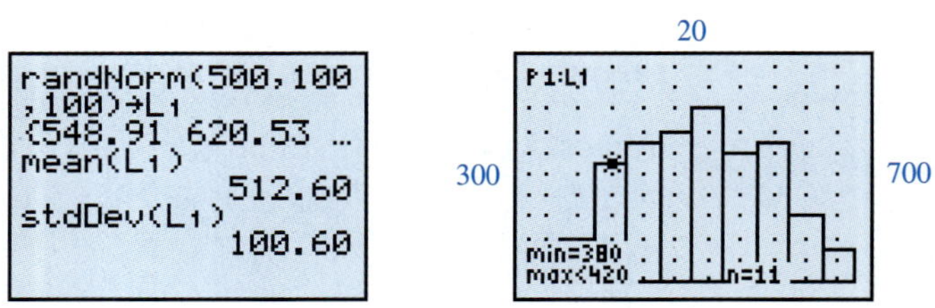

(A) (B)

FIGURE 8

Several important properties of a continuous random variable with normal distribution are listed in the box below:

Properties of a Normal Probability Distribution

1. $P(a \le x \le b) =$ Area under the normal curve from a to b
2. $P(-\infty < x < \infty) = 1 =$ Total area under the normal curve
3. $P(x = c) = 0$

In Example 2, what is the probability of a light bulb chosen at random having a life of exactly 621 hours? The area above 621 and below the normal curve at $x = 621$ is 0 (a line has no width). Thus, the probability of a light bulb chosen at random having a life of exactly 621 hours is 0. However, if the number 621 is the result of rounding a number between 620.5 and 621.5 (which is most likely the case), then the answer to the question is

$$P(620.5 \leq x \leq 621.5) = \text{Area under the normal curve from 620.5 to 621.5}$$

The area is found using the procedures outlined in Example 2.

We have just pointed out an important distinction between a continuous random variable and a discrete random variable: For a probability distribution of a continuous random variable, the probability of x assuming a single value is always 0. On the other hand, for a probability distribution of a discrete random variable, the probability of x assuming a particular value from the set of permissible values is usually a positive number between 0 and 1. Continuous probability distributions are covered in greater depth in a course in calculus.

Approximating a Binomial Distribution with a Normal Distribution

You no doubt found in some of the problems in Exercise 7-5 that when a binomial random variable assumes a large number of values (that is, when n is large), the use of the probability distribution formula

$$P(x \text{ successes in } n \text{ trials}) = C_{n,x}p^x q^{n-x}$$

became very tedious. It would be very helpful if there was an easily computed approximation of this distribution for large n. Such a distribution is found in the form of an appropriately selected normal distribution.

To clarify ideas and relationships, let us consider an example of a normal distribution approximation of a binomial distribution with a relatively small value of n. Then we will consider an example with a large value of n.

Example 3 **Market Research** A credit card company claims that their card is used by 40% of the people buying gasoline in a particular city. A random sample of 20 gasoline purchasers is made. If the company's claim is correct, what is the probability that:

(A) From 6 to 12 people in the sample use the card?
(B) Fewer than 4 people in the sample use the card?

Solution We begin by drawing a normal curve with the same mean and standard deviation as the binomial distribution (Fig. 9, page 516). A histogram superimposed on this normal curve can be used to approximate the histogram for the binomial distribution. The mean and standard deviation of the binomial distribution are

$$\mu = np = (20)(.4) = 8 \qquad n = \text{Sample size}$$
$$\sigma = \sqrt{npq} = \sqrt{(20)(.4)(.6)} \approx 2.19 \qquad p = .4 \text{ (from the 40\% claim)}$$

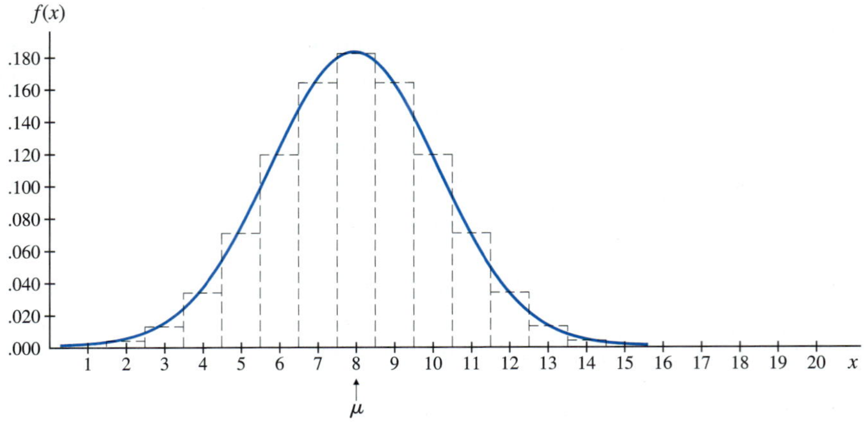

FIGURE 9

(A) To approximate the probability that 6 to 12 people in the sample use the credit card, we find the area under the normal curve from 5.5 to 12.5. We use 5.5 rather than 6, because the rectangle in the histogram corresponding to 6 extends from 5.5 to 6.5; and, reasoning in the same way, we use 12.5 instead of 12. To use Table I in Appendix C, we split the area into two parts: A_1 to the left of the mean and A_2 to the right of the mean. The sketch in Figure 10 is helpful. Areas A_1 and A_2 are found as follows:

$$z_1 = \frac{x - \mu}{\sigma} = \frac{5.5 - 8}{2.19} \approx -1.14 \qquad A_1 = .3729$$

$$z_2 = \frac{x - \mu}{\sigma} = \frac{12.5 - 8}{2.19} \approx 2.05 \qquad A_2 = .4798$$

$$\text{Total area} = A_1 + A_2 = .8527$$

Thus, the approximate probability that the sample will contain between 6 and 12 users of the credit card is .85 (assuming the firm's claim is correct).

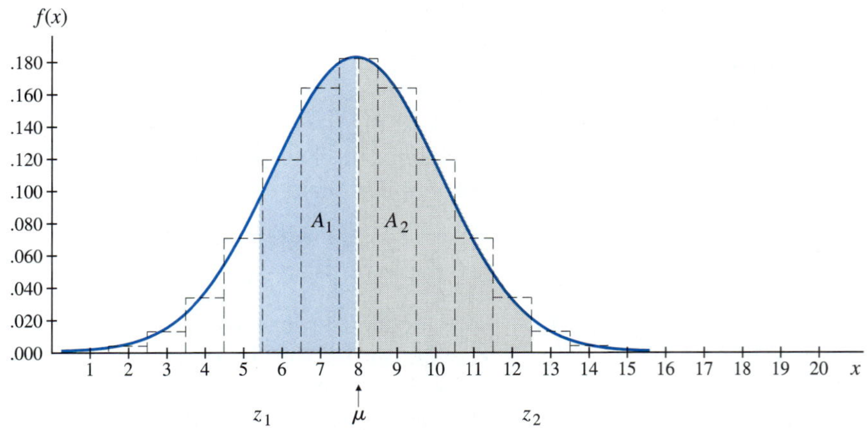

FIGURE 10

(B) To use the normal curve to approximate the probability that the sample contains fewer than 4 users of the credit card, we must find the area A_1 under the normal curve to the left of 3.5. The sketch in Figure 11 is useful. Since the total area under either half of the normal curve is .5, we first use Table I in Appendix C to find the area A_2 under the normal curve from 3.5 to the mean 8, and then subtract A_2 from .5:

$$z = \frac{x - \mu}{\sigma} = \frac{3.5 - 8}{2.19} \approx -2.05 \qquad A_2 = .4798$$
$$A_1 = .5 - A_2 = .5 - .4798 = .0202$$

Thus, the approximate probability that the sample contains fewer than 4 users of the credit card is approximately .02 (assuming the company's claim is correct).

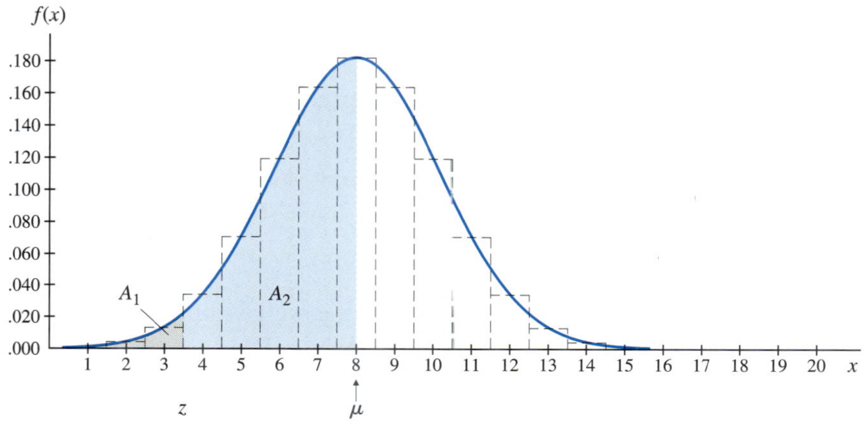

FIGURE 11

Matched Problem 3 In Example 3 use the normal curve to approximate the probability that in the sample there are:

(A) From 5 to 9 users of the credit card
(B) More than 10 users of the card

You no doubt are wondering how large n should be before a normal distribution provides an adequate approximation for a binomial distribution. Without getting too involved, the following rule-of-thumb provides a good test:

Rule-of-Thumb Test

Use a normal distribution to approximate a binomial distribution only if the interval $[\mu - 3\sigma, \mu + 3\sigma]$ lies entirely in the interval from 0 to n.

Note that in Example 3 the interval $[\mu - 3\sigma, \mu + 3\sigma] = [1.43, 14.57]$ lies entirely within the interval from 0 to 20; hence, the use of the normal distribution was justified.

Explore–Discuss 1

(A) Show that if $n \geq 30$ and $.25 \leq p \leq .75$ for a binomial distribution, then it passes the rule-of-thumb test.

(B) Give an example of a binomial distribution that passes the rule-of-thumb test but does not satisfy the conditions of part (A).

Example 4 **Quality Control** A company manufactures 50,000 ballpoint pens each day. The manufacturing process produces 50 defective pens per 1,000, on the average. A random sample of 400 pens is selected from each day's production and tested. What is the probability that the sample contains:

(A) At least 14 and no more than 25 defective pens?

(B) 33 or more defective pens?

SOLUTION Is it appropriate to use a normal distribution to approximate this binomial distribution? The answer is yes, since the rule-of-thumb test passes with ease:

$$\mu = np = 400(.05) = 20 \qquad p = \frac{50}{1{,}000} = .05$$

$$\sigma = \sqrt{npq} = \sqrt{400(.05)(.95)} \approx 4.36$$

$$[\mu - 3\sigma, \mu + 3\sigma] = [6.92, 33.08]$$

This interval is well within the interval from 0 to 400.

(A) To find the approximate probability of the number of defective pens in a sample being at least 14 and not more than 25, we find the area under the normal curve from 13.5 to 25.5. To use Table I in Appendix C, we split the area into an area to the left of the mean and an area to the right of the mean, as shown in Figure 12.

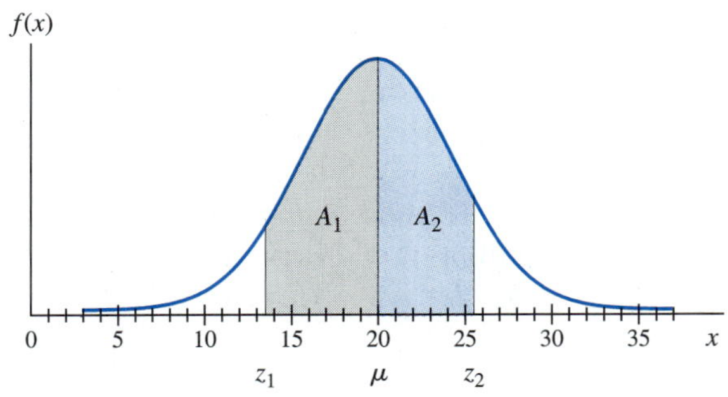

FIGURE 12

$$z_1 = \frac{x - \mu}{\sigma} = \frac{13.5 - 20}{4.36} \approx -1.49 \qquad A_1 = .4319$$

$$z_2 = \frac{x - \mu}{\sigma} = \frac{25.5 - 20}{4.36} \approx 1.26 \qquad A_2 = .3962$$

$$\text{Total area} = A_1 + A_2 = .8281$$

Thus, the approximate probability of the number of defective pens in the sample being at least 14 and not more than 25 is .83.

(B) Since the total area under a normal curve from the mean on is .5, we find the area A_1 (see Fig. 13) from Table I in Appendix C and subtract it from .5 to obtain A_2.

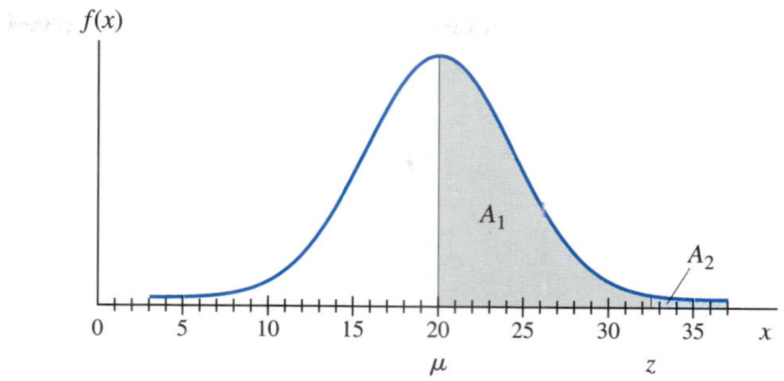

FIGURE 13

$$z = \frac{x - \mu}{\sigma} = \frac{32.5 - 20}{4.36} \approx 2.87 \qquad A_1 = .4979$$
$$A_2 = .5 - A_1 = .5 - .4979 = .0021 \approx .002$$

Thus, the approximate probability of finding 33 or more defective pens in the sample is .002. If a random sample of 400 included more than 33 defective pens, then the management would conclude that either a rare event has happened and the manufacturing process is still producing only 50 defective pens per 1,000, on the average, or something is wrong with the manufacturing process and it is producing more than 50 defective pens per 1,000, on the average. The company might very well have a policy of checking the manufacturing process whenever 33 or more defective pens are found in a sample rather than believing a rare event has happened and that the manufacturing process is still running smoothly.

Matched Problem 4 ➡ Suppose in Example 4 that the manufacturing process produces 40 defective pens per 1,000, on the average. What is the approximate probability that in the sample of 400 pens there are:

(A) At least 10 and no more than 20 defective pens?
(B) 27 or more defective pens?

When to Use the .5 Adjustment

If we are assuming a normal probability distribution for a continuous random variable (such as that associated with heights or weights of people), then we find $P(a \leq x \leq b)$, where a and b are real numbers, by finding the area under the corresponding normal curve from a to b (see Example 2). However, if we use a normal probability distribution to approximate a binomial probability distribution, then we find $P(a \leq x \leq b)$, where a and b are nonnegative integers, by finding the area under the corresponding normal curve from $a - .5$ to $b + .5$ (see Examples 3 and 4).

Explore–Discuss 2

(A) Construct a histogram of the binomial distribution with $n = 8$ and $p = .1$.

(B) Does the binomial distribution of part (A) satisfy the rule-of-thumb test?

C (C) Use a graphing utility to graph the normal distribution that has the same mean and standard deviation as the binomial distribution of part (A). How does the graph compare to the histogram?

(D) Is the normal distribution a good approximation to the binomial distribution in this case? Explain.

Answers to Matched Problems **1.** 49.38% **2.** .3413 **3.** (A) .70 (B) .13 **4.** (A) .83 (B) .004

EXERCISE 7-6

A *Given a normal distribution with mean 50 and standard deviation 10, find the number of standard deviations each of the measurements in Problems 1–8 is from the mean. Express the answer as a positive number.*

1. 65 **2.** 75 **3.** 83 **4.** 79
5. 45 **6.** 38 **7.** 42 **8.** 26

Using the normal distribution described for Problems 1–8 and Table I in Appendix C, find the area under the normal curve from the mean to the indicated measurement in Problems 9–16.

9. 65 **10.** 75 **11.** 83 **12.** 79
13. 45 **14.** 38 **15.** 42 **16.** 26

B *Given a normal distribution with mean 70 and standard deviation 8, find the area under the normal curve above the intervals in Problems 17–24.*

17. 60–80 **18.** 50–90
19. 62–74 **20.** 66–78
21. 88 or larger **22.** 90 or larger
23. 60 or smaller **24.** 56 or smaller

In Problems 25 and 26, discuss the validity of each statement. If the statement is always true, explain why. If not, give a counterexample.

25. (A) All normal distributions have the same shape.
(B) The area above the x axis and below the normal curve is the same for all normal distributions.

26. (A) The distribution of final exam scores from a statistics class of 90 students is a normal distribution.
(B) In a normal distribution, the probability is 0 that a score lies more than 4 standard deviations away from the mean.

In Problems 27–34, use the rule-of-thumb test to check whether a normal distribution (with the same mean and standard deviation as the binomial distribution) is a suitable approximation for the binomial distribution with:

27. $n = 15, p = .7$ **28.** $n = 12, p = .6$
29. $n = 15, p = .4$ **30.** $n = 20, p = .6$
31. $n = 100, p = .05$ **32.** $n = 200, p = .03$
33. $n = 500, p = .05$ **34.** $n = 400, p = .08$

35. A Bernoulli trial has probability of success $p = .1$. Explain how to determine the number of repeated trials necessary to obtain a binomial distribution that passes the rule-of-thumb test for using a normal distribution as a suitable approximation.

36. For a binomial distribution with $n = 100$, explain how to determine the smallest and largest values of p that pass the rule-of-thumb test for using a normal distribution as a suitable approximation.

C *A binomial experiment consists of 500 trials with the probability of success for each trial .4. What is the probability of obtaining the number of successes indicated in Problems 37–44? Approximate these probabilities to two decimal places using a normal curve. (This binomial experiment easily passes the rule-of-thumb test, as you can check. When computing the probabilities, adjust the intervals as in Examples 3 and 4.)*

37. 185–220 **38.** 190–205
39. 210–220 **40.** 175–185
41. 225 or more **42.** 212 or more
43. 175 or less **44.** 188 or less

To graph Problems 45–48, use a graphing utility and refer to the normal probability distribution function with mean μ and standard deviation σ:

$$f(x) = \frac{1}{\sigma\sqrt{2\pi}} e^{-(x-\mu)^2/2\sigma^2} \qquad (1)$$

45. Graph equation (1) with $\sigma = 5$ and:
(A) $\mu = 10$ (B) $\mu = 15$ (C) $\mu = 20$

Graph all three in the same viewing window with Xmin $= -10$, Xmax $= 40$, Ymin $= 0$, and Ymax $= 0.1$.

46. Graph equation (1) with $\sigma = 4$ and:
(A) $\mu = 8$ (B) $\mu = 12$ (C) $\mu = 16$

Graph all three in the same viewing window with Xmin $= -5$, Xmax $= 30$, Ymin $= 0$, and Ymax $= 0.1$.

47. Graph equation (1) with $\mu = 20$ and:
(A) $\sigma = 2$ (B) $\sigma = 4$

Graph both in the same viewing window with Xmin $= 0$, Xmax $= 40$, Ymin $= 0$, and Ymax $= 0.2$.

48. Graph equation (1) with $\mu = 18$ and:
(A) $\sigma = 3$ (B) $\sigma = 6$

Graph both in the same viewing window with Xmin $= 0$, Xmax $= 40$, Ymin $= 0$, and Ymax $= 0.2$.

49. (A) If 120 scores are chosen from a normal distribution with mean 75 and standard deviation 8, how many scores x would be expected to satisfy $67 \leq x \leq 83$?

(B) Use a graphing utility to generate 120 scores from the normal distribution with mean 75 and standard deviation 8. Determine the number of scores x such that $67 \leq x \leq 83$, and compare your results with the answer to part (A).

50. (A) If 250 scores are chosen from a normal distribution with mean 100 and standard deviation 10, how many scores x would be expected to be greater than 110?

(B) Use a graphing utility to generate 250 scores from the normal distribution with mean 100 and standard deviation 10. Determine the number of scores greater than 110, and compare your results with the answer to part (A).

Applications

Business & Economics

51. *Sales.* Salespeople for a business machine company have average annual sales of \$200,000, with a standard deviation of \$20,000. What percentage of the salespeople would be expected to make annual sales of \$240,000 or more? Assume a normal distribution.

52. *Guarantees.* The average lifetime for a car battery of a certain brand is 170 weeks, with a standard deviation of 10 weeks. If the company guarantees the battery for 3 years, what percentage of the batteries sold would be expected to be returned before the end of the warranty period? Assume a normal distribution.

53. *Quality control.* A manufacturing process produces a critical part of average length 100 millimeters, with a standard deviation of 2 millimeters. All parts deviating by more than 5 millimeters from the mean must be rejected. What percentage of the parts must be rejected, on the average? Assume a normal distribution.

54. *Quality control.* An automated manufacturing process produces a component with an average width of 7.55 centimeters, with a standard deviation of 0.02 centimeter. All components deviating by more than 0.05 centimeter from the mean must be rejected. What percentage of the parts must be rejected, on the average? Assume a normal distribution.

55. *Marketing claims.* A company claims that 60% of the households in a given community use their product. A competitor surveys the community, using a random sample of 40 households, and finds only 15 households out of the 40 in the sample using the product. If the company's claim is correct, what is the probability of 15 or fewer households using the product in a sample of 40? Conclusion? Approximate a binomial distribution with a normal distribution.

56. *Labor relations.* A union representative claims 60% of the union membership will vote in favor of a particular settlement. A random sample of 100 members is polled, and out of these, 47 favor the settlement. What is the approximate probability of 47 or fewer in a sample of 100 favoring the settlement when 60% of all the membership favor the settlement? Conclusion? Approximate a binomial distribution with a normal distribution.

Life Sciences

57. *Medicine.* The average healing time of a certain type of incision is 240 hours, with standard deviation of 20 hours. What percentage of the people having this incision would heal in 8 days or less? Assume a normal distribution.

58. *Agriculture.* The average height of a hay crop is 38 inches, with a standard deviation of 1.5 inches. What percentage of the crop will be 40 inches or more? Assume a normal distribution.

59. *Genetics.* In a family with 2 children, the probability that both children are girls is approximately .25. In a random sample of 1,000 families with 2 children, what is the approximate probability that 220 or fewer will have 2 girls? Approximate a binomial distribution with a normal distribution.

60. *Genetics.* In Problem 59, what is the approximate probability of the number of families with 2 girls in the sample being at least 225 and not more than 275? Approximate a binomial distribution with a normal distribution.

Social Sciences

61. *Testing.* Scholastic Aptitude Tests are scaled so that the mean score is 500 and the standard deviation is 100. What percentage of the students taking this test should score 700 or more? Assume a normal distribution.

62. *Politics.* Candidate Harkins claims a private poll shows that she will receive 52% of the vote for governor. Her opponent, Mankey, secures the services of another pollster, who finds that 470 out of a random sample of 1,000 registered voters favor Harkins. If Harkins' claim is correct, what is the probability that only 470 or fewer will favor her in a random sample of 1,000? Conclusion? Approximate a binomial distribution with a normal distribution.

63. *Grading on a curve.* An instructor grades on a curve by assuming the grades on a test are normally distributed. If the average grade is 70 and the standard deviation is 8, find the test scores for each grade interval if the instructor wishes to assign grades as follows: 10% A's, 20% B's, 40% C's, 20% D's, and 10% F's.

64. *Psychology.* A test devised to measure aggressive–passive personalities was standardized on a large group of people. The scores were normally distributed, with a mean of 50 and a standard deviation of 10. If we want to designate the highest 10% as aggressive, the next 20% as moderately aggressive, the middle 40% as average, the next 20% as moderately passive, and the lowest 10% as passive, what ranges of scores will be covered by these five designations?

Important Terms and Symbols

7-1 *Graphing Data.* Bar graph; broken-line graph; pie graph

7-2 *Graphing Quantitative Data.* Quantitative data; frequency table; class interval; data range; class frequency; frequency distribution; relative frequency; probability distribution; histogram; frequency polygon; frequency curve; cumulative frequency table; cumulative frequency polygon; ogive

7-3 *Measures of Central Tendency.* Arithmetic average; mean; sample mean (grouped and ungrouped sample data); population mean (grouped and ungrouped population data); summation symbol; weighted average; expected value; median; mode

$\bar{x}$ = Sample mean; μ = Population mean

Mean (ungrouped data):

$$\frac{\sum_{i=1}^{n} x_i}{n} = \frac{x_1 + x_2 + \cdots + x_n}{n}$$

Mean (grouped data):

$$\frac{\sum_{i=1}^{k} x_i f_i}{n} = \frac{x_1 f_1 + x_2 f_2 + \cdots + x_k f_k}{n}$$

where $n = \sum_{i=1}^{k} f_i$ = Total number of measurements

7-4 *Measures of Dispersion.* Range; deviation; sample variance (ungrouped and grouped sample data); sample standard deviation (ungrouped and grouped sample data); population variance (ungrouped and grouped population data); population standard deviation (ungrouped and grouped population data); law of large numbers

s^2 = Sample variance;
s = Sample standard deviation;
σ^2 = Population variance;
σ = Population standard deviation;

Sample standard deviation (ungrouped data):

$$s = \sqrt{\frac{\sum_{i=1}^{n} (x_i - \bar{x})^2}{n - 1}}$$

Sample standard deviation (grouped data):

$$s = \sqrt{\frac{\sum_{i=1}^{k} (x_i - \bar{x})^2 f_i}{n - 1}}$$

where $n = \sum_{i=1}^{k} f_i$ = Total number of measurements

(If the data set is the whole population, then by replacing s with σ, $\bar{x}$ with μ, and $n - 1$ with n, the above two formulas will yield the population standard deviation for ungrouped and grouped data, respectively.)

7-5 *Bernoulli Trials and Binomial Distributions.* Theoretical probability distribution; Bernoulli trial; success; failure; sequence of Bernoulli trials; binomial experiment; binomial formula; binomial distribution

$$(a + b)^n = C_{n,0}a^n + C_{n,1}a^{n-1}b + C_{n,2}a^{n-2}b^2 + \cdots + C_{n,n}b^n;$$

$$P(x) = P(x \text{ successes in } n \text{ trials}) = C_{n,x}p^x q^{n-x}, \quad x \in \{0, 1, \ldots, n\};$$

$$\text{Mean} = \mu = np; \quad \text{Standard deviation} = \sigma = \sqrt{npq}$$

7-6 *Normal Distributions.* Normal curve; normal probability distribution; discrete random variable; continuous random variable; z value; rule-of-thumb test; approximating a binomial distribution with a normal distribution

$$z = \frac{x - \mu}{\sigma}$$

Review Exercise

Work through all the problems in this chapter review and check your answers in the back of the book. Answers to all review problems are there along with section numbers in italics to indicate where each type of problem is discussed. Where weaknesses show up, review appropriate sections in the text.

A

1. Graph the following data using a bar graph and a broken-line graph:

United Nations Peacekeeping

Year	Number of Active Field Operations
1988	9
1989	12
1990	14
1991	17
1992	21
1993	28
1994	17

2. Graph the data in the following table using two pie graphs, one for men and one for women:

Living Arrangements of the Elderly (65 Years and Over)

	Men (%)	Women (%)
Alone	16.0	42.0
With spouse	74.3	39.7
With relatives	7.4	16.2
With nonrelatives	2.3	2.1

3. (A) Draw a histogram for the binomial distribution

$$P(x) = C_{3,x}(.4)^x(.6)^{3-x}$$

(B) What are the mean and standard deviation?

4. For the set of sample measurements 1, 1, 2, 2, 2, 3, 3, 4, 4, 5, find the:

(A) Mean (B) Median (C) Mode

(D) Standard deviation

5. If a normal distribution has a mean of 100 and a standard deviation of 10, then:

(A) How many standard deviations is 118 from the mean?

(B) What is the area under the normal curve between the mean and 118?

B

6. Given the sample of 25 quiz scores listed in the table below from a class of 500 students:

(A) Construct a frequency table using a class interval of width 2 starting at 9.5.

(B) Construct a histogram.

(C) Construct a frequency polygon.

(D) Construct a cumulative frequency and relative cumulative frequency table.

(E) Construct a cumulative frequency polygon.

Quiz Scores

14	13	16	15	17
19	15	14	17	15
15	13	12	14	14
12	14	13	11	15
16	14	16	17	14

7. For the set of grouped sample data given in the table:
(A) Find the mean.
(B) Find the standard deviation.
(C) Find the median.

INTERVAL	FREQUENCY
0.5–3.5	1
3.5–6.5	5
6.5–9.5	7
9.5–12.5	2

8. (A) Construct a histogram for the binomial distribution

$$P(x) = C_{6,x}(.5)^x(.5)^{6-x}$$

(B) What are the mean and standard deviation?

9. What are the mean and standard deviation for a binomial distribution with $p = .6$ and $n = 1{,}000$?

In Problems 10 and 11, discuss the validity of each statement. If the statement is always true, explain why. If not, give a counterexample.

10. (A) If the data set $x_1, x_2, \ldots, x_n$ has mean $\bar{x}$, then the data set $x_1 + 5, x_2 + 5, \ldots, x_n + 5$ has mean $\bar{x} + 5$.
(B) If the data set $x_1, x_2, \ldots, x_n$ has standard deviation s, then the data set $x_1 + 5, x_2 + 5, \ldots, x_n + 5$ has standard deviation $s + 5$.

11. (A) If X represents a binomial random variable with mean μ, then $P(X \geqslant \mu) = .5$.
(B) If X represents a normal random variable with mean μ, then $P(X \geqslant \mu) = .5$.
(C) The area of a histogram of a binomial distribution is equal to the area above the x axis and below a normal curve.

12. If the probability of success in a single trial of a binomial experiment with 1,000 trials is .6, what is the probability of obtaining at least 550 and no more than 650 successes in 1,000 trials? [*Hint:* Approximate with a normal distribution.]

13. Given a normal distribution with mean 50 and standard deviation 6, find the area under the normal curve:
(A) Between 41 and 62
(B) From 59 on

14. A data set is formed by recording the sums when a pair of dice is rolled 100 times. A second data set is formed by again rolling a pair of dice 100 times, but recording the product, not the sum, of the two numbers.
(A) Which of the two data sets would you expect to have the smaller standard deviation? Explain.
C (B) To obtain evidence for your answer to part (A), use a graphing utility to simulate both experiments, and compute the standard deviations of each of the two data sets.

C

15. For the sample quiz scores in Problem 6 above, find the mean and standard deviation using the data:
(A) Without grouping
(B) Grouped, with class interval of width 2 starting at 9.5

16. A fair die is rolled five times. What is the probability of rolling:
(A) Exactly three 6's? (B) At least three 6's?

17. Two dice are rolled three times. What is the probability of getting a sum of 7 at least once?

18. Ten students take an exam worth 100 points.
(A) Construct a hypothetical set of exam scores for the ten students in which both the median and the mode are 30 points higher than the mean.
(B) Could the median and the mode both be 50 points higher than the mean? Explain.

19. In the last presidential election, 39% of the registered voters in a certain city actually cast ballots.
(A) In a random sample of 20 registered voters from that city, what is the probability that exactly 8 voted in the last presidential election?
(B) Verify by the rule-of-thumb test that the normal distribution with mean 7.8 and standard deviation 2.18 is a good approximation of the binomial distribution with $n = 20$ and $p = .39$.
(C) For the normal distribution of part (B), $P(x = 8) = 0$. Explain the discrepancy between this result and your answer from part (A).

20. A random variable represents the number of wins in a 12 game season for a football team that has a probability of .9 of winning any of its games.
(A) Find the mean and standard deviation of the random variable.
(B) Find the probability that the team wins each of its 12 games.
C (C) Use a graphing utility to simulate 100 repetitions of the binomial experiment associated with the random variable, and compare the empirical probability of a perfect season with the answer to part (B).

APPLICATIONS

Business & Economics

21. *Retail sales.* The daily number of bad checks received by a large department store in a random sample of 10 days out of the past year were 15, 12, 17, 5, 5, 8, 13, 5, 16, and 4. Find the:
(A) Mean (B) Median (C) Mode
(D) Standard deviation

22. *Preference survey.* Find the mean, median, and/or mode, whichever are applicable, for the following employee cafeteria service survey:

DRINK ORDERED WITH MEAL	NUMBER
Coffee	435
Tea	137
Milk	298
Soft drink	522
Milk shake	392

23. *Plant safety.* The weekly record of reported accidents in a large auto assembly plant in a random sample of 35 weeks from the past 10 years is listed below:

34	33	36	35	37	31	37
39	34	35	37	35	32	35
33	35	32	34	32	32	39
34	31	35	33	31	38	34
36	34	37	34	36	39	34

(A) Construct a frequency and relative frequency table using class intervals of width 2 and starting at 29.5.
(B) Construct a histogram and frequency polygon.
C (C) Find the mean and standard deviation for the grouped data.

24. *Personnel screening.* The scores on a screening test for new technicians are normally distributed with mean 100 and standard deviation 10. Find the approximate percentage of applicants taking the test who score:
(A) Between 92 and 108 (B) 115 or higher

25. *Market research.* A newspaper publisher claims that 70% of the people in a community read their newspaper. Doubting the assertion, a competitor randomly surveys 200 people in the community. Based on the publisher's claim (and assuming a binomial distribution):
(A) Compute the mean and standard deviation of the binomial distribution.
(B) Determine whether the rule-of-thumb test warrants the use of a normal distribution to approximate this binomial distribution.
(C) Calculate the approximate probability of finding at least 130 and no more than 155 readers in the sample.
(D) Determine the approximate probability of finding 125 or fewer readers in the sample.
C (E) Use a graphing utility to graph the relevant normal distribution.

Life Sciences

26. *Health care.* A small town has three doctors on call for emergency service. The probability that any one doctor will be available when called is .90. What is the probability that at least one doctor will be available for an emergency call?

Group Activity 1 Analysis of Data on Student Lifestyle

(A) Select several quantitative variables related to student life or the economic impact of students on the surrounding community—for example, the number of hours spent studying outside of class per week, the number of long-distance telephone calls made per month, the number of ounces of alcoholic beverages consumed per week, the number of dollars spent on recreational activities per week, and so on. Interview a total of approximately 40 students to obtain data on each of the variables you have selected. Discuss how the sample of students should be selected in order to obtain an approximately random sample. Discuss how the interviews should be conducted in order to obtain reliable information.

(B) Compute the mean, median, and standard deviation for the data set corresponding to each of your quantitative variables. Compute the proportion of the sample that lies within 1, 2, and 3 standard deviations of the mean. Which of your data sets most closely approximates a normal distribution?

(C) Use histograms and/or tables and other graphs to present the results of your study to those outside your group.

Group Activity 2 *Survival Rates for a Heart Transplant*

In recent years approximately 2,400 heart transplant operations have been performed annually in the United States. The American Heart Association reported that the 1 year survival rate for a heart transplant is 82.4%, the 2 year survival rate is 78.2%, and the 3 year rate is 74.6% (www.amhrt.org).

Ten patients are currently awaiting a heart transplant at St. Luke's Hospital. Assume that each of the ten undergoes the transplant surgery.

(A) Construct probability distribution tables for the random variables X_1, X_2, and X_3, where X_k represents the number from among the ten patients who survive for at least k years. (Assume X_k is binomial.)

(B) What is the probability that eight or more of the ten heart transplant recipients will survive at least 1 year? 2 years? 3 years?

(C) Use a graphing utility to simulate 100 repetitions of the binomial experiments associated with X_1, X_2, and X_3, and compare the results of the simulations with the answers to part (B).

(D) If 2,500 heart transplants are performed in the United States this year, what is the probability that at least 2,000 of the recipients will survive at least 1 year? 2 years? 3 years? (Approximate the appropriate binomial distributions with normal distributions.)

CHAPTER 8

GAMES AND DECISIONS

INTRODUCTION

Game theory is a relatively new branch of mathematics designed to help people who are in conflict situations determine the best course of action out of several possible choices. The theory is applicable to some parlor games, but more importantly, it has been applied with moderate success to decision-making problems in economics, business, psychology, sociology, warfare, and political science. Current research in this new branch of mathematics suggests a promising future.

Even though the theory had its beginnings in the 1920's, its greatest advance occurred in 1944, when John von Neumann and Oskar Morgenstern, both at Princeton University, published their landmark book, *Games and Economic Behavior.* Fifty years later, in 1994, the Nobel prize in economic science was awarded to John Nash of Princeton University, John Harsanyi of the University of California at Berkeley, and Reinhard Selten of the University of Bonn for their work in game theory.

Game theory provides an excellent review of many of the topics studied in the preceding chapters. Linear systems, matrices, probability, expected value, and linear programming are all used in the development of this subject. The general theory of games is mathematically complex. In this book we will consider only the simplest kinds of games, that is, games involving:

1. Only two players
2. A payoff of some amount after each play such that one player's win is the other player's loss (no fee goes to a third party such as a gambling house for playing the game)

Games of this type are called **two-person zero-sum games.** With the indicated restrictions, we will be able to determine the best (optimal) strategies of play for each person.

SECTION 8-1

Strictly Determined Games

- STRICTLY DETERMINED MATRIX GAMES
- NONSTRICTLY DETERMINED MATRIX GAMES

The best way to start this discussion is with an example. Out of this example will evolve basic definitions, theorems, and methodology.

Consider two stores that sell stereo equipment, store R and store C. These are the only two stereo stores in a shopping mall, and each is trying to decide how to price its most competitive stereo package. A market research firm supplies the following information:

$$\begin{array}{cc} & \text{Store } C \\ & \begin{array}{cc} \$200 & \$225 \end{array} \\ \text{Store } R \begin{array}{c} \$200 \\ \$225 \end{array} & \begin{bmatrix} 55\% & 70\% \\ 40\% & 55\% \end{bmatrix} \end{array} \tag{1}$$

The entries in the matrix indicate the percentage of the business store R will receive. That is, if both stores price their packages at \$200, then store R will receive 55% of all the business (store C will lose 55% of the business but will get 45%). If store R chooses a price of \$200 and store C chooses \$225, then store R will receive 70% of the business (store C will lose 70% of the business but will get 30%); and so on. Each store can choose its own price but cannot control the price of the other. The object is for each store to determine a price that will ensure the maximum possible business in this competitive situation.

This marketing competition may be viewed as a game between store R and store C. A single play of the game requires store R to choose (play) Row 1 or Row 2 in matrix (1) (that is, price its stereo package at either \$200 or \$225) and simultaneously requires store C to choose (play) Column 1 or Column 2 (that is, price its stereo package at either \$200 or \$225). It is common to designate the person(s) choosing the rows by R, for **row player,** and the person(s) choosing the columns by C, for **column player.** Each entry in matrix (1) is called the **payoff** for a particular pair of moves by R and C. Matrix (1) is called a **matrix game,** or a **payoff matrix.** This game is a two-person zero-sum game, since each store may be considered a person, and R wins the same amount that C loses, and vice versa.

STRICTLY DETERMINED MATRIX GAMES

Actually, **any $m \times n$ matrix may be considered a two-person zero-sum matrix game** in which player R chooses (plays) any one of m rows and player C simultaneously chooses (plays) any one of n columns. For example, the 3×4 matrix

$$\begin{bmatrix} 0 & 6 & -2 & -4 \\ 5 & 2 & 1 & 3 \\ -8 & -1 & 0 & 20 \end{bmatrix} \tag{2}$$

may be viewed as a matrix game where R has three moves and C has four moves. If R plays Row 2 and C plays Column 4, then R wins 3 units. If, however, R plays Row 3 and C plays Column 1, then R "wins" -8 units; that is, C wins 8 units.

Negative entries in the payoff matrix indicate a win for C, and positive entries indicate a win for R.

[Except when entries are percentages as in matrix (1).]

How should R and C play in matrix game (2)? If R is a little greedy, Row 3 may be chosen because of the payoff of 20 units. But we assume C is not stupid and would be likely to avoid Column 4 (not wanting to lose 20 units). If C plays Column 1, 8 units might be won. But then R might play Row 2, anticipating C's thinking, and C would not win after all. Is there a best play for each? To help us unravel this problem, we now state a **fundamental principle of game theory:**

Fundamental Principle of Game Theory

1. A matrix game is played repeatedly.
2. Player R tries to maximize winnings.
3. Player C tries to minimize losses.

Player R, deciding to be conservative, thinks in terms of the worst that could happen for each row choice, and chooses a row that has the largest minimum payoff. This provides a **security level** that is guaranteed irrespective of C's choices. Returning to matrix game (2), we find that the worst that could happen in Row 1 is a 4 unit loss, in Row 2 a 1 unit gain, and in Row 3 an 8 unit loss. Each of these values is circled below:

$$\begin{bmatrix} 0 & 6 & -2 & \enclose{circle}{-4} \\ 5 & 2 & \enclose{circle}{1} & 3 \\ \enclose{circle}{-8} & -1 & 0 & 20 \end{bmatrix}$$

The best approach (**strategy**) for R is to select the row with the largest of these minimum values—that is, Row 2. With this choice, a win of at least 1 unit is guaranteed for R no matter what C does!

Similarly, C puts a square around the maximum value in each column to identify the worst situation that could happen for each column choice. (Remember that a positive number indicates a loss for C.)

$$\begin{bmatrix} 0 & \boxed{6} & -2 & \enclose{circle}{-4} \\ \boxed{5} & 2 & \boxed{\enclose{circle}{1}} & 3 \\ \enclose{circle}{-8} & -1 & 0 & \boxed{20} \end{bmatrix}$$

Circles mark the minimum value in each row, and squares mark the maximum value in each column.

We see that C's best approach (strategy) is to select the column with the smallest of these maximum values—that is, Column 3. By choosing Column 3, C establishes a security level of a loss of 1 unit irrespective of R's choices, and this is the best that C can do if R continues to make the best moves.

The entry 1 in the second row and third column (enclosed by a square and a circle) is both the largest of the row minimums and the smallest of the column maximums. This means that R should always play the second row and C should always play the third column. The result is a win of 1 unit for R every time.

These are the best *(optimum)* strategies for both R and C. If C keeps playing the third column and R decides to change from the second row, then R's wins cannot increase. Similarly, if R continues to play the second row and C deviates from the third column, C's losses cannot decrease. The game is said to be *strictly determined* in that R must always play Row 2 to maximize winnings and C must always play Column 3 to minimize losses.

The payoff value surrounded by both a circle and a square is called a *saddle value.* To see why, drop all the elements in the matrix above except those in the second row and third column to obtain

$$\begin{bmatrix} & & -2 & \\ 5 & 2 & \boxed{\textcircled{1}} & 3 \\ & & 0 & \end{bmatrix}$$

In one direction 1 is a minimum, and in the other direction 1 is a maximum—the form is characteristic of a horse saddle. Saddle values play an important role in game theory, so we now formalize our discussion and state some important definitions and a theorem.

Strictly Determined Matrix Games

A matrix game is said to be **strictly determined** if a payoff value is simultaneously a row minimum and a column maximum. If such a value exists, it is called a **saddle value.** In a strictly determined game, **optimal strategies** are:

R should choose any row containing a saddle value.
C should choose any column containing a saddle value.

A saddle value is called the **value** of a strictly determined game. The game is **fair** if its value is zero. [*Note:* In a strictly determined game (assuming both players play their optimal strategy), knowledge of an opponent's move provides no advantage, since the payoff will always be a saddle value.]

Locating Saddle Values

Step 1. Circle the minimum value in each row (it may occur in more than one place).

Step 2. Place squares around the maximum value in each column (it may occur in more than one place).

Step 3. Any entry with both a circle and a square around it is a saddle value.

We state Theorem 1 without proof.

Theorem 1 ■■ EQUALITY OF SADDLE VALUES

If a game matrix has two or more saddle values, then they are equal. ■■

Example 1 **Finding Saddle Values and Optimal Strategies** Two large shopping centers have two competing home improvement discount stores, store R in one center and store C in the other. Each week, each store chooses one, and only one, of the following means of promotion: TV, radio, newspaper, or a blanket mailing. A marketing research company provided the following payoff matrix, which indicates the percentage of market gain or loss for each choice of action by R and by C (we assume any gain by R is a loss by C, and vice versa):

Store C

Store R	TV	Radio	Paper	Mail
TV	0	−2	−2	2
Radio	1	2	1	3
Paper	−1	0	0	1
Mail	1	2	1	2

(A) Find all saddle values. (B) Find optimal strategies for R and C.
(C) Find the value of the game.

SOLUTION (A) Follow the three steps listed in the box for locating saddle values:

Store C

Store R	TV	Radio	Paper	Mail
TV	0	(−2)	(−2)	2
Radio	[1]	[2]	[1]	[3]
Paper	(−1)	0	0	1
Mail	[1]	[2]	[1]	2

There are four saddle values, and all are equal to 1.

(B) Optimal strategies for store R and store C are to choose rows and columns, respectively, that contain a saddle value. Thus,

Optimal strategy for R: Choose radio or mail each week.
Optimal strategy for C: Choose TV or newspaper each week.

(C) The value of the game is 1, and store R has the advantage. (When both stores use optimal strategies, store R will gain 1% of the market at store C's expense of losing 1%.)

Matched Problem 1 Repeat Example 1 for the stereo game matrix discussed at the beginning of this section:

Store C

Store R	\$200	\$225
\$200	55%	70%
\$225	40%	55%

Explore–Discuss 1

Suppose that a and k are both saddle values of the 3×4 matrix

$$A = \begin{bmatrix} a & b & c & d \\ e & f & g & h \\ i & j & k & l \end{bmatrix}$$

(A) Show that a must equal k after explaining why each of the following must be true: $a \leqslant c$, $c \leqslant k$, $a \geqslant i$, and $i \geqslant k$.

(B) Show that c and i also must be equal to k.

■ Nonstrictly Determined Matrix Games

Not all matrix games are strictly determined; that is, many matrix games do not have saddle values. Consider the classic penny-matching game: We have two players, player R and player C. Each player has a penny, and they simultaneously choose to show the side of the coin of their choice (H = heads, T = tails). If the pennies match, then R wins (C loses) 1¢. If the pennies do not match, then R loses (C wins) 1¢. In terms of a game matrix, we have

$$\begin{array}{cc} & \text{Player } C \\ & \begin{array}{cc} \text{H} & \text{T} \end{array} \\ \text{Player } R \; \begin{array}{c} \text{H} \\ \text{T} \end{array} & \begin{bmatrix} 1 & -1 \\ -1 & 1 \end{bmatrix} \end{array}$$

Testing this game matrix for saddle values, we obtain

$$\begin{array}{cc} & \text{Player } C \\ & \begin{array}{cc} \text{H} & \text{T} \end{array} \\ \text{Player } R \; \begin{array}{c} \text{H} \\ \text{T} \end{array} & \begin{bmatrix} \boxed{1} & \enclose{circle}{-1} \\ \enclose{circle}{-1} & \boxed{1} \end{bmatrix} \end{array}$$

No entry is enclosed in both a circle and a square; hence, there are no saddle values and the game is called **nonstrictly determined.** A minimum of -1 occurs in each row; thus, R could play either. Similarly, a maximum of 1 occurs in each column, and C could play either. In a nonstrictly determined game, knowledge of the other player's move would certainly be very useful. For example, if C knew that R was going to play Row 1 (heads), then C would obviously play Column 2 (tails) to win 1¢.

In this game, is there an optimum strategy for each player? In general, for nonstrictly determined two-person zero-sum games, are there optimum strategies for each player? The answer, surprisingly, turns out to be yes—this is the subject of Sections 8-2 through 8-4.

Example 2 ➡ **Distinguishing between Strictly and Nonstrictly Determined Games**

Determine whether the following matrix games are strictly determined:

$$A = \begin{bmatrix} -1 & 2 & -3 & 4 \\ 5 & -2 & 3 & 0 \end{bmatrix} \qquad B = \begin{bmatrix} 2 & -1 & -8 \\ 4 & 8 & 10 \\ -5 & -3 & 0 \end{bmatrix}$$

SOLUTION Using the circle and square technique discussed above, we obtain

$$A = \begin{bmatrix} -1 & \boxed{2} & \enclose{circle}{-3} & \boxed{4} \\ \boxed{5} & \enclose{circle}{-2} & \boxed{3} & 0 \end{bmatrix} \qquad B = \begin{bmatrix} 2 & -1 & \enclose{circle}{-8} \\ \boxed{\enclose{circle}{4}} & \boxed{8} & \boxed{10} \\ \enclose{circle}{-5} & -3 & 0 \end{bmatrix}$$

Matrix game A has no saddle values and thus is nonstrictly determined. Matrix game B has a saddle value (4 in Row 2 and Column 1) and hence is strictly determined.

Matched Problem 2 Determine which of the matrix games below are nonstrictly determined:

$$A = \begin{bmatrix} -1 & 3 \\ 0 & 2 \\ -2 & -4 \\ -1 & 0 \end{bmatrix} \qquad B = \begin{bmatrix} 2 & -1 & 4 \\ -3 & 0 & 5 \\ -6 & 10 & -8 \end{bmatrix}$$

Explore–Discuss 2

There are sixteen 2×2 matrices with all entries that are either 1 or -1. Which of them are matrices of strictly determined matrix games? Nonstrictly determined matrix games?

Answers to Matched Problems

1. (A) The 55% in the first row and first column is the only saddle value.
(B) Optimal strategy for R: Play Row 1 ($200 price).
Optimal strategy for C: Play Column 1 ($200 price).
(C) The value of the game is 55%; thus, store R has the advantage. (If the value had been 50%, then neither store would have an advantage; if it had been 45%, then store C would have the advantage.)

2. B is nonstrictly determined.

EXERCISE 8-1

In Problems 1–18, for each matrix game that is strictly determined (if it is not strictly determined, say so), indicate:

(A) All saddle values
(B) Optimal strategies for R and C
(C) The value of the game

A

1. $\begin{bmatrix} 3 & 2 \\ 2 & -1 \end{bmatrix}$ **2.** $\begin{bmatrix} -5 & 8 \\ -1 & 0 \end{bmatrix}$ **3.** $\begin{bmatrix} -2 & 5 \\ 3 & 0 \end{bmatrix}$

4. $\begin{bmatrix} 6 & 2 \\ -1 & 3 \end{bmatrix}$ **5.** $\begin{bmatrix} -3 & 0 \\ 4 & 1 \end{bmatrix}$ **6.** $\begin{bmatrix} 0 & 1 \\ -3 & 0 \end{bmatrix}$

7. $\begin{bmatrix} 1 & 1 \\ 1 & 1 \end{bmatrix}$ **8.** $\begin{bmatrix} -3 & -3 \\ -3 & -3 \end{bmatrix}$

B

9. $\begin{bmatrix} 2 & 2 \\ 2 & 5 \end{bmatrix}$ **10.** $\begin{bmatrix} 4 & 4 \\ -2 & 4 \end{bmatrix}$

11. $\begin{bmatrix} 2 & -1 & -5 \\ 1 & 0 & 3 \\ -3 & -7 & 8 \end{bmatrix}$ **12.** $\begin{bmatrix} 1 & 0 & 3 & 1 \\ -5 & -2 & 4 & -3 \end{bmatrix}$

13. $\begin{bmatrix} 3 & -2 \\ 1 & 5 \\ -4 & 0 \\ 5 & -3 \end{bmatrix}$ **14.** $\begin{bmatrix} 1 & -3 & 5 \\ -2 & 1 & 6 \\ 3 & -4 & 0 \end{bmatrix}$

15. $\begin{bmatrix} 3 & -1 & 4 & -7 \\ 1 & 0 & 2 & 3 \\ 5 & -2 & -3 & 0 \\ 3 & 0 & 1 & 5 \end{bmatrix}$ **16.** $\begin{bmatrix} 1 & -2 & 0 & 3 \\ -5 & 0 & -1 & 8 \\ 4 & 1 & 1 & 2 \end{bmatrix}$

17. $\begin{bmatrix} 0 & 4 & -8 & -3 \\ 2 & 5 & 3 & 2 \\ 1 & -3 & -2 & -9 \\ 2 & 4 & 7 & 2 \end{bmatrix}$ **18.** $\begin{bmatrix} -1 & 9 & -1 & -1 \\ -2 & 4 & -3 & -2 \\ -1 & 5 & -1 & -1 \\ -3 & 0 & -2 & -4 \end{bmatrix}$

19. For the matrix game of Problem 17, would you rather be player R or player C? Explain.

20. For the matrix game of Problem 18, would you rather be player R or player C? Explain.

C

21. Is there a value of m such that the following is not a strictly determined matrix game? Explain.

$$\begin{bmatrix} -3 & m \\ 0 & 1 \end{bmatrix}$$

22. If M is a 2×2 matrix game and both entries in one row are the same, try to find values for the other row so that the game is not strictly determined. What is your conclusion?

Applications

Business & Economics

23. *Price war.* A small town on a major highway has only two service stations: station R, a major brand station, and station C, an independent. A market research firm provided the payoff matrix below, where each entry indicates the percentage of customers who go to station R for the indicated prices per gallon of unleaded gasoline. Find saddle values and optimum strategies for each company.

		Station C	
		\$1.00	\$1.05
Station R	\$1.05	50%	70%
	\$1.10	40%	50%

24. *Investment.* Suppose you have \$10,000 to invest for a period of 5 years. After some investigation and advice from a financial advisor, you come up with the game matrix shown below, where you (R) are playing against the economy (C). Each entry in the matrix is the expected payoff (in dollars) after 5 years for an investment of \$10,000 in the corresponding row designation with the future state of the economy in the corresponding column designation. (The economy is regarded as a rational player who can make decisions against the investor—in any case, the investor would like to do the best possible irrespective of what happens to the economy.) Find saddle values and optimal strategies for each player.

		Economy C		
		Fall	No change	Rise
Investor R	Savings and loan, 5 year note	5,870	5,870	5,870
	Blue chip stock	−2,000	4,000	7,000
	Speculative stock	−5,000	2,000	10,000

25. *Store location.* Two equally competitive pet shops want to locate stores at Lake Tahoe, where there are currently no pet shops. The figure shows the percentages of the total Tahoe Basin population serviced by each of the three main business centers. If both shops locate in the same business center, they split all the business equally; if they locate in two different centers, they each get all the business in the center in which they locate plus half the business in the third center. Where should the two pet shops locate? Set up a game matrix and solve.

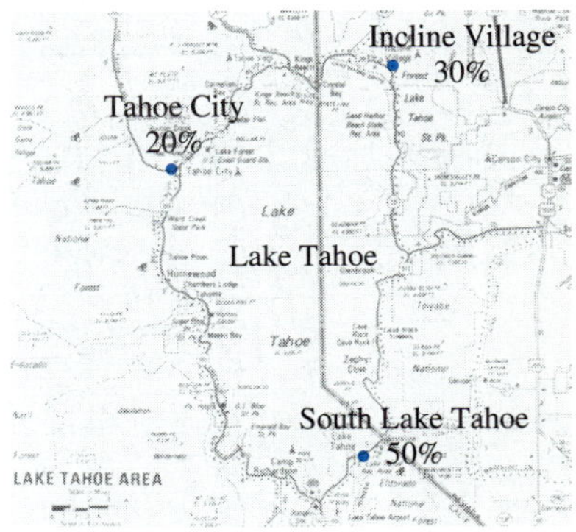

Figure for 25

26. *Store location.* Two competing automobile parts companies (R and C) are trying to decide among three small towns (E, F, and G) for new store locations (see the figure). All three towns have the same business potential. If both companies locate in the same town, then they split the business evenly (payoff is 0 to both). If, however, they locate in different towns, the store that is closer to the third town will get all of that town's business. For example, if R locates in E and C in F, then the payoff is 1 (R has gained one town). If, on the other hand, R locates in F and C in E, then the payoff is -1 (R has lost one town to C). Write the payoff matrix, find all saddle values, and indicate optimum strategies for both stores.

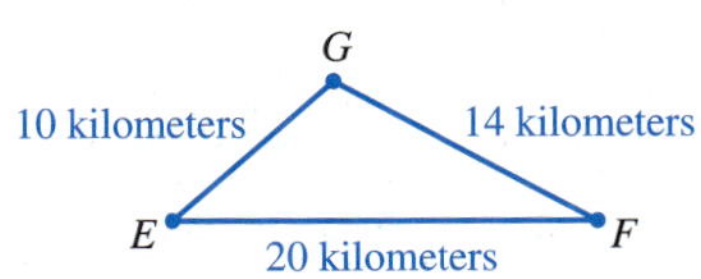

Figure for 26

SECTION 8-2

Mixed Strategy Games

- Nonstrictly Determined Games—An Example
- Pure and Mixed Strategies
- Expected Value of a Game
- Fundamental Theorem of Game Theory
- A Solution to a 2 × 2 Matrix Game
- Recessive Rows and Columns

Two-person zero-sum games can be divided into two classes:

1. Strictly determined games
2. Nonstrictly determined games

In the preceding section we found that in a strictly determined game, both players always choose rows and columns that contain a saddle value. The game is completely open, in the sense that knowledge of an opponent's strategy will be of no advantage to either player.

Nonstrictly Determined Games—An Example

We now consider the more interesting nonstrictly determined games (where we will find some unexpected results). As before, we will start with an example and generalize from our observations.

There are many variations of the **two-finger Morra game.** Basically, the game involves two players who simultaneously show one or two fingers and agree to a payoff for each particular combined outcome. Let us consider the following variation of the game, where the payoff (in dollars) is the sum of the fingers to R if the sum is even and the sum of the fingers to C if the sum is odd:

$$\begin{array}{cc} & \text{Player } C \\ & \begin{array}{cc} \text{1 finger} & \text{2 fingers} \end{array} \\ \text{Player } R \begin{array}{r} \text{1 finger} \\ \text{2 fingers} \end{array} & \begin{bmatrix} 2 & -3 \\ -3 & 4 \end{bmatrix} \end{array}$$

Would you rather be a row player or a column player, or does it matter? (Think about this for a moment before proceeding.)

To answer this question, we start by checking to see whether the game is strictly determined. Using the circle and square method described in the preceding section, we have

$$\begin{bmatrix} \boxed{2} & \textcircled{-3} \\ \textcircled{-3} & \boxed{4} \end{bmatrix}$$

Since no payoff has both a circle and a square around it, there are no saddle values; hence, the game is not strictly determined. How should the players play?

If player R continued to play Row 2 (because of the large payoff of \$4), it would not take C long to detect R's strategy, and C would obviously play Column 1. Then R would play Row 1 and C would have to shift to Column 2, and so on. It appears that neither player should continue playing the same row or same column, but should play each row and each column in some mixed pattern unknown to the other player.

How should a mixed pattern be chosen? Von Neumann discovered that the best way to choose a mixed pattern was to use a probability distribution and a chance device that would produce this distribution. For example, R might choose Row 1 with probability $\frac{1}{4}$ and Row 2 with probability $\frac{3}{4}$. This could be accomplished by placing 1 white marble and 3 black marbles in an urn and drawing one at random, letting white represent Row 1 and black represent Row 2. (The drawn marble would be replaced after each play.) In this way, neither the player nor the opponent would know which row was to be played until a marble was drawn. But, in the long run, R would play Row 1 one-fourth of the time and Row 2 three-fourths of the time. Similarly, player C might choose Column 1 with probability $\frac{3}{5}$ and Column 2 with probability $\frac{2}{5}$ by random drawings from an urn containing 3 white and 2 black marbles.

Pure and Mixed Strategies

It is useful at this time to bring together the ideas expressed above as well as those expressed in Section 8-1 by carefully defining the word *strategy.*

Strategies for R and C

Given the game matrix

$$M = \begin{bmatrix} a & b \\ c & d \end{bmatrix}$$

R's **strategy** is denoted by a probability row matrix:

$$P = [p_1 \quad p_2] \qquad \begin{aligned} p_1 &\geq 0 \\ p_2 &\geq 0 \\ p_1 + p_2 &= 1 \end{aligned}$$

C's **strategy** is denoted by a probability row matrix:

$$Q = \begin{bmatrix} q_1 \\ q_2 \end{bmatrix} \qquad \begin{aligned} q_1 &\geq 0 \\ q_2 &\geq 0 \\ q_1 + q_2 &= 1 \end{aligned}$$

In the introductory example, R would be using the strategy

$$P = \begin{bmatrix} \frac{1}{4} & \frac{3}{4} \end{bmatrix} \qquad \text{Row matrix}$$

and C would be using the strategy

$$Q = \begin{bmatrix} \frac{3}{5} \\ \frac{2}{5} \end{bmatrix} \qquad \text{Column matrix}$$

The reasons for using row and column matrices for strategies will be made clear shortly.

If one of the elements in P (or Q) is 1 and the other is 0, then the strategy is called a **pure strategy.** If a strategy is not a pure strategy, then it is called a **mixed strategy.** Thus, $P = [0 \quad 1]$ is a pure strategy, and R would play Row 2 every play. On the other hand, $P = \begin{bmatrix} \frac{1}{4} & \frac{3}{4} \end{bmatrix}$ is a mixed strategy, indicating that R plays Row 1 with probability $\frac{1}{4}$ and Row 2 with probability $\frac{3}{4}$.

In the two-finger Morra game described above, we found that pure strategies were not the best choice (pure strategies are always used in strictly determined games, and the game above is not strictly determined). This indicates that mixed strategies should be used. But what mixed strategies? Are there optimal mixed strategies for both players? The answer happens to be yes, and we now discuss a way to find them.

Expected Value of a Game

We will use the idea of **expected value** discussed in Section 6-7. What is the expected value of the two-finger Morra game for R if strategies

$$P = \begin{bmatrix} p_1 & p_2 \end{bmatrix} = \begin{bmatrix} .25 & .75 \end{bmatrix} \quad \text{and} \quad Q = \begin{bmatrix} q_1 \\ q_2 \end{bmatrix} = \begin{bmatrix} .6 \\ .4 \end{bmatrix}$$

are used by R and C, respectively? There are four possible outcomes (payoffs) for R, and each has a probability of occurrence that can be determined. Starting with

$$M = \begin{bmatrix} a & b \\ c & d \end{bmatrix} = \begin{bmatrix} 2 & -3 \\ -3 & 4 \end{bmatrix}$$

we see that the probability of payoff a occurring is the probability associated with R playing Row 1 and C playing Column 1. Since R's and C's plays are independent (neither knows what the other will play), the probability of a occurring is simply the product p_1q_1. Similarly, the probabilities of payoffs b, c, and d are p_1q_2, p_2q_1, and p_2q_2, respectively. Thus, using the definition of expected value from Section 6-7, we see that the expected value of the game for R, denoted by $E(P, Q)$, is given by

$$\begin{aligned} E(P, Q) &= ap_1q_1 + bp_1q_2 + cp_2q_1 + dp_2q_2 \\ &= (2)(.25)(.6) + (-3)(.25)(.4) + (-3)(.75)(.6) + (4)(.75)(.4) \\ &= -0.15 \end{aligned} \tag{1}$$

This means that in the long run, with R and C using the indicated strategies, R will lose 15¢ per game, on the average, and C will win 15¢ per game, on the average. Is this the best that R and C can do?

Player R would really like to find a strategy that would give the largest expected value that would hold irrespective of C's choice of strategies. On the other hand, C would like to find a strategy that would give R the smallest expected value that would hold irrespective of R's choice of strategies.

Before we indicate a solution to this problem, here is a convenient way of denoting the expected value $E(P, Q)$—this will clarify the reason we represented P and Q as row and column matrices, respectively:

Theorem 1 ■■ Expected Value of a Matrix Game for R

For the matrix game

$$M = \begin{bmatrix} a & b \\ c & d \end{bmatrix}$$

(continued)

Theorem 1 (*continued*) and strategies

$$P = \begin{bmatrix} p_1 & p_2 \end{bmatrix} \qquad Q = \begin{bmatrix} q_1 \\ q_2 \end{bmatrix}$$

for R and C, respectively, the expected value of the game for R is given by

$$E(P, Q) = PMQ$$

To show that $E(P, Q) = PMQ$, multiply PMQ to obtain

$$PMQ = ap_1q_1 + bp_1q_2 + cp_2q_1 + dp_2q_2$$

But from equation (1),

$$E(P, Q) = ap_1q_1 + bp_1q_2 + cp_2q_1 + dp_2q_2$$

Hence, $E(P, Q) = PMQ$. (The details of expanding PMQ are left to Problem 19 in Exercise 8-2.)

REMARKS

1. After multiplying, PMQ is a 1×1 matrix, which is usually written without brackets.
2. The relation $E(P, Q) = PMQ$ holds for arbitrary $m \times n$ matrix games (including strictly determined games) as well, where P is a $1 \times m$ row matrix and Q is an $n \times 1$ column matrix.

FUNDAMENTAL THEOREM OF GAME THEORY

We have now arrived at the heart of this discussion. We state, without proof, the **fundamental theorem of game theory.**

Theorem 2 FUNDAMENTAL THEOREM OF GAME THEORY[†]

For every $m \times n$ matrix game M, there exist strategies P^* and Q^* (not necessarily unique) for R and C, respectively, and a unique number v such that

$$P^*MQ \geqslant v \tag{2}$$

for every strategy Q of C and

$$PMQ^* \leqslant v \tag{3}$$

for every strategy P of R.

The number v is called the **value of the game** and is the *security level for both R and C.* If $v = 0$, the game is said to be **fair.** Furthermore, any strategies P^* and Q^* that satisfy (2) and (3) are called **optimal strategies** for R and C, respectively. It can be shown that v is the largest guaranteed expectation that R can obtain, irrespective of C's play. Surprisingly, v is also the smallest expectation that C can allow for R, irrespective of R's play.

Essentially, the fundamental theorem states that *every* matrix game has optimal strategies for R and C and a unique game value v. Finding these opti-

[†]Since Theorem 2 applies to *every* $m \times n$ matrix game M, it applies to strictly determined games as well. For the latter, v is a saddle value, and P^* and Q^* are pure strategies.

mal strategies and the corresponding value of a matrix game is called **solving the game.** The triplet (v, P^*, Q^*) is called a **solution of the game.**

An immediate consequence of the fundamental theorem is the fact that the expected value of the game for R, when both R and C use optimal strategies, is v. This is why we call v the value of the game. We now state this as a theorem (the proof is left to Problem 20, Exercise 8-2).

Theorem 3 EXPECTED VALUE FOR OPTIMAL STRATEGIES

$$E(P^*, Q^*) = P^*MQ^* = v$$

A SOLUTION TO A 2 × 2 MATRIX GAME

Now that we know what a solution to a matrix game is, how do we find it? For a 2 × 2 matrix game that is not strictly determined (we already know how to solve a strictly determined matrix game), there exists a set of formulas for the solution in terms of the payoff values in the matrix for the game. Although such simple formulas do not exist for the solution of $m \times n$ matrix games in general, in Sections 8-3 and 8-4 we will show that the problem of solving a matrix game can be converted to an equivalent linear programming problem. We will then use the tools developed in Chapter 5 to obtain a solution to the matrix game.

The following solution of a 2 × 2 nonstrictly determined matrix game is established by showing that inequalities (2) and (3) in Theorem 2 hold (see Problem 21, Exercise 8-2):

Theorem 4 SOLUTION TO A 2 × 2 NONSTRICTLY DETERMINED MATRIX GAME

For the nonstrictly determined game

$$M = \begin{bmatrix} a & b \\ c & d \end{bmatrix}$$

the optimal strategies P^* and Q^* and the value of the game are given by

$$P^* = [p_1^* \quad p_2^*] = \begin{bmatrix} \dfrac{d-c}{D} & \dfrac{a-b}{D} \end{bmatrix} \qquad Q^* = \begin{bmatrix} q_1^* \\ q_2^* \end{bmatrix} = \begin{bmatrix} \dfrac{d-b}{D} \\ \dfrac{a-c}{D} \end{bmatrix}$$

$$v = \frac{ad - bc}{D}$$

where $D = (a + d) - (b + c)$.

[*Note:* Under the assumption that M is not strictly determined, it can be shown that $D = (a + d) - (b + c)$ will never be 0.]

Let us now consider some examples where we can make use of these formulas.

Example 1 **Solving a 2 × 2 Nonstrictly Determined Matrix Game** Solve the two-finger Morra game introduced at the beginning of this section. (Have you decided which player you would rather be yet?)

SOLUTION

$$M = \begin{bmatrix} a & b \\ c & d \end{bmatrix} = \begin{bmatrix} 2 & -3 \\ -3 & 4 \end{bmatrix}$$

We first compute D: $D = (a + d) - (b + c) = 6 - (-6) = 12$

Then the solution of the game is:

$$P^* = \begin{bmatrix} \frac{d - c}{D} & \frac{a - b}{D} \end{bmatrix} = \begin{bmatrix} \frac{7}{12} & \frac{5}{12} \end{bmatrix}$$

$$Q^* = \begin{bmatrix} \frac{d - b}{D} \\ \frac{a - c}{D} \end{bmatrix} = \begin{bmatrix} \frac{7}{12} \\ \frac{5}{12} \end{bmatrix}$$

$$v = \frac{ad - bc}{D} = \frac{8 - 9}{12} = -\frac{1}{12}$$

What does all of this mean? It means that R's optimal strategy is

$$P^* = \begin{bmatrix} \frac{7}{12} & \frac{5}{12} \end{bmatrix}$$

That is, using a random process (such as colored marbles in an urn or a spinner), R should choose Row 1 with a probability of $\frac{7}{12}$ and Row 2 with a probability of $\frac{5}{12}$. The optimal strategy for C is

$$Q^* = \begin{bmatrix} \frac{7}{12} \\ \frac{5}{12} \end{bmatrix}$$

Thus, using a random process, C should choose Column 1 with a probability of $\frac{7}{12}$ and Column 2 with a probability of $\frac{5}{12}$. If optimal strategies are used by both R and C, then the expected value of the game for R is $-\frac{1}{12}$. In the long run, R's average loss will be $v = -\frac{1}{12}$ of a dollar per game, while C's average gain per game will be $\frac{1}{12}$ of a dollar. The game is not fair, since $v = -\frac{1}{12} \neq 0$. It would be better to be player C. You probably did not guess this, since the game matrix M appears to be fair.

Matched Problem 1 ➠ Solve the following version of the two-finger Morra game (this is equivalent to the penny-matching game discussed in Section 8-1):

$$M = \begin{bmatrix} 1 & -1 \\ -1 & 1 \end{bmatrix}$$

Explore–Discuss 1

Let: $M = \begin{bmatrix} a & b \\ c & d \end{bmatrix}$

(A) Show that if the row minima belong to the same column, then at least one of them is a saddle point.

(B) Show that if the column maxima belong to the same row, then at least one of them is a saddle point.

(C) Show that if $(a + d) - (b + c) = 0$, then M has at least one saddle point (that is, M is strictly determined).

(D) Explain why part (C) implies that the denominator D in Theorem 4 will never be 0.

Recessive Rows and Columns

Some strictly determined and nonstrictly determined higher-dimensional matrix games may be reduced to lower-dimensional games because intelligent players would never play certain rows or columns. Consider the following matrix game:

$$M = \begin{bmatrix} -3 & 0 & 1 \\ 3 & -1 & 2 \\ -2 & 1 & 1 \end{bmatrix}$$

Let us first note that this is not a strictly determined game. Now, from R's point of view, there is a row that should never be played. Can you determine which one? If we compare Row 1 with Row 3, we find that for each column choice by C, the payoffs in Row 3 are greater than or equal to those in Row 1. Thus, there would never be an advantage for R to play Row 1, so we delete it to obtain

$$M_1 = \begin{bmatrix} 3 & -1 & 2 \\ -2 & 1 & 1 \end{bmatrix}$$

Player R should choose a mixed strategy for the remaining rows.

We now turn to C. There is a column in M_1 that C should never play. Can you find it? Comparing Columns 2 and 3, we see that for each row choice by R, the payoffs in Column 2 are less than or equal to those in Column 3 for C. Thus, there would never be an advantage for C to play Column 3, so we delete it to obtain

$$M_2 = \begin{bmatrix} 3 & -1 \\ -2 & 1 \end{bmatrix}$$

The game is now 2×2 and can be solved using the formulas in Theorem 4.

We summarize our discussion as follows:

Recessive Rows and Columns

Recessive rows: A row in a game matrix is said to be **recessive** and may be deleted if there exists a **(dominant) row** with corresponding elements *greater than or equal to* those in the given row.

Recessive columns: A column in a game matrix is said to be **recessive** and may be deleted if there exists a **(dominant) column** with corresponding elements *less than or equal to* those in the given column.

In short, a row (column) is recessive if it is never better than another row (column). Optimal behavior will never require the use of a recessive row or column; hence, such a row or column may be discarded.

Example 2 **Deleting Recessive Rows and Columns** Solve the matrix game:

$$M = \begin{bmatrix} -1 & -3 & 1 & 0 \\ 0 & -2 & 4 & 0 \\ 3 & 1 & -3 & 2 \end{bmatrix}$$

SOLUTION **Step 1.** Is M strictly determined? No. (If the answer were yes, we would find a saddle value and corresponding pure strategies for R and C, and then we would be finished.)

Step 2. Delete recessive rows and columns if present.

Row 2 dominates Row 1 — Column 2 dominates Columns 1 and 4

$$\begin{bmatrix} \cancel{1} & \cancel{3} & \cancel{1} & \cancel{0} \\ 0 & -2 & 4 & 0 \\ 3 & 1 & -3 & 2 \end{bmatrix} \rightarrow \begin{bmatrix} \cancel{0} & -2 & 4 & \cancel{0} \\ \cancel{3} & 1 & -3 & \cancel{2} \end{bmatrix}$$

$$\rightarrow \begin{bmatrix} -2 & 4 \\ 1 & -3 \end{bmatrix}$$

Step 3. Solve the remaining matrix game:

$$M_1 = \begin{bmatrix} a & b \\ c & d \end{bmatrix} = \begin{bmatrix} -2 & 4 \\ 1 & -3 \end{bmatrix}$$

$$D = (a + d) - (b + c) = (-5) - (5) = -10$$

$$P^* = \begin{bmatrix} \dfrac{d-c}{D} & \dfrac{a-b}{D} \end{bmatrix} = \begin{bmatrix} \dfrac{-4}{-10} & \dfrac{-6}{-10} \end{bmatrix} = \begin{bmatrix} \dfrac{2}{5} & \dfrac{3}{5} \end{bmatrix}$$

$$Q^* = \begin{bmatrix} \dfrac{d-b}{D} \\ \dfrac{a-c}{D} \end{bmatrix} = \begin{bmatrix} \dfrac{-7}{-10} \\ \dfrac{-3}{-10} \end{bmatrix} = \begin{bmatrix} \dfrac{7}{10} \\ \dfrac{3}{10} \end{bmatrix}$$

$$v = \frac{ad - bc}{D} = \frac{6 - 4}{-10} = \frac{2}{-10} = -\frac{1}{5}$$

The game is not fair and favors C. Note that in terms of the original game M, we would say that R's optimal strategy is

$$P^* = \begin{bmatrix} 0 & \frac{2}{5} & \frac{3}{5} \end{bmatrix}$$

which means that player R, using a random process, should choose Row 1 with a probability of 0, Row 2 with a probability of $\frac{2}{5}$, and Row 3 with a probability of $\frac{3}{5}$. The optimal strategy for C is

$$Q^* = \begin{bmatrix} 0 \\ \frac{7}{10} \\ \frac{3}{10} \\ 0 \end{bmatrix}$$

Player C, using a random process, should choose Column 1 with a probability of 0, Column 2 with a probability of $\frac{7}{10}$, Column 3 with a probability of $\frac{3}{10}$, and Column 4 with a probability of 0.

If both players follow their optimal strategies, then the expected value of the game for R is $-\frac{1}{5}$ and for C is $\frac{1}{5}$.

Matched Problem 2 Solve the matrix game: $M = \begin{bmatrix} -3 & -1 & 2 \\ 3 & 2 & -4 \\ -1 & -1 & 3 \end{bmatrix}$

Even though dominance is applicable to strictly determined games, its primary use is in the reduction of mixed strategy games. Mixed strategy games are much easier to solve if they can be reduced first.

Explore–Discuss 2

(A) Using Theorem 4, give conditions on a, b, c, and d that guarantee that the nonstrictly determined matrix game

$$M = \begin{bmatrix} a & b \\ c & d \end{bmatrix}$$

is fair.

(B) Construct the matrix of payoffs for a two-finger Morra game that is nonstrictly determined and fair.

(C) How many such matrices are there? Explain.

Answers to Matched Problems

1. $P^* = \begin{bmatrix} \frac{1}{2} & \frac{1}{2} \end{bmatrix}$, $Q^* = \begin{bmatrix} \frac{1}{2} \\ \frac{1}{2} \end{bmatrix}$, $v = 0$

2. Eliminate recessive rows and columns to obtain: $M_1 = \begin{bmatrix} 2 & -4 \\ -1 & 3 \end{bmatrix}$

Then: $P^* = \begin{bmatrix} \frac{2}{5} & \frac{3}{5} \end{bmatrix}$, $Q^* = \begin{bmatrix} \frac{7}{10} \\ \frac{3}{10} \end{bmatrix}$

For the original game M: $P^* = \begin{bmatrix} 0 & \frac{2}{5} & \frac{3}{5} \end{bmatrix}$, $Q^* = \begin{bmatrix} 0 \\ \frac{7}{10} \\ \frac{3}{10} \end{bmatrix}$, $v = \frac{1}{5}$

Exercise 8-2

Solve the matrix games in Problems 1–14, indicating optimal strategies P^* and Q^* for R and C, respectively, and the value v of the game. (Both strictly and nonstrictly determined games are included, so check for this first.)

A

1. $\begin{bmatrix} -1 & 2 \\ 2 & -4 \end{bmatrix}$ **2.** $\begin{bmatrix} 2 & -3 \\ -1 & 2 \end{bmatrix}$ **3.** $\begin{bmatrix} -1 & 2 \\ 1 & 0 \end{bmatrix}$

4. $\begin{bmatrix} 2 & -1 \\ -2 & 1 \end{bmatrix}$ **5.** $\begin{bmatrix} 4 & -6 \\ -2 & 3 \end{bmatrix}$ **6.** $\begin{bmatrix} -1 & 3 \\ 2 & -6 \end{bmatrix}$

B

7. $\begin{bmatrix} 5 & -1 \\ 4 & 1 \end{bmatrix}$ **8.** $\begin{bmatrix} 3 & 0 \\ 1 & -4 \end{bmatrix}$

9. $\begin{bmatrix} 0 & 2 & -1 & 2 \\ 1 & -2 & 1 & -1 \end{bmatrix}$ **10.** $\begin{bmatrix} 3 & -6 \\ 4 & -6 \\ -2 & 3 \\ -3 & 0 \end{bmatrix}$

11. $\begin{bmatrix} 1 & -4 & -1 \\ 2 & -3 & 0 \\ -1 & 2 & 2 \end{bmatrix}$ **12.** $\begin{bmatrix} -1 & 2 & 2 \\ 2 & -4 & -2 \\ 2 & -5 & 0 \end{bmatrix}$

13. $\begin{bmatrix} 2 & 1 & 2 \\ 3 & 0 & -5 \\ 1 & -2 & 7 \end{bmatrix}$ **14.** $\begin{bmatrix} -1 & 1 & -5 \\ 2 & 2 & 3 \\ 0 & -2 & 4 \end{bmatrix}$

In Problems 15 and 16, discuss the validity of each statement. If the statement is always true, explain why. If not, give a counterexample.

15. Consider the matrix game: $M = \begin{bmatrix} a & b \\ c & d \end{bmatrix}$

(A) If $a = b$, then M is strictly determined.
(B) If $a = c$, then M is strictly determined.
(C) If $a = d$, then M is strictly determined.

16. (A) There exists a 2×2 nonstrictly determined matrix game M for which the optimal strategy for player R is mixed and the optimal strategy for player C is pure.

(B) There exists a 2×2 nonstrictly determined matrix game M for which the optimal strategy for player R is to choose either row with probability $\frac{1}{2}$ and the optimal strategy for player C is to choose either column with probability $\frac{1}{2}$.

C

17. You (R) and a friend (C) are playing the matrix game shown below, where the entries indicate your winnings from C in dollars. In order to encourage your friend to play, since you cannot lose as the matrix is written, you pay her \$3 before each game.

$$\begin{array}{c} \\ R \end{array} \overset{C}{\begin{bmatrix} 0 & 2 & 1 & 0 \\ 4 & 3 & 5 & 4 \\ 0 & 2 & 6 & 1 \\ 0 & 1 & 0 & 3 \end{bmatrix}}$$

(A) If selection of a particular row or a particular column is made by use of the spinner shown, what is your expected value? (Each number on the spinner is equally likely.)

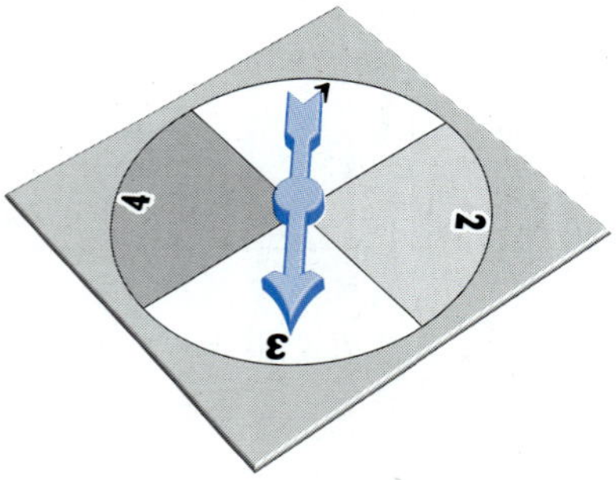

(B) If you make whatever row choice you wish and your opponent continues to use the spinner for her column choice, what is your expected value, assuming you optimize your choice?

(C) If you both disregard the spinner and make your own choices, what is your expected value, assuming you both optimize your choices?

18. You (R) and a friend (C) are playing the matrix game shown at the top of the next column, where the entries indicate your winnings from C in dollars. In order to encourage your friend to play, you pay her \$4 before each game. The jack, queen, king, and ace from the hearts, spades, and diamonds are taken from a standard deck of cards. The game is based on a random draw of a single card from these 12 cards. The play is indicated at the top and side of the matrix.

$$R\;\begin{array}{c} J \\ Q \\ K \\ A \end{array} \overset{\displaystyle C}{\overset{\begin{array}{ccc} H & S & D \end{array}}{\begin{bmatrix} 1 & 0 & 6 \\ 2 & 2 & 1 \\ 4 & 4 & 7 \\ 1 & 2 & 6 \end{bmatrix}}}$$

(A) If you select a row by drawing a single card and your friend selects a column by drawing a single card (after replacement), what is your expected value of the game?

(B) If your opponent chooses an optimum strategy (ignoring the cards) and you make your row choice by drawing a card, what is your expected value?

(C) If you both disregard the cards and make your own choices, what is your expected value, assuming you both choose optimum strategies?

19. For

$$M = \begin{bmatrix} a & b \\ c & d \end{bmatrix}$$

$$P = \begin{bmatrix} p_1 & p_2 \end{bmatrix} \qquad Q = \begin{bmatrix} q_1 \\ q_2 \end{bmatrix}$$

show that $PMQ = E(P, Q)$.

20. Prove, using the fundamental theorem of game theory, that

$$P^*MQ^* = v$$

which is the same as $E(P^*, Q^*) = v$.

21. Show that the solution formulas (Theorem 4) for a 2×2 nonstrictly determined matrix game meet the conditions for a solution stated in the fundamental theorem (Theorem 2).

22. Show that if a 2×2 matrix game has a saddle value, then either one row is recessive or one column is recessive.

23. Explain how to construct a 2×2 matrix game M for which the optimal strategies are

$$P^* = \begin{bmatrix} .9 & .1 \end{bmatrix} \quad \text{and} \quad Q^* = \begin{bmatrix} .3 \\ .7 \end{bmatrix}$$

24. Explain how to construct a 2×2 matrix game M for which the optimal strategies are

$$P^* = \begin{bmatrix} .6 & .4 \end{bmatrix} \quad \text{and} \quad Q^* = \begin{bmatrix} .8 \\ .2 \end{bmatrix}$$

Applications

Business & Economics

25. *Promotion.* A town has only two banks, bank R and bank C, and both compete about equally for the town's business. Each week each bank decides on the use of one, and only one, of the following means of promotion: TV, radio, newspaper, and mail. A marketing research firm provided the following payoff matrix, which indicates the percentage of market gain or loss for each choice of action by R and by C (we assume any gain by R is a loss by C, and vice versa):

		C			
		TV	Radio	Paper	Mail
R	TV	0	−1	−1	0
	Radio	1	2	−1	−1
	Paper	0	−1	0	1
	Mail	−1	−1	−1	0

(A) Find optimum strategies for bank R and bank C. What is the value of the game?

(B) What is the expected value of the game for R if bank R always chooses TV and bank C uses its optimum strategy?

(C) What is the expected value of the game for R if bank C always chooses radio and bank R uses its optimum strategy?

(D) What is the expected value of the game for R if both banks always use the paper?

26. *Viewer ratings.* A city has two closely competitive television stations, station R and station C. Each month each station makes exactly one choice for the Thursday 8–9 PM time slot from the program categories shown above and to the left of the matrix given below. Each entry in the matrix is an average viewer index rating gain (or loss) established by a rating firm using data collected over the past 5 years. (Any gain for station R is a loss for station C, and vice versa.)

		C			
		Nature films	Talk shows	Sports events	Movies
R	Travel	0	2	−1	0
	News	1	0	−1	−2
	Sit coms	2	3	−1	1
	Soaps	1	−2	0	0

(A) Find the optimum strategies for station R and station C. What is the value of the game?

(B) What is the expected value of the game for R if station R always chooses travel and station C uses its optimum strategy?

(C) What is the expected value of the game for R if station C always chooses movies and station R uses its optimum strategy?

(D) What is the expected value of the game for R if station R always chooses sit coms and station C always chooses sports events?

27. *Investment.* You have inherited $10,000 just prior to a presidential election and wish to invest it in solar energy and oil stocks. An investment advisor provides you with a payoff matrix that indicates your probable 4 year gains, depending on which party comes into office. How should you invest your money so that you would have the largest expected gain irrespective of how the election turns out?

		Player C (fate)	
		Republican	Democrat
Player R (you)	Solar energy	$1,000	$4,000
	Oil	$5,000	$3,000

[*Note:* For a one-time play (investment), you would split your investment proportional to the entries in your optimal strategy matrix. Assume fate is a very clever player; then if fate deviates from its optimal strategy, you know you will not do any worse than the value of the game, and may do better.]

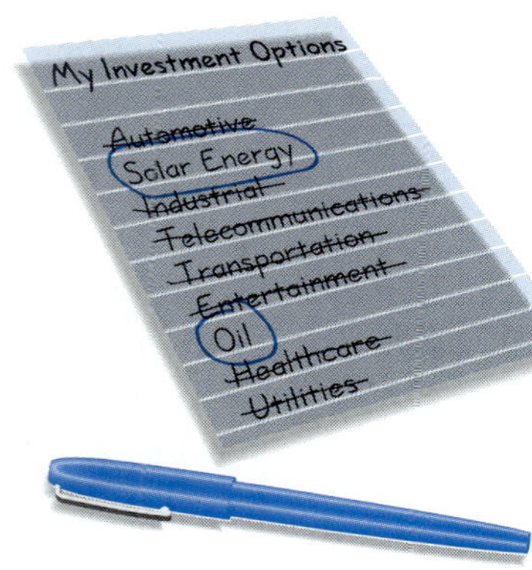

28. *Corporate farming.* A large corporate farm grows both wheat and rice. Rice does better in wetter years and wheat does better in normal or drier years. Based on records over the past 20 years and current market prices, a consulting firm provides the corporate farm management with the following payoff matrix, where the entries are in millions of dollars:

		Weather (fate)		
		Wet	Normal	Dry
Corporate farm	Wheat	−2	8	2
	Rice	7	3	−3

The management would like to determine their best strategy against the weather's "best strategy" to destroy them. Then, no matter what the weather does, the company will do no worse than the value of the game, and may do a lot better. This information could be very useful to the company when applying for loans. [*Note:* For each year the payoff matrix holds, the company can split the planting between wheat and rice proportional to the size of the entries in its optimal strategy matrix.]

(A) Find the optimal strategies for the company and the weather and the value of the game.

(B) What is the expected value of the game for the corporation if the weather (fate) chooses to play the pure strategy "wet" for many years and the company continues to play its optimal strategy?

(C) Answer part (B), replacing "wet" with "normal."

(D) Answer part (B), replacing "wet" with "dry."

SECTION 8-3 Linear Programming and 2 × 2 Games—Geometric Approach

In Section 8-2 we said that formula solutions for the more general $m \times n$ matrix game do not exist. There is, however, a systematic way to solve a nonstrictly determined matrix game without recessive rows or columns. All such games can be converted into linear programming problems, which can be solved using the techniques described in Chapter 5.

We will first describe the linear programming process geometrically using a 2×2 matrix game, where ideas and procedures do not get too involved. (A brief review of Section 5-2 may prove helpful.) In the next section, using the simplex method and the dual from Sections 5-4 and 5-5, we will develop a general procedure for arbitrary $m \times n$ matrix games.

Consider a nonstrictly determined matrix game with *all positive payoffs:*

$$\text{Player } R \overset{\text{Player } C}{\begin{bmatrix} a & b \\ c & d \end{bmatrix}} = M \qquad a, b, c, d > 0 \tag{1}$$

The *all positive payoff* requirement is necessary to guarantee that v (the value of the game) is positive, a fact we will need shortly. The requirement of all positive payoffs may seem too restrictive; however, it causes little trouble because of the following theorem, which we state without proof (see Problems 11 and 12 in Exercise 8-3):

Theorem 1 ■■ INVARIANT OPTIMAL STRATEGIES

Optimal strategies of a matrix game do not change if a constant value k is added to each payoff. If v is the value of the original game, then $v + k$ is the value of the new game. ■■

For example, if we start with the matrix game

$$M = \begin{bmatrix} 2 & -3 \\ -1 & 0 \end{bmatrix}$$

all we have to do is add some number, say 4, to each payoff to obtain the positive matrix

$$M_1 = \begin{bmatrix} 6 & 1 \\ 3 & 4 \end{bmatrix}$$

We find P^*, Q^*, and v_1 for this new game; then P^*, Q^*, and $v = v_1 - 4$ will be the solution of the original game.

Explore–Discuss 1

(A) Suppose a 2×2 matrix game M has all positive payoffs. Explain why the value of the game is positive.
(B) Suppose the value of a 2×2 matrix game M is positive. Does M have all positive payoffs? Explain.
(C) Repeat parts (A) and (B) for an $m \times n$ matrix game.

Recall from Section 8-2 that we introduced the idea of expected value when looking for optimal strategies for R and C. And we found that for a matrix game M, the expected value of the game for R is given by

$$E(P, Q) = PMQ \tag{2}$$

where

$$P = \begin{bmatrix} p_1 & p_2 \end{bmatrix} \text{ is } R\text{'s strategy} \qquad \text{and} \qquad Q = \begin{bmatrix} q_1 \\ q_2 \end{bmatrix} \text{ is } C\text{'s strategy}$$

Naturally, R is interested in finding a strategy P^*, along with the largest possible number v, such that

$$E(P^*, Q) = P^*MQ \geqslant v \tag{3}$$

for any choice of Q by C.

Since (3) is to hold for all Q, it must hold in particular for

$$Q_1 = \begin{bmatrix} 1 \\ 0 \end{bmatrix} \qquad \text{and} \qquad Q_2 = \begin{bmatrix} 0 \\ 1 \end{bmatrix}$$

Using the positive game matrix (1), and successively substituting Q_1 and Q_2 into (3), we obtain the two inequalities

$$\begin{bmatrix} p_1 & p_2 \end{bmatrix} \begin{bmatrix} a & b \\ c & d \end{bmatrix} \begin{bmatrix} 1 \\ 0 \end{bmatrix} \geqslant v \qquad \begin{bmatrix} p_1 & p_2 \end{bmatrix} \begin{bmatrix} a & b \\ c & d \end{bmatrix} \begin{bmatrix} 0 \\ 1 \end{bmatrix} \geqslant v$$

Multiplying (a good exercise for you), we have

$$ap_1 + cp_2 \geqslant v \qquad\qquad bp_1 + dp_2 \geqslant v \tag{4}$$

These inequalities suggest a linear programming problem, but each involves three variables, p_1, p_2, and v, instead of two. We remedy this by dividing through by v. (The sense of the inequalities will not change, since we know v is positive by the assumption that each payoff in M is positive.) The inequalities (4) now become

$$a\frac{p_1}{v} + c\frac{p_2}{v} \geq 1 \qquad b\frac{p_1}{v} + d\frac{p_2}{v} \geq 1 \tag{5}$$

To simplify notation, we introduce two new variables,

$$x_1 = \frac{p_1}{v} \quad \text{and} \quad x_2 = \frac{p_2}{v} \quad \text{where } x_1 \geq 0, x_2 \geq 0 \tag{6}$$

The inequalities (5) can now be written in the simpler form

$$\begin{aligned} ax_1 + cx_2 &\geq 1 \\ bx_1 + dx_2 &\geq 1 \end{aligned} \qquad x_1, x_2 \geq 0 \tag{7}$$

Our problem now is to maximize v subject to the constraints in (7). But how do we express v as a linear function in terms of x_1 and x_2? Adding the two equations in (6) produces the following useful result:

$$\begin{aligned} x_1 + x_2 &= \frac{p_1}{v} + \frac{p_2}{v} \\ &= \frac{p_1 + p_2}{v} \\ &= \frac{1}{v} \qquad \text{Since } p_1 + p_2 = 1 \text{ (Why?)} \end{aligned}$$

Thus,

$$x_1 + x_2 = \frac{1}{v} \quad \text{or} \quad v = \frac{1}{x_1 + x_2} \tag{8}$$

So, to maximize v, we minimize $1/v$ instead (as v gets larger, $1/v$ gets smaller). We now have the following linear programming problem:

$$\begin{aligned} \text{Minimize} \quad & y = \frac{1}{v} = x_1 + x_2 \\ \text{Subject to} \quad & ax_1 + cx_2 \geq 1 \\ & bx_1 + dx_2 \geq 1 \\ & x_1, x_2 \geq 0 \end{aligned} \tag{9}$$

Using the methods discussed in Section 5-2, we solve (9) geometrically and then use (8) and (6) to find v and P^*.

Now we turn to player C, who is naturally interested in finding a strategy Q^* and the smallest possible value v such that

$$E(P, Q^*) = PMQ^* \leq v \tag{10}$$

for any choice of P by R.

Since (10) is to hold for all P, it must hold in particular for

$$P_1 = \begin{bmatrix} 1 & 0 \end{bmatrix} \quad \text{and} \quad P_2 = \begin{bmatrix} 0 & 1 \end{bmatrix}$$

Substituting P_1 and P_2 into (10) and multiplying (another good exercise for the reader), we obtain

$$\begin{aligned} aq_1 + bq_2 &\leq v \\ cq_1 + dq_2 &\leq v \end{aligned} \tag{11}$$

Dividing through by the positive number v and letting

$$z_1 = \frac{q_1}{v} \quad \text{and} \quad z_2 = \frac{q_2}{v} \quad \text{where } z_1 \geq 0, z_2 \geq 0 \tag{12}$$

we obtain

$$\begin{aligned} az_1 + bz_2 &\leq 1 \\ cz_1 + dz_2 &\leq 1 \end{aligned} \qquad z_1, z_2 \geq 0 \tag{13}$$

Thus, we are interested in minimizing v subject to the constraints (13). As before, we note that

$$\begin{aligned} z_1 + z_2 &= \frac{q_1}{v} + \frac{q_2}{v} \\ &= \frac{q_1 + q_2}{v} \\ &= \frac{1}{v} \end{aligned}$$

Thus,

$$z_1 + z_2 = \frac{1}{v} \quad \text{or} \quad v = \frac{1}{z_1 + z_2} \tag{14}$$

Minimizing v is the same as maximizing $1/v$ (decreasing v increases $1/v$). Thus, we have a second linear programming problem:

$$\begin{aligned} \text{Maximize} \quad & y = \frac{1}{v} = z_1 + z_2 \\ \text{Subject to} \quad & az_1 + bz_2 \leq 1 \\ & cz_1 + dz_2 \leq 1 \\ & z_1, z_2 \geq 0 \end{aligned} \tag{15}$$

After (15) is solved geometrically, we find v and Q^* using (14) and (12).

From Section 5-5, you should recognize (15) as the dual of (9). And by Theorem 1 in that section, (9) and (15) must have the same optimal value. Thus, the v found using (9) and the v found using (15) must be the same.

The above process of converting a 2×2 matrix game into a linear programming problem generalizes to $m \times n$ matrix games, which we will consider in the next section. We summarize the above results for a 2×2 matrix game in the next box for convenient reference.

2 × 2 Matrix Games and Linear Programming—Geometric Approach

Given the nonstrictly determined matrix game

$$M = \begin{bmatrix} a & b \\ c & d \end{bmatrix}$$

to find $P^* = [p_1 \quad p_2]$, $Q^* = \begin{bmatrix} q_1 \\ q_2 \end{bmatrix}$, and v, proceed as follows:

Step 1. If M is not a positive matrix (one with all entries positive), convert it into a positive matrix M_1 by adding a suitable positive constant k to each element. Let M_1, the new positive matrix, be represented as follows:

$$M_1 = \begin{bmatrix} e & f \\ g & h \end{bmatrix} \quad \begin{matrix} e = a + k & f = b + k \\ g = c + k & h = d + k \end{matrix}$$

This new matrix game, M_1, has the same optimal strategies P^* and Q^* as M. However, if v_1 is the value of the game M_1, then

$$v = v_1 - k$$

is the value of the original game M.

Step 2. Set up the two corresponding linear programming problems:

(A) Minimize $y = x_1 + x_2$
Subject to
$$ex_1 + gx_2 \geq 1$$
$$fx_1 + hx_2 \geq 1$$
$$x_1, x_2 \geq 0$$

(B) Maximize $y = z_1 + z_2$
Subject to
$$ez_1 + fz_2 \leq 1$$
$$gz_1 + hz_2 \leq 1$$
$$z_1, z_2 \geq 0$$

Step 3. Solve each linear programming problem geometrically.

Step 4. Use the solutions in step 3 to find the value v_1 for game M_1 and the optimal strategies and value v for the original game M:

$$v_1 = \frac{1}{y} = \frac{1}{x_1 + x_2} \quad \text{or} \quad v_1 = \frac{1}{y} = \frac{1}{z_1 + z_2}$$

$$P^* = [p_1 \quad p_2] = [v_1x_1 \quad v_1x_2] \qquad Q^* = \begin{bmatrix} q_1 \\ q_2 \end{bmatrix} = \begin{bmatrix} v_1z_1 \\ v_1z_2 \end{bmatrix}$$

$$v = v_1 - k$$

Step 5. A further check of the solution is provided by showing that

$$P^*MQ^* = v$$ See Theorem 3 of Section 8-2.

REMARK

In step 2 in the box, note that part (B) is the dual of part (A), and consequently, both must have the same optimal value (Theorem 1 in Section 5-5). In the next section, using the simplex method and properties of the dual, we will see that solving part (B) will automatically produce the solution for part (A). In this section we are restricting our attention to the geometric approach; hence, we must solve each part as a separate problem.

Example 1 ➡ **Solving 2 × 2 Matrix Games Using Geometric Methods** Solve the following matrix game using geometric methods to solve the corresponding linear programming problems (see Section 5-2):

$$M = \begin{bmatrix} -2 & 4 \\ 1 & -3 \end{bmatrix}$$

SOLUTION **Step 1.** Convert M into a positive matrix (one with all entries positive) by adding 4 to each payoff. We denote the modified matrix by M_1:

$$M_1 = \begin{bmatrix} 2 & 8 \\ 5 & 1 \end{bmatrix} = \begin{bmatrix} e & f \\ g & h \end{bmatrix} \qquad k = 4$$

Step 2. Set up the two corresponding linear programming problems:

(A) Minimize $y = x_1 + x_2$
Subject to $2x_1 + 5x_2 \geq 1$
$8x_1 + x_2 \geq 1$
$x_1, x_2 \geq 0$

(B) Maximize $y = z_1 + z_2$
Subject to $2z_1 + 8z_2 \leq 1$
$5z_1 + z_2 \leq 1$
$z_1, z_2 \geq 0$

Step 3. Solve each linear programming problem geometrically:

(A)

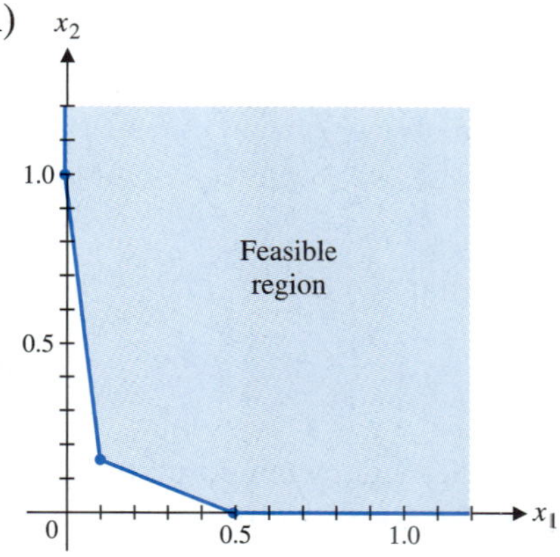

(B)

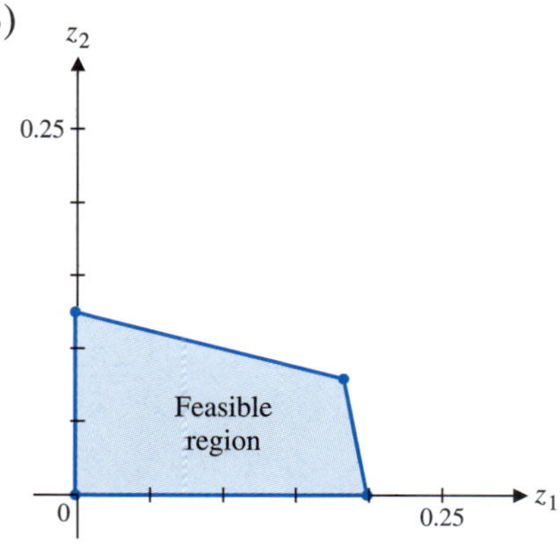

Theorems 1 and 2 in Section 5-2 imply that each problem has a solution that must occur at a corner point.

(A)

CORNER POINTS	MINIMIZE $y = x_1 + x_2$
$(0, 1)$	1
$\left(\frac{2}{19}, \frac{3}{19}\right)$	$\frac{5}{19}$
$\left(\frac{1}{2}, 0\right)$	$\frac{1}{2}$

Min y occurs at

$$x_1 = \tfrac{2}{19} \quad \text{and} \quad x_2 = \tfrac{3}{19}$$

(B)

CORNER POINTS	MAXIMIZE $y = z_1 + z_2$
$(0, 0)$	0
$\left(0, \frac{1}{8}\right)$	$\frac{1}{8}$
$\left(\frac{7}{38}, \frac{3}{38}\right)$	$\frac{5}{19}$
$\left(\frac{1}{5}, 0\right)$	$\frac{1}{5}$

Max y occurs at

$$z_1 = \tfrac{7}{38} \quad \text{and} \quad z_2 = \tfrac{3}{38}$$

Step 4. Use the solutions in step 3 to find the value v_1 for the game M_1 and the optimal strategies and value v for the original game M:

(A) $v_1 = \dfrac{1}{x_1 + x_2} = \dfrac{1}{\frac{2}{19} + \frac{3}{19}} = \dfrac{19}{5}$

$p_1 = v_1 x_1 = \frac{19}{5} \cdot \frac{2}{19} = \frac{2}{5}$

$p_2 = v_1 x_2 = \frac{19}{5} \cdot \frac{3}{19} = \frac{3}{5}$

(B) $v_1 = \dfrac{1}{z_1 + z_2} = \dfrac{1}{\frac{7}{38} + \frac{3}{38}} = \dfrac{19}{5}$

$q_1 = v_1 z_1 = \frac{19}{5} \cdot \frac{7}{38} = \frac{7}{10}$

$q_2 = v_1 z_2 = \frac{19}{5} \cdot \frac{3}{38} = \frac{3}{10}$

[*Note:* v_1 found in part (A) should always be the same as v_1 found in part (B).]

Optimal strategies are the same for both games M and M_1. Thus,

$$P^* = [p_1 \quad p_2] = [\tfrac{2}{5} \quad \tfrac{3}{5}] \qquad Q^* = \begin{bmatrix} q_1 \\ q_2 \end{bmatrix} = \begin{bmatrix} \frac{7}{10} \\ \frac{3}{10} \end{bmatrix}$$

and the value of the original game is

$$v = v_1 - k = \tfrac{19}{5} - 4 = -\tfrac{1}{5}$$

Step 5. A further check of the solution is provided by showing that

$$P^*MQ^* = v$$ See Theorem 3 in Section 8-2.

This check is left to the reader.

Matched Problem 1 Solve the following matrix game using geometric linear programming methods:

$$M = \begin{bmatrix} 2 & -4 \\ -1 & 3 \end{bmatrix}$$

Explore–Discuss 2

Show that

$$M = \begin{bmatrix} 1 & -1 \\ -3 & -2 \end{bmatrix}$$

is a strictly determined matrix game. Nevertheless, apply the geometric procedure given for nonstrictly determined matrix games to M. Does something go wrong? Do you obtain the correct optimal strategies? Explain.

Answers to Matched Problem

1. $P^* = [\frac{2}{5} \quad \frac{3}{5}]$, $Q^* = \begin{bmatrix} \frac{7}{10} \\ \frac{3}{10} \end{bmatrix}$, $v = \frac{1}{5}$

Exercise 8-3

A *In Problems 1–6, solve the matrix games using a geometric linear programming approach.*

1. $\begin{bmatrix} 2 & -3 \\ -1 & 2 \end{bmatrix}$

2. $\begin{bmatrix} 2 & -1 \\ -2 & 1 \end{bmatrix}$

3. $\begin{bmatrix} -1 & 3 \\ 2 & -6 \end{bmatrix}$

4. $\begin{bmatrix} 4 & -6 \\ -2 & 3 \end{bmatrix}$

5. $\begin{bmatrix} -2 & -1 \\ 5 & 6 \end{bmatrix}$

6. $\begin{bmatrix} 6 & 2 \\ -1 & 1 \end{bmatrix}$

B

7. Is there a better way to solve the matrix game in Problem 5 than the geometric linear programming approach? Explain.

8. Is there a better way to solve the matrix game in Problem 6 than the geometric linear programming approach? Explain.

In Problems 9 and 10, remove recessive rows and columns; then solve using geometric linear programming techniques.

9. $\begin{bmatrix} 3 & -6 \\ 4 & -6 \\ -2 & 3 \\ -3 & 0 \end{bmatrix}$

10. $\begin{bmatrix} -1 & 2 & 2 \\ 2 & -4 & -2 \\ 2 & -5 & 0 \end{bmatrix}$

C

11. (A) Let P and Q be strategies for the 2×2 matrix game M. Let k be a constant, and let J be the 2×2 matrix with all 1's as entries. Show that the matrix product $P(kJ)Q$ equals the 1×1 matrix k.

(B) Generalize part (A) to the situation where M and J are $m \times n$ matrices.

12. Use properties of matrix addition and multiplication to deduce from Problem 11 that if P^* and Q^* are optimal strategies for the game M with value v, then they are also optimal strategies for the game $M + kJ$ with value $v + k$.

APPLICATIONS

Business & Economics

Solve the matrix games in Problems 13–16 by using geometric linear programming methods.

13. *Promotion.* Problem 25A, Exercise 8-2

14. *Viewer ratings.* Problem 26A, Exercise 8-2

15. *Investment.* Problem 27, Exercise 8-2

16. *Corporate farming.* Problem 28A, Exercise 8-2

SECTION 8-4 Linear Programming and $m \times n$ Games—Simplex Method and the Dual

In this section we will generalize the process of solving 2×2 matrix games to $m \times n$ matrix games. The simplex method, using the dual, will play a central role in the solution process.

The methods outlined in Section 8-3 for converting a 2×2 matrix game into a pair of linear programming problems, each the dual of the other, generalize completely to $m \times n$ matrix games. We will restate the procedure for a 2×3 game, illustrate the procedure with an example, and solve the example using simplex methods and the dual. The fact that the two linear programming problems are mutually dual provides the advantage that the solution of one automatically gives the solution of the other (see Section 5-5). Based on the analysis of a 2×3 game, you should be able to convert any $m \times n$ matrix game into a pair of linear programming problems, one the dual of the other, and solve them using the methods of Section 5-5.

2 × 3 Matrix Games and Linear Programming—Simplex Method and the Dual

Given the nonstrictly determined matrix game M, free of recessive rows and columns,

$$M = \begin{bmatrix} r_1 & r_2 & r_3 \\ s_1 & s_2 & s_3 \end{bmatrix}$$

to find $P^* = [p_1 \quad p_2]$, $Q^* = \begin{bmatrix} q_1 \\ q_2 \\ q_3 \end{bmatrix}$, and v, proceed as follows:

Step 1. If M is not a positive matrix, convert it into a positive matrix M_1 by adding a suitable positive constant k to each element. Let M_1, the new positive matrix, be represented as follows:

$$M_1 = \begin{bmatrix} a_1 & a_2 & a_3 \\ b_1 & b_2 & b_3 \end{bmatrix}$$

If v_1 is the value of the game M_1, then the value of the original game M is given by $v = v_1 - k$.

Step 2. Set up the two linear programming problems (the maximization problem is always the dual of the minimization problem):

(A) Minimize $y = x_1 + x_2$
Subject to
$$\begin{aligned} a_1x_1 + b_1x_2 &\geqslant 1 \\ a_2x_1 + b_2x_2 &\geqslant 1 \\ a_3x_1 + b_3x_2 &\geqslant 1 \\ x_1, x_2 &\geqslant 0 \end{aligned}$$

(B) Maximize $y = z_1 + z_2 + z_3$
Subject to
$$\begin{aligned} a_1z_1 + a_2z_2 + a_3z_3 &\leqslant 1 \\ b_1z_1 + b_2z_2 + b_3z_3 &\leqslant 1 \\ z_1, z_2, z_3 &\geqslant 0 \end{aligned}$$

Step 3. Solve the maximization problem, part (B), the dual of part (A), using the simplex method as modified in Section 5-5. [You will automatically get the solution of the minimization problem, part (A), as well, by following this process.]

Step 4. Use the solutions in step 3 to find the value v_1 for game M_1 and the optimal strategies and value v for the original game M:

$$v_1 = \frac{1}{y} = \frac{1}{x_1 + x_2} \qquad \text{or} \qquad v_1 = \frac{1}{y} = \frac{1}{z_1 + z_2 + z_3}$$

$$P^* = [p_1 \quad p_2] = [v_1x_1 \quad v_1x_2] \qquad Q^* = \begin{bmatrix} q_1 \\ q_2 \\ q_3 \end{bmatrix} = \begin{bmatrix} v_1z_1 \\ v_1z_2 \\ v_1z_3 \end{bmatrix}$$

$$v = v_1 - k$$

Step 5. A further check of the solution is provided by showing that

$$P^*MQ^* = v$$

Example 1 **Investment Analysis** An investor wishes to invest \$10,000 in bonds and gold. He knows that the return on the investments will be affected by changes in interest rates. After some analysis, he estimates that the return (in thousands of dollars) at the end of a year will be as indicated in the following payoff matrix:

Change in interest rates (fate)

$$\begin{array}{l} \\ \text{Bonds} \\ \text{Gold} \end{array} \begin{array}{c} \begin{array}{ccc} 0\% & +1\% & -3\% \end{array} \\ \begin{bmatrix} 1 & -1 & 6 \\ -1 & 2 & -3 \end{bmatrix} \end{array}$$

(A) We assume fate is a very clever player and will play to reduce the investor's return as much as possible. Find optimal strategies for both the investor and for "fate." What is the value of the game?

(B) Find the expected values of the game if the investor continues with his optimal strategy and fate "switches" to the following pure strategies: (*1*) Play only 0% change. (*2*) Play only +1% change. (*3*) Play only −3% change.

SOLUTION (A) Let: $M = \begin{bmatrix} 1 & -1 & 6 \\ -1 & 2 & -3 \end{bmatrix}$

Step 1. Convert M into a positive matrix M_1 by adding 4 to each entry in M:

$$M_1 = \begin{bmatrix} 5 & 3 & 10 \\ 3 & 6 & 1 \end{bmatrix} \qquad k = 4$$

Step 2. Set up the two corresponding linear programming problems:

(A) Minimize $y = x_1 + x_2$
Subject to
$$\begin{aligned} 5x_1 + 3x_2 &\geq 1 \\ 3x_1 + 6x_2 &\geq 1 \\ 10x_1 + x_2 &\geq 1 \\ x_1, x_2 &\geq 0 \end{aligned}$$

(B) Maximize $y = z_1 + z_2 + z_3$
Subject to
$$\begin{aligned} 5z_1 + 3z_2 + 10z_3 &\leq 1 \\ 3z_1 + 6z_2 + z_3 &\leq 1 \\ z_1, z_2, z_3 &\geq 0 \end{aligned}$$

Step 3. Solve the maximization problem, part (B), the dual of part (A), using the simplex method as modified in Section 5-5. We introduce slack variables x_1 and x_2 to obtain

$$\begin{aligned} 5z_1 + 3z_2 + 10z_3 + x_1 \qquad\qquad &= 1 \\ 3z_1 + 6z_2 + z_3 \qquad + x_2 \qquad &= 1 \\ -z_1 - z_2 - z_3 \qquad\qquad + y &= 0 \end{aligned}$$

Write the simplex tableau and identify the first pivot element:

$$\begin{array}{c} \begin{array}{cccccc} z_1 & z_2 & z_3 & x_1 & x_2 & y \end{array} \\ \left[\begin{array}{cccccc|c} 5 & 3 & 10 & 1 & 0 & 0 & 1 \\ 3 & 6 & 1 & 0 & 1 & 0 & 1 \\ \hline -1 & -1 & -1 & 0 & 0 & 1 & 0 \end{array}\right] \end{array} \begin{array}{l} \frac{1}{5} \text{ (minimum)} \\ \frac{1}{3} \\ \end{array}$$

Since the most negative indicator in the third row appears in each of the first three columns, we can choose any of these as a pivot column.

We choose the first column, and then find that the first row is the pivot row. We now use row operations to pivot on 5:

$$
\begin{array}{c}
\\ x_1 \\ x_2 \\ y
\end{array}
\begin{array}{c}
\begin{array}{cccccc|c} z_1 & z_2 & z_3 & x_1 & x_2 & y & \end{array} \\
\left[\begin{array}{cccccc|c}
\textcircled{5} & 3 & 10 & 1 & 0 & 0 & 1 \\
3 & 6 & 1 & 0 & 1 & 0 & 1 \\
\hline
-1 & -1 & -1 & 0 & 0 & 1 & 0
\end{array}\right]
\end{array}
\quad \tfrac{1}{5}R_1 \to R_1
$$

$$
\sim \left[\begin{array}{cccccc|c}
\textcircled{1} & \frac{3}{5} & 2 & \frac{1}{5} & 0 & 0 & \frac{1}{5} \\
3 & 6 & 1 & 0 & 1 & 0 & 1 \\
\hline
-1 & -1 & -1 & 0 & 0 & 1 & 0
\end{array}\right]
\quad \begin{array}{l} (-3)R_1 + R_2 \to R_2 \\ R_1 + R_3 \to R_3 \end{array}
$$

$$
\sim \begin{array}{c} z_1 \\ x_2 \\ y \end{array}
\left[\begin{array}{cccccc|c}
1 & \frac{3}{5} & 2 & \frac{1}{5} & 0 & 0 & \frac{1}{5} \\
0 & \frac{21}{5} & -5 & -\frac{3}{5} & 1 & 0 & \frac{2}{5} \\
\hline
0 & -\frac{2}{5} & 1 & \frac{1}{5} & 0 & 1 & \frac{1}{5}
\end{array}\right]
\quad \begin{array}{l} \frac{1}{5} \div \frac{3}{5} = \frac{1}{3} \\ \frac{2}{5} \div \frac{21}{5} = \frac{2}{21} \text{ (minimum)} \end{array}
$$

Now select the next pivot element:

$$
\begin{array}{c}
\\ z_1 \\ x_2 \\ y
\end{array}
\begin{array}{c}
\begin{array}{cccccc|c} z_1 & z_2 & z_3 & x_1 & x_2 & y & \end{array} \\
\left[\begin{array}{cccccc|c}
1 & \frac{3}{5} & 2 & \frac{1}{5} & 0 & 0 & \frac{1}{5} \\
0 & \textcircled{\frac{21}{5}} & -5 & -\frac{3}{5} & 1 & 0 & \frac{2}{5} \\
\hline
0 & -\frac{2}{5} & 1 & \frac{1}{5} & 0 & 1 & \frac{1}{5}
\end{array}\right]
\end{array}
\quad \tfrac{5}{21}R_2 \to R_2
$$

$$
\sim \left[\begin{array}{cccccc|c}
1 & \frac{3}{5} & 2 & \frac{1}{5} & 0 & 0 & \frac{1}{5} \\
0 & \textcircled{1} & -\frac{25}{21} & -\frac{1}{7} & \frac{5}{21} & 0 & \frac{2}{21} \\
\hline
0 & -\frac{2}{5} & 1 & \frac{1}{5} & 0 & 1 & \frac{1}{5}
\end{array}\right]
\quad \begin{array}{l} (-\frac{3}{5})R_2 + R_1 \to R_1 \\ \\ \frac{2}{5}R_2 + R_3 \to R_3 \end{array}
$$

$$
\sim \begin{array}{c} z_1 \\ z_2 \\ y \end{array}
\left[\begin{array}{cccccc|c}
1 & 0 & \frac{19}{7} & \frac{2}{7} & -\frac{1}{7} & 0 & \frac{1}{7} \\
0 & 1 & -\frac{25}{21} & -\frac{1}{7} & \frac{5}{21} & 0 & \frac{2}{21} \\
\hline
0 & 0 & \frac{11}{21} & \frac{1}{7} & \frac{2}{21} & 1 & \frac{5}{21}
\end{array}\right]
$$

Max $y = z_1 + z_2 + z_3 = \frac{5}{21}$ occurs at $z_1 = \frac{1}{7}$, $z_2 = \frac{2}{21}$, $z_3 = 0$, $x_1 = 0$, $x_2 = 0$.

The solution to the minimization problem, part (A), can be read from the bottom row of the final simplex tableau for the dual problem above. Thus, from the row

$$
\begin{array}{ccccccc}
 & & & x_1 & x_2 & & \\
[0 & 0 & \frac{11}{21} & \frac{1}{7} & \frac{2}{21} & 1 & \frac{5}{21}]
\end{array}
$$

we conclude that the solution to the minimization problem, part (A), is Min $y = x_1 + x_2 = \frac{5}{21}$ at $x_1 = \frac{1}{7}$, $x_2 = \frac{2}{21}$.

Step 4. Use the solutions in step 3 to find the value v_1 for the game M_1 and the optimal strategies and value v for the original game M:

$$v_1 = \frac{1}{y} = \frac{1}{\frac{5}{21}} = \frac{21}{5}$$

$$P^* = [p_1 \quad p_2] = [v_1x_1 \quad v_1x_2] = [\tfrac{3}{5} \quad \tfrac{2}{5}]$$

$$Q^* = \begin{bmatrix} q_1 \\ q_2 \\ q_3 \end{bmatrix} = \begin{bmatrix} v_1z_1 \\ v_1z_2 \\ v_1z_3 \end{bmatrix} = \begin{bmatrix} \frac{3}{5} \\ \frac{2}{5} \\ 0 \end{bmatrix}$$

$$v = v_1 - k = \tfrac{21}{5} - 4 = \tfrac{1}{5}$$

Step 5. A further check is provided by showing that

$$P^*MQ^* = v$$

This we leave to the reader.

Conclusion: If the investor splits the \$10,000 proportional to the numbers in his optimal strategy, \$6,000 ($\frac{3}{5}$ of \$10,000) in bonds and \$4,000 ($\frac{2}{5}$ of \$10,000) in gold, then no matter which strategy fate chooses for interest rate changes (as long as the payoff matrix remains unchanged), the investor will be guaranteed a return of \$200 ($\frac{1}{5}$ of \$1,000). If fate plays other than the optimal column strategy, then the investor can do no worse than a \$200 return, and may do quite a bit better.

(B) Recall that the expected value of the game M using strategies P and Q is

$$E(P, Q) = PMQ$$

Thus, if the investor continues to use his optimal strategy P^* and fate "switches" to the pure strategies (*1*) 0% change, (*2*) +1% change, and (*3*) −3% change, the expected values are:

$$(1)\quad \overset{P^*}{\begin{bmatrix} \frac{3}{5} & \frac{2}{5} \end{bmatrix}} \overset{M}{\begin{bmatrix} 1 & -1 & 6 \\ -1 & 2 & -3 \end{bmatrix}} \overset{Q}{\begin{bmatrix} 1 \\ 0 \\ 0 \end{bmatrix}} = \frac{1}{5} \quad \text{(or \$200)}$$

$$(2)\quad \overset{P^*}{\begin{bmatrix} \frac{3}{5} & \frac{2}{5} \end{bmatrix}} \overset{M}{\begin{bmatrix} 1 & -1 & 6 \\ -1 & 2 & -3 \end{bmatrix}} \overset{Q}{\begin{bmatrix} 0 \\ 1 \\ 0 \end{bmatrix}} = \frac{1}{5} \quad \text{(or \$200)}$$

$$(3)\quad \overset{P^*}{\begin{bmatrix} \frac{3}{5} & \frac{2}{5} \end{bmatrix}} \overset{M}{\begin{bmatrix} 1 & -1 & 6 \\ -1 & 2 & -3 \end{bmatrix}} \overset{Q}{\begin{bmatrix} 0 \\ 0 \\ 1 \end{bmatrix}} = \frac{12}{5} \quad \text{(or \$2,400)}$$

Notice that in none of these three cases did the investor earn less than \$200, and in the third case he earned quite a bit more.

Explore–Discuss 1

(A) Use graphical techniques to solve the minimization problem in step 2 of Example 1.

(B) Could you similarly use graphical techniques to solve the maximization problem in step 2 of Example 1? Explain.

Matched Problem 1 Suppose the investor in Example 1 wishes to invest \$10,000 in long-term and short-term bonds, as well as in gold, and he is concerned about inflation. After some analysis he estimates that the return (in thousands of dollars) at the end of a year will be as indicated in the following payoff matrix:

$$\begin{array}{l} \\ \text{Gold} \\ \text{Long-term bonds} \\ \text{Short-term bonds} \end{array} \overset{\begin{array}{cc} \text{Inflation (fate)} \\ \text{Up 3\%} & \text{Down 3\%} \end{array}}{\begin{bmatrix} 3 & -3 \\ -3 & 2 \\ -1 & 1 \end{bmatrix}}$$

Again, assume fate is a very good player that will attempt to reduce the investor's return as much as possible. Find the optimal strategies for both the investor and for fate. What is the value of the game?

Explore–Discuss 2

Outline a procedure for solving the 4 × 5 matrix game

$$M = \begin{bmatrix} 2 & -2 & -1 & 6 & -1 \\ -3 & -6 & 4 & -1 & -7 \\ 5 & 3 & 6 & 0 & 4 \\ -4 & 2 & 7 & 3 & -5 \end{bmatrix}$$

without actually solving the game.

Answers to Matched Problem **1.** $P^* = [\frac{1}{4} \quad 0 \quad \frac{3}{4}], Q^* = \begin{bmatrix} \frac{1}{2} \\ \frac{1}{2} \end{bmatrix}, v = 0$

EXERCISE 8-4

In Problems 1–4, solve each matrix game.

A

1. $\begin{bmatrix} 1 & 4 & 0 \\ 0 & -1 & 2 \end{bmatrix}$

2. $\begin{bmatrix} 1 & -1 \\ 2 & -2 \\ 0 & 1 \end{bmatrix}$

B

3. $\begin{bmatrix} 0 & 1 & -2 \\ -1 & 0 & 3 \\ 2 & -3 & 0 \end{bmatrix}$

4. $\begin{bmatrix} 1 & 2 & 0 \\ 0 & 1 & 2 \\ 2 & 0 & 1 \end{bmatrix}$

In Problems 5–8, outline a procedure for solving the matrix game; then solve it.

5. $\begin{bmatrix} 4 & 2 & 2 & 1 \\ 0 & 1 & -1 & 3 \\ -2 & -1 & -3 & 2 \end{bmatrix}$

6. $\begin{bmatrix} -5 & -6 & -7 & 4 \\ -5 & 3 & 2 & 3 \\ -4 & -2 & -6 & 7 \end{bmatrix}$

7. $\begin{bmatrix} -2 & -1 & 3 & -1 \\ 1 & 2 & 4 & 0 \\ -1 & 1 & -1 & 1 \\ 0 & 1 & -1 & 2 \end{bmatrix}$

8. $\begin{bmatrix} 2 & -5 & -3 & -1 \\ 0 & 1 & 2 & -2 \\ -1 & 0 & 1 & 3 \\ 2 & -3 & -2 & 0 \end{bmatrix}$

C

9. *Scissors, paper, stone game.* This game, believed to have originated in the Far East, seems to be well-known in many parts of the world. Two players simultaneously present a hand in one of three positions: an open hand (paper), a closed fist (stone), or two open fingers (scissors). The payoff is 1 unit according to the rule "Paper covers stone, stone breaks scissors, and scissors cut paper." If both players present the same form, the payoff is 0.
(A) Set up the payoff matrix for the game.
(B) Solve the game using the simplex method discussed in this section. (Remove any recessive rows and columns, if present, before you start the simplex method.)

10. Player R has a \$2, a \$5, and a \$10 bill. Player C has a \$1, a \$5, and a \$10 bill. Each player selects and shows (simultaneously) one of their three bills. If the total value of the two bills shown is even, then R wins C's bill; if the value is odd, then C wins R's bill. (Which player would you rather be?)

(A) Set up the payoff matrix for the game.

(B) Solve the game using the simplex method discussed in this section. (Remove any recessive rows and columns, if present, before you start the simplex method.)

Applications

Business & Economics

11. *Product mix.* A large department store chain is about to order deluxe, standard, and economy grade video cassette recorders (VCR's) for next year's inventory. The state of the nation's economy (fate) during the year will be an important factor on sales for that year. Records over the past 5 years indicate that if the economy is up, the company will net 2, 1, and 0 million dollars, respectively, on sales of deluxe, standard, and economy grade VCR's; if the economy is down, the company will net −1, 1, and 4 million dollars, respectively, on sales of deluxe, standard, and economy grade VCR's.

(A) Set up a payoff matrix for this problem.

(B) Find optimal strategies for both the company and fate (the economy). What is the value of the game?

(C) How should the company's VCR budget be allocated to each grade of VCR to maximize their return irrespective of what the economy does the following year?

(D) What is the expected value of the game to the company if they order only deluxe VCR's and fate plays the strategy "Down"? If the company plays its optimal strategy and fate plays the strategy "Down"? Discuss these and other possible scenarios.

12. *Product mix.* A tour agency organizes special standard and luxury tours for the following year. Once the agency has committed to these tours, they cannot be changed. The state of the economy during the following year has a direct effect on tour sales. From past records the company has established the following payoff matrix (in millions of dollars):

Economy (fate)

	Down	No change	Up
Standard	1	2	0
Luxury	0	1	3

(A) Find optimal strategies for both the company and fate (the economy). What is the value of the game?

(B) What proportion of each type of tour should be arranged for in advance in order for the company to maximize its return irrespective of what the economy does the following year?

(C) What is the expected value of the game to the company if they organize only luxury tours and fate plays the strategy "Down"? If the company plays its optimal strategy and fate plays the strategy "No change"? Discuss these and other possible scenarios.

Important Terms and Symbols

8-1 *Strictly Determined Games.* Two-person zero-sum games; row player; column player; payoff; matrix game; payoff matrix; fundamental principle of game theory; security level; optimum strategies; strictly determined game; saddle value; value of a game; fair game; nonstrictly determined game

8-2 *Mixed Strategy Games.* Two-finger Morra game; strategy; pure strategy; mixed strategy; expected value; fundamental theorem of game theory; value of a game; fair game; optimal strategies; solution of a game; recessive rows and columns; dominant rows and columns

8-3 *Linear Programming and 2 × 2 Games—Geometric Approach.* Matrix games and corresponding linear programming problems

8-4 *Linear Programming and m × n Games—Simplex Method and the Dual.* Solution by the simplex method using the dual

Review Exercise

Work through all the problems in this chapter review and check your answers in the back of the book. Answers to all review problems are there along with section numbers in italics to indicate where each type of problem is discussed. Where weaknesses show up, review appropriate sections in the text.

A *In Problems 1–4, for each matrix game that is strictly determined (if it is not strictly determined, say so), indicate:*
(A) All saddle values
(B) Optimal strategies for R and C
(C) The value of the game

1. $\begin{bmatrix} -4 & 6 \\ -3 & 1 \end{bmatrix}$

2. $\begin{bmatrix} -5 & 2 \\ 3 & -1 \end{bmatrix}$

3. $\begin{bmatrix} -3 & -1 & 5 & -8 \\ 1 & 0 & 0 & 2 \\ 3 & 0 & 1 & 0 \\ 6 & -2 & -4 & 2 \end{bmatrix}$

4. $\begin{bmatrix} 1 & -2 & 3 \\ -1 & 2 & 0 \\ 3 & 0 & -4 \end{bmatrix}$

5. Delete as many recessive rows and columns as possible; then write the reduced matrix game:

$$\begin{bmatrix} -2 & 3 & 5 \\ -1 & -3 & 0 \\ 0 & -1 & 1 \end{bmatrix}$$

B *Problems 6–9 refer to the matrix game:*

$$M = \begin{bmatrix} -2 & 1 \\ 0 & -1 \end{bmatrix}$$

6. Solve M using formulas from Section 8-2.

7. Write the two linear programming problems corresponding to M after adding 3 to each payoff.

8. Solve the matrix game M using linear programming and a geometric approach.

9. Solve the matrix game M using linear programming and the simplex method.

In Problems 10 and 11, discuss the validity of each statement. If the statement is always true, explain why. If not, give a counterexample.

10. (A) If a matrix game is strictly determined, then it is fair.
(B) If a and b are saddle values of a game matrix, then $a = b$.

11. (A) If the entries of a row in a game matrix are less than or equal to the corresponding entries in another row, then there is an optimal strategy for R in which the given row is never chosen.
(B) If the entries of a column in a game matrix are less than or equal to the corresponding entries in another column, then there is an optimal strategy for C in which the given column is never chosen.

In Problems 12–16, solve each matrix game (first check for saddle values, recessive rows, and recessive columns).

12. $\begin{bmatrix} -1 & 2 & 8 \\ 0 & 2 & -4 \\ 0 & 1 & 3 \end{bmatrix}$

13. $\begin{bmatrix} -1 & 5 & -3 & 7 \\ -4 & -3 & 2 & -2 \\ 3 & 0 & 2 & 1 \end{bmatrix}$

14. $\begin{bmatrix} 0 & 3 & -1 \\ -1 & -2 & 1 \end{bmatrix}$

15. $\begin{bmatrix} 2 & 6 & -4 & -7 \\ 4 & 7 & -3 & -5 \\ 3 & 3 & 9 & 8 \end{bmatrix}$

16. $\begin{bmatrix} -1 & 1 & -2 \\ 0 & -2 & 2 \\ -3 & 2 & -1 \end{bmatrix}$

C

17. Does every strictly determined 2×2 matrix game have a recessive row or column? Explain.

18. Does every strictly determined 3×3 matrix game have a recessive row or column? Explain.

19. *Finger game.* Consider the following finger game between Ron (rows) and Cathy (columns): Each points either 1 or 2 fingers at the other. If they match, Ron pays Cathy \$2. If Ron points 1 finger and Cathy points 2, then Cathy pays Ron \$3. If Ron points 2 fingers and Cathy points 1, then Cathy pays Ron \$1.
(A) Set up a payoff matrix for this game.
(B) Solve the matrix game by the three different methods discussed in this chapter. What are the optimal strategies for Ron and for Cathy?
(C) What is the expected value of the game for Ron? For Cathy?

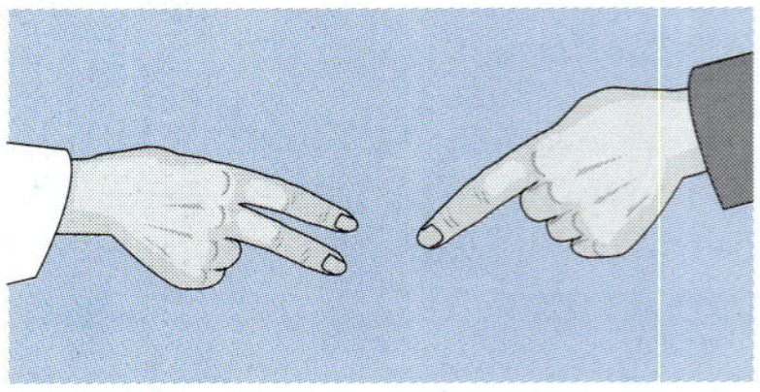

Group Activity 1 *Baseball Strategy*

Two baseball franchises compete for fans in the same large metropolitan area. In order to increase attendance each club decides each year to designate a percentage of profits to one, and only one, of the following: advertising, player salaries, stadium improvements, or special promotions at the ballpark.

Club *R* has finished high in the standings in recent years and has several players with expensive long-term contracts. A stadium renovation project that included the construction of luxury boxes was recently completed. The owner is reluctant to spend more money on player salaries, will not spend any more on stadium improvements, but suspects the club could do a better job of promoting its successful team through advertising and special promotions at the ballpark.

Club *C* has been less successful on the field and plays its games in an old stadium badly in need of modernization. Its owner vacillates between designating profits for player salaries, hoping to sign a couple of free agents, or using the money for stadium improvements. He tends to doubt the effectiveness of additional advertising or special promotions.

A consulting firm provides the following payoff matrix, which indicates the percentage of market gain or loss for each choice of action by *R* or *C*, assuming that any gain by *R* is a loss for *C*, and vice versa (Adv = Advertising, Sal = Player salaries, Sta = Stadium improvements, Pro = Special promotions):

		Club C			
		Adv	Sal	Sta	Pro
	Adv	0	−1	1	2
Club R	Sal	1	2	3	−4
	Pro	1	−3	−2	−1

(A) Find optimum strategies for each club. Is the game fair? What is its value?

(B) How do the optimum strategies correspond with the owners' inclinations?

(C) What is the expected value of the game if *R* plays its optimum strategy but *C* always chooses stadium improvements? Always chooses player salaries?

(D) What is the expected value of the game if *C* plays its optimum strategy but *R* chooses advertising, player salaries, and special promotions, each with probability $\frac{1}{3}$?

Group Activity 2 *Simulation of Matrix Games*

If a matrix game were played just a few times, it would be unusual for player *R*'s winnings per game to average v, the value of the game. Instead, the value of the game is a long-run average of player *R*'s winnings per game, taken over many repetitions of the game.

A graphing utility may be used to simulate repeated plays of a matrix game. For the game matrix

$$M = \begin{bmatrix} 9 & 1 \\ 4 & 6 \end{bmatrix}$$

and strategies

$$P^* = [.2 \quad .8] \qquad \text{and} \qquad Q^* = \begin{bmatrix} .5 \\ .5 \end{bmatrix}$$

the graphing utility commands in Figure 1 simulate 100 repetitions of the game.

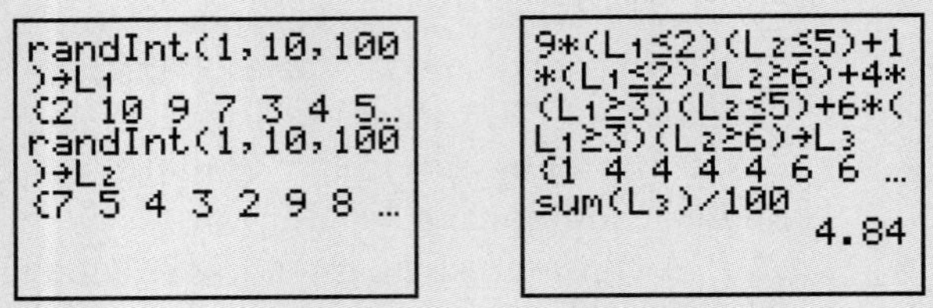

(A) (B)

FIGURE 1

The list L_1 in Figure 1A, which determines the row selected by player *R*, contains 100 integers from 1 to 10 chosen at random. Player *R*, in accordance with the strategy $P^* = [.2 \quad .8]$, selects Row 1 if the number lies in the range 1–2, and selects Row 2 if the number lies in the range 3–10. (Note from Figure 1, for example, that *R* selects Row 1 in the first game, but Row 2 in the second and third games.)

The list L_2, which determines the column selected by player *C*, also contains 100 integers from 1 to 10 chosen at random. Player *C*, in accordance with the strategy

$$Q^* = \begin{bmatrix} .5 \\ .5 \end{bmatrix}$$

selects Column 1 if the number lies in the range 1–5, and selects Column 2 if the number lies in the range 6–10. (Note from Figure 1, for example, that *C* selects Column 2 in the first game, but Column 1 in the second and third games.)

The list L_3 in Figure 1B, constructed by using the appropriate graphing utility test commands (see your user's manual), contains *R*'s winnings for each of the 100 games. Note from Figure 1B that the average of *R*'s winnings per game is 4.84.

(A) Compute the theoretical value of the matrix game *M*, and compare with *R*'s average winnings of 4.84 in the simulation above.

(B) Show that the strategies

$$P^* = [.2 \quad .8] \qquad \text{and} \qquad Q^* = \begin{bmatrix} .5 \\ .5 \end{bmatrix}$$

are optimal strategies for *R* and *C*.

(C) Use a graphing utility to simulate 100 repetitions of the matrix game M if R chooses either row with equal probability and C chooses either column with equal probability. Compare R's average winnings per game with the theoretical value of the game.

(D) How much does R lose by failing to employ the optimal strategy in part (C)? Explain.

(E) Use a graphing utility to simulate 100 repetitions of the matrix game M if R uses the strategy $P = [.5 \quad .5]$ and C uses the strategy

$$Q = \begin{bmatrix} .9 \\ .1 \end{bmatrix}$$

Compare R's average winnings per game with the theoretical value of the game.

CHAPTER 9

MARKOV CHAINS

INTRODUCTION

In this chapter we consider a mathematical model that combines probability and matrices to analyze a *stochastic process,* which consists of a sequence of trials that satisfy certain conditions. The sequence of trials is called a *Markov chain* after the Russian mathematician Andrei Markov (1856–1922), who is credited with supplying much of the foundation work in stochastic processes. Many of the early applications of Markov chains were in the physical sciences. More recent applications of Markov chains involve a wide variety of topics, including finance, market research, genetics, medicine, demographics, psychology, and political science.

In the first section we introduce the basic properties of Markov chains. In the remaining sections we discuss the long-term behavior of two different types of Markov chains.

SECTION 9-1

Properties of Markov Chains

- INTRODUCTION
- TRANSITION AND STATE MATRICES
- POWERS OF TRANSITION MATRICES
- APPLICATION

INTRODUCTION

In this section we are going to be interested in physical *systems* and their possible *states.* To understand what this means, consider the following examples:

1. A stock listed on the New York Stock Exchange either increases, decreases, or does not change in price each day the exchange is open. The stock can be thought of as a physical system with three possible states: increase, decrease, or no change.

2. A commuter, relative to a rapid transit system, can be thought of as a physical system with two states, a user or a nonuser.
3. A voting precinct casts a simple majority vote during each congressional election for a Republican candidate, a Democratic candidate, or a third-party candidate. The precinct, relative to all congressional elections past, present, and future, constitutes a physical system that can be thought of as being in one (and only one) of three states after each election: Republican, Democratic, or other.

If a system evolves from one state to another in such a way that chance elements are involved in progressing from one state to the next, then the progression of the system through a sequence of states is called a **stochastic process** (*stochos* is the Greek word for "guess"). We will now consider a simple example of a stochastic process in detail, and out of it will arise further definitions and methodology.

A toothpaste company markets a product (brand A) that currently has 10% of the toothpaste market. The company hires a market research firm to estimate the percentage of the market it might acquire in the future if it launches an aggressive sales campaign. The research firm uses test marketing and extensive surveys to predict the effect of the campaign. They find that if a person is using brand A, then the probability is .8 that this person will buy it again when he or she runs out of toothpaste. On the other hand, a person using another brand will switch to brand A with a probability of .6 when he or she runs out of toothpaste. Thus, each toothpaste consumer can be considered to be in one of two possible states:

$$A = \text{Uses brand } A \qquad \text{and} \qquad A' = \text{Uses another brand}$$

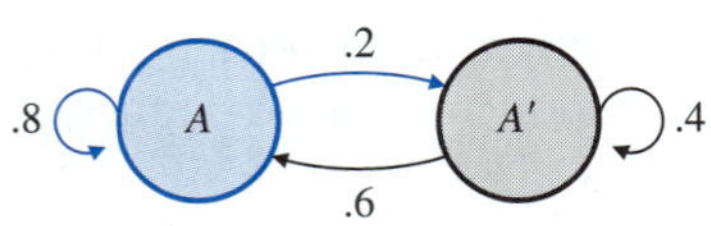

FIGURE 1
Transition diagram

The probabilities determined by the market research firm can be represented graphically in a **transition diagram** as shown in Figure 1.

We can also represent the information from the market research firm numerically in a **transition probability matrix:**

$$\begin{array}{c} \\ \text{Current state} \end{array} \begin{array}{c} \\ A \\ A' \end{array} \overset{\displaystyle\text{Next state}}{\overset{\begin{array}{cc} A & A' \end{array}}{\begin{bmatrix} .8 & .2 \\ .6 & .4 \end{bmatrix}}} = P$$

Explore–Discuss 1

(A) Refer to the transition diagram in Figure 1. What is the probability that a person using brand A will switch to another brand when he or she runs out of toothpaste?

(B) Refer to transition probability matrix P. What is the probability that a person who is not using brand A will not switch to brand A when he or she runs out of toothpaste?

(C) In Figure 1, the sum of the probabilities on the arrows leaving each state is 1. Will this be true for any transition diagram? Explain your answer.

(D) In transition probability matrix P, the sum of the probabilities in each row is 1. Will this be true for any transition probability matrix? Explain your answer.

The toothpaste company's 10% share of the market at the beginning of the sales campaign can be represented as an **initial-state distribution matrix:**

$$S_0 = \begin{matrix} A & A' \\ [.1 & .9] \end{matrix}$$

If a person is chosen at random, then the probability that this person uses brand A (state A) is .1, and the probability that this person does not use brand A (state A') is .9. Thus, S_0 also can be interpreted as an **initial-state probability matrix.**

What are the probabilities of a person being in state A or A' on the first purchase after the start of the sales campaign? Let us look at the probability tree given below. [*Note:* A_0 represents state A at the beginning of the campaign, A_1' represents state A' on the first purchase after the campaign, and so on.]

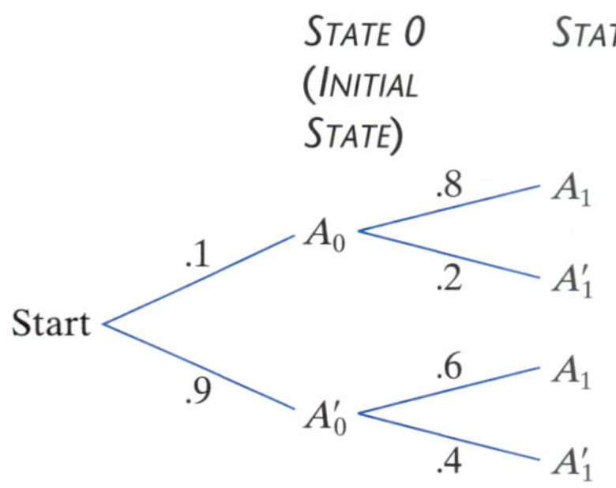

Proceeding as in Chapter 6, we can read the required probabilities directly from the tree:

$$\begin{aligned} P(A_1) &= P(A_0 \cap A_1) + P(A_0' \cap A_1) \\ &= (.1)(.8) + (.9)(.6) = .62 \end{aligned}$$

$$\begin{aligned} P(A_1') &= P(A_0 \cap A_1') + P(A_0' \cap A_1') \\ &= (.1)(.2) + (.9)(.4) = .38 \end{aligned}$$

[*Note:* $P(A_1) + P(A_1') = 1$, as expected.]

Thus, the **first-state matrix** is

$$S_1 = \begin{matrix} A & A' \\ [.62 & .38] \end{matrix}$$

This matrix gives us the probabilities of a randomly chosen person being in state A or A' on the first purchase after the start of the campaign. We see that brand A's market share has increased from 10% to 62%.

Now, if you were asked to find the probabilities of a person being in state A or state A' on the tenth purchase after the start of the campaign, you might start to draw additional branches on the probability tree, but you would soon become discouraged—and for good reason, because the number of branches doubles for each successive purchase. By the tenth purchase, there would be $2^{11} = 2{,}048$ branches! Fortunately, we can convert the summing of branch products to matrix multiplication. In particular, if we multiply the initial-state matrix S_0 by the transition matrix P, we obtain the first-state matrix S_1:

$$\underset{\text{Initial state}}{S_0P = \overset{\quad A \quad\; A'}{[.1 \quad .9]}}\underset{\text{Transition matrix}}{\begin{bmatrix} .8 & .2 \\ .6 & .4 \end{bmatrix}} = \underbrace{[(.1)(.8) + (.9)(.6) \qquad (.1)(.2) + (.9)(.4)]}_{\text{Compare with the tree computations above}} = \underset{\text{First state}}{\overset{\quad A \quad\; A'}{[.62 \quad .38]}} = S_1$$

As you might guess, we can get the second-state matrix S_2 (for the second purchase) by multiplying the first-state matrix by the transition matrix:

$$S_1P = \underset{\text{First state}}{\overset{\quad A \quad\; A'}{[.62 \quad .38]}}\begin{bmatrix} .8 & .2 \\ .6 & .4 \end{bmatrix} = \underset{\text{Second state}}{\overset{\quad A \quad\; A'}{[.724 \quad .276]}} = S_2$$

The third-state matrix S_3 is computed in a similar manner:

$$S_2P = \underset{\text{Second state}}{\overset{\quad A \quad\; A'}{[.724 \quad .276]}}\begin{bmatrix} .8 & .2 \\ .6 & .4 \end{bmatrix} = \underset{\text{Third state}}{\overset{\quad A \quad\; A'}{[.7448 \quad .2552]}} = S_3$$

Examining the values in the first three state matrices, we see that brand A's market share increases after each toothpaste purchase. Will the market share for brand A continue to increase until it approaches 100%, or will it level off at some value less than 100%? These questions will be answered in the next section when we develop techniques for determining the long-run behavior of state matrices.

Transition and State Matrices

The sequence of trials (toothpaste purchases) with the constant transition matrix P described above is a special kind of stochastic process called a *Markov chain.* In general, a **Markov chain,** or **process,** is a sequence of experiments, trials, or observations such that the transition probability matrix from one state to the next is constant. A Markov process has no memory. The various matrices associated with a Markov chain are defined in the next box.

Markov Chains

Given a Markov chain with n states, a ***k*th-state matrix** is a matrix of the form

$$S_k = [s_{k1} \quad s_{k2} \quad \ldots \quad s_{kn}]$$

Each entry s_{ki} is the proportion of the population that is in state i after the kth trial, or, equivalently, the probability of a randomly selected element of the population being in state i after the kth trial. The sum of all the entries in the kth state matrix S_k must be 1.

A **transition matrix** is a constant square matrix P of order n such that the entry in the ith row and jth column indicates the probability of the system moving from the ith state to the jth state on the next observation or trial. The sum of the entries in each row must be 1.

REMARKS

1. Since the entries in a kth-state matrix or a transition matrix are probabilities, they must be real numbers between 0 and 1, inclusive.
2. Rearranging the various states and corresponding transition probabilities in a transition matrix will produce a different, but equivalent, transition matrix. For example, both of the following matrices are transition matrices for the toothpaste manufacturer discussed earlier:

$$P = \begin{array}{c} \\ A \\ A' \end{array}\!\!\begin{array}{c} \begin{array}{cc} A & A' \end{array} \\ \begin{bmatrix} .8 & .2 \\ .6 & .4 \end{bmatrix} \end{array} \qquad P' = \begin{array}{c} \\ A' \\ A \end{array}\!\!\begin{array}{c} \begin{array}{cc} A' & A \end{array} \\ \begin{bmatrix} .4 & .6 \\ .2 & .8 \end{bmatrix} \end{array}$$

Such rearrangements will affect the form of the matrices used in the solution of a problem, but will not affect any of the information obtained from these matrices. In Section 9-3 we will encounter situations where it will be helpful to select a transition matrix that has a special form. For now, you can choose any order for the states in a transition matrix.

As we indicated in the preceding discussion, matrix multiplication can be used to compute the various state matrices of a Markov chain, as described in the following box.

Computing State Matrices for a Markov Chain

If S_0 is the initial-state matrix and P is the transition matrix for a Markov chain, then the subsequent state matrices are given by:

$$\begin{aligned} S_1 &= S_0P && \text{First-state matrix} \\ S_2 &= S_1P && \text{Second-state matrix} \\ S_3 &= S_2P && \text{Third-state matrix} \\ &\vdots \\ S_k &= S_{k-1}P && k\text{th-state matrix} \end{aligned}$$

Example 1 **Insurance** An insurance company found that on the average, over a period of 10 years, 23% of the drivers in a particular community who were involved in an accident one year were also involved in an accident the following year. They also found that only 11% of the drivers who were not involved in an accident one year were involved in an accident the following year. Use these percentages as approximate empirical probabilities for the following:

(A) Draw a transition diagram.
(B) Find the transition matrix P.
(C) If 5% of the drivers in the community are involved in an accident this year, what is the probability that a driver chosen at random from the community will be involved in an accident next year? Year after next?

SOLUTION (A)

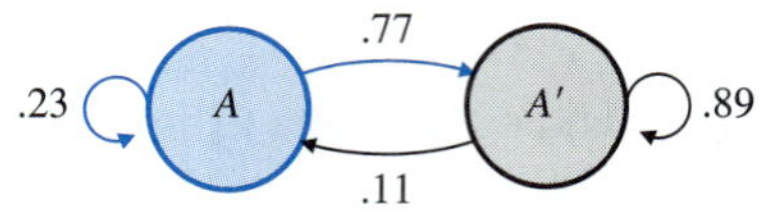

A = Accident
A′ = No accident

(B)

$$\begin{array}{c} \text{Next year} \\ \begin{array}{cc} \qquad\quad A & A' \end{array} \end{array}$$

$$\begin{matrix}\text{This} \\ \text{year}\end{matrix}\ \begin{matrix} A \\ A' \end{matrix}\begin{bmatrix} .23 & .77 \\ .11 & .89 \end{bmatrix} = P \quad \text{Transition matrix}$$

(C) The initial-state matrix S_0 is

$$S_0 = \overset{\begin{matrix}A & \quad A'\end{matrix}}{[.05 \quad .95]} \quad \text{Initial-state matrix}$$

Thus,

$$S_0P = \overset{\begin{matrix}A & \quad A'\end{matrix}}{[.05 \quad .95]}\begin{bmatrix} .23 & .77 \\ .11 & .89 \end{bmatrix} = \overset{\begin{matrix}A & \quad A'\end{matrix}}{[.116 \quad .884]} = S_1$$

This year (initial state) — Next year (first state)

$$S_1P = \overset{\begin{matrix}A & \quad A'\end{matrix}}{[.116 \quad .884]}\begin{bmatrix} .23 & .77 \\ .11 & .89 \end{bmatrix} = \overset{\begin{matrix}A & \quad A'\end{matrix}}{[.12392 \quad .87608]} = S_2$$

Next year (first state) — Year after next (second state)

The probability of a driver chosen at random from the community having an accident next year is .116 and having an accident year after next is .12392. That is, it is expected that 11.6% of the drivers in the community will have an accident next year and 12.392% the year after.

Matched Problem 1 An insurance company classifies drivers as low-risk if they are accident-free for 1 year. Past records indicate that 98% of the drivers in the low-risk category (L) one year will remain in that category the next year, and 78% of the drivers that are not in the low-risk category (L') one year will be in the low-risk category the next year.

(A) Draw a transition diagram.
(B) Find the transition matrix P.
(C) If 90% of the drivers in the community are in the low-risk category this year, what is the probability that a driver chosen at random from the community will be in the low-risk category next year? Year after next?

POWERS OF TRANSITION MATRICES

We will now investigate the effective use of the powers of a transition matrix.

Explore–Discuss 2

Given the transition matrix P and initial-state matrix S_0, where

$$P = \begin{matrix} & \begin{matrix} A & A' \end{matrix} \\ \begin{matrix} A \\ A' \end{matrix} & \begin{bmatrix} .9 & .1 \\ .7 & .3 \end{bmatrix} \end{matrix} \quad \text{and} \quad S_0 = \begin{matrix} \begin{matrix} A & A' \end{matrix} \\ [.5 \quad .5] \end{matrix}$$

(A) Find S_2 and S_4.
(B) Find P^2 and P^4. [Recall that $P^2 = PP$ and $P^4 = P^2P^2$.]
(C) Find S_0P^2 and S_0P^4.
(D) Compare the results of parts (A) and (C). What interpretation of the entries in P^2 and P^4 does this suggest?

The state matrices for a Markov chain are defined **recursively;** that is, each state matrix is defined in terms of the preceding state matrix. For example, to find the fourth-state matrix S_4, it is necessary to compute the preceding three state matrices:

$$S_1 = S_0P \qquad S_2 = S_1P \qquad S_3 = S_2P \qquad S_4 = S_3P$$

Is there any way to compute a given state matrix directly without first computing all the preceding state matrices? If we substitute the equation for S_1 in the equation for S_2, substitute this new equation for S_2 in the equation for S_3, and so on, a definite pattern emerges:

$$\begin{aligned} S_1 &= S_0P \\ S_2 &= S_1P = (S_0P)P = S_0P^2 \\ S_3 &= S_2P = (S_0P^2)P = S_0P^3 \\ S_4 &= S_3P = (S_0P^3)P = S_0P^4 \\ &\vdots \end{aligned}$$

In general, it can be shown that the kth-state matrix is given by $S_k = S_0P^k$. We summarize this important result in Theorem 1.

Theorem 1 ■■ **POWERS OF A TRANSITION MATRIX**

If P is the transition matrix and S_0 is an initial-state matrix for a Markov chain, then the kth-state matrix is given by

$$S_k = S_0P^k$$

The entry in the ith row and jth column of P^k indicates the probability of the system moving from the ith state to the jth state in k observations or trials. The sum of the entries in each row of P^k is 1. ■■

Example 2 ➡ **Using P^k to Compute S_k** Find P^4 and use it to find S_4 for

$$P = \begin{matrix} & \begin{matrix} A & A' \end{matrix} \\ \begin{matrix} A \\ A' \end{matrix} & \begin{bmatrix} .1 & .9 \\ .6 & .4 \end{bmatrix} \end{matrix} \quad \text{and} \quad S_0 = \begin{matrix} \begin{matrix} A & A' \end{matrix} \\ [.2 \quad .8] \end{matrix}$$

SOLUTION

$$P^2 = PP = \begin{bmatrix} .1 & .9 \\ .6 & .4 \end{bmatrix}\begin{bmatrix} .1 & .9 \\ .6 & .4 \end{bmatrix} = \begin{bmatrix} .55 & .45 \\ .3 & .7 \end{bmatrix}$$

$$P^4 = P^2P^2 = \begin{bmatrix} .55 & .45 \\ .3 & .7 \end{bmatrix}\begin{bmatrix} .55 & .45 \\ .3 & .7 \end{bmatrix} = \begin{bmatrix} .4375 & .5625 \\ .375 & .625 \end{bmatrix}$$

$$S_4 = S_0P^4 = [.2 \quad .8]\begin{bmatrix} .4375 & .5625 \\ .375 & .625 \end{bmatrix} = [.3875 \quad .6125]$$

Matched Problem 2 Find P^4 and use it to find S_4 for

$$P = \begin{matrix} & \begin{matrix} A & A' \end{matrix} \\ \begin{matrix} A \\ A' \end{matrix} & \begin{bmatrix} .8 & .2 \\ .3 & .7 \end{bmatrix} \end{matrix} \quad \text{and} \quad S_0 = \begin{matrix} \begin{matrix} A & A' \end{matrix} \\ [.8 \quad .2] \end{matrix}$$

If a graphing calculator or a computer is available for computing matrix products and powers of a matrix, then finding state matrices for any number of trials becomes a routine calculation.

Example 3 **Using a Graphing Utility and P^k to Compute S_k** Use P^8 and a graphing utility to find S_8 for P and S_0 as given in Example 2. Round values in S_8 to six decimal places.

SOLUTION After storing the matrices P and S_0 in the graphing utility's memory, use the equation

$$S_8 = S_0P^8$$

to compute S_8. Figure 2 shows the result on a typical graphing utility. Thus, we see that (to six decimal places)

$$S_8 = [.399219 \quad .600781]$$

FIGURE 2

Matched Problem 3 Use P^8 and a graphing utility to find S_8 for P and S_0 as given in Matched Problem 2. Round values in S_8 to six decimal places.

APPLICATION

The next example illustrates the use of Theorem 1 in an applied problem.

Example 4 **University Enrollment** Part-time students admitted to an MBA program in a university are considered to be first-year students until they successfully complete 15 credits. Then they are classified as second-year students and may begin to take more advanced courses and to work on the thesis required for graduation. Past records indicate that at the end of each year 10% of the first-year students (F) drop out of the program (D) and 30% become second-year students (S). Also, 10% of the second-year students drop out of the program and 40% graduate (G) each year. Students that graduate or drop out never return to the program.

(A) Draw a transition diagram.
(B) Find the transition matrix P.
(C) What is the probability that a first-year student graduates within 4 years? Drops out within 4 years?

SOLUTION (A) If 10% of the first-year students drop out and 30% become second-year students, the remaining 60% must continue as first-year students for another year (see the diagram). Similarly, 50% of the second-year students must continue as second-year students for another year. Since students who drop out never return, all the students in state D in one year will continue in that state the next year. We indicate this by placing a 1 on the arrow from D back to D. State G is labeled in the same manner.

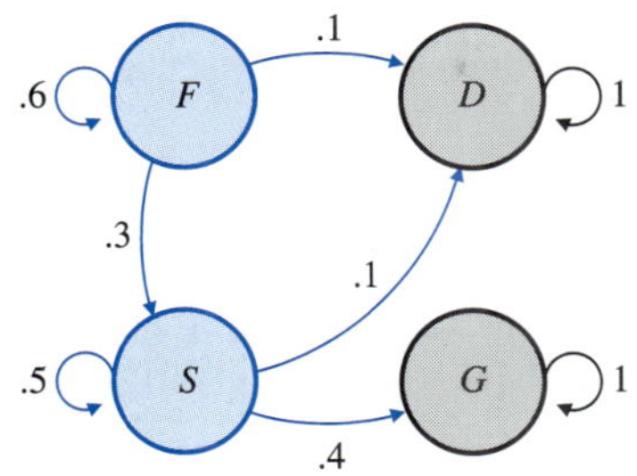

Transition diagram

(B)

$$P = \begin{array}{c} \\ F \\ D \\ S \\ G \end{array} \begin{array}{c} \begin{array}{cccc} F & D & S & G \end{array} \\ \begin{bmatrix} .6 & .1 & .3 & 0 \\ 0 & 1 & 0 & 0 \\ 0 & .1 & .5 & .4 \\ 0 & 0 & 0 & 1 \end{bmatrix} \end{array}$$

Transition matrix

(C) The probability that a first-year student moves from state F to state G within 4 years is the entry in Row 1 and Column 4 of P^4 (Theorem 1). Hand computation of P^4 requires two multiplications:

$$P^2 = \begin{bmatrix} .6 & .1 & .3 & 0 \\ 0 & 1 & 0 & 0 \\ 0 & .1 & .5 & .4 \\ 0 & 0 & 0 & 1 \end{bmatrix} \begin{bmatrix} .6 & .1 & .3 & 0 \\ 0 & 1 & 0 & 0 \\ 0 & .1 & .5 & .4 \\ 0 & 0 & 0 & 1 \end{bmatrix} = \begin{bmatrix} .36 & .19 & .33 & .12 \\ 0 & 1 & 0 & 0 \\ 0 & .15 & .25 & .6 \\ 0 & 0 & 0 & 1 \end{bmatrix}$$

$$P^4 = P^2P^2 = \begin{bmatrix} .36 & .19 & .33 & .12 \\ 0 & 1 & 0 & 0 \\ 0 & .15 & .25 & .6 \\ 0 & 0 & 0 & 1 \end{bmatrix} \begin{bmatrix} .36 & .19 & .33 & .12 \\ 0 & 1 & 0 & 0 \\ 0 & .15 & .25 & .6 \\ 0 & 0 & 0 & 1 \end{bmatrix}$$

$$= \begin{bmatrix} .1296 & .3079 & .2013 & .3612 \\ 0 & 1 & 0 & 0 \\ 0 & .1875 & .0625 & .75 \\ 0 & 0 & 0 & 1 \end{bmatrix}$$

Thus, the probability that a first-year student has graduated within 4 years is .3612. Similarly, the probability that a first-year student has dropped out within 4 years is .3079 (the entry in Row 1 and Column 2 of P^4).

Matched Problem 4 Refer to Example 4. At the end of each year the faculty examines the progress each second-year student has made in writing the required thesis. Past records indicate that 30% of the second-year students (S) have their theses approved (A) and 10% of the students are dropped from the program for insufficient progress (D), never to return. The remaining students continue to work on their theses.

(A) Draw a transition diagram.
(B) Find the transition matrix P.
(C) What is the probability that a second-year student completes the thesis requirement within 4 years? Is dropped from the program for insufficient progress within 4 years?

Explore–Discuss 3

Refer to Example 4. States D and G are referred to as *absorbing states,* because a student who enters either one of these states never leaves it. Absorbing states are discussed in detail in Section 9-3.

(A) How can absorbing states be recognized from a transition diagram? Draw a transition diagram with two states, one that is absorbing and one that is not, to illustrate.
(B) How can absorbing states be recognized from a transition matrix? Write the transition matrix for the diagram you drew in part (A) to illustrate.

Answers to Matched Problems

1. (A)

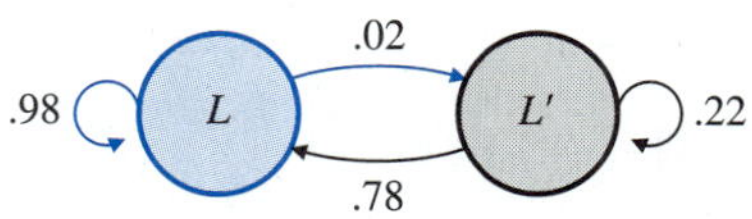

L = Low-risk
L' = Not low-risk

(B)
$$\begin{array}{c} \\ \text{This} \\ \text{year} \end{array} \begin{array}{c} \\ L \\ L' \end{array} \overset{\text{Next year}}{\overset{\begin{array}{cc} L & L' \end{array}}{\begin{bmatrix} .98 & .02 \\ .78 & .22 \end{bmatrix}}} = P$$

(C) Next year: .96; year after next: .972

2. $P^4 = \begin{bmatrix} .625 & .375 \\ .5625 & .4375 \end{bmatrix}$; $S_4 = [.6125 \quad .3875]$ **3.** $S_8 = [.600781 \quad .399219]$

4. (A)

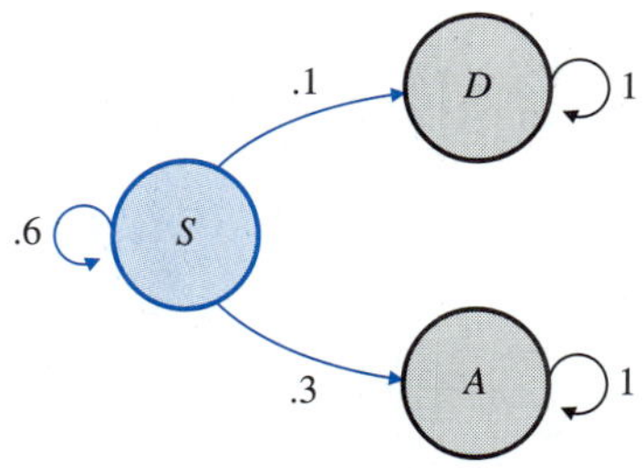

(B)
$$P = \begin{array}{c} S \\ A \\ D \end{array} \overset{\begin{array}{ccc} S & A & D \end{array}}{\begin{bmatrix} .6 & .3 & .1 \\ 0 & 1 & 0 \\ 0 & 0 & 1 \end{bmatrix}}$$

(C) .6528; .2176

EXERCISE 9-1

A *Problems 1–8 refer to the following transition matrix:*

$$P = \begin{array}{c} A \\ B \end{array} \overset{\begin{array}{cc} A & B \end{array}}{\begin{bmatrix} .8 & .2 \\ .4 & .6 \end{bmatrix}}$$

In Problems 1–4, find S_1 for the indicated initial-state matrix S_0, and interpret with the aid of a tree diagram.

1. $S_0 = [1 \quad 0]$
2. $S_0 = [0 \quad 1]$
3. $S_0 = [.5 \quad .5]$
4. $S_0 = [.3 \quad .7]$

In Problems 5–8, find S_2 for the indicated initial-state matrix S_0, and explain what it represents.

5. $S_0 = [1 \quad 0]$
6. $S_0 = [0 \quad 1]$
7. $S_0 = [.5 \quad .5]$
8. $S_0 = [.3 \quad .7]$

In Problems 9–14, is there a unique way of filling in the missing probabilities in the transition diagram? If so, complete the transition diagram and write the corresponding transition matrix. If not, explain why.

9.

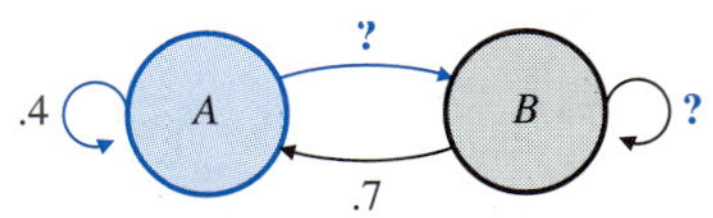

10.

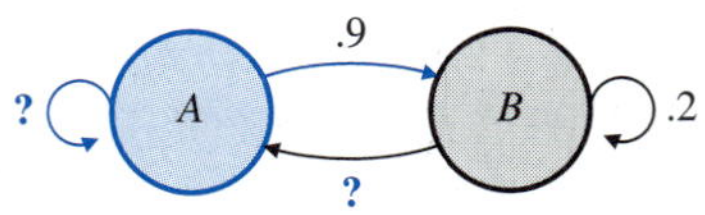

11.

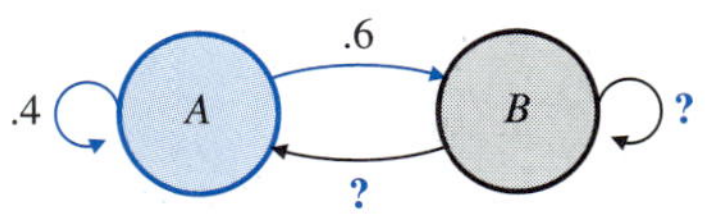

12.

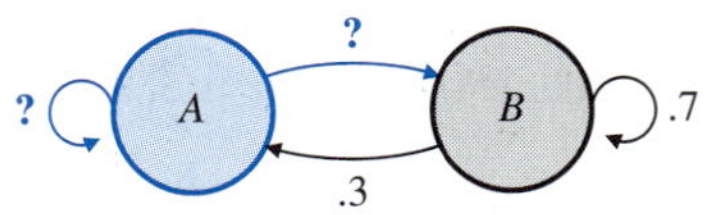

13.

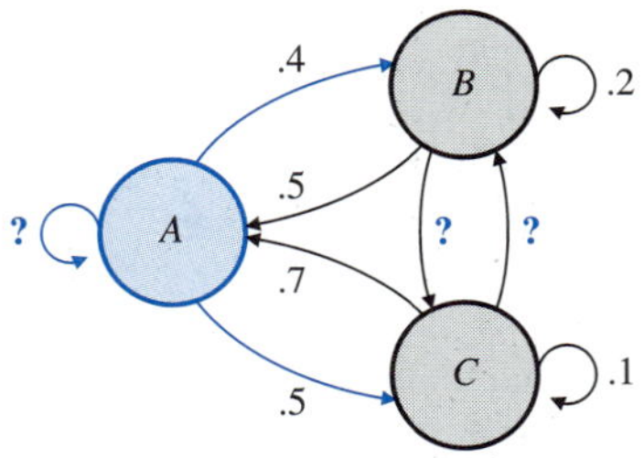

14.

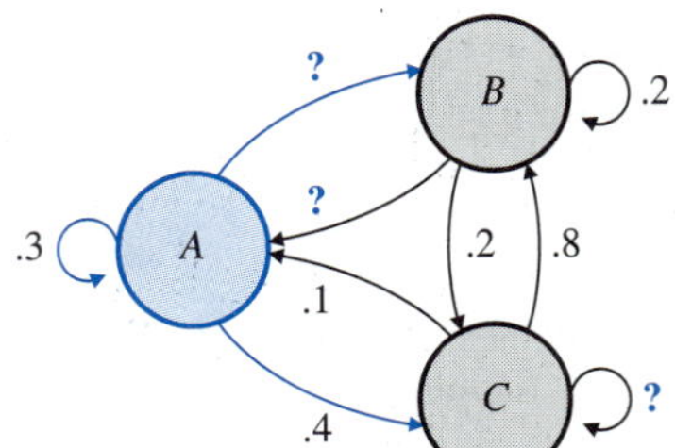

In Problems 15–20, are there unique values of a, b, and c that will make P a transition matrix? If so, complete the transition matrix and draw the corresponding transition diagram. If not, explain why.

15. $P = \begin{array}{c} \\ A \\ B \\ C \end{array}\begin{array}{c} \begin{array}{ccc} A & B & C \end{array} \\ \begin{bmatrix} 0 & .5 & a \\ b & 0 & .4 \\ .2 & c & .1 \end{bmatrix} \end{array}$

16. $P = \begin{array}{c} \\ A \\ B \\ C \end{array}\begin{array}{c} \begin{array}{ccc} A & B & C \end{array} \\ \begin{bmatrix} a & 0 & .9 \\ .2 & .3 & b \\ .6 & c & 0 \end{bmatrix} \end{array}$

17. $P = \begin{array}{c} \\ A \\ B \\ C \end{array}\begin{array}{c} \begin{array}{ccc} A & B & C \end{array} \\ \begin{bmatrix} 0 & a & .3 \\ 0 & b & 0 \\ c & .8 & 0 \end{bmatrix} \end{array}$

18. $P = \begin{array}{c} \\ A \\ B \\ C \end{array}\begin{array}{c} \begin{array}{ccc} A & B & C \end{array} \\ \begin{bmatrix} 0 & 1 & a \\ 0 & 0 & b \\ c & .5 & 0 \end{bmatrix} \end{array}$

19. $P = \begin{array}{c} \\ A \\ B \\ C \end{array}\begin{array}{c} \begin{array}{ccc} A & B & C \end{array} \\ \begin{bmatrix} .2 & .1 & .7 \\ a & .4 & c \\ .5 & b & .4 \end{bmatrix} \end{array}$

20. $P = \begin{array}{c} \\ A \\ B \\ C \end{array}\begin{array}{c} \begin{array}{ccc} A & B & C \end{array} \\ \begin{bmatrix} a & .8 & .1 \\ .3 & b & .4 \\ .6 & .5 & c \end{bmatrix} \end{array}$

B *In Problems 21–24, use the given information to draw the transition diagram and to find the transition matrix.*

21. A Markov process has two states, A and B. The probability of going from state A to state B in one trial is .7, and the probability of going from state B to state A in one trial is .9.

22. A Markov process has two states, A and B. The probability of going from state A to state A in one trial is .6, and the probability of going from state B to state B in one trial is .2.

23. A Markov chain has three states, A, B, and C. The probability of going from state A to state B in one trial is .1, the probability of going from state A to state C in one trial is .3, the probability of going from state B to state A in one trial is .2, the probability of going from state B to state C in one trial is .5, and the probability of going from state C to state C in one trial is 1.

24. A Markov chain has three states, A, B, and C. The probability of going from state A to state B in one trial is 1, the probability of going from state B to state A in one trial is .5, the probability of going from state B to state C in one trial is .5, and the probability of going from state C to state A in one trial is 1.

Problems 25–34 refer to the transition matrix P and the powers of P given below:

$$P = \begin{array}{c} \\ A \\ B \\ C \end{array}\begin{array}{c} \begin{array}{ccc} A & B & C \end{array} \\ \begin{bmatrix} .6 & .3 & .1 \\ .2 & .5 & .3 \\ .1 & .2 & .7 \end{bmatrix} \end{array} \qquad P^2 = \begin{array}{c} \\ A \\ B \\ C \end{array}\begin{array}{c} \begin{array}{ccc} A & B & C \end{array} \\ \begin{bmatrix} .43 & .35 & .22 \\ .25 & .37 & .38 \\ .17 & .27 & .56 \end{bmatrix} \end{array}$$

$$P^3 = \begin{array}{c} \\ A \\ B \\ C \end{array}\begin{array}{c} \begin{array}{ccc} A & B & C \end{array} \\ \begin{bmatrix} .35 & .348 & .302 \\ .262 & .336 & .402 \\ .212 & .298 & .49 \end{bmatrix} \end{array}$$

25. Find the probability of going from state A to state B in two trials.

26. Find the probability of going from state B to state C in two trials.

27. Find the probability of going from state C to state A in three trials.

28. Find the probability of going from state B to state B in three trials.

29. Find S_2 for $S_0 = [1 \quad 0 \quad 0]$, and explain what it represents.

30. Find S_2 for $S_0 = [0 \quad 1 \quad 0]$, and explain what it represents.

31. Find S_3 for $S_0 = [0 \quad 0 \quad 1]$, and explain what it represents.

32. Find S_3 for $S_0 = [1 \quad 0 \quad 0]$, and explain what it represents.

33. Using a graphing utility to compute powers of P, find the smallest positive integer n such that the corresponding entries in P^n and P^{n+1} are equal to two decimal places.

34. Using a graphing utility to compute powers of P, find the smallest positive integer n such that the corresponding entries in P^n and P^{n+1} are equal to three decimal places.

In Problems 35–38, given the transition matrix P and initial-state matrix S_0, find P^4 and use P^4 to find S_4.

35. $$P = \begin{array}{c|cc} & A & B \\ \hline A & .1 & .9 \\ B & .6 & .4 \end{array}; \quad S_0 = [.8 \quad .2]$$

36. $$P = \begin{array}{c|cc} & A & B \\ \hline A & .8 & .2 \\ B & .3 & .7 \end{array}; \quad S_0 = [.4 \quad .6]$$

37. $$P = \begin{array}{c|ccc} & A & B & C \\ \hline A & 0 & .4 & .6 \\ B & 0 & 0 & 1 \\ C & 1 & 0 & 0 \end{array}; \quad S_0 = [.2 \quad .3 \quad .5]$$

38. $$P = \begin{array}{c|ccc} & A & B & C \\ \hline A & 0 & 1 & 0 \\ B & .8 & 0 & .2 \\ C & 1 & 0 & 0 \end{array}; \quad S_0 = [.4 \quad .2 \quad .4]$$

39. A Markov process with two states has transition matrix P. If the initial-state matrix is $S_0 = [1 \quad 0]$, discuss the relationship between the entries in the kth-state matrix and the entries in the kth power of P.

40. Repeat Problem 39 if the initial-state matrix is $S_0 = [0 \quad 1]$.

C

41. Given the transition matrix

$$P = \begin{array}{c|cccc} & A & B & C & D \\ \hline A & .2 & .2 & .3 & .3 \\ B & 0 & 1 & 0 & 0 \\ C & .2 & .2 & .1 & .5 \\ D & 0 & 0 & 0 & 1 \end{array}$$

(A) Find P^4.
(B) Find the probability of going from state A to state D in four trials.
(C) Find the probability of going from state C to state B in four trials.
(D) Find the probability of going from state B to state A in four trials.

42. Repeat Problem 41 for the transition matrix

$$P = \begin{array}{c|cccc} & A & B & C & D \\ \hline A & .5 & .3 & .1 & .1 \\ B & 0 & 1 & 0 & 0 \\ C & 0 & 0 & 1 & 0 \\ D & .1 & .2 & .3 & .4 \end{array}$$

A matrix is called a ***probability matrix*** *if all its entries are real numbers between 0 and 1, inclusive, and the sum of the entries in each row is 1. Thus, transition matrices are square probability matrices and state matrices are probability matrices with one row.*

43. If

$$P = \begin{bmatrix} a & 1-a \\ 1-b & b \end{bmatrix}$$

is a probability matrix, show that P^2 is a probability matrix.

44. If

$$P = \begin{bmatrix} a & 1-a \\ 1-b & b \end{bmatrix} \quad \text{and} \quad S = [c \quad 1-c]$$

are probability matrices, show that SP is a probability matrix.

Use a graphing utility and the formula $S_k = S_0P^k$ (Theorem 1) to compute the required state matrices in Problems 45–48.

45. The transition matrix for a Markov chain is

$$P = \begin{bmatrix} .4 & .6 \\ .2 & .8 \end{bmatrix}$$

(A) If $S_0 = [0 \quad 1]$, find $S_2, S_4, S_8, \ldots$. Can you identify a state matrix S that the matrices S_k seem to be approaching?
(B) Repeat part (A) for $S_0 = [1 \quad 0]$.
(C) Repeat part (A) for $S_0 = [.5 \quad .5]$.
(D) Find SP for any matrix S you identified in parts (A)–(C).
(E) Write a brief verbal description of the long-term behavior of the state matrices of this Markov chain based on your observations in parts (A)–(D).

46. Repeat Problem 45 for $P = \begin{bmatrix} .9 & .1 \\ .4 & .6 \end{bmatrix}$.

47. Refer to Problem 45. Find P^k for $k = 2, 4, 8, \ldots$. Can you identify a matrix Q that the matrices P^k are approaching? If so, how is Q related to the results you discovered in Problem 45?

48. Refer to Problem 46. Find P^k for $k = 2, 4, 8, \ldots$. Can you identify a matrix Q that the matrices P^k are approaching? If so, how is Q related to the results you discovered in Problem 46?

Applications

Business & Economics

49. *Scheduling.* An outdoor restaurant in a summer resort closes only on days that it rains. From past records it is found that from May through September, when it rains one day, the probability of rain for the next day is .4; when it does not rain one day, the probability of rain the next day is .06.
(A) Draw a transition diagram.
(B) Write the transition matrix.
(C) If it rains on Thursday, what is the probability that the restaurant will be closed on Saturday? On Sunday?

50. *Scheduling.* Repeat Problem 49 if the probability of rain following a rainy day is .6 and the probability of rain following a nonrainy day is .1.

51. *Advertising.* A vigorous television advertising campaign is conducted during the football season to promote a well-known brand X shaving cream. For each of several weeks, a survey is made, and it is found that each week 80% of those using brand X continue to use it and 20% switch to another brand. It is also found that of those not using brand X, 20% switch to brand X while the other 80% continue using another brand.
(A) Draw a transition diagram.
(B) Write the transition matrix.
(C) If 20% of the people are using brand X at the start of the advertising campaign, what percentage will be using it 1 week later? 2 weeks later?

52. *Car rental.* A car rental agency has rental and return facilities at both Kennedy and LaGuardia airports, two of the principal airports in the New York City area. Assume a car rented at either airport must be returned to one or the other airport. If a car is rented at LaGuardia, the probability that it will be returned there is .8; if a car is rented at Kennedy, the probability that it will be returned there is .7. Assume the company rents all its 100 cars each day and each car is rented (and returned) only once a day. If we start with 50 cars at each airport:
(A) What is the expected distribution the next day?
(B) What is the expected distribution 2 days later?

53. *Homeowners insurance.* The market for homeowners insurance in a particular city is dominated by two companies, National Property and United Family. Currently, National Property insures 50% of the homes in the city, United Family insures 30%, and the remainder are insured by a collection of smaller companies. United Family decides to offer rebates to increase its market share. This has the following effects on insurance purchases for the next several years: each year 25% of National Property's customers switch to United Family and 10% switch to other companies; 10% of United Family's customers switch to National Property and 5% switch to other companies; and 15% of the customers of other companies switch to National Property and 35% switch to United Family.
(A) Draw a transition diagram.
(B) Write the transition matrix.
(C) What percentage of the homes will be insured by National Property next year? The year after next?
(D) What percentage of the homes will be insured by United Family next year? The year after next?

54. *Service contracts.* A small community has two heating services that offer annual service contracts for home heating systems, Alpine Heating and Badger Furnaces. Currently, 25% of the homeowners have service contracts with Alpine, 30% have service contracts with Badger, and the remainder do not have service contracts. Both companies launch aggressive advertising campaigns to attract new customers, with the following effects on service contract purchases for the next several years: each year 35% of the homeowners with no current service contract decide to purchase a contract from Alpine and 40% decide to purchase one from Badger. In addition, 10% of the previous customers at each company decide to switch to the other company, and 5% decide they do not want a service contract.
(A) Draw a transition diagram.
(B) Write the transition matrix.
(C) What percentage of the homes will have service contracts with Alpine next year? The year after next?
(D) What percentage of the homes will have service contracts with Badger next year? The year after next?

55. *Employee training.* A nationwide chain of travel agencies maintains a training program for new travel agents. Initially, all new employees are classified as beginning agents requiring extensive supervision. Every 6 months the performance of each agent is reviewed. Past records indicate that after each semiannual review, 40% of the beginning agents are promoted to intermediate agents requiring only minimal supervision, 10% are terminated for unsatisfactory performance, and the remainder continue as beginning agents. Furthermore, 30% of the intermediate agents are promoted to qualified travel agents requiring no supervision, 10% are terminated for unsatisfactory performance, and the remainder continue as intermediate agents.
(A) Draw a transition diagram.
(B) Write the transition matrix.
(C) What is the probability that a beginning agent is promoted to qualified agent within 1 year? Within 2 years?

56. *Employee training.* All welders in a factory begin as apprentices. Every year the performance of each apprentice is reviewed. Past records indicate that after each review, 10% of the apprentices are promoted to professional welder, 20% are terminated for unsatisfactory performance, and the remainder continue as apprentices.
   (A) Draw a transition diagram.
   (B) Write the transition matrix.
   (C) What is the probability that an apprentice is promoted to professional welder within 2 years? Within 4 years?

Life Sciences

57. *Health insurance.* A midwestern university offers its employees three choices for health care: a clinic-based health maintenance organization (HMO), a preferred provider organization (PPO), and a traditional fee-for-service program (FFS). Each year the university designates an open enrollment period during which employees may change from one health plan to another. Prior to the last open enrollment period, 20% of the employees were enrolled in the HMO, 25% in the PPO, and the remainder in the FFS. During the open enrollment period, 15% of the employees in the HMO switched to the PPO and 5% switched to the FFS; 20% of the employees in the PPO switched to the HMO and 10% to the FFS; and 25% of the employees in the FFS switched to the HMO and 30% switched to the PPO.
   (A) Write the transition matrix.
   (B) What percentage of employees were enrolled in each plan after the last open enrollment period?
   (C) If this trend continues, what percentage of employees will be enrolled in each program after the next open enrollment period?

58. *Dental insurance.* Refer to Problem 57. During the open enrollment period, employees at the university can switch between the two available dental care programs, the low-option plan (LOP) and the high-option plan (HOP). Prior to the last open enrollment period, 40% of the employees were enrolled in the LOP and 60% in the HOP. During the open enrollment program, 30% of the employees in the LOP switched to the HOP and 10% of the employees in the HOP switched to the LOP.
   (A) Write the transition matrix.
   (B) What percentage of employees were enrolled in each plan after the last open enrollment period?
   (C) If this trend continues, what percentage of employees will be enrolled in each program after the next open enrollment period?

Social Sciences

59. *Housing trends.* Home ownership in the United States declined during the 1980's, ending a 40 year period of growth. The 1980 census reported that 45.1% of the households in a particular city were homeowners and the remainder were renters. During the next decade, 12% of the homeowners became renters and the rest continued to be homeowners. Likewise, 5% of the renters became homeowners and the rest continued to rent.
   (A) Write the appropriate transition matrix.
   (B) According to this transition matrix, what percentage of households were homeowners in 1990?
   (C) If the transition matrix remains the same, what percentage of the households will be homeowners in 2000?

60. *Housing trends.* The 1980 census reported that 73.3% of the households in a suburban county were homeowners. During the next decade, 5% of the homeowners became renters and the rest continued to be homeowners. Likewise, 3% of the renters became homeowners and the rest continued to rent.
   (A) Write the appropriate transition matrix.
   (B) According to this transition matrix, what percentage of households were homeowners in 1990?
   (C) If the transition matrix remains the same, what percentage of the households will be homeowners in 2000?

SECTION 9-2 Regular Markov Chains

- STATIONARY MATRICES
- REGULAR MARKOV CHAINS
- APPLICATIONS
- C GRAPHING UTILITY APPROXIMATIONS

Given a Markov chain with transition matrix P and initial-state matrix S_0, the entries in the state matrix S_k are the probabilities of being in the corresponding states after k trials. What happens to these probabilities as the number of trials k increases? In this section we establish conditions on the transition matrix P that enable us to determine the long-run behavior of both the state matrices S_k and the powers of the transition matrix P^k.

Stationary Matrices

We begin by considering a concrete example—the toothpaste company discussed in the preceding section. Recall that the transition matrix was given by

$$P = \begin{matrix} & \begin{matrix} A & A' \end{matrix} \\ \begin{matrix} A \\ A' \end{matrix} & \begin{bmatrix} .8 & .2 \\ .6 & .4 \end{bmatrix} \end{matrix} \qquad \begin{matrix} A = \text{Uses brand } A \text{ toothpaste} \\ A' = \text{Uses another brand} \end{matrix}$$

Initially, this company had a 10% share of the toothpaste market. If the probabilities in the transition matrix P remain valid over a long period of time, what will happen to the company's market share? Examining the first several state matrices will give us some insight into this situation (matrix multiplication details are omitted):

$$\begin{aligned} S_0 &= [.1 \quad .9] \\ S_1 = S_0P &= [.62 \quad .38] \\ S_2 = S_1P &= [.724 \quad .276] \\ S_3 = S_2P &= [.7448 \quad .2552] \\ S_4 = S_3P &= [.74896 \quad .25104] \\ S_5 = S_4P &= [.749792 \quad .250208] \\ S_6 = S_5P &= [.7499584 \quad .2500416] \end{aligned}$$

It appears that the state matrices are getting closer and closer to $S = [.75 \quad .25]$ as we proceed to higher states. Let us multiply the matrix S (the matrix the other state matrices appear to be approaching) by the transition matrix:

$$SP = [.75 \quad .25]\begin{bmatrix} .8 & .2 \\ .6 & .4 \end{bmatrix} = [.75 \quad .25] = S$$

No change occurs! The matrix $[.75 \quad .25]$ is called a **stationary matrix.** If we reach this state or are very close to it, the system is said to be at steady-state—that is, later states either will not change or will not change very much. In terms of this example, this means that in the long run a person will purchase brand A with a probability of .75; that is, the company can expect to capture 75% of the market, assuming the transition matrix does not change.

The general definition of a stationary matrix is given in the next box.

Stationary Matrix for a Markov Chain

The state matrix $S = [s_1 \quad s_2 \quad \ldots \quad s_n]$ is a **stationary matrix** for a Markov chain with transition matrix P if

$$SP = S$$

where $s_i \geq 0, i = 1, \ldots, n$, and $s_1 + s_2 + \cdots + s_n = 1$.

Explore–Discuss 1

(A) Suppose the toothpaste company started with only 5% of the market instead of 10%. Write the initial-state matrix, find the next six state matrices, and discuss the behavior of these state matrices as you proceed to higher states.
(B) Repeat part (A) if the company started with 90% of the toothpaste market.

Regular Markov Chains

Does every Markov chain have a unique stationary matrix? And if a Markov chain has a unique stationary matrix, will the successive state matrices always approach this stationary matrix? Unfortunately, the answer to both these questions is no. (See Problems 31–34 in Exercise 9-2.) However, there is one important type of Markov chain for which both questions always can be answered in the affirmative. These are called *regular Markov chains.*

Regular Markov Chains

A transition matrix P is **regular** if some power of P has only positive entries. A Markov chain is a **regular Markov chain** if its transition matrix is regular.

Example 1 **Recognizing Regular Matrices** Which of the following matrices are regular?

(A) $P = \begin{bmatrix} .8 & .2 \\ .6 & .4 \end{bmatrix}$ (B) $P = \begin{bmatrix} 0 & 1 \\ 1 & 0 \end{bmatrix}$ (C) $P = \begin{bmatrix} .5 & .5 & 0 \\ 0 & .5 & .5 \\ 1 & 0 & 0 \end{bmatrix}$

SOLUTION (A) This is the transition matrix for the toothpaste company. Since all the entries in P are positive, we can immediately conclude that P is regular.
(B) P has two 0 entries, so we must examine higher powers of P:

$$P^2 = \begin{bmatrix} 1 & 0 \\ 0 & 1 \end{bmatrix} \quad P^3 = \begin{bmatrix} 0 & 1 \\ 1 & 0 \end{bmatrix} \quad P^4 = \begin{bmatrix} 1 & 0 \\ 0 & 1 \end{bmatrix} \quad P^5 = \begin{bmatrix} 0 & 1 \\ 1 & 0 \end{bmatrix}$$

Since the powers of P oscillate between P and I, the 2×2 identity, all powers of P will contain 0 entries. Hence, P is not regular.
(C) Again, we examine higher powers of P:

$$P^2 = \begin{bmatrix} .25 & .5 & .25 \\ .5 & .25 & .25 \\ .5 & .5 & 0 \end{bmatrix} \quad P^3 = \begin{bmatrix} .375 & .375 & .25 \\ .5 & .375 & .125 \\ .25 & .5 & .25 \end{bmatrix}$$

Since all the entries in P^3 are positive, P is regular.

Matched Problem 1 Which of the following matrices are regular?

(A) $P = \begin{bmatrix} .3 & .7 \\ 1 & 0 \end{bmatrix}$ (B) $P = \begin{bmatrix} 1 & 0 \\ 1 & 0 \end{bmatrix}$ (C) $P = \begin{bmatrix} 0 & 1 & 0 \\ .5 & 0 & .5 \\ .5 & 0 & .5 \end{bmatrix}$

Explore–Discuss 2

Consider the toothpaste company (discussed at the beginning of this section) with regular transition matrix P and stationary matrix S, where

$$P = \begin{bmatrix} .8 & .2 \\ .6 & .4 \end{bmatrix} \quad \text{and} \quad S = [.75 \quad .25]$$

Compare P^2, P^4, and P^8. Discuss any apparent relationships between P^k and S.

The relationships among successive state matrices, powers of the transition matrix, and the stationary matrix for a regular Markov chain are given in Theorem 1. The proof of this theorem is left to more advanced courses.

Theorem 1 PROPERTIES OF REGULAR MARKOV CHAINS

Let P be the transition matrix for a regular Markov chain.

(A) There is a unique stationary matrix S that can be found by solving the equation

$$SP = S$$

(B) Given any initial-state matrix S_0, the state matrices S_k approach the stationary matrix S.

(C) The matrices P^k approach a **limiting matrix** $\overline{\boldsymbol{P}}$**,** where each row of $\overline{P}$ is equal to the stationary matrix S.

Example 2 **Finding the Stationary Matrix** The transition matrix for a Markov chain is

$$P = \begin{bmatrix} .7 & .3 \\ .2 & .8 \end{bmatrix}$$

(A) Find the stationary matrix S.

(B) Discuss the long-run behavior of S_k and P^k.

SOLUTION (A) Since P is regular, the stationary matrix S must exist. To find it, we must solve the equation $SP = S$. Let

$$S = [s_1 \quad s_2]$$

and write

$$[s_1 \quad s_2]\begin{bmatrix} .7 & .3 \\ .2 & .8 \end{bmatrix} = [s_1 \quad s_2]$$

After multiplying the left side, we obtain

$$[(.7s_1 + .2s_2) \quad (.3s_1 + .8s_2)] = [s_1 \quad s_2]$$

which is equivalent to the system

$$\begin{matrix} .7s_1 + .2s_2 = s_1 & \text{or} & -.3s_1 + .2s_2 = 0 \\ .3s_1 + .8s_2 = s_2 & \text{or} & .3s_1 - .2s_2 = 0 \end{matrix} \qquad (1)$$

System (1) is dependent and has an infinite number of solutions. However, we are looking for a solution that is also a state matrix. This gives us another equation that we can add to (1) to obtain a system with a unique solution.

$$\begin{aligned} -.3s_1 + .2s_2 &= 0 \\ .3s_1 - .2s_2 &= 0 \\ s_1 + s_2 &= 1 \end{aligned} \qquad (2)$$

System (2) can be solved using matrix methods or elimination to obtain

$$s_1 = .4 \qquad \text{and} \qquad s_2 = .6$$

Thus,

$$S = [.4 \quad .6]$$

is the stationary matrix.

CHECK

$$SP = [.4 \quad .6]\begin{bmatrix} .7 & .3 \\ .2 & .8 \end{bmatrix} = [.4 \quad .6] = S$$

(B) Given any initial-state matrix S_0, Theorem 1 guarantees that the state matrices S_k will approach the stationary matrix S. Furthermore,

$$P^k = \begin{bmatrix} .7 & .3 \\ .2 & .8 \end{bmatrix}^k \quad \text{approaches the limiting matrix} \quad \overline{P} = \begin{bmatrix} .4 & .6 \\ .4 & .6 \end{bmatrix}$$

Matched Problem 2 The transition matrix for a Markov chain is

$$P = \begin{bmatrix} .6 & .4 \\ .1 & .9 \end{bmatrix}$$

Find the stationary matrix S and the limiting matrix $\overline{P}$.

APPLICATIONS

Example 3 **Insurance** Refer to Example 1 in Section 9-1, where we found the following transition matrix for an insurance company:

$$P = \begin{matrix} & \begin{matrix} A & A' \end{matrix} \\ \begin{matrix} A \\ A' \end{matrix} & \begin{bmatrix} .23 & .77 \\ .11 & .89 \end{bmatrix} \end{matrix} \qquad \begin{aligned} A &= \text{Accident} \\ A' &= \text{No Accident} \end{aligned}$$

If these probabilities remain valid over a long period of time, what percentage of drivers can the company expect to have an accident during any given year?

SOLUTION To determine what happens in the long run, we find the stationary matrix by solving the following system:

$$[s_1 \quad s_2]\begin{bmatrix} .23 & .77 \\ .11 & .89 \end{bmatrix} = [s_1 \quad s_2] \qquad \text{and} \qquad s_1 + s_2 = 1$$

which is equivalent to

$$\begin{aligned} .23s_1 + .11s_2 &= s_1 \\ .77s_1 + .89s_2 &= s_2 \\ s_1 + s_2 &= 1 \end{aligned} \qquad \text{or} \qquad \begin{aligned} -.77s_1 + .11s_2 &= 0 \\ .77s_1 - .11s_2 &= 0 \\ s_1 + s_2 &= 1 \end{aligned}$$

Solving this system, we obtain

$$s_1 = .125 \qquad \text{and} \qquad s_2 = .875$$

The stationary matrix is $[.125 \quad .875]$, which means, in the long run, assuming the transition matrix does not change, about 12.5% of the drivers in the community will have an accident during any given year.

Matched Problem 3 ➡ Refer to Matched Problem 1 in Section 9-1, where we found the following transition matrix for an insurance company:

$$P = \begin{matrix} & \begin{matrix} L & L' \end{matrix} \\ \begin{matrix} L \\ L' \end{matrix} & \begin{bmatrix} .98 & .02 \\ .78 & .22 \end{bmatrix} \end{matrix} \qquad \begin{matrix} L = \text{Low-risk} \\ L' = \text{Not low-risk} \end{matrix}$$

If these probabilities remain valid over a long period of time, what percentage of drivers can the company expect to be in the low-risk category during any given year?

Example 4 ➡ **Employee Evaluation** A company rates every employee as below-average, average, or above-average. Past performance indicates that each year 10% of the below-average employees will raise their rating to average and 25% of the average employees will raise their rating to above-average. On the other hand, 15% of the average employees will lower their rating to below-average, and 15% of the above-average employees will lower their rating to average. Company policy prohibits rating changes from below-average to above-average, or conversely, in a single year. Over the long run, what percentage of employees will receive below-average ratings? Average ratings? Above-average ratings?

Solution First, we find the transition matrix:

$$\begin{matrix} & & \text{Next year} \\ & & \begin{matrix} A^- & A & A^+ \end{matrix} \\ \text{This year} & \begin{matrix} A^- \\ A \\ A^+ \end{matrix} & \begin{bmatrix} .9 & .1 & 0 \\ .15 & .6 & .25 \\ 0 & .15 & .85 \end{bmatrix} \end{matrix} \qquad \begin{matrix} A^- = \text{Below-average} \\ A = \text{Average} \\ A^+ = \text{Above-average} \end{matrix}$$

To determine what happens over the long run, we find the stationary matrix by solving the following system:

$$[s_1 \quad s_2 \quad s_3] \begin{bmatrix} .9 & .1 & 0 \\ .15 & .6 & .25 \\ 0 & .15 & .85 \end{bmatrix} = [s_1 \quad s_2 \quad s_3] \qquad \text{and} \qquad s_1 + s_2 + s_3 = 1$$

which is equivalent to

$$\begin{aligned} .9s_1 + .15s_2 &= s_1 \\ .1s_1 + .6s_2 + .15s_3 &= s_2 \\ .25s_2 + .85s_3 &= s_3 \\ s_1 + s_2 + s_3 &= 1 \end{aligned} \qquad \text{or} \qquad \begin{aligned} -.1s_1 + .15s_2 &= 0 \\ .1s_1 - .4s_2 + .15s_3 &= 0 \\ .25s_2 - .15s_3 &= 0 \\ s_1 + s_2 + s_3 &= 1 \end{aligned}$$

Using Gauss–Jordan elimination to solve this system of four equations with three variables, we obtain

$$s_1 = .36 \qquad s_2 = .24 \qquad s_3 = .4$$

Thus, in the long run, 36% of the employees will be rated as below-average, 24% as average, and 40% as above-average.

Matched Problem 4 ➡ A mail order company classifies its customers as preferred, standard, or infrequent, depending on the number of orders placed in a year. Past records indicate that each year 5% of the preferred customers are reclassified as standard and 12% as infrequent; 5% of the standard customers are reclassified as preferred and 5% as infrequent; and 9% of the infrequent customers are reclassified as preferred and 10% as standard. Assuming these percentages remain valid, what percentage of customers can the company expect to have in each category in the long run?

■ GRAPHING UTILITY APPROXIMATIONS

If P is the transition matrix for a regular Markov chain, then the powers of P approach the limiting matrix $\overline{P}$, where each row of $\overline{P}$ is equal to the stationary matrix S (Theorem 1C). We can use this result to approximate S by computing P^k for sufficiently large values of k. The next example illustrates this approach on a graphing utility.

Example 5 ➡ **Approximating the Stationary Matrix** Compute powers of the transition matrix P to approximate $\overline{P}$ and S to four decimal places. Check the approximation in the equation $SP = S$.

$$P = \begin{bmatrix} .5 & .2 & .3 \\ .7 & .1 & .2 \\ .4 & .1 & .5 \end{bmatrix}$$

SOLUTION To approximate $\overline{P}$ to four decimal places we enter P in a graphing utility (Fig. 1A), set the decimal display to four places, and compute powers of P until all three rows of P^k are identical. Examining the output in Figure 1B, we conclude that

$$\overline{P} = \begin{bmatrix} .4943 & .1494 & .3563 \\ .4943 & .1494 & .3563 \\ .4943 & .1494 & .3563 \end{bmatrix} \quad \text{and} \quad S = [.4943 \quad .1494 \quad .3563]$$

Entering S in the graphing utility and computing SP shows that these matrices are correct to four decimal places (see Fig. 1C).

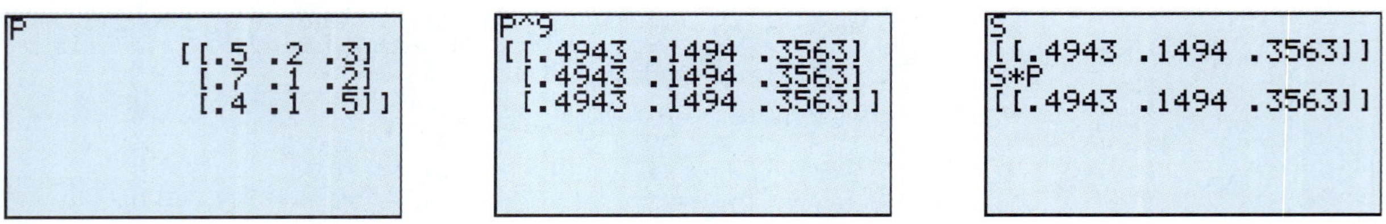

(A) P (B) P^9 (C) Check: $SP = S$

FIGURE 1

Matched Problem 5 ➡ Repeat Example 5 for: $P = \begin{bmatrix} .3 & .6 & .1 \\ .2 & .3 & .5 \\ .1 & .2 & .7 \end{bmatrix}$

REMARKS

1. We used a relatively small value of k to approximate $\overline{P}$ in Example 5. Many graphing utilities will compute P^k for large values of k almost as rapidly as for small values. However, round-off errors can occur in these calculations. A safe procedure is to start with a relatively small value of k, such as $k = 8$, and then keep doubling k until the rows of P^k are identical to the specified number of decimal places.
2. If any of the entries of P^k are approaching 0, the graphing utility may use scientific notation to display these entries as very small numbers. Figure 2 shows the 100th power of a transition matrix P. The entry in Row 2 and Column 2 of P^{100} is approaching 0, but the graphing utility displays it as 5.1538×10^{-53}. If this occurs, simply change this value to 0 in the corresponding entry in $\overline{P}$. Thus, from the output in Figure 2 we conclude that

$$P^k = \begin{bmatrix} 1 & 0 \\ .7 & .3 \end{bmatrix}^k \quad \text{approaches} \quad \overline{P} = \begin{bmatrix} 1 & 0 \\ 1 & 0 \end{bmatrix}$$

FIGURE 2

Answers to Matched Problems

1. (A) Regular (B) Not regular (C) Regular
2. $S = [.2 \quad .8]$; $\overline{P} = \begin{bmatrix} .2 & .8 \\ .2 & .8 \end{bmatrix}$ 3. 97.5%
4. 28% preferred, 43% standard, 29% infrequent
5. $\overline{P} = \begin{bmatrix} .1618 & .2941 & .5441 \\ .1618 & .2941 & .5441 \\ .1618 & .2941 & .5441 \end{bmatrix}$; $S = [.1618 \quad .2941 \quad .5441]$

EXERCISE 9-2

A *In Problems 1–14, determine whether each transition matrix is regular.*

1. $P = \begin{bmatrix} .1 & .9 \\ .7 & .3 \end{bmatrix}$
2. $P = \begin{bmatrix} .8 & .2 \\ .5 & .5 \end{bmatrix}$
3. $P = \begin{bmatrix} 1 & 0 \\ 0 & 1 \end{bmatrix}$
4. $P = \begin{bmatrix} .5 & .5 \\ .5 & .5 \end{bmatrix}$
5. $P = \begin{bmatrix} .2 & .8 \\ 1 & 0 \end{bmatrix}$
6. $P = \begin{bmatrix} .9 & .1 \\ 0 & 1 \end{bmatrix}$
7. $P = \begin{bmatrix} 1 & 0 \\ .6 & .4 \end{bmatrix}$
8. $P = \begin{bmatrix} 0 & 1 \\ .3 & .7 \end{bmatrix}$
9. $P = \begin{bmatrix} .3 & .4 & .3 \\ 0 & 0 & 1 \\ .4 & .2 & .4 \end{bmatrix}$
10. $P = \begin{bmatrix} 0 & 0 & 1 \\ .1 & .1 & .8 \\ .5 & .3 & .2 \end{bmatrix}$
11. $P = \begin{bmatrix} 0 & 1 & 0 \\ .4 & 0 & .6 \\ 0 & 1 & 0 \end{bmatrix}$
12. $P = \begin{bmatrix} .1 & .3 & .6 \\ 0 & 0 & 1 \\ 1 & 0 & 0 \end{bmatrix}$
13. $P = \begin{bmatrix} 0 & 0 & 1 \\ .8 & .1 & .1 \\ 0 & 1 & 0 \end{bmatrix}$
14. $P = \begin{bmatrix} 0 & .5 & .5 \\ 1 & 0 & 0 \\ 1 & 0 & 0 \end{bmatrix}$

B *For each transition matrix P in Problems 15–22, solve the equation $SP = S$ to find the stationary matrix S and the limiting matrix $\overline{P}$.*

15. $P = \begin{bmatrix} .1 & .9 \\ .6 & .4 \end{bmatrix}$
16. $P = \begin{bmatrix} .8 & .2 \\ .3 & .7 \end{bmatrix}$
17. $P = \begin{bmatrix} .5 & .5 \\ .3 & .7 \end{bmatrix}$
18. $P = \begin{bmatrix} .9 & .1 \\ .7 & .3 \end{bmatrix}$
19. $P = \begin{bmatrix} .5 & .1 & .4 \\ .3 & .7 & 0 \\ 0 & .6 & .4 \end{bmatrix}$
20. $P = \begin{bmatrix} .4 & .1 & .5 \\ .2 & .8 & 0 \\ 0 & .5 & .5 \end{bmatrix}$
21. $P = \begin{bmatrix} .8 & .2 & 0 \\ .5 & .1 & .4 \\ 0 & .6 & .4 \end{bmatrix}$
22. $P = \begin{bmatrix} .2 & .8 & 0 \\ .6 & .1 & .3 \\ 0 & .9 & .1 \end{bmatrix}$

In Problems 23 and 24, discuss the validity of each statement. If the statement is always true, explain why. If not, give a counterexample.

23. (A) If two entries of a 2×2 transition matrix P are 0, then P is not regular.
(B) If three entries of a 3×3 transition matrix P are 0, then P is not regular.

24. (A) If P is an $n \times n$ transition matrix for a Markov chain and S is a $1 \times n$ matrix such that $SP = S$, then S is a stationary matrix.
(B) If a transition matrix P for a Markov chain has a stationary matrix S, then P is regular.

In Problems 25–28, approximate the stationary matrix S for each transition matrix P by computing powers of the transition matrix P. Round matrix entries to four decimal places.

25. $P = \begin{bmatrix} .51 & .49 \\ .27 & .73 \end{bmatrix}$

26. $P = \begin{bmatrix} .68 & .32 \\ .19 & .81 \end{bmatrix}$

27. $P = \begin{bmatrix} .5 & .5 & 0 \\ 0 & .5 & .5 \\ .8 & .1 & .1 \end{bmatrix}$

28. $P = \begin{bmatrix} .2 & .2 & .6 \\ .5 & 0 & .5 \\ .5 & 0 & .5 \end{bmatrix}$

C

29. A red urn contains 2 red marbles and 3 blue marbles, and a blue urn contains 1 red marble and 4 blue marbles. A marble is selected from an urn, the color is noted, and the marble is returned to the urn from which it was drawn. The next marble is drawn from the urn whose color is the same as the marble just drawn. Thus, this is a Markov process with two states: draw from the red urn or draw from the blue urn.
(A) Draw a transition diagram for this process.
(B) Write the transition matrix.
(C) Find the stationary matrix and describe the long-run behavior of this process.

30. Repeat Problem 29 if the red urn contains 5 red and 3 blue marbles, and the blue urn contains 1 red and 3 blue marbles.

31. Given the transition matrix:

$$P = \begin{bmatrix} 0 & 1 \\ 1 & 0 \end{bmatrix}$$

(A) Discuss the behavior of the state matrices $S_1, S_2, S_3, \ldots$ for the initial-state matrix $S_0 = [.2 \quad .8]$.
(B) Repeat part (A) for $S_0 = [.5 \quad .5]$.
(C) Discuss the behavior of P^k, $k = 2, 3, 4, \ldots$.
(D) Which of the conclusions of Theorem 1 are not valid for this matrix? Why is this not a contradiction?

32. Given the transition matrix:

$$P = \begin{bmatrix} 0 & 1 & 0 \\ 0 & 0 & 1 \\ 1 & 0 & 0 \end{bmatrix}$$

(A) Discuss the behavior of the state matrices $S_1, S_2, S_3, \ldots$ for the initial-state matrix $S_0 = [.2 \quad .3 \quad .5]$.
(B) Repeat part (A) for $S_0 = [\frac{1}{3} \quad \frac{1}{3} \quad \frac{1}{3}]$.
(C) Discuss the behavior of P^k, $k = 2, 3, 4, \ldots$.
(D) Which of the conclusions of Theorem 1 are not valid for this matrix? Why is this not a contradiction?

33. The transition matrix for a Markov chain is

$$P = \begin{bmatrix} 1 & 0 & 0 \\ .2 & .2 & .6 \\ 0 & 0 & 1 \end{bmatrix}$$

(A) Show that $R = [1 \quad 0 \quad 0]$ and $S = [0 \quad 0 \quad 1]$ are both stationary matrices for P. Explain why this does not contradict Theorem 1A.
(B) Find another stationary matrix for P. [*Hint:* Consider $T = aR + (1 - a)S$, where $0 < a < 1$.]
(C) How many different stationary matrices does P have?

34. The transition matrix for a Markov chain is

$$P = \begin{bmatrix} .7 & 0 & .3 \\ 0 & 1 & 0 \\ .2 & 0 & .8 \end{bmatrix}$$

(A) Show that $R = [.4 \quad 0 \quad .6]$ and $S = [0 \quad 1 \quad 0]$ are both stationary matrices for P. Explain why this does not contradict Theorem 1A.
(B) Find another stationary matrix for P. [*Hint:* Consider $T = aR + (1 - a)S$, where $0 < a < 1$.]
(C) How many different stationary matrices does P have?

Problems 35 and 36 require the use of a graphing utility.

35. Refer to the transition matrix P in Problem 33. What matrix $\overline{P}$ do the powers of P appear to be approaching? Are the rows of $\overline{P}$ stationary matrices for P?

36. Refer to the transition matrix P in Problem 34. What matrix $\overline{P}$ do the powers of P appear to be approaching? Are the rows of $\overline{P}$ stationary matrices for P?

37. The transition matrix for a Markov chain is

$$P = \begin{bmatrix} .1 & .5 & .4 \\ .3 & .2 & .5 \\ .7 & .1 & .2 \end{bmatrix}$$

Let M_k denote the maximum entry in the second column of P^k. Note that $M_1 = .5$.

C (A) Find M_2, M_3, M_4, and M_5 to three decimal places.
(B) Explain why $M_k \geqslant M_{k+1}$ for all positive integers k.

38. The transition matrix for a Markov chain is

$$P = \begin{bmatrix} 0 & .2 & .8 \\ .3 & .3 & .4 \\ .6 & .1 & .3 \end{bmatrix}$$

Let m_k denote the minimum entry in the third column of P^k. Note that $m_1 = .3$.

C (A) Find m_2, m_3, m_4, and m_5 to three decimal places.
(B) Explain why $m_k \leqslant m_{k+1}$ for all positive integers k.

Applications

Business & Economics

39. *Transportation.* Most railroad cars are owned by individual railroad companies. When a car leaves its home railroad's trackage, it becomes part of a national pool of cars and can be used by other railroads. The rules governing the use of these pooled cars are designed to eventually return the car to the home trackage. A particular railroad found that each month 11% of its box cars on the home trackage left to join the national pool and 29% of its box cars in the national pool were returned to the home trackage. If these percentages remain valid for a long period of time, what percentage of its box cars can this railroad expect to have on its home trackage in the long run?

40. *Transportation.* The railroad in Problem 39 also has a fleet of tank cars. If 14% of the tank cars on the home trackage enter the national pool each month and 26% of the tank cars in the national pool are returned to the home trackage each month, what percentage of its tank cars can the railroad expect to have on its home trackage in the long run?

41. *Labor force.* Table 1 gives the percentage of the female population of the United States who were members of the civilian labor force in the indicated years:

Table 1

Year	Percent
1970	43.3
1980	51.5
1990	57.5

Source: U.S. Bureau of Labor Statistics

The following transition matrix P is proposed as a model for the data, where L represents females who are in the labor force and L' represents females who are not in the labor force:

Next decade

$$\begin{array}{ccc} & & \begin{array}{cc} L & L' \end{array} \\ \text{Current decade} & \begin{array}{c} L \\ L' \end{array} & \begin{bmatrix} .93 & .07 \\ .2 & .8 \end{bmatrix} = P \end{array}$$

(A) Let $S_0 = [.433 \quad .567]$, and find S_1 and S_2. (Compute both matrices exactly and then round entries to three decimal places.)
(B) Construct a new table comparing the results from part (A) with the data in Table 1.
(C) According to this transition matrix, what percentage of the female population will be in the labor force in the long run?

42. *Market research.* Table 2 gives the percentage of households that owned a dog over an 8 year period:

Table 2

Year	Percent
1983	42.5
1987	38.2
1991	36.5

Source: American Veterinary Medical Association

The following transition matrix P is proposed as a model for the data, where D represents the households that own a dog:

Next 4 year period

$$\begin{array}{ccc} & & \begin{array}{cc} D & D' \end{array} \\ \text{Current 4 year period} & \begin{array}{c} D \\ D' \end{array} & \begin{bmatrix} .61 & .39 \\ .21 & .79 \end{bmatrix} = P \end{array}$$

(A) Let $S_0 = [.425 \quad .575]$, and find S_1 and S_2. (Compute both matrices exactly and then round entries to three decimal places.)
(B) Construct a new table comparing the results from part (A) with the data in Table 2.
(C) According to this transition matrix, what percentage of the households will own a dog in the long run?

43. *Market share.* Consumers in a certain state can choose between three long-distance telephone services: GTT, NCJ, and Dash. Aggressive marketing by all three companies results in a continual shift of customers among the three services. Each year, GTT loses 5% of its customers to NCJ and 20% to Dash, NCJ loses 15% of its customers to GTT and 10% to Dash, and Dash loses 5% of its customers to GTT and 10% to NCJ. Assuming these percentages remain valid over a long period of time, what is each company's expected market share in the long run?

44. *Market share.* Consumers in a certain area can choose between three package delivery services: APS, GX, and WWP. Each week, APS loses 10% of its customers to GX and 20% to WWP, GX loses 15% of its customers to APS and 10% to WWP, and WWP loses 5% of its customers to APS and 5% to GX. Assuming these percentages remain valid over a long period of time, what is each company's expected market share in the long run?

45. *Insurance.* An auto insurance company classifies its customers in three categories: poor, satisfactory, and preferred. Each year, 40% of those in the poor category are moved to satisfactory, and 20% of those in the satisfactory category are moved to preferred. Also, 20% in the preferred category are moved to the satisfactory category, and 20% in the satisfactory category are moved to the poor category. Customers are never moved from poor to preferred, or conversely, in a single year. Assuming these percentages remain valid over a long period of time, how many customers can the company expect to have in each category in the long run?

46. *Insurance.* Repeat Problem 45 if 40% of the preferred customers are moved to the satisfactory category each year and all other information remains the same.

Problems 47 and 48 require the use of a graphing utility.

47. *Market share.* Acme Soap Company markets one brand of soap, called Standard Acme (SA), and Best Soap Company markets two brands, Standard Best (SB) and Deluxe Best (DB). Currently, Acme has 40% of the market, and the remainder is equally divided between the two Best brands. Acme is considering the introduction of a second brand to get a larger share of the market. A proposed new brand, called brand X, was test-marketed in several large cities, producing the following transition matrix for the consumers' weekly buying habits:

$$P = \begin{array}{c} \\ SB \\ DB \\ SA \\ X \end{array} \begin{array}{c} \begin{array}{cccc} SB & DB & SA & X \end{array} \\ \begin{bmatrix} .4 & .1 & .3 & .2 \\ .3 & .2 & .2 & .3 \\ .1 & .2 & .2 & .5 \\ .3 & .3 & .1 & .3 \end{bmatrix} \end{array}$$

Assuming that P represents the consumers' buying habits over a long period of time, use this transition matrix and the initial-state matrix $S_0 = [.3 \quad .3 \quad .4 \quad 0]$ to compute successive state matrices in order to approximate the elements in the stationary matrix correct to two decimal places. If Acme decides to market this new soap, what is the long-run expected total market share for their two soaps?

48. *Market share.* Refer to Problem 47. The chemists at Acme Soap Company have developed a second new soap, called brand Y. Test-marketing this soap against the three established brands produces the following transition matrix:

$$P = \begin{array}{c} \\ SB \\ DB \\ SA \\ Y \end{array} \begin{array}{c} \begin{array}{cccc} SB & DB & SA & Y \end{array} \\ \begin{bmatrix} .3 & .2 & .2 & .3 \\ .2 & .2 & .2 & .4 \\ .2 & .2 & .4 & .2 \\ .1 & .2 & .3 & .4 \end{bmatrix} \end{array}$$

Proceed as in Problem 47 to approximate the elements in the stationary matrix correct to two decimal places. If Acme decides to market brand Y, what is the long-run expected total market share for Standard Acme and brand Y? Should Acme market brand X or brand Y?

Life Sciences

49. *Genetics.* A given plant species has red, pink, or white flowers according to the genotypes RR, RW, and WW, respectively. If each of these genotypes is crossed with a pink-flowering plant (genotype RW), then the transition matrix is

$$\begin{array}{c} \text{This} \\ \text{generation} \end{array} \begin{array}{c} \\ \text{Red} \\ \text{Pink} \\ \text{White} \end{array} \begin{array}{c} \text{Next generation} \\ \begin{array}{ccc} \text{Red} & \text{Pink} & \text{White} \end{array} \\ \begin{bmatrix} .5 & .5 & 0 \\ .25 & .5 & .25 \\ 0 & .5 & .5 \end{bmatrix} \end{array}$$

Assuming the plants of each generation are crossed only with pink plants to produce the next generation, show that regardless of the makeup of the first generation, the genotype composition will eventually stabilize at 25% red, 50% pink, and 25% white. (Find the stationary matrix.)

50. *Gene mutation.* Suppose a gene in a chromosome is of type A or type B. Assume the probability that a gene of type A will mutate to type B in one generation is 10^{-4} and that a gene of type B will mutate to type A is 10^{-6}.
(A) What is the transition matrix?
(B) After many generations, what is the probability that the gene will be of type A? Of type B? (Find the stationary matrix.)

Social Sciences

51. *Rapid transit.* A new rapid transit system has just started operating. In the first month of operation, it is found that 25% of the commuters are using the system, while 75% still travel by automobile. The following transition matrix was determined from records of other rapid transit systems:

		Next month: Rapid transit	Next month: Automobile
Current month	Rapid transit	.8	.2
	Automobile	.3	.7

(A) What is the initial-state matrix?
(B) What percentage of the commuters will be using the new system after 1 month? After 2 months?
(C) Find the percentage of commuters using each type of transportation after it has been in service for a long time.

52. *Politics—filibuster.* The Senate is in the middle of a floor debate and a filibuster is threatened. Senator Hanks, who is still vacillating, has a probability of .1 of changing his mind during the next 5 minutes. If this pattern continues for each 5 minutes that the debate continues, and if a 24 hour filibuster takes place before a vote is taken, what is the probability that Senator Hanks will cast a yes vote? A no vote?
(A) First, complete the following transition matrix:

		Next 5 minutes: Yes	Next 5 minutes: No
Current 5 minutes	Yes	.9	.1
	No		

(B) Find the stationary matrix and answer the two questions.
(C) What is the stationary matrix if the probability of Senator Hanks changing his mind (.1) is replaced with an arbitrary probability p?

The center of population of the 48 contiguous states of the United States is the point where a flat, rigid map of the contiguous states would balance if the location of each person was represented on the map by a weight of equal measure. In 1790 the population center was 23 miles east of Baltimore, Maryland. By 1990, the center had shifted about 800 miles west and 100 miles south to a point in southeast Missouri. To study this shifting population, the U.S. Bureau of the Census divides the states into the four regions shown in the figure. Problems 53 and 54 deal with population shifts among some of these regions.

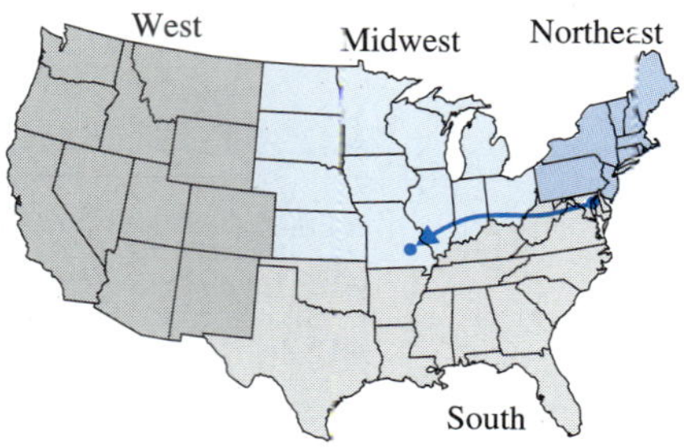

Figure for 53 and 54:
Regions of the United States and the center of population

53. *Population shifts.* Table 3 gives the percentage of the U.S. population living in the south region of the United States in the indicated years:

Table 3

YEAR	PERCENT
1970	30.9
1980	33.3
1990	34.4

Source: U.S. Bureau of the Census

The following transition matrix P is proposed as a model for the data, where S represents the population that lives in the south region:

		Next decade: S	Next decade: S'
Current decade	S	.61	.39
	S'	.21	.79

$$= P$$

(A) Let $S_0 = [.309 \quad .691]$, and find S_1 and S_2. (Compute both matrices exactly and then round entries to three decimal places.)

(B) Construct a new table comparing the results from part (A) with the data in Table 3.

(C) According to this transition matrix, what percentage of the population will live in the south region in the long run?

54. *Population shifts.* Table 4 gives the percentage of the U.S. population living in the northeast region of the United States in the indicated years:

Table 4

YEAR	PERCENT
1970	24.1
1980	21.7
1990	20.4

Source: U.S. Bureau of the Census

The following transition matrix P is proposed as a model for the data, where N represents the population that lives in the northeast region:

$$\begin{array}{c c} & \text{Next decade} \\ & \begin{array}{cc} N & N' \end{array} \\ \text{Current decade} \begin{array}{c} N \\ N' \end{array} & \begin{bmatrix} .61 & .39 \\ .09 & .91 \end{bmatrix} = P \end{array}$$

(A) Let $S_0 = [.241 \quad .759]$, and find S_1 and S_2. (Compute both matrices exactly and then round entries to three decimal places.)

(B) Construct a new table comparing the results from part (A) with the data in Table 4.

(C) According to this transition matrix, what percentage of the population will live in the northeast region in the long run?

SECTION 9-3 Absorbing Markov Chains

- Absorbing States and Absorbing Chains
- Standard Form
- Limiting Matrix
- C Graphing Utility Approximations

In Section 9-2, we saw that the powers of a regular transition matrix always approach a limiting matrix. Not all transition matrices have this property. In this section we discuss another type of Markov chain called an *absorbing Markov chain.* Although regular and absorbing Markov chains have some differences, they have one important similarity: the powers of the transition matrix for an absorbing Markov chain also approach a limiting matrix. After introducing basic concepts, we will develop methods for finding the limiting matrix and discuss the relationship between the states in the Markov chain and the entries in the limiting matrix.

Absorbing States and Absorbing Chains

A state in a Markov chain is called an **absorbing state** if once the state is entered, it is impossible to leave.

Example 1 **Recognizing Absorbing States** Identify any absorbing states for the following transition matrices:

$$\text{(A)}\quad P = \begin{array}{c} A \\ B \\ C \end{array}\overset{\begin{array}{ccc} A & B & C \end{array}}{\begin{bmatrix} 1 & 0 & 0 \\ .5 & .5 & 0 \\ 0 & .5 & .5 \end{bmatrix}} \qquad \text{(B)}\quad P = \begin{array}{c} A \\ B \\ C \end{array}\overset{\begin{array}{ccc} A & B & C \end{array}}{\begin{bmatrix} 0 & 0 & 1 \\ 0 & 1 & 0 \\ 1 & 0 & 0 \end{bmatrix}}$$

SOLUTION (A) The probability of going from state A to state A is 1, and the probability of going from state A to either state B or state C is 0. Thus, once state A is entered, it is impossible to leave; hence, A is an absorbing state. Since the probability of going from state B to state A is nonzero, it is possible to leave B, and B is not an absorbing state. Similarly, the probability of going from state C to state B is nonzero, so C is not an absorbing state.

(B) Reasoning as before, the 1 in Row 2 and Column 2 indicates that state B is an absorbing state. The probability of going from state A to state C and the probability of going from state C to state A are both nonzero. Hence, A and C are not absorbing states.

Matched Problem 1 Identify any absorbing states for the following transition matrices:

$$\text{(A)}\quad P = \begin{array}{c} \\ A \\ B \\ C \end{array}\!\!\begin{array}{c} \begin{array}{ccc} A & B & C \end{array} \\ \begin{bmatrix} .5 & 0 & .5 \\ 0 & 1 & 0 \\ 0 & .5 & .5 \end{bmatrix} \end{array} \qquad \text{(B)}\quad P = \begin{array}{c} \\ A \\ B \\ C \end{array}\!\!\begin{array}{c} \begin{array}{ccc} A & B & C \end{array} \\ \begin{bmatrix} 0 & 1 & 0 \\ 1 & 0 & 0 \\ 0 & 0 & 1 \end{bmatrix} \end{array}$$

The reasoning used in Example 1 to identify absorbing states is generalized in Theorem 1.

Theorem 1 ABSORBING STATES AND TRANSITION MATRICES

A state in a Markov chain is **absorbing** if and only if the row of the transition matrix corresponding to the state has a 1 on the main diagonal and 0's elsewhere.

The presence of an absorbing state in a transition matrix does not guarantee that the powers of the matrix approach a limiting matrix nor that the state matrices in the corresponding Markov chain approach a stationary matrix. For example, if we square the matrix P from Example 1B, we obtain

$$P^2 = \begin{bmatrix} 0 & 0 & 1 \\ 0 & 1 & 0 \\ 1 & 0 & 0 \end{bmatrix}\begin{bmatrix} 0 & 0 & 1 \\ 0 & 1 & 0 \\ 1 & 0 & 0 \end{bmatrix} = \begin{bmatrix} 1 & 0 & 0 \\ 0 & 1 & 0 \\ 0 & 0 & 1 \end{bmatrix} = I$$

Since $P^2 = I$, the 3×3 identity matrix, it follows that

$$P^3 = PP^2 = PI = P \qquad \text{Since } P^2 = I$$
$$P^4 = PP^3 = PP = I \qquad \text{Since } P^3 = P \text{ and } PP = P^2 = I$$

In general, the powers of this transition matrix P oscillate between P and I and do not approach a limiting matrix.

Explore–Discuss 1

(A) For the initial-state matrix $S_0 = [a \quad b \quad c]$, find the first four state matrices, S_1, S_2, S_3, and S_4, in the Markov chain with transition matrix

$$P = \begin{bmatrix} 0 & 0 & 1 \\ 0 & 1 & 0 \\ 1 & 0 & 0 \end{bmatrix}$$

(B) Do the state matrices appear to be approaching a stationary matrix? Discuss.

To ensure that transition matrices for Markov chains with one or more absorbing states have limiting matrices, it is necessary to require the chain to satisfy one additional condition, as stated in the following definition.

Absorbing Markov Chains

A Markov chain is an **absorbing chain** if:

1. There is at least one absorbing state.
2. It is possible to go from each nonabsorbing state to at least one absorbing state in a finite number of steps.

As we saw earlier, absorbing states are easily identified by examining the rows of a transition matrix. It is also possible to use a transition matrix to determine whether a Markov chain is an absorbing chain, but this can be a difficult task, especially if the matrix is large. A transition diagram is often a more appropriate tool for determining whether a Markov chain is absorbing. The next example illustrates this approach for the two matrices discussed in Example 1.

Example 2 **Recognizing Absorbing Markov Chains** Use a transition diagram to determine whether P is the transition matrix for an absorbing Markov chain.

$$\text{(A)}\quad P = \begin{matrix} & \begin{matrix} A & B & C \end{matrix} \\ \begin{matrix} A \\ B \\ C \end{matrix} & \begin{bmatrix} 1 & 0 & 0 \\ .5 & .5 & 0 \\ 0 & .5 & .5 \end{bmatrix} \end{matrix} \qquad \text{(B)}\quad P = \begin{matrix} & \begin{matrix} A & B & C \end{matrix} \\ \begin{matrix} A \\ B \\ C \end{matrix} & \begin{bmatrix} 0 & 0 & 1 \\ 0 & 1 & 0 \\ 1 & 0 & 0 \end{bmatrix} \end{matrix}$$

SOLUTION

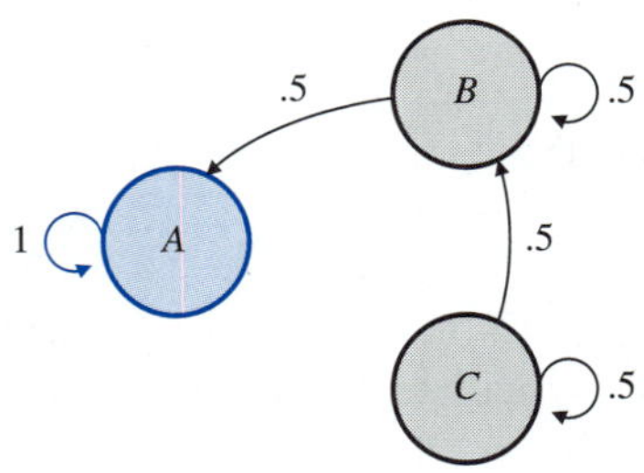

(A) From Example 1A, we know that A is the only absorbing state. The second condition in the definition of an absorbing Markov chain is satisfied if we can show that it is possible to go from the nonabsorbing states B and C to the absorbing state A in a finite number of steps. This is easily determined by drawing a transition diagram. Examining the diagram in the margin, we see that it is possible to go from state B to the absorbing state A in one step and from state C to the absorbing state A in two steps. Thus, P is the transition matrix for an absorbing Markov chain.

(B) Again, we draw the transition diagram for P, as shown. From this diagram it is clear that it is impossible to go from either state A or state C to the absorbing state B. Hence, P is not the transition matrix for an absorbing Markov chain.

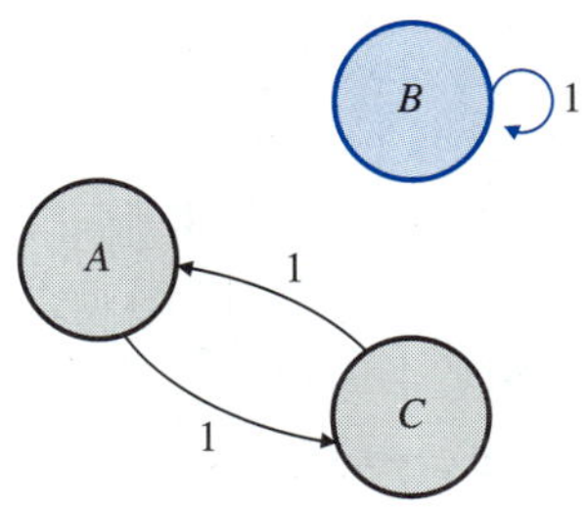

Matched Problem 2 ➡ Use a transition diagram to determine whether P is the transition matrix for an absorbing Markov chain.

$$\text{(A)}\quad P = \begin{array}{c} \\ A \\ B \\ C \end{array}\begin{array}{c} \begin{array}{ccc} A & B & C \end{array} \\ \begin{bmatrix} .5 & 0 & .5 \\ 0 & 1 & 0 \\ 0 & .5 & .5 \end{bmatrix} \end{array} \qquad \text{(B)}\quad P = \begin{array}{c} \\ A \\ B \\ C \end{array}\begin{array}{c} \begin{array}{ccc} A & B & C \end{array} \\ \begin{bmatrix} 0 & 1 & 0 \\ 1 & 0 & 0 \\ 0 & 0 & 1 \end{bmatrix} \end{array}$$

Explore–Discuss 2

Determine whether each statement is true or false. Use examples and verbal arguments to support your conclusions.

(A) A Markov chain with two states, one nonabsorbing and one absorbing, is always an absorbing chain.
(B) A Markov chain with two states, both of which are absorbing, is always an absorbing chain.
(C) A Markov chain with three states, one nonabsorbing and two absorbing, is always an absorbing chain.

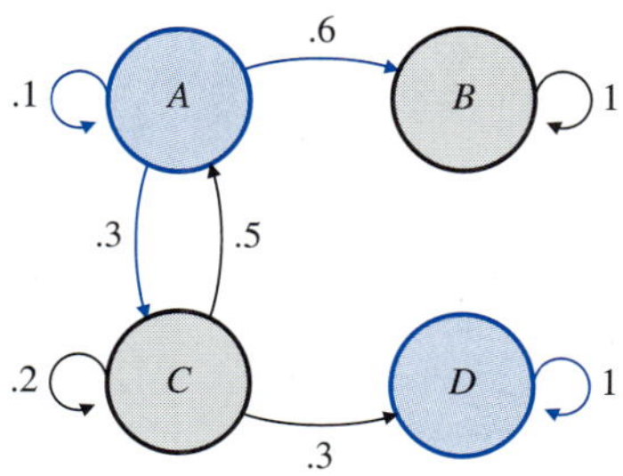

FIGURE 1

Standard Form

The transition matrix for a Markov chain is not unique. Consider the transition diagram in Figure 1. Since there are 4! = 24 different ways to arrange the four states in this diagram, there are 24 different ways to write a transition matrix. (Some of these matrices may have identical entries, but all are different when the row and column labels are taken into account.) For example, matrices M, N, and P shown below are three different transition matrices for this diagram.

$$M = \begin{array}{c} \\ A \\ B \\ C \\ D \end{array}\begin{array}{c} \begin{array}{cccc} A & B & C & D \end{array} \\ \begin{bmatrix} .1 & .6 & .3 & 0 \\ 0 & 1 & 0 & 0 \\ .5 & 0 & .2 & .3 \\ 0 & 0 & 0 & 1 \end{bmatrix} \end{array} \qquad N = \begin{array}{c} \\ D \\ B \\ C \\ A \end{array}\begin{array}{c} \begin{array}{cccc} D & B & C & A \end{array} \\ \begin{bmatrix} 1 & 0 & 0 & 0 \\ 0 & 1 & 0 & 0 \\ .3 & 0 & .2 & .5 \\ 0 & .6 & .3 & .1 \end{bmatrix} \end{array} \qquad P = \begin{array}{c} \\ B \\ D \\ A \\ C \end{array}\begin{array}{c} \begin{array}{cccc} B & D & A & C \end{array} \\ \begin{bmatrix} 1 & 0 & 0 & 0 \\ 0 & 1 & 0 & 0 \\ .6 & 0 & .1 & .3 \\ 0 & .3 & .5 & .2 \end{bmatrix} \end{array} \qquad (1)$$

In matrices N and P, notice that all the absorbing states precede all the nonabsorbing states. A transition matrix written in this form is said to be a *standard form*. We will find standard forms very useful in determining limiting matrices for absorbing Markov chains. The general definition of standard form is given in the next box.

Standard Forms for Absorbing Markov Chains

A transition matrix for an absorbing Markov chain is a **standard form** if the rows and columns are labeled so that all the absorbing states precede all the nonabsorbing states. (There may be more than one standard form.) Any standard form can always be partitioned into four submatrices:

$$\begin{array}{c} \\ A \\ N \end{array}\begin{array}{c} \begin{array}{cc} A & N \end{array} \\ \left[\begin{array}{c|c} I & 0 \\ \hline R & Q \end{array}\right] \end{array} \qquad \begin{array}{l} A = \text{All absorbing states} \\ N = \text{All nonabsorbing states} \end{array}$$

where I is an identity matrix and 0 is a zero matrix.

Referring to the matrix P in (1), we see that the submatrices in this standard form are

$$I = \begin{bmatrix} 1 & 0 \\ 0 & 1 \end{bmatrix} \qquad 0 = \begin{bmatrix} 0 & 0 \\ 0 & 0 \end{bmatrix} \qquad P = \begin{array}{c} \\ B \\ D \\ A \\ C \end{array}\!\!\begin{array}{c} \begin{array}{cccc} B & D & A & C \end{array} \\ \left[\begin{array}{cc:cc} 1 & 0 & 0 & 0 \\ 0 & 1 & 0 & 0 \\ \hdashline .6 & 0 & .1 & .3 \\ 0 & .3 & .5 & .2 \end{array}\right] \end{array}$$

$$R = \begin{bmatrix} .6 & 0 \\ 0 & .3 \end{bmatrix} \qquad Q = \begin{bmatrix} .1 & .3 \\ .5 & .2 \end{bmatrix}$$

Explore–Discuss 3

We used the diagram in Figure 1 to find the transition matrices M, N, and P in (1). Devise a procedure for transforming matrix M into standard forms N and P by interchanging rows and columns in M without reference to the transition diagram. Discuss the relative merits of using transition diagrams versus using row and column rearrangements to find standard forms.

Limiting Matrix

We are now ready to discuss the long-run behavior of absorbing Markov chains. We begin with an application.

Example 3 **Real Estate Development** Two competing real estate companies are trying to buy all the farms in a particular area for future housing development. Each year, 20% of the farmers decide to sell to company A, 30% decide to sell to company B, and the rest continue to farm their land. Neither company ever sells any of the farms they purchase.

(A) Draw a transition diagram for this Markov process and determine whether the associated Markov chain is absorbing.

(B) Write a transition matrix that is in standard form.

(C) If neither company owns any farms at the beginning of this competitive buying process, estimate the percentage of farms that each company will purchase in the long run.

(D) If company A buys 50% of the farms before company B enters this competitive buying process, estimate the percentage of farms that each company will purchase in the long run.

Solution (A)

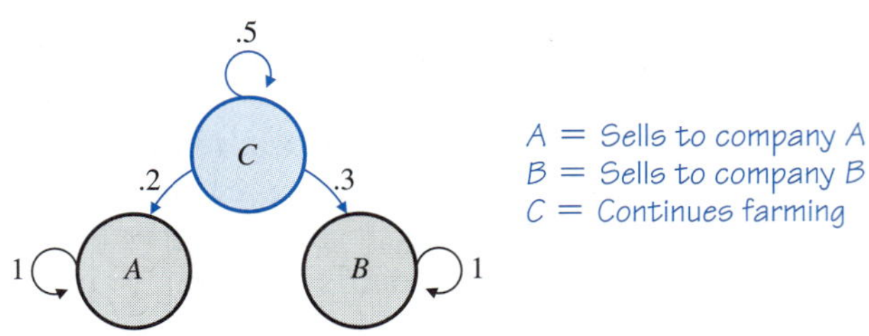

The associated Markov chain is absorbing, since there are two absorbing states, A and B, and it is possible to go from the nonabsorbing state C to either A or B in one step.

(B) We use the transition diagram to write a transition matrix that is in standard form:

$$P = \begin{matrix} & \begin{matrix} A & B & C \end{matrix} \\ \begin{matrix} A \\ B \\ C \end{matrix} & \begin{bmatrix} 1 & 0 & 0 \\ 0 & 1 & 0 \\ .2 & .3 & .5 \end{bmatrix} \end{matrix} \quad \text{Standard form}$$

(C) At the beginning of the competitive buying process all the farmers are in state C (own a farm). Thus, $S_0 = [0 \quad 0 \quad 1]$. The successive state matrices are (multiplication details omitted):

$$\begin{aligned}
S_1 &= S_0P = [.2 \quad .3 \quad .5] \\
S_2 &= S_1P = [.3 \quad .45 \quad .25] \\
S_3 &= S_2P = [.35 \quad .525 \quad .125] \\
S_4 &= S_3P = [.375 \quad .5625 \quad .0625] \\
S_5 &= S_4P = [.3875 \quad .58125 \quad .03125] \\
S_6 &= S_5P = [.39375 \quad .590625 \quad .015625] \\
S_7 &= S_6P = [.396875 \quad .5953125 \quad .0078125] \\
S_8 &= S_7P = [.3984375 \quad .59765625 \quad .00390625] \\
S_9 &= S_8P = [.39921875 \quad .598828125 \quad .001953125]
\end{aligned}$$

It appears that these state matrices are approaching the matrix

$$S = \begin{matrix} \begin{matrix} A & B & C \end{matrix} \\ [.4 \quad .6 \quad 0] \end{matrix}$$

This indicates that in the long run, company A will acquire approximately 40% of the farms and company B will acquire the remaining 60%.

(D) This time, at the beginning of the competitive buying process 50% of the farmers are already in state A and the rest are in state C. Thus, $S_0 = [.5 \quad 0 \quad .5]$. The successive state matrices are (multiplication details omitted):

$$\begin{aligned}
S_1 &= S_0P = [.6 \quad .15 \quad .25] \\
S_2 &= S_1P = [.65 \quad .225 \quad .125] \\
S_3 &= S_2P = [.675 \quad .2625 \quad .0625] \\
S_4 &= S_3P = [.6875 \quad .28125 \quad .03125] \\
S_5 &= S_4P = [.69375 \quad .290625 \quad .015625] \\
S_6 &= S_5P = [.696875 \quad .2953125 \quad .0078125] \\
S_7 &= S_6P = [.6984375 \quad .29765625 \quad .00390625] \\
S_8 &= S_7P = [.69921875 \quad .298828125 \quad .001953125]
\end{aligned}$$

These state matrices approach a matrix different from the one in part (A):

$$S' = \begin{matrix} \begin{matrix} A & B & C \end{matrix} \\ [.7 \quad .3 \quad 0] \end{matrix}$$

Because of its head start, company A will now acquire approximately 70% of the farms and company B will acquire the remaining 30%.

Matched Problem 3 Repeat Example 3 if 10% of the farmers sell to company A each year, 40% sell to company B, and the remainder continue farming.

Recall from Theorem 1 in Section 9-2 that the successive state matrices of a regular Markov chain always approach a stationary matrix. Furthermore, this stationary matrix is unique. That is, changing the initial-state matrix does not change the stationary matrix. The successive state matrices for an absorbing Markov chain also approach a stationary matrix, but this matrix is not unique. To confirm this, consider the transition matrix P and the state matrices S and S' from Example 3:

$$P = \begin{matrix} & \begin{matrix} A & B & C \end{matrix} \\ \begin{matrix} A \\ B \\ C \end{matrix} & \begin{bmatrix} 1 & 0 & 0 \\ 0 & 1 & 0 \\ .2 & .3 & .5 \end{bmatrix} \end{matrix} \qquad S = \begin{matrix} \begin{matrix} A & B & C \end{matrix} \\ [.4 \quad .6 \quad 0] \end{matrix} \qquad S' = \begin{matrix} \begin{matrix} A & B & C \end{matrix} \\ [.7 \quad .3 \quad 0] \end{matrix}$$

It turns out that S and S' are both stationary matrices, as the following multiplications verify:

$$SP = [.4 \quad .6 \quad 0]\begin{bmatrix} 1 & 0 & 0 \\ 0 & 1 & 0 \\ .2 & .3 & .5 \end{bmatrix} = [.4 \quad .6 \quad 0] = S$$

$$S'P = [.7 \quad .3 \quad 0]\begin{bmatrix} 1 & 0 & 0 \\ 0 & 1 & 0 \\ .2 & .3 & .5 \end{bmatrix} = [.7 \quad .3 \quad 0] = S'$$

In fact, this absorbing Markov chain has an infinite number of stationary matrices (see Problems 41 and 42 in Exercise 9-3).

Thus, changing the initial-state matrix for an absorbing Markov chain can cause the successive state matrices to approach a different stationary matrix.

In Section 9-2 we used the unique stationary matrix for a regular Markov chain to find the limiting matrix $\overline{P}$. Since an absorbing Markov chain can have many different stationary matrices, we cannot expect this approach to work for absorbing chains. However, it turns out that transition matrices for absorbing chains do have limiting matrices, and they are not very difficult to find. Theorem 2 gives us the necessary tools. The proof of this theorem is left for more advanced courses.

Theorem 2 ■■ LIMITING MATRICES FOR ABSORBING MARKOV CHAINS

If a standard form P for an absorbing Markov chain is partitioned as

$$P = \left[\begin{array}{c|c} I & 0 \\ \hline R & Q \end{array}\right]$$

then P^k approaches a limiting matrix $\overline{P}$ as k increases, where

$$\overline{P} = \left[\begin{array}{c|c} I & 0 \\ \hline FR & 0 \end{array}\right]$$

The matrix F is given by $F = (I - Q)^{-1}$ and is called the **fundamental matrix** for P.

The identity matrix used to form the fundamental matrix F must be the same size as the matrix Q. ■■

Example 4 **Finding the Limiting Matrix**

(A) Find the limiting matrix $\overline{P}$ for the standard form P found in Example 3.
(B) Use $\overline{P}$ to find the limit of the successive state matrices for $S_0 = [0 \quad 0 \quad 1]$.
(C) Use $\overline{P}$ to find the limit of the successive state matrices for $S_0 = [.5 \quad 0 \quad .5]$.

SOLUTION (A) From Example 3, we have

$$P = \left[\begin{array}{cc:c} 1 & 0 & 0 \\ 0 & 1 & 0 \\ \hdashline .2 & .3 & .5 \end{array}\right] \quad \left[\begin{array}{c:c} I & 0 \\ \hdashline R & Q \end{array}\right]$$

where

$$I = \begin{bmatrix} 1 & 0 \\ 0 & 1 \end{bmatrix} \qquad 0 = \begin{bmatrix} 0 \\ 0 \end{bmatrix} \qquad R = [.2 \quad .3] \qquad Q = [.5]$$

If $I = [1]$ is the 1×1 identity matrix, then $I - Q$ is also a 1×1 matrix, and $F = (I - Q)^{-1}$ is simply the multiplicative inverse of the single entry in $I - Q$. Thus,

$$\begin{aligned} F &= ([1] - [.5])^{-1} = [.5]^{-1} = [2] \\ FR &= [2][.2 \quad .3] = [.4 \quad .6] \end{aligned}$$

and the limiting matrix is

$$\overline{P} = \begin{array}{c} \\ A \\ B \\ C \end{array}\!\!\begin{array}{c} \begin{array}{ccc} A & B & C \end{array} \\ \begin{bmatrix} 1 & 0 & 0 \\ 0 & 1 & 0 \\ .4 & .6 & 0 \end{bmatrix} \end{array} \quad \left[\begin{array}{c:c} I & 0 \\ \hdashline FR & 0 \end{array}\right]$$

(B) Since the successive state matrices are given by $S_k = S_0P^k$ (Theorem 1 in Section 9-1) and P^k approaches $\overline{P}$, it follows that S_k approaches

$$S_0\overline{P} = [0 \quad 0 \quad 1]\begin{bmatrix} 1 & 0 & 0 \\ 0 & 1 & 0 \\ .4 & .6 & 0 \end{bmatrix} = [.4 \quad .6 \quad 0]$$

which agrees with the results in part (C) of Example 3.

(C) This time, the successive state matrices approach

$$S_0\overline{P} = [.5 \quad 0 \quad .5]\begin{bmatrix} 1 & 0 & 0 \\ 0 & 1 & 0 \\ .4 & .6 & 0 \end{bmatrix} = [.7 \quad .3 \quad 0]$$

which agrees with the results in part (D) of Example 3.

Matched Problem 4 Repeat Example 4 for the standard form P found in Matched Problem 3.

Recall that the limiting matrix for a regular Markov chain contains the long-run probabilities of going from any state to any other state. This is also true for the limiting matrix of an absorbing Markov chain. Let's compare the transition matrix P and its limiting matrix $\overline{P}$ from Example 4:

$$P = \begin{matrix} & \begin{matrix} A & B & C \end{matrix} \\ \begin{matrix} A \\ B \\ C \end{matrix} & \begin{bmatrix} 1 & 0 & 0 \\ 0 & 1 & 0 \\ .2 & .3 & .5 \end{bmatrix} \end{matrix} \quad \text{approaches} \quad \overline{P} = \begin{matrix} & \begin{matrix} A & B & C \end{matrix} \\ \begin{matrix} A \\ B \\ C \end{matrix} & \begin{bmatrix} 1 & 0 & 0 \\ 0 & 1 & 0 \\ .4 & .6 & 0 \end{bmatrix} \end{matrix}$$

The rows of P and $\overline{P}$ corresponding to the absorbing states A and B are identical. That is, if the probability of going from state A to state A is 1 at the beginning of the chain, this probability will remain 1 for all trials in the chain and for the limiting matrix. The entries in the third row of $\overline{P}$ give the long-run probabilities of going from the nonabsorbing state C to states A, B, or C.

The fundamental matrix F provides some additional information about an absorbing chain. Recall from Example 4 that $F = [2]$. It can be shown that the entries in F determine the average number of trials it takes to go from a given nonabsorbing state to an absorbing state. In the case of Example 4, the single entry 2 in F indicates that it will take an average of 2 years for a farmer to go from state C (owns a farm) to one of the absorbing states (sells the farm). Some will reach an absorbing state in 1 year, some will take more than 2 years, but the average will be 2 years. These observations are summarized in Theorem 3, which we state below without proof.

Theorem 3 PROPERTIES OF THE LIMITING MATRIX $\overline{P}$

If P is a transition matrix in standard form for an absorbing Markov chain, F is the fundamental matrix, and $\overline{P}$ is the limiting matrix, then:

(A) The entry in Row i and Column j of $\overline{P}$ is the long-run probability of going from state i to state j. For the nonabsorbing states, these probabilities are also the entries in the matrix FR used to form $\overline{P}$.

(B) The sum of the entries in each row of the fundamental matrix F is the average number of trials it will take to go from each nonabsorbing state to some absorbing state.

[Note that the rows of both F and FR correspond to the nonabsorbing states in the order given in the standard form P.]

REMARKS

1. The zero matrix in the lower right corner of the limiting matrix $\overline{P}$ in Theorem 2 indicates that the long-run probability of going from any nonabsorbing state to any other nonabsorbing state is always 0. That is, in the long run, all elements in an absorbing Markov chain end up in one of the absorbing states.
2. If the transition matrix for an absorbing Markov chain is not a standard form, it is still possible to find a limiting matrix (see Problems 37 and 38 in Exercise 9-3). However, it is customary to use a standard form when investigating the limiting behavior of an absorbing chain.

Now that we have developed the necessary tools for analyzing the long-run behavior of an absorbing Markov chain, we apply these tools to an application we considered earlier (see Example 4 in Section 9-1).

Example 5 **University Enrollment** The transition diagram for part-time students enrolled in an MBA program in a university is shown below:

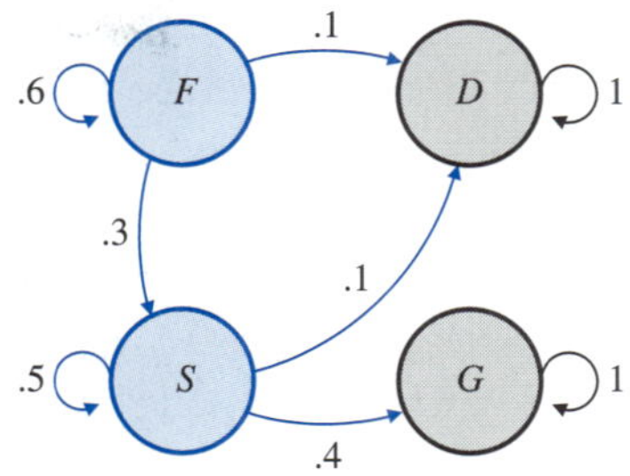

F = First-year students
S = Second-year students
D = Dropouts
G = Graduates

(A) In the long run, what percentage of first-year students will graduate? What percentage of second-year students will not graduate?

(B) What is the average number of years a first-year student will remain in this program? A second-year student?

SOLUTION (A) First, notice that this is an absorbing Markov chain with two absorbing states, state D and state G. A standard form for this absorbing chain is

$$P = \begin{array}{c} \\ D \\ G \\ F \\ S \end{array} \begin{array}{c} \begin{array}{cccc} D & G & F & S \end{array} \\ \left[\begin{array}{cc:cc} 1 & 0 & 0 & 0 \\ 0 & 1 & 0 & 0 \\ \hdashline .1 & 0 & .6 & .3 \\ .1 & .4 & 0 & .5 \end{array}\right] \end{array} \qquad \left[\begin{array}{c:c} I & 0 \\ \hdashline R & Q \end{array}\right]$$

The submatrices in this partition are

$$I = \begin{bmatrix} 1 & 0 \\ 0 & 1 \end{bmatrix} \qquad 0 = \begin{bmatrix} 0 & 0 \\ 0 & 0 \end{bmatrix} \qquad R = \begin{bmatrix} .1 & 0 \\ .1 & .4 \end{bmatrix} \qquad Q = \begin{bmatrix} .6 & .3 \\ 0 & .5 \end{bmatrix}$$

Thus,

$$F = (I - Q)^{-1} = \left(\begin{bmatrix} 1 & 0 \\ 0 & 1 \end{bmatrix} - \begin{bmatrix} .6 & .3 \\ 0 & .5 \end{bmatrix}\right)^{-1}$$

$$= \begin{bmatrix} .4 & -.3 \\ 0 & .5 \end{bmatrix}^{-1}$$

Use row operations to find this matrix inverse.

$$= \begin{bmatrix} 2.5 & 1.5 \\ 0 & 2 \end{bmatrix}$$

and

$$FR = \begin{bmatrix} 2.5 & 1.5 \\ 0 & 2 \end{bmatrix} \begin{bmatrix} .1 & 0 \\ .1 & .4 \end{bmatrix} = \begin{bmatrix} .4 & .6 \\ .2 & .8 \end{bmatrix}$$

The limiting matrix is

$$\bar{P} = \begin{array}{c} \\ D \\ G \\ F \\ S \end{array} \begin{array}{c} \begin{array}{cccc} D & G & F & S \end{array} \\ \begin{bmatrix} 1 & 0 & 0 & 0 \\ 0 & 1 & 0 & 0 \\ .4 & .6 & 0 & 0 \\ .2 & .8 & 0 & 0 \end{bmatrix} \end{array} \qquad \left[\begin{array}{c:c} I & 0 \\ \hdashline FR & 0 \end{array}\right]$$

From this limiting form, we see that in the long run 60% of the first-year students will graduate and 20% of the second-year students will not graduate.

(B) The sum of the entries in the first row of the fundamental matrix F is $2.5 + 1.5 = 4$. According to Theorem 3, this indicates that a first-year student will spend an average of 4 years in the transient states F and S before reaching one of the absorbing states, D or G. The sum of the entries in the second row of F is $0 + 2 = 2$. Thus, a second-year student spends an average of 2 years in the program before either graduating or dropping out.

Matched Problem 5 Repeat Example 5 for the following transition diagram:

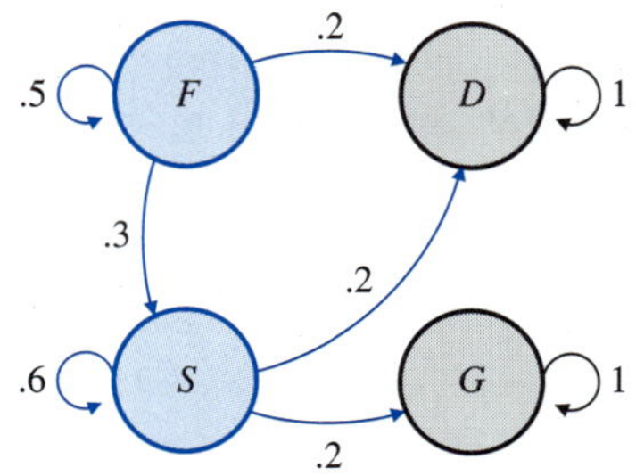

■ GRAPHING UTILITY APPROXIMATIONS

Just as was the case for regular Markov chains, the limiting matrix $\overline{P}$ for an absorbing Markov chain with transition matrix P can be approximated by computing P^k on a graphing utility for sufficiently large values of k. For example, computing P^{50} for the standard form P in Example 5 produces the following results:

$$P^{50} = \begin{bmatrix} 1 & 0 & 0 & 0 \\ 0 & 1 & 0 & 0 \\ .1 & 0 & .6 & .3 \\ .1 & .4 & 0 & .5 \end{bmatrix}^{50} = \begin{bmatrix} 1 & 0 & 0 & 0 \\ 0 & 1 & 0 & 0 \\ .4 & .6 & 0 & 0 \\ .2 & .8 & 0 & 0 \end{bmatrix} = \overline{P}$$

where once again we have replaced very small numbers displayed in scientific notation with 0 (see Remark 2 on page 585 in Section 9-2).

CAUTION

Before you use P^k to approximate $\overline{P}$, be certain to determine that $\overline{P}$ does in fact exist. If you attempt to approximate a limiting matrix when none exists, the results can be misleading. For example, consider the transition matrix

$$P = \begin{bmatrix} 1 & 0 & 0 & 0 & 0 \\ .2 & .2 & 0 & .3 & .3 \\ 0 & 0 & 0 & .5 & .5 \\ 0 & 0 & 1 & 0 & 0 \\ 0 & 0 & 1 & 0 & 0 \end{bmatrix}$$

Computing P^{50} on a graphing utility produces the following matrix:

$$P^{50} = \begin{bmatrix} 1 & 0 & 0 & 0 & 0 \\ .25 & 0 & .625 & .0625 & .0625 \\ 0 & 0 & 1 & 0 & 0 \\ 0 & 0 & 0 & .5 & .5 \\ 0 & 0 & 0 & .5 & .5 \end{bmatrix} \tag{2}$$

It is tempting to stop at this point and conclude that the matrix in (2) must be a good approximation for $\overline{P}$. But to do so would be incorrect! If P^{50} approximates a limiting matrix $\overline{P}$, then P^{51} should also approximate the same matrix. However, computing P^{51} produces quite a different matrix:

$$P^{51} = \begin{bmatrix} 1 & 0 & 0 & 0 & 0 \\ .25 & 0 & .125 & .3125 & .3125 \\ 0 & 0 & 0 & .5 & .5 \\ 0 & 0 & 1 & 0 & 0 \\ 0 & 0 & 1 & 0 & 0 \end{bmatrix} \tag{3}$$

Computing additional powers of P shows that the even powers of P approach the matrix in (2) while the odd powers approach the matrix in (3). Thus, the transition matrix P does not have a limiting matrix.

A graphing utility also can be used to perform the matrix calculations necessary to find $\overline{P}$ exactly, as illustrated in Figure 2 for the transition matrix P from Example 5. This approach has the advantage of producing the fundamental matrix F whose row sums provide additional information about the long-run behavior of the chain.

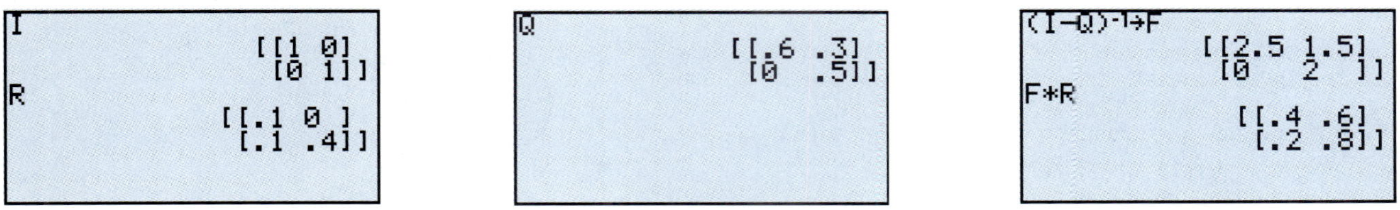

(A) Store I and R in the graphing utility memory

(B) Store Q in the graphing utility memory

(C) Compute F and FR

FIGURE 2
Matrix calculations

Answers to Matched Problems

1. (A) State B is absorbing. (B) State C is absorbing.

2. (A) Absorbing Markov chain (B) Not an absorbing Markov chain

3. (A)

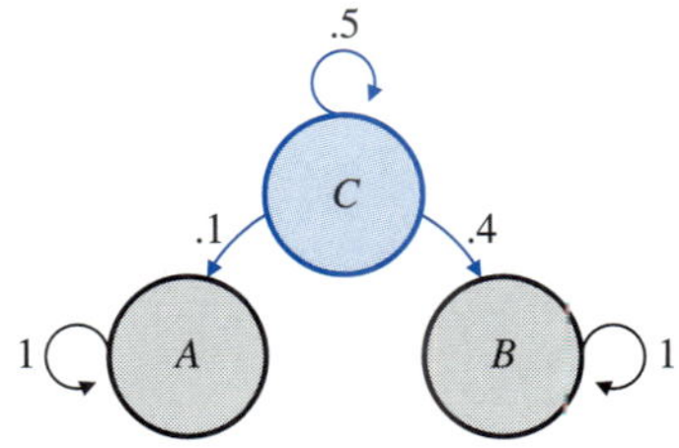

(B) $P = \begin{matrix} & \begin{matrix} A & B & C \end{matrix} \\ \begin{matrix} A \\ B \\ C \end{matrix} & \begin{bmatrix} 1 & 0 & 0 \\ 0 & 1 & 0 \\ .1 & .4 & .5 \end{bmatrix} \end{matrix}$

(C) Company A will purchase 20% of the farms and company B will purchase 80%.

(D) Company A will purchase 60% of the farms and company B will purchase 40%.

4. (A) $\overline{P} = \begin{matrix} & A & B & C \\ A & 1 & 0 & 0 \\ B & 0 & 1 & 0 \\ C & .2 & .8 & 0 \end{matrix}$ (B) [.2 .8 0] (C) [.6 .4 0]

5. (A) 30% of the first-year students will graduate; 50% of the second-year students will not graduate.

(B) A first-year student will spend an average of 3.5 years in the program; a second-year student will spend an average of 2.5 years in the program.

Exercise 9-3

A *In Problems 1–6, identify the absorbing states in the indicated transition matrix.*

1. $P = \begin{matrix} & A & B & C \\ A & .6 & .3 & .1 \\ B & 0 & 1 & 0 \\ C & 0 & 0 & 1 \end{matrix}$

2. $P = \begin{matrix} & A & B & C \\ A & 0 & 1 & 0 \\ B & .3 & .2 & .5 \\ C & 0 & 0 & 1 \end{matrix}$

3. $P = \begin{matrix} & A & B & C \\ A & 0 & 0 & 1 \\ B & 1 & 0 & 0 \\ C & 0 & 1 & 0 \end{matrix}$

4. $P = \begin{matrix} & A & B & C \\ A & 1 & 0 & 0 \\ B & .3 & .4 & .3 \\ C & 0 & 0 & 1 \end{matrix}$

5. $P = \begin{matrix} & A & B & C & D \\ A & 1 & 0 & 0 & 0 \\ B & 0 & 0 & 1 & 0 \\ C & .1 & .1 & .5 & .3 \\ D & 0 & 0 & 0 & 1 \end{matrix}$

6. $P = \begin{matrix} & A & B & C & D \\ A & 0 & 1 & 0 & 0 \\ B & 1 & 0 & 0 & 0 \\ C & .1 & .2 & .3 & .4 \\ D & .7 & .1 & .1 & .1 \end{matrix}$

8.

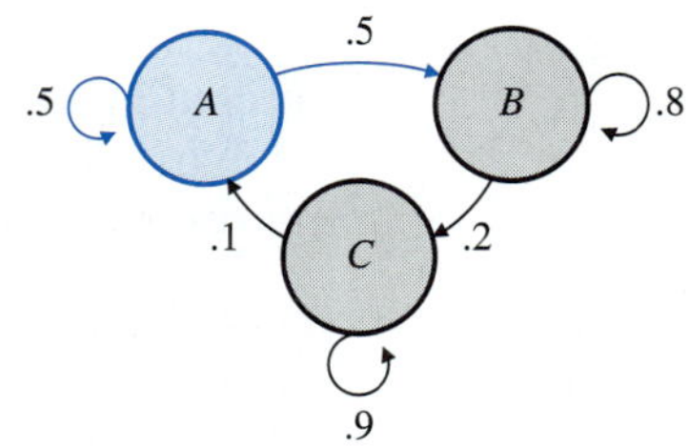

9.

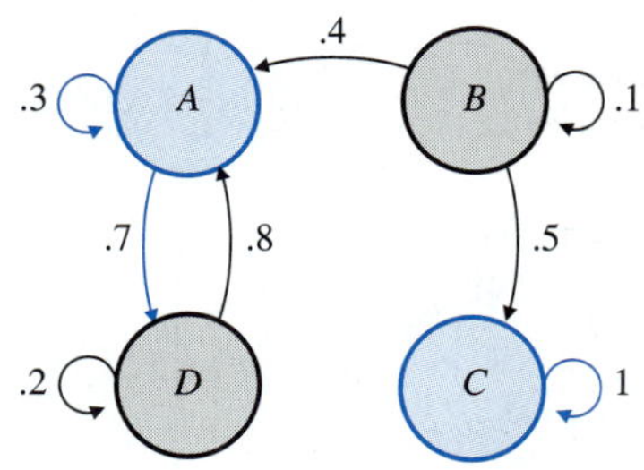

10.

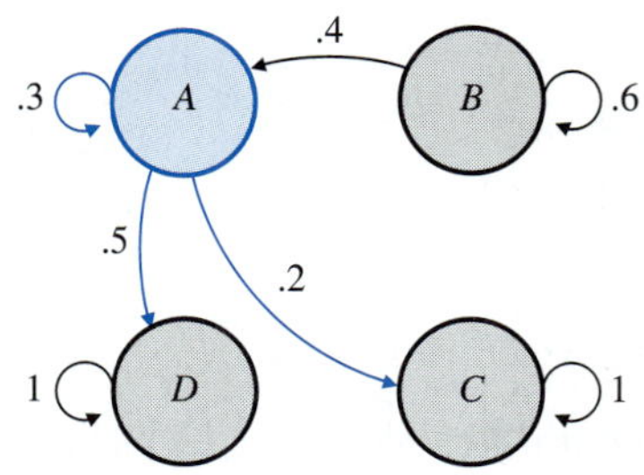

In Problems 7–10, identify the absorbing states for each transition diagram, and determine whether the diagram represents an absorbing Markov chain.

7.

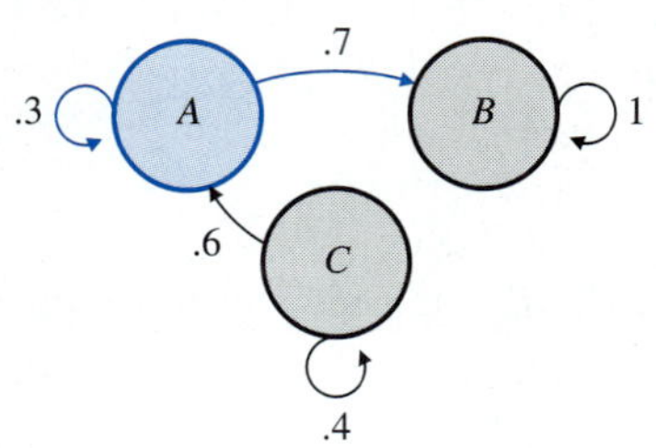

B *In Problems 11–14, find a standard form for the absorbing Markov chain with the indicated transition diagram.*

11.

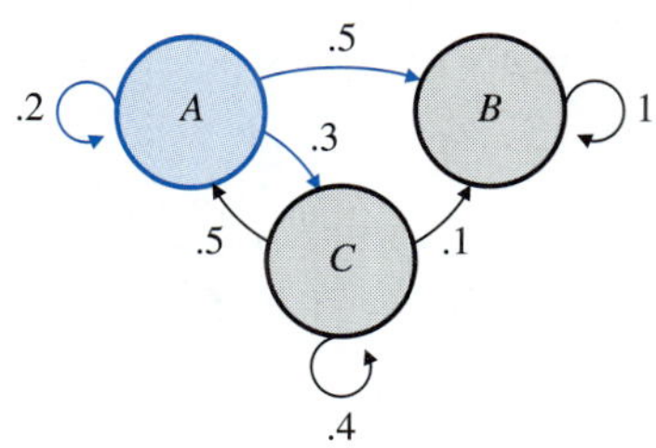

12.

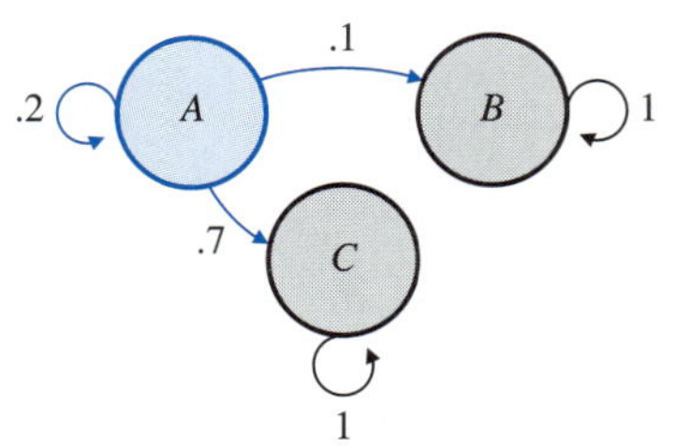

13.

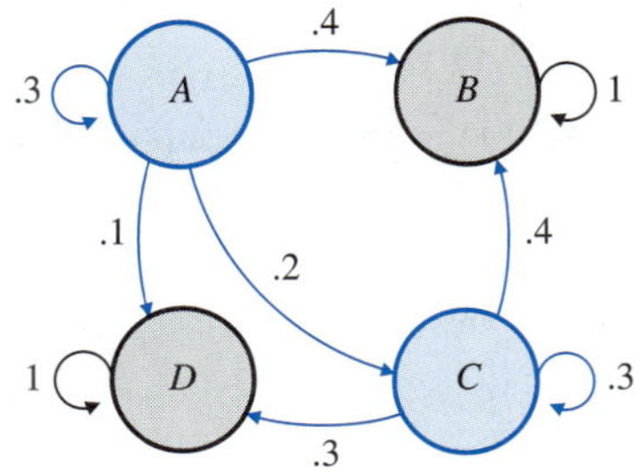

14.

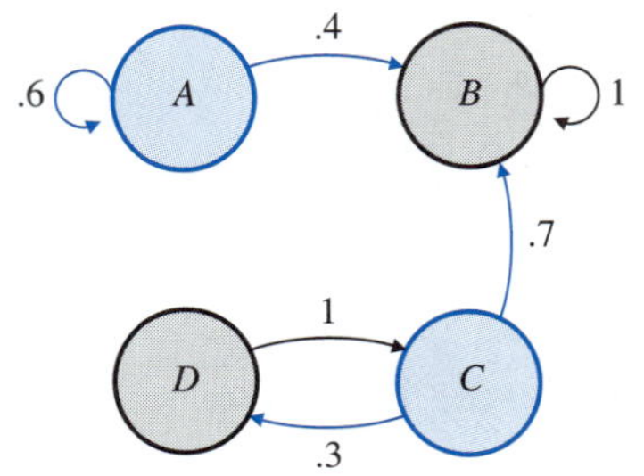

In Problems 15–18, find a standard form for the absorbing Markov chain with the indicated transition matrix.

15. $P = \begin{array}{c} \\ A \\ B \\ C \end{array}\begin{array}{c} \begin{array}{ccc} A & B & C \end{array} \\ \begin{bmatrix} .2 & .3 & .5 \\ 1 & 0 & 0 \\ 0 & 0 & 1 \end{bmatrix} \end{array}$

16. $P = \begin{array}{c} \\ A \\ B \\ C \end{array}\begin{array}{c} \begin{array}{ccc} A & B & C \end{array} \\ \begin{bmatrix} 0 & 0 & 1 \\ 0 & 1 & 0 \\ .7 & .2 & .1 \end{bmatrix} \end{array}$

17. $P = \begin{array}{c} \\ A \\ B \\ C \\ D \end{array}\begin{array}{c} \begin{array}{cccc} A & B & C & D \end{array} \\ \begin{bmatrix} .1 & .2 & .3 & .4 \\ 0 & 1 & 0 & 0 \\ .5 & .2 & .2 & .1 \\ 0 & 0 & 0 & 1 \end{bmatrix} \end{array}$

18. $P = \begin{array}{c} \\ A \\ B \\ C \\ D \end{array}\begin{array}{c} \begin{array}{cccc} A & B & C & D \end{array} \\ \begin{bmatrix} 0 & .3 & .3 & .4 \\ 0 & 1 & 0 & 0 \\ 0 & 0 & 1 & 0 \\ .8 & .1 & .1 & 0 \end{bmatrix} \end{array}$

In Problems 19–24, find the limiting matrix for the indicated standard form. Find the long-run probability of going from each nonabsorbing state to each absorbing state and the average number of trials needed to go from each nonabsorbing state to an absorbing state.

19. $P = \begin{array}{c} \\ A \\ B \\ C \end{array}\begin{array}{c} \begin{array}{ccc} A & B & C \end{array} \\ \begin{bmatrix} 1 & 0 & 0 \\ 0 & 1 & 0 \\ .1 & .4 & .5 \end{bmatrix} \end{array}$

20. $P = \begin{array}{c} \\ A \\ B \\ C \end{array}\begin{array}{c} \begin{array}{ccc} A & B & C \end{array} \\ \begin{bmatrix} 1 & 0 & 0 \\ 0 & 1 & 0 \\ .3 & .2 & .5 \end{bmatrix} \end{array}$

21. $P = \begin{array}{c} \\ A \\ B \\ C \end{array}\begin{array}{c} \begin{array}{ccc} A & B & C \end{array} \\ \begin{bmatrix} 1 & 0 & 0 \\ .2 & .6 & .2 \\ .4 & .2 & .4 \end{bmatrix} \end{array}$

22. $P = \begin{array}{c} \\ A \\ B \\ C \end{array}\begin{array}{c} \begin{array}{ccc} A & B & C \end{array} \\ \begin{bmatrix} 1 & 0 & 0 \\ .1 & .6 & .3 \\ .2 & .2 & .6 \end{bmatrix} \end{array}$

23. $P = \begin{array}{c} \\ A \\ B \\ C \\ D \end{array}\begin{array}{c} \begin{array}{cccc} A & B & C & D \end{array} \\ \begin{bmatrix} 1 & 0 & 0 & 0 \\ 0 & 1 & 0 & 0 \\ .1 & .2 & .6 & .1 \\ .2 & .2 & .3 & .3 \end{bmatrix} \end{array}$

24. $P = \begin{array}{c} \\ A \\ B \\ C \\ D \end{array}\begin{array}{c} \begin{array}{cccc} A & B & C & D \end{array} \\ \begin{bmatrix} 1 & 0 & 0 & 0 \\ 0 & 1 & 0 & 0 \\ .1 & .1 & .7 & .1 \\ .3 & .1 & .4 & .2 \end{bmatrix} \end{array}$

Problems 25–30 refer to the matrices in Problems 19–24, as indicated. Use the limiting matrix $\overline{P}$ found for each transition matrix P in Problems 19–24 to determine the long-run behavior of the successive state matrices for the indicated initial-state matrices.

25. For matrix P from Problem 19 with:
(A) $S_0 = [0 \quad 0 \quad 1]$ (B) $S_0 = [.2 \quad .5 \quad .3]$

26. For matrix P from Problem 20 with:
(A) $S_0 = [0 \quad 0 \quad 1]$ (B) $S_0 = [.2 \quad .5 \quad .3]$

27. For matrix P from Problem 21 with:
(A) $S_0 = [0 \quad 0 \quad 1]$ (B) $S_0 = [.2 \quad .5 \quad .3]$

28. For matrix P from Problem 22 with:
(A) $S_0 = [0 \quad 0 \quad 1]$ (B) $S_0 = [.2 \quad .5 \quad .3]$

29. For matrix P from Problem 23 with:
(A) $S_0 = [0 \quad 0 \quad 0 \quad 1]$ (B) $S_0 = [0 \quad 0 \quad 1 \quad 0]$
(C) $S_0 = [0 \quad 0 \quad .4 \quad .6]$ (D) $S_0 = [.1 \quad .2 \quad .3 \quad .4]$

30. For matrix P from Problem 24 with:
(A) $S_0 = [0 \quad 0 \quad 0 \quad 1]$ (B) $S_0 = [0 \quad 0 \quad 1 \quad 0]$
(C) $S_0 = [0 \quad 0 \quad .4 \quad .6]$ (D) $S_0 = [.1 \quad .2 \quad .3 \quad .4]$

In Problems 31 and 32, discuss the validity of each statement. If the statement is always true, explain why. If not, give a counterexample.

31. (A) If every state of a Markov chain is absorbing, then it is an absorbing chain.
(B) If a Markov chain has absorbing states, then it is an absorbing chain.

32. (A) In an absorbing Markov chain, if a nonabsorbing state is exited, it can never be entered again.
(B) An absorbing Markov chain can have an infinite number of stationary matrices.

In Problems 33–36, use a graphing utility to approximate the limiting matrix for the indicated standard form.

33. $$P = \begin{array}{c} \\ A \\ B \\ C \\ D \end{array} \begin{array}{c} \begin{array}{cccc} A & B & C & D \end{array} \\ \begin{bmatrix} 1 & 0 & 0 & 0 \\ 0 & 1 & 0 & 0 \\ .5 & .3 & .1 & .1 \\ .6 & .2 & .1 & .1 \end{bmatrix} \end{array}$$

34. $$P = \begin{array}{c} \\ A \\ B \\ C \\ D \end{array} \begin{array}{c} \begin{array}{cccc} A & B & C & D \end{array} \\ \begin{bmatrix} 1 & 0 & 0 & 0 \\ 0 & 1 & 0 & 0 \\ .1 & .1 & .5 & .3 \\ 0 & .2 & .3 & .5 \end{bmatrix} \end{array}$$

35. $$P = \begin{array}{c} \\ A \\ B \\ C \\ D \\ E \end{array} \begin{array}{c} \begin{array}{ccccc} A & B & C & D & E \end{array} \\ \begin{bmatrix} 1 & 0 & 0 & 0 & 0 \\ 0 & 1 & 0 & 0 & 0 \\ 0 & .4 & .5 & 0 & .1 \\ 0 & .4 & 0 & .3 & .3 \\ .4 & .4 & 0 & .2 & 0 \end{bmatrix} \end{array}$$

36. $$P = \begin{array}{c} \\ A \\ B \\ C \\ D \\ E \end{array} \begin{array}{c} \begin{array}{ccccc} A & B & C & D & E \end{array} \\ \begin{bmatrix} 1 & 0 & 0 & 0 & 0 \\ 0 & 1 & 0 & 0 & 0 \\ .5 & 0 & 0 & 0 & .5 \\ 0 & .4 & 0 & .2 & .4 \\ 0 & 0 & .1 & .7 & .2 \end{bmatrix} \end{array}$$

C

37. The following matrix P is a nonstandard transition matrix for an absorbing Markov chain:

$$P = \begin{array}{c} \\ A \\ B \\ C \\ D \end{array} \begin{array}{c} \begin{array}{cccc} A & B & C & D \end{array} \\ \begin{bmatrix} .2 & .2 & .6 & 0 \\ 0 & 1 & 0 & 0 \\ .5 & .1 & 0 & .4 \\ 0 & 0 & 0 & 1 \end{bmatrix} \end{array}$$

To find a limiting matrix for P, follow the steps outlined below:

Step 1. Using a transition diagram as an aid, rearrange the columns and rows of P to produce a standard form for this chain.

Step 2. Find the limiting matrix for this standard form.

Step 3. Using a transition diagram as an aid, reverse the process used in step 1 to produce a limiting matrix for the original matrix P.

38. Repeat Problem 37 for

$$P = \begin{array}{c} \\ A \\ B \\ C \\ D \end{array} \begin{array}{c} \begin{array}{cccc} A & B & C & D \end{array} \\ \begin{bmatrix} 1 & 0 & 0 & 0 \\ .3 & .6 & 0 & .1 \\ .2 & .3 & .5 & 0 \\ 0 & 0 & 0 & 1 \end{bmatrix} \end{array}$$

39. Verify the results in Problem 37 by computing P^k on a graphing utility for large values of k.

40. Verify the results in Problem 38 by computing P^k on a graphing utility for large values of k.

41. Show that $S = [x \quad 1 - x \quad 0], 0 \leq x \leq 1$, is a stationary matrix for the transition matrix

$$P = \begin{array}{c} \\ A \\ B \\ C \end{array} \begin{array}{c} \begin{array}{ccc} A & B & C \end{array} \\ \begin{bmatrix} 1 & 0 & 0 \\ 0 & 1 & 0 \\ .1 & .5 & .4 \end{bmatrix} \end{array}$$

Discuss the generalization of this result to any absorbing Markov chain with two absorbing states and one nonabsorbing state.

42. Show that $S = [x \quad 1 - x \quad 0 \quad 0], 0 \leq x \leq 1$, is a stationary matrix for the transition matrix

$$P = \begin{array}{c} \\ A \\ B \\ C \\ D \end{array} \begin{array}{c} \begin{array}{cccc} A & B & C & D \end{array} \\ \begin{bmatrix} 1 & 0 & 0 & 0 \\ 0 & 1 & 0 & 0 \\ .1 & .2 & .3 & .4 \\ .6 & .2 & .1 & .1 \end{bmatrix} \end{array}$$

Discuss the generalization of this result to any absorbing Markov chain with two absorbing states and two nonabsorbing states.

43. An absorbing Markov chain has the following matrix P as a standard form:

$$P = \begin{array}{c} \\ A \\ B \\ C \\ D \end{array} \begin{array}{c} \begin{array}{cccc} A & B & C & D \end{array} \\ \begin{bmatrix} 1 & 0 & 0 & 0 \\ .2 & .3 & .1 & .4 \\ 0 & .5 & .3 & .2 \\ 0 & .1 & .6 & .3 \end{bmatrix} \end{array} \quad \left[\begin{array}{c|c} I & 0 \\ \hline R & Q \end{array}\right]$$

Let w_k denote the maximum entry in Q^k. Note that $w_1 = .6$.

(A) Find w_2, w_4, w_8, w_{16}, and w_{32} to three decimal places.

(B) Describe Q^k when k is large.

44. Refer to the matrices P and Q of Problem 43. For k a positive integer, let $T_k = I + Q + Q^2 + \cdots + Q^k$.

(A) Explain why $T_{k+1} = T_k Q + I$.

(B) Using a graphing utility and part (A) to quickly compute the matrices T_k, discover and describe the connection between $(I - Q)^{-1}$ and T_k when k is large.

Applications

Business & Economics

45. *Loans.* A credit union classifies automobile loans into one of four categories: the loan has been paid in full (F), the account is in good standing (G) with all payments up to date, the account is in arrears (A) with one or more missing payments, or the account has been classified as a bad debt (B) and sold to a collection agency. Past records indicate that each month 10% of the accounts in good standing pay the loan in full, 80% remain in good standing, and 10% become in arrears. Furthermore, 10% of the accounts in arrears are paid in full, 40% become accounts in good standing, 40% remain in arrears, and 10% are classified as bad debts.

(A) In the long run, what percentage of the accounts in arrears will pay their loan in full?

(B) In the long run, what percentage of the accounts in good standing will become bad debts?

(C) What is the average number of months an account in arrears will remain in this system before it is either paid in full or classified as a bad debt?

46. *Employee training.* A national chain of automobile muffler and brake repair shops maintains a training program for its mechanics. All new mechanics begin training in muffler repairs. Every 3 months the performance of each mechanic is reviewed. Past records indicate that after each quarterly review, 30% of the muffler repair trainees are rated as qualified to repair mufflers and begin training in brake repairs, 20% are terminated for unsatisfactory performance, and the remainder continue as muffler repair trainees. Also, 30% of the brake repair trainees are rated as fully qualified mechanics requiring no further training, 10% are terminated for unsatisfactory performance, and the remainder continue as brake repair trainees.

(A) In the long run, what percentage of the muffler repair trainees will become fully qualified mechanics?

(B) In the long run, what percentage of the brake repair trainees will be terminated?

(C) What is the average number of quarters a muffler repair trainee will remain in the training program before being either terminated or promoted to fully qualified mechanic?

47. *Marketing.* Three electronics firms are aggressively marketing their graphing calculators to high school and college mathematics departments by offering volume discounts, complimentary display equipment, and assistance with curriculum development. Due to the amount of equipment involved and the necessary curriculum changes, once a department decides to use a particular calculator in their courses, they never switch to another brand or stop using calculators. Each year, 6% of the departments decide to use calculators from company A, 3% decide to use calculators from company B, 11% decide to use calculators from company C, and the remainder decide not to use any calculators in their courses.

(A) In the long run, what is the market share of each company?

(B) On the average, how many years will it take a department to decide to use calculators from one of these companies in their courses?

48. *Pensions.* Once a year employees at a company are given the opportunity to join one of three pension plans, A, B, or C. Once an employee decides to join one of these plans, the employee cannot drop the plan or switch to another plan. Past records indicate that each year 4% of the employees elect to join plan A, 14% elect to join plan B, 7% elect to join plan C, and the remainder do not join any plan.

(A) In the long run, what percentage of the employees will elect to join plan A? Plan B? Plan C?

(B) On the average, how many years will it take an employee to decide to join a plan?

Life Sciences

49. *Medicine.* After bypass surgery, patients are placed in an intensive care unit (ICU) until their condition stabilizes. Then they are transferred to a cardiac care ward (CCW) where they remain until they are released from the hospital. In a particular metropolitan area, a study of hospital records produced the following data: each day 2% of the patients in the ICU died, 52% were transferred to the CCW, and the remainder stayed in the ICU. Furthermore, each day 4% of the patients in the CCW developed complications and were returned to the ICU, 1% died while in the CCW, 22% were released from the hospital, and the remainder stayed in the CCW.

(A) In the long run, what percentage of the patients in the ICU are released from the hospital?

(B) In the long run, what percentage of the patients in the CCW die without ever being released from the hospital?

(C) What is the average number of days a patient in the ICU will stay in the hospital?

50. *Medicine.* The study discussed in Problem 49 also produced the following data for patients who underwent aortic valve replacements: each day 2% of the patients in the ICU died, 60% were transferred to the CCW, and the remainder stayed in the ICU. Furthermore, each day 5% of the patients in the CCW developed complications and were returned to the ICU, 1% died while in the

CCW, 19% were released from the hospital, and the remainder stayed in the CCW.

(A) In the long run, what percentage of the patients in the CCW are released from the hospital?

(B) In the long run, what percentage of the patients in the ICU die without ever being released from the hospital?

(C) What is the average number of days a patient in the CCW will stay in the hospital?

Social Sciences

51. *Psychology.* A rat is placed in room F or room B of the maze shown in the figure. The rat wanders from room to room until it enters one of the rooms containing food, L or R. Assume that the rat chooses an exit from a room at random and that once it enters a room with food it never leaves.

(A) What is the long-run probability that a rat placed in room B ends up in room R?

(B) What is the average number of exits a rat placed in room B will choose until it finds food?

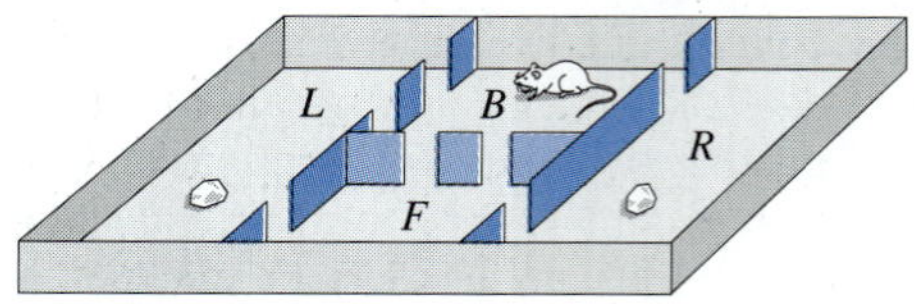

Figure for 51 and 52

52. *Psychology.* Repeat Problem 51 if the exit from room B to room R is blocked.

Important Terms and Symbols

9-1 *Properties of Markov Chains.* Stochastic process; transition diagram; transition probability matrix; initial-state distribution matrix; initial-state probability matrix; first-state matrix; Markov chain or process; kth-state matrix; recursive definition of state matrices; powers of a transition matrix

$$S_k = S_{k-1}P = S_0P^k$$

9-2 *Regular Markov Chains.* Stationary matrix; regular transition matrix; regular Markov chain; limiting matrix

$$SP = S; \quad P^k \text{ approaches } \bar{P}$$

9-3 *Absorbing Markov Chains.* Absorbing state; absorbing Markov chain; standard form; limiting matrix; fundamental matrix

$$P = \left[\begin{array}{c|c} I & 0 \\ \hline R & Q \end{array}\right]; \quad \bar{P} = \left[\begin{array}{c|c} I & 0 \\ \hline FR & 0 \end{array}\right];$$

$$F = (I - Q)^{-1}$$

Review Exercise

Work through all the problems in this chapter review and check your answers in the back of the book. Answers to all review problems are there along with section numbers in italics to indicate where each type of problem is discussed. Where weaknesses show up, review appropriate sections in the text.

A

1. Given the transition matrix P and initial-state matrix S_0 shown below, find S_1 and S_2 and explain what each represents:

$$P = \begin{matrix} & \begin{matrix} A & B \end{matrix} \\ \begin{matrix} A \\ B \end{matrix} & \begin{bmatrix} .6 & .4 \\ .2 & .8 \end{bmatrix} \end{matrix} \qquad S_0 = [.3 \quad .7]$$

In Problems 2–6, P is a transition matrix for a Markov chain. Identify any absorbing states and classify the chain as regular, absorbing, or neither.

2. $P = \begin{matrix} & \begin{matrix} A & B \end{matrix} \\ \begin{matrix} A \\ B \end{matrix} & \begin{bmatrix} 1 & 0 \\ .7 & .3 \end{bmatrix} \end{matrix}$

3. $P = \begin{matrix} & \begin{matrix} A & B \end{matrix} \\ \begin{matrix} A \\ B \end{matrix} & \begin{bmatrix} 0 & 1 \\ .7 & .3 \end{bmatrix} \end{matrix}$

4. $P = \begin{matrix} & \begin{matrix} A & B \end{matrix} \\ \begin{matrix} A \\ B \end{matrix} & \begin{bmatrix} 0 & 1 \\ 1 & 0 \end{bmatrix} \end{matrix}$

5. $$P = \begin{array}{c|ccc} & A & B & C \\ \hline A & .8 & 0 & .2 \\ B & 0 & 1 & 0 \\ C & 0 & 0 & 1 \end{array}$$

6. $$P = \begin{array}{c|cccc} & A & B & C & D \\ \hline A & 1 & 0 & 0 & 0 \\ B & 0 & 1 & 0 & 0 \\ C & 0 & 0 & .3 & .7 \\ D & 0 & 0 & .6 & .4 \end{array}$$

In Problems 7–10, write a transition matrix for the indicated transition diagram, identify any absorbing states, and classify each Markov chain as regular, absorbing, or neither.

7.

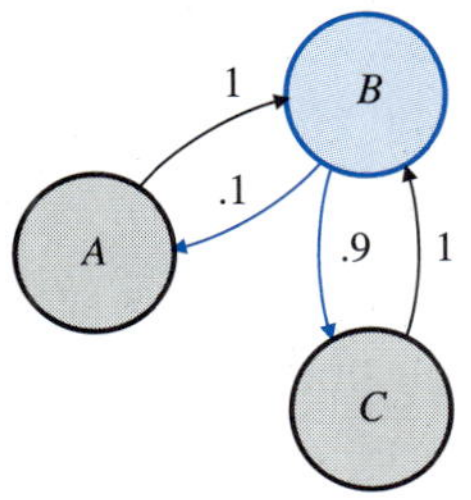

8.

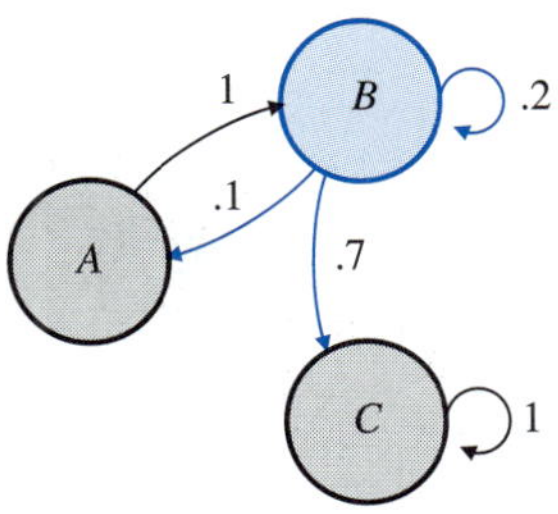

9.

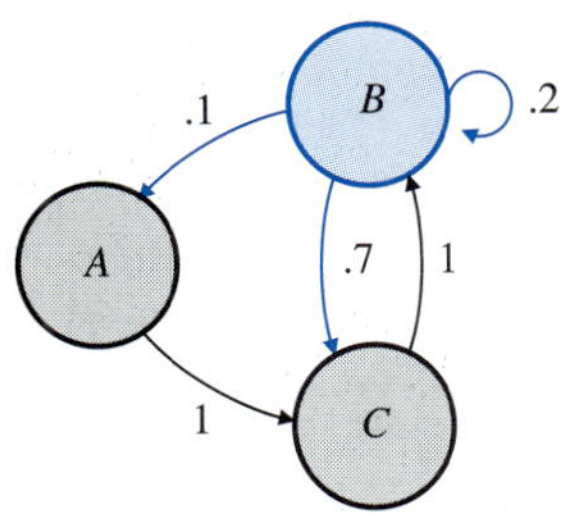

10.

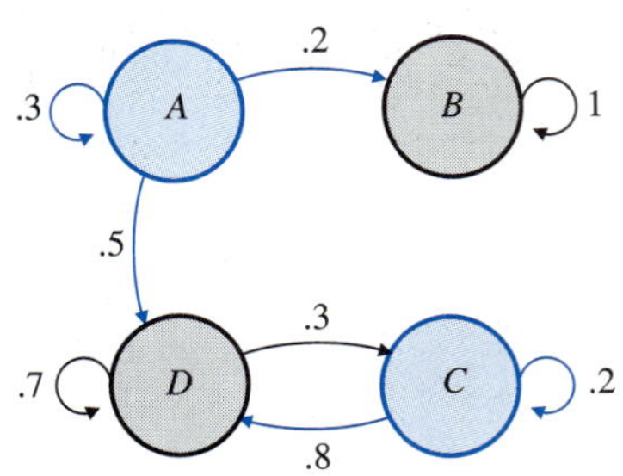

B

11. A Markov chain has three states, *A*, *B*, and *C*. The probability of going from state *A* to state *B* in one trial is .2, the probability of going from state *A* to state *C* in one trial is .5, the probability of going from state *B* to state *A* in one trial is .8, the probability of going from state *B* to state *C* in one trial is .2, the probability of going from state *C* to state *A* in one trial is .1, and the probability of going from state *C* to state *B* in one trial is .3. Draw a transition diagram and write a transition matrix for this chain.

12. Given the transition matrix

$$P = \begin{array}{c|cc} & A & B \\ \hline A & .4 & .6 \\ B & .9 & .1 \end{array}$$

find the probability of:

(A) Going from state *A* to state *B* in two trials.
(B) Going from state *B* to state *A* in three trials.

In Problems 13 and 14, solve the equation $SP = S$ to find the stationary matrix S and the limiting matrix $\bar{P}$.

13. $$P = \begin{array}{c|cc} & A & B \\ \hline A & .4 & .6 \\ B & .2 & .8 \end{array}$$

14. $$P = \begin{array}{c|ccc} & A & B & C \\ \hline A & .4 & .6 & 0 \\ B & .5 & .3 & .2 \\ C & 0 & .8 & .2 \end{array}$$

In Problems 15 and 16, find the limiting matrix for the indicated standard form. Find the long-run probability of going from each nonabsorbing state to each absorbing state and the average number of trials needed to go from each nonabsorbing state to an absorbing state.

15. $$P = \begin{array}{c|ccc} & A & B & C \\ \hline A & 1 & 0 & 0 \\ B & 0 & 1 & 0 \\ C & .3 & .1 & .6 \end{array}$$

16. $$P = \begin{array}{c|cccc} & A & B & C & D \\ \hline A & 1 & 0 & 0 & 0 \\ B & 0 & 1 & 0 & 0 \\ C & .1 & .5 & .2 & .2 \\ D & .1 & .1 & .4 & .4 \end{array}$$

In Problems 17–20, use a graphing utility to approximate the limiting matrix for the indicated transition matrix.

17. Matrix *P* from Problem 13

18. Matrix *P* from Problem 14

19. Matrix *P* from Problem 15

20. Matrix *P* from Problem 16

21. Find a standard form for the absorbing Markov chain with transition matrix

$$P = \begin{array}{c} \\ A \\ B \\ C \\ D \end{array}\begin{array}{c} \begin{array}{cccc} A & B & C & D \end{array} \\ \begin{bmatrix} .6 & .1 & .2 & .1 \\ 0 & 1 & 0 & 0 \\ .3 & .2 & .3 & .2 \\ 0 & 0 & 0 & 1 \end{bmatrix} \end{array}$$

In Problems 22 and 23, determine the long-run behavior of the successive state matrices for the indicated transition matrix and initial-state matrices.

22. $P = \begin{array}{c} \\ A \\ B \\ C \end{array}\begin{array}{c} \begin{array}{ccc} A & B & C \end{array} \\ \begin{bmatrix} 0 & 1 & 0 \\ 0 & 0 & 1 \\ .2 & .6 & .2 \end{bmatrix} \end{array}$

(A) $S_0 = [0 \quad 0 \quad 1]$ (B) $S_0 = [.5 \quad .3 \quad .2]$

23. $P = \begin{array}{c} \\ A \\ B \\ C \end{array}\begin{array}{c} \begin{array}{ccc} A & B & C \end{array} \\ \begin{bmatrix} 1 & 0 & 0 \\ 0 & 1 & 0 \\ .2 & .6 & .2 \end{bmatrix} \end{array}$

(A) $S_0 = [0 \quad 0 \quad 1]$ (B) $S_0 = [.5 \quad .3 \quad .2]$

24. Let P be a 2×2 transition matrix for a Markov chain. Can P be regular if two of its entries are 0? Explain.

25. Let P be a 3×3 transition matrix for a Markov chain. Can P be regular if three of its entries are 0? If four of its entries are 0? Explain.

C

26. A red urn contains 2 red marbles, 1 blue marble, and 1 green marble. A blue urn contains 1 red marble, 3 blue marbles, and 1 green marble. A green urn contains 6 red marbles, 3 blue marbles, and 1 green marble. A marble is selected from an urn, the color is noted, and the marble is returned to the urn from which it was drawn. The next marble is drawn from the urn whose color is the same as the marble just drawn. Thus, this is a Markov process with three states: draw from the red urn, draw from the blue urn, or draw from the green urn.

(A) Draw a transition diagram for this process.
(B) Write the transition matrix P.
(C) Determine whether this chain is regular, absorbing, or neither.
(D) Find the limiting matrix $\overline{P}$, if it exists, and describe the long-run behavior of this process.

27. Repeat Problem 26 if the blue and green marbles are removed from the red urn.

28. Show that $S = [x \quad y \quad z \quad 0]$, where $x + y + z = 1$, $0 \le x \le 1, 0 \le y \le 1, 0 \le z \le 1$, is a stationary matrix for the transition matrix

$$P = \begin{array}{c} \\ A \\ B \\ C \\ D \end{array}\begin{array}{c} \begin{array}{cccc} A & B & C & D \end{array} \\ \begin{bmatrix} 1 & 0 & 0 & 0 \\ 0 & 1 & 0 & 0 \\ 0 & 0 & 1 & 0 \\ .1 & .3 & .4 & .2 \end{bmatrix} \end{array}$$

Discuss the generalization of this result to any absorbing chain with three absorbing states and one nonabsorbing state.

In Problems 29–35, either give an example of a Markov chain with the indicated properties or explain why no such chain can exist.

29. A regular Markov chain with an absorbing state.

30. An absorbing Markov chain that is regular.

31. A regular Markov chain with two different stationary matrices.

32. An absorbing Markov chain with two different stationary matrices.

33. A Markov chain with no limiting matrix.

34. A regular Markov chain with no limiting matrix.

35. An absorbing Markov chain with no limiting matrix.

In Problems 36 and 37, use a graphing utility to approximate the entries (to three decimal places) of the limiting matrix, if it exists, of the indicated transition matrix.

36. $P = \begin{array}{c} \\ A \\ B \\ C \\ D \end{array}\begin{array}{c} \begin{array}{cccc} A & B & C & D \end{array} \\ \begin{bmatrix} .2 & .3 & .1 & .4 \\ 0 & 0 & 1 & 0 \\ 0 & .8 & 0 & .2 \\ 0 & 0 & 1 & 0 \end{bmatrix} \end{array}$

37. $P = \begin{array}{c} \\ A \\ B \\ C \\ D \end{array}\begin{array}{c} \begin{array}{cccc} A & B & C & D \end{array} \\ \begin{bmatrix} .1 & 0 & .3 & .6 \\ .2 & .4 & .1 & .3 \\ .3 & .5 & 0 & .2 \\ .9 & .1 & 0 & 0 \end{bmatrix} \end{array}$

Applications

Business & Economics

38. *Product switching.* A company's brand (X) has 20% of the market. A market research firm finds that if a person uses brand X, the probability is .7 that he or she will buy it next time. On the other hand, if a person does not use brand X (represented by X'), the probability is .5 that he or she will switch to brand X the next time.

(A) Draw a transition diagram.
(B) Write a transition matrix.
(C) Write the initial-state matrix.
(D) Find the first-state matrix and explain what it represents.
(E) Find the stationary matrix.
(F) What percentage of the market will brand X have in the long run if the transition matrix does not change?

39. *Marketing.* Recent technological advances have led to the development of three new milling machines, brand A, brand B, and brand C. Due to the extensive retooling and startup costs, once a company converts its machine shop to one of these new machines, it never switches to another brand. Each year 6% of the machine shops convert to brand A machines, 8% convert to brand B machines, 11% convert to brand C machines, and the remainder continue to use their old machines.

(A) In the long run, what is the market share of each brand?
(B) What is the average number of years a company waits before converting to one of the new milling machines?

40. *Market research.* Table 1 gives the percentage of households that owned a VCR in the indicated years:

Table 1

Year	Percent
1984	10.6
1988	58.0
1992	72.5

Source: Television Bureau of Advertising, Inc.

The following transition matrix P is proposed as a model for the data, where V represents the households that own a VCR:

Four years later (V, V'); Current year (V, V'):

$$\begin{array}{c c} & \begin{array}{cc} V & V' \end{array} \\ \begin{array}{c} V \\ V' \end{array} & \begin{bmatrix} .853 & .147 \\ .553 & .447 \end{bmatrix} \end{array} = P$$

(A) Let $S_0 = [.106 \quad .894]$, and find S_1 and S_2. (Compute both matrices exactly and then round entries to three decimal places.)
(B) Construct a new table comparing the results from part (A) with the data in Table 1.
(C) According to the transition matrix, what percentage of the households will own a VCR in the long run?

41. *Employee training.* In order to become a Fellow of the Society of Actuaries, an individual must pass a series of ten examinations given by the Society. Passage of the first two preliminary exams is a prerequisite for employment as a trainee in the actuarial department of a large insurance company. Each year 15% of the trainees complete the next three exams in the program and become associates of the Society of Actuaries, 5% leave the company, never to return, and the remainder continue as trainees. Furthermore, each year 17% of the associates complete the remaining five exams and become fellows of the Society of Actuaries, 3% leave the company, never to return, and the remainder continue as associates.

(A) In the long run, what percentage of the trainees will become fellows?
(B) In the long run, what percentage of the associates will leave the company?
(C) What is the average number of years a trainee remains in this program before either becoming a fellow or being discharged?

Life Sciences

42. *Genetics.* A given plant species has red, pink, or white flowers according to the genotypes RR, RW, and WW, respectively. If each of these genotypes is crossed with a red-flowering plant, then the transition matrix is

Next generation (Red, Pink, White); This generation (Red, Pink, White):

$$\begin{array}{c c} & \begin{array}{ccc} \text{Red} & \text{Pink} & \text{White} \end{array} \\ \begin{array}{c} \text{Red} \\ \text{Pink} \\ \text{White} \end{array} & \begin{bmatrix} 1 & 0 & 0 \\ .5 & .5 & 0 \\ 0 & 1 & 0 \end{bmatrix} \end{array}$$

If each generation of the plant is crossed only with red plants to produce the next generation, show that eventually all the flowers produced by the plants will be red. (Find the limiting matrix.)

Social Sciences

43. *Farm population.* Table 2 gives the percentage of the U.S. population living on farms in the indicated years:

Table 2

Year	Percent
1980	2.8
1985	2.3
1990	1.9

Source: U.S. Bureau of the Census

The following transition matrix P is proposed as a model for the data, where F represents the population that lives on a farm:

$$\begin{array}{cc} & \text{Five years later} \\ & \begin{array}{cc} F & F' \end{array} \\ \text{Current year} \begin{array}{c} F \\ F' \end{array} & \begin{bmatrix} .751 & .249 \\ .001 & .999 \end{bmatrix} = P \end{array}$$

(A) Let $S_0 = [.028 \quad .972]$, and find S_1 and S_2.
(B) Construct a new table comparing the results from part (A) with the data in Table 2.
(C) According to this transition matrix, what percentage of the population will live on a farm in the long run?

Group Activity 1 Social Mobility

The government of a developing country uses annual income to classify its citizens into three classes: lower (L), middle (M), and upper (U). A Markov chain with transition matrix P is used to model the mobility of the population among the three classes:

$$P = \begin{array}{c} \\ L \\ M \\ U \end{array} \begin{array}{c} \begin{array}{ccc} L & M & U \end{array} \\ \begin{bmatrix} .9 & .1 & 0 \\ .15 & .8 & .05 \\ 0 & .1 & .9 \end{bmatrix} \end{array}$$

Each year, for example, 10% of the population moves from the lower to the middle class, 5% from the middle to the upper class, and so on.

(A) If 75% of the population is now in the lower class, 20% in the middle class, and 5% in the upper class, predict the proportions in each class after 1, 2, and 3 years.
(B) Explain why P is regular, and find the stationary and limiting matrices.

One of the goals of the government is to increase the size of the middle class. Consequently, the government has directed its ministries of economics, education, and human services to promote policies that will increase the movement from the lower to the middle class—that is, to increase the value of a in the transition matrix P':

$$P' = \begin{array}{c} \\ L \\ M \\ U \end{array} \begin{array}{c} \begin{array}{ccc} L & M & U \end{array} \\ \begin{bmatrix} 1-a & a & 0 \\ .15 & .8 & .05 \\ 0 & .1 & .9 \end{bmatrix} \end{array}$$

(C) Find the stationary matrix for P' if $a = .2$. If $a = .3$.
(D) Find the value of a so that in the long run the middle class will be three times the size of the lower class.
(E) Can the lower class be eliminated entirely by increasing a? Explain.

Group Activity 2 Gambler's Ruin

Games of chance between two players that continue until one player goes broke are often referred to as **gambler's ruin problems.** For example, suppose two people who have a total of \$4 between them are playing a simple coin-tossing game. A fair coin is tossed. If the coin comes up heads, player A pays player B \$1. If the coin comes up tails, player B pays player A \$1. The game continues until one player has all the money and the other player is broke, or ruined, hence the name *gambler's ruin.*

This game can be analyzed as an absorbing Markov chain, where the states are the amount of money player A has at each stage of the game. Thus, the possible states are \$0, \$1, \$2, \$3, and \$4. The game ends when player A has \$0 or \$4, so these two states are absorbing. The transition diagram and a standard form for the transition matrix are shown below.

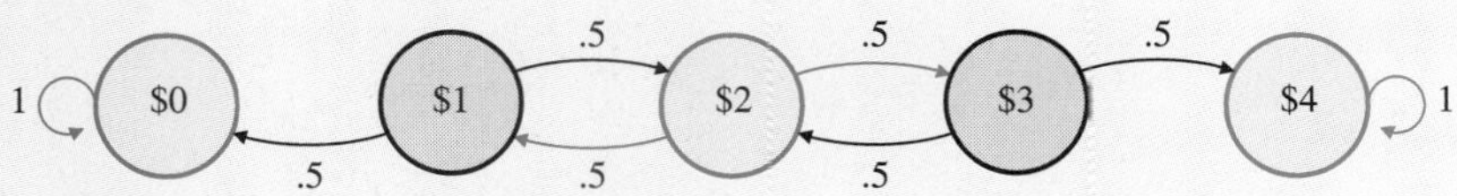

$$P = \begin{array}{c} \\ \$0 \\ \$4 \\ \$1 \\ \$2 \\ \$3 \end{array} \begin{array}{c} \begin{array}{ccccc} \$0 & \$4 & \$1 & \$2 & \$3 \end{array} \\ \begin{bmatrix} 1 & 0 & 0 & 0 & 0 \\ 0 & 1 & 0 & 0 & 0 \\ .5 & 0 & 0 & .5 & 0 \\ 0 & 0 & .5 & 0 & .5 \\ 0 & .5 & 0 & .5 & 0 \end{bmatrix} \end{array}$$

(A) Find the fundamental matrix F and the limiting matrix $\overline{P}$. What is the probability that player A wins all the money if player A starts with \$1? With \$2? With \$3? What is the average number of coin tosses until the game ends if player A starts with \$1? With \$2? With \$3?

(B) Repeat part (A), but assume the players start with a total of \$5 between them. Draw the corresponding transition diagram, and find a standard form for the transition matrix. Be certain to consider all the different amounts of money player A could have at the beginning of the game.

(C) Roulette wheels in Nevada generally have 38 equally spaced slots numbered 00, 0, 1, 2, . . . , 36. The numbers from 1 to 36 are evenly divided between red and black. A player who bets \$1 on black wins \$1 (and gets the bet back) if the ball comes to rest on black; otherwise (if the ball lands on red, 0, or 00), the \$1 bet is lost. A player at a roulette wheel starts with \$3 and plans to continue to make \$1 wagers on black until he either doubles his money or goes broke. Find the probability that the player goes broke. Find the average number of wagers the player will make before either going broke or doubling his money. Solve using Markov chain techniques.

APPENDIX A

Basic Algebra Review

INTRODUCTION

Appendix A reviews some important basic algebra concepts usually studied in earlier courses. The material may be studied systematically before commencing with the rest of the book or reviewed as needed. The Self-Test on Basic Algebra that precedes Section A-1 may be taken to locate areas of weakness. All the answers to the self-test are in the answer section in the back of the book and are keyed to the sections in Appendix A where the related topics are discussed.

SELF-TEST ON BASIC ALGEBRA

Work through all the problems in this self-test and check your answers in the back of the book. All answers are there and are keyed to relevant sections in Appendix A. Where weaknesses show up, review appropriate sections in the appendix.

1. Indicate true (T) or false (F):
(A) $7 \notin \{4, 6, 8\}$ (B) $\{8\} \subset \{4, 6, 8\}$
(C) $\varnothing \in \{4, 6, 8\}$ (D) $\varnothing \subset \{4, 6, 8\}$

2. Replace each question mark with an appropriate expression that will illustrate the use of the indicated real number property:
(A) Commutative ($\cdot$): $x(y + z) = ?$
(B) Associative (+): $2 + (x + y) = ?$
(C) Distributive: $(2 + 3)x = ?$

Problems 3–7 refer to the following polynomials:
(A) $3x - 4$
(B) $x + 2$
(C) $3x^2 + x - 8$
(D) $x^3 + 8$

3. Add all four.

4. Subtract the sum of (A) and (C) from the sum of (B) and (D).

5. Multiply (C) and (D).

6. What is the degree of (D)?

7. What is the coefficient of the second term in (C)?

In Problems 8–13, perform the indicated operations and simplify.

8. $5x^2 - 3x[4 - 3(x - 2)]$

9. $(2x + y)(3x - 4y)$

10. $(2a - 3b)^2$

11. $(2x - y)(2x + y) - (2x - y)^2$

12. $(m^2 + 2mn - n^2)(m^2 - 2mn - n^2)$

13. $(x - 2y)^3$

14. Write in scientific notation:
(A) 4,065,000,000,000 (B) 0.0073

15. Write in standard decimal form:
(A) 2.55×10^8 (B) 4.06×10^{-4}

16. If $U = \{2, 4, 5, 6, 8\}$, $M = \{2, 4, 5\}$, and $N = \{5, 6\}$, find:
(A) $M \cup N$ (B) $M \cap N$
(C) $(M \cup N)'$ (D) $M \cap N'$

17. Given the Venn diagram shown in the figure, how many elements are in each of the following sets:
(A) $A \cup B$ (B) $A \cap B$
(C) $(A \cup B)'$ (D) $A \cap B'$

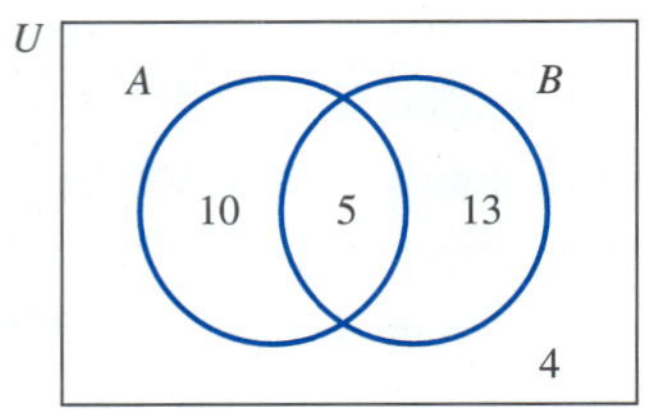

18. Indicate true (T) or false (F):
(A) A natural number is a rational number.
(B) A number with a repeating decimal expansion is an irrational number.

19. Give an example of an integer that is not a natural number.

Simplify Problems 20–28 and write answers using positive exponents only. All variables represent positive real numbers.

20. $6(xy^3)^5$

21. $\dfrac{9u^8v^6}{3u^4v^8}$

22. $(2 \times 10^5)(3 \times 10^{-3})$

23. $(x^{-3}y^2)^{-2}$

24. $u^{5/3}u^{2/3}$

25. $(9a^4b^{-2})^{1/2}$

26. $\dfrac{5^0}{3^2} + \dfrac{3^{-2}}{2^{-2}}$

27. $(x^{1/2} + y^{1/2})^2$

28. $(3x^{1/2} - y^{1/2})(2x^{1/2} + 3y^{1/2})$

Write Problems 29–34 in completely factored form relative to the integers. If a polynomial cannot be factored further relative to the integers, say so.

29. $12x^2 + 5x - 3$

30. $8x^2 - 18xy + 9y^2$

31. $t^2 - 4t - 6$

32. $6n^3 - 9n^2 - 15n$

33. $(4x - y)^2 - 9x^2$

34. $2x^2 + 4xy - 5y^2$

In Problems 35–40, perform the indicated operations and reduce to lowest terms. Represent all compound fractions as simple fractions reduced to lowest terms.

35. $\dfrac{2}{5b} - \dfrac{4}{3a^3} - \dfrac{1}{6a^2b^2}$

36. $\dfrac{3x}{3x^2 - 12x} + \dfrac{1}{6x}$

37. $\dfrac{x}{x^2 - 16} - \dfrac{x + 4}{x^2 - 4x}$

38. $\dfrac{y - 2}{y^2 - 4y + 4} \div \dfrac{y^2 + 2y}{y^2 + 4y + 4}$

39. $\dfrac{\dfrac{1}{7 + h} - \dfrac{1}{7}}{h}$

40. $\dfrac{x^{-1} + y^{-1}}{x^{-2} - y^{-2}}$

41. Each statement illustrates the use of one of the following real number properties or definitions. Indicate which one.

Commutative $(+, \cdot)$	Associative $(+, \cdot)$	Distributive
Identity $(+, \cdot)$	Inverse $(+, \cdot)$	Subtraction
Division	Negatives	Zero

(A) $(-7) - (-5) = (-7) + [-(-5)]$
(B) $5u + (3v + 2) = (3v + 2) + 5u$
(C) $(5m - 2)(2m + 3) = (5m - 2)2m + (5m - 2)3$
(D) $9 \cdot (4y) = (9 \cdot 4)y$
(E) $\dfrac{u}{-(v - w)} = -\dfrac{u}{v - w}$
(F) $(x - y) + 0 = (x - y)$

42. In a freshman class of 100 students, 70 are taking English, 45 are taking math, and 25 are taking both English and math.
(A) How many students are taking either English or math?
(B) How many students are taking English and not math?

43. Change to rational exponent form:
$6\sqrt[5]{x^2} - 7\sqrt[4]{(x - 1)^3}$

44. Change to radical form: $2x^{1/2} - 3x^{2/3}$

45. Write in the form $ax^p + bx^q$, where a and b are real numbers and p and q are rational numbers:

$$\frac{4\sqrt{x} - 3}{2\sqrt{x}}$$

In Problems 46 and 47, rationalize the denominator.

46. $\dfrac{3x}{\sqrt{3x}}$

47. $\dfrac{x-5}{\sqrt{x}-\sqrt{5}}$

In Problems 48 and 49, rationalize the numerator.

48. $\dfrac{\sqrt{x-5}}{x-5}$

49. $\dfrac{\sqrt{u+h}-\sqrt{u}}{h}$

Solve Problems 50–54 for x.

50. $\dfrac{x}{12}-\dfrac{x-3}{3}=\dfrac{1}{2}$

51. $x^2 = 5x$

52. $3x^2 - 21 = 0$

53. $x^2 - x - 20 = 0$

54. $2x = 3 + \dfrac{1}{x}$

In Problems 55–57, solve and graph on a real number line.

55. $2(x + 4) > 5x - 4$

56. $1 - \dfrac{x-3}{3} \leq \dfrac{1}{2}$

57. $-2 \leq \dfrac{x}{2} - 3 < 3$

In Problems 58 and 59, solve for y in terms of x.

58. $2x - 3y = 6$

59. $xy - y = 3$

60. If $A \cap B = A$, then is it always true that $A \subset B$? Explain.

Applications

61. *Economics.* If the Gross Domestic Product (GDP) was $5,951,000,000,000 for the United States in 1992 and the population was 255,100,000, determine the GDP per person using scientific notation. Express the answer in scientific notation and in standard decimal form to the nearest dollar.

62. *Investment.* An investor has $60,000 to invest. If part is invested at 8% and the rest at 14%, how much should be invested at each rate to yield 12% on the total amount?

63. *Break-even analysis.* A producer of educational videos is producing an instructional video. The producer estimates it will cost $72,000 to shoot the video and $12 per unit to copy and distribute the tape. If the wholesale price of the tape is $30, how many tapes must be sold for the producer to break even?

Section A-1 Sets

- Set Properties and Set Notation
- Set Operations
- Application

In this section we will review a few key ideas from set theory. Set concepts and notation not only help us talk about certain mathematical ideas with greater clarity and precision, but are indispensable to a clear understanding of probability.

Set Properties and Set Notation

We can think of a **set** as any collection of objects specified in such a way that we can tell whether any given object is or is not in the collection. Capital letters, such as *A*, *B*, and *C*, are often used to designate particular sets. Each object in a set is called a **member,** or **element,** of the set. Symbolically:

$a \in A$	means	"a is an element of set A"
$a \notin A$	means	"a is not an element of set A"

A set without any elements is called the **empty,** or **null, set.** For example, the set of all people over 20 feet tall is an empty set. Symbolically:

$\varnothing$	represents	"the empty, or null, set"

A set is usually described either by listing all its elements between braces { } or by enclosing a rule within braces that determines the elements of the set. Thus, if $P(x)$ is a statement about x, then

$S = \{x \mid P(x)\}$	means	"S is the set of all x such that $P(x)$ is true"

Recall that the vertical bar within the braces is read "such that." The following example illustrates the rule and listing methods of representing sets.

Example 1 **Representing Sets**

Rule		Listing
$\{x \mid x \text{ is a weekend day}\}$	=	{Saturday, Sunday}
$\{x \mid x^2 = 4\}$	=	$\{-2, 2\}$
$\{x \mid x \text{ is an odd positive counting number}\}$	=	$\{1, 3, 5, \ldots\}$

The three dots (. . .) in the last set given in Example 1 indicate that the pattern established by the first three entries continues indefinitely. The first two sets in Example 1 are **finite sets** (we intuitively know that the elements can be counted, and there is an end); the last set is an **infinite set** (we intuitively know that there is no end in counting the elements). When listing the elements in a set, we do not list an element more than once, and the order in which the elements are listed does not matter.

Matched Problem 1 Let G be the set of all numbers such that $x^2 = 9$.

(A) Denote G by the rule method. (B) Denote G by the listing method.
(C) Indicate whether the following are true or false: $3 \in G$; $9 \notin G$.

If each element of a set A is also an element of set B, we say that A is a **subset** of B. For example, the set of all women students in a class is a subset of the whole class. Note that the definition allows a set to be a subset of itself. If set A and set B have exactly the same elements, then the two sets are said to be **equal.** Symbolically:

$A \subset B$	means	"A is a subset of B"
$A = B$	means	"A and B have exactly the same elements"
$A \not\subset B$	means	"A is not a subset of B"
$A \neq B$	means	"A and B do not have exactly the same elements"

It can be proved that:

$\varnothing$ is a subset of every set.

Example 2 ➡ **Set Notation** If $A = \{-3, -1, 1, 3\}$, $B = \{3, -3, 1, -1\}$, and $C = \{-3, -2, -1, 0, 1, 2, 3\}$, then each of the following statements is true:

$$\begin{array}{lll} A = B & A \subset C & A \subset B \\ C \neq A & C \not\subset A & B \subset A \\ \varnothing \subset A & \varnothing \subset C & \varnothing \notin A \end{array}$$

Matched Problem 2 ➡ Given $A = \{0, 2, 4, 6\}$, $B = \{0, 1, 2, 3, 4, 5, 6\}$, and $C = \{2, 6, 0, 4\}$, indicate whether the following relationships are true (T) or false (F):

(A) $A \subset B$ (B) $A \subset C$ (C) $A = C$
(D) $C \subset B$ (E) $B \not\subset A$ (F) $\varnothing \subset B$

Example 3 ➡ **Subsets** List all the subsets of the set $\{a, b, c\}$.

SOLUTION $\{a, b, c\}$, $\{a, b\}$, $\{a, c\}$, $\{b, c\}$, $\{a\}$, $\{b\}$, $\{c\}$, $\varnothing$

Matched Problem 3 ➡ List all the subsets of the set $\{1, 2\}$.

Explore–Discuss 1

Which of the following statements are true?

(A) $\varnothing \subset \varnothing$ (B) $\varnothing \in \varnothing$ (C) $\varnothing = \{0\}$ (D) $\varnothing \subset \{0\}$

Set Operations

The **union** of sets A and B, denoted by $A \cup B$, is the set of all elements formed by combining all the elements of A and all the elements of B into one set. Symbolically:

Union

$$A \cup B = \{x | x \in A \textbf{ or } x \in B\}$$

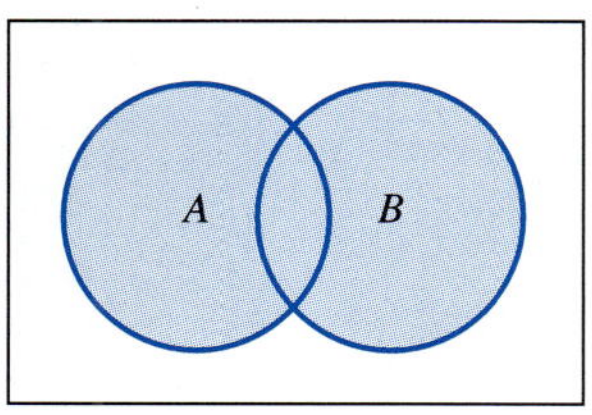

FIGURE 1
$A \cup B$ is the shaded region.

Here we use the word **or** in the way it is always used in mathematics; that is, x may be an element of set A or set B or both.

Venn diagrams are useful in visualizing set relationships. The union of two sets can be illustrated as shown in Figure 1. Note that

$$A \subset A \cup B \quad \text{and} \quad B \subset A \cup B$$

The **intersection** of sets A and B, denoted by $A \cap B$, is the set of elements in set A that are also in set B. Symbolically:

Intersection

$$A \cap B = \{x | x \in A \textbf{ and } x \in B\}$$

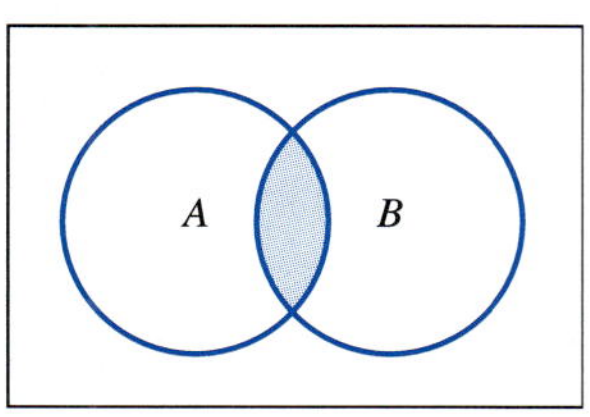

FIGURE 2
$A \cap B$ is the shaded region.

This relationship is easily visualized in the Venn diagram shown in Figure 2. Note that

$$A \cap B \subset A \quad \text{and} \quad A \cap B \subset B$$

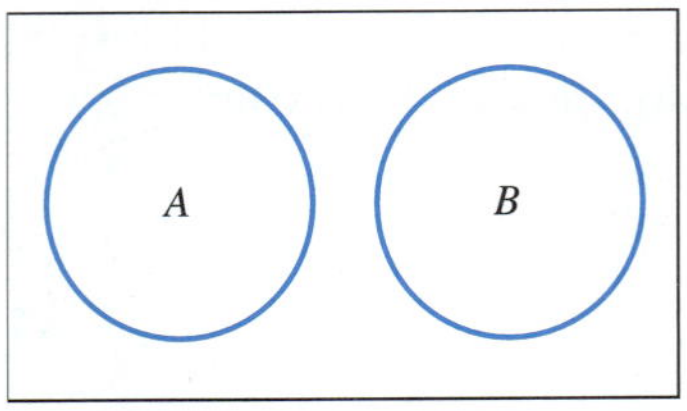

FIGURE 3
$A \cap B = \varnothing$; A and B are disjoint.

If $A \cap B = \varnothing$, then the sets A and B are said to be **disjoint;** this is illustrated in Figure 3.

The set of all elements under consideration is called the **universal set** U**.** Once the universal set is determined for a particular case, all other sets under discussion must be subsets of U.

We now define one more operation on sets, called the *complement*. The **complement** of A (relative to U), denoted by A', is the set of elements in U that are not in A (see Fig. 4). Symbolically:

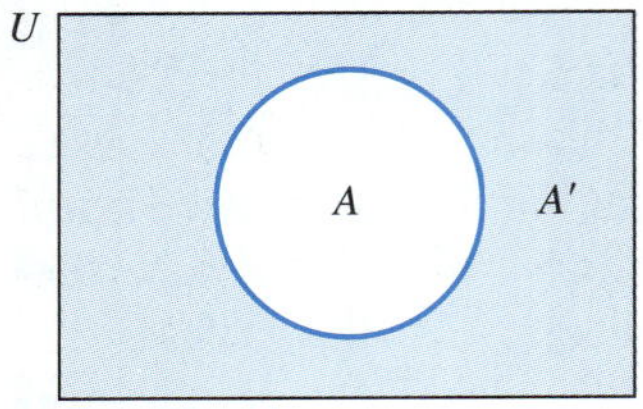

FIGURE 4
The complement of A is A'.

Complement

$$A' = \{x \in U | x \notin A\}$$

Example 4 **Union, Intersection, and Complement** If $A = \{3, 6, 9\}$, $B = \{3, 4, 5, 6, 7\}$, $C = \{4, 5, 7\}$, and $U = \{1, 2, 3, 4, 5, 6, 7, 8, 9\}$, then

$$A \cup B = \{3, 4, 5, 6, 7, 9\}$$
$$A \cap B = \{3, 6\}$$
$$A \cap C = \varnothing \qquad \textit{A and C are disjoint}$$
$$B' = \{1, 2, 8, 9\}$$

Matched Problem 4 If $R = \{1, 2, 3, 4\}$, $S = \{1, 3, 5, 7\}$, $T = \{2, 4\}$, and $U = \{1, 2, 3, 4, 5, 6, 7, 8, 9\}$, find:

(A) $R \cup S$ (B) $R \cap S$ (C) $S \cap T$ (D) S'

■ Application

Example 5 **Marketing Survey** In a survey of 100 randomly chosen students, a marketing questionnaire included the following three questions:

1. Do you own a TV?
2. Do you own a car?
3. Do you own a TV and a car?

Seventy-five answered yes to 1, 45 answered yes to 2, and 35 answered yes to 3.

(A) How many students owned either a car or a TV?
(B) How many students did not own either a car or a TV?

Solution Venn diagrams are very useful for this type of problem. If we let

U = Set of students in sample (100)
T = Set of students who own TV's (75)
C = Set of students who own cars (45)
$T \cap C$ = Set of students who own cars and TV's (35)

then:

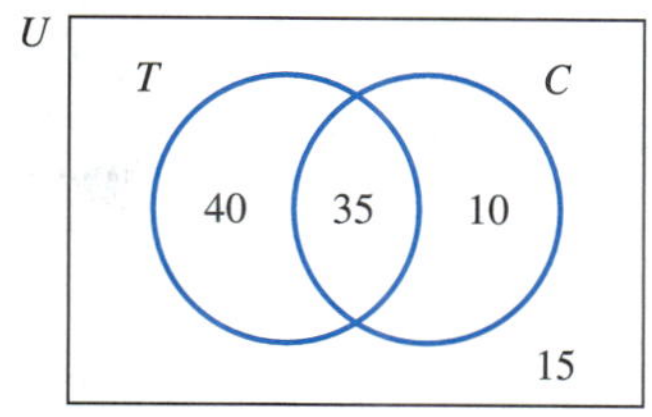

Place the number in the intersection first; then work outward:

$40 = 75 - 35$
$10 = 45 - 35$
$15 = 100 - (40 + 35 + 10)$

(A) The number of students who own either a car or a TV is the number of students in the set $T \cup C$. You might be tempted to say that this is just the number of students in T plus the number of students in C, $75 + 45 = 120$, but this sum is larger than the sample we started with! What is wrong? We have actually counted the number in the intersection (35) twice. The correct answer, as seen in the Venn diagram, is

$$40 + 35 + 10 = 85$$

(B) The number of students who do not own either a car or a TV is the number of students in the set $(T \cup C)'$; that is, 15.

Matched Problem 5 Referring to Example 5:

(A) How many students owned a car but not a TV?
(B) How many students did not own both a car and a TV?

Note in Example 5 and Matched Problem 5 that the word **and** is associated with intersection and the word **or** is associated with union.

Explore–Discuss 2

In Example 5, find the number of students in the set $(T \cup C) \cap C'$. Describe this set verbally and with a Venn diagram.

Answers to Matched Problems

1. (A) $\{x \mid x^2 = 9\}$ (B) $\{-3, 3\}$ (C) True; True
2. All are true **3.** $\{1, 2\}, \{1\}, \{2\}, \varnothing$
4. (A) $\{1, 2, 3, 4, 5, 7\}$ (B) $\{1, 3\}$ (C) $\varnothing$ (D) $\{2, 4, 6, 8, 9\}$
5. (A) 10, the number in $T' \cap C$ (B) 65, the number in $(T \cap C)'$

Exercise A-1

A *Indicate true (T) or false (F) in Problems 1–8.*

1. $4 \in \{2, 3, 4\}$
2. $6 \notin \{2, 3, 4\}$
3. $\{2, 3\} \subset \{2, 3, 4\}$
4. $\{3, 2, 4\} = \{2, 3, 4\}$
5. $\{3, 2, 4\} \subset \{2, 3, 4\}$
6. $\{3, 2, 4\} \in \{2, 3, 4\}$
7. $\varnothing \subset \{2, 3, 4\}$
8. $\varnothing = \{0\}$

In Problems 9–20, write the resulting set using the listing method.

9. $\{1, 3, 5\} \cup \{2, 3, 4\}$
10. $\{3, 4, 6, 7\} \cup \{3, 4, 5\}$
11. $\{1, 3, 4\} \cap \{2, 3, 4\}$
12. $\{3, 4, 6, 7\} \cap \{3, 4, 5\}$
13. $\{1, 5, 9\} \cap \{3, 4, 6, 8\}$
14. $\{6, 8, 9, 11\} \cap \{3, 4, 5, 7\}$

B

15. $\{x \mid x - 2 = 0\}$

16. $\{x \mid x + 7 = 0\}$

17. $\{x \mid x^2 = 49\}$

18. $\{x \mid x^2 = 100\}$

19. $\{x \mid x \text{ is an odd number between 1 and 9, inclusive}\}$

20. $\{x \mid x \text{ is a month starting with } M\}$

21. For $U = \{1, 2, 3, 4, 5\}$ and $A = \{2, 3, 4\}$, find A'.

22. For $U = \{7, 8, 9, 10, 11\}$ and $A = \{7, 11\}$, find A'.

Problems 23–34 refer to the Venn diagram below. How many elements are in each of the indicated sets?

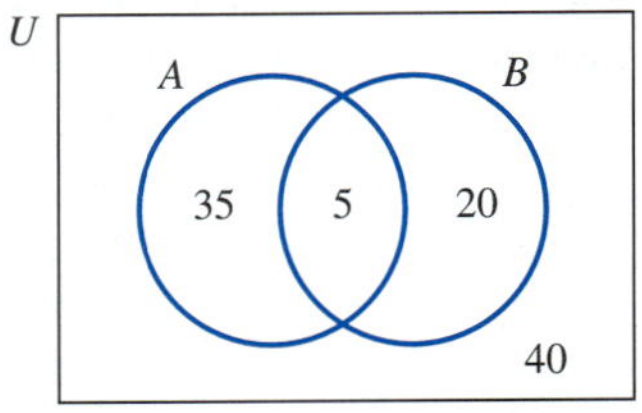

23. A

24. U

25. A'

26. B'

27. $A \cup B$

28. $A \cap B$

29. $A' \cap B$

30. $A \cap B'$

31. $(A \cap B)'$

32. $(A \cup B)'$

33. $A' \cap B'$

34. U'

35. If $R = \{1, 2, 3, 4\}$ and $T = \{2, 4, 6\}$, find:
(A) $\{x \mid x \in R \textbf{ or } x \in T\}$ (B) $R \cup T$

36. If $R = \{1, 3, 4\}$ and $T = \{2, 4, 6\}$, find:
(A) $\{x \mid x \in R \textbf{ and } x \in T\}$ (B) $R \cap T$

37. For $P = \{1, 2, 3, 4\}$, $Q = \{2, 4, 6\}$, and $R = \{3, 4, 5, 6\}$, find $P \cup (Q \cap R)$.

38. For P, Q, and R in Problem 37, find $P \cap (Q \cup R)$.

C *Venn diagrams may be of help in Problems 39–44.*

39. If $A \cup B = B$, can we always conclude that $A \subset B$?

40. If $A \cap B = B$, can we always conclude that $B \subset A$?

41. If A and B are arbitrary sets, can we always conclude that $A \cap B \subset B$?

42. If $A \cap B = \varnothing$, can we always conclude that $B = \varnothing$?

43. If $A \subset B$ and $x \in A$, can we always conclude that $x \in B$?

44. If $A \subset B$ and $x \in B$, can we always conclude that $x \in A$?

45. How many subsets does each of the following sets have? Also, try to discover a formula in terms of n for a set with n elements.
(A) $\{a\}$ (B) $\{a, b\}$
(C) $\{a, b, c\}$ (D) $\{a, b, c, d\}$

46. How do the sets $\varnothing$, $\{\varnothing\}$, and $\{0\}$ differ from each other?

Applications

Business & Economics

Marketing survey. Problems 47–58 refer to the following survey: A marketing survey of 1,000 car commuters found that 600 answered yes to listening to the news, 500 answered yes to listening to music, and 300 answered yes to listening to both. Let

N = *Set of commuters in the sample who listen to news*
M = *Set of commuters in the sample who listen to music*

Following the procedures in Example 5, find the number of commuters in each set described below.

47. $N \cup M$

48. $N \cap M$

49. $(N \cup M)'$

50. $(N \cap M)'$

51. $N' \cap M$

52. $N \cap M'$

53. Set of commuters who listen to either news or music

54. Set of commuters who listen to both news and music

55. Set of commuters who do not listen to either news or music

56. Set of commuters who do not listen to both news and music

57. Set of commuters who listen to music but not news

58. Set of commuters who listen to news but not music

59. *Committee selection.* The management of a company, a president and three vice-presidents, denoted by the set $\{P, V_1, V_2, V_3\}$, wish to select a committee of 2 people from among themselves. How many ways can this committee be formed? That is, how many 2-person subsets can be formed from a set of 4 people?

60. *Voting coalition.* The management of the company in Problem 59 decides for or against certain measures as follows: The president has 2 votes and each vice-president has 1 vote. Three favorable votes are needed to pass a measure. List all minimal winning coalitions; that is, list all subsets of $\{P, V_1, V_2, V_3\}$ that represent exactly 3 votes.

Life Sciences

Blood types. *When receiving a blood transfusion, a recipient must have all the antigens of the donor. A person may have one or more of the three antigens A, B, and Rh, or none at all. Eight blood types are possible, as indicated in the following Venn diagram, where U is the set of all people under consideration:*

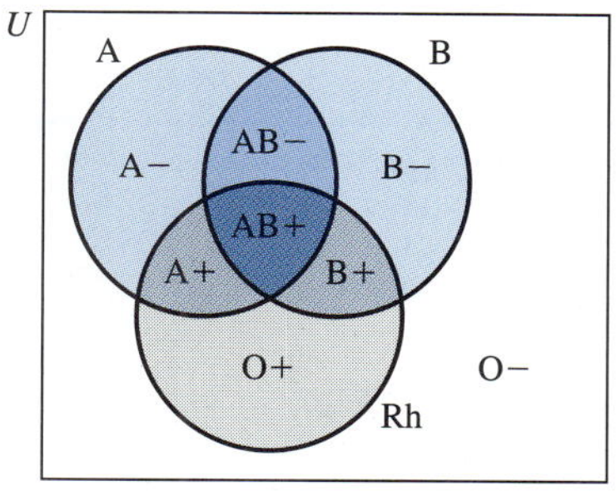

Blood Types

A−
A+
B−
B+
AB−
AB+
O−
O+

An A− person has A antigens but no B or Rh; an O+ person has Rh but neither A nor B; an AB− person has A and B antigens but no Rh; and so on.

Using the Venn diagram, indicate which of the eight blood types are included in the sets in Problems 61–68.

61. $A \cap Rh$ **62.** $A \cap B$ **63.** $A \cup Rh$

64. $A \cup B$ **65.** $(A \cup B)'$ **66.** $(A \cup B \cup Rh)'$

67. $A' \cap B$ **68.** $Rh' \cap A$

Social Sciences

Group structures. *R. D. Luce and A. D. Perry, in a study on group structure (Psychometrika, 1949, 14:95–116), used the idea of sets to formally define the notion of a clique within a group. Let G be the set of all persons in the group and let $C \subset G$. Then C is a clique provided that:*

1. *C contains at least 3 elements.*
2. *For every $a, b \in C$, $a \mathbf{R} b$ and $b \mathbf{R} a$.*
3. *For every $a \notin C$, there is at least one $b \in C$ such that $a \not\mathbf{R} b$ or $b \not\mathbf{R} a$ or both.*

[Note: Interpret "$a \mathbf{R} b$" to mean "a relates to b," "a likes b," "a is as wealthy as b," and so on. Of course, "$a \not\mathbf{R} b$" means "a does not relate to b," and so on.]

69. Translate statement 2 into ordinary English.

70. Translate statement 3 into ordinary English.

SECTION A-2

Algebra and Real Numbers

- The Set of Real Numbers
- The Real Number Line
- Basic Real Number Properties
- Further Properties
- Fraction Properties

The rules for manipulating and reasoning with symbols in algebra depend, in large measure, on properties of the real numbers. In this section we will look at some of the important properties of this number system. To make our discussions here and elsewhere in the text clearer and more precise, we will occasionally make use of simple *set* concepts and notation. Refer to Section A-1 if you are not yet familiar with the basic ideas concerning sets.

The Set of Real Numbers

What number system have you been using most of your life? The *real number system.* Informally, a **real number** is any number that has a decimal representation. Table 1 describes the set of real numbers and some of its important subsets. Figure 1 illustrates how these sets of numbers are related.

Table 1
THE SET OF REAL NUMBERS

SYMBOL	NAME	DESCRIPTION	EXAMPLES
N	Natural numbers	Counting numbers (also called positive integers)	$1, 2, 3, \ldots$
Z	Integers	Natural numbers, their negatives, and 0	$\ldots, -2, -1, 0, 1, 2, \ldots$
Q	Rational numbers	Numbers that can be represented as a/b, where a and b are integers and $b \neq 0$; decimal representations are repeating or terminating	$-4, 0, 1, 25, \frac{-3}{5}, \frac{2}{3}, 3.67, -0.33\overline{3}, 5.272\ 7\overline{27}$*
I	Irrational numbers	Numbers that can be represented as nonrepeating and nonterminating decimal numbers	$\sqrt{2}, \pi, \sqrt[3]{7}, 1.414\ 213\ldots, 2.718\ 281\ 82\ldots$
R	Real numbers	Rational and irrational numbers	

*The overbar indicates that the number (or block of numbers) repeats indefinitely. The space after every third digit is used to help keep track of the number of decimal places.

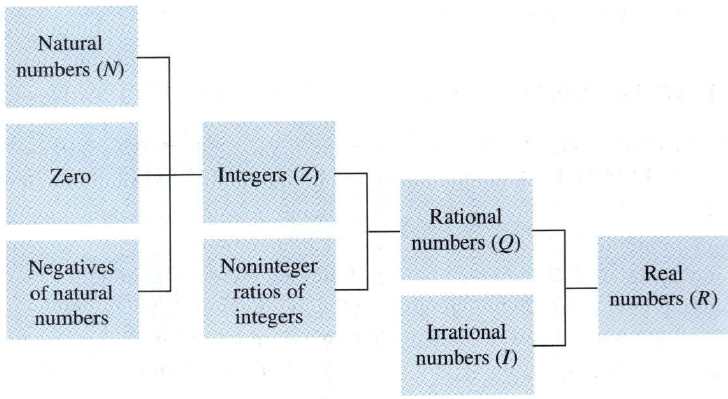

FIGURE 1
Real numbers and important subsets

The set of integers contains all the natural numbers and something else—their negatives and 0. The set of rational numbers contains all the integers and something else—noninteger ratios of integers. And the set of real numbers contains all the rational numbers and something else—irrational numbers.

■ THE REAL NUMBER LINE

A one-to-one correspondence exists between the set of real numbers and the set of points on a line. That is, each real number corresponds to exactly one point, and each point corresponds to exactly one real number. A line with a real number associated with each point, and vice versa, as shown in Figure 2, is called a **real number line**, or simply a **real line.** Each number associated with a point is called the **coordinate** of the point.

The point with coordinate 0 is called the **origin.** The arrow on the right end of the line indicates a positive direction. The coordinates of all points to the right of the origin are called **positive real numbers,** and those to the left of the origin are called **negative real numbers.** The real number 0 is neither positive nor negative.

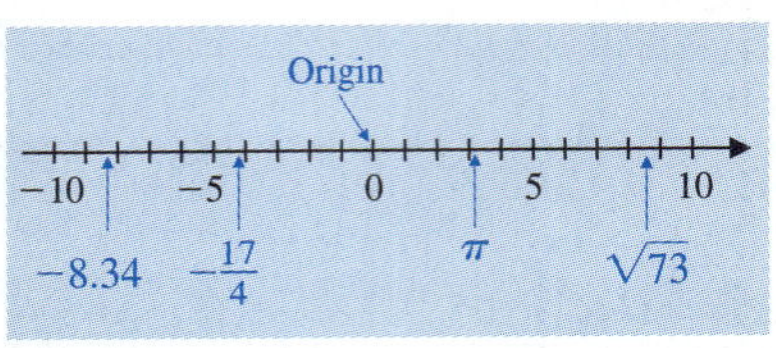

FIGURE 2
The real number line

Basic Real Number Properties

We now take a look at some of the basic properties of the real number system that enable us to convert algebraic expressions into *equivalent forms*. These assumed properties become operational rules in the algebra of real numbers.

Basic Properties of the Set of Real Numbers

Let a, b, and c be arbitrary elements in the set of real numbers R.

Addition Properties

Associative: $(a + b) + c = a + (b + c)$

Commutative: $a + b = b + a$

Identity: 0 is the additive identity; that is, $0 + a = a + 0 = a$ for all a in R, and 0 is the only element in R with this property.

Inverse: For each a in R, $-a$ is its unique additive inverse; that is, $a + (-a) = (-a) + a = 0$, and $-a$ is the only element in R relative to a with this property.

Multiplication Properties

Associative: $(ab)c = a(bc)$

Commutative: $ab = ba$

Identity: 1 is the multiplicative identity; that is, $(1)a = a(1) = a$ for all a in R, and 1 is the only element in R with this property.

Inverse: For each a in R, $a \neq 0$, $1/a$ is its unique multiplicative inverse; that is, $a(1/a) = (1/a)a = 1$, and $1/a$ is the only element in R relative to a with this property.

Distributive Properties

$$a(b + c) = ab + ac \qquad (a + b)c = ac + bc$$

Do not be intimidated by the names of these properties. Most of the ideas presented here are quite simple. In fact, you have been using many of these properties in arithmetic for a long time.

You are already familiar with the **commutative properties** for addition and multiplication. They indicate that the order in which the addition or multiplication of two numbers is performed does not matter. For example,

$$7 + 2 = 2 + 7 \quad \text{and} \quad 3 \cdot 5 = 5 \cdot 3$$

Is there a commutative property relative to subtraction or division? That is, does $a - b = b - a$ or does $a \div b = b \div a$ for all real numbers a and b (division by 0 excluded)? The answer is no, since, for example,

$$8 - 6 \neq 6 - 8 \quad \text{and} \quad 10 \div 5 \neq 5 \div 10$$

When computing

$$3 + 2 + 6 \quad \text{or} \quad 3 \cdot 2 \cdot 6$$

why do we not need parentheses to indicate which two numbers are to be added or multiplied first? The answer is to be found in the **associative properties.** These properties allow us to write

$$(3 + 2) + 6 = 3 + (2 + 6) \qquad \text{and} \qquad (3 \cdot 2) \cdot 6 = 3 \cdot (2 \cdot 6)$$

so it does not matter how we group numbers relative to either operation. Is there an associative property for subtraction or division? The answer is no, since, for example,

$$(12 - 6) - 2 \neq 12 - (6 - 2) \qquad \text{and} \qquad (12 \div 6) \div 2 \neq 12 \div (6 \div 2)$$

Evaluate each side of each equation to see why.

Relative to addition, commutativity and associativity permit us to change the order of addition at will and insert or remove parentheses as we please. The same is true for multiplication, but not for subtraction and division.

What number added to a given number will give that number back again? What number times a given number will give that number back again? The answers are 0 and 1, respectively. Because of this, 0 and 1 are called the **identity elements** for the real numbers. Hence, for any real numbers a and b,

$$0 + 5 = 5 \qquad \text{and} \qquad (a + b) + 0 = a + b$$
$$1 \cdot 4 = 4 \qquad \text{and} \qquad (a + b) \cdot 1 = a + b$$

We now consider **inverses.** For each real number a, there is a unique real number $-a$ such that $a + (-a) = 0$. The number $-a$ is called the **additive inverse** of a, or the **negative** of a. For example, the additive inverse (or negative) of 7 is -7, since $7 + (-7) = 0$. The additive inverse (or negative) of -7 is $-(-7) = 7$, since $-7 + [-(-7)] = 0$. It is important to remember that:

$-a$ is not necessarily a negative number; it is positive if a is negative and negative if a is positive.

For each nonzero real number a, there is a unique real number $1/a$ such that $a(1/a) = 1$. The number $1/a$ is called the **multiplicative inverse** of a, or the **reciprocal** of a. For example, the multiplicative inverse (or reciprocal) of 4 is $\frac{1}{4}$, since $4(\frac{1}{4}) = 1$. (Also note that 4 is the multiplicative inverse of $\frac{1}{4}$.) The number 0 has no multiplicative inverse.

We now turn to the **distributive properties,** which involve both multiplication and addition. Consider the following two computations:

$$5(3 + 4) = 5 \cdot 7 = 35 \qquad 5 \cdot 3 + 5 \cdot 4 = 15 + 20 = 35$$

Thus,

$$5(3 + 4) = 5 \cdot 3 + 5 \cdot 4$$

and we say that multiplication by 5 *distributes* over the sum $(3 + 4)$. In general, **multiplication distributes over addition** in the real number system. Two more illustrations are

$$9(m + n) = 9m + 9n \qquad (7 + 2)u = 7u + 2u$$

Example 1 **Real Number Properties** State the real number property that justifies the indicated statement.

STATEMENT	PROPERTY ILLUSTRATED
(A) $x(y + z) = (y + z)x$	Commutative ($\cdot$)
(B) $5(2y) = (5 \cdot 2)y$	Associative ($\cdot$)
(C) $2 + (y + 7) = 2 + (7 + y)$	Commutative (+)
(D) $4z + 6z = (4 + 6)z$	Distributive
(E) If $m + n = 0$, then $n = -m$.	Inverse (+)

Matched Problem 1 State the real number property that justifies the indicated statement.

(A) $8 + (3 + y) = (8 + 3) + y$
(B) $(x + y) + z = z + (x + y)$
(C) $(a + b)(x + y) = a(x + y) + b(x + y)$
(D) $5xy + 0 = 5xy$
(E) If $xy = 1$, $x \neq 0$, then $y = 1/x$.

FURTHER PROPERTIES

Subtraction and *division* can be defined in terms of addition and multiplication, respectively:

Subtraction and Division

For all real numbers a and b:

Subtraction: $a - b = a + (-b)$ $\qquad$ $7 - (-5) = 7 + [-(-5)] = 7 + 5 = 12$

Division: $a \div b = a\left(\frac{1}{b}\right)$, $b \neq 0$ $\qquad$ $9 \div 4 = 9\left(\frac{1}{4}\right) = \frac{9}{4}$

Thus, to subtract b from a, add the negative (the additive inverse) of b to a. To divide a by b, multiply a by the reciprocal (the multiplicative inverse) of b. Note that division by 0 is not defined, since 0 does not have a reciprocal. Thus:

0 can never be used as a divisor!

The following properties of negatives can be proved using the preceding assumed properties and definitions.

Properties of Negatives

For all real numbers a and b:

1. $-(-a) = a$
2. $(-a)b = -(ab) = a(-b) = -ab$
3. $(-a)(-b) = ab$
4. $(-1)a = -a$
5. $\frac{-a}{b} = -\frac{a}{b} = \frac{a}{-b}$, $b \neq 0$
6. $\frac{-a}{-b} = -\frac{-a}{b} = -\frac{a}{-b} = \frac{a}{b}$, $b \neq 0$

We now state two important properties involving 0:

Zero Properties

For all real numbers a and b:

1. $a \cdot 0 = 0$ $\quad 0 \cdot 0 = 0 \quad (-35)(0) = 0$
2. $ab = 0$ if and only if $a = 0$ or $b = 0$
   If $(3x + 2)(x - 7) = 0$, then either $3x + 2 = 0$ or $x - 7 = 0$.

Explore–Discuss 1

In general, a set of numbers is closed under an operation if performing the operation on numbers in the set always produces another number in the set. For example, the real numbers R are closed under addition, multiplication, subtraction, and division, excluding division by 0. Replace each **?** in the following tables with T (true) or F (false), and illustrate each false statement with an example. (See Table 1 for the definitions of the sets N, Z, Q, I, and R.)

	Closed under addition	Closed under multiplication
N	?	?
Z	?	?
Q	?	?
I	?	?
R	T	T

	Closed under subtraction	Closed under division*
N	?	?
Z	?	?
Q	?	?
I	?	?
R	T	T

*Excluding division by 0.

Fraction Properties

Recall that the quotient $a \div b\ (b \neq 0)$ written in the form a/b is called a **fraction.** The quantity a is called the **numerator,** and the quantity b is called the **denominator.**

Fraction Properties

For all real numbers a, b, c, d, and k (division by 0 excluded):

1. $\dfrac{a}{b} = \dfrac{c}{d}$ if and only if $ad = bc$ $\quad \dfrac{4}{6} = \dfrac{6}{9}$ since $4 \cdot 9 = 6 \cdot 6$

2. $\dfrac{ka}{kb} = \dfrac{a}{b}$ $\quad \dfrac{7 \cdot 3}{7 \cdot 5} = \dfrac{3}{5}$

3. $\dfrac{a}{b} \cdot \dfrac{c}{d} = \dfrac{ac}{bd}$ $\quad \dfrac{3}{5} \cdot \dfrac{7}{8} = \dfrac{3 \cdot 7}{5 \cdot 8}$

4. $\dfrac{a}{b} \div \dfrac{c}{d} = \dfrac{a}{b} \cdot \dfrac{d}{c}$ $\quad \dfrac{2}{3} \div \dfrac{5}{7} = \dfrac{2}{3} \cdot \dfrac{7}{5}$

5. $\dfrac{a}{b} + \dfrac{c}{b} = \dfrac{a + c}{b}$ $\quad \dfrac{3}{6} + \dfrac{5}{6} = \dfrac{3 + 5}{6}$

6. $\dfrac{a}{b} - \dfrac{c}{b} = \dfrac{a - c}{b}$ $\quad \dfrac{7}{8} - \dfrac{3}{8} = \dfrac{7 - 3}{8}$

7. $\dfrac{a}{b} + \dfrac{c}{d} = \dfrac{ad + bc}{bd}$ $\quad \dfrac{2}{3} + \dfrac{3}{5} = \dfrac{2 \cdot 5 + 3 \cdot 3}{3 \cdot 5}$

Answers to Matched Problem **1.** (A) Associative (+) (B) Commutative (+) (C) Distributive (D) Identity (+) (E) Inverse (•)

Exercise A-2

All variables represent real numbers.

A *In Problems 1–6, replace each question mark with an appropriate expression that will illustrate the use of the indicated real number property.*

1. Commutative property (•): $uv = ?$

2. Commutative property (+): $x + 7 = ?$

3. Associative property (+): $3 + (7 + y) = ?$

4. Associative property (•): $x(yz) = ?$

5. Identity property (•): $1(u + v) = ?$

6. Identity property (+): $0 + 9m = ?$

In Problems 7–26, indicate true (T) or false (F).

7. $5(8m) = (5 \cdot 8)m$ **8.** $a + cb = a + bc$

9. $5x + 7x = (5 + 7)x$

10. $uv(w + x) = uvw + uvx$

11. $7 - 11 = 7 + (-11)$ **12.** $8 \div (-5) = 8(\frac{1}{-5})$

13. $(x + 3) + 2x = 2x + (x + 3)$

14. $(4x + 3) + (x + 2) = 4x + [3 + (x + 2)]$

15. $\dfrac{2x}{-(x + 3)} = -\dfrac{2x}{x + 3}$ **16.** $-\dfrac{2x}{-(x - 3)} = \dfrac{2x}{x - 3}$

17. $(-3)(\frac{1}{-3}) = 1$ **18.** $(-0.5) + (0.5) = 0$

19. $-x^2y^2 = (-1)x^2y^2$

20. $[-(x + 2)](-x) = (x + 2)x$

21. $\dfrac{a}{b} + \dfrac{c}{d} = \dfrac{a + c}{b + d}$ **22.** $\dfrac{k}{k + b} = \dfrac{1}{1 + b}$

23. $(x + 8)(x + 6) = (x + 8)x + (x + 8)6$

24. $u(u - 2v) + v(u - 2v) = (u + v)(u - 2v)$

25. If $(x - 2)(2x + 3) = 0$, then either $x - 2 = 0$ or $2x + 3 = 0$.

26. If either $x - 2 = 0$ or $2x + 3 = 0$, then $(x - 2)(2x + 3) = 0$.

B

27. If $uv = 1$, does either u or v have to be 1? Explain.

28. If $uv = 0$, does either u or v have to be 0? Explain.

29. Indicate whether the following are true (T) or false (F):
(A) All integers are natural numbers.
(B) All rational numbers are real numbers.
(C) All natural numbers are rational numbers.

30. Indicate whether the following are true (T) or false (F):
(A) All natural numbers are integers.
(B) All real numbers are irrational.
(C) All rational numbers are real numbers.

31. Give an example of a real number that is not a rational number.

32. Give an example of a rational number that is not an integer.

33. Given the sets of numbers N (natural numbers), Z (integers), Q (rational numbers), and R (real numbers), indicate to which set(s) each of the following numbers belongs:
(A) 8 (B) $\sqrt{2}$ (C) -1.414 (D) $\frac{-5}{2}$

34. Given the sets of numbers N, Z, Q, and R (see Problem 33), indicate to which set(s) each of the following numbers belongs:
(A) -3 (B) 3.14 (C) π (D) $\frac{2}{3}$

35. Indicate true (T) or false (F), and for each false statement find real number replacements for a, b, and c that will provide a counterexample. For all real numbers a, b, and c:
(A) $(a + b) + c = a + (b + c)$
(B) $(a - b) - c = a - (b - c)$
(C) $a(bc) = (ab)c$
(D) $(a \div b) \div c = a \div (b \div c)$

36. Indicate true (T) or false (F), and for each false statement find real number replacements for a and b that will provide a counterexample. For all real numbers a and b:
(A) $a + b = b + a$
(B) $a - b = b - a$
(C) $ab = ba$
(D) $a \div b = b \div a$

C

37. If $c = 0.151\ 515\ldots$, then $100c = 15.151\ 5\ldots$ and

$$\begin{aligned} 100c - c &= 15.151\ 5\ldots - 0.151\ 515\ldots \\ 99c &= 15 \\ c &= \tfrac{15}{99} = \tfrac{5}{33} \end{aligned}$$

Proceeding similarly, convert the repeating decimal 0.090 909... into a fraction. (All repeating decimals are rational numbers, and all rational numbers have repeating decimal representations.)

38. Repeat Problem 37 for 0.181 818....

Use a calculator to express each number in Problems 39 and 40 as a decimal to the capacity of your calculator. Observe the repeating decimal representation of the rational numbers and the nonrepeating decimal representation of the irrational numbers.

39. (A) $\frac{13}{6}$ (B) $\sqrt{21}$ (C) $\frac{7}{16}$ (D) $\frac{29}{111}$

40. (A) $\frac{8}{9}$ (B) $\frac{3}{11}$ (C) $\sqrt{5}$ (D) $\frac{11}{8}$

SECTION A-3

Operations on Polynomials

- NATURAL NUMBER EXPONENTS
- POLYNOMIALS
- COMBINING LIKE TERMS
- ADDITION AND SUBTRACTION
- MULTIPLICATION
- COMBINED OPERATIONS

This section covers basic operations on *polynomials,* a mathematical form that is encountered frequently. Our discussion starts with a brief review of natural number exponents. Integer and rational exponents and their properties will be discussed in detail in subsequent sections. (Natural numbers, integers, and rational numbers are important parts of the real number system; see Table 1 and Figure 1 in Appendix A-2.)

NATURAL NUMBER EXPONENTS

We define a **natural number exponent** as follows:

> **Natural Number Exponent**
>
> For n a natural number and b any real number,
>
> $b^n = b \cdot b \cdot \cdots \cdot b$ n factors of b
>
> $3^5 = 3 \cdot 3 \cdot 3 \cdot 3 \cdot 3$ 5 factors of 3
>
> where n is called the **exponent** and b is called the **base.**

Along with this definition, we state the **first property of exponents:**

Theorem 1 FIRST PROPERTY OF EXPONENTS

For any natural numbers m and n, and any real number b:

$$b^m b^n = b^{m+n} \qquad (2t^4)(5t^3) = 2 \cdot 5t^{4+3} = 10t^7$$

POLYNOMIALS

Algebraic expressions are formed by using constants and variables and the algebraic operations of addition, subtraction, multiplication, division, raising to powers, and taking roots. Special types of algebraic expressions are called *polynomials.* A **polynomial in one variable** x is constructed by adding or subtracting constants and terms of the form ax^n, where a is a real number and n is a natural number. A **polynomial in two variables** x and y is constructed by adding and subtracting constants and terms of the form $ax^m y^n$, where a is a real number and m and n are natural numbers. Polynomials in three and more variables are defined in a similar manner.

POLYNOMIALS		NOT POLYNOMIALS	
8	0	$\frac{1}{x}$	$\frac{x-y}{x^2+y^2}$
$3x^3 - 6x + 7$	$6x + 3$		
$2x^2 - 7xy - 8y^2$	$9y^3 + 4y^2 - y + 4$	$\sqrt{x^3 - 2x}$	$2x^{-2} - 3x^{-1}$
$2x - 3y + 2$	$u^5 - 3u^3v^2 + 2uv^4 - v^4$		

Polynomial forms are encountered frequently in mathematics. For the efficient study of polynomials it is useful to classify them according to their *degree.* If a term in a polynomial has only one variable as a factor, then the **degree of the term** is the power of the variable. If two or more variables are present in a term as factors, then the **degree of the term** is the sum of the powers of the variables. The **degree of a polynomial** is the degree of the nonzero term with the highest degree in the polynomial. Any nonzero constant is defined to be a **polynomial of degree 0.** The number 0 is also a polynomial but is not assigned a degree.

Example 1 **Degree**

(A) The degree of the first term in $5x^3 + \sqrt{3}x - \frac{1}{2}$ is 3, the degree of the second term is 1, the degree of the third term is 0, and the degree of the whole polynomial is 3 (the same as the degree of the term with the highest degree).

(B) The degree of the first term in $8u^3v^2 - \sqrt{7}uv^2$ is 5, the degree of the second term is 3, and the degree of the whole polynomial is 5.

Matched Problem 1

(A) Given the polynomial $6x^5 + 7x^3 - 2$, what is the degree of the first term? The second term? The third term? The whole polynomial?

(B) Given the polynomial $2u^4v^2 - 5uv^3$, what is the degree of the first term? The second term? The whole polynomial?

In addition to classifying polynomials by degree, we also call a single-term polynomial a **monomial,** a two-term polynomial a **binomial,** and a three-term polynomial a **trinomial.**

COMBINING LIKE TERMS

The concept of *coefficient* plays a central role in the process of combining *like terms.* A constant in a term of a polynomial, including the sign that precedes it, is called the **numerical coefficient,** or simply, the **coefficient,** of the term. If a constant does not appear, or only a + sign appears, the coefficient is understood to be 1. If only a − sign appears, the coefficient is understood to be −1. Thus, given the polynomial

$$5x^4 - x^3 - 3x^2 + x - 7 = 5x^4 + (-1)x^3 + (-3)x^2 + 1x + (-7)$$

the coefficient of the first term is 5, the coefficient of the second term is −1, the coefficient of the third term is −3, the coefficient of the fourth term is 1, and the coefficient of the fifth term is −7.

The following distributive properties are fundamental to the process of combining *like terms.*

Distributive Properties of Real Numbers

1. $a(b + c) = (b + c)a = ab + ac$
2. $a(b - c) = (b - c)a = ab - ac$
3. $a(b + c + \cdots + f) = ab + ac + \cdots + af$

Two terms in a polynomial are called **like terms** if they have exactly the same variable factors to the same powers. The numerical coefficients may or may not be the same. Since constant terms involve no variables, all constant terms are like terms. If a polynomial contains two or more like terms, these terms can be combined into a single term by making use of distributive properties. The following example illustrates the reasoning behind the process:

$$\begin{aligned} 3x^2y - 5xy^2 + x^2y - 2x^2y &= 3x^2y + x^2y - 2x^2y - 5xy^2 \\ &= (3x^2y + 1x^2y - 2x^2y) - 5xy^2 \\ &= (3 + 1 - 2)x^2y - 5xy^2 \\ &= 2x^2y - 5xy^2 \end{aligned}$$

Note the use of distributive properties.

It should be clear that free use is made of the real number properties discussed in Appendix A-2. The steps shown in the dashed box are usually done mentally, and the process is quickly mechanized as follows:

Like terms in a polynomial are combined by adding their numerical coefficients.

How can we simplify expressions such as $4(x - 2y) - 3(2x - 7y)$? We clear the expression of parentheses using distributive properties, and combine like terms:

$$\begin{aligned} 4(x - 2y) - 3(2x - 7y) &= 4x - 8y - 6x + 21y \\ &= -2x + 13y \end{aligned}$$

Example 2 **Removing Parentheses** Remove parentheses and simplify:

(A)
$$\begin{aligned} 2(3x^2 - 2x + 5) + (x^2 + 3x - 7) &= 2(3x^2 - 2x + 5) + 1(x^2 + 3x - 7) \\ &= 6x^2 - 4x + 10 + x^2 + 3x - 7 \\ &= 7x^2 - x + 3 \end{aligned}$$

(B) $(x^3 - 2x - 6) - (2x^3 - x^2 + 2x - 3)$
$$\begin{aligned} &= 1(x^3 - 2x - 6) + (-1)(2x^3 - x^2 + 2x - 3) \\ &= x^3 - 2x - 6 - 2x^3 + x^2 - 2x + 3 \\ &= -x^3 + x^2 - 4x - 3 \end{aligned}$$

Be careful with the sign here.

(C)
$$\begin{aligned} [3x^2 - (2x + 1)] - (x^2 - 1) &= [3x^2 - 2x - 1] - (x^2 - 1) \\ &= 3x^2 - 2x - 1 - x^2 + 1 \\ &= 2x^2 - 2x \end{aligned}$$

Remove inner parentheses first.

Matched Problem 2 Remove parentheses and simplify:

(A) $3(u^2 - 2v^2) + (u^2 + 5v^2)$
(B) $(m^3 - 3m^2 + m - 1) - (2m^3 - m + 3)$
(C) $(x^3 - 2) - [2x^3 - (3x + 4)]$

ADDITION AND SUBTRACTION

Addition and subtraction of polynomials can be thought of in terms of removing parentheses and combining like terms, as illustrated in Example 2. Horizontal and vertical arrangements are illustrated in the next two examples. You should be able to work either way, letting the situation dictate your choice.

Example 3 **Adding Polynomials** Add horizontally and vertically:

$$x^4 - 3x^3 + x^2, \quad -x^3 - 2x^2 + 3x, \quad \text{and} \quad 3x^2 - 4x - 5$$

SOLUTION Add horizontally:

$$\begin{aligned}(x^4 - 3x^3 + x^2) &+ (-x^3 - 2x^2 + 3x) + (3x^2 - 4x - 5)\\ &= x^4 - 3x^3 + x^2 - x^3 - 2x^2 + 3x + 3x^2 - 4x - 5\\ &= x^4 - 4x^3 + 2x^2 - x - 5\end{aligned}$$

Or vertically, by lining up like terms and adding their coefficients:

$$\begin{array}{rrrrr} x^4 & -\ 3x^3 & +\ x^2 & & \\ & -\ x^3 & -\ 2x^2 & +\ 3x & \\ & & 3x^2 & -\ 4x & -\ 5 \\ \hline x^4 & -\ 4x^3 & +\ 2x^2 & -\ x & -\ 5 \end{array}$$

Matched Problem 3 Add horizontally and vertically:

$$3x^4 - 2x^3 - 4x^2, \quad x^3 - 2x^2 - 5x, \quad \text{and} \quad x^2 + 7x - 2$$

Example 4 **Subtracting Polynomials** Subtract $4x^2 - 3x + 5$ from $x^2 - 8$, both horizontally and vertically.

SOLUTION

$$\begin{aligned}(x^2 - 8) &- (4x^2 - 3x + 5)\\ &= x^2 - 8 - 4x^2 + 3x - 5\\ &= -3x^2 + 3x - 13\end{aligned}$$

or

$$\begin{array}{rrr} x^2 & & -\ 8 \\ -4x^2 & +\ 3x & -\ 5 \\ \hline -3x^2 & +\ 3x & -\ 13 \end{array}$$

← Change signs and add.

Matched Problem 4 Subtract $2x^2 - 5x + 4$ from $5x^2 - 6$, both horizontally and vertically.

MULTIPLICATION

Multiplication of algebraic expressions involves the extensive use of distributive properties for real numbers, as well as other real number properties.

Example 5 **Multiplying Polynomials** Multiply: $(2x - 3)(3x^2 - 2x + 3)$

SOLUTION

$$\begin{aligned}(2x - 3)(3x^2 - 2x + 3) &= 2x(3x^2 - 2x + 3) - 3(3x^2 - 2x + 3)\\ &= 6x^3 - 4x^2 + 6x - 9x^2 + 6x - 9\\ &= 6x^3 - 13x^2 + 12x - 9\end{aligned}$$

Or, using a vertical arrangement,

$$\begin{array}{r}
3x^2 - 2x + 3 \\
\underline{2x - 3} \\
6x^3 - 4x^2 + 6x \\
\underline{-9x^2 + 6x - 9} \\
6x^3 - 13x^2 + 12x - 9
\end{array}$$

Matched Problem 5 Multiply: $(2x - 3)(2x^2 + 3x - 2)$

Thus, to multiply two polynomials, multiply each term of one by each term of the other, and combine like terms.

Products of binomial factors occur frequently, so it is useful to develop procedures that will enable us to write down their products by inspection. To find the product $(2x - 1)(3x + 2)$, we proceed as follows:

$$\begin{aligned}(2x - 1)(3x + 2) &= 6x^2 + 4x - 3x - 2 \\ &= 6x^2 + x - 2\end{aligned}$$

The inner and outer products are like terms, so combine into a single term.

To speed the process, we do the step in the dashed box mentally.

Products of certain binomial factors occur so frequently that it is useful to learn formulas for their products. The following formulas are easily verified by multiplying the factors on the left:

Special Products

1. $(a - b)(a + b) = a^2 - b^2$
2. $(a + b)^2 = a^2 + 2ab + b^2$
3. $(a - b)^2 = a^2 - 2ab + b^2$

Explore–Discuss 1

(A) Explain the relationship between special product formula 1 and the areas of the rectangles in the figure.

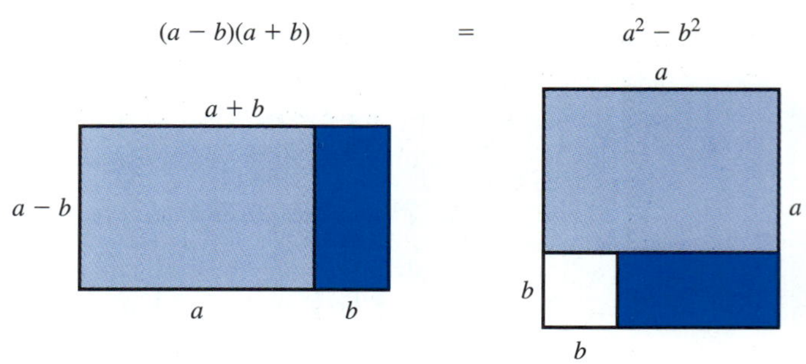

(B) Construct similar figures to provide geometric interpretations for special product formulas 2 and 3.

Example 6 **Special Products** Multiply mentally where possible

(A) $(2x - 3y)(5x + 2y)$ (B) $(3a - 2b)(3a + 2b)$
(C) $(5x - 3)^2$ (D) $(m + 2n)^3$

SOLUTION

(A) $(2x - 3y)(5x + 2y) = 10x^2 + 4xy - 15xy - 6y^2$
$= 10x^2 - 11xy - 6y^2$

(B) $(3a - 2b)(3a + 2b) = (3a)^2 - (2b)^2$
$= 9a^2 - 4b^2$

(C) $(5x - 3)^2 = (5x)^2 - 2(5x)(3) + 3^2$
$= 25x^2 - 30x + 9$

(D)
$$\begin{aligned}(m + 2n)^3 &= (m + 2n)^2(m + 2n)\\ &= (m^2 + 4mn + 4n^2)(m + 2n)\\ &= m^2(m + 2n) + 4mn(m + 2n) + 4n^2(m + 2n)\\ &= m^3 + 2m^2n + 4m^2n + 8mn^2 + 4mn^2 + 8n^3\\ &= m^3 + 6m^2n + 12mn^2 + 8n^3\end{aligned}$$

Matched Problem 6 Multiply mentally where possible:

(A) $(4u - 3v)(2u + v)$ (B) $(2xy + 3)(2xy - 3)$
(C) $(m + 4n)(m - 4n)$ (D) $(2u - 3v)^2$
(E) $(2x - y)^3$

Combined Operations

We complete this section by considering several examples that use all the operations just discussed. Note that in simplifying, we usually remove grouping symbols starting from the inside. That is, we remove parentheses ( ) first, then brackets [ ], and finally braces { }, if present. Also:

Multiplication and division precede addition and subtraction, and taking powers precedes multiplication and division.

Example 7 **Combined Operations** Perform the indicated operations and simplify:

(A)
$$\begin{aligned}3x - \{5 - 3[x - x(3 - x)]\} &= 3x - \{5 - 3[x - 3x + x^2]\}\\ &= 3x - \{5 - 3x + 9x - 3x^2\}\\ &= 3x - 5 + 3x - 9x + 3x^2\\ &= 3x^2 - 3x - 5\end{aligned}$$

(B)
$$\begin{aligned}(x - 2y)(2x + 3y) - (2x + y)^2 &= 2x^2 - xy - 6y^2 - (4x^2 + 4xy + y^2)\\ &= 2x^2 - xy - 6y^2 - 4x^2 - 4xy - y^2\\ &= -2x^2 - 5xy - 7y^2\end{aligned}$$

Matched Problem 7 Perform the indicated operations and simplify:

(A) $2t - \{7 - 2[t - t(4 + t)]\}$
(B) $(u - 3v)^2 - (2u - v)(2u + v)$

Answers to Matched Problems

1. (A) 5, 3, 0, 5 (B) 6, 4, 6
2. (A) $4u^2 - v^2$ (B) $-m^3 - 3m^2 + 2m - 4$ (C) $-x^3 + 3x + 2$
3. $3x^4 - x^3 - 5x^2 + 2x - 2$ **4.** $3x^2 + 5x - 10$ **5.** $4x^3 - 13x + 6$
6. (A) $8u^2 - 2uv - 3v^2$ (B) $4x^2y^2 - 9$ (C) $m^2 - 16n^2$
(D) $4u^2 - 12uv - 9v^2$ (E) $8x^3 - 12x^2y + 6xy^2 - y^3$
7. (A) $-2t^2 - 4t - 7$ (B) $-3u^2 - 6uv + 10v^2$

EXERCISE A-3

A *Problems 1–8 refer to the following polynomials:*
(A) $2x - 3$ *(B)* $2x^2 - x + 2$ *(C)* $x^3 + 2x^2 - x + 3$

1. What is the degree of (C)?

2. What is the degree of (A)?

3. Add (B) and (C).

4. Add (A) and (B).

5. Subtract (B) from (C).

6. Subtract (A) from (B).

7. Multiply (B) and (C).

8. Multiply (A) and (C).

In Problems 9–30, perform the indicated operations and simplify.

9. $2(u - 1) - (3u + 2) - 2(2u - 3)$

10. $2(x - 1) + 3(2x - 3) - (4x - 5)$

11. $4a - 2a[5 - 3(a + 2)]$

12. $2y - 3y[4 - 2(y - 1)]$

13. $(a + b)(a - b)$

14. $(m - n)(m + n)$

15. $(3x - 5)(2x + 1)$

16. $(4t - 3)(t - 2)$

17. $(2x - 3y)(x + 2y)$

18. $(3x + 2y)(x - 3y)$

19. $(3y + 2)(3y - 2)$

20. $(2m - 7)(2m + 7)$

21. $(3m + 7n)(2m - 5n)$

22. $(6x - 4y)(5x + 3y)$

23. $(4m + 3n)(4m - 3n)$

24. $(3x - 2y)(3x + 2y)$

25. $(3u + 4v)^2$

26. $(4x - y)^2$

27. $(a - b)(a^2 + ab + b^2)$

28. $(a + b)(a^2 - ab + b^2)$

29. $(4x + 3y)^2$

30. $(3x + 2)^2$

B *In Problems 31–44, perform the indicated operations and simplify.*

31. $m - \{m - [m - (m - 1)]\}$

32. $2x - 3\{x + 2[x - (x + 5)] + 1\}$

33. $(x^2 - 2xy + y^2)(x^2 + 2xy + y^2)$

34. $(2x^2 + x - 2)(x^2 - 3x + 5)$

35. $(3a - b)(3a + b) - (2a - 3b)^2$

36. $(2x - 1)^2 - (3x + 2)(3x - 2)$

37. $(m - 2)^2 - (m - 2)(m + 2)$

38. $(x - 3)(x + 3) - (x - 3)^2$

39. $(x - 2y)(2x + y) - (x + 2y)(2x - y)$

40. $(3m + n)(m - 3n) - (m + 3n)(3m - n)$

41. $(u + v)^3$

42. $(x - y)^3$

43. $(x - 2y)^3$

44. $(2m - n)^3$

45. Subtract the sum of the last two polynomials from the sum of the first two: $2x^2 - 4xy + y^2$, $3xy - y^2$, $x^2 - 2xy - y^2$, $-x^2 + 3xy - 2y^2$

46. Subtract the sum of the first two polynomials from the sum of the last two: $3m^2 - 2m + 5$, $4m^2 - m$, $3m^2 - 3m - 2$, $m^3 + m^2 + 2$

C *In Problems 47–50, perform the indicated operations and simplify.*

47. $(2x - 1)^3 - 2(2x - 1)^2 + 3(2x - 1) + 7$

48. $2(x - 2)^3 - (x - 2)^2 - 3(x - 2) - 4$

49. $2\{(x - 3)(x^2 - 2x + 1) - x[3 - x(x - 2)]\}$

50. $-3x\{x[x - x(2 - x)] - (x + 2)(x^2 - 3)\}$

51. If you are given two polynomials, one of degree *m* and the other of degree *n*, where *m* is greater than *n*, what is the degree of their product?

52. What is the degree of the sum of the two polynomials in Problem 51?

53. How does the answer to Problem 51 change if the two polynomials can have the same degree?

54. How does the answer to Problem 52 change if the two polynomials can have the same degree?

55. Show by example that, in general, $(a + b)^2 \neq a^2 + b^2$. Discuss possible conditions on *a* and *b* that would make this a valid equation.

56. Show by example that, in general, $(a - b)^2 \neq a^2 - b^2$. Discuss possible conditions on *a* and *b* that would make this a valid equation.

Applications

Business & Economics

57. *Investment.* You have $10,000 to invest, part at 9% and the rest at 12%. If x is the amount invested at 9%, write an algebraic expression that represents the total annual income from both investments. Simplify the expression.

58. *Investment.* A person has $100,000 to invest. If $$x$ are invested in a money market account yielding 7% and twice that amount in certificates of deposit yielding 9%, and if the rest is invested in high-grade bonds yielding 11%, write an algebraic expression that represents the total annual income from all three investments. Simplify the expression.

59. *Gross receipts.* Four thousand tickets are to be sold for a musical show. If x tickets are to be sold for $10 each and three times that number for $30 each, and if the rest are sold for $50 each, write an algebraic expression that represents the gross receipts from ticket sales, assuming all tickets are sold. Simplify the expression.

60. *Gross receipts.* Six thousand tickets are to be sold for a concert, some for $9 each and the rest for $15 each. If x is the number of $9 tickets sold, write an algebraic expression that represents the gross receipts from ticket sales, assuming all tickets are sold. Simplify the expression.

Life Sciences

61. *Nutrition.* Food mix A contains 2% fat, and food mix B contains 6% fat. A 10 kilogram diet mix of foods A and B is formed. If x kilograms of food A are used, write an algebraic expression that represents the total number of kilograms of fat in the final food mix. Simplify the expression.

62. *Nutrition.* Each ounce of food M contains 8 units of calcium, and each ounce of food N contains 5 units of calcium. A 160 ounce diet mix is formed using foods M and N. If x is the number of ounces of food M used, write an algebraic expression that represents the total number of units of calcium in the diet mix. Simplify the expression.

Section A-4 Factoring Polynomials

- Common Factors
- Factoring by Grouping
- Factoring Second-Degree Polynomials
- Special Factoring Formulas
- Combined Factoring Techniques

A polynomial is written in factored form if it is written as the product of two or more polynomials. The following polynomials are written in factored form:

$$4x^2y - 6xy^2 = 2xy(2x - 3y) \qquad 2x^3 - 8x = 2x(x - 2)(x + 2)$$
$$x^2 - x - 6 = (x - 3)(x + 2) \qquad 5m^2 + 20 = 5(m^2 + 4)$$

Unless stated to the contrary, we will limit our discussion of factoring of polynomials to polynomials with integer coefficients.

A polynomial with integer coefficients is said to be **factored completely** if each factor cannot be expressed as the product of two or more polynomials with integer coefficients, other than itself or 1. All the polynomials above, as we will see by the conclusion of this section, are factored completely.

Writing polynomials in completely factored form is often a difficult task. But accomplishing it can lead to the simplification of certain algebraic expressions and to the solution of certain types of equations and inequalities. The distributive properties for real numbers are central to the factoring process.

Common Factors

Generally, a first step in any factoring procedure is to factor out all factors common to all terms.

Example 1 **Common Factors** Factor out all factors common to all terms:

(A) $3x^3y - 6x^2y^2 - 3xy^3$ (B) $3y(2y + 5) + 2(2y + 5)$

SOLUTION (A)

$$3x^3y - 6x^2y^2 - 3xy^3 = \mathbf{(3xy)}x^2 - \mathbf{(3xy)}2xy - \mathbf{(3xy)}y^2$$
$$= 3xy(x^2 - 2xy - y^2)$$

(B)

$$3y(2y + 5) + 2(2y + 5) = 3y\mathbf{(2y + 5)} + 2\mathbf{(2y + 5)}$$
$$= (3y + 2)(2y + 5)$$

Matched Problem 1 Factor out all factors common to all terms:

(A) $2x^3y - 8x^2y^2 - 6xy^3$ (B) $2x(3x - 2) - 7(3x - 2)$

Factoring by Grouping

Occasionally, polynomials can be factored by grouping terms in such a way that we obtain results that look like Example 1B. We can then complete the factoring following the steps used in that example. This process will prove useful in the next subsection, where an efficient method is developed for factoring a second-degree polynomial as the product of two first-degree polynomials, if such factors exist.

Example 2 **Factoring by Grouping** Factor by grouping:

(A) $3x^2 - 3x - x + 1$ (B) $4x^2 - 2xy - 6xy + 3y^2$
(C) $y^2 + xz + xy + yz$

SOLUTION (A)

$$3x^2 - 3x - x + 1$$
$$= (3x^2 - 3x) - (x - 1)$$ Group the first two and the last two terms.
$$= 3x\mathbf{(x - 1)} - \mathbf{(x - 1)}$$ Factor out any common factors from each group. The common factor $(x - 1)$ can be taken out, and the factoring is complete.
$$= \mathbf{(x - 1)}(3x - 1)$$

(B)

$$4x^2 - 2xy - 6xy + 3y^2 = (4x^2 - 2xy) - (6xy - 3y^2)$$
$$= 2x\mathbf{(2x - y)} - 3y\mathbf{(2x - y)}$$
$$= \mathbf{(2x - y)}(2x - 3y)$$

(C) If we group the first two terms and the last two terms of $y^2 + xz + xy + yz$, as in parts (A) and (B), no common factor can be taken out of each group to complete the factoring. However, if the two middle terms are reversed, we can proceed as before:

$$y^2 + xz + xy + yz = y^2 + xy + xz + yz$$
$$= (y^2 + xy) + (xz + yz)$$
$$= y(y + x) + z(x + y)$$
$$= y\mathbf{(x + y)} + z\mathbf{(x + y)}$$
$$= \mathbf{(x + y)}(y + z)$$

Matched Problem 2 ➡ Factor by grouping:

(A) $6x^2 + 2x + 9x + 3$ (B) $2u^2 + 6uv - 3uv - 9v^2$
(C) $ac + bd + bc + ad$

Factoring Second-Degree Polynomials

We now turn our attention to factoring second-degree polynomials of the form

$$2x^2 - 5x - 3 \qquad \text{and} \qquad 2x^2 + 3xy - 2y^2$$

into the product of two first-degree polynomials with integer coefficients. Since many second-degree polynomials with integer coefficients cannot be factored in this way, it would be useful to know ahead of time that the factors we are seeking actually exist. The factoring approach we use, involving the *ac test,* determines at the beginning whether first-degree factors with integer coefficients do exist. Then, if they exist, the test provides a simple method for finding them.

ac Test for Factorability

If in polynomials of the form

$$ax^2 + bx + c \qquad \text{or} \qquad ax^2 + bxy + cy^2 \tag{1}$$

the product ac has two integer factors p and q whose sum is the coefficient b of the middle term; that is, if integers p and q exist so that

$$pq = ac \qquad \text{and} \qquad p + q = b \tag{2}$$

then the polynomials have first-degree factors with integer coefficients. If no integers p and q exist that satisfy (2), then the polynomials in (1) will not have first-degree factors with integer coefficients.

If integers p and q exist that satisfy (2) in the *ac* test, then the factoring always can be completed as follows: Using $b = p + q$, split the middle terms in (1) to obtain

$$ax^2 + bx + c = ax^2 + px + qx + c$$
$$ax^2 + bxy + cy^2 = ax^2 + pxy + qxy + cy^2$$

Complete the factoring by grouping the first two terms and the last two terms as in Example 2. This process always works, and it does not matter if the two middle terms on the right are interchanged.

Several examples should make the process clear. After a little practice, you will perform many of the steps mentally and will find the process fast and efficient.

Example 3 ➡ **Factoring Second-Degree Polynomials** Factor, if possible, using integer coefficients:

(A) $4x^2 - 4x - 3$ (B) $2x^2 - 3x - 4$ (C) $6x^2 - 25xy + 4y^2$

SOLUTION (A) $4x^2 - 4x - 3$

Step 1. Use the ac test to test for factorability. Comparing $4x^2 - 4x - 3$ with $ax^2 + bx + c$, we see that $a = 4, b = -4$, and $c = -3$. Multiply a and c to obtain

$$ac = (4)(-3) = -12$$

pq
$(1)(-12)$
$(-1)(12)$
$(2)(-6)$
$(-2)(6)$
$(3)(-4)$
$(-3)(4)$

All factor pairs of $-12 = ac$

List all pairs of integers whose product is -12, as shown in the margin. These are called **factor pairs** of -12. Then try to find a factor pair that sums to $b = -4$, the coefficient of the middle term in $4x^2 - 4x - 3$. (In practice, this part of step 1 is often done mentally and can be done rather quickly.) Notice that the factor pair 2 and -6 sums to -4. Thus, by the ac test, $4x^2 - 4x - 3$ has first-degree factors with integer coefficients.

Step 2. Split the middle term, using $b = p + q$, and complete the factoring by grouping. Using $-4 = 2 + (-6)$, we split the middle term in $4x^2 - 4x - 3$ and complete the factoring by grouping:

$$\begin{aligned} 4x^2 - 4x - 3 &= 4x^2 + 2x - 6x - 3 \\ &= (4x^2 + 2x) - (6x + 3) \\ &= 2x(2x + 1) - 3(2x + 1) \\ &= (2x + 1)(2x - 3) \end{aligned}$$

The result can be checked by multiplying the two factors to obtain the original polynomial.

(B) $2x^2 - 3x - 4$

Step 1. Use the ac test to test for factorability:

$$ac = (2)(-4) = -8$$

pq
$(-1)(8)$
$(1)(-8)$
$(-2)(4)$
$(2)(-4)$

All factor pairs of $-8 = ac$

Does -8 have a factor pair whose sum is -3? None of the factor pairs listed in the margin sums to $-3 = b$, the coefficient of the middle term in $2x^2 - 3x - 4$. According to the ac test, we can conclude that $2x^2 - 3x - 4$ does not have first-degree factors with integer coefficients, and we say that the polynomial is **not factorable.**

(C) $6x^2 - 25xy + 4y^2$

Step 1. Use the ac test to test for factorability:

$$ac = (6)(4) = 24$$

Mentally checking through the factor pairs of 24, keeping in mind that their sum must be $-25 = b$, we see that if $p = -1$ and $q = -24$, then

$$pq = (-1)(-24) = 24 = ac$$

and

$$p + q = (-1) + (-24) = -25 = b$$

Thus, the polynomial is factorable.

Step 2. Split the middle term, using $b = p + q$, and complete the factoring by grouping. Using $-25 = (-1) + (-24)$, we split the middle term in $6x^2 - 25xy + 4y^2$ and complete the factoring by grouping:

$$\begin{aligned} 6x^2 - 25xy + 4y^2 &= 6x^2 - xy - 24xy + 4y^2 \\ &= (6x^2 - xy) - (24xy - 4y^2) \\ &= x(6x - y) - 4y(6x - y) \\ &= (6x - y)(x - 4y) \end{aligned}$$

The check is left to the reader.

Matched Problem 3 ➡ Factor, if possible, using integer coefficients:

(A) $2x^2 + 11x - 6$ (B) $4x^2 + 11x - 6$ (C) $6x^2 + 5xy - 4y^2$

Special Factoring Formulas

The factoring formulas listed below will enable us to factor certain polynomial forms that occur frequently. These formulas can be established by multiplying the factors on the right.

Special Factoring Formulas

Perfect square:	1. $u^2 + 2uv + v^2 = (u + v)^2$
Perfect square:	2. $u^2 - 2uv + v^2 = (u - v)^2$
Difference of squares:	3. $u^2 - v^2 = (u - v)(u + v)$
Difference of cubes:	4. $u^3 - v^3 = (u - v)(u^2 + uv + v^2)$
Sum of cubes:	5. $u^3 + v^3 = (u + v)(u^2 - uv + v^2)$

Caution

Notice that $u^2 + v^2$ is not included in the list of special factoring formulas. In general,

$$u^2 + v^2 \neq (au + bv)(cu + dv)$$

for any choice of real number coefficients a, b, c, and d.

Example 4 ➡ **Factoring** Factor completely:

(A) $4m^2 - 12mn + 9n^2$ (B) $x^2 - 16y^2$ (C) $z^3 - 1$
(D) $m^3 + n^3$ (E) $a^2 - 4(b + 2)^2$

Solution (A) $4m^2 - 12mn + 9n^2 = (2m - 3n)^2$

(B) $x^2 - 16y^2 = x^2 - (4y)^2 = (x - 4y)(x + 4y)$

(C) $z^3 - 1 = (z - 1)(z^2 + z + 1)$

(D) $m^3 + n^3 = (m + n)(m^2 - mn + n^2)$

(E) $a^2 - 4(b + 2)^2 = [a - 2(b + 2)][a + 2(b + 2)]$

Matched Problem 4 ➡ Factor completely:

(A) $x^2 + 6xy + 9y^2$ (B) $9x^2 - 4y^2$ (C) $8m^3 - 1$
(D) $x^3 + y^3z^3$ (E) $9(m - 3)^2 - 4n^2$

Explore–Discuss 1

(A) Verify the following factoring formulas for $u^4 - v^4$:

$$\begin{aligned} u^4 - v^4 &= (u - v)(u + v)(u^2 + v^2) \\ &= (u - v)(u^3 + u^2v + uv^2 + v^3) \end{aligned}$$

(B) Discuss the pattern in the following formulas:

$$\begin{aligned} u^2 - v^2 &= (u - v)(u + v) \\ u^3 - v^3 &= (u - v)(u^2 + uv + v^2) \\ u^4 - v^4 &= (u - v)(u^3 + u^2v + uv^2 + v^3) \end{aligned}$$

(C) Use the pattern you discovered in part (B) to write similar formulas for $u^5 - v^5$ and $u^6 - v^6$. Verify your formulas by multiplication.

Combined Factoring Techniques

We complete this section by considering several factoring problems that involve combinations of the preceding techniques. Generally speaking: **when factoring a polynomial, we first take out all factors common to all terms, if they are present.** Then we continue, using techniques discussed above, until the polynomial is in a completely factored form.

Example 5 **Combined Factoring Techniques** Factor completely:

(A) $3x^3 - 48x$ (B) $3u^4 - 3u^3v - 9u^2v^2$ (C) $3m^4 - 24mn^3$
(D) $3x^4 - 5x^2 + 2$

SOLUTION
(A) $3x^3 - 48x = 3x(x^2 - 16) = 3x(x - 4)(x + 4)$
(B) $3u^4 - 3u^3v - 9u^2v^2 = 3u^2(u^2 - uv - 3v^2)$
(C) $3m^4 - 24mn^3 = 3m(m^3 - 8n^3) = 3m(m - 2n)(m^2 + 2mn + 4n^2)$
(D) $3x^4 - 5x^2 + 2 = (3x^2 - 2)(x^2 - 1) = (3x^2 - 2)(x - 1)(x + 1)$

Matched Problem 5 Factor completely:

(A) $18x^3 - 8x$ (B) $4m^3n - 2m^2n^2 + 2mn^3$ (C) $2t^4 - 16t$
(D) $2y^4 - 5y^2 - 12$

Answers to Matched Problems

1. (A) $2xy(x^2 - 4xy - 3y^2)$ (B) $(2x - 7)(3x - 2)$
2. (A) $(3x + 1)(2x + 3)$ (B) $(u + 3v)(2u - 3v)$ (C) $(a + b)(c + d)$
3. (A) $(2x - 1)(x + 6)$ (B) Not factorable (C) $(3x + 4y)(2x - y)$
4. (A) $(x + 3y)^2$ (B) $(3x - 2y)(3x + 2y)$
(C) $(2m - 1)(4m^2 + 2m + 1)$ (D) $(x + yz)(x^2 - xyz + y^2z^2)$
(E) $[3(m - 3) - 2n][3(m - 3) + 2n]$
5. (A) $2x(3x - 2)(3x + 2)$ (B) $2mn(2m^2 - mn + n^2)$
(C) $2t(t - 2)(t^2 + 2t + 4)$ (D) $(2y^2 + 3)(y - 2)(y + 2)$

EXERCISE A-4

A *In Problems 1–8, factor out all factors common to all terms.*

1. $6m^4 - 9m^3 - 3m^2$
2. $6x^4 - 8x^3 - 2x^2$
3. $8u^3v - 6u^2v^2 + 4uv^3$
4. $10x^3y + 20x^2y^2 - 15xy^3$
5. $7m(2m - 3) + 5(2m - 3)$
6. $5x(x + 1) - 3(x + 1)$
7. $a(3c + d) - 4b(3c + d)$
8. $2w(y - 2z) - x(y - 2z)$

In Problems 9–18, factor by grouping.

9. $2x^2 - x + 4x - 2$
10. $x^2 - 3x + 2x - 6$
11. $3y^2 - 3y + 2y - 2$
12. $2x^2 - x + 6x - 3$
13. $2x^2 + 8x - x - 4$
14. $6x^2 + 9x - 2x - 3$
15. $wy - wz + xy - xz$
16. $ac + ad + bc + bd$
17. $am - bn - bm + an$
18. $ab + 6 + 2a + 3b$

B *In Problems 19–56, factor completely. If a polynomial cannot be factored, say so.*

19. $3y^2 - y - 2$
20. $2x^2 + 5x - 3$
21. $u^2 - 2uv - 15v^2$
22. $x^2 - 4xy - 12y^2$
23. $m^2 - 6m - 3$
24. $x^2 + x - 4$
25. $w^2x^2 - y^2$
26. $25m^2 - 16n^2$
27. $9m^2 - 6mn + n^2$
28. $x^2 + 10xy + 25y^2$
29. $y^2 + 16$
30. $u^2 + 81$
31. $4z^2 - 28z + 48$
32. $6x^2 + 48x + 72$
33. $2x^4 - 24x^3 + 40x^2$
34. $2y^3 - 22y^2 + 48y$
35. $4xy^2 - 12xy + 9x$
36. $16x^2y - 8xy + y$
37. $6m^2 - mn - 12n^2$
38. $6s^2 + 7st - 3t^2$
39. $4u^3v - uv^3$
40. $x^3y - 9xy^3$
41. $2x^3 - 2x^2 + 8x$
42. $3m^3 - 6m^2 + 15m$
43. $r^3 - t^3$
44. $m^3 + n^3$
45. $a^3 + 1$
46. $c^3 - 1$

C

47. $(x + 2)^2 - 9y^2$
48. $(a - b)^2 - 4(c - d)^2$
49. $5u^2 + 4uv - 2v^2$
50. $3x^2 - 2xy - 4y^2$
51. $6(x - y)^2 + 23(x - y) - 4$
52. $4(A + B)^2 - 5(A + B) - 6$
53. $y^4 - 3y^2 - 4$
54. $m^4 - n^4$
55. $27a^2 + a^5b^3$
56. $s^4t^4 - 8st$

SECTION A-5 Operations on Rational Expressions

- REDUCING TO LOWEST TERMS
- MULTIPLICATION AND DIVISION
- ADDITION AND SUBTRACTION
- COMPOUND FRACTIONS

We now turn our attention to fractional forms. A quotient of two algebraic expressions (division by 0 excluded) is called a **fractional expression.** If both the numerator and the denominator are polynomials, the fractional expression is called a **rational expression.** Some examples of rational expressions are

$$\frac{1}{x^3 + 2x} \qquad \frac{5}{x} \qquad \frac{x + 7}{3x^2 - 5x + 1} \qquad \frac{x^2 - 2x + 4}{1}$$

In this section we will discuss basic operations on rational expressions, including multiplication, division, addition, and subtraction.

Since variables represent real numbers in the rational expressions we will consider, the properties of real number fractions summarized in Appendix A-2 will play a central role in much of the work that we will do.

Even though not always explicitly stated, we always assume that variables are restricted so that division by 0 is excluded.

For example, given the rational expression

$$\frac{2x + 5}{x(x + 2)(x - 3)}$$

the variable x is understood to be restricted from being 0, -2, or 3, since these values would cause the denominator to be 0.

Reducing to Lowest Terms

Central to the process of reducing rational expressions to *lowest terms* is the *fundamental property of fractions,* which we restate here for convenient reference:

Fundamental Property of Fractions

If a, b, and k are real numbers with $b, k \neq 0$, then

$$\frac{ka}{kb} = \frac{a}{b} \qquad \frac{5 \cdot 2}{5 \cdot 7} = \frac{2}{7} \qquad \frac{x(x + 4)}{2(x + 4)} = \frac{x}{2}, \quad x \neq -4$$

Using this property from left to right to eliminate all common factors from the numerator and the denominator of a given fraction is referred to as **reducing a fraction to lowest terms.** We are actually dividing the numerator and denominator by the same nonzero common factor.

Using the property from right to left—that is, multiplying the numerator and denominator by the same nonzero factor—is referred to as **raising a fraction to higher terms.** We will use the property in both directions in the material that follows.

Example 1 **Reducing to Lowest Terms** Reduce each rational expression to lowest terms.

(A) $$\frac{6x^2 + x - 1}{2x^2 - x - 1} = \frac{(2x + 1)(3x - 1)}{(2x + 1)(x - 1)}$$ Factor numerator and denominator completely.

$$= \frac{3x - 1}{x - 1}$$ Divide numerator and denominator by the common factor $(2x + 1)$.

(B) $$\frac{x^4 - 8x}{3x^3 - 2x^2 - 8x} = \frac{x(x - 2)(x^2 + 2x + 4)}{x(x - 2)(3x + 4)}$$

$$= \frac{x^2 + 2x + 4}{3x + 4}$$

Matched Problem 1 Reduce each rational expression to lowest terms.

(A) $\dfrac{x^2 - 6x + 9}{x^2 - 9}$ (B) $\dfrac{x^3 - 1}{x^2 - 1}$

Multiplication and Division

Since we are restricting variable replacements to real numbers, multiplication and division of rational expressions follow the rules for multiplying and dividing real number fractions summarized in Appendix A-2.

Multiplication and Division

If a, b, c, and d are real numbers, then:

1. $\dfrac{a}{b} \cdot \dfrac{c}{d} = \dfrac{ac}{bd}, \quad b, d \neq 0$ $\qquad \dfrac{3}{5} \cdot \dfrac{x}{x+5} = \dfrac{3x}{5(x+5)}$

2. $\dfrac{a}{b} \div \dfrac{c}{d} = \dfrac{a}{b} \cdot \dfrac{d}{c}, \quad b, c, d \neq 0$ $\qquad \dfrac{3}{5} \div \dfrac{x}{x+5} = \dfrac{3}{5} \cdot \dfrac{x+5}{x}$

Explore–Discuss 1

Write a verbal description of the process of multiplying two rational expressions. Do the same for the quotient of two rational expressions.

Example 2 **Multiplication and Division** Perform the indicated operations and reduce to lowest terms.

(A) $\dfrac{10x^3y}{3xy + 9y} \cdot \dfrac{x^2 - 9}{4x^2 - 12x}$ Factor numerators and denominators. Then divide any numerator and any denominator with a like common factor.

$$= \frac{\overset{5x^2}{\cancel{10x^3y}}}{\underset{3 \cdot 1}{\cancel{3y(x+3)}}} \cdot \frac{\overset{1 \cdot 1}{\cancel{(x-3)(x+3)}}}{\underset{2 \cdot 1}{\cancel{4x(x-3)}}}$$

$$= \frac{5x^2}{6}$$

(B) $\dfrac{4 - 2x}{4} \div (x - 2) = \dfrac{\overset{1}{\cancel{2}}(2 - x)}{\underset{2}{\cancel{4}}} \cdot \dfrac{1}{x - 2}$ $\qquad x - 2 = \dfrac{x - 2}{1}$

$$= \frac{2 - x}{2(x - 2)} = \frac{\overset{-1}{\cancel{-(x-2)}}}{2\underset{1}{\cancel{(x-2)}}}$$ $b - a = -(a - b)$, a useful change in some problems

$$= -\frac{1}{2}$$

Matched Problem 2 Perform the indicated operations and reduce to lowest terms.

(A) $\dfrac{12x^2y^3}{2xy^2 + 6xy} \cdot \dfrac{y^2 + 6y + 9}{3y^3 + 9y^2}$ (B) $(4 - x) \div \dfrac{x^2 - 16}{5}$

Addition and Subtraction

Again, because we are restricting variable replacements to real numbers, addition and subtraction of rational expressions follow the rules for adding and subtracting real number fractions.

Addition and Subtraction

For a, b, and c real numbers:

1. $\dfrac{a}{b} + \dfrac{c}{b} = \dfrac{a + c}{b}$, $b \neq 0$ $\quad \dfrac{x}{x+5} + \dfrac{8}{x+5} = \dfrac{x+8}{x+5}$
2. $\dfrac{a}{b} - \dfrac{c}{b} = \dfrac{a - c}{b}$, $b \neq 0$ $\quad \dfrac{x}{3x^2y^2} - \dfrac{x+7}{3x^2y^2} = \dfrac{x-(x+7)}{3x^2y^2}$

Thus, we add rational expressions with the same denominators by adding or subtracting their numerators and placing the result over the common denominator. If the denominators are not the same, we raise the fractions to higher terms, using the fundamental property of fractions to obtain common denominators, and then proceed as described.

Even though any common denominator will do, our work will be simplified if the *least common denominator (LCD)* is used. Often, the LCD is obvious, but if it is not, the steps in the next box describe how to find it.

The Least Common Denominator (LCD)

The LCD of two or more rational expressions is found as follows:

1. Factor each denominator completely, including integer factors.
2. Identify each different factor from all the denominators.
3. Form a product using each different factor to the highest power that occurs in any one denominator. This product is the LCD.

Example 3 **Addition and Subtraction** Combine into a single fraction and reduce to lowest terms.

(A) $\dfrac{3}{10} + \dfrac{5}{6} - \dfrac{11}{45}$ (B) $\dfrac{4}{9x} - \dfrac{5x}{6y^2} + 1$ (C) $\dfrac{1}{x-1} - \dfrac{1}{x} - \dfrac{2}{x^2-1}$

SOLUTION (A) To find the LCD, factor each denominator completely:

$$\left.\begin{aligned} 10 &= 2 \cdot 5 \\ 6 &= 2 \cdot 3 \\ 45 &= 3^2 \cdot 5 \end{aligned}\right\} \text{LCD} = 2 \cdot 3^2 \cdot 5 = 90$$

Now use the fundamental property of fractions to make each denominator 90:

$$\frac{3}{10} + \frac{5}{6} - \frac{11}{45} = \frac{9\cdot 3}{9\cdot 10} + \frac{15\cdot 5}{15\cdot 6} - \frac{2\cdot 11}{2\cdot 45}$$
$$= \frac{27}{90} + \frac{75}{90} - \frac{22}{90}$$
$$= \frac{27 + 75 - 22}{90} = \frac{80}{90} = \frac{8}{9}$$

(B) $\left.\begin{aligned} 9x &= 3^2x \\ 6y^2 &= 2\cdot 3y^2 \end{aligned}\right\}$ LCD $= 2\cdot 3^2xy^2 = 18xy^2$

$$\frac{4}{9x} - \frac{5x}{6y^2} + 1 = \frac{2y^2\cdot 4}{2y^2\cdot 9x} - \frac{3x\cdot 5x}{3x\cdot 6y^2} + \frac{18xy^2}{18xy^2}$$
$$= \frac{8y^2 - 15x^2 + 18xy^2}{18xy^2}$$

(C) $\dfrac{1}{x-1} - \dfrac{1}{x} - \dfrac{2}{x^2-1}$

$$= \frac{1}{x-1} - \frac{1}{x} - \frac{2}{(x-1)(x+1)} \qquad \text{LCD} = x(x-1)(x+1)$$
$$= \frac{x(x+1) - (x-1)(x+1) - 2x}{x(x-1)(x+1)}$$
$$= \frac{x^2 + x - x^2 + 1 - 2x}{x(x-1)(x+1)}$$
$$= \frac{1 - x}{x(x-1)(x+1)}$$
$$= \frac{\overset{-1}{-\cancel{(x-1)}}}{x\underset{1}{\cancel{(x-1)}}(x+1)} = \frac{-1}{x(x+1)}$$

Matched Problem 3 Combine into a single fraction and reduce to lowest terms.

(A) $\dfrac{5}{28} - \dfrac{1}{10} + \dfrac{6}{35}$ (B) $\dfrac{1}{4x^2} - \dfrac{2x+1}{3x^3} + \dfrac{3}{12x}$

(C) $\dfrac{2}{x^2 - 4x + 4} + \dfrac{1}{x} - \dfrac{1}{x-2}$

Explore–Discuss 2

What is the value of $\dfrac{\;16\;}{\frac{4}{2}}$?

What is the result of entering $16 \div 4 \div 2$ on a calculator?

What is the difference between $16 \div (4 \div 2)$ and $(16 \div 4) \div 2$?

How could you use fraction bars to distinguish between these two cases when writing $\dfrac{16}{\frac{4}{2}}$?

Compound Fractions

A fractional expression with fractions in its numerator, denominator, or both is called a **compound fraction.** It is often necessary to represent a compound fraction as a **simple fraction**—that is (in all cases we will consider), as the quotient of two polynomials. The process does not involve any new concepts. It is a matter of applying old concepts and processes in the correct sequence.

Example 4 **Simplifying Compound Fractions** Express as a simple fraction reduced to lowest terms:

(A) $\dfrac{\dfrac{1}{5+h} - \dfrac{1}{5}}{h}$ (B) $\dfrac{\dfrac{y}{x^2} - \dfrac{x}{y^2}}{\dfrac{y}{x} - \dfrac{x}{y}}$

Solution We will simplify the expressions in parts (A) and (B) using two different methods—each is suited to the particular type of problem.

(A) We simplify this expression by combining the numerator into a single fraction and using division of rational forms.

$$\frac{\dfrac{1}{5+h} - \dfrac{1}{5}}{h} = \left[\frac{1}{5+h} - \frac{1}{5}\right] \div \frac{h}{1}$$

$$= \frac{5-5-h}{5(5+h)} \cdot \frac{1}{h}$$

$$= \frac{-h}{5(5+h)h} = \frac{-1}{5(5+h)}$$

(B) The method used here makes effective use of the fundamental property of fractions in the form

$$\frac{a}{b} = \frac{ka}{kb} \qquad b, k \neq 0$$

Multiply the numerator and denominator by the LCD of all fractions in the numerator and denominator—in this case, x^2y^2:

$$\frac{x^2y^2\left(\dfrac{y}{x^2} - \dfrac{x}{y^2}\right)}{x^2y^2\left(\dfrac{y}{x} - \dfrac{x}{y}\right)} = \frac{x^2y^2\dfrac{y}{x^2} - x^2y^2\dfrac{x}{y^2}}{x^2y^2\dfrac{y}{x} - x^2y^2\dfrac{x}{y}} = \frac{y^3 - x^3}{xy^3 - x^3y}$$

$$= \frac{\overset{1}{\cancel{(y-x)}}(y^2 + xy + x^2)}{xy\underset{1}{\cancel{(y-x)}}(y+x)}$$

$$= \frac{y^2 + xy + x^2}{xy(y+x)} \quad \text{or} \quad \frac{x^2 + xy + y^2}{xy(x+y)}$$

Matched Problem 4 Express as a simple fraction reduced to lowest terms:

(A) $\dfrac{\dfrac{1}{2+h} - \dfrac{1}{2}}{h}$ (B) $\dfrac{\dfrac{a}{b} - \dfrac{b}{a}}{\dfrac{a}{b} + 2 + \dfrac{b}{a}}$

Answers to Matched Problems

1. (A) $\dfrac{x-3}{x+3}$ (B) $\dfrac{x^2+x+1}{x+1}$ **2.** (A) $2x$ (B) $\dfrac{-5}{x+4}$

3. (A) $\dfrac{1}{4}$ (B) $\dfrac{3x^2-5x-4}{12x^3}$ (C) $\dfrac{4}{x(x-2)^2}$

4. (A) $\dfrac{-1}{2(2+h)}$ (B) $\dfrac{a-b}{a+b}$

Exercise A-5

A *In Problems 1–18, perform the indicated operations and reduce answers to lowest terms.*

1. $\dfrac{d^5}{3a} \div \left(\dfrac{d^2}{6a^2} \cdot \dfrac{a}{4d^3}\right)$

2. $\left(\dfrac{d^5}{3a} \div \dfrac{d^2}{6a^2}\right) \cdot \dfrac{a}{4d^3}$

3. $\dfrac{x^2}{12} + \dfrac{x}{18} - \dfrac{1}{30}$

4. $\dfrac{2y}{18} - \dfrac{-1}{28} - \dfrac{y}{42}$

5. $\dfrac{4m-3}{18m^3} + \dfrac{3}{4m} - \dfrac{2m-1}{6m^2}$

6. $\dfrac{3x+8}{4x^2} - \dfrac{2x-1}{x^3} - \dfrac{5}{8x}$

7. $\dfrac{x^2-9}{x^2-3x} \div (x^2-x-12)$

8. $\dfrac{2x^2+7x+3}{4x^2-1} \div (x+3)$

9. $\dfrac{2}{x} - \dfrac{1}{x-3}$

10. $\dfrac{3}{m} - \dfrac{2}{m+4}$

11. $\dfrac{3}{x^2-1} - \dfrac{2}{x^2-2x+1}$

12. $\dfrac{1}{a^2-b^2} + \dfrac{1}{a^2+2ab+b^2}$

13. $\dfrac{x+1}{x-1} - 1$

14. $m - 3 - \dfrac{m-1}{m-2}$

15. $\dfrac{3}{a-1} - \dfrac{2}{1-a}$

16. $\dfrac{5}{x-3} - \dfrac{2}{3-x}$

17. $\dfrac{2x}{x^2-16} - \dfrac{x-4}{x^2+4x}$

18. $\dfrac{m+2}{m^2-2m} - \dfrac{m}{m^2-4}$

B *In Problems 19–30, perform the indicated operations and reduce answers to lowest terms. Represent any compound fractions as simple fractions reduced to lowest terms.*

19. $\dfrac{x^2}{x^2+2x+1} + \dfrac{x-1}{3x+3} - \dfrac{1}{6}$

20. $\dfrac{y}{y^2-y-2} - \dfrac{1}{y^2+5y-14} - \dfrac{2}{y^2+8y+7}$

21. $\dfrac{2-x}{2x+x^2} \cdot \dfrac{x^2+4x+4}{x^2-4}$

22. $\dfrac{9-m^2}{m^2+5m+6} \cdot \dfrac{m+2}{m-3}$

23. $\dfrac{c+2}{5c-5} - \dfrac{c-2}{3c-3} + \dfrac{c}{1-c}$

24. $\dfrac{x+7}{ax-bx} + \dfrac{y+9}{by-ay}$

25. $\dfrac{1+\dfrac{3}{x}}{x-\dfrac{9}{x}}$

26. $\dfrac{1-\dfrac{y^2}{x^2}}{1-\dfrac{y}{x}}$

27. $\dfrac{\dfrac{1}{2(x+h)} - \dfrac{1}{2x}}{h}$

28. $\dfrac{\dfrac{1}{x+h} - \dfrac{1}{x}}{h}$

29. $\dfrac{\dfrac{x}{y} - 2 + \dfrac{y}{x}}{\dfrac{x}{y} - \dfrac{y}{x}}$

30. $\dfrac{1 + \dfrac{2}{x} - \dfrac{15}{x^2}}{1 + \dfrac{4}{x} - \dfrac{5}{x^2}}$

In Problems 31–38, imagine that the indicated "solutions" were given to you by a student whom you were tutoring in this class.
(A) Is the solution correct? If the solution is incorrect, explain what is wrong and how it can be corrected.
(B) Show a correct solution for each incorrect solution.

31. $\dfrac{x^2 + 4x + 3}{x + 3} = \dfrac{x^2 + 4x}{x} = x + 4$

32. $\dfrac{x^2 - 3x - 4}{x - 4} = \dfrac{x^2 - 3x}{x} = x - 3$

33. $\dfrac{(x + h)^2 - x^2}{h} = (x + 1)^2 - x^2 = 2x + 1$

34. $\dfrac{(x + h)^3 - x^3}{h} = (x + 1)^3 - x^3 = 3x^2 + 3x + 1$

35. $\dfrac{x^2 - 3x}{x^2 - 2x - 3} + x - 3 = \dfrac{x^2 - 3x + x - 3}{x^2 - 2x - 3} = 1$

36. $\dfrac{2}{x - 1} - \dfrac{x + 3}{x^2 - 1} = \dfrac{2x + 2 - x - 3}{x^2 - 1} = \dfrac{1}{x + 1}$

37. $\dfrac{2x^2}{x^2 - 4} - \dfrac{x}{x - 2} = \dfrac{2x^2 - x^2 - 2x}{x^2 - 4} = \dfrac{x}{x + 2}$

38. $x + \dfrac{x - 2}{x^2 - 3x + 2} = \dfrac{x + x - 2}{x^2 - 3x + 2} = \dfrac{2}{x - 2}$

C *Represent the compound fractions in Problems 39–42 as simple fractions reduced to lowest terms.*

39. $\dfrac{\dfrac{1}{3(x + h)^2} - \dfrac{1}{3x^2}}{h}$

40. $\dfrac{\dfrac{1}{(x + h)^2} - \dfrac{1}{x^2}}{h}$

41. $1 - \dfrac{1}{1 - \dfrac{1}{1 - \dfrac{1}{x}}}$

42. $2 - \dfrac{1}{1 - \dfrac{2}{a + 2}}$

SECTION A-6 Integer Exponents and Scientific Notation

- INTEGER EXPONENTS
- SCIENTIFIC NOTATION

We now review basic operations on integer exponents and scientific notation and its use.

INTEGER EXPONENTS

Definitions for **integer exponents** are listed below.

Definition of a^n

For n an integer and a a real number:

1. For n a positive integer,

$$a^n = a \cdot a \cdot \cdots \cdot a \qquad n \text{ factors of } a \qquad 5^4 = 5 \cdot 5 \cdot 5 \cdot 5$$

2. For $n = 0$,

$$a^0 = 1 \qquad a \neq 0 \qquad 12^0 = 1$$

0^0 is not defined.

3. For n a negative integer,

$$a^n = \frac{1}{a^{-n}} \qquad a \neq 0 \qquad a^{-3} = \frac{1}{a^{-(-3)}} = \frac{1}{a^3}$$

[If n is negative, then $(-n)$ is positive.]
Note: It can be shown that for *all* integers n,

$$a^{-n} = \frac{1}{a^n} \quad \text{and} \quad a^n = \frac{1}{a^{-n}} \qquad a \neq 0 \qquad a^5 = \frac{1}{a^{-5}}, \quad a^{-5} = \frac{1}{a^5}$$

The following integer exponent properties are very useful in manipulating integer exponent forms.

Theorem 1 ■■ EXPONENT PROPERTIES

For n and m integers and a and b real numbers:

1. $a^m a^n = a^{m+n}$ $\qquad a^8 a^{-3} = a^{8+(-3)} = a^5$
2. $(a^n)^m = a^{mn}$ $\qquad (a^{-2})^3 = a^{3(-2)} = a^{-6}$
3. $(ab)^m = a^m b^m$ $\qquad (ab)^{-2} = a^{-2}b^{-2}$
4. $\left(\dfrac{a}{b}\right)^m = \dfrac{a^m}{b^m} \quad b \neq 0$ $\qquad \left(\dfrac{a}{b}\right)^5 = \dfrac{a^5}{b^5}$
5. $\dfrac{a^m}{a^n} = a^{m-n} = \dfrac{1}{a^{n-m}} \quad a \neq 0$ $\qquad \dfrac{a^{-3}}{a^7} = \dfrac{1}{a^{7-(-3)}} = \dfrac{1}{a^{10}}$ ■■

Explore–Discuss 1

Property 1 in Theorem 1 can be expressed verbally as follows:

To find the product of two exponential forms with the same base, add the exponents and use the same base.

Express the other properties in Theorem 1 verbally. Decide which you find easier to remember—a formula or a verbal description.

Exponent forms are frequently encountered in algebraic applications. You should sharpen your skills in using these forms by reviewing the above basic definitions and properties and the examples that follow.

Example 1 **Simplifying Exponent Forms** Simplify, and express the answers using positive exponents only.

(A) $(2x^3)(3x^5) = 2 \cdot 3x^{3+5} = 6x^8$

(B) $x^5x^{-9} = x^{-4} = \dfrac{1}{x^4}$

(C) $\dfrac{x^5}{x^7} = x^{5-7} = x^{-2} = \dfrac{1}{x^2}$

or $\dfrac{x^5}{x^7} = \dfrac{1}{x^{7-5}} = \dfrac{1}{x^2}$

(D) $\dfrac{x^{-3}}{y^{-4}} = \dfrac{y^4}{x^3}$

(E) $(u^{-3}v^2)^{-2} = (u^{-3})^{-2}(v^2)^{-2} = u^6v^{-4} = \dfrac{u^6}{v^4}$

(F) $\left(\dfrac{y^{-5}}{y^{-2}}\right)^{-2} = \dfrac{(y^{-5})^{-2}}{(y^{-2})^{-2}} = \dfrac{y^{10}}{y^4} = y^6$

(G) $\dfrac{4m^{-3}n^{-5}}{6m^{-4}n^3} = \dfrac{2m^{-3-(-4)}}{3n^{3-(-5)}} = \dfrac{2m}{3n^8}$

Matched Problem 1 Simplify, and express the answers using positive exponents only.

(A) $(3y^4)(2y^3)$ (B) m^2m^{-6} (C) $(u^3v^{-2})^{-2}$

(D) $\left(\dfrac{y^{-6}}{y^{-2}}\right)^{-1}$ (E) $\dfrac{8x^{-2}y^{-4}}{6x^{-5}y^2}$

Example 2 **Converting to a Simple Fraction** Write as a simple fraction with positive exponents:

$$\frac{1 - x}{x^{-1} - 1}$$

SOLUTION First note that

$$\frac{1 - x}{x^{-1} - 1} \neq \frac{x(1 - x)}{-1} \qquad \text{A common error}$$

The original expression is a complex fraction, and we proceed to simplify it as follows:

$$\begin{aligned} \frac{1 - x}{x^{-1} - 1} &= \frac{1 - x}{\dfrac{1}{x} - 1} \qquad \text{Multiply numerator and denominator by } x \text{ to clear internal fractions.} \\ &= \frac{x(1 - x)}{x\left(\dfrac{1}{x} - 1\right)} \\ &= \frac{x(1 - x)}{1 - x} = x \end{aligned}$$

Matched Problem 2 Write as a simple fraction with positive exponents: $\dfrac{1 + x^{-1}}{1 - x^{-2}}$

Scientific Notation

In the real world, one often encounters very large numbers. For example:

The public debt in the United States in 1992, to the nearest billion dollars, was

$4,065,000,000,000

The world population in the year 2000, to the nearest million, is projected to be

6,166,000,000

Very small numbers are also encountered:

The sound intensity of a normal conversation is

0.000 000 000 316 watt per square centimeter*

It is generally troublesome to write and work with numbers of this type in standard decimal form. The first and last example cannot even be entered into many calculators as they are written. But with exponents

---

*We write 0.000 000 000 316 in place of 0.000000000316, because it is then easier to keep track of the number of decimal places. We follow this convention when there are more than five decimal places to the right of the decimal.

defined for all integers, we can now express any finite decimal form as the product of a number between 1 and 10 and an integer power of 10; that is, in the form

$$a \times 10^n \qquad 1 \le a < 10, \quad a \text{ in decimal form}, \quad n \text{ an integer}$$

A number expressed in this form is said to be in **scientific notation.** The following are some examples of numbers in standard decimal notation and in scientific notation:

DECIMAL AND SCIENTIFIC NOTATION

$$\begin{aligned} 7 &= 7 \times 10^0 & 0.5 &= 5 \times 10^{-1} \\ 67 &= 6.7 \times 10 & 0.45 &= 4.5 \times 10^{-1} \\ 580 &= 5.8 \times 10^2 & 0.0032 &= 3.2 \times 10^{-3} \\ 43{,}000 &= 4.3 \times 10^4 & 0.000\ 045 &= 4.5 \times 10^{-5} \\ 73{,}400{,}000 &= 7.34 \times 10^7 & 0.000\ 000\ 391 &= 3.91 \times 10^{-7} \end{aligned}$$

Note that the power of 10 used corresponds to the number of places we move the decimal to form a number between 1 and 10. The power is positive if the decimal is moved to the left and negative if it is moved to the right. Positive exponents are associated with numbers greater than or equal to 10; negative exponents are associated with positive numbers less than 1; and a zero exponent is associated with a number that is 1 or greater, but less than 10.

Example 3 ➡ **Scientific Notation**

(A) Write each number in scientific notation:

$$7{,}320{,}000 \qquad \text{and} \qquad 0.000\ 000\ 54$$

(B) Write each number in standard decimal form:

$$4.32 \times 10^6 \qquad \text{and} \qquad 4.32 \times 10^{-5}$$

SOLUTION (A)

$$7{,}320{,}000 = 7.320\ 000. \times 10^6 = 7.32 \times 10^6$$

6 places left
Positive exponent

$$0.000\ 000\ 54 = 0.000\ 000\ 5.4 \times 10^{-7} = 5.4 \times 10^{-7}$$

7 places right
Negative exponent

(B)

$$4.32 \times 10^6 = 4{,}320{,}000 \qquad 4.32 \times 10^{-5} = \frac{4.32}{10^5} = 0.000\ 043\ 2$$

6 places right; 5 places left

Positive exponent 6; Negative exponent −5

Matched Problem 3 ➡

(A) Write each number in scientific notation: 47,100; 2,443,000,000; 1.45
(B) Write each number in standard decimal form: 3.07×10^8; 5.98×10^{-6}

Explore–Discuss 2

Scientific and graphing calculators can calculate in either standard decimal mode or scientific notation mode. If the result of a computation in decimal mode is either too large or too small to be displayed, then most calculators will automatically display the answer in scientific notation. Read the manual for your calculator and experiment with some operations on very large and very small numbers in both decimal mode and scientific notation mode. For example, show that:

$$\frac{216{,}700{,}000{,}000}{0.000\ 000\ 000\ 000\ 078\ 8} = 2.75 \times 10^{24}$$

Answers to Matched Problems

1. (A) $6y^7$ (B) $\frac{1}{m^4}$ (C) $\frac{v^4}{u^6}$ (D) y^4 (E) $\frac{4x^3}{3y^6}$ **2.** $\frac{x}{x-1}$

3. (A) 4.7×10^4; 2.443×10^9; 1.45×10^0 (B) 307,000,000; 0.000 005 98

Exercise A-6

A *In Problems 1–14, simplify and express answers using positive exponents only. Variables are restricted to avoid division by 0.*

1. $2x^{-9}$ **2.** $3y^{-5}$ **3.** $\frac{3}{2w^{-7}}$

4. $\frac{5}{4x^{-9}}$ **5.** $2x^{-8}x^5$ **6.** $3c^{-9}c^4$

7. $\frac{w^{-8}}{w^{-3}}$ **8.** $\frac{m^{-11}}{m^{-5}}$ **9.** $5v^8v^{-8}$

10. $7d^{-4}d^4$ **11.** $(a^{-3})^2$ **12.** $(b^4)^{-3}$

13. $(x^6y^{-3})^{-2}$ **14.** $(a^{-3}b^4)^{-3}$

Write each number in Problems 15–20 in scientific notation.

15. 82,300,000,000 **16.** 5,380,000

17. 0.783 **18.** 0.019

19. 0.000 034 **20.** 0.000 000 007 832

Write each number in Problems 21–28 in standard decimal notation.

21. 4×10^4 **22.** 9×10^6

23. 7×10^{-3} **24.** 2×10^{-5}

25. 6.171×10^7 **26.** 3.044×10^3

27. 8.08×10^{-4} **28.** 1.13×10^{-2}

B *In Problems 29–38, simplify and express answers using positive exponents only.*

29. $(22 + 31)^0$ **30.** $(2x^3y^4)^0$

31. $\frac{10^{-3} \cdot 10^4}{10^{-11} \cdot 10^{-2}}$ **32.** $\frac{10^{-17} \cdot 10^{-5}}{10^{-3} \cdot 10^{-14}}$

33. $(5x^2y^{-3})^{-2}$ **34.** $(2m^{-3}n^2)^{-3}$

35. $\frac{8 \times 10^{-3}}{2 \times 10^{-5}}$ **36.** $\frac{18 \times 10^{12}}{6 \times 10^{-4}}$

37. $\frac{8x^{-3}y^{-1}}{6x^2y^{-4}}$ **38.** $\frac{9m^{-4}n^3}{12m^{-1}n^{-1}}$

In Problems 39–42, write each expression in the form $ax^p + bx^q$ or $ax^p + bx^q + cx^r$, where a, b, and c are real numbers and p, q, and r are integers. For example,

$$\frac{2x^4 - 3x^2 + 1}{2x^3} = \frac{2x^4}{2x^3} - \frac{3x^2}{2x^3} + \frac{1}{2x^3} = x - \frac{3}{2}x^{-1} + \frac{1}{2}x^{-3}$$

39. $\frac{7x^5 - x^2}{4x^5}$ **40.** $\frac{5x^3 - 2}{3x^2}$

41. $\frac{3x^4 - 4x^2 - 1}{4x^3}$ **42.** $\frac{2x^3 - 3x^2 + x}{2x^2}$

Write each expression in Problems 43–46 with positive exponents only, and as a single fraction reduced to lowest terms.

43. $\dfrac{3x^2(x-1)^2 - 2x^3(x-1)}{(x-1)^4}$

44. $\dfrac{5x^4(x+3)^2 - 2x^5(x+3)}{(x+3)^4}$

45. $2x^{-2}(x-1) - 2x^{-3}(x-1)^2$

46. $2x(x+3)^{-1} - x^2(x+3)^{-2}$

In Problems 47–50, convert each number to scientific notation and simplify. Express the answer in both scientific notation and in standard decimal form.

47. $\dfrac{9{,}600{,}000{,}000}{(1{,}600{,}000)(0.000\ 000\ 25)}$

48. $\dfrac{(60{,}000)(0.000\ 003)}{(0.0004)(1{,}500{,}000)}$

49. $\dfrac{(1{,}250{,}000)(0.000\ 38)}{0.0152}$

50. $\dfrac{(0.000\ 000\ 82)(230{,}000)}{(625{,}000)(0.0082)}$

51. What is the result of entering 2^{3^2} on a calculator?

52. Refer to Problem 51. What is the difference between $2^{(3^2)}$ and $(2^3)^2$? Which agrees with the value of 2^{3^2} obtained with a calculator?

53. If $n = 0$, then property 1 in Theorem 1 implies that $a^m a^0 = a^{m+0} = a^m$. Explain how this helps motivate the definition of a^0.

54. If $m = -n$, then property 1 in Theorem 1 implies that $a^{-n}a^n = a^0 = 1$. Explain how this helps motivate the definition of a^{-n}.

C *Write the fractions in Problems 55–58 as simple fractions reduced to lowest terms.*

55. $\dfrac{u+v}{u^{-1}+v^{-1}}$

56. $\dfrac{x^{-1}-y^{-1}}{x-y}$

57. $\dfrac{b^{-2}-c^{-2}}{b^{-3}-c^{-3}}$

58. $\dfrac{xy^{-2}-yx^{-2}}{y^{-1}-x^{-1}}$

Applications

Business & Economics

Problems 59 and 60 refer to Table 1.

Table 1

Assets of the Five Largest U.S. Commercial Banks, 1992

Bank	Assets ($)
Citicorp, New York	213,701,000,000
Bank of America, San Francisco	180,646,000,000
Chemical Banking, New York	139,655,000,000
J. P. Morgan, New York	102,941,000,000
Chase Manhattan, New York	95,862,000,000

59. *Financial assets*
(A) Write Citicorp's assets in scientific notation.
(B) After converting to scientific notation, determine the ratio of the assets of Citicorp to the assets of Chase Manhattan. Write the answer in standard decimal form to four decimal places.
(C) Repeat part (B) with the banks reversed.

60. *Financial assets*
(A) Write Bank of America's assets in scientific notation.
(B) After converting to scientific notation, determine the ratio of the assets of Bank of America to the assets of J. P. Morgan. Write the answer in standard decimal form to four decimal places.
(C) Repeat part (B) with the banks reversed.

Problems 61 and 62 refer to Table 2.

Table 2

U.S. Public Debt, Interest on Debt, and Population

Year	Public Debt ($)	Interest on Debt ($)	Population
1982	1,142,000,000,000	117,400,000,000	231,500,000
1992	4,065,000,000,000	292,300,000,000	255,100,000

61. *Public debt.* Carry out the following computations using scientific notation, and write final answers in standard decimal form.
(A) What was the per capita debt in 1992 (to the nearest dollar)?
(B) What was the per capita interest paid on the debt in 1992 (to the nearest dollar)?
(C) What was the percentage interest paid on the debt in 1992 (to two decimal places)?

62. *Public debt.* Carry out the following computations using scientific notation, and write final answers in standard decimal form.
(A) What was the per capita debt in 1982 (to the nearest dollar)?
(B) What was the per capita interest paid on the debt in 1982 (to the nearest dollar)?
(C) What was the percentage interest paid on the debt in 1982 (to two decimal places)?

Life Sciences

Air pollution. Air quality standards establish maximum amounts of pollutants considered acceptable in the air. The amounts are frequently given in parts per million (ppm). A standard of 30 ppm also can be expressed as follows:

$$30 \text{ ppm} = \frac{30}{1{,}000{,}000} = \frac{3 \times 10}{10^6}$$
$$= 3 \times 10^{-5} = 0.000\ 03 = 0.003\%$$

In Problems 63 and 64, express the given standard:
(A) In scientific notation
(B) In standard decimal notation
(C) As a percent

63. 9 ppm, the standard for carbon monoxide, when averaged over a period of 8 hours

64. 0.03 ppm, the standard for sulfur oxides, when averaged over a year

Social Sciences

65. *Crime.* In 1992, the United States had a violent crime rate of 757.5 per 100,000 people and a population of 255.1 million people. How many violent crimes occurred that year? Compute the answer using scientific notation and convert the answer to standard decimal form (to the nearest thousand).

66. *Population density.* The United States had a 1992 population of 255.1 million people and a land area of 3,539,000 square miles. What was the population density? Compute the answer using scientific notation and convert the answer to standard decimal form (to one decimal place).

SECTION A-7 Rational Exponents and Radicals

- *n*TH ROOTS OF REAL NUMBERS
- RATIONAL EXPONENTS AND RADICALS
- PROPERTIES OF RADICALS

Square roots may now be generalized to *nth roots,* and the meaning of exponent may be generalized to include all rational numbers.

*n*TH ROOTS OF REAL NUMBERS

Consider a square of side r with area 36 square inches. We can write

$$r^2 = 36$$

and conclude that side r is a number whose square is 36. We say that r is a **square root** of b if $r^2 = b$. Similarly, we say that r is a **cube root** of b if $r^3 = b$. And, in general:

*n*th Root

For any natural number n,

r is an ***n*th root** of b if $r^n = b$

Thus, 4 is a square root of 16, since $4^2 = 16$, and -2 is a cube root of -8, since $(-2)^3 = -8$. Since $(-4)^2 = 16$, we see that -4 is also a square root of 16. It can be shown that any positive number has two real square roots, two real 4th roots, and, in general, two real nth roots if n is even. Negative numbers have no real square roots, no real 4th roots, and, in general, no real nth roots if n is even. The reason is that no real number raised to an even power can be negative. For odd roots the situation is simpler. Every real number has exactly one real cube root, one real 5th root, and, in general, one real nth root if n is odd.

Additional roots can be considered in the *complex number system.* But in this book we restrict our interest to *real roots of real numbers,* and "root" will always be interpreted to mean "real root."

Rational Exponents and Radicals

We now turn to the question of what symbols to use to represent nth roots. For n a natural number greater than 1, we use

$$b^{1/n} \quad \text{or} \quad \sqrt[n]{b}$$

to represent a **real *n*th root of *b*.** The exponent form is motivated by the fact that $(b^{1/n})^n = b$ if exponent laws are to continue to hold for rational exponents. The other form is called an ***n*th root radical.** In the expression below, the symbol $\sqrt{}$ is called a **radical,** n is the **index** of the radical, and a is the **radicand:**

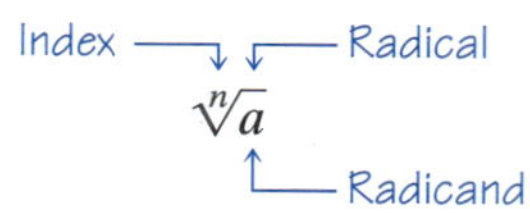

When the index is 2, it is usually omitted. That is, when dealing with square roots, we simply use $\sqrt{b}$ rather than $\sqrt[2]{b}$. If there are two real nth roots, both $b^{1/n}$ and $\sqrt[n]{b}$ denote the positive root, called the **principal *n*th root.**

Example 1 **Finding *n*th Roots** Evaluate each of the following:

(A) $4^{1/2}$ and $\sqrt{4}$ (B) $-4^{1/2}$ and $-\sqrt{4}$ (C) $(-4)^{1/2}$ and $\sqrt{-4}$

(D) $8^{1/3}$ and $\sqrt[3]{8}$ (E) $(-8)^{1/3}$ and $\sqrt[3]{-8}$ (F) $-8^{1/3}$ and $-\sqrt[3]{8}$

SOLUTION

(A) $4^{1/2} = \sqrt{4} = 2 \quad (\sqrt{4} \neq \pm 2)$ (B) $-4^{1/2} = -\sqrt{4} = -2$

(C) $(-4)^{1/2}$ and $\sqrt{-4}$ are not real numbers

(D) $8^{1/3} = \sqrt[3]{8} = 2$ (E) $(-8)^{1/3} = \sqrt[3]{-8} = -2$

(F) $-8^{1/3} = -\sqrt[3]{8} = -2$

Matched Problem 1 Evaluate each of the following:

(A) $16^{1/2}$ (B) $-\sqrt{16}$ (C) $\sqrt[3]{-27}$ (D) $(-9)^{1/2}$ (E) $(\sqrt[4]{81})^3$

Common Error

The symbol $\sqrt{4}$ represents the single number 2, not ± 2. Do not confuse $\sqrt{4}$ with the solutions of the equation $x^2 = 4$, which are usually written in the form $x = \pm\sqrt{4} = \pm 2$.

We now define b^r for any rational number $r = m/n$.

Rational Exponents

If m and n are natural numbers without common prime factors, b is a real number, and b is nonnegative when n is even, then

$$b^{m/n} = \begin{cases} (b^{1/n})^m = (\sqrt[n]{b})^m \\ (b^m)^{1/n} = \sqrt[n]{b^m} \end{cases} \qquad \begin{aligned} 8^{2/3} &= (8^{1/3})^2 = (\sqrt[3]{8})^2 = 2^2 = 4 \\ 8^{2/3} &= (8^2)^{1/3} = \sqrt[3]{8^2} = \sqrt[3]{64} = 4 \end{aligned}$$

and

$$b^{-m/n} = \frac{1}{b^{m/n}} \quad b \neq 0 \qquad 8^{-2/3} = \frac{1}{8^{2/3}} = \frac{1}{4}$$

Note that the two definitions of $b^{m/n}$ are equivalent under the indicated restrictions on m, n, and b.

All the properties listed for integer exponents in Theorem 1 in Section A-6 also hold for rational exponents, provided b is nonnegative when n is even. Unless stated to the contrary, all variables in the rest of the discussion represent positive real numbers.

Example 2 **From Rational Exponent Form to Radical Form and Vice Versa**
Change rational exponent form to radical form.

(A) $x^{1/7} = \sqrt[7]{x}$
(B) $(3u^2v^3)^{3/5} = \sqrt[5]{(3u^2v^3)^3}$ or $(\sqrt[5]{3u^2v^3})^3$ The first is usually preferred.

(C) $y^{-2/3} = \dfrac{1}{y^{2/3}} = \dfrac{1}{\sqrt[3]{y^2}}$ or $\sqrt[3]{y^{-2}}$ or $\sqrt[3]{\dfrac{1}{y^2}}$

Change radical form to rational exponent form.

(D) $\sqrt[5]{6} = 6^{1/5}$ (E) $-\sqrt[3]{x^2} = -x^{2/3}$
(F) $\sqrt{x^2 + y^2} = (x^2 + y^2)^{1/2}$ Note that $(x^2 + y^2)^{1/2} \neq x + y$. Why?

Matched Problem 2 Convert to radical form.

(A) $u^{1/5}$ (B) $(6x^2y^5)^{2/9}$ (C) $(3xy)^{-3/5}$

Convert to rational exponent form.

(D) $\sqrt[4]{9u}$ (E) $-\sqrt[7]{(2x)^4}$ (F) $\sqrt[3]{x^3 + y^3}$

Example 3 **Working with Rational Exponents** Simplify each and express answers using positive exponents only. If rational exponents appear in final answers, convert to radical form.

(A) $(3x^{1/3})(2x^{1/2}) = 6x^{1/3+1/2} = 6x^{5/6} = 6\sqrt[6]{x^5}$
(B) $(-8)^{5/3} = [(-8)^{1/3}]^5 = (-2)^5 = -32$

(C) $(2x^{1/3}y^{-2/3})^3 = 8xy^{-2} = \dfrac{8x}{y^2}$

(D) $\left(\dfrac{4x^{1/3}}{x^{1/2}}\right)^{1/2} = \dfrac{4^{1/2}x^{1/6}}{x^{1/4}} = \dfrac{2}{x^{1/4-1/6}} = \dfrac{2}{x^{1/12}} = \dfrac{2}{\sqrt[12]{x}}$

Matched Problem 3 Simplify each and express answers using positive exponents only. If rational exponents appear in final answers, convert to radical form.

(A) $9^{3/2}$ (B) $(-27)^{4/3}$ (C) $(5y^{1/4})(2y^{1/3})$ (D) $(2x^{-3/4}y^{1/4})^4$

(E) $\left(\dfrac{8x^{1/2}}{x^{2/3}}\right)^{1/3}$

Explore–Discuss 1

In each of the following, evaluate both radical forms:

$$16^{3/2} = \sqrt{16^3} = (\sqrt{16})^3$$
$$27^{2/3} = \sqrt[3]{27^2} = (\sqrt[3]{27})^2$$

Which radical conversion form is easier to use if you are performing the calculations by hand?

Example 4 **Working with Rational Exponents** Multiply, and express answers using positive exponents only.

(A) $3y^{2/3}(2y^{1/3} - y^2)$ (B) $(2u^{1/2} + v^{1/2})(u^{1/2} - 3v^{1/2})$

SOLUTION (A) $3y^{2/3}(2y^{1/3} - y^2) = 6y^{2/3+1/3} - 3y^{2/3+2}$
$= 6y - 3y^{8/3}$

(B) $(2u^{1/2} + v^{1/2})(u^{1/2} - 3v^{1/2}) = 2u - 5u^{1/2}v^{1/2} - 3v$

Matched Problem 4 Multiply, and express answers using positive exponents only.

(A) $2c^{1/4}(5c^3 - c^{3/4})$ (B) $(7x^{1/2} - y^{1/2})(2x^{1/2} + 3y^{1/2})$

Example 5 **Working with Rational Exponents** Write the following expression in the form $ax^p + bx^q$, where a and b are real numbers and p and q are rational numbers:

$$\frac{2\sqrt{x} - 3\sqrt[3]{x^2}}{2\sqrt[3]{x}}$$

SOLUTION

$$\frac{2\sqrt{x} - 3\sqrt[3]{x^2}}{2\sqrt[3]{x}} = \frac{2x^{1/2} - 3x^{2/3}}{2x^{1/3}} \quad \text{Change to rational exponent form.}$$
$$= \frac{2x^{1/2}}{2x^{1/3}} - \frac{3x^{2/3}}{2x^{1/3}} \quad \text{Separate into two fractions.}$$
$$= x^{1/6} - 1.5x^{1/3}$$

Matched Problem 5 Write the following expression in the form $ax^p + bx^q$, where a and b are real numbers and p and q are rational numbers:

$$\frac{5\sqrt[3]{x} - 4\sqrt{x}}{2\sqrt{x^3}}$$

Properties of Radicals

Changing or simplifying radical expressions is aided by several properties of radicals that follow directly from the properties of exponents considered earlier.

Properties of Radicals

If c, n, and m are natural numbers greater than or equal to 2, and if x and y are positive real numbers, then:

1. $\sqrt[n]{x^n} = x$ $\quad \sqrt[3]{x^3} = x$
2. $\sqrt[n]{xy} = \sqrt[n]{x}\sqrt[n]{y}$ $\quad \sqrt[5]{xy} = \sqrt[5]{x}\sqrt[5]{y}$
3. $\sqrt[n]{\frac{x}{y}} = \frac{\sqrt[n]{x}}{\sqrt[n]{y}}$ $\quad \sqrt[4]{\frac{x}{y}} = \frac{\sqrt[4]{x}}{\sqrt[4]{y}}$

Example 6 **Applying Properties of Radicals** Simplify using properties of radicals:

(A) $\sqrt[4]{(3x^4y^3)^4}$ (B) $\sqrt[4]{8}\sqrt[4]{2}$ (C) $\sqrt[3]{\frac{xy}{27}}$

Solution

(A) $\sqrt[4]{(3x^4y^3)^4} = 3x^4y^3$ Property 1

(B) $\sqrt[4]{8}\sqrt[4]{2} = \sqrt[4]{16} = \sqrt[4]{2^4} = 2$ Properties 2 and 1

(C) $\sqrt[3]{\frac{xy}{27}} = \frac{\sqrt[3]{xy}}{\sqrt[3]{27}} = \frac{\sqrt[3]{xy}}{3}$ or $\frac{1}{3}\sqrt[3]{xy}$ Properties 3 and 1

Matched Problem 6 Simplify using properties of radicals:

(A) $\sqrt[7]{(x^3 + y^3)^7}$ (B) $\sqrt[3]{8y^3}$ (C) $\frac{\sqrt[3]{16x^4y}}{\sqrt[3]{2xy}}$

Explore–Discuss 2

Multiply:

$$(\sqrt{a} - \sqrt{b})(\sqrt{a} + \sqrt{b})$$

How can this product be used to simplify radical forms?

A question arises regarding the best form in which a radical expression should be left. There are many answers, depending on what use we wish to make of the expression. In deriving certain formulas, it is sometimes useful to clear either a denominator or a numerator of radicals. The process is referred to as **rationalizing** the denominator or numerator. Examples 7 and 8 illustrate the rationalizing process.

Example 7 **Rationalizing Denominators** Rationalize each denominator:

(A) $\frac{6x}{\sqrt{2x}}$ (B) $\frac{6}{\sqrt{7} - \sqrt{5}}$ (C) $\frac{x - 4}{\sqrt{x} + 2}$

Solution

(A) $\frac{6x}{\sqrt{2x}} = \frac{6x}{\sqrt{2x}} \cdot \frac{\sqrt{2x}}{\sqrt{2x}} = \frac{6x\sqrt{2x}}{2x} = 3\sqrt{2x}$

$$\text{(B)}\quad \frac{6}{\sqrt{7}-\sqrt{5}} = \frac{6}{\sqrt{7}-\sqrt{5}} \cdot \frac{\sqrt{7}+\sqrt{5}}{\sqrt{7}+\sqrt{5}} = \frac{6(\sqrt{7}+\sqrt{5})}{2} = 3(\sqrt{7}+\sqrt{5})$$

$$\text{(C)}\quad \frac{x-4}{\sqrt{x}+2} = \frac{x-4}{\sqrt{x}+2} \cdot \frac{\sqrt{x}-2}{\sqrt{x}-2} = \frac{(x-4)(\sqrt{x}-2)}{x-4} = \sqrt{x}-2$$

Matched Problem 7 Rationalize each denominator:

$$\text{(A)}\ \frac{12ab^2}{\sqrt{3ab}} \qquad \text{(B)}\ \frac{9}{\sqrt{6}+\sqrt{3}} \qquad \text{(C)}\ \frac{x^2-y^2}{\sqrt{x}-\sqrt{y}}$$

Example 8 **Rationalizing Numerators** Rationalize each numerator:

$$\text{(A)}\ \frac{\sqrt{2}}{2\sqrt{3}} \qquad \text{(B)}\ \frac{3+\sqrt{m}}{9-m} \qquad \text{(C)}\ \frac{\sqrt{2+h}-\sqrt{2}}{h}$$

SOLUTION

$$\text{(A)}\quad \frac{\sqrt{2}}{2\sqrt{3}} = \frac{\sqrt{2}}{2\sqrt{3}} \cdot \frac{\sqrt{2}}{\sqrt{2}} = \frac{2}{2\sqrt{6}} = \frac{1}{\sqrt{6}}$$

$$\text{(B)}\quad \frac{3+\sqrt{m}}{9-m} = \frac{3+\sqrt{m}}{9-m} \cdot \frac{3-\sqrt{m}}{3-\sqrt{m}} = \frac{9-m}{(9-m)(3-\sqrt{m})} = \frac{1}{3-\sqrt{m}}$$

$$\text{(C)}\quad \frac{\sqrt{2+h}-\sqrt{2}}{h} = \frac{\sqrt{2+h}-\sqrt{2}}{h} \cdot \frac{\sqrt{2+h}+\sqrt{2}}{\sqrt{2+h}+\sqrt{2}} = \frac{h}{h(\sqrt{2+h}+\sqrt{2})} = \frac{1}{\sqrt{2+h}+\sqrt{2}}$$

Matched Problem 8 Rationalize each numerator:

$$\text{(A)}\ \frac{\sqrt{3}}{3\sqrt{2}} \qquad \text{(B)}\ \frac{2-\sqrt{n}}{4-n} \qquad \text{(C)}\ \frac{\sqrt{3+h}-\sqrt{3}}{h}$$

Answers to Matched Problems

1. (A) 4 (B) -4 (C) -3 (D) Not a real number (E) 27

2. (A) $\sqrt[5]{u}$ (B) $\sqrt[9]{(6x^2y^5)^2}$ or $(\sqrt[9]{6x^2y^5})^2$ (C) $1/\sqrt[5]{(3xy)^3}$
(D) $(9u)^{1/4}$ (E) $-(2x)^{4/7}$ (F) $(x^3+y^3)^{1/3}$ (not $x+y$)

3. (A) 27 (B) 81 (C) $10y^{7/12} = 10\sqrt[12]{y^7}$ (D) $16y/x^3$
(E) $2/x^{1/18} = 2/\sqrt[18]{x}$

4. (A) $10c^{13/4} - 2c$ (B) $14x + 19x^{1/2}y^{1/2} - 3y$ **5.** $2.5x^{-7/6} - 2x^{-1}$

6. (A) $x^3 + y^3$ (B) $2y$ (C) $2x$

7. (A) $4b\sqrt{3ab}$ (B) $3(\sqrt{6}-\sqrt{3})$ (C) $(x+y)(\sqrt{x}+\sqrt{y})$

8. (A) $\dfrac{1}{\sqrt{6}}$ (B) $\dfrac{1}{2+\sqrt{n}}$ (C) $\dfrac{1}{\sqrt{3+h}+\sqrt{3}}$

EXERCISE A-7

A *Change each expression in Problems 1–6 to radical form. Do not simplify.*

1. $6x^{3/5}$ **2.** $7y^{2/5}$ **3.** $(4xy^3)^{2/5}$
4. $(7x^2y)^{5/7}$ **5.** $(x^2 + y^2)^{1/2}$ **6.** $x^{1/2} + y^{1/2}$

Change each expression in Problems 7–12 to rational exponent form. Do not simplify.

7. $5\sqrt[4]{x^3}$ **8.** $7m\sqrt[5]{n^2}$ **9.** $\sqrt[5]{(2x^2y)^3}$
10. $\sqrt[9]{(3m^4n)^2}$ **11.** $\sqrt[3]{x} + \sqrt[3]{y}$ **12.** $\sqrt[3]{x^2 + y^3}$

In Problems 13–24, find rational number representations for each, if they exist.

13. $25^{1/2}$ **14.** $64^{1/3}$ **15.** $16^{3/2}$
16. $16^{3/4}$ **17.** $-36^{1/2}$ **18.** $-32^{3/5}$
19. $(-36)^{1/2}$ **20.** $(-32)^{3/5}$ **21.** $(\frac{4}{25})^{3/2}$
22. $(\frac{8}{27})^{2/3}$ **23.** $9^{-3/2}$ **24.** $8^{-2/3}$

In Problems 25–34, simplify each expression and write answers using positive exponents only. All variables represent positive real numbers.

25. $x^{4/5}x^{-2/5}$ **26.** $y^{-3/7}y^{4/7}$ **27.** $\dfrac{m^{2/3}}{m^{-1/3}}$
28. $\dfrac{x^{1/4}}{x^{3/4}}$ **29.** $(8x^3y^{-6})^{1/3}$ **30.** $(4u^{-2}v^4)^{1/2}$
31. $\left(\dfrac{4x^{-2}}{y^4}\right)^{-1/2}$ **32.** $\left(\dfrac{w^4}{9x^{-2}}\right)^{-1/2}$ **33.** $\dfrac{8x^{-1/3}}{12x^{1/4}}$
34. $\dfrac{6a^{3/4}}{15a^{-1/3}}$

Simplify each expression in Problems 35–40 using properties of radicals. All variables represent positive real numbers.

35. $\sqrt[5]{(2x + 3)^5}$ **36.** $\sqrt[3]{(7 + 2y)^3}$ **37.** $\sqrt{18x^3}\sqrt{2x^3}$
38. $\sqrt{2a^3}\sqrt{32a^5}$ **39.** $\dfrac{\sqrt{6x}\sqrt{10}}{\sqrt{15x}}$ **40.** $\dfrac{\sqrt{8}\sqrt{12y}}{\sqrt{6y}}$

B *In Problems 41–48, multiply, and express answers using positive exponents only.*

41. $3x^{3/4}(4x^{1/4} - 2x^8)$

42. $2m^{1/3}(3m^{2/3} - m^6)$

43. $(3u^{1/2} - v^{1/2})(u^{1/2} - 4v^{1/2})$

44. $(a^{1/2} + 2b^{1/2})(a^{1/2} - 3b^{1/2})$

45. $(5m^{1/2} + n^{1/2})(5m^{1/2} - n^{1/2})$

46. $(2x^{1/2} - 3y^{1/2})(2x^{1/2} + 3y^{1/2})$

47. $(3x^{1/2} - y^{1/2})^2$

48. $(x^{1/2} + 2y^{1/2})^2$

Write each expression in Problems 49–54 in the form $ax^p + bx^q$, where a and b are real numbers and p and q are rational numbers.

49. $\dfrac{\sqrt[3]{x^2} + 2}{2\sqrt[3]{x}}$ **50.** $\dfrac{12\sqrt{x} - 3}{4\sqrt{x}}$ **51.** $\dfrac{2\sqrt[4]{x^3} + \sqrt[3]{x}}{3x}$
52. $\dfrac{3\sqrt[3]{x^2} + \sqrt{x}}{5x}$ **53.** $\dfrac{2\sqrt[3]{x} - \sqrt{x}}{4\sqrt{x}}$ **54.** $\dfrac{x^2 - 4\sqrt{x}}{2\sqrt[3]{x}}$

Rationalize the denominators in Problems 55–60.

55. $\dfrac{12mn^2}{\sqrt{3mn}}$ **56.** $\dfrac{14x^2}{\sqrt{7x}}$ **57.** $\dfrac{2}{\sqrt{x} - 2}$
58. $\dfrac{3}{\sqrt{x} + 4}$ **59.** $\dfrac{7(x - y)^2}{\sqrt{x} - \sqrt{y}}$ **60.** $\dfrac{3a - 3b}{\sqrt{a} + \sqrt{b}}$

Rationalize the numerators in Problems 61–66.

61. $\dfrac{\sqrt{5xy}}{5x^2y^2}$ **62.** $\dfrac{\sqrt{3mn}}{3mn}$
63. $\dfrac{\sqrt{x + h} - \sqrt{x}}{h}$ **64.** $\dfrac{\sqrt{2(a + h)} - \sqrt{2a}}{h}$
65. $\dfrac{\sqrt{t} - \sqrt{x}}{t - x}$ **66.** $\dfrac{\sqrt{x} - \sqrt{y}}{\sqrt{x} + \sqrt{y}}$

Problems 67–70 illustrate common errors involving rational exponents. In each case, find numerical examples that show that the left side is not always equal to the right side.

67. $(x + y)^{1/2} \neq x^{1/2} + y^{1/2}$

68. $(x^3 + y^3)^{1/3} \neq x + y$

69. $(x + y)^{1/3} \neq \dfrac{1}{(x + y)^3}$

70. $(x + y)^{-1/2} \neq \dfrac{1}{(x + y)^2}$

C *In Problems 71–76, simplify by writing each expression as a simple or single fraction reduced to lowest terms and without negative exponents.*

71. $-\frac{1}{2}(x - 2)(x + 3)^{-3/2} + (x + 3)^{-1/2}$

72. $2(x - 2)^{-1/2} - \frac{1}{2}(2x + 3)(x - 2)^{-3/2}$

73. $\dfrac{(x - 1)^{1/2} - x(\frac{1}{2})(x - 1)^{-1/2}}{x - 1}$

74. $\dfrac{(2x - 1)^{1/2} - (x + 2)(\frac{1}{2})(2x - 1)^{-1/2}(2)}{2x - 1}$

75. $\dfrac{(x + 2)^{2/3} - x(\frac{2}{3})(x + 2)^{-1/3}}{(x + 2)^{4/3}}$

76. $\dfrac{2(3x - 1)^{1/3} - (2x + 1)(\frac{1}{3})(3x - 1)^{-2/3}(3)}{(3x - 1)^{2/3}}$

In Problems 77–82, evaluate using a calculator. (Refer to the instruction book for your calculator to see how exponential forms are evaluated.)

77. $22^{3/2}$ **78.** $15^{5/4}$ **79.** $827^{-3/8}$

80. $103^{-3/4}$ **81.** $37.09^{7/3}$ **82.** $2.876^{8/5}$

In Problems 83 and 84, evaluate each expression on a calculator and determine which pairs have the same value. Verify these results algebraically.

83. (A) $\sqrt{3} + \sqrt{5}$ (B) $\sqrt{2 + \sqrt{3}} + \sqrt{2 - \sqrt{3}}$
(C) $1 + \sqrt{3}$ (D) $\sqrt[3]{10 + 6\sqrt{3}}$
(E) $\sqrt{8 + \sqrt{60}}$ (F) $\sqrt{6}$

84. (A) $2\sqrt[3]{2 + \sqrt{5}}$ (B) $\sqrt{8}$
(C) $\sqrt{3} + \sqrt{7}$ (D) $\sqrt{3 + \sqrt{8}} + \sqrt{3 - \sqrt{8}}$
(E) $\sqrt{10 + \sqrt{84}}$ (F) $1 + \sqrt{5}$

SECTION A-8 Linear Equations and Inequalities in One Variable

- Linear Equations
- Linear Inequalities
- Applications

The equation

$$3 - 2(x + 3) = \frac{x}{3} - 5$$

and the inequality

$$\frac{x}{2} + 2(3x - 1) \geq 5$$

are both first-degree in one variable. In general, a **first-degree, or linear, equation** in one variable is any equation that can be written in the form

$$\textbf{Standard form:} \quad ax + b = 0 \qquad a \neq 0 \tag{1}$$

If the equality symbol, =, in (1) is replaced by $<, >, \leq$, or $\geq$, then the resulting expression is called a **first-degree, or linear, inequality.**

A **solution** of an equation (or inequality) involving a single variable is a number that, when substituted for the variable, makes the equation (or inequality) true. The set of all solutions is called the **solution set.** When we say that we **solve an equation** (or inequality), we mean that we find its solution set.

Knowing what is meant by the solution set is one thing; finding it is another. We start by recalling the idea of equivalent equations and equivalent inequalities. If we perform an operation on an equation (or inequality) that produces another equation (or inequality) with the same solution set, then the two equations (or inequalities) are said to be **equivalent.** The basic idea in solving equations and inequalities is to perform operations on these forms that produce simpler equivalent forms, and to continue the process until we obtain an equation or inequality with an obvious solution.

Linear Equations

Linear equations are generally solved using the following equality properties:

> **Equality Properties**
>
> An equivalent equation will result if:
>
> 1. The same quantity is added to or subtracted from each side of a given equation.
> 2. Each side of a given equation is multiplied by or divided by the same nonzero quantity.

Several examples should remind you of the process of solving equations.

Example 1 **Solving a Linear Equation** Solve and check:

$$8x - 3(x - 4) = 3(x - 4) + 6$$

Solution

$$\begin{aligned} 8x - 3(x - 4) &= 3(x - 4) + 6 \\ 8x - 3x + 12 &= 3x - 12 + 6 \\ 5x + 12 &= 3x - 6 \\ 2x &= -18 \\ x &= -9 \end{aligned}$$

Check

$$\begin{aligned} 8x - 3(x - 4) &= 3(x - 4) + 6 \\ 8(-9) - 3[(-9) - 4] &\stackrel{?}{=} 3[(-9) - 4] + 6 \\ -72 - 3(-13) &\stackrel{?}{=} 3(-13) + 6 \\ -33 &\stackrel{\checkmark}{=} -33 \end{aligned}$$

Matched Problem 1 Solve and check: $3x - 2(2x - 5) = 2(x + 3) - 8$

Explore–Discuss 1

According to equality property 2, multiplying both sides of an equation by a nonzero number always produces an equivalent equation. What is the smallest positive number that you could use to multiply both sides of the following equation in order to produce an equivalent equation without fractions?

$$\frac{x + 1}{3} - \frac{x}{4} = \frac{1}{2}$$

Example 2 **Solving a Linear Equation** Solve and check: $\dfrac{x + 2}{2} - \dfrac{x}{3} = 5$

Solution What operations can we perform on

$$\frac{x + 2}{2} - \frac{x}{3} = 5$$

to eliminate the denominators? If we can find a number that is exactly divisible by each denominator, then we can use the multiplication property of equality to clear the denominators. The LCD (least common denominator) of the fractions, 6, is exactly what we are looking for! Actually, any common denominator will do, but the LCD results in a simpler equivalent equation. Thus, we multiply both sides of the equation by 6:

$$6\left(\frac{x+2}{2} - \frac{x}{3}\right) = 6 \cdot 5$$

$$\overset{3}{\cancel{6}} \cdot \frac{(x+2)}{\underset{1}{\cancel{2}}} - \overset{2}{\cancel{6}} \cdot \frac{x}{\underset{1}{\cancel{3}}} = 30$$

$$3(x+2) - 2x = 30$$
$$3x + 6 - 2x = 30$$
$$x = 24$$

CHECK

$$\frac{x+2}{2} - \frac{x}{3} = 5$$
$$\frac{24+2}{2} - \frac{24}{3} \stackrel{?}{=} 5$$
$$13 - 8 \stackrel{?}{=} 5$$
$$5 \stackrel{\checkmark}{=} 5$$

Matched Problem 2 Solve and check: $\frac{x+1}{3} - \frac{x}{4} = \frac{1}{2}$

In many applications of algebra, formulas or equations must be changed to alternative equivalent forms. The following examples are typical.

Example 3 **Solving a Formula for a Particular Variable** Solve the amount formula for simple interest, $A = P + Prt$, for:

(A) r in terms of the other variables
(B) P in terms of the other variables

SOLUTION (A)

$A = P + Prt$ Reverse equation.
$P + Prt = A$ Now isolate r on the left side.
$Prt = A - P$ Divide both members by Pt.
$r = \frac{A - P}{Pt}$

(B)

$A = P + Prt$ Reverse equation.
$P + Prt = A$ Factor out P (note the use of the distributive property).
$P(1 + rt) = A$ Divide by $(1 + rt)$.
$P = \frac{A}{1 + rt}$

Matched Problem 3 Solve $M = Nt + Nr$ for:
(A) t (B) N

Linear Inequalities

Before we start solving linear inequalities, let us recall what we mean by $<$ (less than) and $>$ (greater than). If a and b are real numbers, then we write

$a < b$ a is less than b

if there exists a positive number p such that $a + p = b$. Certainly, we would expect that if a positive number was added to any real number, the sum would be larger than the original. That is essentially what the definition states. If $a < b$, we may also write

$$b > a \quad b \text{ is greater than } a$$

Example 4 **Inequalities**

(A) $3 < 5$ Since $3 + 2 = 5$
(B) $-6 < -2$ Since $-6 + 4 = -2$
(C) $0 > -10$ Since $-10 < 0$

Matched Problem 4 Replace each question mark with either $<$ or $>$.

(A) 2 **?** 8 (B) −20 **?** 0 (C) −3 **?** −30

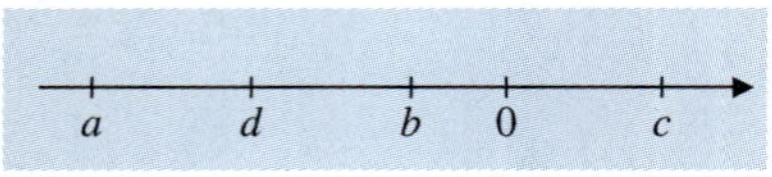

FIGURE 1
$a < b, c > d$

The inequality symbols have a very clear geometric interpretation on the real number line. If $a < b$, then a is to the left of b on the number line; if $c > d$, then c is to the right of d (Fig. 1). Check this geometric property with the inequalities in Example 4.

Explore–Discuss 2

Replace **?** with $<$ or $>$ in each of the following:

(A) -1 **?** 3 and $2(-1)$ **?** $2(3)$
(B) -1 **?** 3 and $-2(-1)$ **?** $-2(3)$
(C) 12 **?** -8 and $\frac{12}{4}$ **?** $\frac{-8}{4}$
(D) 12 **?** -8 and $\frac{12}{-4}$ **?** $\frac{-8}{-4}$

Based on these examples, describe verbally the effect of multiplying both sides of an inequality by a number.

Now let us turn to the problem of solving linear inequalities in one variable. Recall that a **solution** of an inequality involving one variable is a number that, when substituted for the variable, makes the inequality true. The set of all solutions is called the **solution set.** When we say that we **solve an inequality,** we mean that we find its solution set. The procedures used to solve linear inequalities in one variable are almost the same as those used to solve linear equations in one variable but with one important exception, as noted in property 3 below.

Inequality Properties

An equivalent inequality will result and the **sense will remain the same** if each side of the original inequality:

1. Has the same real number added to or subtracted from it.
2. Is multiplied or divided by the same positive number.

(continued)

Inequality Properties *(continued)*

An equivalent inequality will result and the **sense will reverse** if each side of the original inequality:

3. Is multiplied or divided by the same negative number.

Note: Multiplication by 0 and division by 0 are not permitted.

Thus, we can perform essentially the same operations on inequalities that we perform on equations, with the exception that **the sense of the inequality reverses if we multiply or divide both sides by a negative number.** Otherwise, the sense of the inequality does not change. For example, if we start with the true statement

$$-3 > -7$$

and multiply both sides by 2, we obtain

$$-6 > -14$$

and the sense of the inequality stays the same. But if we multiply both sides of $-3 > -7$ by -2, then the left side becomes 6 and the right side becomes 14, so we must write

$$6 < 14$$

to have a true statement. Thus, the sense of the inequality reverses.

If $a < b$, the double inequality $a < x < b$ means that $\boldsymbol{x > a}$ **and** $\boldsymbol{x < b}$; that is, x is between a and b. Other variations, as well as a useful interval notation, are given in Table 1. Note that an end point on a line graph has a square bracket through it if it is included in the inequality and a parenthesis through it if it is not.

Table 1

Interval Notation	Inequality Notation	Line Graph
$[a, b]$	$a \leq x \leq b$	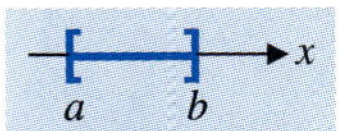
$[a, b)$	$a \leq x < b$	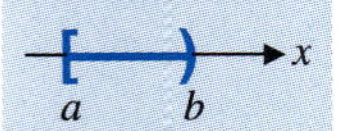
$(a, b]$	$a < x \leq b$	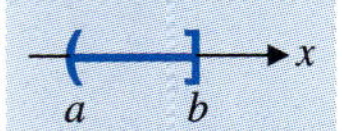
(a, b)	$a < x < b$	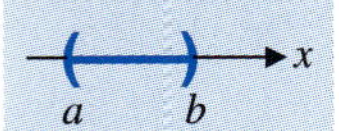
$(-\infty, a]$	$x \leq a$	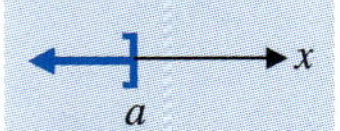
$(-\infty, a)$	$x < a$	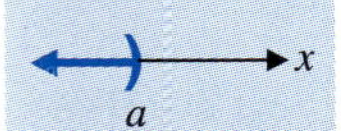
$[b, \infty)$*	$x \geq b$	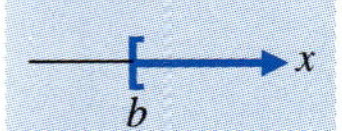
(b, ∞)	$x > b$	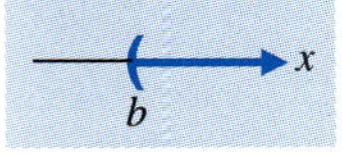

*The symbol ∞ (read "infinity") is not a number. When we write $[b, \infty)$, we are simply referring to the interval starting at b and continuing indefinitely to the right. We would never write $[b, \infty]$.

Example 5 **Interval and Inequality Notation, and Line Graphs**

(A) Write $[-2, 3)$ as a double inequality and graph.
(B) Write $x \geqslant -5$ in interval notation and graph.

SOLUTION (A) $[-2, 3)$ is equivalent to $-2 \leqslant x < 3$.

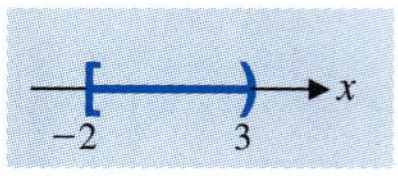

(B) $x \geqslant -5$ is equivalent to $[-5, \infty)$.

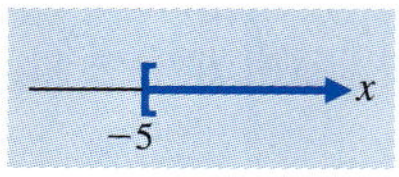

Matched Problem 5 (A) Write $(-7, 4]$ as a double inequality and graph.
(B) Write $x < 3$ in interval notation and graph.

Explore–Discuss 3

The solution to Example 5B shows the graph of the inequality $x \geqslant -5$. What is the graph of $x < -5$? What is the corresponding interval? Describe the relationship between these sets.

Example 6 **Solving a Linear Inequality** Solve and graph:

$$2(2x + 3) < 6(x - 2) + 10$$

SOLUTION

$$\begin{aligned} 2(2x + 3) &< 6(x - 2) + 10 \\ 4x + 6 &< 6x - 12 + 10 \\ 4x + 6 &< 6x - 2 \\ -2x + 6 &< -2 \\ -2x &< -8 \\ x &> 4 \quad \text{or} \quad (4, \infty) \end{aligned}$$

Notice that the sense of the inequality reverses when we divide both sides by −2.

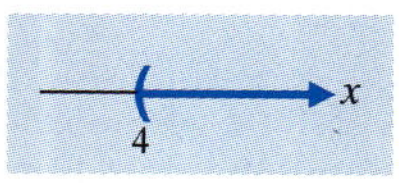

Notice that in the graph of $x > 4$, we use a parenthesis through 4, since the point 4 is not included in the graph.

Matched Problem 6 Solve and graph: $3(x - 1) \leqslant 5(x + 2) - 5$

Example 7 **Solving a Double Inequality** Solve and graph: $-3 < 2x + 3 \leqslant 9$

SOLUTION We are looking for all numbers x such that $2x + 3$ is between −3 and 9, including 9 but not −3. We proceed as above except that we try to isolate x in the middle:

$$-3 < 2x + 3 \leqslant 9$$

$$-3 - \mathbf{3} < 2x + 3 - \mathbf{3} \leqslant 9 - \mathbf{3}$$

$$-6 < 2x \leqslant 6$$

$$\frac{-6}{2} < \frac{2x}{2} \leqslant \frac{6}{2}$$

$$-3 < x \leqslant 3 \quad \text{or} \quad (-3, 3]$$

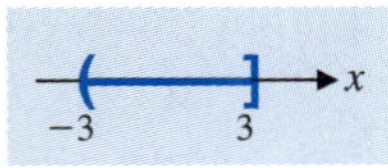

Matched Problem 7 ➡ Solve and graph: $-8 \leqslant 3x - 5 < 7$

Note that a linear equation usually has exactly one solution, while a linear inequality usually has infinitely many solutions.

Applications

To realize the full potential of algebra, we must be able to translate real-world problems into mathematical forms. In short, we must be able to do word problems.

The first example below involves the important concept of **break-even analysis,** which is encountered in several places in this text. Any manufacturing company has **costs, *C*,** and **revenues, *R*.** The company will have a **loss** if $R < C$, will **break even** if $R = C$, and will have a **profit** if $R > C$. Costs involve **fixed costs,** such as plant overhead, product design, set-up, and promotion; and **variable costs,** which are dependent upon the number of items produced at a certain cost per item.

Example 8 ➡ **Break-Even Analysis** A recording company produces compact disks (CD's). One-time fixed costs for a particular CD are \$24,000, which includes costs such as recording, album design, and promotion. Variable costs amount to \$6.20 per CD and include the manufacturing, distribution, and royalty costs for each disk actually manufactured and sold to a retailer. The CD is sold to retail outlets at \$8.70 each. How many CD's must be manufactured and sold for the company to break even?

Solution Let x = Number of CD's manufactured and sold
C = Cost of producing x CD's
R = Revenue (return) on sales of x CD's

The company breaks even if $R = C$, with

$$\begin{aligned} C &= \text{Fixed costs} + \text{Variable costs} \\ &= \$24{,}000 + \$6.20x \\ R &= \$8.70x \end{aligned}$$

Find x such that $R = C$; that is, such that

$$\begin{aligned} 8.7x &= 24{,}000 + 6.2x \\ 2.5x &= 24{,}000 \\ x &= 9{,}600 \text{ CD's} \end{aligned}$$

Check For $x = 9{,}600$,

$$\begin{aligned} C &= 24{,}000 + 6.2x \\ &= 24{,}000 + 6.2(9{,}600) \\ &= \$83{,}520 \end{aligned} \qquad \begin{aligned} R &= 8.7x \\ &= 8.7(9{,}600) \\ &= \$83{,}520 \end{aligned}$$

Matched Problem 8 ➡ What is the break-even point in Example 8 if fixed costs are \$18,000, variable costs are \$5.20 per CD, and the CD's are sold to retailers for \$7.60 each?

Algebra has many different types of applications—so many, in fact, that no single approach applies to all. However, the following suggestions may help you get started:

Suggestions for Solving Word Problems

1. Read the problem very carefully.
2. Write down important facts and relationships.
3. Identify unknown quantities in terms of a single letter, if possible.
4. Write an equation (or inequality) relating the unknown quantities and the facts in the problem.
5. Solve the equation (or inequality).
6. Write all solutions requested in the original problem.
7. Check the solution(s) in the original problem.

Example 9 ➡ **Consumer Price Index** The Consumer Price Index (CPI) is a measure of the average change in prices over time from a designated reference period, which equals 100. The index is based on prices of basic consumer goods and services, and is published at regular intervals by the Bureau of Labor Statistics. Table 2 lists the CPI for several years from 1960 to 1992. What net annual salary in 1992 would have the same purchasing power as a net annual salary of \$13,000 in 1960? Compute the answer to the nearest dollar.

Table 2

CPI (1982–1984 = 100)

YEAR	INDEX
1960	29.6
1970	38.8
1980	82.4
1990	130.7
1992	140.3

SOLUTION To have the same purchasing power, the ratio of a salary in 1992 to a salary in 1960 would have to be the same as the ratio of the CPI in 1992 to the CPI in 1960. Thus, if x is the net annual salary in 1992, we solve the equation

$$\frac{x}{13{,}000} = \frac{140.3}{29.6}$$

$$x = 13{,}000 \cdot \frac{140.3}{29.6}$$

$$= \$61{,}618 \text{ per year}$$

Matched Problem 9 ➡ What net annual salary in 1970 would have had the same purchasing power as a net annual salary of \$75,000 in 1992? Compute the answer to the nearest dollar.

Answers to Matched Problems

1. $x = 4$ **2.** $x = 2$ **3.** (A) $t = \frac{M - Nr}{N}$ (B) $N = \frac{M}{t + r}$

4. (A) $<$ (B) $<$ (C) $>$

5. (A) $-7 < x \leq 4$;

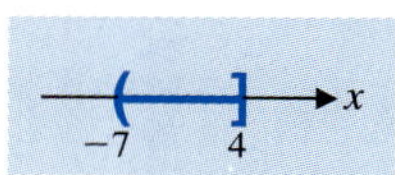

(B) $(-\infty, 3)$

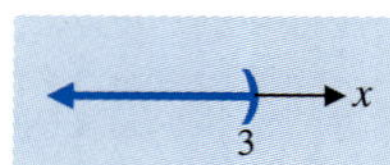

6. $x \geq -4$ or $[-4, \infty)$ 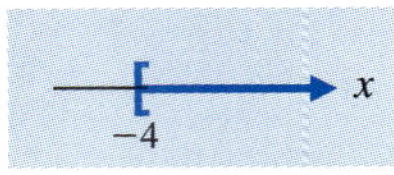

7. $-1 \leq x < 4$ or $[-1, 4)$ 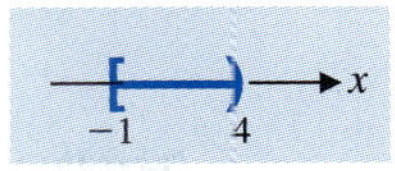

8. 7,500 CD's **9.** \$20,741

EXERCISE A-8

A *Solve Problems 1–6.*

1. $2m + 9 = 5m - 6$

2. $3y - 4 = 6y - 19$

3. $x + 5 < -4$

4. $x - 3 > -2$

5. $-3x \geq -12$

6. $-4x \leq 8$

Solve Problems 7–10 and graph.

7. $-4x - 7 > 5$

8. $-2x + 8 < 4$

9. $2 \leq x + 3 \leq 5$

10. $-3 < y - 5 < 8$

Solve Problems 11–26.

11. $\frac{y}{7} - 1 = \frac{1}{7}$

12. $\frac{m}{5} - 2 = \frac{3}{5}$

13. $\frac{x}{3} > -2$

14. $\frac{y}{-2} \leq -1$

15. $\frac{y}{3} = 4 - \frac{y}{6}$

16. $\frac{x}{4} = 9 - \frac{x}{2}$

B

17. $10x + 25(x - 3) = 275$

18. $-3(4 - x) = 5 - (x + 1)$

19. $3 - y \leq 4(y - 3)$

20. $x - 2 \geq 2(x - 5)$

21. $\frac{x}{5} - \frac{x}{6} = \frac{6}{5}$

22. $\frac{y}{4} - \frac{y}{3} = \frac{1}{2}$

23. $\frac{m}{5} - 3 < \frac{3}{5} - m$

24. $u - \frac{2}{3} > \frac{u}{3} + 2$

25. $0.1(x - 7) + 0.05x = 0.8$

26. $0.4(u + 5) - 0.3u = 17$

Solve Problems 27–30 and graph.

27. $2 \leq 3x - 7 < 14$

28. $-4 \leq 5x + 6 < 21$

29. $-4 \leq \frac{9}{5}C + 32 \leq 68$

30. $-1 \leq \frac{2}{3}t + 5 \leq 11$

C *Solve Problems 31–38 for the indicated variable.*

31. $3x - 4y = 12$; for y

32. $y = -\frac{2}{3}x + 8$; for x

33. $Ax + By = C$; for y $(B \neq 0)$

34. $y = mx + b$; for m

35. $F = \frac{9}{5}C + 32$; for C

36. $C = \frac{5}{9}(F - 32)$; for F

37. $A = Bm - Bn$; for B

38. $U = 3C - 2CD$; for C

Solve Problems 39 and 40 and graph.

39. $-3 \leq 4 - 7x < 18$

40. $-1 < 9 - 2u \leq 5$

41. What can be said about the signs of the numbers a and b in each case?

(A) $ab > 0$ (B) $ab < 0$

(C) $\frac{a}{b} > 0$ (D) $\frac{a}{b} < 0$

42. What can be said about the signs of the numbers a, b, and c in each case?

(A) $abc > 0$ (B) $\frac{ab}{c} < 0$

(C) $\frac{a}{bc} > 0$ (D) $\frac{a^2}{bc} < 0$

43. Replace each question mark with < or >, as appropriate:

(A) If $a - b = 2$, then a **?** b.

(B) If $c - d = -1$, then c **?** d.

44. For what c and d is $c + d < c - d$?

45. If both a and b are positive numbers and b/a is greater than 1, then is $a - b$ positive or negative?

46. If both a and b are negative numbers and b/a is greater than 1, then is $a - b$ positive or negative?

Applications

Business & Economics

47. *Puzzle.* A jazz concert brought in \$165,000 on the sale of 8,000 tickets. If the tickets sold for \$15 and \$25 each, how many of each type of ticket were sold?

48. *Puzzle.* An all-day parking meter takes only dimes and quarters. If it contains 100 coins with a total value of \$14.50, how many of each type of coin are in the meter?

49. *Investing.* You have \$12,000 to invest. If part is invested at 10% and the rest at 15%, how much should be invested at each rate to yield 12% on the total amount?

50. *Investing.* An investor has \$20,000 to invest. If part is invested at 8% and the rest at 12%, how much should be invested at each rate to yield 11% on the total amount?

51. *Inflation.* If the price change of cars parallels the change in the CPI (see Table 2 in Example 9), what would a car sell for (to the nearest dollar) in 1992 if a comparable model sold for \$5,000 in 1970?

52. *Inflation.* If the price change in houses parallels the CPI (see Table 2 in Example 9), what would a house valued at \$200,000 in 1992 be valued at (to the nearest dollar) in 1960?

53. *Break-even analysis.* A publisher for a promising new novel figures fixed costs (overhead, advances, promotion, copy editing, typesetting, and so on) at \$55,000, and variable costs (printing, paper, binding, shipping) at \$1.60 for each book produced. If the book is sold to distributors for \$11 each, what is the break-even point for the publisher?

54. *Break-even analysis.* The publisher of a new book called *Muscle-Powered Sports* figures fixed costs at \$92,000 and variable costs at \$2.10 for each book produced. If the book is sold to distributors for \$15 each, how many must be sold for the publisher to break even?

55. *Break-even analysis.* The publisher in Problem 53 finds that rising prices for paper increase the variable costs to \$2.10 per book.

(A) Discuss possible strategies the company might use to deal with this increase in costs.

(B) If the company continues to sell the books for \$11, how many books must they sell now to make a profit?

(C) If the company wants to start making a profit at the same production level as before the cost increase, how much should they sell the book for now?

56. *Break-even analysis.* The publisher in Problem 54 finds that rising prices for paper increase the variable costs to \$2.70 per book.

(A) Discuss possible strategies the company might use to deal with this increase in costs.

(B) If the company continues to sell the books for \$15, how many books must they sell now to make a profit?

(C) If the company wants to start making a profit at the same production level as before the cost increase, how much should they sell the book for now?

Life Sciences

57. *Wildlife management.* A naturalist for a fish and game department estimated the total number of rainbow trout in a certain lake using the popular capture–mark–recapture technique. He netted, marked, and released 200 rainbow trout. A week later, allowing for thorough mixing, he again netted 200 trout and found 8 marked ones among them. Assuming that the proportion of marked fish in the second sample was the same as the proportion of all marked fish in the total population, estimate the number of rainbow trout in the lake.

58. *Ecology.* If the temperature for a 24 hour period at an Antarctic station ranged between −49°F and 14°F (that is, $-49 \leq F \leq 14$),what was the range in degrees Celsius? [*Note:* $F = \frac{9}{5}C + 32$.]

Social Sciences

59. *Psychology.* The IQ (intelligence quotient) is found by dividing the mental age (MA), as indicated on standard tests, by the chronological age (CA) and multiplying by 100. For example, if a child has a mental age of 12 and a chronological age of 8, the calculated IQ is 150. If a 9-year-old girl has an IQ of 140, compute her mental age.

60. *Anthropology.* In their study of genetic groupings, anthropologists use a ratio called the *cephalic index.*

This is the ratio of the width of the head to its length (looking down from above) expressed as a percentage. Symbolically,

$$C = \frac{100W}{L}$$

where C is the cephalic index, W is the width, and L is the length. If an Indian tribe in Baja California (Mexico) had an average cephalic index of 66 and the average width of their heads was 6.6 inches, what was the average length of their heads?

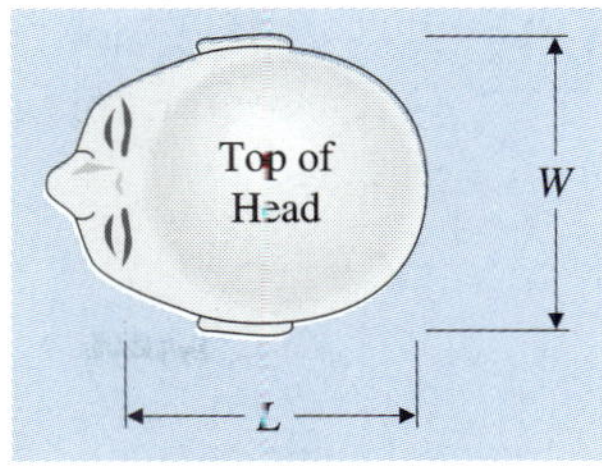

Figure for 60

SECTION A-9 Quadratic Equations

- SOLUTION BY SQUARE ROOT
- SOLUTION BY FACTORING
- QUADRATIC FORMULA
- THE QUADRATIC FORMULA AND FACTORING
- APPLICATION: SUPPLY AND DEMAND

A **quadratic equation** in one variable is any equation that can be written in the form

$$ax^2 + bx + c = 0 \qquad a \neq 0$$

where x is a variable and a, b, and c are constants. We will refer to this form as the **standard form.** The equations

$$5x^2 - 3x + 7 = 0 \qquad \text{and} \qquad 18 = 32t^2 - 12t$$

are both quadratic equations, since they are either in the standard form or can be transformed into this form.

We will restrict our review to finding real solutions to quadratic equations.

SOLUTION BY SQUARE ROOT

The easiest type of quadratic equation to solve is the special form where the first-degree term is missing:

$$ax^2 + c = 0 \qquad a \neq 0$$

The method of solution of this special form makes direct use of the square root property:

Square Root Property

If $a^2 = b$, then $a = \pm\sqrt{b}$.

Explore–Discuss 1

Determine whether each of the following pairs of equations are equivalent. Explain.

(A) $x^2 = 4$ and $x = 2$
(B) $x^2 = 4$ and $x = -2$
(C) $x = \sqrt{4}$ and $x = 2$
(D) $x = \sqrt{4}$ and $x = -2$
(E) $x = -\sqrt{4}$ and $x = -2$

Example 1 **Square Root Method** Use the square root property to solve each equation.

(A) $x^2 - 7 = 0$ (B) $2x^2 - 10 = 0$ (C) $3x^2 + 27 = 0$
(D) $(x - 8)^2 = 9$

SOLUTION (A)
$$x^2 - 7 = 0$$
$$x^2 = 7 \quad \text{What real number squared is 7?}$$
$$x = \pm\sqrt{7} \quad \text{Short for } \sqrt{7} \text{ and } -\sqrt{7}$$

(B)
$$2x^2 - 10 = 0$$
$$2x^2 = 10$$
$$x^2 = 5 \quad \text{What real number squared is 5?}$$
$$x = \pm\sqrt{5}$$

(C)
$$3x^2 + 27 = 0$$
$$3x^2 = -27$$
$$x^2 = -9 \quad \text{What real number squared is } -9\text{?}$$

No real solution, since no real number squared is negative.

(D)
$$(x - 8)^2 = 9$$
$$x - 8 = \pm\sqrt{9}$$
$$x - 8 = \pm 3$$
$$x = 8 \pm 3 = 5 \quad \text{or} \quad 11$$

Matched Problem 1 Use the square root property to solve each equation.

(A) $x^2 - 6 = 0$ (B) $3x^2 - 12 = 0$ (C) $x^2 + 4 = 0$
(D) $(x + 5)^2 = 1$

SOLUTION BY FACTORING

If the left side of a quadratic equation when written in standard form can be factored, then the equation can be solved very quickly. The method of solution by factoring rests on the following important property of real numbers (see Appendix A-2):

If a and b are real numbers, then $ab = 0$ if and only if $a = 0$ or $b = 0$ (or both).

Example 2 **Factoring Method** Solve by factoring using integer coefficients, if possible.

(A) $3x^2 - 6x - 24 = 0$ (B) $3y^2 = 2y$ (C) $x^2 - 2x - 1 = 0$

SOLUTION (A)

$$3x^2 - 6x - 24 = 0$$ Divide both sides by 3, since 3 is a factor of each coefficient.

$$x^2 - 2x - 8 = 0$$ Factor the left side, if possible.

$$(x - 4)(x + 2) = 0$$

$$x - 4 = 0 \quad \text{or} \quad x + 2 = 0$$

$$x = 4 \quad \text{or} \quad x = -2$$

(B)

$$3y^2 = 2y$$

$$3y^2 - 2y = 0$$ We lose the solution $y = 0$ if both sides are divided by y ($3y^2 = 2y$ and $3y = 2$ are not equivalent).

$$y(3y - 2) = 0$$

$$y = 0 \quad \text{or} \quad 3y - 2 = 0$$

$$3y = 2$$

$$y = \tfrac{2}{3}$$

(C) $x^2 - 2x - 1 = 0$

This equation cannot be factored using integer coefficients. We will solve this type of equation by another method, considered below.

Matched Problem 2 Solve by factoring using integer coefficients, if possible.

(A) $2x^2 + 4x - 30 = 0$ (B) $2x^2 = 3x$ (C) $2x^2 - 8x + 3 = 0$

Note that an equation such as $x^2 = 25$ can be solved by either the square root or the factoring method, and the results are the same (as they should be). Solve this equation both ways and compare.

Also, note that the factoring method can be extended to higher-degree polynomial equations. Consider the following:

$$x^3 - x = 0$$

$$x(x^2 - 1) = 0$$

$$x(x - 1)(x + 1) = 0$$

$$x = 0 \quad \text{or} \quad x - 1 = 0 \quad \text{or} \quad x + 1 = 0$$

Solution: $x = 0, 1, -1$

Check these solutions in the original equation.

The factoring and square root methods are fast and easy to use when they apply. However, there are quadratic equations that look simple but cannot be solved by either method. For example, as was noted in Example 2C, the polynomial in

$$x^2 - 2x - 1 = 0$$

cannot be factored using integer coefficients. This brings us to the well-known and widely used *quadratic formula*.

QUADRATIC FORMULA

There is a method called *completing the square* that will work for all quadratic equations. After briefly reviewing this method, we will then use it to develop the famous quadratic formula—a formula that will enable us to solve any quadratic equation quite mechanically.

Explore–Discuss 2

Replace **?** in each of the following with a number that makes the equation valid.

(A) $(x + 1)^2 = x^2 + 2x + \textbf{?}$ (B) $(x + 2)^2 = x^2 + 4x + \textbf{?}$
(C) $(x + 3)^2 = x^2 + 6x + \textbf{?}$ (D) $(x + 4)^2 = x^2 + 8x + \textbf{?}$

Replace **?** in each of the following with a number that makes the trinomial a perfect square.

(E) $x^2 + 10x + \textbf{?}$ (F) $x^2 + 12x + \textbf{?}$ (G) $x^2 + bx + \textbf{?}$

The method of **completing the square** is based on the process of transforming a quadratic equation in standard form,

$$ax^2 + bx + c = 0$$

into the form

$$(x + A)^2 = B$$

where A and B are constants. Then, this last equation can be solved easily (if it has a real solution) by the square root method discussed above.

Consider the equation from Example 2C:

$$x^2 - 2x - 1 = 0 \tag{1}$$

Since the left side does not factor using integer coefficients, we add 1 to each side to remove the constant term from the left side:

$$x^2 - 2x = 1 \tag{2}$$

Now we try to find a number that we can add to each side to make the left side a square of a first-degree polynomial. Note the following square of a binomial:

$$(x + m)^2 = x^2 + 2mx + m^2$$

We see that the third term on the right is the square of one-half the coefficient of x in the second term on the right. To complete the square in equation (2), we add the square of one-half the coefficient of x, $(-\frac{2}{2})^2 = 1$, to each side. (This rule works only when the coefficient of x^2 is 1, that is, $a = 1$.) Thus,

$$x^2 - 2x \mathbf{+ 1} = 1 \mathbf{+ 1}$$

The left side is the square of $x - 1$, and we write

$$(x - 1)^2 = 2$$

What number squared is 2?

$$x - 1 = \pm\sqrt{2}$$
$$x = 1 \pm \sqrt{2}$$

And equation (1) is solved!

Let us try the method on the general quadratic equation

$$ax^2 + bx + c = 0 \qquad a \neq 0 \tag{3}$$

and solve it once and for all for x in terms of the coefficients a, b, and c. We start by multiplying both sides of (3) by $1/a$ to obtain

$$x^2 + \frac{b}{a}x + \frac{c}{a} = 0$$

Add $-c/a$ to both sides:

$$x^2 + \frac{b}{a}x = -\frac{c}{a}$$

Now we complete the square on the left side by adding the square of one-half the coefficient of x, that is, $(b/2a)^2 = b^2/4a^2$, to each side:

$$x^2 + \frac{b}{a}x + \frac{b^2}{4a^2} = \frac{b^2}{4a^2} - \frac{c}{a}$$

Writing the left side as a square and combining the right side into a single fraction, we obtain

$$\left(x + \frac{b}{2a}\right)^2 = \frac{b^2 - 4ac}{4a^2}$$

Now we solve by the square root method:

$$x + \frac{b}{2a} = \pm\sqrt{\frac{b^2 - 4ac}{4a^2}}$$

$$x = -\frac{b}{2a} \pm \frac{\sqrt{b^2 - 4ac}}{2a}$$ Since $\pm\sqrt{4a^2} = \pm 2a$ for any real number a

When this is written as a single fraction, it becomes the **quadratic formula:**

Quadratic Formula

If $ax^2 + bx + c = 0, a \neq 0$, then

$$x = \frac{-b \pm \sqrt{b^2 - 4ac}}{2a}$$

This formula is generally used to solve quadratic equations when the square root or factoring methods do not work. The quantity $b^2 - 4ac$ under the radical is called the **discriminant,** and it gives us the useful information about solutions listed in Table 1.

Table 1

$b^2 - 4ac$	$ax^2 + bx + c = 0$
Positive	Two real solutions
Zero	One real solution
Negative	No real solutions

Example 3 **Quadratic Formula Method** Solve $x^2 - 2x - 1 = 0$ using the quadratic formula.

SOLUTION

$$x^2 - 2x - 1 = 0$$

$$x = \frac{-b \pm \sqrt{b^2 - 4ac}}{2a} \qquad a = 1, b = -2, c = -1$$

$$= \frac{-(-2) \pm \sqrt{(-2)^2 - 4(1)(-1)}}{2(1)}$$

$$= \frac{2 \pm \sqrt{8}}{2} = \frac{2 \pm 2\sqrt{2}}{2} = 1 \pm \sqrt{2} \approx -0.414 \quad \text{or} \quad 2.414$$

CHECK

$$x^2 - 2x - 1 = 0$$

When $x = 1 + \sqrt{2}$,

$$(1 + \sqrt{2})^2 - 2(1 + \sqrt{2}) - 1 = 1 + 2\sqrt{2} + 2 - 2 - 2\sqrt{2} - 1 = 0$$

When $x = 1 - \sqrt{2}$,

$$(1 - \sqrt{2})^2 - 2(1 - \sqrt{2}) - 1 = 1 - 2\sqrt{2} + 2 - 2 + 2\sqrt{2} - 1 = 0$$

Matched Problem 3 Solve $2x^2 - 4x - 3 = 0$ using the quadratic formula.

If we try to solve $x^2 - 6x + 11 = 0$ using the quadratic formula, we obtain

$$x = \frac{6 \pm \sqrt{-8}}{2}$$

which is not a real number. (Why?)

THE QUADRATIC FORMULA AND FACTORING

As in Section A-4, we restrict our interest in factoring to polynomials with integer coefficients. If a polynomial cannot be factored as a product of lower-degree polynomials with integer coefficients, we say that the polynomial is **not factorable in the integers.**

Suppose you were asked to factor

$$x^2 - 19x - 372 \tag{4}$$

The larger the coefficients, the more difficult the process of applying the *ac* test discussed in Section A-4. The quadratic formula provides a simple and efficient method of factoring a second-degree polynomial with integer coefficients as the product of two first-degree polynomials with integer coefficients, if the factors exist. We illustrate the method using equation (4), and generalize the process from this experience.

We start by solving the corresponding quadratic equation using the quadratic formula:

$$x^2 - 19x - 372 = 0$$

$$x = \frac{-(-19) \pm \sqrt{(-19)^2 - 4(1)(-372)}}{2}$$

$$= \frac{19 \pm \sqrt{1,849}}{2}$$

$$= \frac{19 \pm 43}{2} = -12 \quad \text{or} \quad 31$$

Now, we write

$$x^2 - 19x - 372 = [x - (-12)](x - 31) = (x + 12)(x - 31)$$

Multiplying the two factors on the right produces the second-degree polynomial on the left.

What is behind this procedure? The following two theorems justify and generalize the process:

Theorem 1 ■■ FACTORABILITY THEOREM

A second-degree polynomial, $ax^2 + bx + c$, with integer coefficients can be expressed as the product of two first-degree polynomials with integer coefficients if and only if $\sqrt{b^2 - 4ac}$ is an integer. ■■

Theorem 2 ■■ FACTOR THEOREM

If r_1 and r_2 are solutions to $ax^2 + bx + c = 0$, then

$$ax^2 + bx + c = a(x - r_1)(x - r_2)$$ ■■

Example 4 **Factoring with the Aid of the Discriminant** Factor, if possible, using integer coefficients:

(A) $4x^2 - 65x + 264$ (B) $2x^2 - 33x - 306$

SOLUTION (A) $4x^2 - 65x + 264$

Step 1. Test for factorability:

$$\sqrt{b^2 - 4ac} = \sqrt{(-65)^2 - 4(4)(264)} = 1$$

Since the result is an integer, the polynomial has first-degree factors with integer coefficients.

Step 2. Factor, using the factor theorem. Find the solutions to the corresponding quadratic equation using the quadratic formula:

$$4x^2 - 65x + 264 = 0$$

$$x = \frac{-(-65) \pm 1}{2 \cdot 4} = \frac{33}{4} \quad \text{or} \quad 8$$

From step 1

Thus,

$$4x^2 - 65x + 264 = 4\left(x - \frac{33}{4}\right)(x - 8)$$
$$= (4x - 33)(x - 8)$$

(B) $2x^2 - 33x - 306$

Step 1. Test for factorability:

$$\sqrt{b^2 - 4ac} = \sqrt{(-33)^2 - 4(2)(-306)} = \sqrt{3{,}537}$$

Since $\sqrt{3{,}537}$ is not an integer, the polynomial is not factorable in the integers.

Matched Problem 4 Factor, if possible, using integer coefficients:

(A) $3x^2 - 28x - 464$ (B) $9x^2 + 320x - 144$

Application: Supply and Demand

Supply and demand analysis is a very important part of business and economics. In general, producers are willing to supply more of an item as the price of an item increases, and less of an item as the price decreases. Similarly, buyers are willing to buy less of an item as the price increases, and more of an item as the price decreases. Thus, we have a dynamic situation where the price, supply, and demand fluctuate until a price is reached at which the supply is equal to the demand. In economic theory, this point is called the **equilibrium point**—if the price increases from this point, the supply will increase and the demand will decrease; if the price decreases from this point, the supply will decrease and the demand will increase.

Example 5 **Supply and Demand** At a large beach resort in the summer, the weekly supply and demand equations for folding beach chairs are

$$p = \frac{x}{140} + \frac{3}{4} \quad \text{Supply equation}$$
$$p = \frac{5{,}670}{x} \quad \text{Demand equation}$$

The supply equation indicates that the supplier is willing to sell x units at a price of p dollars per unit. The demand equation indicates that consumers are willing to buy x units at a price of p dollars per unit. How many units are required for supply to equal demand? At what price will supply equal demand?

Solution Set the right side of the supply equation equal to the right side of the demand equation and solve for x:

$$\frac{x}{140} + \frac{3}{4} = \frac{5{,}670}{x} \quad \text{Multiply by 140x, the LCD.}$$
$$x^2 + 105x = 793{,}800 \quad \text{Write in standard form.}$$
$$x^2 + 105x - 793{,}800 = 0 \quad \text{Use the quadratic formula.}$$
$$x = \frac{-105 \pm \sqrt{105^2 - 4(1)(-793{,}800)}}{2}$$
$$x = 840 \text{ units}$$

The negative root is discarded, since a negative number of units cannot be produced or sold. Substitute $x = 840$ back into either the supply equation or the demand equation to find the equilibrium price (we use the demand equation):

$$p = \frac{5{,}670}{x} = \frac{5{,}670}{840} = \$6.75$$

Thus, at a price of \$6.75 the supplier is willing to supply 840 chairs and consumers are willing to buy 840 chairs during a week.

Matched Problem 5 Repeat Example 5 if near the end of summer the supply and demand equations are

$$p = \frac{x}{80} - \frac{1}{20} \quad \text{Supply equation}$$

$$p = \frac{1{,}264}{x} \quad \text{Demand equation}$$

Answers to Matched Problems

1. (A) $\pm\sqrt{6}$ (B) ± 2 (C) No real solution (D) $-6, -4$
2. (A) $-5, 3$ (B) $0, \frac{3}{2}$ (C) Cannot be factored using integer coefficients
3. $(2 \pm \sqrt{10})/2$
4. (A) Cannot be factored using integer coefficients (B) $(9x - 4)(x + 36)$
5. 320 chairs at \$3.95 each

EXERCISE A-9

Find only real solutions in the problems below. If there are no real solutions, say so.

A *Solve Problems 1–4 by the square root method.*

1. $2x^2 - 22 = 0$
2. $3m^2 - 21 = 0$
3. $(x - 1)^2 = 4$
4. $(x + 2)^2 = 9$

Solve Problems 5–8 by factoring.

5. $2u^2 - 8u - 24 = 0$
6. $3x^2 - 18x + 15 = 0$
7. $x^2 = 2x$
8. $n^2 = 3n$

Solve Problems 9–12 by using the quadratic formula.

9. $x^2 - 6x - 3 = 0$
10. $m^2 + 8m + 3 = 0$
11. $3u^2 + 12u + 6 = 0$
12. $2x^2 - 20x - 6 = 0$

B *Solve Problems 13–30 by using any method.*

13. $2x^2 = 4x$
14. $2x^2 = -3x$
15. $4u^2 - 9 = 0$
16. $9y^2 - 25 = 0$
17. $8x^2 + 20x = 12$
18. $9x^2 - 6 = 15x$
19. $x^2 = 1 - x$
20. $m^2 = 1 - 3m$
21. $2x^2 = 6x - 3$
22. $2x^2 = 4x - 1$
23. $y^2 - 4y = -8$
24. $x^2 - 2x = -3$
25. $(x + 4)^2 = 11$
26. $(y - 5)^2 = 7$
27. $\frac{3}{p} = p$
28. $x - \frac{7}{x} = 0$
29. $2 - \frac{2}{m^2} = \frac{3}{m}$
30. $2 + \frac{5}{u} = \frac{3}{u^2}$

In Problems 31–38, factor, if possible, as the product of two first-degree polynomials with integer coefficients. Use the quadratic formula and the factor theorem.

31. $x^2 + 40x - 84$
32. $x^2 - 28x - 128$
33. $x^2 - 32x + 144$
34. $x^2 + 52x + 208$
35. $2x^2 + 15x - 108$
36. $3x^2 - 32x - 140$
37. $4x^2 + 241x - 434$
38. $6x^2 - 427x - 360$

C

39. Solve $A = P(1 + r)^2$ for r in terms of A and P; that is, isolate r on the left side of the equation (with coefficient 1) and end up with an algebraic expression on the right side involving A and P but not r. Write the answer using positive square roots only.

40. Solve $x^2 + mx + n = 0$ for x in terms of m and n.

41. Consider the quadratic equation

$$x^2 + 4x + c = 0$$

where c is a real number. Discuss the relationship between the values of c and the three types of roots listed in Table 1.

42. Consider the quadratic equation

$$x^2 - 2x + c = 0$$

where c is a real number. Discuss the relationship between the values of c and the three types of roots listed in Table 1.

Applications

Business & Economics

43. *Supply and demand.* A company wholesales a certain brand of shampoo in a particular city. Their marketing research department established the following weekly supply and demand equations:

$$p = \frac{x}{450} + \frac{1}{2} \quad \text{Supply equation}$$

$$p = \frac{6{,}300}{x} \quad \text{Demand equation}$$

How many units are required for supply to equal demand? At what price per bottle will supply equal demand?

44. *Supply and demand.* An importer sells a certain brand of automatic camera to outlets in a large metropolitan area. During the summer, the weekly supply and demand equations are

$$p = \frac{x}{6} + 9 \quad \text{Supply equation}$$

$$p = \frac{24{,}840}{x} \quad \text{Demand equation}$$

How many units are required for supply to equal demand? At what price will supply equal demand?

45. *Interest rate.* If P dollars is invested at $100r$ percent compounded annually, at the end of 2 years it will grow to $A = P(1 + r)^2$. At what interest rate will \$100 grow to \$144 in 2 years? [*Note:* If $A = 144$ and $P = 100$, find r.]

46. *Interest rate.* Using the formula in Problem 45, determine the interest rate that will make \$1,000 grow to \$1,210 in 2 years.

Life Sciences

47. *Ecology.* An important element in the erosive force of moving water is its velocity. To measure the velocity v (in feet per second) of a stream, we position a hollow L-shaped tube with one end under the water pointing upstream and the other end pointing straight up a couple of feet out of the water. The water will then be pushed up the tube a certain distance h (in feet) above the surface of the stream. Physicists have shown that $v^2 = 64h$. Approximately how fast is a stream flowing if $h = 1$ foot? If $h = 0.5$ foot?

Social Sciences

48. *Safety research.* It is of considerable importance to know the least number of feet d in which a car can be stopped, including reaction time of the driver, at various speeds v (in miles per hour). Safety research has produced the formula $d = 0.044v^2 + 1.1v$. If it took a car 550 feet to stop, estimate the car's speed at the moment the stopping process was started.

APPENDIX B

Special Topics

SECTION B-1

Sequences, Series, and Summation Notation

- SEQUENCES
- SERIES AND SUMMATION NOTATION

If someone asked you to list all natural numbers that are perfect squares, you might begin by writing

$$1,\ 4,\ 9,\ 16,\ 25,\ 36$$

But you would soon realize that it is impossible to actually list all the perfect squares, since there are an infinite number of them. However, you could represent this collection of numbers in several different ways. One common method is to write

$$1, 4, 9, \ldots, n^2, \ldots \qquad n \in N$$

where N is the set of natural numbers. A list of numbers such as this is generally called a *sequence*. Sequences and related topics form the subject matter of this section.

SEQUENCES

Consider the function f given by

$$f(n) = 2n + 1 \tag{1}$$

where the domain of f is the set of natural numbers N. Note that

$$f(1) = 3, \quad f(2) = 5, \quad f(3) = 7, \quad \ldots$$

The function f is an example of a sequence. In general, a **sequence** is a function with domain a set of successive integers. Instead of the standard function notation used in equation (1), sequences are usually defined in terms of a special notation.

The range value $f(n)$ is usually symbolized more compactly with a symbol such as a_n. Thus, in place of equation (1), we write

$$a_n = 2n + 1$$

and the domain is understood to be the set of natural numbers unless something is said to the contrary or the context indicates otherwise. The elements in the range are called **terms of the sequence;** a_1 is the first term, a_2 is the second term, and a_n is the ***n*th term,** or **general term.**

$$a_1 = 2(1) + 1 = 3 \quad \text{First term}$$
$$a_2 = 2(2) + 1 = 5 \quad \text{Second term}$$
$$a_3 = 2(3) + 1 = 7 \quad \text{Third term}$$
$$\vdots$$
$$a_n = 2n + 1 \quad \text{General term}$$

The ordered list of elements

$$3, 5, 7, \ldots, 2n + 1, \ldots$$

obtained by writing the terms of the sequence in their natural order with respect to the domain values is often informally referred to as a sequence. A sequence also may be represented in the abbreviated form $\{a_n\}$, where a symbol for the nth term is written within braces. For example, we could refer to the sequence $3, 5, 7, \ldots, 2n + 1, \ldots$ as the sequence $\{2n + 1\}$.

If the domain of a sequence is a finite set of successive integers, then the sequence is called a **finite sequence.** If the domain is an infinite set of successive integers, then the sequence is called an **infinite sequence.** The sequence $\{2n + 1\}$ discussed above is an infinite sequence.

Example 1 **Writing the Terms of a Sequence** Write the first four terms of each sequence:

(A) $a_n = 3n - 2$ (B) $\left\{\frac{(-1)^n}{n}\right\}$

SOLUTION (A) 1, 4, 7, 10 (B) $-1, \frac{1}{2}, \frac{-1}{3}, \frac{1}{4}$

Matched Problem 1 Write the first four terms of each sequence:

(A) $a_n = -n + 3$ (B) $\left\{\frac{(-1)^n}{2^n}\right\}$

Explore–Discuss 1

(A) A multiple-choice test question asked for the next term in the sequence

$$2, 4, 8, \ldots$$

and gave the following choices:

(1) 16 (2) 14 (3) $\frac{25}{2}$

Which is the correct answer?

(B) Compare the first four terms of the following sequences:

(1) $a_n = 2^n$ (2) $b_n = n^2 - n + 2$ (3) $c_n = 5n + \frac{6}{n} - 9$

Now, which of the choices in part (A) appears to be correct?

Now that we have seen how to use the general term to find the first few terms in a sequence, we consider the reverse problem. That is, can a sequence be defined just by listing the first three or four terms of the sequence? And can we then use these initial terms to find a formula for the nth term? In general, without other information, the answer to the first question is no. As Explore–Discuss 1 illustrates, many different sequences may start off with the same terms. Simply listing the first three terms (or any other finite number of terms) does not specify a particular sequence. In fact, it can be shown that given any list of m numbers, there are an infinite number of sequences whose first m terms agree with these given numbers.

What about the second question? That is, given a few terms, can we find the general formula for at least one sequence whose first few terms agree with the given terms? The answer to this question is a qualified yes. If we can observe a simple pattern in the given terms, then we usually can construct a general term that will produce that pattern. The next example illustrates this approach.

Example 2 **Finding the General Term of a Sequence** Find the general term of a sequence whose first four terms are:

(A) $3, 4, 5, 6, \ldots$ (B) $5, -25, 125, -625, \ldots$

SOLUTION (A) Since these terms are consecutive integers, one solution is $a_n = n$, $n \geq 3$. If we want the domain of the sequence to be all natural numbers, then another solution is $b_n = n + 2$.

(B) Each of these terms can be written as the product of a power of 5 and a power of -1:

$$\begin{aligned} 5 &= (-1)^0 5^1 = a_1 \\ -25 &= (-1)^1 5^2 = a_2 \\ 125 &= (-1)^2 5^3 = a_3 \\ -625 &= (-1)^3 5^4 = a_4 \end{aligned}$$

If we choose the domain to be all natural numbers, then a solution is

$$a_n = (-1)^{n-1} 5^n$$

Matched Problem 2 Find the general term of a sequence whose first four terms are:

(A) $3, 6, 9, 12, \ldots$ (B) $1, -2, 4, -8, \ldots$

In general, there is usually more than one way of representing the nth term of a given sequence (see the solution of Example 2A). However, unless something is stated to the contrary, we assume the domain of the sequence is the set of natural numbers N.

Series and Summation Notation

If $a_1, a_2, a_3, \ldots, a_n, \ldots$ is a sequence, then the expression

$$a_1 + a_2 + a_3 + \cdots + a_n + \cdots$$

is called a **series.** If the sequence is finite, the corresponding series is a **finite series.** If the sequence is infinite, the corresponding series is an

infinite series. We will consider only finite series in this section. For example,

$$1, 3, 5, 7, 9 \qquad \text{Finite sequence}$$
$$1 + 3 + 5 + 7 + 9 \qquad \text{Finite series}$$

Notice that we can easily evaluate this series by adding the five terms:

$$1 + 3 + 5 + 7 + 9 = 25$$

Series are often represented in a compact form called **summation notation.** Consider the following examples:

$$\begin{aligned}\sum_{k=3}^{6} k^2 &= 3^2 + 4^2 + 5^2 + 6^2\\ &= 9 + 16 + 25 + 36 = 86\end{aligned}$$

$$\begin{aligned}\sum_{k=0}^{2} (4k + 1) &= (4\cdot 0 + 1) + (4\cdot 1 + 1) + (4\cdot 2 + 1)\\ &= 1 + 5 + 9 = 15\end{aligned}$$

In each case, the terms of the series on the right are obtained from the expression on the left by successively replacing the **summing index *k*** with integers, starting with the number indicated below the **summation sign** Σ and ending with the number that appears above Σ. The summing index may be represented by letters other than k and may start at any integer and end at any integer greater than or equal to the starting integer. Thus, if we are given the finite sequence

$$\frac{1}{2}, \frac{1}{4}, \frac{1}{8}, \ldots, \frac{1}{2^n}$$

the corresponding series is

$$\frac{1}{2} + \frac{1}{4} + \frac{1}{8} + \cdots + \frac{1}{2^n} = \sum_{j=1}^{n} \frac{1}{2^j}$$

where we have used j for the summing index.

Example 3 **Summation Notation** Write

$$\sum_{k=1}^{5} \frac{k}{k^2 + 1}$$

without summation notation. Do not evaluate the sum.

SOLUTION

$$\begin{aligned}\sum_{k=1}^{5} \frac{k}{k^2 + 1} &= \frac{1}{1^2 + 1} + \frac{2}{2^2 + 1} + \frac{3}{3^2 + 1} + \frac{4}{4^2 + 1} + \frac{5}{5^2 + 1}\\ &= \frac{1}{2} + \frac{2}{5} + \frac{3}{10} + \frac{4}{17} + \frac{5}{26}\end{aligned}$$

Matched Problem 3 Write

$$\sum_{k=1}^{5} \frac{k + 1}{k}$$

without summation notation. Do not evaluate the sum.

Explore–Discuss 2

(A) Find the smallest value of n for which the value of the series

$$\sum_{k=1}^{n} \frac{k}{k^2 + 1}$$

is greater than 3.

(B) Find the smallest value of n for which the value of the series

$$\sum_{j=1}^{n} \frac{1}{2^j}$$

is greater than 0.99. Greater than 0.999.

If the terms of a series are altenately positive and negative, we call the series an **alternating series.** The next example deals with the representation of such a series.

Example 4 **Summation Notation** Write the alternating series

$$\frac{1}{2} - \frac{1}{4} + \frac{1}{6} - \frac{1}{8} + \frac{1}{10} - \frac{1}{12}$$

using summation notation with:

(A) The summing index k starting at 1
(B) The summing index j starting at 0

SOLUTION (A) $(-1)^{k+1}$ provides the alternation of sign, and $1/(2k)$ provides the other part of each term. Thus, we can write

$$\frac{1}{2} - \frac{1}{4} + \frac{1}{6} - \frac{1}{8} + \frac{1}{10} - \frac{1}{12} = \sum_{k=1}^{6} \frac{(-1)^{k+1}}{2k}$$

(B) $(-1)^j$ provides the alternation of sign, and $1/[2(j + 1)]$ provides the other part of each term. Thus, we can write

$$\frac{1}{2} - \frac{1}{4} + \frac{1}{6} - \frac{1}{8} + \frac{1}{10} - \frac{1}{12} = \sum_{j=0}^{5} \frac{(-1)^j}{2(j + 1)}$$

Matched Problem 4 Write the alternating series

$$1 - \frac{1}{3} + \frac{1}{9} - \frac{1}{27} + \frac{1}{81}$$

using summation notation with:

(A) The summing index k starting at 1
(B) The summing index j starting at 0

Summation notation provides a compact notation for the sum of any list of numbers, even if the numbers are not generated by a formula. For example, suppose the results of an examination taken by a class of 10 students are given in the following list:

87, 77, 95, 83, 86, 73, 95, 68, 75, 86

If we let $a_1, a_2, a_3, \ldots, a_{10}$ represent these 10 scores, then the average test score is given by

$$\frac{1}{10}\sum_{k=1}^{10} a_k = \frac{1}{10}(87 + 77 + 95 + 83 + 86 + 73 + 95 + 68 + 75 + 86)$$
$$= \frac{1}{10}(825) = 82.5$$

More generally, in statistics, the **arithmetic mean** $\bar{a}$ of a list of n numbers $a_1, a_2, \ldots, a_n$ is defined as

$$\bar{a} = \frac{1}{n}\sum_{k=1}^{n} a_k$$

Example 5 **Arithmetic Mean** Find the arithmetic mean of 3, 5, 4, 7, 4, 2, 3, and 6.

SOLUTION $\bar{a} = \frac{1}{8}\sum_{k=1}^{8} a_k = \frac{1}{8}(3 + 5 + 4 + 7 + 4 + 2 + 3 + 6) = \frac{1}{8}(34) = 4.25$

Matched Problem 5 Find the arithmetic mean of 9, 3, 8, 4, 3, and 6.

Answers to Matched Problems

1. (A) 2, 1, 0, −1 (B) $\frac{-1}{2}, \frac{1}{4}, \frac{-1}{8}, \frac{1}{16}$
2. (A) $a_n = 3n$ (B) $a_n = (-2)^{n-1}$ **3.** $2 + \frac{3}{2} + \frac{4}{3} + \frac{5}{4} + \frac{6}{5}$
4. (A) $\sum_{k=1}^{5} \frac{(-1)^{k-1}}{3^{k-1}}$ (B) $\sum_{j=0}^{4} \frac{(-1)^j}{3^j}$ **5.** 5.5

EXERCISE B-1

A *Write the first four terms for each sequence in Problems 1–6.*

1. $a_n = 2n + 3$ **2.** $a_n = 4n - 3$

3. $a_n = \frac{n+2}{n+1}$ **4.** $a_n = \frac{2n+1}{2n}$

5. $a_n = (-3)^{n+1}$ **6.** $a_n = (-\frac{1}{4})^{n-1}$

7. Write the 10th term of the sequence in Problem 1.

8. Write the 15th term of the sequence in Problem 2.

9. Write the 99th term of the sequence in Problem 3.

10. Write the 200th term of the sequence in Problem 4.

In Problems 11–16, write each series in expanded form without summation notation, and evaluate.

11. $\sum_{k=1}^{6} k$ **12.** $\sum_{k=1}^{5} k^2$ **13.** $\sum_{k=4}^{7} (2k - 3)$

14. $\sum_{k=0}^{4} (-2)^k$ **15.** $\sum_{k=0}^{3} \frac{1}{10^k}$ **16.** $\sum_{k=1}^{4} \frac{1}{2^k}$

Find the arithmetic mean of each list of numbers in Problems 17–20.

17. 5, 4, 2, 1, and 6 **18.** 7, 9, 9, 2, and 4

19. 96, 65, 82, 74, 91, 88, 87, 91, 77, and 74

20. 100, 62, 95, 91, 82, 87, 70, 75, 87, and 82

B *Write the first five terms of each sequence in Problems 21–26.*

21. $a_n = \frac{(-1)^{n+1}}{2^n}$ **22.** $a_n = (-1)^n(n-1)^2$

23. $a_n = n[1 + (-1)^n]$ **24.** $a_n = \frac{1 - (-1)^n}{n}$

25. $a_n = \left(-\frac{3}{2}\right)^{n-1}$ **26.** $a_n = \left(-\frac{1}{2}\right)^{n+1}$

In Problems 27–42, find the general term of a sequence whose first four terms agree with the given terms.

27. −2, −1, 0, 1, . . . **28.** 4, 5, 6, 7, . . .

29. 4, 8, 12, 16, . . . **30.** −3, −6, −9, −12, . . .

31. $\frac{1}{2}, \frac{3}{4}, \frac{5}{6}, \frac{7}{8}, \ldots$

32. $\frac{1}{2}, \frac{2}{3}, \frac{3}{4}, \frac{4}{5}, \ldots$

33. $1, -2, 3, -4, \ldots$

34. $-2, 4, -8, 16, \ldots$

35. $1, -3, 5, -7, \ldots$

36. $3, -6, 9, -12, \ldots$

37. $1, \frac{2}{5}, \frac{4}{25}, \frac{8}{125}, \ldots$

38. $\frac{4}{3}, \frac{16}{9}, \frac{64}{27}, \frac{256}{81}, \ldots$

39. $x, x^2, x^3, x^4, \ldots$

40. $1, 2x, 3x^2, 4x^3, \ldots$

41. $x, -x^3, x^5, -x^7, \ldots$

42. $x, \frac{x^2}{2}, \frac{x^3}{3}, \frac{x^4}{4}, \ldots$

Write each series in Problems 43–50 in expanded form without summation notation. Do not evaluate.

43. $\sum_{k=1}^{5} (-1)^{k+1}(2k - 1)^2$

44. $\sum_{k=1}^{4} \frac{(-2)^{k+1}}{2k + 1}$

45. $\sum_{k=2}^{5} \frac{2^k}{2k + 3}$

46. $\sum_{k=3}^{7} \frac{(-1)^k}{k^2 - k}$

47. $\sum_{k=1}^{5} x^{k-1}$

48. $\sum_{k=1}^{3} \frac{1}{k} x^{k+1}$

49. $\sum_{k=0}^{4} \frac{(-1)^k x^{2k+1}}{2k + 1}$

50. $\sum_{k=0}^{4} \frac{(-1)^k x^{2k}}{2k + 2}$

Write each series in Problems 51–54 using summation notation with:
(A) The summing index k starting at $k = 1$
(B) The summing index j starting at $j = 0$

51. $2 + 3 + 4 + 5 + 6$

52. $1^2 + 2^2 + 3^2 + 4^2$

53. $1 - \frac{1}{2} + \frac{1}{3} - \frac{1}{4}$

54. $1 - \frac{1}{3} + \frac{1}{5} - \frac{1}{7} + \frac{1}{9}$

Write each series in Problems 55–58 using summation notation with the summing index k starting at $k = 1$.

55. $2 + \frac{3}{2} + \frac{4}{3} + \cdots + \frac{n + 1}{n}$

56. $1 + \frac{1}{2^2} + \frac{1}{3^2} + \cdots + \frac{1}{n^2}$

57. $\frac{1}{2} - \frac{1}{4} + \frac{1}{8} - \cdots + \frac{(-1)^{n-1}}{2^n}$

58. $1 - 4 + 9 - \cdots + (-1)^{n+1}n^2$

C *Some sequences are defined by a* ***recursion formula****—that is, a formula that defines each term of the sequence in terms of one or more of the preceding terms. For example, if $\{a_n\}$ is defined by*

$$a_1 = 1 \quad \text{and} \quad a_n = 2a_{n-1} + 1 \quad \text{for } n \geq 2$$

then

$$a_2 = 2a_1 + 1 = 2 \cdot 1 + 1 = 3$$
$$a_3 = 2a_2 + 1 = 2 \cdot 3 + 1 = 7$$
$$a_4 = 2a_3 + 1 = 2 \cdot 7 + 1 = 15$$

and so on. In Problems 59–62, write the first five terms of each sequence.

59. $a_1 = 2$ and $a_n = 3a_{n-1} + 2$ for $n \geq 2$

60. $a_1 = 3$ and $a_n = 2a_{n-1} - 2$ for $n \geq 2$

61. $a_1 = 1$ and $a_n = 2a_{n-1}$ for $n \geq 2$

62. $a_1 = 1$ and $a_n = -\frac{1}{3}a_{n-1}$ for $n \geq 2$

If A is a positive real number, then the terms of the sequence defined by

$$a_1 = \frac{A}{2} \quad \text{and} \quad a_n = \frac{1}{2}\left(a_{n-1} + \frac{A}{a_{n-1}}\right) \quad \text{for } n \geq 2$$

can be used to approximate $\sqrt{A}$ to any decimal place accuracy desired. In Problems 63 and 64, compute the first four terms of this sequence for the indicated value of A, and compare the fourth term with the value of $\sqrt{A}$ obtained from a calculator.

63. $A = 2$

64. $A = 6$

SECTION B-2 Arithmetic and Geometric Sequences

- Arithmetic and Geometric Sequences
- *n*th-Term Formulas
- Sum Formulas for Finite Arithmetic Series
- Sum Formulas for Finite Geometric Series
- Sum Formula for Infinite Geometric Series
- Applications

For most sequences it is difficult to sum an arbitrary number of terms of the sequence without adding term by term. But particular types of sequences—*arithmetic sequences* and *geometric sequences*—have certain properties that lead to convenient and useful formulas for the sums of the corresponding *arithmetic series* and *geometric series*.

Arithmetic and Geometric Sequences

The sequence $5, 7, 9, 11, 13, \ldots, 5 + 2(n - 1), \ldots$, where each term after the first is obtained by adding 2 to the preceding term, is an example of an arithmetic sequence. The sequence $5, 10, 20, 40, 80, \ldots, 5(2)^{n-1}, \ldots$, where each term after the first is obtained by multiplying the preceding term by 2, is an example of a geometric sequence.

Arithmetic Sequence

A sequence of numbers

$$a_1, a_2, a_3, \ldots, a_n, \ldots$$

is called an **arithmetic sequence** if there is a constant d, called the **common difference,** such that

$$a_n - a_{n-1} = d$$

That is,

$$a_n = a_{n-1} + d \qquad \text{for every } n > 1$$

Geometric Sequence

A sequence of numbers

$$a_1, a_2, a_3, \ldots, a_n, \ldots$$

is called a **geometric sequence** if there exists a nonzero constant r, called a **common ratio,** such that

$$\frac{a_n}{a_{n-1}} = r$$

That is,

$$a_n = ra_{n-1} \qquad \text{for every } n \geqslant 1$$

Explore–Discuss 1

(A) Describe verbally all arithmetic sequences with common difference 2.

(B) Describe verbally all geometric sequences with common ratio 2.

Example 1 **Recognizing Arithmetic and Geometric Sequences** Which of the following can be the first four terms of an arithmetic sequence? Of a geometric sequence?

(A) $1, 2, 3, 5, \ldots$ (B) $-1, 3, -9, 27, \ldots$
(C) $3, 3, 3, 3, \ldots$ (D) $10, 8.5, 7, 5.5, \ldots$

SOLUTION (A) Since $2 - 1 \neq 5 - 3$, there is no common difference, so the sequence is not an arithmetic sequence. Since $2/1 \neq 3/2$, there is no common ratio, so the sequence is not geometric either.
(B) The sequence is geometric with common ratio -3. It is not arithmetic.
(C) The sequence is arithmetic with common difference 0, and is also geometric with common ratio 1.
(D) The sequence is arithmetic with common difference -1.5. It is not geometric.

Matched Problem 1 Which of the following can be the first four terms of an arithmetic sequence? Of a geometric sequence?

(A) $8, 2, 0.5, 0.125, \ldots$ (B) $-7, -2, 3, 8, \ldots$
(C) $1, 5, 25, 100, \ldots$

*n*TH-TERM FORMULAS

If $\{a_n\}$ is an arithmetic sequence with common difference d, then

$$a_2 = a_1 + d$$
$$a_3 = a_2 + d = a_1 + 2d$$
$$a_4 = a_3 + d = a_1 + 3d$$

This suggests that:

nth Term of an Arithmetic Sequence

$$a_n = a_1 + (n - 1)d \qquad \text{for all } n > 1 \qquad (1)$$

Similarly, if $\{a_n\}$ is a geometric sequence with common ratio r, then

$$a_2 = a_1 r$$
$$a_3 = a_2 r = a_1 r^2$$
$$a_4 = a_3 r = a_1 r^3$$

This suggests that:

nth Term of a Geometric Sequence

$$a_n = a_1 r^{n-1} \qquad \text{for all } n > 1 \qquad (2)$$

Example 2 **Finding Terms in Arithmetic and Geometric Sequences**

(A) If the 1st and 10th terms of an arithmetic sequence are 3 and 30, respectively, find the 40th term of the sequence.
(B) If the 1st and 10th terms of a geometric sequence are 3 and 30, find the 40th term to three decimal places.

SOLUTION (A) First use formula (1) with $a_1 = 3$ and $a_{10} = 30$ to find d:

$$\begin{aligned} a_n &= a_1 + (n-1)d \\ a_{10} &= a_1 + (10-1)d \\ 30 &= 3 + 9d \\ d &= 3 \end{aligned}$$

Now find a_{40}:

$$a_{40} = 3 + 39 \cdot 3 = 120$$

(B) First use formula (2) with $a_1 = 3$ and $a_{10} = 30$ to find r:

$$\begin{aligned} a_n &= a_1 r^{n-1} \\ a_{10} &= a_1 r^{10-1} \\ 30 &= 3r^9 \\ r^9 &= 10 \\ r &= 10^{1/9} \end{aligned}$$

Now find a_{40}:

$$a_{40} = 3(10^{1/9})^{39} = 3(10^{39/9}) = 64{,}633.041$$

Matched Problem 2 (A) If the 1st and 15th terms of an arithmetic sequence are -5 and 23, respectively, find the 73rd term of the sequence.

(B) Find the 8th term of the geometric sequence

$$\frac{1}{64}, \frac{-1}{32}, \frac{1}{16}, \ldots$$

Sum Formulas for Finite Arithmetic Series

If $a_1, a_2, a_3, \ldots, a_n$ is a finite arithmetic sequence, then the corresponding series $a_1 + a_2 + a_3 + \cdots + a_n$ is called a *finite arithmetic series.* We will derive two simple and very useful formulas for the sum of a finite arithmetic series. Let d be the common difference of the arithmetic sequence $a_1, a_2, a_3, \ldots, a_n$ and let S_n denote the sum of the series $a_1 + a_2 + a_3 + \cdots + a_n$. Then

$$S_n = a_1 + (a_1 + d) + \cdots + [a_1 + (n-2)d] + [a_1 + (n-1)d]$$

Reversing the order of the sum, we obtain

$$S_n = [a_1 + (n-1)d] + [a_1 + (n-2)d] + \cdots + (a_1 + d) + a_1$$

Something interesting happens if we combine these last two equations by addition (adding corresponding terms on the right sides):

$$2S_n = [2a_1 + (n-1)d] + [2a_1 + (n-1)d] + \cdots + [2a_1 + (n-1)d] + [2a_1 + (n-1)d]$$

All the terms on the right side are the same, and there are n of them. Thus,

$$2S_n = n[2a_1 + (n-1)d]$$

and we have the following general formula:

Sum of a Finite Arithmetic Series—First Form

$$S_n = \frac{n}{2}[2a_1 + (n-1)d] \tag{3}$$

Replacing

$$[a_1 + (n - 1)d] \quad \text{in} \quad \frac{n}{2}[a_1 + a_1 + (n - 1)d]$$

by a_n from equation (1), we obtain a second useful formula for the sum:

Sum of a Finite Arithmetic Series—Second Form

$$S_n = \frac{n}{2}(a_1 + a_n) \tag{4}$$

Example 3 **Finding a Sum** Find the sum of the first 30 terms in the arithmetic sequence:

3, 8, 13, 18, . . .

SOLUTION Use formula (3) with $n = 30$, $a_1 = 3$, and $d = 5$:

$$S_{30} = \frac{30}{2}[2 \cdot 3 + (30 - 1)5] = 2{,}265$$

Matched Problem 3 Find the sum of the first 40 terms in the arithmetic sequence:

15, 13, 11, 9, . . .

Example 4 **Finding a Sum** Find the sum of all the even numbers between 31 and 87.

SOLUTION First, find n using equation (1):

$$\begin{aligned} a_n &= a_1 + (n - 1)d \\ 86 &= 32 + (n - 1)2 \\ n &= 28 \end{aligned}$$

Now find S_{28} using formula (4):

$$\begin{aligned} S_n &= \frac{n}{2}(a_1 + a_n) \\ S_{28} &= \frac{28}{2}(32 + 86) = 1{,}652 \end{aligned}$$

Matched Problem 4 Find the sum of all the odd numbers between 24 and 208.

Sum Formulas for Finite Geometric Series

If $a_1, a_2, a_3, \ldots, a_n$ is a finite geometric sequence, then the corresponding series $a_1 + a_2 + a_3 + \cdots + a_n$ is called a *finite geometric series.* As with arithmetic series, we can derive two simple and very useful formulas for the sum of a finite geometric series. Let r be the common ratio of the geometric sequence $a_1, a_2, a_3, \ldots, a_n$ and let S_n denote the sum of the series $a_1 + a_2 + a_3 + \cdots + a_n$. Then

$$S_n = a_1 + a_1r + a_1r^2 + \cdots + a_1r^{n-2} + a_1r^{n-1}$$

If we multiply both sides by r, we obtain

$$rS_n = a_1r + a_1r^2 + a_1r^3 + \cdots + a_1r^{n-1} + a_1r^n$$

Now combine these last two equations by subtraction to obtain

$$\begin{aligned} rS_n - S_n &= (a_1r + a_1r^2 + a_1r^3 + \cdots + a_1r^{n-1} + a_1r^n) - (a_1 + a_1r + a_1r^2 + \cdots + a_1r^{n-2} + a_1r^{n-1}) \\ (r-1)S_n &= a_1r^n - a_1 \end{aligned}$$

Notice how many terms drop out on the right side. Solving for S_n, we have:

Sum of a Finite Geometric Series—First Form

$$S_n = \frac{a_1(r^n - 1)}{r - 1} \qquad r \neq 1 \qquad (5)$$

Since $a_n = a_1r^{n-1}$, or $ra_n = a_1r^n$, formula (5) also can be written in the form:

Sum of a Finite Geometric Series—Second Form

$$S_n = \frac{ra_n - a_1}{r - 1} \qquad r \neq 1 \qquad (6)$$

Example 5 **Finding a Sum** Find the sum of the first ten terms of the geometric sequence:

$$1, 1.05, 1.05^2, \ldots$$

SOLUTION Use formula (5) with $a_1 = 1$, $r = 1.05$, and $n = 10$:

$$\begin{aligned} S_n &= \frac{a_1(r^n - 1)}{r - 1} \\ S_{10} &= \frac{1(1.05^{10} - 1)}{1.05 - 1} \\ &\approx \frac{0.6289}{0.05} \approx 12.58 \end{aligned}$$

Matched Problem 5 Find the sum of the first eight terms of the geometric sequence:

$$100, 100(1.08), 100(1.08)^2, \ldots$$

Sum Formula for Infinite Geometric Series

Explore–Discuss 2

(A) For any n, the sum of a finite geometric series with $a_1 = 5$ and $r = \frac{1}{2}$ is given by [see formula (5)]

$$S_n = \frac{5[(\frac{1}{2})^n - 1]}{\frac{1}{2} - 1} = 10 - 10\left(\frac{1}{2}\right)^n$$

Discuss the behavior of S_n as n increases.

(B) Repeat part (A) if $a_1 = 5$ and $r = 2$.

Given a geometric series, what happens to the sum S_n of the first n terms as n increases without stopping? To answer this question, let us write formula (5) in the form

$$S_n = \frac{a_1 r^n}{r - 1} - \frac{a_1}{r - 1}$$

It is possible to show that if $-1 < r < 1$, then r^n will approach 0 as n increases. Thus, the first term above will approach 0 and S_n can be made as close as we please to the second term, $-a_1/(r - 1)$ [which can be written as $a_1/(1 - r)$], by taking n sufficiently large. Thus, if the common ratio r is between -1 and 1, we define the sum of an infinite geometric series to be:

Sum of an Infinite Geometric Series

$$S_\infty = \frac{a_1}{1 - r} \qquad -1 < r < 1 \qquad (7)$$

If $r \leqslant -1$ or $r \geqslant 1$, then an infinite geometric series has no sum.

Applications

Example 6 **Loan Repayment** A person borrows \$3,600 and agrees to repay the loan in monthly installments over a period of 3 years. The agreement is to pay 1% of the unpaid balance each month for using the money and \$100 each month to reduce the loan. What is the total cost of the loan over the 3 years?

Solution Let us look at the problem relative to a time line:

	\$3,600	\$3,500	\$3,400	. . .	\$200	\$100		Unpaid balance
0	1	2	3	. . .	34	35	36	Months
	0.01(3,600) = 36	0.01(3,500) = 35	0.01(3,400) = 34	. . .	0.01(300) = 3	0.01(200) = 2	0.01(100) = 1	1% of unpaid balance

The total cost of the loan is

$$1 + 2 + \cdots + 34 + 35 + 36$$

The terms form a finite arithmetic series with $n = 36$, $a_1 = 1$, and $a_{36} = 36$, so we can use formula (4):

$$S_n = \frac{n}{2}(a_1 + a_n)$$

$$S_{36} = \frac{36}{2}(1 + 36) = \$666$$

And we conclude that the total cost of the loan over the period of 3 years is \$666.

Matched Problem 6 Repeat Example 6 with a loan of \$6,000 over a period of 5 years.

Example 7 **Economy Stimulation** The government has decided on a tax rebate program to stimulate the economy. Suppose you receive \$600 and you spend 80% of this, and each of the people who receive what you spend also spend 80% of what they receive, and this process continues without end. According to the **multiplier principle** in economics, the effect of your \$600 tax rebate on the economy is multiplied many times. What is the total amount spent if the process continues as indicated?

SOLUTION We need to find the sum of an infinite geometric series with the first amount spent being $a_1 = (0.8)(\$600) = \$480$ and $r = 0.8$. Using formula (7), we obtain

$$S_\infty = \frac{a_1}{1 - r} = \frac{\$480}{1 - 0.8} = \$2{,}400$$

Thus, assuming the process continues as indicated, we would expect the \$600 tax rebate to result in about \$2,400 of spending.

Matched Problem 7 Repeat Example 7 with a tax rebate of \$1,000.

Answers to Matched Problems

1. (A) The sequence is geometric with $r = \frac{1}{4}$. It is not arithmetic.
(B) The sequence is arithmetic with $d = 5$. It is not geometric.
(C) The sequence is neither arithmetic nor geometric.

2. (A) 139 (B) -2 **3.** -960 **4.** 10,672

5. 1,063.66 **6.** \$1,830 **7.** \$4,000

EXERCISE B-2

A *In Problems 1 and 2, determine whether the indicated sequence can be the first three terms of an arithmetic or geometric sequence, and, if so, find the common difference or common ratio and the next two terms of the sequence.*

1. (A) $-11, -16, -21, \ldots$ (B) $2, -4, 8, \ldots$
(C) $1, 4, 9, \ldots$ (D) $\frac{1}{2}, \frac{1}{6}, \frac{1}{18}, \ldots$

2. (A) $5, 20, 100, \ldots$ (B) $-5, -5, -5, \ldots$
(C) $7, 6.5, 6, \ldots$ (D) $512, 256, 128, \ldots$

B *Let $a_1, a_2, a_3, \ldots, a_n, \ldots$ be an arithmetic sequence. In Problems 3–8, find the indicated quantities.*

3. $a_1 = 7; d = 4; a_2 = ?; a_3 = ?$

4. $a_1 = -2; d = -3; a_2 = ?; a_3 = ?$

5. $a_1 = 2; d = 4; a_{21} = ?; S_{31} = ?$

6. $a_1 = 8; d = -10; a_{15} = ?; S_{23} = ?$

7. $a_1 = 18; a_{20} = 75; S_{20} = ?$

8. $a_1 = 203; a_{30} = 261; S_{30} = ?$

Let $a_1, a_2, a_3, \ldots, a_n, \ldots$ be a geometric sequence. In Problems 9–18, find the indicated quantities.

9. $a_1 = 3; r = -2; a_2 = ?; a_3 = ?; a_4 = ?$

10. $a_1 = 32; r = -\frac{1}{2}; a_2 = ?; a_3 = ?; a_4 = ?$

11. $a_1 = 1; a_7 = 729; r = -3; S_7 = ?$

12. $a_1 = 3; a_7 = 2{,}187; r = 3; S_7 = ?$

13. $a_1 = 100; r = 1.08; a_{10} = ?$

14. $a_1 = 240; r = 1.06; a_{12} = ?$

15. $a_1 = 100; a_9 = 200; r = ?$

16. $a_1 = 100; a_{10} = 300; r = ?$

17. $a_1 = 500$; $r = 0.6$; $S_{10} = ?$; $S_\infty = ?$

18. $a_1 = 8{,}000$; $r = 0.4$; $S_{10} = ?$; $S_\infty = ?$

19. $S_{41} = \sum_{k=1}^{41} (3k + 3) = ?$

20. $S_{50} = \sum_{k=1}^{50} (2k - 3) = ?$

21. $S_8 = \sum_{k=1}^{8} (-2)^{k-1} = ?$

22. $S_8 = \sum_{k=1}^{8} 2^k = ?$

23. Find the sum of all the odd integers between 12 and 68.

24. Find the sum of all the even integers between 23 and 97.

25. Find the sum of each infinite geometric sequence (if it exists).
(A) $2, 4, 8, \ldots$ (B) $2, -\frac{1}{2}, \frac{1}{8}, \ldots$

26. Repeat Problem 25 for:
(A) $16, 4, 1, \ldots$ (B) $1, -3, 9, \ldots$

C

27. Find $f(1) + f(2) + f(3) + \cdots + f(50)$ if $f(x) = 2x - 3$.

28. Find $g(1) + g(2) + g(3) + \cdots + g(100)$ if $g(t) = 18 - 3t$.

29. Find $f(1) + f(2) + \cdots - f(10)$ if $f(x) = (\frac{1}{2})^x$.

30. Find $g(1) + g(2) + \cdots + g(10)$ if $g(x) = 2^x$.

31. Show that the sum of the first n odd positive integers is n^2, using appropriate formulas from this section.

32. Show that the sum of the first n even positive integers is $n + n^2$, using formulas in this section.

Applications

Business & Economics

33. *Loan repayment.* If you borrow \$4,800 and repay the loan by paying \$200 per month to reduce the loan and 1% of the unpaid balance each month for the use of the money, what is the total cost of the loan over 24 months?

34. *Loan repayment.* If you borrow \$5,400 and repay the loan by paying \$300 per month to reduce the loan and 1.5% of the unpaid balance each month for the use of the money, what is the total cost of the loan over 18 months?

35. *Economy stimulation.* The government, through a subsidy program, distributes \$5,000,000. If we assume each individual or agency spends 70% of what is received, and 70% of this is spent, and so on, how much total increase in spending results from this government action? (Let $a_1 = \$3{,}500{,}000$.)

36. *Economy stimulation.* Due to reduced taxes, an individual has an extra \$1,200 in spendable income. If we assume that the individual spends 65% of this on consumer goods, and the producers of these goods in turn spend 65% on consumer goods, and that this process continues indefinitely, what is the total amount spent (to the nearest dollar) on consumer goods?

37. *Compound interest.* If \$1,000 is invested at 5% compounded annually, the amount A present after n years forms a geometric sequence with common ratio $1 + 0.05 = 1.05$. Use a geometric sequence formula to find the amount A in the account (to the nearest cent) after 10 years. After 20 years. [*Hint:* Use a time line.]

38. *Compound interest.* If \$$P$ is invested at $100r$% compounded annually, the amount A present after n years forms a geometric sequence with common ratio $1 + r$. Write a formula for the amount present after n years. [*Hint:* Use a time line.]

Section B-3 The Binomial Theorem

- Factorial
- Binomial Theorem—Development

The binomial form

$$(a + b)^n$$

where n is a natural number, appears more frequently than you might expect. The coefficients in the expansion play an important role in probability studies. The *binomial formula,* which we will derive informally, enables us to expand $(a + b)^n$ directly for n any natural number. Since the formula involves *factorials,* we digress for a moment here to introduce this important concept.

Factorial

For n a natural number, ***n* factorial,** denoted by ***n*!,** is the product of the first n natural numbers. **Zero factorial** is defined to be 1. That is:

$$n! = n \cdot (n-1) \cdot \cdots \cdot 2 \cdot 1$$
$$1! = 1$$
$$0! = 1$$

It is also useful to note that:

$$n! = n \cdot (n-1)! \qquad n \geq 1$$

Example 1 **Factorial Forms** Evaluate each:

(A) $5! = 5 \cdot 4 \cdot 3 \cdot 2 \cdot 1 = 120$ (B) $\dfrac{8!}{7!} = \dfrac{8 \cdot 7!}{7!} = 8$

(C) $\dfrac{10!}{7!} = \dfrac{10 \cdot 9 \cdot 8 \cdot \cancel{7!}}{\cancel{7!}} = 720$

Matched Problem 1 Evaluate each:

(A) $4!$ (B) $\dfrac{7!}{6!}$ (C) $\dfrac{8!}{5!}$

The following important formula involving factorials has applications in many areas of mathematics and statistics. We will use this formula to provide a more concise form for the expressions encountered later in this discussion.

For n and r integers satisfying $0 \leq r \leq n$,

$$C_{n,r} = \frac{n!}{r!(n-r)!}$$

Example 2 **Evaluating $C_{n,r}$**

(A) $C_{9,2} = \dfrac{9!}{2!(9-2)!} = \dfrac{9!}{2!7!} = \dfrac{9 \cdot 8 \cdot \cancel{7!}}{2 \cdot \cancel{7!}} = 36$

(B) $C_{5,5} = \dfrac{5!}{5!(5-5)!} = \dfrac{5!}{5!0!} = \dfrac{5!}{5!} = 1$

Matched Problem 2 Find:

(A) $C_{5,2}$ (B) $C_{6,0}$

Binomial Theorem—Development

Let us expand $(a + b)^n$ for several values of n to see if we can observe a pattern that leads to a general formula for the expansion for any natural number n:

$$
\begin{aligned}
(a + b)^1 &= a + b \\
(a + b)^2 &= a^2 + 2ab + b^2 \\
(a + b)^3 &= a^3 + 3a^2b + 3ab^2 + b^3 \\
(a + b)^4 &= a^4 + 4a^3b + 6a^2b^2 + 4ab^3 + b^4 \\
(a + b)^5 &= a^5 + 5a^4b + 10a^3b^2 + 10a^2b^3 + 5ab^4 + b^5
\end{aligned}
$$

Observations

1. The expansion of $(a + b)^n$ has $(n + 1)$ terms.
2. The power of a decreases by 1 for each term as we move from left to right.
3. The power of b increases by 1 for each term as we move from left to right.
4. In each term, the sum of the powers of a and b always equals n.
5. Starting with a given term, we can get the coefficient of the next term by multiplying the coefficient of the given term by the exponent of a and dividing by the number that represents the position of the term in the series of terms. For example, in the expansion of $(a + b)^4$ above, the coefficient of the third term is found from the second term by multiplying 4 and 3, and then dividing by 2 [that is, the coefficient of the third term $= (4 \cdot 3)/2 = 6$].

We now postulate these same properties for the general case:

$$
\begin{aligned}
(a + b)^n &= a^n + \frac{n}{1}a^{n-1}b + \frac{n(n-1)}{1 \cdot 2}a^{n-2}b^2 + \frac{n(n-1)(n-2)}{1 \cdot 2 \cdot 3}a^{n-3}b^3 + \cdots + b^n \\
&= \frac{n!}{0!(n-0)!}a^n + \frac{n!}{1!(n-1)!}a^{n-1}b + \frac{n!}{2!(n-2)!}a^{n-2}b^2 + \frac{n!}{3!(n-3)!}a^{n-3}b^3 + \cdots + \frac{n!}{n!(n-n)!}b^n \\
&= C_{n,0}a^n + C_{n,1}a^{n-1}b + C_{n,2}a^{n-2}b^2 + C_{n,3}a^{n-3}b^3 + \cdots + C_{n,n}b^n
\end{aligned}
$$

And we are led to the formula in the binomial theorem (a formal proof requires mathematical induction, which is beyond the scope of this book):

> **Binomial Theorem**
>
> For all natural numbers n,
>
> $$(a + b)^n = C_{n,0}a^n + C_{n,1}a^{n-1}b + C_{n,2}a^{n-2}b^2 + C_{n,3}a^{n-3}b^3 + \cdots + C_{n,n}b^n$$

Example 3 **Using the Binomial Theorem** Use the binomial formula to expand $(u + v)^6$.

SOLUTION

$$
\begin{aligned}
(u + v)^6 &= C_{6,0}u^6 + C_{6,1}u^5v + C_{6,2}u^4v^2 + C_{6,3}u^3v^3 + C_{6,4}u^2v^4 + C_{6,5}uv^5 + C_{6,6}v^6 \\
&= u^6 + 6u^5v + 15u^4v^2 + 20u^3v^3 + 15u^2v^4 + 6uv^5 + v^6
\end{aligned}
$$

Matched Problem 3 ➡ Use the binomial formula to expand $(x + 2)^5$.

Example 4 ➡ **Using the Binomial Theorem** Use the binomial formula to find the sixth term in the expansion of $(x - 1)^{18}$.

SOLUTION

$$\text{Sixth term} = C_{18,5}x^{13}(-1)^5 = \frac{18!}{5!(18-5)!}x^{13}(-1)$$
$$= -8{,}568x^{13}$$

Matched Problem 4 ➡ Use the binomial formula to find the fourth term in the expansion of $(x - 2)^{20}$.

Explore–Discuss 1

(A) Use the formula for $C_{n,r}$ to find

$$C_{6,0} + C_{6,1} + C_{6,2} + C_{6,3} + C_{6,4} + C_{6,5} + C_{6,6}$$

(B) Write the binomial theorem for $n = 6$, $a = 1$, and $b = 1$, and compare with the results of part (A).

(C) For any natural number n, find

$$C_{n,0} + C_{n,1} + C_{n,2} + \cdots + C_{n,n}$$

Answers to Matched Problems

1. (A) 24 (B) 7 (C) 336 **2.** (A) 10 (B) 1

3. $x^5 + 10x^4 + 40x^3 + 80x^2 + 80x + 32$

4. $-9{,}120x^{17}$

EXERCISE B-3

A *In Problems 1–20, evaluate each expression.*

1. $6!$ **2.** $7!$ **3.** $\frac{10!}{9!}$

4. $\frac{20!}{19!}$ **5.** $\frac{12!}{9!}$ **6.** $\frac{10!}{6!}$

7. $\frac{5!}{2!3!}$ **8.** $\frac{7!}{3!4!}$ **9.** $\frac{6!}{5!(6-5)!}$

10. $\frac{7!}{4!(7-4)!}$ **11.** $\frac{20!}{3!17!}$ **12.** $\frac{52!}{50!2!}$

B

13. $C_{5,3}$ **14.** $C_{7,3}$ **15.** $C_{6,5}$ **16.** $C_{7,4}$

17. $C_{5,0}$ **18.** $C_{5,5}$ **19.** $C_{18,15}$ **20.** $C_{18,3}$

Expand each expression in Problems 21–26 using the binomial formula.

21. $(a + b)^4$ **22.** $(m + n)^5$ **23.** $(x - 1)^6$

24. $(u - 2)^5$ **25.** $(2a - b)^5$ **26.** $(x - 2y)^5$

Find the indicated term in each expansion in Problems 27–32.

27. $(x - 1)^{18}$; 5th term **28.** $(x - 3)^{20}$; 3rd term

29. $(p + q)^{15}$; 7th term **30.** $(p + q)^{15}$; 13th term

31. $(2x + y)^{12}$; 11th term **32.** $(2x + y)^{12}$; 3rd term

C

33. Show that $C_{n,0} = C_{n,n}$.

34. Show that $C_{n,r} = C_{n,n-r}$.

35. The triangle below is called **Pascal's triangle.** Can you guess what the next two rows at the bottom are? Compare these numbers with the coefficients of binomial expansions.

```
        1
      1   1
    1   2   1
  1   3   3   1
1   4   6   4   1
```

APPENDIX C

Tables

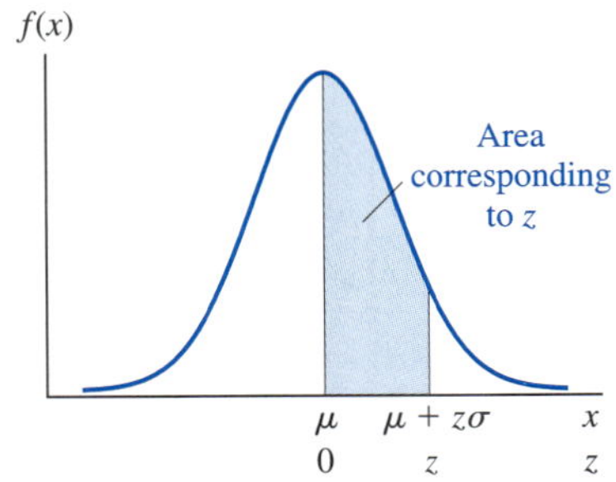

Table I

AREA UNDER THE STANDARD NORMAL CURVE
—*TABLE ENTRIES REPRESENT THE AREA UNDER THE STANDARD NORMAL CURVE FROM 0 TO z, z ⩾ 0*

z	*.00*	*.01*	*.02*	*.03*	*.04*	*.05*	*.06*	*.07*	*.08*	*.09*
0.0	0.0000	0.0040	0.0080	0.0120	0.0160	0.0199	0.0239	0.0279	0.0319	0.0359
0.1	0.0398	0.0438	0.0478	0.0517	0.0557	0.0596	0.0636	0.0675	0.0714	0.0753
0.2	0.0793	0.0832	0.0871	0.0910	0.0948	0.0987	0.1026	0.1064	0.1103	0.1141
0.3	0.1179	0.1217	0.1255	0.1293	0.1331	0.1368	0.1406	0.1443	0.1480	0.1517
0.4	0.1554	0.1591	0.1628	0.1664	0.1700	0.1736	0.1772	0.1808	0.1844	0.1879
0.5	0.1915	0.1950	0.1985	0.2019	0.2054	0.2088	0.2123	0.2157	0.2190	0.2224
0.6	0.2257	0.2291	0.2324	0.2357	0.2389	0.2422	0.2454	0.2486	0.2517	0.2549
0.7	0.2580	0.2611	0.2642	0.2673	0.2704	0.2734	0.2764	0.2794	0.2823	0.2852
0.8	0.2881	0.2910	0.2939	0.2967	0.2995	0.3023	0.3051	0.3078	0.3106	0.3133
0.9	0.3159	0.3186	0.3212	0.3238	0.3264	0.3289	0.3315	0.3340	0.3365	0.3389
1.0	0.3413	0.3438	0.3461	0.3485	0.3508	0.3531	0.3554	0.3577	0.3599	0.3621
1.1	0.3643	0.3665	0.3686	0.3708	0.3729	0.3749	0.3770	0.3790	0.3810	0.3830
1.2	0.3849	0.3869	0.3888	0.3907	0.3925	0.3944	0.3962	0.3980	0.3997	0.4015
1.3	0.4032	0.4049	0.4066	0.4082	0.4099	0.4115	0.4131	0.4147	0.4162	0.4177
1.4	0.4192	0.4207	0.4222	0.4236	0.4251	0.4265	0.4279	0.4292	0.4306	0.4319
1.5	0.4332	0.4345	0.4357	0.4370	0.4382	0.4394	0.4406	0.4418	0.4429	0.4441
1.6	0.4452	0.4463	0.4474	0.4484	0.4495	0.4505	0.4515	0.4525	0.4535	0.4545
1.7	0.4554	0.4564	0.4573	0.4582	0.4591	0.4599	0.4608	0.4616	0.4625	0.4633
1.8	0.4641	0.4649	0.4656	0.4664	0.4671	0.4678	0.4686	0.4693	0.4699	0.4706
1.9	0.4713	0.4719	0.4726	0.4732	0.4738	0.4744	0.4750	0.4756	0.4761	0.4767
2.0	0.4772	0.4778	0.4783	0.4788	0.4793	0.4798	0.4803	0.4808	0.4812	0.4817
2.1	0.4821	0.4826	0.4830	0.4834	0.4838	0.4842	0.4846	0.4850	0.4854	0.4857
2.2	0.4861	0.4864	0.4868	0.4871	0.4875	0.4878	0.4881	0.4884	0.4887	0.4890
2.3	0.4893	0.4896	0.4898	0.4901	0.4904	0.4906	0.4909	0.4911	0.4913	0.4916
2.4	0.4918	0.4920	0.4922	0.4925	0.4927	0.4929	0.4931	0.4932	0.4934	0.4936
2.5	0.4938	0.4940	0.4941	0.4943	0.4945	0.4946	0.4948	0.4949	0.4951	0.4952
2.6	0.4953	0.4955	0.4956	0.4957	0.4959	0.4960	0.4961	0.4962	0.4963	0.4964
2.7	0.4965	0.4966	0.4967	0.4968	0.4969	0.4970	0.4971	0.4972	0.4973	0.4974
2.8	0.4974	0.4975	0.4976	0.4977	0.4977	0.4978	0.4979	0.4979	0.4980	0.4981
2.9	0.4981	0.4982	0.4982	0.4983	0.4984	0.4984	0.4985	0.4985	0.4986	0.4986
3.0	0.4987	0.4987	0.4987	0.4988	0.4988	0.4989	0.4989	0.4989	0.4990	0.4990
3.1	0.4990	0.4991	0.4991	0.4991	0.4992	0.4992	0.4992	0.4992	0.4993	0.4993
3.2	0.4993	0.4993	0.4994	0.4994	0.4994	0.4994	0.4994	0.4995	0.4995	0.4995
3.3	0.4995	0.4995	0.4995	0.4996	0.4996	0.4996	0.4996	0.4996	0.4996	0.4997
3.4	0.4997	0.4997	0.4997	0.4997	0.4997	0.4997	0.4997	0.4997	0.4997	0.4998
3.5	0.4998	0.4998	0.4998	0.4998	0.4998	0.4998	0.4998	0.4998	0.4998	0.4998
3.6	0.4998	0.4998	0.4999	0.4999	0.4999	0.4999	0.4999	0.4999	0.4999	0.4999
3.7	0.4999	0.4999	0.4999	0.4999	0.4999	0.4999	0.4999	0.4999	0.4999	0.4999
3.8	0.4999	0.4999	0.4999	0.4999	0.4999	0.4999	0.4999	0.4999	0.4999	0.4999
3.9	0.5000	0.5000	0.5000	0.5000	0.5000	0.5000	0.5000	0.5000	0.5000	0.5000

Table II

Basic Geometric Formulas

1. Similar Triangles

(A) Two triangles are similar if two angles of one triangle have the same measure as two angles of the other.

(B) If two triangles are similar, their corresponding sides are proportional:

$$\frac{a}{a'} = \frac{b}{b'} = \frac{c}{c'}$$

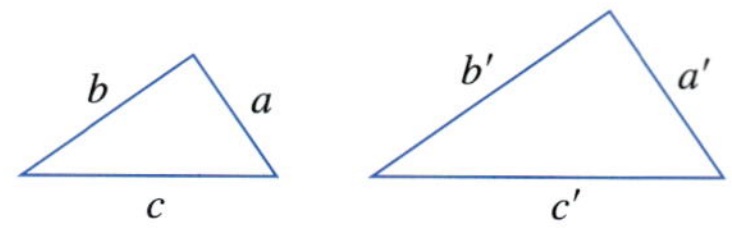

2. Pythagorean Theorem

$$c^2 = a^2 + b^2$$

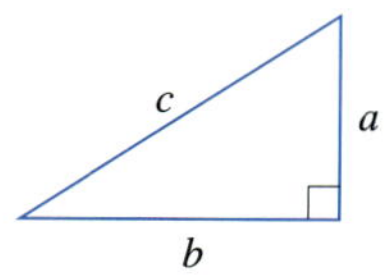

3. Rectangle

$A = ab$ Area

$P = 2a + 2b$ Perimeter

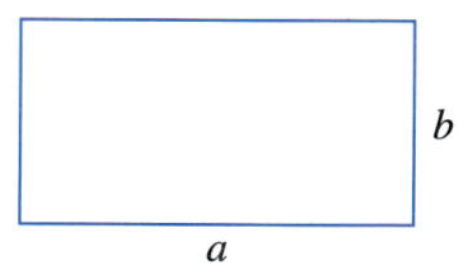

4. Parallelogram

$h =$ Height

$A = ah = ab \sin \theta$ Area

$P = 2a + 2b$ Perimeter

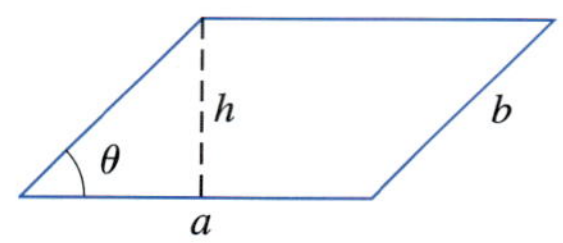

5. Triangle

$h =$ Height

$A = \frac{1}{2}hc$ Area

$P = a + b + c$ Perimeter

$s = \frac{1}{2}(a + b + c)$ Semiperimeter

$A = \sqrt{s(s - a)(s - b)(s - c)}$ Area—Heron's formula

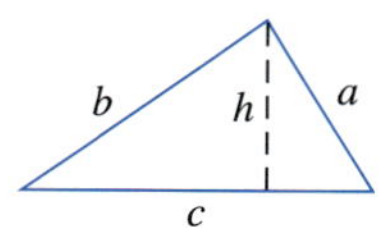

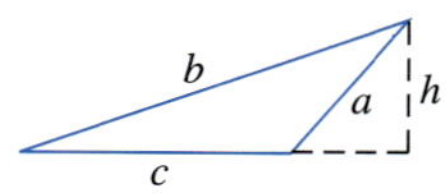

6. Trapezoid

Base a is parallel to base b.

$h =$ Height

$A = \frac{1}{2}(a + b)h$ Area

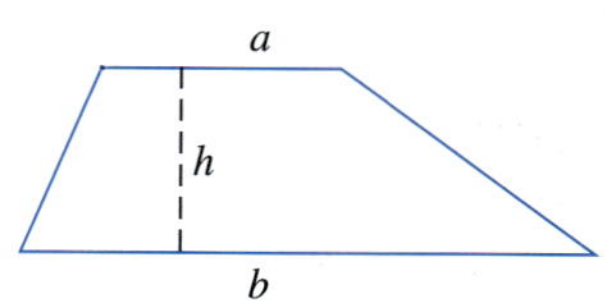

Table II *(Continued)*

7. Circle

$R =$ Radius
$D =$ Diameter
$D = 2R$
$A = \pi R^2 = \frac{1}{4}\pi D^2$ Area
$C = 2\pi R = \pi D$ Circumference
$\frac{C}{D} = \pi$ For all circles
$\pi \approx 3.141\ 59$

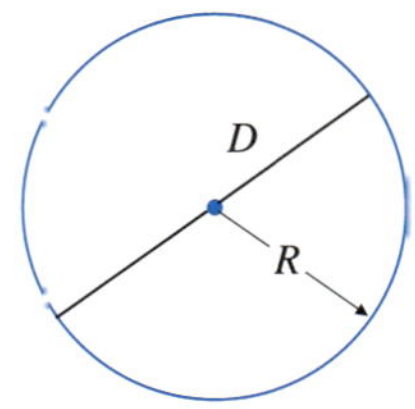

8. Rectangular Solid

$V = abc$ Volume
$T = 2ab + 2ac + 2bc$ Total surface area

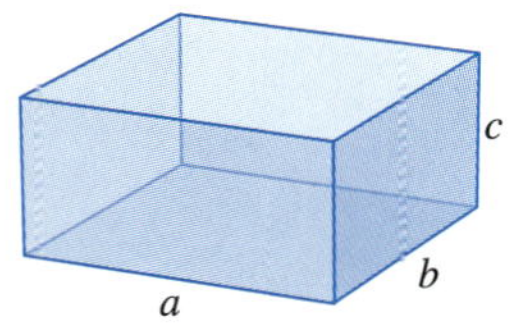

9. Right Circular Cylinder

$R =$ Radius of base
$h =$ Height
$V = \pi R^2 h$ Volume
$S = 2\pi Rh$ Lateral surface area
$T = 2\pi R(R + h)$ Total surface area

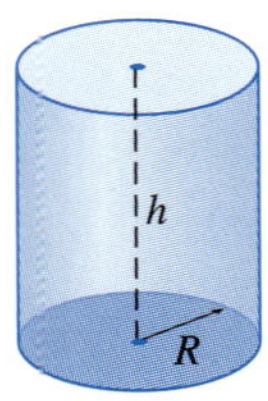

10. Right Circular Cone

$R =$ Radius of base
$h =$ Height
$s =$ Slant height
$V = \frac{1}{3}\pi R^2 h$ Volume
$S = \pi Rs = \pi R\sqrt{R^2 + h^2}$ Lateral surface area
$T = \pi R(R + s) = \pi R(R + \sqrt{R^2 + h^2})$ Total surface area

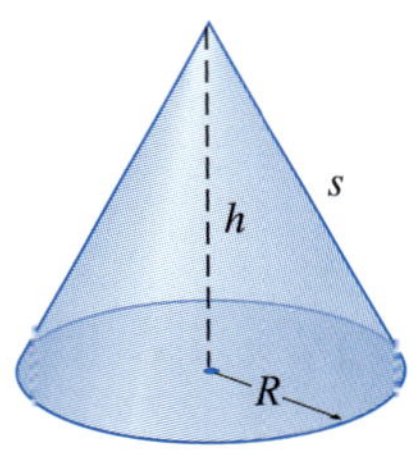

11. Sphere

$R =$ Radius
$D =$ Diameter
$D = 2R$
$V = \frac{4}{3}\pi R^3 = \frac{1}{6}\pi D^3$ Volume
$S = 4\pi R^2 = \pi D^2$ Surface area

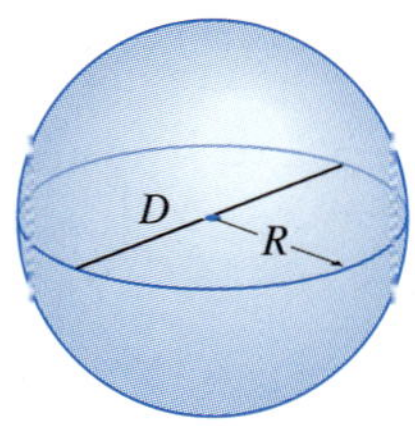

ANSWERS

CHAPTER 1

Exercise 1-1

1. Function **3.** Not a function **5.** Function **7.** Function **9.** Not a function **11.** Function **13.** 4
15. -5 **17.** -6 **19.** -2 **21.** -12 **23.** -1 **25.** -6 **27.** 12 **29.** $\frac{3}{4}$ **31.** $y = 0$ **33.** $y = 4$
35. $x = -5, 0, 4$ **37.** $x = -6$ **39.** All real numbers **41.** All real numbers except -4
43. All real numbers except -4 and 1 **45.** $x \leqslant 7$ **47.** $x < 7$
49. $f(2) = 0$, and 0 is a number; therefore, $f(2)$ exists. On the other hand, $f(3)$ is not defined, since the denominator would be 0; therefore, we say that $f(3)$ does not exist.
51. $g(x) = 2x^3 - 5$ **53.** $G(x) = 2\sqrt{x} - x^2$
55. Function f multiplies the domain element by 2 and subtracts 3 from the result.
57. Function F multiplies the cube of the domain element by 3 and subtracts twice the square root of the domain element from the result.
59. A function with domain R **61.** A function with domain R **63.** Not a function; for example, when $x = 1$, $y = \pm 3$
65. A function with domain all real numbers except $x = 4$ **67.** Not a function; for example, when $x = 4$, $y = \pm 3$
69. 4 **71.** $h - 1$ **73.** 4 **75.** $8a + 4h - 7$ **77.** $3a^2 + 3ah + h^2$ **79.** $\dfrac{1}{\sqrt{a+h} + \sqrt{a}}$
81. $P(w) = 2w + \dfrac{50}{w}, w > 0$ **83.** $A(l) = l(50 - l), 0 \leqslant l \leqslant 50$

85.

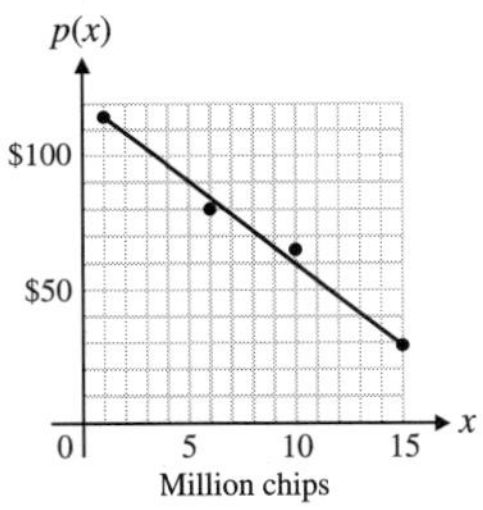

$p(8) = 71$ dollars per chip;
$p(11) = 53$ dollars per chip

87. (A) $R(x) = xp(x) = x(119 - 6x)$ million dollars; Domain: $1 \leqslant x \leqslant 15$

(B)

x (MILLION)	R(x) (MILLION $)
1	113
3	303
6	498
9	585
12	564
15	435

(C)

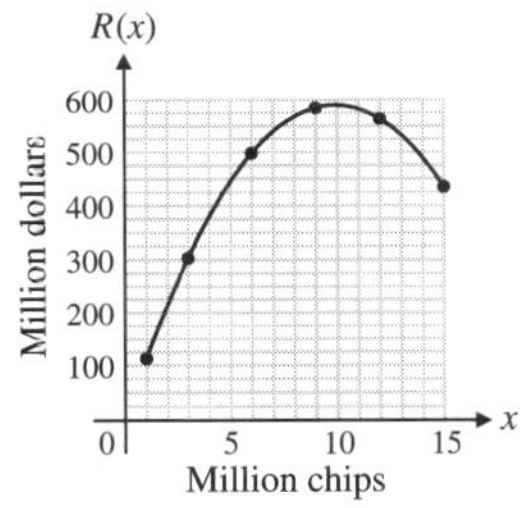

89. (A) $P(x) = R(x) - C(x) = x(119 - 6x) - (234 + 23x)$ million dollars; Domain: $1 \leqslant x \leqslant 15$

(B)

x (MILLION)	P(x) (MILLION $)
1	-144
3	0
6	126
9	144
12	54
15	-144

(C)

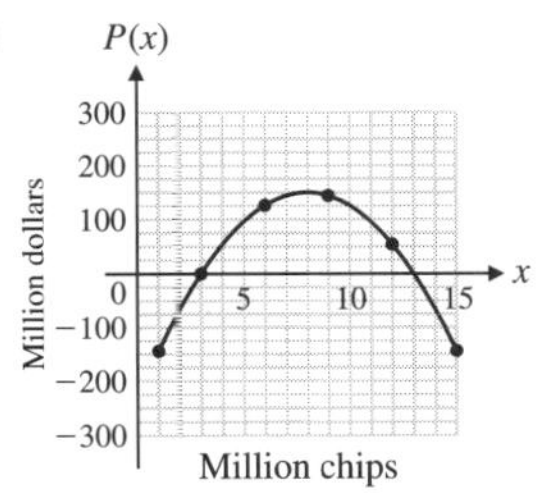

91. (A) $V(x) = x(8 - 2x)(12 - 2x)$
(B) $0 \leqslant x \leqslant 4$

(C)

x	$V(x)$
1	60
2	64
3	36

(D)

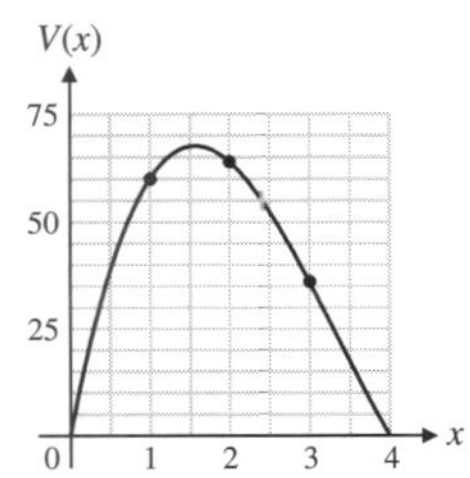

93. (A) The graph indicates that there is a value of x near 2 that will produce a volume of 65. The table shows $x = 1.9$ to one decimal place:

x	1.8	1.9	2
$V(x)$	66.5	65.4	64

(B) $x = 1.93$ to two decimal places

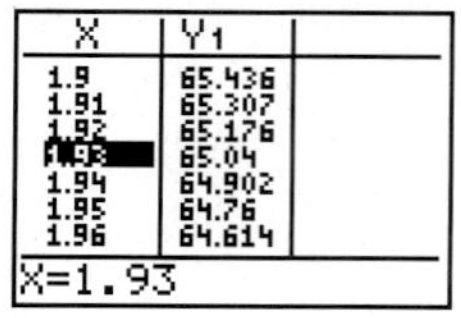

X	Y1
1.9	65.436
1.91	65.307
1.92	65.176
1.93	65.04
1.94	64.902
1.95	64.76
1.96	64.614

X=1.93

95. $v = \dfrac{75 - w}{15 + w}$; 1.9032 cm/sec

Exercise 1-2

1. Domain: All real numbers; Range: All real numbers
3. Domain: All real numbers; Range: $(-\infty, 0]$
5. Domain: $[0, \infty)$; Range: $(-\infty, 0]$
7. Domain: All real numbers; Range: All real numbers

9.

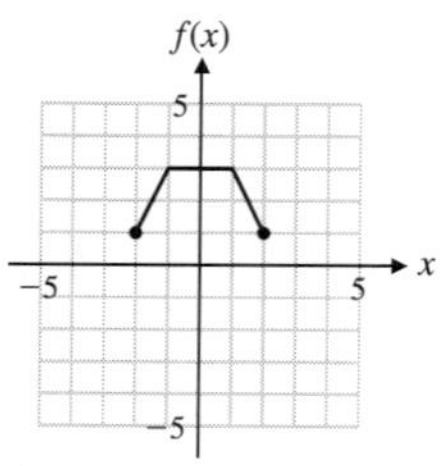

11.

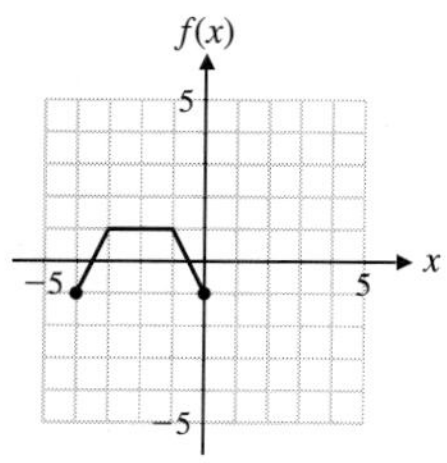

13.

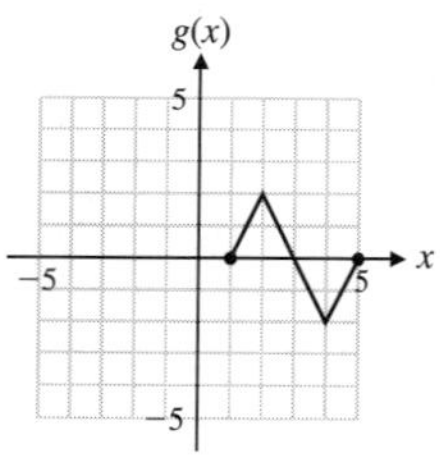

15.

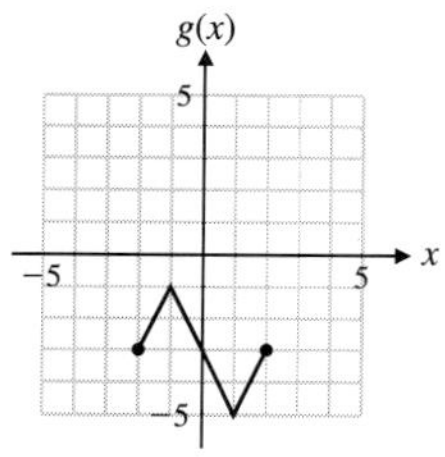

17.

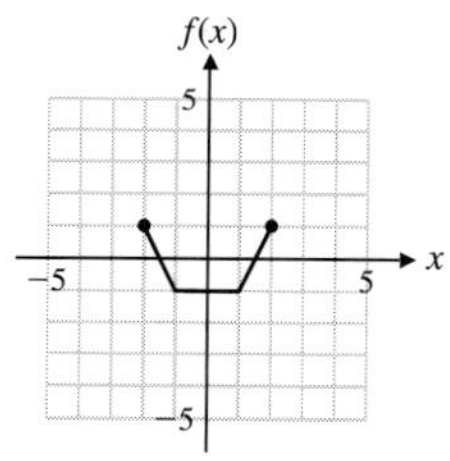

19.

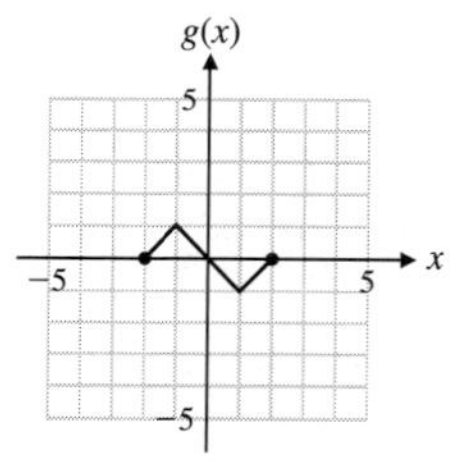

21. The graph of $g(x) = -|x + 3|$ is the graph of $y = |x|$ reflected in the x axis and shifted 3 units to the left.

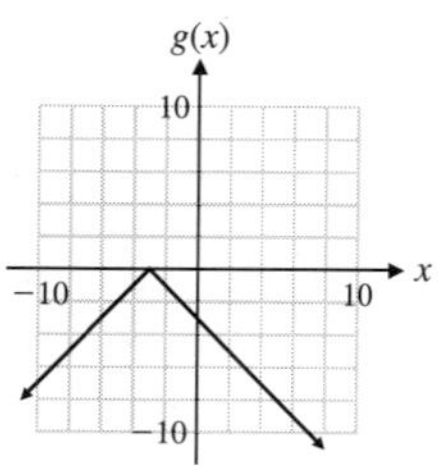

23. The graph of $f(x) = (x - 4)^2 - 3$ is the graph of $y = x^2$ shifted 4 units to the right and 3 units down.

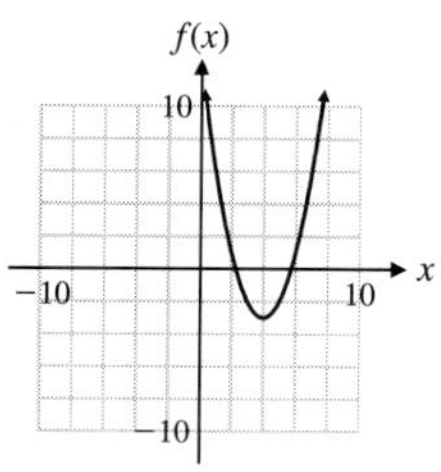

25. The graph of $f(x) = 7 - \sqrt{x}$ is the graph of $y = \sqrt{x}$ reflected in the x axis and shifted 7 units up.

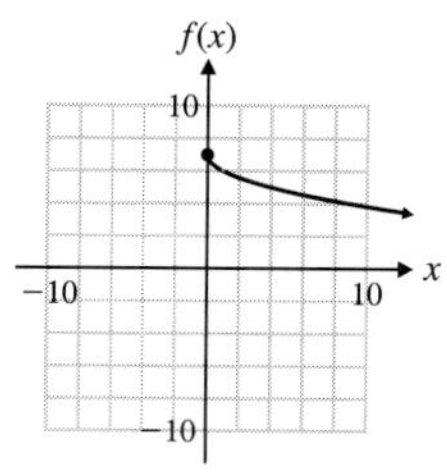

27. The graph of $h(x) = -3|x|$ is the graph of $y = |x|$ reflected in the x axis and vertically expanded by a factor of 3.

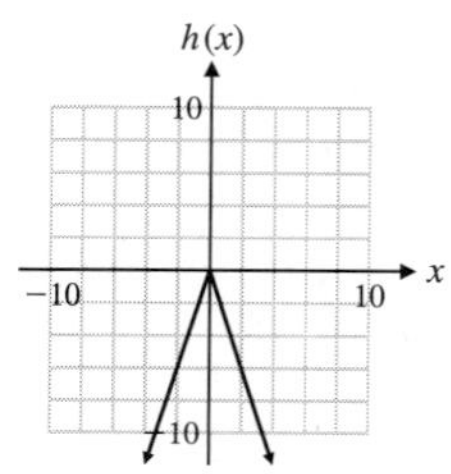

29. The graph of the basic function $y = x^2$ is shifted 2 units to the left and 3 units down. Equation: $y = (x + 2)^2 - 3$.
31. The graph of the basic function $y = x^2$ is reflected in the x axis and shifted 3 units to the right and 2 units up. Equation: $y = 2 - (x - 3)^2$.
33. The graph of the basic function $y = \sqrt{x}$ is reflected in the x axis and shifted 4 units up. Equation: $y = 4 - \sqrt{x}$.
35. The graph of the basic function $y = x^3$ is shifted 2 units to the left and 1 unit down. Equation: $y = (x + 2)^3 - 1$.
37. $g(x) = \sqrt{x - 2} - 3$ **39.** $g(x) = -|x + 3|$ **41.** $g(x) = -(x - 2)^3 - 1$

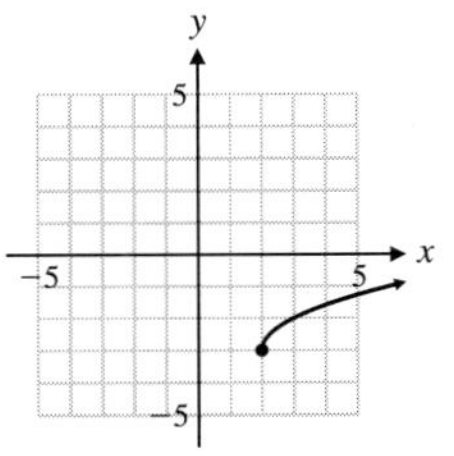

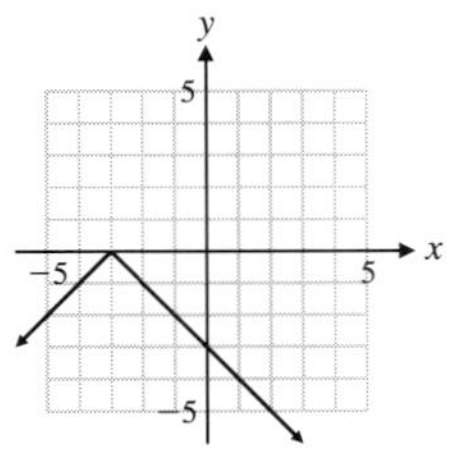

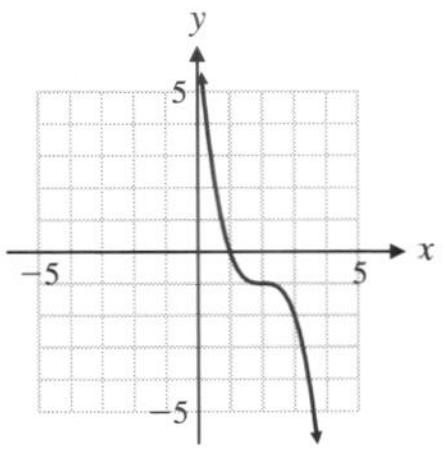

43. Discontinuous at $x = 0$ **45.** No points of discontinuity **47.** Discontinuous at $x = -1$ and $x = 1$

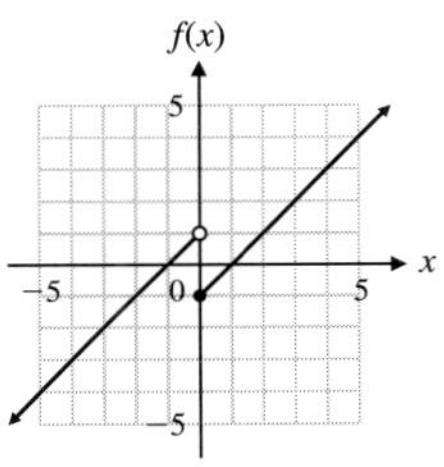

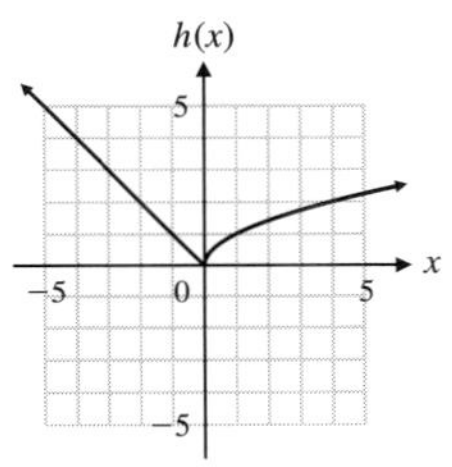

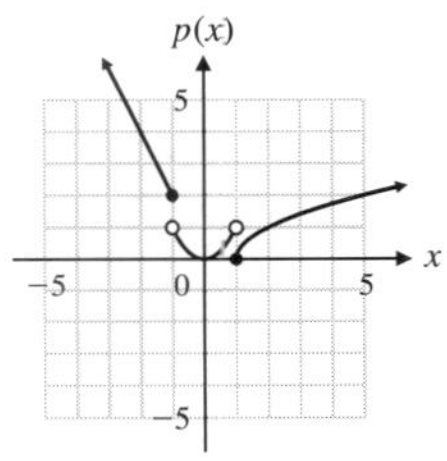

49. The graph of the basic function $y = |x|$ is reflected in the x axis and vertically contracted by a factor of 0.5. Equation: $y = -0.5|x|$.
51. The graph of the basic function $y = x^2$ is reflected in the x axis and vertically expanded by a factor of 2. Equation: $y = -2x^2$.
53. The graph of the basic function $y = \sqrt[3]{x}$ is reflected in the x axis and vertically expanded by a factor of 3. Equation: $y = -3\sqrt[3]{x}$.
55. Reversing the order does not change the result.
57. Reversing the order can change the result.
59. Reversing the order can change the result.
61. (A) The graph of the basic function $y = \sqrt{x}$ is reflected in the x axis, vertically expanded by a factor of 4, and shifted up 115 units.

(B)

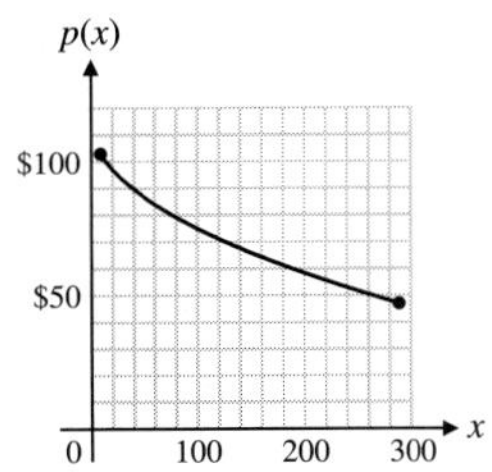

63. (A) The graph of the basic function $y = x^3$ is vertically contracted by a factor of 0.000 48 and shifted right 500 units and up 60,000 units.

(B)

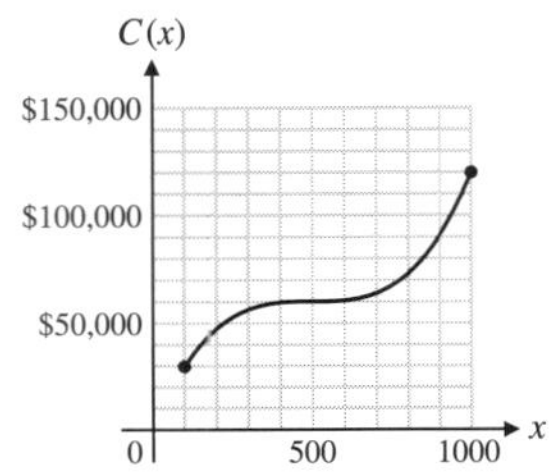

65. (A) $C(x) = \begin{cases} 10.06x & \text{if } 1 \leq x < 4 \\ 8.52x & \text{if } 4 \leq x < 10 \\ 7.31x & \text{if } 10 \leq x \end{cases}$

(B) 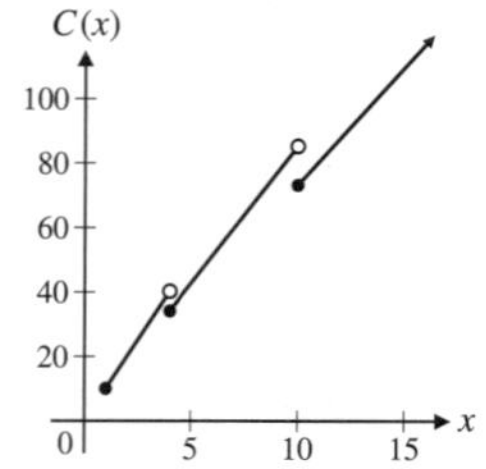

(C) Since $C(3) = \$30.18$ is less than $C(4) = \$34.08$, the customer would have to decide if a fourth box for an additional \$3.90 was desirable. However, since $C(9) = \$76.68$ is greater than $C(10) = \$73.10$, it is always to the customer's advantage to purchase 10 boxes instead of 9 boxes.

67. (A) $C(x) = \begin{cases} 0.5x & \text{if } 0 \leq x \leq 50 \\ 0.35x + 7.5 & \text{if } 50 < x \leq 100 \\ 0.2x + 22.5 & \text{if } 100 < x \end{cases}$

(B) 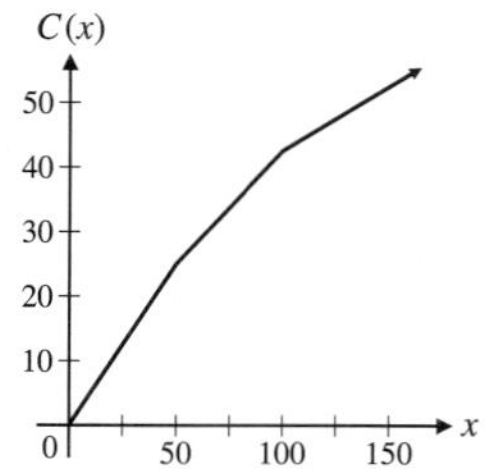

(C) In this case, the discount mileage rates do not apply to previous mileage, so there is no break in the graph at the points where the mileage rate changes.

69. (A) The graph of the basic function $y = x$ is vertically expanded by a factor of 5.5 and shifted down 220 units.

(B) 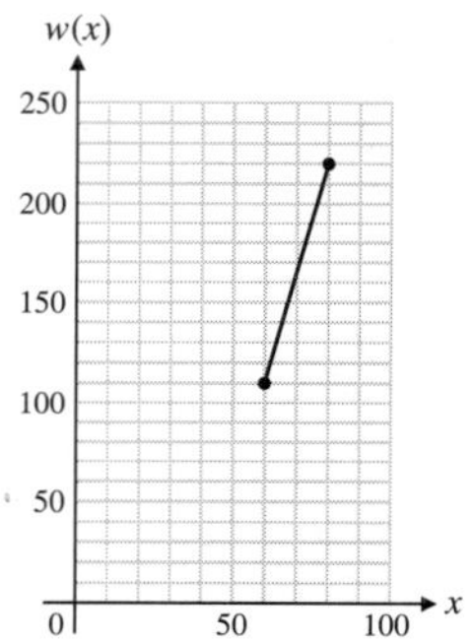

71. (A) The graph of the basic function $y = \sqrt{x}$ is vertically expanded by a factor of 7.08.

(B) 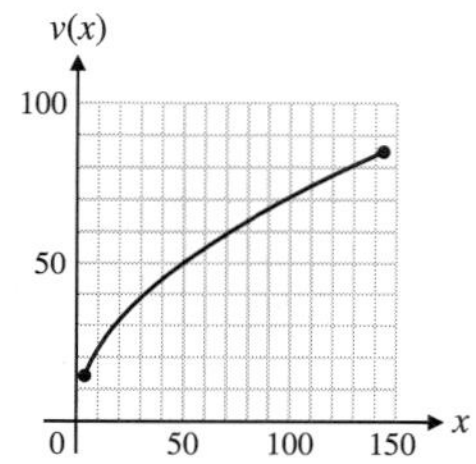

Exercise 1-3

1. (D) **3.** (C); slope is 0

5. 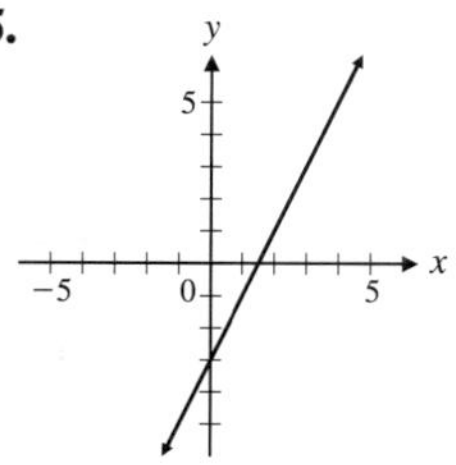

7. 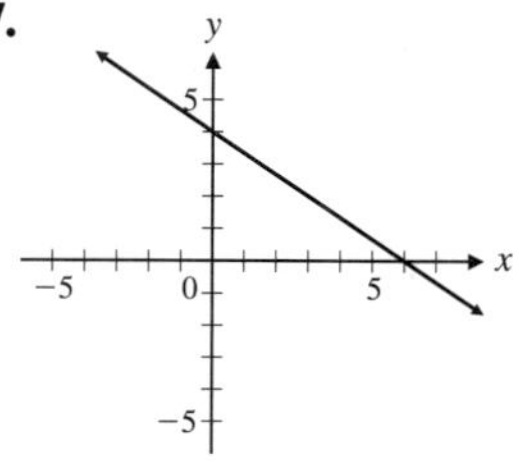

9. Slope = 2; y intercept = −3
11. Slope = $-\frac{2}{3}$; y intercept = 2
13. $y = -2x + 4$
15. $y = -\frac{3}{5}x + 3$

17. 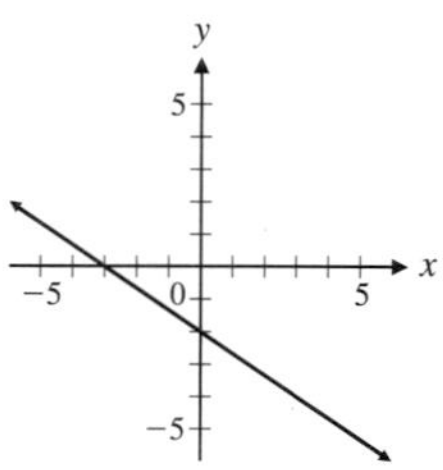

19. 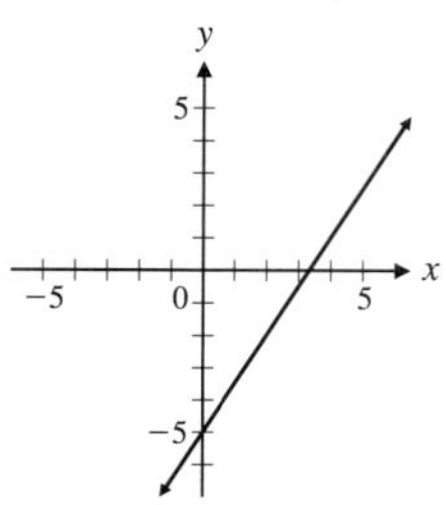

21. 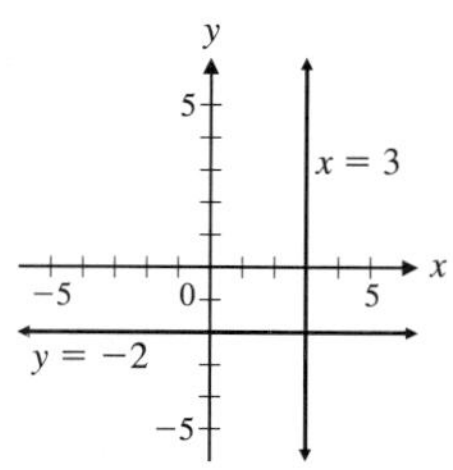

23. $y = -3x + 5$; $m = -3$
25. $y = -\frac{2}{3}x + 4$; $m = -\frac{2}{3}$

27. (A)

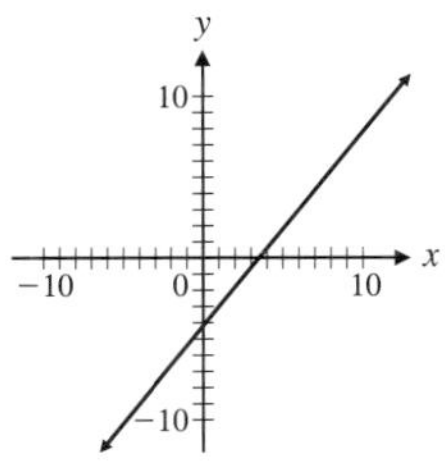

(B) x intercept: 3.5; y intercept: -4.2

(C)

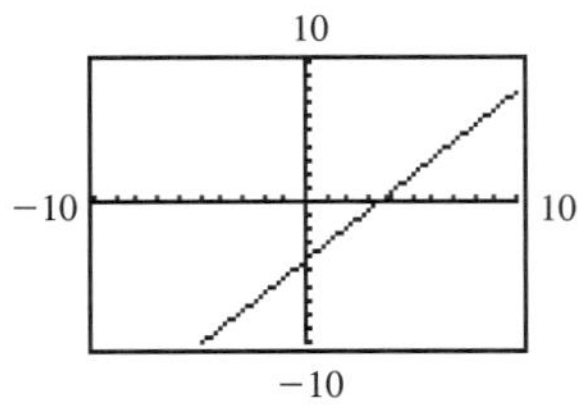

(D) x intercept: 3.5; y intercept: -4.2

(E) $x > 3.5$ or $(3.5, \infty)$

29. $x = 3; y = -5$ **31.** $x = -1; y = -3$

33. $y + 1 = -3(x - 4); y = -3x + 11$

35. $y + 5 = \frac{2}{3}(x + 6); y = \frac{2}{3}x - 1$ **37.** $y = 0x - 5$, or $y = -5$ **39.** $\frac{1}{3}$

41. $-\frac{1}{5}$ **43.** Not defined **45.** 0 **47.** $(y - 3) = \frac{1}{3}(x - 1); x - 3y = -8$

49. $(y + 2) = -\frac{1}{5}(x + 5); x + 5y = -15$ **51.** $x = 2$ **53.** $y = 3$

55. A linear function **57.** Not a function **59.** A constant function

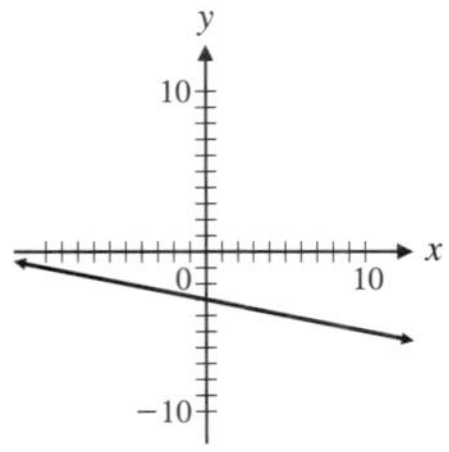

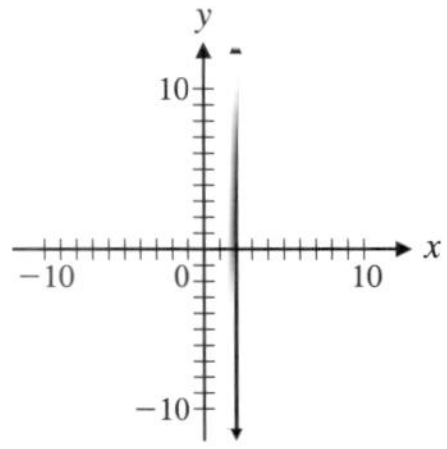

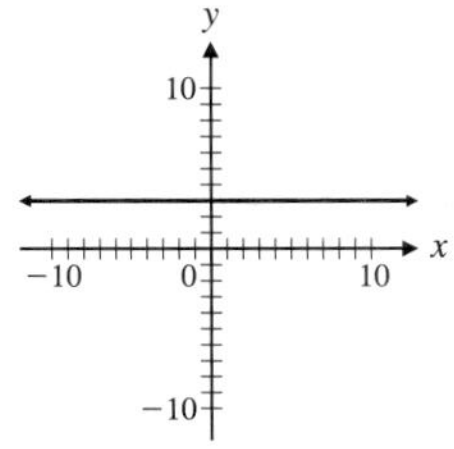

61. The graphs have the same y intercept, $(0, 2)$.

63. (A)

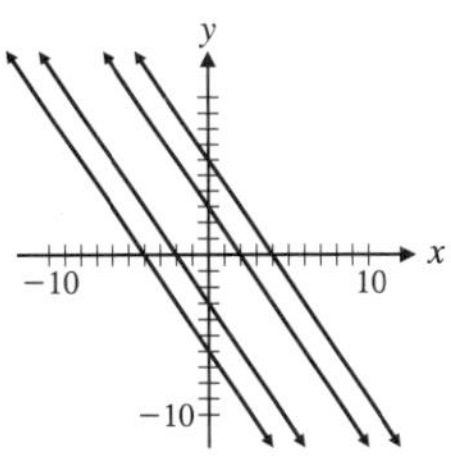

(B) Varying C produces a family of parallel lines. This is verified by observing that varying C does not change the slope of the lines but changes the intercepts.

65. The graph of g is the same as the graph of f for x satisfying $mx + b \geqslant 0$ and the reflection of the graph of f in the x axis for x satisfying $mx + b < 0$. The function g is never linear.

67. (A) \$130; \$220

(B)

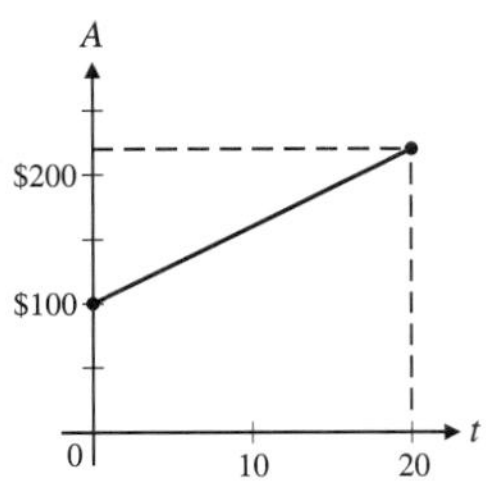

(C) The slope is 6. The amount in the account is growing at the rate of \$6 per year.

69. (A) $C(x) = 180x + 200$

(B) \$2,360

(C)

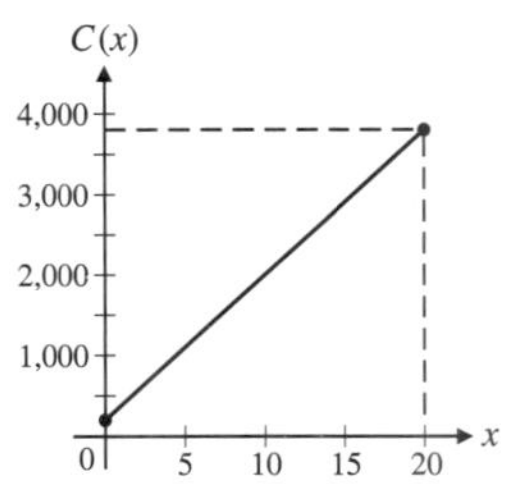

71. (A)

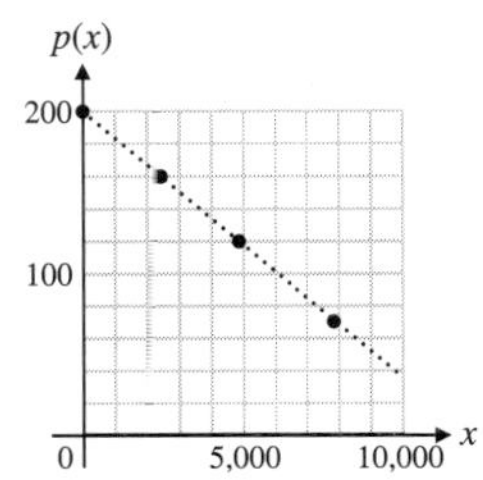

(B) $p(x) = -\frac{1}{60}x + 200$

(C) $p(3{,}000) = \$150$

(D) The slope is $-\frac{1}{60} \approx -0.02$. The price decreases \$0.02 for each unit increase in demand.

73. (A)

x	0	1	2	3	4
SALES	5.9	6.5	7.7	8.6	9.7
$f(x)$	5.7	6.7	7.7	8.6	9.6

(B)

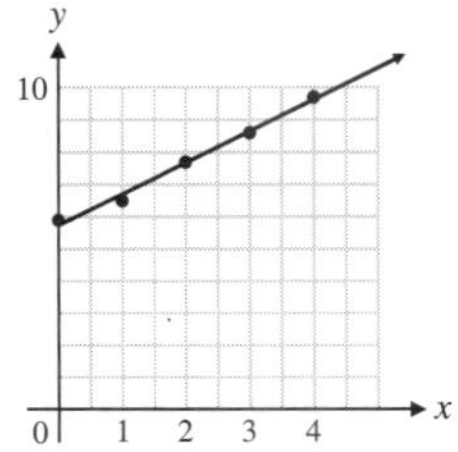

(C) \$10.6 billion; \$17.4 billion

(D) The sales are \$5.9 billion in 1988 and increase at approximately \$0.97 billion per year for the next 4 years.

75. $0.2x + 0.1y = 20$

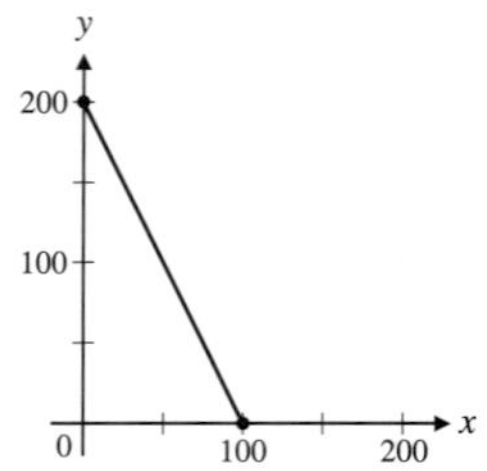

77. (A) 64 g; 35 g

(B)

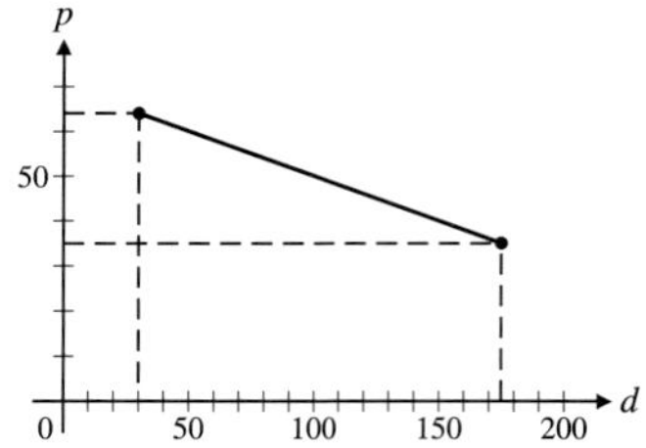

(C) $-\frac{1}{5}$

Exercise 1-4

1. (A), (C), (E), (F) **3.** (A) m (B) g (C) f (D) n

5. (A) x intercepts: 1, 3; y intercept: -3
(B) Vertex: $(2, 1)$
(C) Maximum: 1
(D) Range: $y \leqslant 1$ or $(-\infty, 1]$
(E) Increasing interval: $x \leqslant 2$ or $(-\infty, 2]$
(F) Decreasing interval: $x \geqslant 2$ or $[2, \infty)$

7. (A) x intercepts: $-3, -1$; y intercept: 3
(B) Vertex: $(-2, -1)$
(C) Minimum: -1
(D) Range: $y \geqslant -1$ or $[-1, \infty)$
(E) Increasing interval: $x \geqslant -2$ or $[-2, \infty)$
(F) Decreasing interval: $x \leqslant -2$ or $(-\infty, -2]$

9. (A) x intercepts: 1, 3; y intercept: -3
(B) Vertex: $(2, 1)$
(C) Maximum: 1
(D) Range: $y \leqslant 1$ or $(-\infty, 1]$

11. (A) x intercepts: $-3, -1$; y intercept: 3 (B) Vertex: $(-2, -1)$ (C) Minimum: -1 (D) Range: $y \geqslant -1$ or $[-1, \infty)$

13. $y = -[x - (-2)]^2 + 5$ or $y = -(x + 2)^2 + 5$ **15.** $y = (x - 1)^2 - 3$

17. $f(x) = (x - 4)^2 - 3$ (A) x intercepts: 2.3, 5.7; y intercept: 13 (B) Vertex: $(4, -3)$ (C) Minimum: -3 (D) Range: $y \geqslant -3$ or $[-3, \infty)$

19. $M(x) = -(x + 3)^2 + 10$ (A) x intercepts: $-6.2, 0.2$; y intercept: 1 (B) Vertex: $(-3, 10)$ (C) Maximum: 10 (D) Range: $y \leqslant 10$ or $(-\infty, 10]$

21. $G(x) = 0.5(x - 4)^2 + 2$ (A) x intercepts: none; y intercept: 10 (B) Vertex: $(4, 2)$ (C) Minimum: 2 (D) Range: $y \geqslant 2$ or $[2, \infty)$

23. (A) $-4.87, 8.21$ (B) $-3.44, 6.78$ (C) No solution **25.** The vertex of the parabola is on the x axis.

27. $g(x) = 0.25(x - 3)^2 - 9.25$ (A) x intercepts: $-3.08, 9.08$; y intercept: -7 (B) Vertex: $(3, -9.25)$ (C) Minimum: -9.25 (D) Range: $y \geqslant -9.25$ or $[-9.25, \infty)$

29. $f(x) = -0.12(x - 4)^2 + 3.12$ (A) x intercepts: $-1.1, 9.1$; y intercept: 1.2 (B) Vertex: $(4, 3.12)$ (C) Maximum: 3.12 (D) Range: $y \leqslant 3.12$ or $(-\infty, 3.12]$

31. $x = -5.37, 0.37$ **33.** $-1.37 < x < 2.16$ **35.** $x \leqslant -0.74$ or $x \geqslant 4.19$

37. Axis: $x = 2$; Vertex: $(2, 4)$; Range: $y \geqslant 4$ or $[4, \infty)$; no x intercepts

39. (A)

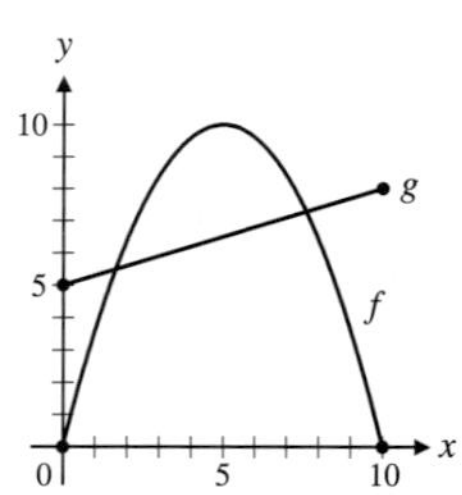

(B) 1.64, 7.61
(C) $1.64 < x < 7.61$
(D) $0 \leqslant x < 1.64$ or $7.61 < x \leqslant 10$

41. (A)

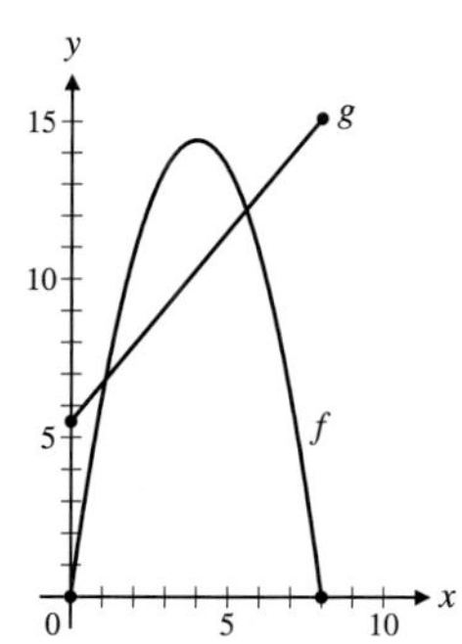

(B) 1.10, 5.57
(C) $1.10 < x < 5.57$
(D) $0 \leqslant x < 1.10$ or $5.57 < x \leqslant 8$

43. $f(x) = x^2 + 1$ and $g(x) = -(x - 4)^2 - 1$ are two examples. The graphs do not cross the x axis.

45. (A)

x	28	30	32	34	36
MILEAGE	45	52	55	51	47
$f(x)$	45.3	51.8	54.2	52.4	46.5

(B)

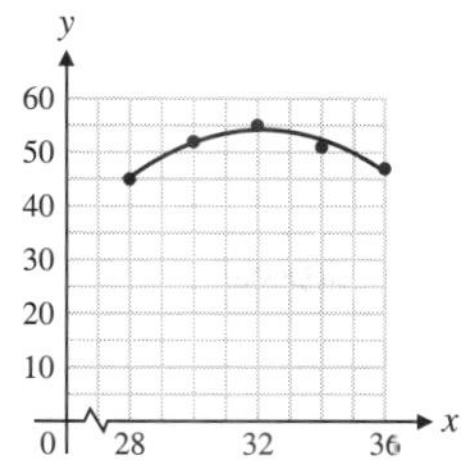

(C) $f(31) = 53.50$ thousand miles; $f(35) = 49.95$ thousand miles

47. (A)

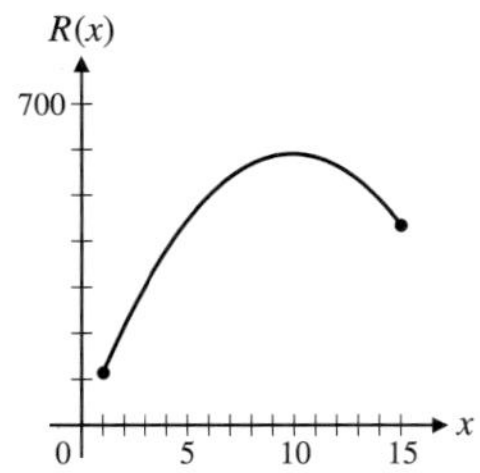

(B) $R(x) = -6(x - 9.917)^2 + 590.042$; 9.917 million chips (9,917,000 chips); 590.042 million dollars ($590,042,000)

(C)

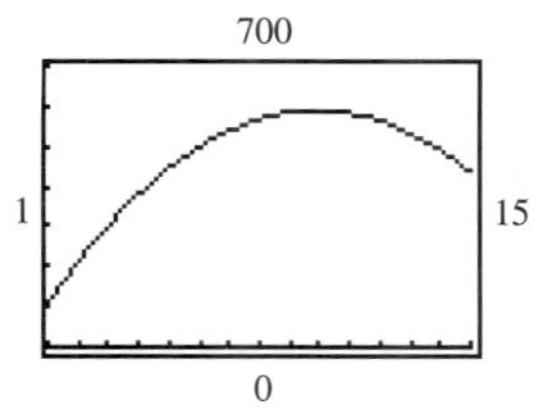

(D) 9.917 million chips (9,917,000 chips); 590.042 million dollars ($590,042,000)

(E) $59

49. (A)

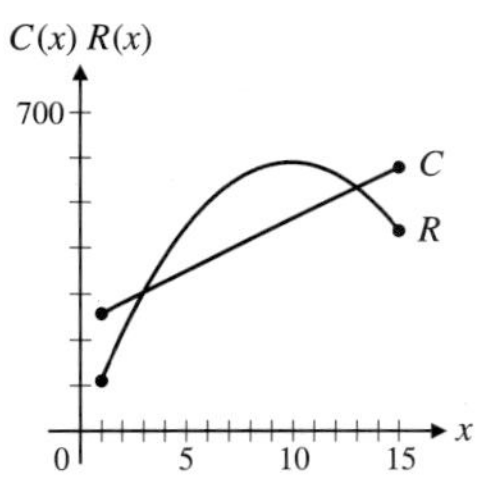

(B) Break even at 3.000 million (3,000,000) and 13.000 million (13,000,000) chip production

(C)

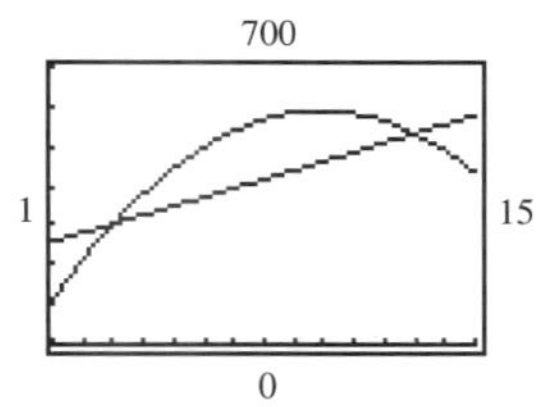

(D) Break even at 3.000 million (3,000,000) and 13.000 million (13,000,000) chip production

(E) Loss: $1 \leq x < 3$ or $13 < x \leq 15$; Profit: $3 < x < 13$

51. (A) $P(x) = R(x) - C(x) = -6x^2 + 96x - 234$

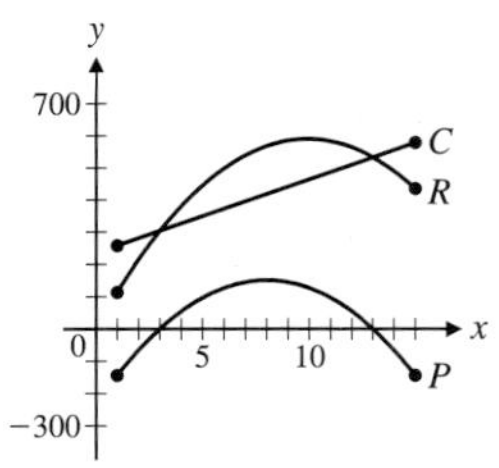

(B)

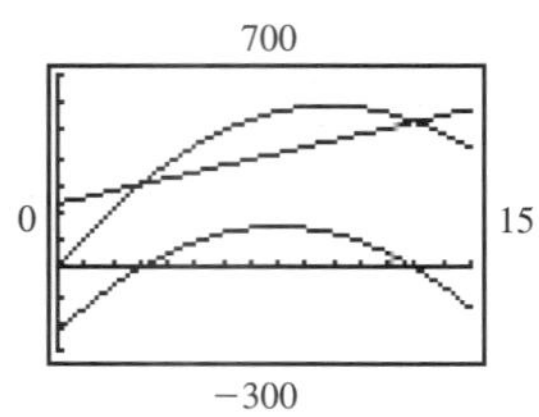

(C) The x coordinates of the intersection points of R and C are the same as the x intercepts of P.
(D) x intercepts and break-even points: 3.000 million (3,000,000) and 13.000 million (13,000,000) chips
(E) Same as (D)
(F) The maximum revenue and the maximum profit do not occur at the same production level; they are not equal. Maximizing profit takes both revenue and cost into account. Maximizing revenue does not consider the production costs.
(G) Maximum profit is 150.000 million dollars ($150,000,000) at an output of 8.000 million (8,000,000) chips. This is much smaller than the maximum revenue of 590.042 million dollars ($590,042,000).
(H) Same as (G).

53. (A) $x = 0.14$ cm (B) $x = 0.14$ cm

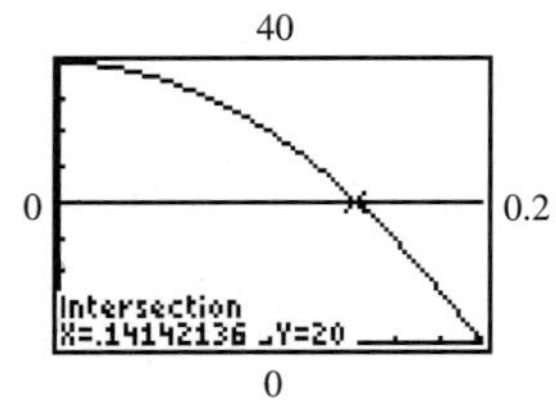

Review Exercise

1. *(1-1)*

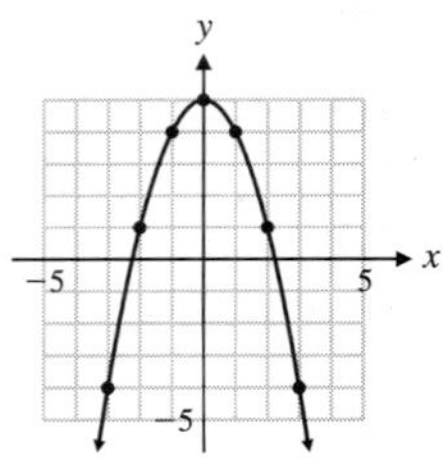

2. (A) Not a function (B) A function (C) A function (D) Not a function *(1-1)*
3. (A) -2 (B) -8 (C) 0 (D) Not defined *(1-1)*
4. (A) $y = 4$ (B) $x = 0$ (C) $y = 1$ (D) $x = -1$ or 1 (E) $y = -2$
(F) $y = -5$ or 5 *(1-1)*

5. (A)

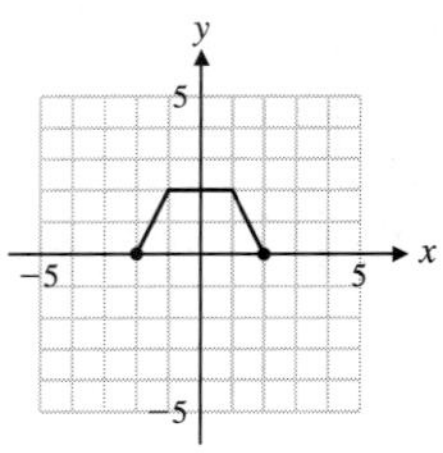

(B)

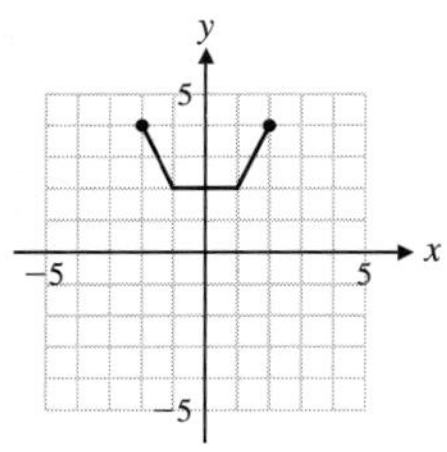

(C)

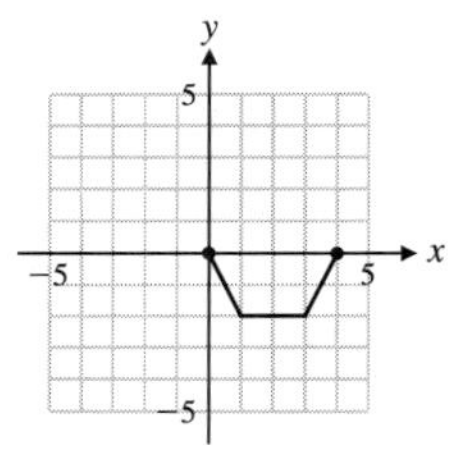

(D)

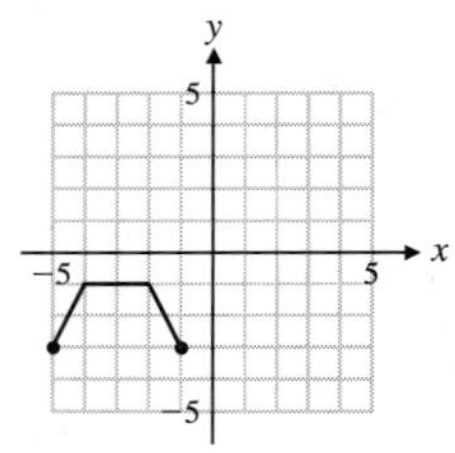

(1-2)

6. (A) n (B) g (C) m; slope is zero (D) f; slope is not defined *(1-3)* **7.** $y = -\frac{2}{3}x + 6$ *(1-3)*
8. Vertical line: $x = -6$; Horizontal line: $y = 5$ *(1-3)*
9. x intercept $= 9$; y intercept $= -6$; Slope $= \frac{2}{3}$ *(1-3)*

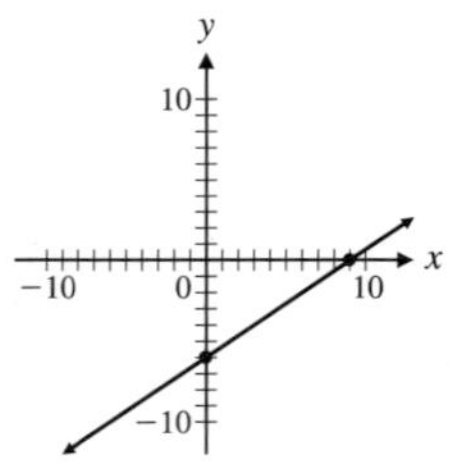

10. (B), (C), (D), (F) *(1-4)*
11. (A) g (B) m (C) n (D) f *(1-2, 1-4)*
12. (A) x intercepts: $-4, 0$; y intercept: 0
(B) Vertex: $(-2, -4)$ (C) Minimum: -4
(D) Range: $y \geqslant -4$ or $[-4, \infty)$
(E) Increasing on $[-2, \infty)$
(F) Decreasing on $(-\infty, -2]$ *(1-4)*
13. Linear functions: (A), (C), (E), (F);
Constant function: (D) *(1-3)*
14. (A) All real numbers except $x = -2$ and 3
(B) $x < 5$ *(1-1)*

15. Function g multiplies a domain element by 2 and then subtracts 3 times the square root of the domain element from the result. *(1-1)*

16. The graph of $x = -3$ is a vertical line with x intercept -3, and the graph of $y = 2$ is a horizontal line with y intercept 2. *(1-3)*

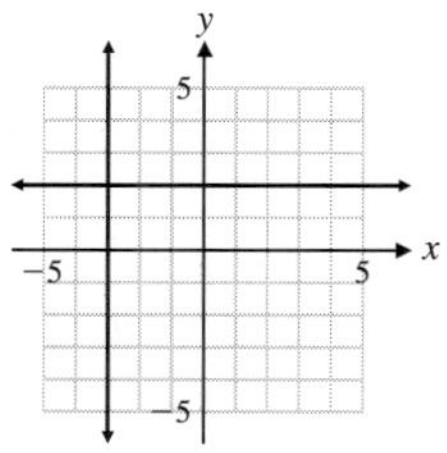

17. (A)

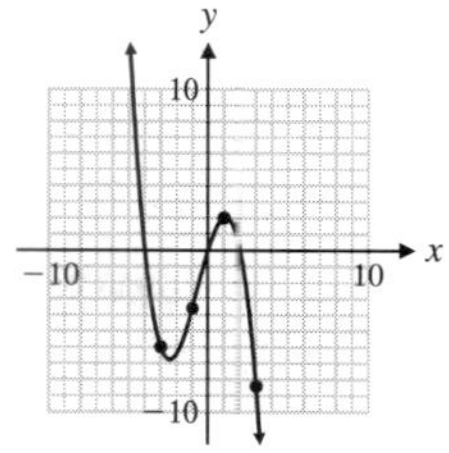

(B) The equation $f(x) = 3$ has one solution; the equation $f(x) = 2$ has two solutions; and the equation $f(x) = 1$ has three solutions.
(C) $f(x) = 3$: $x = -4.28$; $f(x) = 2$: $x = -4.19, 1$; $f(x) = 1$: $x = -4.10, 0.35, 1.75$ *(1-1)*

18. -2 *(1-1)* **19.** $2a + h - 3$ *(1-1)*

20. The graph of function m is the graph of $y = |x|$ reflected in the x axis and shifted to the right 4 units. *(1-2)*

21. The graph of function g is the graph of $y = x^3$ vertically contracted by a factor of 0.3 and shifted up 3 units. *(1-2)*

22. The graph of $y = x^2$ is vertically expanded by a factor of 2, reflected in the x axis, and shifted to the left 3 units. Equation: $y = -2(x + 3)^2$. *(1-2)*

23. $f(x) = 2\sqrt{x + 3} - 1$ *(1-2)*

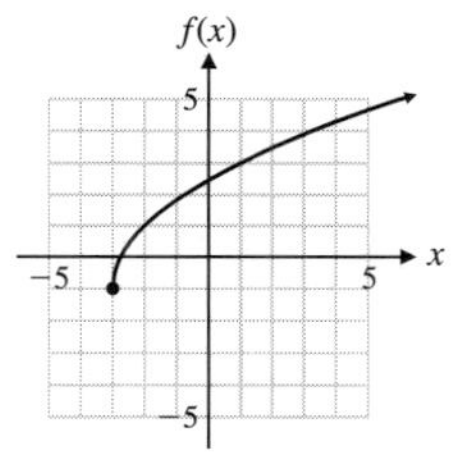

24. (A)

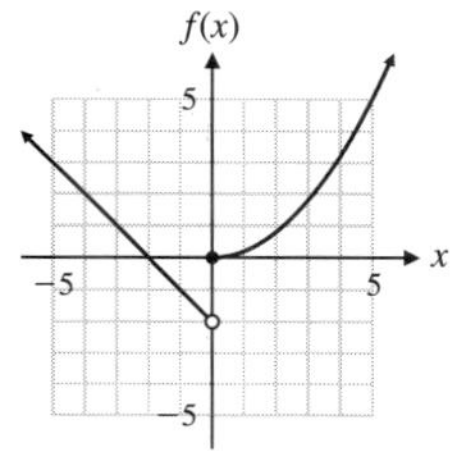

(B) Discontinuous at $x = 0$

(C)

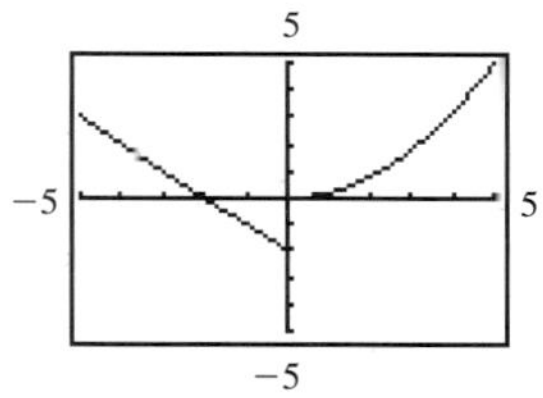

(1-2)

25. (A) $y = -\frac{2}{3}x$ (B) $y = 3$ *(1-3)* **26.** (A) $3x + 2y = 1$ (B) $y = 5$ (C) $x = -2$ *(1-3)*

27. $y = -(x - 4)^2 + 3$ *(1-2, 1-4)*

28. $f(x) = -0.4(x - 4)^2 + 7.6$ (A) x intercepts: $-0.4, 8.4$; y intercept: 1.2 (B) Vertex: (4.0, 7.6) (C) Maximum: 7.6 (D) Range: $x \leq 7.6$ or $(-\infty, 7.6]$ *(1-4)*

29.

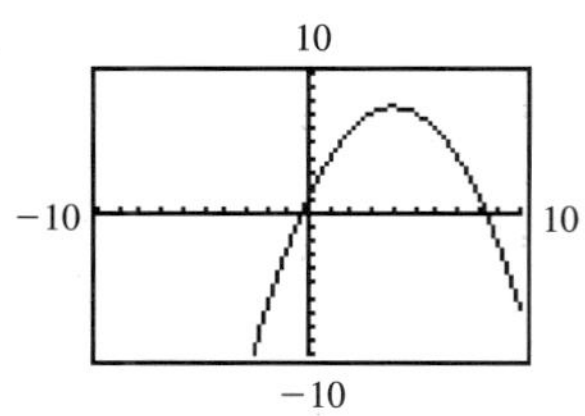

(A) x intercepts: $-0.4, 8.4$; y intercept: 1.2
(B) Vertex: (4.0, 7.6)
(C) Maximum: 7.6
(D) Range: $x \leq 7.6$ or $(-\infty, 7.6]$ *(1-4)*

30. The graph of $y = \sqrt[3]{x}$ is vertically expanded by a factor of 2, reflected in the x axis, and shifted 1 unit left and 1 unit down. Equation: $y = -2\sqrt[3]{x + 1} - 1$. *(1-2)*

31. The graphs appear to be perpendicular to each other. (It can be shown that if the slopes of two slant lines are the negative reciprocals of each other, then the two lines are perpendicular.) *(1-3)*

32. (A) $\dfrac{1}{\sqrt{x + h} + \sqrt{x}}$ (B) $\dfrac{-1}{x(x + h)}$ *(1-2)*

33. $G(x) = 0.3(x + 2)^2 - 8.1$ (A) x intercepts: $-7.2, 3.2$; y intercept: -6.9 (B) Vertex: $(-2, -8.1)$ (C) Minimum: -8.1 (D) Range: $x \geq -8.1$ or $[-8.1, \infty)$ (E) Decreasing: $(-\infty, -2]$; Increasing: $[-2, \infty)$ *(1-4)*

34.

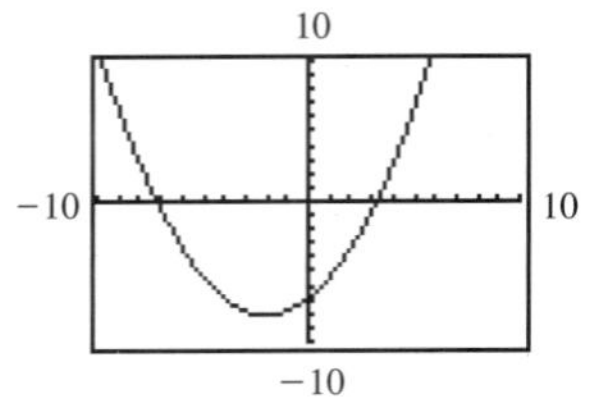

(A) x intercepts: $-7.2, 3.2$;
y intercept: -6.9
(B) Vertex: $(-2, -8.1)$
(C) Minimum: -8.1
(D) Range: $x \geq -8.1$ or $[-8.1, \infty)$
(E) Decreasing: $(-\infty, -2]$;
Increasing: $[-2, \infty)$ *(1-4)*

35. (A) $V(t) = -1{,}250t + 12{,}000, 0 \leq t \leq 8$

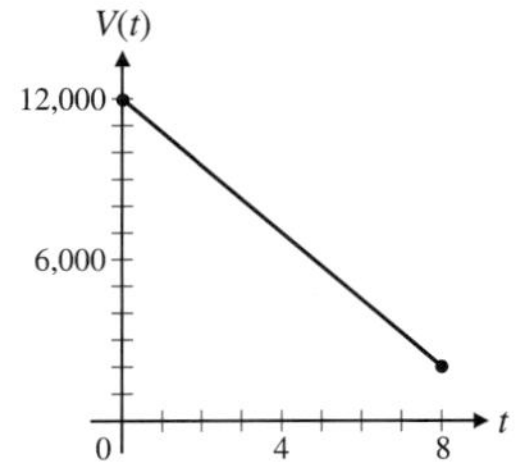

(B) $V(5) = \$5{,}750$ *(1-3)*

36. (A)

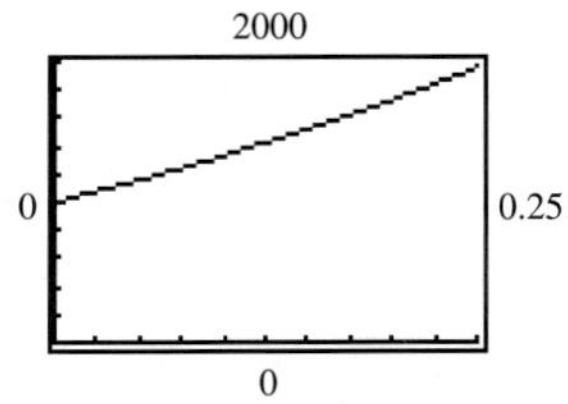

(B) $r = 0.1447$ or 14.47% compounded annually *(1-1, 1-2)*

37. (A) $R = 1.6C$ (B) \$192 (C) \$110 (D) 1.6; the slope indicates the change in retail price per unit change in cost. *(1-3)*

38. (A)

x	0	5	10	15	20
CONSUMPTION	309	276	271	255	233
$f(x)$	303	286	269	252	234

(B)

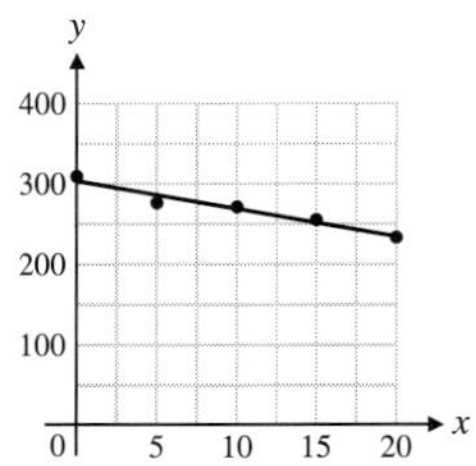

(C) 217 eggs per capita in 1995; 200 eggs per capita in 2000
(D) Per capita egg consumption is dropping by about 17 eggs every 5 years. *(1-3)*

39. (A) $C(x) = \begin{cases} 0.49x & \text{for } 0 \leq x < 36 \\ 0.44x & \text{for } 36 \leq x < 72 \\ 0.39x & \text{for } 72 \leq x \end{cases}$ (B) Discontinuous at $x = 36$ and $x = 72$

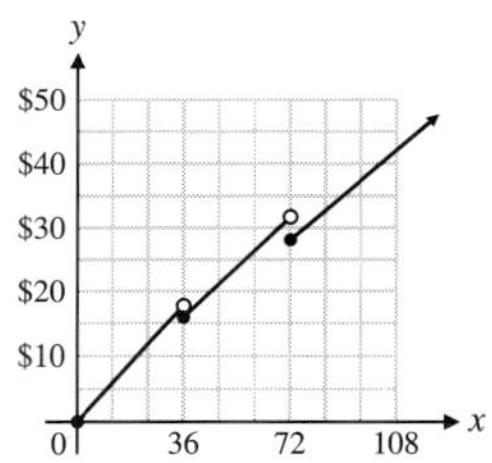

(1-2)

40. (A) $C = 84{,}000 + 15x$; $R = 50x$

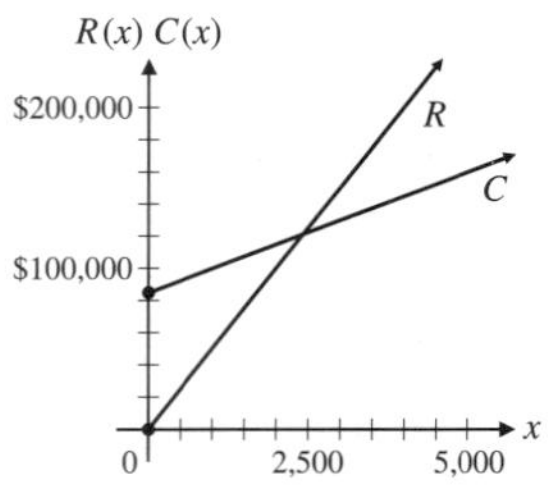

(B) $R = C$ at $x = 2{,}400$ units; $R < C$ for $0 \leq x < 2{,}400$; $R > C$ for $x > 2{,}400$
(C) Same as (B) *(1-3)*

41. (A)

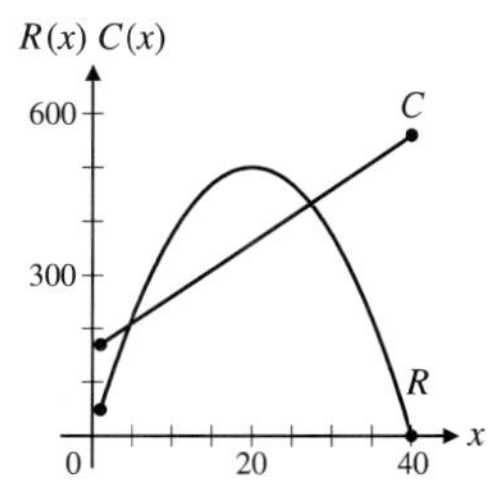

(B) $R = C$ for $x = 4.686$ thousand units (4,686 units) and for $x = 27.314$ thousand units (27,314 units); $R < C$ for $1 \leq x < 4.686$ or $27.314 < x \leq 40$; $R > C$ for $4.686 < x < 27.314$.
(C) Maximum revenue is 500 thousand dollars (\$500,000). This occurs at an output of 20 thousand units (20,000 units). At this output, the wholesale price is $p(20) = \$25$. *(1-3, 1-4)*

42. (A) $P(x) = R(x) - C(x) = x(50 - 1.25x) - (160 + 10x)$

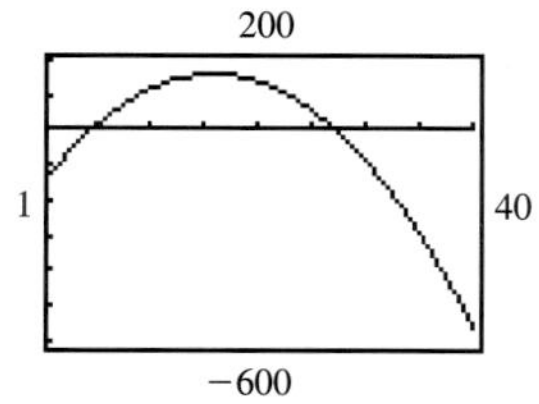

(B) $P = 0$ for $x = 4.686$ thousand units (4,686 units) and for $x = 27.314$ thousand units (27,314 units); $P < 0$ for $1 \leqslant x < 4.686$ or $27.314 < x \leqslant 40$; $P > 0$ for $4.686 < x < 27.314$.
(C) Maximum profit is 160 thousand dollars ($160,000). This occurs at an output of 16 thousand units (16,000 units). At this output, the wholesale price is $p(16) = \$30$. *(1-4)*

43. (A) $A(x) = -\frac{3}{2}x^2 + 420x$
(B) Domain: $0 \leqslant x \leqslant 280$

(C)

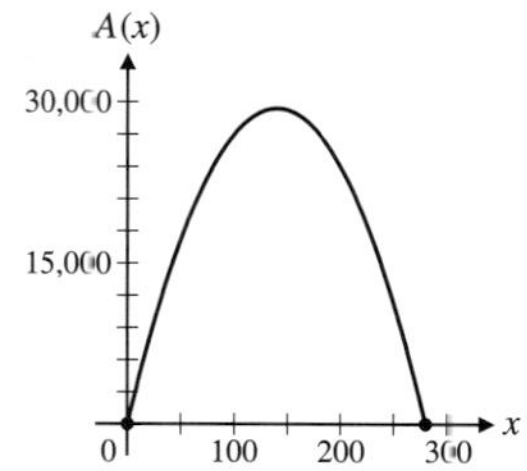

(D) There are two solutions to the equation $A(x) = 25{,}000$, one near 90 and another near 190.
(E) 86 ft; 194 ft
(F) Maximum combined area is 29,400 ft^2. This occurs for $x = 140$ ft and $y = 105$ ft. *(1-4)*

44. (A) $P(x) = 15x + 20$
(B) 95

(C)

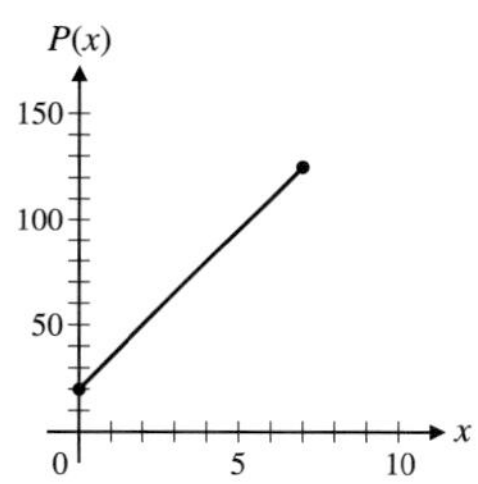

(D) 15 *(1-4)*

45. (A) 1 lb; 3 lb

(B)

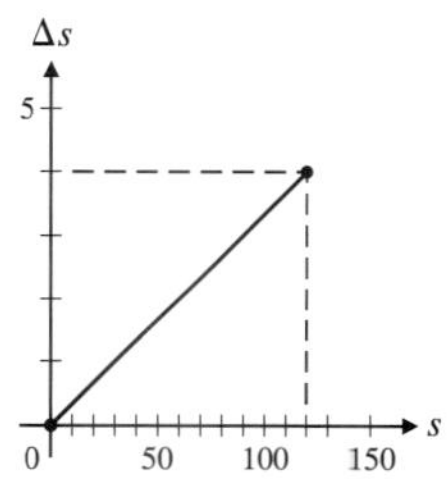

(C) $\frac{1}{30}$ *(1-4)*

Chapter 2

Exercise 2-1

1. (A) 2 (B) 1 (C) 2 (D) 0 (E) 1 (F) 1 **3.** (A) 5 (B) 4 (C) 5 (D) 1 (E) 1 (F) 1
5. (A) 6 (B) 5 (C) 6 (D) 0 (E) 1 (F) 1 **7.** (A) 3 (B) 4 (C) Negative **9.** (A) 4 (B) 5 (C) Negative
11. (A) 0 (B) 1 (C) Negative **13.** (A) 5 (B) 6 (C) Positive
15. (A) x intercept: -2; y intercept: -1 (B) Domain: all real numbers except 2

(C) Vertical asymptote: $x = 2$; horizontal asymptote: $y = 1$ (D)

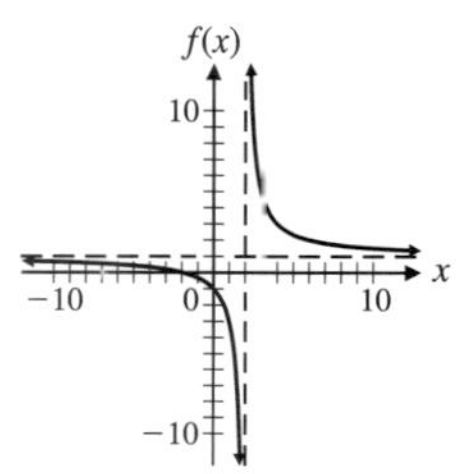

(E)

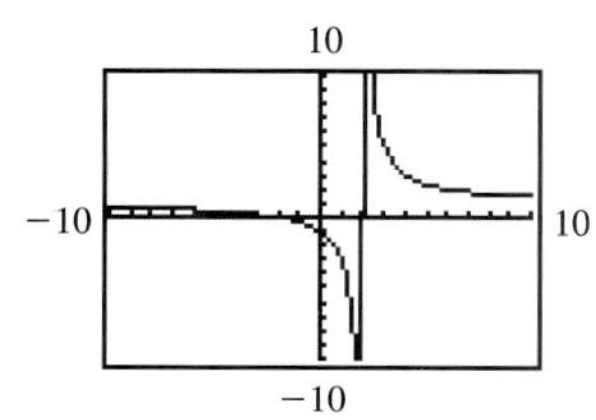

17. (A) x intercept: 0; y intercept: 0 (B) Domain: all real numbers except -2

(C) Vertical asymptote: $x = -2$; horizontal asymptote: $y = 3$ (D)

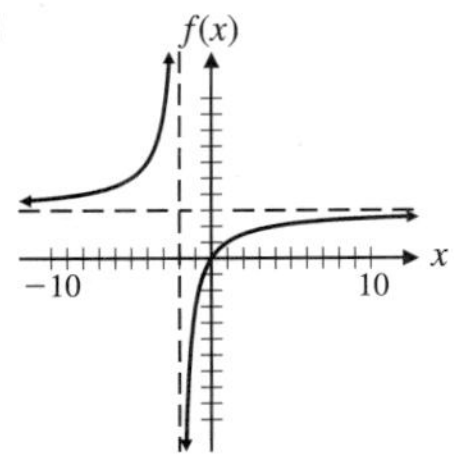

(E)

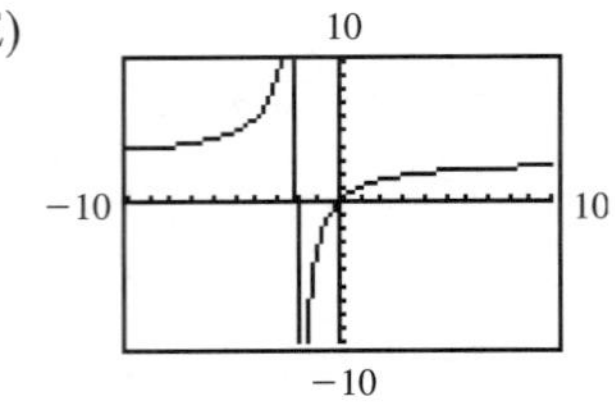

19. (A) x intercept: 2; y intercept: -1 (B) Domain: all real numbers except 4

(C) Vertical asymptote: $x = 4$; horizontal asymptote: $y = -2$ (D)

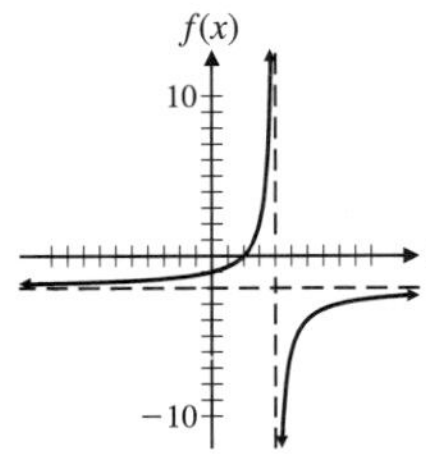

(E)

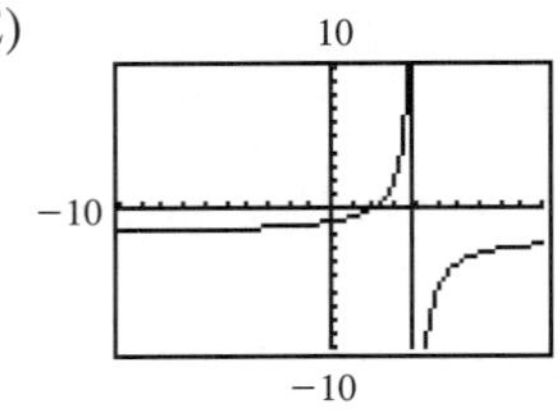

21. The graph will more and more like the graph of $y = 2x^4$. **23.** The graph will look more and more like the graph of $y = -x^5$.

25. (A)

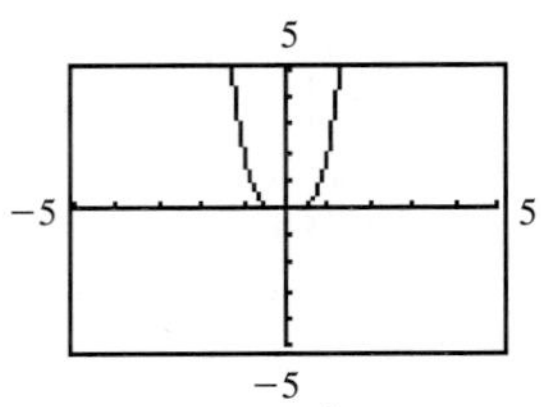

$y = 2x^4$

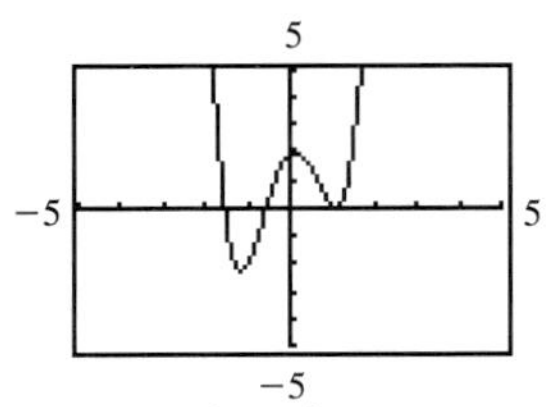

$y = 2x^4 - 5x^2 + x + 2$

(B)

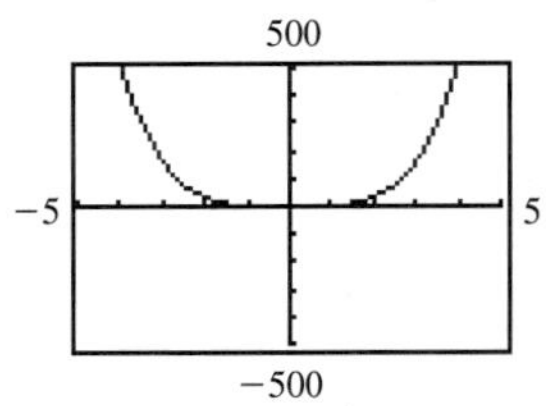

$y = 2x^4$

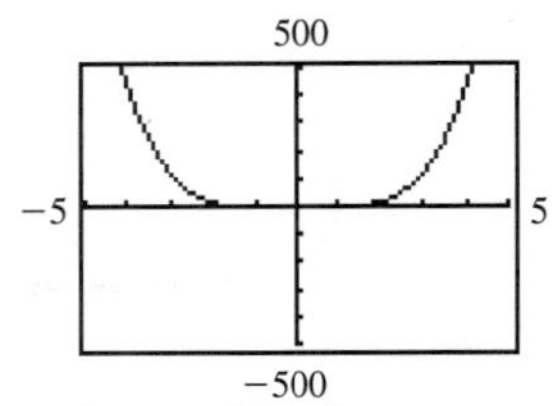

$y = 2x^4 - 5x^2 + x + 2$

27. (A)

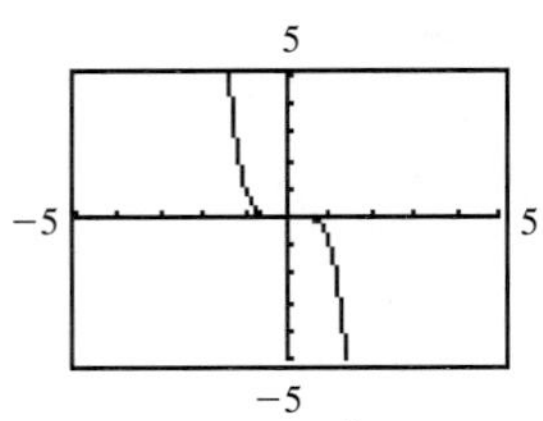

$y = -x^5$

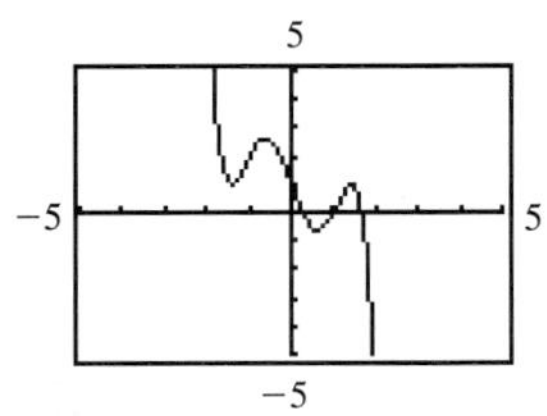

$y = -x^5 + 4x^3 - 4x + 1$

(B)

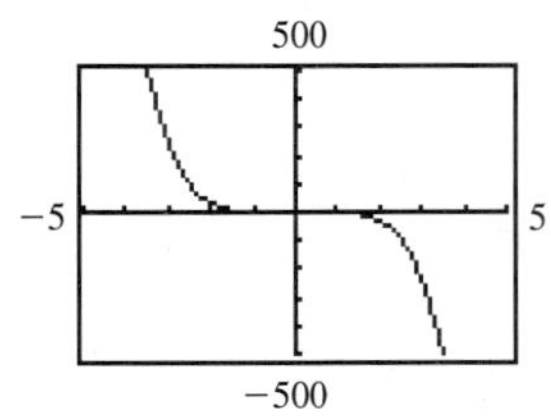

$y = -x^5$

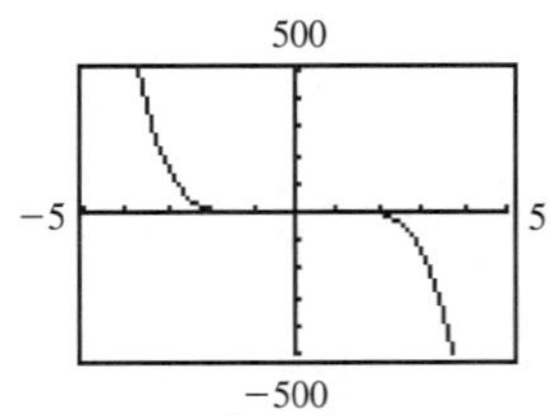

$y = -x^5 + 4x^3 - 4x + 1$

29. $x = 2.58$ **31.** $x = -2.68, -1.16$ **33.** $x = -0.84, 1.26, 11.29$

37. (A) x intercept; 0; y intercept; 0 (B) Vertical asymptotes: $x = -2$, $x = 3$; Horizontal asymptote: $y = 2$

(C)

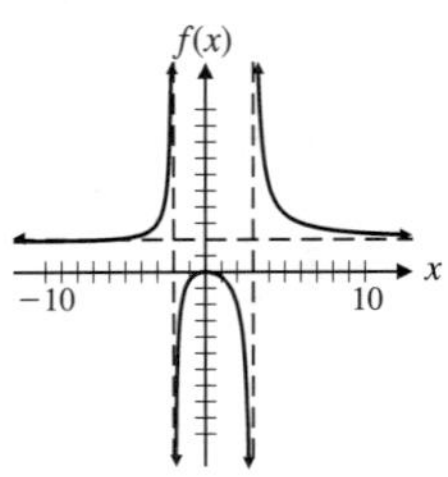

(D)

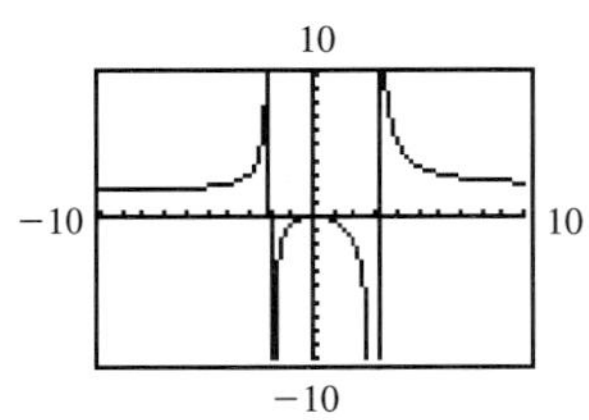

39. (A) x intercept: $\pm\sqrt{3}$; y intercept: $-\frac{2}{3}$ (B) Vertical asymptotes: $x = -3$, $x = 3$; Horizontal asymptote: $y = -2$

(C) 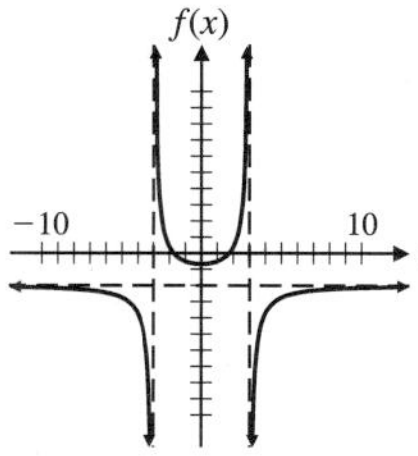

(D) 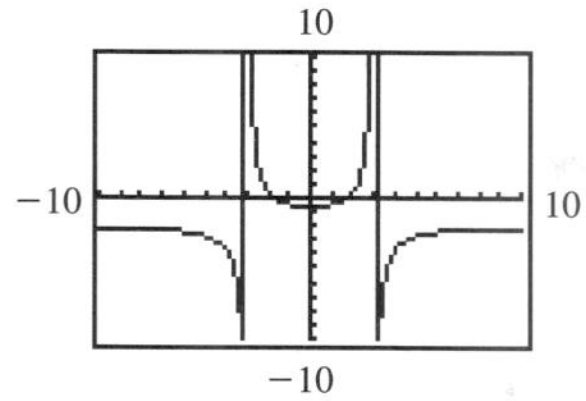

41. (A) x intercept: 0; y intercept: 0 (B) Vertical asymptotes: $x = -3$, $x = 2$; Horizontal asymptote: $y = 0$

(C) 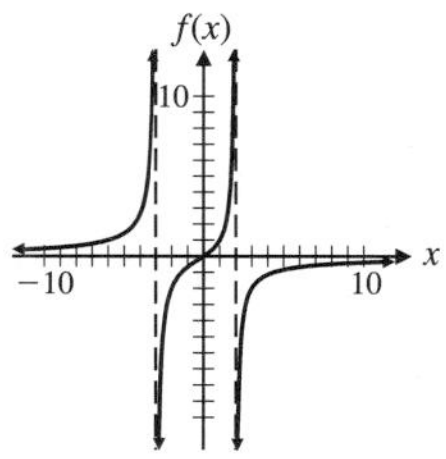

(D) 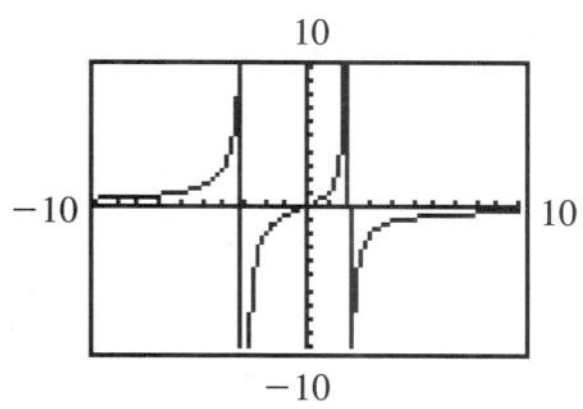

43. $f(x) = x^2 - x - 2$

45. $f(x) = 4x - x^3$

47. (A) $C(x) = 180x + 200$

(B) $\overline{C}(x) = \dfrac{180x + 200}{x}$

(C) 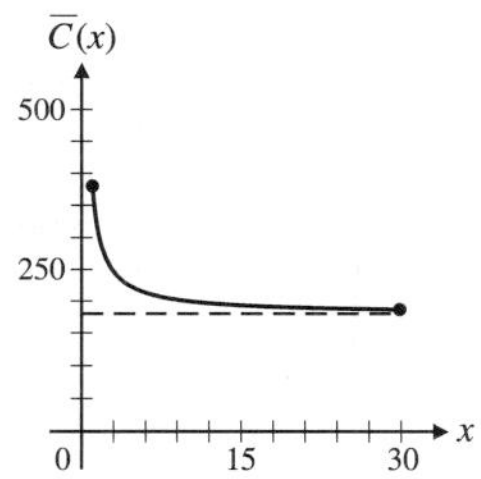

(D) $180 per board

49. (A) $\overline{C}(n) = \dfrac{2{,}500 + 175n + 25n^2}{n}$ (D) 10 yr; $675.00 per yr

(B) 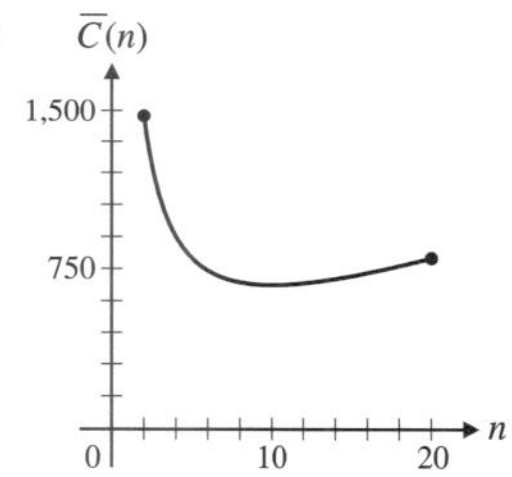

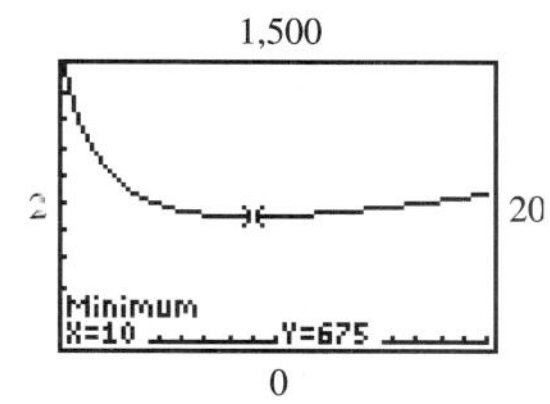

(C) 10 yr; $675.00 per yr

51. (A) $\overline{C}(x) = \dfrac{0.00048(x - 500)^3 + 60{,}000}{x}$

(B) 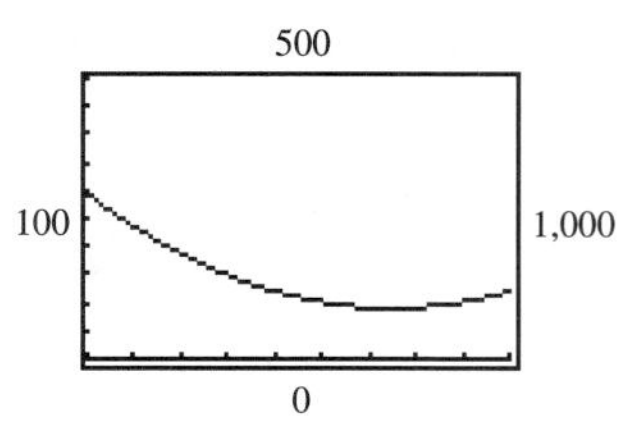

(C) 750 cases per month; $90 per case

53. (A) 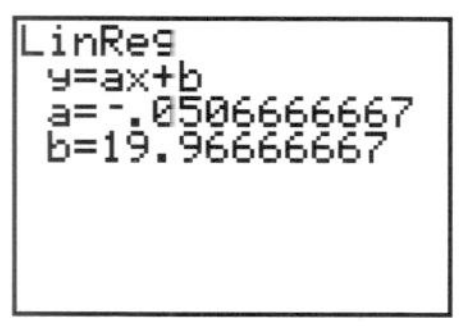

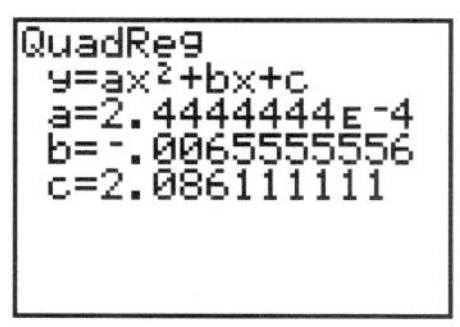

(B) $\overline{x} = 195$; $\overline{p} = \$10.09$

55. (A) 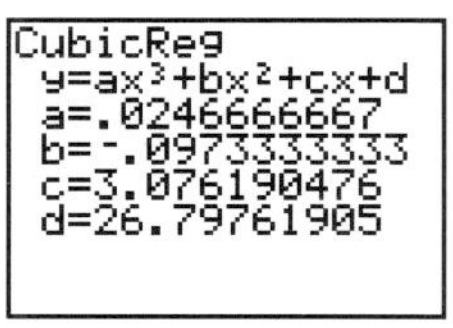

(B) $1,072.8 billion

57. (A) 0.06 cm/sec

(B) 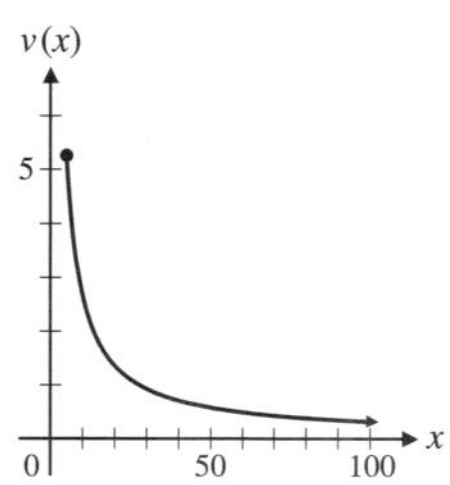

59. (A) 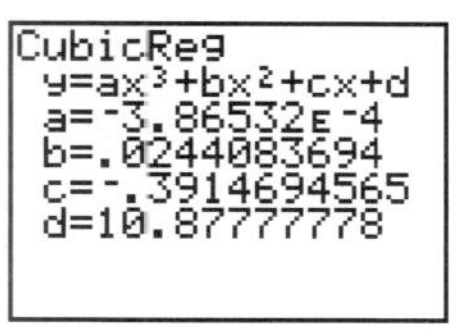

(B) 7.5 marriages per 1,000 population

Exercise 2-2

1.

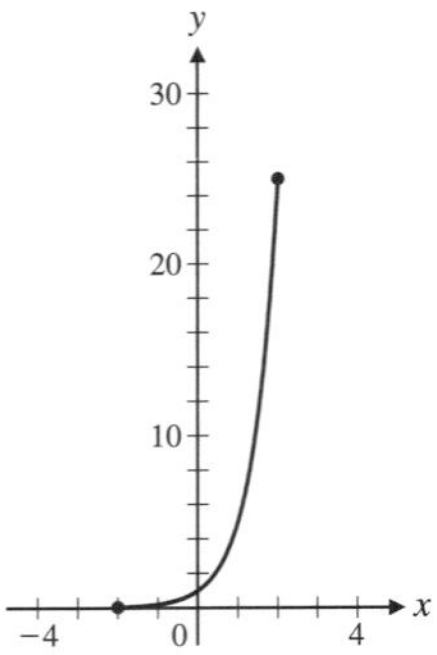

3.

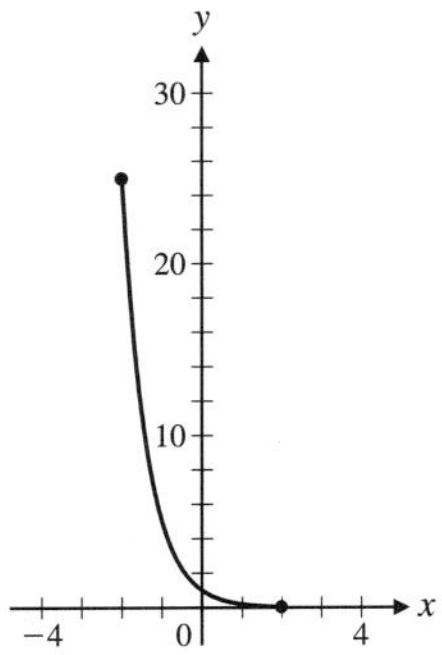

5.

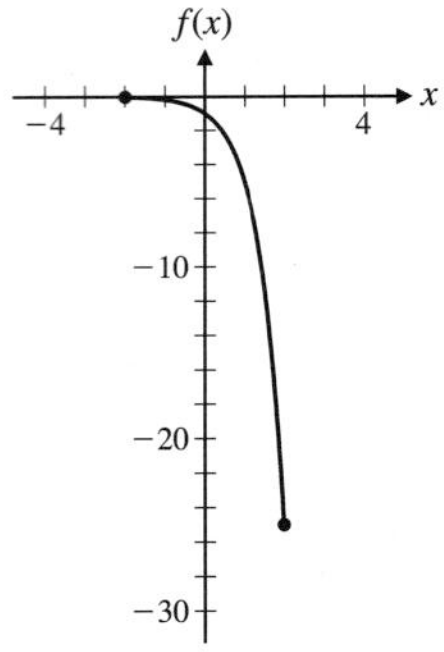

7.

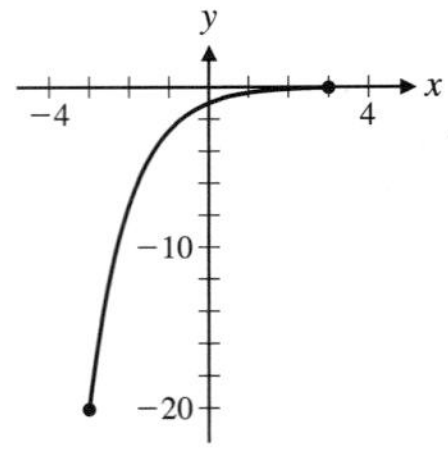

9.

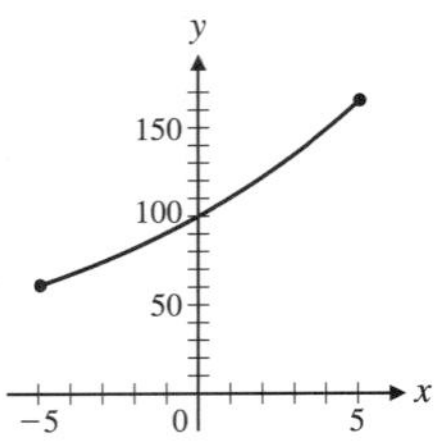

11.

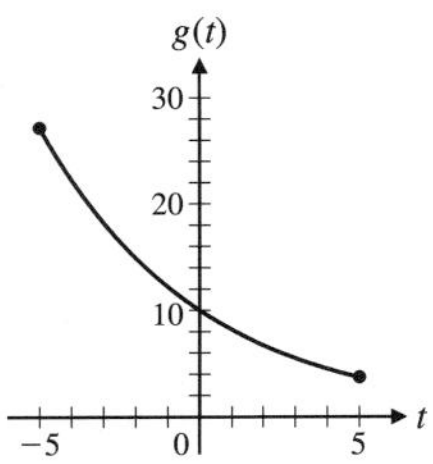

13. 4^{6xy} **15.** e **17.** $8e^{3.6t}$

19. The graph of g is a reflection of the graph *f* in the *x* axis.

21. The graph of *g* is the graph of *f* shifted 1 unit to the left.

23. The graph of *g* is the graph of *f* shifted 1 unit up.

25. The graph of *g* is the graph of *f* vertically expanded by a factor of 2 and shifted to the left 2 units.

27.

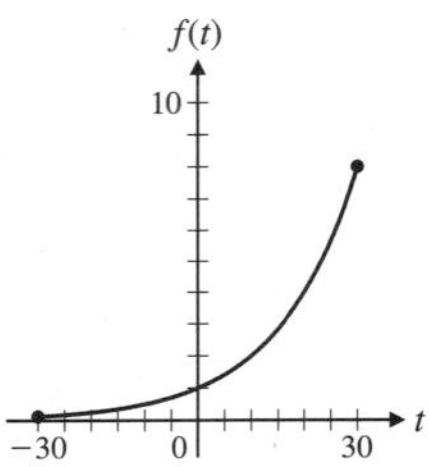

29.

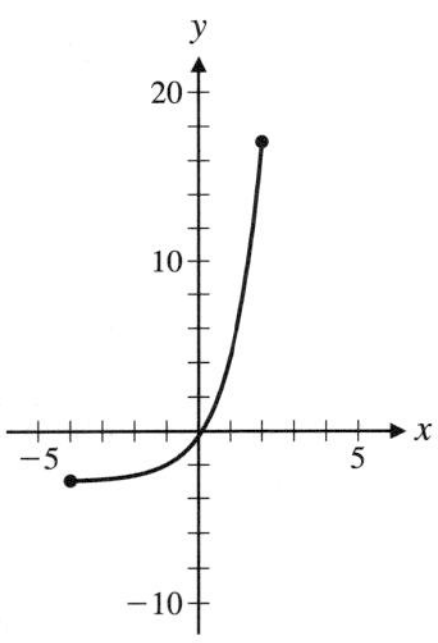

31.

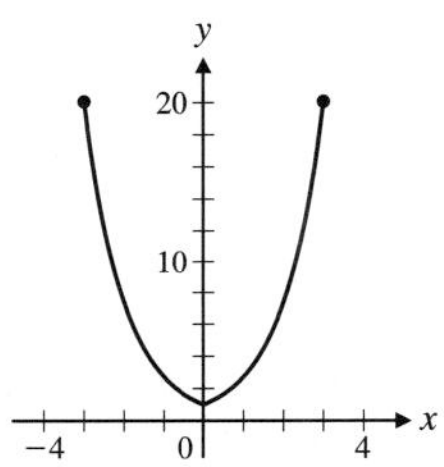

33.

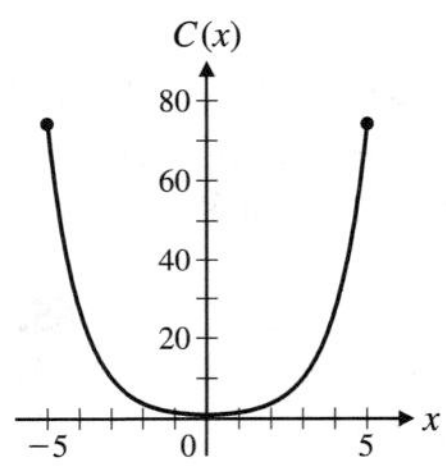

35.

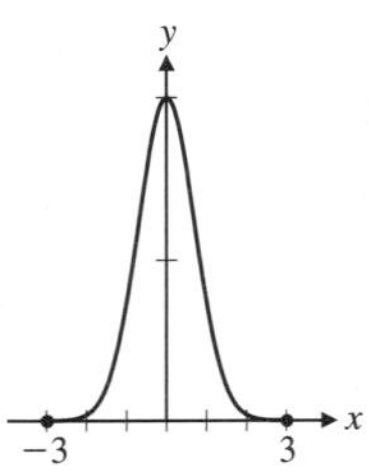

37. $a = 1, -1$ **39.** The top curve belongs to *f;* the bottom curve belongs to *g.*

41. The top curve belongs to *g;* the bottom curve belongs to *f.*

43. $x = 1$ **45.** $x = -1, 6$ **47.** $x = 3$ **49.** $x = 3$ **51.** $x = -3, 0$

53.

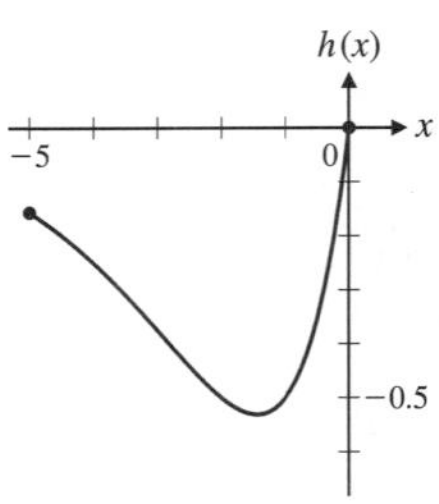

55.

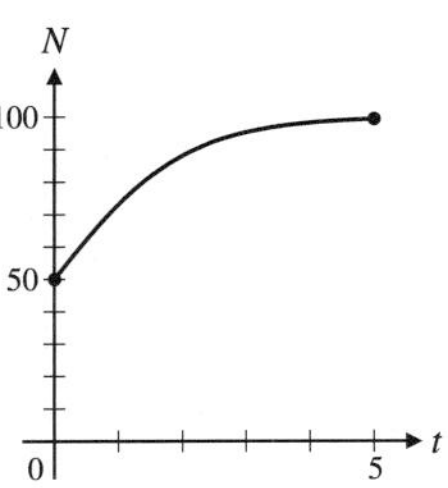

57. $x = 1.40$ **59.** $x = -0.73$

61. (A) \$2,633.56 (B) \$7,079.54 **63.** (A) \$11,871.65 (B) \$20,427.93 **65.** \$9,217

67. (A) \$10,850.88 (B) \$10,838.29 **69.** \$28,847.49

71. N approaches 2 as t increases without bound.

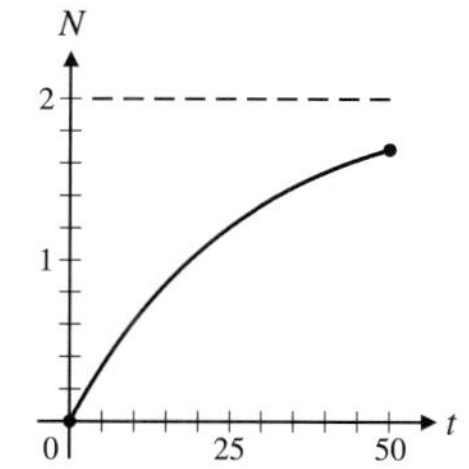

73. (A) 1995: \$673,000; 2010: \$5,162,000

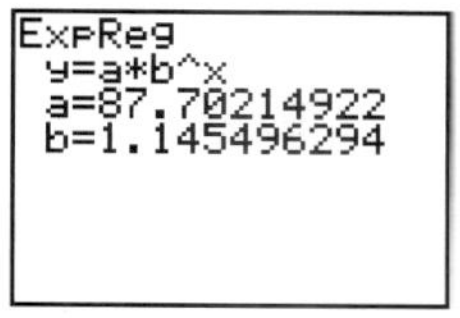

75. (A) 10%
(B) 1%

77. (A) $N = 40{,}000e^{0.21t}$
(B) 215,000; 613,000
(C)

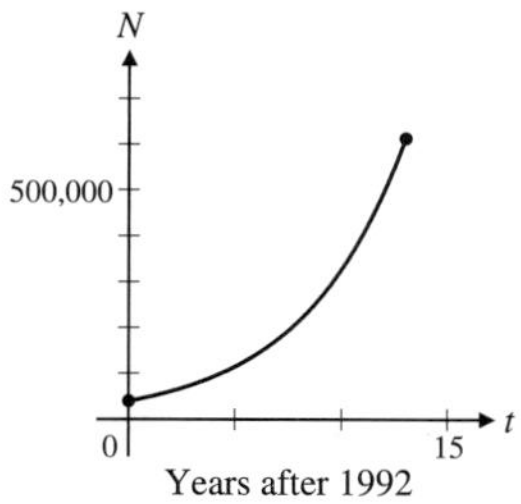

79. (A) $P = 5.7e^{0.0114t}$
(B) 6.8 billion; 8.5 billion
(C)

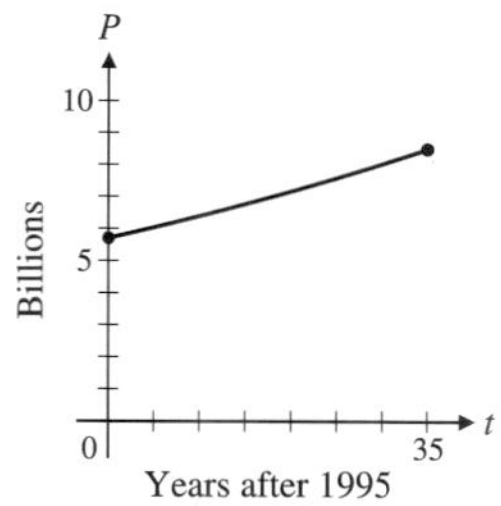

81. (A) 1995: 7,244,000; 1996: 13,507,000

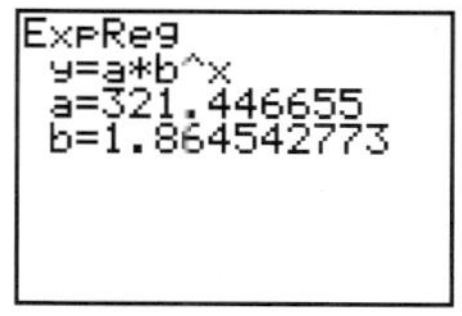

Exercise 2-3

1. $27 = 3^3$ **3.** $10^0 = 1$ **5.** $8 = 4^{3/2}$ **7.** $\log_7 49 = 2$ **9.** $\log_4 8 = \frac{3}{2}$ **11.** $\log_b A = u$ **13.** 0 **15.** 1
17. 1 **19.** 3 **21.** -3 **23.** 3 **25.** $\log_b P - \log_b Q$ **27.** $5 \log_b L$ **29.** $\log_b p - \log_b q - \log_b r - \log_b s$
31. $x = 9$ **33.** $y = 2$ **35.** $b = 10$ **37.** $x = 2$ **39.** $y = -2$ **41.** $b = 100$ **43.** $5 \log_b x - 3 \log_b y$
45. $\frac{1}{3} \log_b N$ **47.** $2 \log_b x + \frac{1}{3} \log_b y$ **49.** $\log_b 50 - 0.2t \log_b 2$ **51.** $\log_b P + t \log_b(1 + r)$ **53.** $\log_e 100 - 0.01t$
55. $x = 2$ **57.** $x = 8$ **59.** $x = 7$ **61.** No solution

63.

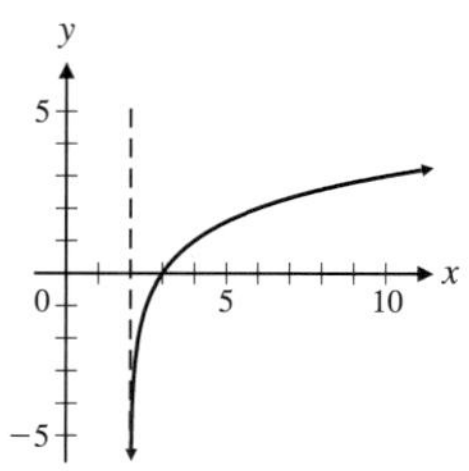

65. The graph of $y = \log_2(x - 2)$ is the graph of $y = \log_2 x$ shifted to the right 2 units.
67. Domain: $(-1, \infty)$; Range: All real numbers
69. (A) 3.547 43 (B) -2.160 32 (C) 5.626 29 (D) -3.197 04
71. (A) 13.4431 (B) 0.0089 (C) 16.0595 (D) 0.1514
73. 1.0792 **75.** 1.4595 **77.** 30.6589 **79.** 18.3559

81. Increasing: $(0, \infty)$

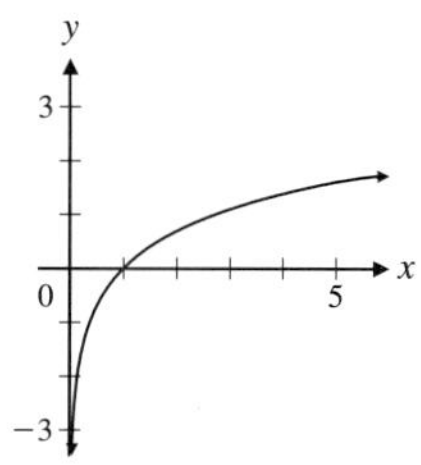

83. Decreasing: $(0, 1]$
Increasing: $[1, \infty)$

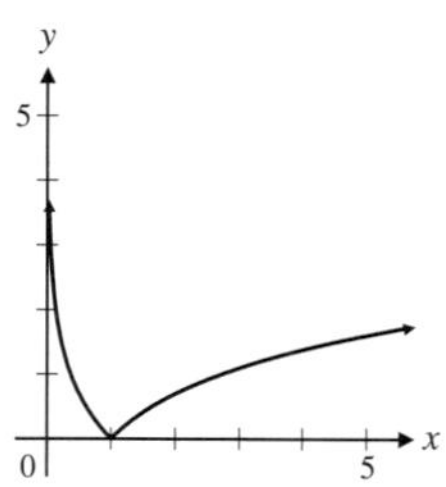

85. Increasing: $(-2, \infty)$

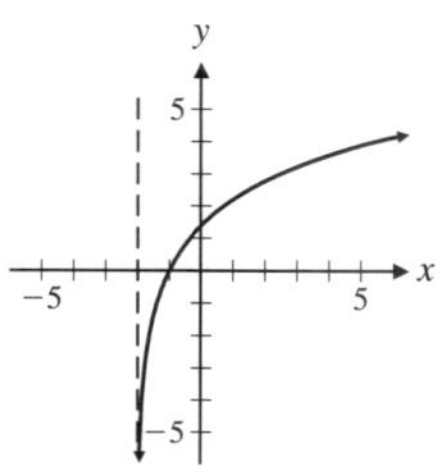

87. Increasing: $(0, \infty)$

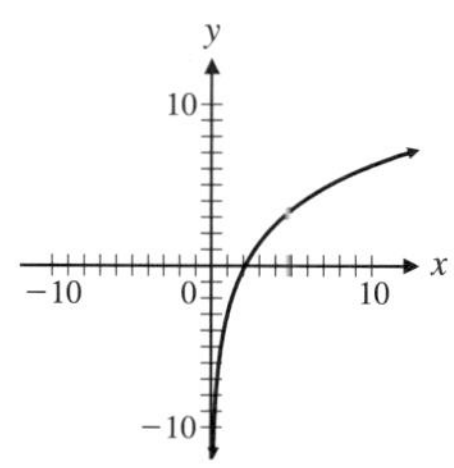

89. The calculator interprets log 13/7 as $\frac{\log 13}{7}$, not as $\log \frac{13}{7}$. **91.** Because $b^0 = 1$ for any permissible base b ($b > 0, b \neq 1$).
93. $y = c \cdot 10^{0.8x}$ **95.** $x > \sqrt{x} > \ln x$ for $1 < x \leq 16$ **97.** 12 yr **99.** 9.87 yr; 9.80 yr
101. 8.664% compounded continuously

103. (A) 5,373

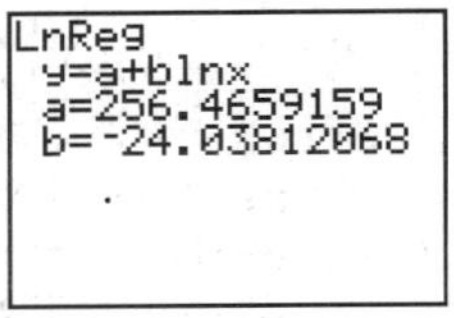

(B) 7,220

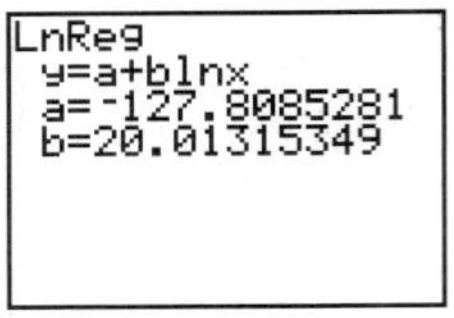

107. (A) 1996: 123.1 bushels/acre; 2010: 141.5 bushels/acre

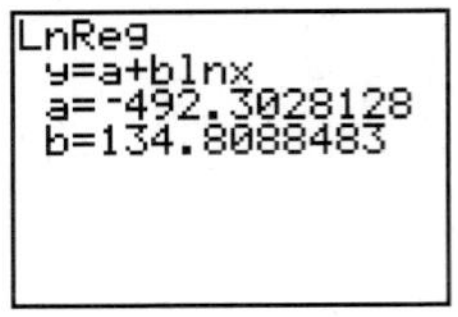

109. 901 yr

Review Exercise

1. $v = \ln u$ *(2-3)* **2.** $y = \log x$ *(2-3)* **3.** $M = e^N$ *(2-3)* **4.** $u = 10^v$ *(2-3)* **5.** 5^{2x} *(2-2)* **6.** e^{2u^2} *(2-2)*
7. $x = 9$ *(2-3)* **8.** $x = 6$ *(2-3)* **9.** $x = 4$ *(2-3)* **10.** $x = 2.157$ *(2-3)* **11.** $x = 13.128$ *(2-3)*
12. $x = 1{,}273.503$ *(2-3)* **13.** $x = 0.318$ *(2-3)* **14.** (A) 3 (B) 2 (C) 3 (D) 1 (E) 1 (F) 1 *(2-1)*
15. (A) 4 (B) 3 (C) 4 (D) 0 (E) 1 (F) 1 *(2-1)* **16.** (A) 2 (B) 3 (C) Positive *(2-1)*
17. (A) 3 (B) 4 (C) Negative *(2-1)*
18. (A) x intercept: -4; y intercept: -2 (B) All real numbers, except $x = 2$ (C) Vertical asymptote: $x = 2$; Horizontal asymptote: $y = 1$

(D)

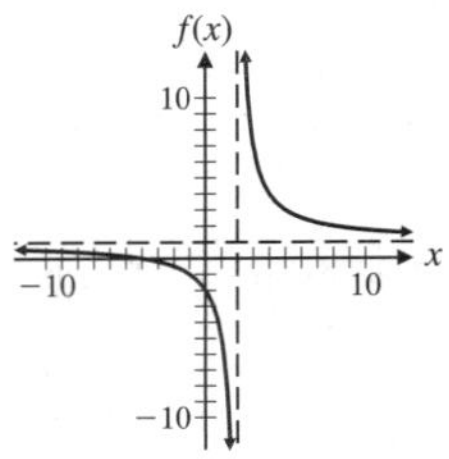

(E)

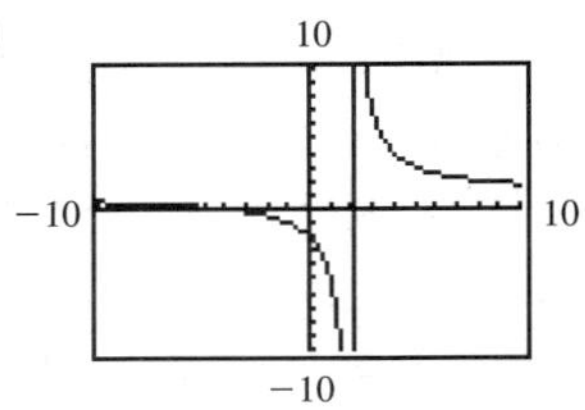

(2-1)

19. (A) x intercept: $\frac{4}{3}$; y intercept: -2 (B) All real numbers, except $x = -2$ (C) Vertical asymptote: $x = -2$; Horizontal asymptote: $y = 3$

(D)

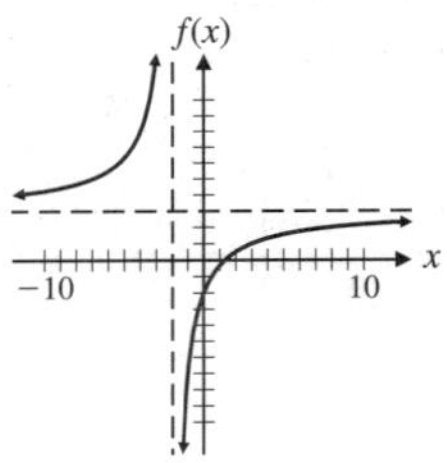

(E)

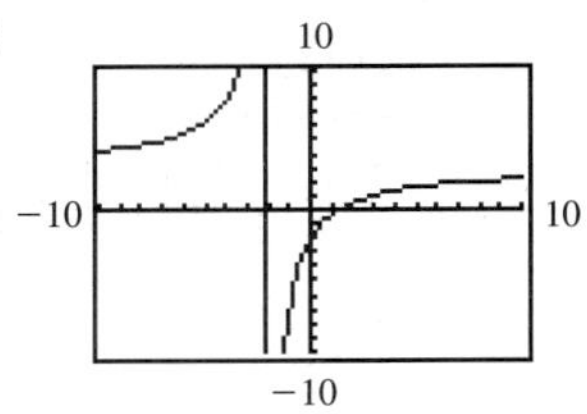

(2-1)

20. $x = 8$ *(2-3)* **21.** $x = 3$ *(2-3)* **22.** $x = 3$ *(2-2)* **23.** $x = -1, 3$ *(2-2)* **24.** $x = 0, \frac{3}{2}$ *(2-2)* **25.** $x = -2$ *(2-3)*
26. $x = \frac{1}{2}$ *(2-3)* **27.** $x = 27$ *(2-3)* **28.** $x = 13.3113$ *(2-3)* **29.** $x = 158.7552$ *(2-3)* **30.** $x = 0.0097$ *(2-3)*
31. $x = 1.4359$ *(2-3)* **32.** $x = 1.4650$ *(2-3)* **33.** $x = 92.1034$ *(2-3)* **34.** $x = 9.0065$ *(2-3)* **35.** $x = 2.1081$ *(2-3)*
36. $x = 2.8074$ *(2-3)* **37.** $x = -1.0387$ *(2-3)* **38.** They look very much alike. *(2-1)*

39. (A)

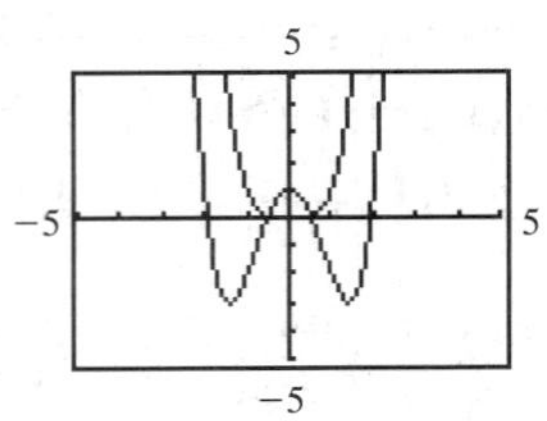

(B)

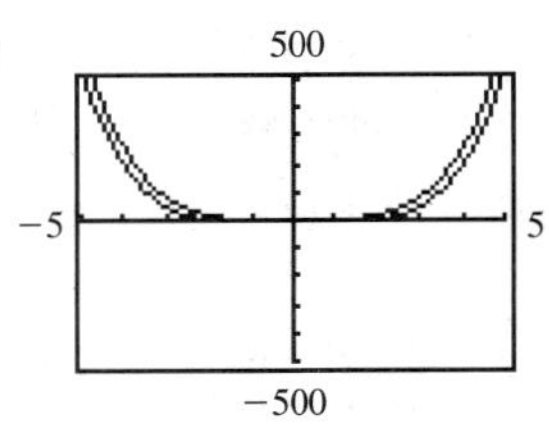

(2-1)

40. $x = -1.14, 6.78$ *(2-1)* **41.** $(-1.54, -0.79)$; $(0.69, 0.99)$ *(2-2, 2-3)* **42.** $1 + 2e^x - e^{-x}$ *(2-2)* **43.** $2e^{-2x} - 2$ *(2-2)*

44. Increasing: $[-2, 4]$ *(2-2)* **45.** Decreasing $[0, \infty)$ *(2-2)* **46.** Increasing: $(-1, 10]$ *(2-3)*

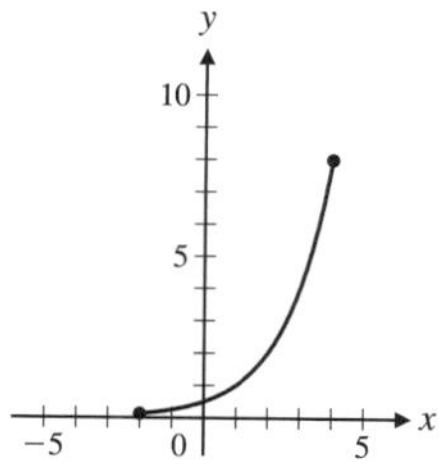

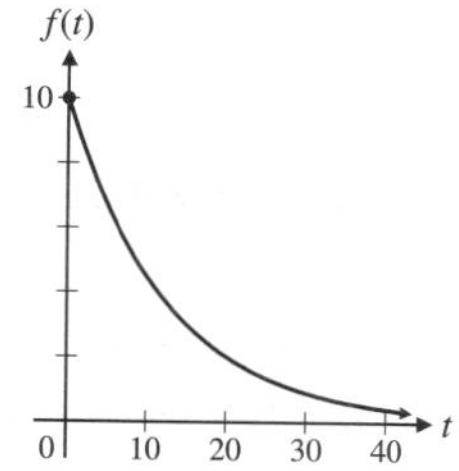

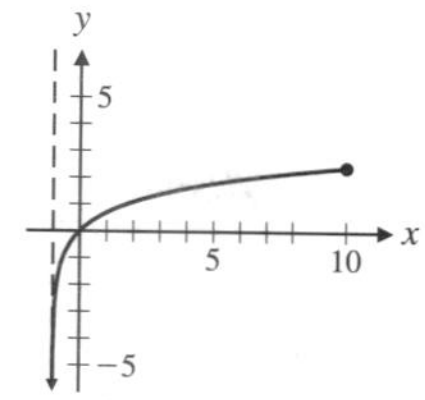

47. $\log 10^{\pi} = \pi$ and $10^{\log\sqrt{2}} = \sqrt{2}$; $\ln e^{\pi} = \pi$ and $e^{\ln\sqrt{2}} = \sqrt{2}$ *(2-3)* **48.** $x = 2$ *(2-3)* **49.** $x = 2$ *(2-3)*
50. $x = 1$ *(2-3)* **51.** $x = 300$ *(2-3)* **52.** $y = ce^{-5t}$ *(2-3)*
53. If $\log_1 x = y$, then $1^y = x$; that is, $1 = x$ for all positive real numbers x, which is not possible. *(2-3)*
54. \$10,263.65 *(2-2)* **55.** \$10,272.17 *(2-2)* **56.** 8 yr *(2-2, 2-3)* **57.** 6.93 yr *(2-2, 2-3)*

58. (A) $C(x) = 40x + 300$; $\overline{C}(x) = \dfrac{40x + 300}{x}$

(B)

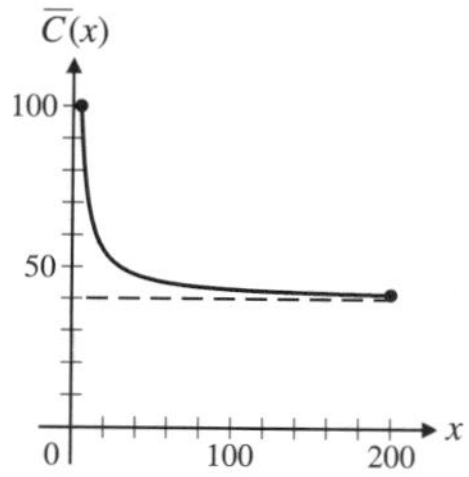

(C) $y = 40$ (D) \$40 per pair *(2-1)*

59. (A) $\overline{C}(x) = \dfrac{20x^3 - 360x^2 + 2{,}300x - 1{,}000}{x}$

(B)

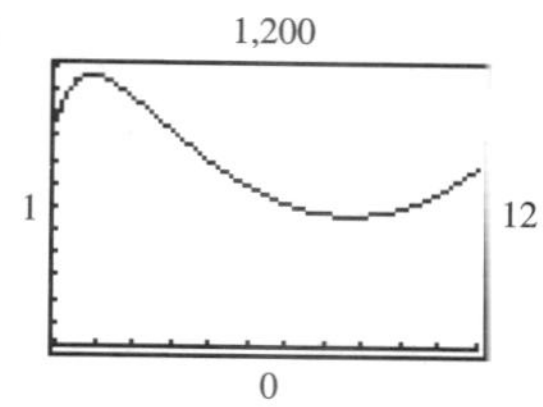

(C) 8.667 thousand cases (8,667)
(D) \$567 per case *(2-1)*

60. (A) 2,833 sets

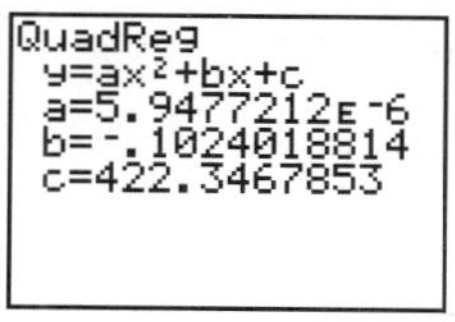

(B) 4,836

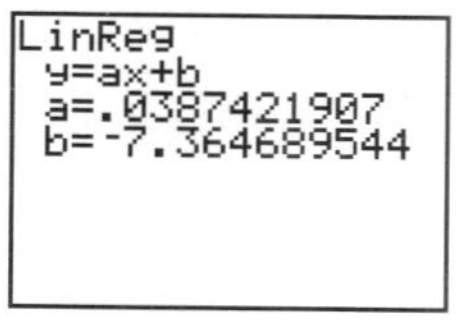

(D) Equilibrium price: \$131.59; Equilibrium quantity: 3,587 cookware sets *(2-1)*
61. (A) $N = 2^{2t}$ or $N = 4^t$ (B) 15 days *(2-2, 2-3)* **62.** $k = 0.009\ 42$; 489 ft *(2-2, 2-3)*
63. (A) 1996: 1,724 million bushels; 2010: 2,426 million bushels. *(2-3)*

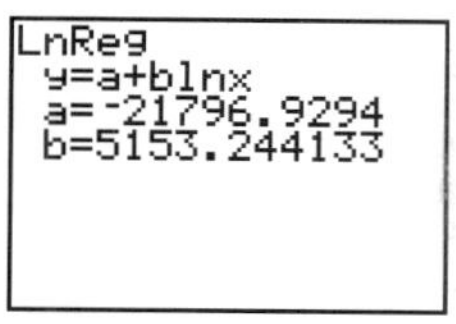

64. 23.4 yr *(2-2, 2-3)* **65.** 23.1 yr *(2-2, 2-3)* **66.** (A) 1996: \$207 billion; 2010: \$886 billion (B) Midway through 2004 *(2-2)*

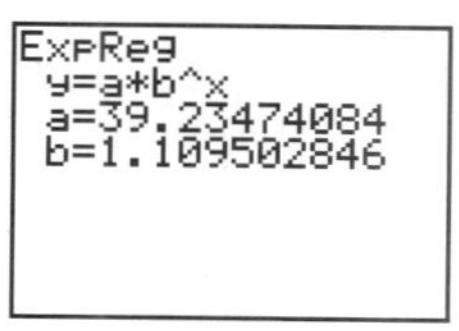

CHAPTER 3

Exercise 3-1

1. 0.095; $\frac{1}{6}$ yr **3.** 18%; $\frac{5}{12}$ yr **5.** $I = \$20$ **7.** $r = 0.08$ or 8% **9.** $A = \$112$ **11.** $P = \$888.89$
13. $r = I/Pt$ **15.** $P = A/(1 + rt)$
17. The graphs are linear, all with y intercept \$1,000; their slopes are 40, 80, and 120, respectively. **19.** \$140 **21.** \$9.23
23. \$7,685.00 **25.** 10.125% **27.** 18% **29.** \$1,680; 36% **31.** 9.126% **33.** \$9,693.91 **35.** 13.538%
37. 21.335% **39.** 11.161%

Exercise 3-2

1. $A = \$112.68$ **3.** $A = \$3{,}433.50$ **5.** $P = \$2{,}419.99$ **7.** $P = \$7{,}351.04$ **9.** 0.75% per month
11. 1.75% per quarter **13.** 9.6% compounded monthly **15.** 9% compounded semiannually
17. (A) \$126.25; \$26.25 (B) \$126.90; \$26.90 (C) \$127.05; \$27.05 **19.** (A) \$7,147.51 (B) \$10,217.39
21. All three graphs are increasing, curve upward, and have the same y intercept; the greater the interest rate, the greater the increase. The amounts at the end of 8 years are \$1,376.40, \$1,892.46, and \$2,599.27, respectively.

23.

PERIOD	INTEREST	AMOUNT
0		\$1,000.00
1	\$97.50	\$1,097.50
2	\$107.01	\$1,204.51
3	\$117.44	\$1,321.95
4	\$128.89	\$1,450.84
5	\$141.46	\$1,592.29
6	\$155.25	\$1,747.54

25. (A) \$6,755.64 (B) \$4,563.87
27. (A) 10.38% (B) 12.68%
29. $5\frac{1}{2}$ yr
31. $n \approx 12$
33. (A) $7\frac{1}{4}$ yr (B) 6 yr
35. \$22,702.60
37. \$196,993.25
39. \$14.26/ft^2/mo
41. 18 yr

43. 9% compounded monthly, since its effective rate is 9.38%; the effective rate of 9.3% compounded annually is 9.3%
45. (A) In 1998, 222 years after the signing, it would be worth \$76,139.26.
(B) If interest were compounded monthly, daily, or continuously, it would be worth \$77,409.05, \$78,033.73, or \$78,055.09, respectively.
(C)

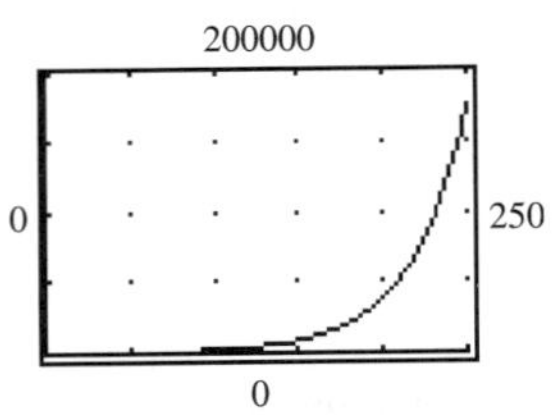

47. 2 yr, 10 mo **49.** \$328,791.70 **51.** 4.952 yr; 4.959 yr

53.

YEARS	EXACT RATE	RULE OF 72
6	12.2	12.0
7	10.4	10.3
8	9.1	9.0
9	8.0	8.0
10	7.2	7.2
11	6.5	6.5
12	5.9	6.0

55. 23 quarters

57. To maximize earnings, choose 10% simple interest for investments lasting less than 11 years and 7% compound interest for investments lasting 11 years or more.
59. 7.14% **61.** $5,935.34 **63.** 9.08% **65.** (A) 8.60% (B) 8.60% (C) 8.57% **67.** 14.48% **69.** 7.95%

Exercise 3-3

1. $FV = \$13{,}435.19$ **3.** $FV = \$60{,}401.98$ **5.** $PMT = \$123.47$ **7.** $PMT = \$310.62$ **9.** $n = 17$ **11.** $i = 0.09$
13. Value: $30,200.99; Interest: $10,200.99 **15.** $20,931.01 **17.** $331.46 **19.** $625.28

21.

PERIOD	AMOUNT	INTEREST	BALANCE
1	$1,000.00	$0.00	$1,000.00
2	$1,000.00	$83.20	$2,083.20
3	$1,000.00	$173.32	$3,256.52
4	$1,000.00	$270.94	$4,527.46
5	$1,000.00	$376.69	$5,904.15

23. First year: $50.76; second year: $168.09; third year: $296.42
25. (A) $413,092 (B) $393,965
27. $177.46; $1,481.92
29. 3 yr, 5 mo
31. 11.58%
33. 7.13%
35. After 11 quarterly payments

Exercise 3-4

1. $PV = \$3{,}458.41$ **3.** $PV = \$4{,}606.09$ **5.** $PMT = \$199.29$ **7.** $PMT = \$586.01$ **9.** $n = 29$ **11.** $i = 0.029$
13. $109,421.92 **15.** $14,064.67; $16,800.00 **17.** (A) $36.59/mo; $58.62 interest (B) $38.28/mo; $89.04 interest
19. $273.69/mo; $7,705.68 interest

21.

PAYMENT NUMBER	PAYMENT	INTEREST	UNPAID BALANCE REDUCTION	UNPAID BALANCE
0				$5,000.00
1	$ 758.05	$ 225.00	$ 533.05	4,466.95
2	758.05	201.01	557.04	3,909.91
3	758.05	175.95	582.10	3,327.81
4	758.05	149.75	608.30	2,719.51
5	758.05	122.38	635.67	2,083.84
6	758.05	93.77	664.28	1,419.56
7	758.05	63.88	694.17	725.39
8	758.03	32.64	725.39	0.00
Totals	$6,064.38	$1,064.38	$5,000.00	

23. First year interest = $625.07; Second year interest = $400.91; Third year interest = $148.46
25. $85,846.38; $128,153.62 **27.** $143.85/mo; $904.80
29. Monthly payment = $841.39
(A) $70,952.33 (B) $55,909.02 (C) $36,813.32
31. (A) Monthly payment = $395.04; Total interest = $64,809.60
(B) 114 mo, or 9.5 yr; Interest saved = $38,375.04
33. (A) 157 (B) 243 (C) The withdrawals continue forever.
35. (A) Monthly withdrawals: $1,229.66; Total interest: $185,338.80
(B) Monthly deposits: $162.65
37. $29,799
39. All three graphs are decreasing, curve downward, and have the same x intercept; the unpaid balances are always in the ratio 2:3:4. The monthly payments are $402.31, $603.47, and $804.62, with total interest amounting to $94,831.60, $142,249.20, and $189,663.20, respectively.
41. 14.45% **43.** 10.21%

Review Exercise

1. $A = \$104.50$ *(3-1)* **2.** $P = \$800$ *(3-1)* **3.** $t = 0.75$ yr, or 9 mo *(3-1)* **4.** $r = 6\%$ *(3-1)*
5. $A = \$1{,}393.68$ *(3-2)* **6.** $P = \$3{,}193.50$ *(3-2)* **7.** $FV = \$69{,}770.03$ *(3-3)* **8.** $PMT = \$115.00$ *(3-3)*
9. $PV = \$33{,}944.27$ *(3-4)* **10.** $PMT = \$166.07$ *(3-4)* **11.** $n \approx 16$ *(3-2)* **12.** $n \approx 41$ *(3-3)*
13. $3,350.00; $350.00 *(3-1)* **14.** $27,551.32 *(3-2)* **15.** $9,422.24 *(3-2)*

16. (A)

PERIOD	INTEREST	AMOUNT
0		$400.00
1	$21.60	$421.60
2	$22.77	$444.37
3	$24.00	$468.36
4	$25.29	$493.65

(3-2)

(B)

PERIOD	INTEREST	PAYMENT	BALANCE
1		$100.00	$100.00
2	$5.40	$100.00	$205.40
3	$11.09	$100.00	$316.49
4	$17.09	$100.00	$433.58

(3-3)

17. To maximize earnings, choose 13% simple interest for investments lasting less than 9 years and 9% compound interest for investments lasting 9 years or more. *(3-2)*
18. $164,402 *(3-2)*
19. 9% compounded quarterly, since its effective rate is 9.31%, while the effective rate of 9.25% compounded annually is 9.25% *(3-2)*
20. $27,971.23; $8,771.23 *(3-3)* **21.** $11.64 *(3-1)* **22.** $10,210.25 *(3-2)* **23.** $6,268.21 *(3-2)* **24.** 15% *(3-1)*
25. $10,318.91; $2,281.09 *(3-4)* **26.** 2 yr, 3 mo *(3-2)* **27.** 5 yr, 10 mo; 3 yr, 11 mo *(3-2)*
28. (A) $571,499 (B) $1,973,277 *(3-3)* **29.** 10.45% *(3-2)* **30.** 20% *(3-1)* **31.** $10,988.22 *(3-4)*
32. (A) $115,573.86 (B) $359.64 (C) $171,228.80 *(3-4)* **33.** $99.85/mo; $396.36 *(3-4)* **34.** $526.28/mo *(3-3)*
35. 6.93 yr; 7.27 yr *(3-2)* **36.** $175.28; $2,516.80 *(3-4)* **37.** 5 yr, 10 mo *(3-3)* **38.** 18 yr *(3-4)* **39.** 28.8% *(3-1)*

40.

PAYMENT NUMBER	PAYMENT	INTEREST	UNPAID BALANCE REDUCTION	UNPAID BALANCE
0				$1,000.00
1	$ 265.82	$25.00	$ 240.82	759.18
2	265.82	18.98	246.84	512.34
3	265.82	12.81	253.01	259.33
4	265.81	6.48	259.33	0.00
Totals	$1,063.27	$63.27	$1,000.00	

(3-4)

41. 1 yr, 1 mo *(3-2)*
42. $55,347.48; $185,830.24 *(3-3)*
43. West Lake S & L: 9.800%; Security S & L: 9.794% *(3-2)*
44. 8.24% *(3-1)*
45. 43 *(3-3)*
46. (A) $1,435.63 (B) $74,397.48 (C) $11,625.04 *(3-4)*
47. The certificate would be worth $81,041.86 when the 360th payment is made. By reducing the principal the loan would be paid off in 238 months. If the monthly payment were then invested at 8.75% compounded monthly, it would be worth $105,167.89 at the time of the 360th payment. *(3-2, 3-3, 3-4)*
48. The lower rate would save $15,485.03 in interest payments. *(3-4)* **49.** $3,176.14 *(3-2)* **50.** 8.37% *(3-2)*
51. $4,744.73 *(3-1)* **52.** $6,697.11 *(3-4)* **53.** 9.38% *(3-2)* **54.** (A) $398,807 (B) $374,204 *(3-3)*
55. (A) $746.79/mo; $896.76/mo (B) $73,558.78; $41,482.19 *(3-4)* **56.** $19,239 *(3-4)* **57.** 33.52% *(3-4)*
58. (A) 10.74% (B) 15 yr; 40 yr *(3-3)*

Chapter 4

Exercise 4-1

1. Graph (B); no solution **3.** Graph (A); $x = -3, y = 1$ **5.** $x = 2, y = 4$ **7.** No solution (parallel lines)
9. $x = 4, y = 5$ **11.** $x = 1, y = 4$ **13.** $u = 2, v = -3$ **15.** $m = 8, n = 6$ **17.** $x = 1, y = -5$
19. No solution (inconsistent) **21.** $x = -\frac{4}{3}, y = 1$ **23.** Infinitely many solutions (dependent) **25.** $x = \frac{4}{7}, x = \frac{3}{7}$
27. $x = 1.1, y = 0.3$ **29.** $x = -\frac{5}{4}, y = \frac{5}{3}$ **31.** $(2.41, 1.12)$ **33.** $(-3.31, -2.24)$

35.

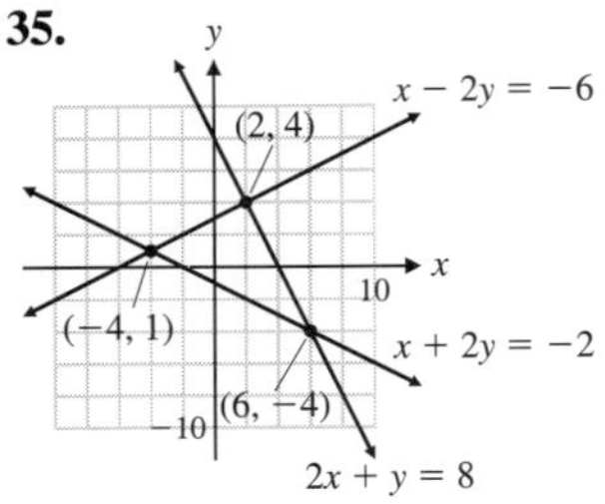

37.

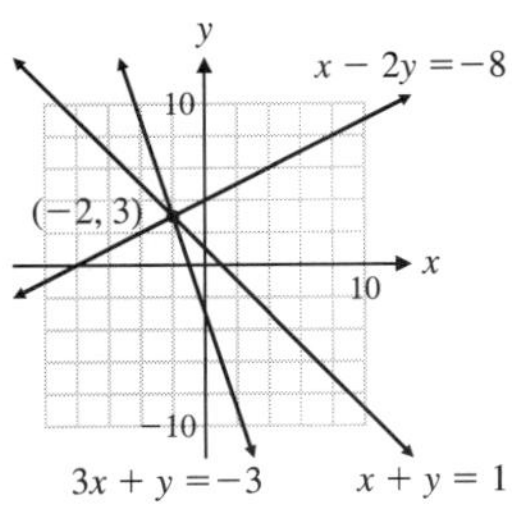

39.

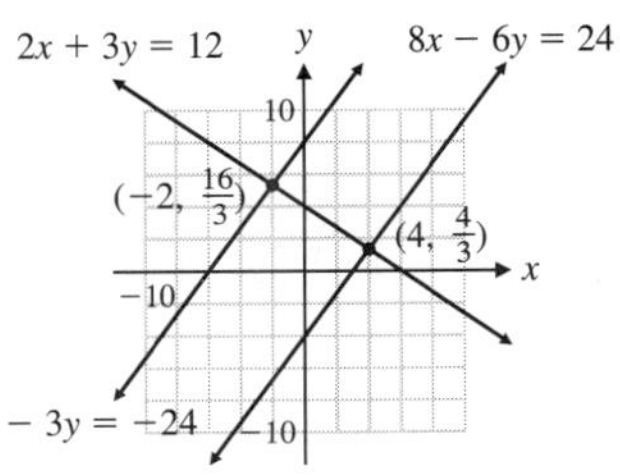

41. (A) $(20, -24)$
(B) $(-4, 6)$
(C) No solution

43. (A) Supply: 143 T-shirts; Demand: 647 T-shirts
(B) Supply: 857 T-shirts; Demand: 353 T-shirts
(C) Equilibrium price = \$6.50; Equilibrium quantity = 500 T-shirts
(D)

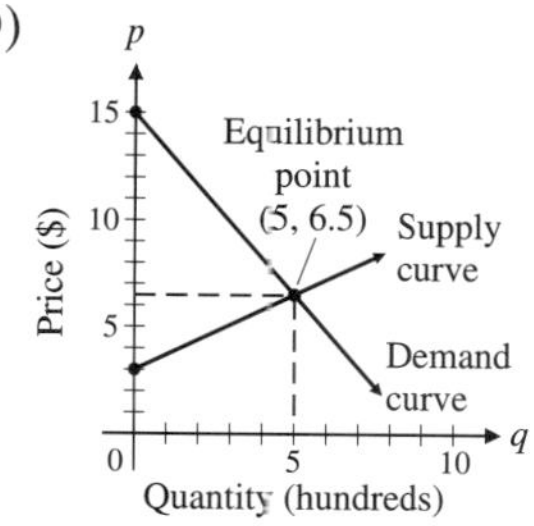

45. (A) $p = 0.001q + 0.15$
(B) $p = -0.002q + 1.74$
(C) Equilibrium price = \$0.68;
Equilibrium quantity = 530
(D)

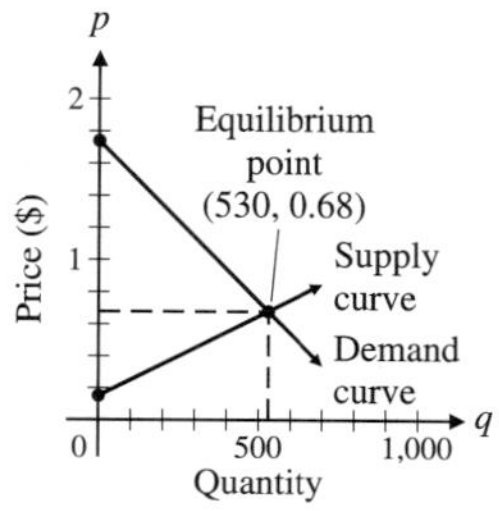

47. (A) For $x = 120$ units,
Cost = \$216,000 = Revenue
(B)

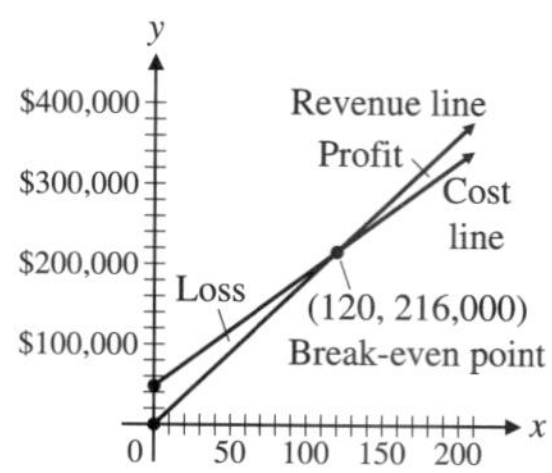

49. (A) For $x = 1{,}920$ tapes,
Cost = \$38,304 = Revenue
(B)

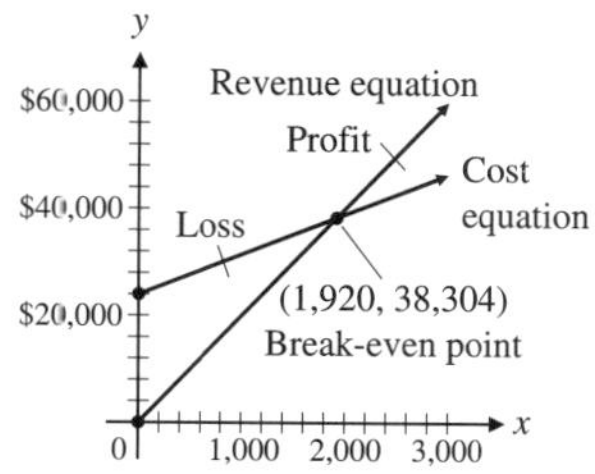

51. Base price = \$17.95; Surcharge = \$2.45/lb
53. 5,720 lb robust blend; 6,160 lb mild blend
55. Mix *A*: 80 g; Mix *B*: 60 g

57. (A)

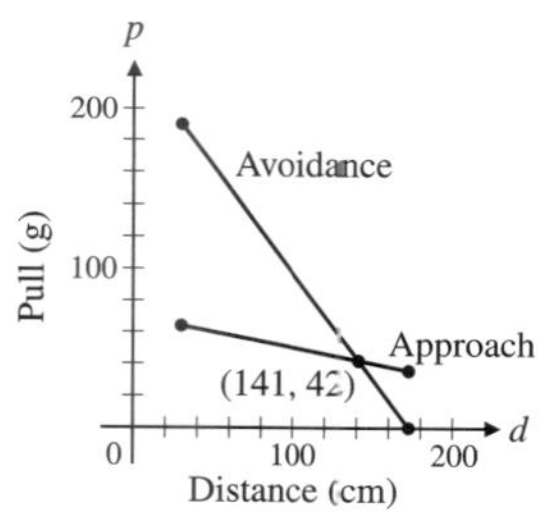

(B) $d = 141$ cm (approx.)
(C) Vacillate

Exercise 4-2

1. *A* is 2×3; *C* is 1×3 **3.** *C* **5.** *B* **7.** $a_{12} = -4$; $a_{23} = -5$ **9.** $-1, 8, 0$
11. (A) 2×4 (B) 1 (C) $e_{23} = 7$; $f_{12} = -6$

13. $\left[\begin{array}{rr|r} 4 & -6 & -8 \\ 1 & -3 & 2 \end{array}\right]$ **15.** $\left[\begin{array}{rr|r} -4 & 12 & -8 \\ 4 & -6 & -8 \end{array}\right]$ **17.** $\left[\begin{array}{rr|r} 1 & -3 & 2 \\ 8 & -12 & -16 \end{array}\right]$

19. $\left[\begin{array}{rr|r} 1 & -3 & 2 \\ 0 & 6 & -16 \end{array}\right]$ **21.** $\left[\begin{array}{rr|r} 1 & -3 & 2 \\ 2 & 0 & -12 \end{array}\right]$ **23.** $\left[\begin{array}{rr|r} 1 & -3 & 2 \\ 3 & -3 & -10 \end{array}\right]$

25. $\frac{1}{3}R_2 \rightarrow R_2$ **27.** $6R_1 + R_2 \rightarrow R_2$ **29.** $\frac{1}{3}R_2 + R_1 \rightarrow R_1$
31. $x_1 = 3$, $x_2 = 2$; each pair of lines has the same intersection point.

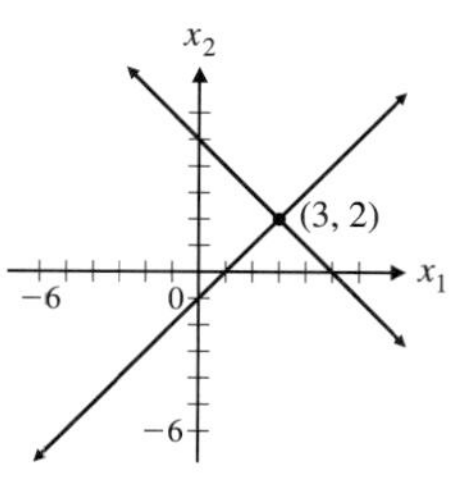

$x_1 + x_2 = 5$
$x_1 - x_2 = 1$

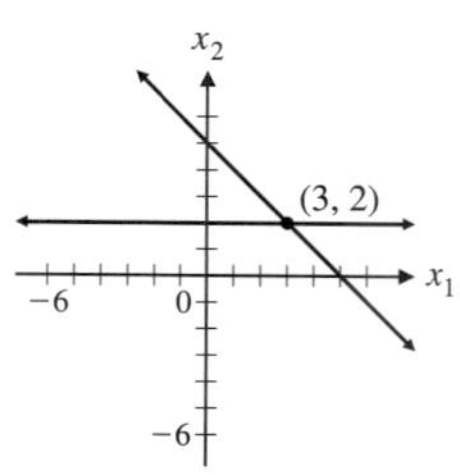

$x_1 + x_2 = 5$
$-2x_2 = -4$

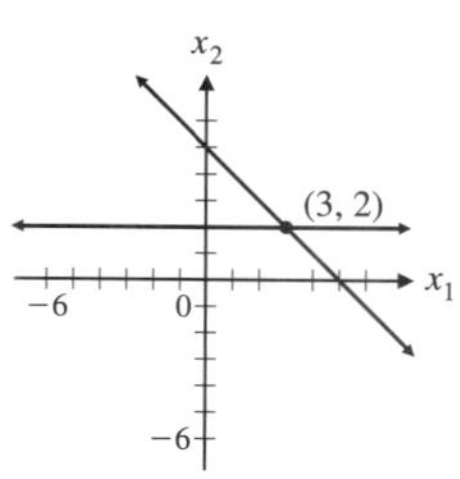

$x_1 + x_2 = 5$
$x_2 = 2$

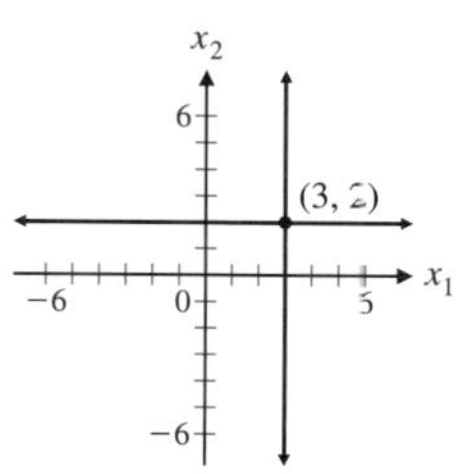

$x_1 = 3$
$x_2 = 2$

33. $x_1 = 3, x_2 = 1$
35. $x_1 = 2, x_2 = 1$
37. $x_1 = 2, x_2 = 4$
39. No solution
41. $x_1 = 1, x_2 = 4$

43. Infinitely many solutions: $x_2 = s$, $x_1 = 2s - 3$ for any real number s
45. Infinitely many solutions; $x_2 = s$, $x_1 = \frac{1}{2}s + \frac{1}{2}$ for any real number s
47. $x_1 = -1, x_2 = 3$ **49.** No solution **51.** Infinitely many solutions: $x_2 = t$, $x_1 = \frac{3}{2}t + 2$ for any real number t
53. $x_1 = 2, x_2 = -1$ **55.** $x_1 = 2, x_2 = -1$ **57.** $x_1 = 1.1, x_2 = 0.3$ **59.** $x_1 = -23.125, x_2 = 7.8125$
61. $x_1 = 3.225, x_2 = -6.9375$

Exercise 4-3

1. Reduced form **3.** Not reduced form; $R_2 \leftrightarrow R_3$ **5.** Not reduced form; $\frac{1}{3}R_2 \rightarrow R_2$
7. Reduced form **9.** Not reduced form; $2R_2 + R_1 \rightarrow R_1$ **11.** $x_1 = -2, x_2 = 3, x_3 = 0$
13. $x_1 = 2t + 3, x_2 = -t - 5, x_3 = t$ for t any real number **15.** No solution
17. $x_1 = 2s + 3t - 5, x_2 = s, x_3 = -3t + 2, x_4 = t$ for s and t any real numbers **19.** $\left[\begin{array}{cc|c} 1 & 0 & -7 \\ 0 & 1 & 3 \end{array}\right]$
21. $\left[\begin{array}{ccc|c} 1 & 0 & 0 & -5 \\ 0 & 1 & 0 & 4 \\ 0 & 0 & 1 & -2 \end{array}\right]$ **23.** $\left[\begin{array}{ccc|c} 1 & 0 & 2 & -\frac{5}{3} \\ 0 & 1 & -2 & \frac{1}{3} \\ 0 & 0 & 0 & 0 \end{array}\right]$ **25.** $x_1 = -2, x_2 = 3, x_3 = 1$ **27.** $x_1 = 0, x_2 = -2, x_3 = 2$
29. $x_1 = 2t + 3, x_2 = t - 2, x_3 = t$ for t any real number **31.** $x_1 = 1, x_2 = 2$ **33.** No solution
35. $x_1 = t - 1, x_2 = 2t + 2, x_3 = t$ for t any real number **37.** $x_1 = -2s + t + 1, x_2 = s, x_3 = t$ for s and t any real numbers
39. No solution **41.** $x_1 = 2.5t - 4, x_2 = t, x_3 = -5$ for t any real number **43.** $x_1 = 1, x_2 = -2, x_3 = 1$
45. (A) Dependent system with two parameters and an infinite number of solutions
(B) Dependent system with one parameter and an infinite number of solutions
(C) Independent system with a unique solution (D) Impossible
47. The system has no solution for $k = -3$ and a unique solution for all other values of k.
49. The system has an infinite number of solutions for $k = 3$ and a unique solution for all other values of k.
51. $x_1 = 2s - 3t + 3, x_2 = s + 2t + 2, x_3 = s, x_4 = t$ for s and t any real numbers
53. $x_1 = -0.5, x_2 = 0.2, x_3 = 0.3, x_4 = -0.4$
55. $x_1 = 2s - 1.5t + 1, x_2 = s, x_3 = -t + 1.5, x_4 = 0.5t - 0.5, x_5 = t$ for s and t any real numbers
57. (A) x_1 = Number of one-person boats
x_2 = Number of two-person boats
x_3 = Number of four-person boats
$$\begin{aligned} 0.5x_1 + x_2 + 1.5x_3 &= 380 \\ 0.6x_1 + 0.9x_2 + 1.2x_3 &= 330 \\ 0.2x_1 + 0.3x_2 + 0.5x_3 &= 120 \end{aligned}$$
20 one-person boats, 220 two-person boats, and 100 four-person boats
(B) $$\begin{aligned} 0.5x_1 + x_2 + 1.5x_3 &= 380 \\ 0.6x_1 + 0.9x_2 + 1.2x_3 &= 330 \end{aligned}$$
$t - 80$ one-person boats, $420 - 2t$ two-person boats, and t four-person boats, where t is an integer satisfying $80 \leq t \leq 210$
(C) $$\begin{aligned} 0.5x_1 + x_2 &= 380 \\ 0.6x_1 + 0.9x_2 &= 330 \\ 0.2x_1 + 0.3x_2 &= 120 \end{aligned}$$
There is no production schedule that will use all the labor-hours in all departments.
59. x_1 = Number of 6,000 gal tank cars
x_2 = Number of 8,000 gal tank cars
x_3 = Number of 18,000 gal tank cars
$$\begin{aligned} x_1 + x_2 + x_3 &= 24 \\ 6{,}000x_1 + 8{,}000x_2 + 18{,}000x_3 &= 250{,}000 \end{aligned}$$
$5t - 29$ 6,000 gal tank cars, $53 - 6t$ 8,000 gal tank cars, and t 18,000 gal tank cars, where $t = 6, 7$, or 8
61. x_1 = Federal income tax
x_2 = State income tax
x_3 = Local income tax
$$\begin{aligned} x_1 + 0.5x_2 + 0.5x_3 &= 3{,}825{,}000 \\ 0.2x_1 + x_2 + 0.2x_3 &= 1{,}530{,}000 \\ 0.1x_1 + 0.1x_2 + x_3 &= 765{,}000 \end{aligned}$$
Tax liability is 57.65%.
63. x_1 = Taxable income of company A
x_2 = Taxable income of company B
x_3 = Taxable income of company C
$$\begin{aligned} x_1 - 0.08x_2 - 0.03x_3 - 0.07x_4 &= 2.272 \\ -0.12x_1 + x_2 - 0.11x_3 - 0.13x_4 &= 2.106 \\ -0.11x_1 - 0.09x_2 + x_3 - 0.08x_4 &= 2.736 \\ -0.06x_1 - 0.02x_2 - 0.14x_3 + x_4 &= 3.168 \end{aligned}$$
Taxable incomes are $2,927,000 for company A, $3,372,000 for company B, $3,675,000 for company C, and $3,926,000 for company D.
65. (A) x_1 = Number of ounces of food A
x_2 = Number of ounces of food B
x_3 = Number of ounces of food C
$$\begin{aligned} 30x_1 + 10x_2 + 20x_3 &= 340 \\ 10x_1 + 10x_2 + 20x_3 &= 180 \\ 10x_1 + 30x_2 + 20x_3 &= 220 \end{aligned}$$
8 oz of food A, 2 oz of food B, and 4 oz of food C
(B) $$\begin{aligned} 30x_1 + 10x_2 &= 340 \\ 10x_1 + 10x_2 &= 180 \\ 10x_1 + 30x_2 &= 220 \end{aligned}$$
There is no combination that will meet all the requirements.
(C) $$\begin{aligned} 30x_1 + 10x_2 + 20x_3 &= 340 \\ 10x_1 + 10x_2 + 20x_3 &= 180 \end{aligned}$$
8 oz of food A, $10 - 2t$ oz of food B, and t oz of food C, where $0 \leq t \leq 5$

67. x_1 = Number of barrels of mix A
x_2 = Number of barrels of mix B
x_3 = Number of barrels of mix C
x_4 = Number of barrels of mix D

$$30x_1 + 30x_2 + 30x_3 + 60x_4 = 900$$
$$50x_1 + 75x_2 + 25x_3 + 25x_4 = 750$$
$$30x_1 + 20x_2 + 20x_3 + 50x_4 = 700$$

$10 - t$ barrels of mix A, $t - 5$ barrels of mix B, $25 - 2t$ barrels of mix C, and t barrels of mix D, where t is an integer satisfying $5 \leq t \leq 10$

69. x_1 = Number of hours for company A
x_2 = Number of hours for company B
$30x_1 + 20x_1 = 600$
$10x_1 + 20x_1 = 400$
Company A: 10 hr; company B: 15 hr

71. (A) 6th St. and Washington Ave.: $x_1 + x_2 = 1{,}200$; 6th St. and Lincoln Ave.: $x_2 + x_3 = 1{,}000$; 5th St. and Lincoln Ave.: $x_3 + x_4 = 1{,}300$
(B) $x_1 = 1{,}500 - t$, $x_2 = t - 300$, $x_3 = 1{,}300 - t$, and $x_4 = t$, where $300 \leq t \leq 1{,}300$
(C) 1,300; 300 (D) Washington Ave.: 500; 6th St.: 700; Lincoln Ave.: 300

Exercise 4-4

1. $\begin{bmatrix} -1 & 0 \\ 5 & -3 \end{bmatrix}$ **3.** Not defined **5.** $\begin{bmatrix} 5 & -7 \\ -5 & 2 \\ 0 & 4 \end{bmatrix}$ **7.** $\begin{bmatrix} 5 & -10 & 0 & 20 \\ -15 & 10 & -5 & 30 \end{bmatrix}$ **9.** $\begin{bmatrix} 5 \\ -3 \end{bmatrix}$

11. $\begin{bmatrix} 2 & 4 \\ 1 & -5 \end{bmatrix}$ **13.** $\begin{bmatrix} 1 & -5 \\ -2 & -4 \end{bmatrix}$ **15.** $[-7]$ **17.** $\begin{bmatrix} -15 & 6 \\ -20 & 8 \end{bmatrix}$ **19.** $[11]$ **21.** $\begin{bmatrix} 3 & -2 & -4 \\ 6 & -4 & -8 \\ -9 & 6 & 12 \end{bmatrix}$

23. $\begin{bmatrix} -12 & 12 & 18 \\ 20 & -18 & -6 \end{bmatrix}$ **25.** Not defined **27.** $\begin{bmatrix} 11 & 2 \\ 4 & 27 \end{bmatrix}$ **29.** $\begin{bmatrix} 6 & 4 \\ 0 & -3 \end{bmatrix}$ **31.** $\begin{bmatrix} -1.3 & -0.7 \\ -0.2 & -0.5 \\ 0.1 & 1.1 \end{bmatrix}$

33. $\begin{bmatrix} -66 & 69 & 39 \\ 92 & -18 & -36 \end{bmatrix}$ **35.** Not defined **37.** $\begin{bmatrix} -18 & 48 \\ 54 & -34 \end{bmatrix}$ **39.** $\begin{bmatrix} -26 & -15 & -25 \\ -4 & -18 & 4 \\ 2 & 43 & -19 \end{bmatrix}$

41. B^n approaches $\begin{bmatrix} 0.25 & 0.75 \\ 0.25 & 0.75 \end{bmatrix}$; AB^n approaches $[0.25 \quad 0.75]$ **43.** $a = -1, b = 1, c = 3, d = -5$ **45.** $x = 2, y = -3$

47. $x = 3, y = 2$ **49.** $a = 3, b = 4, c = 1, d = 2$ **51.** All are true. **53.**

	Guitar	Banjo
Materials	\$33	\$26
Labor	\$57	\$77

55. Markup

	Basic car	Air	AM/FM radio	Cruise control
Model A	\$1,180	\$82	\$54	\$30
Model B	\$1,075	\$98	\$81	\$30
Model C	\$1,715	\$106	\$108	\$36

57. (A) \$11.80 (B) \$30.30 (C) MN gives the labor costs per boat at each plant.

(D) MN =

	Plant I	Plant II
One-person boat	\$11.80	\$13.80
Two-person boat	\$18.50	\$21.60
Four-person boat	\$26.00	\$30.30

59. (A) $A^2 = \begin{bmatrix} 0 & 0 & 2 & 0 & 0 \\ 1 & 0 & 0 & 0 & 1 \\ 0 & 1 & 0 & 2 & 0 \\ 1 & 0 & 0 & 0 & 1 \\ 0 & 0 & 1 & 0 & 0 \end{bmatrix}$ There is one way to travel from Baltimore to Atlanta with one intermediate connection; there are two ways to travel from Atlanta to Chicago with one intermediate connection. In general, the elements in A^2 indicate the number of different ways to travel from the ith city to the jth city with one intermediate connection.

(B) $A^3 = \begin{bmatrix} 2 & 0 & 0 & 0 & 2 \\ 0 & 1 & 0 & 2 & 0 \\ 0 & 0 & 3 & 0 & 0 \\ 0 & 1 & 0 & 2 & 0 \\ 1 & 0 & 0 & 0 & 1 \end{bmatrix}$ There is one way to travel from Denver to Baltimore with two intermediate connections; there are two ways to travel from Atlanta to El Paso with two intermediate connections. In general, the elements in A^3 indicate the number of different ways to travel from the ith city to the jth city with two intermediate connections.

(C) $A + A^2 + A^3 + A^4 = \begin{bmatrix} 2 & 3 & 2 & 5 & 2 \\ 1 & 1 & 4 & 2 & 1 \\ 4 & 1 & 3 & 2 & 4 \\ 1 & 1 & 4 & 2 & 1 \\ 1 & 1 & 1 & 3 & 1 \end{bmatrix}$ It is possible to travel from any origin to any destination with at most three intermediate connections.

61. (A) 70 g (B) 30 g
(C) MN gives the amount (in grams) of protein, carbohydrate, and fat in 20 oz of each mix.

(D) $MN = \begin{bmatrix} 70 & 60 & 50 \\ 380 & 360 & 340 \\ 50 & 40 & 30 \end{bmatrix}$ (columns: Mix X, Mix Y, Mix Z; rows: Protein, Carbohydrate, Fat)

63. (A) \$2,650 (B) \$4,400
(C) NM gives the total cost per town.

(D) $NM = \begin{bmatrix} \$2{,}650 \\ \$4{,}400 \end{bmatrix}$ (Cost per town; rows: Berkeley, Oakland)

(E) $[1 \quad 1]N = [3{,}000 \quad 1{,}300 \quad 13{,}000]$ (Telephone call, House call, Letter)

(F) $N\begin{bmatrix} 1 \\ 1 \\ 1 \end{bmatrix} = \begin{bmatrix} 6{,}500 \\ 10{,}800 \end{bmatrix}$ (Total contacts; rows: Berkeley, Oakland)

65. (A) $\begin{bmatrix} 0 & 0 & 1 & 1 & 1 & 0 \\ 1 & 0 & 0 & 1 & 1 & 0 \\ 0 & 1 & 0 & 1 & 0 & 0 \\ 0 & 0 & 0 & 0 & 0 & 1 \\ 0 & 0 & 1 & 1 & 0 & 1 \\ 1 & 1 & 1 & 0 & 0 & 0 \end{bmatrix}$ (B) $\begin{bmatrix} 0 & 1 & 2 & 3 & 1 & 2 \\ 1 & 0 & 2 & 3 & 2 & 2 \\ 1 & 1 & 0 & 2 & 1 & 1 \\ 1 & 1 & 1 & 0 & 0 & 1 \\ 1 & 2 & 2 & 2 & 0 & 2 \\ 2 & 2 & 2 & 3 & 2 & 0 \end{bmatrix}$ (C) $BC = \begin{bmatrix} 9 \\ 10 \\ 6 \\ 4 \\ 9 \\ 11 \end{bmatrix}$ where $C = \begin{bmatrix} 1 \\ 1 \\ 1 \\ 1 \\ 1 \\ 1 \end{bmatrix}$

(D) Frank, Bart, Aaron and Elvis (tie), Charles, Dan

Exercise 4-5

1. $\begin{bmatrix} 2 & -3 \\ 4 & 5 \end{bmatrix}$ **3.** $\begin{bmatrix} 2 & -3 \\ 4 & 5 \end{bmatrix}$ **5.** $\begin{bmatrix} -2 & 1 & 3 \\ 2 & 4 & -2 \\ 5 & 1 & 0 \end{bmatrix}$ **7.** $\begin{bmatrix} -2 & 1 & 3 \\ 2 & 4 & -2 \\ 5 & 1 & 0 \end{bmatrix}$

9. Yes **11.** No **13.** Yes **15.** No **17.** Yes

19. $\begin{bmatrix} -1 & 0 \\ -3 & 1 \end{bmatrix}$ **21.** $\begin{bmatrix} 3 & -2 \\ -1 & 1 \end{bmatrix}$ **23.** $\begin{bmatrix} 7 & -3 \\ -2 & 1 \end{bmatrix}$ **25.** $\begin{bmatrix} -5 & -12 & 3 \\ -2 & -4 & 1 \\ 2 & 5 & -1 \end{bmatrix}$

27. $\begin{bmatrix} 6 & -2 & -1 \\ -5 & 2 & 1 \\ -3 & 1 & 1 \end{bmatrix}$ **29.** $\begin{bmatrix} -2 & -3 \\ 3 & 4 \end{bmatrix}$ **31.** Does not exist **33.** $\begin{bmatrix} 1.5 & -0.5 \\ -2 & 1 \end{bmatrix}$

35. $\begin{bmatrix} 1 & 2 & 2 \\ -2 & -3 & -4 \\ -1 & -2 & -1 \end{bmatrix}$ **37.** Does not exist **39.** $\begin{bmatrix} -3 & -2 & 1.5 \\ 4 & 3 & -2 \\ 3 & 2 & -1.25 \end{bmatrix}$ **41.** $\begin{bmatrix} -1.75 & -0.375 & 0.5 \\ -5.5 & -1.25 & 1 \\ 0.5 & 0.25 & 0 \end{bmatrix}$

45. M^{-1} exists if and only if all the elements on the main diagonal are nonzero.

47. $A^{-1} = A; A^2 = I$ **49.** $A^{-1} = A; A^2 = I$

51. $\begin{bmatrix} 0.5 & -0.3 & 0.85 & -0.25 \\ 0 & 0.1 & 0.05 & -0.25 \\ -1 & 0.9 & -1.55 & 0.75 \\ -0.5 & 0.4 & -1.3 & 0.5 \end{bmatrix}$ **53.** $\begin{bmatrix} 1.75 & 5.25 & 8.75 & -1 & -18.75 \\ 1.25 & 3.75 & 6.25 & -1 & -13.25 \\ -4.75 & -13.25 & -22.75 & 3 & 48.75 \\ -1.375 & -4.625 & -7.875 & 1 & 16.375 \\ 3.25 & 8.75 & 15.25 & -2 & -32.25 \end{bmatrix}$

55. 36 44 59 86 61 82 68 95 25 37 49 64 63 81 47 66 43 62

57. GONE WITH THE WIND

59. 27 19 82 9 89 50 114 101 70 123 50 100 75 72 89 63 96 99 69 99 52 105 106 56 97 59 127 110 81 132

61. THE GREATEST SHOW ON EARTH

Exercise 4-6

1. $\begin{aligned} 3x_1 + x_2 &= 5 \\ 2x_1 - x_2 &= -4 \end{aligned}$ **3.** $\begin{aligned} -3x_1 + x_2 \quad\quad &= 3 \\ 2x_1 \quad + x_3 &= -4 \\ -x_1 + 3x_2 - 2x_3 &= 2 \end{aligned}$ **5.** $\begin{bmatrix} 3 & -4 \\ 2 & 1 \end{bmatrix}\begin{bmatrix} x_1 \\ x_2 \end{bmatrix} = \begin{bmatrix} 1 \\ 5 \end{bmatrix}$ **7.** $\begin{bmatrix} 1 & -3 & 2 \\ -2 & 3 & 0 \\ 1 & 1 & 4 \end{bmatrix}\begin{bmatrix} x_1 \\ x_2 \\ x_3 \end{bmatrix} = \begin{bmatrix} -3 \\ 1 \\ -2 \end{bmatrix}$

9. $x_1 = -8, x_2 = 2$ **11.** $x_1 = 0, x_2 = 4$ **13.** $x_1 = 3, x_2 = -2$ **15.** $x_1 = 11, x_2 = 4$

17. (A) $x_1 = -3, x_2 = 2$ (B) $x_1 = -1, x_2 = 2$ (C) $x_1 = -8, x_2 = 3$

19. (A) $x_1 = 17, x_2 = -5$ (B) $x_1 = 7, x_2 = -2$ (C) $x_1 = 24, x_2 = -7$

21. (A) $x_1 = 1, x_2 = 0, x_3 = 0$ (B) $x_1 = -7, x_2 = -2, x_3 = 3$ (C) $x_1 = 17, x_2 = 5, x_3 = -7$

23. (A) $x_1 = 8, x_2 = -6, x_3 = -2$ (B) $x_1 = -6, x_2 = 6, x_3 = 2$ (C) $x_1 = 20, x_2 = -16, x_3 = -10$

25. $x_1 = 2t + 2.5, x_2 = t$, t any real number

27. No solution

29. $x_1 = 13t + 3, x_2 = 8t + 1, x_3 = t$, t any real number

31. $X = (A - B)^{-1}C$ **33.** $X = (A + I)^{-1}C$ **35.** $X = (A + B)^{-1}(C + D)$

37. (A) $x_1 = 1, x_2 = 0$ (B) $x_1 = -2{,}000, x_2 = 1{,}000$ (C) $x_1 = 2{,}001, x_2 = -1{,}000$

39. $x_1 = 18.2, x_2 = 27.9, x_3 = -15.2$ **41.** $x_1 = 24, x_2 = 5, x_3 = -2, x_4 = 15$

43. (A) x_1 = Number of \$4 tickets sold
x_2 = Number of \$8 tickets sold
$x_1 + x_2 = 10{,}000$
$4x_1 + 8x_2 = k_1$ Return
Concert 1: 6,000 \$4 tickets and 4,000 \$8 tickets
Concert 2: 5,000 \$4 tickets and 5,000 \$8 tickets
Concert 3: 3,000 \$4 tickets and 7,000 \$8 tickets
(B) No
(C) \$40,000 + 4t, where t is an integer satisfying $0 \leqslant t \leqslant 10{,}000$

45. x_1 = Number of hours plant A operates
x_2 = Number of hours plant B operates
$10x_1 + 8x_2 = k_1$ Number of car frames produced
$5x_1 + 8x_2 = k_2$ Number of truck frames produced
Order 1: 280 hr at plant A and 25 hr at plant B
Order 2: 160 hr at plant A and 150 hr at plant B
Order 3: 80 hr at plant A and 225 hr at plant B

47. x_1 = President's bonus
x_2 = Executive vice-president's bonus
x_3 = Associative vice-president's bonus
x_4 = Assistant vice-president's bonus

$$\begin{aligned} x_1 + 0.03x_2 + 0.03x_3 + 0.03x_4 &= 60{,}000 \\ 0.025x_1 + x_2 + 0.025x_3 + 0.025x_4 &= 50{,}000 \\ 0.02x_1 + 0.02x_2 + x_3 + 0.02x_4 &= 40{,}000 \\ 0.015x_1 + 0.015x_2 + 0.015x_3 + x_4 &= 30{,}000 \end{aligned}$$

President: \$56,600; Executive vice-president: \$47,000; Associate vice-president: \$37,400; Assistant vice-president: \$27,900

49. (A) x_1 = Number of ounces of mix A
x_2 = Number of ounces of mix B
$0.2x_1 + 0.1x_2 = k_1$ Protein
$0.02x_1 + 0.06x_2 = k_2$ Fat
Diet 1: 60 oz mix A and 80 oz mix B
Diet 2: 20 oz mix A and 60 oz mix B
Diet 3: 0 oz mix A and 100 oz mix B
(B) No

Exercise 4-7

1. 40¢ from A; 20¢ from E **3.** $\begin{bmatrix} 0.6 & -0.2 \\ -0.2 & 0.9 \end{bmatrix}; \begin{bmatrix} 1.8 & 0.4 \\ 0.4 & 1.2 \end{bmatrix}$ **5.** $X = \begin{bmatrix} x_1 \\ x_2 \end{bmatrix} = \begin{bmatrix} 16.4 \\ 9.2 \end{bmatrix}$

7. 20¢ from A; 10¢ from B; 10¢ from E **11.** Agriculture: \$18 billion; Building: \$15.6 billion; Energy: \$22.4 billion

13. $\begin{bmatrix} 1.4 & 0.4 \\ 0.6 & 1.6 \end{bmatrix}; \begin{bmatrix} 24 \\ 46 \end{bmatrix}$ **15.** $\begin{bmatrix} 1.58 & 0.24 & 0.58 \\ 0.4 & 1.2 & 0.4 \\ 0.22 & 0.16 & 1.22 \end{bmatrix}; \begin{bmatrix} 38.6 \\ 18 \\ 17.4 \end{bmatrix}$

17. (A) Agriculture: \$80 million; Manufacturing: \$64 million
(B) The final demand for agriculture increases to \$54 million and the final demand for manufacturing decreases to \$38 million.

19. The total output of the energy sector should be 75% of the total output of the mining sector.

21. Each element should be between 0 and 1, inclusive. **23.** Coal: \$28 billion; Steel: \$26 billion

25. Agriculture: \$148 million; Tourism: \$146 million

27. Agriculture: \$40.1 billion; Manufacturing: \$29.4 billion; Energy: \$34.4 billion

29. Year 1: Agriculture: \$65 billion; Energy: \$83 billion; Labor: \$71 billion; Manufacturing: \$88 billion.
Year 2: Agriculture: \$81 billion; Energy: \$97 billion; Labor: \$83 billion; Manufacturing: \$99 billion.
Year 3: Agriculture: \$117 billion; Energy: \$124 billion; Labor: \$106 billion; Manufacturing: \$120 billion.

Review Exercise

1. $x = 4, y = 4$ *(4-1)* **2.** $x = 4, y = 4$ *(4-1)*

3. (A) Not in reduced form; $R_1 \leftrightarrow R_2$ (B) Not in reduced form; $\frac{1}{3}R_2 \rightarrow R_2$
(C) Reduced form (D) Not in reduced form; $(-1)R_2 + R_1 \rightarrow R_1$ *(4-3)*

4. (A) $2 \times 5, 3 \times 2$ (B) $a_{24} = 3, a_{15} = 2, b_{31} = -1, b_{22} = 4$ (C) AB is not defined; BA is defined *(4-2, 4-4)*

5. $x_1 = 8, x_2 = 2$ *(4-6)* **6.** $\begin{bmatrix} 3 & 3 \\ 4 & 2 \end{bmatrix}$ *(4-4)* **7.** Not defined *(4-4)* **8.** $\begin{bmatrix} -3 & 0 \\ 1 & -1 \end{bmatrix}$ *(4-4)* **9.** $\begin{bmatrix} 4 & 3 \\ 7 & 4 \end{bmatrix}$ *(4-4)*

10. Not defined *(4-4)* **11.** $\begin{bmatrix} 5 \\ 5 \end{bmatrix}$ *(4-4)* **12.** $\begin{bmatrix} 2 & 3 \\ 4 & 6 \end{bmatrix}$ *(4-4)* **13.** $[8]$ *(4-4)* **14.** Not defined *(4-4)*

15. $\begin{bmatrix} -2 & 3 \\ 3 & -4 \end{bmatrix}$ *(4-5)* **16.** $x_1 = 9, x_2 = -11$ *(4-1)* **17.** $x_1 = 9, x_2 = -11$ *(4-2)*

18. $x_1 = 9, x_2 = -11; x_1 = 16, x_2 = -19; x_1 = -2, x_2 = 4$ *(4-6)* **19.** Not defined *(4-4)* **20.** $\begin{bmatrix} 10 & -8 \\ 4 & 6 \end{bmatrix}$ *(4-4)*

21. $\begin{bmatrix} -2 & 8 \\ 8 & 6 \end{bmatrix}$ *(4-4)* **22.** $\begin{bmatrix} -2 & -1 & -3 \\ 4 & 2 & 6 \\ 6 & 3 & 9 \end{bmatrix}$ *(4-4)* **23.** $[9]$ *(4-4)* **24.** $\begin{bmatrix} 10 & -5 & 1 \\ -1 & -4 & -5 \\ 1 & -7 & -2 \end{bmatrix}$ *(4-4)*

25. $\begin{bmatrix} -\frac{5}{2} & 2 & -\frac{1}{2} \\ 1 & -1 & 1 \\ \frac{1}{2} & 0 & -\frac{1}{2} \end{bmatrix}$ *(4-5)* **26.** (A) $x_1 = 2, x_2 = 1, x_3 = -1$
(B) $x_1 = -5t - 12, x_2 = 3t + 7, x_3 = t$ for t any real number *(4-3)*

27. $x_1 = 2, x_2 = 1, x_3 = -1; x_1 = 1, x_2 = -2, x_3 = 1; x_1 = -1, x_2 = 2, x_3 = -2$ *(4-6)*

28. The system has an infinite number of solutions for $k = 3$ and a unique solution for any other value of k. *(4-3)*

29. $(I - M)^{-1} = \begin{bmatrix} 1.4 & 0.3 \\ 0.8 & 1.6 \end{bmatrix}; X = \begin{bmatrix} 48 \\ 56 \end{bmatrix}$ *(4-7)* **30.** $x = 3.46, y = 1.69$ *(4-1)* **31.** $\begin{bmatrix} -0.9 & -0.1 & 5 \\ 0.8 & 0.2 & -4 \\ 0.1 & -0.1 & 0 \end{bmatrix}$ *(4-5)*

32. $x_1 = 1{,}400, x_2 = 3{,}200, x_3 = 2{,}400$ *(4-6)* **33.** $x_1 = 1{,}400, x_2 = 3{,}200, x_3 = 2{,}400$ *(4-3)*

34. $(I - M)^{-1} = \begin{bmatrix} 1.3 & 0.4 & 0.7 \\ 0.2 & 1.6 & 0.3 \\ 0.1 & 0.8 & 1.4 \end{bmatrix}; X = \begin{bmatrix} 81 \\ 49 \\ 62 \end{bmatrix}$ *(4-7)*

35. (A) Unique solution
(B) Either no solution or an infinite number of solutions *(4-6)*

36. (A) Unique solution
(B) No solution
(C) Infinite number of solutions *(4-3)*

37. (B) is the only correct solution. *(4-6)*

38. (A) $C = 243{,}000 + 22.45x$; $R = 59.95x$
(B) $x = 6{,}480$ machines; $R = C = \$388{,}476$
(C) Profit occurs if $x > 6{,}480$;
loss occurs if $x < 6{,}480$.

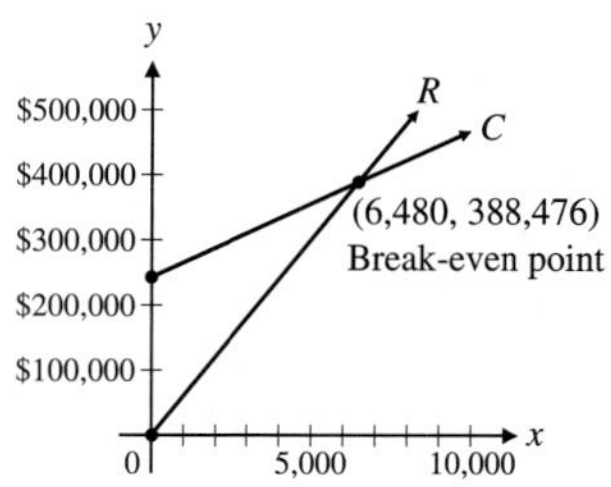

(4-1)

39. x_1 = Number of tons of ore A
x_2 = Number of tons of ore B
$0.01x_1 + 0.02x_2 = 4.5$
$0.02x_1 + 0.05x_2 = 10$
$x_1 = 250$ tons of ore A,
$x_2 = 100$ tons of ore B *(4-3)*

40. (A) $\overset{X}{\begin{bmatrix} x_1 \\ x_2 \end{bmatrix}} = \overset{A^{-1}}{\begin{bmatrix} 500 & -200 \\ -200 & 100 \end{bmatrix}} \begin{bmatrix} 4.5 \\ 10 \end{bmatrix} = \begin{bmatrix} 250 \\ 100 \end{bmatrix}$
$x_1 = 250$ tons of ore A, $x_2 = 100$ tons of ore B

(B) $\overset{X}{\begin{bmatrix} x_1 \\ x_2 \end{bmatrix}} = \overset{A^{-1}}{\begin{bmatrix} 500 & -200 \\ -200 & 100 \end{bmatrix}} \begin{bmatrix} 2.3 \\ 5 \end{bmatrix} = \begin{bmatrix} 150 \\ 40 \end{bmatrix}$ *(4-6)*
$x_1 = 150$ tons of ore A, $x_2 = 40$ tons of ore B

41. x_1 = Number of model *A* trucks purchased
x_2 = Number of model *B* trucks purchased
x_3 = Number of model *C* trucks purchased

$$\begin{aligned} x_1 + x_2 + x_3 &= 12 \\ 18{,}000x_1 + 22{,}000x_2 + 30{,}000x_3 &= 300{,}000 \end{aligned}$$

$2t - 9$ model *A* trucks, $21 - 3t$ model *B* trucks, and t model *C* trucks, where $t = 5, 6,$ or 7 *(4-3)*

42. (A) Elements in *MN* give the cost of materials for each alloy from each supplier.

(B) $MN = \begin{bmatrix} \$7{,}620 & \$7{,}530 \\ \$13{,}880 & \$13{,}930 \end{bmatrix}$ (columns: Supplier A, Supplier B; rows: Alloy 1, Alloy 2)

(C) $[1 \quad 1]MN = [\$21{,}500 \quad \$21{,}460]$ (Supplier A, Supplier B) Total material costs *(4-4)*

43. (A) \$6.35
(B) Elements in *MN* give the total labor costs for each calculator at each plant.
(C) $MN = \begin{bmatrix} \$3.65 & \$3.00 \\ \$6.35 & \$5.20 \end{bmatrix}$ (columns: CA, TX; rows: Model A, Model B) *(4-4)*

44. x_1 = Amount invested at 5%
x_2 = Amount invested at 10%

$$\begin{aligned} x_1 + x_2 &= 5{,}000 \\ 0.05x_1 + 0.1x_2 &= 400 \end{aligned}$$

\$2,000 at 5%, \$3,000 at 10% *(4-3)*

45. \$2,000 at 5% and \$3,000 at 10% *(4-6)*

46. No to both. The annual yield must be between \$250 and \$500 inclusive. *(4-6)*

47. x_1 = Number of \$8 tickets
x_2 = Number of \$12 tickets
x_3 = Number of \$20 tickets

$$\begin{aligned} x_1 + x_2 + x_3 &= 25{,}000 \\ 8x_1 + 12x_2 + 20x_3 &= k_1 \quad \text{Return requested} \\ x_1 - x_3 &= 0 \end{aligned}$$

Concert 1: 5,000 \$8 tickets, 15,000 \$12 tickets, and 5,000 \$20 tickets
Concert 2: 7,500 \$8 tickets, 10,000 \$12 tickets, and 7,500 \$20 tickets
Concert 3: 10,000 \$8 tickets, 5,000 \$12 tickets, and 10,000 \$20 tickets *(4-6)*

48.

$$\begin{aligned} x_1 + x_2 + x_3 &= 25{,}000 \\ 8x_1 + 12x_2 + 20x_3 &= k_1 \quad \text{Return requested} \end{aligned}$$

Concert 1: $2t - 5{,}000$ \$8 tickets, $30{,}000 - 3t$ \$12 tickets, and t \$20 tickets, where t is an integer satisfying $2{,}500 \le t \le 10{,}000$
Concert 2: $2t - 7{,}500$ \$8 tickets, $32{,}500 - 3t$ \$12 tickets, and t \$20 tickets, where t is an integer satisfying $3{,}750 \le t \le 10{,}833$
Concert 3: $2t - 10{,}000$ \$8 tickets, $35{,}000 - 3t$ \$12 tickets, and t \$20 tickets, where t is an integer satisfying $5{,}000 \le t \le 11{,}666$ *(4-3)*

49. (A) Agriculture: \$80 billion; Fabrication: \$60 billion (B) Agriculture: \$135 billion; Fabrication: \$145 billion *(4-7)*

50. GRAPHING UTILITY *(4-5)*

51. (A)
1st & Elm: $x_1 + x_4 = 1{,}300$
2nd & Elm: $x_1 - x_2 = 400$
2nd & Oak: $x_2 + x_3 = 700$
1st & Oak: $x_3 - x_4 = -200$

(B) $x_1 = 1{,}300 - t$, $x_2 = 900 - t$, $x_3 = t - 200$, $x_4 = t$, where $200 \le t \le 900$
(C) 900; 200
(D) Elm St.: 800; 2nd St.: 400; Oak St.: 300 *(4-3)*

52. (A) $M = \begin{bmatrix} 0 & 1 & 0 & 0 \\ 0 & 0 & 1 & 1 \\ 1 & 0 & 0 & 1 \\ 1 & 0 & 0 & 0 \end{bmatrix}$ (rows and columns: A, B, C, D)

(B) $M + M^2 = \begin{bmatrix} 0 & 1 & 1 & 1 \\ 2 & 0 & 1 & 2 \\ 2 & 1 & 0 & 1 \\ 1 & 1 & 0 & 0 \end{bmatrix}$ (rows and columns: A, B, C, D)

Rank: *B, C, A, D (4-4)*

Chapter 5

Exercise 5-1

1.

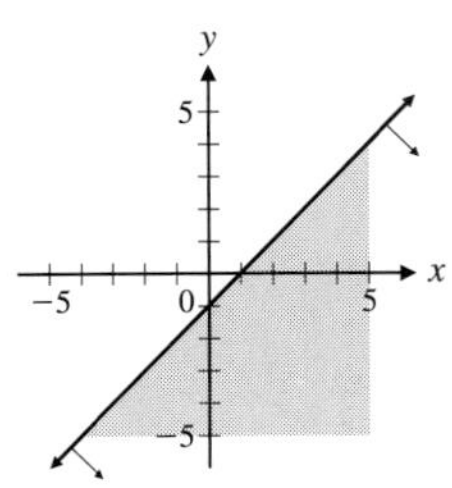

3.

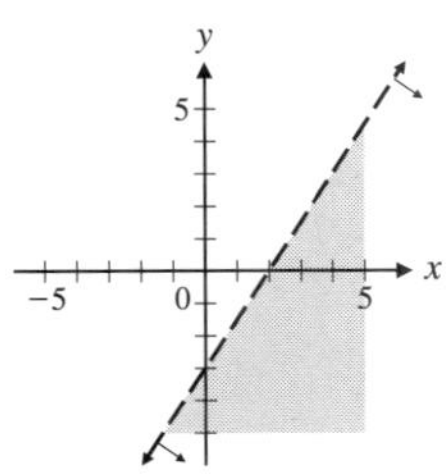

5.

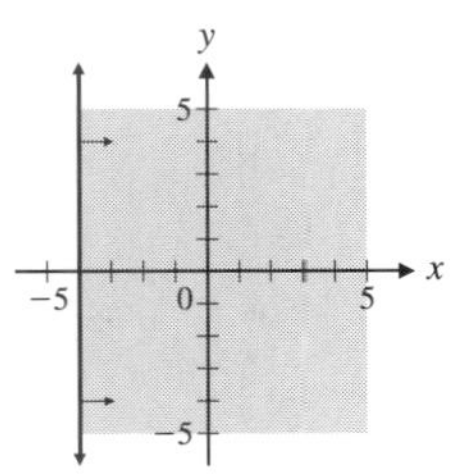

7.

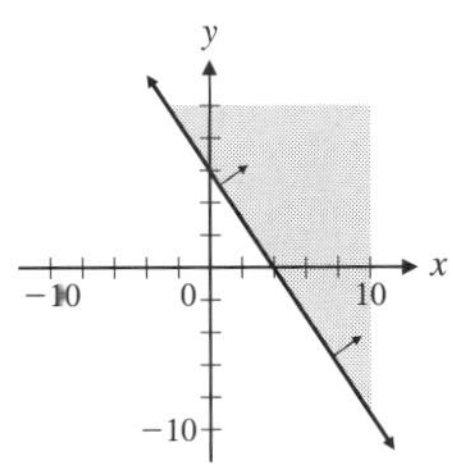

9.

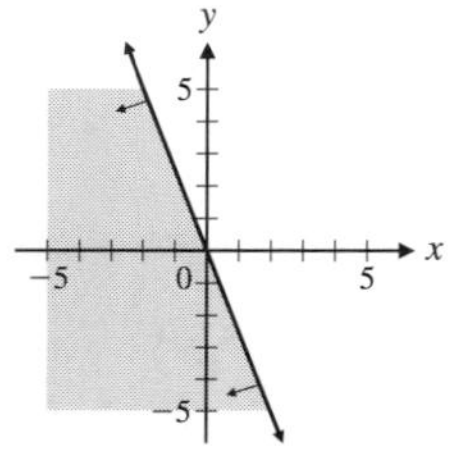

11. IV **13.** I **15.**

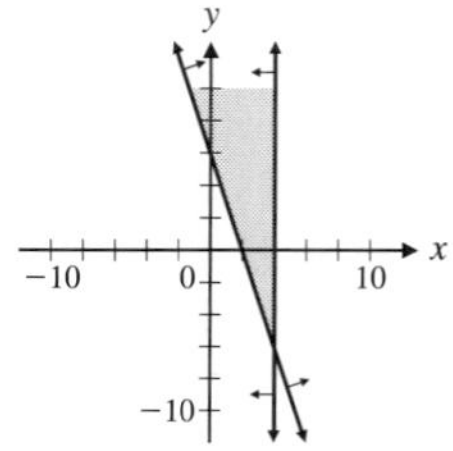

17.

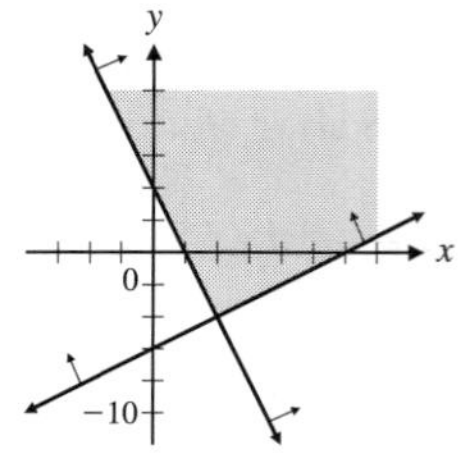

19. (A) Solution region is the double-shaded region.

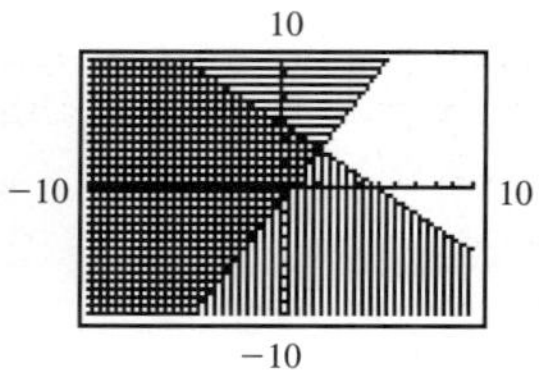

(B) Solution region is the unshaded region.

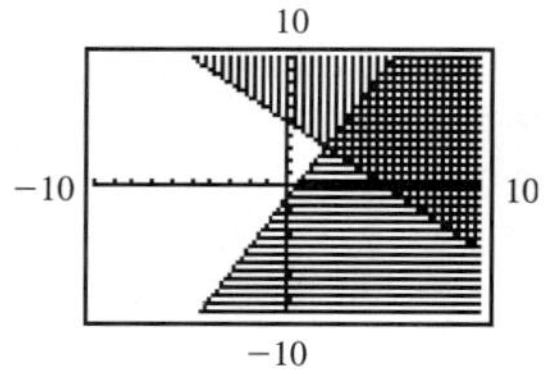

21. (A) Solution region is the double-shaded region.

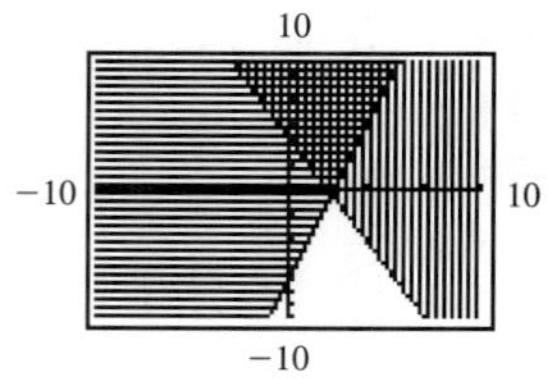

(B) Solution region is the unshaded region.

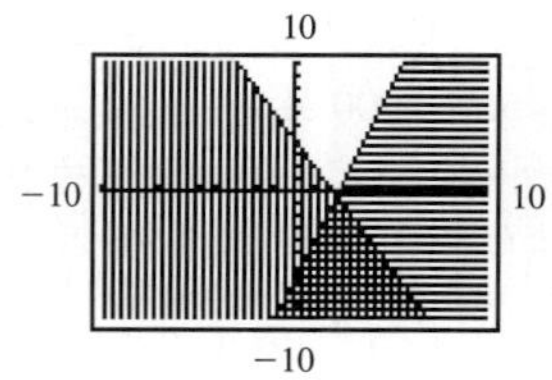

23. IV; (8, 0), (18, 0), (6, 4) **25.** I; (0, 16), (6, 4), (18, 0) **27.** Bounded

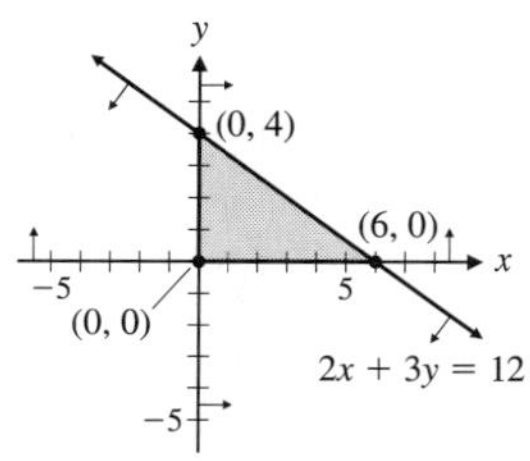

29. Bounded

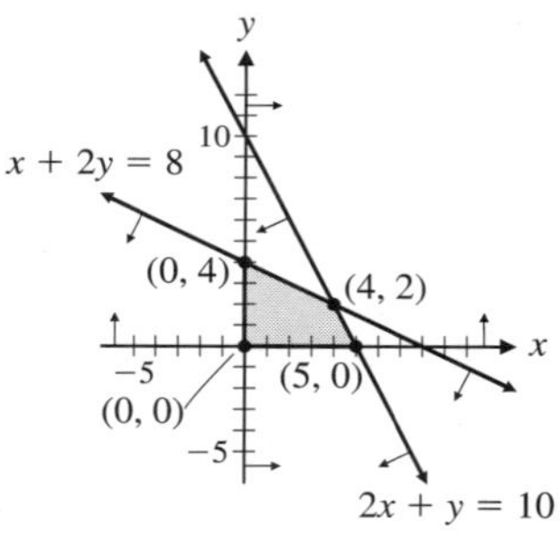

31. Unbounded

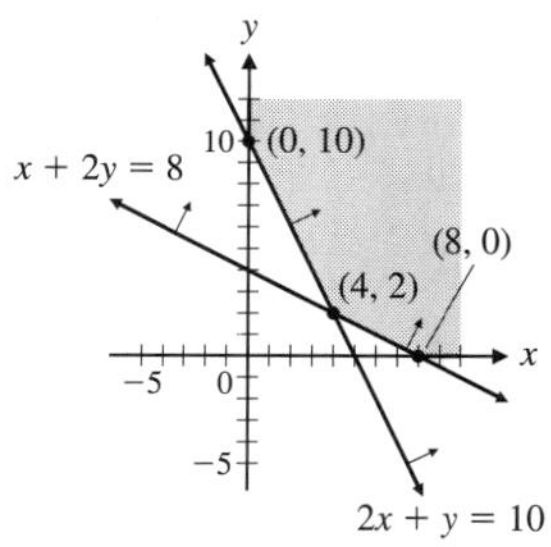

33. Bounded

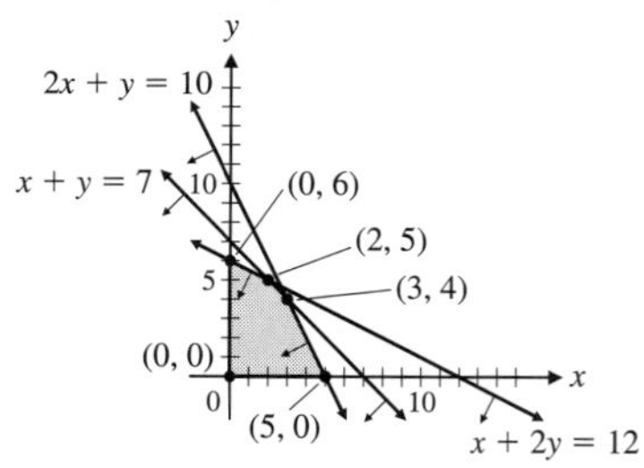

35. Unbounded

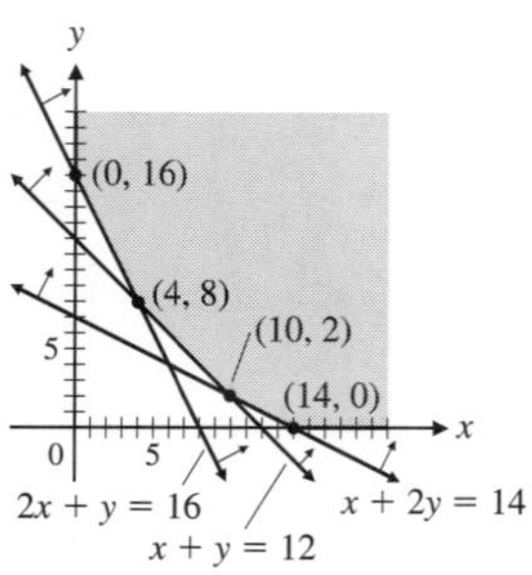

37. Bounded

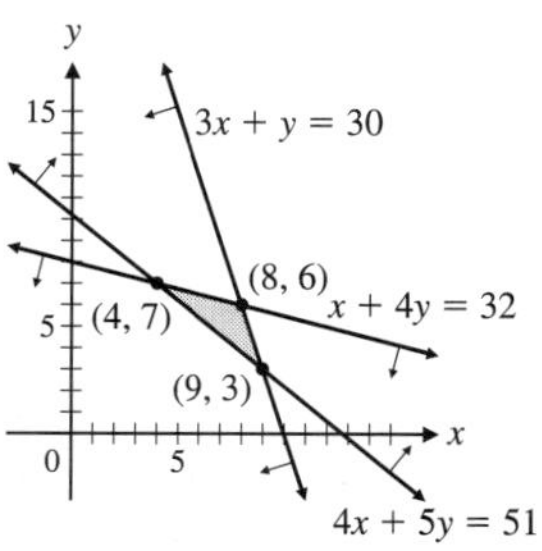

39. Empty

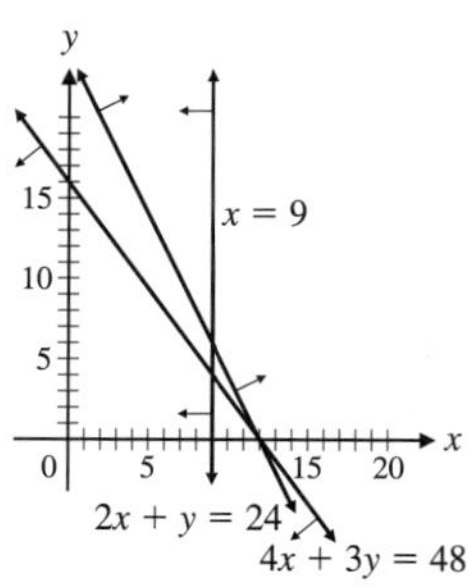

41. Unbounded

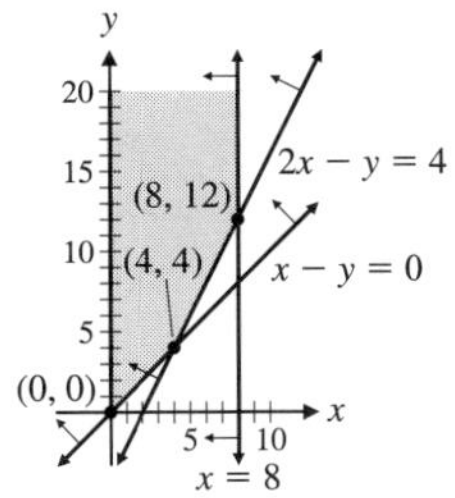

43. Bounded

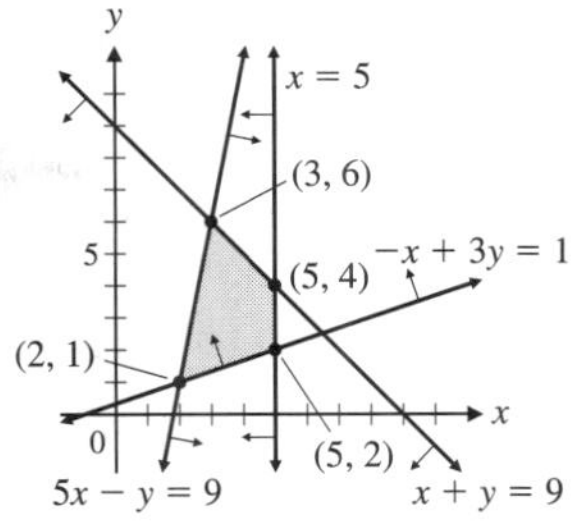

45. Bounded

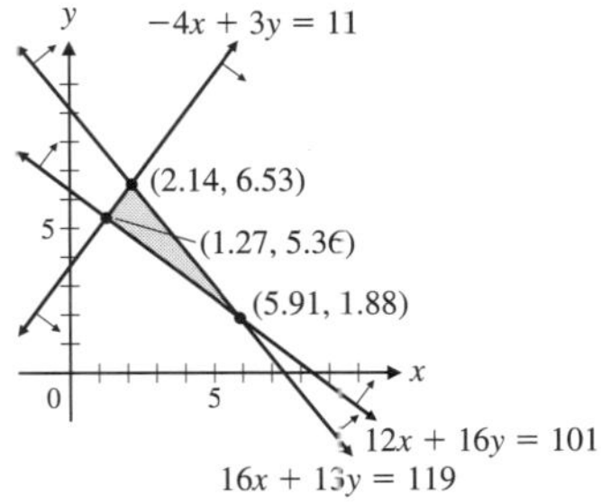

47. (A) $3x + 4y = 36$ and $3x + 2y = 30$ intersect at $(8, 3)$; $3x + 4y = 36$ and $x = 0$ intersect at $(0, 9)$; $3x + 4y = 36$ and $y = 0$ intersect at $(12, 0)$; $3x + 2y = 30$ and $x = 0$ intersect at $(0, 15)$; $3x + 2y = 30$ and $y = 0$ intersect at $(10, 0)$; $x = 0$ and $y = 0$ intersect at $(0, 0)$

(B) $(8, 3), (0, 9), (10, 0), (0, 0)$

49.
$$\begin{aligned} 6x + 4y &\le 108 \\ x + y &\le 24 \\ x &\ge 0 \\ y &\ge 0 \end{aligned}$$

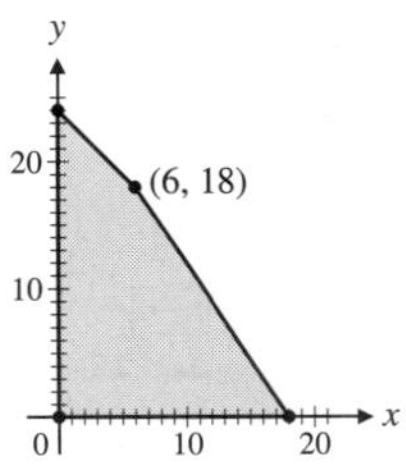

51. (A) All production schedules in the feasible region that are on the graph of $50x + 60y = 1{,}100$ will result in a profit of \$1,100.

(B) There are many possible choices. For example, producing 5 trick skis and 15 slalom skis will produce a profit of \$1,150. All the production schedules in the feasible region that are on the graph of $50x + 60y = 1{,}150$ will result in a profit of \$1,150.

53.
$$\begin{aligned} 20x + 10y &\ge 460 \\ 30x + 30y &\ge 960 \\ 5x + 10y &\ge 220 \\ x &\ge 0 \\ y &\ge 0 \end{aligned}$$

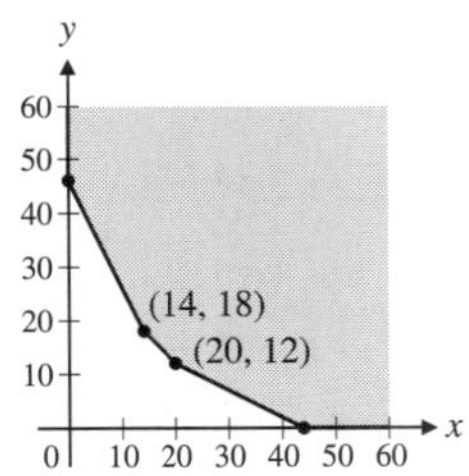

55.
$$\begin{aligned} 10x + 20y &\le 800 \\ 20x + 10y &\le 640 \\ x &\ge 0 \\ y &\ge 0 \end{aligned}$$

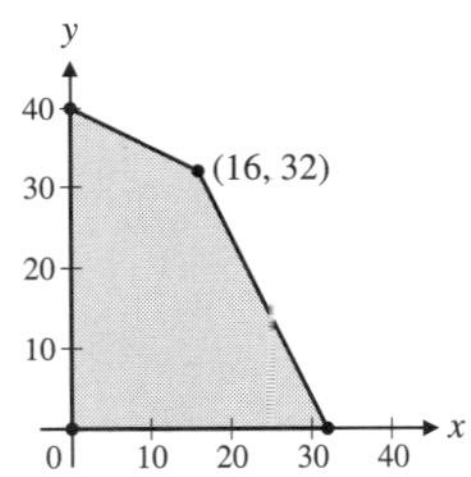

Exercise 5-2

1. 16 **3.** 84 **5.** 32 **7.** 36 **9.** Max $P = 30$ at $x_1 = 4$ and $x_2 = 2$

11. Min $z = 14$ at $x_1 = 4$ and $x_2 = 2$; no max

13. Max $P = 260$ at $x_1 = 2$ and $x_2 = 5$ **15.** Min $z = 140$ at $x_1 = 14$ and $x_2 = 0$; no max

17. Min $P = 20$ at $x_1 = 0$ and $x_2 = 2$; Max $P = 150$ at $x_1 = 5$ and $x_2 = 0$

19. Feasible region empty; no optimal solutions

21. Min $P = 140$ at $x_1 = 3$ and $x_2 = 8$; Max $P = 260$ at $x_1 = 8$ and $x_2 = 10$, at $x_1 = 12$ and $x_2 = 2$, or at any point on the line segment from $(8, 10)$ to $(12, 2)$

23. Max $P = 26{,}000$ at $x_1 = 400$ and $x_2 = 600$ **25.** Max $P = 5{,}507$ at $x_1 = 6.62$ and $x_2 = 4.25$

27. Max $z = 2$ at $x_1 = 4$ and $x_2 = 2$; Min z does not exist

29. (A) $2a < b$ (B) $\frac{1}{3}a < b < 2a$ (C) $b < \frac{1}{3}a$ (D) $b = 2a$ (E) $b = \frac{1}{3}a$

31. (A) Max profit = \$780 when 6 trick skis and 18 slalom skis are produced.

(B) Max profit decreases to \$720 when 18 trick skis and no slalom skis are produced.

(C) Max profit increases to \$1,080 when no trick skis and 24 slalom skis are produced.

33. (A) Plant *A:* 5 days; Plant *B:* 4 days; min cost \$8,600
(B) Plant *A:* 10 days; Plant *B:* 0 days; min cost \$6,000
(C) Plant *A:* 0 days; Plant *B:* 10 days; min cost \$8,000
35. 7 buses, 15 vans; min cost \$9,900
37. \$40,000 in CD's and \$20,000 in mutual fund; max return is \$3,800
39. (A) Max $P =$ \$450 when 750 gal are produced using the old process exclusively.
(B) Max $P =$ \$380 when 400 gal are produced using the old process and 700 gal are produced using the new process.
(C) Max $P =$ \$288 when 1,440 gal are produced using the new process exclusively.
41. (A) 150 bags brand *A*, 100 bags brand *B;* max nitrogen 1,500 lb
(B) 0 bags brand *A*, 250 bags brand *B;* min nitrogen 750 lb
43. 20 yd^3 *A*, 12 yd^3 *B;* \$1,020 **45.** 48; 16 mice, 32 rats

Exercise 5-3

1. (A) 2 (B) 2 basic and 3 nonbasic variables (C) 2 linear equations with 2 variables
3. (A) 5 (B) 4 (C) 5 basic and 4 nonbasic variables (D) 5 linear equations with 5 variables

5.

	NONBASIC	BASIC	FEASIBLE?
(A)	x_1, x_2	s_1, s_2	Yes
(B)	x_1, s_1	x_2, s_2	Yes
(C)	x_1, s_2	x_2, s_1	No
(D)	x_2, s_1	x_1, s_2	No
(E)	x_2, s_2	x_1, s_1	Yes
(F)	s_1, s_2	x_1, x_2	Yes

7.

	x_1	x_2	s_1	s_2	FEASIBLE?
(A)	0	0	50	40	Yes
(B)	0	50	0	−60	No
(C)	0	20	30	0	Yes
(D)	25	0	0	15	Yes
(E)	40	0	−30	0	No
(F)	20	10	0	0	Yes

9.

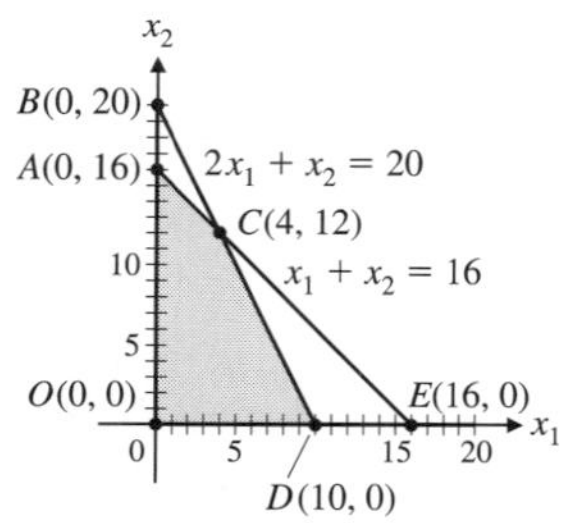

$$\begin{aligned} x_1 + x_2 + s_1 \quad\quad &= 16 \\ 2x_1 + x_2 \quad\quad + s_2 &= 20 \end{aligned}$$

x_1	x_2	s_1	s_2	INTERSECTION POINT	FEASIBLE?
0	0	16	20	*O*	Yes
0	16	0	4	*A*	Yes
0	20	−4	0	*B*	No
16	0	0	−12	*E*	No
10	0	6	0	*D*	Yes
4	12	0	0	*C*	Yes

11.

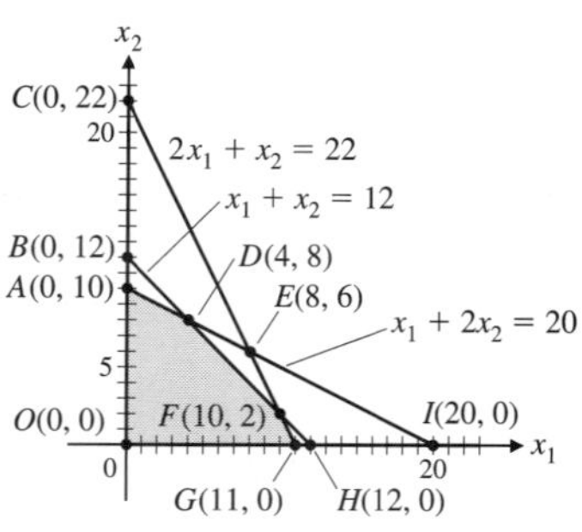

$$\begin{aligned} 2x_1 + x_2 + s_1 \quad\quad\quad\quad &= 22 \\ x_1 + x_2 \quad\quad + s_2 \quad\quad &= 12 \\ x_1 + 2x_2 \quad\quad\quad\quad + s_3 &= 20 \end{aligned}$$

x_1	x_2	s_1	s_2	s_3	INTERSECTION POINT	FEASIBLE?
0	0	22	12	20	*O*	Yes
0	22	0	−10	−24	*C*	No
0	12	10	0	−4	*B*	No
0	10	12	2	0	*A*	Yes
11	0	0	1	9	*G*	Yes
12	0	−2	0	8	*H*	No
20	0	−18	−8	0	*I*	No
10	2	0	0	6	*F*	Yes
8	6	0	−2	0	*E*	No
4	8	6	0	0	*D*	Yes

Exercise 5-4

1. (A) Basic: x_2, s_1, P; nonbasic: x_1, s_2 (B) $x_1 = 0, x_2 = 12, s_1 = 15, s_2 = 0, P = 20$ (C) Additional pivot required

3. (A) Basic: x_2, x_3, s_3, P; nonbasic: x_1, s_1, s_2 (B) $x_1 = 0, x_2 = 15, x_3 = 5, s_1 = 0, s_2 = 0, s_3 = 12, P = 45$
(C) No optimal solution

5. Enter ↓ x_1; Exit → s_1

$$\begin{array}{c} \\ s_1 \\ s_2 \\ P \end{array}\begin{bmatrix} x_1 & x_2 & s_1 & s_2 & P & \\ \textcircled{1} & 4 & 1 & 0 & 0 & 4 \\ 3 & 5 & 0 & 1 & 0 & 24 \\ \hline -8 & -5 & 0 & 0 & 1 & 0 \end{bmatrix}$$

$$\sim \begin{array}{c} x_1 \\ s_2 \\ P \end{array}\begin{bmatrix} 1 & 4 & 1 & 0 & 0 & 4 \\ 0 & -7 & -3 & 1 & 0 & 12 \\ \hline 0 & 27 & 8 & 0 & 1 & 32 \end{bmatrix}$$

7. Enter ↓ x_1; Exit → x_2

$$\begin{array}{c} \\ x_2 \\ s_2 \\ s_3 \\ P \end{array}\begin{bmatrix} x_1 & x_2 & s_1 & s_2 & s_3 & P & \\ \textcircled{2} & 1 & 1 & 0 & 0 & 0 & 4 \\ 3 & 0 & 1 & 1 & 0 & 0 & 8 \\ 0 & 0 & 2 & 0 & 1 & 0 & 2 \\ \hline -4 & 0 & -3 & 0 & 0 & 1 & 5 \end{bmatrix}$$

$$\sim \begin{array}{c} x_1 \\ s_2 \\ s_3 \\ P \end{array}\begin{bmatrix} 1 & \frac{1}{2} & \frac{1}{2} & 0 & 0 & 0 & 2 \\ 0 & -\frac{3}{2} & -\frac{1}{2} & 1 & 0 & 0 & 2 \\ 0 & 0 & 2 & 0 & 1 & 0 & 2 \\ \hline 0 & 2 & -1 & 0 & 0 & 1 & 13 \end{bmatrix}$$

9. (A)
$$\begin{aligned} 2x_1 + x_2 + s_1 \quad &= 10 \\ x_1 + 3x_2 \quad + s_2 &= 10 \\ -15x_1 - 10x_2 \quad + P &= 0 \end{aligned}$$

(B) Enter ↓ x_1; Exit → s_1

$$\begin{array}{c} \\ s_1 \\ s_2 \\ P \end{array}\begin{bmatrix} x_1 & x_2 & s_1 & s_2 & P & \\ \textcircled{2} & 1 & 1 & 0 & 0 & 10 \\ 1 & 3 & 0 & 1 & 0 & 10 \\ \hline -15 & -10 & 0 & 0 & 1 & 0 \end{bmatrix}$$

(C) Max $P = 80$ at $x_1 = 4$ and $x_2 = 2$

11. (A)
$$\begin{aligned} 2x_1 + x_2 + s_1 \quad &= 10 \\ x_1 + 3x_2 \quad + s_2 &= 10 \\ -30x_1 - x_2 \quad + P &= 0 \end{aligned}$$

(B) Enter ↓ x_1; Exit → s_1

$$\begin{array}{c} \\ s_1 \\ s_2 \\ P \end{array}\begin{bmatrix} x_1 & x_2 & s_1 & s_2 & P & \\ \textcircled{2} & 1 & 1 & 0 & 0 & 10 \\ 1 & 3 & 0 & 1 & 0 & 10 \\ \hline -30 & -1 & 0 & 0 & 1 & 0 \end{bmatrix}$$

(C) Max $P = 150$ at $x_1 = 5$ and $x_2 = 0$

13. Max $P = 260$ at $x_1 = 2$ and $x_2 = 5$
15. No optimal solution exists.
17. Max $P = 7$ at $x_1 = 3$ and $x_2 = 5$
19. Max $P = 58$ at $x_1 = 12, x_2 = 0$, and $x_3 = 2$
21. Max $P = 17$ at $x_1 = 4, x_2 = 3$, and $x_3 = 0$
23. Max $P = 22$ at $x_1 = 1, x_2 = 6$, and $x_3 = 0$
25. Max $P = 26{,}000$ at $x_1 = 400$ and $x_2 = 600$
27. Max $P = 450$ at $x_1 = 0, x_2 = 180$, and $x_3 = 30$

29. Max $P = 88$ at $x_1 = 24$ and $x_2 = 8$

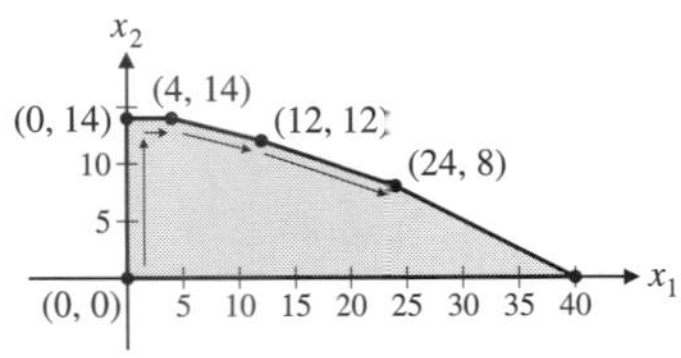

31. Choosing either column produces the same optimal solution: Max $P = 13$ at $x_1 = 3$ and $x_2 = 10$.

33. Choosing Column 1: Max $P = 60$ at $x_1 = 12, x_2 = 8$, and $x_3 = 0$. Choosing Column 2: Max $P = 60$ at $x_1 = 0, x_2 = 20$, and $x_3 = 0$.

35. Let: x_1 = Number of A components
x_2 = Number of B components
x_3 = Number of C components

Maximize $P = 7x_1 + 8x_2 + 10x_3$
Subject to
$$\begin{aligned} 2x_1 + 3x_2 + 2x_3 &\le 1{,}000 \\ x_1 + x_2 + 2x_3 &\le 800 \\ x_1, x_2, x_3 &\ge 0 \end{aligned}$$

200 A components, 0 B components, and 300 C components; Max profit is \$4,400

37. Let: x_1 = Amount invested in government bonds
x_2 = Amount invested in mutual funds
x_3 = Amount invested in money market funds

Maximize $P = 0.08x_1 + 0.13x_2 + 0.15x_3$
Subject to
$$\begin{aligned} x_1 + x_2 + x_3 &\le 100{,}000 \\ -x_1 + x_2 + x_3 &\le 0 \\ x_1, x_2, x_3 &\ge 0 \end{aligned}$$

\$50,000 in government bonds, \$0 in mutual funds, and \$50,000 in money market funds; Max return is \$11,500

39. Let: x_1 = Number of ads placed in daytime shows
x_2 = Number of ads placed in prime-time shows
x_3 = Number of ads placed in late-night shows

$$\begin{array}{ll} \text{Maximize} & P = 14{,}000x_1 + 24{,}000x_2 + 18{,}000x_3 \\ \text{Subject to} & x_1 + x_2 + x_3 \le 15 \\ & 1{,}000x_1 + 2{,}000x_2 + 1{,}500x_3 \le 20{,}000 \\ & x_1, x_2, x_3 \ge 0 \end{array}$$

10 daytime ads, 5 prime-time ads, and 0 late-night ads; Max number of potential customers is 260,000

41. (A) Let: x_1 = Number of colonial houses
x_2 = Number of split-level houses
x_3 = Number of ranch houses

$$\begin{array}{ll} \text{Maximize} & P = 20{,}000x_1 + 18{,}000x_2 + 24{,}000x_3 \\ \text{Subject to} & \frac{1}{2}x_1 + \frac{1}{2}x_2 + x_3 \le 30 \\ & 60{,}000x_1 + 60{,}000x_2 + 80{,}000x_3 \le 3{,}200{,}000 \\ & 4{,}000x_1 + 3{,}000x_2 + 4{,}000x_3 \le 180{,}000 \\ & x_1, x_2, x_3 \ge 0 \end{array}$$

20 colonial, 20 split-level, and 10 ranch houses; Max profit is $1,000,000

(B) 0 colonial, 40 split-level, and 10 ranch houses; Max profit is $960,000; 20,000 labor-hours are not used

(C) 45 colonial, 0 split-level, and 0 ranch houses; Max profit is $1,125,000; 7.5 acres of land and $500,000 of capital are not used

43. (A) Let: x_1 = Number of boxes of assortment I
x_2 = Number of boxes of assortment II
x_3 = Number of boxes of assortment III

$$\begin{array}{ll} \text{Maximize} & P = 4x_1 + 3x_2 + 5x_3 \\ \text{Subject to} & 4x_1 + 12x_2 + 8x_3 \le 4{,}800 \\ & 4x_1 + 4x_2 + 8x_3 \le 4{,}000 \\ & 12x_1 + 4x_2 + 8x_3 \le 5{,}600 \\ & x_1, x_2, x_3 \ge 0 \end{array}$$

200 boxes of assortment I, 100 boxes of assortment II, and 350 boxes of assortment III; Max profit is $2,850

(B) 100 boxes of assortment I, 0 boxes of assortment II, and 550 boxes of assortment III; Max profit is $3,150; 200 fruit-filled candies are not used

(C) 0 boxes of assortment I, 50 boxes of assortment II, and 675 boxes of assortment III; Max profit is $3,525; 400 fruit-filled candies are not used

45. Let: x_1 = Number of grams of food *A*
x_2 = Number of grams of food *B*
x_3 = Number of grams of food *C*

$$\begin{array}{ll} \text{Maximize} & P = 3x_1 + 4x_2 + 5x_3 \\ \text{Subject to} & x_1 + 3x_2 + 2x_3 \le 30 \\ & 2x_1 + x_2 + 2x_3 \le 24 \\ & x_1, x_2, x_3 \ge 0 \end{array}$$

0 g food *A*, 3 g food *B*, and 10.5 g food *C*; Max protein is 64.5 units

47. Let: x_1 = Number of undergraduate students
x_2 = Number of graduate students
x_3 = Number of faculty members

$$\begin{array}{ll} \text{Maximize} & P = 18x_1 + 25x_2 + 30x_3 \\ \text{Subject to} & x_1 + x_2 + x_3 \le 20 \\ & 100x_1 + 150x_2 + 200x_3 \le 3{,}200 \\ & x_1, x_2, x_3 \ge 0 \end{array}$$

0 undergraduate students, 16 graduate students, and 4 faculty members; Max number of interviews is 520

Exercise 5-5

1. $\begin{bmatrix} -5 \\ 0 \\ 3 \\ -1 \\ 8 \end{bmatrix}$ **3.** $\begin{bmatrix} 1 & -2 & 0 & 4 \end{bmatrix}$ **5.** $\begin{bmatrix} 2 & 5 \\ 1 & 2 \\ -6 & 0 \\ 0 & 1 \\ -1 & 3 \end{bmatrix}$ **7.** $\begin{bmatrix} 1 & 0 & 8 & 4 \\ 2 & 2 & 0 & -1 \\ -1 & -7 & 1 & 3 \end{bmatrix}$

9. (A)
$$\begin{array}{ll} \text{Maximize} & P = 4y_1 + 5y_2 \\ \text{Subject to} & y_1 + 2y_2 \le 8 \\ & 3y_1 + y_2 \le 9 \\ & y_1, y_2 \ge 0 \end{array}$$

(B)
$$\begin{array}{l} y_1 + 2y_2 + x_1 = 8 \\ 3y_1 + y_2 + x_2 = 9 \\ -4y_1 - 5y_2 + P = 0 \end{array}$$

(C)
$$\begin{array}{c} \\ \\ \\ \end{array}\begin{matrix} y_1 & y_2 & x_1 & x_2 & P & \\ \end{matrix}$$
$$\left[\begin{array}{ccccc|c} 1 & 2 & 1 & 0 & 0 & 8 \\ 3 & 1 & 0 & 1 & 0 & 9 \\ \hline -4 & -5 & 0 & 0 & 1 & 0 \end{array}\right]$$

11. (A) Max $P = 121$ at $y_1 = 3$ and $y_2 = 5$ (B) Min $C = 121$ at $x_1 = 1$ and $x_2 = 2$

13. (A)
$$\begin{array}{ll} \text{Maximize} & P = 13y_1 + 12y_2 \\ \text{Subject to} & 4y_1 + 3y_2 \le 9 \\ & y_1 + y_2 \le 2 \\ & y_1, y_2 \ge 0 \end{array}$$

(B) Min $C = 26$ at $x_1 = 0$ and $x_2 = 13$

15. (A)
$$\begin{array}{ll} \text{Maximize} & P = 15y_1 + 8y_2 \\ \text{Subject to} & 2y_1 + y_2 \le 7 \\ & 3y_1 + 2y_2 \le 12 \\ & y_1, y_2 \ge 0 \end{array}$$

(B) Min $C = 54$ at $x_1 = 6$ and $x_2 = 1$

17. (A) Maximize $P = 8y_1 + 4y_2$
Subject to $2y_1 - 2y_2 \leqslant 11$
$y_1 + 3y_2 \leqslant 4$
$y_1, y_2 \geqslant 0$
(B) Min $C = 32$ at $x_1 = 0$ and $x_2 = 8$

19. (A) Maximize $P = 6y_1 + 4y_2$
Subject to $-3y_1 + y_2 \leqslant 7$
$y_1 - 2y_2 \leqslant 9$
$y_1, y_2 \geqslant 0$
(B) No optimal solution exists.

21. Min $C = 24$ at $x_1 = 8$ and $x_2 = 0$ **23.** Min $C = 20$ at $x_1 = 0$ and $x_2 = 4$
25. Min $C = 140$ at $x_1 = 14$ and $x_2 = 0$ **27.** Min $C = 44$ at $x_1 = 6$ and $x_2 = 2$
29. Min $C = 43$ at $x_1 = 0$, $x_2 = 1$, and $x_3 = 3$ **31.** No optimal solution exists.
33. 2 variables and 4 problem constraints **35.** 2 constraints and any number of variables
37. No; the dual problem is not a standard maximization problem. **39.** Yes; multiply both sides of the inequality by -1.
41. Min $C = 44$ at $x_1 = 0$, $x_2 = 3$, and $x_3 = 5$ **43.** Min $C = 166$ at $x_1 = 0$, $x_2 = 12$, $x_3 = 20$, and $x_4 = 3$

45. (A) Let: x_1 = Numbers of hours the Cedarburg plant is operated
x_2 = Number of hours the Grafton plant is operated
x_3 = Number of hours the West Bend plant is operated

Minimize $C = 70x_1 + 75x_2 + 90x_3$
Subject to $20x_1 + 10x_2 + 20x_3 \geqslant 300$
$10x_1 + 20x_2 + 20x_3 \geqslant 200$
$x_1, x_2, x_3 \geqslant 0$

Cedarburg plant 10 hr per day, West Bend plant 5 hr per day, Grafton plant not used; Min cost is \$1,150
(B) West Bend plant 15 hr per day, Cedarburg and Grafton plants not used; Min cost is \$1,350
(C) Grafton plant 10 hr per day, West Bend plant 10 hr per day, Cedarburg plant not used; Min cost is \$1,650

47. (A) Let: x_1 = Number of single-sided drives ordered from Associated Electronics
x_2 = Number of double-sided drives ordered from Associated Electronics
x_3 = Number of single-sided drives ordered from Digital Drives
x_4 = Number of double-sided drives ordered from Digital Drives

Minimize $C = 250x_1 + 350x_2 + 290x_3 + 320x_4$
Subject to
$$\begin{aligned} x_1 + x_2 \qquad\qquad &\leqslant 1{,}000 \\ x_3 + x_4 &\leqslant 2{,}000 \\ x_1 \qquad + x_3 \qquad &\geqslant 1{,}200 \\ x_2 \qquad + x_4 &\geqslant 1{,}600 \\ x_1, x_2, x_3, x_4 &\geqslant 0 \end{aligned}$$

1,000 single-sided drives from Associated Electronics, 200 single-sided and 1,600 double-sided drives from Digital Drives; Min cost is \$820,000

49. Let: x_1 = Number of ounces of food L
x_2 = Number of ounces of food M
x_3 = Number of ounces of food N

Minimize $C = 20x_1 + 24x_2 + 18x_3$
Subject to $20x_1 + 10x_2 + 10x_3 \geqslant 300$
$10x_1 + 10x_2 + 10x_3 \geqslant 200$
$10x_1 + 15x_2 + 10x_3 \geqslant 240$
$x_1, x_2, x_3 \geqslant 0$

10 oz *L*, 8 oz *M*, 2 oz *N*; Min cholesterol intake is 428 units

51. Let: x_1 = Number of students bused from North Division to Central
x_2 = Number of students bused from North Division to Washington
x_3 = Number of students bused from South Division to Central
x_4 = Number of students bused from South Division to Washington

Minimize $C = 5x_1 + 2x_2 + 3x_3 + 4x_4$
Subject to
$$\begin{aligned} x_1 + x_2 \qquad\qquad &\geqslant 300 \\ x_3 + x_4 &\geqslant 500 \\ x_1 \qquad + x_3 \qquad &\leqslant 400 \\ x_2 \qquad + x_4 &\leqslant 500 \\ x_1, x_2, x_3, x_4 &\geqslant 0 \end{aligned}$$

300 students bused from North Division to Washington, 400 from South Division to Central, and 100 from South Division to Washington; Min cost is \$2,200

Exercise 5-6

1. (A) Maximize $P = 5x_1 + 2x_2 - Ma_1$
Subject to
$$\begin{aligned} x_1 + 2x_2 + s_1 \qquad\qquad &= 12 \\ x_1 + x_2 \qquad - s_2 + a_1 &= 4 \\ x_1, x_2, s_1, s_2, a_1 &\geqslant 0 \end{aligned}$$

(B)
$$\begin{array}{c} \begin{matrix} x_1 & x_2 & s_1 & s_2 & a_1 & P \end{matrix} \\ \left[\begin{array}{cccccc|c} 1 & 2 & 1 & 0 & 0 & 0 & 12 \\ 1 & 1 & 0 & -1 & 1 & 0 & 4 \\ \hline -M-5 & -M-2 & 0 & M & 0 & 1 & -4M \end{array}\right] \end{array}$$

(C) $x_1 = 12, x_2 = 0, s_1 = 0, s_2 = 8, a_1 = 0, P = 60$ (D) Max $P = 60$ at $x_1 = 12$ and $x_2 = 0$

3. (A) Maximize $P = 3x_1 + 5x_2 - Ma_1$
Subject to
$$\begin{aligned} 2x_1 + x_2 + s_1 \quad &= 8 \\ x_1 + x_2 \quad + a_1 &= 6 \\ x_1, x_2, s_1, a_1 &\geq 0 \end{aligned}$$

(B)

x_1	x_2	s_1	a_1	P	
2	1	1	0	0	8
1	1	0	1	0	6
$-M - 3$	$-M - 5$	0	0	1	$-6M$

(C) $x_1 = 0, x_2 = 6, s_1 = 2, a_1 = 0, P = 30$ (D) Max $P = 30$ at $x_1 = 0$ and $x_2 = 6$

5. (A) Maximize $P = 4x_1 + 3x_2 - Ma_1$
Subject to
$$\begin{aligned} -x_1 + 2x_2 + s_1 \quad &= 2 \\ x_1 + x_2 \quad - s_2 + a_1 &= 4 \\ x_1, x_2, s_1, s_2, a_1 &\geq 0 \end{aligned}$$

(B)

x_1	x_2	s_1	s_2	a_1	P	
-1	2	1	0	0	0	2
1	1	0	-1	1	0	4
$-M - 4$	$-M - 3$	0	M	0	1	$-4M$

(C) No optimal solution exists. (D) No optimal solution exists.

7. (A) Maximize $P = 5x_1 + 10x_2 - Ma_1$
Subject to
$$\begin{aligned} x_1 + x_2 + s_1 \quad &= 3 \\ 2x_1 + 3x_2 \quad - s_2 + a_1 &= 12 \\ x_1, x_2, s_1, s_2, a_1 &\geq 0 \end{aligned}$$

(B)

x_1	x_2	s_1	s_2	a_1	P	
1	1	1	0	0	0	3
2	3	0	-1	1	0	12
$-2M - 5$	$-3M - 10$	0	M	0	1	$-12M$

(C) $x_1 = 0, x_2 = 3, s_1 = 0, s_2 = 0, a_1 = 3, P = -3M + 30$ (D) No optimal solution exists.

9. Min $P = 1$ at $x_1 = 3$ and $x_2 = 5$; Max $P = 16$ at $x_1 = 8$ and $x_2 = 0$ **11.** Max $P = 44$ at $x_1 = 2$ and $x_2 = 8$

13. No optimal solution exists. **15.** Min $C = -9$ at $x_1 = 0$, $x_2 = \frac{7}{4}$, and $x_3 = \frac{3}{4}$

17. Max $P = 32$ at $x_1 = 0$, $x_2 = 4$, and $x_3 = 2$ **19.** Max $P = 65$ at $x_1 = \frac{35}{2}$, $x_2 = 0$, and $x_3 = \frac{15}{2}$

21. Max $P = 120$ at $x_1 = 20$, $x_2 = 0$, and $x_3 = 20$

23. Problem 5—Unbounded feasible region:

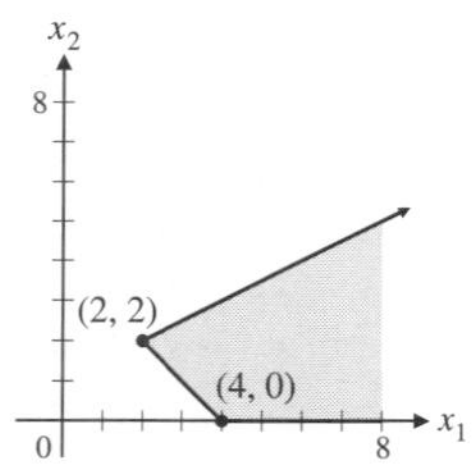

Problem 7—Empty feasible region:

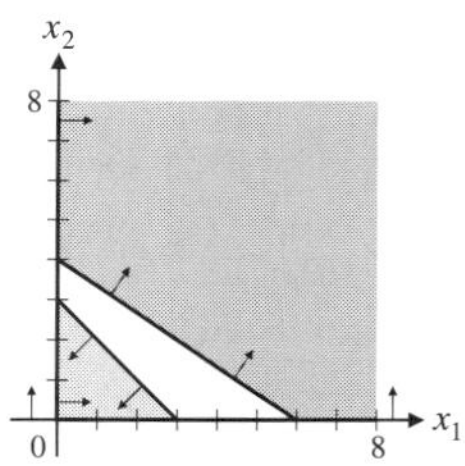

25. Min $C = -30$ at $x_1 = 0$, $x_2 = \frac{3}{4}$, and $x_3 = 0$ **27.** Max $P = 17$ at $x_1 = \frac{49}{5}$, $x_2 = 0$, and $x_3 = \frac{22}{5}$

29. Min $C = \frac{135}{2}$ at $x_1 = \frac{15}{4}$, $x_2 = \frac{3}{4}$, and $x_3 = 0$ **31.** Max $P = 372$ at $x_1 = 28$, $x_2 = 4$, and $x_3 = 0$

33. Let: x_1 = Number of 16k modules manufactured daily
x_2 = Number of 64k modules manufactured daily

Maximize $P = 18x_1 + 30x_2$
Subject to
$$\begin{aligned} 10x_1 + 15x_2 &\leq 2{,}200 \\ 2x_1 + 4x_2 &\leq 500 \\ x_1 \quad &\geq 50 \\ x_1, x_2 &\geq 0 \end{aligned}$$

130 of the 16k modules and 60 of the 64k modules; Max profit is \$4,140

35. Let: x_1 = Number of ads placed in the *Sentinel*
x_2 = Number of ads placed in the *Journal*
x_3 = Number of ads placed in the *Tribune*

Minimize $C = 200x_1 + 200x_2 + 100x_3$
Subject to
$$\begin{aligned} x_1 + x_2 + x_3 &\leq 10 \\ 2{,}000x_1 + 500x_2 + 1{,}500x_3 &\geq 16{,}000 \\ x_1, x_2, x_3 &\geq 0 \end{aligned}$$

2 ads in the *Sentinel,* 0 ads in the *Journal,* 8 ads in the *Tribune;* Min cost is \$1,200

37. Let: x_1 = Number of bottles of brand A
x_2 = Number of bottles of brand B
x_3 = Number of bottles of brand C

Minimize $C = 0.6x_1 + 0.4x_2 + 0.9x_3$
Subject to
$$\begin{aligned} 10x_1 + 10x_2 + 20x_3 &\geq 100 \\ 2x_1 + 3x_2 + 4x_3 &\leq 24 \\ x_1, x_2, x_3 &\geq 0 \end{aligned}$$

0 bottles of A, 4 bottles of B, 3 bottles of C; Min cost is \$4.30

39. Let: x_1 = Number of cubic yards of mix A
x_2 = Number of cubic yards of mix B
x_3 = Number of cubic yards of mix C

$$\begin{aligned} &\text{Maximize} && P = 12x_1 + 16x_2 + 8x_3 \\ &\text{Subject to} && 12x_1 + 8x_2 + 16x_3 \leqslant 700 \\ &&& 16x_1 + 8x_2 + 16x_3 \geqslant 800 \\ &&& x_1, x_2, x_3 \geqslant 0 \end{aligned}$$

25 yd³ A, 50 yd³ B, 0 yd³ C: Max is 1,100 lb

41. Let: x_1 = Number of car frames produced at the Milwaukee plant
x_2 = Number of truck frames produced at the Milwaukee plant
x_3 = Number of car frames produced at the Racine plant
x_4 = Number of truck frames produced at the Racine plant

$$\begin{aligned} &\text{Maximize} && P = 50x_1 + 70x_2 + 50x_3 + 70x_4 \\ &\text{Subject to} && x_1 + x_3 \leqslant 250 \\ &&& x_2 + x_4 \leqslant 350 \\ &&& x_1 + x_2 \leqslant 300 \\ &&& x_3 + x_4 \leqslant 200 \\ &&& 150x_1 + 200x_2 \leqslant 50{,}000 \\ &&& 135x_3 + 180x_4 \leqslant 35{,}000 \\ &&& x_1, x_2, x_3, x_4 \geqslant 0 \end{aligned}$$

43. Let: x_1 = Number of barrels of A used in regular gasoline
x_2 = Number of barrels of A used in premium gasoline
x_3 = Numbers of barrels of B used in regular gasoline
x_4 = Number of barrels of B used in premium gasoline
x_5 = Numbers of barrels of C used in regular gasoline
x_6 = Number of barrels of C used in premium gasoline

$$\begin{aligned} &\text{Maximize} && P = 10x_1 + 18x_2 + 8x_3 + 16x_4 + 4x_5 + 12x_6 \\ &\text{Subject to} && x_1 + x_2 \leqslant 40{,}000 \\ &&& x_3 + x_4 \leqslant 25{,}000 \\ &&& x_5 + x_6 \leqslant 15{,}000 \\ &&& x_1 + x_3 + x_5 \geqslant 30{,}000 \\ &&& x_2 + x_4 + x_6 \geqslant 25{,}000 \\ &&& -5x_1 + 5x_3 + 15x_5 \geqslant 0 \\ &&& -15x_2 - 5x_4 + 5x_6 \geqslant 0 \\ &&& x_1, x_2, x_3, x_4, x_5, x_6 \geqslant 0 \end{aligned}$$

45. Let: x_1 = Percentage invested in high-tech funds
x_2 = Percentage invested in global funds
x_3 = Percentage invested in corporate bonds
x_4 = Percentage invested in municipal bonds
x_5 = Percentage invested in CD's

$$\begin{aligned} &\text{Maximize} && P = 0.11x_1 + 0.1x_2 + 0.09x_3 + 0.08x_4 + 0.05x_5 \\ &\text{Subject to} && x_1 + x_2 + x_3 + x_4 + x_5 = 1 \\ &&& 2.7x_1 + 1.8x_2 + 1.2x_3 + 0.5x_4 \leqslant 1.8 \\ &&& x_5 \geqslant 0.2 \\ &&& x_1, x_2, x_3, x_4, x_5 \geqslant 0 \end{aligned}$$

47. Let: x_1 = Number of ounces of food L
x_2 = Number of ounces of food M
x_3 = Number of ounces of food N

$$\begin{aligned} &\text{Minimize} && C = 0.4x_1 + 0.6x_2 + 0.8x_3 \\ &\text{Subject to} && 30x_1 + 10x_2 + 30x_3 \geqslant 400 \\ &&& 10x_1 + 10x_2 + 10x_3 \geqslant 200 \\ &&& 10x_1 + 30x_2 + 20x_3 \geqslant 300 \\ &&& 8x_1 + 4x_2 + 6x_3 \leqslant 150 \\ &&& 60x_1 + 40x_2 + 50x_3 \leqslant 900 \\ &&& x_1, x_2, x_3 \geqslant 0 \end{aligned}$$

49. Let: x_1 = Number of students from town A enrolled in school I
x_2 = Number of students from town A enrolled in school II
x_3 = Number of students from town B enrolled in school I
x_4 = Number of students from town B enrolled in school II
x_5 = Number of students from town C enrolled in school I
x_6 = Number of students from town C enrolled in school II

$$\begin{aligned} &\text{Minimize} && C = 4x_1 + 8x_2 + 6x_3 + 4x_4 + 3x_5 + 9x_6 \\ &\text{Subject to} && x_1 + x_2 = 500 \\ &&& x_3 + x_4 = 1{,}200 \\ &&& x_5 + x_6 = 1{,}800 \\ &&& x_1 + x_3 + x_5 \leqslant 2{,}000 \\ &&& x_2 + x_4 + x_6 \leqslant 2{,}000 \\ &&& x_1 + x_3 + x_5 \geqslant 1{,}400 \\ &&& x_2 + x_4 + x_6 \geqslant 1{,}400 \\ &&& x_1 \leqslant 300 \\ &&& x_2 \leqslant 300 \\ &&& x_3 \leqslant 720 \\ &&& x_4 \leqslant 720 \\ &&& x_5 \leqslant 1{,}080 \\ &&& x_6 \leqslant 1{,}080 \\ &&& x_1, x_2, x_3, x_4, x_5, x_6 \geqslant 0 \end{aligned}$$

Review Exercise

1. Bounded *(5-1)*

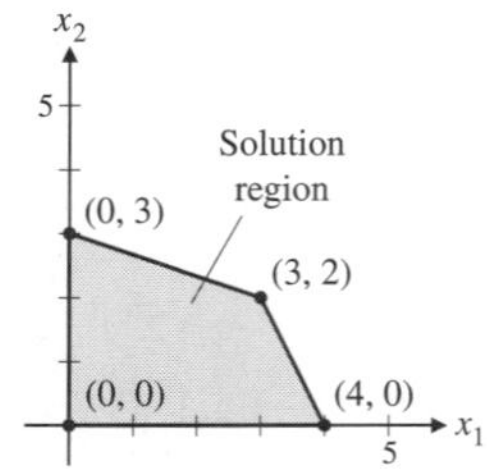

2. Unbounded *(5-1)*

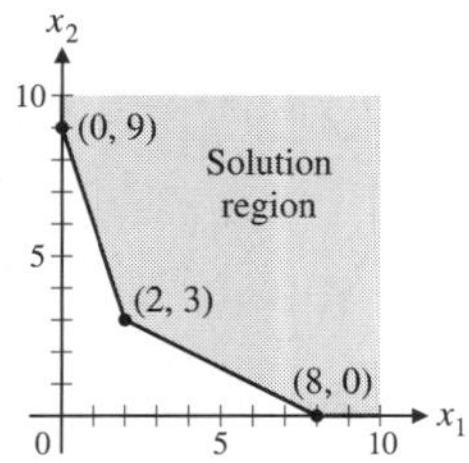

3. Max $P = 24$ at $x_1 = 4$ and $x_2 = 0$ *(5-2)*

4.
$$\begin{aligned} 2x_1 + x_2 + s_1 \quad &= 8 \\ x_1 + 2x_2 \quad + s_2 &= 10 \end{aligned}$$ *(5-3)*

5. 2 basic and 2 nonbasic variables *(5-3)*

6.

x_1	x_2	s_1	s_2	FEASIBLE?
0	0	8	10	Yes
0	8	0	−6	No
0	5	3	0	Yes
4	0	0	6	Yes
10	0	−12	0	No
2	4	0	0	Yes

(5-3)

7. Enter ↓ (column x_1); Exit → (row s_1)

$$\begin{array}{c} \\ s_1 \\ s_2 \\ P \end{array} \begin{array}{c} \begin{array}{cccccc} x_1 & x_2 & s_1 & s_2 & P & \end{array} \\ \left[\begin{array}{ccccc|c} \textcircled{2} & 1 & 1 & 0 & 0 & 8 \\ 1 & 2 & 0 & 1 & 0 & 10 \\ \hline -6 & -2 & 0 & 0 & 1 & 0 \end{array}\right] \end{array}$$ *(5-4)*

8. Max $P = 24$ at $x_1 = 4$ and $x_2 = 0$ *(5-4)*

9. Basic variables: x_2, s_2, s_3, P; nonbasic variables: x_1, x_3, s_1

Enter ↓ (column x_1); Exit → (row s_3)

$$\begin{array}{c|ccccccc|c} & x_1 & x_2 & x_3 & s_1 & s_2 & s_3 & P & \\ \hline x_2 & 2 & 1 & 3 & -1 & 0 & 0 & 0 & 20 \\ s_2 & 3 & 0 & 4 & 1 & 1 & 0 & 0 & 30 \\ s_3 & \textcircled{2} & 0 & 5 & 2 & 0 & 1 & 0 & 10 \\ \hline P & -8 & 0 & -5 & 3 & 0 & 0 & 1 & 50 \end{array} \sim \begin{array}{c|ccccccc|c} & x_1 & x_2 & x_3 & s_1 & s_2 & s_3 & P & \\ \hline x_2 & 0 & 1 & -2 & -3 & 0 & -1 & 0 & 10 \\ s_2 & 0 & 0 & -\frac{7}{2} & -2 & 1 & -\frac{3}{2} & 0 & 15 \\ x_1 & 1 & 0 & \frac{5}{2} & 1 & 0 & \frac{1}{2} & 0 & 5 \\ \hline P & 0 & 0 & 15 & 11 & 0 & 4 & 1 & 90 \end{array}$$ *(5-4)*

10. (A) $x_1 = 0, x_2 = 2, s_1 = 0, s_2 = 5, P = 12$; additional pivoting required
(B) $x_1 = 0, x_2 = 0, s_1 = 0, s_2 = 7, P = 22$; no optimal solution exists
(C) $x_1 = 6, x_2 = 0, s_1 = 15, s_2 = 0, P = 10$; optimal solution *(5-4)*

11. Min $C = 40$ at $x_1 = 0$ and $x_2 = 20$ *(5-2)*

12. Maximize $P = 15y_1 + 20y_2$
Subject to
$$\begin{aligned} y_1 + 2y_2 &\le 5 \\ 3y_1 + y_2 &\le 2 \\ y_1, y_2 &\ge 0 \end{aligned}$$ *(5-5)*

13.
$$\begin{aligned} y_1 + 2y_2 + x_1 \qquad\qquad &= 5 \\ 3y_1 + y_2 \qquad + x_2 \qquad &= 2 \\ -15y_1 - 20y_2 \qquad\qquad + P &= 0 \end{aligned}$$ *(5-5)*

14.
$$\begin{array}{ccccc|c} y_1 & y_2 & x_1 & x_2 & P & \\ \hline 1 & 2 & 1 & 0 & 0 & 5 \\ 3 & 1 & 0 & 1 & 0 & 2 \\ \hline -15 & -20 & 0 & 0 & 1 & 0 \end{array}$$ *(5-5)*

15. Max $P = 40$ at $y_1 = 0$ and $y_2 = 2$ *(5-4)*

16. Min $C = 40$ at $x_1 = 0$ and $x_2 = 20$ *(5-5)*

17. Max $P = 26$ at $x_1 = 2$ and $x_2 = 5$ *(5-2)*

18. Max $P = 26$ at $x_1 = 2$ and $x_2 = 5$ *(5-4)*

19. Min $C = 51$ at $x_1 = 9$ and $x_2 = 3$ *(5-2)*

20. Maximize $P = 10y_1 + 15y_2 + 3y_3$
Subject to
$$\begin{aligned} y_1 + y_2 \qquad &\le 3 \\ y_1 + 2y_2 + y_3 &\le 8 \\ y_1, y_2, y_3 &\ge 0 \end{aligned}$$ *(5-5)*

21. Min $C = 51$ at $x_1 = 9$ and $x_2 = 3$ *(5-5)*

22. No optimal solution exists. *(5-4)*

23. Max $P = 23$ at $x_1 = 4$, $x_2 = 1$, and $x_3 = 0$ *(5-4)*

24. (A) Modified problem:
Maximize $P = x_1 + 3x_2 - Ma_1$
Subject to
$$\begin{aligned} x_1 + x_2 - s_1 + a_1 \qquad &= 6 \\ x_1 + 2x_2 \qquad\qquad + s_2 &= 8 \\ x_1, x_2, s_1, s_2, a_1 &\ge 0 \end{aligned}$$

(B) Preliminary simplex tableau:

$$\begin{array}{cccccc|c} x_1 & x_2 & s_1 & a_1 & s_2 & P & \\ \hline 1 & 1 & -1 & 1 & 0 & 0 & 6 \\ 1 & 2 & 0 & 0 & 1 & 0 & 8 \\ \hline -1 & -3 & 0 & M & 0 & 1 & 0 \end{array}$$

Initial simplex tableau:

$$\begin{array}{cccccc|c} x_1 & x_2 & s_1 & a_1 & s_2 & P & \\ \hline 1 & 1 & -1 & 1 & 0 & 0 & 6 \\ 1 & 2 & 0 & 0 & 1 & 0 & 8 \\ \hline -M-1 & -M-3 & M & 0 & 0 & 1 & -6M \end{array}$$

24. (C) $x_1 = 4, x_2 = 2, s_1 = 0, a_1 = 0, s_2 = 0, P = 10$
(D) Since $a_1 = 0$, the optimal solution to the original problem is Max $P = 10$ at $x_1 = 4$ and $x_2 = 2$ *(5-6)*

25. (A) Modified problem:

$$\begin{aligned} \text{Maximize} \quad & P = x_1 + x_2 - Ma_1 \\ \text{Subject to} \quad & x_1 + x_2 - s_1 + a_1 = 5 \\ & x_1 + 2x_2 + s_2 = 4 \\ & x_1, x_2, s_1, s_2, a_1 \geq 0 \end{aligned}$$

(B) Preliminary simplex tableau:

x_1	x_2	s_1	a_1	s_2	P	
1	1	−1	1	0	0	5
1	2	0	0	1	0	4
−1	−1	0	M	0	1	0

Initial simplex tableau:

x_1	x_2	s_1	a_1	s_2	P	
1	1	−1	1	0	0	5
1	2	0	0	1	0	4
$-M-1$	$-M-1$	M	0	0	1	$-5M$

(C) $x_1 = 4, x_2 = 0, s_1 = 0, s_2 = 0, a_1 = 1, P = -M + 4$
(D) Since $a_1 \neq 0$, the original problem has no optimal solution. *(5-6)*

26.

$$\begin{aligned} \text{Maximize} \quad & P = 2x_1 + 3x_2 + x_3 - Ma_1 - Ma_2 \\ \text{Subject to} \quad & x_1 - 3x_2 + x_3 + s_1 = 7 \\ & x_1 + x_2 - 2x_3 - s_2 + a_1 = 2 \\ & 3x_1 + 2x_2 - x_3 + a_2 = 4 \\ & x_1, x_2, x_3, s_1, s_2, a_1, a_2 \geq 0 \end{aligned}$$

(5-6)

27. The geometric method solves maximization and minimization problems involving two decision variables. If the feasible region is bounded, there are no restrictions on the coefficients or constants. If the feasible region is unbounded and the coefficients of the objective function are positive, then a minimization problem has a solution, but a maximization does not. *(5-2)*

28. The basic simplex method with slack variables solves standard maximization problems involving $\leq$ constraints with nonnegative constants on the right side. *(5-4)*

29. The dual method solves minimization problems with positive coefficients in the objective function. *(5-5)*

30. The big M method solves any linear programming problem. *(5-6)*

31. Max $P = 36$ at $x_1 = 6, x_2 = 8$ *(5-2)*

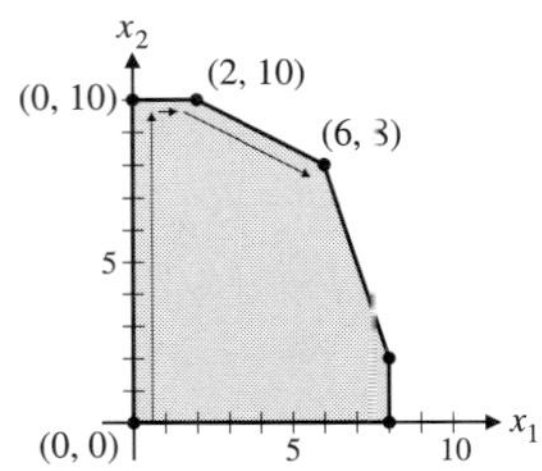

32. Min $C = 15$ at $x_1 = 3$ and $x_2 = 3$ *(5-5)*

33. Min $C = 15$ at $x_1 = 3$ and $x_2 = 3$ *(5-6)*

34. Min $C = 9{,}960$ at $x_1 = 0, x_2 = 240, x_3 = 400$, and $x_4 = 60$ *(5-5)*

35. (A) Let: x_1 = Number of regular sails
x_2 = Number of competition sails

$$\begin{aligned} \text{Maximize} \quad & P = 100x_1 + 200x_2 \\ \text{Subject to} \quad & 2x_1 + 3x_2 \leq 150 \\ & 4x_1 + 10x_2 \leq 380 \\ & x_1, x_2 \geq 0 \end{aligned}$$

Max $P = \$8{,}500$ when 45 regular and 20 competition sails are produced.
(B) Max profit increases to $9,880 when 38 competition and no regular sails are produced.
(C) Max profit decreases to $7,500 when no competition and 75 regular sails are produced. *(5-2)*

36. (A) Let: x_1 = Amount invested in oil stock
x_2 = Amount invested in steel stock
x_3 = Amount invested in government bonds

$$\begin{aligned} \text{Maximize} \quad & P = 0.12x_1 + 0.09x_2 + 0.05x_3 \\ \text{Subject to} \quad & x_1 + x_2 + x_3 \leq 150{,}000 \\ & x_1 \leq 50{,}000 \\ & x_1 + x_2 - x_3 \leq 25{,}000 \\ & x_1, x_2, x_3 \geq 0 \end{aligned}$$

Max return is $12,500 when $50,000 is invested in oil stock, $37,500 is invested in steel stock, and $62,500 in government bonds.
(B) Max return is $13,625 when $87,500 is invested in steel stock and $62,500 in government bonds. *(5-4)*

37. Let: x_1 = Number of motors shipped from factory A to plant X
x_2 = Number of motors shipped from factory A to plant Y
x_3 = Number of motors shipped from factory B to plant X
x_4 = Number of motors shipped from factory B to plant Y

$$\begin{aligned}\text{Minimize} \quad & C = 5x_1 + 8x_2 + 9x_3 + 7x_4 \\ \text{Subject to} \quad & x_1 + x_2 \leqslant 1{,}500 \\ & x_3 + x_4 \leqslant 1{,}000 \\ & x_1 + x_3 \geqslant 900 \\ & x_2 + x_4 \geqslant 1{,}200 \\ & x_1, x_2, x_3, x_4 \geqslant 0\end{aligned}$$

Min C = \$13,100 when 900 motors are shipped from factory A to plant X, 200 motors are shipped from factory A to plant Y, and 1,000 motors are shipped from factory B to plant Y *(5-5)*

38. Let: x_1 = Number of pounds of long-grain rice used in brand A
x_2 = Number of pounds of long-grain rice used in brand B
x_3 = Number of pounds of wild rice used in brand A
x_4 = Number of pounds of wild rice used in brand B

$$\begin{aligned}\text{Maximize} \quad & P = 0.8x_1 + 0.5x_2 - 1.9x_3 - 2.2x_4 \\ \text{Subject to} \quad & 0.1x_1 - 0.9x_3 \leqslant 0 \\ & 0.05x_2 - 0.95x_4 \leqslant 0 \\ & x_1 + x_2 \leqslant 8{,}000 \\ & x_3 + x_4 \leqslant 500 \\ & x_1, x_2, x_3, x_4 \geqslant 0\end{aligned}$$

Max profit is \$3,350 when 1,350 lb long-grain rice and 150 lb wild rice are used to produce 1,500 lb brand A, and 6,650 lb long-grain rice and 350 lb wild rice are used to produce 7,000 lb brand B. *(5-4)*

39. (A) Let: x_1 = Number of grams of mix A
x_2 = Number of grams of mix B

$$\begin{aligned}\text{Minimize} \quad & C = 0.04x_1 + 0.09x_2 \\ \text{Subject to} \quad & 2x_1 + 5x_2 \geqslant 850 \\ & 2x_1 + 4x_2 \geqslant 800 \\ & 4x_1 + 5x_2 \geqslant 1{,}150 \\ & x_1, x_2 \geqslant 0\end{aligned}$$

Min C = \$16.50 when 300 g mix A and 50 g mix B are used.
(B) The minimum cost decreases to \$13.00 when 100 g mix A and 150 g mix B are used.
(C) The minimum cost increases to \$17.00 when 425 g mix A and no mix B are used. *(5-2)*

Chapter 6

Exercise 6-1

1. 115 **3.** 300 **5.** 165 **7.** 210 **9.** 95 **11.** 260

13. (A) 4 ways:

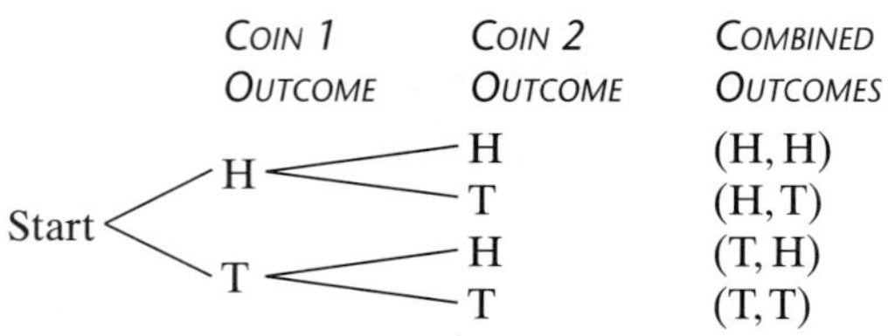

(B) $2 \cdot 2 = 4$ ways

15. (A) 12 combined outcomes:

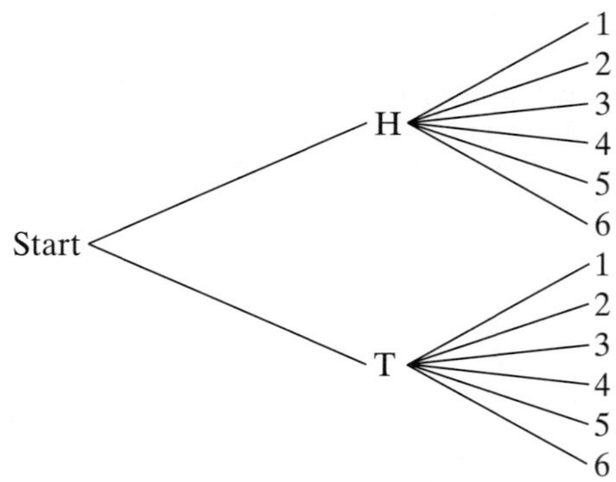

(B) $2 \cdot 6 = 12$ combined outcomes

17. $n(A \cap B') = 60, n(A \cap B) = 20, n(A' \cap B) = 30, n(A' \cap B') = 90$

19. $n(A \cap B') = 5, n(A \cap B) = 20, n(A' \cap B) = 35, n(A' \cap B') = 40$

21.

	A	A'	*Totals*
B	30	60	90
B'	40	70	110
Totals	70	130	200

23.

	A	A'	*Totals*
B	20	35	55
B'	25	20	45
Totals	45	55	100

25. (A) True (B) False
27. $5 \cdot 3 \cdot 4 \cdot 2 = 120$
29. $6 \cdot 5 \cdot 4 \cdot 3 = 360$; $6 \cdot 6 \cdot 6 \cdot 6 = 1{,}296$; $6 \cdot 5 \cdot 5 \cdot 5 = 750$

31. $10 \cdot 9 \cdot 8 \cdot 7 \cdot 6 = 30{,}240$; $10 \cdot 10 \cdot 10 \cdot 10 \cdot 10 = 100{,}000$; $10 \cdot 9 \cdot 9 \cdot 9 \cdot 9 = 65{,}610$
33. $26 \cdot 26 \cdot 26 \cdot 10 \cdot 10 \cdot 10 = 17{,}576{,}000$; $26 \cdot 25 \cdot 24 \cdot 10 \cdot 9 \cdot 8 = 11{,}232{,}000$
35. No, the same 8 combined choices are available either way. **37.** 14 **39.** 14
41. (A) 6 combined outcomes:

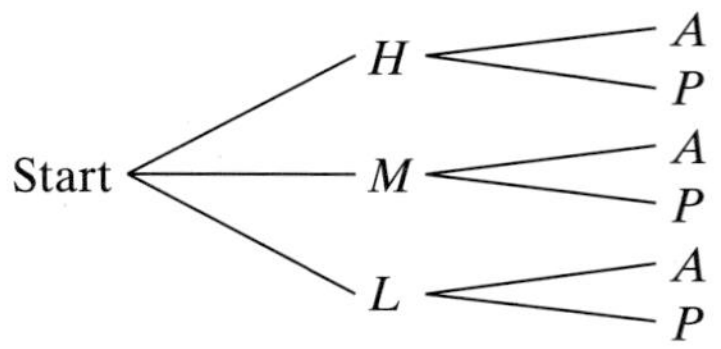

(B) $3 \cdot 2 = 6$

43. 12 **45.** (A) 1,010 (B) 190 (C) 270 **47.** 1,570
49. (A) 102 (B) 689 (C) 1,470 (D) 1,372

51. (A) 12 classifications:

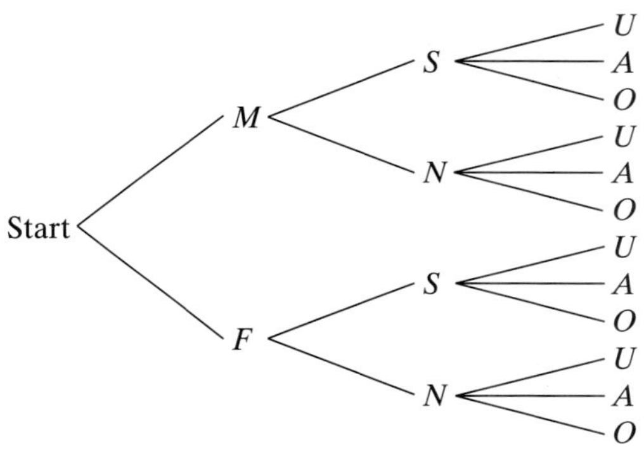

(B) $2 \cdot 2 \cdot 3 = 12$

53. 2,905

Exercise 6-2

1. 24 **3.** 9 **5.** 990 **7.** 10 **9.** 35 **11.** 1 **13.** 60 **15.** 6,497,400 **17.** 10 **19.** 270,725
21. (A) Permutation (B) Combination **23.** $P_{10,3} = 10 \cdot 9 \cdot 8 = 720$ **25.** $C_{7,3} = 35$; $P_{7,3} = 210$
27. The factorial function $x!$ grows much faster than the exponential function 3^x, which in turn grows much faster than the cubic function x^3.
29. $C_{13,5} = 1{,}287$ **31.** $C_{13,5}C_{13,2} = 100{,}386$ **33.** $C_{8,3}C_{10,4}C_{7,2} = 246{,}960$
35. The numbers are the same read up or down, since $C_{n,r} = C_{n,n-r}$.
37. (A) $C_{8,2} = 28$ (B) $C_{8,3} = 56$ (C) $C_{8,4} = 70$ **39.** $P_{5,2} = 20$; $P_{5,3} = 60$; $P_{5,4} = 120$; $P_{5,5} = 120$
41. (A) $P_{8,5} = 6{,}720$ (B) $C_{8,5} = 56$ (C) $2 \cdot C_{6,4} = 30$
43. For many calculators $k = 69$, but your calculator may be different. **45.** $C_{24,12} = 2{,}704{,}156$
47. (A) $C_{24,3} = 2{,}024$ (B) $C_{19,3} = 969$
49. (A) $C_{30,10} = 30{,}045{,}015$ (B) $C_{8,2}C_{12,5}C_{10,3} = 2{,}661{,}120$
51. (A) $C_{6,3}C_{5,2} = 200$ (B) $C_{6,4}C_{5,1} = 75$ (C) $C_{6,5} = 6$ (D) $C_{11,5} = 462$ (E) $C_{6,4}C_{5,1} + C_{6,5} = 81$
53. 336; 512 **55.** $P_{4,2} = 12$

Exercise 6-3

1. Occurrence of E is certain. **3.** $\frac{1}{2}$
5. (A) Reject; no probability can be negative (B) Reject; $P(J) + P(G) + P(P) + P(S) \neq 1$ (C) Acceptable
7. $P(J) + P(P) = .56$ **9.** $\frac{1}{8}$ **11.** $1/P_{10,3} \approx .0014$ **13.** $C_{26,5}/C_{52,5} \approx .025$ **15.** $C_{12,5}/C_{52,5} \approx .000\ 305$
17. $S = \{$All days in a year, 365, excluding leap year$\}$; $\frac{1}{365}$, assuming each day is as likely as any other day for a person to be born

19. $1/P_{5,5} = 1/5! = .008\ 33$ **21.** $\frac{1}{36}$ **23.** $\frac{5}{36}$ **25.** $\frac{1}{6}$ **27.** $\frac{7}{9}$ **29.** 0 **31.** $\frac{1}{3}$ **33.** $\frac{2}{9}$ **35.** $\frac{2}{3}$
37. $\frac{1}{4}$ **39.** $\frac{1}{4}$ **41.** $\frac{3}{4}$
43. (A) Yes (B) Yes, because we would expect, on average, 20 heads in 40 flips; $P(\text{H}) = \frac{37}{40} = .925$; $P(\text{T}) = \frac{3}{40} = .075$
45. $\frac{1}{9}$ **47.** $\frac{1}{3}$ **49.** $\frac{1}{9}$ **51.** $\frac{4}{9}$ **53.** $C_{16,5}/C_{52,5} \approx .001\ 68$ **55.** $48/C_{52,5} \approx .000\ 018\ 5$ **57.** $4/C_{52,5} \approx .000\ 001\ 5$
59. $C_{4,2}C_{4,3}/C_{52,5} \approx .000\ 009$ **61.** (A) $\frac{7}{50} = .14$ (B) $\frac{1}{6} \approx .167$ (C) Answer depends on results of simulation.
63. (A) Represent the outcomes H and T by 1 and 2, respectively, and select 500 random integers from the integers 1 and 2. (B) Answer depends on results of simulation. (C) Each is $\frac{1}{2} = .5$
65. (A) $1/P_{12,4} \approx .000\ 084$ (B) $1/12^4 \approx .000\ 048$
67. (A) $C_{6,3}C_{5,2}/C_{11,5} \approx .433$ (B) $C_{6,4}C_{5,1}/C_{11,5} \approx .162$ (C) $C_{6,5}C_{11,5} \approx .013$ (D) $(C_{6,4}C_{5,1} + C_{6,5})/C_{11,5} \approx .175$
69. (A) $1/P_{8,3} \approx .0030$ (B) $1/8^3 \approx .0020$ **71.** (A) $P_{6,2}/P_{11,2} \approx .273$ (B) $(C_{5,3} + C_{6,1}C_{5,2})/C_{11,3} \approx .424$

Exercise 6-4

1. .997 **3.** (1); $\frac{1}{2}$ **5.** (2); $\frac{7}{10}$ **7.** .4 **9.** .25 **11.** .05 **13.** .2 **15.** .6 **17.** .65 **19.** $\frac{1}{4}$ **21.** $\frac{11}{36}$
23. (A) $\frac{3}{5}; \frac{5}{3}$ (B) $\frac{1}{3}; \frac{3}{1}$ (C) $\frac{2}{3}; \frac{3}{2}$ (D) $\frac{11}{9}; \frac{9}{11}$ **25.** (A) $\frac{3}{11}$ (B) $\frac{11}{18}$ (C) $\frac{4}{5}$ or .8 (D) .49 **27.** (A) False (B) True
29. 1:1 **31.** 7:1 **33.** 2:1 **35.** 1:2 **37.** (A) $\frac{1}{8}$ (B) $8 **39.** (A) .31; $\frac{31}{69}$ (B) .6; $\frac{3}{2}$ **41.** $\frac{11}{26}; \frac{11}{15}$
43. $\frac{7}{13}; \frac{7}{6}$ **45.** .78 **47.** $\frac{250}{1,000} = .25$
49. Either events *A*, *B*, and *C* are mutually exclusive, or events *A* and *B* are not mutually exclusive and the other pairs of events are mutually exclusive.
51. There are fewer calculator steps, and, in addition, 365! produces an overflow error on many calculators, while $P_{365,n}$ does not produce an overflow error for many values of *n*.
53. $P(E) = 1 - \dfrac{12!}{(12 - n)!12^n}$
57. (A) $\frac{10}{50} + \frac{10}{50} = \frac{20}{50} = .4$ (B) $\frac{6}{36} + \frac{5}{36} = \frac{11}{36} \approx .306$ (C) Answer depends on results of simulation.
59. (A) $P(C \cup S) = P(C) + P(S) - P(C \cap S) = .45 + .75 - .35 = .85$ (B) $P(C' \cap S') = .15$
61. (A) $P(M_1 \cup A) = P(M_1) + P(A) - P(M_1 \cap A) = .2 + .3 - .05 = .45$
(B) $P[(M_2 \cap A') \cup (M_3 \cap A')] = P(M_2 \cap A') + P(M_3 \cap A') = .2 + .35 = .55$
63. $P(K' \cap D') = .9$ **65.** .83 **67.** $P(A \cap S) = \frac{50}{1,000} = .05$
69. (A) $P(U \cup N) = .22; \frac{11}{39}$ (B) $P[(D \cap A) \cup (R \cap A)] = .3; \frac{7}{3}$ **71.** $1 - C_{15,3}/C_{20,3} \approx .6$

Exercise 6-5

1. .50 **3.** .20 **5.** .10 **7.** .06 **9.** .50 **11.** .30 **13.** Independent **15.** Dependent
17. (A) $\frac{1}{2}$ (B) $2(\frac{1}{2})^8 \approx .007\ 81$ **19.** (A) $\frac{1}{4}$ (B) Dependent **21.** (A) .18 (B) .26
23. (A) Independent and not mutually exclusive (B) Dependent and mutually exclusive. **25.** $(\frac{1}{2})(\frac{1}{2}) = \frac{1}{4}; \frac{1}{2} + \frac{1}{2} - \frac{1}{4} = \frac{3}{4}$
27. (A) $(\frac{1}{4})(\frac{13}{51}) \approx .0637$ (B) $(\frac{1}{4})(\frac{1}{4}) = .0625$ **29.** (A) $\frac{3}{13}$ (B) Independent **31.** (A) Dependent (B) Independent

33. (A) 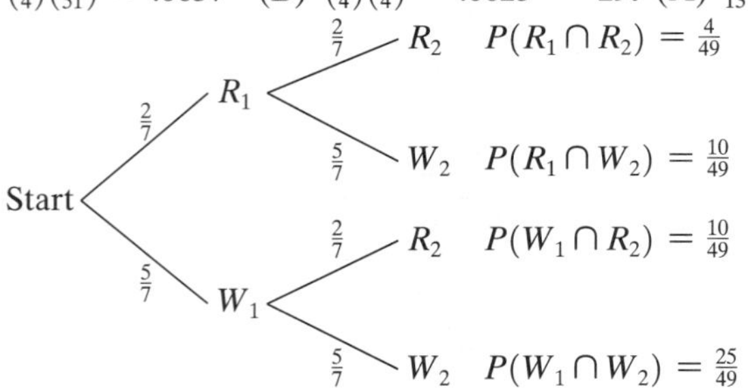

(B) 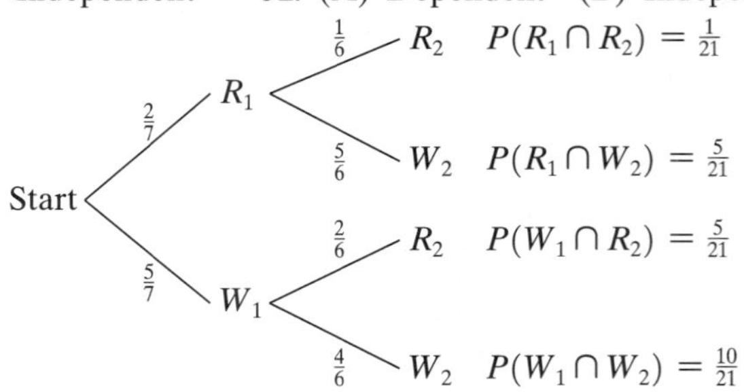

35. (A) $\frac{24}{49}$ (B) $\frac{11}{21}$ **37.** (A) False (B) False **39.** $\frac{5}{18}$ **41.** (A) .167 (B) .25 (C) .25
45. $P(A|A) = P(A \cap A)/P(A) = P(A)/P(A) = 1$ **47.** $P(A)P(B) \neq 0 = P(A \cap B)$

49. (A)

	H	*S*	*B*	*Totals*
Y	.400	.180	.020	.600
N	.150	.120	.130	.400
Totals	.550	.300	.150	1.000

(B) $P(Y|H) = \dfrac{.400}{.550} \approx .727$ (C) $P(Y|B) = \dfrac{.020}{.150} \approx .133$
(D) $P(S) = .300$; $P(S|Y) = .300$ (E) $P(H) = .550$; $P(H|Y) \approx .667$
(F) $P(B \cap N) = .130$ (G) Yes (H) No (I) No

51. (A) .167 (B) .25 (C) .25

53. (A)

	C	C'	*Totals*
R	.06	.44	.50
R'	.02	.48	.50
Totals	.08	.92	1.00

(B) Dependent
(C) $P(C|R) = .12$ and $P(C) = .08$; since $P(C|R) > P(C)$, the red dye should be banned.
(D) $P(C|R) = .04$ and $P(C) = .08$; since $P(C|R) < P(C)$, the red dye should not be banned, since it appears to prevent cancer.

55. (A)

	A	B	C	*Totals*
F	.130	.286	.104	.520
F'	.120	.264	.096	.480
Totals	.250	.550	.200	1.000

(B) $P(A|F) = \dfrac{.130}{.520} = .250$; $P(A|F') = \dfrac{.120}{.480} = .250$
(C) $P(C|F) = \dfrac{.104}{.520} = .200$; $P(C|F') = \dfrac{.096}{.480} = .200$
(D) $P(A) = .250$ (E) $P(B) = .550$; $P(B|F') = .550$
(F) $P(F \cap C) = .104$
(G) No; A, B, and C are independent of F and F'.

Exercise 6-6

1. $(.6)(.8) = .48$ **3.** $(.6)(.8) + (.4)(.3) = .60$ **5.** .80 **7.** .417 **9.** .375 **11.** .222 **13.** .50 **15.** .278

17. .125 **19.** .50 **21.** .375 **23.** Start

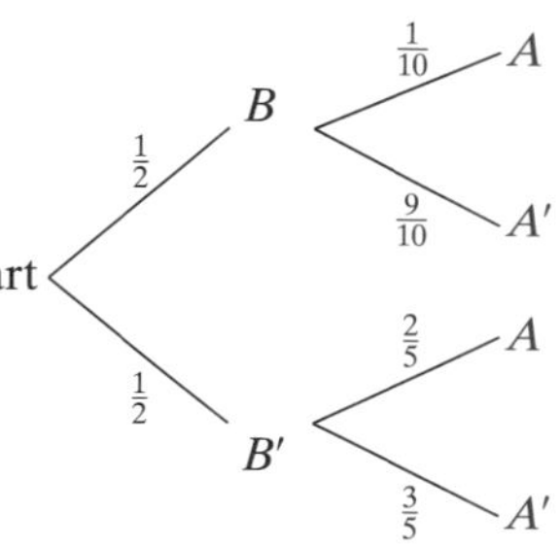

25. .25 **27.** .333 **29.** .50 **31.** .745

33. M and U are independent, which follows from any one of the following: $P(M|U) = c = P(M)$; $P(U|M) = a = P(U)$; $P(M \cap U) = ca = P(M)P(U)$.

35. (A) True (B) True **37.** .235 **39.** $\dfrac{P(U_1 \cap R)}{P(R)} + \dfrac{P(U_1' \cap R)}{P(R)} = \dfrac{P(U_1 \cap R) + P(U_1' \cap R)}{P(R)} = \dfrac{P(R)}{P(R)} = 1$

41. .913; .226 **43.** .091; .545; .364 **45.** .667; .000 412 **47.** .231; .036 **49.** .941; .0588

Exercise 6-7

1. $E(X) = -.1$

3. Probability distribution:

x_i	0	1	2
p_i	$\frac{1}{4}$	$\frac{1}{2}$	$\frac{1}{4}$

$E(X) = 1$

5. Payoff table:

x_i	\$1	−\$1
p_i	$\frac{1}{2}$	$\frac{1}{2}$

$E(X) = 0$; game is fair

7. Payoff table:

x_i	−\$3	−\$2	−\$1	\$0	\$1	\$2
p_i	$\frac{1}{6}$	$\frac{1}{6}$	$\frac{1}{6}$	$\frac{1}{6}$	$\frac{1}{6}$	$\frac{1}{6}$

$E(X) = -50¢$; game is not fair

9. −\$0.50 **11.** −\$0.036; \$0.036

13. \$40. Let x = amount you should lose if a 6 turns up. Set up a payoff table; then set the expected value of the game equal to zero and solve for x.

15. Win \$1 **17.** A_2; \$210

19. Payoff table:

x_i	\$35	−\$1
p_i	$\frac{1}{38}$	$\frac{37}{38}$

$E(X) = -5.26¢$

21. .002

23. Payoff table:

x_i	\$499	\$99	\$19	\$4	−\$1
p_i	.0002	.0006	.001	.004	.9942

$E(X) = -80¢$

25. (A)

x_i	0	1	2
p_i	$\frac{7}{15}$	$\frac{7}{15}$	$\frac{1}{15}$

(B) .60

27. (A)

x_i	−\$5	\$195	\$395	\$595
p_i	.985	.0149	.000 059 9	.000 000 06

(B) $E(X) \approx -\$2$

29. (A) −\$92 (B) The value per game is $\frac{-\$92}{200} = -\$0.46$, compared with an expected value of −\$0.0526.

(C) The simulated gain or loss depends on the results of the simulation; the expected loss is \$26.32.

31. Payoff table:

x_i	\$4,850	−\$150
p_i	.01	.99

$E(X) = -\$100$

33. Site A, with $E(X) = \$3.6$ million

35. 1.54

37. For A_1, $E(X) = \$4$, and for A_2, $E(X) = \$4.80$; A_2 is better

Review Exercise

1. (A) 12 combined outcomes:

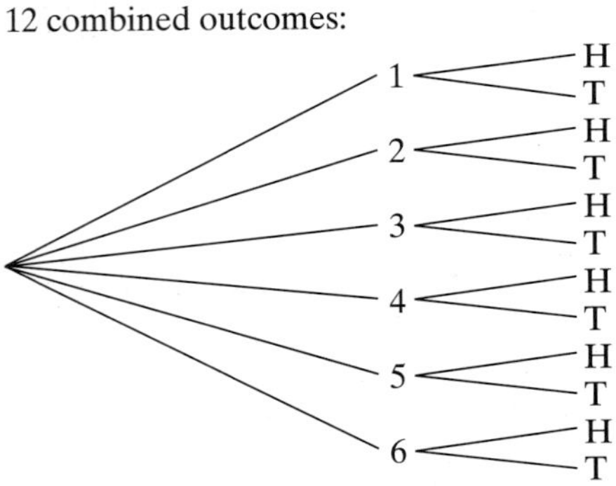

(B) $6 \cdot 2 = 12$ *(6-1)*

2. (A) 65 (B) 75 (C) 35 (D) 105 (E) 150 (F) 85 (G) 115 (H) 45 *(6-1)*

3. 15; 30 *(6-2)* **4.** $6 \cdot 5 \cdot 4 \cdot 3 \cdot 2 \cdot 1 = 720$ *(6-1)*

5. $P_{6,6} = 6! = 720$ *(6-2)*

6. $C_{13,5}/C_{52,5} \approx .0005$ *(6-3)*

7. $1/P_{15,2} \approx .0048$ *(6-3)*

8. $1/P_{10,3} \approx .0014$; $1/C_{10,3} \approx .0083$ *(6-3)*

9. .05 *(6-3)*

10. Payoff table:

x_i	−\$2	−\$1	\$0	\$1	\$2
p_i	$\frac{1}{5}$	$\frac{1}{5}$	$\frac{1}{5}$	$\frac{1}{5}$	$\frac{1}{5}$

$E(X) = 0$; game is fair *(6-7)*

11. (A) .7 (B) .6 *(6-4)* **12.** $P(R \cup G) = .8$; odds for $R \cup G$ are 8 to 2 *(6-4)* **13.** $\frac{5}{11} \approx .455$ *(6-4)*
14. .27 *(6-5)* **15.** .20 *(6-5)* **16.** .02 *(6-5)* **17.** .03 *(6-5)* **18.** .15 *(6-5)* **19.** .1304 *(6-5)* **20.** .1*(6-5)*
21. No, since $P(T|Z) \neq P(T)$ *(6-5)* **22.** Yes, since $P(S \cap X) = P(S)P(X)$ *(6-5)* **23.** .4 *(6-5)* **24.** .2 *(6-5)*
25. .3 *(6-5)* **26.** .08 *(6-5)* **27.** .18 *(6-5)* **28.** .26 *(6-5)* **29.** .31 *(6-6)* **30.** .43 *(6-6)*

31. (A) $\frac{10}{32}$ (B) $\frac{1}{4}$

(C) As the sample in part (A) is increased in size, approximate empirical probabilities should approach the theoretical probabilities. *(6-3)*

32. (A) True (B) False *(6-4)* **33.** (A) False (B) False *(6-5)*

34. Payoff table:

x_i	\$5	−\$4	\$2
p_i	.25	.5	.25

$E(X) = -25¢$; game is not fair *(6-7)*

35. (A) $\frac{1}{3}$ (B) $\frac{2}{9}$ *(6-5)*

36. (A) $\frac{2}{13}$; 2 to 11 (B) $\frac{4}{13}$; 4 to 9 (C) $\frac{12}{13}$; 12 to 1 *(6-4)*

37. (A) 1 to 8 (B) \$8 *(6-4)*

38. (A) $P(2 \text{ heads}) = .21$, $P(1 \text{ head}) = .48$, $P(0 \text{ heads}) = .31$
(B) $P(2 \text{ heads}) = .25$, $P(1 \text{ head}) = .50$, $P(0 \text{ heads}) = .25$
(C) 2 heads, 250; 1 head, 500; 0 heads, 250 *(6-3, 6-7)*

39. 5 children, 15 grandchildren, and 30 great grandchildren, for a total of 50 descendants *(6-1, 6-2)*

40. $\frac{1}{2}$; since the coin has no memory, the 10th toss is independent of the preceding 9 tosses. *(6-5)*

41. (A)

x_i	2	3	4	5	6	7	8	9	10	11	12
p_i	$\frac{1}{36}$	$\frac{2}{36}$	$\frac{3}{36}$	$\frac{4}{36}$	$\frac{5}{36}$	$\frac{6}{36}$	$\frac{5}{36}$	$\frac{4}{36}$	$\frac{3}{36}$	$\frac{2}{36}$	$\frac{1}{36}$

(B) $E(X) = 7$ *(6-7)*

42. $A = \{(1,3), (2,2), (3,1), (2,6), (3,5), (4,4), (5,3), (6,2), (6,6)\}$;
$B = \{(1,5), (2,4), (3,3), (4,2), (5,1), (6,6)\}$; $P(A) = \frac{1}{4}$; $P(B) = \frac{1}{6}$; $P(A \cap B) = \frac{1}{36}$; $P(A \cup B) = \frac{7}{18}$ *(6-4)*

43. (1) Probability of an event cannot be negative; (2) sum of probabilities of simple events must be 1; (3) probability of an event cannot be greater than 1. *(6-3)*

44.

	A	A'	*Totals*
B	15	30	45
B'	35	20	55
Totals	50	50	100

(6-4)

45. (A) .6 (B) $\frac{5}{6}$ *(6-5)* **46.** (A) $\frac{1}{13}$ (B) Independent *(6-5)*
47. 336; 512; 392 *(6-1)* **48.** (A) $P_{6,3} = 120$ (B) $C_{5,2} = 10$ *(6-2)*
49. $C_{25,12} = C_{25,13} = 5{,}200{,}300$ *(6-2)* **50.** (A) $\frac{6}{25}$ (B) $\frac{3}{10}$ *(6-5)*
51. Part (B) *(6-5)* **52.** (A) 1.2 (B) 1.2 *(6-7)*
53. (A) $\frac{3}{5}$ (B) $\frac{1}{3}$ (C) $\frac{7}{15}$ (D) $\frac{9}{14}$ (E) $\frac{5}{8}$ (F) $\frac{3}{10}$ *(6-5, 6-6)* **54.** No *(6-5)*
55. (A) $C_{13,5}/C_{52,5}$ (B) $C_{13,3} \cdot C_{13,2}/C_{52,5}$ *(6-3)* **56.** $C_{8,2}/C_{10,4} = \frac{2}{15}$ *(6-3)* **57.** $N_1 \cdot N_2 \cdot N_3$ *(6-1)*
58. Events S and H are mutually exclusive. Hence, $P(S \cap H) = 0$, while $P(S) \neq 0$ and $P(H) \neq 0$. Therefore, $P(S \cap H) \neq P(S)P(H)$, which implies S and H are dependent. *(6-5)*
59. (A) $\frac{9}{50} = .18$ (B) $\frac{9}{36} = .25$
(C) The empirical probability depends on the results of the simulation; the theoretical probability is $\frac{5}{36} \approx .139$. *(6-3)*
60. The empirical probability depends on the results of the simulation; the theoretical probability is $\frac{2}{52} \approx .038$. *(6-3)*
61. (A) .350 (B) $\frac{3}{8} = .375$ (C) 375 *(6-3)* **62.** $-.0172$; .0172; no *(6-7)*
63. (A) $P_{10,3} = 720$ (B) $P_{6,3}/P_{10,3} = \frac{1}{6}$ (C) $C_{10,3} = 120$ (D) $(C_{6,3} + C_{6,2} \cdot C_{4,1})/C_{10,3} = \frac{2}{3}$ *(6-1, 6-2, 6-3)*
64. 33 *(6-1)* **65.** $1 - C_{7,3}/C_{10,3} = \frac{17}{24}$ *(6-3)* **66.** $\frac{12}{51} \approx .235$ *(6-5)* **67.** $\frac{12}{51} \approx .235$ *(6-5)*
68. (A)

x_i	2	3	4	5	6
p_i	$\frac{9}{36}$	$\frac{12}{36}$	$\frac{10}{36}$	$\frac{4}{36}$	$\frac{1}{36}$

(B) $E(X) = \frac{10}{3}$ *(6-7)*

69. $E(X) \approx -\$0.167$; no; $\$(10/3) \approx \$3.33$ *(6-7)* **70.** $2^5 = 32$; 6 *(6-1)*
71. (A) $\frac{1}{4}$; 1 to 3 (B) \$3 *(6-4, 6-6)* **72.** $1 - 10!/(5!10^5) \approx .70$ *(6-4)*
73. $P_{5,5} = 120$ *(6-1)* **74.** (A) 610 (B) 390 (C) 270 *(6-1)*
75. (A) .8 (B) .2 (C) .5 *(6-1, 6-4)*
76. $P(A \cap P) = P(A)P(P|A) = .34$ *(6-5)*
77. (A) $P(A) = .290$; $P(B) = .290$; $P(A \cap B) = .100$; $P(A|B) = .345$; $P(B|A) = .345$
(B) No, since $P(A \cap B) \neq P(A)P(B)$.
(C) $P(C) = .880$; $P(D) = .120$; $P(C \cap D) = 0$; $P(C|D) = 0$; $P(D|C) = 0$
(D) Yes, since $C \cap D = \varnothing$; dependent, since $P(C \cap D) = 0$ and $P(C)P(D) \neq 0$ *(6-5)*
78. Plan *A:* $E(X) = \$7.6$ million; plan *B:* $E(X) = \$7.8$ million; plan *B* *(6-7)*
79. Payoff table:

x_i	\$270	$-\$30$
p_i	.08	.92

$E(X) = -\$6$ *(6-7)*

80. $1 - (C_{10,4}/C_{12,4}) \approx .576$ *(6-2, 6-4)*
81. (A)

x_i	0	1	2
p_i	$\frac{12}{22}$	$\frac{9}{22}$	$\frac{1}{22}$

(B) $E(X) = \frac{1}{2}$ *(6-7)*

82. .955 *(6-6)* **83.** $\frac{6}{7} \approx .857$ *(6-6)*

Chapter 7

Exercise 7-1

1.

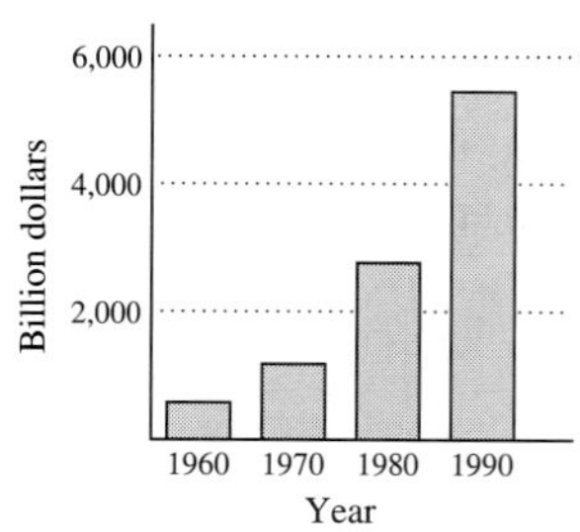

3. Approx. \$2,200 in 1984, approx. \$2,500 in 1994; total taxes; recreation

5.

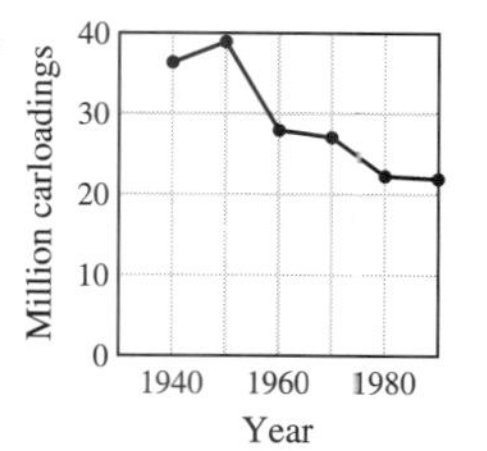

7.

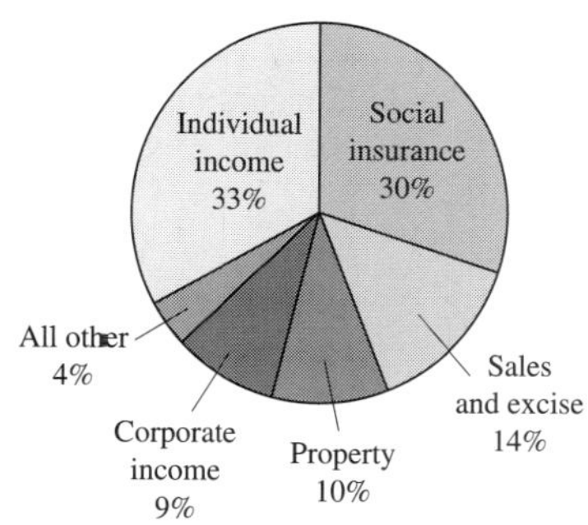

9. Find the percentage of cars imported for each of the five countries; for example, Canada accounted for $1{,}590/(1{,}590 + 1{,}013 + 551 + 234 + 141) = 45.1\%$ of the imports from the five countries. The central angles would be 162.2°, 103.3°, 56.2°, 23.9°, 14.4°.

11.

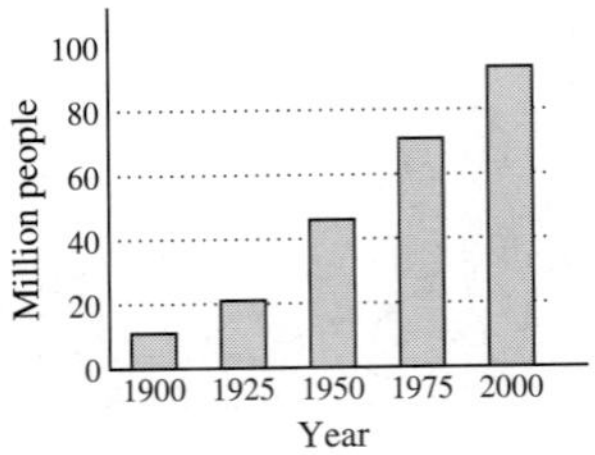

13.

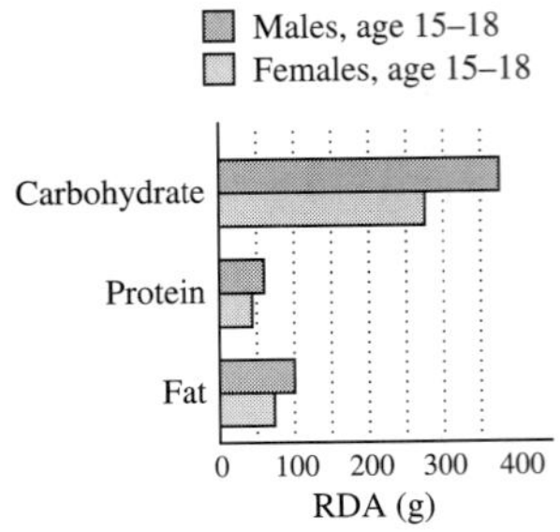

15. (A)

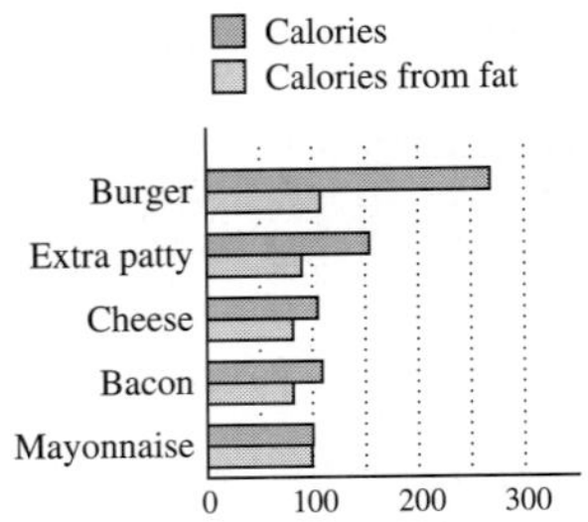

(B) The double bacon cheeseburger with mayo would supply 76.5% of the calories from fat allowed for the entire day.

17.

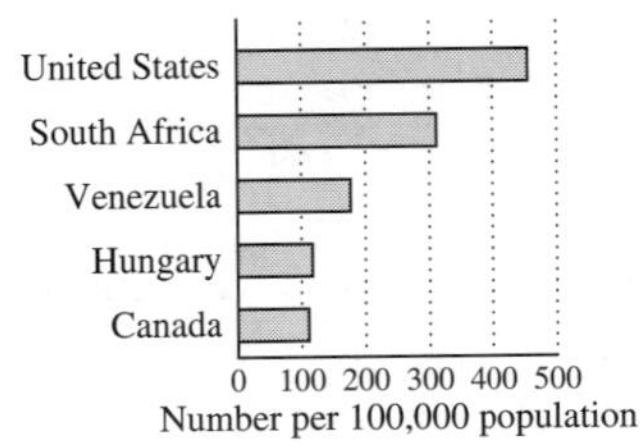

19. Approx. 23 in 1900, approx. 33 in 1990; the median age decreased in the 1950's and 1960's, but increased in the other decades.

Exercise 7-2

1. (A) and (B)

Class interval	Tally	Frequency	Relative frequency
0.5–2.5	|	1	.1
2.5–4.5	|	1	.1
4.5–6.5	||||	4	.4
6.5–8.5	|||	3	.3
8.5–10.5	|	1	.1

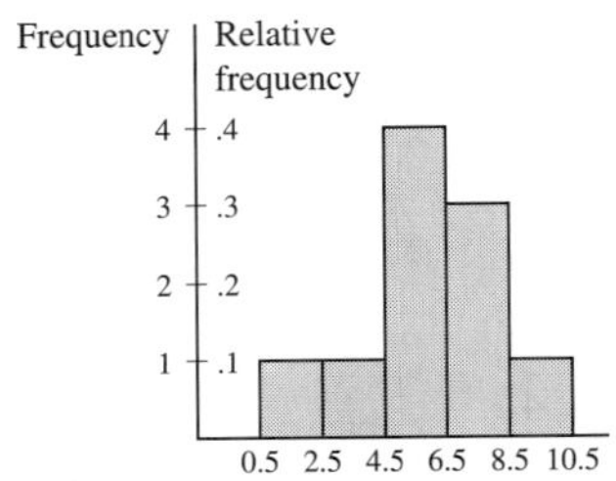

(C) The frequency tables and histograms are identical, but the data set in part (B) is more spread out than that of part (A).

3. (A) Let Xmin = 1.5, Xmax = 25.5, change Xscl from 1 to 2, and multiply Ymax and Yscl by 2; change Xscl from 1 to 4, and multiply Ymax and Yscl by 4.

(B) The shape becomes more symmetrical and more rectangular.

5. (A)

Class interval	Tally	Frequency	Relative frequency
20.5–24.5	|	1	.05
24.5–28.5	|||	3	.15
28.5–32.5	~~||||~~ ||	7	.35
32.5–36.5	~~||||~~ |	6	.30
36.5–40.5	||	2	.10
40.5–44.5	|	1	.05

(B)

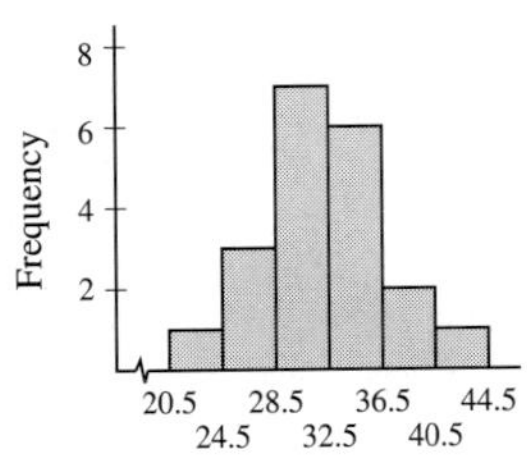

(C) .45; .2 (D)

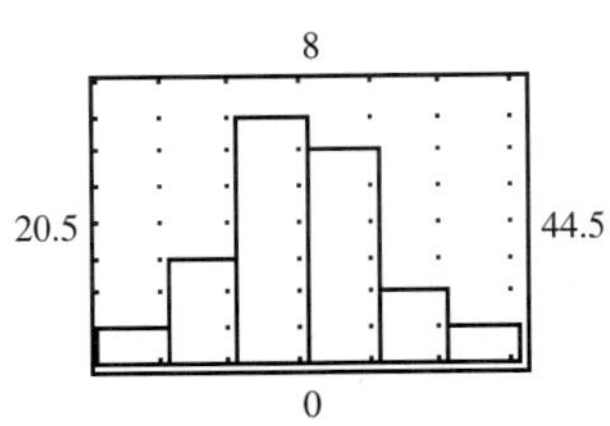

7. (A)

CLASS INTERVAL	FREQUENCY	RELATIVE FREQUENCY
−0.5–4.5	5	.05
4.5–9.5	54	.54
9.5–14.5	25	.25
14.5–19.5	13	.13
19.5–24.5	0	.00
24.5–29.5	1	.01
29.5–34.5	2	.02
	100	1.00

(B)

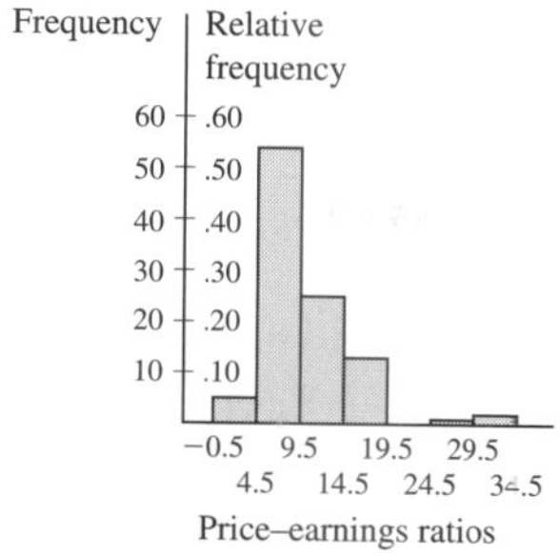

(C)

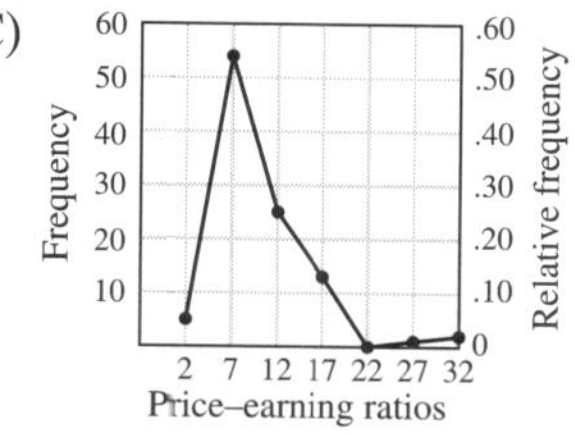

(D)

CLASS INTERVAL	FREQUENCY	CUMULATIVE FREQUENCY	RELATIVE CUMULATIVE FREQUENCY
−0.5–4.5	5	5	.05
4.5–9.5	54	59	.59
9.5–14.5	25	84	.84
14.5–19.5	13	97	.97
19.5–24.5	0	97	.97
24.5–29.5	1	98	.98
29.5–34.5	2	100	1.00

P(PE ratio between 4.5 and 14.5) $= .79$

(E)

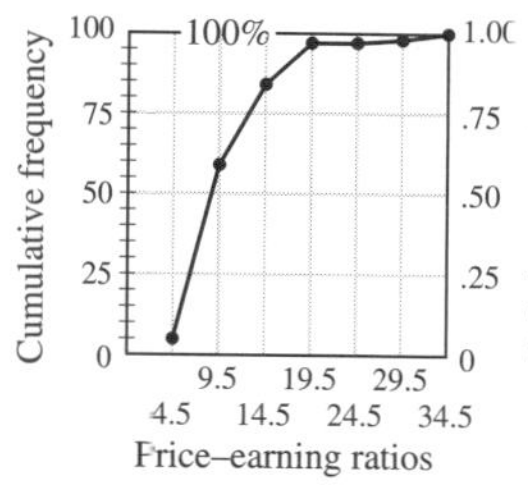

9. (A)

CLASS INTERVAL	FREQUENCY	RELATIVE FREQUENCY
1.95–2.15	21	.21
2.15–2.35	19	.19
2.35–2.55	17	.17
2.55–2.75	14	.14
2.75–2.95	9	.09
2.95–3.15	6	.06
3.15–3.35	5	.05
3.35–3.55	4	.04
3.55–3.75	3	.03
3.75–3.95	2	.02
	100	1.00

(B)

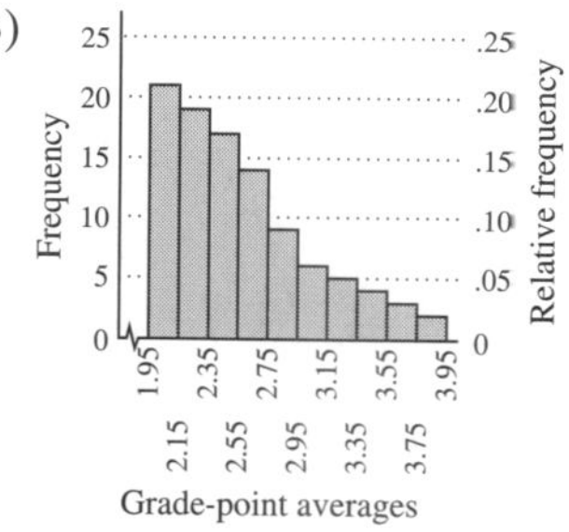

(C)

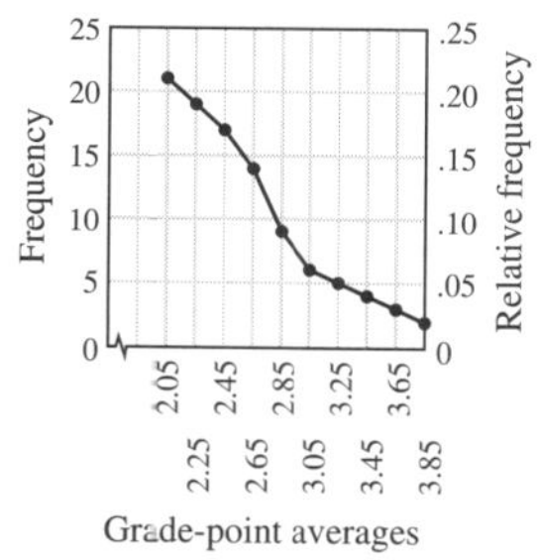

(D)

CLASS INTERVAL	FREQUENCY	CUMULATIVE FREQUENCY	RELATIVE CUMULATIVE FREQUENCY
1.95–2.15	21	21	.21
2.15–2.35	19	40	.40
2.35–2.55	17	57	.57
2.55–2.75	14	71	.71
2.75–2.95	9	80	.80
2.95–3.15	6	86	.86
3.15–3.35	5	91	.91
3.35–3.55	4	95	.95
3.55–3.75	3	98	.98
3.75–3.95	2	100	1.00

$P(\text{GPA} > 2.95) = .2$

(E)

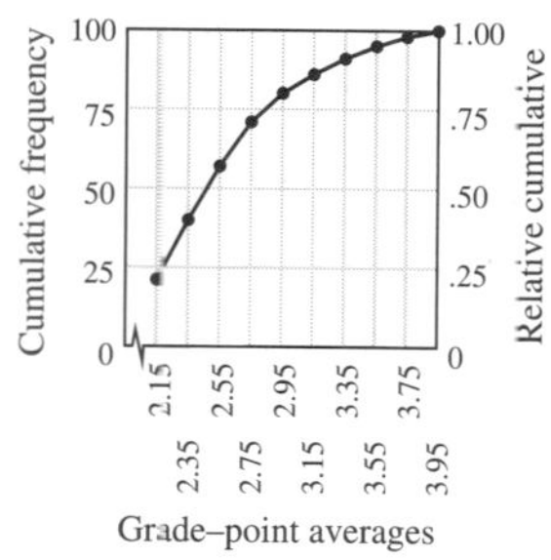

Exercise 7-3

1. Mean = 3; Median = 3; Mode = 3 **3.** Modal preference is chocolate. **5.** Mean = 4.4 **7.** The median

9. (A) Close to 3.5; close to 3.5
(B) Answer depends on results of simulation.

11. (A) 175, 175, 325, 525
(B) Let the four numbers be u, v, w, x, where u and v are both equal to m_3. Choose w so that the mean of w and m_3 is m_2; then choose x so that the mean of u, v, w, and x is m_1.

13. Mean $\approx$ 14.7; Median = 11.5; Mode = 10.1 **15.** Mean = 1,045.5 hr; Median = 1,049.5 hr

17. Mean $\approx$ 1,328,000; Median = 1,290,000; Mode = 1,200,000 **19.** Mean = 50.5 g; Median = 50.55 g

21. Mean = 27,000; Median = 25,000; no mode **23.** Median = 577

Exercise 7-4

1. 1.15 **3.** (A) 70%; 100%; 100% (B) Yes (C)

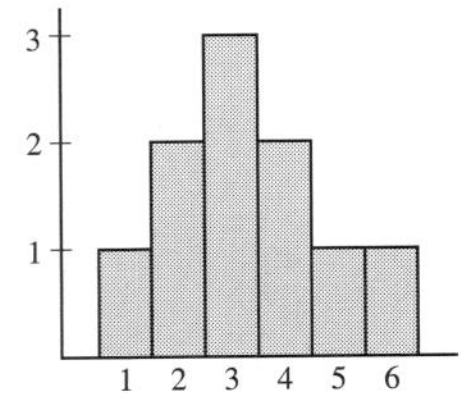

5. 2.5 **7.** (A) False (B) True

9. (A) The first data set. It is more likely that the sum is close to 7, for example, than to 2 or 12.
(B) Answer depends on results of simulation.

11. $\bar{x}$ = \$4.35; s = \$2.45 **13.** $\bar{x}$ = 8.7 hr; s = 0.6 hr **15.** $\bar{x}$ = 5.1 min; s = 0.9 min **17.** $\bar{x}$ = 11.1; s = 2.3

Exercise 7-5

1. $\frac{3}{8}$ = .375

3. $\frac{1}{8}$ = .125

5. .230

7. $C_{3,2}(.5)^2(.5) = .375$

9. $C_{3,0}(.5)^0(.5)^3 = .125$

11. $C_{3,2}(.5)^2(.5) + C_{3,3}(.5)^3(.5)^0 = .500$

13. $\mu = .6$; $\sigma = .65$

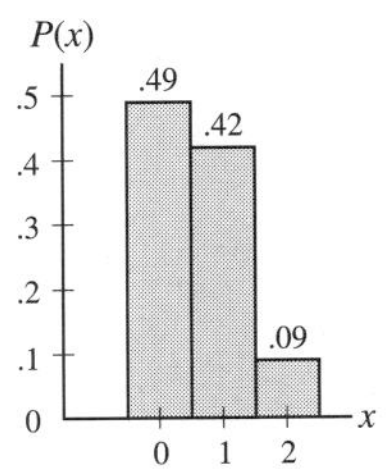

15. $\mu = 2$; $\sigma = 1$

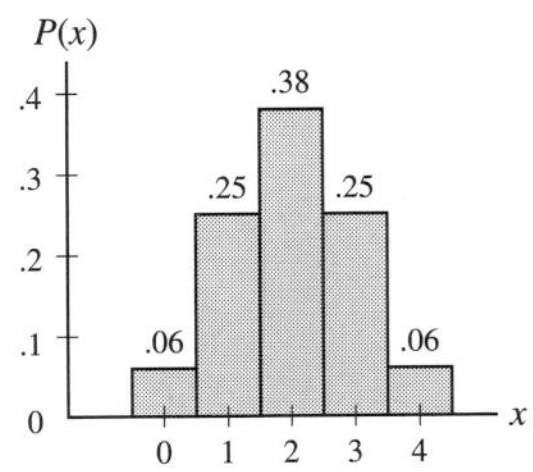

17. $C_{4,3}(\frac{1}{6})^3(\frac{5}{6}) \approx .0154$ **19.** $C_{4,0}(\frac{1}{6})^0(\frac{5}{6})^4 \approx .482$ **21.** $1 - [C_{4,0}(\frac{1}{6})^0(\frac{5}{6})^4] \approx .518$ **23.** (A) .311 (B) .437

25. It is more likely that all answers are wrong (.107) than that at least half are right (.033).

27. $\mu = 2.4$; $\sigma = 1.2$

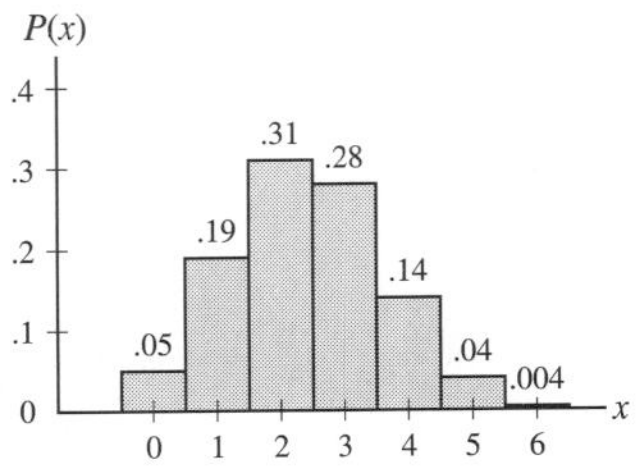

29. $\mu = 2.4$; $\sigma = 1.3$

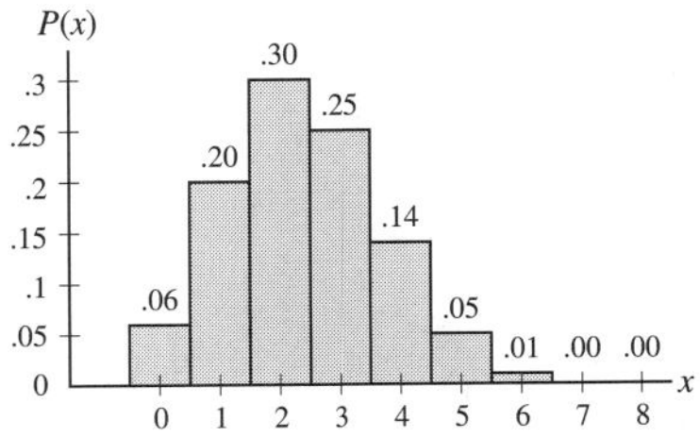

31. (A) $\mu = 17$; $\sigma = 1.597$ (B) .654

33. .238

35. The theoretical probability distribution is given by $P(x) = C_{3,x}(.5)^x(.5)^{3-x} = C_{3,x}(.5)^3$. **37.** $p = .5$

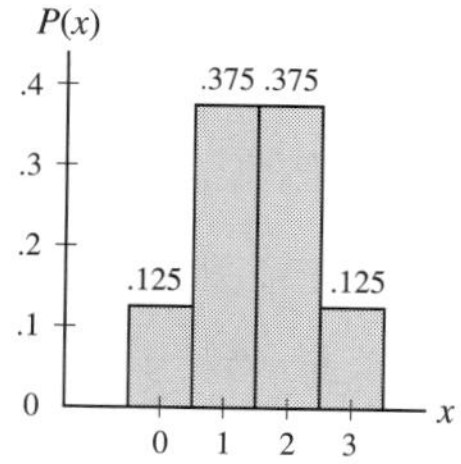

FREQUENCY OF HEADS IN 100 TOSSES OF THREE COINS

NUMBER OF HEADS	THEORETICAL FREQUENCY	ACTUAL FREQUENCY
0	12.5	List your experimental results here.
1	37.5	
2	37.5	
3	12.5	

39. (A) $\mu = 5; \sigma = 1.581$ (B) Answer depends on results of simulation. **41.** (A) .318 (B) .647 **43.** .0188

45. (A) $P(x) = C_{6,x}(.05)^x(.95)^{6-x}$

(B)

x	$P(x)$
0	.735
1	.232
2	.031
3	.002
4	.000
5	.000
6	.000

(C)

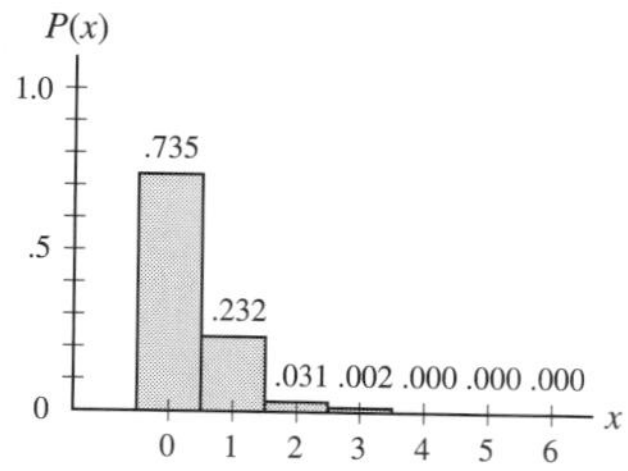

(D) $\mu = .30; \sigma = .53$

47. .998 **49.** (A) .001 (B) .264 (C) .897

51. (A) $P(x) = C_{6,x}(.6)^x(.4)^{6-x}$

(B)

x	$P(x)$
0	.004
1	.037
2	.138
3	.276
4	.311
5	.187
6	.047

(C)

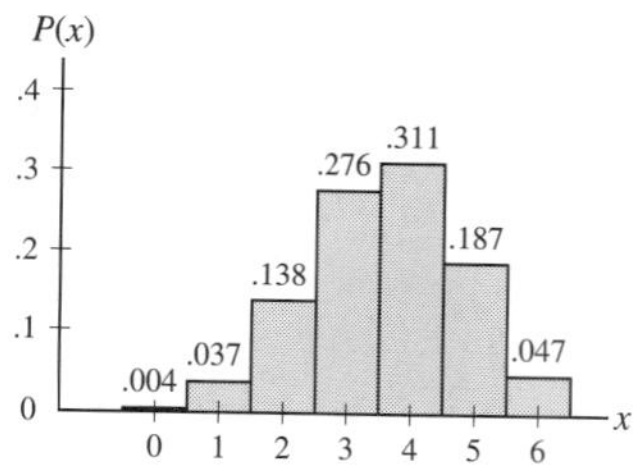

(D) $\mu = 3.6; \sigma = 1.2$

53. .000 864 **55.** (A) $P(x) = C_{5,x}(.2)^x(.8)^{5-x}$

(B)

x	$P(x)$
0	.328
1	.410
2	.205
3	.051
4	.006
5	.000

(C)

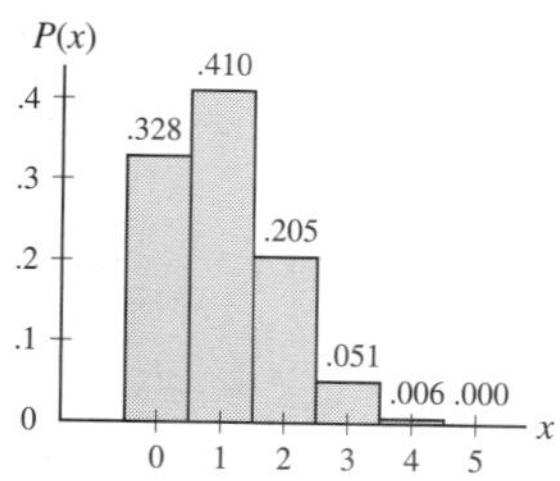

(D) $\mu = 1; \sigma = .89$

57. (A) .0041 (B) .0467 (C) .138 (D) .959

Exercise 7-6

1. 1.5 **3.** 3.3 **5.** 0.5 **7.** 0.8 **9.** .4332 **11.** .4995 **13.** .1915 **15.** .2881 **17.** .7888 **19.** .5328 **21.** .0122 **23.** .1056 **25.** (A) False (B) True **27.** No **29.** Yes **31.** No **33.** Yes **35.** Solve the inequality $np - 3\sqrt{npq} \geq 0$ to obtain $n \geq 81$. **37.** .89

39. .16 **41.** .01 **43.** .01 **45.**

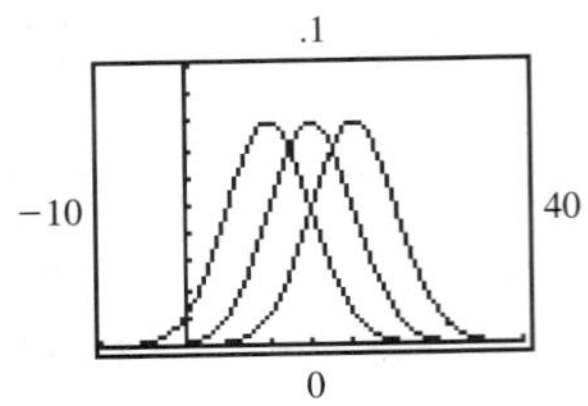

47.

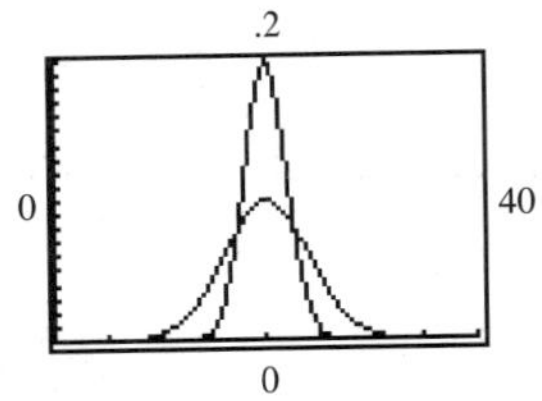

49. (A) Approx. 82 (B) Answer depends on results of simulation. **51.** 2.28% **53.** 1.24% **55.** .0031; either a rare event has happened or the company's claim is false. **57.** 0.82% **59.** .0158 **61.** 2.28% **63.** A's, 80.2 or greater; B's, 74.2–80.2; C's, 65.8–74.2; D's, 59.8–65.8; F's, 59.8 or lower

Review Exercise

1. *(7-1)*

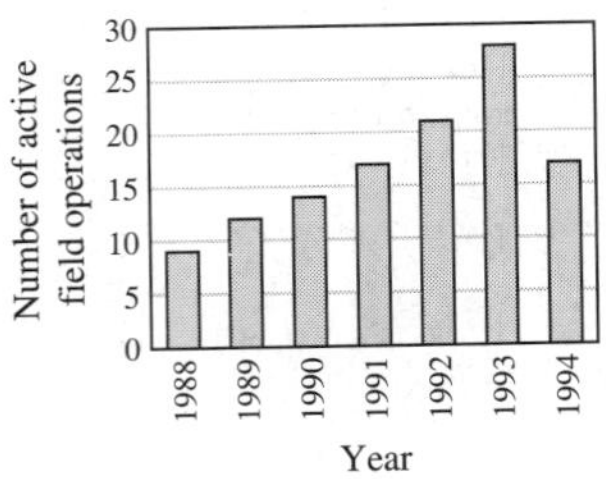

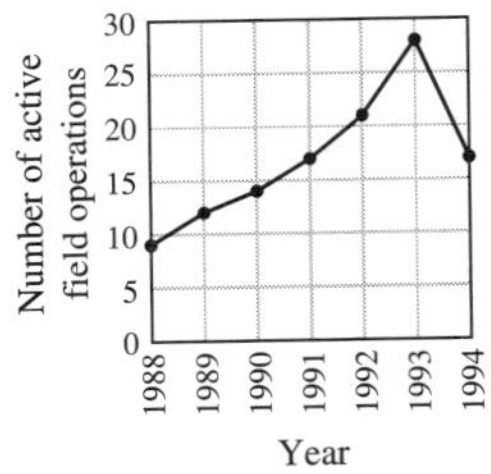

2. *(7-1)*

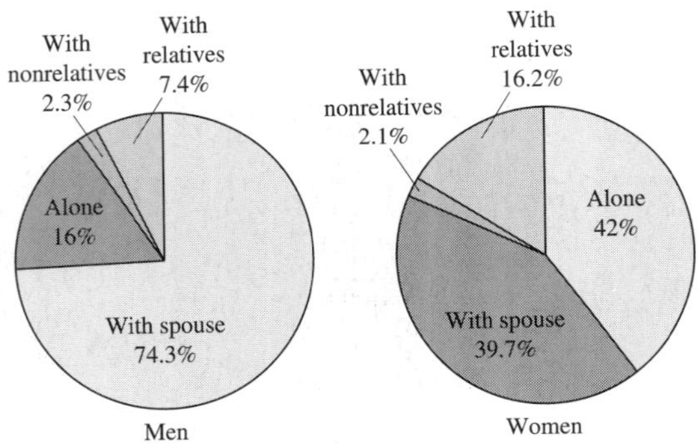

3. (A)

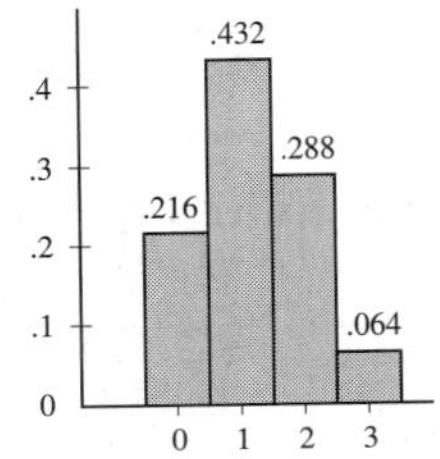

(B) $\mu = 1.2; \sigma = .85$ *(7-5)*

4. (A) $\bar{x} = 2.7$ (B) 2.5 (C) 2 (D) $s = 1.34$ *(7-3, 7-4)* **5.** (A) 1.8 (B) .4641 *(7-6)*

6. (A)

Class interval	Frequency	Relative frequency
9.5–11.5	1	.04
11.5–13.5	5	.20
13.5–15.5	12	.48
15.5–17.5	6	.24
17.5–19.5	1	.04
	25	1.00

(B)

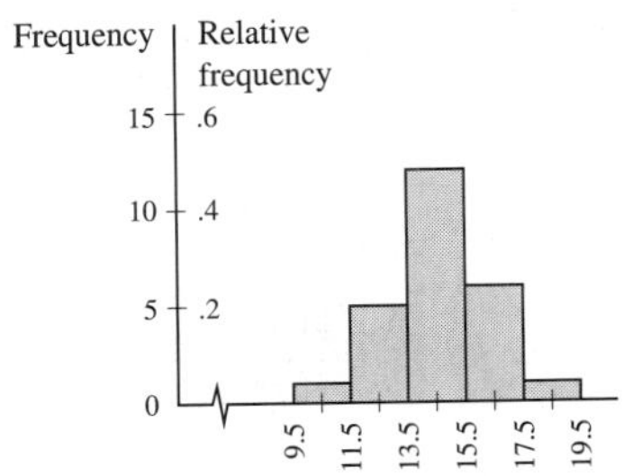

(C)

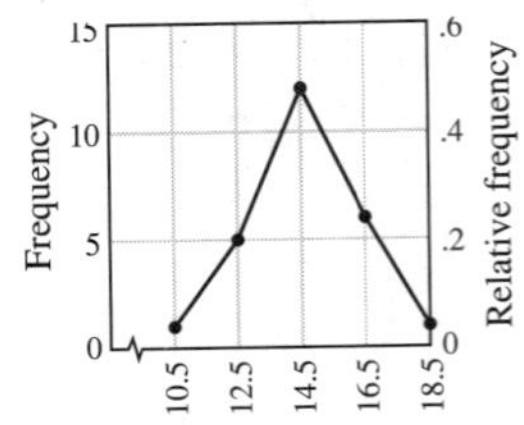

(D)

Class interval	Frequency	Cumulative frequency	Relative cumulative frequency
9.5–11.5	1	1	.04
11.5–13.5	5	6	.24
13.5–15.5	12	18	.72
15.5–17.5	6	24	.96
17.5–19.5	1	25	1.00

(E)

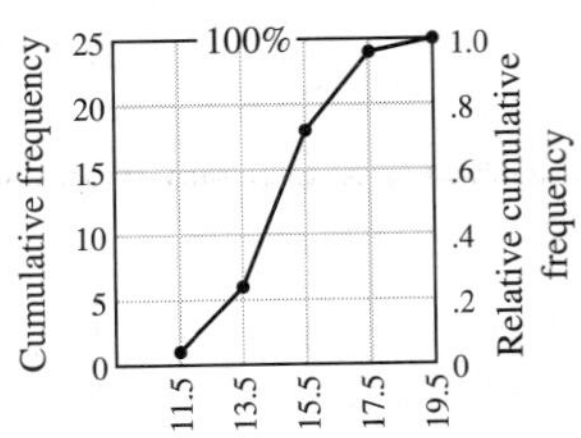

(7-2)

7. (A) $\bar{x} = 7$
(B) $s = 2.45$
(C) 7.14
(7-3, 7-4)

8. (A)

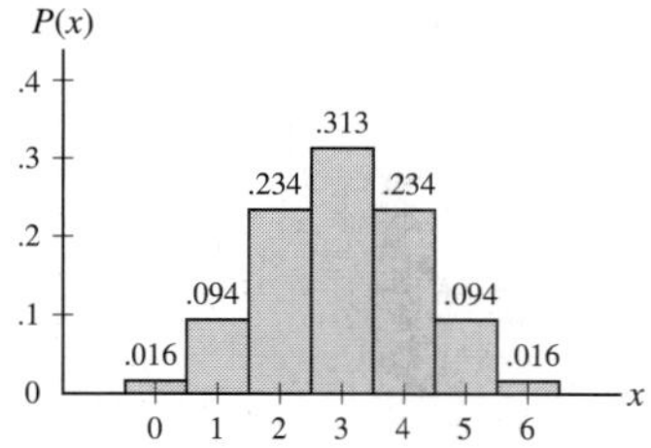

(B) $\mu = 3; \sigma = 1.22$ *(7-5)* **9.** $\mu = 600; \sigma = 15.49$ *(7-5)*

10. (A) True (B) False *(7-3, 7-4)* **11.** (A) False (B) True (C) True *(7-5, 7-6)*
12. .999 *(7-6)* **13.** (A) .9104 (B) .0668 *(7-6)*
14. (A) The first data set. Sums range from 2 to 12, but products range from 1 to 36.
(B) Answers depend on results of simulation. *(7-4)*
15. (A) $\bar{x} = 14.6; s = 1.83$ (B) $\bar{x} = 14.6; s = 1.78$ *(7-3, 7-4)* **16.** (A) .0322 (B) .0355 *(7-5)*
17. .421 *(7-5)* **18.** (A) 10, 10, 20, 20, 90, 90, 90, 90, 90, 90 (B) No *(7-3)*
19. (A) .179
(C) The normal distribution is continuous, not discrete, so the correct analogue of part (A) is $P(7.5 \leqslant x \leqslant 8.5) \approx .18$ (using Table I in Appendix C). *(7-6)*
20. (A) $\mu = 10.8; \sigma = 1.039$ (B) .282 (C) Answer depends on results of simulation. *(7-5)*
21. (A) $\bar{x} = 10$ (B) 10 (C) 5 (D) $s = 5.14$ *(7-3, 7-4)* **22.** Modal preference is soft drink *(7-3)*

23. (A)

Class Interval	Frequency	Relative Frequency
29.5–31.5	3	.086
31.5–33.5	7	.2
33.5–35.5	14	.4
35.5–37.5	7	.2
37.5–39.5	4	.114
	35	1.00

(B)

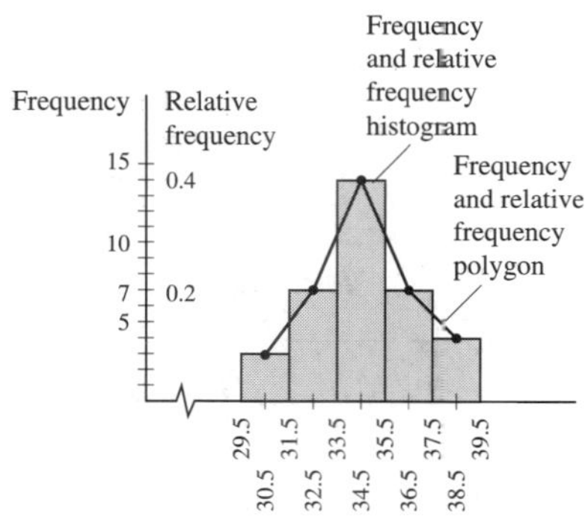

(C) $\bar{x} = 34.61; s = 2.22$ *(7-2, 7-3, 7-4)*

24. (A) 57.62%
(B) 6.68% *(7-6)*

25. (A) $\mu = 140; \sigma = 6.48$
(B) Yes
(C) .939
(D) .0125
(E)

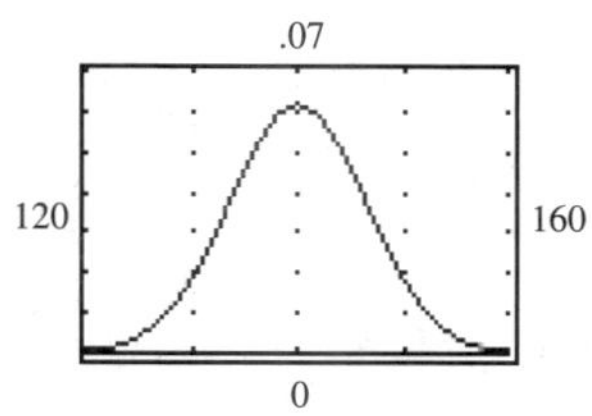

(7-5, 7-6)

26. .999 *(7-5)*

Chapter 8

Exercise 8-1

1. (A) 2 (upper right) (B) *R* plays Row 1, *C* plays column 2 (C) 2 **3.** Not strictly determined
5. (A) 1 (B) *R* plays Row 2, *C* plays Column 2 (C) 1
7. (A) All entries (B) *R* plays either Row 1 or Row 2, *C* plays either Column 1 or Column 2 (C) 1
9. (A) Both 2's in Column 1 (B) *R* plays either Row 1 or Row 2, *C* plays Column 1 (C) 2
11. (A) 0 (B) *R* plays Row 2, *C* plays Column 2 (C) 0 **13.** Not strictly determined
15. (A) Both 0's in Column 2 (B) *R* plays either Row 2 or Row 4, *C* plays Column 2 (C) 0
17. (A) All 2's (B) *R* plays either Row 2 or Row 4, *C* plays either Column 1 or Column 4 (C) 2
19. Player *R*, because the value of the game is positive.
21. No; 0 is a saddle value irrespective of the value of *m*.
23. Saddle value: 50% in upper left; *R* plays Row 1, *C* plays Column 1

25.

Store R \ Store C		Tahoe City	Incline Village	South Lake Tahoe
Store R	Tahoe City	50	45	35
	Incline Village	55	50	40
	South Lake Tahoe	65	60	50

Optimum strategy for both stores is to locate in South Lake Tahoe.

Exercise 8-2

1. $P^* = \begin{bmatrix} \frac{2}{3} & \frac{1}{3} \end{bmatrix}, Q^* = \begin{bmatrix} \frac{2}{3} \\ \frac{1}{3} \end{bmatrix}, v = 0$ **3.** $P^* = \begin{bmatrix} \frac{1}{4} & \frac{3}{4} \end{bmatrix}, Q^* = \begin{bmatrix} \frac{1}{2} \\ \frac{1}{2} \end{bmatrix}, v = \frac{1}{2}$ **5.** $P^* = \begin{bmatrix} \frac{1}{3} & \frac{2}{3} \end{bmatrix}, Q^* = \begin{bmatrix} \frac{3}{5} \\ \frac{2}{5} \end{bmatrix}, v = 0$

7. Strictly determined with $P^* = \begin{bmatrix} 0 & 1 \end{bmatrix}, Q^* = \begin{bmatrix} 0 \\ 1 \end{bmatrix}, v = 1$

9. Eliminate recessive columns to obtain $\begin{bmatrix} 2 & -1 \\ -2 & 1 \end{bmatrix}$; then $P^* = \begin{bmatrix} \frac{1}{2} & \frac{1}{2} \end{bmatrix}, Q^* = \begin{bmatrix} 0 \\ \frac{1}{3} \\ \frac{2}{3} \\ 0 \end{bmatrix}, v = 0$

11. Eliminate recessive rows and columns to obtain $\begin{bmatrix} 2 & -3 \\ -1 & 2 \end{bmatrix}$; then $P^* = \begin{bmatrix} 0 & \frac{3}{8} & \frac{5}{8} \end{bmatrix}, Q^* = \begin{bmatrix} \frac{5}{8} \\ \frac{3}{8} \\ 0 \end{bmatrix}, v = \frac{1}{8}$

13. Strictly determined with $P^* = \begin{bmatrix} 1 & 0 & 0 \end{bmatrix}, Q^* = \begin{bmatrix} 0 \\ 1 \\ 0 \end{bmatrix}, v = 1$

15. (A) True (B) True (C) False **17.** (A) −$1 (B) $1 (C) $0

19. $PMQ = \begin{bmatrix} p_1 & p_2 \end{bmatrix} \begin{bmatrix} a & b \\ c & d \end{bmatrix} \begin{bmatrix} q_1 \\ q_2 \end{bmatrix} = \begin{bmatrix} p_1 & p_2 \end{bmatrix} \begin{bmatrix} aq_1 + bq_2 \\ cq_1 + dq_2 \end{bmatrix} = ap_1q_1 + bp_1q_2 + cp_2q_1 + dp_2q_2 = E(P, Q)$

23. Assign the values of $D \neq 0$ and a arbitrarily, say $D = 10$ and $a = 0$, and solve for b, c, and d in Theorem 4, obtaining

$$M = \begin{bmatrix} 0 & -1 \\ -7 & 2 \end{bmatrix}.$$

25. (A) $P^* = \begin{bmatrix} 0 & \frac{1}{4} & \frac{3}{4} & 0 \end{bmatrix}, Q^* = \begin{bmatrix} 0 \\ \frac{1}{4} \\ \frac{3}{4} \\ 0 \end{bmatrix}, v = -\frac{1}{4}$ (B) −1 (C) $-\frac{1}{4}$ (D) 0

27. $P^* = \begin{bmatrix} \frac{2}{5} & \frac{3}{5} \end{bmatrix}, v = \$3{,}400$ [This means you should invest $(\frac{2}{5})(10{,}000) = \$4{,}000$ in solar energy stocks and $(\frac{3}{5})(10{,}000) = \$6{,}000$ in oil stocks, for an expected gain of $3,400 no matter how the election turns out.]

Exercise 8-3

1. $P^* = \begin{bmatrix} \frac{3}{8} & \frac{5}{8} \end{bmatrix}, Q^* = \begin{bmatrix} \frac{5}{8} \\ \frac{3}{8} \end{bmatrix}, v = \frac{1}{8}$ **3.** $P^* = \begin{bmatrix} \frac{2}{3} & \frac{1}{3} \end{bmatrix}, Q^* = \begin{bmatrix} \frac{3}{4} \\ \frac{1}{4} \end{bmatrix}, v = 0$ **5.** $P^* = \begin{bmatrix} 0 & 1 \end{bmatrix}, Q^* = \begin{bmatrix} 1 \\ 0 \end{bmatrix}, v = 5$

7. Yes. The matrix game is strictly determined, so it may be solved easily by noting that 5 is a saddle value.

9. Eliminate recessive rows to obtain $\begin{bmatrix} 4 & -6 \\ -2 & 3 \end{bmatrix}$; then $P^* = \begin{bmatrix} 0 & \frac{1}{3} & \frac{2}{3} & 0 \end{bmatrix}, Q^* = \begin{bmatrix} \frac{3}{5} \\ \frac{2}{5} \end{bmatrix}, v = 0$

11. (A) $P(kJ)Q = k(PJ)Q = k[p_1 + p_2 \quad p_1 + p_2]Q = k[1 \quad 1]Q = k(q_1 + q_2) = k$
(B) $P(kJ)Q = k(PJ)Q = k[1 \quad 1 \quad \dots \quad 1]Q = k$

13. $P^* = [0 \quad \frac{1}{4} \quad \frac{3}{4} \quad 0], Q^* = \begin{bmatrix} 0 \\ \frac{1}{4} \\ \frac{3}{4} \\ 0 \end{bmatrix}, v = -\frac{1}{4}$ **15.** $P^* = [\frac{2}{5} \quad \frac{3}{5}], v = \$3{,}400$

Exercise 8-4

1. $P^* = [\frac{2}{3} \quad \frac{1}{3}], Q^* = \begin{bmatrix} \frac{2}{3} \\ 0 \\ \frac{1}{3} \end{bmatrix}, v = \frac{2}{3}$ **3.** $P^* = [\frac{1}{2} \quad \frac{1}{3} \quad \frac{1}{6}], Q^* = \begin{bmatrix} \frac{1}{2} \\ \frac{1}{3} \\ \frac{1}{6} \end{bmatrix}, v = 0$

5. Eliminate recessive rows and columns to obtain $\begin{bmatrix} 2 & 1 \\ -1 & 3 \end{bmatrix}$; then $P^* = [\frac{4}{5} \quad \frac{1}{5} \quad 0], Q^* = \begin{bmatrix} 0 \\ 0 \\ \frac{2}{5} \\ \frac{3}{5} \end{bmatrix}, v = \frac{7}{5}$

7. Eliminate recessive rows and columns to obtain $\begin{bmatrix} 1 & 4 & 0 \\ 0 & -1 & 2 \end{bmatrix}$ (the matrix of Problem 1); then

$P^* = [0 \quad \frac{2}{3} \quad 0 \quad \frac{1}{3}], Q^* = \begin{bmatrix} \frac{2}{3} \\ 0 \\ 0 \\ \frac{1}{3} \end{bmatrix}, v = \frac{2}{3}$

9. (A)

	Paper	Stone	Scissors
Paper	0	1	−1
Stone	−1	0	1
Scissors	1	−1	0

(B) $P^* = [\frac{1}{3} \quad \frac{1}{3} \quad \frac{1}{3}], Q^* = \begin{bmatrix} \frac{1}{3} \\ \frac{1}{3} \\ \frac{1}{3} \end{bmatrix}, v = 0$

11. (A)

	Economy (fate) Up	Down
Deluxe	2	−1
Standard	1	1
Economy	0	4

(B) $P^* = [\frac{4}{7} \quad 0 \quad \frac{3}{7}], Q^* = \begin{bmatrix} \frac{5}{7} \\ \frac{2}{7} \end{bmatrix}, v = \frac{8}{7}$ million dollars
(C) $\frac{4}{7}$ of budget on deluxe, 0 standard, $\frac{3}{7}$ of budget on economy
(D) $-1; \frac{8}{7}$

Review Exercise

1. (A) −3 (B) *R* plays Row 2, *C* plays Column 1 (C) −3 *(8-1)* **2.** Not strictly determined *(8-1)*
3. (A) Two 0's in Column 2 (B) *R* plays either Row 2 or Row 3, *C* plays Column 2 (C) 0 *(8-1)*
4. Not strictly determined *(8-1)*

5. $\begin{bmatrix} -2 & 3 \\ 0 & -1 \end{bmatrix}$ *(8-2)* **6.** $P^* = [\frac{1}{4} \quad \frac{3}{4}], Q^* = \begin{bmatrix} \frac{1}{2} \\ \frac{1}{2} \end{bmatrix}, v = -\frac{1}{2}$ *(8-2)*

7. $M_1 = \begin{bmatrix} 1 & 4 \\ 3 & 2 \end{bmatrix}$ (A) Minimize $y = x_1 + x_2$ Subject to $x_1 + 3x_2 \geqslant 1$, $4x_1 + 2x_2 \geqslant 1$, $x_1, x_2 \geqslant 0$ (B) Maximize $y = z_1 + z_2$ Subject to $z_1 + 4z_2 \leqslant 1$, $3z_1 + 2z_2 \leqslant 1$, $z_1, z_2 \geqslant 0$ *(8-3)*

8. $P^* = [\frac{1}{4} \;\; \frac{3}{4}]$, $Q^* = \begin{bmatrix} \frac{1}{2} \\ \frac{1}{2} \end{bmatrix}$, $v = -\frac{1}{2}$ *(8-3)* **9.** Same as in Problem 8 *(8-4)* **10.** (A) False (B) True *(8-1)*

11. (A) True (B) False *(8-2)* **12.** Strictly determined with $P^* = [0 \;\; 0 \;\; 1]$, $Q^* = \begin{bmatrix} 1 \\ 0 \\ 0 \end{bmatrix}$, $v = 0$ *(8-1)*

13. Eliminate recessive rows and columns to obtain $\begin{bmatrix} 5 & -3 \\ 0 & 2 \end{bmatrix}$; then $P^* = [\frac{1}{5} \;\; 0 \;\; \frac{4}{5}]$, $Q^* = \begin{bmatrix} 0 \\ \frac{1}{2} \\ \frac{1}{2} \\ 0 \end{bmatrix}$, $v = 1$ *(8-2)*

14. $P^* = [\frac{2}{3} \;\; \frac{1}{3}]$, $Q^* = \begin{bmatrix} \frac{2}{3} \\ 0 \\ \frac{1}{3} \end{bmatrix}$, $v = -\frac{1}{3}$ *(8-4)*

15. Eliminate recessive rows and columns to obtain $\begin{bmatrix} 4 & -5 \\ 3 & 8 \end{bmatrix}$; then $P^* = [0 \;\; \frac{5}{14} \;\; \frac{9}{14}]$, $Q^* = \begin{bmatrix} \frac{13}{14} \\ 0 \\ 0 \\ \frac{1}{14} \end{bmatrix}$, $v = \frac{47}{14}$ *(8-2)*

16. $P^* = [\frac{1}{2} \;\; \frac{1}{2} \;\; 0]$, $Q^* = \begin{bmatrix} \frac{3}{4} \\ \frac{1}{4} \\ 0 \end{bmatrix}$, $v = -\frac{1}{2}$ *(8-4)* **17.** Yes *(8-1, 8-2)* **18.** No *(8-1, 8-2)*

19. (A) Ron (rows 1, 2), Cathy (columns 1, 2): $\begin{bmatrix} -2 & 3 \\ 1 & -2 \end{bmatrix}$ (B) Ron: $P^* = [\frac{3}{8} \;\; \frac{5}{8}]$; Cathy: $Q^* = \begin{bmatrix} \frac{5}{8} \\ \frac{3}{8} \end{bmatrix}$ (C) Ron: $-\$\frac{1}{8}$; Cathy: $\$\frac{1}{8}$ *(8-1, 8-2, 8-3, 8-4)*

CHAPTER 9

Exercise 9-1

1. $S_1 = \overset{A \quad B}{[.8 \;\; .2]}$; the probability of being in state A after one trial is .8, and the probability of being in state B after one trial is .2.

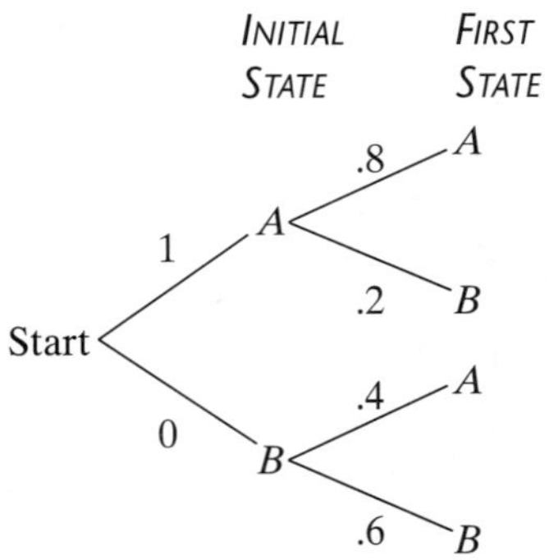

A: $(1)(.8) + (0)(.4) = .8$

B: $(1)(.2) + (0)(.6) = .2$

3. $S_1 = \overset{A \quad B}{[.6 \;\; .4]}$; the probability of being in state A after one trial is .6, and the probability of being in state B after one trial is .4.

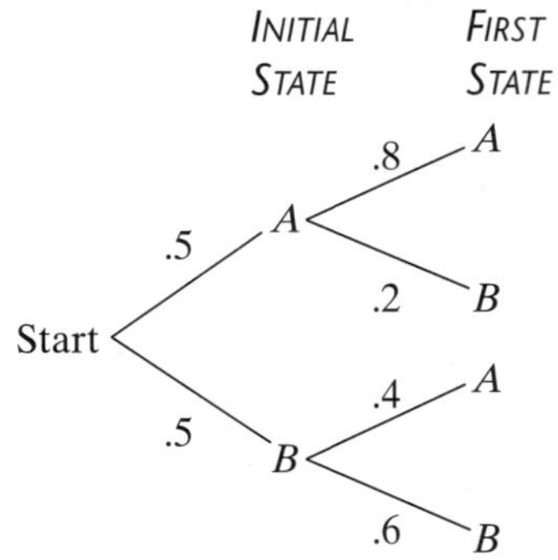

A: $(.5)(.8) + (.5)(.4) = .6$

B: $(.5)(.2) + (.5)(.6) = .4$

5. $S_2 = \overset{\quad A \qquad B}{\begin{bmatrix} .72 & .28 \end{bmatrix}}$; the probability of being in state A after two trials is .72, and the probability of being in state B after two trials is .28.

7. $S_2 = \overset{\quad A \qquad B}{\begin{bmatrix} .64 & .36 \end{bmatrix}}$; the probability of being in state A after two trials is .64, and the probability of being in state B after two trials is .36.

9. 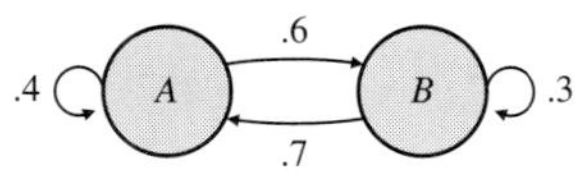

$\begin{array}{c} \\ A \\ B \end{array} \overset{\quad A \quad B}{\begin{bmatrix} .4 & .6 \\ .7 & .3 \end{bmatrix}}$

11. No

13. 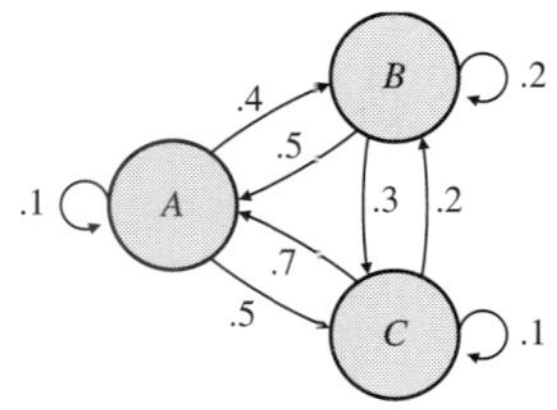

$\begin{array}{c} \\ A \\ B \\ C \end{array} \overset{\quad A \quad B \quad C}{\begin{bmatrix} .1 & .4 & .5 \\ .5 & .2 & .3 \\ .7 & .2 & .1 \end{bmatrix}}$

15. $a = .5, b = .6, c = .7$

17. $a = .7, b = 1, c = .2$

19. No

21. 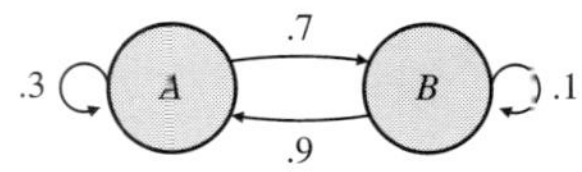

$\begin{array}{c} \\ A \\ B \end{array} \overset{\quad A \quad B}{\begin{bmatrix} .3 & .7 \\ .9 & .1 \end{bmatrix}}$

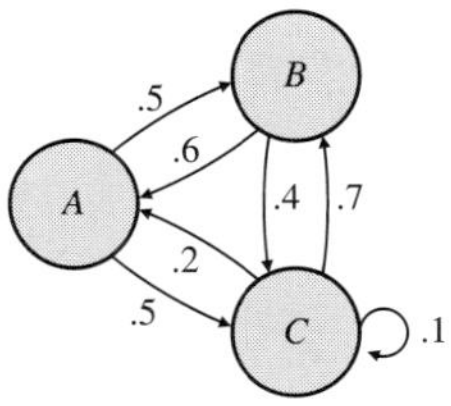

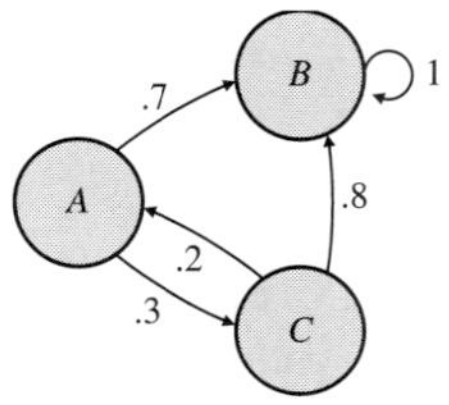

23. 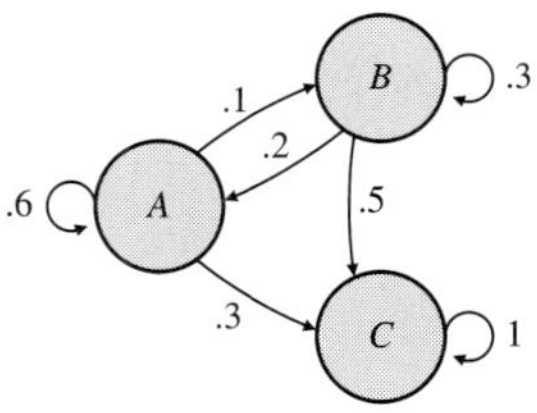

$\begin{array}{c} \\ A \\ B \\ C \end{array} \overset{\quad A \quad B \quad C}{\begin{bmatrix} .6 & .1 & .3 \\ .2 & .3 & .5 \\ 0 & 0 & 1 \end{bmatrix}}$

25. .35

27. .212

29. $S_2 = \overset{\quad A \qquad B \qquad C}{\begin{bmatrix} .43 & .35 & .22 \end{bmatrix}}$; the probabilities of going from state A to states A, B, and C in two trials

31. $S_3 = \overset{\quad A \qquad B \qquad C}{\begin{bmatrix} .212 & .298 & .49 \end{bmatrix}}$; the probabilities of going from state C to states A, B, and C in three trials

33. $n = 9$

35. $P^4 = \begin{array}{c} \\ A \\ B \end{array} \overset{\quad A \qquad B}{\begin{bmatrix} .4375 & .5625 \\ .375 & .625 \end{bmatrix}}$; $S_4 = \overset{\quad A \qquad B}{\begin{bmatrix} .425 & .575 \end{bmatrix}}$

37. $P^4 = \begin{array}{c} \\ A \\ B \\ C \end{array} \overset{\quad A \qquad B \qquad C}{\begin{bmatrix} .36 & .16 & .48 \\ .6 & 0 & .4 \\ .4 & .24 & .36 \end{bmatrix}}$; $S_4 = \overset{\quad A \qquad B \qquad C}{\begin{bmatrix} .452 & .152 & .396 \end{bmatrix}}$

41. (A) $\begin{array}{c} \\ A \\ B \\ C \\ D \end{array} \overset{\quad A \qquad B \qquad C \qquad D}{\begin{bmatrix} .0154 & .3534 & .0153 & .6159 \\ 0 & 1 & 0 & 0 \\ .0102 & .2962 & .0103 & .6833 \\ 0 & 0 & 0 & 1 \end{bmatrix}}$ (B) .6159 (C) .2962 (D) 0

45. (A) $\begin{bmatrix} .25 & .75 \end{bmatrix}$ (B) $\begin{bmatrix} .25 & .75 \end{bmatrix}$ (C) $\begin{bmatrix} .25 & .75 \end{bmatrix}$ (D) $\begin{bmatrix} .25 & .75 \end{bmatrix}$
(E) All the state matrices appear to approach the same matrix, $S = \begin{bmatrix} .25 & .75 \end{bmatrix}$, regardless of the values in the initial-state matrix.

47. $Q = \begin{bmatrix} .25 & .75 \\ .25 & .75 \end{bmatrix}$; the rows of Q are the same as the matrix S from Problem 45.

49. (A) R = Rain, R' = No rain

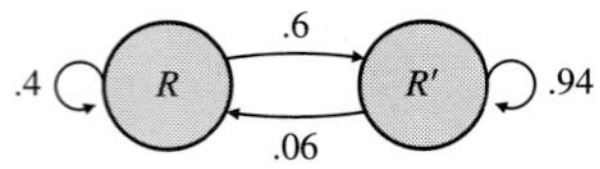

(B)
$$\begin{array}{c} \\ R \\ R' \end{array} \begin{array}{c} \begin{array}{cc} R & R' \end{array} \\ \begin{bmatrix} .4 & .6 \\ .06 & .94 \end{bmatrix} \end{array}$$

(C) Saturday: .196; Sunday: .12664

51. (A)

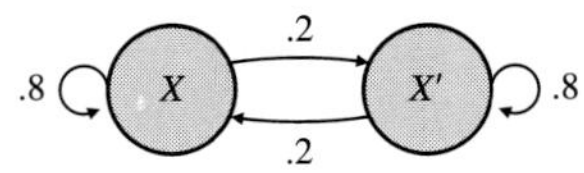

(B)
$$\begin{array}{c} \\ X \\ X' \end{array} \begin{array}{c} \begin{array}{cc} X & X' \end{array} \\ \begin{bmatrix} .8 & .2 \\ .2 & .8 \end{bmatrix} \end{array}$$

(C) 32%; 39.2%

53. (A) N = National Property, U = United Family, O = Other companies

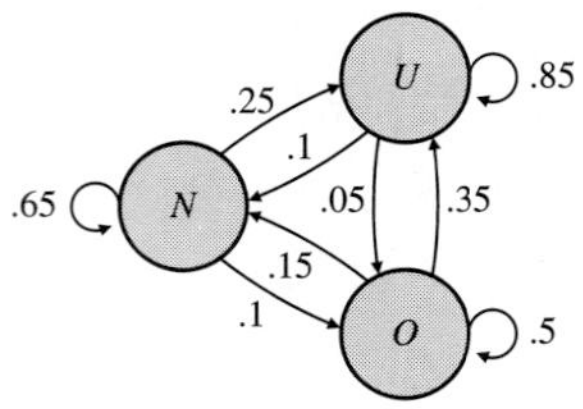

(B)
$$\begin{array}{c} \\ N \\ U \\ O \end{array} \begin{array}{c} \begin{array}{ccc} N & U & O \end{array} \\ \begin{bmatrix} .65 & .25 & .1 \\ .1 & .85 & .05 \\ .15 & .35 & .5 \end{bmatrix} \end{array}$$

(C) 38.5%; 32% (D) 45%; 53.65%

55. (A) B = Beginning agent, I = Intermediate agent, T = Terminated agent, Q = Qualified agent

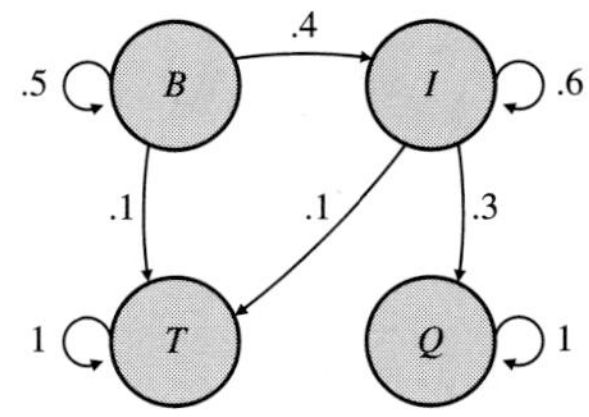

(B)
$$\begin{array}{c} \\ B \\ I \\ T \\ Q \end{array} \begin{array}{c} \begin{array}{cccc} B & I & T & Q \end{array} \\ \begin{bmatrix} .5 & .4 & .1 & 0 \\ 0 & .6 & .1 & .3 \\ 0 & 0 & 1 & 0 \\ 0 & 0 & 0 & 1 \end{bmatrix} \end{array}$$

(C) .12; .3612

57. (A)
$$\begin{array}{c} \\ \text{HMO} \\ \text{PPO} \\ \text{FFS} \end{array} \begin{array}{c} \begin{array}{ccc} \text{HMO} & \text{PPO} & \text{FFS} \end{array} \\ \begin{bmatrix} .8 & .15 & .05 \\ .2 & .7 & .1 \\ .25 & .3 & .45 \end{bmatrix} \end{array}$$

(B) HMO: 34.75%; PPO: 37%; FFS: 28.25%
(C) HMO: 42.2625%; PPO: 39.5875%; FFS: 18.15%

59. (A)
$$\begin{array}{c} \\ H \\ R \end{array} \begin{array}{c} \begin{array}{cc} H & R \end{array} \\ \begin{bmatrix} .88 & .12 \\ .05 & .95 \end{bmatrix} \end{array}$$
H = Homeowner
R = Renter

(B) H: 42.433% (C) H: 40.219%

Exercise 9-2

1. Regular **3.** Not regular **5.** Regular **7.** Not regular **9.** Regular **11.** Not regular **13.** Regular

15. $S = [.4 \quad .6]; \overline{P} = \begin{bmatrix} .4 & .6 \\ .4 & .6 \end{bmatrix}$ **17.** $S = [.375 \quad .625]; \overline{P} = \begin{bmatrix} .375 & .625 \\ .375 & .625 \end{bmatrix}$

19. $S = [.3 \quad .5 \quad .2]; \overline{P} = \begin{bmatrix} .3 & .5 & .2 \\ .3 & .5 & .2 \\ .3 & .5 & .2 \end{bmatrix}$ **21.** $S = [.6 \quad .24 \quad .16]; \overline{P} = \begin{bmatrix} .6 & .24 & .16 \\ .6 & .24 & .16 \\ .6 & .24 & .16 \end{bmatrix}$

23. (A) True (B) False **25.** $S = [.3553 \quad .6447]$ **27.** $S = [.3636 \quad .4091 \quad .2273]$

29. (A)

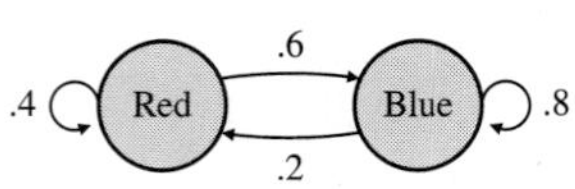

(B)
$$\begin{array}{c} \\ \text{Red} \\ \text{Blue} \end{array} \begin{array}{c} \begin{array}{cc} \text{Red} & \text{Blue} \end{array} \\ \begin{bmatrix} .4 & .6 \\ .2 & .8 \end{bmatrix} \end{array}$$

(C) $[.25 \quad .75]$; in the long run, the red urn will be selected 25% of the time and the blue urn 75% of the time.

31. (A) The state matrices alternate between $[.2 \quad .8]$ and $[.8 \quad .2]$; hence, they do not approach any one matrix.
(B) The state matrices are all equal to S_0; hence, S_0 is a stationary matrix.
(C) The powers of P alternate between P and I, the 2×2 identity; hence, they do not approach a limiting matrix.
(D) Parts (B) and (C) of Theorem 1 are not valid for this matrix. Since P is not regular, this is not a contradiction.
33. (A) Since P is not regular, it may have more than one stationary matrix.
(B) $[.5 \quad 0 \quad .5]$ is another stationary matrix.
(C) P has an infinite number of stationary matrices.

35. $\overline{P} = \begin{bmatrix} 1 & 0 & 0 \\ .25 & 0 & .75 \\ 0 & 0 & 1 \end{bmatrix}$; each row of $\overline{P}$ is a stationary matrix for P.

37. (A) .39; .3; .284; .277
(B) Each entry of the second column of P^{k+1} is the product of a row of P and the second column of P^k. Each entry of the latter is $\leq M_k$, so the product is $\leq M_k$.
39. 72.5%
41. (A) $S_1 = [.516 \quad .484]$; $S_2 = [.577 \quad .423]$

(B)

YEAR	DATA (%)	MODEL (%)
1970	43.3	43.3
1980	51.5	51.6
1990	57.5	57.7

(C) 74.1%
43. GTT: 25%; NCJ: 25%; Dash: 50%
45. Poor: 20%; satisfactory: 40%; preferred: 40%
47. 51%
49. Stationary matrix $= [.25 \quad .50 \quad .25]$
51. (A) $[.25 \quad .75]$ (B) 42.5%; 51.25%
(C) 60% rapid transit; 40% automobile
53. (A) $S_1 = [.334 \quad .666]$; $S_2 = [.343 \quad .657]$

(B)

YEAR	DATA (%)	MODEL (%)
1970	30.9	30.9
1980	33.3	33.4
1990	34.4	34.3

(C) 35%

Exercise 9-3

1. B, C **3.** No absorbing states **5.** A, D **7.** B is an absorbing state; absorbing chain
9. C is an absorbing state; not an absorbing chain

11. $\begin{matrix} & B & A & C \\ B & 1 & 0 & 0 \\ A & .5 & .2 & .3 \\ C & .1 & .5 & .4 \end{matrix}$ **13.** $\begin{matrix} & B & D & A & C \\ B & 1 & 0 & 0 & 0 \\ D & 0 & 1 & 0 & 0 \\ A & .4 & .1 & .3 & .2 \\ C & .4 & .3 & 0 & .3 \end{matrix}$ **15.** $\begin{matrix} & B & C & A \\ B & 1 & 0 & 0 \\ C & 0 & 1 & 0 \\ A & .3 & .5 & .2 \end{matrix}$ **17.** $\begin{matrix} & B & D & A & C \\ B & 1 & 0 & 0 & 0 \\ D & 0 & 1 & 0 & 0 \\ A & .2 & .4 & .1 & .3 \\ C & .2 & .1 & .5 & .2 \end{matrix}$

19. $\overline{P} = \begin{matrix} & A & B & C \\ A & 1 & 0 & 0 \\ B & 0 & 1 & 0 \\ C & .2 & .8 & 0 \end{matrix}$; $P(C \text{ to } A) = .2$; $P(C \text{ to } B) = .8$. It will take an average of 2 trials to go from C to either A or B.

21. $\overline{P} = \begin{matrix} & A & B & C \\ A & 1 & 0 & 0 \\ B & 1 & 0 & 0 \\ C & 1 & 0 & 0 \end{matrix}$; $P(B \text{ to } A) = 1$; $P(C \text{ to } A) = 1$. It will take an average of 4 trials to go from B to A and an average of 3 trials to go from C to A.

23. $\overline{P} = \begin{array}{c|cccc} & A & B & C & D \\ \hline A & 1 & 0 & 0 & 0 \\ B & 0 & 1 & 0 & 0 \\ C & .36 & .64 & 0 & 0 \\ D & .44 & .56 & 0 & 0 \end{array}$; $P(C \text{ to } A) = .36$; $P(C \text{ to } B) = .64$; $P(D \text{ to } A) = .44$; $P(D \text{ to } B) = .56$. It will take an average of 3.2 trials to go from C to either A or B and an average of 2.8 trials to go from D to either A or B.

25. (A) $[.2 \quad .8 \quad 0]$ (B) $[.26 \quad .74 \quad 0]$ **27.** (A) $[1 \quad 0 \quad 0]$ (B) $[1 \quad 0 \quad 0]$

29. (A) $[.44 \quad .56 \quad 0 \quad 0]$ (B) $[.36 \quad .64 \quad 0 \quad 0]$ (C) $[.408 \quad .592 \quad 0 \quad 0]$ (D) $[.384 \quad .616 \quad 0 \quad 0]$

31. (A) True (B) False

33. $\begin{array}{c|cccc} & A & B & C & D \\ \hline A & 1 & 0 & 0 & 0 \\ B & 0 & 1 & 0 & 0 \\ C & .6375 & .3625 & 0 & 0 \\ D & .7375 & .2625 & 0 & 0 \end{array}$ **35.** $\begin{array}{c|ccccc} & A & B & C & D & E \\ \hline A & 1 & 0 & 0 & 0 & 0 \\ B & 0 & 1 & 0 & 0 & 0 \\ C & .0875 & .9125 & 0 & 0 & 0 \\ D & .1875 & .8125 & 0 & 0 & 0 \\ E & .4375 & .5625 & 0 & 0 & 0 \end{array}$ **37.** $\begin{array}{c|cccc} & A & B & C & D \\ \hline A & 0 & .52 & 0 & .48 \\ B & 0 & 1 & 0 & 0 \\ C & 0 & .36 & 0 & .64 \\ D & 0 & 0 & 0 & 1 \end{array}$

43. (A) .370; .297; .227; .132; .045 (B) For large k, all entries of Q^k are close to 0.

45. (A) 75% (B) 12.5% (C) 7.5 months **47.** (A) Company *A*: 30%, company *B*: 15%, company *C*: 55% (B) 5 yr

49. (A) 91.52% (B) 4.96% (C) 6.32 days **51.** (A) .375 (B) 1.75 exits

Review Exercise

1. $S_1 = \overset{\begin{matrix}A & B\end{matrix}}{[.32 \quad .68]}$; $S_2 = \overset{\begin{matrix}A & B\end{matrix}}{[.328 \quad .672]}$. The probability of being in state A after one trial is .32 and after two trials is .328; the probability of being in state B after one trial is .68 and after two trials is .672. *(9-1)*

2. State A is absorbing; chain is absorbing. *(9-2, 9-3)* **3.** No absorbing states; chain is regular. *(9-2, 9-3)*

4. No absorbing states; chain is neither. *(9-2, 9-3)* **5.** States B and C are absorbing; chain is absorbing. *(9-2, 9-3)*

6. States A and B are absorbing; chain is neither. *(9-2, 9-3)*

7. $\begin{array}{c|ccc} & A & B & C \\ \hline A & 0 & 1 & 0 \\ B & .1 & 0 & .9 \\ C & 0 & 1 & 0 \end{array}$; no absorbing states; chain is neither. *(9-1, 9-2, 9-3)*

8. $\begin{array}{c|ccc} & A & B & C \\ \hline A & 0 & 1 & 0 \\ B & .1 & .2 & .7 \\ C & 0 & 0 & 1 \end{array}$; C is absorbing; chain is absorbing. *(9-1, 9-2, 9-3)*

9. $\begin{array}{c|ccc} & A & B & C \\ \hline A & 0 & 0 & 1 \\ B & .1 & .2 & .7 \\ C & 0 & 1 & 0 \end{array}$; no absorbing states; chain is regular. *(9-1, 9-2, 9-3)*

10. $\begin{array}{c|cccc} & A & B & C & D \\ \hline A & .3 & .2 & 0 & .5 \\ B & 0 & 1 & 0 & 0 \\ C & 0 & 0 & .2 & .8 \\ D & 0 & 0 & .3 & .7 \end{array}$; B is absorbing; chain is neither. *(9-1, 9-2, 9-3)*

11. 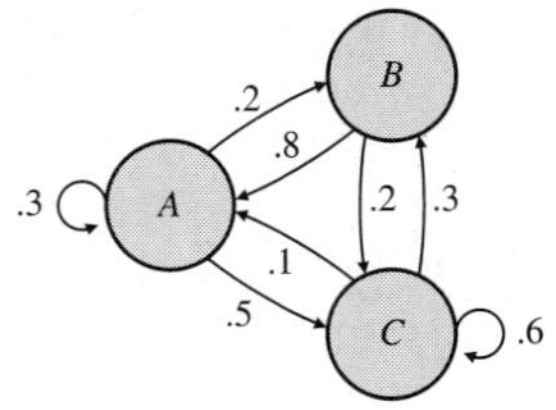

$\begin{array}{c|ccc} & A & B & C \\ \hline A & .3 & .2 & .5 \\ B & .8 & 0 & .2 \\ C & .1 & .3 & .6 \end{array}$ *(9-1)* **12.** (A) .3 (B) .675 *(9-1)*

13. $S = \overset{\begin{matrix}A & B\end{matrix}}{[.25 \quad .75]}$; $\overline{P} = \begin{array}{c|cc} & A & B \\ \hline A & .25 & .75 \\ B & .25 & .75 \end{array}$ *(9-2)* **14.** $S = \overset{\begin{matrix}A & B & C\end{matrix}}{[.4 \quad .48 \quad .12]}$; $\overline{P} = \begin{array}{c|ccc} & A & B & C \\ \hline A & .4 & .48 & .12 \\ B & .4 & .48 & .12 \\ C & .4 & .48 & .12 \end{array}$ *(9-2)*

15. $\begin{array}{c|ccc} & A & B & C \\ \hline A & 1 & 0 & 0 \\ B & 0 & 1 & 0 \\ C & .75 & .25 & 0 \end{array}$; $P(C \text{ to } A) = .75$; $P(C \text{ to } B) = .25$. It takes an average of 2.5 trials to go from C to an absorbing state. *(9-3)*

16. $\begin{array}{c|cccc} & A & B & C & D \\ \hline A & 1 & 0 & 0 & 0 \\ B & 0 & 1 & 0 & 0 \\ C & .2 & .8 & 0 & 0 \\ D & .3 & .7 & 0 & 0 \end{array}$; $P(C \text{ to } A) = .2$; $P(C \text{ to } B) = .8$; $P(D \text{ to } A) = .3$; $P(D \text{ to } B) = .7$. It takes an average of 2 trials to go from C to an absorbing state and an average of 3 trials to go from D to an absorbing state. *(9-3)*

21. $\begin{array}{c|cccc} & B & D & A & C \\ \hline B & 1 & 0 & 0 & 0 \\ D & 0 & 1 & 0 & 0 \\ A & .1 & .1 & .6 & .2 \\ C & .2 & .2 & .3 & .3 \end{array}$ *(9-3)*   **22.** (A) $\begin{array}{ccc} A & B & C \\ [.1 & .4 & .5] \end{array}$ (B) $\begin{array}{ccc} A & B & C \\ [.1 & .4 & .5] \end{array}$ *(9-3)*   **23.** (A) $\begin{array}{ccc} A & B & C \\ [.25 & .75 & 0] \end{array}$ (B) $\begin{array}{ccc} A & B & C \\ [.55 & .45 & 0] \end{array}$ *(9-3)*

24. No. Each row of P would contain a 0 and a 1, but none of the four matrices with this property is regular. *(9-2)*

25. Yes; for example, $P = \begin{bmatrix} 0 & 0 & 1 \\ 0 & 0 & 1 \\ .2 & .3 & .5 \end{bmatrix}$ is regular. *(9-2)*

26. (A)

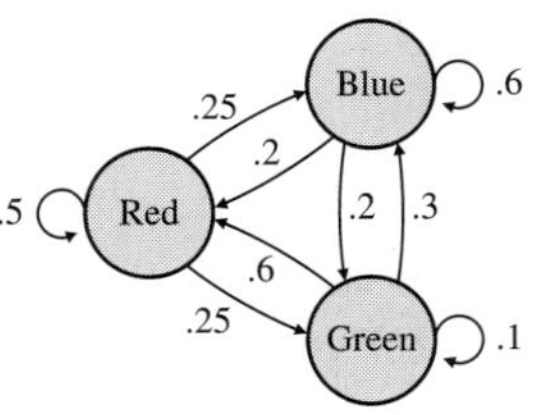

(B) $\begin{array}{c|ccc} & R & B & G \\ \hline R & .5 & .25 & .25 \\ B & .2 & .6 & .2 \\ G & .6 & .4 & .1 \end{array}$ (C) Regular (D) $\begin{array}{c|ccc} & R & B & G \\ \hline R & .4 & .4 & .2 \\ B & .4 & .4 & .2 \\ G & .4 & .4 & .2 \end{array}$ In the long run, the red urn will be selected 40% of the time, the blue urn 40% of the time, and the green urn 20% of the time. *(9-2)*

27. (A)

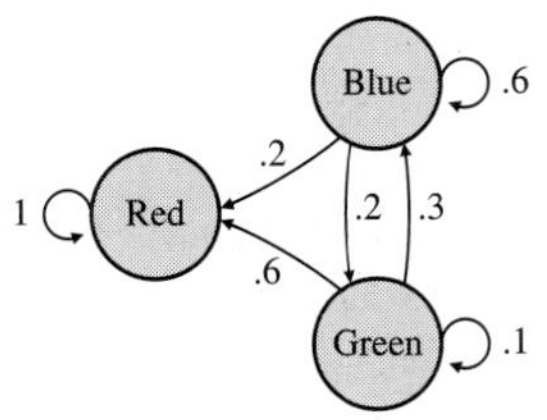

(B) $\begin{array}{c|ccc} & R & B & G \\ \hline R & 1 & 0 & 0 \\ B & .2 & .6 & .2 \\ G & .6 & .3 & .1 \end{array}$ (C) Absorbing (D) $\begin{array}{c|ccc} & R & B & G \\ \hline R & 1 & 0 & 0 \\ B & 1 & 0 & 0 \\ G & 1 & 0 & 0 \end{array}$ Once the red urn is selected, the blue and green urns will never be selected again. It will take an average of 3.67 trials to reach the red urn from the blue urn and an average of 2.33 trials to reach the red urn from the green urn. *(9-3)*

29. No such chain exists. *(9-2, 9-3)*   **30.** No such chain exists. *(9-2, 9-3)*   **31.** No such chain exists. *(9-2)*

32. $S = [1 \quad 0 \quad 0]$ and $S' = [0 \quad 1 \quad 0]$ are both stationary matrices for $P = \begin{array}{c|ccc} & A & B & C \\ \hline A & 1 & 0 & 0 \\ B & 0 & 1 & 0 \\ C & .6 & .3 & .1 \end{array}$ *(9-3)*

33. $P = \begin{array}{c|cc} & A & B \\ \hline A & 0 & 1 \\ B & 1 & 0 \end{array}$ *(9-2, 9-3)*   **34.** No such chain exists. *(9-2)*   **35.** No such chain exists. *(9-3)*

36. No limiting matrix *(9-2, 9-3)* **37.** $P = \begin{array}{c} \\ A \\ B \\ C \\ D \end{array}\begin{array}{c} \begin{array}{cccc} A & B & C & D \end{array} \\ \begin{bmatrix} .392 & .163 & .134 & .311 \\ .392 & .163 & .134 & .311 \\ .392 & .163 & .134 & .311 \\ .392 & .163 & .134 & .311 \end{bmatrix} \end{array}$ *(9-2)*

38. (A) 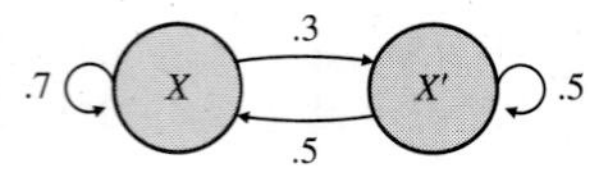

(B) $\begin{array}{c} \\ X \\ X' \end{array}\begin{array}{c} \begin{array}{cc} X & X' \end{array} \\ \begin{bmatrix} .7 & .3 \\ .5 & .5 \end{bmatrix} \end{array}$ (C) $\begin{array}{cc} X & X' \\ [.2 & .8] \end{array}$

(D) $\begin{array}{cc} X & X' \\ [.54 & .46] \end{array}$; 54% of the consumers will purchase brand X on the next purchase. (E) $\begin{array}{cc} X & X' \\ [.625 & .375] \end{array}$ (F) 62.5% *(9-2)*

39. (A) Brand *A:* 24%, brand *B:* 32%, brand *C:* 44% (B) 4 yr *(9-3)*

40. (A) $S_1 = \begin{array}{cc} V & V' \\ [.585 & .415] \end{array}$; $S_2 = \begin{array}{cc} V & V' \\ [.728 & .272] \end{array}$ (B)

YEAR	DATA (%)	MODEL (%)
1984	10.6	10.6
1988	58.0	58.5
1992	72.5	72.8

(C) 79% *(9-2)*

41. (A) 63.75% (B) 15% (C) 8.75 yr *(9-3)* **42.** $\overline{P} = \begin{array}{c} \\ \text{Red} \\ \text{Pink} \\ \text{White} \end{array}\begin{array}{c} \begin{array}{ccc} \text{Red} & \text{Pink} & \text{White} \end{array} \\ \begin{bmatrix} 1 & 0 & 0 \\ 1 & 0 & 0 \\ 1 & 0 & 0 \end{bmatrix} \end{array}$ *(9-3)*

43. (A) $S_1 = \begin{array}{cc} F & F' \\ [.022 & .978] \end{array}$; $S_2 = \begin{array}{cc} F & F' \\ [.0175 & .9825] \end{array}$ (B)

YEAR	DATA (%)	MODEL (%)
1980	2.8	2.8
1985	2.3	2.2
1990	1.9	1.75

(C) 0.4% *(9-2)*

APPENDIX A

Self-Test on Basic Algebra

1. (A) T (B) T (C) F (D) T *(A-1)* **2.** (A) $(y + z)x$ (B) $(2 + x) + y$ (C) $2x + 3x$ *(A-2)*
3. $x^3 + 3x^2 + 5x - 2$ *(A-3)* **4.** $x^3 - 3x^2 - 3x + 22$ *(A-3)* **5.** $3x^5 + x^4 - 8x^3 + 24x^2 + 8x - 64$ *(A-3)*
6. 3 *(A-3)* **7.** 1 *(A-3)* **8.** $14x^2 - 30x$ *(A-3)* **9.** $6x^2 - 5xy - 4y^2$ *(A-3)* **10.** $4a^2 - 12ab + 9b^2$ *(A-3)*
11. $4xy - 2y^2$ *(A-3)* **12.** $m^4 - 6m^2n^2 + n^2$ *(A-3)* **13.** $x^3 - 6x^2y + 12xy^2 - 8y^3$ *(A-3)*
14. (A) 4.065×10^{12} (B) 7.3×10^{-3} *(A-6)* **15.** (A) 255,000,000 (B) 0.000 406 *(A-6)*
16. (A) {2, 4, 5, 6} (B) {5} (C) {8} (D) {2, 4} *(A-1)* **17.** (A) 28 (B) 5 (C) 4 (D) 10 *(A-1)*
18. (A) T (B) F *(A-2)* **19.** 0 and −3 are two examples of infinitely many. *(A-2)* **20.** $6x^5y^{15}$ *(A-6)*
21. $3u^4/v^2$ *(A-6)* **22.** 6×10^2 *(A-6)* **23.** x^6/y^4 *(A-6)* **24.** $u^{7/3}$ *(A-7)* **25.** $3a^2/b$ *(A-7)* **26.** $\frac{5}{9}$ *(A-6)*
27. $x + 2x^{1/2}y^{1/2} + y$ *(A-7)* **28.** $6x + 7x^{1/2}y^{1/2} - 3y$ *(A-7)* **29.** $(3x - 1)(4x + 3)$ *(A-4)*
30. $(4x - 3y)(2x - 3y)$ *(A-4)* **31.** Not factorable relative to the integers *(A-4)*
32. $3n(2n - 5)(n + 1)$ *(A-4)* **33.** $(x - y)(7x - y)$ *(A-4)*
34. Not factorable relative to the integers *(A-4)* **35.** $\dfrac{12a^3b - 40b^3 - 5a}{30a^3b^2}$ *(A-5)* **36.** $\dfrac{7x - 4}{6x(x - 4)}$ *(A-5)*
37. $\dfrac{-8(x + 2)}{x(x - 4)(x + 4)}$ *(A-5)* **38.** $\dfrac{y + 2}{y(y - 2)}$ *(A-5)* **39.** $\dfrac{-1}{7(7 + h)}$ *(A-5)* **40.** $\dfrac{xy}{y - x}$ *(A-7)*

41. (A) Subtraction (B) Commutative (+) (C) Distributive (D) Associative (•) (E) Negatives (F) Identity (+) *(A-2)*
42. (A) 90 (B) 45 *(A-1)* **43.** $6x^{2/5} - 7(x-1)^{3/4}$ *(A-7)* **44.** $2\sqrt{x} - 3\sqrt[3]{x^2}$ *(A-7)* **45.** $2 - \frac{3}{2}x^{-1/2}$ *(A-7)*
46. $\sqrt{3x}$ *(A-7)* **47.** $\sqrt{x} + \sqrt{5}$ *(A-7)* **48.** $\dfrac{1}{\sqrt{x}-5}$ *(A-7)* **49.** $\dfrac{1}{\sqrt{u+h}+\sqrt{u}}$ *(A-7)* **50.** $x = 2$ *(A-8)*
51. $x = 0, 5$ *(A-9)* **52.** $x = \pm\sqrt{7}$ *(A-9)* **53.** $x = -4, 5$ *(A-9)* **54.** $x = (3 \pm \sqrt{17})/4$ *(A-9)*

55. $x < 4$ or $(-\infty, 4)$ 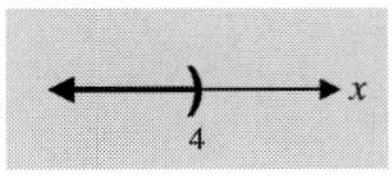

(A-8) **56.** $x \geq \frac{9}{2}$ or $[\frac{9}{2}, \infty)$ 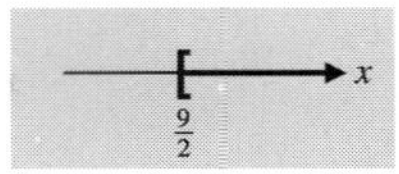

(A-8)

57. $2 \leq x < 12$ or $[2, 12)$ 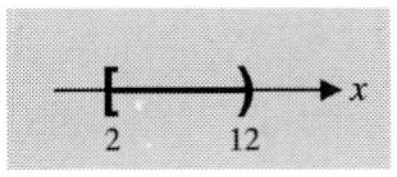

(A-8) **58.** $y = \frac{2}{3}x - 2$ *(A-8)* **59.** $y = 3/(x - 2)$ *(A-8)* **60.** Yes *(A-1)*

61. $2.3328 \times 10^4 = \$23{,}328$ per person *(A-6)* **62.** \$20,000 at 8%; \$40,000 at 14% *(A-8)* **63.** 4,000 tapes *(A-8)*

Exercise A-1

1. T **3.** T **5.** T **7.** T **9.** $\{1, 2, 3, 4, 5\}$ **11.** $\{3, 4\}$ **13.** $\varnothing$ **15.** $\{2\}$ **17.** $\{-7, 7\}$
19. $\{1, 3, 5, 7, 9\}$ **21.** $A' = \{1, 5\}$ **23.** 40 **25.** 60 **27.** 60 **29.** 20 **31.** 95 **33.** 40
35. (A) $\{1, 2, 3, 4, 6\}$ (B) $\{1, 2, 3, 4, 6\}$ **37.** $\{1, 2, 3, 4, 6\}$ **39.** Yes **41.** Yes **43.** Yes
45. (A) 2 (B) 4 (C) 8 (D) 16; 2^n **47.** 800 **49.** 200 **51.** 200 **53.** 800 **55.** 200 **57.** 200
59. 6 **61.** A+, AB+ **63.** A−, A+, B+, AB− AB+, O+ **65.** O+, O− **67.** B−, B+
69. Everybody in the clique relates to each other.

Exercise A-2

1. vu **3.** $(3 + 7) + y$ **5.** $u + v$ **7.** T **9.** T **11.** T **13.** T **15.** T **17.** T **19.** T **21.** F
23. T **25.** T **27.** No **29.** (A) F (B) T (C) T **31.** $\sqrt{2}$ and π are two examples of infinitely many.
33. (A) *N, Z, Q, R* (B) *R* (C) *Q, R* (D) *Q, R*
35. (A) T (B) F, since, for example, $(8 - 4) - 2 \neq 8 - (4 - 2)$. (C) T
(D) F, since, for example, $(8 \div 4) \div 2 \neq 8 \div (4 \div 2)$.
37. $\frac{1}{11}$ **39.** (A) 2.166 666 666... (B) 4.582 575 69... (C) 0.437 500 000... (D) 0.261 261 261...

Exercise A-3

1. 3 **3.** $x^3 + 4x^2 - 2x + 5$ **5.** $x^3 + 1$ **7.** $2x^5 + 3x^4 - 2x^3 + 11x^2 - 5x + 6$ **9.** $-5u + 2$
11. $6a^2 + 6a$ **13.** $a^2 - b^2$ **15.** $6x^2 - 7x - 5$ **17.** $2x^2 + xy - 6y^2$ **19.** $9y^2 - 4$ **21.** $6m^2 - mn - 35n^2$
23. $16m^2 - 9n^2$ **25.** $9u^2 + 24uv + 16v^2$ **27.** $a^3 - b^3$ **29.** $16x^2 + 24xy + 9y^2$ **31.** 1 **33.** $x^4 - 2x^2y^2 + y^4$
35. $5a^2 + 12ab - 10b^2$ **37.** $-4m + 8$ **39.** $-6xy$ **41.** $u^3 + 3u^2v + 3uv^2 + v^3$
43. $x^3 - 6x^2y + 12xy^2 - 8y^3$ **45.** $2x^2 - 2xy + 3y^2$ **47.** $8x^3 - 20x^2 + 1$ **49.** $4x^3 - 14x^2 + 8x - 6$
51. $m + n$ **53.** No change **55.** $(1 + 1)^2 \neq 1^2 + 1^2$; either *a* or *b* must be 0
57. $0.09x + 0.12(10{,}000 - x) = 1{,}200 - 0.03x$ **59.** $10x + 30(3x) + 50(4{,}000 - x - 3x) = 200{,}000 - 100x$
61. $0.02x + 0.06(10 - x) = 0.6 - 0.04x$

Exercise A-4

1. $3m^2(2m^2 - 3m - 1)$ **3.** $2uv(4u^2 - 3uv + 2v^2)$ **5.** $(7m + 5)(2m - 3)$ **7.** $(a - 4b)(3c + d)$
9. $(2x - 1)(x + 2)$ **11.** $(y - 1)(3y + 2)$ **13.** $(x + 4)(2x - 1)$ **15.** $(w + x)(y - z)$ **17.** $(a - b)(m + n)$
19. $(3y + 2)(y - 1)$ **21.** $(u - 5v)(u + 3v)$ **23.** Not factorable **25.** $(wx - y)(wx + y)$ **27.** $(3m - n)^2$
29. Not factorable **31.** $4(z - 3)(z - 4)$ **33.** $2x^2(x - 2)(x - 10)$ **35.** $x(2y - 3)^2$ **37.** $(2m - 3n)(3m + 4n)$
39. $uv(2u - v)(2u + v)$ **41.** $2x(x^2 - x + 4)$ **43.** $(r - t)(r^2 + rt + t^2)$ **45.** $(a + 1)(a^2 - a + 1)$
47. $[(x + 2) - 3y][(x + 2) + 3y]$ **49.** Not factorable **51.** $(6x - 6y - 1)(x - y + 4)$
53. $(y - 2)(y + 2)(y^2 + 1)$ **55.** $a^2(3 + ab)(9 - 3ab + a^2b^2)$

Exercise A-5

1. $8d^6$ **3.** $\dfrac{15x^2 + 10x - 6}{180}$ **5.** $\dfrac{15m^2 + 14m - 6}{36m^3}$ **7.** $\dfrac{1}{x(x-4)}$ **9.** $\dfrac{x-6}{x(x-3)}$ **11.** $\dfrac{x-5}{(x-1)^2(x+1)}$ **13.** $\dfrac{2}{x-1}$ **15.** $\dfrac{5}{a-1}$ **17.** $\dfrac{x^2+8x-16}{x(x-4)(x+4)}$ **19.** $\dfrac{7x^2-2x-3}{6(x+1)^2}$ **21.** $-\dfrac{1}{x}$ **23.** $\dfrac{-17c+16}{15(c-1)}$ **25.** $\dfrac{1}{x-3}$ **27.** $\dfrac{-1}{2x(x+h)}$ **29.** $\dfrac{x-y}{x+y}$ **31.** (A) Incorrect (B) $x + 1$ **33.** (A) Incorrect (B) $2x + h$ **35.** (A) Incorrect (B) $\dfrac{x^2-x-3}{x+1}$ **37.** (A) Correct **39.** $\dfrac{-2x-h}{3(x+h)^2x^2}$ **41.** x

Exercise A-6

1. $2/x^9$ **3.** $3w^7/2$ **5.** $2/x^3$ **7.** $1/w^5$ **9.** 5 **11.** $1/a^6$ **13.** y^6/x^{12} **15.** 8.23×10^{10} **17.** 7.83×10^{-1} **19.** 3.4×10^{-5} **21.** 40,000 **23.** 0.007 **25.** 61,710,000 **27.** 0.000 808 **29.** 1 **31.** 10^{14} **33.** $y^6/25x^4$ **35.** 4×10^2 **37.** $4y^3/3x^5$ **39.** $\frac{7}{4} - \frac{1}{4}x^{-3}$ **41.** $\frac{3}{4}x - \frac{1}{4}x^{-1} - \frac{1}{4}x^{-3}$ **43.** $\dfrac{x^2(x-3)}{(x-1)^3}$ **45.** $\dfrac{2(x-1)}{x^3}$ **47.** 2.4×10^{10}; 24,000,000,000 **49.** 3.125×10^4; 31,250 **51.** 64 **55.** uv **57.** $\dfrac{bc(c+b)}{c^2+bc+b^2}$ **59.** (A) 2.13701×10^{11} (B) 2.2293 (C) 0.4486 **61.** (A) \$15,935 (B) \$1,146 (C) 7.19% **63.** (A) 9×10^{-6} (B) 0.000 009 (C) 0.0009% **65.** 1,932,000

Exercise A-7

1. $6\sqrt[5]{x^3}$ **3.** $\sqrt[5]{(4xy^3)^2}$ **5.** $\sqrt{x^2+y^2}$ (not $x + y$) **7.** $5x^{3/4}$ **9.** $(2x^2y)^{3/5}$ **11.** $x^{1/3} + y^{1/3}$ **13.** 5 **15.** 64 **17.** -6 **19.** Not a rational number (not even a real number) **21.** $\frac{8}{125}$ **23.** $\frac{1}{27}$ **25.** $x^{2/5}$ **27.** m **29.** $2x/y^2$ **31.** $xy^2/2$ **33.** $2/3x^{7/12}$ **35.** $2x + 3$ **37.** $6x^3$ **39.** 2 **41.** $12x - 6x^{35/4}$ **43.** $3u - 13u^{1/2}v^{1/2} + 4v$ **45.** $25m - n$ **47.** $9x - 6x^{1/2}y^{1/2} + y$ **49.** $\frac{1}{2}x^{1/3} + x^{-1/3}$ **51.** $\frac{2}{3}x^{-1/4} + x^{-2/3}$ **53.** $\frac{1}{2}x^{-1/6} - \frac{1}{4}$ **55.** $4n\sqrt{3mn}$ **57.** $\dfrac{2\sqrt{x-2}}{x-2}$ **59.** $7(x-y)(\sqrt{x}+\sqrt{y})$ **61.** $\dfrac{1}{xy\sqrt{5xy}}$ **63.** $\dfrac{1}{\sqrt{x+h}+\sqrt{x}}$ **65.** $\dfrac{1}{\sqrt{t}+\sqrt{x}}$ **67.** $x = y = 1$ is one of many choices. **69.** $x = y = 1$ is one of many choices. **71.** $\dfrac{x+8}{2(x+3)^{3/2}}$ **73.** $\dfrac{x-2}{2(x-1)^{3/2}}$ **75.** $\dfrac{x+6}{3(x+2)^{5/3}}$ **77.** 103.2 **79.** 0.0805 **81.** 4,588 **83.** (A) and (E); (B) and (F); (C) and (D)

Exercise A-8

1. $m = 5$ **3.** $x < -9$ **5.** $x \le 4$ **7.** $x < -3$ or $(-\infty, -3)$

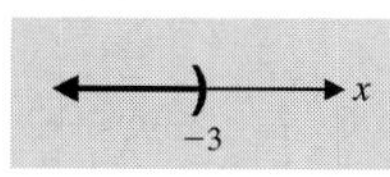

9. $-1 \le x \le 2$ or $[-1, 2]$

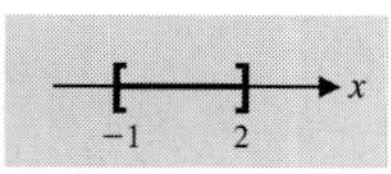

11. $y = 8$ **13.** $x > -6$ **15.** $y = 8$ **17.** $x = 10$ **19.** $y \ge 3$ **21.** $x = 36$ **23.** $m < 3$ **25.** $x = 10$ **27.** $3 \le x < 7$ or $[3, 7)$

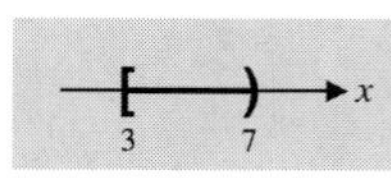

29. $-20 \le C \le 20$ or $[-20, 20]$

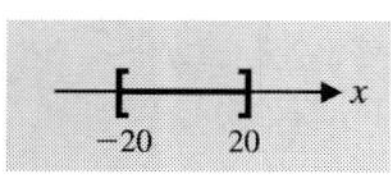

31. $y = \frac{3}{4}x - 3$ **33.** $y = -(A/B)x + (C/B) = (-Ax + C)/B$ **35.** $C = \frac{5}{9}(F - 32)$ **37.** $B = A/(m - n)$

39. $-2 < x \leq 1$ or $(-2, 1]$ 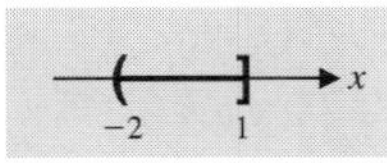

41. (A) and (C): $a > 0$ and $b > 0$, or $a < 0$ and $b < 0$ (B) and (D): $a > 0$ and $b < 0$, or $a < 0$ and $b > 0$
43. (A) $>$ (B) $<$ **45.** Negative **47.** 5,000 \$15 tickets; 3,000 \$25 tickets **49.** \$7,200 at 10%; \$4,800 at 15%
51. \$18,080 **53.** 5,851 books **55.** (B) 6,180 books (C) At least \$11.50 **57.** 5,000 **59.** 12.6 yr

Exercise A-9

1. $\pm\sqrt{11}$ **3.** $-1, 3$ **5.** $-2, 6$ **7.** $0, 2$ **9.** $3 \pm 2\sqrt{2}$ **11.** $-2 \pm \sqrt{2}$ **13.** $0, 2$ **15.** $\pm\frac{3}{2}$
17. $\frac{1}{2}, -3$ **19.** $(-1 \pm \sqrt{5})/2$ **21.** $(3 \pm \sqrt{3})/2$ **23.** No real solution **25.** $-4 \pm \sqrt{11}$ **27.** $\pm\sqrt{3}$
29. $-\frac{1}{2}, 2$ **31.** $(x - 2)(x + 42)$ **33.** Not factorable in the integers **35.** $(2x - 9)(x + 12)$
37. $(4x - 7)(x + 62)$ **39.** $r = \sqrt{A/P} - 1$
41. If $c < 4$, there are two distinct real roots; if $c = 4$, there is one real double root; and if $c > 4$, there are no real roots.
43. 1,575 bottles at \$4 each **45.** 0.2, or 20% **47.** 8 ft/sec; $4\sqrt{2}$ or 5.66 ft/sec

Appendix B

Exercise B-1

1. 5, 7, 9, 11 **3.** $\frac{3}{2}, \frac{4}{3}, \frac{5}{4}, \frac{6}{5}$ **5.** $9, -27, 81, -243$ **7.** 23 **9.** $\frac{101}{100}$ **11.** $1 + 2 + 3 + 4 + 5 + 6 = 21$
13. $5 + 7 + 9 + 11 = 32$ **15.** $1 + \frac{1}{10} + \frac{1}{100} + \frac{1}{1{,}000} = \frac{1{,}111}{1{,}000}$ **17.** 3.6 **19.** 82.5 **21.** $\frac{1}{2}, -\frac{1}{4}, \frac{1}{8}, -\frac{1}{16}, \frac{1}{32}$
23. 0, 4, 0, 8, 0 **25.** $1, -\frac{3}{2}, \frac{9}{4}, -\frac{27}{8}, \frac{81}{16}$ **27.** $a_n = n - 3$ **29.** $a_n = 4n$ **31.** $a_n = (2n - 1)/2n$
33. $a_n = (-1)^{n+1}n$ **35.** $a_n = (-1)^{n+1}(2n - 1)$ **37.** $a_n = (\frac{2}{5})^{n-1}$ **39.** $a_n = x^n$ **41.** $a_n = (-1)^{n+1}x^{2n-1}$
43. $1 - 9 + 25 - 49 + 81$ **45.** $\frac{4}{7} + \frac{8}{9} + \frac{16}{11} + \frac{32}{13}$ **47.** $1 + x + x^2 + x^3 + x^4$ **49.** $x - \dfrac{x^3}{3} + \dfrac{x^5}{5} - \dfrac{x^7}{7} + \dfrac{x^9}{9}$
51. (A) $\sum_{k=1}^{5} (k + 1)$ (B) $\sum_{j=0}^{4} (j + 2)$ **53.** (A) $\sum_{k=1}^{4} \dfrac{(-1)^{k+1}}{k}$ (B) $\sum_{j=0}^{3} \dfrac{(-1)^j}{j + 1}$ **55.** $\sum_{k=1}^{n} \dfrac{k + 1}{k}$ **57.** $\sum_{k=1}^{n} \dfrac{(-1)^{k+1}}{2^k}$
59. 2, 8, 26, 80, 242 **61.** 1, 2, 4, 8, 16 **63.** $1, \frac{3}{2}, \frac{17}{12}, \frac{577}{408}$; $a_4 = \frac{577}{408} \approx 1.414\,216$, $\sqrt{2} \approx 1.414\,214$

Exercise B-2

1. (A) Arithmetic, with $d = -5$; $-26, -31$ (B) Geometric, with $r = 2$; $-16, 32$ (C) Neither
(D) Geometric, with $r = \frac{1}{3}$; $\frac{1}{54}, \frac{1}{162}$
3. $a_2 = 11, a_3 = 15$ **5.** $a_{21} = 82, S_{31} = 1{,}922$ **7.** $S_{20} = 930$ **9.** $a_2 = -6, a_3 = 12, a_4 = -24$ **11.** $S_7 = 547$
13. $a_{10} = 199.90$ **15.** $r = 1.09$ **17.** $S_{10} = 1{,}242, S_\infty = 1{,}250$ **19.** 2,706 **21.** -85 **23.** 1,120
25. (A) Does not exist (B) $S_\infty = \frac{8}{5} = 1.6$ **27.** 2,400 **29.** 0.999
31. Use $a_1 = 1$ and $d = 2$ in $S_n = (n/2)[2a_1 + (n - 1)d]$. **33.** \$48 + \$46 + ⋯ + \$4 + \$2 = \$600
35. About \$11,670,000 **37.** \$1,628.89; \$2,653.30

Exercise B-3

1. 720 **3.** 10 **5.** 1,320 **7.** 10 **9.** 6 **11.** 1,140 **13.** 10 **15.** 6 **17.** 1 **19.** 816
21. $C_{4,0}a^4 + C_{4,1}a^3b + C_{4,2}a^2b^2 + C_{4,3}ab^3 + C_{4,4}b^4 = a^4 + 4a^3b + 6a^2b^2 + 4ab^3 + b^4$
23. $x^6 - 6x^5 + 15x^4 - 20x^3 + 15x^2 - 6x + 1$ **25.** $32a^5 - 80a^4b + 80a^3b^2 - 40a^2b^3 + 10ab^4 - b^5$ **27.** $3{,}060x^{14}$
29. $5{,}005p^9q^6$ **31.** $264x^2y^{10}$ **33.** $C_{n,0} = \dfrac{n!}{0!n!} = 1$; $C_{n,n} = \dfrac{n!}{n!0!} = 1$ **35.** 1 5 10 10 5 1; 1 6 15 20 15 6 1

INDEX

Applications Index

Business & Economics